HISTOPHYSIOLOGY OF THE IMMUNE SYSTEM

The Life History, Organization, and Interactions of Its Cell Populations

---

Recent Volumes in this Series

Volume 230
CELL AND MOLECULAR BIOLOGY OF THE UTERUS
Edited by Wendell W. Leavitt

Volume 231
ADVANCES IN POST-TRANSLATIONAL MODIFICATIONS OF PROTEINS AND AGING
Edited by Vincenzo Zappia, Patrizia Galletti, Raffaele Porta, and Finn Wold

Volume 232
CELLULAR Ca^{2+} REGULATION
Edited by Douglas R. Pfeiffer, Jeanie B. McMillin, and Steve Little

Volume 233
CANCER METASTASIS
Edited by Giorgio Prodi, Lance A. Liotta, Pier-Luigi Lollini, Spiridione Garbisa, Sergio Gorini, and Kurt Hellmann

Volume 234
BIOLOGY OF GROWTH FACTORS: Molecular Biology, Oncogenes, Signal Transduction, and Clinical Implications
Edited by Jeffrey E. Kudlow, David H. MacLennan, Alan Bernstein, and Avrum I. Gotlieb

Volume 235
CENTRAL D_1 DOPAMINE RECEPTORS
Edited by Menek Goldstein, Kjell Fuxe, and Irving Tabachnick

Volume 236
NEURORECEPTORS AND SIGNAL TRANSDUCTION
Edited by Shozo Kito, Tomio Segawa, Kinya Kuriyama, Masaya Tohyama, and Richard W. Olsen

Volume 237
HISTOPHYSIOLOGY OF THE IMMUNE SYSTEM: The Life History, Organization, and Interactions of Its Cell Populations
Edited by Sigbjørn Fossum and Bent Rolstad

HISTOPHYSIOLOGY OF THE IMMUNE SYSTEM

The Life History, Organization, and Interactions of Its Cell Populations

Edited by

Sigbjørn Fossum and
Bent Rolstad

Anatomical Institute
Oslo, Norway

PLENUM PRESS • NEW YORK AND LONDON

Library of Congress Cataloging in Publication Data

International Conference on Lymphatic Tissues and Germinal Centers in Immune Reactions (9th: 1987: Oslo, Norway)
Histophysiology of the immune system: the life history, organization, and interactions of its cell populations / edited by Sigbjørn Fossum and Bent Rolstad.
p. cm. — (Advances in experimental medicine and biology; v. 237)
"Proceedings of the Ninth International Conference on Lymphatic Tissues and Germinal Centers in Immune Reactions, held August 10–14, 1987, in Oslo, Norway" — T.p. verso.
Includes bibliographies and index.

ISBN 978-1-4684-5537-3 ISBN 978-1-4684-5535-9 (eBook)
DOI 10.1007/978-1-4684-5535-9

1. Immune system — Physiology — Congresses. 2. Immune system — Histology — Congresses. 3. Immunocompetent cells — Congresses. I. Fossum, Sigbjørn. II. Rolstad, Bent. III. Title. IV. Series.
[DNLM: 1. Cell Communication — congresses. 2. Cell Differentiation — congresses. 3. Immune System — anatomy & histology — congresses. 4. Immune System — physiology — congresses. W1 AD559 v. 237 / QW 504 I59175 1987h]
QR180.3.I53 1987
616.07′9 — dc19
DNLM/DLC 88-18005
for Library of Congress CIP

Proceedings of the Ninth International Conference on
Lymphatic Tissues and Germinal Centers in Immune Reactions,
held August 10–14, 1987, in Oslo, Norway

Softcover reprint of the hardcover 1st edition 1988
A Division of Plenum Publishing Corporation
233 Spring Street, New York, N.Y. 10013

J.H. HUMPREY

This volume is dedicated to the memory of John H. Humphrey. He was one of the pioneers of modern cellular immunology. Already in the 1950s he had become a central figure in cellular immunology, and he continued to contribute to most areas of this rapidly expanding and diversifying field during the following decades. He realized the importance of looking at the immune system as a whole and was a regular participant as well as a leading force in the Germinal Centre Conferences. Despite the explosive growth of immunology, he continued to have a unique grasp of the subject. He was therefore heavily in demand as a chairman or as an invited speaker in the Germinal Centre Conferences as well as in many other national or international immunological meetings. At the Germinal Centre Conference in Groningen, 1981, he was presented with the "7th Germinal Centre Conference Medal", especially designed to acknowledge his outstanding contributions to this series of conferences as well as to the immunological society as a whole. John Humphrey was also a humanist, deeply engaged in basic political and ethical problems concerning the future of mankind. He will be remembered by his many friends and colleagues with respect and gratitude.

S. Fossum and B. Rolstad

INTRODUCTION

The 9th International Conference on Lymphatic Tissues and Germinal Centres in Immune Reactions was held in Oslo, 9-14 August, 1987. These conferences, by the regular devotees just referred to as the germinal centre conferences or GCC, have been held regularly at roughly three-year intervals since 1966.

The credo of these conferences is "in vivo veritas", signifying that investigating components, like molecules and cells, only gives partial truth. The components must ultimately be explored in their natural context, to see how they interact with other parts and are integrated to a whole.

To the biologist it is obvious that the world must be investigated at many different levels of organization. At each level the patterns observed represent just some of many possible ways of putting together the elements of the lower levels. These patterns are not predetermined, but the results of evolution, i.e. of a history in which stochastic processes play a major role. The organic world can therefore not be torn apart and then reconstructed from basic principles alone. This realization is often expressed as the whole being more than the sum of its parts.

A corollary is that properties of the components depend upon the pattern of which they are parts, i.e. by their interactions. Just as the sentence gets its meaning from the words, the words get their meaning from the sentence. These ideas are evident to biologists. They will therefore reject the recent claims that holism should represent an entirely new way of perceiving the world, but will also be reluctant to discredit the analytical approach, or reductionism. In biology, reductionism and holism, analysis and integration, structures and interactions are complementary.

The GCC-meetings and their proceedings exemplify this complementarity and illustrate the necessity to simultaneously study several levels of organization. Thus, although particular emphasis traditionally has been laid upon the tissue level, i.e. upon cellular behaviour and interactions, in other words upon higher levels of organization, it is no contradiction that the conference this time had added a symposium devoted to leukocyte surface molecules, i.e. the molecular level.

This wide range of analytical levels, from the molecular, via the cellular and tissue levels to organ systems and their interactions, characterizes the germinal centre conferences. They have therefore become established as the main international forum for the ventilation of relationships between structure and function in lymphoid tissues.

It was a privilege and a pleasure to arrange this conference in Oslo. The meeting was successful, thanks to the two hundred devoted scientists from all five continents. We will in particular extend our thanks to dr. Alan Williams, Dunn School of Pathology, Oxford, UK, for his invaluable assistance in arranging the symposium on surface molecules in leukocyte interactions.

Prof. Per Brandtzaeg, Department of Pathology, prof. Lasse Braathen, Department of Dermatology, Rikshospitalet, Oslo, Norway, dr. William E. Bowers, Cooperstown, New York, USA, dr. Gerry G.B. Klaus, Mill Hill, London, UK, and prof. Paul Nieuwenhuis, Groningen, The Netherlands, shared with us the responsibility of the scientific committee and their contributions are hereby gratefully acknowledged.

Finally, we would like to express our gratitude to J.E. Isberg's Fund, Sigval Bergesen D.Y. and Wife Nanki's Foundation of Public Utility, Anders Jahres Fund for the Promotion of Science, Norwegian Cancer Association, The Norwegian Department of Health, The Royal Norwegian Ministry of Development Cooperation (NORAD), The Norwegian Research Council for Science and the Humanities (NAVF), Norwegian Association for Fighting Cancer, Den norske Creditbank, International Business Machines A/S (IBM), Nycomed AS, Merck Sharp & Dohme A/S (MSD), Dynal A.S., Skatron A/S, plus a number of other foundations and companies for their generous financial support, without which it would not have been possible to arrange this conference.

Oslo, February 1988

Sigbjørn Fossum   Bent Rolstad
organizers

Tove Eliassen
secretary

Anatomical Institute
University of Oslo

CONTENTS

THE ROLE OF FOLLICULAR DENDRITIC CELLS AND GERMINAL CENTERS IN IMMUNE REACTIONS

THYMUS, T-CELL DIFFERENTIATION AND THE T-CELL SYSTEM

CELL TYPES INVOLVED IN NON-ADAPTIVE OR NATURAL IMMUNITY

CELL TRAFFIC IN THE IMMUNE SYSTEM

SECRETORY IMMUNITY

ANTIGEN PRESENTATION AND CELLULAR INTERACTIONS IN INITIATION AND REGULATION OF IMMUNE RESPONSES

A. SYMPOSIUM ON SURFACE MOLECULES IN LEUKOCYTE INTERACTIONS

THE LEUKOCYTE-COMMON ANTIGEN (L-CA) FAMILY

A. Neil Barclay, David I. Jackson, Anthony C. Willis[+]
and Alan F. Williams

MRC Cellular Immunology Unit, Sir William Dunn School
of Pathology, University of Oxford
South Parks Road, Oxford OX1 3RE

[+]MRC Immunochemistry Department, Department of Biochemistry
University of Oxford
South Parks Road, Oxford OX1 3QU

INTRODUCTION

The leukocyte common antigen (L-CA, T200, Ly-5 or CD45) consists of a family of abundant membrane glycoproteins of apparent Mr 180-240k present on all lymphoid and myeloid cells but absent from other tissues (1-4). The L-CA shows cell type specific heterogeneity in apparent Mr, glucosylation and antigenicity. L-CA from rat thymocytes gives a single major band on SDS PAGE of 180k Mr, whilst T cells give four bands of 180, 190, 200 and 220k Mr and B cells one broad band at 240k (5) and similar patterns are found on mouse and human leukocytes (2,6). Differences in glycosylation in these forms have been shown with anti-carbohydrate antibodies (7) and by lectin binding (8). Antigenic heterogeneity is revealed by antibodies of the higher Mr bands of L-CA seen on SDS PAGE; e.g. the monoclonal antibody MRC OX-33 is a marker for B cells and the MRC OX-22 monoclonal antibody recognizes B cells, T cytotoxic cells and about 2/3 of T helper cells but not the remainder of the helper cells nor thymocytes (5). The division of T helper cells using this antibody is of particular interest as these subpopulations have important functional differences (9,10). In addition the MRC OX-22 antigenic determinant is stable to trypsin digestion enabling an antigenically active peptide to be purified by antibody affinity chromatography and sequenced (1).

The sequence of the L-CA has now been determined by analysis of cDNA clones in rat (12,13), mouse (14,17) and man(18). An unusual feature is the exceptionally large cytoplasmic domain of 705 amino acids which is separated from the extracellular part by a single transmembrane spanning segment. Much of the extracellular part of thymocyte L-CA has also been sequenced at the protein level showing 13 N-linked glycosylation positions with some '0' linked glycosylation confined to the NH_2-terminal region (13).

It is now clear that some of the heterogeneity in L-CA is due to differences in polypeptide sequence and in this article we review the evidence for polypeptide heterogeneity of rat L-CA by analysis of cDNA and genomic clones.

MATERIALS AND METHODS

Genomic clones for rat L-CA were isolated from a genomic library obtained from Clontech Laboratories. All methods were as in (13).

RESULTS AND DISCUSSION

Four variants of L-CA, DNA clones were characterized from thymocyte and lymph node cDNA libraries. These sequences differed only by the insertion of extra segments of sequence 6 residues after the NH_2-terminus of the smallest form of 1118 amino acids (Fig. 1). The longest form (variant 4 figure 1) contains all 3 segments and its full sequence is shown in figure 2 whilst the other forms have either extra segment II or III. All four forms were found in cDNA libraries from lymph nodes but only the two smallest forms found in thymocytes. The smallest sequence of 1118 would have an Mr of 127,133 which gives an estimate Mr of 169,000 for the processed polypeptide given that thymocyte L-CA contains 25% by weight of carbohydrate (19). This corresponds quite closely to the apparent Mr of 180,000 for thymocyte L-CA. Variant 2 contains an extra 41 amino acids which would give an extra protein Mr of 4122 but carbohydrate could easily bring this to an extra 10,000 apparent Mr. Thus this form could represent a component of the 190k band present on T cells but only a minor component on thymocytes. It is not possible to assign particular bands to these variants and a total of 8 combinations utilizing these three extra segments are possible. However, the partial sequence of the MRC OX-22 antigenic peptide (G A D T Q - L S S E A D L; se ref 11) is found within extra segment II and this segment is only found in lymph node cDNA clones. It is likely that these variants code for the larger forms of the L-CA of T or B cells as MRC OX22 does not react with thymocyte L-CA (180k) but reacts with the B cell form and a subfraction of the 190, 200 and 220k T cell forms.

	COMMON	EXTRA SEGMENTS OF SEQUENCE			COMMON SEQUENCE	cDNA FROM
1	QGTTDD				GANLTT.....PSS*	Thymus, Lymph node
2	QGTTDD			III	GANLTT.....PSS*	Thymus, Lymph node
3	QGTTDD		II		GANLTT.....PSS*	Lymph node
4	DD	I	II	III	GANLTT.....PSS*	Lymph node
	6	42	49	41	1112 amino acids	

Fig. 1. Schematic organization of variants deduced from L-CA cDNA clones. Four different types of variants were found containing extra segments I, II, III. The cDNA clone for variant 4 only extended into the common 6 amino acids whilst the remainder contained the common 6 amino acids together with 4 or 5 amino acids of the leader sequence. The position of cleavage of the signal peptide was based on similarities of this sequence to the mouse and human L-CA leade sequences and on peptide data from rat thymocyte L-CA (13).

Analysis of cDNA clones for mouse and human L-CA shows the presence of similar extra segments of sequence as shown in figure 3. The three segments are quite similar between the three species although the extra segment I is somewhat larger in human L-CA than in rodents. The longest form containing all three segments was identified in a human B cell line (18) and a mouse pre B cell line (17) but another mouse B cell line contained a different variant containing segments corresponding to II and III in figure 1 (16). Thus summarizing the data from the three species indicates 5 out of 8 possible forms have been found so far at the cDNA level.

```
     ↓ ←————————————————— I ——————————————————→↓←——————————
QGTTDDGLDTTEIVLLPQTDPLPARTTEFTPPSISERGNGSSETTYLPGFSSTLMPHLTP
——————————— II ——————————————————————→↓←——————————— III ————
QPDSQTPSARGADTQTLSSQADLTTLTAAPSGETDPPGVPEESTVPETFPGGTPILARNS
———————————————→↓
TAPSPTHTSNVSTTDISSGANLTTPAPSTLGFASNTTTSTEIATPQTKPSCDEKFGNVTV
RYIYDDSSKNFNANLEGDKKPKCEYTDCEKELKNLPECSQKNVTLSNGSCTPDKIINLDV
PPGTHNFNLTNCTPDIEANTSICLEWKIKNKFTCDIQKISYNFRCTPEMKTFALDKHGTL
WLHNLTVRTNYTCAAEVLYNNVILLKQDRRVQTDFGTPEMLPHVQCKNSTNSTTLVSWAE
PASKHHGYILCYKKTPSEKCENLANDVNSFEVKNLRPYTEYTVSLFAYVIGRVQRNGPAK
DCNFRTKAARPGKVNGMKTSRASDNSINVTCNSPYEINGPEARYILEVKSGGSLVKTFNQ
STCKFVVDNLYYSTDYEFLVYFYNGEYLGDPEIKPQSTSYNSKALIIFLVFLIIVTSIAL
LVVLYKIYDLRKKRSSNLDEQQELVERDEEKQLINVDPIHSDLLLETYKRKIADEGRLFL
AEFQSIPRVFSKFPIKDARKSQNQNKNRYVDILPYDYNRVELSEINGDAGSTYINASYID
GFKEPRKYIAAQGPRDETVDDFWKMIWEQKATVIVMVTRCEEGNRNKCAEYWPCMEEGTR
TFRDVVVTINDHKRCPDYIIQKLSIAHKKEKATGREVTHIQFTSWPDHGVPEDPHLLLKL
RRRVNAFSNFFSGPIVVHCSAGVGRTGTYIGIDAMLESLEAEGKVDVYGYVVNLRRQRCL
MVQVEAQYILIHQALVEYNQFGETEVNLSELHSCLQNLKKRDPPSDPSPLEAEYQRLPSY
RSWRTQHIGNQEENKKKNRSSNVVPYDFNRVPLKHELEMSKESEAESDESSDEDSDSEET
SKYINASFVMSYWKPEMMIAAQGPLKETIGDFWQMIFQRKVKVIVMLTELMSGDQEVCAQ
YWGEGKQTYGDMEVMLKDTNKSSAYILRAFELRHSKRKEPRTVYQYQCTTWKGEELPAEP
KDLVTLIQNIKQKLPKSGSEGMKYHKHASILVHCRDGSQQTGLFCALFNLLESAETEDVV
DVFQVVKSLRKARPGMVGSFEQYQFLYDIMASIYPTQNGQVKKANSQDKIEFHNEVDGAK
QDANCVQPADPLNKAQEDSKEVGASEPASGSEEPEHSANGPMSPALTPSS
```

Fig. 2. Sequence of the longest form of rat L-CA (variant 4 in Fig. 1). The sequence shown is for the mature protein and is derived from sequence of several cDNA clones (13). The three extra segments that are present in some variants (Fig. 1) are indicated.

The most likely method for generation of these variants is by alternative splicing of mRNA. Analysis of genomic clones for rat L-CA shows that each of the three extra segments of sequence is coded for by separate exons bounded by typical splice sites (Fig. 4). This has also been shown for two of extra segments in mouse L-CA (16,17). There are no striking sequence features in the immediate flanking regions that distinguish the alternative spliced exons from the remainder. One unusual feature of the mouse gene is that the leader exon is separated by about 50kb from the first exon of the mature polypeptide (16).

The function of L-CA is not known but it has interesting structural features including an extremely large cytoplasmic domain of 80,000 Mr which is known to be phosphorylated (29) and seems likely to interact

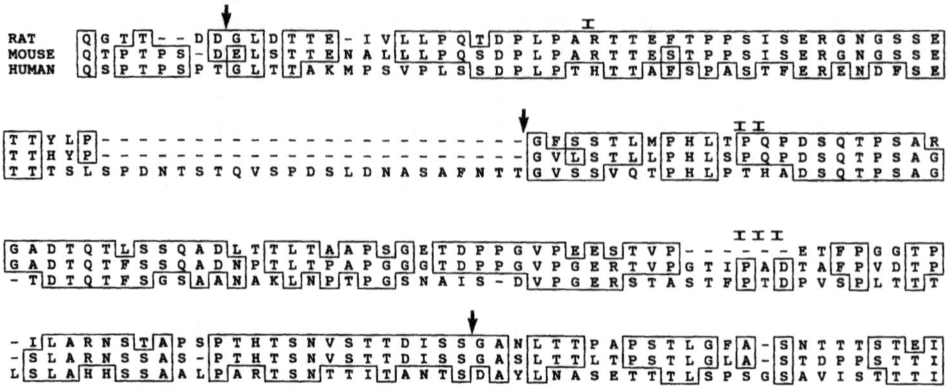

Fig. 3. Comparison of the NH_2-terminal sequence of the longest forms of L-CA from rat, mouse and human. The positions of the three extra segments in rat L-CA are aligned with those from mouse (17) and human (18).

```
SEGMENT I
                G L D T T E I V L L P Q T D P L P A R T T E F T P
ccattaacatgacagGACTGGACACAACAGAGATTGTTCTTCTGCCTCAAACTGACCCTTTACCTGCTCGCACCACTGAATTCACACC

 P S I S E R G N G S S E T T Y L P
CCCAAGCATCTCTGAAAGAGGAAATGGCTCCTCAGAGACCACATATCTTCCAGgtgaccactcaagcc

SEGMENT II
                G F S S T L M P H L T P Q P D S Q T P S A R G A D
actttctaacctcagGTTTTTCATCCACGCTGATGCCTCACCTGACCCCACAGCCTGACTCGCAGACCCCCTCTGCCAGAGGAGCTGA

 T Q T L S S Q A D L T T L T A A P S G E T D P P
CACTCAGACACTCAGCAGCCAAGCTGACCTCACCACACTCACGGCTGCTCCCAGCGGCGAGACTGACCCACCAGgtggtttgggggct

SEGMENT III
                G V P E E S T V P E T F P G G T P I L A R N S T A
cccttctgattgcagGTGTGCCCGAAGAGAGTACTGTACCGGAGACCTTTCCTGGAGGCACCCCCATCCTTGCACGCAACAGCACTGC

 P S P T H T S N V S T T D I S S
TCCCTCACCTACACACACTTCCAATGTCAGCACCACAGATATCTCCTCAGgtgtgaccattatgt
```

Fig. 4. Three exons encode the extra segments of sequence present in rat L-CA. The coding region of each exon is shown in upper case and flanking regions in lower case; the boundaries of the exons contain typical RNA splice sites.

with cytoskeletal components. There is evidence for interactions between L-CA and fodrin (21). The extracellular part is highly glycosylated and the thymocyte form was visualized in the electron microscope as a rod of about 28nm together with a globular domain of diameter 12nm containing the cytoplasmic sequence (22). The extra sequences in the variants differ from the majority of the extracellular sequence in containing an unusually high content of Ser and Thr (33%), Pro (14% and lacking Cys. This region is likely to have an extended structure and is probably extensively 0-glycosylated. Given the abundance of this molecule and the fact that the expression of variants is conserved across species these regions of extra sequence are presumably involved in cell type specific interactions.

REFERENCES

1. J. W. Fabre and A. F. Williams, Transplantation 23:349 (1977).
2. J. Michaelson, M. Scheid, and E. A. Boyse, Immunogenetics 9:193 (1979).
3. R. Dalchau, J. Kirkely and J. W. Fabre, Eur. J. Immunol. 10:737 (1980).
4. I. S. Trowbridge, P. Ralph, and M. J. Bevan, Proc. Natl. Acad. Sci. USA, 72:157 (1975).
5. G. R. Woollett, A. N. Barclay, M. Puklavec, and A. F. Williams, Eur. J. Immunol. 15:168 (1985).
6. I. S. Trowbridge, J. Exp. Med. 148:313 (1978).
7. R. A. Childs, R. Dalchau, P. Scudder, E. F. Hounsell, J. W. Fabre, and T. Feizi, Biochem. Biophys. Res. Comm. 110:424 (1983).
8. W. R. A. Brown and A. F. Williams, Immunology 46:713 (1982).
9. G. P. Spickett, M. R. Brandon, D. W. Mason, A. F. Williams, and G. R. Woollett, J. Exp. Med. 158:795 (1983).
10. R. P. Arthur and D. W. Mason, J. Exp. Med. 163:774 (1986).
11. P. Jonson, A. F. Williams, and G. R. Woollett, in: "Hybridoma Technology in the Biosciences and Medicine" Chapter 9, T. A. Springer, ed., Plenum Press (1985).
12. M. L. Thomas, A. N. Barclay, J. Gagnon, and A. F. Williams, Cell 41:83 (1985),
13. A. N. Barclay, D. I. Jackson, A. C. Willis, and A. F. Williams, EMBO J. 6:1259 (1987).
14. F.-W. Shen, Y. Saga, G. Litman, G. Freeman, J.-S. Tung, H. Cantor and E. A. Boyse, Proc. Natl. Avad. Sci. USA 82:7360 (1985).
15. Y. Saga, J.-S. Tung, F.-W. Shen, and E. A. Boyse, Proc. Natl. Acad. Sci. USA 83:6940 (1986).
16. Y. Saga, J.-S. Tung, F.-W. Shen, and E. A. Boyse, Proc. Natl. Acad. Sci. USA 84:5364 (1987).
17. M. L. Thomas, P. J. Reynolds, A. Chain, Y. Ben-Neriah, and I. S. Trowbridge, Proc. Natl. Acad. Sci. USA 84:5360 (1987).
18. S. J. Ralph, M. L. Thomas, C. C. Morton, and I. S. Trowbridge, EMBO J. 6:1251 (1987).
19. W. R. A. Brown, A. N. Barclay, C. A. Sunderland, and A. F. Williams, Nature 289:456 (1981).
20. D. A. Shackelford and I. S. Trowbridge, J. Biol. Chem. 261:8334 (1986).
21. L. Y. W. Bourguignon, S. J. Suchard, M. L. Nagpal, and J. R. Glenny Jr, J. Cell Biol. 101:477 (1985).
22. G. R. Woollett, A. F. Williams, and D. M. Shotton, EMBO J. 4:2827 (1985).

THE POLY-Ig RECEPTOR - FUNCTIONAL ASPECTS OF SECRETORY COMPONENT EXPRESSION

P. Brandtzaeg, D. Kvale, L.M. Sollid and P.S. Thrane

Laboratory for Immunohistochemistry and Immunopathology (LIIPAT), Institute of Pathology, University of Oslo, Rikshospitalet, Oslo, Norway

INTRODUCTION

Secretory tissues are quantitatively the most important mediator organ of humoral immunity - with the gut as the major contributor. There are about 10^{10} Ig-producing cells per metre of human small bowel (20). Most of these immunocytes produce J-chain-containing dimers or larger polymers of IgA (21). Polymeric IgA (pIgA) can be transported through the crypt epithelium along with pentameric IgM by combining with a pIg receptor protein called the "secretory component" or SC (19). More IgA is translocated to the gut lumen as secretory IgA (SIgA) every day than the total daily production of IgG, which is approximately 30 mg/kg (27). The gut epithelium is responsible for most of the external IgA transport as the human bile contributes only about 1 mg/kg/day to the luminal IgA pool (27).

A comprehensive literature has accumulated about the properties of SIgA, which is considered to function as a highly stabilized antibody that performs immune exclusion (Fig. 1) at mucosal surfaces in co-operation with innate immune mechanisms (16,38). The small amounts of SIgM and IgG (and minor traces of IgD and IgE) that normally appear in external secretions may likewise participate in immunological antigen exclusion (Fig. 1), although these isotypes are less stable than SIgA. In a pure glandular fluid, with minimum contamination by leakage of tissue fluid through surface epithelia, the IgG-to-IgA concentration ratio is reduced to about 0.2% of that in serum, and the IgG-to-IgM ratio may be reduced to less than 2% (2,28). This signifies preferential external translocation of both IgA and IgM compared with IgG, as first reported by our laboratory in 1970 (24).

The high proportion of IgA in external secretions cannot be ascribed to the mere dominance of immunocytes producing this isotype in secretory tissues. The concentration of IgG in interstitial tissue fluid is in fact higher than that of IgA because about 50% of serum IgG is extravascularly distributed; but virtually no IgG is taken up by normal secretory epithelium (20). Only J-chain-containing Ig polymers can be transported by the selective SC-dependent mechanism (19).

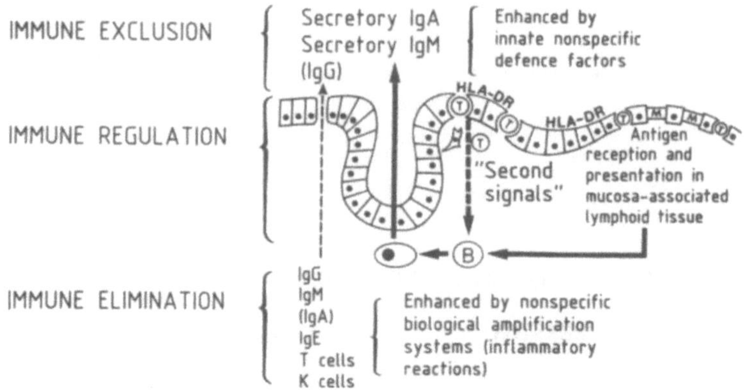

Fig. 1. Schematic representation of the three main components of mucosal immune defense: immune exclusion, immune regulation and immune elimination. Antigen stimulation in mucosa-associated lymphoid tissue provides "first signals" to B cells which migrate to secretory sites via lymph and peripheral blood. "Second signals" (dashed heavy arrow) induce local proliferation and terminal differentiation of the extravasated B cells. Most plasma cells generated in this way produce J-chain-containing polymeric IgA, which is translocated to the lumen as stabilised secretory IgA. Also some IgM is actively transported by the secretory epithelial cells, whereas small amounts of IgG normally reach the lumen by passive diffusion (dashed thin arrow).

Local Immune Regulation

The efferent limb of secretory immunity thus depends on an intimate cooperation of the B-cell system and the secretory epithelium by their production of the two key factors, J chain and SC, respectively (19). But it is still obscure how the particular immunoregulatory requirements of local immunity are normally fullfilled at the various mucosal surfaces.

Immune regulation takes place both in mucosa-associated lymphoid tissue such as the Peyer's patches and at secretory tissue sites (18). Antigen is primarily taken up by the former structures through the follicle-associated epithelium (FAE) which contains "membrane" (M) cells particularly designed for sampling and inward transport (26). Antigen may be processed by subepithelial macrophages and presented to sub- and intraepithelial T lymphocytes by dendritic cells and perhaps also by FAE, both of which express MHC class II (HLA-DR) determinants required for this function (1,18). Stimulated T cells release immunoregulatory factors which act on the local B lymphocytes. After thus receiving their "first signals", the stimulated B cells will migrate through lymph and peripheral blood and, perhaps after receiving "second signals" in mesenteric lymph nodes and spleen, finally seed distant secretory tissues such as the gut lamina propria (Fig. 1). Here further "second signals", modulated by MHC class II-expressing histiocytes and epithelial cells along with T lymphocytes, cause proliferation and terminal differentiation of the extravasated B cells (18). Most of them are of a clonal maturation stage that is compatible with J-chain production, regardless of concomitant cytoplasmic heavy-chain isotype - IgA normally being predominant (19).

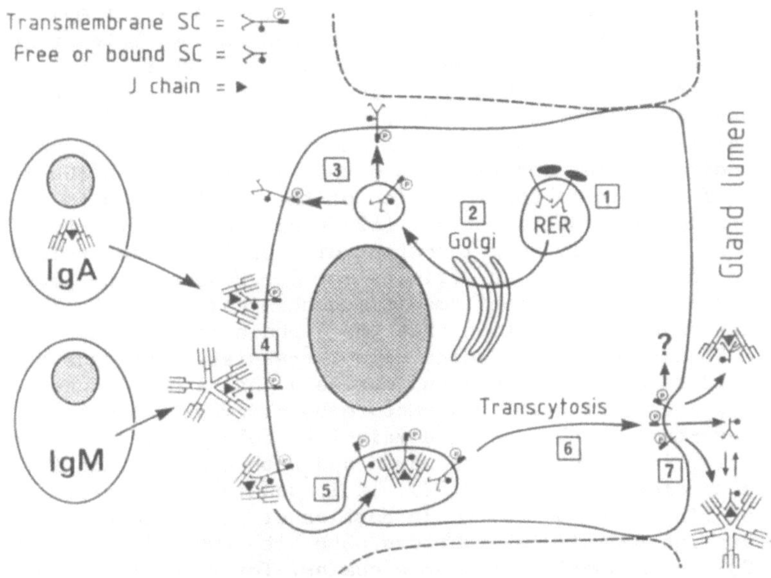

Fig. 2. Model for SC-mediated epithelial transport of SIgA and SIgM. [1] Synthesis and core glycosylation (—) of transmembrane SC in rough endoplasmic reticulum (RER) of epithelial cell. [2] Terminal glycosylation (—•) in Golgi complex. [3] Phosphorylation (p) at some later step. [4] Complexing of SC with J-chain-containing dimeric IgA and pentameric IgM on basolateral cell membrane. [5] Endocytosis of complexes and excess of SC. [6] Transcytosis of vesicles. [7] Cleavage and release of SIgA, SIgM, and free SC. The cleavage mechanism and the fate of the cytoplasmic tail of transmembrane SC are unknown (?). During the external translocation, covalent stabilization of the IgA-SC complexes regularly occurs (two disulphide bridges indicated in SIgA), whereas an excess of free SC in the secretion serves to stabilize the non-covalent IgM-SC complexes (dynamic equilibrium indicated for SIgM).

A COMMON TRANSPORT MODEL FOR POLYMERIC IMMUNOGLOBULINS

The model proposed by our laboratory in 1973-74, suggested that both pIgA and pIgM are transported through secretory epithelia by SC-mediated endocytosis (2-5). J chain and SC were taken to represent the "lock and key" in this process (Fig. 2). We had at that time obtained evidence indicating that the presence of J chain in the pIg molecules was decisive for their specific SC-binding site. Our model depicted that ligand-receptor complexes were subjected to vesicular transport through the cytoplasm outside the Golgi complex, but it was still not possible to suggest how SC could reach the cell surface. However, the model implied that SC was an integral membrane protein when functioning as a receptor.

In vitro Interactions between pIg and SC

It is not clear why the J chain is mandatory for spontaneous non-covalent interactions between Ig polymers and SC. The first evidence for such a crucial role of the J chain was obtained when we noted that a J-chain deficient IgM polymer failed to bind SC (30). We subsequently quantified immunochemically the reductive release of J chain in twenty-four IgA preparations: monomers and polymers that contained only 0.2-0.8 mg J chain/100 mg IgA showed 6-12% SC-binding capacity compared with 69-82% for polymers that contained more than 4.0 mg J chain/100 mg (12).

A role of J chain in the formation of the SC-binding site of pIg was recently questioned by Schiff et al. (62) when they observed that some IgA dimers with an apparently normal content of J chain (as determined electrophoretically) showed no affinity for SC. However, the presence of SC in these dimers was not confirmed immunochemically. The balance of evidence thus favours the notion that the J chain is necessary for an appropriate SC-binding site. Recent computer-assisted analyses based on the amino acid sequence of human and murine J chains have suggested that, despite a low degree of primary structure homology with Ig chains, the J chain may nevertheless show an Ig domain-like folding of its secondary structure (71). Importantly, both human and rabbit SC contain five Ig domain-like homology regions (29,52). It is possible, therefore, that the non-covalent binding of SC to the Fc portion of pIg is determined by complementary interdomain interactions both with the J chain and the constant (C)-region domains of α or μ chains. These possibilities have been discussed in detail by Mestecky and McGhee (48) and by Pumphrey (56).

It is the so-called I ("inaccessible") region of SC that is chiefly involved in the non-covalent binding (11,34); this region apparently corresponds to the NH_2-terminal domain (31). The interactions with pIgA seem to take place both with the Cα2 and the Cα3 domain (33). However, the covalent stabilization that occurs selectively for SIgA (13), may be caused by disulphide exchange reaction between the fifth domain of SC and the Cα2 domain of one IgA subunit (48).

There is normally much stronger non-covalent interactions between pIgM and SC than between pIgA and SC (6), and the apparent equilibrium constant of association (K_a) is several times higher for pIgM than for pIgA (13,65). It remains to be clarified whether a "bonus" effect of a higher molar J-chain content in pIgM than in pIgA explains the different properties of their SC-binding sites. Immunochemical quantitations (7,36) have challenged the notion that there is only one J chain per pIg molecule regardless of the polymer size (41,48). According to estimates obtained in our laboratory there are, on an average, two J chains per normal (polyclonal) IgA dimer and at least three per IgM pentamer. Also, the tendency for spontaneous dimerization shown by isolated J chain, obtained from dimeric IgA by mild reduction without further denaturation, suggests that the J chain exists as a dimer in its native bound form (8). This possibility may explain the marginal but significant affinity that purified J-chain dimers show for free SC (9).

Three independent research groups have determined the affinity (K_a) of SC for pIg to be about $10^8 M^{-1}$ (13,64,44). This value is similar to that of a variety of antigen-antibody reactions, but it is probably an underestimate when it comes to interactions with SC-expressing epithelial cells. Based on binding experiments in dilute solution and kinetic measurements with anti-Ig Sepharose to separate pIg-SC complexes, a $K_a \sim 10^{10} M^{-1}$ was obtained (61). Moreover, experiments with dimeric IgA and dispersed rabbit mammary cells (43) indicated that the association and dissociation rates were faster and K_a higher ($10^9 M^{-1}$) than the comparable values obtained with SC in solution. A similarly raised K_a was obtained for the binding of pIgA to SC-expressing cells of the human colonic HT-29 cell line (68).

Distribution of Ig and SC in Glandular Epithelia

As reviewed previously (14,40), earlier literature contains highly discordant reports on the cellular origin of SC. In our laboratory it has been distinctly localized by immunohistochemistry in the intestinal columnar crypts cells but we have been unable to detect it in goblet cells, regardless of the tissue preparation method used (23). This does not exclude that mucous secretory cells may show some capacity for SC synthesis. Occasional faint staining was in fact seen in mucous acini of salivary and respiratory glands, especially related to the periphery of the cells (14,40). Immunoelectron-microscopic studies have substantiated that not only do serous secretory cells synthesize SC, as demonstrated by its localization in the rough endoplasmic reticulum (RER), but they do also clearly contain SC in the basolateral plasma membrane (25,35,37,39, 59).

IgA can likewise be traced both in the cytoplasm and in relation to the basolateral borders of serous-type secretory epithelial cells. As for SC, the latter feature is especially marked in respiratory and salivary glands. The epithelial distribution of IgM is similar to that of IgA (10,25), but the IgM staining is usually fainter and may be almost undetectable beyond the gut.

Our laboratory originally suggested that the epithelial staining noted for IgA and IgM at the lateral cell borders identified these isotypes associated with the plasma membrane (4,5). Subsequent ultrastructural studies confirmed such a localization for both IgA and IgM in intestinal columnar cells (25,39) and pancreatic ducts (59), and for IgA in bronchial glands (35) and non-ciliated cells of the bronchioles (39). Moreover, these studies demonstrated SC and IgA (and IgM) in endocytic-appearing invaginations of the basolateral plasmalemma and in cytoplasmic vesicles.

Although the epithelial distributions of SC, IgA and IgM appeared to be similar (except for the quantitative differences just described), the detailed staining patterns revealed by paired immunofluorescence were not completely congruent. SC was selectively concentrated in a granular pattern corresponding to the Golgi zone. This feature was best revealed in intestinal glands (4,5) but could be seen also in respiratory (5,14) and lacrimal glands (unpublished). Ultrastructural studies of rabbit mammary and human intestinal and bronchial gland cells have confirmed these immunofluorescence observations; SC was found virtually alone in the Golgi elements whereas both IgA and SC occurred in more apically located vesicles (25,35,42). Moreover, it has been shown that the antigenic properties of SC signify its free form in the Golgi zone and its bound form apically in the cytoplasm (5,57).

Immunofluorescence studies of isolated human colonic columnar cells showed directly that J-chain-containing IgA and IgM had combined with SC on the surface of secretory epithelial cells in vivo (15). A large fraction of the isolated cells contained free SC and SIgA in an intracellular distribution that corresponded to the immunohistochemical results described above. The same cells showed on their surface a patchy distribution of SC in association with pIgA but only traces of free SC as revealed by staining for the I determinant (15).

SECRETORY COMPONENT AS A RECEPTOR

Cell biology studies first substantiated in the rabbit that SC is produced as a transmembrane protein family 25-30 kD larger than secreted free SC (45,53). It was proposed that the relatively large cytoplasmic extension (~15 kD) might function as an effector domain that contains information for guiding SC on its complex migration through the cell; this notion has recently been supported by experiments with mutant transmembrane SC (51). Studies of rat SC has shown that the endodomain is phosphorylated (47), which may be of importance for the receptor function. However, this possibility remains to be examined.

SC is synthesized as a core glycosylated transmembrane glycoprotein on the RER and needs 30-60 min for maturation in the Golgi apparatus (Fig. 2); the subsequent migration to the basolateral plasma membrane and the transport of pIgA to the luminal cell face apparently take place very rapidly, and this route seems to be followed by at least 90% of the SC (50,66).

Experiments with the HT-29 human adenocarcinoma cell line have demonstrated that also human SC is produced as a transmembrane precursor (~95 kD) which, by addition of sugars, becomes about 20 kD larger than the secreted form; the latter, therefore, is generated from the NH_2-terminal non-cytoplasmic five-domain portion of the molecule (49). Also transmembrane SC depends on incorporation of J chain into pIgA and pIgM to perform specific epithelial uptake of these polymers (22). Their transcytosis apparently starts with ligand-induced active clustering of the pIg-SC complexes (15,32), but the transport of SC does not depend on ligand binding (50,54).

MODULATION OF SECRETORY COMPONENT EXPRESSION

When cDNA of rabbit SC was cloned and sequenced, it was found that the five extracellular domains showed remarkable homology to each other and to Ig variable or constant regions (52). This observation is relevant to the pIg receptor function of SC; it is apparently encoded by a gene belonging to the Ig-gene superfamily that seems to be responsible for a variety of cell-surface recognition structures. The amino acid sequence of human free SC suggested the existence of similar homology regions (29). Several members of the Ig-gene superfamily show a striking responsiveness to immunoregulatory cytokines, and it was of interest to examine if the same holds true for SC.

Effect of Cytokines

Experiments were first performed with the HT-29 cell line and recombinant interferon-γ (IFN-γ). After incubation with IFN-γ for 48 h, there was a dose-dependent increase in the proportion of vital cells with surface expression of SC (68). Studies of the kinetics showed increased positivity up to day 8. Membrane expression of MHC class II determinants was induced even more briskly by IFN-γ - from nil to almost 100% positivity.

When we measured secretion of SC in supernatant fluids from cell cultures incubated for 48 h with IFN-γ, a dose-dependent relationship appeared in keeping with that observed for expression of membrane SC. Moreover, the intracellular pool of SC was increased about 50 times by IFN-γ. We also determined whether IFN-γ induced functionally active receptors and found a dose-dependent increase in the capacity of the cells to bind ^{125}I-pIgA. Because IFN-γ is secreted by T cells during immune responses, our observations suggested that activated T cells may be able to promote the external transport of pIgA and pIgM and thereby enhance the efferent limb of the secretory immune system (Fig. 3).

We have recently found that also another recombinant cytokine, namely tumor necrosis factor-α (TNF-α), can enhance the production, secretion and surface membrane expression of functional SC in a subclone of the HT-29 cell line (46). The expression of MHC class I molecules was likewise upregulated. Conversely, MHC class II molecules, which are normally absent on these cells (67), were not induced by TNF-α. This cytokine has previously been shown to induce transcriptional and posttranscriptional upregulation of constitutive MHC class I and class II molecules (55,60). It is interesting, therefore, that SC and class I molecules, which are constitutively expressed by the HT-29 cells, could be enhanced by TNF-α. The possibility remains that the modulation of SC by IFN-γ and TNF-α involves different regulatory mechanisms.

TFN-α is produced by several cell types although activated macrophages are the best recognized source (58). Such cells occur beneath all mucosal surfaces, particularly in the gastrointestinal (64) and respiratory tract (17) along with T lymphocytes. Both IFN-γ and TNF-α may thus participate in the upregulation of SC (Fig. 3). This possibility has been suggested by our immunohistochemical studies of inflammatory lesions in secretory tissues. Signs of increased SC expression and enhanced uptake of IgA was observed in fundic and antral glands in gastritis specimens with dense lymphoid infiltration (70), and increased epithelial expression of both MHC class II molecules and SC was seen in coeliac disease (63) and in inflamed salivary glands (69).

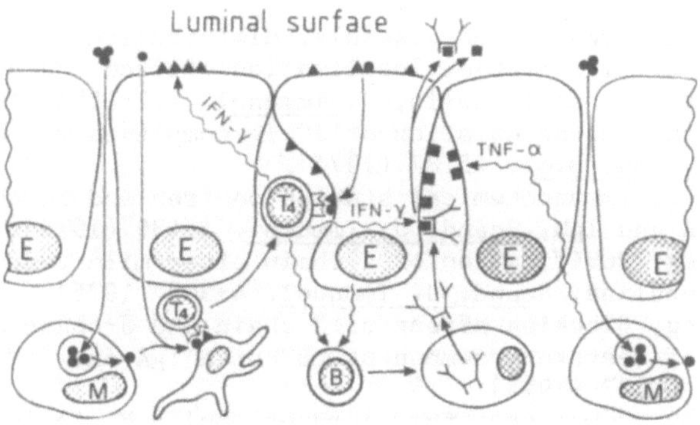

Fig. 3. Schematic depiction of putative mechanisms for upregulation of SC-dependent epithelial transport of dimeric IgA. Antigen (●) penetrates epithelial lining and becomes processed by macrophage (M) before presentation to T4 lymphocyte by dendritic cell. Antigen can also be processed in the lumen or on passage through epithelial cell (E). After activation, T4 cell migrates into epithelium and receives secondary signals by epithelial presentation of antigen or by mere autoreactive stimulation by MHC class II determinants (▲). IFN-γ released from activated T4 cell will stimulate secretory epithelium to enhanced expression of both class II determinants and SC (■). Activated macrophages release TNF-α that contributes to enhanced SC expression. In addition, B cells may be stimulated to J-chain and IgA production by both IFN-γ and undefined epithelial factors.

CONCLUSIONS

Humoral immunity at mucosal surfaces depends on complex interactions between the B-cell system and the secretory epithelium. The "lock and key" factors in this co-operation are the J chain produced by the local Ig-forming cells and the SC expressed by the epithelial cells. The obvious biological significance of the striking J-chain production shown by gland-associated immunocytes, is that IgA and IgM polymers are generated locally and become readily available for SC-mediated external transport. Cytokines such as IFN-γ and TNF-α released from activated mononuclear cells, are probably involved in immune response-associated upregulation of this epithelial transport mechanism. External translocation of pIgA and pIgM may thus be adjusted to increased local production of these polymers. Mucosal T cells and macrophages may in this way play a role in enhancing the efferent limb of the secretory immune system in addition to participating in local immune regulation conducive to the induction of J-chain-expressing B-cell clones.

REFERENCES

1. K. Bjerke and P. Brandtzaeg, Lack of relation between expression of HLA-DR and secretory component (SC) in follicle-associated epithelium of human Peyer's patches, submitted (1987).
2. P. Brandtzaeg, Structure, synthesis and external transfer of mucosal immunoglobulins, Ann. Immunol. (Inst. Pasteur) 124C:417 (1973).
3. P. Brandtzaeg, Two types of IgA immunocytes in man, Nature, New Biol. 243:142 (1973).
4. P. Brandtzaeg, Mucosal and glandular distribution of immunoglobulin components. Immunohistochemistry with a cold ethanol-fixation technique, Immunology 26:1101 (1974).
5. P. Brandtzaeg, Mucosal and glandular distribution of immunoglobulin components. Differential localization of free and bound SC in secretory epithelial cells, J. Immunol. 112:1553 (1974).
6. P. Brandtzaeg, Characteristics of SC-Ig complexes formed in vitro, Adv. Exp. Med. Biol. 45:87 (1974).
7. P. Brandtzaeg, Immunochemical studies on free and bound J chain of human IgA and IgM, Scand. J. Immunol. 4:439 (1975).
8. P. Brandtzaeg, Purification of J chain after mild reduction of human immunoglobulins, Scand. J. Immunol. 4:309 (1975).
9. P. Brandtzaeg, Blocking effect of J chain and J-chain antibody on the binding of secretory component to human IgA and IgM, Scand. J. Immunol. 4:837 (1975).
10. P. Brandtzaeg, Human secretory immunoglobulin M. An immunochemical and immunohistochemical study, Immunology 29:559 (1975).
11. P. Brandtzaeg, Human secretory component - IV. Aggregation and fragmentation of free secretory component, Immunochemistry 12:877 (1975).
12. P. Brandtzaeg, Complex formation between secretory component and human immunoglobulins related to their content of J chain, Scand. J. Immunol. 5:411 (1976).
13. P. Brandtzaeg, Human secretory component. VI. Immunoglobulin-binding properties, Immunochemistry 14:179 (1977).
14. P. Brandtzaeg, Immunohistochemical studies on various aspects of glandular immunoglobulin transport in man, Histochem. J. 9:553 (1977).
15. P. Brandtzaeg, Polymeric IgA is complexed with secretory component (SC) on the surface of human intestinal epithelial cells, Scand. J. Immunol. 8:39 (1978).
16. P. Brandtzaeg, Immune functions of human nasal mucosa and tonsils in health and disease, pp. 28-95, in: "Immunology of the Lung and Upper Respiratory Tract," J. Bienenstock, ed., McGraw-Hill, New York (1984).
17. P. Brandtzaeg, Cells producing immunoglobulins and other immune factors in human nasal mucosa, Protids Biol. Fluids 32:363 (1985).
18. P. Brandtzaeg, Research in gastrointestinal immunology: state of the art, Scand. J. Gastroenterol. 20 (Suppl. 114):137 (1985).

19. P. Brandtzaeg, Role of J chain and secretory component in receptor-mediated glandular and hepatic transport of immunoglobulins in man, Scand. J. Immunol. 22:111 (1985).
20. P. Brandtzaeg and K. Baklien, Immunohistochemical studies of the formation and epithelial transport of immunoglobulins in normal and diseased human intestinal mucosa, Scand. J. Gastroenterol 11 (Suppl. 36):1 (1976).
21. P. Brandtzaeg and F. R. Korsrud, Significance of different J chain profiles in human tissues: generation of IgA and IgM with binding site for secretory component is related to the J chain expressing capacity of the total local immunocyte population, including IgG and IgD producing cells, and depends on the clinical state of the tissue. Clin. Exp. Immunol. 58:709 (1984).
22. P. Brandtzaeg and H. Prydz, Direct evidence for an integrated function of J chain and secretory component in epithelial transport of immunoglobulins, Nature 311:71 (1984).
23. P. Brandtzaeg and T. O. Rognum, Evaluation of nine different fixatives. 1. Preservation of immunoglobulin isotypes, J chain, and secretory component in human tissues, Pathol. Res. Pract. 179:250 (1984).
24. P. Brandtzaeg, I. Fjellanger, and S. T. Gjeruldsen, Human secretory immunoglobulins. I. Salivary secretions from individuals with normal or low levels of serum immunoglobulins, Scand. J. Haematol. Suppl. 12, pp. 1-83 (1970).
25. W. R. Brown, Y. Isobe, and P. K. Nakane, Studies on translocation of immunoglobulins across intestinal epithelium. II. Immunoelectron-microscopic localization of immunoglobulins and secretory component in human intestinal mucosa, Gastroenterology 71:985 (1976).
26. J. R. Carlson and R. L. Owen, Structure and functional role of Peyer's patches, pp. 21-40, in: "Immunopathology of the Small Intestine," M. N. March, ed., John Wiley & Sons, Chichester (1987).
27. D. L. Delacroix, "The human immunoglobulin A system: its vascular compartment," pp.1-111, Thesis, Universite Catholique de Louvain, European Medical Press, Brugge (1985).
28. D. L. Delacroix and J.-P. Vaerman, Function of the human liver in IgA homeostasis in plasma, Ann. N.Y. Acad. Sci. 409:383 (1983).
29. H. Eiffert, E. Quentin, J. Decker, S. Hillemeir, M. Hufschmidt, D. Klingmuller, M. H. Weber, and N. Hilschmann, Die Primärstruktur der menschlichen freien Sekretkomponente und die Anordnung der Disulfidbrücken, Hoppe-Seyler's Z. Physiol. Chem. 365:1489 (1984).
30. T. Eskeland and P. Brandtzaeg, Does J chain mediate the combination of 19S IgM and dimeric IgA with the secretory component rather than being necessary for their polymerization?, Immunochemistry 11:161 (1974).
31. S. Frutiger, G. J. Hughes, W. C. Hanly, M. Kingzette, and J. C. Jaton, The amino-terminal domain of rabbit secretory component is responsible for noncovalent binding to immunoglobulin A dimers, J. Biol. Chem. 261:16673 (1986).
32. R. Gebhart and H. Robenek, Ligand-dependent redistribution of the IgA receptor on cultured rat hepatocytes and its disturbance by cytochalasin B, J. Histochem. Cytochem. 35:301 (1987).
33. C. Geneste, S. Iscaki, R. Mangalo, and J. Pillot, Both Fcα domains of human IgA are involved in in vitro interaction between secretory component and dimeric IgA, Immunol. Letters 13:221 (1986).
34. C. Geneste, R. Mangalo, S. Iscaki, and J. Pillot, Human secretory component. IV. Antigenic regions involved in in vitro binding to dimeric IgA, Immunol. Letters 13:121 (1986).

35. M. R. Goodman, D. W. Link, W. R. Brown, and P. K. Nakane, Ultrastructural evidence of transport of secretory IgA across bronchial epithelium, Am. Rev. respir. Dis. 123:115 (1981).
36. A. O. Grubb, Quantitation of J chain in human biological fluids by a simple immunochemical procedure, Acta. Med. Scand. 204:453 (1978).
37. H. Haimoto, H. Nagura, M. Imaizumi, K. Watanabe, and S. Iijima, Immunoelectron-microscopic study on the transport of secretory IgA in the lower respiratory tract and alveoli, Virchows Arch. A (Pathol. Anat.) 404:369 (1984).
38. L. Å. Hanson and P. Brandtzaeg, The mucosal defense system, pp. 137-162, in: "Immunological Disorders in Infants and Children," E. R. Stiehem and W. A. Fulginiti, eds., Saunders, Philadelphia (1980).
39. J. Jos, F. Labbe, B. Geny, and C. Griscelli, Immunoelectron-microscopic localization of immunoglobulin A and secretory component in jejunal mucosa from children with coeliac disease, Scand. J. Immunol. 9:441 (1979).
40. F. R. Korsrud and P. Brandtzaeg, Characterization of epithelial elements in human major salivary glands by functional markers: localization of amylase, lactoferrin, lysozyme, secretory component, and secretory immunoglobulins by paired immunofluorescence staining, J. Histochem. Cytochem. 30:657 (1982).
41. M. E. Koshland, The coming of age of the immunoglobulin J chain, Ann. Rev. Immunol. 3:425 (1985).
42. J. P. Kraehenbuhl, L. Racine, and R. E. Galardy, Localization of secretory IgA, secretory component, and α heavy chain in the mammary gland of lactating rabbits by immunoelectron microscopy, Ann. N.Y. Acad. Sci. 254:190 (1975).
43. L. Kühn and J. P. Kraehenbuhl, Role of secretory component, a secreted glycoprotein, in the specific uptake of IgA dimer by epithelial cells, J. Biol. Chem. 254:11072 (1979).
44. L. Kühn and J. P. Kraehenbuhl, Interaction of rabbit secretory component with rabbit IgA dimer, J. Biol. Chem. 254:11066 (1979).
45. L. C. Kühn and J. P. Kraehenbuhl, The membrane receptor for polymeric immunoglobulin is structurally related to secretory component. Isolation and characterization of membrane secretory component from rabbit liver and mammary gland, J. Biol. Chem. 256:12490 (1981).
46. D. Kvale, D. Lövhaug, L. M. Sollid, and P. Brandtzaeg, Tumour necrosis factor-α upregulates expression of secretory component, the epithelial receptor for polymeric immunoglobulins, submitted (1987).
47. J. M. Larkin, E. S. Sztul, and G. E. Palade, Phosphorylation of the rat hepatic polymeric IgA receptor, Proc. Natl. Acad. Sci. USA 83:4759 (1986).
48. J. Mestecky and J. R. McGhee, Immunoglobulin A (IgA): molecular and cellular interactions involved in IgA biosynthesis and immune response, Advan. Immunol. 40:153 (1987).
49. K. E. Mostov and G. Blobel, A transmembrane precursor of secretory component. The receptor for transcellular transport of polymeric immunoglobulins, J. Biol. Chem. 257:11816 (1982).
50. K. E. Mostov and J. L. Deitcher, Polymeric immunoglobulin receptor expressed in MDCK cells transcytoses IgA, Cell 46:613 (1986).
51. K. E. Mostov, A. de Bruyn Kops, and D. L. Deitcher, Deletion of the cytoplasmic domain of the polymeric immunoglobulin receptor prevents basolateral localization and endocytosis, Cell 47:359 (1986).
52. K. E. Mostov, M. Friedlander, and G. Blobel, The receptor for transepithelial transport of IgA and IgM contains multiple immunoglobulin-like domains, Nature 308:37 (1984).

53. K. E. Mostov, J. P. Kraehenbuhl, and G. Blobel, Receptor-mediated transcellular transport of immunoglobulin: synthesis of secretory component as multiple and larger transmembrane forms, Proc. Natl. Acad. Sci. USA 77:7257 (1980).
54. H. Nagura, P. K. Nakane, and W. R. Brown, Secretory component in immunoglobulin deficiency: an immunoelectron microscopic study of intestinal epithelium, Scand. J. Immunol. 12:359 (1980).
55. K. Pfizenmaier, P. Scheurich, C. Schluter, and M. Kronke, Tumor necrosis factor enhances HLA-A, B, C, and HLA-DR gene expression in human tumor cells, J. Immunol. 138:975 (1987).
56. R. S. H. Pumphrey, Computer models of the human immunoglobulins. Binding sites and molecular interactions, Immunol. Today 7:206, (1986).
57. M. E. Poger and M. E. Lamm, Localization of free and bound secretory component in human intestinal epithelial cells. A model for the assembly of secretory IgA, J. Exp. Med. 139:629 (1974).
58. N. H. Ruddle, Tumor necrosis factor and related cytokines. Immunol. Today 8:129 (1987).
59. H. Saito, T. Kasajima, and H. Nagura, An immunocytochemical study on secretory mechanisms of IgA in human pancreas, Acta Pathol. Jpn. 35:87 (1985).
60. P. Scheurich, M. Kronke, C. Schluter, U. Ucer, and K. Pfizenmaier, Noncytocidal mechanisms of action of tumor necrosis factor-α on human tumor cells: enhancement of HLA gene expression synergistic with interferon-γ, Immunobiology 172:291 (1986).
61. J. M. Schiff, K. J. Dorrington, and B. J. Underdown, Secretory component and immunoglobulin transport, 4th International Congress of Immunology, Abstr. No. 7.5.29 (1980).
62. J. M. Schiff, M. M. Fisher, and B. J. Underdown, Secretory component as the mucosal transport receptor: separation of physicochemically analogous human IgA fractions with different receptor-binding capacities, Molec. Immunol. 23:45 (1986).
63. H. Scott, P. Brandtzaeg, B. G. Solheim, and E. Thorsby, Relation between HLA-DR-like antigens and secretory component (SC) in jejunal epithelium of patients with coeliac disease or dermatitis herpetiformis, Clin. Exp. Immunol. 44:233 (1981).
64. W. S. Selby, L. W. Poulter, S. Hobbs, D. P. Jewell, and G. Janossy, Heterogeneity of HLA-DR.positive histiocytes in human intestinal lamina propria: a combined histochemical and immunohistological analysis, J. Clin. Pathol. 36:379 (1983).
65. D. J. Socken and B. J. Underdown, Comparison of human bovine and rabbit secretory component-immunoglobulin interactions, Immunochemistry 15:499 (1978).
66. R. Solari and J.-P. Kraehenbuhl, Biosynthesis of the IgA antibody receptor: a model for the transepithelial sorting of a membrane glycoprotein, Cell 36:61 (1984).
67. L. M. Sollid, G. Gaudernack, G. Markussen, D. Kvale, P. Brandtzaeg, and E. Thorsby, Induction of various HLA class II molecules in a human colonic adenocarcinoma cell line, Scand. J. Immunol. 25:175 (1987).
68. L. M. Sollid, D. Kvale, P. Brandtzaeg, G. Markussen, and E. Thorsby, Interferon-γ enhances expression of secretory component, the epithelial receptor for polymeric immunoglobulins, J. Immunol. 138:4303 (1987).
69. P. S. Thrane, L. M. Sollid, P. Brandtzaeg, and H. R. Haanes, Increased expression of secretory component (SC) and HLA-DR in salivary gland epithelium from patients with Sjögren's syndrome, submitted (1987).

70. K. Valnes, P. Brandtzaeg, L. Elgjo, and R. Stave, Specific and nonspecific humoral defense factors in the epithelium of normal and inflamed gastric mucosa. Immunohistochemical localization of immunoglobulins, secretory component, lysozyme, and lactoferrin, Gastroenterology 86:402 (1984).
71. J. Zikan, J. Novotny, T. L. Trapane, M. E. Koshland, D. W. Urry, J. C. Bennett, and J. Mestecky, Secondary structure of the immunoglobulin J chain, Proc. Natl. Acad. Sci. USA 82:5905 (1985).

LYMPHOCYTE MIGRATION MOLECULES

Sirpa Jalkanen*, Robert Bargatze+, Markku Jalkanen•, David Lewinsohn+, Philip Streeter+, Ellen Lakey+, and Eugene C. Butcher+

Department of Medical Microbiology* and Biochemistry•, Turku University, Turku, Finland, and Department of Pathology+, Stanford University Medical Center, Stanford University, Stanford, California 94305

INTRODUCTION

Recirculating lymphocytes leave the blood by selectively binding to specialized high endothelial venules (HEV) in the lymphoid organs of the body[1]. This binding is mediated by specific lymphocyte surface molecules called homing receptors[2]. Functionally distinct homing receptors control lymphocyte migration to peripheral lymph nodes, to mucosa-associated lymphatic tissues (appendix, Peyer's patches) or to inflamed joint tissue (synovium)[3,4]. Most virgin lymphocytes appear to express all three different types of receptors and can, therefore, migrate throughout the body in search of antigen. Following antigenic activation the lymphocytes go through a nonmigratory phase after which many of them, as memory and effector cells, appear to express a single homing receptor class that controls their migration to particular tissues of the body similar to the site where they first encountered their cognate antigen. Thus, organ-specific homing receptors are thought to play an essential role in determining the characteristics of local immune responses[2,5].

In this paper we will review our studies on the biochemical structure of lymphocyte homing receptors and their relation to other leukocyte receptors for endothelium. Also, we will briefly discuss preliminary studies of tissue-specific endothelial cell surface molecules involved in lymphocyte homing.

AN 85-95 kDa CLASS OF PUTATIVE GLYCOPROTEIN HOMING RECEPTORS

Several monoclonal antibodies against mouse, rat and human lymphocyte surface molecules involved in lymphocyte-endothelial cell interaction have been produced[6-9]. MEL-14, against a putative mouse homing receptor for peripheral lymph node HEV, was the first antibody isolated, and it indicated the existence of specific receptor molecules mediating lymphocyte binding to HEV[6]. MEL-14 inhibits mouse lymphocyte homing to peripheral lymph nodes, and precipitates an 80-90 kDa glycoprotein from the lymphocyte surface[6]. In later studies MEL-14 has also been shown to react with 90 kDa putative homing receptors in humans, as defined by monoclonal antibody Hermes-1. In high concentrations MEL-14 inhibits human lymphocyte binding to peripheral lymph node HEV[4,10]. Two dimensional gel analyses in Figure 1 show the remarkable similarity between the human Hermes-1 antigen and the mouse MEL-14 antigen, suggesting substantial conservation of the structure of these putative homing receptors during evolution.

In addition to Hermes-1, itself a non-blocking antibody, we have produced a monoclonal antibody, Hermes-3, that precipitates the same 90 kDa glycoprotein as Hermes-1[10] but binds to a distinct epitope and functionally inhibits human lymphocyte binding to mucosal HEV. Furthermore, a polyclonal antiserum (anti-gp90) against the affinity purified Hermes-1 antigen inhibits lymphocyte binding to peripheral lymph node, mucosal and synovial HEV. The data gained from these functional blocking studies and immunoprecipitation analyses suggest that human lymphocyte homing receptors for peripheral lymph nodes, mucosal lymphoid tissues and inflamed synovium belong to a family of related but functionally independent molecules. Theoretically it is possible that the Hermes-1 defined 90 kDa glycoprotein class is merely associated both physically and functionally with the actual, as yet unidentified, homing receptor molecules. However, this seems unlikely because no associated molecules have been identified in biochemical/immunoprecipitation analyses.

In the rat a monoclonal antibody capable of blocking lymphocyte binding to mucosal HEV precipitates a molecule similar in molecular weight (80 kD) to the human Hermes-1 antigen[8]. However, Rasmussen et al.[7] have reported a monoclonal antibody, A.11, that inhibits rat lymphocytes binding to lymph node but not to mucosal HEV, and precipitates several molecular species, of 40, 63 and 135 kDa. Their findings suggest that other molecules, in addition to the gp80-90 class, might be involved in lymphocyte binding to HEV. Strong evidence of involvement of the LFA-1 molecular complex in a non-organ-specific fashion in lymphocyte-HEV recognition is presented in this volume (see Pals et al, and Hamann et al).

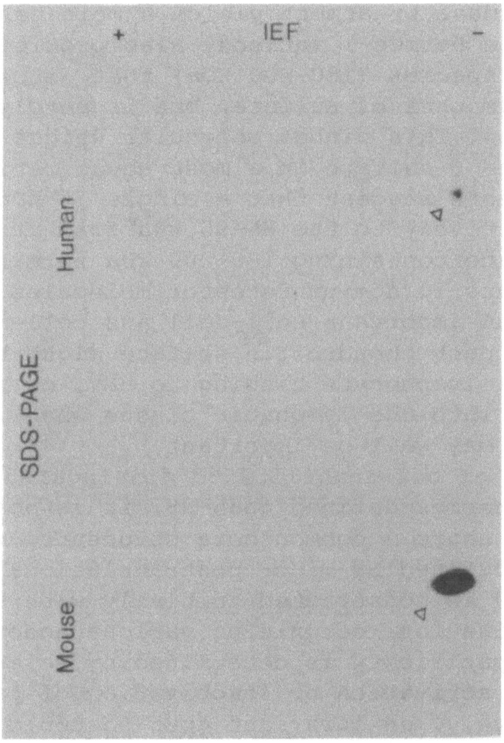

Fig. 1 Two-dimensional gel analyses of the Hermes-1 antigen from human peripheral blood lymphocytes and of the MEL-14 antigen from mouse lymphocytes. Antigens were immunoprecipitated from lysates of surface-iodinated cells. The approximate pI of the receptors is 4.2

BIOCHEMICAL PROPERTIES OF THE HERMES-DEFINED GLYCOPROTEIN

Hermes-1 and anti-gp90 serum were used for biochemical characterization of the Hermes antigen on human lymphocytes. No discrete variants corresponding to distinct receptor specificities were found. Thus, several functional receptor specificities are associated with a highly glycosylated, acidic (pI 4.2), 85-95 kDa molecule or molecules. The gp85-95 contain 3-4 N-linked glycans as well as O-linked carbohydrates[11]. The acidity of the molecules is at least partly due to sialic acid, because neuraminidase treatment yields a more basic pI. Interestingly, the Hermes-1 antibody also precipitates a higher molecular weight species (180-200 kDa) that incorporates relatively high amounts of sulfate, but is hardly visible after surface iodination. This higher molecular weight form contains chondroitin-4- and 6-sulfate in almost equal ratio[11]. Metabolic labeling experiments suggest that a single 76 kDa precursor is processed alternatively to the 85-95 kDa form or to the chondroitin sulfate-containing 180-200 kDa form. The role of chondroitin sulfate in homing receptor molecules is not known at present, but as an important cell-cell and cell-matrix adhesive interaction component chondroitin sulfate might be important either in initial lymphocyte binding to HEV, or in subsequent lymphocyte entry into the lymphatic tissue when interactions with matrix molecules may well be important.

On the basis of our inability to distinguish discrete variants of the Hermes-defined gp85-95, it is possible that human homing receptors share a common core structure, with HEV-specificity defined by minor postranslational modifications; or that each gp90 is constructed initially with several independent domains for recognizing various endothelial cell types, and the specificity is determined by selective postranslational activation or inactivation of particular binding sites. Alternatively, an invariant gp85-95 could be one component of a complex of cell surface molecules that together define receptor specificity.

The studies of Siegelman et al[12] suggest that the mouse lymphocyte surface gp90 recognized by MEL-14 contain one or more covalently linked ubiquitin molecules. Indeed, St. John et al[13] presented evidence that ubiquitin is involved in formation of the epitope recognized by MEL-14. The crossreactivity of MEL-14 with both 90 kDa and 180-200 kDa forms of the Hermes antigen therefore suggests that these putative human homing receptors are ubiquinated as well[11]. Ubiquitin is frequently found in the nucleus and cytoplasm of most or all eukaryotic cells, and it has been reported to be involved in degradation of cytoplasmic and nuclear proteins[14-16]. Therefore, ubiquitin might play some role in the cleavage of homing receptors, perhaps by specific proteases responsible for release of migrating lymphocytes from endothelial cells during entry into lymphoid organs.

METABOLIC REQUIREMENTS FOR LYMPHOCYTE BINDING TO HEV

Metabolic requirements for homing receptor-mediated interaction were assayed by using an in vitro HEV assay[17,18] These studies showed that human lymphocyte binding to HEV requires live, metabolically active cells, because formalin fixed cells could not bind to HEV, and sodium azide inhibited lymphocyte binding to HEV. However, the effect of sodium azide was reversible, and the cells regained their binding ability after the metabolic inhibitor was washed off[18]. Calcium is essential for the binding of human lymphocytes, whereas magnesium does not seem to have any marked effect on the binding as shown in Table 1. These results are similar to those reported for murine systems[19], and in combination with the maintenance of organ-specific lymphocyte-HEV recognition in xenogeneic combinations (N. Wu, personal communication), support the general conservation of lymphocyte-HEV binding mechanisms during evolution.

Table 1. Effect of Ca^{++} and Mg^{++} on lymphocyte binding to HEV

Pretreatment (30')[a] of the cells	Present during assay	Relative adherence ratio[b]
none	0.5 mM Edta	0
0.5 mM Edta	" + 0.75 mM Mg	0.05
0.5 mM Edta	" + 1.0 mM Mg	0.08
0.5 mM Edta	" + 1.5 mM Mg	0.12
0.5 mM Edta	" + 3.0 mM Mg	0.07
0.5 mM Edta	" + 0.75 mM Ca	0.49
0.5 mM Edta	" + 1.0 mM Ca	0.93
0.5 mM Edta	" + 1.5 mM Ca	1.09
0.5 mM Edta	" + 3.0 mM Ca	0.98
0.5 mM Edta	" + 1.5 mM Mg + 3.0 mM Ca	unity
none	1.0 mM Mg + 2.5 mM Ca	1.7

[a]Peripheral blood lymphocytes were pretreated with calcium and magnesium free HBSS medium containing 0.5 mM EDTA, after which the indicated amount of $MgCl_2$, $CaCl_2$, or both were added.
[b]The average number of sample cells bound per HEV divided by the average number of control cells (EDTA pretreated cells supplemented with physiological amount of Ca and Mg) per HEV.

RELATED GLYCOPROTEINS INVOLVED IN LEUKOCYTE BINDING TO ENDOTHELIUM

Lymphocyte homing receptors seem to be members of a common system of leukocyte receptors for endothelium, in that Hermes-1 and MEL-14-defined molecules are also expressed on nonlymphoid cells such as neutrophils, monocytes and eosinophils[20]. In the case of neutrophils these receptors are roughly 10 kDa larger than on lymphocytes, and they have been shown to mediate lymphocyte extravasation to sites of inflammation[20]. Direct structure/function comparisons between the receptor molecules on lymphoid and nonlymphoid cells are required to determine if they play a role in functional differences in endothelial cell interactions with different leukocyte classes.

For example, expression of the neutrophil receptor molecules for endothelium allows neutrophils to bind to HEV of noninflamed lymphoid tissues in an in vitro frozen section assay, but it is not sufficient for neutrophils to enter noninflamed lymphoid organs *in vivo*[20]. These findings indicate that receptors for organ-specific endothelial determinants may be important for the binding of leukocytes to endothelium in sites of acute and chronic inflammation, but that some additional signals or molecules (or unsuspected subtlety in the homing receptor class) may be needed for subsequent neutrophil extravasation. One of these additional molecules is probably the LFA-1 complex, since antibodies to the LFA-1/Mac-1 complex partially inhibit neutrophil interactions with endothelium *in vitro*[21] and *in vivo*[22].

POSSIBLE ENDOTHELIAL CELL LIGANDS FOR LYMPHOCYTE HOMING RECEPTORS

Rosen and his coworkers[23,24] have studied the possible involvement of different carbohydrates in lymphocyte-HEV interaction, and found that mannose-6-phosphate at mM concentrations, as well as mannose-6-phosphate-rich mannan core structure from *Hanensula holstii*, block mouse lymphocyte binding to lymph node HEV. The involvement of mannose-6-phosphate in lymphocyte binding is further supported by the fact that the binding of mannan to lymphocytes is specifically inhibited by MEL-14[25]. Sialic acid may also play a role in lymphocyte binding to HEV, in that neuraminidase treatment of lymph node and Peyer's patch sections completely blocks lymphocyte binding to lymph node HEV, without effecting binding to Peyer's patch HEV[24]. Lymphocyte binding to Peyer's patch HEV might also be dependent on carbohydrate moieties, since periodate treatment of the frozen sections prevents lymphocyte binding to both Peyer's patch and peripheral lymph node HEV[24]. These findings suggest that the lymphocyte receptors may act as mammalian lectins interacting with specific oligosaccharide structures on high endothelium.

Recently monoclonal antibodies against mouse and human HEV differentiation antigens have been produced[26-28]. One of these monoclonal antibodies, MECA-367, seems to recognize HEV determinants that are involved in lymphocyte migration in the mouse[27] (see Streeter et al, this volume). MECA-367 stains brightly all HEV in Peyer's patches. Venules in the lamina propria are also stained, but less intensely. In contrast, HEV in peripheral lymph nodes are negative. In an in vitro assay MECA-367 inhibits lymphocyte binding to Peyer's patch HEV, but does not have any effect on their binding to peripheral lymph node HEV. Furthermore, intravenous administration of MECA-367 blocks lymphocyte migration to Peyer's patches *in vivo*[27]. Thus, the MECA-367 antigen may be the ligand for the mouse homing receptor for mucosa-associated lymphatic tissues. Structural analyses of the 58-66 kDa molecule recognized by MECA-367 should help to resolve the specific elements involved in lymphocyte binding to HEV.

ACKNOWLEDGEMENTS

This work was supported by NIH grants AI-19957 and GM37734, the Yrjö Jahnsson Foundation and the Finnish Cancer Union, and the Veterans Administration. DL is a predoctoral fellow of the Cancer Biology Program, supported by USPHS grant 5T32 CA09302-09. PS is a fellow of the American Cancer Society, California Division, and EL is a fellow of the American Cancer Society, New York. SJ was a Jane Coffin Childs fellow, Stanford University 1984-1986. ECB is an Established Investigator of the American Heart Association.

REFERENCES

1. J.L. Gowans, and E.J. Knight, The route of recirculation of lymphocytes in the rat, Proc. Roy. Soc. Ser. B. 159:257 (1964).
2. E.C. Butcher, The regulation of lymphocyte traffic, Curr. Top. Microbiol. Immunol. 128:85 (1986).
3. E.C. Butcher, R.G. Scollay, and I.L. Weissman, Organ specificity of lymphocyte migration: Mediation by highly selective lymphocyte interaction with organ-specific determinants on high endothelial venules, Eur. J. Immunol. 10:556 (1980).
4. S. Jalkanen, A.C. Steere, R.I Fox, and E.C. Butcher, A distinct endothelial cell recognition system controlling lymphocyte traffic into inflamed synovium, Science 233:556 (1986).
5. S. Jalkanen, R.A. Reichert, W.M. Gallatin, R.F. Bargatze, I.L. Weissman, and E.C. Butcher, Homing receptors and the control of lymphocyte migration, Immunol. Rev. 91:39 (1986).
6. W.M. Gallatin, I.L. Weissman, and E.C. Butcher, A cell surface molecule involved in organ-specific homing of lymphocytes. Nature 303:30 (1983).
7. R.A. Rasmussen, Y-H Chin, J.J. Woodruff, and T.G. Easton, Lymphocyte recognition of lymph node high endothelium. VII. Cell surface proteins involved in adhesion defined by monoclonal anti-HEBFLN (A.11) antibody, J. Immunol. 135:19 (1985).

8. Y-H. Chin, R.A. Rasmussen, J.J. Woodruff, and T.G. Easton, A monoclonal anti-HEBF antibody with specificity for lymphocyte surface molecules mediating adhesion to Peyer's patch high endothelium of the rat, J. Immunol. 136:1 (1986).
9. S. Jalkanen, R. Bargatze, L. Herron, and E.C. Butcher, A lymphoid cell surface glycoprotein involved in endothelial recognition and lymphocyte homing in man, Eur. J. Immunol. 16:1195 (1986).
10. S. Jalkanen, R.F. Bargatze, J. de los Toyos, and E.C. Butcher. Lymphocyte recognition of high endothelium: antibodies to distinct epitopes of an 85-95 kD glycoprotein antigen differ- entially inhibit lymphocyte binding to lymph node, mucosal, or synovial endothelial cells, J. Cell. Biol. 105:983 (1987).
11. S. Jalkanen, R. Bargatze, M. Jalkanen, M. Tammi, and E.C. Butcher, Biochemical properties of glycoproteins involved in lymphocyte recognition of high endothelial venules in man, submitted.
12. M. Siegelman, M. Bond, W.M. Gallatin, T. St. John, H. Smith, V. Fried, and I.L. Weissman, A putative lymphocyte homing receptor is a ubiquitinated branched-chain glycoprotein: Additional cell surface proteins also appear ubiquitinated, Science 231:823 (1986).
13. T. St. John, W.M. Gallatin, M. Siegelman, H.T. Smith, V.A. Fried, and I.L. Weissman, Expression cloning of a lymphocyte homing receptor cDNA: Ubiquitin is the reactive species, Science 231:845 (1986).
14. A. Hershko, Ubiquitin: Roles in protein modification and breakdown, Cell 34:11 (1983).
15. A. Ciechanover, D. Finley, and A. Varshavsky, The ubiquitin-mediated proteolytic pathway and mechanisms of energy dependent intracellular protein degradation, J. Cell. Biochem. 24: 27 (1984).
16. A. Bachmeir, D. Finley, and A. Varshavsky, In vivo half-life of a protein is a function of its amino-terminal residue, Science 234:179 (1986).
17. H.B. Stamper Jr, and J.J. Woodruff, Lymphocyte homing into lymph nodes: In vitro demonstration of the selective affinity of recirculating lymphocytes for high endothelial venules, J. Exp. Med. 144:828 (1976).
18. S. Jalkanen, and E.C. Butcher, In vitro analysis of the homing properties of human lymphocytes: Developmental regulation of functional receptors for high endothelial venules, Blood 66:577 (1985).
19. J.J. Woodruff, M. Katz, E. Lucas, and H.B. Stamber Jr, An in vitro model of lymphocyte homing.II.Membrane and cytoplasmic events involved in lymphocyte adherence to specialized high-endothelial venules of lymph nodes, J. Immunol.119:1603 (1977).
20. D.M. Lewinsohn, R.F. Bargatze, and E.C. Butcher, Leukocyte-endothelial cell recognition: evidence of a common molecular

mechanism shared by neutrophils, lymphocytes, and other leukocytes, J. Immunol. 138:4313 (1987).

21. J.M. Harlan, P.D. Killen, F.M. Senecal, B.R. Schwartz, E.K. Yee, R.F. Taylor, P.G. Beatty, T.H. Price, and H.D. Ochs, The role of neutrophil membrane glycoprotein GP-150 in neutrophil adherence to endothelium in vitro, Blood 69:338 (1985).
22. K.E. Arfors, C. Lundberg, L. Lindbom, K. Lundberg, P.G. Beatty, A monoclonal antibody to the membrane glycoprotein complex CD18 inhibits polymorphonuclear leukocyte accumulation and plasma leakage in vivo, Blood 69:338 (1987)
23. L.M. Stoolman, T.S. Tenforde, and S.D. Rosen, Phosphomannosyl receptors may participate in the adhesive interaction between lymphocytes and high endothelial venules, J. Cell Biol. 99: 1535 (1984).
24. S.D. Rosen, M.S. Singer, T.A. Yednock, and L.M. Stoolman, Involvement of sialic acid on endothelial cells in organ-specific lymphocyte recirculation, Science 228:1005 (1985).
25. T.A. Yednock, E.C. Butcher, L.M. Stoolman, S.D. Rosen, Lymphocyte homing receptors: relationship between MEL-14 antigen and a carbohydrate-binding receptor, J. Cell. Biol. 104:725 (1987).
26. A.D. Duijvestijn, A. Schreiber, and E.C. Butcher, Interferon-gamma induces an antigen specific for endothelium involved in lymphocyte homing, PNAS 83:9114 (1986).
27. P. Streeter, E. Lakey, R.F. Bargatze, B.T. Rouse, and E.C. Butcher, A tissue-specific endothelial cell molecule involved in lymphocyte homing to mucosal tissues, submitted.
28. A.D. Duijvestijn, E. Horst, S.T. Pals, B.T. Rouse, A. Steere, L. Picker, C.J.M. Meijer, and E.C. Butcher, High endothelial differentiation in human lymphoid and inflammatory tissue defined by monoclonal antibody HECA-452. Am. J. Path., in press.

THE BINDING OF ANTIGEN PRESENTING CELLS TO T LYMPHOCYTES

Ralph M. Steinman and Kayo Inaba

The Rockefeller University and Irvington House Institute, N.Y. N.Y.10021, and Dep't of Zoology, Kyoto University, Kyoto 606

Clonal selection theory predicted, and subsequent experiments have shown, that B and T lymphocytes express individualized receptors that recognize specific antigens. The receptors are heterodimers in which the amino terminal domains of each chain are highly variable and suitably diverse for antigen recognition.

Ig receptors on B cells recognize epitopes on soluble molecules and small particulates. Therefore it is easy to visualize how antigen gains access to, and selects, specific clones of B lymphocytes. It is less easy to understand how antigen selects specific clones of T lymphocytes. The antigen is not free in solution but on the surface of other antigen presenting cells or APC[1]. In transplantation immunity, which is the most powerful T cell-mediated immune response, the antigens are located on transmembrane class I and II MHC products. For other antigens, the T cell recognizes epitopes that most likely are created by the association of antigen and self-MHC molecules. How do MHC molecules on a presenting cell and T cell receptor (TcR) heterodimers on a lymphocyte find each other? Are there mechanisms other than antigen presentation and recognition that temporarily bind APC to T cells to facilitate the alignment of complementary MHC and TcR molecules? How are interactions generated _in situ_ where APC must find infrequent clones of antigen-specific lymphocytes?

These questions are highlighted by experiments on the properties of dendritic cells _in vitro_ and _in vivo_. In several responses, immunization is greatly enhanced by administering the antigen on dendritic cells, a distinct subset of APC (1). One way in which dendritic cells are specialized as APC is the capacity to form stable contacts with antigen-specific T lymphocytes. Two tissue culture systems will be reviewed. The first is the binding of dendritic cells to antigen-specific T cells to form multicellular aggregates early in the _afferent limb_ of immunity. The aggregates are the site for T cell-sensitization in the MLR (2,3) and T-

---

[1]Abbreviations: APC, Antigen presenting cell; TcR, T cell receptor for antigen; MLR, Mixed leukocyte reaction; LC, Langerhans cells; MHC, Major histocompatibility complex; TcR, T cell receptor for antigen/MHC; IL-2, Interleukin-2; GM-CSF, Granulocyte-macrophage colony stimulating factor.

dependent antibody response (4-6). By "sensitization," we mean the development of resting T cells into lymphoblasts which a) respond to T cell growth factors and b) produce lymphokines or in the case of killer cells, the proteins required for target cell lysis. A second cell-cell contact system is the binding of dendritic cells and other APC to freshly sensitized T lymphoblasts (7). This type of binding may be responsible for the diverse manifestations of the efferent limb of T cell-mediated immunity. Examples are the capacities of $CD4^+$ T blasts to help antibody-secreting B cells, and $CD8^+$ T blasts to kill specific targets. We will then try to relate tissue culture findings to APC-T interactions in situ.

INTERACTIONS OF UNPRIMED LYMPHOCYTES AND DENDRITIC CELLS

Immune responses in tissue culture frequently occur in cell aggregates

When antigen and small numbers of dendritic cells are applied to suspensions of lymphocytes, spherical cellular aggregates 5-20 cells in diameter are evident within the first day of culture. The aggregates can be separated from nonclustered cells by velocity sedimentation. One then monitors the composition and function of isolated clusters, nonclusters, and mixtures of the two fractions.

In several responses, immunity develops almost exclusively in clusters. The clusters function when cultured under sparse conditions ($2\text{-}3 \times 10^4$ cells /6 mm diameter microtest well). Nonclusters have little or no activity, even at higher cell doses ($1\text{-}4 \times 10^5$/well). Mixtures of clusters and nonclusters are comparable or a little more active than clusters alone. The role of clusters as a microenvironment for T cell-mediated immunity has been evident in T cell proliferation to soluble (5) and alloantigens (2,3), production of several lymphokines (3,8), development of antibody-secreting cells to foreign erythrocytes and hapten-carrier conjugates (4,6), and generation of cytolytic T lymphocytes (3,9).

Identification of dendritic cells in clusters

The movement of dendritic cells into clusters has been monitored in several ways. One is to identify dendritic cells in cluster and noncluster fractions with cytologic and cell surface markers (2-4). Stellate cells that stain brightly with anti-class II MHC antibodies, but not with anti-monocyte or anti-lymphocyte mAb, are found primarily in the clusters. A second approach is to label the applied dendritic cell inoculum with a vital dye. Fluorescent carbocyanine dyes remain dendritic cell-associated for days and have allowed us to monitor dendritic cell numbers more easily and effectively. Again most of the dendritic cells are in the clusters (10,11). A third approach is underway. It is to observe live "unlabelled" dendritic cells by phase contrast and time lapse. Dendritic cells, because of their size and very active cell surface, are easy to identify. One then can observe interactions with lymphocytes directly.

Clustering of antigen-reactive lymphocytes

Rechallenge experiments in the MLR illustrate the efficiency with which specific T cells cluster with dendritic cells. Lymphoblasts derived from the clusters respond to first (antigen-specific) but not third party (nonspecific) stimulators, while nonclustered T cells respond primarily to third party (2,3). As stated above, the response of isolated clusters approaches that of the cluster-noncluster mix. T cells also have been sensitized to carrier proteins in syngeneic cultures. The cluster-derived blasts are highly reponsive to antigen rechallenge and are MHC-restricted (5). These findings indicate that dendritic cells can solve the question we sited initially, i.e., MHC products on dendritic cells successfully select antigen-specific T lymphocytes even though this process requires the interaction of cell surface constituents on two infrequent cell types (dendritic cells and specific lymphocyte clones).

Two aspects of clustering further illustrate specificity. One has to do with the syngeneic MLR. With murine cells, it is hard to distinguish a syngeneic and allogeneic MLR in the first day since comparable numbers of clusters are often observed. Syngeneic MLR clusters subsequently proliferate much less actively. We do not understand why so many clusters develop in the syngeneic murine system, but the finding raises the possibility of an element of nonspecificity perhaps related to exposure to fetal calf serum. However, we do not encounter clustering in the human syngeneic MLR, which is performed in human serum (2).

A second finding has to do with B lymphocytes. In antibody responses to certain antigens, e.g., hapten-carrier conjugates, some antigens on heterologous red cells, the antigen-specific B cells cluster with dendritic and helper T cells (4,6). For other antigens, e.g., T-independent antigens like TNP- or DNP-Ficoll, or red cells plus soluble T cell replacing factors, most antibody-producing cells develop from the noncluster fraction (6). These latter antibody responses are known not to require direct T-B contact. We think that MHC-restricted, helper T lymphocytes are first activated by dendritic cells and then bind those B cells that have been selected by antigen and are presenting carrier determinants (see below). After clustering, B cells become responsive to antigen-nonspecific but essential lymphokines (4,6). This has been demonstrated by isolating the clusters, removing the T lymphocytes, and inducing the formation of antibody-secreting cells with factors that are present in medium that has been conditioned by dendritic-T cell clusters.

HOW DO DENDRITIC-LYMPHOCYTE CLUSTERS FORM?

The weak activity of macrophages and B lymphocytes in primary responses

The key finding is that presentation of antigen does not in itself lead to the formation of clusters. This is particularly evident in the MLR. In both mouse and man, macrophages and B lymphocytes cluster allogeneic resting T cells weakly if at all. These APC do not cluster even when the level of Ia on APC is upregulated by appropriate cytokines or by LPS (11). As will be discussed below, B cells and macrophages can bind and stimulate sensitized T lymphoblasts in an antigen-specific manner. However resting T cells seem to ignore alloantigens on these leukocytes.

Effects of blocking monoclonal antibodies

A number of surface molecules on T cells, especially CD4 and LFA-1, are considered to be "accessory molecules" that enhance the "avidity" of APC-T binding (12,13). mAb to LFA-1 and CD4 markedly inhibit stimulation of the MLR by dendritic cells (11). However, the mAb do not inhibit DC-T binding but block the function of DC-T clusters after they have formed. This suggests that moieties other than LFA-1 and CD4 are required for dendritic-T cell cluster formation (11).

Hypothesis

Given the above information, we suspect that dendritic cells bind and survey T lymphocytes in an antigen-independent fashion. If antigen is present, the binding is stabilized and subsequent events in T cell activation can ensue. The capacity of dendritic cells to survey clustered lymphocytes and vice versa may be critical for allowing MHC and TcR molecules on opposing cell surfaces to align. This hypothesis is being tested in ongoing experiments on the human syngeneic and allogeneic MLR's. In both, we observe the dendritic cell sending out processes and binding lymphocytes for periods of several minutes. In the allogeneic MLR, T cells are occasionally retained and the cluster begins to form. Therefore we are proposing that lymphocytes do not recognize antigen on APC as the first step in T-dependent immunity. Instead the first step is that dendritic cells bind to T cells by an antigen-independent, reversible, mechanism. Then it becomes feasible for TcR molecules to align with the large numbers of MHC molecules that are found on this class of APC.

ANTIGEN-DEPENDENT BINDING OF APC TO SENSITIZED T LYMPHOBLASTS

Two stages in T-dependent immune responses

The idea that most APC bind to sensitized T blasts originated in studies of the antibody response. Dendritic cells are essential accessory cells for antibody formation to heterologous red cells and to hapten-carrier conjugates (4,14,15). The antibody forming cells arise almost exclusively from clusters (4-6). However neither dendritic cells nor resting helper T cells cluster B lymphocytes directly. Therefore a two-stage mechanism seems likely. Dendritic cells first bind and activate carrier-specific helper T lymphocytes, and then the primed T cells become capable of interacting with hapten-specific B lymphocytes.

To test the T-B component of the above sequence, one requires a population of antigen-specific primed T cells. The requirement can be met with dendritic-T cell clusters as the source. In the MLR and in cultures primed to protein carriers, dendritic-T cell clusters generate large numbers of antigen-specific T blasts (3,5). When these blasts are challenged with B cells as APC, stimulation is apparent (3,5). Dendritic cells are 5-30 times more active as APC, but B cells are very active nonetheless. The B-T interaction is MHC restricted, and B cell growth and Ig secretion ensue (3,5). Current experiments are testing if B cells must first be activated by antigen prior to becoming capable of binding to T blasts.

Binding of B cells to sensitized T blasts

The capacity of B cells to act as APC for T blasts is correlated with their ability to bind T blasts in rapid binding assays. 10 min APC-T binding assays have become feasible using T blasts are the test cell (7). The T blasts are labeled with a fluorescent dye and are sedimented with graded doses of APC. 10 min later, single nonclustered T cells are enumerated. Dendritic cells prove to be 3-10 times more active than populations of macrophages and B cells in rapid binding assays. Yet the differences are quantitative in nature and not qualitative, as in the total lack of B cell stimulating activity in primary responses.

The above assays are very different in design from prior experiments on the interaction of macrophages and primed lymph node T cells (16-18). The latter studies: utilized bulk and not antigen-specific populations, required prolonged incubation times, and were associated with high levels of nonspecific binding, often 50-80% of the specific binding.

When T lymphoblasts are "rested" in the absence of APC or lymphokines, the "memory" T cells bind and respond primarily to dendritic cells rather than B cells or macrophages (5,7,19). The memory cells are antigen-specific, as assessed by challenge with dendritic cells, and express surface TcR molecules as defined by the KJ 16 anti-TcR mAb (unpublished). So again it seems that something in addition to expression of MHC and TcR molecules is required to obtain APC-T binding. Perhaps a molecule like LFA-1 contributes to the binding of T blasts to APC, but the LFA-1 is only functional in the lymphoblast. Wright et al have shown that another member of the LFA-1 family, the C3bi receptor, undergoes changes in binding and functional properties without significant changes in receptor numbers (20).

Our studies of B cells as APC have been performed with bulk, Sephadex G10 nonadherent populations. Current experiments are addressing the relative contributions of resting and activated B cells as APC.

Observations on the CD8$^+$ subset

Another example of the distinct binding features of resting and sensitized T lymphocytes occurs with cytotoxic CD8$^+$ T cells. We have confirmed the work of Sprent and coworkers (21,22) that CD8$^+$ cells can be stimulated directly in the MLR, that is in the apparent absence of CD4$^+$ helpers (9). The experiments rely on the anti-CD4 reagent, GK 1.5, which is very effective in eliminating CD4$^+$ cells with rabbit complement and in blocking function (should any CD4$^+$ cells persist). Dendritic cells cluster the responding CD8$^+$ cells early in the MLR and in the presence of blocking levels of GK 1.5. The clusters are similar to those observed in the CD4$^+$ response.

Primary macrophages from spleen and peritoneal cavity do not induce a CD8$^+$ MLR (9) even though the phagocytes express readily detectable class I MHC products (23,24). However, the CTL that are induced by dendritic cells are assayed using macrophages as the ^{51}Cr-labeled targets (9). Therefore macrophages fail to cluster or activate resting CD8$^+$ cells but do interact with sensitized CTL. A similar situation occurs when lectin-induced T blasts are the source of class I MHC antigens. In unpublished studies, we note that ConA blasts are inactive as stimulators of the MLR. Yet con A blasts serve as standard targets for sensitized CTL.

ANTIGEN-INDEPENDENT BINDING OF DENDRITIC CELLS AND T LYMPHOBLASTS

Results from rapid binding assays

As summarized above, the binding of B cells and macrophages to sensitized T lymphoblasts is antigen-specific. This is the case whether the binding assay is done at 4° or 37°. When dendritic cells are studied at 4°, the formation of dendritic-T blast conjugates is also antigen-specific, but at 37° large aggregates form in the absence of antigen (7). Dendritic cells also bind T cells in unprimed populations, although we cannot discern if the bound T cells have already been activated in situ. On a per cell basis, T blasts and memory cells are much more efficient in antigen-independent clustering than "unprimed" populations (19). Dendritic cells do not bind B cells or most thymocytes.

The big difference between the dendritic-T cell conjugates that form in the absence and presence of antigen is that antigen-independent conjugates disassemble within a few hours, and there is no IL-2 release or cell growth. Antigen-dependent conjugates release IL-2 within the first 2-5h of culture (11).

The question then arises: is the antigen-independent mechanism which mediates the binding of dendritic and T cells in 10 min assays at 37°C the same as the proposed antigen-independent binding utilized for the formation of clusters with unprimed T cells (see above)? We are trying to answer this question by generating mAb that selectively block the rapid binding of dendritic cells to T blasts and then testing if the same mAb would block clustering at the onset of an MLR or T-dependent antibody response.

Observations on epidermal Langerhans cells

There is one situation where both types of antigen-independent binding seem to go hand in hand. It involves epidermal Langerhans cells (LC). A relationship of LC do dendritic cells has become apparent in the last two years (25-27). The relationship is evident only after the LC have been cultured 1-3 days, since freshly isolated LC have some striking phenotypic and functional differences from dendritic cells. For example, dose response curves indicate that fresh LC are at least 10-20 times less active than dendritic cells in stimulating the MLR even though their levels of class II MHC products are comparable (25,27,28). When LC are tested after 1-3d in culture, accessory function in the primary MLR and T-dependent antibody response rise sharply. If anything, LC are even more active than lymphoid dendritic cells.

When LC are tested in rapid T cell clustering assays, it is apparent that fresh LC are very efficient in antigen-dependent binding but inactive in antigen-independent binding. The latter develops after culture. Fresh LC also are weak in generating clusters during the primary MLR. We have concluded that the capacity of LC to initiate primary responses is more closely tied to its capacity to form clusters with requisite T cells, and that this may require the development of an antigen-independent clustering mechanism (27).

Effects of monoclonal antibodies

Several mAb have been tested as blockers of clustering in rapid binding assays. When binding of dendritic cells to $CD4^+$ T blasts is monitored, anti-Ia markedly blocks antigen-dependent clustering at 4° but has little effect at 37° in the presence or absence of antigen. Anti-CD4 and anti-LFA-1 do not block either form of clustering, although these mAb have marked effects on conjugates once they have formed. The anti-CD4 conjugates do not make IL-2, and the anti-LFA-1 conjugates are readily disassembled by gentle pipetting (11). These blocking studies now need to be performed with other types of APC such as B cells and macrophages. However, it seems that CD4 and LFA-1 are both not needed for the initial contact between dendritic cells and T blasts.

Are dendritic cells unique in binding and activating T lymphocytes?

The answer is a qualified "no." Several alternatives are apparent, although their physiologic relevance has to be ascertained. Effective accessory function is observed with: a) EBV-transformed B cell lines, and the A.20 mouse B lymphoma, which are are active MLR stimulators and generate clusters of lymphoblasts. It is not clear how the behaviour of these B cell lines pertains to primary B lymphocytes, since the latter are weak or inactive in inducing the MLR (see above). b) Macrophages do not cluster unprimed T cells in antigen-specific responses, but efficient clustering occurs with mitogens. With sodium periodate, clusters form on day 1 but then disassemble and there is no proliferation (J.Austyn, submitted). With conA and anti-CD3 mAb, both clustering and proliferation occur (unpublished). c) Highly purified T cells can respond to mitogens and PMA in the apparent absence of accessory cells (29,30), and the proliferation occurs in clusters (unpublished).

Therefore there are instances in which one can bypass the need for dendritic cells in the formation of clusters and in the activation of T lymphocytes. On the otherhand, dendritic cells represent a highly specialized primary population for inducing antigen-specific responses. The study of accessory dendritic cells should also be important experimentally. When one bypasses the dendritic cell with a mAb or with PMA, one then must decide how and when to remove the stimulus. It is likely that APC and T cells communicate via several different ligand-receptor interactions during cell-mediated immunity, and that each occurs at a specific time interval. For example, T cell molecules like TcR, LFA-1 and CD4 may play separate and sequential roles, each of which involves an interaction with specific constituents on the APC (11). Another reason to study dendritic cells as accessory cells is to acquire more knowledge about APC-T cell interactions in vivo where it is not feasible to use such agents as PMA, lectins, and transformed cell lines.

APC-T CELL INTERACTIONS IN SITU

Interdigitating cells

Since the first micrographs of Veldman and others (31,32), it has been evident that the T-dependent regions of lymphoid organs contain "interdigitating cells" which bear some resemblance to dendritic cells (33,34). Both have extensive cytoplasmic processes and irregularly shaped nuclei. The cytoplasm lacks the many phagolysosomes and vacuoles that are typical of the macrophages found in other regions of the lymphoid organs such as the red pulp, medulla, and subcapsular sinus. It is also difficult to load interdigitating and dendritic cells with exogenous endocytosed materials, although evidence for endocytosis of lymphocytes in situ has been obtained (35). When surface markers are compared, both dendritic cells in vitro and presumptive interdigitating cells in situ express abundant Ia (36,37). Both have little or no reactivity with several mAb that react strongly with macrophages including 2.4G2 anti-Fc receptor, M1/70 anti-C3bi receptor, and F4/80 anti-macrophage (37,38). Birbeck granules have been noted in interdigitating cells, suggesting an origin from Langerhans cells (39), which are related to dendritic cells (see above).

New and more direct evidence for a relationship of interdigitating and dendritic cells comes from Austyn et al (submitted). They have succeeded in labeling splenic dendritic cells with a vital, fluorescent, DNA-binding dye called Hoechst 33342. Labeling does not interfere with dendritic cell function in culture. When the cells are adoptively transferred i.v., and the spleens are taken 24h later for sectioning, the fluorescent cells are distributed throughout the T-dependent areas. Transfer into the footpad seeds the lymph node. These experiments provide the first direct information for homing of dendritic cells to the T area and for a relationship between isolated dendritic cells and interdigitating cells.

Taken together, the data indicate that dendritic cells are localized precisely in the pathway of recirculating lymphocytes in situ. This provides a ready opportunity for establishing contacts with antigen-specific T cells. Conceivably some of the many hours that are required for lymphocyte recirculation through the T area reflects the time that dendritic cells and T lymphocytes survey one another to facilitate the formation of antigen-receptor interactions across two cell surfaces. It seems to us that the more demanding unknown is not whether dendritic cells and antigen-specific T cell clones can make contact in situ, because the known structure of lymphoid tissues seems designed towards this end, but how antigen gains access to this locale. It is not easy to demonstrate antigen accumulation in the T-dependent region during cell-mediated immunity.

Dendritic cells in nonlymphoid tissues

Identification of dendritic cells in nonlymphoid tissues is an area of increasing interest. The immunohistological findings of Fabre (40) have pointed out that Ia^+ dendritic leukocytes, which lack macrophage and B cell traits, are found in all organs but brain. A priority is to isolate these cells and determine their functional capacities directly. Klinkert et al find dendritic cells in liver suspensions (41), and recent studies reveal comparable cells in lung (42). If the study of epidermal LC bear upon dendritic cells in other nonlymphoid tissues, it seems that a maturation process is required before tissue dendritic cells exhibit the full functional and phenotypic properties of their lymphoid counterparts.

Central vs. peripheral pathways for T cell sensitization

Central sensitization means that lymphocytes are sensitized in lymphoid organs that drain the site of antigen administration. Antigens and/or APC move via the blood to the spleen, or via lymphatics to the lymph node. This pathway seems to be important for the development of sensitivity to skin allografts and contact allergens, since immunization does not occur if draining lymphatics are severed (43,44). It has been proposed that the lymphatics act as a conduit for antigen-bearing Langerhans cells or "veiled" cells (39,46-50), most likely to the T areas of the draining lymphoid tissue. Lymphoblast formation in the T area, comparable to that seen in tissue culture with dendritic cells as APC, is a prominent cytologic feature of T-dependent allograft responses, contact sensitivity, and antibody formation.

In the central pathway, antigens and/or dendritic cells gain access to the T-dependent regions. Antigen-specific T cells are selected from the recirculating lymphocyte pool and induced to become growing lymphoblasts with the capacity to make the products required for cell-mediated immunity, i.e., lymphokines and cytolysins. The T blasts leave via the efferent lymph, gain access to the blood stream, and home to inflammatory sites (51,52). Because the T cells are freshly sensitized blasts, effective interactions with APC should then ensue in the inflammatory site. Therefore many of the in vitro findings reviewed above can be applied in vivo. A major unknown is the mechanism whereby antigen is carried, or gains access to, the T area.

In peripheral sensitization, a phenomenon initially observed with kidney grafts (53), T cells would begin to grow and differentiate locally at the site of antigen deposition. One unknown is whether T cells encounter tissue APC with the stimulating properties of dendritic cells. Perhaps the deposition of antigens leads to the local mobilization and maturation of dendritic cells that are comparable to cultured epidermal LC (see above). GM-CSF may be the limiting factor, as the studies of murine LC are beginning to indicate (28), and GM-CSF can be produced by many cell types including macrophages, epithelial cells, keratinocytes, and smooth muscle. Sufficient numbers of antigen-specific T cells would also have to gain access to the tissue site, but this may be feasible as part of the inflammatory response.

A frequently sited alternative to the tissue dendritic cell is the antigen-presenting endothelial cell. Unfortunately the study of endothelial cells during in vitro primary responses remains limited. It also is not clear how endothelial cells, which often may require lymphokines to express Ia (54), would function early on in the sensitization of $CD4^+$ helper lymphocytes.

Two recent studies have visualized "peripheral" dendritic-T cell clusters in situ. One is the clustering of T cells and dermal Langerhans cells during delayed type hypersensitivity reactions to PPD (55). The other is the binding of T cells to interstitial dendritic cells during cardiac allograft rejection (56). New approaches should shed light on the functional properties of these aggregates. These include antibody and nucleic acid probes to activation antigens and to specific cytokines.

SUMMARY

In the introduction, we asked how MHC molecules on the surfaces of APC make contact with antigen-specific receptors on the surfaces of T cells. We have reviewed two models in which antigen-specific contacts occur with primary leukocyte populations in vitro. One system involves resting T cells; the other sensitized T lymphoblasts. At the onset of a primary immune response, dendritic cells seem critical for binding and activating both $CD4^+$ and $CD8^+$ subsets. Given the current evidence, we suggest that dendritic cells literally find the right T cell clone, and not vice versa, and that dendritic cells do so by first binding and surveying T cells by an antigen-independent mechanism. In the efferent limb of immunity, other types of APC including B cells and macrophages bind and stimulate freshly sensitized T lymphoblasts. Freshly isolated epidermal Langerhans cells do not cluster T cells by an antigen-independent mechanism but acquire this capacity during epidermal suspension culture. Under the control of GM-CSF, the Langerhans cell becomes a powerful accessory cell for the primary or sensitization limb of T-dependent immune responses like the MLR and primary antibody response. Isolated lymphoid dendritic cells have many features in common with interdigitating cells in lymphoid T areas, and may be related to some of the irregularly-shaped Ia^+ cells in certain epithelia and interstitial regions. Contact with dendritic cells may be important in both central and peripheral pathways of T cell sensitization in situ.

ACKNOWLEDGEMENT

The experiments reviewed here were performed in collaboration with Jon Austyn, Nina Bhardwaj, Eve Flechner, Peter Freudenthal, Angela Granelli-Piperno, Sumi Koide, Lisa Lau, Gerold Schuler, Jay Valinsky, David Weinstein, Margit Witmer-Pack, James Young, Shigeru Muramatsu, and Zanvil Cohn.

REFERENCES

1. Steinman, R.M., Inaba, K., Schuler, G. & Witmer, M.D. In: Mechanisms of host resistance to infectious agents, tumors, and allografts., Steinman, R.M. & North, R.J. (ed.) pp. 71-97.. The Rockefeller University Press, New York (1986).
2. Flechner, E., Freudenthal, P., Kaplan, G. & Steinman, R.M. Cell.Immunol. In Press.
3. Inaba, K. & Steinman, R.M. J. Exp. Med. **160**, 1717-1735 (1984).
4. Inaba, K., Witmer, M.D. & Steinman, R.M. J. Exp. Med. **160**, 858-876 (1984).
5. Inaba, K. & Steinman, R.M. Science **229**, 475-479 (1985).

6. Inaba, K. & Steinman, R.M. Cell. Immunol. **105**, 432-442 (1987).
7. Inaba, K. & Steinman, R.M. J. Exp. Med. **163**, 247-261 (1986).
8. Inaba, K., Granelli-Piperno, A. & Steinman, R.M. J. Exp. Med. **158**, 2040-2057 (1983).
9. Inaba, K., Young, J.W. & Steinman, R.M. J. Exp. Med. **166**, 182-194 (1987).
10. Koide, S.L., Inaba, K. & Steinman, R.M. J. Exp. Med. **165**, 515-530 (1987).
11. Inaba, K. & Steinman, R.M. J. Exp. Med. **165**, 1403-1417 (1987).
12. Marrack, P., Endres, R., Shimonkevitz, R., Zlotnik, A., Dialynas, D., Fitch, F. & Kappler, J. J. Exp. Med. **158**, 1077-1091 (1983).
13. Gougeon, M.-L., Bismuth, G. & Theze, J. Cell. Immunol. **95**, 75-83 (1985).
14. Inaba, K., Steinman, R.M., Van Voorhis, W.C. & Muramatsu, S. Proc. Natl. Acad. Sci. USA **80**, 6041-6045 (1983).
15. Komatsubara, S., Hirayama, Y., Inaba, K., Naito, K., Yoshida, K., Kawai, J. & Muramatsu, S. Cell. Immunol. **95**, 288-296 (1985).
16. Nielson, M.H., Jensen, H.J., Braendstrup, O. & Werdelin, O. J. Exp. Med. **140**, 1266-1272 (1974).
17. Lipsky, P.E. & Rosenthal, A.S. J. Exp. Med. **141**, 138-154 (1975).
18. Werdelin, O., Braendstrup, O. & Shevach, E.M. J. Immunol. **123**, 1755-1762 (1979).
19. Inaba, K., Koide, S. & Steinman, R.M. Proc. Natl. Acad. Sci. USA **82**, 7686-7690 (1985).
20. Wright, S.D., Detmers, P.A., Jong, M.T.C. & Meyer, B.C. J. Exp. Med. **163**, 1245-1259 (1986).
21. Sprent, J. & Schaefer, M. Nature **322**, 541-544 (1986).
22. Sprent, J. & Schaefer, M. J. Exp. Med. **162**, 2068-2088 (1985).
23. Steinman, R.M., Nogueira, N., Witmer, M.D., Tydings, J.D. & Mellman, I.S. J. Exp. Med. **152**, 1248-1261 (1980).
24. Nussenzweig, M.C., Steinman, R.M., Unkeless, J.C., Witmer, M.D., Gutchinov, B. & Cohn, Z.A. J. Exp. Med. **154**, 168-187 (1981).
25. Schuler, G. & Steinman, R.M. J. Exp. Med. **161**, 526-546 (1985).
26. Schuler, G., Romani, N. & Steinman, R.M. J. Invest. Dermatol. **85**, 99s-106s (1985).
27. Inaba, K., Schuler, G., Witmer, M.D., Valinsky, J., Atassi, B. & Steinman, R.M. J. Exp. Med. **164**, 605-613 (1986).
28. Witmer-Pack, M.D., Olivier, W., Valinsky, J., Schuler, G. & Steinman, R.M. J. Exp. Med. In Press.
29. Rosenstreich, D.L., & Mizel,S.B. J. Immunol. 123: 1749-1754 (1973).
30. Hara, T. & Fu, S.M. J. Exp. Med. **161**, 641-656 (1985).
31. Heusermann, U., Stutte, H.J. & Muller-Hermelink, H.K. Cell. Tissue. Res. **153**, 415-417 (1974).
32. Veldman, J.E. & Kaiserling, E. In: The Reticuloendothelial System.Morphology, Carr, I. & Daems, W.T. (ed.) pp. 381-416. Plenum Publishing Corp., New York (1980).
33. Steinman, R.M., Adams, J.C. & Cohn, Z.A. J. Exp. Med. **141**, 804-820 (1975).
34. Tew, J.G., Thorbecke, J. & Steinman, R.M. J. Reticuloendothelial Soc. **31**, 371-380 (1982).
35. Fossum, S. & Rolstad, B. Eur. J. Immunol. **16**, 440-450 (1986).
36. Dijkstra, C.D. J. Reticuloendothel. Soc. **32**, 167-178 (1982).
37. Witmer, M.D. & Steinman, R.M. Am. J. Anat. **170**, 465-481 (1984).
38. Hume, D.A. & Gordon, G. J. Exp. Med. **158**, 1522-1536 (1983).
39. Hoefsmit, E.C.M, Duijvestijn, A.M. & Kamperdijk, W.A. Immunobiol. **161**, 255-265 (1982).
40. Hart, D.N.J. & Fabre, J.W. J. Exp. Med. **154**, 347-361 (1981).
41. Klinkert, W.E.F., LaBadie, J.H. & Bowers, W.E. J. Exp. Med. **156**, 1-19 (1982).
42. Sertl, K., Takemura, T., Tschachler, E., Ferrans, V.J., Kaliner, M.A. & Shevach, E.M. J. Exp. Med. **163**, 436-451 (1986).
43. Barker, C.F. & Billingham, R.E. J. Exp. Med. **128**, 197-221 (1968).

44. Landsteiner, K. & Chase, M.W. J. Exp. Med. **69**, 767-784 (1939).
45. Macatonia, S., Edwards, A., Griffiths, S., Fryer, P., & Knight, S. J. Exp. Med. In Press.
46. Knight, S.C., Balfour, B.M., O'Brien, J., Buttifant, L., Sumerska, T. & Clark, J. Eur. J. Immunol. **12**, 1057-1060 (1982).
47. Mason, D.W., Pugh, C.W. & Webb, M. Immunol. **44**, 75-87 (1981).
48. Silberberg, I.R., Baer, L. & Rosenthal, S.A. J. Invest. Dermatol. **66**, 210-217 (1976).
49. Mayrhofer, P., Holt, P.G. & Papadimitriou, J.M. Immunol. **58**, 379-387 (1986).
50. Knight, S.C., Krejci, J., Malkovsky, M., Colizzi, V., Gautam, A. & Asherson, G.L. Cell. Immunol. **94**, 427-434 (1985).
51. Ottaway, C.A. & Parrott, D.M.V. J. Exp. Med. **150**, 218-230 (1979).
52. Koster, F.T., McGregor, D.D. & Mackaness, G.B. J. Exp. Med. **133**, 400-409 (1971).
53. Strober, S. & Gowans, J.L. J. Exp. Med. **122**, 347 (1965).
54. Pober, J.S., Gimbrone, M.A., Cotran, R.S., Reiss, C.S., Burakoff, S.J., Fiers, W. & Ault, K.A. J. Exp. Med. **157**, 1339-1353 (1983).
55. Kaplan, G., Nusrat, A., Witmer, M.D., Nath, I. & Cohn, Z.A. J. Exp. Med. **165**, 763-776 (1987).
56. Forbes, R.D.C., Parfrey, N.A., Gomersall, M., Darden, A.G. & Guttmann, R.D. J. Exp. Med. **164**, 1239-1258 (1986).

B. 9th GERMINAL CENTRE CONFERENCE

BONE MARROW, B-CELL DIFFERENTIATION AND THE B-CELL SYSTEM

IN VIVO LOCALIZATION OF B LYMPHOCTYE PROGENITOR CELLS

IN MOUSE BONE MARROW

D.G. Osmond, K. Jacobsen, Y-H Park[1]
and L. Lamontagne[2]

Department of Anatomy, McGill University
Montreal, Quebec, Canada H3A 2B2

INTRODUCTION

The microenvironmental organisation of primary B lymphocytopoiesis in the bone marrow remains poorly understood.

In previous studies we have developed techniques to identify B lineage cells in tissue sections by the cell surface binding of radiolabeled antibodies administered in vivo (1). In mouse femoral bone marrow, the distribution of B lymphocytes bearing surface μ heavy chains of IgM (su) has been visualized by the binding of ^{125}I-anti μ antibodies using both light microscope and electron microscope radioautography (1,2).

The present work aimed to localize B lineage cells bearing the B lineage-associated B220 glycoprotein (3,4), as revealed by the binding of the monoclonal antibody (mAb) 14.8 (5,6), administered systematically to young mice. Immunofluorescence labeling has shown that mAb 14.8 binding characterises not only $s\mu^+$ B lymphocytes but also a wide range of B lymphcyte progenitor cells (5-8). The latter include pre-B cells which bear cytoplasmic μ heavy chains ($c\mu$) without $s\mu$, as well as large u-cells, putative early B progenitor cells before the expression of u chains (6-8). Some of these $14.8^+\mu^-$ early progenitors also contain terminal deoxynucleotidyl transferase (TdT) (7), and intranuclear enzyme apparently involved in generating diversity during u chain gene rearrangement (9). Some putative earlier TdT^+ cells are both 14.8^- and μ^- (7,8). Whereas the 14.8^+u^- cells and large $14.8^+c\mu^+s\mu^-$ pre-B cells are actively proliferating, small $14.8^+c\mu^+s\mu^-$ cells are post mitotic pre-B cells which give rise to su^+ B lymphocytes without further division (7,8,10,11). In addition to the in situ radiolabeling of 14.8^+-binding cells the present work also compares the prevalence of the various putative B lymphocyte progenitor cell populations in 2 types of bone marrow sample, medullary cells flushed from the medullary cavity and bone-associated cells removed by curetting the inner surface of the surrounding cortical bone.

Present addresses:
[1]Kitano Hospital, 3-13 Kamiyama, Kita-ku, Osaka, Japan
[2]Département des Sciences biologiques, Université de Quéec à Montréal

MATERIALS AND METHODS

Animals

C3H/HeJ mice (Jackson Laboratories, Bar Harbor, ME) were used at 4-5 wk and 7-9 wk of age.

Antibodies and radioiodination

Rat mAb 14.8 (5,6) pooled from hybridoma cell cultures was purified on an anti-rat K light chain sepharose column and coupled to carrier-free Na-^{125}I (specific activity 1.5x10^7 cCi/µg) by the chloramine T method (2).

For immunofluorescence studies, fluorescein isothiocyanate (FITC)-conjugated goat anti-rat IgG (Cooper Biomedical Inc., Malvern, PA), rabbit anti-TdT (Supertechs Inc., Bethesda, MD), rhodamine isothiocyanate (TRITC)-conjugated goat anti-rabbit IgG F(ab')$_2$ (Jackson Immunoresearch Laboratories Inc., Mississauga, Ontario), and TRITC- and FITC-conjugated affinity-purified goat antibodies to mouse µ chains (Kirkegaard and Perry Laboratories, Gaithersburg, MD) were used, as detailed (7,11).

In vivo labeling with ^{125}I-mAb 14.8

A 100 ul injection of ^{125}I-mAb 14.8 (5 µg, 50x10^6 cpm) was given to 4-5 wk old mice via the external jugular vein under chloral hydrate anaesthesia (1,2). After a period of 3 m, unbound ^{125}I-mAb 14.8 was washed out of the entire body by perfusing cold 10% newborn calf serum (NCS) in Eagle's minimal essential medium (MEM) through a cannula in the left ventricle (3ml/min for 10 m), followed by a perfusion of fixative (cold 2.5% glutaraldehyde, 2% acrolein; 10 m), prior to taking tissues for histological processing. Alternatively, tissues were removed 3 m after the ^{125}I-mAb injection for preparation of single cell suspensions.

Radioautography of tissue sections and cell suspensions

Femurs were removed, immersed in fixative (2.5% glutaraldehyde, 2% acrolein at 4°C) overnight, washed in cacodylate buffer, postfixed in 1% buffered osmium tetroxide for 3 h at 4°C, dehydrated in acetone, and embedded in epon. Transverse sections were cut through the mid diaphysis. For light microscopy, 1 µ sections were cut, stained with iron hematoxylin, processed for radioautography (12) and exposed for 7-28 d. For electron microscopy 70-85 nm-thick sections were carbon-coated, processed for radioautography, exposed for 4-12 wk and developed with a fine grain technique (13). For the analysis of single cells, samples of femoral bone marrow, thymus and spleen were suspended in 10% NCS-MEM as described (7), washed by centrifugation in 6% bovine serum albumin, processed for radiography and exposed for 1-7 d.

Immunofluorescence labeling of bone associated B lineage cells

Femurs were taken from 7-9 wk old mice, in groups of 3 mice, the epiphyses were removed and the diaphyses were split longitudinally. Bone marrow cells were removed in 2 samples. 1) Cells were flushed off with 10% NES-MEM and other loosely adherent cells were removed (medullary cells). 2) The internal aspect of the cortical bone was firmly scraped with a dental curette to remove residual cells (bone-associated cells). In each case, single cell suspensions were prepared, washed, and subjected to double immunofluorescence labeling procedures for su, cu, 14.8 and TdT in various combinations to detect and quantitate B lymphocytes (su), pre-B cells (cµ$^+$sµ$^-$) and early B precursors (14.8$^+$µ$^-$; TdT$^+$14.8$^-$; TdT$^+$14.8$^+$), as detailed previously (7,11).

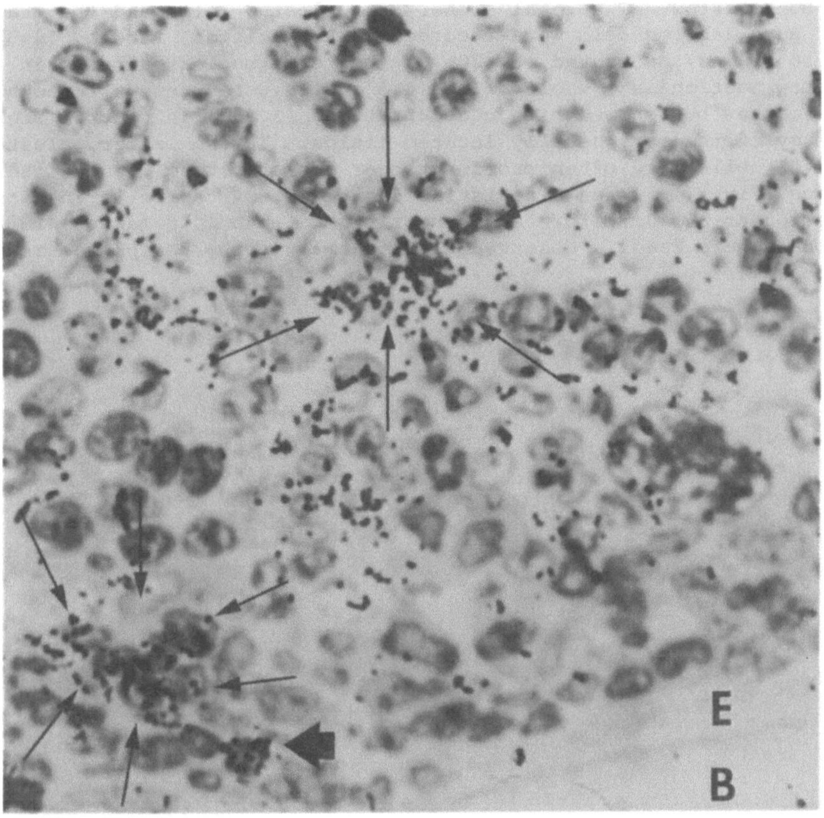

Fig. 1. Light microscope radioautograph of bone marrow section showing labeled 14.8^{+} cells as single cells (large arrows) and as clusters (small arrows) near the endosteum (E) and cortical bone (B) (X1,600).

RESULTS

In vivo labeling of B lymphocyte progenitor cells

To verify the effectiveness of the in vivo labeling procedure, single cell suspensions of femoral bone marrow spleen and thymus were examined after the iv administration of ^{125}I-mAb 14.8. Well-labeled cells were found in MacNeal's tetrachrome-stained radioautographic smear preparations from each tissue. Analysis of cells with lymphoid morphology showed that the proportion of small lymphocytes with surface labeling (> 10 grains) by ^{125}I-mAb 14.8 was 80% in the bone marrow, 54% in the spleen and 0.2% in the thymus. Corresponding values for large lymphoid cells were 71% in the bone marrow, 73% in the spleen and 1.4% in the thymus. No other morphologic cell types were labeled.

Light microscope radioautography of bone marrow sections revealed labeled 14.8^{+} cells in the extravascular compartment, both as single cells and in clusters of 6-8 cells closely grouped together (Fig. 1). The spatial distribution of 14.8^{+} cells was analysed by examining successive 400 μm^2 microscopic fields throughout the entire transverse section of femoral bone marrow. The total number of nucleated cells and the grain count of each labeled cell were recorded in each field. Labeled 14.8^+ cells (> 6 grains) accounted for 20% of the total nucleated cells in the bone marrow section as a whole. The 14.8^+ cells were present throughout the bone marrow but showed high concentrations (> 25%) in central fields and in many of the peripherally located fields. In the latter, clusters of large 14.8^+ cells were often prominent near the endosteum linging the cortical bone (Fig. 1). When vincristine was given to stop cells in mitosis for a period of 2h 40m before administering ^{125}I-mAb 14.8 large 14.8^+ cells in metaphase were readily observed particularly in the peripheral regions of the bone marrow.

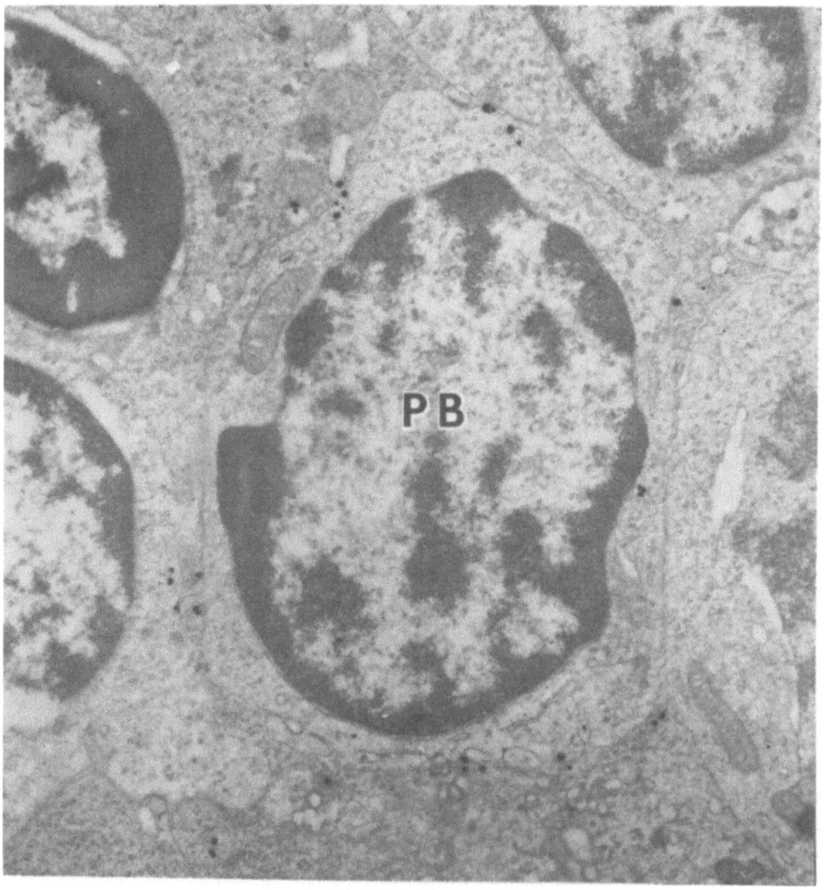

Fig. 2. Electron microscope radioautograph of labeled 14.8^+ progenitor B cell (PB) in bone marrows (x56,100).

The clusters of 14.8^+ cells were well shown by electron microscopy, groups of 4-8 morphologically similar 14.8^+ cells being found together in individual electron microscope fields (Fig. 3).

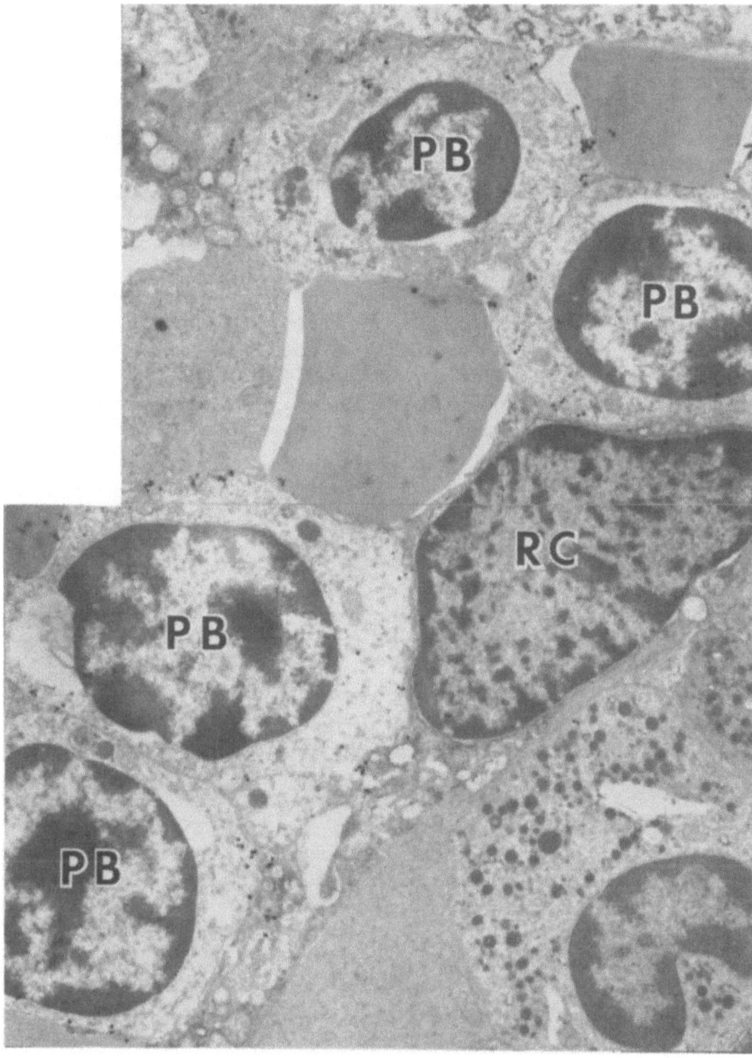

Fig. 3. Cluster of 4 labeled 14.8^+ progenitor B cells (PB) associated with the processes of a reticular cell (RC) (X41,910).

Electron microscope radioautography revealed details of the morphology and cellular associations of the labeled 14.8^+ cells. Radioautographic grains were well localised over the surface membrane profile of labeled cells (Fig. 2). Background grains were virtually absent. The morphology of the labeled cells ranged from typical small lymphocytes with densely heterochromatic nuclei and small cytoplasmic volume (2) to larger cells with greater euchromatic nuclear areas and more extensive cytoplasm characteristic of B lymphocyte precursors (Fig. 2). No other cell type was labeled.

Labeled 14.8^+ cells were in apparently direct contact with each other, with other haemopoietic cell lineages and with complex processes from reticular (stromal) cells which formed irregular profiles in section. The close association with reticular cells was well shown in some clusters of 14.8^+ cells, in which all the cells of the cluster appeared to be in contact with the surface of a single irregularly shaped reticular cell (Fig. 3).

Bone-associated B lymphocyte progenitor cells

Immunofluorescence labeling revealed B lymphocyte progenitors not only in medullary samples of femoral bone marrow obtained by flushing the medullary cavity but also in further samples collected by curetting the bone. Histological sections revealed that the bone-associated samples were recovered from small indentations in the bone removed with the surface of the bone by curettage. The incidences of early B progenitors ($14.8^{+}\mu^{-}$) and pre-B cells ($c\mu^{+}s\mu^{-}$) were actually higher in bone associated samples (9.1±0.3%; 11.4±3.0%) than in medullary samples (4.8±0.1%; 8.1±0.2%). The pre-B cells were larger and had a higher mitotic index in bone-asociated samples (5.3±2.2%) than in medullary samples (2.5±0.8%). Other putative early progenitor cells containing intranuclear TdT were also present and tended to be more frequent in bone-associated samples (1.6%) than in medullary samples (1.1%).

DISCUSSION

In defining the microenvironment and the cellular interactions involved in B lymphocyte genesis, two major questions concern the tissue distribution of B lineage cells and the nature of possible associations between B lymphocyte progenitors and other cells in the bone marrow. Both questions were addressed by the present techniques.

The labeling of cells recovered in suspensions of the femoral bone marrow, spleen and thymus after iv administration of ^{125}I-mAb 14.8 correspond to the incidences of B lineage cells previously revealed in these tissues by exposing cell suspensions to ^{125}I-mAb 14.8 in vitro. This result, together with the absence of labeling of any non-lymphoid cell, either in suspensions or tissue sections, demonstrated the effectiveness and specificity of the in vivo labeling method.

The high concentrations of 14.8^{+} cells found in peripheral areas of bone marrow sections as well as in central areas are in contrast with previous studies of $s\mu^{+}$ B lymphocytes in mouse bone marrow which, although present to some extent in all regions, showed particularly high incidences in the central areas around the longitudinal venous sinusoid (1). These findings suggest that the peripherally located 14.8^{+} cells are progenitor B cells, as indicated also by the observation of many notably large mitotic 14.8^{+} cells in the subendosteal region of the bone marrow and of TdT^{+} cells, $14.8^{+}\mu^{-}$ early B progenitors and $c\mu^{+}s\mu^{-}$pre-B cells closely associated with the surface of the surrounding bone. Presumptively, each of the clusters of 14.8^{+} cells found in peripheral regions of the bone marrow may represent the progeny of a single primitive B lineage progenitor cell. The present findings accord with evidence that the peripheral region of bone marrow near the surrounding bone apparently represents a supportive environment for early proliferative progenitors of hemopoietic lineages in general (14-16) from which the differentiating progeny may migrate centrally to be released into the blood stream. Various factors contributing to such a postulated centripetal gradient of proliferative activity could include gradients of pH or oxygen tension reflecting the radial organisation of the bone marrow blood circulation (2), bone-derived growth factors, and local stromal cell interactions.

Electron microscopy reveals clearly that clusters of 14.8^{+} B lymphocyte progenitors can be intimately associated with individual reticular cells and their complex processes under conditions of normal B cell genesis in situ. While the significance of these associations remains to be verified the findings suggest that the interactions between certain non-macrophage stromal cells in long term bone marrow cultures (LTBMC) which appear to be essential to support the activity of B lymphocyte progenitors (17-19) may have a normal in vivo counterpart. Following the present demonstration of B lineage associatedreticular cells in vivo, further studies will aim to characterize these cells and to determine whether they are identical with the subset of bone marrow stromal cells derived from LTBMC which can produce B lineage stimulatory factors.

ACKNOWLEDGEMENTS

This work was supported by the Medical Research Council of Canada.

REFERENCES

1. S. J. Batten and D. G. Osmond, J. Immunol. Meth. 7:381 (1984).
2. D. G. Osmond and S. J. Batten, Amer. J. Anat. 170:349 (1984).
3. R. L. Coffman and I. L. Weissman, J. Exp. Med. 153:269 (1981).
4. R. L. Coffman, Immunol. Rev. 69:5 (1983).
5. P. W. Kincade, G. Lee, T. Watanabe, L. Sun, and M. P. Scheid, J. Immunol. 127:2262 (1981).
6. K. S. Landreth, P. W. Kincade, G. Lee, and E. S. Medlock, J. Immunol. 131:572 (1983).
7. Y-H. Park and D. G. Osmond, J. Exp. Med. 165:444 (1987).
8. D. G. Osmond and Y-H. Park, Int. Rev. Immnol. 2:241 (1987).
9. T. A. Kunkel, K. P. Gopinathan, D. K. Dube, E. T. Snow, and L. A. Loeb, Proc. Natl. Acad. Sci. USA 83:1967 (1986).
10. K. S. Landreth, C. Rosse, and J. Clagett, J. Immunol. 127-2027 (1981).
11. D. Opstelten and D. G. Osmond, J. Immunol. 131:2635 (1983).
12. B. M. Kopriwa and C. P. Leblond, J. Histochem. Cytochem. 10:269 (1962).
13. B. M. Kopriwa, Histochemie 44:201 (1975).
14. S. E. Shackney, S. S. Ford, and A. B. Witting, Cell Tissue Kinet. 8:505 (1975).
15. J. K. Gong, Science 199:1443 (1978).
16. D. G. Osmond, M. T. E. Fahlman, G. M. Fulop, and D. M. Rahal, in: "Microenvironments in Haemopoietic and Lymphoid Differentiation", Pitman Medical, London, CIBA Foundation Symposium (1981).
17. K. Dorshkind, L. Schouest, and W. H. Fletcher, Cell Tiss. Res. 239-375 (1985).
18. P. L. Witte, M. Robinson, M. G. Low, D. L. Stiers, S. Perkins, R. A. Fleischman, and P. W. Kincade, Eur. J. Immunol. (1987) (in press).
19. Z. X. Song, R. W. Shadduck, D. J. Innes, A. Waheed, and P. J. Quesenbery, Blood 66:273 (1985).

PATTERN OF DISTRIBUTION OF EARLY B LINEAGE CELLS IN RAT BONE MARROW

Mirjam Hermans[1], Hans Hartsuiker[1] and Davine Opstelten

Department of Histology
University of Groningen
Oostersingel 69/1, 9713 EZ Groningen, the Netherlands

INTRODUCTION

In mammals bone marrow (BM) is the primary site of B lymphocyte genesis in postnatal life, providing precursors for antibody forming cells for humoral immune responses. The B progenitor cells are situated in the BM within a three dimensional network formed by stromal cells, vascular elements and erythroid, myeloid and lymphoid cells. Little is known about their spatial organisation.

Of the various stages of development of B lymphocytes present in the BM, the final stages that are correlated with immunoglobulin (Ig) expression are best defined. Newly formed B lymphocytes carry both Ig μ heavy chains and light chains on their cell surface; their immediate precursors are pre-B cells, synthesising μ heavy chains detectable in their cytoplasm only (1,2). Less well established are the stages of B lymphopoiesis preceding the expression of μ chains.

Part of the preceding stages may be represented by terminal deoxynucleotidyl transferase (TdT)-containing cells in BM. TdT is a mutagenic enzyme which supposedly contributes to somatic diversification of antigen receptors (3). In rats most BM TdT^+ cells are thought to be B precursor cells as they carry B lineage-associated antigens (4), and partly express μ chains in their cytoplasm (5). Furthermore, Ig heavy chain gene rearrangements have been detected in TdT^+ cells enriched by short term BM culture comparable to those found in pre-B and B cells (6).

In this study we report the location of B, pre-B and TdT^+ cells detected in frozen sections of rat femurs stained for k light chain, μ heavy chain and TdT using the immunofluorescence technique.

MATERIALS AND METHODS

Animals

Male 5-6 weeks old (AO/G x BN/G) F1 hybrid rats (RT1 u/n) were obtained from stocks held at the Department of Histology, University of Groningen, and maintained under conventional conditions.

---

[1]Supported by a grant from the Dutch Foundation for Medical and Health Research MEDIGON, which is subsidised by the Netherlands Organisation for the Advancement of Pure Research (ZWO)(grant 900-505-125).

Antibodies

Rabbit anti rat μ chain (7) and horse anti rabbit Ig conjugated to tetramethylrhodamine isothiocyanate (TRITC) (7) were a kind gift from Dr. A.F. Williams (MRC Cellular Immunology Unit, Sir William Dunn School of Pathology, Oxford, England). Rabbit anti TdT and goat anti rabbit Ig(H+L) conjugated to fluorescein isothiocyanate (FITC) were from Supertechs Inc. (Bethesda, MD, USA). Affinity purified goat anti mouse Ig(H+L)-FITC was from Kirkegaard and Perry Laboratories (Gaithersburg, MD, USA) and goat anti mouse IgG_1-TRITC was from Southern Biotechnology Associates (Birmingham, Ala, USA). HIS 8 (mouse IgG1 anti rat k light chain (8)) and HIS 43 (mouse IgG1 anti rat sinus endothelium) were from the Department of Histology, University of Groningen. HIS 43 was produced according to the immunization protocol described (9); rat BM cells were used to immunize BALB/c mice, and SP2-0 myeloma cells served as fusion partner. Hybridoma culture supernatants were screened on BM cytocentrifuge preparations and femoral frozen sections using indirect immunofluorescent staining procedures. All first stage antibodies were diluted to the appropriate working dilution with phosphate buffered saline (PBS), all second stage antibodies in PBS containing 2-5% (v/v) normal rat serum.

Cryostat sectioning of femurs

Rats were killed by an overdose of ether. The femurs were dissected out and immersed in precooled PBS (0^oC). The outer layer of the bone was removed by using a dental drill equipped with a diamond disc (10). The femurs were frozen in liquid freon and sectioned (7 μm) transversely to the longitudinal axis in a cryostat (at -20^oC to -30^oC). Sections were transferred from the knife onto room temperature glass slides and quickly dried under a room temperature fan.

Fluorescence staining procedures

The sections were fixed in precooled (0^oC) 5% (v/v) glacial acetic acid in absolute ethanol for 12-15 min on ice, and washed five times in ice-cold PBS. The fixed sections were incubated with first stage antibody (60 min, room temperature, humidified chamber), washed three times in PBS (10 min, room temperature), incubated with the appropriate second stage antibody (30 min) and finally washed three times in PBS. For double immunofluorescence the fixed sections were stained sequentially by two sets of 1° plus 2° stage antibody.

Sections were mounted in 90% (v/v) glycerol (Merck, Darmstadt, FRG) in PBS containing 2.5% (w/v) 1,4 diazabicyclo-[2,2,2] octane (Sigma, St. Louis, MO 63178 USA), pH 8.6.

RESULTS

All sections showed a rather disperse pattern of B cells, pre-B cells and TdT^+ cells. The relative number of B cell precursors in the subendosteal area o the BM appeared considerably higher than in the central area. Therefore, for further quantitative analysis, we subdivided the BM in a peripheral region (0-144 μm distance from the endosteum) and a central region (over 500 μm distance from the endosteum).

Cell frequencies in the centre and the periphery of the BM

μ^+ B and pre-B cells : In sections double stained for μ heavy chain and k light chain, the vast majority of the μ^+k^+ B cells was strongly μ^+, while μ^+k^- pre-B cells were weakly μ^+. The μ^+ B cell frequency in the centre of the BM was 9% of all nucleated cells, similar to the 8% μ^+ B cells in the periphery of the

BM. For pre-B cells a difference in frequency in the two compartments was evident. In the centre they composed 13% of the cells, while the overall frequency in the periphery was 29%.

TdT$^+$ cells : The enzyme TdT was located mostly within the cell nuclei. The cells equaled the pre-B cells in their spatial distribution, 2% in the centre of the BM and 4% in the peripheral BM compartment.

Distribution in parenchyma versus sinuses

HIS 43 was used to analyse intrasinus location of cells. This monoclonal antibody stained BM endothelium strongly, and the reticular network of stromal cells weakly. Screening HIS 43 for binding to the thymus, lymph node, spleen, Peyer's patches, heart, tongue, liver, pancreas, intestine, bladder, ovary, vagina, kidney, adrenal gland, pituitary gland, brain and nerve tissue, using immunohistology of frozen sections revealed that specificity was confined to fibrocytes, fibroblasts, smooth muscle cells (apparently their cell membrane), and the endothelium of all blood vessels. In BM sections double stained for HIS 43 and μ or TdT we analysed the parenchymal versus sinus lumen location of B lineage cell subsets.

In the anti-μ stained sections radially distributed extended clusters of strongly μ$^+$ cells, situated within the BM sinuses, were visible. Weakly μ$^+$ pre-B cells were almost exclusively seen in the parenchyma. In the centre of the BM the proportion of all strongly μ$^+$ cells that were located within sinuses was slightly lower than in the periphery (26% and 42%, respectively).

The number of TdT$^+$ cells located within BM sinuses was very low: in the 2 sections analysed completely we encountered 1 and 2, respectively.

DISCUSSION

By developing a technique to obtain high quality frozen sections of femoral marrow we were able to reveal the location of B, pre-B and TdT$^+$ cells in rat BM.

Within the femoral cross sections an uneven distribution of B cell precursors was evident. Both pre-B and TdT$^+$ cells showed a gradient with a high frequency in the subendosteal area declining gradually towards the centre of the marrow. B cells did not show such a gradient. Our present data conform with the idea of a radial developmental sequence in which the data of hemopoietic progenitors proliferate actively in the peripheral regions before bone marrow and their progeny differentiates and migrates simultaneously, before being discharged through the sinus walls into the blood stream (11).

Within BM sinuses we found accumulations of strongly μ$^+$ cells, consisting of up to 20 cells; they probably represent newly formed B cells, recently discharged from the parenchyma (11). TdT$^+$ and pre-B cells were rarely detected in the sinus lumens, suggesting no significant emigration of those cells from the BM.

All three cell subsets studied occurred in a rather disperse pattern in the BM parenchyma. This in contrast to mature B lymphocytes in the peripheral lymphoid organs where many are found highly organised in follicles. Also in comparison with erythropoiesis, where clear cut erythroblastic islands around macrophages can be seen, B lymphopoiesis in BM appears less spatially structured. However, this does not exclude the existence of specific B lymphopoietic microenvironments in BM.

REFERENCES

1 M.C.Raff, M. Megson, J.J.T. Owen, and M.D. Cooper. Early production of intracellular IgM by B-lymphocyte precursors in mouse. Nature. 259:224 (1976).

2 D. Levitt, and M.D. Cooper. Mouse pre-B cells synthesize and secrete μ heavy chains but not light chains. Cell 19:617 (1980).

3 S.V. Desiderio, G.D. Yancopoulos, M. Paskind, E. Thomas, M.A. Boss, N. Landau, F.W. Alt, and D. Baltimore. Insertions of N regions into heavy-chain genes is correlated with expression of terminal deoxytransferase in B cells. Nature 311:752 (1984).

4 D. Opstelten, G.J. Deenen, J. Rozing, and S.V. Hunt. B lymphocyte-associated antigens on terminal deoxynucleotidyl transferase-positive cells and pre-B cells in bone marrow of the rat. J. Immunol. 137:76 (1986).

5 G.J. Deenen, S.V. Hunt, and D. Opstelten. A stathmokinetic study of B-lymphocytopoiesis in rat bone marrow: Proliferation of cells containing cytoplasmic μ chains, terminal deoxynucleotidyl transferase and carrying His 24 antigen. J. Immunol. 139:702 (1987).

6 S.V. Hunt, E.S. Medlock, D.L. Greiner, I. Goldschneider, and D. Opstelten. Rat immunoglobulin genes have comparable patterns of J_H rearrangements in normal peripheral B cells and in pre-B and cultured TdT^+ cells from bone marrow. This volume.

7 A.F. Williams. IgG2 and other immunoglobulin classes on the cell surface of rat lymphoid cells. Eur. J. Immunol. 5:883 (1975).

8 F.G.M. Kroese, D. Opstelten, A.S. Wubbena, G.J. Deenen, J. Aten, E.H. Schwander, L. de Ley, and P. Nieuwenhuis. Monoclonal antibodies to rat B lymphocyte (sub-) populations. In: Microenvironments in the lymphoid system. G.G.B. Klaus, ed. Plenum Press, New York. P. 81. (1985).

9 F.G.M. Kroese, A.S. Wubbena, D. Opstelten, G.J. Deenen, E.H. Schwander, L. de Leij, H. Vos, S. Poppema, J. Volberda, and P. Nieuwenhuis. B lymphocyte differentiation in the rat : production and characterization of monoclonal antibodies to B lineage associated antigens. Eur. J. Immunol. 17:921 (1987).

10 S.A. Bentley, J.M. Foidart, and H.K. Kleinman. Connective tissue elements in rat bone marrow: Immunofluorescent visualization of the hematopoietic microenvironment. J. Histochem. Cytochem. 32:114 (1984).

11 D.G. Osmond, and Y.H. Park. B lymphocyte progenitors in mouse bone marrow. Intern. Rev. Immunol. 2:241 (1987).

B CELL PRECURSOR POPULATIONS IN FETAL AND NEONATAL RAT LIVER: FREQUENCY, TOPOGRAPHY AND ANTIGENIC PHENOTYPE

G.J. Veldhuis and D. Opstelten

Department of Histology
State University Groningen
The Netherlands

INTRODUCTION

In rodents and man B lymphocyte development starts in fetal liver, and gradually shifts to the bone marrow where it is maintained throughout life (1). The immediate precursors of newly formed surface IgM bearing B cells are pre-B cells defined by the presence of μ heavy chains (but not light chains) in the cytoplasm only (1).
In rat, B cell precursors have been characterised extensively in adult bone marrow (2-4). Pre-B cells are abundantly present, and candidate precursors before μ chain expression have been identified by the presence of B lineage surface markers, and the enzyme terminal deoxynucleotidyl transferase (TdT) supposedly involved in diversifying antigen receptors during gene rearrangement (5).
In contrast to mouse (6-8) and man (9), little is known about the ontogeny of B lymphocytopoiesis in rat. Functional studies have shown the fetal liver to be the strongest in B cell generating capacity, located in the Thy-1$^+$ subset (10). The fact that TdT$^+$ cells have been reported absent from rat fetal and neonatal liver (11) indicated to us that B cell development early in life may differ from that in adults. Therefore, we searched for the presence of B, pre-B, TdT$^+$ and B lineage marker bearing cells in fetal and neonatal liver *in situ*, allowing detection of very low frequencies as well as information on their spatial organisation.

MATERIALS AND METHODS

Animals: Fetal and neonatal (AO/G x BN/G) F1 hybrid rats (day 14 of gestation - day 10 after birth) were obtained from stocks held at the Department of Histology, State University of Groningen and maintained under conventional conditions.
Tissue sampling and (immuno) histology: Timed matings were scored by vaginal smearing (appearance of sperm in smear = day 0). Livers were removed by using fine catarrac knifes and a stereo-microscope. Part of the tissue was frozen in liquid freon. Cryostat sections (5 μm) were fixed in ethanol/acetic acid, washed in PBS and stained with the markers summarized in Table 1 by double immunofluorescence as described (2,4). For proper detection of TdT it was important to quickly dry sections upon removal from the cryostat.

The other part of tissue was fixed and paraffin embedded, sectioned and stained with methylgreen-pyronin for general histology.

Table 1. Antibodies used in this study

(mouse monoclonal antibody)	source	isotype	specificity	reference
HIS8	1	IgG1	k light chain	12
HIS40	1	IgG1	μ heavy chain	3
HIS24	1	IgG2b	200kD B-lineage associated membrane molecule; on bone marrow B, pre-B and TdT^+ cells but not on CFU-S, B/T stem cells	2,3,13
HIS14	1		B lineage surface antigen in adult rats	2,13
86.4.2G4	1	IgM	erythroid lineage	a)
86.4.2D9	1	IgM	myeloid lineage	a)
86.4.1C5	1	IgG2a	Thy-1, strongly on stem cells and TdT^+ cells, weakly on pre-B and B cells	a)
Rabbit α Rat μ	2		μ heavy chain	14
Rabbit α TdT	3		TdT	
Goat α Mouse Ig-FITC/TRITC	4		mouse Ig (H+L), IgH isotypes (μ, γ_{2a}, γ_{2b})	
Goat α Rabbit Ig-FITC/TRITC	3,4		rabbit Ig (H+L)	
Horse α Rabbit-TRITC	2		rabbit Ig (H+L)	14

1) Department of Histology, State University of Groningen
2) MRC Cellular Immunology Unit (Dr. A.F. Williams), Sir William Dunn School of Pathology, Oxford, U.K.
3) Supertechs Inc., Bethesda, MD, U.S.A.
4) Southern Biotechnology Associates, Birmingham, Al, U.S.A.
a) M. Hermans et al., unpublished.

RESULTS

Paraffin sections showed that hemopoiesis in fetal liver mainly concerned erythroid and myeloid elements, but some cells of lymphoid morphology were observed as well. With increasing age the non-hemopoietic area of the liver developed rapidly; at 6 days after birth only some small colonies of erythroid, myeloid and lymphoid cells were observed. The proportion of lymphoid cells related to other hemopoietic cells increased with increasing age. At day 10 after birth hemopoiesis in liver was absent.

Pre-B and B cells, number and localisation

Pre-B cells (μ^+k^-) were first observed on day 17 of gestation; they were weakly μ^+ and varied in cell size. Their frequency was very low, 1 to 3 per section. The following days their number steadily increased (see Figure 1). Peak levels of μ^+k^- cells per unit area of liver section were reached 2 days after birth.
μ^+k^+ B cells, which were stronger μ^+ than pre-B cells, appeared later than pre-B cells and were first observed on day 19 of gestation. At that time, they were 3% of total μ^+ cells. This proportion increased

throughout the period studied and was 60% on day 6 after birth (Fig. 1). Pre-B cells were often observed in clusters of two or more cells (Fig. 2). Two days after birth about 70% of the μ^+k^- cells were in clusters. B cells were mostly solitary.

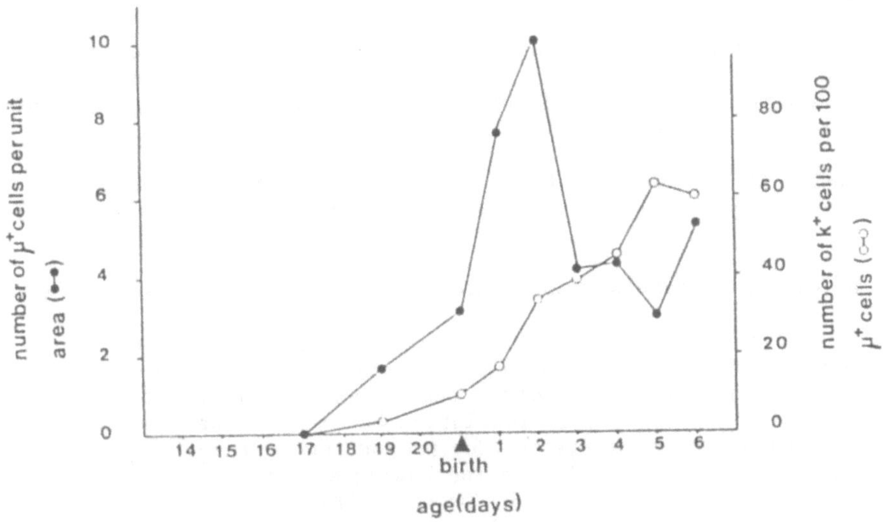

Figure 1. μ^+ B lineage cells in sections of fetal and neonatal liver. In sections stained for μ (TRITC) and k (FITC), the number of positive cells per unit area were scored using a rectangular eye piece graticule; one unit corresponds to 80.000 μm^2.

Figure 2. Pre-B cells in liver of a 3 days old rat. In this frozen section of liver, stained for μ and k, a cluster of 4 weakly μ^+ cells was observed. They were k^- (not shown).

Surface antigens of μ^+ B lineage cells

HIS24: On day 19 of gestation some strongly μ^+ cells showed a weak expression of HIS24. After birth HIS24 expression was also observed on some clusters of weakly μ^+ cells. The number of μ^+ cells which expressed HIS24-antigen increased with increasing age, but remained lower than the number of μ^+HIS24$^-$ cells. The intensity of HIS24-binding increased with age but was much lower than that of adult BM μ^+ cells.

HIS14: The first μ^+HIS14$^+$ cells were observed on day 21 of gestation, they were strongly μ^+ solitary cells. Clusters of weakly μ^+ cells were HIS14$^-$. The number of μ^+HIS14$^+$ increased with increasing age, but the HIS14 intensity remained lower than in adult BM.

1C5 (Thy-1): All μ^+ cells in fetal and neonatal liver sections were Thy-1$^+$; weakly μ^+ cells expressed more Thy-1 than strongly μ^+ cells.

Thy-1$^+\mu^-$ cells: a precursor population before μ chain expression

Thy-1$^+$ (1C5+) cells were present in fetal liver at 14 days of gestation. It concerned medium to large sized cells, mostly solitary, which appeared distributed randomly throughout the liver parenchyma. With increasing age the Thy-1$^+$ population became more heterogenous in cell size and staining intensity. The proportion of Thy-1$^+\mu^-$ cells among Thy-1$^+$ cells decreased with increasing age (Table 2), reaching a plateau two days after birth. All Thy-1$^+$ cells were negative for 2G4 and 2D9, the erythroid and myeloid marker respectively, at all ages studied.

Table 2

Proportion of μ^- cells among Thy-1$^+$ liver cells as a function of age

age (in days)	fetal 14	17	19	neonatal 1	2	3	6
fraction of μ^- cells among Thy-1$^+$ cells	1	1	.85	.34	.40	.37	.42

TdT$^+$ cells

When screening fetal and neonatal liver sections for the presence of TdT, frozen sections of adult rat BM and thymus served as positive controls. No cells were found, at any time, that showed the staining pattern typical for TdT$^+$ cells in adult rat (see M. Hermans et al., this volume). Occasionally some weak cytoplasmic fluorescence was observed in the vicinity of eosiniphilic granulocytes; presumably this was non-specific uptake of the reagents.

DISCUSSION

The first μ^+k^- pre-B cells detected in rat liver, at day 17 of gestation, were randomly distributed throughout the liver parenchyma. With time these cells were more and more observed in clusters of two or more cells suggesting that one or more cell divisions may take place in this compartment. This accords with observations in mice, where the clonal nature of pre-B cell foci has been demonstrated (8).

The earliest μ^+ cells expressed the B lineage epitopes HIS24 and HIS14 poorly. Thus, they differ from μ^+ cells in adult rats in antigenic phenotype. In mouse, the B lineage antigen B220 is not present either on the earliest μ^+ pre-B cells observed (15). Apparently these epitopes have no function in the mere genesis of B lymphocytes. Thy-1, however was on all μ^+ cells observed in the liver as is the case for μ^+ cells in adult bone marrow (2).

The Thy-$1^+\mu^-$ cells detected must include the population of hemopoietic stem cells (10). They were large, strongly Thy-1^+ cells, in accord with data from day 15 - 19 fetal liver cell suspensions (16). No clustering of Thy-$1^+\mu^-$ cells was evident at any time, suggesting that either few or no cell divisions occur before μ chain expression, or that their progeny dissiminates rapidly.

In the present study on $TdT^{\mp}$ cells were observed in sections fetal and neonatal rat liver. There is no obvious technical explanation for the absence of TdT. In cell suspensions of both rat and mouse fetal and neonatal liver TdT has been reported absent as well (11). In humans, however, TdT^+ cells that are phenotypically similar to pre-B and B cells have been detected in fetal liver and bone marrow at 8-20 weeks of gestation (9).
The present data suggests that in contrast to later in life B lymphocytopoiesis in fetal and neonatal rodents proceeds without the action of TdT. Considering the possible role of TdT, this may contribute to the lower Ig diversity in neonates (17).

REFERENCES

1. J.J.T. Owen. Progress in Immunology IV, p. 304 (1980).

2. D. Opstelten, G.J. Deenen, J. Rozing, and S.V. Hunt. J. Immunol. 137:76 (1987).

3. G.J. Deenen, S.V. Hunt, and D. Opstelten. J. Immunol. 139:702 (1987).

4. M. Hermans, H. Hartsuiker, and D. Opstelten, this volume.

5. S.V. Desiderio, G.D. Yancopolos, M. Paskind, E. Thomas, M.A. Boss, N. Landau, F.W. Alt and D. Baltimore. Nature 311:752 (1984).

6. W.A. Kamps and M.D. Cooper. J.Immunol. 129:526 (1982).

7. A. Velardi and M.D. Cooper. J.Immunol. 133:672 (1984).

8. J. Rossant, K.M. Vijh, C.E. Grossi, and M.D. Cooper. Nature 319:519 (1986).

9. G.E.M. Asma, R. Langlois van den Bergh, and J.M. Vossen. Clin. Exp. Immunol. 64:356 (1986).

10. S.V. Hunt. In: Immunobiology of bone marrow transplantation. Springer-Verlag, Berlin. p. 193 (1980).

11. K.E. Gregoire, I. Goldschneider, R.W. Barton, and F.J. Bollum. J.Immunol. 123:1347 (1979).

12. F.G.M. Kroese, D. Opstelten, A.S. Wubbena, G.J. Deenen, J. Aten, E.H. Schwander, L. de Ley, and P. Nieuwenhuis. In: Microenvironments in the lymphoid system. (G.G.B. Klaus, e.d.) Plenum Press. New York. p. 81 (1985).

13. F.G.M. Kroese, A.S. Wubbena, D. Opstelten, G.J. Deenen, E.H. Schwander, L. de Leij, H. Vos, S. Poppema, J. Volberda, and P. Nieuwenhuis. Eur. J. Immunol. 17:921 (1987).

14. A.F. Williams. Eur. J. Immunol. 5:883 (1975).

15. E.S. Medlock, K.S. Landreth, and P.W. Kincade. Dev.& Comp. Immunol. 8:877 (1984).

16. M.A. Ritter, L.K. Gordon, and I. Goldschneider. J. Immunol. 121:2463 (1978).

17. N.R. Klinman, and J.L. Press. Fed. Proc. 34:47 (1975).

RAT IMMUNOGLOBULIN GENES HAVE COMPARABLE PATTERNS OF J_H REARRANGEMENT IN NORMAL PERIPHERAL B CELLS AND IN PRE-B AND CULTURED TdT^+ CELLS FROM BONE MARROW

S.V. Hunt[1], E.S. Medlock[2], D.L. Greiner[2], I. Goldschneider[2], and D. Opstelten[3,4]

[1]Sir William Dunn School of Pathology, University of Oxford South Parks Road, OXFORD OX1 3RE, U.K.
[2]Department of Pathology, University of Connecticut Health Center, Farmington, U.S.A.
[3]Histologisch Laboratorium, Rijksuniversiteit Groningen The Netherlands
[4]Supported in part by a grant from the Dutch Foundation for Medical and Health Research MEDIGON, which is subsidised by the Netherlands Organisation for the Advancement of Pure Research (ZWO, grant 900-505-125)

SUMMARY

Pre-B cells from normal rat bone marrow were isolated by first rosette-depleting granulocytes, T cells and stem cells with W3/13 mAb and mature B cells with anti-Ig mAb, and then in some samples enriching by FACS-sorting $HIS24^+$ cells. Presumptive precursors of these pre-B cells possessing terminal deoxy-nucleotidyl transferase were prepared by culture of normal marrow. The extent of rearrangement of kappa and heavy chain genes was estimated on both these types of cells and on mature peripheral B cells by measuring the disappearance of the germ-line restriction fragments encoding J segments (other than J_H1) relative to a non-rearranging C_K fragment. Pre-B cells and B cells showed similar degrees of rearrangement of their kappa genes, leaving roughly half in germ-line form; the TdT^+-enriched cells showed approximately similar heavy chain rearrangement as the peripheral B cells. Some non-germ-line restriction fragments hybridising with J_H in sufficient amount to register as discrete bands on Southern blots were found in common between these different cell populations taken from different individual donors. These may represent preferred $D\text{-}J_H$ or perhaps $V\text{-}D\text{-}J_H$ rearrangement arising repetitively in independent clones, or may be due to dominance by a few preferred clones.

INTRODUCTION

The B-lineage-restricted cell surface antigen recognised by the mAb HIS24 is found on pre-B cells of rat bone marrow[1]. A minor marrow lymphocyte population bears HIS24 but lacks u; this group may therefore include precursors of pre-B cells. It can be subdivided into those that stain for terminal deoxynucleotidyl transferase (TdT^+) and those that lack it. All TdT^+ bear HIS 24 in normal marrow. This itself is evidence that many TdT^+ belong to the B cell lineage, as is the presence of μ chains in some of the TdT^+ cells[2]. Further clues to the relationships between these populations might be gained by studying the state of arrangement of immunoglobulin

Light and Heavy chain genes. We have chosen to examine cells from normal animals, rather then virus-transformed or fusion-immortalised lines which may be selective in their transformation targets and may continue to rearrange their genes in culture. To get enough TdT$^+$ cells for assay, short-term cultures of rat marrow on mouse feeders[3] were grown. HIS24$^+\mu^-$TdT$^-$ cells were not available in sufficient quantity for assay.

Our aim was therefore to determine the nature and extent of rearrangement of Ig kappa and Heavy chain genes in DNA from normal populations of mature B cells and their presumptive precursors, pre-B and TdT$^+$. The theory of the assay is to measure independently for kappa and Heavy genes the reduction in the amount of the germ-line restriction fragment cut out by a particular enzyme to contain the 5'-most J, since this fragment is bound to be altered in size by incoming V or D segments. In practice PstI, the enzyme used here because of its good resolution of J_K, C_K and J_H, does not permit the detection of the J_H1 gene segment but only of some of the other J_H segments. A spectrum of unpredictably diverse fragment sizes should be produced, each one in amount too small to detect. If new, non-germ-line bands are found, this indicates a predominant rearrangement of some kind surfacing above this undetectable background.

METHODS

B cells prepared from pooled cervical and mesenteric lymph nodes (10^8 cells) of unprimed PVG rats by depletion of rosettes with W3/13-coated erythrocytes to remove T cells[4]. Purity was >89% by FACS re-analysis.

Pre-B cells were prepared from bone marrow (10^9 cells) of 3 adolescent unprimed PVG rats by depletion of rosettes with W3/13- and anti-Ig-(OX12)-coated erythrocytes to remove granulocytes, T cells and mature B cells. This fraction (lanes 6,7, Fig. 1) contained 83% HIS24$^+$ cells. A portion of these cells were stained by indirect immunofluorescence (saturating HIS24 tissue culture supernatant followed by FITC-Rabbit anti-mouse Ig) and sorted for HIS24$^+$ lymphocytes by FACS. Purity was 96% (lane 8, Fig. 1). In one experiment HIS24$^+$ cells were size-selected by forward light scatter to exclude large cells (lane 9, Fig. 1).

TdT$^+$ cells were prepared from normal unprimed LEW rats by culture on marrow stromal cell feeder layers from C57Bl/6 mice as described[3]. The cultures were fractionated into non-adherent and adherent fractions, yielding 10-30 x 10^6 rat lymphoid cells, and the mouse feeder layer cells were removed by labelling with anti-H2^b (ATCC HB31) and panning on anti-mouse Ig coated plastic dishes. See legend to Fig. 1 for % TdT$^+$ cells in each sample.

High molecular weight DNA from replicate samples of these preparations and control DNA from 5 LEW thymomas[5], LEW liver and PVG spleen was extracted by detergent lysis and CsCl centrifugation[6]. Probes were the 5'-most J_K($J_{4/5}$ of ref 6 and a J_H-containing Bam-Xba fragment of a genomic sub-clone spanning the region half-way between DQ_{52} and the presumed enhancer region[7] (to detect rearrangments) and C_K[6] (as non-rearranging internal standard to estimate DNA loaded per lane), labelled to >2x10^8 cpm per ug by nick-translation.

RESULTS

Germline genomic restriction fragments were observed at 7.8 kb (C_K), 6.2 kb (J_K) as expected[6] and 2.1 kb (J_H) (Fig. 1). No differences were seen between PVG and LEW fragments and the results from these two strains are considered together.

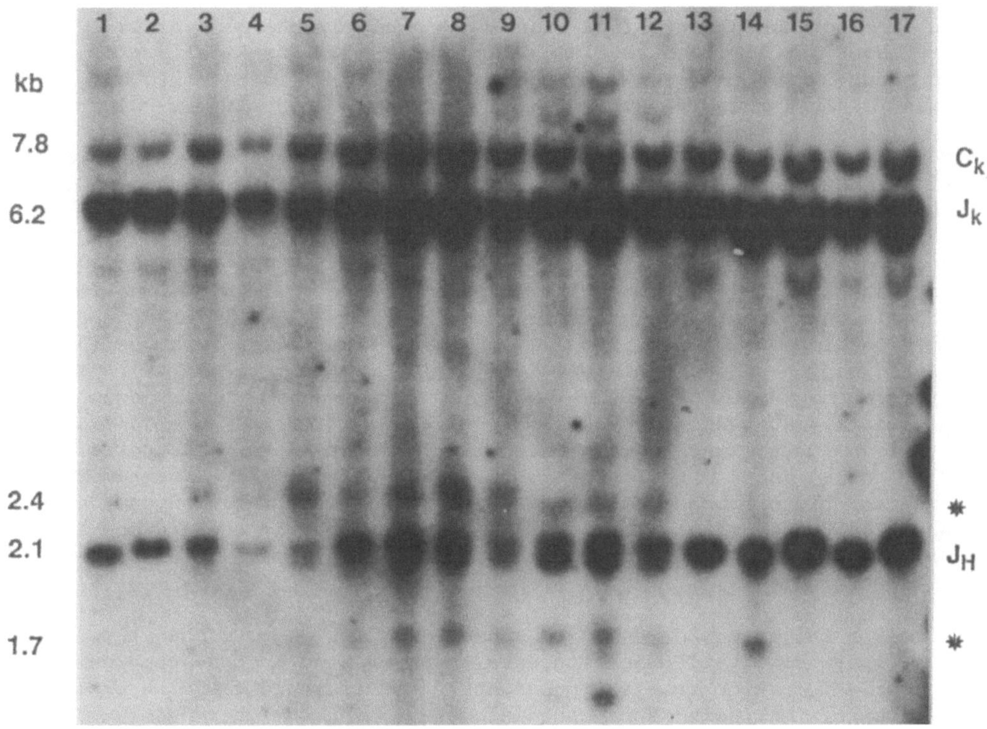

Fig 1. Southern blot of DNA samples cut with PstI and probed for J_k, J_H, and C_k. Lanes 1-5 are standards with known proportions of germ-line and B cell DNA. Lanes 6,7: pre-B without FACS separation. Lanes 8,9 sorted for $HIS24^+$(lane 9 selecting for small cells only). Lanes 10, 12: non-adherent fraction of cultured BM (56% and 19% TdT^+)(10 day and 14 day cultures, respectively); lane 11: adherent fraction (68% TdT^+)(10 day culture). 5 - 10 µg DNA loaded per lane. Blotting and hybridisation as described[6]. Washing: 3 x 30 minutes, 55^0C, 0.4x SSPE. Autoradiographic exposure: 19 days (pre-flashed, 2 intensifying screens).

Table 1. Densitometry of J_k, J_H and C_k genomic DNA Pst1 restriction fragments. An LKB Gelscan Densitometer measured the full length of each lane. Band densities were computed having set the background manually to correspond to the density of the region immediately adjacent to each band. The data were transformed first by determining the ratio of J_k or J_H to C_k for each lane, then by setting these ratios relative to the germ-line standard in lane 2 (100%) (figures underlined in Table). The J_H data for lanes 5-12 integrate the sum of the split 2.1/2.0 kb bands together.

		Standards				BM					BM						
Sample	Liver	Sperm ------>			B cell	pre-B --->		pre-B --->		<---	TdT^+	--->	<-----		Thymomas		----->
Sperm		100	50	25	0	W3/13⁻sIg⁻		$HIS24^+$		A	B	C	35	45	46	47	75
B cell		0	50	75	100												
Strain	LEW	<------				PVG			------>	<------			LEW				------>
C_k	0.45	0.60	0.98	0.36	1.07	1.36	2.58	2.35	1.19	1.48	1.55	0.84	0.99	1.09	1.18	0.65	1.10
J_k	2.29	3.00	3.27	1.54	2.34	4.38	5.68	5.80	2.55	5.37	6.98	4.82	4.98	6.30	6.61	4.35	6.07
J_H	0.69	0.70	0.85	0.25	0.51*	1.87*	2.63*	1.99*	0.78*	1.60*	2.09*	1.34*	1.58	1.57	3.02	2.10	3.07
RATIOS																	
J_k/C_k	5.08	5.00	3.33	4.28	2.19	3.22	2.20	2.47	2.14	3.63	4.50	5.74	5.03	5.78	5.60	6.70	5.52
ratio to lane 2	102	100	67	86	44	64	44	49	43	73	90	114	100	115	112	134	110
J_H/C_k	1.53	1.16	0.86	0.69	0.48	1.38	1.02	0.85	0.66	1.08	1.35	1.60	1.60	1.44	2.56	3.23	2.79
ratio to lane 2	131	100	74	59	41	117	87	72	56	92	115	136	136	123	218	276	240

In the measurements of kappa rearrangment of peripheral B cells J_K/C_K fell to about 45% of its germ-line value (Table 1). Pre-B cells also showed about 50% of the germ-line intensity. TdT^+-enriched cells and the thymomas retained roughly 100%.

For J_H rearrangement the 2.1 kb band in all the pre-B and TdT^+ preparations and in other samples of peripheral B cells (data not shown) split to give two just-resolvable sub-bands, which were too close to be distinguished by the densitometer (about 100 bp apart). Therefore the densitometer data in Table 1 integrate these sub-bands as one peak and overestimate the residual germ-line component by up to 2-fold. Qualitatively, additional weak bands can be seen at 2.4 and 1.7 kb in all B, pre-B and TdT^+ populations (lanes 5 to 12) but not in the sperm, liver or thymoma samples (except a solo unexplained band at 1.7 kb for thymoma #45). Sample densitometer traces are presented in Fig. 2 to demonstrate these extra bands. The 2.4 kb band represents very roughly 20% of the germ-line intensity that would be expected from the C_K internal standard. The split 2.1/2.0 and 2.4 and 1.7 kb bands are due to J_H because they have never been seen in many blots probed for J_K and C_K, but they were revealed on a comparable blot probed solely for J_H. A band at 1.4 kb was seen in one sample only of TdT^+ (lane 11) but it is not known to which of the three probes this was attributable.

Three of the thymomas (#46,47,75) had J_H/C_K ratios more than twice those of sperm or liver, which suggests an alteration in gene copy number. Since the J_K/C_K ratios remain approximately normal, this could be explained by the loss of the whole kappa-containing region on one chromosome, or by the duplication of the H chain genome.

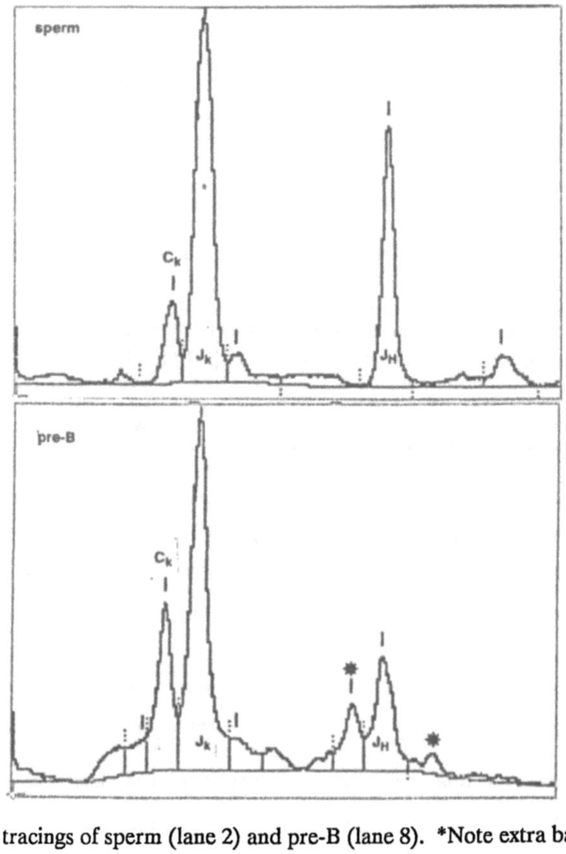

Fig 2. Densitometer tracings of sperm (lane 2) and pre-B (lane 8). *Note extra bands at 2.4 and 1.7 kb

DISCUSSION

Kappa Rearrangement

In peripheral B cells the residual germ-line component found, just under half, agrees substantially with previous determinations in the mouse and rat[6,8]. Some cells rearrange both J_K genes, with one non-productively rearranged and the other homologue productively rearranged; rather more cells rearrange only one of the pair and leave the other in germ-line form. In the four pre-B cell samples there was as much rearrangement as in the B cells; former determinations in the mouse showed virtually no kappa rearrangement[9]. This suggests that in the rat, most pre-B cells are more mature, despite their lack of kappa protein. This might fit with the higher frequency of pre-B cells in rat bone marrow than in the mouse (about double), and it hints at a greater delay in rats between kappa rearrangement and protein expression than in mice.

In the TdT^+ samples, there was some variation in the measured kappa rearrangement. With the single result on the purer preparation of non-adherent cells (lane 10) about three-quarters remained germ-line. This single value showing a little rearrangement evidently needs confirmation with other samples. Even this tentative part-rearrangement was less than in either B or pre-B and some allowance may have to be made for possible comtaminating pre-B cells in the cultures. The lesser rearrangement in the TdT^+ samples compared with pre-Bs along with their lack of cytoplasmic Ig substantiate the idea that, assuming they are on the B-lineage (see below), TdT^+ cells precede pre-B cells developmentally.

Heavy Chain Rearrangement

The present assay used a J_H probe which should detect the earliest rearrangements, D to J, but on its own cannot distinguish D-J from V-D-J joins. A Pst site falls within the J_H region downstream of J_H1, so the assay does not measure J_H1 rearrangements. Incoming V_H segments will be expected to alter any D-J fragments superimposed on any alteration due initially to D-J integration. Measurement of the loss of the germ-line J_H component was complicated by the fact that two close sub-bands appeared at the characteristic 2.1 kb germ-line Pst I J_H position (we are currently using digests with other enzymes to improve the resolution). The quantitation of the residual germ-line band by densitometry does not therefore help to assign the maturation stage of pre-B and TdT^+ cells. However, qualitatively it is clear that all of the samples, B, pre-B and TdT^+, show similar restriction patterns, with 2.4, split 2,1/2.0 and 1.7 kb bands, while none of the thymomas nor the sperm nor liver samples show this. This is evidence that they all belong to the same (B) lineage.

The new bands at 2.4 and 1.7 and one of the split bands at 2.1/2.0 suggest predominant rearrangements. The most likely explanation is that they independently occur in many different clones due to some bias inherent in the mechanisms of rearrangement. Such preferred rearrangements have been found in the mouse for V_H7183 and V_H OX2 genes [10,11] and if they can occur similarly at the earlier D-J- stage also then this could explain how a few predominant bands stand out more obviously than if further diversification occurred due to V_H incursion. Less likely might be a very distorted clonal proliferation with the bias reflecting an antigen-independent early paucicloncal expansion. This is less reasonable because the phenomenon recurs independently in different groups and strains of rats. A comparable bias does not show itself in kappa rearrangement. A full analysis of gene segment usage would require characterisation of cloned DNA from the 2.4 and 1.7 kb regions.

REFERENCES

1. D. Opstelten, G. J. Deenen, J. Rozing, and S. V. Hunt, B lymphocyte associated antigens on terminal deoxynucleotidyl transferase-positive cells and pre-B cells in bone marrow of the rat, J. Immunol. 137:76 (1986).
2. G. J. Deenen, S. V. Hunt, and D. Opstelten, A stathmokinetic study of B lymphocytopoiesis in rat bone marrow: proliferation of cells containing cytoplasmic mu chains, terminal deoxynucleotidyl transferase and carrying HIS24antigen, J. Immunol. 139:702 (1987).
3. J. Hayashi, E. S. Medlock, and I. Goldschneider, A selective culture system for generating terminal deoxynucleotidyl-transferase positive (TdT^+) lymphoid precursor cells in vitro. I. Description of the culture system, J. Exp. Med. 160:1622 (1984).
4. S. V. Hunt, Preparative immunoselection of lymphocyte populations, in: "Handbook of Experimental Immunology", D. M. Weir, L. A. Herzenberg, and C. Blackwell, eds., Blackwell Scientific Publications (4th ed., 1986).
5. R. W. Barton, D. L. Greiner, E. S. Medlock, and I, Goldschneider, Phenotypic heterogeneity of Gross-virus induced thymic lymphomas in the rat: cellular origin and migratory properties, Cell. Immunol. 94:113 (1985).
6. A. Tsukamoto, I. L. Weissman, and S. V. Hunt, Allelic exclusion in rat kappa immunoglobulin chains; extent of J_K rearrangement in normal B lymphocytes, EMBO J. 3:975 (1984).
7. M. Bruggemann, J. Free, A. Diamond, J. C. Howard, S. Cobbold, and H. Waldmann, Immunoglobulin heavy chain locus of the rat: striking homology to mouse antibody genes, Proc. Natl. Acad. Sci. USA 83:6075 (1986).
8. R. Joho, C. Nottenburg, R. L. Coffman, and I. L. Weissman, Immunoglobulin gene rearrangement and expression during lymphocyte development, Curr. Top. Dev. Biol. 18:16 (1983).
9. R. L. Coffman and I. L. Weismann, Immunoglobulin gene rearrangement during pre-B cell differantiation, J. mol.cell. Immunol. 1:31 (1983).
10. A. M. Lawler, P. S. Lin, and P. J. Gearhart, Adult B-cell repertoire is biased toward two heavy-chain variable-region genes that rearrange frequently in fetal pre-B cells, Proc. Natl. Acad. Sci. USA 84:2454 (1987).
11. B. A. Malynn, J. E. Berman, G. D. Yancopoulos, C. A. Bona, and F. W. Alt, Expression of the immunoglobulin heavy-chain variable gene repertoire, Curr. Top Microbiol. Immunol. 135:75 (1987).

CHICKEN B LYMPHOCYTE DIFFERENTIATION: BURSAL MICROENVIRONMENT AND DIFFERENCES IN ONTOGENY BETWEEN NORMAL AND SPF BIRDS

Helen C. Ramm, Kathy Mitrangas, Trevor J. Wilson, Richard L. Boyd and Harry A. Ward

Department of Pathology and Immunology, Monash University Melbourne, Victoria, Australia

INTRODUCTION

The avian bursa of Fabricius provides a unique model for the investigation of B lymphocytopoiesis since its involvement in this process is essential (Grossi et al., 1976; Eerola et al., 1982; Boyd et al., 1983). This organ contributes significantly to the clonal expansion of B cells, the majority of follicles being derived from 2-5 stem cells committed prior to proliferation within the bursa (Pink et al., 1985b; Ratcliffe et al., 1986). Close to hatching, these differentiated cells acquire the ability to migrate to peripheral tissues (Boyd and Ward, 1984). A common precursor to the follicular cortex and medulla has been described (Pink et al., 1985a) and it appears less differentiated lymphocytes occur in the cortex (Boyd and Ward, 1978). It is also established that the bursal epithelium is receptive to stem cell influx between 8 and 15 days of embryogenesis (Le Douarin, 1986).

It is apparent from bursal studies that close interaction between stromal elements and B lymphocytes is essential during B cell differentiation. Whilst it is recognized that these provide a specific microenvironment, the nature of stromal cells and their products remains elusive. Previous experiments (St. Pierre and Ackerman, 1965; Dent et al., 1968; Heller and Friedman, 1979; Ekino et al., 1985) also suggest possible important influences of environmental antigens on B cell differentiation. Close associations between intestinal contents (Bockman and Cooper, 1976), bursal reticular epithelial (REp) cells and developing B lymphocytes have additionally been observed (Boyd et al., 1983; Boyd and Ward, 1984).

To investigate the role of environmental antigenic stimulation in early B lymphocyte differentiation, an ontogenic study was made in normal and specific pathogen-free (SPF) bursae. A panel of M.Abs was also used to relate B cell development to in situ associations with distinct stromal components involved in the regulation of intra-bursal B cell differentiation.

MATERIALS AND METHODS

Animals

Chickens and embryos (Australorp x White Leghorn F1 hybrids) were obtained from Research Poultry Farm, Research, Victoria, Australia. Specific pathogen-free (SPF) birds and eggs were obtained from C.S.I.R.O. Experimental Station, Maribyrnong, Victoria.

Preparation and staining of cell suspensions

Bursae removed from birds and embryos freshly killed by chloroform overdose and ice immersion respectively were pressed through a fine nylon mesh sieve into HBSS (0-4°C). Resultant cell suspensions were washed (x3) in HBSS (containing 0.01% NaN_3). For fixation, cells were resuspended in 1 ml 5% (v/v) FCS in HBSS, and 4 ml (w/v) formaldehyde was added dropwise with constant shaking for 20 min at 0-4°C. The cells were then pelleted (400 g_{max}, 5 min) resuspended in 5 ml 0.13% (v/v) Triton X-100 [Sigma, St. Louis, USA] in HBSS (R.T., 3 min), washed in HBSS and finally resuspended in 5% FCS in HBSS (0-4°C).

Approximately 10^6 cells were incubated with appropriate M.Ab supernatant (R.T., 30 min). Suspensions were then washed (x3) in HBSS and incubated with FITC-labelled goat anti-mouse IgG [Tago, Burlingame, California] (R.T., 30 min). Suspensions were washed as before and air-dried on slides. These were then fixed in absolute ethanol (30 s) and mounted in buffered glycerol/p-phenylenediamine. The preparations were examined with a Leitz Dialux 20 immunofluorescence microscope using narrow-band blue epi-illumination. The M.Abs used, were MUI-36 which displays non-Ig, non-BL B lymphocyte reactivity (Ward et al., in preparation) comparable to L22 or Bu-1a (Pink et al., 1985a), and a M.Ab detecting μ chains kindly provided by Professor M.D. Cooper.

In ovo injections

13-14 day old embryonated eggs were candled and small "windows" (~4 mm x 8 mm) cut in the shell over an area of underlying blood vesels. The embryos were injected intravenously with 0.1 ml LPS (100 µg/ml) [W.E. coli 0111:B4, Difco Labs, Detroit, USA]. Controls received 0.1 ml PBS. Approximately 12 embryos were injected in each group. The eggs were sealed with Parafilm and cell suspensions were prepared at 18 days of embryogenesis.

Preparation and staining of tissue sections

Bursae were removed from freshly killed birds and embryos. The procedure for immunofluorescence staining of frozen sections has been described in detail previously (Boyd et al., 1987b). For polyester wax sections, the tissue was fixed in absolute ethanol/glacial acetic acid (19:1 v/v) for 12-24 h depending on size (2 changes, 4°C), processed in absolute ethanol for 3 h (2 changes, 4°C), then infiltrated with polyester wax (BDH) at 35kPa for 3 h (3 changes, 37°C). Tissue was finally embedded in fresh wax and stored at 4°C. Sections (3 µm) were floated out onto a cold waterbath, rapidly picked up on gelatinized slides and dried for 2-3 h at 37°C before storing at 4°C. Immediately preceding staining, the sections were dewaxed by immersion in absolute ethanol (5-10 min), 90% ethanol (5 min), 70% ethanol (5 min) and then washed in PBS (1 min, 5 min).

Immunoperoxidase staining of sections was as follows. Endogenous peroxidase activity was blocked by immersion of slides in 3% H_2O_2 /70% ethanol (R.T., 10-30 min) during the dewaxing procedure. Sections were then incubated with either M.Ab or NS-1 conditioned medium and washed with PBS (1 min, 2x5 min, gentle shaking). The specificities and methods of production of the M.Abs to bursal stromal components have been described previously (Boyd et al., 1987a). Sections were next incubated with peroxidase-labelled rabbit anti-mouse Igs [Dakopatts, Copenhagen, Denmark] (R.T., 30 min) and washed as before. Slides were finally immersed in freshly prepared substrate solution (0.05% DAB, 0.01% H_2O_2 in PBS) for 8 min. Sections were then rinsed in H_2O, counterstained - if desired - in Harris' haematoxylin and, following dehydration mounted in coverslipping resin [Lab Tek Division, Miles Labs, Napperville, Illinois].

RESULTS

Ontogenic development of cu^+, su^+ and MUI-36$^+$ cells in normal and SPF bursae

No significant difference in percentage of cu^+ cells between these birds was observed until 18 days of incubation. Thereafter until 1 day post-hatching, a significantly higher proportion ($P<0.01$) of cu^+ cells was detected in normal birds, reaching a maximum at 20 days of embryogenesis. The corresponding peak in SPF chickens was not detected until 3 days post-hatching. This retarded development was reversible by intravenous injection of 13-14 day SPF embryos with 10 µg LPS - a statistically significant increase ($P<0.01$) from ~ 15% to 21% being observed in the percentage of cu^+ cells in 18 day embryos.

Significantly fewer ($P<0.02$) $s\mu^+$ and MUI-36$^+$ cells were observed in 18 day SPF embryos, after which similar distributions in both types of bird were recorded.

Table 1. A developmental comparison between normal and SPF bursae of % $c\mu^+$, su^+ and MUI-36$^+$ cells

AGE days	Normal chickens			SPF chickens		
	cµ	sµ	MUI-36	cµ	sµ	MUI-36
12e	1.2 ± 0.6	-	1.6 ± 0.7	1.5 ± 0.7	-	1.3 ± 0.7
14e	2.9 ± 1.0	-	1.3 ± 0.6	2.1 ± 0.8	-	0.7 ± 0.5
16e	5.0 ± 1.3	-	13.4 ± 2.2	4.4 ± 1.2	-	10.1 ± 1.8
18e	19.9 ± 2.4	12.9 ± 2.0	55.0 ± 4.3	14.7 ± 2.2	3.4 ± 1.0	53.0 ± 4.2
20e	22.8 ± 2.5	9.9 ± 1.8	77.1 ± 4.5	13.6 ± 2.0	12.6 ± 2.1	59.3 ± 4.9
1	22.3 ± 2.6	32.0 ± 3.2	79.3 ± 4.8	13.1 ± 2.0	19.9 ± 2.6	77.2 ± 4.6
3	21.4 ± 2.6	36.3 ± 3.4	95.4 ± 5.6	24.7 ± 2.8	31.6 ± 3.2	81.8 ± 5.1
5	13.8 ± 2.1	36.3 ± 3.5	94.1 ± 5.2	19.7 ± 2.5	30.4 ± 3.2	89.0 ± 5.4
7	12.3 ± 2.0	46.4 ± 3.8	93.3 ± 5.4	16.2 ± 2.2	45.0 ± 3.7	83.8 ± 5.2

- not detected; e denotes days of embryogenesis
Each data point has been calculated from a total cell count of at least 600 cells.
Results were obtained from 2 experiments; within each experiment the tests were performed in duplicate.
Tissues were pooled from 3-9 embryos or 2-3 post-hatch chickens.

Distribution of $c\mu^+$ and MUI-36$^+$ cells in normal and SPF bursae

At 12 days of embryogenesis, MUI-36$^+$ cells were predominantly located in the tunica propria but some were in perifollicular areas. Globular cytoplasmic-staining μ^+ cells were detected within follicles in 14 day embryos. Concomitant with follicle development was a marked expansion of both cell populations. Diffuse cytoplasmic-staining μ^+ cells whilst appearing within normal follicles at 18 days of embryogenesis, occurred in SPF birds 3 days post-hatching. Thereafter $c\mu^+$ cells were localized to the medulla. MUI-36$^+$ cells were present throughout the follicles; whilst a globular staining pattern was initially observed this antigen became restricted to the plasma membrane upon development.

Ontogenic development of stromal antigens

This ontogenic study revealed a sequential development of stromal antigens. M.Abs reactive with adult bursal surface epithelium represented the earliest antigens detected, appearing at 10 days of embryogenesis. Most M.Abs reacting with the adult bursa cortex stained only the tunica propria in 12 days embryos. Some M.Abs specific to the adult medulla were restricted to this region throughout development, while the majority of these M.Abs reacted with embryonic surface epithelium. BL (MUI-78) initially identified in the tunica propria at 12 days of embryogenesis was subsequently predominant in outer follicular areas.

DISCUSSION

B lymphocyte differentiation appears retarded in SPF birds hatched into a normal environment, the percentage of bursal $c\mu^+$ cells reaching a peak at 3 days in contrast to a maximum just prior to hatching in normal birds. The latter corresponds with the large intestine unfolding (Schaffner et al., 1974), enabling contact between developing B cells and both intestinal and yolk sac antigens. Histological investigations revealed a corresponding retardation in appearance of diffuse cytoplasmic-staining μ^+ cells, possibly indicating a different stage of μ chain synthesis within follicles. Since during later stages of development, $c\mu^+$ cells represent both plasma and pre-B cells, this delay may be attributed to inadequate stimulation of differentiated cells. This would result in fewer pre-B cells required to maintain the smaller pool of responding B lymphocytes and inevitably less plasma cells developing. A similar proposal has been suggested (Opstelten and Osmond, 1985) as pre-B cell proliferation in mouse bone marrow increased following intraperitoneal SRBC injection, indicating susceptibility of this population possibly mediated indirectly to extrinsic stimulation. However the $c\mu^+$ cells detected within the present study were not unequivocally shown to be classical pre-B cells. This requires verification by revealing an absence of light chains in double-labelling experiments.

A significantly lower percentage of $s\mu^+$ cells was observed in 18 day embryos, possibly again due to inadequate microenvironmental stimulation. Thereafter, a rapid rise in proportion of these cells was observed. These findings differ noticeably from previous investigations where cIgM$^+$sIgM$^+$ bursal cells (<0.1%) were seen in 12 day embryos, reaching adult values (>85%) in 18 day embryos (Cooper et al., 1976; Grossi et al., 1976). However since the predominant variable in these studies was the use of an anti-μ M.Ab in the current investigation, and chicken IgM being polymorphic (Ivanyi, 1975; Ratcliffe and Ivanyi, 1979; Pink et al., 1985b; Ratcliffe, 1985; Ratcliffe et al., 1986), it is suggested that this M.Ab recognized only 1 IgM allotype. Previous experiments (Ratcliffe and Ivanyi, 1979) detected ~45% sIgM-1a+ and ~46% sIgM-1b+ cells in 1 day old birds, and earlier attempts were unsuccessful.

A comparative investigation of the expression of another B cell antigen - MUI-36 - indicated similar distributions and staining patterns throughout ontogeny. Thus the percentage of cu+ cells gradually decreases possibly as a consequence of su expression upon maturation and the proportion of MUI-36+ cells increases with age as the bursa develops.

Recent studies in mice have revealed normal marrow B cell generation whilst regulated by endogenous factors can be increased through exogenous stimulation (Osmond et al., 1981; Fulop and Osmond, 1983a; Fulop and Osmond, 1983b; Opstelten and Osmond, 1985). Later experiments have suggested initial mediation by splenic macrophages (Osmond et al., 1985; Pietrangeli and Osmond, 1985). In a more detailed study (Ramm et al., 1987, in preparation) we report no effect on the chicken marrow B cell lineage after LPS stimulation, but a statistically significant rise in the proportion of bursal cu+ cells was detected; this is compatible with the murine results since the bone marrow is suggested to be the bursal equivalent in mammals.

Pronounced diversity in stromal components was also observed throughout ontogeny. Current double-labelling experiments using a selection of anti-stroma M.Abs, MUI-36 and the anti-μ M.Ab will enable clearer interpretation of stromal associations with B cells during development. It will be of interest to determine whether B cell precursors invade the surface epithelium where follicles are developing, or in conjunction with mesenchymal cells (Oláh et al., 1986), induce this event. Recently a subset of epithelial cells was detected overlying the area where follicles bulge into the lumen (Houssaint et al., 1986). These may be crucially involved in antigenic stimulation of developing B cells since this follicle-associated epithelium is capable of transporting particles (Beezhold et al., 1985). The altered staining pattern observed in this study, from embryonic surface epithelium to adult medulla upon maturation, agrees with the developmental relationship suggested between these regions (Le Dourain et al., 1975; Houssaint et al., 1976). A similar relationship is proposed for the adult cortex and tunica propria (Le Douarin et al., 1975; Boyd et al., 1976; Boyd and Ward, 1984).

Therefore past findings and the results obtained in this investigation, suggest environmental antigens contribute in the regulation of early intrabursal B cell differentiation; such an effect may indirectly cause a rise in cμ+ cell proliferation resulting in amplification of the responding pool of su+ lymphocytes. This study also revealed the extreme heterogeneity of the bursal stroma during development, indicating significant changes throughout ontogeny, possibly providing specific microenvironments for distinct phases of B cell development.

ACKNOWLEDGEMENTS

This work was supported by grants from the National Health and Medical Research Council, Australia (grant No. 860396), the Anti-Cancer Council of Victoria (grant No. 13165) and the Australian Research Grants Scheme (grant No. D28315655).

REFERENCES

Beezhold, D. H., Sachs, H. G., and Van Alten, P.J., 1985, Dev. Comp. Immunol. 9:351.

Bockman, D. E., and Cooper, M. D., 1976, in: Proc. International Symposium on the Phylogeny of T and B cells, R. K. Wright, and E. L. Cooper, eds., Elsevier/North Holland Biomedical Press, Amsterdam, p. 277.

Boyd, R. L., Mitrangas, K., Ramm, H. C., Wilson, T. J., Fahey, K. J., and Ward, H. A., 1987b, in: Progress in Clinical Biological Research, Vol. 238 "Avian Immunology", W. T. Weber, and D. L. Ewert, eds., Alan R. Liss Inc., New York, p. 41.

Boyd, R. L., and Ward, H. A., 1978, Immunology 34:9.

Boyd, R. L., and Ward, H. A., 1984, Dev. Comp. Immunol. 8:149.

Boyd, R. L., Ward, H. A., and Muller, H. K., 1976, Int. Arch. Allergy Appl. Immunol. 50:129.

Boyd, R. L., Ward, H. A., and Muller, H. K., 1983, J. Retic. Soc. 34:383.

Boyd, R. L., Wilson, T. J., Mitrangas, K., and Ward, H. A., 1987a, in: Progress in Clinical Biological Research, Vol. 238 "Avian Immunology", W. T. Weber, and D. L. Ewert, eds., Alan R. Liss Inc., New York, p. 29.

Cooper, M. D., Kearney, J. F., Lydyard, P. M., Grossi, C. E., and Lawton, A. R., 1976, Cold Spr. Harb. Symp. Quant. Biol. 41:139.

Dent, P. B., Perey, D. Y. E., Cooper, M. D., and Good, R. A., 1968, J. Immunol. 101:799.

Eerola, E., Lassila, O., Gilmour, D. G., and Toivanen, A., 1982, J. Immunol. 128:2652.

Ekino, S., Suginohara, K., Urano, T., Fujii, H., Matsuno, K., and Kotani, M., 1985, immunology 55:405.

Fulop, G. M., and Osmond, D. G., 1983a, Cell Immunol. 75:80.

Fulop, G. M., and Osmond, D. G., 1983b, Cell Immunol. 75:91.

Grossi, C. E., Lydyard, P. M., and Cooper, M. D., 1976, Ann. Immunol. (Inst. Pasteur) 127:931.

Heller, D. E., and Friedman, A. R., 1979, Dev. Comp. Biol. 3:667.

Houssaint, E., Belo, M., and Le Douarin, N. M., 1976, Dev. Biol. 53:250.

Houssaint, E., Diez, E., and Hallet, M., 1986, Immnology 58:43.

Ivanyi, J., 1975, J. Immunogenetics 2:69.

Le Douarin, N., 1986, Dev. Biol. 20:291.

Le Douarin, N. M., Houssaint, E., Jotereau, F. V., and Belo, M., 1975, Proc. Natl. Acad. Sci. USA 72:2701.

Oláh, I., Glick, B., and Toro, I., 1986, Poultry Science 65:574.

Opstelten, D., and Osmond, D. G., 1985, Eur. J. Immunol. 15:599.

Osmond, D. G., Fahlman, M. T. E., Fulop, G. M., and Rahal, D. M., 1981, Ciba Foundation Symposium 84, Pitman Medical, London, p.68.

Osmond, D. G., Fulop, G. M., Opstelten, D., and Pietrangeli, C., 1985, Adv. Exp. Med. Biol. 186:35.

Pietrangeli, C. E., and Osmond, D. G., 1985, Cell Immunol. 94:147.

Pink, J. R., Ratcliffe, M. J. H., and Vainio, O., 1985b, Eur. J. Immunol. 15:617.

Pink, J. R., Vainio, O., and Rijnbeek, A., 1985a, Eur. J. Immunol. 15:83.

Ratcliffe, M. J. H., 1985, Immunol. Today 6:223.

Ratcliffe, M. J. H., and Ivanyi, J., 1979, Immunogenetics 9:149.

Ratcliffe, M. J. H., Lassila, O., Pink, J. R. L., and Vainio, O., 1986, Eur. J. Immnol. 16:129.

Schaffner, T., Hess, M. W., and Cottier, H., 1974, Ser. Haemat. VII(4):568.

St. Pierre, R. L. and Ackerman, G. A., 1965, Science 147:1307.

RESPONSE OF THE CHICKEN BURSAL STROMA TO TREATMENT WITH CYCLOPHOSPHAMIDE AND IBD VIRUS

T.J. Wilson, K. Mitrangas, H.C. Ramm, R.L. Boyd and H.A. Ward

Department of Pathology and Immunology, Monash University Medical School, Prahran, 3181, Victoria, Australia

INTRODUCTION

Pluripotential stem cells appear to require discrete microenvironmental factors to differentiate along a particular developmental pathway. In the chicken, the avian bursa of Fabricius provides a complex collection of cells and factors (1), which enable blood borne precursor cells to differentiate into functional B-lymphocytes (2). The exact nature of the stromal components affecting B-cell differentiation, however, remains poorly defined.

Injection of cyclophosphamide (Cy) into chickens destroys bursal lymphocytes resulting in severe immunodeficiency. Transplantation of stem cells into these chickens restores B-cell function (3), indicating that the essential stromal elements are Cy resistant. Infectious bursal disease virus (IBDV) is a naturally occurring double stranded RNA virus which induces humoral immunodepression in chickens (4). The mechanisms underlying the IBDV effect however, have not been adequately investigated. There is evidence of viral infection of IgM positive B-cells (5,6) but this does not exclude infection of bursal non-lymphoid cells as a possible cause of the disease. Thus Cy-treated and IBDV infected chickens are ideal models for the study of bursal microenvironmental factors influencing B-cell differentiation.

We have previously described a panel of monoclonal antibodies (M.Ab's) reactive with the non-lymphoid components of the chicken bursa and thymus (7). In the present study we have used two models, Cy-treatment and IBDV infection, to gain insight into the relative importance of the different antigens.

Following IBDV infection, in addition to the loss of bursal lymphocytes, there was a substantial reduction in the expression of antigens on the surface epithelium (S.Ep), medullary epithelium and basement membrane associated epithelium. Cy, however, induced an increased expression of two medullary antigens while other determinants were expressed on the follicle associated epithelium (FAE) which are normally present only during ontogeny.

MATERIALS AND METHODS

Chickens: Australorp X White Leghorn F1 hybrid embryos were obtained from Research Poultry Farm (Research, Victoria, Australia). Specific pathogen-free (SPF) birds were obtained from CSIRO, Division of Animal Health (Parkville, Victoria, Australia).

Cyclophosphamide Treatment: Commencing on the day of hatching, chicks were given i.p. injections of 3mg Cy (Cycloblastin, Farmitalia Carlo Erba, Hawthorn, Victoria, Australia) on four consecutive days (8). Seven, 14, and 21 days after the final injection bursae, thymuses and spleens were removed from injected and age-matched control animals, and were snap frozen in a liquid nitrogen-isopentane slurry.

IBDV Infection: SPF chickens were infected intra-ocularly with a purified Australian isolate of IBDV (9). The thymus, bursa and spleen were removed from birds at 24, 48, 72 and 96 hours post-infection and immediately snap frozen. Duplicate uninfected chickens served as controls.

Monoclonal Antibodies and Indirect Immunofluorescence: Monoclonal antibodies (M.Ab) which react with various components of the bursal and thymic stroma were prepared and characterized as described elsewhere (7). A panel of 30 of these, reacting with distinct antigens, was chosen for this study, and staining visualized by standard indirect immunofluorescence procedures (7). Briefly, 4um cryostat sections were incubated with M.Ab or NS-1 culture supernatants (20min) and washed 3 times with PBS (1x1min, 2x5min). Finally they were incubated with an anti-mouse Ig FITC conjugate (Silenus Labs, Melbourne, Australia) for 20min and washed as before. The sections were examined on a Leitz Diavert Fluorescence Microscope (narrow band blue illumination) and photos were taken using a Leitz Photoautomat camera and Kodak Ektachrome 1600 ASA film.

RESULTS

Cyclophosphamide Treatment

Following Cy treatment the bursae demonstrated severe atrophy in conjunction with depletion of lymphoid cells and showed no reconstitution during the 3 weeks post treatment. Of the thirty M.Ab's screened on frozen sections of these bursae, 22 demonstrated no significant alteration in the distribution of their antigens (Table 1). The major components of the bursa including the cortex, medulla, surface epithelium (S.Ep), follicle associated epithelium (FAE), basement membrane (BM) and basement membrane associated epithelium (BMAE) all expressed one or more antigens which were not sensitive to Cy. The determinant recognized by MUI-78, which detects a B-L-like molecule, was also resistant. The remaining 8 M.Ab identified modifications in the treated bursae concomitant with a loss of B-lymphocytes (Table 2).

MUI-36 identifies a marker on the B-cell lineage in addition to a subpopulation of bursal and thymic medullary stromal cells (Ward et al. manuscript in preparation). The non-lymphoid cells bearing this marker were resistant to the Cy treatment although the MUI-36 positive bursal lymphocytes were depleted within one week of the Cy injections. Corresponding to the loss of lymphocytes, an influx of macrophages (MO), identified by MUI-79, was observed in both the cortex and medulla. This M.Ab stains monocytes in peripheral blood and MO-like cells in many tissues, whereas MUI-66, a M.Ab which identifies MO-like cells in the bursal cortex and medulla, was confluent in the medulla post-Cy treatment.

Table I. Monoclonal Antibodies which do not alter their Specificity on Chicken Bursa Post-Cyclophosphamide Treatment

Specificity	Monoclonal
S.Ep, especially FAE.	MUI-60.
S.Ep and Med Ep.	MUI-55, 62, 64, 65, 71 & 80.
S.Ep and BM.	MUI-70.
Isol. in S.Ep and med.	MUI-77.
FAE. and med Ep.	MUI-61.
S.Ep and BMAE.	MUI-53.
Cort. cells and med. Ep.	MUI-78, 57, 63.
Med. Ep.	MUI-67, 68 & 69.
Med MO.	MUI-81.
T-cells or Thymic Stroma.	MUI-56, 82, 83.

Abbreviations:- cort:-cortex; med:-medulla; isol:-isolated; Ep:-epithelium; S.Ep:-surface epithelium; BM:-basement membrane; BMAE:-basement membrane associated epithelium; FAE:-follicle associated epithelium; MO:-macrophage-like cells.

The bursal FAE was also modified within one week of Cy treatment. Two M.Ab's (MUI-51 and MUI-58) identify antigens expressed on the S.Ep and BMAE in the control chickens. Following Cy treatment however, there was an increased expression of these antigens on the FAE similar to their distribution during ontogeny where they are identified on the FAE and epithelial buds. The antigen detected by M.Ab MUI-52, normally exclusively expressed on isolated stellate cells in the thymic cortex, was also expressed on the FAE post Cy treatment.

M.Ab's MUI-54 and MUI-72 both demonstrated extensive, increased antigenic expression in the bursal medulla from two weeks post-treatment. In the control chicks MUI-54 stained isolated medullary epithelial cells in addition to S.Ep and BMAE. MUI-72, however, detected cells in the bursal cortex, with only rare fibres seen in the medulla.

Thymus and spleen were also examined from treated birds and controls. In the thymus there was an alteration in a number of antigens, particularly in the cortex, one week after the Cy injections. By two weeks there was no observed difference between treated and control thymic tissue. No alterations were observed in the treated spleen other than the depletion of B-cells.

Effect of IBDV Infection

In chickens infected with IBDV (Fig.1, Table III), there was a rapid, progressive loss of over 90% MUI-36 positive cells in the bursa, spleen and peripheral blood over the 4 days. As observed with Cy treatment there was a resistant population of MUI-36 antigen bearing cells in the outer medulla. The viral infection had a varied effect on bursal stromal antigens. Similar to that demonstrated with Cy the vast majority of antigens showed no significant alteration, including all those expressed in the cortex. Whereas on the S.Ep, MUI-58 and MUI-64 showed a loss of staining by 48 hours, and MUI-54 and 80 by 96 hours. In the medulla four antigens, detected by MUI-61, 62, 63 and 65, were lost by 96 hours after infection. The loss of medullary antigens, particularly MUI-61, was gradual, closely parallelling that of the MUI-36 positive cells.

Table II. Alterations in the Staining of Chicken Bursa Following Cyclophosphamide Treatment

Normal Staining Pattern	Post-Cyclophosphamide Staining Pattern	Monoclonal
Lymphocytes.	Isol. cells in outer med.	MUI-36.
+ S.Ep, BMAE.	+ S.Ep, BMAE, FAE ++.	MUI-51 & 58.
+ Cort. & med.	+ Cort. & med., FAE ++.	MUI-52.
S.Ep, BMAE, isol med Ep.	S.Ep, BMAE, extensive med.	MUI-54.
+++ isol. cort. cells, rare med fibres.	++ isol. cort. cells, extensive med fibres.	MUI-72
+++ cort. MO, + med MO.	+++ cort. MO, med.	MUI-66.
few MO cort/interstit.	MO throughout.	MUI-79.

Abbreviations:- cort:-cortex; med:-medulla; isol:-isolated; S.Ep:-surface epithelium; BMAE:-basement membrane associated epithelium; FAE:-Follicle Associated Epithelium; MO:-macrophage-like cells.

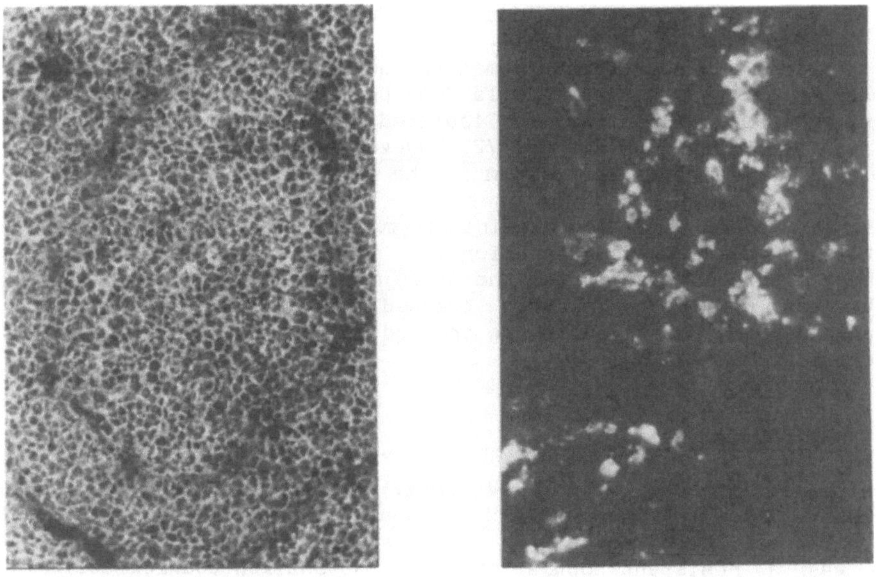

1(a) 1(b)

Figure 1. Immunofluorescent staining of adult chicken bursa:- MUI-36 on (a) Normal; (b) 96 hours after IBDV infection (x250).

Table III. Alterations in Monoclonal Antibody Staining of Chicken Bursa Following Inoculation with Infectious Bursal Disease Virus (IBDV)

Normal Staining Pattern	Specificity Post IBDV Infection	Monoclonals
Lymphocytes.	Isol outer med cells	MUI-36
BMAE., S.Ep.	Neg by 48 hrs.	MUI-58,64.
Med Ep., S.Ep.	Neg by 96 hrs.	MUI-54,80.
Med Ep.,S.Ep.	Isol. fibres by 96 hrs.	MUI-61,62, 63 & 65.

Abbreviations:- cort:-cortex; med:-medulla; isol:-isolated; BMAE:-basement membrane associated epithelium; S.Ep:-Surface Epithelium; neg:-negative; Ep:-Epithelium.

DISCUSSION

Cy-treatment and IBDV infection both deplete B-cells in chickens. Simultaneously with the loss of lymphocytes we have demonstrated an alteration in a number of stromal antigens with an increase in the expression of some antigens on the FAE and bursal medullary epithelium following Cy-injection and a decrease in the same regions post-IBDV infection.

Cy is believed to deplete lymphocytes by the crosslinking of DNA and thus fast dividing cells are primarily effected. Therefore bursae post-Cy treatment are depleted of lymphoid cells while the stroma is left intact (8). Similarly, bursal lymphoid cells, particularly those expressing surface IgM have been shown to be the targets for IBDV infection (5,6). Thus if the effect on the bursal stroma is secondary in both these conditions, we would expect to detect similar alterations in the bursal non-lymphoid components.

In the bursa cortex, no alteration was observed in the antigenic distribution in either condition. In the medulla however, two antigens detected by MUI-54 and 72 showed increased expression following Cy-treatment while post-IBDV infection 6 antibodies detecting medullary epithelial antigens (MUI-54, 61, 62, 63, 65 & 80) demonstrated substantially reduced staining. IBDV infection also destroyed antigens expressed on the S.Ep, with determinants identified by MUI-54, 58, 64 and 80 all depleted by 96 hours post-infection. Conversely after Cy-treatment, the FAE re-expressed three antigens recognised by MUI-51, 52 and 58 which are normally expressed only *in ovo*.

The majority of stromal antigens were unaltered following Cy-treatment and therefore they may have potential roles in B-cell differentiation. The expression of some antigens, however, increased on the FAE and bursal medulla and this may indicate that they have a function which is normally regulated by lymphoid cells.

Our data also indicate that IBD virus has a more direct influence on the bursal epithelium. IBDV has been shown to infect reticular epithelial cells, macrophages and other stromal cells of the bursa, in addition to lymphocytes (10). Stromal cells may be necessary for the development of the immunodeficient state in infected animals as peripheral lymphocytes are poor targets for IBDV infection (5), while surgically bursectomized chickens did not develop the clinical symptoms of the disease post-infection (11).

We propose that the immunodepression of chickens infected with IBDV is not caused solely by the direct viral infection of B-cells, but also by the depletion of some important stromal elements in the bursal microenvironment. Thus these antigens which we have demonstrated to have altered distribution post-IBDV infection may potentially have important functions in B-cell differentiation. The investigation of these particular stromal molecules, using further M.Ab to inhibit reconstitution of bursae with stem cells during ontogeny and after Cy-treatment, is currently in progress.

ACKNOWLEDGEMENTS

This work was supported by grants from the National Health and Medical Research Council, Australia (grant No. 860396), the Anti-Cancer Council of Victoria (grant No. 131658) and the Australian Research Grants Scheme (grant No. D28315655).

REFERENCES

1. B. Glick, Adv. Vet. Sci. Comp. Med. 30:67 (1985).

2. N.M. Le Douarin, E. Houssaint, F.U. Jotereau and M. Belo, Proc. Natl. Acad. Sci. USA. 72:2701 (1975)

3. A. Toivanen, J. Eskola and P. Toivanen, Ann. Immunol. (Inst. Pasteur) 127C:923 (1976)

4. L. Hudson, M. Pattison and N. Thantrey, Eur. J. Immunol. 5:675 (1975)

5. K. Hirai and R.W. Calnek, Infect. Immun. 25:964 (1979)

6. T. Nakai and K. Hirai, Avian Dis. 25:831 (1981)

7. R.L. Boyd, T.J. Wilson, K. Mitrangas and H.A. Ward, in: Progress in Clinical Biological Research Vol. 238:"Avian Immunology," W.T. Weber and D.L. Ewert, ed., Alan R. Liss Inc., New York, p29 (1987)

8. P. Toivanen and A. Toivanen, Eur. J. Immunol. 3:585 (1973)

9. K.J. Fahey, I.J. O'Donell and A.A. Azad, J. Gen. Virol. 66:1479 (1985)

10. I. Kaufer and E. Weiss, Avian Dis. 20:483 (1976)

11. I. Kaufer and E. Weiss, Infect. Immune. 27:364 (1980)

AGE-DEPENDENCE OF LYMPHOCYTE PRODUCTION IN PEYER'S PATCH FOLLICLES IN CONTRAST TO THE OTHER PEYER'S PATCH COMPARTMENTS AND THE THYMUS

H.J. Rothkötter, M. Geist, F.J. Fritz and R. Pabst

Centre of Anatomy, Medical School of Hannover
D-3000 Hannover 61, West Germany

INTRODUCTION

The gut-associated lymphoid tissue consists of intraepithelial and lamina propria lymphocytes and organized lymphoid tissue, the Peyer's patches (PP). In pigs there are two types of PP, discrete ones in the jejunum (jejPP) and a continuous PP in the ileum (ilPP). Their role in the pig's lymphoid system is not known: probably antigen from the gut lumen crossing the epithelium of the PP stimulates their lymphocytopoiesis.

Differences have been shown in the homing pattern of blood lymphocytes to jejPP and ilPP (1). Morphological differences have also been described, especially in the dome epithelium (2). Both types of PP develop during gestation, but a division of follicles and interfollicular area was first observed after birth (3). In a previous study using pigs of 13 weeks of age, lymphocyte production was observed not only in the follicles but also in the interfollicular area of PP and no differences were found between jejPP and ilPP (4). It is not known, however, whether differences occur earlier in their postnatal development.

In the present study 5 groups of pigs at different ages were used, the youngest being 1 day old. The lymphocyte production in jejPP and ilPP follicles, interfollicular area and dome was determined by the metaphase arrest technique using vincristine. The lymphocyte production in the cortex and medulla of the thymus was also studied. An attempt was made to answer the following questions:

1. Is lymphocyte production in the PP of the pig age-dependent?
2. Is there a difference in lymphocyte production between jejPP and ilPP in general or at different ages?
3. Do follicles, dome and interfollicular area differ in lymphocyte production?
4. Is there a difference in lymphocyte production between the PP and the thymus?

MATERIALS AND METHODS

Forty-one young pigs of the German Landrace, which were conventionally reared, were used in this study. Their age was 1 day (n=12), 5 days (n=4), 12 days (n=9), 40 days (n=12) and 13 weeks (n=6).

The metaphase arrest was carried out using Vincristine (VCR) at a dose of 0.25 mg/kg body weight. The animals were treated in two different ways: A. Animals of 1, 5 and 12 days of age received the VCR injection intraperitoneally, those of 40 days of age intravenously. In these four groups at least two animals were sacrificed by an overdose of barbiturate 1, 2, 2.5, 3.5 h after the VCR injection. Two pigs of each group were sacrificed after injection of NaCl as a negative control. In each animal, samples were taken from the jejPP, ilPP and thymus. B. The 13 week old animals were anesthetized by intravenous injections of thiopental. In each pig, samples were taken from the jejPP, ilPP and thymus before and 1, 2, 2.5, 3.5 h after injecting VCR using operative techniques.

All samples were fixed in Schaffer's solution, embedded in glycolmethacrylate, cut at 1 µm or 4 µm and stained with Giemsa solution. On histological sections at least 2000 lymphocytes and the number of mitoses were counted in the dome, interfollicular area and follicles of the PP and in the cortex and medulla of the thymus.

The rate of cells entering mitosis per hour is called r_m, and it was calculated by linear regression analysis using the scientific tables of Geigy. The increase in the line of regression equals the increase in the mitotic rate per hour. The mean and standard deviation are given.

RESULTS

Lymphocyte production was observed in each compartment of the PP and thymus studied here. No anaphases were detected.

Peyer's patches

In PP follicles the mitotic rate per hour increased dramatically from 1 d to 13 weeks (Fig. 1). There was no significant difference between jejPP and ilPP. In jejPP the rate was 0.02 ± 0.07%/h on d 1 and increased to 4.35 ± 0.16%/h at 13 weeks. In ilPP the mitotic rate per hour increased from 0.24 ± 0.14%/h on d 1 to 3.05 ± 0.16%/h at 13 weeks. In the interfollicular area only a small increase in the mitotic rate per hour was observed during postnatal development: in jejPP from 0.11 ± 0.05%/h on d 1 to 0.43 ± 0.05%/h at 13 weeks. This was comparable to the ilPP (0.10 ± 0.04%/h on d 1 to 0.53 ± 0.09%/h at 13 weeks). In the dome the mitotic rate per hour also showed a small increase: in jejPP from 0.12 ± 0.06%/h on d 1 to 0.52 ± 0.14%/h on d 40, and in ilPP from 0.08 ± 0.04%/h on d 1 to 0.61 ± 0.19%/h on d 40. Again, no significant difference was observed between jejPP and ilPP (Fig. 1).

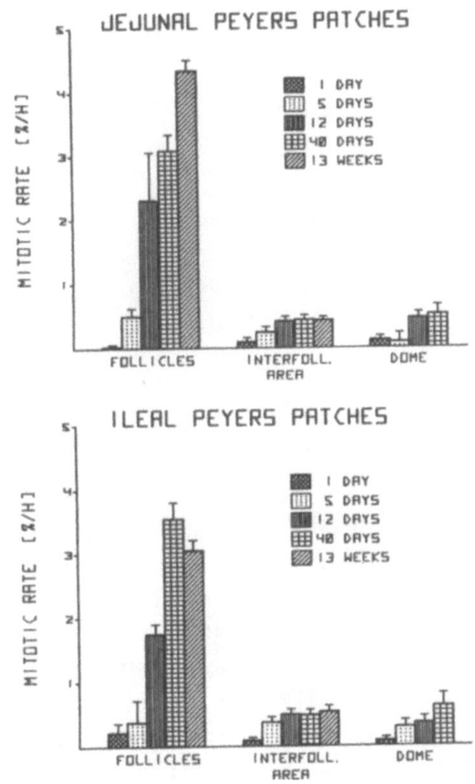

Fig. 1 Comparison of the increase in mitotic rate per hour after vincristine in the three compartments of jejunal and ileal PP. The mitotic rate was calculated by linear regression analysis.

Thymus

The mitotic rate per hour in the cortex and the medulla of the thymus showed small differences at the different ages. In the thymic cortex the mitotic rate per hour ranged from 0.26 ± 0.09%/h (d 1) to 0.43 ± 0.03%/h (13 weeks). In the medulla the mitotic rate per hour varied between 0.07 ± 0.05%/h (d 12) and 0.32 ± 0.03%/h (13 weeks) (Fig. 2).

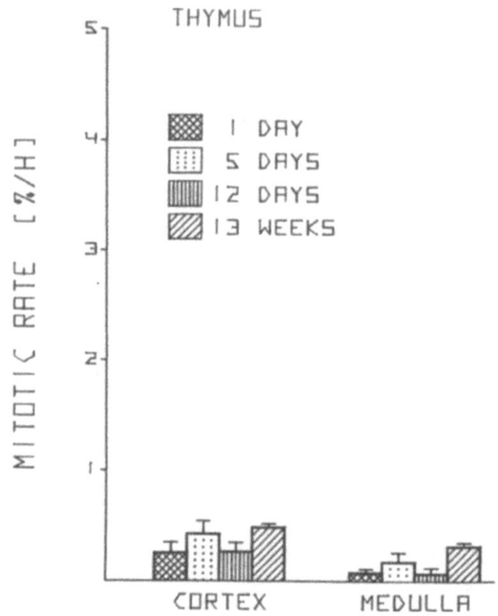

Fig. 2 Mitotic rate per hour in the cortex and medulla of the thymus after vincristine.

DISCUSSION

Lymphocytopoiesis is one functional aspect of lymphoid organs and can be studied by the metaphase arrest technique. This technique enables a histotopographical localization of the lymphocyte production in the different compartments of a lymphoid organ. A particular problem with the technique is the correct dose of VCR. In the present experiments the animals received a dose of 0.25 mg/kg body weight. After the same dose in a previous study neither degenerated metaphases nor anaphases were observed (4).

Lymphocytes are produced in all compartments of PP, but during early postnatal growth the production only increases in follicles. This might be an effect of antigen in the gut lumen. Immediately after birth the gut-associated lymphoid tissue is exposed to nutritional and microbial antigens. The question is, however, why the dome and interfollicular area only show little lymphocyte production, which is not age-dependent.

In lambs there were no differences between fetal and postnatal lymphocyte production in PP (5). The mitotic rate per hour in jejPP follicles was 4.8 $\pm$ 0.2%/h in fetal lambs and 5.4 $\pm$ 0.6%/h in 6 week old animals. The results in ilPP follicles were similar: 5.2 $\pm$ 0.6%/h in fetal and 5.4 $\pm$ 0.8%/h in 6 week old animals (5). Thus, lymphocytopoiesis in PP of lambs seems to be antigen-independent as in a "primary lymphoid organ". This is obviously not the case in pigs.

For comparison, the lymphocyte production of the thymus was determined in this study. The cortex and the medulla showed an age-independent mitotic rate. These results are similar to those in lambs (5).

The present study provides no evidence that jejPP and ilPP differ in lymphocytopoiesis at any time point, which is similar to data on fetal and postnatal lambs (5).

The fate of the newly formed lymphocytes in the follicles, interfollicular area and dome is not clear. How many of these cells emigrate to other lymphoid organs? This question could be answered by selectively labelling the PP with DNA precursors, a method used previously in lambs (6). However, it is impossible to locate the compartment of the PP where the newly formed lymphocytes come from.

ACKNOWLEDGEMENTS

This study was supported by a grant from the Deutsche Forschungsgemeinschaft (Pa 240/5).

REFERENCES

1. R.M. Binns, and S.T. Licence, in: "Microenvironments in the Lymphoid System," p. 661, G.G.B. Klaus, ed., Plenum Press, New York, (1985).
2. R.M. Chu, and C.H. Liu. Vet. Immunol. Immunopathol. 6:391 (1984).
3. H.A. Chapman, J.S. Johnson, and M.D. Cooper. J. Immunol. 112:555 (1974).
4. R. Pabst, and F.J. Fritz. Cell Tissue Res. 245:423 (1986).
5. J.D. Reynolds. Eur. J. Immunol. 17:503 (1987).
6 R. Pabst, and J.D. Reynolds. J. Immunol. 136:2011 (1986).

MIGRATION OF NEWLY-PRODUCED VIRGIN B CELLS FROM BONE MARROW TO SECONDARY LYMPHOID ORGANS

J.E. Lortan, S. OLdfield, C.A. Roobottom and I.C.M. MacLennan

Dept. of Immunology, the Medical School
University of Birmingham, England

If marked B cells from thoracic duct lymph are transferred to syngeneic recipients they characteristically migrate via high endothelial venules or marginal zone blood sinusoids to the follicles of secondary lymphoid organs (1,2). Less is known of the capacity of virgin B cells, recently produced in the bone marrow, to migrate to secondary lymphoid organs. Under normal conditions the rate of primary B lymphopoiesis greatly exceeds the rate at which B cells are lost from the peripheral B cell pool (3). It follows that most newly-produced virgin B cells can only have a short life span. We have identified two pathways by which newly-produced virgin B cells are recruited to the peripheral pool. One appears to be antigen-independent and recruits B cells in sufficient numbers to replenish loss from the peripheral pool (4,5). If the peripheral pool is depleted selectively by cytotoxic agents a high proportion of the bone marrow output is recruited until the peripheral pool is replenished. The second mechanism resulting in virgin B cell recruitment appears to be antigen-dependent. In this situation newly produced virgin B cells are activated by T cell-dependent antigens and the memory B cells generated may become part of the peripheral pool. This process appears to occur only in periods immediately following exposure to antigen (6). The established phase of T cell-dependent responses seems to be maintained by sequential cycles of memory B cell activation with little new virgin B cell recruitment (3,5,6).

The present study investigates the capacity of newly-produced virgin B cells from the marrow to migrate to secondary lymphoid organs. A full report of this study is being published elsewhere (7).

Materials and Methods

Animals

Congenic strains of rats which differ in their kappa immunoglobulin light chain allotype were bred in the animal house, Department of Immunology, University of Birmingham. The strains are PVG/Ciu and PVG-RI-A^{a} (DA) (8) which have k1a and k1b allotypes respectively, and will be referred to as PVG-k1a and PVG-k1b. Adult rats were used of between 3 and 6 months of age.

Cell collection and transfer

Normal, intact PVG-k1b recipients received 10^8, PVG-k1a thoracic duct lymphocytes (TDL) or 10^8 bone marrow (B.M.) mononuclear cells suspended in 0.5 ml RPMI-1640, by slow intravenous injection. TDL were collected from PVG-k1a animals (9). Bone marrow cells depleted of recirculating cells were obtained from PVG-k1a rats which had undergone continuous thoracic duct drainage for 48 hrs.

Antisera

Monoclonal mouse anti-rat k1a (MARK-3) (10) and polyclonal rabbit anti-rat IgM and rabbit anti-rat IgD (11) were kind gifts from Professor H. Bazin, Institute D'Immunolgie Experimentale, University of Louvain, Brussels. Sheep anti-mouse IgG (heavy and light chains) and a horseradish peroxidase-conjugate of sheep anti-mouse IgG were obtained from I.D.R.L., The Medical School, Birmingham. Swine immunoglobulins to rabbit immunoglobulins, soluble complex of horseradish peroxidase and rabbit anti-horseradish peroxidase (rabbit PAP); soluble complex of alkaline phosphatase and mouse anti-alkaline phosphatase (APAAP) were purchased from Dakopatts, Denmark. Fluorescein isothiocyanate-conjugated sheep anti-mouse IgG (heavy and light chains) was kindly provided by Dr G.D. Johnson, Department of Immunology, The Medical School, Birmingham.

Immunofluorescence of cell suspensions

The proportion of nucleated cells with surface membrane immunoglobulin in normal rat TDL and BM was determined by staining in suspension (12).

Immunohistology

PVG-k1b recipient animals were sacrificed at 2, 8, 24 or 72 hours after receiving 10^8 donor TDL or B.M. cells by intravenous injection. Portions of spleen and cervical lymph nodes were snap frozen in liquid nitrogen. Frozen sections were stained by a double immunoenzyme technique using the PAP system to demonstrate either surface IgM or IgD (13) and the APAAP system to identify the transferred PVG-k1a B lymphocytes (14). This allowed the location of the transferred B lymphocytes to be seen in the same section in relation to the lymphoid compartments of the spleen and lymph node.

Quantitative microscopy

The number of transferred donor k1a cells was determined in unit areas of the tissues using a light microscope, and according to the point counting method described by Weibel (15). Results are expressed as the number of k1a-positive B cells per 10^3 line intercepts on a 1cm graticule which was placed in one eyepiece.

Results

Experimental design

B cell transfers were carried out between rats of congenic strains which differ in kappa immunoglobulin light chain allotype. This allows trasferred cells to be identified in tissue sections with immunohistological techniques using anti-allotype antibodies. Peripheral B cells were obtained from thoracic duct lymph, while newly-produced virgin B cells were obtained from the bone marrow of rats which had been depleted of recirculating B cells by chronic thoracic duct drainage. In both cases B cells were transferred by slow intravenous injection. At intervals after transfer groups of rats were killed and their secondary lymphoid organs examined for the presence and distribution of transferred B cells.

The B cell content of donor preparations

Using immunofluorescence it was determined that a mean of 42.9% (SD±13) of nucleated cells in thoracic duct lymph from PVG-k1a donor rats were k1a positive, and a mean of 5.4% (SD±2.26) of nucleated bone marrow cells bear surface membrane immunoglobulin.

Distribution of donor B cells within cervical lymph nodes

This is shown in Figure 1a.

Thoracic Duct B cells: Two hours after transfer, most of the B cells of thoracic duct origin were situated in extrafollicular areas of the lymph nodes; particularly, the perilymphatic regions of the paracortex and the corticomedullary junction. By 8 hours after transfer the majority of transferred B cells were seen within follicles (Figures 2a and 2b) and this distribution pattern remained constant throughout the period of observation.

B cells from bone marrow depleted of recirculating cells: As seen with thoracic duct B cells, at 2 hours most transferred cells were found in the perilymphatic and corticomedullary areas of the paracortex. However, at 8 and 24 hours after transfer these cells were still predominantly outside follicles (Figure 2c and 2d). It was only by 3 days that the proportion of these cells seen within follicles increased.

Distribution of donor B cells within the spleen

This is illustrated in Figure 1b.

Thoracic duct B cells: Two hours after intravenous transfer approximately two thirds of the donor cells were found outside follicles within the periarteriolar lymphocytic sheath and red pulp. By 8 hours, as seen with lymph nodes, more than two thirds of the cells were now within follicles. This pattern persisted at 24 and 72 hours after transfer.

B cells from bone marrow depleted of recirculating cells: At 2 hours after injection most donor B cells were extrafollicular. In sharp contrast to the picture seen with thoracic duct B cells, 8 and 24 hours after transfer more than two thirds of these cells remained outside follicles. By 3 days, the number of cells within follicles had increased but still only represented 40% of the transferred cells identified within the spleen .

Discussion

The present study confirms the conclusion derived from functional studies, that newly-produced virgin B cells can gain access to antigen in secondary lymphoid organs. However, this access is mainly confined to those areas rich in interdigitating cells and T cells. There is a considerable body of evidence which supports the view that interdigitating

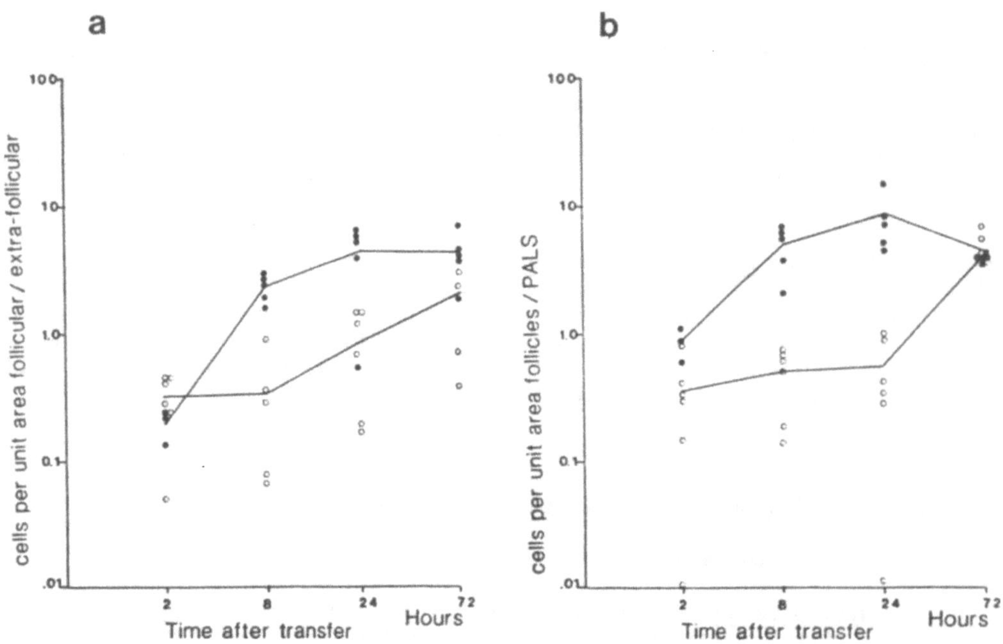

Figure 1. Ratios of donor B cells a) in follicles of lymph node to those in extrafollicular areas and b) in follicles of spleen to those in the periarteriolar lymphocytic sheath at intervals following transfer of k allotype-marked B cells from: (O) B.M. depleted of recirculating cells and (●) thoracic duct lymph.

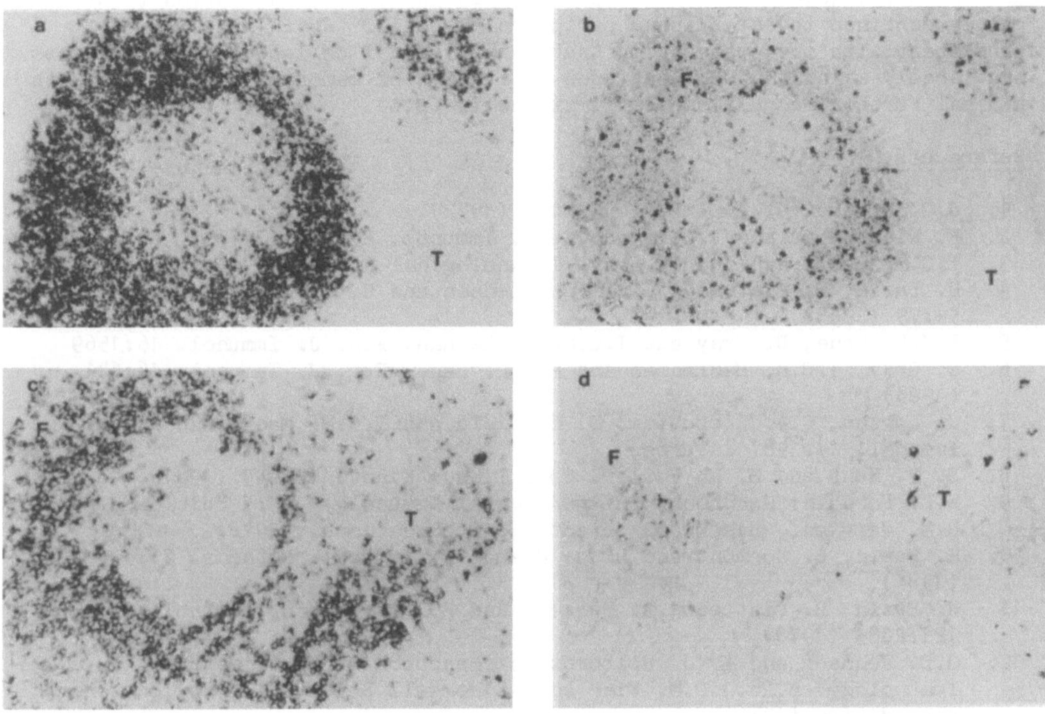

Figure 2. Location of donor cells, within lymph nodes, from TDL (a and b) or B.M. (c and d) 8 h after transfer. Sections were double stained to demonstrate both surface membrane IgD and transferred B cells using rabbit anti-rat IgD and a peroxidase system together with MARK-3 anti-rat k1a followed by an alkaline phosphatase system. The peroxidase staining was developed using diaminobenzidine to give a brown product and alkaline phosphatase staining was developed with naphthol AS-MX phosphate and fast blue BB salt to give a blue product. Photomicrographs a and c were taken through a blue filter transmitting light of wavelength <510 nm to emphasise the brown, sIgD-bearing cells. Photomicrographs c and d were taken through a red filter transmitting light of wavelength >590 nm to show the blue, k1a-positive, donor B cells.

F = follicle, T = T cell-rich area

cells can present antigen to T cells and assist in T cell-dependent B cell activation (16,17). However, the turnover rate of interdigitating cells appears to be relatively rapid so it is unlikely that antigen taken up by these cells can persist in interdigitating cell-rich areas for extended periods. This is consistent with the observation that activation of newly-produced virgin B cells by T cell-dependent antigens is mainly confined to periods immediately following administration of antigen (3,6). In contrast, antigen taken up by follicular dendritic cells can persist in follicles for months (18). As newly-produced virgin-B cells have limited access to follicles one might expect follicular B cell activation to be mainly confined to established peripheral B cells. Again this is consistent with the observation that established T cell-dependent responses are largely maintained by continued activation of memory B cell clones with minimal further recruitment of virgin cells (3,6).

References

1. J.G. Howard. J. Exp. Med. 135:185 (1972).
2. P. Niewenhuis and W.L. Ford. Cell. Immunol. 23:254 (1976).
3. I.C.M. MacLennan and D. Gray. Immunological Reviews. 91:61 (1986).
4. H. Bazin, B. Platteau, I.C.M. MacLennan and G. D. Johnson. Immunol. 54:79 (1985).
5. P.J.L. Lane, D. Gray and I.C.M. MacLennan. Eur. J. Immunol. 16:1569
6. D. Gray, I.C.M. MacLennan and P.J.L. Lane. Eur. J. Immunol. 16:641 (1986)
7. J. Lortan, C.A. Roobottam, S. Oldfield and I.C.M. MacLennan. Eur J. Immunol. 17:1987 in press
8. S. V. Hunt and M. H. Fowler. Cell Tissue Kinet. 14:445 (1981).
9. W.L. Ford in:"Handbook of Experimental Immunolgy". 3rd Edition p.23.1, D.M. Weir ed., Blackwell Scientific Publications. (1978).
10. H. Bazin, L. Cormont and L. de Clerq. J. Immunol. Methods. 71:9 (1984)
11. H. Bazin, B. Platteau, A. Beckers and R. Pauwels. J. Immunol. 121:2083 (1978).
12. G.D. Johnson and E. J. Holborow in "Handbook of Experimental Immunology" p.28.1 D.M. Wier ed., Blackwell Scientific publications 4th Edition (1986).
13. G. Hoffman-Feazer, H. Rodt, M. Eulitz and S. J. Thierfelder. J. Immunol. Methods. 13:261 (1976).
14. J. L. Cordell, B. Falini, W. Euber, A.K. Ghosh, Z. Abdulaziz, S. MacDonald, K. Pulford, H Stein and D.Y. Mason. J. Histochem. Cytochem. 32:219 (1984).
15. E. R. Weibel. Lab. Invest. 12:131 (1963).
16. S. C. Knight, C. M. Balfour, J. O'Brien, L. Battifaust, T. Sumersk, and J. Clarke. Eur. J. Immunol. 12:1057 (1982).
17. K. Inaba, M.D. Witmer and R.M. Syeinman. J. Exp. Med. 160:858 (1984).
18. T.E. Mandel, R.P. Phipps, A.P. Abbot and J.G. Tew. Immunology 43:353 (1981).

MEMORY B CELLS GENERATED IN T CELL-DEPENDENT ANTIBODY RESPONSES COLONISE THE SPLENIC MARGINAL ZONE

Susan Oldfield, Liu Yong-Jun, Martin Beaman and C.M. MacLennan

Dept. of Immunology, University of Birmingham
Birmingham, U.K.

Introduction

The marginal zones are a major B cell compartment in both rat and human spleen. These B cells possess a distinct phenotype in that they express surface membrane IgM (sIgM) with little surface membrane IgD (sIgD). Although they do not recirculate as such, they are derived from recirculating precursors (1). They appear in normal numbers in congenitally athymic animals and much indirect evidence suggests they can be activated by thymus-independent type 2 (TI-2) antigens (2). However, experiments presented here indicate that some memory B cells generated during established T cell-dependent antibody responses colonise the marginal zones and take up the characteristic phenotype of marginal zone B cells. These cells have been revealed using a method developed by Claasen & Van Rooijen (3) in which haptenated enzymes are used to study the appearance of specific cells producing anti-hapten antibodies. These workers used this technique to identify antibody-secreting cells. However, during the course of the experiments reported it became clear that hapten-specific surface immunoglobulin can also be identified by this technique.

Materials and Methods

Animals. Wistar rats between 3 and 6 months of age were used in all experiments except those involving cell transfer, where PVG/k1a and PVG/k1b congenic pairs were used. These differ in the allotype of their kappa immunoglobulin light chain: PVG/k1a rats were used as donors and PVG/k1b rats as recipients.

Antigens and antisera

Spider crab (Maia Squida) haemocyanin (MSH) and DNP-hydroxyethyl starch were gifts of Professor Humphrey, Royal Postgraduate Medical School, London. Dinitrophenylated MSH (DNP-MSH) was prepared according to Gray et al (4) and oxazalone-MSH according to Mäkelä et al (5). Preparation of oxazalone-Ficoll was based on the method of Inman (6). The monoclonal antibody HIS 14 was a gift from Dr. Kroese and Professor Nieuwenhuis,

University of Groningen, The Netherlands and rabbit anti-rat IgM and rabbit anti-rat IgD were gifts from Professor Bazin, University of Louvain, Brussels, Belgium. Sheep anti-mouse IgG labelled with peroxidase was obtained from I.D.R.L., The Medical School, University of Birmingham. Swine anti-rabbit immunoglobulin and peroxidase/anti-peroxidase complexes were purchased from Dakopatts, Denmark.

Immunisations

Primary immunisations with 50μg alum-precipitated MSH or dinitrophenylated haemocyanin (DNP-MSH) together with 10^9 chemically killed Bordatella pertussis were given intraperitoneally. Secondary immunisations with 50μg DNP-MSH, 50μg oxazalone-MSH (Ox-MSH), 20μg DNP-hydroxyethyl starch (DNP-HES) or 150μg oxazalone-Ficoll (Ox-Ficoll) were given either subcutaneously into the footpad or intravenously.

Cell transfer

Thoracic duct lymphocytes were collected into ice-cold RPMI 1640 containing heparin (20 units/ml) (7). Cells were harvested after 24 hours, 3×10^8 lymphocytes resuspended in 0.5ml RPMI 1640 and injected intravenously via the tail vein.

Immunohistology

Spleens were snap frozen in liquid nitrogen and stored at -70°C. 5μm frozen sections were double stained with DNP-alkaline phosphatase and oxazalone-peroxidase (3) to demonstrate specific antibody-containing cells. Alternatively sections were stained with DNP-alkaline phosphatase or oxazalone-alkaline phosphatase together with anti-rat IgM or anti-rat IgD, using a peroxidase anti-peroxidase technique (8) or with the monoclonal antibody HIS 14 which was revealed with a sheep anti-mouse peroxidase conjugate.

Results

Production of specific plasma cells in the spleen

Rats were primed with MSH and boosted with DNP-MSH and Ox-MSH in the left and right footpads respectively. Many specific antibody-containing cells were seen in the red pulp of the spleen and around the outer edge of the periarteriolar lymphocytic sheath four days after immunisation as previously observed by Claasen et al (9). These cells stained densely with hapten-enzyme conjugates and had the appearance of plasma cells. Double-staining with DNP-alkaline phosphatase and oxazalone-peroxidase demonstrated that almost all the specific antibody-containing cells bound either one or the other hapten but not both. The numbers of specific plasma cells was maximal on day four after immunisation and then fell rapidly to low levels. Serum antibody concentrations remained high and it is likely that production of antibody shifted to the bone marrow (10). Following immunisation with the thymus-independent type 2 antigens, plasma cells were again seen in the red pulp and outer edge of the periarteriolar lymphocytic sheath (Figure 1). They persisted longer than splenic plasma cells in the TD reponse and were seen both at 7 and 12 days after immunisation.

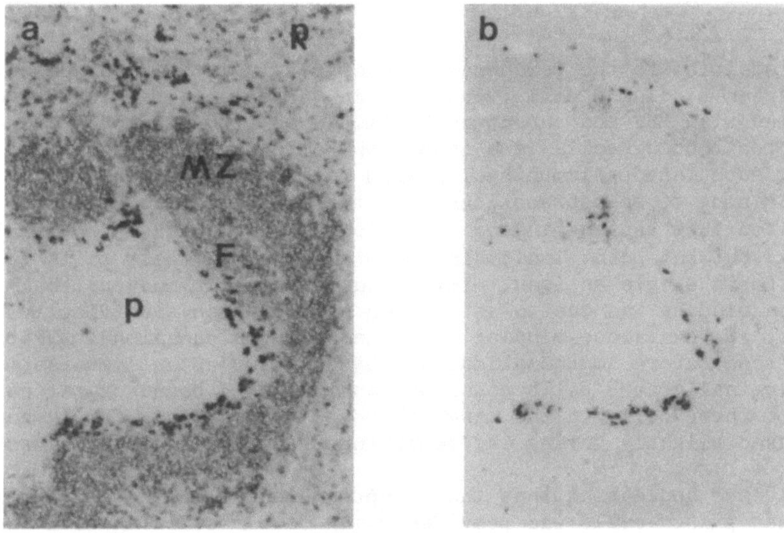

Figure 1. Anti-DNP antibody-secreting cells in the periarteriolar lymphocytic sheath seven days after immunisation with DNP-HES. a) shows a section of spleen stained with HIS 14 and DNP-alkaline phosphatase b) shows the same section photographed through a red filter to highlight blue, DNP-binding cells. MZ = marginal zone, F = follicle, P = periarteriolar lymphocytic sheath, R = red pulp.

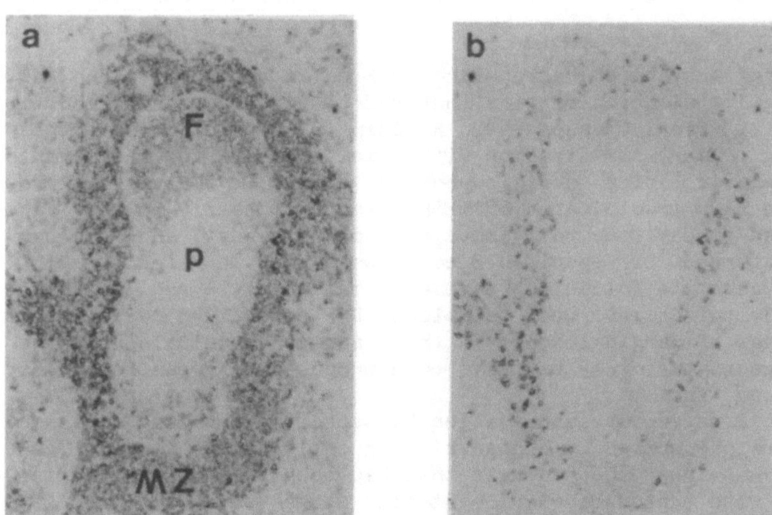

Figure 2. DNP-binding cells in the splenic marginal zone seven days after immunisation with T-cell dependent antigens. Sections were stained and photographed as in Fig 1a & b.

Appearance of antigen-binding B cells in the marginal zone following TD immunisation

In addition to the presence of specific plasma cells as described above, hapten-binding B cells were seen in the splenic marginal zones of rats primed with MSH and subsequently immunised with DNP-MSH and Ox-MSH. These hapten-binding cells were seen both following intravenous and following subcutaneous immunisation with hapten-MSH (Figure 2). Double-staining of spleen sections from rats that received both antigens demonstrated that hapten-binding cells in the marginal zone bound either DNP or Ox, but not both, indicating that antibody on these cells is specific for a single antigen. This finding makes it most unlikely that the hapten binding was due to antibodies adsorbed non-specifically by Fc receptors. No oxazalone-binding cells and only an occasional DNP-binding cell was seen before immunisation, but by day four after immunisation up to 10% of marginal zone B cells bound DNP and up to 5% bound oxazalone. The numbers of these cells in the marginal zone then remained fairly constant or increased slightly during the remaining 3 weeks of the experiments.

Evidence indicating that these hapten-binding cells were B cells rather than plasma cells was provided using double-staining with haptenated alkaline phosphatase and HIS 14, a monoclonal antibody that binds to B and pre-B cells but not to plasma cells (11). The hapten-binding cells in the marginal zones were found to be His 14 +ve. Double-staining to demonstrate hapten-binding together with IgM or IgD, demonstrated that the hapten-binding B cells had the characteristic marginal zone phenotype bearing sIgM but not sIgD. Immunisation with the TI-2 antigens, DNP-HES and Ox-Ficoll, does not lead to the accumulation of hapten-binding B cells within the marginal zone, although a good serum antibody response was seen and plasma cells were present in the spleen (Figure 1).

Hapten-binding B cells in the marginal zone are not associated with primary responses

Large numbers of hapten-binding B cells are seen in the marginal zone following immunisation of rats that had previously been primed with carrier. However, Claasen & Van Rooijen (9) did not observe these cells following primary immunisation with trinitrophenylated haemocyanin by intravenous injection on six successive days. We immunised a group of ten rats with alum-precipitated DNP-MSH intraperitoneally. Three rats were sacrificed on day four post immunisation, 3 on day 7 and 4 on day twelve. Small numbers of DNP-specific B cells were seen in a few areas of marginal zone, adjacent to follicles, in one animal at day 4 and two animals at day 7 after immunisation. Only occasional DNP-specific B cells were seen in other areas of marginal zones in these and other animals at these times. This suggests that they may have been produced in local follicles.

At 12 days post immunisation DNP-specific B cells were seen in all 4 rats. The proportion of hapten-binding cells in the marginal zones, however, was lower than in rats that had been primed with carrier before boosting with hapten-carrier. At this stage they were now scattered randomly throughout the marginal zone .

Inability to transfer hapten-binding B cells into the marginal zone from thoracic duct lymph

Experiments were performed to investigate whether adoptive transfer of thoracic duct lymphocytes would lead to the appearance of hapten-binding B cells within the marginal zones of recipient animals. 3×10^8 thoracic duct lymphocytes from PVG/k1a rats, primed with MSH and reimmunised subcutaneously with DNP-MSH, were transferred into normal PVG/k1b animals. Recipient animals were killed 10 days later and frozen sections of spleen stained to demonstrate DNP-binding cells together with donor B cells. Very few DNP-binding cells were seen in the marginal zone and those which were positive were of host origin. The numbers seen were comparable to the levels found in animals which had not been immunised. Donor animals had good numbers of hapten-binding cells in their splenic marginal zones after 48 hrs thoracic duct drainage.

In a second experiment, 3 x 10^8 thoracic duct lymphocytes from PVG/k1a rats that had been primed and boosted with DNP-MSH, were transferred into PVG/κ1b rats that had received 8Gy γ-irradiation with shielding of the hind limbs to preserve the bone marrow. Two recipients were immunised with DNP-MSH five days later and the serum response measured. Both animals had made specific antibody of donor allotype indicating that memory cells had indeed been transferred. All animals were killed 10 days after transfer. No DNP-binding cells were found in the marginal zones of animals that had not been immunised although most marginal zone B cells were of donor origin. Hapten-binding cells were found in the two recipient rats which had been immunised. These data indicate that recirculating memory B cells do not colonise the marginal zones. It may be that hapten-binding cells migrate directly to marginal zones from adjacent follicles, as suggested by Nieuwenhuis and Keuning (12).

Conclusion

Antigen-specific memory B cells colonise the splenic marginal zone following immunisation with T cell-dependent, but not TI-2 antigens. These cells have the characteristic marginal zone phenotype, are generated locally and are probably independent of the recirculating memory B cell pool.

This is unlikely to be the only source of marginal zone B cells as marginal zone develops normally both in congenitally athymic and germ-free rats.

References

1) D.S. Kumararatne, and I.C.M. MacLennan. Eur. J. Immunol. 11: 865 (1981).

2) P.J.L. Lane, D. Gray, S. Oldfield, and I.C.M. MacLennan. Eur. J. Immunol. 16: 1569 (1986).

3) E. Claasen, and N. Van Rooijen. J. Histochem. and Cytochem. 33: 840 (1985).

4) D. Gray, D. Chassoux, I.C.M. MacLennan, and H. Bazin. Clin. exp. Immunol. 60: 70 (1985).

5) O. Mäkelä, M. Kartinen, J.L.T. Pelkonen, and K. Karjalainen. J. Exp. Med. 148: 1645 (1978).

6) J.K. Inman. J. Immunol. 114: 704 (1975).

7) W.L. Ford, in: "Handbook of Experimental Immunology", D.M. Weir, ed., Blackwell Scientific Publications, 3rd edition, Oxford (1978).

8) G. Hoffmann-Fezer, H. Rodt, M. Eulitz, S. Thierfelder. J. Immunol. Methods 13: 261 (1976).

9) E. Claasen, N. Kors, C.D. Dijkstra, and N. Van Rooijen. Immunol. 57: 399 (1985).

10) R. Benner, W. Hijmans, and J.J. Haaijman. Clin. Exp. Immunol. 46: 1 (1981).

11) F.G.M. Kroese, D. Opstelten, A.S. Wubbena, F.J. Deener, J. Ater, E.H. Schwander, L. de Leij, and P. Nieuwenhuis. Adv. Exp. Med. Biol. 186: 81 (1985).

12) P. Nieuwenhius, and F.J. Keuning. Immunol. 26: 509 (1974).

MARGINAL ZONE B CELLS AND THE LOCALISATION OF ANTIGEN ON FOLLICULAR DENDRITIC CELLS

S. Oldfield, J.E. Lortan, M.A.J. Hyatt and I.C.M. MacLennan

Department of Immunology, The Medical School
University of Birmingham, Birmingham, United Kingdom

The mechanism of uptake of antigen by follicular dendritic cells (FDC)is poorly understood, but complement and preformed antibody have been shown to be necessary for this to process (1). Brown et al (2) suggested that lymphocytes may be involved in the active transport of antigen into follicles. More recent studies (3,4) have led to the suggestion that it may be marginal zone B cells which transport antigen into splenic follicles.

B cells in the marginal zone of the spleen do not recirculate. They are larger than recirculating follicular B cells, and express surface membrane IgM in the absence of easily detected surface IgD (5). In the rat they have been shown to have Fc receptors and receptors for C3b, C3bi and C3d (6). In addition, marginal zone B lymphocytes are stimulated to move into follicles following intravenous injection of gram-negative bacteria or endotoxin (3,7). This migration probably occurs via an intrasplenic route (3). Thirty minutes after intravenous injection, large molecular weight antigens like aggregated human IgG, are demonstrable within the splenic marginal zone, both within macrophages and on the surface of lymphoid cells (2,8). Twenty four hours later the antigen is almost exclusively within follicles. Thus B lymphocytes in the marginal zone are well placed to bind antigen in the form of immune complexes and transport it into follicles.

External irradiation of rats with 8 Gy, competely depletes secondary lymphoid organs of lymphocytes without destroying FDC or displacing previously localised immune complex. If primary B lymphopoietic capacity is retained by hind limb shielding, follicles are completely repopulated from bone marrow precursor cells by 7 days, whereas it takes about 3 weeks for marginal zone reconstitution (9). Using this model, it is possible to investigate whether localisation of immune complex on FDC can take place in rats that have follicular B cells but which are deficient in marginal zone B cells.

Materials and Methods

Animals

Normal adult male Lou/C rats were bred under standard laboratory conditions in the animal house of the Department of Immunology, University of Birmingham. They were used at 3 to 6 months of age and were matched for age within each group.

Total body irradiation

An ACEL beam ^{60}Co source in the department of Radiation Physics, University of Birmingham, was used to deliver 8 Gy γ-irradiation over 5 minutes. The animals were anaesthetised with Hypnorm (Janssen Pharmaceuticals) and lead blocks were positioned to shield their hind limbs from the irradiation beam. This conserves 10-20% of the rats' haemopoietic capacity.

Aggregated fluorescein-conjugated human IgG

Human IgG (10) was conjugated to fluorescein-isothiocyanate (isomer 1, 10% on Celite from BDH chemicals, Poole, UK) (11). The conjugate was then aggregated by heating (12). Each animal received 2 mg of the aggregated fluorescein-conjugated human IgG by intravenous injection, before or 3,7 or 21 days after, irradiation.

Collection and preparation of tissue

Twenty four hours after the injection of the aggregated flourescein-conjugated human IgG the animals were sacrificed and the spleens snap-frozen in liquid nitrogen. Double immunofluorescence staining (11) of frozen sections was used to demonstrate simultaneous antigen localisation and the level of reconstitution of lymphoid compartments.

Antisera

Polyclonal rabbit anti-rat IgM was a kind gift from Professor H. Bazin, Institute d'Immunologie Experimentale, University of Louvain, Brussels. This was conjugated to Biotin N-hydroxysuccinimide ester (Miles-Yeda, Slough, G.B.) (13). Mouse monoclonal antibody, KiM4R which recognises rat follicular dendritic cells was a kind gift from Dr Waker and Prof Parawaresh, Kiel, FRG. Rhodamine 600- conjugated avidin D was purchased from Seralab Ltd., Crawley Down, G.B. and was used to identify the anti-rat IgM. Rhodamine-conjugated rabbit anti-mouse immunoglobulin was obtained from DAKO, High Wycombe, G.B. and was use to identify KiM4R.

Results

Four matched groups of adult rats were used in these experiments. Three of these groups received 8Gy whole body irradiation with hind limb shielding to preserve haemopoietic activity including a capacity for primary B lymphopoiesis. These irradiated groups respectively received intravenous, fluorescein-conjugated, aggregated IgG by intravenous injections 3, 7 or 21 days later and were killed 24 hours after these injections. The non-irradiated group also received aggregated IgG and was examined 24 hours later. The level of localisation of immune complex in splenic follicles; the number of B cells in the follicles and marginal zones of the spleen and the presence of follicular dendritic cells were then assessed for each animal. The results are shown in the table 1 and figures 1 and 2. Irradiation did not result in loss of FDC as assessed by KiM4R in any of the animals examined. However, immune complex only localised in those rats which had normal numbers of marginal zone B cells i.e. in the non-irradiated group and in those rats examined 3 weeks after

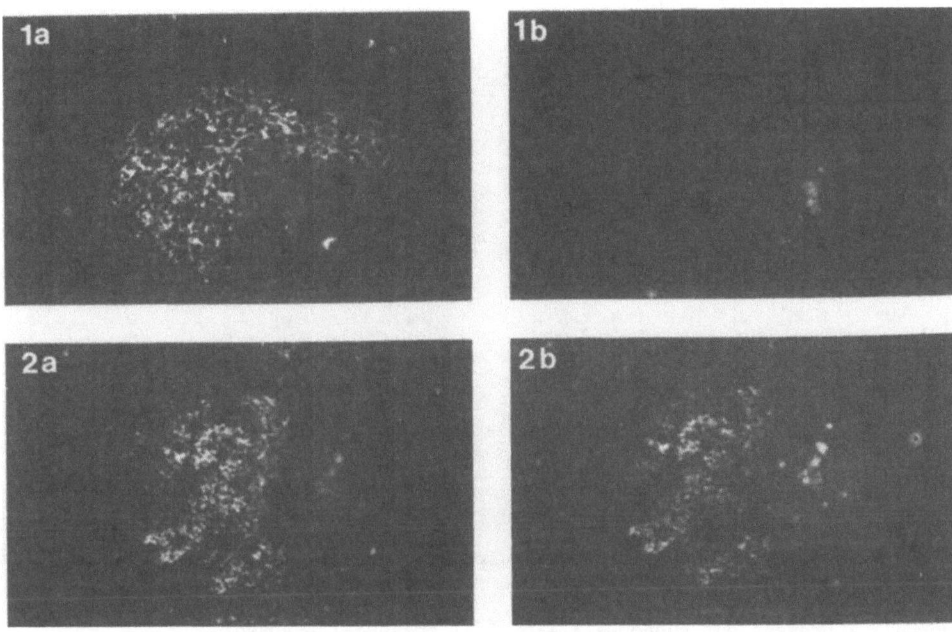

Figure 1

Frozen sections of spleen from rats irradiated with hind limb shielding.
1. from a rat 7 days post irradiation and 2. from a rat 21 days post irradiation. Both rats had received fluorescein-conjugated aggregated IgG by intravenous injection 24 hrs before sacrifice.

Photomicrographs are of sections stained by double immunofluorescence to reveal FDC using kiM4R (1a and 2a) and localisation of IgG aggregates (1b and 2b).

irradiation. At 7 days post irradiation there were good levels of follicular B cells but no localisation of immune complex was achieved at this stage.

Table 1

Localisation of Aggregated Fluorescein-IgG within Wollicles and the Presence of Cells in Lymphoid Compartments of the Spleen after Whole Body Irradiation with Hind Limb Shielding

Time after irradiation	Follicular Dendritic cells	Follicular B cells	T cells in T areas	Marginal zone B cells	Aggregated IgG localisation in FDC
No irradiation	++	++	++	++	++
	++	++	++	++	++
	++	++	++	++	++
3 days	++	+	+	-	-
	++	+	+	-	-
	++	+	+	-	-
7 days	++	++	+	±	-
	++	++	+	±	-
	++	++	+	±	-
21 days	++	++	+	++	++
	++	++	+	++	++
	++	++	+	++	++

All rats received fluorescein-conjugated IgG aggregates by intravenous injection 24 hrs before sacrifice

++=normal; +=present but subnormal; ±=minimal presence; -=absent.

A further set of experiments were performed in which two groups of rats received 8Gy total body irradiation with hind limb shielding. The following day the rats of one group were each reconstituted with $3x10^8$ syngeneic thoracic duct lymphocytes. All rats were injected intravenously with fluorescein- conjugated aggregated IgG 7 days after irradiation and assessed for lymphocyte reconstitution and localisation of the fluorescein-labelled aggregated IgG 24 hours after injection. The results are summarised in table 2. Both groups of irradiated rats had follicles well filled with B cells and FDC. However, only those irradiated rats which had been reconstituted with thoracic duct lymphocytes had substantial numbers of marginal zone cells, but in only one of these reconstituted rats did the number of marginal zone B cells approach normal levels. No localization of aggregated IgG was seen in the irradiated but not reconstituted group. The irradiated and reconstituted group had localised aggregated IgG although the amount of aggregate found in FDC in these rats was less than that seen in the control rats. These data indicate that the FDC of irradiated rats are able to accept IgG aggregates providing thay have had a radiation sensitive component replaced; this replacement being achieved with thoracic duct lymphocytes.

Table 2

The Ability of Thoracic Duct Lymphocytes to Reconstitute Marginal Zone B Cells in Irradiated Rats and the Capacity for Localisation of Aggregated IgG on FDC

Rat treatment		Immunohistological finding		
8Gy TBI hind limb shielding day 0	3×10^8 Syngeneic Thoracic duct lymphocytes day 1	Follicular B cells	Marginal zone B cells	Aggregated IgG localization on FDC day 7
Yes	No	++	±	-
		++	±	-
		++	±	-
Yes	Yes	++	++	+
		++	+	+
		++	±	±
No	No	++	++	++
		++	++	++
		++	++	++

Fluorescein conjugated IgG aggregates were injected i.v. in all rats 24 hours sacrifice before

++=normal; +=present but subnormal; ±=minimal presence; -=absent.

Discussion

Efficient localisation of immune complexes or aggregated IgG on FDC *in vivo* appears to require a radiosensitive population of cells which can be restored one week following transfer of syngeneic thoracic duct lymphocytes. Follicular B cells do not appear to have this capacity, for rats irradiated with hind limb shielding have good follicles within 7 days of irradiation but do not localise aggregated IgG. The present study and that of Kroese et al (4) show that failure of FDC to accept immune complexes is not due to irradiation damage. For rats reconstituted with thoracic duct lymphocytes recover the capacity for immune complex localisation within a week of irradiation. The data in the present study add to the indirect evidence which indicate that it is marginal zone B cells which are responsible for active transport of immune complexes or IgG aggregates to FDC. These studies provide direct evidence that follicular B cells are unable to act in this way.

References

1. G.G.B. Klaus and J. H. Humphrey. Immunology 33:31 (1977).

2. J.C. Brown, G. Harris, M. Papamichail, V.S. Sljivic, E. J. Holborow. Immunology 24:955 (1973).

3. D. Gray, D.S. Kumararatne, J. E. Lortan, M. Khan and I.C.M. MacLennan. Immunology 52:659 (1984).

4. F.G.M. Kroese, A.S. Wubbena and P. Niewenhuis. Immunology 57:99 (1986).

5. D. Gray, I.C.M. MacLennan, H. Bazin and M. Khan. Eur. J. Immunol. 12:564 (1982).

6. D. Gray, I. McConnell, D.S. Kumararatne, I.C. M. MacLennan, J. H. Humphrey and H. Bazin. Eur. J. Immunol. 14:47 (1984).

7. J.C. Pettersen, D.F. Borgen, K.C. Graupner. Amer. J. Anat. 121:305 (1967).

8. A.J.P. Veerman and N. van Rooijen. Cell Tissue Res. 161:211 (1975).

9. P.J.L. Lane, D.Gray and I.C.M.MacLennan. Eur J. Immunol (1986).

10. L. Hudson and F.C. Hay "Practical Immunology" p.1. Blackwell Scientific Pyublications 2nd Edition (1980).

11. G. D. Johnson and E. J. Holborow *in*: "Handbook of Experimental Immunology" p.28.1, D.M.Weir ed., Blackwell Scientific Publications 4th Edition (1986).

12. M.R. Haeney *in*: "Techniques in Clinical Immunology "p.173. R. A. Thompson ed., Blackwell Scientific Publications 2nd Edition (1981).

13. E. A. Bayer, E. Skutelsky and M. Wilcheck. Methods in Enzymology 62:308 (1979).

CBA/N MICE HAVE MARGINAL ZONE B CELLS WITH NORMAL SURFACE IMMUNOGLOBULIN PHENOTYPE

Liu Yong-Jun, Jennifer E. Lortan, Susan Oldfield and Ian C.M. MacLennan

Department of Immunology, University of Birmingham
Birmingham, U.K.

Introduction

Antigens can be broadly classified into those which require T cell help to evoke an immune response and those which do not (1). However, the T cell-independent (TI) antigens can be further subdivided on the basis of time during ontogeny when they can evoke antibody respones (2). In man the TI-1 group of antigens can activate B cells in vivo before birth while TI-2 antigens fail to evoke normal responses until some months after birth. Another criterion which allows the TI-1 and TI-2 antigens to be distinguished is the relative capacity of these two groups of antigens to generate antibody responses in CBA/N mice (3). TI-1 antigens evoke responses in these mice while TI-2 antigens do not. TI-2 antigens are mostly based upon a core structure of polysaccharide. Haptens linked to polysaccharides such as Ficoll or hydroxyethyl starch, behave as TI-2 antigens.

Several lines of evidence implicate splenic marginal zone B cells in responses to TI-2 antigens (4). These include a) the presence of dendritic cells, situated in the marginal zones of the spleen, which localise polysaccharides (5,6); b) association of responsiveness to TI-2 antigens with the presence of marginal zone B cells but not the presence of follicular B cells, in experiments in vivo where B cells are depleted and selectively reconstituted (4); c) the selective depressive effect of adult splenectomy on the capacity of rodents to respond to TI-2 antigens (7).

CBA/N mice carry on X-linked immunodeficiency attributable to the x-id gene. It has been shown that the profile of surface immunoglobulin expression of splenic B cells in these mice is abnormal. They are in particular deficient in B cells which have high levels of surface IgD with relatively low levels of surface IgM (8). Paradoxically splenic marginal zone B cells express IgM on their surface with relatively little IgD (9). CBA/N mice appear to have normal numbers of splenic B cells with this pattern of surface immunoglobulin expression. This suggests that CBA/N mice might have normal numbers of marginal zone B cells. However, recent studies by Forrester et al indicate that there is active B lymphopoiesis in the spleens of young CBA/N mice (10). Some of the newly-produced virgin B cells at that site might be expected to be IgM+ve, IgD-ve.

The object of the present study has been to examine the spleens of CBA/N mice using immunohistology to quantitate the size of B cell areas and determine the phenotype of B cells in these areas. CBA/Ca mice which lack the x-id gene were used as controls. Young CBA/N mice were found to have a marked reduction in the size of their follicles and some reduction in the size of their marginal zones. No qualitative abnormality in the surface immunoglobulin phenotype of B cells in these areas was apparent.

Methods

Animals CBA/N x-id/x-id female mice and CBA/Ca female mice which lack the x-id gene were bred in the Department of Zoology animal house of the University of Edinburgh and were kindly donated by Dr H S Micklem.

Antisera Rabbit anti-mouse IgD was a kind gift of Dr R Coico, Department of Pathology, New York University Medical Centre. Sheep anti-mouse IgM, sheep anti-mouse IgG3, sheep-anti mouse IgA, rhodamine-conjugated pig anti-sheep immunoglobulin, horseradish peroxidase-conjugated pig anti-sheep immunoglobulin and avidin-conjugated horseradish peroxidase were all produced by BDS biologicals, Birmingham, England and were a kind gift of Mr R Drew. Biotinylated mouse monoclonal anti-Thy 1-2 was purchased from Becton and Dickinson. Fluorescein conjugated mouse anti-mouse IgD^a allotype AMS15.1, was a kind gift from Dr A M Stall, Department of Genetics, Stanford University Medical School, California (11).

Immunohistology Serial 5μm frozen sections of spleen were stained to demonstrate IgM, IgD and Thy 1-2, respectively, using immunoperoxidase methods. Sections were incubated with sheep anti-mouse IgM, rabbit anti-mouse IgD or biotinylated mouse monoclonal anti-Thy 1-2 followed by horseradish peroxidase conjugates of pig anti-sheep immunoglobulin, pig anti-rabbit immunoglobulin and avidin respectively. Peroxidase was developed with diaminobenzidine (5 mg/10ml Tris-buffered saline to which 1 drop of 100 vol H_2O_2 was added immediately before use) and the sections counterstained with haematoxylin.

To demonstrate sIgD together with sIgM, sections were examined by double immunofluorescence using fluorescein-labelled mouse anti-mouse IgD^a allotype and sheep anti-mouse IgM followed by rhodamine-labelled pig anti-sheep immunoglobulin.

Measurement of relative areas of splenic compartments

Stained sections were examined by light microscopy at 250x magnification. A 1 cm^2 graticule was placed in one eyepiece. This graticule is divided into vertical and horizontal lines set at 1mm intervals. The proportion of intercepts between the vertical and horizontal lines falling on each zone in a tissue section is taken as a measure of the area of that zone. Sections were examined by serial sweeps from top to bottom, until the entire section had been covered.

Results

CBA/N mice have marginal zone and follicular B cells but their numbers in young mice are significantly reduced compared with controls

Groups of female CBA/N (x-id/x-id) and CBA/Ca mice were killed either at 4 or 13 weeks of life. These were then examined using immunohistology to detect: a) IgM-bearing B cells b) IgD-bearing B cells and c) Thy 1-2-bearing T cells. For qualitative interpretation double immunofluorescence was carried out. This showed that CBA/N mice had both marginal zones and follicles (fig. 1). Quantitative analysis was carried out on spleen sections stained using immunoperoxidase. The relative areas of marginal zones and follicles to a) the T cell-rich periarteriolar lymphocytic sheath (PALS) and b) the red pulp was determined and these results are shown in figure 2. At 4 weeks of age the follicles of CBA/N mice are about 30% of the size of those found in the spleens of CBA/Ca

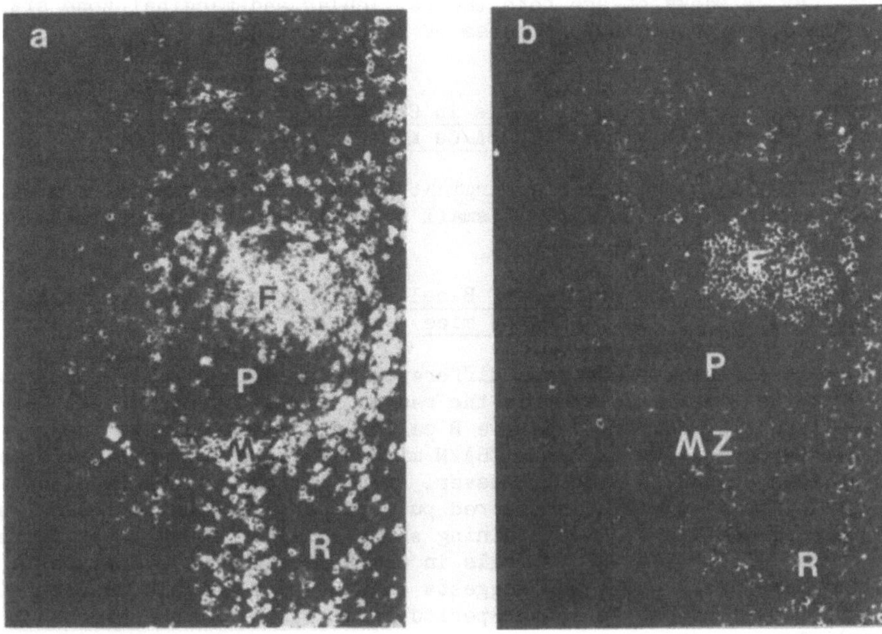

Figure 1. Double immunofluorescence staining of spleen to demonstrate a) sIgM, b) sIgD
F = follicle, MZ = marginal zone, P = periarteriolar lymphocytic sheath, R = red pulp

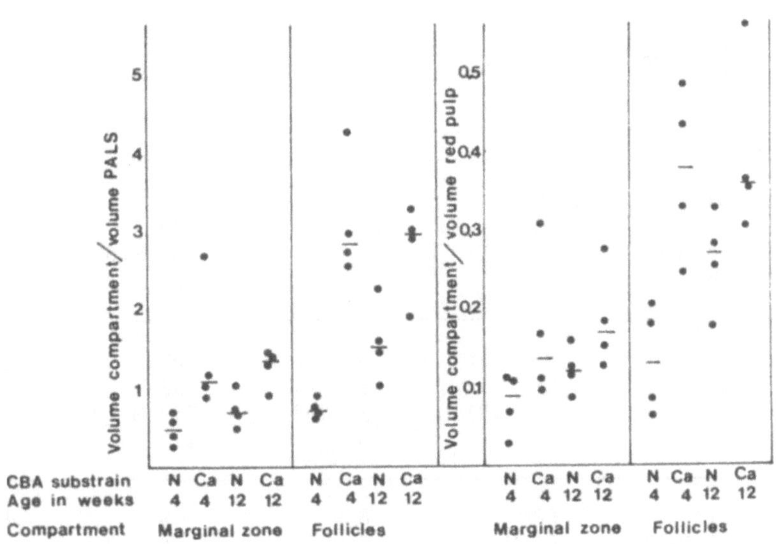

Figure 2. Relative areas of marginal zone and follicles in spleen sections from CBA/N and normal mice.

mice, while the marginal zones are about half the size in the CBA/N mice of this age. By 13 weeks of age both the follicular and marginal zone sizes of CBA/N mice are around 60% of those of the controls.

The phenotype of marginal zone cells in CBA/N mice was not found to be obviously different from that of CBA/Ca mice

Most B cells in the splenic marginal zones of both strains of mice expressed IgM with little IgD. A small proportion of these cells also expressed IgG3.

Major differences in the profile of B cell phenotypes in the red pulp of the spleen between CBA/N and CBA/Ca mice

The most striking phenotypic difference in splenic B cells was that observed in the B cells located in the red pulp. At 4 weeks of age CBA/N mice have a similar number of IgM+ve B cells in the red pulp to their CBA/Ca counterparts. By 13 weeks CBA/N mice have more IgM+ve B cells in this area than do CBA/Ca mice. However, at both times, CBA/N mice are deficient in IgD+ve B cells in the red pulp compared to the controls (Table 1). Double immunofluorescence staining shows this to be due to an increase in the number of IgM+ve, IgD-ve cells in CBA/N mice. This phenotype is typical of immature B cells and suggests that B lymphopoiesis is occurring at a higher level and for a longer period after birth in the spleens of these animals.

Table 1 Sections of spleen were stained to demonstrate sIgM and sIgD by double immunofluorescence. The numbers of sIgM+ve and sIgD+ve B cells per 100 intercept area of red pulp at 400 x magnification is given. Individual counts on 4 animals in each group, together with the median (M) are given.

	Total IgM	Total IgD	IgM^+ IgD^+	IgM^+ IgD^-	IgM^- IgD^+
	92	6	2	95	4
	99 M=78	14 M=9	5 M=3	94 M=79	9 M=7
CBA/N	64	9	1	63	8
4 weeks	59	8	3	56	5
	34	15	6	30	11
CBA/Ca	80	35	11	79	24
4 weeks	98 M=59	42 M=31	17 M=9	82 M=60	25 M=23
	38	27	6	32	21
	92	37	5	87	32
CBA/N	104	25	5	99	20
13 weeks	280 M=142	40 M=39	21 M=7	259 M=136	19 M=26
	180	60	8	172	52
	11	62	8	3	54
CBA/Ca	9 M=11	61 M=62	5 M=8	4 M=4	56 M=56
13 weeks	30	109	10	20	99

Lack of IgM and IgG3 but not IgA plasma cells in the red pulp of CBA/N mice

Sections of spleen were stained by indirect immunofluorescence to demonstrate IgM, IgG3 and IgA and the number of plasma cells secreting these isotypes was enumerated (Table 2). At 4 weeks of age CBA/N mice have less than 2% of the number of IgM and IgG3 plasma cells seen in CBA/Ca mice of the same age. By 13 weeks of age the numbers have increased slightly, but are still only around 10% of control numbers. In contrast numbers of IgA plasma cells are normal at 4 weeks and are relatively increased by 13 weeks.

Table 2. Number and isotype distribution of plasma cells in the splenic red pulp. Number of cells per 1000 intercepts tissue section at 400x magnification are given (M = median).

	IgM		IgG3		IgA	
CBA/N	8		<1		20	
4 weeks	7	M=8	<1	M=≤1	15	M=23
	3		<1		29	
	21		<1		26	
CBA/Ca	540		87		38	
4 weeks	320	M=500	136	M=176	13	M=37
	560		216		37	
	610		372		36	
CBA/N	16		6		142	
13 weeks	31	M=8	12	M=9	171	M=161
	24		3		151	
	34		17		181	
CBA/Ca	324		106		84	
13 weeks	315	M=320	48	M=90	44	M=47
	337		102		50	
	175		77		41	

Discussion

This study has shown, using immunohistological techniques, that CBA/N mice do possess splenic marginal zones containing B cells which are sIgM positive, sIgD negative; the characteristic surface immunoglobulin phenotype of normal marginal zone B cells. However, over recent years, marginal zones have been implicated in the response to TI-2 antigens (4), but CBA/N mice cannot respond to these antigens (3). There seem to be three possible explanations for this.

Firstly, the marginal zone may have no role in TI-2 responses. This is unlikely, but all the evidence associating marginal zones with TI-2 responses is indirect and there is no formal proof of the connection. For example, correlation of the responsiveness of rats to TI-2 antigens with the recovery of marginal zone but not with follicular recovery following depletion of B cells and subsequently reconstitution (4) may be coincidental.

Secondly, immunohistology shows the presence of marginal zones and can be used to delineate the phenotype of the constituent cells but cannot give any information on their functional integrity. Phenotypically normal marginal zone B cells may still have a defect in their ability to respond to TI-2 antigens. For example they might lack a receptor for a growth factor produced by polysaccharide-localising cells (5).

Thirdly, there appears to be two routes of entry of B cells into the marginal zone. We have shown that memory B cells in thymus-dependent responses colonise the marginal zones (12). These cells do not appear in the marginal zones following responses to TI-2 antigens. In contrast marginal zones acquire normal B cell numbers in congenitally athymic, nude rats and in germ-free, antigen-deprived rats. Also recirculating B cells recolonise depleted marginal zones at a rate and with a consistency which suggests this is an antigen-independent process. It may be that these cells which leave the recirculating B cell pool to colonise the marginal zones are the main responders to TI-2 antigens. It has previously been shown that newly produced virgin B cells are not able to colonise the marginal zones directly and that colonisation is from recirculating B cells (4). Also we have shown that newly produced virgin B cells are recruited to the recirculating B cell pool to replenish depleted recirculating B cell numbers by an antigen-independent mechanism (13). If this mechanism for replacing recirculating cells is absent or deficient one would expect animals to have reduced numbers of SIgM and IgD+ve follicular B cells as is seen in CBA/N mice. B cell recruitment to the recirculating pool in this situation, would be confined to recruitment of memory B cells generated following activation by T cell-dependent antigens. In T cell-deprived animals, which lack the capacity for antigen-independent recruitment of newly produced virgin B cells to the recirculating pool, one would expect to see normal B lymphopoiesis but a total lack of follicular or marginal zone B cells. This is precisely the situation observed in CBA/N nu/nu mice (14,15).

References

1) D.E. Mosier and B. Subbarao. Immunology Today. 3: 217 (1982).
2) J.J. Mond, A. Ahmed, I. Scher and W.E. Paul. Immunol. Rev. 37: 89 (1977).
3) J.J. Mond, I. Scher, D.E. Mosier, M. Baese and W.E. Paul. Eur. J. Immunol. 8: 459 (1978).
4) P.J.L. Lane, D. Gray, S. Oldfield and I.C.M. MacLennan. Eur. J. Immunol. 16: 1569 (1986).
5) J.H. Humphrey and D. Grennan. Eur. J. Immunol. 11: 212 (1981).
6) D. Gray, I. McConnell, D.S. Kumararatne, I.C.M. MacLennan, J.H. Humphrey and H. Bazin. Eur. J. Immunol. 14: 47 (1984).
7) D. Gray, D. Chassoux, I.C.M. MacLennan, and H. Bazin. Clin. exp. Immunol. 60: 78 (1985).

8) D. Finkelman, A.H. Smith, I. Scher and W.E. Paul. J. Exp. Med. 142: 1316 (1975).
9) D. Gray, I.C.M. MacLennan, H. Bazin and M. Khan, Eur. J. Immunol. 12: 564.
10) L.M. Forrester, J.D. Ansell and H.S. Michlem. J. Exp. Med. 165 949 (1987)
11) A.M. Stall and M.R. Lohen. J. Immunol. 132: 787 (1984).
12) S. Oldfield, Y-J. Liu, M. Beaman and I.C.M. MacLennan. This volume.
13) H. Bazin, B. Platteau, I.C.M. MacLennan and G.D. Johnson. Immunol. 54:79 (1985)
14) H.H. Wortis, L. Birkly, D. Hughes, S. Roschelle and G. Wanech. J. Exp. Med. 155: 903 (1982).
15) J.J. Mond, J. Scher, J. Gossman, S. Kessler, P.K.A. Mongiri, C. Hansen, F.D. Finkelman and W.E. Paul. J. Exp. Med. 155: 924 (1982).

THYMUS-INDEPENDENT TYPE 2 RESPONSES IN LYMPH NODES

Liu Yong-Jun, Susan Oldfield and Ian C.M. MacLennan
Department of Immunology, University of Birmingham
Birmingham, U.K.

Introduction

Splenectomy in adult rats produces a profound and prolonged impairment of their capacity to respond to the thymus-independent type 2 (TI-2) antigen, dinitrophenylated hydroxyethyl starch (DNP-HES) but has no effect on their response to the TD antigen, DNP-haemocyanin (1). However, this impairment is not complete suggesting that some other organs have some capacity to respond to TI-2 antigens. In this study we show the ability of peripheral lymph nodes to produce limited responses to these antigens.

Materials and Methods

Animals

Wistar rats were used throughout. These animals were bred and maintained under standard laboratory conditions in the animal house of the Department of Immunology, University of Birmingham and were used at between 3 and 6 months of age.

Splenectomy was carried out by a standard technique. The blood vessels in the splenic peduncle were tied off with cotton thread and the spleen removed. The wound was stitched and skin clips applied. These animals were immunised 14 days after the operation.

Antigens and Immunisations

Dinitrophenylated hydroxyethyl starch and fluorescein-labelled Ficoll (FITC-Ficoll) were kind gifts from Professor J H Humphrey, Royal Postgraduate Medical School, London. Oxazalone-Ficoll was prepared by a method based on that of Inman (2). 10 or 20μg DNP-HES in aqueous solution was injected either intravenously into the tail vein, or subcutaneously into the hind footpad.
150μg oxazalone-Ficoll in aqueous solution was injected into the hind footpad.

Antisera

Monoclonal mouse anti-rat IgD (MARD3) and polyclonal rabbit anti-rat class and subclass antibodies were a gift of Professor H Bazin, University of Louvain, Brussels. Sheep anti-rabbit IgG was obtained from I.D.R.L., the Medical School, University of Birmingham and rhodamine-labelled rabbit anti-mouse immunoglobulin was purchased from Dakopatts, Copenhagen, Denmark. The mouse anti-rat B cell monoclonal antibodies, HIS 14 and HIS 24 were a gift from Dr. Kroese and Professor Nieuwenhuis, University of Groningen, The Netherlands.

Immunohistology

Spleens were snap frozen in liquid nitrogen and stored at -70°C. Five μm frozen sections were cut and fixed in acetone for 20 mins at 4°C. To demonstrate specific antibody-containing cells, sections were double-stained with DNP-alkaline phosphatase and oxazalone-peroxidase (3). Alkaline phosphatase was developed with naphthol AS-MX phosphate and Fast Blue BB salt to give a blue product (4) and peroxidase then developed with 3-amino 9-ethylcarbazole to give a reddish product (5).

Double immunofluorescence staining to demonstrate IgM and IgD, HIS 24 or HIS 14 was by the indirect technique using rabbit anti-rat IgM together with a mouse monoclonal anti-rat IgD (MARD3) or HIS 14 or HIS 24 in the first layer, followed by a mixture of sheep anti-rabbit immunoglobulin labelled with fluorescein and rabbit anti-mouse immunoglobulin labelled with rhodamine. The procedure is described by Johnson and Holborow (6).

Radioimmunoassay

Serum anti-DNP titres were measured using a solid phase radioimmunoassay as described by Gray et al (1).

Results

Immunohistology of lymph node responses

Groups of rats were immunised with two TI-2 antigens simultaneously; oxazalone-Ficoll in the right footpad and DNP-HES in the left footpad. The number of specific hapten-binding plasma cells in each popliteal lymph node was assessed by immunohistology at seven and twelve days after immunisation. At both times many hapten-specific plasma cells were seen in the draining nodes with little or no response in the contralateral nodes (Table 1). No hapten-binding plasma cells were seen in control rats that had not been immunised.

Table 1. Hapten-binding plasma cells in the medulla of popliteal lymph nodes from rats immunised with DNP-HES in the left footpad and ozaxalone-Ficoll in the right footpad. Number of cells per 0.5 mm^2 area of tissue section is given.

Day	Animal	anti-DNP		anti-Ox		anti-Ox / anti-DNP	
		left	right	left	right	left	right
7	1	20	0	7	19	0.4	19/0
	2	15	2	0	9	<0.1	4.5
	3	64	5	10	52	0.2	10.4
	4	60	3	3	44	<0.1	14.7
12	5	3	5	0	41	<0.1	8.2
	6	6	1	3	5	<0.1	5
	7	6	0	3	4	0.5	4/0
	8	8	0	38	6	4.8	6/0

Comparison of serum responses following intravenous or subcutaneous immunisation with DNP-HES in splenectomised and control rats

Groups of rats were subjected to splenectomy or a sham splenectomy operation. Two weeks later they were immunised with DNP-HES either intravenously or by subcutaneous injection into each hind footpad. Titres of antibody in the serum on day 9 following immunisation are shown in figure 1. The results confirm that splenectomy impairs serum antibody response following intravenous immunisation and show that subcutaneous immunisation failed to improve titres of antibody in these rats. Both control and splenectomised rats showed a poor response to subcutaneous immunisation.

"Marginal zones" in lymph nodes

The popliteal lymph nodes from rats immunised with DNP-HES and oxazalone-Ficoll in the footpads were also stained by double immunofluorescence to demonstrate IgM together with IgD or HIS 14 (a monoclonal antibody recognising all rat B cells [7]) or HIS 24 (a monoclonal antibody specific for follicular and germinal centre B cells). In seven out of thirty lymph nodes examined, patches of cells larger than follicular B cells, were seen in the subcapsular region. These cells showed strong surface IgM staining with little or no surface IgD staining (Figure 2). They were also stained by HIS 14, but not by HIS 24. In the spleen, this phenotype is characteristic of the marginal zone.

FITC-Ficoll localisation in popliteal lymph nodes

Humphrey and Grennan (8) showed that TI-2 antigens labelled with fluorescein were localised *in vivo* in macrophage-like cells in the marginal zone of spleen and in the subcapsular area, intranodal lymphatics and medulla of lymph nodes. In preliminary experiments, 20μg of FITC-Ficoll was injected into the hind footpads of groups of rats which were sacrificed 2 or 21 days later. FITC-Ficoll was found located predominantly in the medullary cords and lymphatic channels of the popliteal nodes, with smaller amounts in some subcapsular areas which open into the medulla. Anti-IgM and anti-IgD staining has not identified marginal zone type B cells in these areas. There was no decrease in fluorescence within the node between day 2 and day 21 and no fluorescence was visible in the spleen, indicating that the draining lymph nodes are efficient at trapping and retaining the antigen.

Discussion

B cell activation by TI-2 antigens does occur in lymph nodes, but the rate of activation in this site is small compared to that achieved in the spleen and contributes little towards the serum antibody response. Several lines of indirect evidence suggest that the marginal zone is involved in the splenic response to TI-2 antigens (9). Using double immunofluorescence techniques, B cells with the marginal zone phenotype were found in areas deep to the subcapsular sinus in lymph nodes of rats. The marginal zones comprise the largest identifiable B cell compartment in the spleen of rats (10). By contrast, the number of B cells with a phenotype similar to that of marginal zone B cells, in the areas deep to the subcapsular sinus vary from a few patches of cells to a few single, scattered cells. It has previously been shown that the proportion of B cells that are IgM^{+} IgD^{-} in lymph nodes is much less than that in the spleen (11) and that a

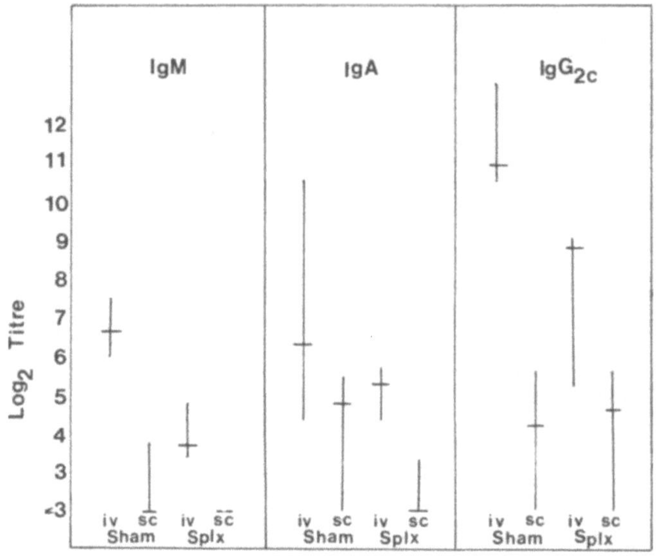

Figure 1. Anti-DNP antibody titres 9 days after immunisation with DNP-HES either intravenously (i.v.) or subcutaneously (s.c.) in splenectomised (splx) or sham splenectomised rats (Sham).

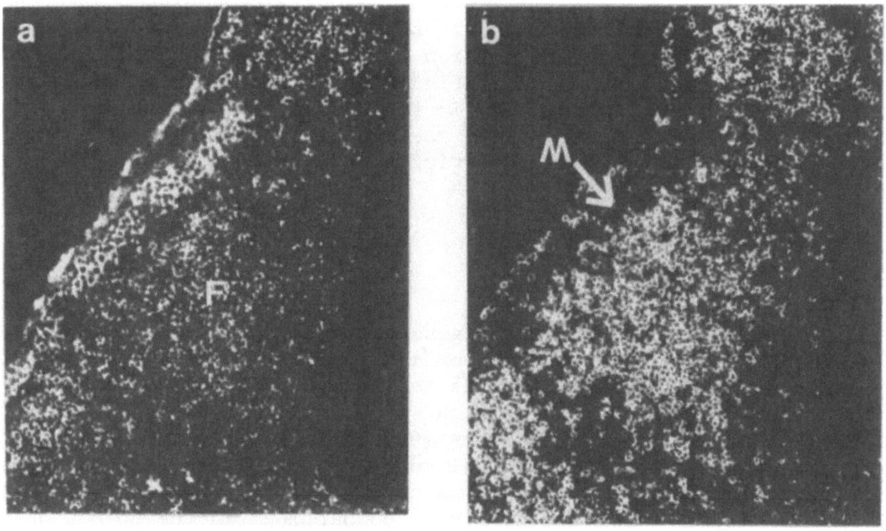

Figure 2. Double immunofluorescent staining of a section of popliteal lymph node to demonstrate surface IgM (a) together with surface IgD (b). F = follicle. M = marginal zone-like area.

substantial number of these are situated within follicles. This indicates that development of marginal zone B cells in peripheral lymph nodes is poor compared with that of the marginal zones in the spleen. This might account for the limited response of lymph nodes to TI-2 antigens. This is reflected in the poorer systemic response to TI-2 antigens following local immunisation compared with intravenous immunisation. In this situation most of the antigen may be retained by the draining popliteal, inguinal and femoral nodes, so the spleen doesn't receive sufficient antigen to mount a good response. At the same time the deficient marginal zone B cell presence in the extra-splenic peripheral lymphoid tissues may limit.

References

1) D. Gray, D. Chassoux, I.C.M. MacLennan, and H. Bazin Clin. exp. Immunol. 60: 70 (1985).

2) J.K. Inman. J. Immunol. 114: 704 (1975).

3) E. Claasen, and N. Van Rooijen. J. Histochem. & Cytochem. 33: 840 (1985).

4) A.G.E. Pearse. Histochemistry. Theoretical and Applied, vol. 1 & 2, Churchill Livingstone, Edinburgh (1968).

5) R.C. Graham, U. Lundholm, N.J. Karnovsky. J. Histochem. Cytochem. 13: 150 (1965).

6) G.D. Johnson, and E.J. Holborow. In "Handbook of Experimental Immunology", Blackwell Scientific Publications, Oxford (1984).

7) F.G.M. Kroese, D. Opstelten, A.S. Wubbena, G.J. Deenen, J. Ater, E.H. Schwander, L. de Leij, and P. Nieuwenhuis. Adv. Exp. Med. Biol. 186: 81 (1985).

8) J.H. Humphrey, and D. Grennan. Eur. J. Immunol. 11: 221 (1981).

9) P.J.L. Lane, D. Gray, S. Oldfield, and I.C.M. MacLennan. Eur. J. Immunol. 16: 1569 (1986).

10) D.S. Kumararatne, H. Bazin, and I.C.M. MacLennan. Eur. J. Immunol. 11: 858 (1981).

11) D. Gray, I.C.M. MacLennan, H. Bazin and M. Khan. Eur. J. Immunol. 12: 564 (1982).

BOTH LY-1 B CELLS AND CONVENTIONAL B CELLS MAKE AUTOANTIBODIES TO BROMELAIN-TREATED AUTOLOGOUS ERYTHROCYTES

E. Andrew, W. Annis and R.N. Maini

Kennedy Institute of Rheumatology
Bute Gardens
London W6 7DW

INTRODUCTION

In mice a minor subpopulation of B cells has been described which has a different phenotype and function from the majority of B cells[1 2]and which belongs to a separate B cell lineage[3]. These B cells are characterized by the expression of about 5-fold more membrane IgM and 10-fold less IgD than cells of the conventional B lineage[4]. In addition, many of them express a surface molecule, Ly-1 (CD5), previously thought to be present only on T cells[4 5]. These Ly-1 B cells occur at low frequency (1-2%) in adult spleen, are not detectable in lymph nodes or blood[1], but account for up to 50% of the B cells in the peritoneal cavity[3].

It has been reported that Ly-1 B cells are responsible for all spontaneous IgM secretion in normal mice and for the production of IgM autoantibodies to DNA, thymocytes and bromelain-treated red blood cells (Brm-rbc)[2]. Autoimmune NZB mice and their F_1 hybrids, (NZB x NZW) (BW), which spontaneously produce these autoantibodies, have increased percentages of splenic Ly-1 B cells[1]. These findings have led to the suggestion that Ly-1 B cells play an important role in autoimmune disease. However, the pathogenic autoantibodies in NZB and BW mice are thought to be of the IgG class[6], whereas the IgM autoantibodies mentioned above (which are also produced by normal non-autoimmune strains) are probably not pathogenic and may function to aid the removal of cells damaged as a result of infection or injury. Thus the role of Ly-1 B cells in the pathogenesis and maintenance of autoimmunity is unclear and needs to be evaluated carefully. In attempting to determine the relationship between the cells producing pathogenic IgG autoantibodies and those producing non-pathogenic IgM autoantibodies, we were unable to substantiate the initial observations that only Ly-1 B cells secrete these IgM autoantibodies. Instead, a small but significant number of cells of the conventional B cell phenotype (IgD-bright, IgM-dull and Ly-1-negative) produced autoantibodies to Brm-rbc. The relationship between these cells and Ly-1 B cells is now under investigation.

MATERIALS AND METHODS

Spleen cell suspensions pooled from NZB or BW female mice aged 3-5 months were stained with fluorescent antibodies and separated by flow cytometry on a Becton Dickinson FACStar. The following two combinations of antibodies were used; a fluoresceinated goat-anti mouse IgM antibody specific for μ heavy chains (FITC-GaMμ) together with a biotinylated rat anti-mouse Ly-1 antibody (53-7.3), or FITC-GaMμ together with a biotinylated mouse anti-IgD antibody (10.4.22). The biotinylated antibodies were followed by streptavidin-phycoerythrin. Immediately after separation on the FACS the number of IgM-secreting cells per group was assessed by an ELISA spot technique (Elispot) using an affinity-purified goat anti-mouse μ heavy chain antibody to coat the ELISA plates and an alkaline phosphatase-conjugated rabbit anti-mouse IgM antibody to reveal secreted IgM[7]. The number of plaque-forming cells to bromelain-treated red blood cells (Brm-PFC) was determined by a modification of the Cunningham plaque assay[8].

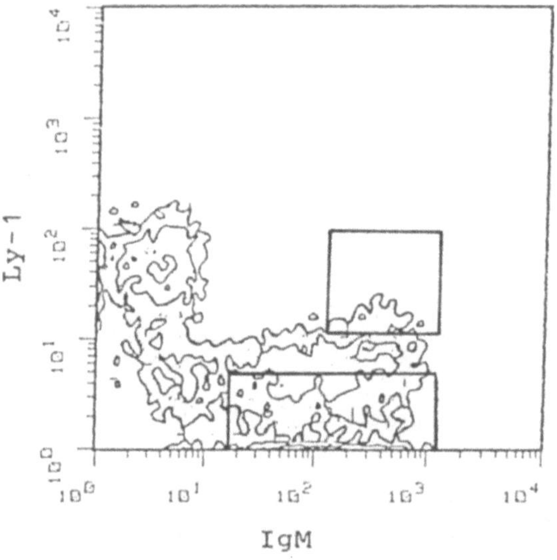

Figure 1. FACS profile of BW spleen cells co-stained for IgM and Ly-1. The boxes indicate the position of the sort windows selected for Ly-1 B cells (upper) and conventional B cells (lower).

RESULTS AND DISCUSSION

In the first series of experiments spleen cells were fractionated on the FACS into Ly-1 B-enriched (IgM-bright/Ly-1-dull) and Ly-1 B-depleted (IgM-positive/Ly-1-negative) groups (Fig.1). The number of spontaneous Brm-PFC and IgM-secreting cells in the Ly-1 B-enriched population was 3-4 fold more than in the Ly-1 B-depleted group (Table 1a). However, the Ly-1 B-enriched group represented only 1.43% of the spleen cells, whereas the Ly-1 B-depleted group accounted for 10.72%. This meant in fact that in the spleen only half as many Brm-PFC and IgM-secreting cells were produced by the Ly-1 B-enriched group as by the Ly-1 B-depleted group (Table 1b).

Table 1. FACS-separated Ly-1 B-enriched and Ly-1 B-depleted spleen cells spontaneously secreting IgM and autoantibodies to Brm-rbc.

(a) No. of IgM-secreting and Brm-PFC per 10^7 unseparated or separated spleen cells

Cell population	Percentage of B cells	No. of IgM-secreting cells/10^7 cells	No. of Brm-PFC/ 10^7 cells
Unstained unsorted	46.56	78.0 ± 12.0*	9.0 ± 0
Stained unsorted	46.56	92.8 ± 16.4	8.0 ± 0.7
Ly-1 B-enriched	100.00	142.6 ± 21.2	23.2 ± 1.5
Ly-1 B-depleted	100.00	40.1 ± 5.2	5.0 ± 0.7

* ± = s.e

(b) Total no. of IgM-secreting and Brm-PFC per spleen

B cell population	Percentage of total spleen cells	No. of IgM-secreting cells/spleen	No. of Brm-PFC/ spleen
unstained unsorted	46.56	49.4×10^3	56.9×10^2
stained unsorted	46.56	58.8×10^3	50.6×10^2
Ly-1 B-enriched	1.43	2.8×10^3	3.3×10^2
Ly-1 B-depleted	10.72	5.8×10^3	5.4×10^2

These results differ from previously published findings that the majority of spontaneous IgM-secreting cells and all Brm-PFC belong to the Ly-1 B cell lineage [2 3]. There are two possibilities to account for this. One is that the FACS did not distinguish between the (IgM-bright/Ly-1-dull) Ly-1 B cells and the (IgM-positive/Ly-1-negative) conventional B cells. From the FACS profile (Fig.1) it can be seen that the Ly-1 B cells form a continuum with the conventional B cells, rather than appearing as a discrete population. For the FACS this makes separation more difficult and may result in Ly-1 B cells contaminating the Ly-1 B-depleted group.An alternative explanation for the difference is that the Brm-PFC detected in the Ly-1 B-depleted group may have come from the Ly-1-negative "sister" population described by Herzenberg[9]. This has an identical immunoglobulin phenotype to Ly-1 B cells (IgM-bright/IgD-dull) but lacks the Ly-1 antigen. Spleen cells co-stained with anti-IgM and anti-IgD do segregate into discrete populations (Fig.2), with the Ly-1 B cells falling in the IgM-bright/IgD-dull group. Thus if cells were selected for intensity of IgM and IgD expression (Fig.2), all Ly-1 B cells (IgM-bright/IgD-dull) should be excluded from the conventional B cell (IgM-dull/IgD-bright) population.

It was found that after separating the spleen cells into IgM-bright/ IgD dull (Ly-1 B containing) and IgM-dull/IgD-bright (conventional B) groups, almost all of the spontaneous IgM-secreting cells were in the Ly-1 B-containing population (Table 2), indicating that one of the two possibilities above was correct. However of the few cells secreting IgM in the conventional B cell group, all made antibodies to Brm-rbc. These cells were not Ly-1 B cells by definition of their membrane immunoglobulin phenotype, and did not fall within the activated B cell (IgM-dull/IgD-dull) group either. Their identity is still unknown, but whether they are related to Ly-1 B cells and/or are the cells which switch isotype to produce pathogenic IgG autoantibodies is now under investigation.

Table 2. FACS-separated IgM-bright/IgD-dull (Ly-1 B-containing) and IgM-dull/IgD-bright (Ly-1 B-negative) spleen cells spontaneously secreting IgM and autoantibodies to Brm-rbc.

(a) No. of IgM-secreting and Brm-PFC per 10^7 unseparated or separated spleen cells

Cell population	Percentage of B cells	No. of IgM-secreting cells/10^7 cells	No. of Brm-PFC/ 10^7 cells
unstained unsorted	40.49	590.0 ± 67.1*	25.7 ± 2.6
stained unsorted	40.49	336.7 ± 23.7	14.0 ± 1.1
IgM-bright/IgD-dull	100.00	383.3 ± 5.3	22.5 ± 3.1
IgM-dull/IgD-bright	100.00	5.8 ± 0.3	10.3 ± 1.6

* ±= s.e.

(b) Total no. of IgM-secreting and Brm-PFC per spleen

B cell population	Percentage of total spleen cells	No. of IgM-secreting cells/spleen	No. of Brm-PFC/ spleen
unstained unsorted	40.49	26.28 x 10^4	1.14 x 10^4
stained unsorted	40.49	14.99 x 10^4	0.62 x 10^4
IgM-bright/IgD-dull	11.50	4.85 x 10^4	0.28 x 10^4
IgM-dull/IgD-bright	15.49	0.10 x 10^4	0.17 x 10^4

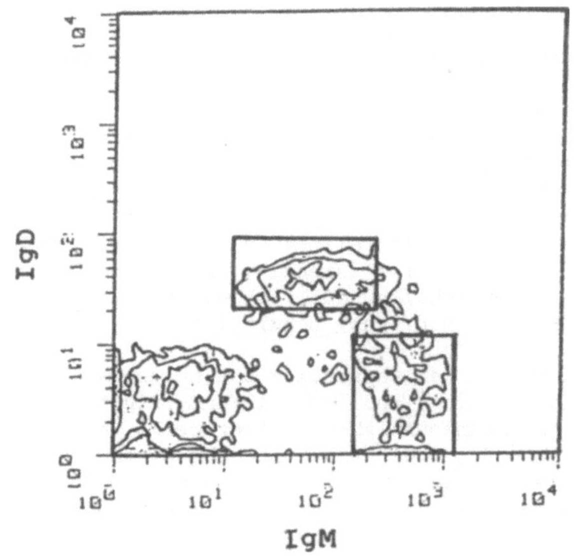

Figure 2. FACS profile of BW spleen cells co-stained for IgM and IgD. The boxes indicate the position of the sort windows selected for Ly-1 B-containing cells (lower right) and Ly-1 B-negative cells (upper left)

REFERENCES

1. Hayakawa K, Hardy RR, Parks DR & Herzenberg LA (1983) J.Exp.Med.157: 202
2. Hayakawa K, Hardy RR, Herzenberg LA & Herzenberg LA (1984) Proc. Natl. Acad. Sci. USA 81: 2492
3. Hayakawa K, Hardy RR, Herzenberg LA & Herzenberg LA(1985) J.Exp.Med. 161: 1554
4. Hardy RR & Hayakawa K (1986). Immunol.Rev. 93: 53
5. Manohar V, Brown E, Leiserson WM & Chused TM (1982) J.Immunol. 129:532
6. Theofilopoulos AG & Dixon FJ (1985) Adv.Immunol. 37; 269
7. Sedgwick JD & Holt PG (1983) J.Immunol.Meth. 57: 301
8. Cunningham AJ (1974) Nature 252:749
9. Herzenberg LA, Stall AM, Lalor PA et al (1986) Immunol.Rev. 93:81

IgG SUBCLASS DISTRIBUTION IN JUVENILE HUMAN TONSIL: IgG3 AND IgG4 RESULTS OF SPECIFIC ANTIBODY PRODUCTION USING SYNTHETIC PEPTIDES

W.J.A. Boersma, C. Deen, N.D. Zegers, J. Radl*, and J.J. Haaijman

Medical Biological Lab. TNO, P.O. Box 45, 2280 AA Rijswijk
The Netherlands
*Institute for Experimental Gerontology TNO
P.O. Box 5815, 2280 HV Rijswijk, The Netherlands

SUMMARY

A decapeptide with a sequence corresponding to a part of the hinge region of human IgG3 was used to prepare a mouse monoclonal antibody (Mab 330-2.2). The Mab recognized IgG3 in ELISA only when the IgG3 was denatured by mild heat. Mab specificity in immunohistochemistry was calibrated with a number of reference Mabs obtained through a WHO/IUIS collaborative study. Finally, Mab 330-2.2 was used in conjunction with a set of other isotype specific Mabs to study the (sub)class distribution of cytoplasmic Ig containing cells in palatine tonsil. IgG3 within the IgG class was strongly over-represented in our tonsil material if compared to the representation of IgG3 in serum.

INTRODUCTION

The primary and tertiary structures of human IgG (HuIgG) molecules have been shown to be highly similar (1). With respect to their amino acid sequences, for example, a homology of approximately 95% has been demonstrated (2). Despite such structural similarities, successful production of specific monoclonal antibodies (Mabs) for the IgG3 and IgG4 subclasses of HuIgG has been accomplished, both in our laboratory as well as those of others, using classical techniques (3). However, results from a recent study sponsored by the World Health Organization (WHO) have demonstrated that classical methods do not produce generally applicable Mabs for the IgG1 and IgG2 subclasses of human IgG (3). Thus, studies were undertaken to develop more suitable techniques for the production of specific Mabs for the IgG subclasses.

Because different IgG subclasses share similar immunodominant epitopes, efforts were made to develop a technique which would circumvent the intramolecular immunodominance of class specific determinants. In the present study, synthetic peptides (SP) comprised of subclass-specific amino acid substitutions were synthesized and utilized for the production of IgG3-specific Mabs. IgG3 was chosen as the first target because reference Mabs were available. The antibodies to be raised with SPs could thus be compared and contrasted with regard to their specificity and performance in different immunoassays.

The IgG3 molecule has an extended hinge region which consists, in part, of a repeated amino acid sequence (2). This repeat in the hinge region contains several prolines which can be expected to contribute to a relatively high intramolecular flexibility of the molecule, an asset allegedly contributing to antigenicity (4,5). Further, most polyclonal sera and HuIgG3-specific Mabs recognize a determinant in this hinge region (3). Thus, a decapeptide with the same repeat sequence as the IgG3 hinge was synthesized and made immunogenic in a determinant carrier system. Carriers were necessary to provide an immunogenic backbone for the decapeptide since such peptides themselves do not possess high immunogenic properties.

An IgG3-specific Mab (330-2.2) with unique properties was produced which proved particularly useful in experiments utilizing enzymoimmunohistochemical techniques. The IgG3-specific Mab, together with Mabs specific for IgG4 and IgG (common determinant), were subsequently utilized to investigate IgG subclass distribution in juvenile human, palatine tonsil.

MATERIALS AND METHODS

Decapeptides were synthesized with a Biosearch Sam II peptide synthesizer using the Merrifield solid phase technique with tBOC-protected amino acids. SPs were purified by chromatography on Sephadex G-15. The hinge region amino acid sequence of IgG3 synthesized was: H2N-ASP-THR-PRO-PRO-PRO-CYS-PRO-ARG-CYS-PRO-COOH. Conjugates for the IgG3-specific peptide were made with chicken gamma globulin (CGG), using glutaraldehyde. The conjugate was precipitated on to Al(OH)3 gel. BALB/c mice (8-12 weeks-old) received for immunization a primary injection intraperitoneally of 100 µg precipitated conjugate, followed by two booster injections of 50 µg each. Three days prior to fusion the mice received a final booster of 100 µg. Spleen cells were fused with SP2/0 plasmacytoma cells in the presence of polyethyleneglycol, using routine protocols as described by Haaijman et al. (6). The fusion mixture was plated out in microcultures such that most likely cultures were obtained with single clones. Supernates were tested in ELISA, cytoplasmic and membrane immunofluorescence as well as immunoperoxidase staining as described (6).

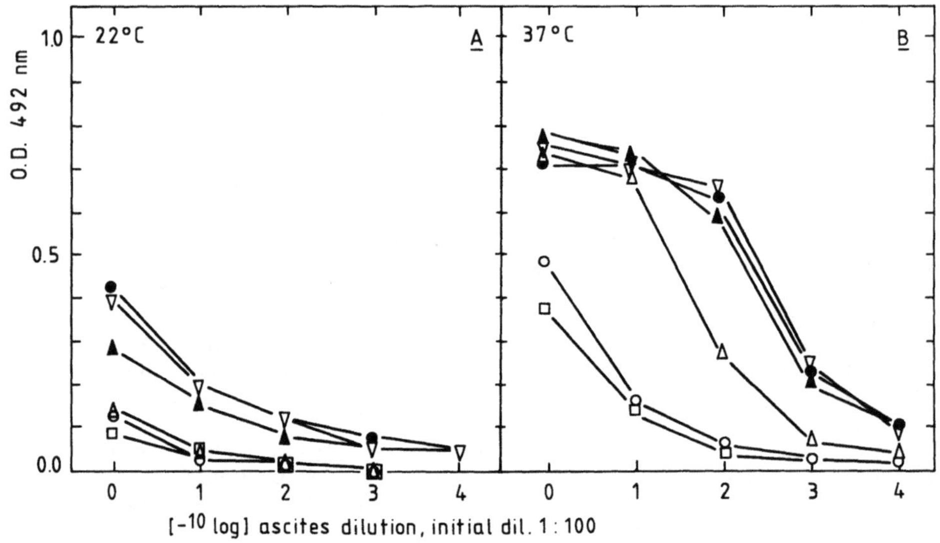

Fig.1. Evaluation of Mab 330-2.2 specificity in ELISA at two different temperatures (A: 22°C; B: 37°C). The following antigens were used for coating: IgG(o), IgG3χ(□), IgG3λ(Δ), and three IgG3 's(∇, • and▲). The coated antigens were incubated with various dilutions of 330-2.2 ascitic fluid, followed by peroxidase labelled rabbit anti-mouse/Ig. Ortho-phenylene diamine functioned as the substrate and the reaction product was read at 492 nm.

Several ELISA protocols were used. Wells of 96 well polyvinylchloride microtitration plates were prepared either by direct coating with the antigen or via presentation of the antigen via polyclonal rabbit anti-human IgG Fc (RAHuIgG/Fc) and rabbit anti-human IgG Fab (RAHuIgG/Fab) antibodies. The technique for cytoplasmic staining was described previously (6). The reactivity of anti-SP Mabs was compared and contrasted using a double enzymatic staining technique with that of reference Mabs obtained from the WHO collaborative study (3). Double staining was also used in the determination of the Ig(sub)class distribution of plasma cells in tonsil.

RESULTS

Immunization of mice and goats with the CGG-SP conjugate resulted in anti-sera that reácted specifically with HuIgG3. Mabs derived after immunization with CGG-SP, however, showed various specificity profiles. Some were SP specific and did not recognize native IgG3, some recognized IgG3 but did not recognize the SP in a bovine serum albumin-SP conjugate and some reacted with both SP and IgG3. Of the latter, clone 330-2.2 was selected for further experiments in ELISA and immunocytochemistry.

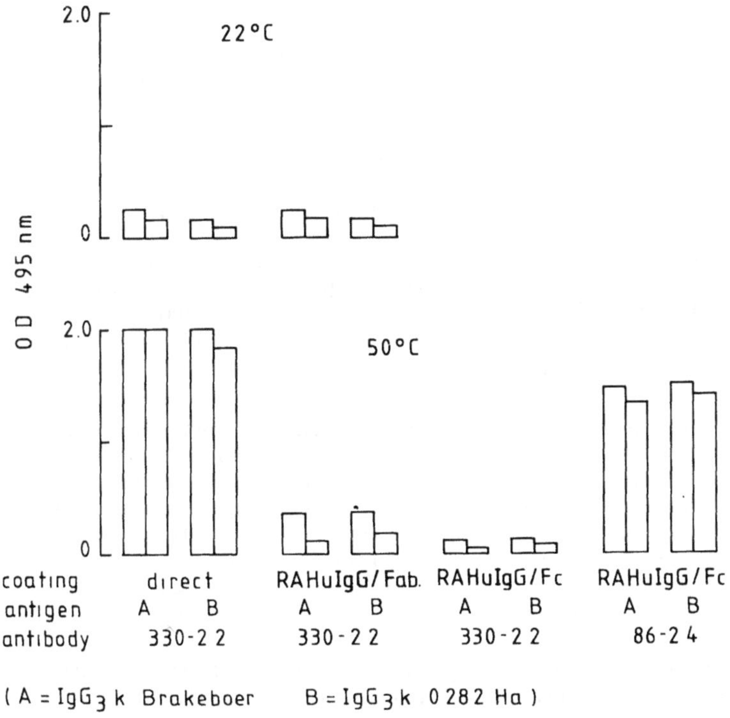

Fig.2. Temperature dependency of ELISA reactivity of anti-HuIgG3, in relation to "in assay" antigen presentation.

Mab 330-2.2 gave variable results in ELISA. Accidentally it was revealed, that temperature played a crucial role. A strong ELISA signal with IgG3 coated onto the plate was obtained only when the coating and/or the antigen-antibody interaction were allowed to proceed at elevated temperature (Fig.1). The temperature as such, however, did not influence the antigen structure. When the antigen was presented "in assay" via anti-HuIgG/Fab or anti-HuIgG/Fc antibodies, the IgG3 was not recognized by 330-2.2 even when the procedure was performed at 50°C. Another anti-HuIgG3 (clone 86-2.4, ref 3, renamed HP6080), was used as a control. This antibody

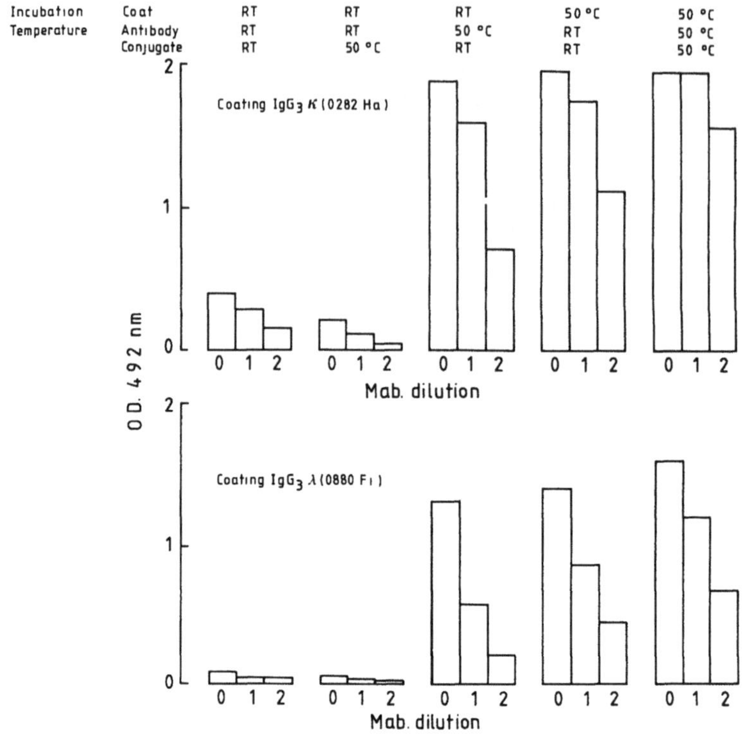

Fig.3. Temperature dependent antigen presentation in ELISA. Coating: IgG3 5 µg/ml. Antibody: 330-2.2 ascites. Conjugate: rabbit anti-mouse-peroxidase. Ordinate: optical density; abscissa: different Mab dilutions 0=1:500, 1=1:5,000, 2=1:50,000. At the top of the figure the temperature is shown used during the various steps of the procedure.

was able to recognize IgG3 when presented via anti-HuIgG/Fc at normal as well as at elevated temperature (Fig.2). Subsequently the various steps in the ELISA procedure were investigated with respect to their temperature dependency. The results shown in Fig. 3 indicate that an elevated temperature, either during the coating of antigen or during those steps in which antigen/antibody interaction took place was required for a strong ELISA signal. Since Mab 330-2.2 did not readily recognize (heated) IgG3 in solution, it appears that adsorption is necessary to stabilize the conformational change.

The antibodies produced by clone 330-2.2 strongly stained in the immunoperoxidase technique. Fig. 4 presents an example of juvenile human tonsil showing the distribution of IgG3 positive cells. Because of the many pitfalls which may be encountered in the use of Mabs (8), the performance of Mab 330-2.2 was evaluated in a comparison with a set of putative anti-HuIgG3 Mabs which were obtained by the courtesy of a working party under auspices of WHO/IUIS (3). Double staining was performed on human tonsil material and the frequency of cells stained with 330-2.2 and the coded WHO/IUIS antibodies was determined. The specificity patterns are shown schematically in Fig. 5. Results indicated that the different Mabs did not recognize one population of cells. However, a complete overlap was obtained using Mabs HP6050 and 330-2.2. At present, too little is known regarding the expression of Ig epitopes under denaturing conditions, such as those present in a histological slide to explain the molecular basis of the observed microheterogeneity in staining performance. Assuming that the reactivity profile of 330-2.2 was sufficiently well characterized, we proceeded to use the Mab as a diagnostic reagent.

Human juvenile tonsils were obtained from six individuals. Serial slides were used in single and double staining techniques. Plasma cell class and subclass frequencies were determined using methods similar to those described by Hijmans et al. (9). In brief, separate slides were stained with the following combinations of Mabs: 1. anti-HuIgM (clone 152-7.4,ref.8) + anti-HuIgA1 (clone 69-11.4,ref.8). 2. anti-HuIgD (clone 158-9R19,ref.8) + anti-HuIgA1. 3. anti-HuIgG (clone HP6017,ref.3) + anti-HuIgA1. 4. anti-HuIgA2 (clone 512-H5,ref.8) + anti-HuIgA1. 5. anti-HuIgG3 (clone 330-2.2) + anti-HuIgG. 6. anti-HuIgG4 (clone 315-1.1, unpubl.) + anti-HuIgG.

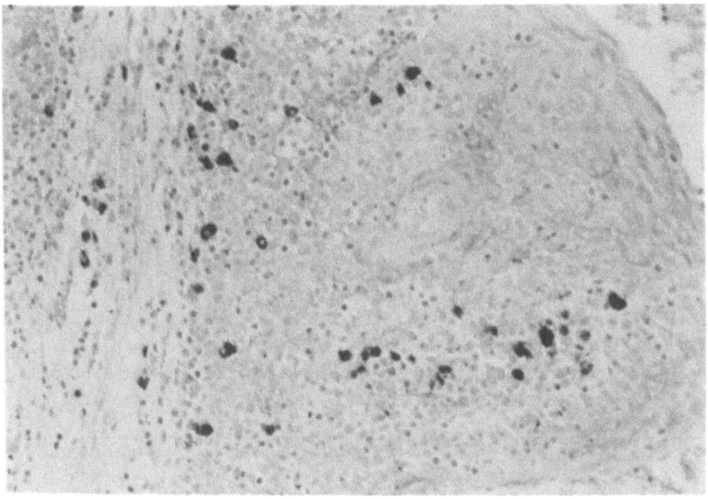

Fig.4. Mab 330-2.2 strongly stains IgG3 containing plasma cells (C-IgG3 cells) in juvenile human, palatine tonsil.

Fig.5. In a double enzymatic cytochemical procedure Mab 330-2.2 was compared with and contrasted to monoclonal anti-HuIgG3's from the WHO/IUIS working group on IgG subclass-specific monoclonal antibodies. The scheme indicates the differences and overlap in reactivity patterns.

In each slide the number of cytoplasmic Ig containing cells (C-Ig cells) reacting with either of the staining reagents was counted . Slides made, using combinations 1-4, yielded the following relative distribution of Ig-containing cells: C-IgM:C-IgD:C-IgG:C-IgA1:C-IgA2. Slides prepared using combinations 5 and 6 gave the proportion of C-IgG3 and C-IgG4 cells, relative to the total number of C-IgG cells. Compartments were distinguished according to Brandtzaeg et al. (10).

C-IgM cells formed the second largest population in the follicles (34%) but were of limited importance in the parafollicular area (6%). For C-IgD cells, the opposite held true with 20% of C-Ig cells positive for IgD in the parafollicular area and 7% in the follicles.

C-IgG cells made up 38% of all parafollicular C-Ig cells and 47% of those cells in the follicles. Further, C-IgG3 cells formed almost half of the C-IgG cells both in the parafollicular area (45%) as well as in the follicles (57%). C-IgG4 cells, although a minor population in the follicles (2%), formed 11% of the parafollicular cells. For IgG1 and IgG2 specific Mabs were not available.

C-IgA cells formed the second largest population (36%) in the parafollicular area. Within the IgA class, the C-IgA1 cells constituted the majority of cells in both the follicular (83%) as well as in the parafollicular area (63%) (Table 1).

Table I

Ig (SUB)CLASS DISTRIBUTION OF JUVENILE HUMAN TONSIL
N=6, 3-6 slides per tonsil, average % ± s.e.m.

		Parafollicular Area	Follicles
IgM		6 ± 1	34 ± 4
IgD		20 ± 3	7 ± 2
IgG		38 ± 2	47 ± 4
	IgG3 (% of IgG)	45	57
	IgG4 (% of IgG)	11	2
IgA		36 ± 2	12 ± 11
	IgA1 (% of IgA)	63	83
	IgA2 (% of IgA)	37	17

DISCUSSION

The present study demonstrates that synthetic linear determinants can be used for the production of useful diagnostic anti-Ig reagents. While immunization with native IgG3 leads mainly to the production of class-specific antibodies, the use of the synthetic peptides allows one, in principle, to circumvent the phenomenon related to immunodominance with a resulting greater specificity. The decapeptides used here were not very immunogenic, thus necessitating a carrier protein. However, it is necessary to strike a balance between the immunogenic character of the carrier, with its own immunodominant determinants, and the immune reactivity that is directed towards the synthetic determinants. The choice of the amino acid sequence in the peptide and the conformation of the SP in the conjugate during the antigen presentation are likely to be of prime importance for the efficient production of specific Mabs. Although the method presented here was successful in producing a specific Mab, it is understood that many more studies are needed to enable general guidelines to be drawn up for the use of SPs in the preparation of diagnostics. The results indicate, moreover, that in the application of Mabs, detailed performance testing is needed. As in the case reported here application may be confined to only one type of immunoassay.

According to Brandtzaeg et al. (10) most C-Ig cells in human tonsil are found in the reticular epithelial area (64%), the extra-follicular area (25%) and the germinal centers (8%). In our study, only the first and third area were scored and designated as parafollicular and follicular area. Brandtzaeg and collaborators studied both "normal" tonsils (10) and tonsils in various states of hyperplasia (11). Histopathological grading was not available for our material, therefore the C-Ig data are compared to the normal values.

Compared to previously published values (10), the frequencies of C-IgM and C-IgD cells found in the present study were elevated in both the follicular and parafollicular areas with a concomittant reduced frequency of C-IgG cells. C-IgA cell frequencies were not much different from those observed by Brandtzaeg (10).

Tonsillitis appears to be accompanied by high serum levels of IgG but decreased numbers of C-IgM, C-IgG and C-IgA cells in tonsil (11). Interestingly, the C-IgD cell frequencies are increased. Our data implicate, at least in part, a hyperplastic nature of the tonsils studied in this paper.

C-IgG3 cells form the most prominent IgG subclass in the specimens investigated, forming approximately 50% of all C-IgG cells. To our knowledge, no data on the distribution of IgG subclasses in lymphoid tissue other than bone marrow has yet been published. Since specific monoclonal anti-subclass antibodies applicable in immunohistochemistry became available only recently, comparison to other results awaits further investigation. More studies are needed to evaluate the possible function of the C-IgG3 cells in tonsils. Complete subclass analysis (including G1 and G2) of other lymphoid tissues could yield important clues on this respect. It may be argued that the relatively high C-IgG3 cell frequency reflects in part the decrease in the other subclasses. Alternatively, however, a specific function of IgG3 in inflammatory processes cannot be ruled out.

Sixty to eighty per cent of the C-IgA cells belonged to the IgA1 subclass. This finding is consistent with earlier observations (12) showing a preponderance of IgA1 representation in the proximal portions of the digestive and respiratory tracts. The relationship between this subclass distribution and immunobiological function remains yet unclear.

The results described here substantiate the need for the production of diagnostic reagents specific for human Ig and notably IgG subclasses. It has been shown here that a method employing synthetic peptides may form a reasonable alternative to conventional approaches. It allows the circumvention of the phenomenon of class specific immunodominant epitopes. Experiments are in progress to produce monoclonal antibodies directed against HuIgG1 and HuIgG2, using this approach.

ACKNOWLEDGEMENT

The supply of Mabs from the WHO/IUIS working party by Dr. C.B. Reimer, CDC, Atlanta, USA and the supply of purified Ig preparations by Dr. F. Skvaril, Bern, Switzerland is gratefully acknowledged. Part of this research was made possible by a grant from Nordic Immunological Laboratories, Tilburg, The Netherlands.

REFERENCES

1. Pumphrey R., Immunology Today, 7:14 (1986)
2. Kabat E.A., Wu T.T., Bilofsky H., Reid-Miller M., Perry H., Sequences of proteins of immunological interest. U.S. Department of Health and Human Services (1983)
3. Jefferis R. et al., Immunology Letters 10:223 (1985)

4. Westhof E., Altschuh D., Moras D., Bloomer A.C., Mondragon A., Klug A., van Regenmortel M.H.V., Nature 311:123 (1984)
5. Tainer J.A., Getzoff E.D., Alexander H., Houghten R.A., Olson A.J., Lerner R.A. Hendrickson W.A., Nature 312:127 (1984)
6. Haaijman J.J., Coolen J., Deen C., Kröse C.J.M., Zijlstra J.J., Radl J., In: Pal, S.B. (ed) Reviews on immunoassay technology. The MacMillan Press Ltd. London, U.K., 1987, in press
7. Haaijman J.J. Coolen J., Kröse C.J.M., Pronk G.J., Ming Z.F., Histochemistry 84:363 (1986)
8. Haaijman J.J., Deen C., Kröse C.J.M., Zijlstra J.J., Coolen J., Radl J., Immunology Today 5:56 (1984)
9. Hijmans W., Schuit H.R.E., Hulsing-Hesselink E., Ann. N.Y. Acad. Sci. 177:290, (1971)
10. Brandtzaeg P., Surjan L., Berdal P., Clin. Exp. Immunol. 31:367 (1978)
11. Surjan L., Brandtzaeg P., Berdal P., Clin. Exp. Immunol. 31:382 (1978)
12. Kett K., Brandtzaeg P., Radl J., Haaijman J.J., J. Immunol. 136:3681 (1986)

INTERRELATIONSHIPS OF CELL-SUBSTRATE AND CELL-CELL ADHESION STRUCTURES IN B-CHRONIC LYMPHOCYTIC LEUKEMIA (B-CLL) CELLS[1]

F. Caligaris-Cappio, L. Bergui, L. Tesio, M. Riva, M. Schena, G. Corbascio, and P.C. Marchisio.

Dipartimento di Scienze Biomediche e Oncologia Umana
Cattedra di Clinica Medica A e Cattedra di Istologia
Universita' di Torino, Torino, Italy

The investigation of cell-substrate adhesion structures and cell-cell interaction molecules may be crucial to understanding the traffic and distribution of malignant lymphoid cells. In this respect, B-chronic lymphocytic leukemia (B-CLL) is a useful model because of three reasons. First, malignant B-CLL cells relentlessly accumulate in lymphoid organs and the bone marrow, but fail, in large part, to recirculate (1), suggesting the existence of adhesion abnormalities. Second, the cytoskeletal organization of B-CLL cells differs from that of normal B lymphocytes (2). B-CLL cells adhere spontaneously in vitro utilizing, instead of the conventional "adhesion plaques", dot-shaped close contact sites similar to the podosomes described in cells of monocyte-macrophage lineage and in class I oncogene-transformed cells (3-5). Third, the cytoskeleton function of B-CLL cells may be abnormal as suggested by the observation that different ligands, including surface immunoglobulins (sIg), are capped sluggishly if at all (6,7).

In this work, we aimed at defining the relationships between podosomes, the microfilament-matrix interaction structures vinculin and talin (8-10) and the trans-membrane heterodimeric integrins (reviewed in ref. 11) which are central to the processes of both cell-substrate adhesion and cell-cell interaction.

MATERIAL AND METHODS

Cells

Peripheral blood lymphocytes (PBL) from 14 patients with B-CLL and 3 patients with follicular lymphomas in leukemic phase were separated on a Ficoll-Hypaque (FH) gradient. Normal B lymphocytes were obtained from tonsils.

Antibodies

Rabbit (R) antisera to human (Hu) IgM, IgD, k and light chains directly conjugated with tetraethylrhodamine isothiocyanate (TRITC) from Dakopatts were used in direct immunofluorescence (IF). CD5 (RFT1, ref. 12), CD19 (B4, Coulter), CD20 (B1, Coulter) MoAbs were used in indirect IF and revealed with R-anti-mouse (M)-Ig conjugated with fluorescein isothiocyanate (FITC) from Dakopatts.

F-actin was detected with fluorescein- or rhodamine-tagged phalloidin (F-PHD and R-PHD; ref. 13). Vinculin was evaluated with a monoclonal antibody (MoAb) from Bio Yeda, talin with a rabbit Ab to chicken gizzard talin cross-reacting with the mammalian form (14).

The different classes of integrins were tested using specific MoAbs revealed with R-anti-M-Ig-FITC or -TRITC. $Beta_1$ was detected with the MoAb, clone A1A5 (15), $beta_2$ with a MoAb specific for the beta chain of the LFA1 complex (CD18; ref. 16), beta $_3$ with MoAbs to the human vitronectin receptor (17) and the platelet glycoprotein complex gpIIb-IIIa (18). The different alpha chains of the CD11 complex were analysed with anti-LFA1 alfa (CD11a; ref. 19), MO1 (CD11b, Coulter) and anti-p150,95 (SHCL3, CD11c; ref. 20).

Cell culture and staining

B-CLL and normal B cells (2 x 10^6 cells/ml) were cultured in RPMI medium supplemented with 10% fetal calf serum at 37°C in a humidified atmosphere with 5% CO_2 for 72 hrs. Cultures were set up in duplicate as controls and in the presence of 1.6 x 10^{-8} M 12-O-tetradecanoyl phorbol-13-acetate (TPA).

The surface features of cultured cells were analysed on harvested cells resuspended at the concentration of 1 x 10^6 cells/50 ul phosphate buffered saline (PBS) and stained with MoAbs (and R-anti-Hu-Ig-TRITC) for 10 min at room temperature (RT). The cells were washed, reacted with G-anti-M-Ig-FITC for 10 min at RT and analysed under a coveslip with a Zeiss epifluorescence microscope equipped with a Zeiss planapochromat 63 x/1.4 oil immersion lens.

For IF analysis of cytoskeletal structures, cells were cultured on glass coverslips, fixed and permeabilized as previously described (4). Coverslips were simultaneously reacted with the relevant Abs (revealed by FITC- or TRITC-G-anti-M-Ig) and F- or R-PHD in order to colocalize F-actin and different cytoskeletal or trans-membrane antigens. Cells were analysed with a Leitz Diavert microscope equipped with epifluorescence and interference reflection optics.

RESULTS

Surface expression of LFA1 molecules by normal and B-CLL cells

The common beta chain (CD18) as well as the LFA1 alpha chain (CD11a) were expressed by > 95% normal B cells, while, as expected, CD11b and CD11c were not observed. The TPA treatment did not produce any appreciable modification in the expression of CD11 molecules. B-CLL cells as well as PBL from leukemic follicular lymphomas had a different pattern. CD18 was, as expected, present on the surface of >95% monoclonal B cells, but CD11a was surprisingly absent, or extremely faint. In the same samples, the residual T lymphocytes were instead strongly CD11a+. The TPA treatment of malignant B cells did not produce a significant increase in the proportion of monoclonal B cells coexpressing CD11a: the percentage of CD11a+ cells raised from <5% to 10 + 5%. As already shown (21), >60% B-CLL cells expressed CD11c after TPA.

Cytoskeletal co-localization of F-Actin, Vinculin, Talin and Integrins

F-actin was concentrated in multiple dots scattered over the ventral membrane (podosomes; see ref. 2). The F-actin core of each single podosome was surrounded by a ring containg vinculin and talin. Both F-actin and the vinculin and talin rings consensually clustered in the centre of the cell upon TPA treatment producing a honeycomb pattern. We next tested the reactivity of the different classes of integrins. The most relevant observation was that the $beta_2$ integrin (CD18) co-localized with vinculin and talin in rings surrounding the F-actin core of podosomes. CD11a, CD11b and CD11c could not be observed, confirming cell suspension studies. After TPA, CD11c was diffuse over the whole surface with no obvious sign of podosome localization.
$Beta_1$ integrin precisely co-localized with the F-actin core, while $beta_3$ integrin was not expressed by malignant B cells.

Normal B cells did not adhere spontaneously to coverslips even after TPA treatment and were thus glued onto poli-L-lysin coated slides. When tested with the same reagents employed on B-CLL cells, some homogeneous fluorescence was seen beneath the plasma membrane or diffusely on the surface. No dot- or ring-like pattern were seen.

DISCUSSION

The lymphocyte receptor functions mediated by surface molecules are closely associated with the cellular contractile system (22,23). It is therefore relevant to investigate the relationships between the cytoskeleton, the molecules which control the microfilament/matrix interactions (8-10) and the structures which are involved in cell recognition events (CD11; ref. 24,25). The rationale for comparing B-CLL to normal B lymphocytes is provided by the peculiar cytoskeleton organization of B-CLL cells which adhere in vitro through dot-shaped close contact sites that concentrate the bulk of F-actin (podosomes). Our first observation is that at least four different members of the complex controlling cell-matrix adhesion and cell-cell interactions are specifically related to podosomes in malignant B-CLL cells. The microanatomy of podosomes indicates that they are composed of a core of F-actin and $beta_1$ integrin surrounded by a ring where vinculin, talin and $beta_2$ integrin (common beta chain of LFA1, CD18) co-localize. This cytoskeleton organization is not shared by normal B cells, indicating that B-CLL cells have developed new adhesion properties. The second observation is that B-CLL cells as well as circulating malignant lymphocytes from patients with follicular lymphomas in leukemic phase do not express the alpha chain of LFA1 (CD11a), again differing from normal B lymphocytes. This unexpected finding suggests either that a yet unidentified alpha chain is linked to LFA1 beta chain in B-CLL cells or that in these malignant cells the alpha chain has undergone molecular modifications which prevent its recognition by known antibodies.

REFERENCES

1. H.D. Flad, C. Huber, K. Bremer, H.D. Menne, and H. Huber. Eur. J. Immunol. 3: 688 (1973).

2. F. Caligaris-Cappio, L. Bergui, L. Tesio, G. Corbascio, F. Tousco, and P.C. Marchisio. Blood 67:233 (1986).

3. P.C. Marchisio, O. Capasso, L. Nitsch, R. Cancedda, and E. Gionti. Exp. Cell Res. 151:332 (1984).

4. P.C. Marchisio, D. Cirillo, L. Naldini, M.V. Primavera, A. Tetl, and A. Zambonin-Zallone. J. Cell Biol. 99:1696 (1984).

5. P.C. Marchisio, D. Cirillo, A. Teti, A. Zambonin-Zallone, and G. Tarone. Exp. Cell Res. 169:202 (1987).

6. H.J. Cohen. J. Clin. Invest. 55:84 (1975).

7. U. Mintz, L. Sachs. Int. J. Cancer 15:253 (1975).

8. I.I. Singer, and P.R. Paradiso. Cell 24:481 (1981).

9. B. Geiger. Biochim. Biophys. Acta 737:305 (1983).

10. K. Burridge, and P. Mangeat. Nature 308:744 (1984).

11. R.O. Hynes. Cell 48:549 (1987).

12. F. Caligaris-Cappio, M. Gobbi, M. Bofill, and G. Janossy. J. Exp. Med. 155:623 (1982).

13. E. Wulf, A. Deboben, F.A. Bautz, H. Faulstich, and T. Wieland. Proc. Natl. Acad. Sci. U.S.A. 76:4498 (1979).

14. K. Burridge, and L. Connell. J. Cell Biol. 97:359 (1983).

15. M.E. Hemler, C. Huang, and L. Schwarz. J. Biol. Chem. 262:3300 (1987).

16. F. Miedema, P.A.T. Tetteroo, W.G. Hesselink, G. Werner, H. Spits, and C.J.M. Melief. Eur. J. Immunol. 14:518 (1984).

17. R. Pytela, M.D. Pierschbacher, and E. Ruoslahti. Proc. Natl. Acad. Sci. U.S.A. 82:5766 (1985).

18. R. Pytela, M.D. Pierschbacher, M.H. Ginsberg, E.F. Plow, and E. Ruoslahti. Science 231:1559 (1986).

19. J.E.K. Hildreth, F.M. Gotch, P.D.K. Hildreth, and A.J. McMichael. Eur. J. Immunol. 13:202 (1983).

20. R. Schwarting, H. Stein, and C.Y. Wang. Blood 65:974 (1985).

21. F. Caligaris-Cappio, G. Pizzolo, M. Chilosi, L Bergui, G. Semenzato, L. Tesio, L. Morittu, F. Malavasi, M. Gobbi, R. Schwarting, D. Campana, and G. Janossy. Blood 66:1035 (1985).

22. J. Braun. Immunol. Today 1:21 (1983).

23. J. Braun, and E.R. Unanue. Semin. Hematol. 20:322 (1983).

24. F. Sanchez-Madrid, A.M. Krensky, C.F. Ware, E. Robbins, J.L. Strominger, S.J.Burakoff, and T.A. Springer. Proc. Natl. Acad. Sci. U.S.A. 79:7489 (1982).

25. A. Fischer, A. Durandy, G. Sterkers, and C. Griscelli. J. Immunol. 136:3198 (1986).

1. This work was supported by Progetto Finalizzato Oncologia, CNR and by A.I.R.C. L.B. is a recipient of a A.I.R.C. fellowship. L.T. is a recipient of a fellowship of Regione Piemonte. The authors are indebted to all the colleagues who made this investigation possible by generously providing samples of their antibodies.

HETEROGENEITY IN THE EXPRESSION OF B-CELL ANTIGEN IN NON-HODGKIN'S LYMPHOMA

D. B. Jones, K. Moore* and D. H. Wright

University Department of Pathology and *Regional Immunology Laboratory, Southampton General Hospital Tremona Road, Southampton S09 4XY

INTRODUCTION

The value of monoclonal antibodies (MoAbs) for the identification of neoplastic B-cells is now undisputed. Studies of normal B-cell populations with large series of MoAbs have demonstrated that distinct markers are acquired and lost during ontogeny.[1,2] Workshops on human leucocyte differentiating antigens have described a universal system for MoAb classification, MoAbs falling into the same cluster of differentiation, or CD group, bind to the same B-cell monoclonal antigen but may bind to different epitopes on that molecule.[3] The majority of pan-B reagents identify the bulk of B-cell malignancies, although the efficiency of detection varies in different histological types of tumour and with different CD specificities.[4,5,6] In this presentation we have reviewed the performance of a range of B-cell antibodies in our routine diagnostic profiles. We report on variation in the staining obtained with different reagents and also present new data on the efficacy of two locally produced antibodies, WR14[7] and WR17[8] in the diagnosis of B-non-Hodgkin's lymphoma (B-NHL).

MATERIALS AND METHODS

The tissues employed in this study represent routine diagnostic biopsies obtained from Southampton General and associated hospitals. Lymph nodes were bisected and snap frozen for immunohistological analysis. Additional material was fixed in formol saline and subjected to routine histological analysis according to the Kiel system. In the majority of cases of B-NHL monoclonality was confirmed by light chain staining.

Table 1 lists the antibodies employed for immunohistochemical staining. In addition to reagents which have been assiged a CD number, the monoclonal antibody KB61 is of restricted B-cell distribution and in reactive tonsil and lymph node sections strongly stains mantle B-cells but spares

Table 1. Antibodies Employed in this Study

Antibody	Designation
WR17,[8] HD28[17]	CD37
SHCl-1,[16] Töl5[4]	CD22
HD37[18]	CD19
WR14[7]	T, restricted B
KB61[9]	restricted B

follicle centres.[9] WR14 stains all T-lymphocytes in the circulation but fails to stain circulating B-cells. Its reactivity with malignant B-cells is similar to that described for CD5 cluster reagents.[7] We have demonstrated that WR14 binds a molecular species of 120 kd molecular weight. This observation shows WR14 to be distinct from CD5 antibodies which precipitate a molecule of 65 kd.

Immunostaining: Immunoperoxidase staining was performed by a modification of the method of Stein et al.[10] Briefly, fresh tissue was frozen in liquid nitrogen and 6mu sections were fixed in dry acetone, washed in tris buffered saline and incubated with monoclonal antibody for 30 minutes. They were subsequently washed 3 tims in Tris Buffered Saline pH 7.6, incubated with peroxidase conjugated rabbit anti-mouse Ig (Dako) and washed 3 times further before developing the reaction product using 3,3-diaminobenzidine.[11]

RESULTS

In general, the staining of tumour cell populations with all of the monoclonal antibodies was of lower intensity than that seen in reactive proliferations. Most antibodies stained neoplastic cells with a pattern that was either membrane only or both membrane and cytoplasmic. In some of the large cell tumours unusual patterns of staining were seen with individual monoclonal reagents. These were granular and cytoplasmic, or focal and apparently restricted to the Golgi region of the cell. In tumours containing mixed populations it was usual to find that the small cell component was more strongly stained than the large cell component. Staining of tissue histocytes was seen with the CD37 antibodies and also with the antibody KB61 but did not complicate the interpretation of tumour cell populations.

The relative efficiencies of staining of the different reagents with different histological types of B-NHL is given below. For clarity, we have listed the staining patterns obtained with the major subtypes of B-NHL identified by the Kiel classification.

Lymphocytic Lymphoma: Thirty-four cases of lymphocytic lymphoma were stained. All were positive with MoAbs identifying CD37, CD22 and CD19 antigen groups. WR14 identified 30 of the 34 cases. KB61 was positive on 14 of 20 cases stained. All cases showed uniform strong expression of individual antigens. Occasionally, proliferation centres could be distinguished within lymphocytic lymphoma by intensity of monoclonal antibody staining.

Follicle Centre Cell Lymphoma/Centroblastic/Centrocytic, Follicular and Diffuse: Tumours of this type formed the bulk of the material stained (Table 2). In all, 67 lymphomas were investigated. All cases were positive for the CD37 antigen, although in 2 cases WR17 was positive whilst the HD28 antibody was negative. Sixty-three of 67 were positive for CD22, although the intensity of staining

Table 2. Antigen Expression in FCC

CD19	61/67
CD22	63/67
CD37	67/67
WR14	5/67
KB61*	22/67**

*reactive mantle cells frequently positive
**weak staining.

with individual antibodies identifying this antigen varied greatly between tumours. CD19 was positive on 61 of 67 cases. WR14 was positive on only 5, whilst KB61 showed weak positivity on approximately one-third of the follicle centre cell tumours. The intensity of staining seen with KB61 was much lower than that seen in lymphocytic lymphoma. It was often possible to identify the residue of reactive mantles surrounding neoplastic follicles with KB61. This interpretation was confirmed by inspection of the kappa and lambda staining pattern on the same lymph node biopsy.

Malignant Lymphoma/Centrocytic/Diffuse: Twelve cases presented during the period of this study. CD37, CD22 and CD19 antibodies stained the majority of cases, although heterogeneity was seen between individual reagents. In general, CD37 gave the strongest staining. The intensity of staining varied greatly in centrocytic diffuse lymphoma, from a level equivalent to lymphocytic lymphoma to extremely weak membrane or cytoplasmic positivity. The antibody WR14 stained half the cases of centrocytic/diffuse lymphoma. In each case the pattern of staining was that of membrane positivity.

Malignant Lymphoma/Centroblastic/Diffuse: Fifteen cases presented during the period of study. The majority were identified with CD19 and CD22 reagents, although marked heterogeneity was seen in the intensity of staining.

WR17 stained all of the cases, although the range of intensity of staining was considerable and HD28 similarly stained all cases. In general, WR17 was the stronger reagent, although in 2 cass HD28 gave stronger staining than the alternative CD37 antibody. K.B61 stained 2 of the centroblastic lymphomas, whilst WR14 was positive in 3 cases.

Large Cell B-NHL: During the period of investigation we were also able to stain 5 cases of large cell B-NHL. All were positive for the CD37 reagent, WR17, although the patterns of staining were abnormal and often intracellular. Three of the 5 were positive with CD22 and in 2 of these cases the staining pattern was cytoplasmic. Three of the lymphomas were positive for the CD30 activation antigen, BERH2.[12] In these cases a histological diagnosis of Ki-1 lymphoma (B-cell) would be appropriate.

DISCUSSION

The monoclonal antibody, WR17, has been shown to identify the CD37 gp40-45 kd antigen complex.[8] The general reactivity of this antibody is that of a pan-B-cell reagent. The investigation of neoplastic B-cells reported, confirms that this reagent is of great value in the diagnosis of B-NHL. In all cases studied, WR17 was positive. The CD37 positivity seen with WR17 was stronger than that observed with either CD22 or CD19 reagents. The value of WR17 in immunodiagnosis is underlined by the observation that all B-cell lymphoblastic lymphomas investigated were positive with this antibody and also that 2 cases of lymphoplasmacytic lymphoma stained with WR17 (data not shown). This suggests that the CD37 antibody appears early in the B-cell lineage and is present until late in the B-cell differentiation.

The staining patterns of WR14 and KB61 in different histological types of B-NHL are of interest. Although initially identified as T-cell antibody, WR14 clearly stained a proportion of neoplastic B-cells. It is clear from Table 3

Table 3. Relative Staining with WR14 and KB61

	WR14	KB61
CLL/LL	30/34	14/20
FCC	5/67	22/67 (weak)
CB:D	3/15	2/15
CC:D	6/12	7/15

that WR14 positivity is closely associated with a histological diagnosis of lymphocytic lymphoma or of centrocytic/diffuse lymphoma.[13] In this respect, WR14 resembles CD5 reagents. Whilst the 120 Kd molecule, identified by WR14, is distinct from that identified by CD5 antibodies, it clearly behaves in a manner which closely resembles the 65 kd CD5 molecule during B-cell differentiation.[13]

Heterogeneity of staining was seen with the CD22 reagents. Most antibodies stained the majority of tumor cells although, in individual cases, there was considerable heterogeneity in the intensity of staining seen with individual CD22 MoAbs. This variation could not be attributed to the avidity of the MoAb as the range of heterogeneity seen varied in individual cases. Epitope mapping studies have shown that whilst all CD22 monoclonal antibodies precipitate an identical molecule they may identify individual non-overlapping epitopes present on that molecule.[8] This heterogeneity of antigen expression could lead to diagnostic problems where a single CD22 MoAb is employed as the sole means of identifying B-cell lineage.

Heterogeneity was also seen between the 2 CD37 antibodies. In general, WR17 stained tumour cells more strongly than HD28, although in a small number of B-NHL this was not the case. Further, in a minority of cases WR17 was positive whilst HD28 was negative. Studies, employing endoglycosidases, have shown that monoclonal antibodies identifying the CD37 antigen bind to a carbohydrate epitope on the molecule[14] and glycosilation differences presumably account for the variation in positivity seen between WR17 and HD28.

Heterogeneity in antigen expression is well established in the node-based T-cell lymphomas.[15] The data presented indicates that B-NHL show similar heterogeneity in their lineage antigen expression. The greatest variation is seen in large cell B-NHL showing a highly pleomorphic tumour component. Overall, the data presented suggests that whilst MoAbs to B-cell lineage antigens are a reliable tool for the diagnosis of B-cell lymphoma the level of heterogeneity shown is such that reagents representative of more than one CD antigen group should be employed for reliable diagnosis.

References

1. L M Nadler, K C Anderson, M Bates, E Park, B L Slaughenhoupt and S F Schlossman. In: 'Leucocyte Typing'. Springer-Verlag: Berlin: 354 (1984).

2. L M Nadler, P Stashenko, R Hardy, A van Agthoven, C Terhorst and S F Schlossman. J Immunol 126: 1941 (1983).

3. J M S Williamson, I Grigor, M E F Smith, C S Holgate, P Quirke, C C Bird, D L Alison and J A Child. J Pathol 150: 51 (1986).

4. H Stein, J Gerdes and D Y Mason. Clin Haematol 11: 531 (1982).

5. L M Nadler, P Stashenko, J Ritz, R Hardy, J M Pesando, S F Schlossman. J Clin Invest 67: 134 (1981).

6. T Al Saati, G Laurent, P Cavariviere, F Rigalard, G Delsol. Clin Exp Immunol 58: 631 (1984).

7. K Moore, S A Cooper, A M Nesbitt and D B Jones. Submitted for publication.

8. K Moore, S A Cooper and D B Jones. J Pathol 152: 13 (1987).

9. K Pulford, E Raflkiar, S M MacDonald, W Erber, B Falini, K Gatter, D Y Mason. Immunology 57: 71 (1986).

10. H Stein, A Bonk, G Tolksdorf, K Lennert, H Rodt and J Gerdes. J Histochem Cytochem 28: 746 (1980).

11. D B Jones, K M Britten and D H Wright. In: Leucocyte TypingII, Vol 2. Springer-Verlag: Berlin: 299 (1986).

12. H Stein, D Y Mason, J Gerdes, N O'Connor, J Wainscoat, G Pallesen, K Gatter, B Falini, G Delsol, H Lemke, R Schwarting and K Lennert. Blood 66: 848 (1985).

13. F Calgaris-Cappio, M Gobbi and L Bergui. J Haematol 56: 655 (1984).

14. R Schwartz. Personal communication.

15. D B Jones, D H Wright, F K Paul and J L Smith. Hem Oncol 4: 219 (1986).

16. R Schwarting, H Stein and C Y Wang. Blood 65: 974 (1985).

17. B Dorken, G Moldenhauer, A Pezzutto, R Schwartz and S Krescal. In: Leucocyte Typing II, Vol 2. Springer-Verlag: Berlin: 47 (1986).

18. A Pezzutto, A Dorken, G Feller, R Moldenhauer, R Schwartz. In: Leucocyte Typing II, Vol 2. Springer-Verlag: Berlin: 47 (1986).

GROWTH REQUIREMENTS OF SJL LYMPHOMAS *IN VITRO*: EFFECT OF BCGF II

Jennifer L. Lasky and G. Jeanette Thorbecke

Dept. of Pathology
NYU School of Medicine
New York, NY 10016

INTRODUCTION

Reticulum cell sarcomas (RCS) arise spontaneously in > 90% of aging SJL mice[1]. The tumors are reported to arise in germinal centers of mesenteric lymph node (LN) and Peyer's patches[2]. When injected intravenously into SJL recipients, labeled tumor cells localize in germinal centers of lymphoid tissue[3]. Thus these "RCS" may in fact represent some form of murine follicular center B cell lymphomas[4].

Recently we have established an *in vitro* RCS cell line, cRCS-X, from the transplantable RCS-X line. cRCS-X maintains all of the properties of spontaneous and transplantable RCS, many of which are consistent with a B cell derivation[5]. Primary RCS do not appear in SJL mice which have been injected with anti-μ antibodies from birth[6]. Despite the fact that primary RCS[7], transplantable RCS lines and cRCS-X have rearranged immunoglobulin (Ig) genes (J. Stavnezer, J.L. Lasky, N.M. Ponzio and G.J. Thorbecke, manuscript in preparation), they do not have detectable Ig on their cell surface. The cells are negative for expression of the T cell markers Thy 1.2, Ly 1.2, Ly 2.2 and L3T4, for the macrophage surface antigen Mac 1, and for the natural killer cell marker, NK 1.1. Both *in vivo* and *in vitro* maintained RCS bear the B cell marker 14.8[8], I-A^s [9], and H-$2K^s$ but no detectable H-$2D^s$ [5,10]. Studies with the *in vivo* RCS lines have shown that I-A^s molecules on RCS resemble those on LPS-activated SJL B cell blasts with respect to their electrophoretic migration patterns in 2 dimensional gels[11].

One of the most striking characteristics of both *in vivo* RCS and cRCS-X is their ability to induce strong proliferation in syngeneic T cells and their profound dependence on such syngeneic T cells for their growth[9]. Transplantable RCS and cRCS-X will not grow in SJL mice that have been immunocompromised by γ-irradiation[12,3,9]. Thymectomized SJL mice have a lower incidence of generalized primary RCS[13]. cRCS-X is dependent on the addition of γ-irradiated LN cells (LNγ) to the cultures every 3 weeks for it's continued growth *in vitro*. Adherent SJL spleen cells do not support cRCS-X growth. The Lyl^+, $Ly2^-$, $L3T4^+$ T cell proliferative response induced by RCS is accompanied by the production of a number of lymphokines including IL-2, IFN-γ [14] and T cell replacing factor[15]. We have recently shown that supernatants from cultures of cRCS-X grown with LNγ not only contain high titers of IL-2, but also both anti-μ (IL-4) and dextran sulfate (DxS) costimulator activities when assayed on normal murine splenic B cells[5]. These findings suggest that one or more of these lymphokines produced by T helper cells in response to RCS is necessary for RCS growth both in vivo and in vitro.

Human B cell growth factor (BCGF)[16] has activity in the murine B cell DxS costimulator assay[17-19]. In the present studies we have determined whether several recombinant lymphokines and BCGF can serve as growth factors for cRCS-X . The results show that only BCGF induces dose dependent proliferation of cRCS-X. In addition, this lymphokine has much stronger costimulatory activity with DxS on PNA^+ than on PNA^- LN B cells, suggesting that germinal center B cells also use BCGF as a major growth factor.

MATERIALS AND METHODS

Mice and Tumors. SJL/J mice were obtained from the Jackson Laboratory, Bar Harbor, ME. cRCS-X was derived from tumorous LN containing RCS-X by complement mediated cytolysis following treatment with an anti-T cell cocktail containing monoclonal anti-Thy 1.2 (6.80), anti-L3T4 (GK 1.5)[20], anti-Ly 2.2 (HO-2.2)[21] and anti-Ly 1.2 (C3PO)[22]. One x 10^7 purified RCS-X cells were plated with or without equal numbers of γ-irradiated SJL LN cells in 24 well plates (Linbro, New Haven, CN). The cells were cultured in 2 ml of Iscove's modified Dulbecco's medium per well, supplemented with 15% FCS (Hyclone, Logan, UT). The cells are maintained at a density of 5 x 10^5 cells/ml with equal numbers of fresh 3500R γ-irradiated SJL LN (LNγ) added approximately every 3 weeks.
B Cell Growth Factor Assay. One x 10^5 B cells from SJL LN were cultured in 96 well plates (Linbro, New Haven, CN) for 3 days in RPMI-1640 with 10% FCS at 37° C in 5% CO_2. The cells were cultured with either 10 μg/ml goat anti-μ antibodies (provided by Dr. F. Finkelman, Bethesda, MD) conjugated to sepharose CL-4B (Pharmacia, Piscataway, NJ) or 10 μg/ml DxS (Sigma Chemical Co., St. Louis, MO) as costimulators in the presence of serial dilutions of lymphokines. Cultures were labeled with ^{3}H-thymidine for the last 6-8 hours of incubation and harvested as described[23]. Results are expressed as counts per minute (cpm).
Colony Formation by cRCS-X Cells. Colony formation in soft agar by cRCS-X cells was assayed by a modification of the method described by Kincade et al.[24]. Briefly, 1-2 x 10^5 cRCS-X cells were cultured in 1 ml Iscove's medium, supplemented with 1% Nutridoma-SP (Boehringer Mannheim Biochem., Indianapolis, IN) which contained 0.8% Sea Prep ultra low gelling temperature agarose (FMC Corp., Rockland, ME). The cells were cultured in 35 x 100 mm tissue culture dishes (Falcon 3001, Becton Dickinson and Co., Oxnard, CA) which were rimmed with 25 μl of 20% sheep red blood cells (Colorado Serum Co., Denver, CO) in saline before the cells in medium containing agarose were added. After solidifying the agarose at 4° C for 1 hour, cultures were incubated for 7 days at 37° C in 5% CO_2. The numbers of colonies (CFU) of cRCS-X were determined by visualization in an inverted microscope.
Enrichment of PNA^+ Germinal Center Cells by Panning. SJL mice were immunized in the front footpads with 3 μg trinitrophenylated B. abortus (TNP-BA). Seven days later B cells were purified from the draining LN by removal of adherent cells and T cells as described above. Bacteriological grade petri dishes were coated for 18 hours with 40 μg peanut agglutinin (PNA)/ml. Three x 10^7 B cells in phosphate buffered saline (PBS) with 2% FCS were added to each PNA coated dish, incubated at 4° C for 1 hour, and nonadherent (PNA^-) cells were harvested. High affinity PNA binding (PNA^+) cells were eluted from the plates with 0.2 M D-galactose in PBS/2% FCS.

Lymphokines. Purified human BCGF (0.6-2.69 U/ml was purchased from Cellular Products Inc., Buffalo, NY). Recombinant murine IL-4 was provided by Dr. S. Gillis (Immunex Corporation, Seattle, WA) and recombinant IFN-γ was a gift from Genentech Corporation (San Fransisco, CA). Ultrapure human IL-1 (100 U/ml), recombinant murine IL-2 (10^5 U/ml) and murine IL-3 (5×10^4 U/ml) were obtained from Genzyme Corporation (Boston, MA).

Response of cRCS-X to Lymphokines. cRCS-X cells were cultured at 5×10^4 cells/well in Iscove's medium without FCS and supplemented with 1% Nutridoma-SP for 2 days in the presence of serial 2 fold dilutions of various lymphokines at 37° C in 5% CO_2. Proliferation was measured by ^{3}H-thymidine incorporation.

RESULTS AND DISCUSSION

Factors responsible for cRCS-X growth were characterized by examining the effect of purified factors in ^{3}H-thymidine incorporation assays. Dose response curves for BCGF (Fig.1), IL-1 (Table 1), IFN-γ (Table 1), IL-2, IL-3 and IL-4 were determined. The background proliferation was much higher in the presence than in the absence of FCS; therefore, only the dose response curves generated in the absence of FCS are shown. The only lymphokine tested which, by itself, significantly increased cRCS-X proliferation in the presence or absence of FCS, was BCGF. Human serum albumin, which was the major contaminant in the preparation of BCGF used was found to have only a negligeable effect on cRCS-X proliferation (not shown). BCGF at concentrations of 0.005-0.01 U/ml significantly increased cRCS-X proliferation and caused a linear increase in ^{3}H-thymidine uptake when added at concentrations up to 0.05-0.1 U/ml (Fig.1). IL-2, IL-3, and IL-4 did not induce RCS proliferation (data not shown).

Although IL-1 and IFN-γ when used alone also had no detectable effect, they were able to synergize with BCGF in promoting cRCS-X growth. In the presence of 0.05 U/ml of BCGF each of these cytokines caused a significant increase in cRCS-X proliferation which plateaued at more than twice the level of ^{3}H-thymidine incorporation caused by BCGF alone. This effect was obtained with 0.3 U/ml IL-1 and with 300 U/ml IFN-γ (Table 1).

In view of the marked proliferation of cRCS-X induced by BCGF, this lymphokine was tested to determine whether it, like LNγ, could be used to maintain cRCS-X growth indefinitely. It was indeed found that cRCS-X could be maintained for at least 10 months in medium containing 0.1 U of BCGF per ml without the addition of LNγ. In medium containing FCS alone the cells died within approximately 40 days. It is clear that BCGF is the essential ingredient which causes cRCS-X to continue to grow _in vitro_. Moreover, the findings suggest a similarity of cRCS-X to human B cell lines which also require 0.1 U/ml of these BCGF preparations for their maintenance[25].

Another short term assay of the effect of factors on cRCS-X was developed which made use of the ability of cRCS-X cells to produce colonies in soft agar (Fig.2). No colonies were seen in serum free medium unless BCGF was added. 0.1 U/ml of BCGF induced approximately 800 colonies per 10^5 cRCS-X cells, the 100% values used in Figs. 2a and 2b. Addition of 10% FCS or 1% human serum caused an equivalent number of colonies to grow as did BCGF alone. Addition of both BCGF and serum had a synergistic effect on colony formation such that more than 1% of plated cRCS-X cells produced colonies and larger colonies were produced. IL-2 and IFN-γ did not induce colony formation when added alone, but IFN-γ

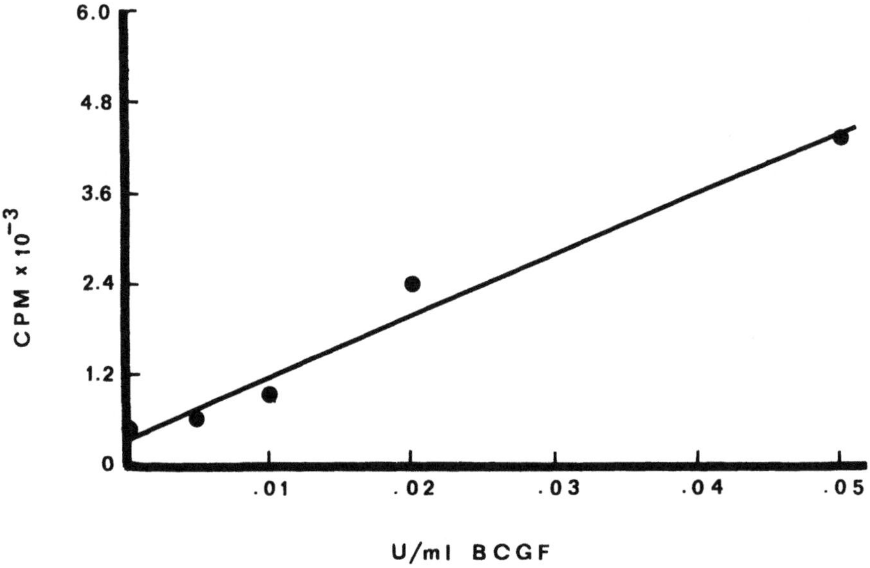

Fig. 1. Response of cRCS-X cells to BCGF. The linear portion of the dose response curve from a representative assay is shown here. The BCGF preparation used contained 2.69 U/ml BCGF. Maximum stimulation obtained in this assay was 12,845 cpm at 0.3 U/ml BCGF. Background proliferation in the absence of BCGF was 452 cpm.

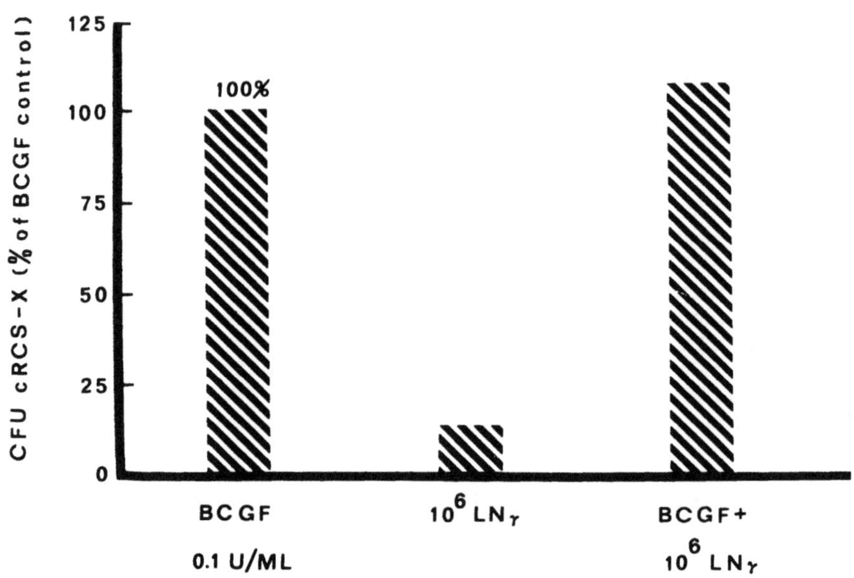

Fig. 2. Synergy between BCGF and sera for the induction of cRCS-X colony forming units. Final concentration of IFNγ used was 7.6 x 10^3 U/ml. IL-2 was added at 25 U/ml. Background colony formation by cRCS-X cells in serum free medium was 30 colonies per 10^5 cRCS-X cells (2a) and 0 colonies per 10^5 cRCS-X cells (2b).

Table 1

SYNERGISTIC EFFECT ON cRCS-X PROLIFERATION OF IL-1 AND IFNγ WITH BCGF

U/ml	Synergizing Agent	WITHOUT BCGF		WITH 0.05 U/ml BCGF	
		cpm	Δcpm*	cpm	Δcpm
5	IL-1	4897	3825	20286	11811
1.25		2612	1540	14392	5971
0.3		1323	251	12028	3553
19.0×10^3	IFNγ	2481	1409	21494	13019
5.0×10^3		2478	1406	18366	9891
2.0×10^3		1028	0	11751	3276
0.3×10^3		833	0	9939	1464
0	None	1072	0	8475	0

* Δ cpm = cpm induced by addition of the synergizing agent.

appeared to increase the effect of BCGF. In the absence of FCS and BCGF, the addition of LNγ caused a small increase in colony formation by cRCS-X cells (Fig.3) and in the presence of FCS, LNγ caused a much larger increase (not shown). However, in the presence of BCGF, the addition of LNγ did not cause a further increase. The fact that sera are able to synergize with BCGF in the induction of cRCS-X growth suggests that their effect is not due to the presence of BCGF in these sera. In contrast the absence of synergy between LNγ and BCGF indicates that LN cells contribute BCGF itself. The factor(s) responsible for the effect of serum have not been identified. It is unlikely that IFNγ is present in serum and, even though IL-1 could be present, alone it is not nearly as effective in supporting cRCS-X growth as are either human or fetal calf sera.

The fact that cRCS-X survive (but do not grow) in vitro for 44-56 days in medium containing FCS, in the absence of LNγ[5], suggests that RCS cells may be able to produce autocrine factors. The production of factors by B cells, affecting B cell differentiation[26] and proliferation[27-29], has been reported. However, the effect of any factor produced by cRCS-X cells themselves would have to be limited, since the cells cannot be maintained indefinitely without addition of LNγ or exogenous BCGF.

The most prominent lymphokine in the human BCGF preparation used in the present studies is a 12 Kd molecule[16]. This lymphokine, previously called BCGF II by investigators working with mouse cells, has recently been cloned [30] and shown to have additional activities besides promoting B cell growth. It induces proliferation and IgM secretion in BCL_1 cells and also promotes differentiation and IgG production by primed B cells in the presence of antigen[30]. In addition, this factor has eosinophil differentiation-inducing activity[31], and it upregulates IL-2 receptors on murine B cells[32]. It has been proposed that this lymphokine be named IL-5[30]. Preliminary results indicate that recombinant murine IL-5 (kindly provided by Dr. T. Honjo) stimulates cRCS-X proliferation to the same extent as does human BCGF.

cRCS-X may represent a good cell line for the assay of BCGF. In short term proliferation assays of ^{3}H-thymidine incorporation there is a linear relationship between BCGF concentration and cRCS-X proliferation.

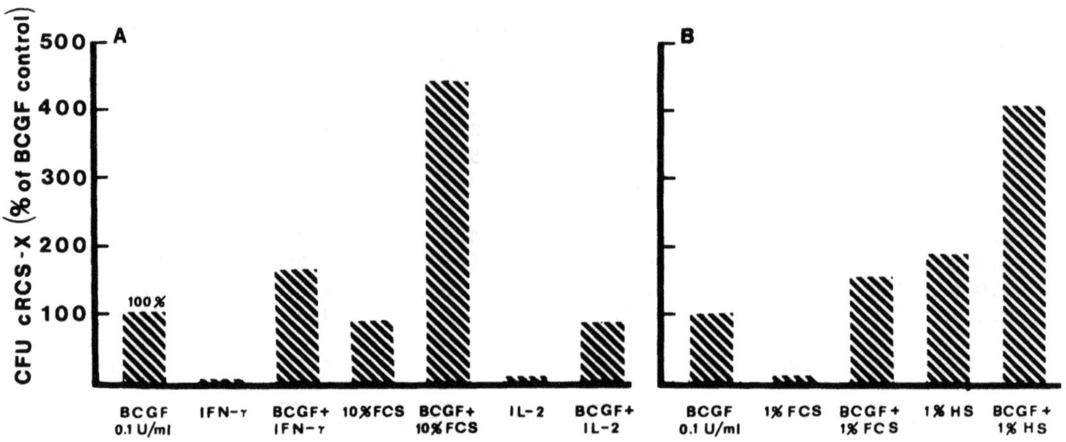

Fig. 3. Absence of synergy between BCGF and LNγ. 10^5 cRCS-X cells were cultured with or without 10^6 LNγ. Background colony formation in serum free medium was 4 colonies per 10^5 cRCS-X cells plated.

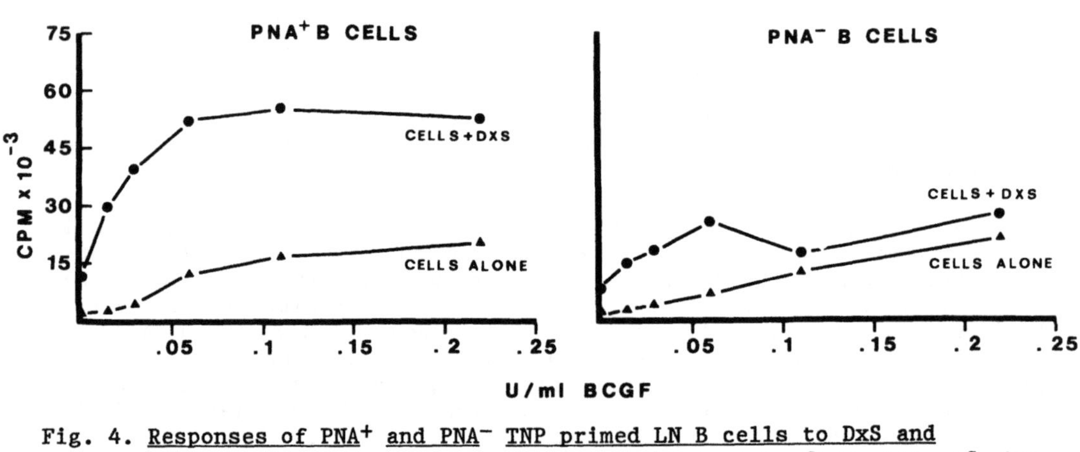

Fig. 4. Responses of PNA+ and PNA- TNP primed LN B cells to DxS and BCGF. The dose response curves shown are from one of two representative experiments with identical results. Cells were cultured with or without 10 μg/ml DxS in the presence of increasing concentrations of BCGF.

In this respect it is particularly interesting that the cRCS-X cells do not respond to recombinant IL-4, whereas the BCL_1 tumor cell line, commonly used to assay BCGF II activity does[30].

Originally our laboratory began to study the RCS model in SJL mice because these lymphomas appear to originate in germinal centers. Like germinal centers[33], RCS growth is T cell dependent. Therefore, we were interested to see if BCGF, which so strongly stimulates RCS growth in vitro, also causes proliferation of isolated germinal center B cells. We employed the PNA fractionation technique developed in our laboratory[34] to isolate PNA^+ germinal center B cells from brachial LN in which germinal centers had been induced by immunization with TNP-BA. PNA^+ and PNA^- B cells were assayed for their ability to proliferate in response to BCGF in the presence and absence of DxS as a costimulator (Fig. 4). Both fractions responded similarly to BCGF alone. The responses of PNA^+ and PNA^- cells to 10 μg/ml of DxS were 11,481 cpm and 7,573 cpm respectively, also not significantly different. However, in the presence of both BCGF and DxS the PNA^+ germinal center B cells showed markedly higher proliferation than did the PNA^- B cells.

These findings suggest that BCGF II (IL-5) is a growth factor both for germinal center cells and for lymphoma cells originating in germinal centers. PNA^+ B cells require a costimulator, such as the DxS used here, in order to proliferate maximally in response to BCGF. Such a stimulus in the germinal centers in vivo might be derived from C3 containing immune complexes[35]. It is also interesting to note that germinal centers have been proposed as sites for isotype switching in B cells[36], and memory cell proliferation[34]. A relatively higher sensitivity of germinal center cells than of other B cells to IL-5 would be in concordance with the augmenting effects on IgG and IgA production attributed to IL-5[37] and with the proliferative effect of BCGF on germinal center cells described here.

ACKNOWLEDGEMENTS

We would like to thank Dr. U. Hammerling (Sloan-Kettering Institute, NY) for kindly providing the 6.80 monoclonal antibody and Dr. F. Fitch (University of Chicago, IL) for the generous gift of the GK1.5 producing hybridoma. Supported by USPHS grants CA-14462 and AG-04980.

1. E.D. Murphy. SJL/J, A new inbred strain of mouse with a high, early incidence of reticulum-cell neoplasms, Proc. Amer. Assoc. Cancer Res. 4:46 (1963).
2. R. Siegler and M. A. Rich. Pathogenesis of reticulum cell sarcoma in mice, J. Natl. Cancer Inst. 41:125-143 (1968).
3. E.A. Carswell, S.P. Lerman, and G.J. Thorbecke. Properties of reticulum cell sarcomas in SJL/J mice II. Fate of labeled tumor cells in normal and irradiated syngeneic mice, J. Immunol. 23:39 (1976).
4. P.K. Pattengale and C.R. Taylor. Experimental Models of Lymphoproliferative Disease. The Mouse as a model for human non-Hodgkin's lymphomas and related leukemias, Amer. J. Path. 113:237 (1983).
5. J.L. Lasky, G.S. Erianne, N.M. Ponzio, and G.J. Thorbecke. Characterization and growth factor requirements of SJL lymphomas. I. Development of an in vitro reticulum cell sarcoma (RCS) cell line (cRCS-X), Fed. Proc. 46:1506 (1987).
6. I.R. Katz, R. Asofsky, and G.J. Thorbecke. Suppression of spontaneous reticulum cell sarcoma development and of syngeneic stimulator cells by anti-μ treatment of SJL/J mice, J. Immunol. 125:1355 (1980).
7. H. Osada, H. Nakauchi, and K. Okumura. Molecular evidence

that SJL reticulum cell sarcoma is B cell derived, Fed. Proc. 46:1056 (1987).

8. M.P. Scheid, J.M. Chapman-Alexander, T. Hayama, I.R. Katz, S.P. Lerman, C. Nagler, N.M. Ponzio, and G.J. Thorbecke. Identification of transplantable reticulum cell sarcomas in SJL/J mice as belonging to the B cell lineage, In: "B Lymphocytes in the Immune Response: Functional, Developmental, and Interactive Properties," N.R. Klinman, D.E. Mosier, I. Scher and E.S. Vitetta, eds. Elsevier, North Holland (1981).
9. N.M. Ponzio, P.H. Brown and G.J. Thorbecke. Host-tumor interactions in the SJL lymphoma model, Intern. Rev. Immunol. 1:273 (1986).
10. K.W. Beisel, and S.P. Lerman. Progressive loss of H-$2D^s$ antigens during passage of a spontaneous SJL/J lymphoma, Transplant. Proc. 13:1811 (1981).
11. P.H. Brown, and G.J. Thorbecke. Characterization of the molecules on SJL/J lymphomas which stimulate syngeneic T cells, J. Immunol. 135:3572 (1985).
12. S.P. Lerman, N.M. Ponzio, E.A. Carswell and G.J. Thorbecke, Properties of transplantable reticulum cell sarcomas in SJL/J mice, in: "Immune Reactivity of Lymphocytes," Feldman and Globerson, eds., Plenum Publishing Corporation, New York (1976).
13. I.R. Katz, J. Chapman-Alexander, and G.J. Thorbecke. Growth of SJL/J-derived transplantable reticulum cell sarcoma as related to its ability to induce T-cell proliferation in the host. IV. Effect of thymectomy on primary lymphoma incidence, Cell. Immunol. 74:394 (1982).
14. N.M. Ponzio, T. Hayama, C. Nagler, I.R. Katz, M.K. Hoffmann, K. Gilbert, J. Vilcek, and G.J. Thorbecke. Ia-restricted interaction of normal lymphoid cells and SJL lymphoma (reticulum cell sarcoma) leading to lymphokine production. II. Rapid production of antibody-enhancing factor, interleukin 2, and immune interferon, J Natl. Cancer Inst. 72:311 (1984).
15. T. Hayama, C. Nagler, D.T. Umetsu, J. Chapman-Alexander, and G.J. Thorbecke. Ia-Restricted interaction between normal lymphoid cells and SJL lymphoma (RCS) leading to lymphokine production, Cell. Immunol. 79:134 (1983).
16. S.R. Mehta, D. Conrad, R. Sandler, J. Mortan, R. Montagna, and A.L. Maizel. Purification of human B cell growth factor, J. Immunol. 135:3298 (1985).
17. S. Swain. Role of BCGF II in the differentiation to antibody secretion of normal and tumor B cells, J. Immunol. 134: 3934 (1985).
18. K. Yoshizaki, T. Nakagawa, K. Fukunaga, T. Kaieda, S. Maruyama, S. Kishimoto, Y. Yamamura, and T. Kishimoto. Characterization of human B cell growth factor (BCGF) from cloned T cells or mitogen-stimulated T cells, J. Immunol. 130:1241 (1983).
19. K. Shimizu, T. Hirano, K. Ishibashi, N. Nakano, T. Taga, K. Sugamura, Y. Yamamura and T. Kishimoto. Immortalization of BGDF (BCGF II)- and BCDF-producing T cells by human T cell leukemia virus (HTLV) and characterization of human BGDF (BCGF II), J. Immunol. 134:1728 (1985).
20. D.P. Dialynas, Z.S. Quan, K.A. Wall, A. Pierres, J. Quintans, M.R. Loken, M. Pierres, and F.W. Fitch. Characterization of the murine T cell surface molecule, designated L3T4, identified by monoclonal antibody GK1.5: Similarity of L3T4 to the human Leu-3/T4 molecule, J. Immunol. 131:2445 (1983).
21. D.H. Raulet, P.D. Gottlieb, and M.J. Bevan. Fractionation of lymphocyte populations with monoclonal antibodies specific for Lyt-2.2 and Lyt 3.1, J. Immunol. 125:1136 (1980).
22. C. Mark, F. Figueroa, Z.A. Nagy, and J. Klein. Cytotoxic monoclonal antibody specific for the Lyt-1.2 antigen, Immunogenetics. 16:95 (1982).

23. I.R. Katz, S.P. Lerman, N.M. Ponzio, D.C. Schreffler and G.J. Thorbecke. Growth of SJL/J derived transplantable reticulum cell sarcoma as related to its ability to induce T-cell proliferation in the host. I. Dominant negative genetic influences of other parent haplotype in F_1 hybrids of SJL/J mice, J. Exp. Med. 151:347 (1980).
24. P.W. Kincade, P. Ralph, and M.A.S. Moore. Growth of B-lymphocyte clones in semisolid culture is mitogen dependent, J. Exp. Med. 143:1265 (1976).
25. A.L. Maizel, J.W. Morgan, S.R. Mehta, N.M. Kouttab, J.M. Bator, and C.G. Sahasrabuddhe. Long-term growth of human B cells and their use in a microassay for B-cell growth factor, Proc. Natl. Acad. Sci. 80:5047 (1983).
26. K. Yoshizáki, T. Nakagawa, K. Fukunaga, L.T. Tseng, Y. Yamamura, and T. Kishimoto. Isolation and characterization of B cell differentiation factor (BCDF) secreted from a human B lymphoblastoid cell line, J. Immunol. 132:2948 (1984).
27. K. Nakajima, T. Hirano, F. Takatsuki, N. Sakaguchi, N. Yoshida, and T. Kishimoto. Physicochemical and functional properties of murine B cell-derived B cell growth factor II (WEHI-231-BCGF-II), J. Immunol. 135:1207 (1985).
28. K.H. Brooks, J.W. Uhr, and E.S. Vitetta. A B cell growth factor-like activity is secreted by cloned, neoplastic B cells, J. Immunol. 133:3133 (1984).
29. G. Scala, A. Shaw, C.I. Pastore F. Alfinito, G. Morrone, and S. Venuta. Autocrine growth function of human interleukin-1 molecules, Fed. Proc. 46:1344 (1987).
30. T. Kinashi, N. Harada, E. Severinson, T. Tanabe, P. Sideras, M. Konishi, C. Azuma, A. Tominaga, S. Bergstedt-Lindqvist, M. Takahashi, F. Matsuda, Y. Yaoita, K. Takatsu, and T. Honjo. Cloning of complementary DNA encoding T-cell replacing factor and identity with B-cell growth factor II, Nature 324:70 (1986).
31. A. O'Garra, D.J. Warren, M. Holman, A.M. Popham, C.J. Sanderson, and G.G.B. Klaus. Interleukin 4 (B-cell growth factor II/eosinophil differentiation factor) is a mitogen and differentiation factor for preactivated murine B lymphocytes, Proc. Natl. Acad. Sci. 83:5228 (1986).
32. M. S. Loughnan, K. Takatsu, N. Harada, and G.J.V. Nossal. T-cell-replacing factor (interleukin 5) induces expression of interleukin 2 receptors on murine splenic B cells, Proc. Natl. Acad. Sci. 84:5399 (1987).
33. E.B. Jacobson, L.H. Caporale and G.J. Thorbecke. Effect of thymus cell injections on germinal center formation in lymphoid tissues of nude (thymusless) mice, Cell. Immunol. 13:416 (1974).
34. R.F. Coico, B.S. Bhogal and G.J. Thorbecke. Relationship of germinal centers in lymphoid tissue to immunologic memory. VI. Transfer of B cell memory with lymph node cells fractionated according to their receptors for peanut agglutinin, J. Immunol. 131:2254 (1983).
35. M. Papamichail, C. Gutierrez, P. Embling, P. Johnson and E.J. Holborow. Complement dependency of localization of aggregated IgG in germinal centres, Scand. J. Immunol. 4:343 (1975).
36. G. Kraal, I.L. Weissman and E.C. Butcher. Germinal centre B cells: antigen specificity and changes in heavy chain class expression, Nature 298:377 (1982).
37. T. Yokota, R.L. Coffman, H. Hagiwara, D.M. Rennick, Y. Takebe, K. Yokota, L. Gemmell, B. Shrader, G.Yang, P. Meyerson, J. Luh, P. Hoy, J. Pene, F. Briere, J. Banchereau, J. De Vries, F.D. Lee, N. Arai, and K. Arai. Proc. Natl. Acad. Sci. USA, in press (1987).

THE ROLE OF FOLLICULAR DENDRITIC CELLS AND GERMINAL CENTERS IN IMMUNE REACTIONS

MORPHOLOGICL DIVERSITY OF DRC-1 POSPITIVE CELLS:

HUMAN FOLLICULAR DENDRITIC CELLS AND THEIR RELATIVES

Kazuo Terashma, Katsuaki Ukai, Katsushi Tajima, Fumiaki Yuda, and Yutaka Imai

Department of Pathology, School of Medicine, Yamagata Univ.
Zao-Iida, Yamagata, Japan 990-23

INTRODUCTION

Follicular dendritic cells(FDC) have been regarded as antigen-trapping cells playing roles in immune amplification especially in the secondary reponse.[4,7,9] However their morphological spectrum has not been enough -clarified since their cytoplasmic extensions are too complicatedly interwoven to observe in banal electron microsopy. The purpose of this report is to analyze their morphological profile and spectrum of human FDC with a monoclonal antibody DRC-1(DAKO)[2] specific to FDC and their relatives, in a combination with an isolation technique of FDC according to Tew et al.[6] Surveying FDC-precursors the authors much remarked extrafollicular DRC -1^+ cells and first observed DRC-1^+ cells proliferating in culture, which were originally obtained from the peripheral blood of autoimmune disease -patients. These morphological analysis certainly will give some new informations for FDC-study.

MATERIALS and METHODS

The tonsils were surgically extirpated from children or young adults suffering from hyperplastic tonsillitis. Some lymph nodes and granulation tissues were also surgically obtained. One third of these was snap-frozen in dry ice-aceton bath for cryostatt sections. One third was fixed in PLP-fixative to make 50u sections with Microslicer for immunoelectron microscopy. One third was enzymatically digested with collagenase according to Tew et al.to isolate DRC-1^+ cells, followed by BSA-gradient fractionation and Cytobucket or Cytospin centrifuge preparation.[6] Mononuclear cells in the peripheral blood of the normal and autoimmune disease-patients including SLE and RA were isolated with Conray Hypaque and similarly treated as described above. In some cases mononuclear cells were cultivated in ISCOV medium with 20% FCS under 7%CO_2. In immunochemistry and immunoelectron micoscopy,after dipping in 0.5% HIO_4/PBS at 4 C for 15mins. to inhibit endogenous peroxidase, frozen or sliced sections and cells centrifuged were treated with DRC-1 monoclonal antibody(DAKO Co, M 709) diluted x100 in PBS, followed with HRP-labelled sheep $F(ab')_2$ anti -mouse immunoglobulins(Amersham Ltd, diluted x50) enoughly absorbed with heat-coagulated human plasma. Peroxidase reaction was done in Karnovsky's method with DAB+H_2O_2. Double stainings were performed with HRP-labelled anti-mouse IgM for DRC-1 and alkaline phosphatase-labelled anti-mouse IgG

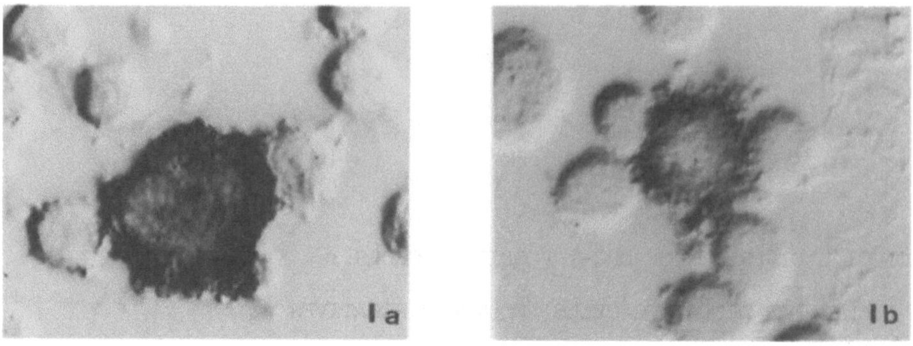

Fig. 1 DRC-1^{+} cells prepared by Cytobucket apparatus(x1000).

for other monoclonal antibodies of IgG subclass. The former was stained with DAB and the latter was done with Fast blue BB.

RESULT and DISCUSSION

In frozen or Microslicer-sections of the lymphoid tissue DRC-1 intensively stained the germinal center especially of the light zone in a lacy pattern, finely the corona and the primary follicle in a similar pattern. In addition, a few DRC-1^{+} cells singularly sparsed could be confirmed outside the follicles, in the interfollicular or paracortex and medulla. Immunoelectron microscopy of Microslicer-sections and cells collagenase -treated yielded three different types A, B, and C of DRC-1^{+} cells as shown in Table 1. In addition to organellae each type was characterized by cytoplasmic projections varying in their developmental stage. They were grouped into the primary, secondary and teritiary projections. The primary(P) is a fundamental blunt extension. The secondary(S) is a branch of the primary. The teritiary(T) is a bud-like or worm-like projection sprouting from the secondary and composes labyrinth like structures(LBS).

DRC-1^{+} cells of type A in tissue is a polygonal to spindle-shaped typical FDC bearing well-developed LBS, predominating in the light zone of the germinal centers. In the Cytobucket-preparations, enzymatically

Table 1. Three Types of DRC-1 Positive Cells

Type	General Profiles	Cyotoplasmic Extensions Primary	Secondary	Tertiary	LBS	Predominating in
		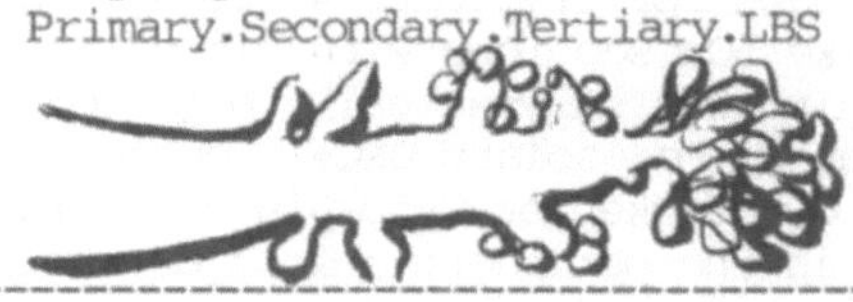				
A	Large mono/multinuclear., Sea anemone-like	+	+	+	+++	Light zone of G.C.
B	Large mononuclear, oligodendritic, stellate	+	+	+/-	-	Corona,Dark zone of GC, Primary FOL
C	Individually isolated, Medium-sized, Lymphoid	+	+/-	-	-/+	Outside FOL Paracortex,Medulla

LBS;labyrinth-like structures. GC;germinal center, FOL;follicle

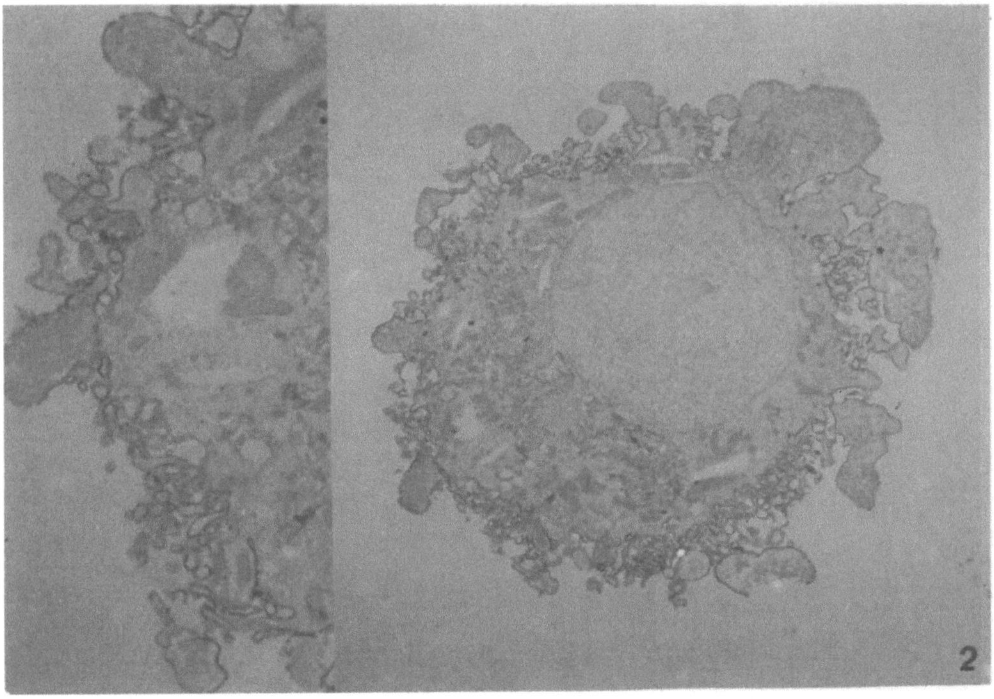

Fig. 2 A DRC-1+ FDC of typ A with a nucleolated euchromatic round nucleus and tactile-like cytoplasmic extensions(primary :P, secondary:S, and teritiary:T), giving a profile of a sea anemone.(x3000, x5000

isolated DRC-1+ cells of type A is a large mono- to binucleated round cell ruffled by various cytoplasmic extensions of P, S and Ts fully developed to form abundant LBS. On occasion P-extentions bluntly extruded intervening between S- and T-ones. The nucleus is often blastic with euchromatin and nucleoli. Binucleated cells were fascile to obsesrve often with a cytoplasm deeply infolded or invaginated. The outer-surface of the infolded cytoplasm was also decorated with LBSs.

Type B were spindle-shaped stellate cells predominating in the corona and primary follicles, extending their primary projections among lymphoid cells. In enzyme-treated cell-suspension, they accorded with oligipolar large mononuclear round cells bearing a few non-lamified primary cytoplasmic extensions. They were aligned with twig or root tubercle-shaped secondary projections but rarely with the tertiary projections.

Type C were singularly confirmed outside follicles in a profile of a medium-sized lymphoid or mononuclear cells. In some instances they located close to vessels especially of the high endothelial vessels. Their cytoplasmic extensions resemble blunt pseudopodes. which was serrated in some instances to form a miniature of LBSs

The Cytobucket-preparations of isolated FDCs revised out such sturbborn image of us for FDC as flattened spindle-shaped fibroblastic or reticulum cells observed in ordinary-processing electron microscopy. A majority of them were large round, rather euchromatic mono- to polykaryonic giant cells, entirely marginated with varying cytoplasmic projections, in a fashion of sea-anemone or jelly fish abundant in tactiles. In gradient separation methods the specific gravity to collect FDCs curiously varied according to researchers.[2,3,5,6] This may be explained by different enzyme-digestion technique,[6] which reflect upon a cell number of an isolated unit. Tew et al' method[6] is superior in a sense of a smaller number of

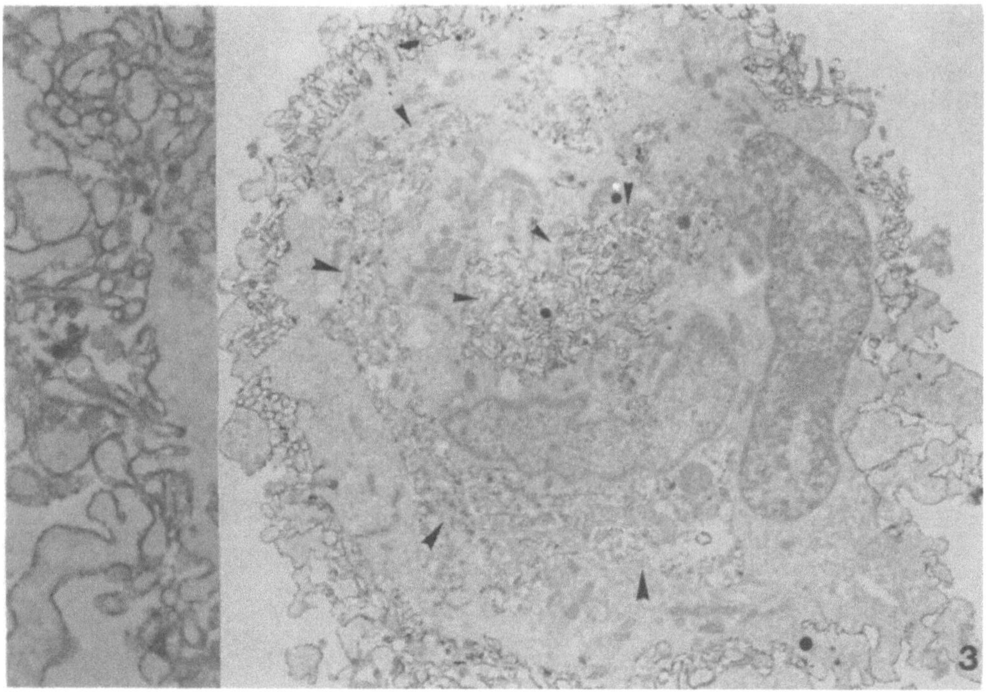

Fig. 3. Another FDC of type A, binucleated, is marginated by DRC-1$^+$ primary, secondary and tertiary projections composing LBS. Note pronounced cytoplasmic infoldings(▼▼) giving an profile of jelly fish (Cells suspended with collagenase, x3000 et x5000).

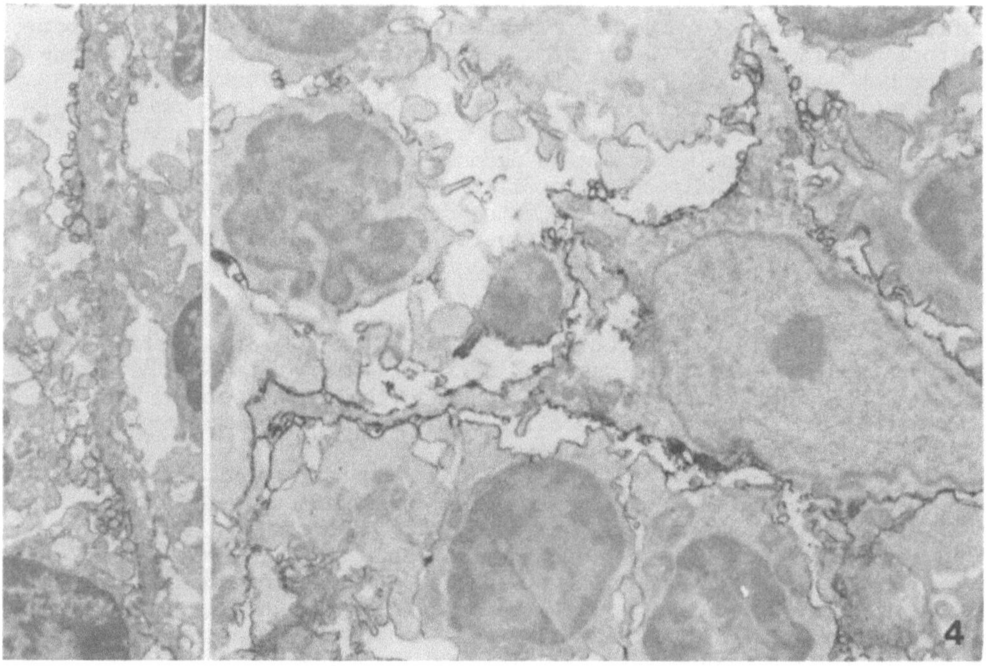

Fig. 4 A type B stellate dendritic cell in the corona with DRC-1$^+$ elongated primary and twig-like secondary projections (Micro-slicer section, x3000 et x5000)

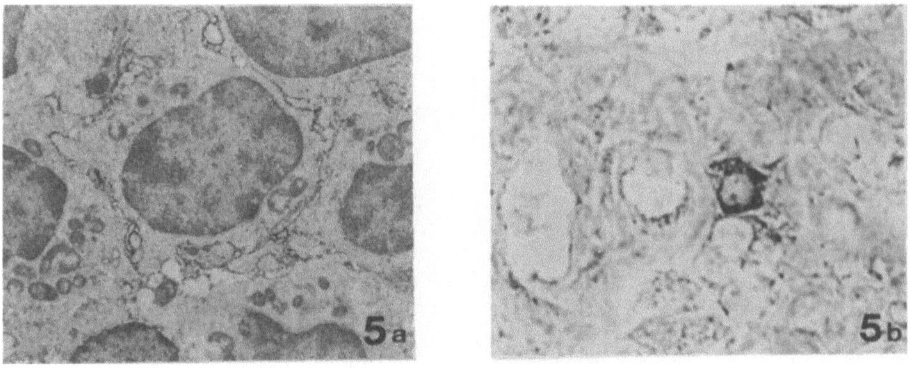

Fig. 5. A mononucleated DRC-1$^+$ cell separately located in the paracortex of the lymph node(a. x1500) and a large mononuclear round DRC-1$^+$ cells near the vessel in a granultion tissue (b. frozen-section, x400)

aggregated or clustered FDCs to those in others' digesting methods. Typical FDC or type A individually separated were preferably recognized in the interface between 1.063 and 1.030 in a similar manner to murine FDC.[6] At the same time a small portion of them dispersed even in the lighter and heavier fractions. Type B and possibly C were in the heavier fraction. The clustered or aggregated DRC-1$^+$ cells were confirmed in the bottom embracing lymphocytes.[3]

Human FDC or type A DRC-1$^+$ cells, isolated, in a "sea anemone" or" jelly fish" profile were very different in shape from "sun-burst"[6] or "arachid sea star"-shaped FDC according to Tew et al.[6,8] of mice immunized against horseradish peroxidase in the secondary response. In our material, the latter rather resemble type B DRC-1$^+$ cells. This difference might come from different species and/or different stimulating conditions. Furthermore the tertiary or bud-like cytoplasmic projections of type A cells resemble "ICOSOM" designated by Tew et al.[9] as antigen -antibody complex trapping-beaded sites of the cytoplasmic projections The ICOSOM were assumed to be intaken by germinal center cells to allow antigen presentation. In human materials it requires further analysis, with some defined antigens including streptococcus group A antigen whether the tertiary projections were analogous to ICOSOM or not. On the other hand the DRC-1$^+$FDC did not definitely resemble "nurse cell" in the thymus and interdigitating cells in the thymus dependant areas.

Double staining of DRC-1$^+$ cells Cytobucket-prepared disclosed C3bR^{++}, CR2$^+$, CR3$^+$, LCA$^+$, LeuM2$^{+/-}$, LeuM3$^-$, LeuM5$^-$, MB1$^-$, MT1$^-$, HLA DR$^{+/-}$, HLA DQ$^+$, Desmin$^-$, Actin$^-$, *S-100$^-$, *NSE$^-$,

In some granulation tissues, round mononuclear DRC-1$^+$ cells were individually confirmed occasionally near high endothelial venules. On the subject of germinal center- or lymph follicle-neoformation, it remained open to questions whether a skeletal construction of the follicle with FDC or its relatives preceeded an aggregation of lymphoid cells or vice versa.[1,5] The granulation-specimen disclosed a possibility of DRC-1$^+$ precursors for FDC to settle in loco, waiting for lymphocytic assembling to form an extranodal follicle. Although DRC-1$^+$ cells were disclosed to share with LCA(leukocyte common antigen) as presented above, it interested the authors whether the DRC-1$^+$ cells were fixed or circulating. As far as we examined, DRC-1$^+$ cells were not identified in the peripheral blood of 10 healthy persons. Howver, they were confirmed in 4 SLE or RA patients out of 30 autoimmune disease patients. The appeared very tiny in occurence (0.01-0.05% of nucleated cells). Most of them resembled confit-shaped medium-sized lymphoid cells and mononuclear cells.

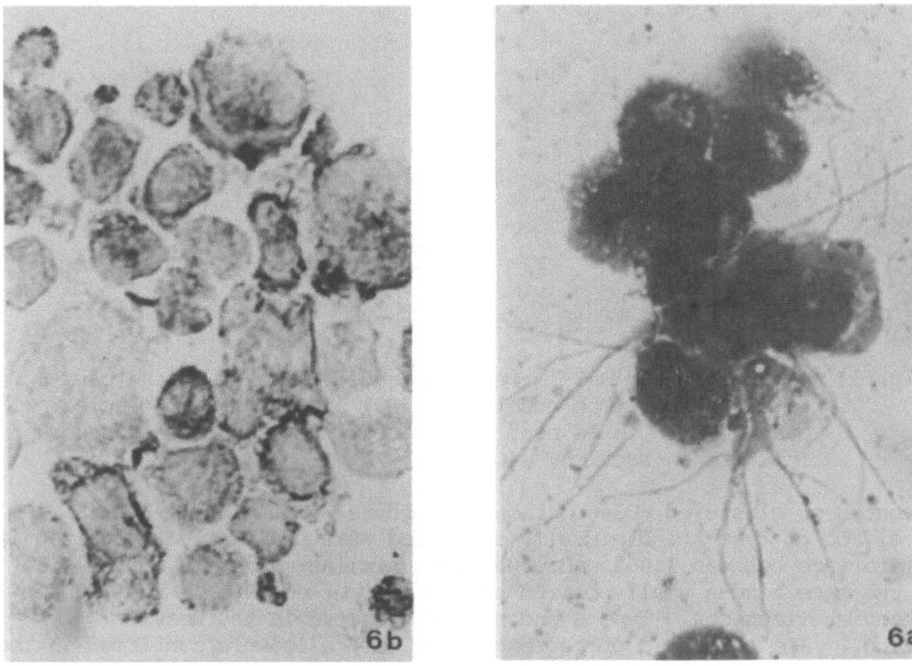

Fig. 6 A cell-cluster originated from the peripheral blood of a rheumatoid arthritis patient(T. O.) showing prominent cytoplasmic projections (a. May-Grunwald x 1000) and a scattering of DRC-1+ cells in a colony of an autoimmune disease-patient (S. O.) (b. DRC-1 staining x1000).

Furthermore in vitro culture system of the mononuclear cells of the peripheral blood(MNCPB) the present authors recognized DRC-1$^+$ cells spontaneously proliferate in 5 cases out of about 50 cases of autoimmune diseases. One case(T.O.) of RA showed a vivid proliferation of IgG kappa-bearing B cells, making floating round colonies or clusters up to the diameter of 2 to 3 mm, during 8 months. DRC-1$^+$ cells counted 2 to 20 in percentage. Some cells showed many cytoplasmic projections radiating in a sea-anemone or stellate fashion Another case (S.O), suspected of lupoid hepatitis, proliferated with no colony formation during 6 months. DRC-1$^+$ cells counted more than 50%. These proliferating series are not completely homogenious but more or less heterogenious in a similar fashion to an long-lasting proliferation of B cells infected by EB-virus. In our preliminary experiment a high frequency of DRC-1$^+$ cells was also confirmed in EB virus-induced lasting lines of MNCPB Double staining analysis with other monoclonal antibodies of IgG class and a limiting dilution culture will be reported to estimate the significance of these DRC-1$^+$ series. Furthermore the present authors recognized an remarkable increment of clustering DRC-1$^+$ cells in the lymph nodes of IBL(immunoblastic lymph adenopathy)-like T cell lymphopmas and in the bone marrow of a chronic lymphocytic leukemia case. As a precursor or candidate of FDC, various sorts of cells have hereto been counted including fibroblastic reticulum cells,[1,4] pericytes, endothelia,[10] and mononuclear cells in the marginal sinus[6] and marginal zone cells.[10] The present study gave it due consideration that some precursors of FDC, LCA(Leukocyte common antigen)-positive, were blood born[5] and they engageed in a neogenesis of germinal centers after anchoring in tissue as a DRC-1$^+$ cells of type C under various chronic inflammation including autoimmune diseases.

SUMMARY

Follicular dendritic cells (FDC) and their relatives of the human tonsils and lymph nodes were immunoelectron microscopically analyzed with a monoclonal antibody DRC-1(DAKO) in a combination with an enzymatic isolation technique. DRC-1+ cells were largely grouped into three type A (large mono- to multinucleated, sea anemone-fashioned, bearing labyrinth structures, in germinal center), type B (oligodendritic or stellate, medium-sized, in the corona and the primary follicle) and type C (individually located, medium-sized. lymphoid cell-fashioned, outside follicles). Type C were found in hyperplastic lymphoid tissue as well as in some granulation tissue, The DRC-1+ mononuclear cells were curiously encountered in the peripheral blood of some autoimmune disease patients in contrast with no identification in healthy persons. Furthermore some DRC-1+ cells obtained from autoimmune patient were confirmed to proliferate in culture.

REFERENCES

1. Dobashi,M, Terashima K et Imai Y : Postnatal development of dendrtic reticulum cells and their immune complex trapping ability. Histol. Histopath.1,19,1986
2. Gerdes,J., Naiem,M., Mason, D.Y.and Stein, H.:Human complement (C3b) receptors defined by a mouse monoclonal antibody. Immunology,45,45,1982
3. Heinen, E., Lilet-Leclercq,Ch., Mason,DY., Stein, H., Boniever, J. et al.: Isolation of follicular dendritic cells from human tonsils and adenoids. II Immunocytochemicl characterizatioln. Eur. J. Immunol. 14, 267-273, 1984.
4. Imai, Y., Terashima, K., Matsuda, M., Dobashi, M., Maeda, K.et Kasajima T.: Reticulum cell and dendritic cells-origin and function Rec. Adv. RES. 21, 51, 1983
5. Parwaresch, M.R., Radzun,H.J., Hanmann, M.L. et Peters, K.P.:Monoclonal antibody Ki-M4 specifically recognizes human dendritic cells(follicular dendritic clls) and their possible precursors in blood. Blood, 72, 585, 1985
6. Schnizlen, CT., Kosco, MH., Szakal, AK. et Tew, JG.:Follicuar dendritic cells in suspension enrichment, and initial characterization indicating immune complex trapping and lack of adherence and phagocytic activity. J. Immunol. 134, 1360, 1985
7. Terashima,.K., Imai, Y., Kasajima, T., Tsunoda, R., Takahashi, K. et Kojima, M.: An ultrastructural study on antibody roduction of the lymph nodes of rats with special reference to the role of germinal centers. Acta Pathol. Jpn, 27, 1, 1977
8. Terashima, K., Ukai, K. et Imai Y: Heterogenity of DRC-1 positve cells in Abstract n24 in international conference of "Antigen Presenting Cells-Diversity, differentiation and regulation" Richmond Virginia, USA 1987
9. Tew, JG: Role of follicular dendritic cells in the "alternative antigen pathway leading from immune complex formation to antigen presentation by germinal center B cells. in Abstract n2 in International Conference of "Antigen Presenting Cells-Diversity,differtiation and regulation" in Richmond, Virginia, USA, 1987
10 Humphery, JH, Grennan, D. and Sundaram, V.: The origin of follicular dendrritic cells in the mouse and the mechanism of trapping of immune complex. Eur. J. Immunol. 14, 854-864, 1984.

IDENTIFICATION OF ALKALINE PHOSPHATASE POSITIVE CELLS IN HUMAN GERMINAL CENTRES AS FOLLICULAR DENDRITIC CELLS

L.H.P.M. Rademakers, R.A. De Weger, and P.J.M. Roholl

Pathologisch Instituut, Rijksuniversiteit Utrecht en Academisch Ziekenhuis, Pasteurstraat 2, 3511 HX Utrecht, The Netherlands

INTRODUCTION

Follicular dendritic cells (FDC) are exclusively found in germinal centers, where they form an extensive meshwork between the lymfoid cells. Many controversies exist about their origin. Ultrastructural studies on their ontogeny (1-3) suggest that FDC transform from fibroblastic reticulum cells. Marker studies have shown that FDC have antigens which are common to macrophages (4,5) and a monocyte-macrophage lineage has been suggested (6,7). Humphrey and Sundaram (8), however, excluded in their chimaera studies a bone marrow derivation of FDC. In a previous enzyme-cytochemical study on FDC isolated from human tonsils (9) we have shown that FDC have an enzyme pattern that is different from macrophages. At least a proportion of FDC is alkaline phosphatase-positive (AF^+). Thus far it is unclear whether the alkaline phosphatase activity of FDC is an expression of a specialized activity of FDC or a "rudimental marker" of precursor cells of FDC. To study the nature and "in situ" localization of AF^+ FDC we examined AF^+ cells in germinal centers at the light and electron microscopical level.

MATERIALS AND METHODS

Cytochemistry

Fresh human tonsils were snap frozen in liquid nitrogen. Alkaline phosphatase positive cells were characterized by simultaneous immunostaining with various antibodies. FDC were stained with monoclonaal antibody DRC-1 (Dakopatts) using the biotin-avidin-peroxidase method; peroxidase was visualized by 3',5'-diaminobenzidine (DAB) and H_2O_2. Subsequently the frozen sections were fixed in acid formalin and incubated for endogenous alkaline phosphatase (AF) in tetra-nitro-blue-tetrazolium (TNBT)/5-bromo-4-chloro-3-indolyl phosphate (BCIP) medium (10). Counterstaining was done by Methyl Green.

Electron microscopy

Slices of tonsils were fixed either for conventional transmission electronmicroscopy (11) or for ultrastructural alkaline phosphatase cytochemistry. For demonstration of AF activity, vibratome sections of tissue, fixed in cacodylate buffered 1% glutaraldehyde and 2% paraformaldehyde solution, were incubated in a modified Na-beta-glycerophosphate/cerium chloride medium (12). The tissue was postfixed with 1% OsO_4 and embedded in epon. Ceriumphosphate precipitates were visualized lightmicroscopicallly in one micron sections by conversion into lead sulphide using a magnetron device (13). Ultrathin sections of germinal centres containing AF-positive cells were stained with lead citrate and uranyl acetate.

RESULTS AND DISCUSSION

The staining for DRC-1 monoclonal antibody clearly demarcated the germinal centres and visualized FDC in the dark, light and mantle zones. The staining pattern for alkaline phosphatase was mostly restricted to the interfollicular areas, blood vessels throughout the tonsils and small lymphocytes and branched cells in the mantle zones. In seven out of the 46 tonsils studied FDC could be recognized in a number of germinal centres, having both reactivity with the DRC-1 antibody and for endogenous AF. AF^+ FDC (Fig. 1) were in particular located around the capillaries present in the light zone, dark zone and in the mantle zone of the follicles. In addition, solitary AF^+ FDC were observed at the periphery of the germinal centres.

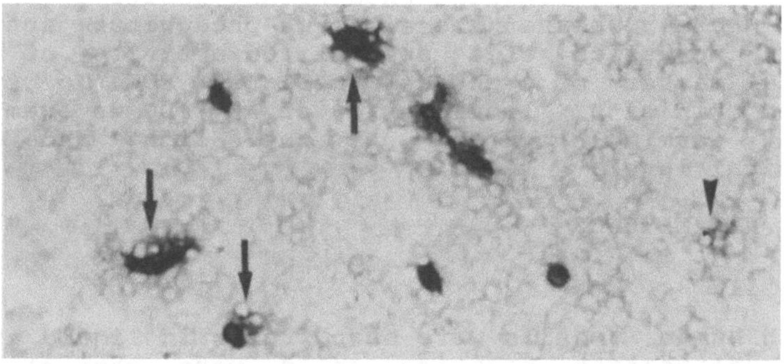

Fig. 1 Germinal centre; DRC-1/AF staining. Anti-FDC peroxidase staining was visualized as faint homogenous dendritic staining pattern. Capillaries are strongly AF^+. Some capillaries are surrounded by dendritic cells (arrows) showing both AF- and DRC-1 reactivity. Arrowhead: AF^+ FDC at the periphery of the germinal centre.

Electronmicroscopy

Ultrastructurally cerium phosphate deposits, indicating alkaline phosphatase activity (Fig. 2), were present on the surface of follicular dendritic cells which were recognized according to their morphological characteristics. Mononuclear as well as binuclear forms of AF+ FDC were encountered solitarly and around capillaries. AF+ FDC broader cytoplasmic extensions and generally do not show an extensive labyrinth of villous processes which is common for the majority of FDC in the light zones of germinal centres (10). AF+ FDC may bear kinocilia. Within the cytoplasm cisterns of the rough

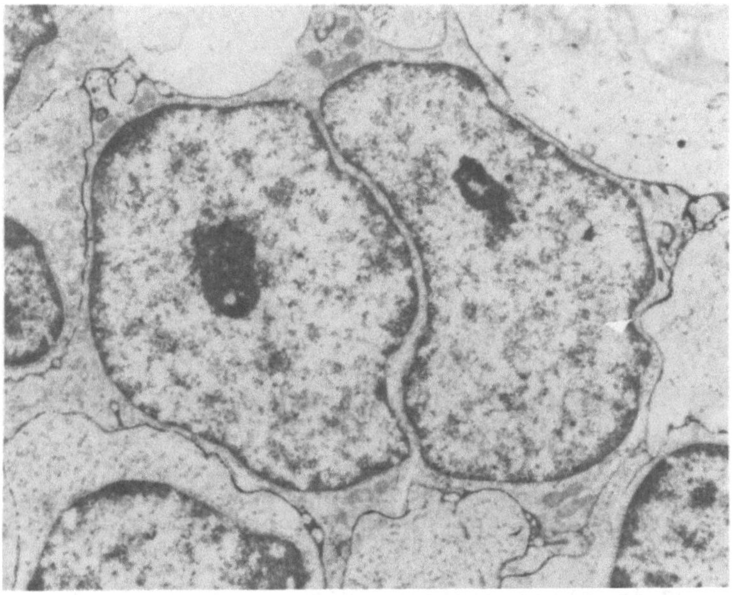

Fig. 2. Binucleated AF+ FDC with ectomembranous cerium phosphate depostis (arrow). Few villous extensions are visible (arrowhead).

endoplasmic reticulum were scarce, the Golgi complex was not obvious, and polyribosomes were abundant. Occasionally bundles of filaments were observed. This ultrastructure corresponds to undifferentiated FDC present in the dark zone and clearly differs from that of the well developed FDC in the light zone. AF+ FDC in the mantle zone also showed small accumulations of villous extentions and more pronounced bundles of filaments. In the interfollicular areas pericyte-like cells, surrounding blood vessels and fibroblastic reticulum cells (14) were AF+ as well.

Ultrastructural examination of the germinal centres consistently shows that FDC extensions surround the capillaries and border directly the perivascular space. At this site parts of an external lamina and intracellular electron-dense condensations on the plasma membrane may be present (Fig. 3). Occasionally FDC were found in direct contact with cells present within the perivascular space by

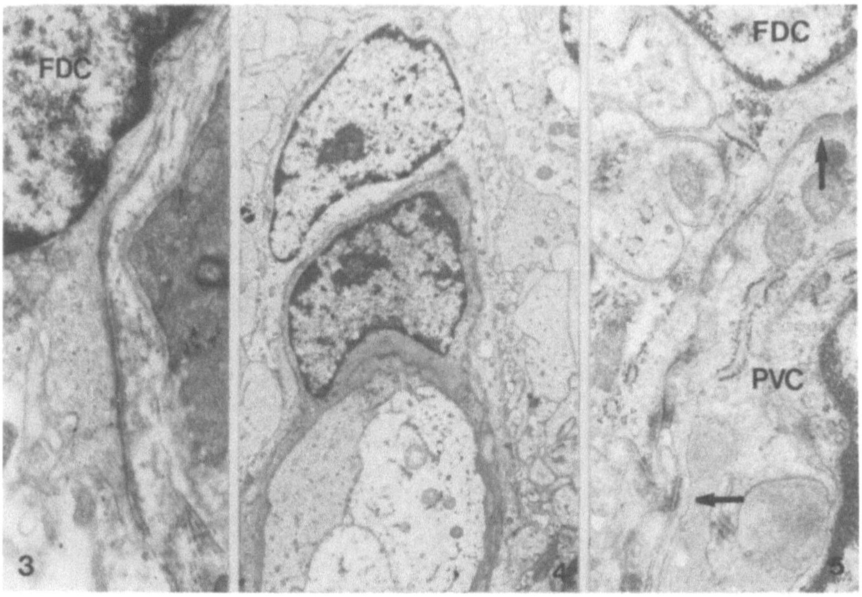

Fig. 3. Pericapillary FDC with remnants of an external lamina opposing the perivascular space (arrowheads). Arrows: internal plasma membrane condensations; asterisk: endothelial cell.

Fig. 4. FDC in close contact with a perivascular space. At site of contact, the perivascular matrix is absent (arrows). Arrowhead: Villous extension of FDC.

Fig. 5. Desmosome-like contacts (arrows) between FDC and perivascular cells (PVC).

desmosome-like attachments. These cells also have an external membrane (Figs. 4, 5). They are partly surrounded by the connective tissue matrix of the perivascular space. The perivascular cells have moderate amounts of RER, local accumulations of intermediate filaments and show some pinocytic activity. Their ultrastructure is compatible with that described for pericytes (15) which, in addition have been electronmicroscopically identified as AF+ cells (16).

The results show that AF$^+$ FDC are ultrastructurally poorly differentiated and represent a minor fraction of the FDC population in the germinal centers. This suggest that the AF-activity of FDC might be a property of mesenchymal precursor cells which is lost during further maturation of FDC. Thus far there are no indications that FDC can divide (17). It is presumed that FDC transforms from primitive stromal mesenchymal cells, present in lymph nodes (18). However the striking location of AF$^+$ FDC around blood vessels and the structural connection of the FDC meshwork with cells, present in the capillary wall, suggests that FDC may transform from pericytes. Among other functions described, pericytes are believed to form a reserve pool of mesenchymal cells from which other mesenchymal cell types develop (15). Pericytes are able to proliferate and differentiate quickly and may contribute to the increase of FDC during the rapid expansion of the germinal centres during the immune response.

REFERENCES

1. U. Heusermann, K.H. Zurborn, L. Schroeder, M.J. Stutte. Cell Tissue Res. 209: 279 (1980).
2. P. Groscurth. Pathol. Res. Pract. 169: 235 (1980).
3. C.D. Dijkstra, E.W.A. Kamperdijk, E.A. Dopp. Cell Tissue Res. 236: 203 (1984).
4. E. Heinen, Ch. Lilet-Leclercq, D.Y. Mason, H. Stein et al. Eur. J. Immunol. 14: 267 (1984).
5. Z.N. Nemes, R. Adany, V. Thomazy. Virch. Arch. (Cell Pathol) 52: 453 (1987).
6. J. Gerdes, D.Y. Mason, A. Zugler. Virch. Arch. (Cell Pathol) 42: 235 (1983).
7. M.R. Parwaresh, H.J. Radzun, A.C. Feller, K.P. Peters, M.L. Hansmann. J. Immunol. 131: 2719 (1983).
8. J.H. Humphrey, V. Sundaram. Adv. Exp. Med. Biol. 186: 167 (1985).
9. L.H.P.M. Rademakers, J.P.J. Peters, D.M.D.S. Go, R.A. De Weger, J.A.M. van Unnik. Cell. Biol. Int. Rep. 8: 185 (1984).
10. A.S.H. De Jong, R. Van Kessel-van Vonk, A.K. Raap. Histochem. J. 17: 1119 (1985).
11. L.H.P.M. Rademakers, J.P.J. Peters, J.A.M. Van Unnik. Virch. Arch. (Cell Pathol) 44: 85 (1983).
12. C.E. Hulstaert, D. Kalicharan, M.J. Hardonk. Histochemistry 78: 71 (1983).
13. D. Kalicharan, C.E. Hulstaert. Eur. J. Cell. Biol. 39, supl. 11: 24 (1985).
14. H.K. Muller-Hermelink, U. Heusermann, E. Kaiserling, H.J. Stutte. Adv. Exp. Med. Biol. 666: 177 (1976).
15. D.E. Sims. Tissue Cell 18: 153 (1986).
16. S. Mori, M. Nagano. Histochemistry 82: 225 (1985).
17. J.P.J. Peters, L.H.P.M. Rademakers, J.J.M. Roelofs, D. De Jong, J.A.M. Van Unnik. Virch. Arch. (Cell Pathol) 46: 215 (1984).
18. B. Severi, G. Biagini, E. Govoni et al. Acta anat. 120: 164 (1984).

ISOLATION OF FOLLICULAR DENDRITIC CELLS FROM HUMAN TONSILS AND ADENOIDS. IN VITRO CULTURE

N. Cormann, E. Heinen, C. Kinet-Denoel, R. Tsunoda, and L.J. Simar

Institute of human histology, University of Liège, 20, rue de Pitteurs, B-4020 Liège, Belgium

Germinal centers of lymphoid follicles are made of lymphoid cells and of accessory cells. Among these are the follicular dendritic cells or FDC's. They surround lymphoid cells with fine membrane extensions and we have shown that those closed relations are preserved after isolation of the cells from lymphoid tissues (1,2,3,4). To analyze the interactions that may take place between FDC and lymphocytes in such associations, we wanted keep them alive in vitro. For this purpose, we tried to maintain in culture entire human follicles obtained after dissection under biomicroscope and, secondly, to cultivate isolated human FDC. On an other hand we hoped to maintain pure FDC in culture, that means FDC populations without any accompanying lymphocyte. For this purpose we used a culture technic described by Jordan et al. (5). They obtained a rapid degeneration of mice thymic lymphocytes when the cultures were performed at 24° C.
In each culture condition, we observed after various periods of time the morphological features of FDC at ultrastructural level and checked the presence of their C3b and Fc receptors using homologous antibodies linked to colloidal gold particles or specific monoclonal antibodies. Tritiated thymidine uptake was tested by autoradiography.

1. Organotypic cultures of entire follicles

Using the Maximov culture technique and a fetal calf serum enriched culture medium, we found at 37°c strong cell alterations and degeneration after more than 3 days of culture. Paradoxically cultures maintained at 24° C show living cells until day 7, especially in germinal centers. Lymphocytes outside the centers are pycnotic whereas those enveloped by FDC cytoplasmic extensions remain alive (Fig.1).

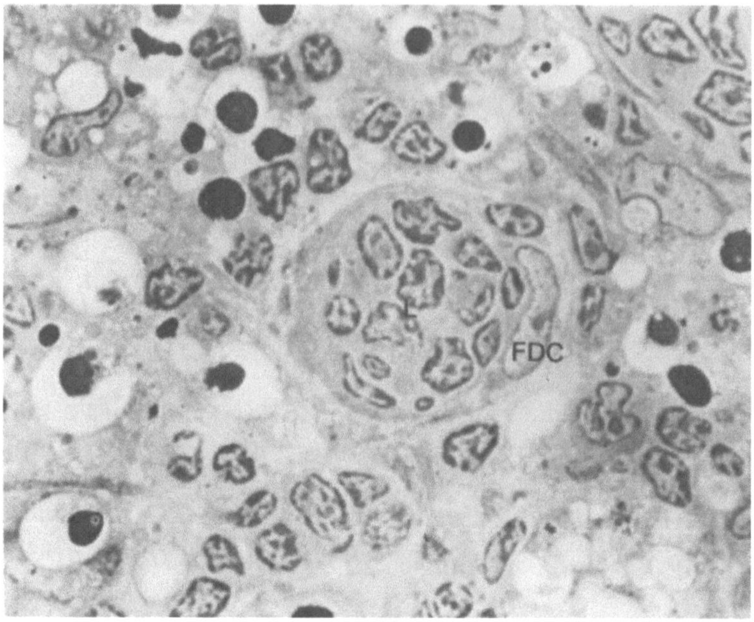

Fig. 1. Lymphoid follicle observed after a 7 days culture at 24°C: the lymphoid cells (L) are pycnotic except when they are in close contact with FDC.x1 650.

2. <u>Culture of isolated FDC</u>

The spheric configuration of FDC is preserved during at least 24 H. During this time the size even seems to enlarge because of the aggregation of other lymphocytes at the periphery. Some FDC are degenerating, others appear in contact with lymphocytes without forming a sperical arrangement. This last aspect is observed more frequently when the cells are allowed to settle down on the bottom of the culture flask. After 48 H. FDC can hardly be detected, only few spherical clusters are still found. After 72 H. Of culture. no FDC can be recognized by their morphological features.

2. <u>Fc and C3b receptors on isolated and cultured FDC</u>

Freshly isolated FDC stained intensely with human anti-tetanus toxoid antibodies bound to colloidal gold particles (α-FcR) and with anti-C3b receptor antibodies. After 24 H. they remain positive but staining is diminished (Fig 2). Later no antibody fixation can be found.

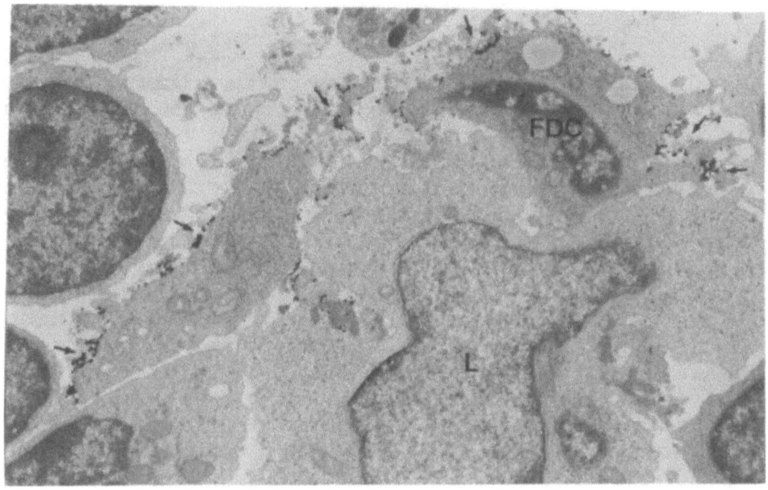

Fig. 2. 24 H. after isolation, FDC labeled with immune complexes bound to gold particles (↦), preserve their contact with lymphoid cells (L), but not always in form of spherical clusters.x6 600.

4. Tritiated thymidine incorporation

Freshly prepared or 24 H. cultured FDC are incubated with tritiated thymidine, fixed and examined on sections prepared for electron microscopy. Immediately after isolation, 8.8% of the cells incorporate the precursor, 4,4% are positive after 24. H. Only lymphocytes, free in suspension or enfolded in FDC, are positive (Fig. 3). No FDC incorporate thymidine visibly.

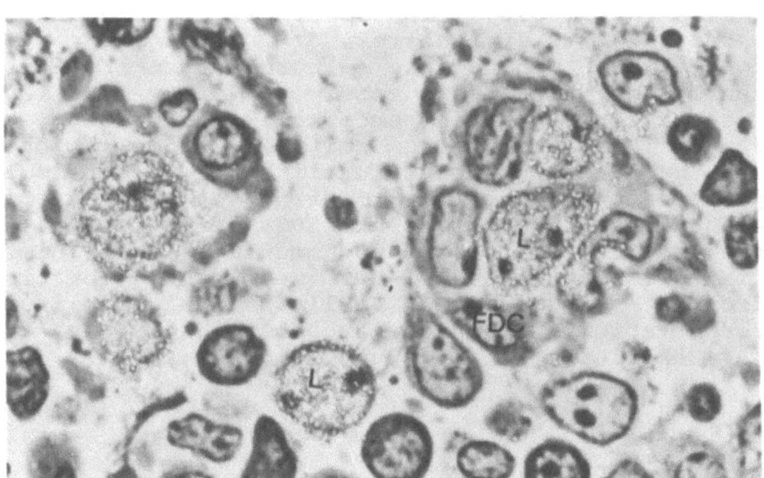

FIg. 3. Lymphoid cells (L), free or surrounded with FDC cytoplasmic extensions are able to incorporate tritiated thymidine. x1 900.

DISCUSSION

Isolated FDC appear as spherical clusters enclosing lymphoid cells by cytoplasmic extensions. In culture conditions, this arrangement can be found for the first 24 H. Thereafter FDC could less and less be identified morphologically or with gold labelled antibodies. Two hypothesis can be proposed: FDC may dedifferentiate in our culture conditions and loose their receptors, so that even present they could not be detected with our methods; or they may degenerate and disappear. In our opinion, both phenomena occur. Indeed we found degenerating FDC as well as cells fixing less labelled antibodies. These cells are poorly characterized morphologically and might be dedifferentiated FDC. After isolation of FDC, Schnizlein et al. (6) obtained free cells which have morphological and cytochemical features in common with FDC; these cells would divide in vitro and preserve their capacity to fix immune complexes. However, until now formal ultrastructural and immunocytochemical observations are lacking to identify the actual nature of these cells.

In organotypic cultures of follicles, FDC remain alive for at least 3 days at 37° C and 7 days at 24° C. In this last case, culturing at 24° C may slow down metabolic activities and even cell death. But despite this fact a culture independent phenomenon occurs, lymphocyte degenerate more rapidly than FDC except when they are within the FDC cytoplasmic extensions. FDC seem thus to create a microenvironment favorable to lymphocytes and perhaps exert a trophic action. In those cultured follicles, the lymphocytes which are still alive form with the FDC spherical clusters similar to those we obtained after an enzymatic digestion. Thus our isolation procedure apparently maintains cells preserving the relationships they have in situ. The spherical appearance we always obtain after isolation may be a consequence of resuspending cells in a liquid culture medium. When FDC settle down in the culture flasks they loose this spherical configuration but remain in contact with lymphocytes.

In conclusion, organotypic cultures of entire follicles are apparently not the best model to study the FDC. In contrast, short term cultures performed with FDC separated by mild enzymatic digestion are suitable since lymphocytes incorporate thymidine at least until 24 H. after isolation and FDC preserve their receptors during a short period of time. Recent results show that FDC effectively exert a mitogenic effect on lymphocytes and modulate their differentiation in plasmocytes (7). FDC seem thus to create a microenvironment specific for B cell survival, proliferation and differentiation which is probably analogous to that observed by Wekerle et al. (8) in thymic nurse cells.

REFERENCES

1. Lilet-Leclercq C., Radoux D., Heinen E., Kinet-Denoël C., Defraigne J.O., Houben-Defresne M.P., Simar L.J. J. Immunol. Meth. 66:235 (1984).

2. Heinen E. , Lilet-Leclercq C., Mason D.Y., Stein H., Bonvier J., Radoux D., Simar L.J., Eur. J. Immunol. 14: 267, (1984).

3. Heinen E., Kinet-Denoël C., Radoux D., Simar, L.J., Adv. Exp. Med. Biol. 186:171 (1985a).

4. Heinen E, Radoux D, Kinet-Denoël C., Moeremans M., De Mey J., Simar L.J. Immunol. 54:777 (1985b).

5. Jordan R.K., Robinson J.H., House K.C., Hopkinson N.A. Abstract 8th International Conf. on Lymphatic Tissues and Germinal Centers in Immune Reactions. Cambridge (1984).

6. Schnitzlein C.T., Kosco M.H., Szakal A.K., Tew J.G. J. Immunol. 134:1360 (1985).

7. Cormann N., Lesage F., Heinen E., Schaaf-Lafontaine N., Kinet-Denoël C., Simar L.J. Immunol. Lett. 14:29 (1986).

8. Wekerle H., Ketelsen U.P. Nature 283:402 (1980)

IMMUNOHISTOCHEMICAL STUDY ON CULTURED FDC-C ENRICHED LYMPHOID CELL POPULATIONS

R. Tsunoda, N. Cormann, E. Heinen, F. Lesage, C. Kinet-Denoel, and L.J. Simar

Institute of Human Histology, University of Liège
20, rue de Pitteurs, B-4020 Liège, Belgium

The follicular dendritic cell-clusters (FDC-C), which are obtained by mild enzymic digestion from lymph follicles, are now considered as the structural minimum units of the germinal centers in situ (1, 2). One isolated FDC forms a globular structure, engulfing 8 to 9 germinal center B cells. Cultured FDC-C exhibit an accessory function to lymphoid cell proliferation but reduce B-cell differentiation (3). To investigate this accessory function morphologically, we tried to clear the cellular phenomenons happening in the cultured FDC-C enriched populations, especially inside and at the periphery of the FDC-C, which an immunohistochemical method and electron microscopy.

The FDC-C were obtained after enzymatic digestion of the enucleated human tonsillar lymphoid follicles (4). For control, two populations, mixed T & B-tonsillar cells and T, B-cells & macrophages populations were also cultivated. The cultured cells were adjusted in 5 x 10^5/ml in RPMI 1640 + 10% FCS + 1% PWM. The reverse-positioned culture system using Terasaki-plate, which is sometimes recommended for the thymic nurse cells, was used. After fixation with aceton, cytocentrifuged cells were stained with the avidin-biotin indirect immunostaining method.

After 24 hrs of culture, FDC-C could hardly be detected under light microscopy. Their antigenicities to anti DRC-1 (anti-human FDC antibody, Dako) and anti C3b-receptor (anti-C3bR, Dako) were lower than those observed at time 0. Besides, many lymphoid cells began to adhere on the periphery of the FDC-C (5), and the size of FDC-C became slightly enlarged. Electron microscopical observations revealed that FDC-C structure were still well-preserved and there is also possibility of an active communication between lymphoid cells located inside or outside the clusters. After 48 hrs culture, FDC themselves and the structure of the FDC-C seemed to be destroyed, only some rare DRC-1 positive cells could be detected.

Among the free lymphoid cells outside the FDC-C, lots of blastic cells were found especially after 72 hrs of culture and their number gradually increased thereafter. In FDC-C enriched populations, the proportion of living lymphoid cells was always higher in comparison with the control cultures composed of B + T alone or B + T + macrophages (Fig. 1). However, the fraction of surviving lymphocytes decreased with the duration of the culture,

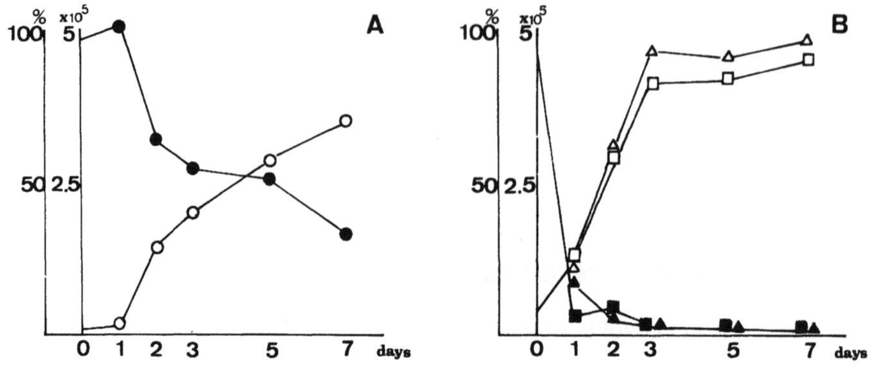

Fig. 1. Evolution of living or death cells during culture. Open marks; percentage of death cells; black marks; number of living cells (o—o FDC-C enriched; △—△ BT; □—□ BTM). Each value represents the mean of five experiments.

apparently parallelly with the disappearance of FDC-C. As previously described (3), the FDC-C enriched lymphoid population takes up actively ^{3}H-thymidine. This means that large proportion of the surviving cells are able to divide. Anyway, any marked increase of the number of blastic lymphoid cells was not found in the two control culture populations. Originally, the FDC-C enriched population contained few % of cytoplasmic IgG (CIgG)-containing cells, and less than 1% cytoplasmic IgM (CIgM)-containing cells. The ratio of the CIgM-containing cells increased up to 8% during 72-152 hrs of culture, but, in contrast, that of the CIgG-containing cells decreased gradually (Fig. 2). Almost all of those CIgM-containing cells were enlarged blastoid cells, but the CIgG-containing cells were smaller. Previous ELISA study (3) suggested that those blastic CIg-containing cells were non-secretory. Any distinct positivity against anti pan-B or pan-T was not revealed to these lymphoid blast in cultured FDC-C enriched populations.

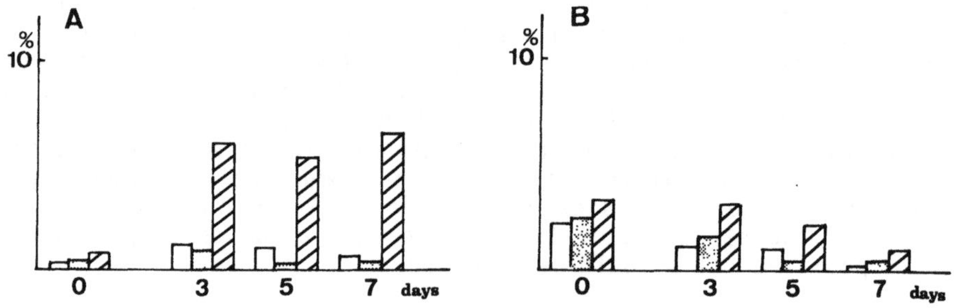

Fig. 2. CIg-containing cells among lymphoid cells in culture. A: CIgM-containing cells, B: CIgG-containing cells. (slashed bar: FDC-C enriched; open bar: BT; dotted bar: BTM). Each percentage is the mean value of five experiments.

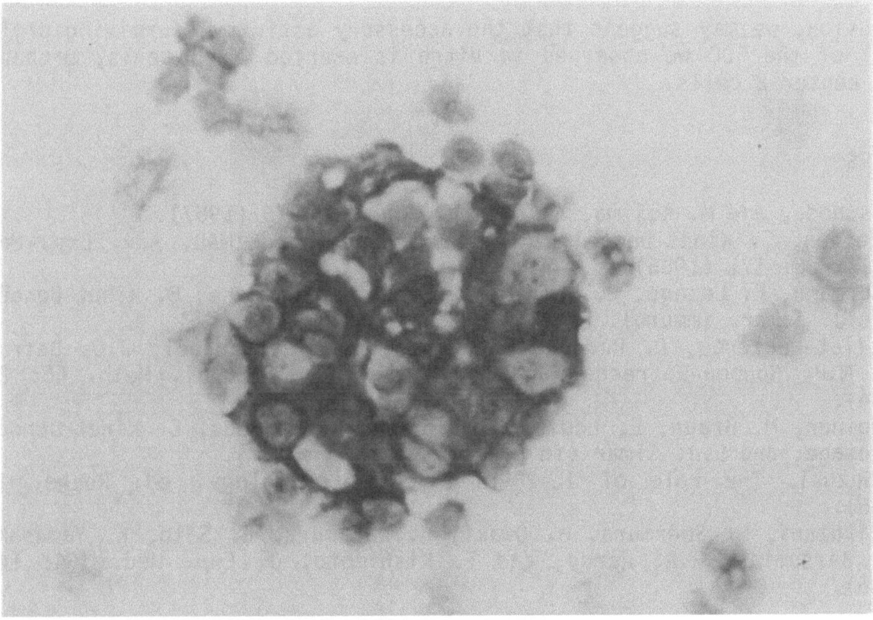

Fig 3. Isolated FDC-C, strongly positive against MoAb-Anti Fc R (x 50).

The blastic cells reacted with the anti IL-2 receptor antibody. The lymphoid cell proliferation in the FDC-C enriched fraction could be inhibited by the addition of an antibody directed against IL-2 receptor (N. Cormann, in preparation). According to the opinion of Suzuki (6) that the enhanced proliferation of T cells was more difficult to inhibit with anti-Tac 1 antibody than that of B cells, our proliferating lymphoid cells seemed to arise from non-T cells. The Fc receptor (CD23) (generous gift of Drs. Kikutani and Kishimoto, Osaka, Japan) was only expressed on the coronal $SIgD^+$, $SIgM^+$ B cells and reacted strongly with FDC in the follicles (Fig. 3). Any blastic lymphoid cells in the FDC-C enriched fraction didn't react clearly with anti-Fc receptor antibody during cultivation. Since Kikutani et al. (7) suggested that the Fc receptor must be lost before lymphoid cells can perform their isotype switch, the proliferating cells we observed in our cultures may arise from germinal center B cells as those ones are Fc receptor negative.

In conclusion, we may suggest that the accessory activity (surviving,proliferation) of the FDC we observed in vitro is exerted on B cells, probably germinal center B cells.

REFERENCES

1. R. Tsunoda, and M. Kojima. Acta Path. Jpn, 37: 575 (1987).
2. E. Heinen, C. Kinet-Denoël, D. Radoux, and L.J. SIMAR. Adv. Exp. Med. Biol., 186: 171 (1985).
3. N. Cormann, F. Lesage, E. Heinen, N. Schaaf-Lafontaine, C. Kinet-Denoël, and L.J. Simar. Immunol. Lett. 14: 29 (1986).
4. C. Lilet-Leclercq, D. Radoux, E. Heinen, C. Kinet-Denoël, J.O. Defraigne, M.P. Houben-Defresne, and L.J. Simar. J. Immunol. Meth. 66: 235 (1984).
5. E. Heinen, M. Braun, E. Louis, N. Cormann, R. Tsunoda, C. Kinet-Denoël, F. Lesage, and L.J. Simar (in press, see this volum).
6. T. Suzuki. The role of IL-2 in B cell physiology. ed. Roche, 145 (1985).
7. H. Kikutani, M. Suermura, H. Owaki, H. Nakamura, R. Sato, K. Yamasaki, E.L. Barsumian, R.R. Hardy, and T. Kishimoto. J. Exp. Med. 164: 1455 (1986).

INTERACTIONS BETWEEN FOLLICULAR DENDRITIC CELLS AND LYMPHOID CELLS

E. Heinen, M. Braun, E. Louis, N. Cormann, R. Tsunoda,
C. Kinet-Denoel, F. Lesage, and L.J. Simar

Institute of Human Histology, University of Liège
20, rue de Pitteurs, B-4020 Liège, Belgium.

Circulating B or T cells may home to lymph follicles either into their corona or their germinal center. Little is known about this migration. B cells settling in the follicles are either memory cells or newly formed lymphocytes arising from the bone marrow. Mel 14+ and mostly IgD+ cells accumulate in the corona whereas Mel 14- and IgD- cells are preferently found in the germinal centers where about 5% T helper cells are also present (1).

Opstelten et al. (2) showed that lymphocytes prepared from the appendix dome localize in the corona, some of them, apparently blast cells, can migrate towards the germinal centers.

The migration and the evolution of all these cells in the follicles is poorly understood that is why we choose to study the follicular dendritic cells (FDC) which form a peculiar microenvironment inside the germinal centers (3). FDC represent there about 2% of the cells (4). They surround lymphoid cells by long and complex cytoplasmic extensions and retain huge amounts of immune complexes via C3b or Fc receptors (5-9).

Our group succeeded in isolating FDC from human or murine lymph organs (4, 10). FDC then appeared as clusters enveloping lymphoid cells, reacted with specific antibodies directed against FDC and were able to retain immune complexes (Fig.1) (11).

In the present study we examine morphological and functional aspects of the interactions existing between lymphoid cells and FDC either in situ or after isolation.

1. Ultrastructural aspects of interactions between FDC and lymphoid cells

The relationships between FDC and lymphoid cells appear clearly at the ultrastructural level: the complex dendritic network of FDC is intermingled with the short microvilli of lymphoid cells, evenmore, FDC send out fine cytoplasmic extensions surrounding lymphoid cells.

These associations are preserved after isolation when delicate cell separation procedures are used (10, 12); the cell clusters obtained by this technic appear to be fragments of germinal centers (1).

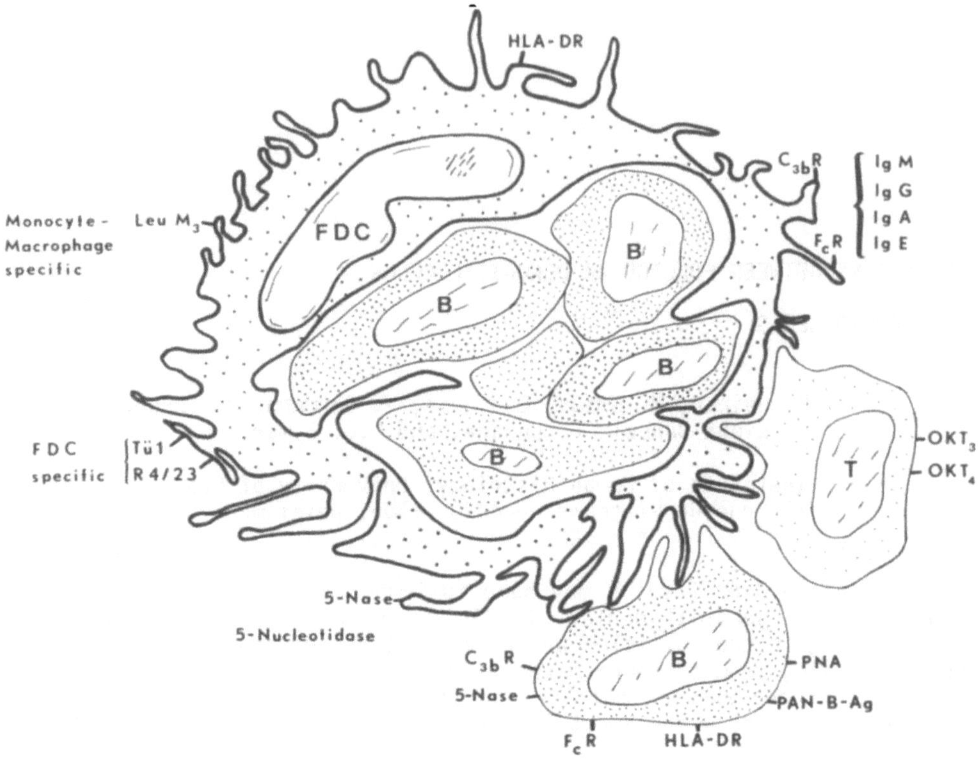

Fig. 1.Schematic representation of an isolated FDC. The surface features of the different cells are precised.

2. Lymphoid cells esthablish close contacts with FDC

When isolated FDC are incubated in presence of FITC-labeled lymphoid cells, some of these last then adhere on the periphery of FDC (table I). The number of adhering lymphocytes increases with time. B cells as well as T cells esthablish such contacts.

Immune complexes added to the incubating medium induce no modifications of the attachment of lymphoid cells to FDC, however when these immune complexes bear C3b fractions, the adherence is reduced (table I).

Antibodies against FDC specific antigens (DRC1, Dako), against 150/95 glycoprotein (Leu M5, Dako) or against CR3 complement receptor (OKM1) do not influence the frequency of lymphocyte attachment to FDC. Antibody 12B1, prepared by Farace et al. (13) and recognizing apparently IL1 secreting cells, inhibits lymphoid cell adherence to FDC.

TABLE I

Incubation time (min.)	Number of labeled lymphocytes per FDC		
	Control	+ IC	+ IC + C3b
30	5,3 ± 2,1	4,8 ± 3,2	3,5 ± 2,6
120	8,3 ± 3	9,1 ± 2,7	6,9 ± 3,7

3. Lymphoid cells transfer immune complexes to FDC

In vivo experiments indicate that lymphoid cells, shortly after sub-cutaneous injection of gold labeled immune complexes, can fix those immune complexes and migrate from the subcapsular sinus of draining lymph nodes towards the lymph follicles (14). We demonstrated in vitro experiments that lymphoid cells bearing immune complexes can adhere on isolated FDC and transfer the immune complexes to these FDC (14). Apparently, B cells can better handle immune complexes to FDC than T cells and this transfer depends on the nature of the antibodies isotypes forming those complexes (15).

4. FDC can influence the survival, the proliferation and the differentiation of lymphoid cells

Lymph follicles, dissected under biomicroscopes from human tonsils and cultured in vitro, exhibit soon heavy cell death in the corona but not in the germinal centers. In this part of the cultured lymph follicles, lymphoid cells survive during several days when they are in close contact with FDC. These last thus appear to exert a trophic action on lymphoid cells (16).This effect on the lymphod cell survival was also observed by Tsunoda et al. (17).

When isoltad FDC mixed to lymphoid cells are cultured in presence of adequate lectins, lymphoid cells divide. We previously showed that lymphoid cells proliferate much better in these conditions than when cultured alone or in presence of macrophages (18). In these culture conditions differentiation of lymphoid cells in Ig secreting cells is inhibited, indeed, the Ig content of the supernatants of FDC containing cultures was much lower than in absence of FDC (18).

FDC thus favourize survival, proliferation but inhibit the differentiation of lymphoid cells in Ig secreting cells. This is apparently achieved by the secretion of products by FDC: we demonstrated the presence of PGE2 (18), IL1 and TNF (unpublished data) in the supernatant of FDC containing suspensions.

CONCLUSIONS

In germinal centers cell contacts exist between FDC and lymphoid cells (B cells and T cells). During these interactions exchanges occur: lymphoid cells can transfer immune complexes whereas FDC can influence the behavior of the lymphoid cells in secreting cytokines.

REFERENCES

1. R. Tsunoda, and M. Kojima. Acta Path. Jpn, 37: 575 (1987).
2. D. Opstelten, G.J. Deenen, J. Bos, and P. Nieuwenhuis. Adv. Exp. Med. Biol., 149: 757 (1982).
3. E. Heinen, and R. Tsunoda. Immunol. Today, 8: 142 (1987).
4. E. Heinen, C. Kinet-Denoël, D. Radoux, and L.J. Simar. Adv. Exp. Med. Biol., 186: 171 (1985).
5. Z.L. Herd, and G.L. Ada. Aust. J. Exp. Biol. Med. Sci. 47: 63 (1969).
6. G.G.B. Klaus, J.H. Humphrey, A. Kunkl, and D.W. Dongworth. Immunol. Rev. 53: 28 (1980).
7. T.E. Mandel, R.P. Phipps, A. Abbot, and J.G. Tew. Immunol. Rev. 53: 30 (1980).
8. E. Heinen, D. Radoux, C. Kinet-Denoël, M. Moeremans, J. De Mey, and L.J. Simar. Immunology 54, 777 (1985).
9. D. Radoux, C. Kinet-Denoël, E. Heinen, M. Moeremans, J. De Mey, and L.J. Simar. Scand. J. Immunol. 21: 345 (1985).

10. C. Lilet-Leclercq, D. Radoux, E. Heinen, C. Kinet-Denoël, J.O. Defraigne, M.P. Houben-Defresne, and L.J. Simar. J. Immunol. Meth. 66: 235 (1984).
11. E. Heinen, C. Lilet-Leclercq, D.Y. Mason, H. Stein, J. Boniver, D. Radoux, C. Kinet-Denoël, and L.J. Simar. Eur. J. Immunol. 14: 267 (1984).
12. C. Kinet-Denoël, E. Heinen, and L.J. Simar. Cell Separation Meth. Select. Appl. 4: 238 (1986).
13. F. Farace, N. Kieffer, B. Caillou, W. Vainschenker, T. Turaz, and M.C. Dokkelar. Europ. J. Immunol. 16: 1521 (1986).
14. E. Heinen, M. Braun, P.G. Coulie, J. Van Snick, M. Moeremans, N. Cormann, C. Kinet-Denoël, and L.J. Simar. Europ. J. Immunol. 16: 167 (1986).
15. M. Braun, E. Heinen, N. Cormann, C. Kinet-Denoël, and L.J. Simar. Cell. Immunol. 107: 99 (1987).
16. N. Cormann, E. Heinen, C. Kinet-Denoël, M. Braun, et L.J. Simar. C.R.Sc. Soc. Biol., 180: 218 (1986).
17. R.Tsunoda, N.Cormann, E.Heinen, F.Lesage, C.Kinet-Denoël and L.J.Simar (in press, see this volume).
18. N. Cormann, F. Lesage, E. Heinen, N. Schaaf-Lafontaine, C. Kinet-Denoël, and L.J. Simar. Immunol. Lett. 14: 29 (1986).
19. E. Heinen, N. Cormann, M. Braun, C. Kinet-Denoël, J. Vanderschelden, and L.J. Simar. Ann. Inst. Pasteur, Immunol. 137: 369 (1987).

DEPOSITS OF TERMINAL COMPLEMENT COMPLEX AND S-PROTEIN ON FOLLICULAR DENDRITIC CELLS IN HUMAN LYMPHOID TISSUE

T.S. Halstensen, T.E. Mollnes, P. Garred and P. Brandtzaeg

Laboratory for Immunohistochemistry and Immunopathology (LIIPAT), Institute of Pathology, and Institute of Immunology and Rheumatology, Rikshospitalet, N-0027 Oslo 1, Norway

INTRODUCTION

Follicular dendritic cells (FDC) retain C3-containing immune complexes (IC) (1,2) mainly by their complement receptors (CR) (3). Theoretically, membrane-bound IC might lead to C activation and eventually membrane damage of FDC in a type II hypersensitivity reaction. Complement activation splits C3 into the C3a and C3b. The latter binds covalently to target surfaces and initiate terminal pathway activation by splitting C5 into C5a and C5b. On biological membranes, the membranolytic terminal complement complex (TCC) C5b-9(m) is formed (4). By contrast, if C activation takes place in the extracellular fluid phase, S-protein binds to the terminal components and the hydrophilic non-lytic SC5b-9 is formed (5). Both C5b-9(m) and SC5b-9 expose TCC neoepitopes, which are not present in the native components. S-protein is produced by the liver and is a multi-functional 65-75 Kd protein with a serum concentration of 0.14-0.70 g/l (6). In addition, S-protein, now known to be identical to vitronectin (7,8), is one of many proteins capable of promoting cell adhesion through the arg-gly-asp sequence and is found in various human tissues (9-11). In this study FDC of human lymphoid tissue were examined by immunohistochemistry for terminal pathway activation and S-protein distribution.

METHODS

Histologically normal tissue specimens obtained from forensic autopsy cases, comprised five axillary and five mesenteric lymph nodes and five spleens. In addition, five samples of Peyer's patches, appendices and colonic tissue were collected from each of five renal graft donors. Tissue specimens with follicular lymphoid hyperplasia consisted of five tonsils and five cervical lymph nodes.

Saline-extracted, ethanol-fixed and paraffin-embedded tissue specimens (12) were incubated for 20 hours at room temperature (13) with optimaly diluted mouse monoclonal antibodies (MoAbs) in phosphate buffer containing 12.5% bovine serum albumin. Antibodies with the following specificities were included: FDC (DRC1, 1:350; Dakopatts, Glostrup, Denmark); a C9 neoepitope in TCC (aE11, 1:2000; 14); S-protein (1:500;

Cytotech, San Diego, USA); IgM(μ chain, 1:10; Dakopatts); and a C3 neoepitope located on C3b, iC3b and C3c (bH6, 1:100; 15). In addition, we used rabbit antisera to C3d (1:500; Dakopatts), C5 (1:100; Dakopatts), and C9 (1:500; Behring). Biotinylated horse anti-mouse IgG, rhodamine-conjugated swine anti-rabbit IgG, and fluorescein-conjugated avidin were applied in an indirect paired immunofluorescence staining method as described previously (16). We also used the alkaline phosphatase-anti-alkaline phosphatase (APAAP) staining method (17) with primary mouse or rabbit antibodies as detailed elsewhere (18).

RESULTS

Germinal centers stained in a similar manner for C5, C9, TCC neoepitope and S-protein and the pattern was identical to that for FDC (Figs. 1 and 2). The staining intensity for TCC (C5, C9, TCC neoepitope) and S-protein varied from negative to bright. In histologically normal active lymphoid tissue (Peyer's patches, appendices and lymphoid follicles of the colon), approximately half of the germinal centers produced positive staining for TCC and S-protein, but only some were positive for the C3 neoepitope. Immunologically resting lymphoid tissue (axillary and mesenteric lymph nodes and spleens) contained very few or no well developed secondary follicles, but small DRC1-positive follicles with some TCC positive material could be seen. By contrast, in tissue specimens with follicular hyperplasia (tonsils and lymph nodes) virtually all germinal centers stained brightly for C3d, TCC and S-protein, but most centers were weak or negative for the C3 neoepitope.

In half of the axillary and mesenteric lymph nodes, fibrotic areas stained intensely for C3d, TCC and S-protein. Some DRC1 positivity could be demonstrated in the same areas. The fibrous capsule and trabeculae of spleen stained positively for C3d, TCC and S-protein in a fibrillary manner, but there was no positivity for DRC1. Deposits of C3d, TCC and S-protein were frequently seen in blood vessel walls of all organs.

DISCUSSION

The identical staining pattern observed for C5, C9 and TCC on FDC suggested that the terminal pathway activation had been completed. Moreover, TCC was probably in the form of SC5b-9, because the staining for TCC and S-protein varied in parallel. Serum concentration of SC5b-9 is under normal conditions low (19), but it rises during intra- or extravascular fluid phase terminal pathway activation (20). In our study we used washed tissue specimens, from which extracellular diffusible proteins had been extracted. Therefore, we believe that the SC5b-9 detected was in a tissue-bound form on FDC.

The cell adhesion sequence (arg-gly-asp) in S-protein (7) could be responsible for the binding of preformed SC5b-9 to FDC. However, FDC might not express any S-protein (vitronectin) receptor because no S-protein staining was observed in the absence of other C deposits. In addition, the parallel staining for TCC, S-protein and C3d in vessel walls, the capsule and fibrous trabeculae of the spleen is difficult to explain on the basis of isolated SC5b-9 deposition; therefore, other mechanisms might be involved in SC5b-9 deposition.

Membrane-bound C3b-containing IC on FDC may induce terminal pathway activation, but the generation of non-lytic SC5b-9 instead of C5b-9(m) probably protects FDC against membrane damage. Nevertheless, this protection may not be sufficient under all circumstances since SC5b-9 deposits in fibrotic parts of lymph nodes sometimes showed positive staining with

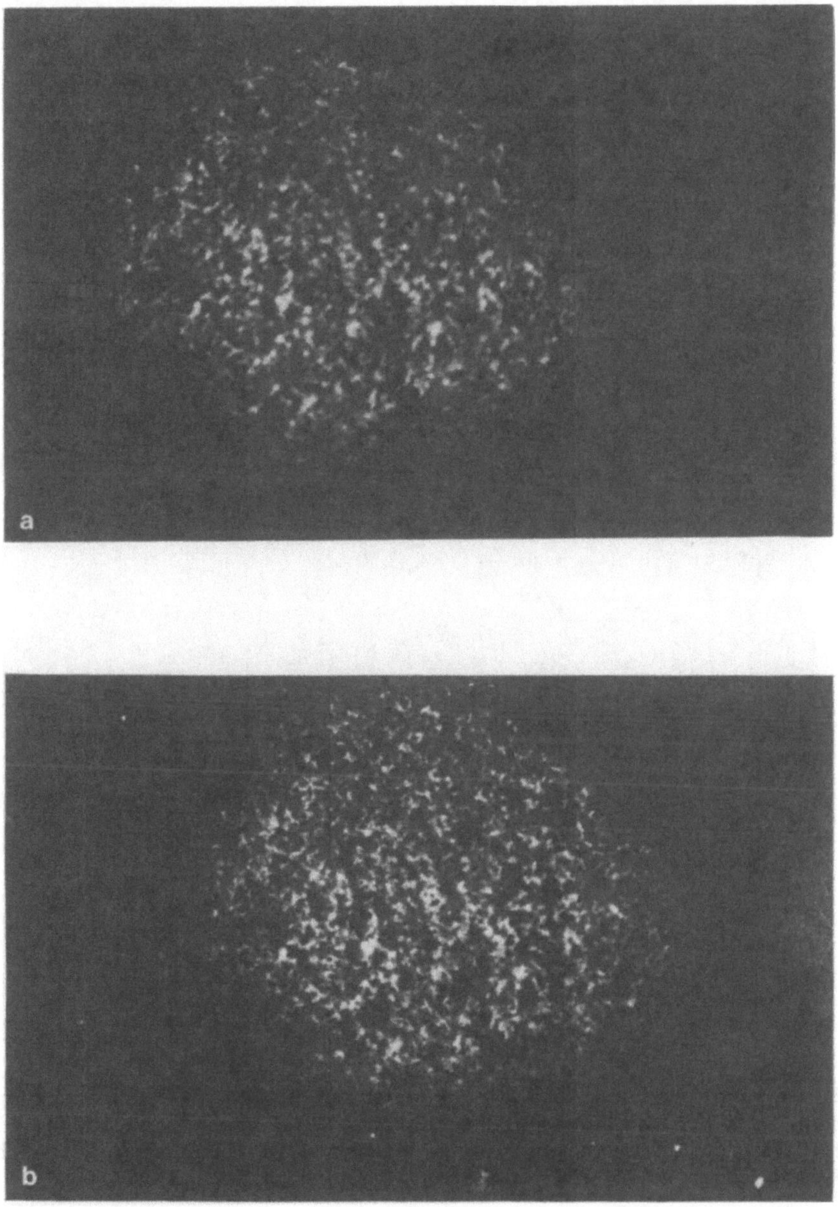

Fig. 1. Paired immunofluorescence staining in same field with (a) MoAb DRC1 decorating FDC in a typical dendritic pattern and (b) antiserum to C9. Note identical staining pattern, suggesting that C9 is deposited on FDC. Washed ethanol-fixed tissue specimens from tonsil with follicular hyperplasia. (Magnification: X 200).

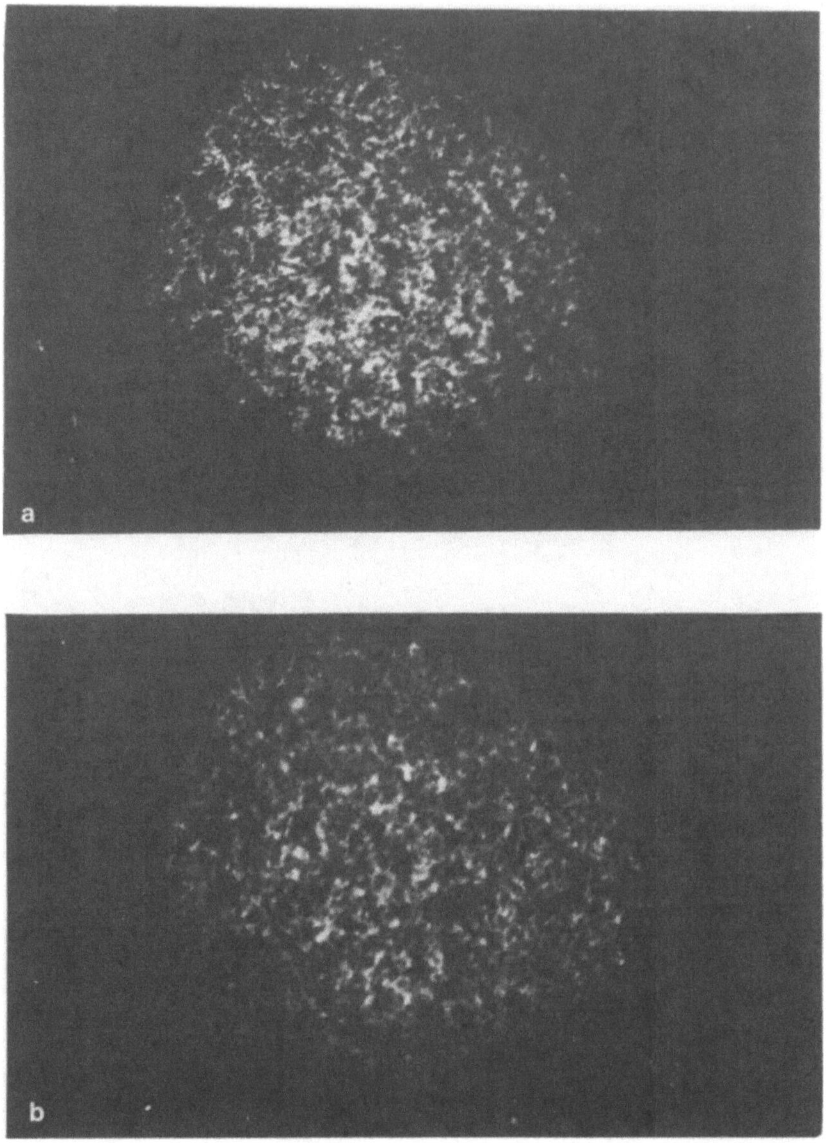

Fig. 2. Immunofluorescence staining with (a) MoAb to a TCC neoepitope and (b) MoAb to S-protein in adjacent sections from washed and ethanol-fixed tonsil. The follicles show identical staining patterns suggesting that TCC might be located on FDC as SC5b-9. (Magnification: X 200).

MoAb DRC1. Such areas might represent burned-out follicles with remnants of FDC, although it should be noted that MoAb DRC1 is not completely specific to FDC but produces a weak staining of B-lymphocytes and endothelial cells. It seems relevant, though, that a recent immunoelectronmicroscopic study of normal renal tissue showed TCC deposits on cell remnants entrapped in the extracellular matrix (21).

The biological implications of our findings are uncertain, but in situ terminal pathway activation on FDC may play a role in B-cell stimulation.

REFERENCES

1. G. J. V. Nossal, A. Abbot, J. Mitchell and Z. Lummus, J. Exp. Med. 127:277 (1986).
2. T. E. Mandel, R. P. Phippes and J. G. Tew, Immunological rev. 53:29 (1980).
3. M. Reynes, J. P. Aubert, J. H. M. Cohen, J. Audouin, V. Tricottet, J. Diebold and M. D. Kazatchkine, J. Immunol. 135:2687 (1985).
4. S. Bahakdi and J. Tranum-Jensen, Biochemica et Biophysia Acta 737:343 (1983).
5. E. R. Podack and H. J. Müller-Eberhard, J. Biol. Chem. 254:9908 (1979).
6. E. R. Podack, K. T. Preisener and H. J. Müller-Eberhard, Acta Pathol. Microbiol. Immunol. Scand. Ser. C, Suppl. 284, 92:89 (1984).
7. D. Jenne and K. K Stanley, The EMBO journal 4:3153 (1985).
8. S. Suzuki, A. Oldberg, E. G. Haymann, M. D. Pierschlater and E. Ruoslathi, The EMBO journal 10:2519 (1985).
9. E. G. Hayman, M. D. Piersbacher, Y. Öhgren and E. Ruoslathi, Proc. Natl. Acad. Sci. USA 80:4003 (1983).
10. K. Dahlbäck, H. Lofberg and B. Dahlbäck, Acta Dermatol. Venerol. 66:461 (1986).
11. R. J. Falck, E. Podack, P. A. Dalmasso and J. C. Jennette, Am. J. Pathol. 127:182 (1987).
12. P. Brandtzaeg, Immunology 26:1101 (1974).
13. P. Brandtzaeg, J. Histochem. Cytochem. 29:1302 (1981).
14. T. E. Mollnes, T. Lea, M. Harboe and J. Tschopp, Scand. J. Immunol. 22:183 (1985).
15. P. Garred, T. E. Mollnes, T. Lea and E. Fischer, Scand. J. Immunol., submitted (1987).
16. P. Brandtzaeg and T. O. Rognum, Histochem J. 15:655 (1983).
17. D. Y. Mason, Immunocytochemistry, vol 3. edited by G. R. Bullock, P. Petrusz. London, Academic Press, pp 25 (1985).
18. P. Brandtzaeg, I. Dale and M. Fagerhol, Am. J. Clin. Pathol. 87:681 (1987).
19. T. E. Mollnes, T. Lea, S. S. Frøland and M. Harboe, Scand. J. Immunol. 22:197 (1985).
20. T. E. Mollnes, Complement 2:156 (1985).
21. N. Hinglais, M. D. Kazatchkine, S. Bachdi, M. D. Appay, C. Mandet, J. Grossetete and J. Barnety, Kidney International 30:399 (1986).

TRANSPORT OF IMMUNE COMPLEXES FROM THE SUBCAPSULAR SINUS INTO THE LYMPH NODE FOLLICLES OF THE RAT

E.W.A. Kamperdijk, E.A. Döpp, and C.D. Dijkstra

Department of Cell Biology, Medical Faculty, Free University
P.O. Box 7161, NL-1007 MC Amsterdam, The Netherlands

Immune complexes localize in lymphoid follicles of different lymphoid organs. It has been shown that immune complex trapping in lymphoid follicles depends on the presence of complement factor C3 (1,2). The complexes are retained in follicular centers for a long time, although they may be replaced gradually by new complexes from the circulation (3). Trapping and retaining of immune complexes by follicular dendritic cells (FDC) may play an important role in the generation of immunological memory, feedback inhibition of the immune response or in both of these processes (4).

In a previous study (5) we showed that not only FDC in follicular centers, but also fibroblastic reticulum cells (FRC) in primary follicles trap horse radish peroxidase (HRP)-anti HRP complexes. These observations together with the description of transitional forms between both cell types during the ontogenetic development (6-8), strongly suggest that FDC are specialized FRC and that these cells form the stroma of the follicular area (9). However, it has also been suggested that FDC are immigrating cells which belong to the mononuclear phagocyte system since they share phenotypical characteristics with cells of the monocyte/macrophage lineage, such as Fc and C3 receptors, MHC class II antigens and some macrophage related antigens (10). Another unresolved question which is directly related to that of the origin of the FDC, concerns the mode of entry of immune complexes into the follicular area of lymphoid tissues and its contact with FDC. It has been suggested that in spleen marginal zone B cells are involved in transporting immune complexes into the follicles followed by deposition on FDC (11,12). However, Tew et al. (13), using ultrastructural techniques, suggested that in the subcapsular sinus of mouse lymph node immune complexes are trapped by dendritic cells (14) and transported to the follicular areas. The mechanism of transport might involve migration of these "pre-FDC: with a concomitant maturation into FDC.

To get more information about the mode of transport of immune complexes into the follicles we injected HRP-anti HRP complexes in rat footpads and studied the distribution pattern of these complexes on ultrastructural level in the outer cortex including the follicles of the draining popliteal lymph nodes. Control animals only received HRP.

MATERIALS AND METHODS

Footpads of random bred Wistar rats, aged between 8 and 10 weeks and weighing 180-200 g, were injected with 0.2 ml of a solution of HRP-anti HRP complexes (PAP, Dako, Denmark). Control animals were only given 0.2 ml of a solution of HRP. After 6 and 20 min, 2 and 24 hrs, the draining popliteal lymph nodes were fixed by perfusion fixation (at least 3 animals per time interval, i.e. 6 lymph nodes), and the outer cortex with the follicular area selectively trimmed out.

Visualisation of HRP and HRP-anti HRP complexes occurred using the method earlier described (5). Hereby the HRP and HRP-anti HRP complexes are visible as electron dense material.

RESULTS

Control animals

Only 6 and 20 min after injection of HRP electron dense material (HRP) was visible in the subcapsular sinus freely or associated with some macrophages, lymphocytes and stromal cells. This antigen was also present in relatively small amounts between the cells bordering the sinus. However, HRP never penetrated the deeper parts of the outer cortex including the follicles (Fig. 1).

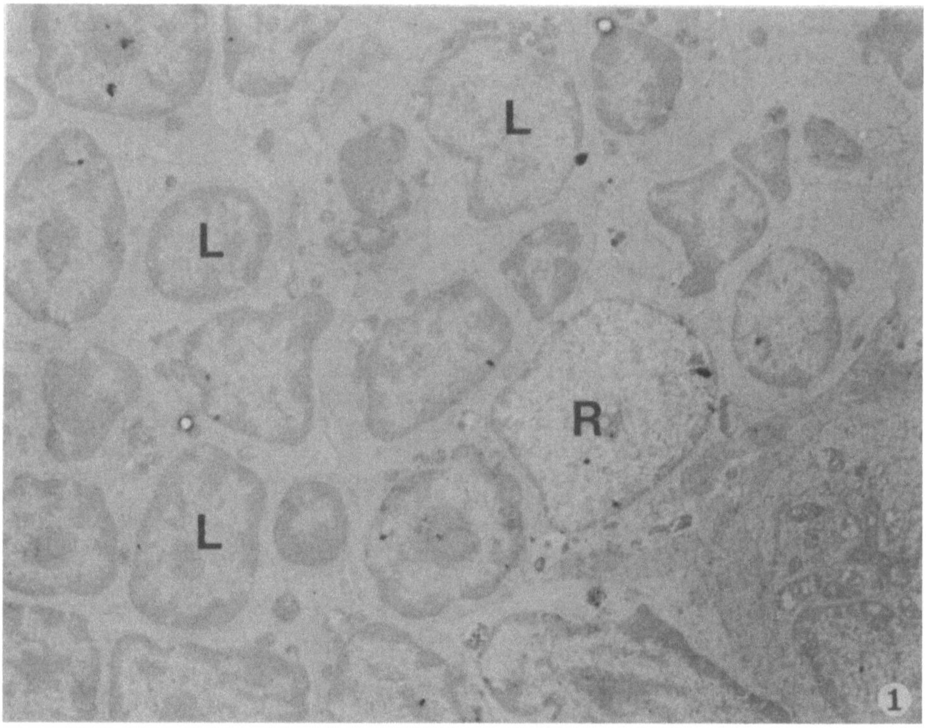

Fig. 1. Outer cortex lymph node, 20 min after HRP injection. No electron dense material (i.e. HRP) is present between the lymphocytes (L). R: reticulum cell.

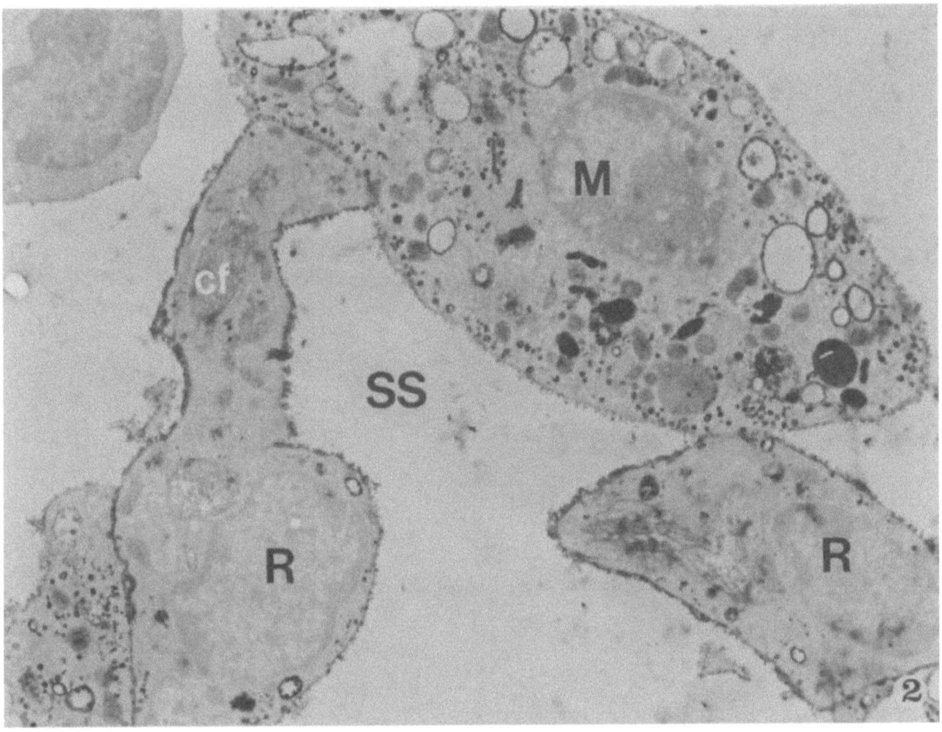

Fig. 2. Subcapsular sinus (ss) 6 min after HRP-anti HRP injection. Electron dense material (complexes) cover the plasma membrane of a macrophage (M) and reticulum cells (R) traversing the sinus. The lymphocyte (L) is negative. cf: collagen fibres.

HRP-anti HRP complex localization and distribution pattern

During the experiments the localization and distribution pattern of the injected complexes changed. The complexes never showed a distinct association with cells showing the morphological characteristics of dendritic cells (14).

Six minutes after injection of HRP-anti HRP electron dense material (i.e. complexes) was observed in the subcapsular sinus freely or attached to the plasma membrane of macrophages, lymphocytes and even bound to reticulum cells traversing the sinus and sinus lining cells (Fig. 2). They were also localized in the vicinity of the sinus.

Twenty minutes after injection complexes had reached the deeper parts of the outer cortex. They always were scattered between lymphoid and non-lymphoid cells (Fig. 3). Macrophages bordering the sinus frequently contained complexes in vacuoles or in phagolysosomes (immune complex phagocytosis).

Two hours after injection complexes were already found in the periphery of relatively small follicles, diffusely localized between lymphoid and non-lymphoid cells or in enclosed spaces between these different cell populations. On this moment PDC in the follicular area never showed any relation with the injected immune complexes.

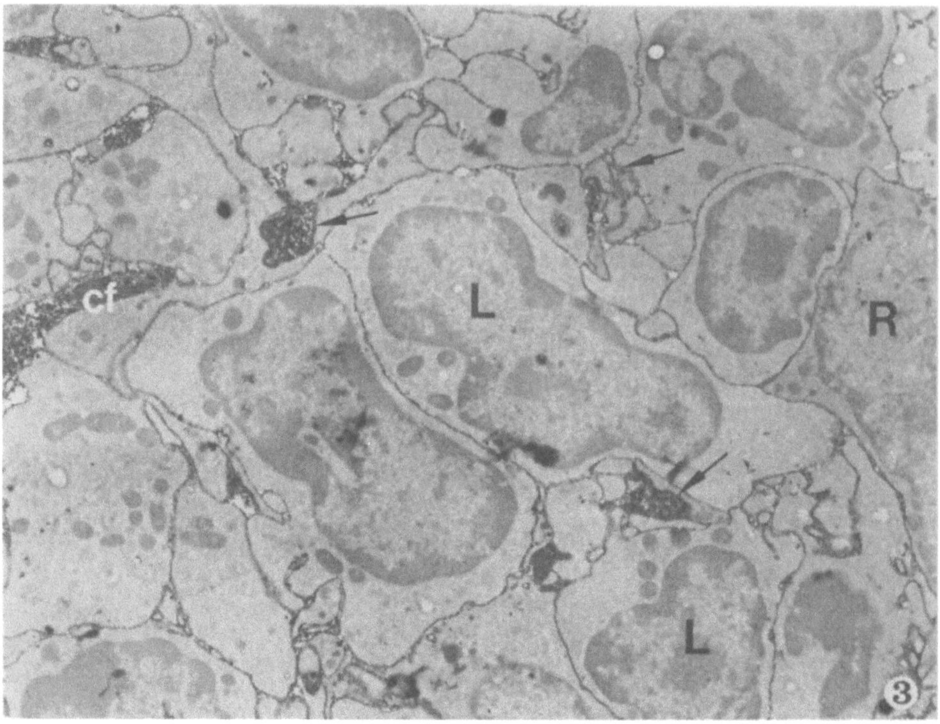

Fig. 3. Outer cortex lymph node, 20 min after HRP-anti HRP injection. Electron dense material (complexes) are diffusely scattered between lymphocytes (arrows). R: reticulum cell; cf: collagen fibres.

Twentyfour hours after injection immune complexes were present in the follicles. They were found in vacuoles or phagolysosomes of macrophages or trapped by FDC (Fig. 4).

DISCUSSION

Our results showed that after footpad injection of HRP-anti HRP complexes first appeared in the subcapsular sinus, freely or on the plasma membrane of lymphocytes, macrophages and reticulum cells, i.e. cells forming the stroma of this area. Within 2 hours the complexes penetrated towards the follicles, they were always seen in a random distribution between lymphoid and non-lymphoid cells. These results do not support the suggestion that well defined lymphocyte subpopulations are involved in transport of immune complexes into the follicles as suggested for the marginal zone lymphocytes in the spleen (12,15) and in particular the IgM positive, IgD negative cells (11). We suggest that the immune complexes are passively transported through the narrow intercellular spaces between the lymphoid and non-lymphoid cells (16). This means that except for macrophages, during transfer, complexes are not strongly bound to the plasma membrane of the different cell populations. Once in the follicles the complexes

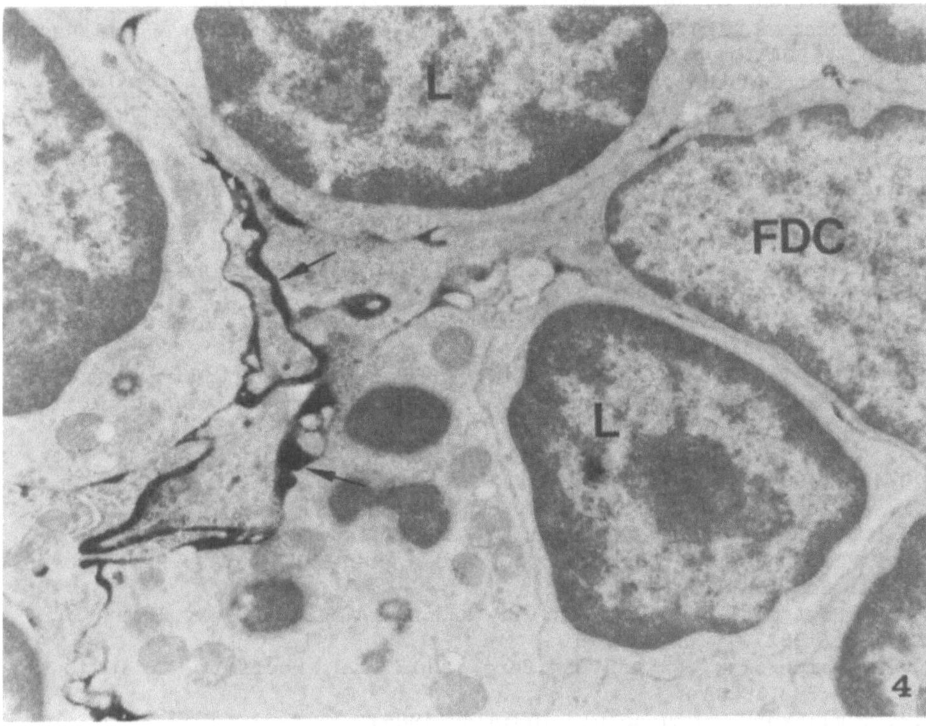

Fig. 4. FDC, 24 hours after HRP-anti HRP injection. Electron dense material (complexes) is present in the intercellular space (arrows). L: lymphocyte.

were trapped and retained by FDC. This migration appeared to be immune complex specific since HRP failed to penetrate into the deeper parts of the outer cortex including the follicles.

Tew et al. (13) suggest that dendritic cells (14) i.e. cells which belong to the cell lineage Langerhans cell, veiled cell and interdigitating cell (17,18) transport immune complexes from the subcapsular sinus towards the follicles by which they transform into FDC. However, in our experiments dendritic cells did not show any association with the complexes. Moreover, it has been shown that dendritic cells are bone marrow derived (19) whereas FDC are not (20). In addition FDC are ultrastructurally different from dendritic cells, express Fc and C3 receptors (21) and do not demonstrate acid phosphatase activity (22).

We conclude that FDC are not related to dendritic cells or macrophages but represent stromal cells characteristic for the follicular area.

REFERENCES

1. M. Papamichail, C. Guttierrez, P. Embling, P. Johnson, E.J. Holborow, and M.B. Pepys. Scand. J. Immunol. 4:343 (1975).
2. G.G.B. Klaus. Immunology 34:643 (1978).
3. N. van Rooijen. Int. Arch. Allergy Appl. Immunol. 49:754 (1975).

4. N. van Rooijen. J. Reticuloendothel. Soc. 21:143 (1977).
5. C.D. Dijkstra, E.W.A. Kamperdijk, and E.A. Döpp. Cell Tissue Res. 236:203 (1984).
6. P. Groscurth. Pathol. Res. Pract. 169:235 (1980).
7. U. Heusermann, K.H. Zurborn, L. Schroeder, and M.J. Stutte. Cell Tissue Res. 209:279 (1980).
8. A. Villena, A. Zapata, J.M. Rivera-Pomar, M.G. Barrutia, and J. Fongua. Cell Tissue Res. 229:219 (1983).
9. E.W.A. Kamperdijk, E.M. Raaijmakers, J.H.S. de Leeuw, and E.C.M. Hoefsmit. Cell Tissue Res. 192:1 (1978).
10. J. Gerdes, H. Stein, D.Y. Mason, and A. Ziegler. Virchows Arch. B. Cell Pathol. 42:161 (1983).
11. D. Gray, D.S. Kumararatne, J. Lortan, M. Khan, and I.C.M. MacLennan. Immunology 52:659 (1984).
12. E. Heinen, M. Braun, P.G. Coulie, J. van Snick, M. Moeremans, N. Cormann, C. Kinet-Denoel, and L.J. Simar. Eur. J. Immunol. 16:167 (1986).
13. J.G. Tew, T.E. Mandel, R.P. Phipps, and A.K. Szakal. Am. J. Anat. 170:407 (1984).
14. R.M. Steinman, and Z.A. Cohn. J. Exp. Med. 139:380 (1974).
15. A.J.P. Veerman, and N. van Rooijen. Cell Tissue Res. 161:211 (1975).
16. P.H.P. Groeneveld, P. Eikelenboom, and N. van Rooijen. J. Reticuloendothel. Soc. 33:109 (1983).
17. E.C.M. Hoefsmit, A.M. Duijvestijn, and E.W.A. Kamperdijk. Immunobiol. 161:255 (1982).
18. E.W.A. Kamperdijk, M. van den Berg, and E.C.M. Hoefsmit. Cell Tissue Res. 237:39 (1984).
19. W.E.F. Klinkert. Immunobiol. 168:414 (1984).
20. J.H. Humphrey, and V. Sundaram. Adv. Exp. Med. Biol. 186:167 (1985).
21. D. Radoux, C. Kinet-Denoel, E. Heinen, M. Moeremans, J. de Mey, and L.J. Simar. Scand. J. Immunol. 21:345 (1985).

FDC-ICCOSOME MEDIATED ANTIGEN DELIVERY TO GERMINAL CENTER B CELLS, ANTIGEN PROCESSING AND PRESENTATION TO T CELLS

Andras K. Szakal, Marie H. Kosco, and John G. Tew

Department of Anatomy, Division of Immunobiology, Medical College of Virginia, VCU, Richmond VA 23298, USA

Department of Microbiology and Immunology, Medical College of Virginia, VCU, Richmond VA 23298, USA

INTRODUCTION

Antigen trapped by follicular dendritic cells (FDC) is retained for long periods as intact molecules (1). An antigen must be internalized, partially degraded and reexpressed on the surface of an antigen presenting cell to be recognized by T cells (2). Consequently, it was difficult to explain how antigen retained by FDC could be presented to T cells, which are required for germinal center (GC) development (3) and can be demonstrated in GCs (4,5). Recently, antigen specific B cells were shown to process and present antigen in a highly efficient manner _in vitro_ (6-8). Germinal center B cells have the requisite features of good antigen presenting cells, i.e., they are metabolically active, rich in Ia, and include cells specific for the FDC-trapped antigen. Thus, GC B cells could present antigen obtained from FDC to GC T cells.

Scanning electron microscopy (SEM) showed two types of FDCs, one with long filiform dendrites and one with "beaded" dendrites (see Fig. 1, Day 1)(9). Intermediate forms suggesting maturation of the first to the second type were also seen. Cell culturing supported this maturation sequence. Both types of dendrites were shown to retain immune complexes on their surfaces. These 0.3-0.7 um diameter beads are now termed "iccosomes" (immune complex coated bodies). Due to the apparent fragility of connections between them, we suggested that iccosomes may function as a mechanism to increase efficiency in follicular dispersion of antigenic information (9). Using immune mice after antigenic challenge with horseradish peroxidase (HRP) as the histochemically detectable antigen, we demonstrated by light (LM) and electron (EM) microscopy the cell-mediated transport of immune complexes from the afferent lymph to the site of developing popliteal GCs during a period between 1 min and 24 hr for retention by FDCs (10). We extended these studies to examine the interaction of the FDC trapped antigen with follicular lymphocytes and to determine if iccosomes existed in GCs _in vivo_.

MATERIALS AND METHODS

In this study, 8-12 week old HRP-immune C3H mice were challenged with 8ug of HRP per hind-footpad, perfused with cold paraformaldehyde (1%)/ glutaraldehyde (0.9%) for LM and EM histochemical localization of the antigen, HRP with DAB/H_2O_2, at days 1,3,5,8, and 14 after antigen injection. For controls, HSA immune mice were challenged with HSA and treated histochemically the same way as HRP injected mice. For technical details see (11).

RESULTS

Light Microscopy

At day 1, localization of HRP-anti-HRP complexes in the ARR of follicles was identical to that described previously (10). At day 3 follicles increased in size and the PO activity was more diffuse. From day 5 through 14 HRP localization became more condensed. As confirmed by controls, especially at day 5, large numbers of TBM contained HRP. This was confirmed by LM with HSA controls and by EM with gold conjugated antigen. For details refer to (11).

Electron Microscopy

Day 1. Predominantly FDC with their typical convoluted filiform (non-beaded) type dendritic processes were observable in follicles. A smaller FDC subpopulation represented FDC with "beaded" dendrites, which coexisted with FDC of the filiform type in the same follicle. The "beads" or iccosomes at day 1 were found primarily near the cell body connected to one another. Most of the lymphocytes seen were of small to medium size interphase cells.

Day 3. At this time after HRP injection, the most outstanding feature in the kinetics of the germinal center response was the apparent dispersion of iccosomes among the germinal center lymphocytes (Fig. 1, Day 3). Lymphocytes appeared to respond to the attachment of iccosomes to their surfaces by surrounding them with their microvillous processes, which gave the impression of impending endocytic events (Fig. 1, Day 3; arrows). Occasional lymphocytes were also seen with a single PO+ granule in their cytoplasm (Fig. 1, Day 3; open arrow). Similar observations were made with respect to endocytosis of iccosomes on TBM. Since no lymphocytes with PO+ granules were seen in HSA controls, it was concluded that the PO activity in the granules represented endocytosed HRP.

Day 5. By this time, iccosomes were only rarely seen among the lymphocytes. Convolutions of filiform dendrites with their trapped PO+ immune complexes extended for long distances away from the FDC cell body. The most impressive observation at day 5 was the presence of one to several PO+ small inclusions in nearly all germinal center lymphocytes (Fig. 1, Day 5a). Since the majority of germinal center lymphocytes contained these PO+ inclusions, it may be safely argued that these cells represent B lymphocytes. The assumption that the PO+ inclusions represent endocytosed antigen was confirmed in a parallel experiment in which passively immunized animals were injected with gold conjugated goat IgG as the antigen. Five days after antigen injection and indirect PO labeling of Ia cell surface determinants, the gold conjugated antigen was localized in the cytoplasm of Ia+ B lymphocytes in membrane bound endocytic vesicles (Fig. 1, Day 5b; arrow).

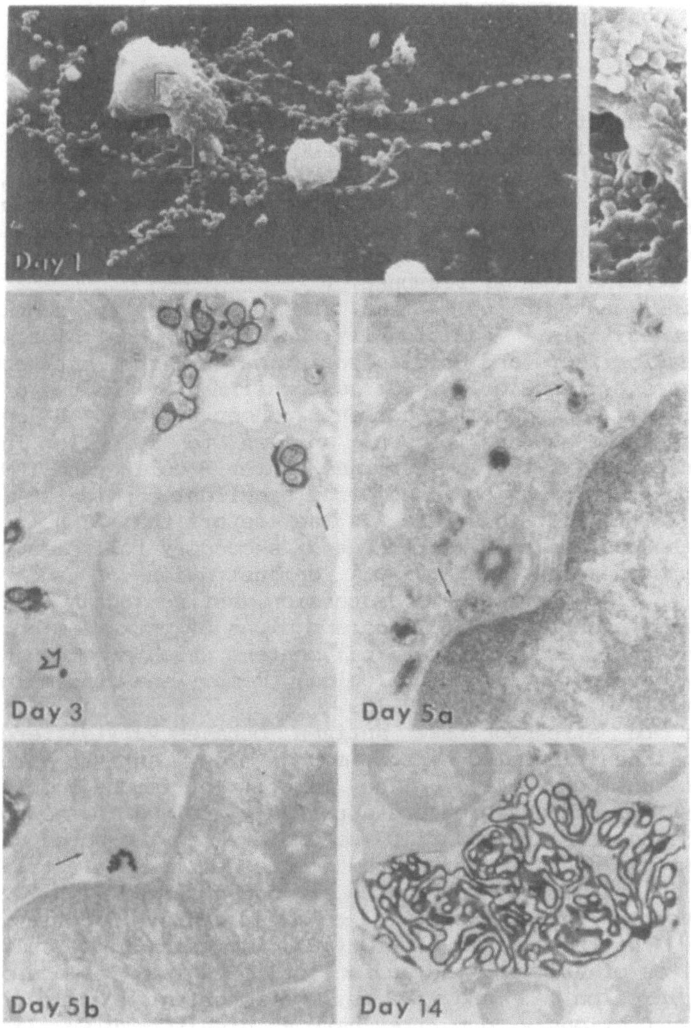

Figure 1. Day 1: Isolated FDC with "beaded" (iccosome) dendrites, X2,800; Right: High magnification of iccosomes, X8,000; Day 3: Iccosome dispersion among germinal center B cells; Note, attached iccosomes surrounded by microvilli (arrows), X9,600; Day 5a: Germinal center B cell with HRP (black) in endocytic vesicles, apparently being degraded (arrows); Note association with Golgi, X23,000; Day 5b: Gold conjugated antigen in germinal center B cell endocytic vesicle, X22,300; Day 14: Tightly convoluted HRP+ filiform FDC dendrites, X7,300.

Day 8 - 14. At day 8, morphological changes in the cellular components of germinal centers appeared to be intermediate between that of day 5 and day 14. By day 14, the ARR was represented predominantly by tightly convoluted bundles of filiform dendrites of FDC (Fig. 1, Day 14) and these were located distantly from the cell bodies. By this time only a few lymphocytes contained PO+ inclusions in their cytoplasm.

DISCUSSION

This study demonstrated a cellular mechanism in immune mice for the delivery of antigen, via iccosomes, from FDC to germinal center B lymphocytes and to tingible body macrophages (TBM). In view of the efficiency of germinal center B cells in antigen presentation (see report by Kosco et al.; this volume) and the unavailability of information on the antigen presenting capacity of TBM, this report focuses primarily on the role of B cells in the germinal center response.

Both germinal center B cells and TBM responded to contact with iccosomes by endocytosis. Additionally, B cells demonstrated apparent lysosomal processing steps and Golgi-association of the antigen. The uptake of antigen by these cell types was confirmed by the use of a gold conjugated antigen. According to current concepts, both B cells and macrophages can process and present antigen to T cells in vitro (2,6-8,14). Consequently, the Ia+ germinal center B cells and TBM, which were also Ia positive, may be considered excellent candidates for this role in vivo. Critical to this idea is the report that T helper cells are found in the corona and light zone of secondary follicles (5). The present study provides the first visual demonstration of such antigen driven events in vivo as iccosome dispersion, antigen uptake by B cells and TBM, antigen fragmentation and apparent signs of reexpression. It is significant that this occurs in germinal centers which are thought to be integral to the development of immunological memory and the maintenance of the secondary immune response (12,13).

It should be stressed that antigen behaves very differently in nonimmune mice. When injected into immune mice, antigen immediately complexes with antibody (10). The majority of immune complexes are phagocytized by macrophages which are apparently drained to the popliteal lymph node where they temporarily accumulate in the subcapsular and medullary sinuses. The remainder of the immune complexes are transported by ATC to follicles (10). Lymphocytes are not observed with PO+ immune complexes on their surfaces. However, germinal center B cells bind to immune complexes on the surface of ATC (10), including those coating the filiform dendrites of FDC, and the surface of iccosomes. The conformation of immune complexes on ATC and FDC may be important for binding by B cells. As shown previously, dendrites have a highly regular 440-490 A° periodic spiral immune complex binding pattern (9). This periodicity and spiral binding pattern may relate to why B cells bind immune complexes on iccosomes and endocytose the antigen so efficiently. This interpretation is consistent with the report that in order to bind and directly stimulate B cells, an immunogen (e.g., Dnp-polymer) must have a certain number of epitopes spaced 120-670 A° apart (15).

Apparently, the endocytosis of antigen must, however, await the maturation of certain FDC to the beaded type and the subsequent dispersion of iccosomes. The conclusion that iccosomes are endocytosed by lymphocytes is based on the sequential association of iccosomes with the surface of lymphocytes at day 3 and the occurrence of the PO+ inclusions in most germinal center lymphocytes by day 5. Iccosomes were attached apparently via the PO+ immune complex to the surface of lymphocytes. Such an attachment suggests binding of the HRP-anti-HRP complexes by B cell surface receptors (e.g. surface Ig, Fc, C3b). Microvillous processes tended to surround the attached iccosomes giving the impression of an impending endocytic event. The mechanism of the event following iccosome attachment is not clear. It is possible that the iccosome is only a means for the dissemination, delivery and attachment of the immune complexes, resulting in the endocytosis of the antigen and not the entire iccosome. More significant is the observation that the gold conjugated antigen was found in endocytic vesicles.

Close examination showed that the antigen in the secondary lysosome-like granules of germinal center B cells appeared to be in various stages of breakdown, as indicated by the different degrees of PO reactivity. Some of these were closely associated with the Golgi lamellae and appeared to be in the process of fusion with Golgi vesicles (see Fig. Day 5a). These observations suggest that fragmented antigen may be emptied into Golgi vesicles or cisternae. The Golgi apparatus, along with other endomembrane components is involved in the biogenesis and modification of the plasma membrane (16) and may provide a mechanism in the B lymphocyte for the reexpression of processed antigen. That germinal center B cells efficiently present antigen obtained by them under these circumstances *in vivo* was confirmed *in vitro* using isolated germinal center B cells and an antigen specific, IL 2 producing, T cell hybridoma. This *in vitro* study fully supports the *in vivo* kinetics of iccosome mediated release, endocytosis and antigen processing by B cells [See Table 1, by Kosco, Szakal and Tew (this volume)].

It is well known that macrophages internalize, partially degrade and reexpress antigen on their surface in a form recognizable by T cells (2). Grey and coworkers (17) showed that nonspecifically bound antigen can be processed by B cell lymphomas and that antigen captured by B cell Ig receptors can be presented efficiently to T cells. Most recently Lanzavecchia (6) conclusively demonstrated that the Ig receptor is involved in the uptake and concentration of the antigen which is then processed and presented with high efficiency to T cells in an MHC restricted manner. Consistent with these *in vitro* findings and concepts, we observed the germinal center B lymphocyte: 1. sequestering the antigen (iccosomes) on their surface for endocytosis; 2. the apparent degradation of the antigen in secondary lysosome-like granules and 3. the association of the antigen with the Golgi apparatus, which may provide a mechanism for reexpression of antigen fragments. These observations suggested the B cell to be the candidate antigen presenting cell in the germinal center.

REFERENCES

1. Tew, J. G., T. E. Mandel, and A. W. Burgess. 1979. Retention of intact HSA for prolonged periods in the popliteal lymph nodes of specifically immunized mice. Cell Immunol., 45:207.

2. Unanue, E. R. 1984. Antigen presenting function of the macrophage. Ann. Rev. Immunol., 2:395.

3. Thorbecke, G. and S. P. Lerman. 1976. Germinal centers and their role in immune responses. *In*: The Reticuloendothelial System in Health and

Disease: Functions and Characteristics. S. M. Reichard, M. R. Escobar, and H. Friedman, eds. Plenum Press, New York. P. 83.

4. Rouse, R. V., J. A. Ledbetter, and I. L. Weissman. 1982. Mouse lymph node germinal centers contain a selected subset of T cells - the helper phenotype. J. Immunol., 128:2243.

5. Nieuwenhuis, P. and D. Opstelten. 1984. Functional anatomy of germinal centers. Am. J. Anat., 170:421.

6. Lanzavecchia, A. 1975. Antigen-specific interaction between T and B cells. Nature, 314:537.

7. Malynn, B. A., D. T. Romeo, and H. H. Wortis. 1985. Antigen-specific B cells effectively present low doses of antigen for induction of T cell proliferation. J. Immunol., 135:980.

8. Abbas, A. K., S. Haber, and K. L. Rock. Antigen presentation by hapten-specific B lymphocytes. II. Specificity and properties of antigen presenting B lymphocytes, and function of immunoglobulin receptors. J. Immunol., 135:1661.

9. Szakal, A. K., R. L. Gieringer, M. H. Kosco, and J. G. Tew. 1985. Isolated follicular dendritic cells: Cytochemical antigen localization, Nomarski, SEM, and TEM morphology. J. Immunol., 134:1349.

10. Szakal, A. K., K. L. Holmes and J. G. Tew. 1983. Transport of immune complexes from the subcapsular sinus to lymph node follicles on the surface of nonphagocytic cells, including cells with dendritic morphology. J. Immunol., 131:1714.

11. Szakal, A. K., M. H. Kosco, and J. G. Tew. 1987. A novel _in vivo_ follicular dendritic cell dependent iccosome mediated mechanism for delivery of antigen to antigen processing cells. J. Immunol. (Accepted).

12. Klaus, G. G. B., J. H. Humphrey, A. Kunkl, and W. Dongworth. 1980. The follicular dendritic cell: its role in antigen presentation in the generation of immunological memory. Immunol. Rev., 53:3.

13. Tew, J. G., R. P. Phipps, and T. E. Mandel. 1980. The maintenance and regulation of the humoral immune response: persisting antigen and the role of follicular antigen-binding dendritic cells. Immunol. Rev. 53:173.

14. Grey, H. M. and R. Chesnut. 1985. Antigen processing and presentation to T cells. Immunology Today. 6:101.

15. Dintzis, R. Z., M. H. Middleton and H. M. Dintzis. 1983. Studies on the immunogenecity and tolerogeneciti of T-independent antigens. J. Immunol. 131:2196.

16. Whaley, W.G., and M. Dauwalder. 1979. The Golgi apparatus, the plasma membrane, and functional integration. In Int. Rev. Cytol., 58:199.

17. Chesnut, R. W. and H.M. Grey. 1981. Studies on the capacity of B cells to serve as antigen-presenting cells. J. Immunol., 126:1075.

IS THE SURVIVAL OF MEMORY B CELLS DEPENDENT ON THE PERSISTENCE OF ANTIGEN?

David Gray

Basel Institute for Immunology
Grenzacherstrasse 487
CH-4058, Basel
Switzerland

INTRODUCTION

The immunisation of chimaeras constructed by transferring kappa allotype-marked thoracic duct B cells from donors primed with DNP-haemocyanin (DNP-MSH) into irradiated (femoral bone marrow shielded) recipients elicits high serum levels of anti-DNP antibodies lasting for at least 6 months (1). The majority of this antibody derives from the donor population even though this comprises only 10-20% of the total B cell pool at this time (1). Continued antibody production for such long periods may be the consequence of secretion by long-lived plasma cells in the bone marrow (2) but it seems more likely to result from repeated stimulation of B cells within memory clones.

We know that antigen remains within lymphoid tissues in the form of immune complex for periods of up to 1 year (3). Established immune responses may be maintained over long periods by activation of memory clones by antigen on follicular dendritic cells (4,5). If this is the case then a memory clone will contain activated B cells and B cells that have returned to the resting state. It is not clear if the life span of resting memory B cells is significantly different from the average life span of B cells in the recirculating pool, variously measured at 4 weeks (6) or 6 weeks (7). In other words, does a memory B cell survive for very long periods in the absence of further antigenic stimulation. Experiments by Askonas and Williamson (8,9) demonstrate that transfer of dominant memory clones without subsequent immunisation results in the loss of these clones.

In the experiments reported here thoracic duct lymphocytes from primed (DNP-MSH) κ-allotype distinct donor rats have been transferred into sublethally irradiated recipients. These chimaeras have been immunised at various times after transfer such that an estimate of clonal half-life in the absence of antigen can be made.

MATERIALS AND METHODS

Animals

PVG/Ciu rats and a PVG-RI-a^{a} (DA) κ.allotype distinct congenic line were bred and maintained in the Institute for Biomedical Reseach, Füllinsdorf, Switzerland. All rats were aged between 10 and 16 weeks at the start of experiments.

Donor immunisation

Klb (PVG-RI-a^{a}) rats were immunised first i.p. with 50 μg alum precipitated DNP-MSH (MSH = *Maia squinada* - spider crab haemocyanin) plus 10^{9} *B. pertusis* and then 1 month later with 50 μg DNP-MSH i.p. in soluble form. These rats were cannulated for thoracic duct drainage 1-2 months later.

Adoptive transfer

Kla (PVG/Ciu) recipient rats received 5.5 Gy γ-irradiation (Gammacell 40 ^{137}Cs source, Atomic Energy Ltd. Canada) 1 day prior to injection of 10^{8} thoracic duct lymphocytes from DNP-MSH primed donors. This is a sublethal dose of irradiation; regeneration of host B cells takes place very quickly (~1 week). Recipients were immunized with 50 μg DNP-MSH plus 50 μg OX-MSH either in soluble form (iv) or alum precipitated + *B. pertussis* (ip) at various times after transfer.

Detection of donor and recipient antibody responses

Anti-DNP and anti-OX responses were assayed in a solid phase radioimmunoassay. DNP-BSA (Bovine Serum Albumin) and OX-BSA were coated onto polystyrene plates at 10 μg/ml (50 μl) in 0.1M carbonate buffer pH9.6 at 4°C overnight and then the plates were blocked with 100 μl 1% BSA in PBS for 30 min at room temperature. Kla (host) and Klb (donor) serum antibody responses were detected using the monoclonal antibodies MARK-3 and FH8, respectively. The radioimmunoassay and quantitation are exactly as described in (1).

RESULTS AND DISCUSSION

The survival of DNP-specific memory B cell clones in the absence of antigen was investigated by immunising chimaeras with soluble DNP-MSH 1 week, 3 weeks and 11 weeks after transfer. The responses derived from donor (DNP-specific memory cells) and host (virgin cells that repopulate the peripheral lymphoid tissues) B cells in these chimaeras are shown in Figure 1. Immunisation 1 week after transfer elicits equivalent early responses (0-7 days) from both host and donor (memory) cells. After this the donor (memory) response becomes predominant and maintains high serum antibody levels. The host response falls to zero by 28 days and exhibits a primary response profile. The transient nature of the host (virgin) response has been noted previously in bone marrow chimaeras that received primed donor B cells (1) and probably reflects the higher affinity for antigen of memory B cells but may also be related to relocalisation of antigen onto follicular dendritic cells (4,10).

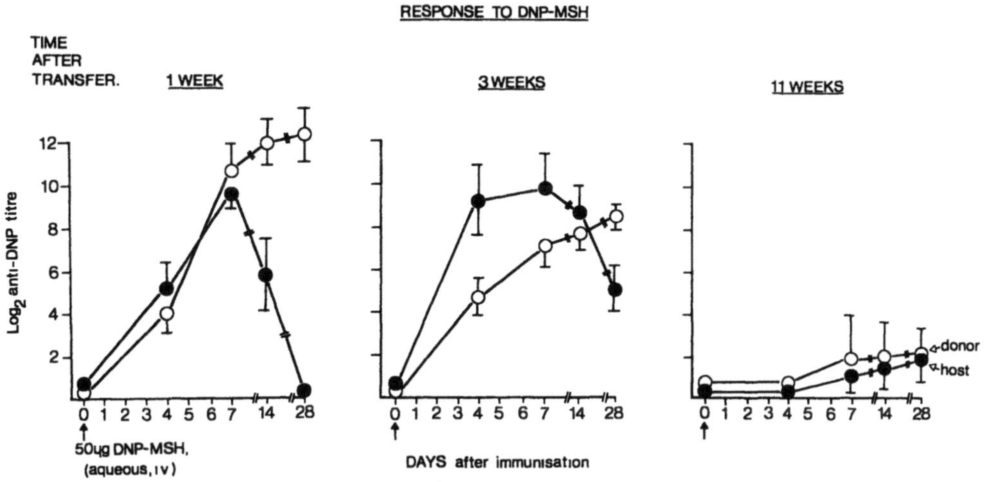

Figure 1 The survival of DNP-specific memory B cells in the absence of antigen. Chimaeras were immunised 1 week, 3 weeks or 11 weeks after transfer with 50 µg of soluble DNP-MSH (iv). Serum anti-DNP titres ($\log_2$) are expressed as mean values (± standard deviation) of groups of 3-4 rats.

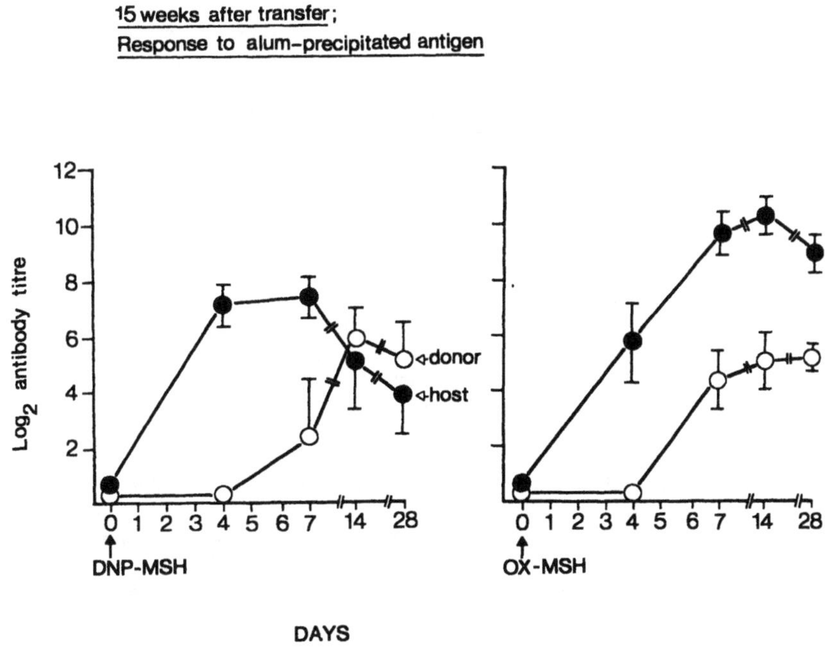

Figure 2. Response of host and donor (DNP-MSH primed) B cells to immunisation with alum-precipitated DNP-MSH + OX-MSH, 15 weeks after transfer. Serum anti-hapten titres ($\log_2$) are expressed as mean values (± SD) of 4 rats.

Immunisation 3 weeks after transfer shows that already there has been a decay of the memory B cell population. The donor response only begins to predominate between days 14 and 28. It seems likely that expansion of memory clones is necessary between days 0-14. The fact that the host response wanes while the donor response continues to rise at day 28 suggests that the donor antibody is derived from higher affinity memory B cells.

Immunisation 11 weeks after transfer evokes negligible host or donor response. Memory to DNP appears to be largely lost. The lack of host response may indicate that T cell help is no longer available. If this is so then the half-life of T cell memory is relatively short in the absence of antigen. The results of the experiment shown in Figure 2 suggest that this is not the case. Chimaeras were immunised 15 weeks after transfer of DNP-MSH-primed cells, with DNP-MSH + OX-MSH in alum precipitated form (+ B. pertussis). In these animals good host responses are made (Fig. 2) that predominate throughout. The rapidity of this host response indicates that T cell memory is still present and can be stimulated by antigen in this form. It is clear that the response to soluble antigen requires the presence of T cell help prior to immunisation (iv. injection of soluble DNP-MSH into naive rats elicits no response), however the processing and presentation of soluble antigen to T cells may be carried out by antigen-specific activated B cells (11). By 11 weeks after transfer of memory B cells, even if significant numbers survive, in the absence of antigen they are probably not in an activated state. The cellular requirements of presentation to memory T cells are currently under test. Figure 2 also illustrates the lack of donor, memory responses 15 weeks after transfer. The similar kinetics of donor anti-DNP and anti-OX responses indicate that the anti-DNP antibodies are not derived from a high affinity memory population (donor cells were from rats primed with DNP-MSH but not OX-MSH).

CONCLUSION

We have shown that the survival of memory B cells is dependent upon the presence of antigen within lymphoid tissues. In the absence of antigen the half-life of memory B cell populations is in the order of a few weeks. We suggest that memory B cells are not long-lived resting B cells but turnover at a similar rate to the majority of peripheral, recirculating B cells (6,7).

In the presence of antigen, memory B cell clones appear to undergo repeated stimulation and may be very long-lived (1,10). The data presented also suggest that soluble antigen can only elicit T cell help if activated or memory B cells are present; it may be that processing and presentation of antigen in this form is only efficiently carried out by antigen-specific B cells and not other types of antigen presenting cell.

ACKNOWLEDGEMENTS

I would like to thank Lena Skarval for technical assistance, Judie Hossmann and Carolyn Harley for typing the manuscripts and Drs. Ivan Lefkovits and Ton Rolink for critical reading. The Basel Institute for Immunology was founded and is supported by F. Hoffmann-La Roche & Co. Ltd., Basel, Switzerland.

REFERENCES

1. D Gray, ICM MacLennan and PJL Lane. Eur. J. Immunol. 16 641 (1986).
2. F Ho, JE Lortan, ICM MacLennan and M Khan. Eur. J. Immunol. 16 1297 (1986).
3. TE Mandel, RP Phipps, A Abbot and JG Tew. Immunol. Rev. 53 29 (1980).
4. D Gray. This volume.
5. JG Tew, RP Phipps and TE Mandel. Immunol. Rev. 53 175 (1980).
6. D Gray, manuscript submitted.
7. J Sprent and A Basten. Cell. Immunol. 7 40 (1973).
8. BA Askonas, AR Williamson and BEG Wright. Proc. Nat. Acad. Sci. 67 1398 (1970).
9. BA Askonas and AR Williamson. Eur. J. Immunol. 2 487 (1972).
10. ICM MacLennan and D Gray. Immunol. Rev. 91 61 (1986).
11. JI Krieger, SF Grammer, HM Grey and RW Chesnut. J. Immunol. 135 2937 (1985).

MEMORY B CELLS BUT NOT VIRGIN B CELLS ARE ACTIVATED IN GERMINAL CENTERS

David Gray

Basel Institute for Immunology
Grenzacherstrasse 487
CH-4058, Basel
Switzerland

INTRODUCTION

In chimaeras constructed by transferring κ-allotype distinct DNP-haemocyanin-primed thoracic duct lymphocytes (TDL) into irradiated (hind limb bone marrow shielded) recipients the activation of virgin (host allotype) B cells is often transient falling to negligible levels within 14 days (1). The lower affinity of antigen receptors on virgin B cells compared to memory B cells precludes their continued activation as antigen becomes limiting. However, as antibody levels rise antigen is re-localised into follicles in the form of immune complex (2,3). Recently it has been shown that virgin B cells after transfer do not migrate into follicular structures, in contrast to transferred TDL (4). This raises the possibility that virgin and memory B cell activation occurs at different sites. When the supply of antigen is high (soon after immunisation) both virgin and memory B cells are activated at extra-follicular sites but as antigen levels fall and re-localisation onto follicular dendritic cells (FDC) occurs, only memory B cells gain access to the antigen supply.

To test this hypothesis chimaeras similar to those already described (1) were immunised with antigens in the form of immune complex. In this way antigen rapidly localises into follicles. Differences in activation of virgin and memory B cells at this site may only reflect the ability of memory cells to compete for complexed antigen because of their higher affinity receptors. To avoid this possibility we have used complexes in which haptenic epitopes are not masked.

MATERIALS AND METHODS

Animals

PVG/Ciu rats and a PVG-RI-a^a (DA) κ-allotype distinct congenic line were bred and maintained in the Institute for Biomedical Research, Füllinsdorf, Switzerland. All recipient animals were 10-16 weeks old at the time of transfer.

Chimaeras

Chimaeras were constructed as described in (1). Briefly this involved irradiating Kla recipient rats while shielding their hind limbs (8 Gy γ irradiation from a Gamma cell 40 ^{137}Cs source, Atomic Energy Ltd. Canada). One day later these rats were reconstituted with 2×10^8 thoracic duct lymphocytes (TDL) from Klb donor rats primed with DNP-MSH (MSH = Maia squinada - spider crab haemocyanin) and ovalbumin. Donor rats were primed with 50 μg DNP-MSH (alum precipitate + B. pertussis i.p.) and then boosted one month later with soluble antigen. TDL was obtained by cannulating 1-2 months after boosting. This protocol produces chimaeras that almost immediately after reconstitution are exporting Kla virgin B cells into a peripheral pool of Klb B cells (including primed 'memory' populations). For experimental validation see (1,5).

Immunisation

All immunisations were with pre-formed antigen-antibody complexes. Anti-DNP-MSH and anti-OX-MSH sera were from rats primed and boosted (2x) with these antigens. Immune complexes were formed in slight antibody excess such that few DNP epitopes were left unbound. A volume of anti-DNP serum was reacted with 10 μg DNP-MSH such that residual antibody activity was still retained (assessed by radioimmunoassay). Complexes were spun at 100,000g for 20 minutes before intravenous injection. Complexes in which haptenic epitopes were not masked were made in a similar way; OX-MSH was reacted with anti-DNP-MSH or DNP-MSH was reacted with anti-OX-MSH.

Chimaeras were immunised with pre-formed immune complex 4 days after reconstitution with TDL. This protocol was shown to result in rapid localisation of antigen into follicles by injecting complex containing fluorescein (FITC) labelled MSH. No FITC-MSH was observed outside follicular structures in spleen or lymph nodes.

Detection of donor and recipient antibody responses

Anti-DNP and anti-OX responses were measured in a solid phase radioimmunoassay. DNP-BSA (Bovine serum albumin) and OX-BSA were coated onto polystyrene plates at 10 μg/ml (50 μl) in 0.1M carbonate buffer pH 9.6 at 4°C overnight and then the plates were blocked with 100 μl 1% BSA in PBS for 30 minutes at room temperature. Kla (host) and Klb (donor) serum antibody responses were detected using the monoclonal antibodies MARK-3 and FH8 respectively. The radioimmunoassay and quantitation are exactly as described in (1).

RESULTS AND DISCUSSION

Immunisation of chimaeras with DNP-MSH anti-DNP-MSH complex results in the potent stimulation of memory (donor) B cells (Figure 1). In contrast, most chimaeras (6 out of 10) make very little virgin (host) B cell response. However, 4 out of 10 chimaeras do make host anti-DNP responses (albeit ~50 fold lower than donor responses); while antigen-antibody complex is not detected in extra-follicular sites (by injection of FITC-MSH anti-MSH complex) we cannot rule out the possibility that antigen is still present outside follicles and is available to activate virgin B cells. The results in Figure 1 demonstrate that, in general, memory B cells carry higher affinity antigen receptors than virgin B cells and hence compete more effectively for epitopes on complexed antigen.

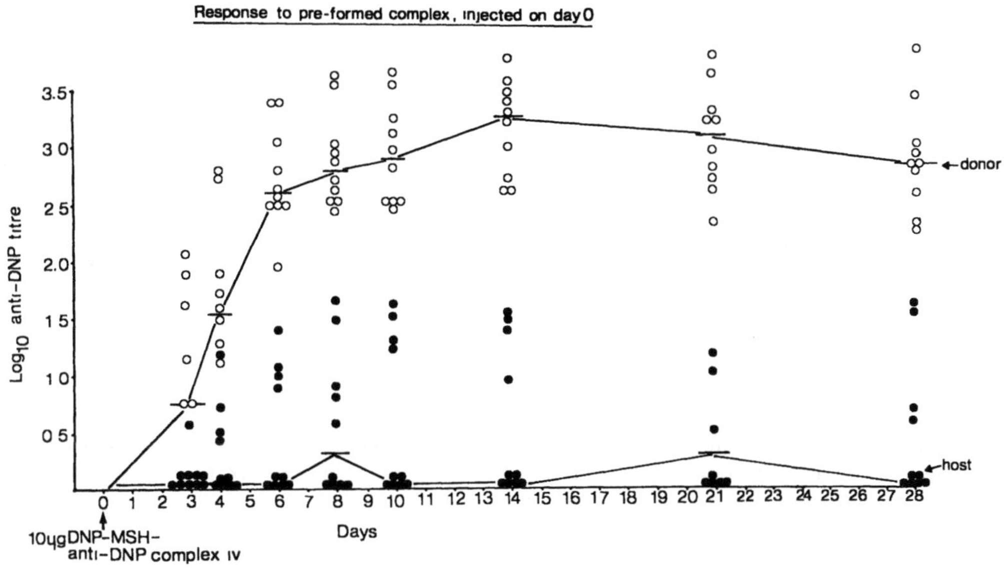

Figure 1. Donor and host-allotype anti-DNP response evoked by immune complex in which both hapten and carrier determinants may be masked: DNP-MSH anti-DNP-MSH. Each dot represents the serum from a single chimaeric rat (donor and host). Medians are plotted.

In a second experiment similar chimaeras (again receiving DNP-MSH primed donor cells) were immunised with immune complex containing OX-MSH anti-MSH. While this localizes onto FDC in follicles the oxazolone determinants are not masked. Again in this situation (Figure 2) donor B cells make the majority of the serum antibody; it is important to remember that in this experiment the donor cells are not actively primed memory B cells but are part of the mature recirculating pool of B cells. We will discuss later the role of antigen in selecting cells for incorporation into this pool. Activation of virgin (host) B cells is at a very low level during the first 10 days of the response. The difference in activation of host and donor cells cannot be explained by the relative affinities of receptors as neither host nor donor cells have been exposed to oxazolone previously.

It is possible that the number of OX-responsive B cells is higher in the donor population than in the host virgin B cell pool at the time of immunisation. To exclude this we have analysed the relative amounts of donor and host serum antibodies after chimaeras are immunised with non-complexed soluble antigen. Donors in this case were primed with ovalbumin (OVA) and chimaeras immunised iv. with 10 μg soluble OX-OVA. Figure 3 shows that comparable responses are elicited from both host and donor cells. The ratio of donor:host anti-OX titres is compared at day 7 and day 10 with the response to complexed oxazolone (Table 1).

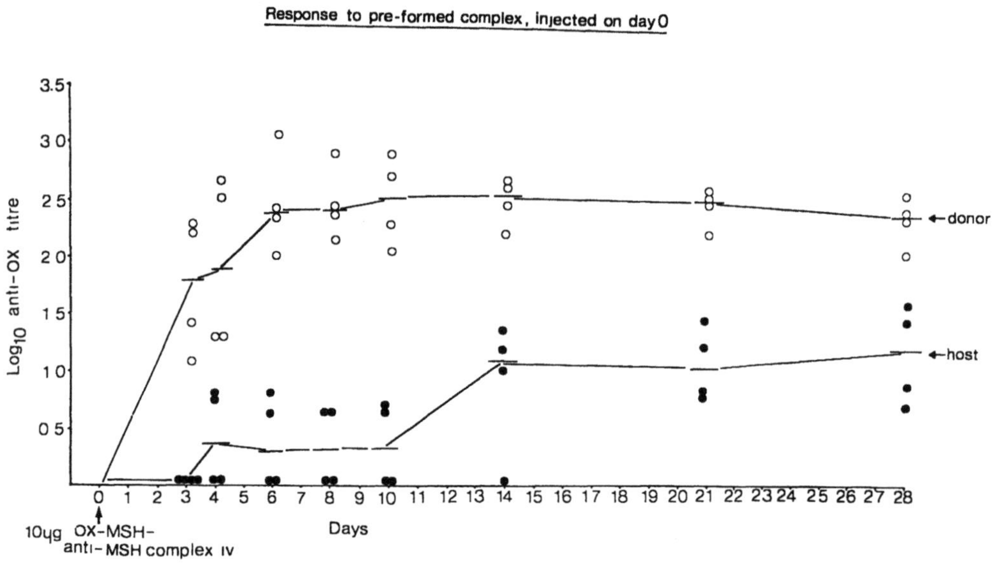

Figure 2. Donor and host-allotype anti-OX response evoked by immune complex in which hapten determinants are not masked: OX-MSH anti-MSH. See legend to Figure 1 for other details.

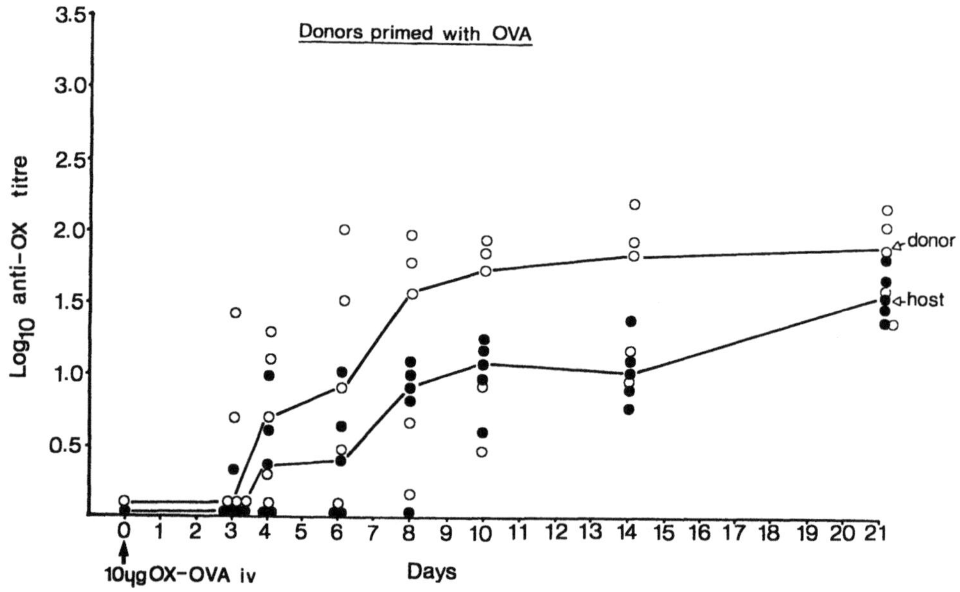

Figure 3. Donor and host-allotype anti-OX resonse evoked by iv. immunisation with soluble OX-OVA. See legend to Figure 1 for other details.

Table 1

Response to:	Ratio of donor:host anti-OX titre	
	day 7	day 10
OX-OVA (soluble) (Fig. 3)	3.9	4.5
OX-MSH anti-MSH (Fig. 2) (complex)	125.6	126.4

A 4:1 ratio can be explained in terms of the proportion of host cells in peripheral tissues at the time of immunisation as this varies from 10 to 30% of total B cells. A ratio of 125:1 cannot be explained in this way.

We are left with the conclusion that once antigen is complexed and becomes localised onto FDC it is no longer available for the activation of virgin B cells as they emerge into the peripheral lymphoid tissues from the marrow. If virgin B cells are stimulated by antigen before this re-localisation occurs then they become part of the recirculating pool and enter follicles. The expansion of virgin clones stimulated in extra-follicular sites may explain the rise in the host response to OX after 10 days (Figure 2). The response of unprimed donor cells to oxazolone (Figure 2) suggests that virgin B cells can be incorporated into the recirculating pool by means other than by specific activation by antigen (1,7): However, it is possible that interaction required for selection of new B cells is a low affinity one (i.e. the anti-OX response may be derived from B cells that were initially stimulated by an antigen that crossreacts with oxazolone).

CONCLUSION

Virgin B cells in the first few days after leaving the bone marrow are not efficiently activated by antigen-antibody complexes localised on FDC. In contrast, activation of virgin B cells does occur as soon as they emerge from the marrow if the antigen is not complexed and is in extra-follicular sites (1). This does not appear to be due to lower affinity of receptors or lower precursor frequency in virgin B cell populations in these chimaeras. Antigen in follicles is not available to virgin B cells because they do not migrate through these areas (4).

Memory B cells are very effectively activated by immune complexes localised on FDC. Activation of memory clones in these sites may continue for many months (1,6) and result in long-term antibody responses.

The stimulation of virgin B cells seems to occur in extra-follicular sites only early in a response and constitutes one mechanism by which B cells newly-emerged from the bone marrow are selected for incorporation into the mature, peripheral, recirculating pool.

REFERENCES

1. D Gray, ICM MacLennan and PJL Lane. Eur. J. Immunol. 16 641 (1986).
2. RG White, DC Henderson, MB Eslami and KH Nielsen. Immunology 28 1 (1975).
3. M Papamichail, C Gutierrez, P Embling, P Johnson, EJ Holborrow and MB Pepys. Scand. J. Immunol. 4 343 (1975).
4. JE Lortan, CA Roobottom, S Oldfield and ICM MacLennan. Eur. J. Immunol. (1987) in press.
5. PJL Lane, D Gray, S Oldfield and ICM MacLennan. Eur. J. Immunol. 16 1569 (1986).
6. ICM MacLennan and D Gray. Immunol. Rev. 91 61 (1986).
7. D Gray. Population of kinetics of rat peripheral B cells (submitted) (1987).

A SUBPOPULATION OF GERMINAL CENTER B CELLS DIFFERENTIATE DIRECTLY INTO ANTIBODY FORMING CELLS UPON SECONDARY IMMUNIZATION

John G. Tew, Greg F. Burton, Andras K. Szakal, and Marie H. Kosco

Department of Microbiology and Immunology, Medical College of Virginia, VCU, Richmond, VA 23298, USA

Department of Anatomy and Immunobiology, Medical College of Virginia, VCU, Richmond, VA 23298, USA

INTRODUCTION

Considerable evidence implicates germinal center (GC) B cells in the development of the B memory cell pool (1,2). However, the ability of GC B cells to differentiate directly into antibody forming cells (AFC) has not been well established although the idea that GC B cells may be involved in production of antibody is not new. Plasma cells near follicular dendritic cells (FDC) in the GCs have been observed (3) and electron microscopic (EM) studies have shown specific AFC in GCs after antigen challenge (4,5). The objective of the present study was to determine if GC B cells are capable of differentiating directly into AFCs. The data reported here confirm that: 1- Some GC B cells differentiate into AFCs in the GC. 2- By day five after secondary immunization large numbers of GC B cells appeared to migrate into the medullary cords where they may fully differentiate into plasma cells. 3- GC B cells isolated at various times after immunization produced specific antibody in culture in the absence of any *in vitro* stimulation by antigen or T cells. This was dramatic in GC B cells isolated 4 days after booster immunization and had declined by day 7. Examination of these cells indicated that many of the cultured cells differentiated into plasma cells in culture. We conclude that at least a subpopulation of GC B cells receive all the stimulation *in vivo* by antigen and T cells needed to fully differentiate into plasma cells.

MATERIALS AND METHODS

The methods used have been described previously: 1- Staining tissue sections with PNA (2). 2- Isolation of the PNA positive GC B cells (2,6). 3- Radioimmunoassay (7). 4- Detection of AFCs *in situ* using EM and horseradish peroxidase (HRP) as the antigen (4).

RESULTS AND DISCUSSION

In a recent study we followed the development of the PNA positive GC B cells over a ten day period after secondary challenge. We noted that at day 5 after secondary challenge large numbers of PNA positive cells were migrating from the GC into the medullary cords. The fact that large numbers of plasma cells accumulate in the medullary cords during the secondary response caused us to wonder whether the GC could be a site for initiation of plasma cell differentiation. Clearly very few of the GC cells were producing antibody in GC. However, we reasoned that GC B cells could be stimulated toward terminal differentiation in the GC and mature as they migrated into the medullary cords.

Support for the hypothesis that GC B cells differentiate into plasma cells included the following: We know that the GC B cells interact with antigen on FDC processes and on immune complex coated bodies (iccosomes) released by the FDCs (8). Binding of antigen by the immunoglobulin receptors delivers an important differentiative signal (9). The iccosomes released by the FDCs are endocytosed by the B cells and this antigen is processed by the GC B cells and is capable of being presented to T cells with great efficiency (6,8). It is known that T cells of the helper T cell phenotype are present in the GCs (10,11) and the GC B cells could serve as the antigen presenting cell. The appropriate helper T cells could then release the T cell growth and differentiation factors needed to convert GC B cells into plasma cells (9).

To assess whether GCs contain cells differentiated to the AFC stage, we actively immunized C3H mice with HRP. After secondary challenge with small quantities of HRP (8 ug/footpad), we observed the iccosome mediated delivery of antigen to GC B cells and its endocytosis by GC B cells between day 3 and 5 after antigen challenge. Previous work indicated that the day 5 GC B cells bind PNA and, when isolated by Percoll gradient separation followed by panning for PNA binding cells, these PNA+ GC B cells can present the *in vivo* obtained antigen to T cells *in vitro* with remarkable efficiency (6). We used immunoperoxidase techniques applied previously by others to look for AFCs in the period 4 to 7 days after HRP challenge (4,5). This technique allowed the localization of the anti-HRP antibodies within cells of the plasmacytic series. Briefly the lymph nodes were obtained after perfusion of mice with fixative. The popliteal lymph nodes were excised, further fixed, sectioned using a vibratome, incubated in a solution of HRP to allow the antibody being made in HRP specific AFCs to bind HRP, and the HRP labeling was developed. The sections were examined using a light microscope and the GC areas were cut out, osmicated, and processed for EM. The localization of GCs with EM was facilitated by the presence of HRP-anti-HRP complexes on the convoluted processes of FDCs. In the relative proximity of FDCs and deeper in the dark zone of GCs, some of the GC B cells were found to contain intracytoplasmic peroxidase positive material (due to the binding of HRP to anti-HRP) in the cisterna of the nuclear envelope and in developing rough endoplasmic reticula (Figure 1). These patterns of antibody localization were characteristic of the various stages of plasma cell differentiation (12). The series of AFCs observed in the GC started with the most primitive proplasmacytes producing only a small amount of antibody which was localized in the nuclear envelope and included cells that represented almost fully differentiated plasma cells.

In view of the cells in the GC already differentiated to the point where antibody formation was detected, and the apparent migration and differentiation of GC B cells, we reasoned that isolated GC B cells might be able to differentiate directly into plasma cells *in vitro*. If GC B cells have encountered antigen *in vivo* and the T cell growth and differentiation factors may have been produced by the T cells in the GC, the differentiative process should continue *in vitro*. We tested this by isolating GC B cells from ovalbumin (OVA) immune mice at various times using the combination of percoll gradient separation followed by PNA panning (6). The purified B cells were cultured in the absence of antigen and with and without a source of T cell growth and differentiation factors. The cells were cultured for five days and the supernatants were monitored for production of anti-OVA using an RIA. The results are tabulated below:

Table 1. Production of Anti-Ovalbumin by PNA+ GC B cell from Ovalbumin immune Mice *in vitro* after Secondary Immunization.

	Days in culture		
Culture Groups	Day 1 *Ab ng/ml*	Day 3 *Ab ng/ml*	Day 5 *Ab ng/ml*
Medium Alone D-4*	400±273@	570±40	715±60
Medium Alone D-7	27±1.1	45±3.8	56±1.4
Medium + Con A D-7 Supernatant	82±8	180±8	270±38

* GC B cells taken 4 days (D-4) after booster and D-7 means GC B cells taken 7 days after booster.
@ Mean of three cultures ± standard error of the mean.

Clearly the purified GC B cells taken 4 days after booster immunization could produce specific antibody in culture in the absence of antigen or T cell factors. Examination of the day 4 cells after five days in culture revealed that many cells had fully differentiated into large plasma cells. The spontaneous production was down considerably at day 7 after booster. A larger portion of the day 7 cells than the day 4 may to be destined to become B memory cell. However, the addition of T cell factors present

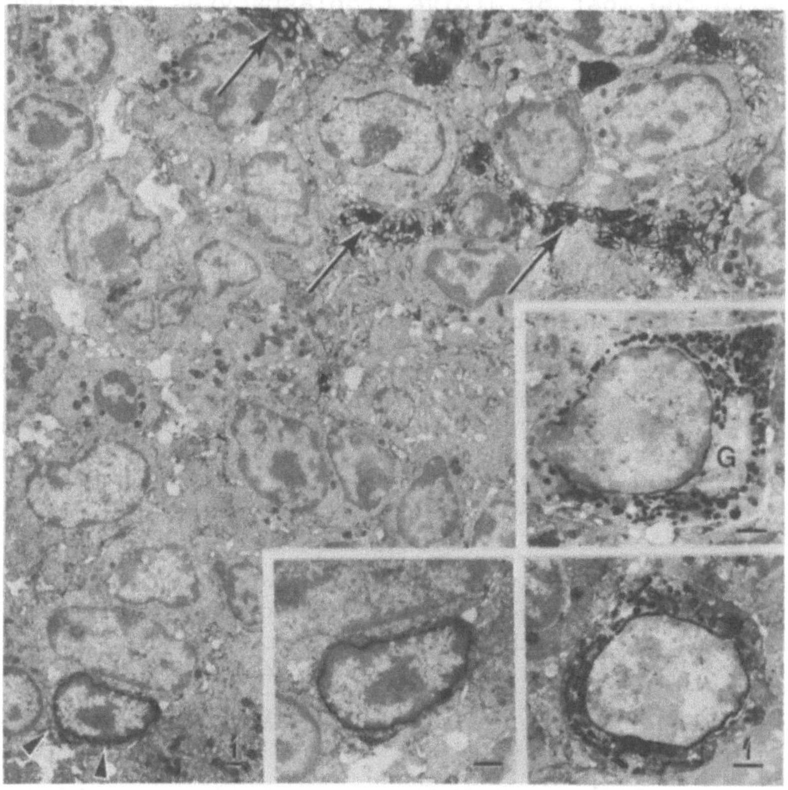

Figure 1. Electron micrographs showing immunoperoxidase localization of anti-HRP antibodies in germinal center cells of the popliteal lymph node 4 to 7 days after challenge in the hind footpad. Note: The arrows (top right) indicate the immune complex coated dendritic processes of FDCs. On bottom left arrow heads indicate the beginning differentiation of a cell of the plasmocyte series. This cell shows the presence of antibodies in its nuclear envelope. Inset a., same cell enlarged. Insets b. and c. show a proplasmocyte and a plasmocyte (G, golgi) respectively. Observe the increasing development of the rER as reflected by the immunoperoxidase localization of specific antibody (black). All of these cells were found in the popliteal lymph node germinal centers in the 4 to 7 day period after booster immunization. Horizontal bars represent the indicated multiple of micrometers.

in a Con A supernatant improved the anti-OVA response of the Day 7 cells considerably indicating the cells maintain the capacity to differentiate into antibody forming cells. A similar effect could be obtained using a supernatant from the EL-4 thymoma. A second significant feature of the response was that antibody production was apparent on day one and an induction period was not required. In contrast, induction of an *in vitro* secondary responses typically requires a two to three day lag period before antibody synthesis begins (13). These results support the idea that these GC cells have received the inductive signal by antigen and many of the T cell signals *in vivo* and are capable of producing antibody in culture immediately.

SUMMARY AND CONCLUSIONS

In the present study we sought to test the hypothesis that GC B cells may be stimulated to differentiate into AFCs during the induction of the secondary antibody response. To test this we gave HRP immune mice a booster immunization and looked for AFCs in the GCs making anti-HRP specific antibody. Within 3 days cells in the GCs were making anti-HRP. Most of the cells were just starting to make immunoglobulin in the rough endoplasmic reticulum but some were nearly mature plasma cells like those in the medullary cords. Furthermore, we isolated GC B cells after a boost with OVA and tested their ability to differentiate directly into AFCs. Data obtained using an OVA specific RIA indicated that purified GC B cells would differentiate and produce specific antibody in the absence of added T cells or antigen. Examination of the cells after 5 days in culture indicated that many of them had differentiated into large plasma cells. These results obtained both *in vivo* and *in vitro* demonstrate that shortly after booster immunization GC B cells receive the appropriate antigen and T cell signals *in vivo* to differentiate and become AFCs. (Supported by NIH Grant AI 17142).

ACKNOWLEDGEMENTS

This work was supported by NIH Grant # AI 17142.

REFERENCES

1. Klaus, G.G.B., Humphrey J.H., Kukkl A. and Dongworth D.W. The follicular dendritic cell: its role in antigen presentation in the generation of immunological memory. Immunol. Rev. 53 (1980) 3-28.

2. Coico, R.F., Bhogal, B.S. and Thorbecke, G.J. Relationship of germinal centers in lymphoid tissue to immunologic memory. VI, Transfer of B cell memory with lymph node cells fractionated according to their receptors for peanut agglutinin. J. Immunol. 131 (1983) 2254-2257.

3. Hanna, M.G., and Szakal, A.K. Localization of ^{125}I-labeled antigen in germinal centers of mouse spleen. Histological and ultrastructrual autoradiographic studies of the secondary immune reaction. J. Immunol. 101 (1968) 949-962.

4. Sordat, B., Sordat M., Hess M.W., Stoner R.D., Cottiere H. Specific antibody within lymphoid germinal center Cells of mice after primary immunization with horseradish peroxidase: A light and electron microscopic study. J. Exp Med. 131 (1970) 77-92.

5. Terashima, K. Imai, Y. Kasajima, T. Tsunoda R. Takahashi K. and Kojima M. An ultrastructural study on antibody production of the lymph nodes of rats with special references to the role of germinal centers. Acta Path. Jap. 27 (1977) 1-24.

6. Kosco, M.H., Szakal, A.K., and Tew, J.G. *In vivo* obtained antigen presented by GC B cell to T cells *in vitro*. J. Immunol (1987) (in press).

7. Tew, J.G., Marshall, D.R., Moore, W.E.C., Best, A.M., Palcanis, K.G., and Ranney, R.R. Serum antibody reactive with predominant organisms the subgingival flora of young adults with generalized severe periodontitis. Infection and Immunity 48 (1985) 303-311.

8. Szakal, A.K., Kosco, M.H., and Tew, J.G. A novel *in vivo* follicular dendritic cell dependent iccosome mediated mechanism for delivery of antigen to antigen processing cells. J. Immunol (1987) (in press).

9. Howard, M., and Paul, W.E. Regulation of B cell growth and differentiation by soluble factors. Ann Rev Immunol 1 (1983) 307-333.

10. Nieuwenhuis, P, and Opstelten, D. Functional anatomy of germinal centers. Am J. Anat. 170 (1984) 421-435.

11. Rouse, R.V., Ledbetter, J.A., and Weissman, I.L. Mouse lymph node germinal centers contain a selected subset of T cells - the helper phenotype. J. Immunol 128 (1982) 2243-2246.

12. Leduc, E.H., Avrames, S., and Bouteille, M. Ultrastructural localization of antibody in differentiating plasma cells. J. Exp. Med. 127 (1968) 109-117.

13. Tew, J.G. and T.E. Mandel. The maintenance and regulation of serum antibody levels: Evidence indicating a role for antigen retained in lymphoid follicles. J. Immunol. 120 (1978) 1063-1069.

SPECIFIC ANTIBODY FORMING CELLS IN LYMPHOID FOLLICLES OF SPLEEN: AN ASPECT OF THE LATE IMMUNE RESPONSE

N. van Rooijen and N. Kors

Immunocytochemistry Unit, Medical Faculty, Department of Histology, Free University, P.O. Box 7161, 1007 MC Amsterdam, The Netherlands

Previously we described the pathway followed by antigen-reactive B cells that are differentiating into antibody-forming cells during the early primary and secondary immune response in the spleen (1). B cells following this 'differentiation pathway' start in the marginal zone (MZ) and enter the outer periarteriolar lymphocyte sheath (PALS). There they migrate along the border with the MZ to the sites where small sheaths of lymphoid tissue around the terminal arterioles traverse the MZ. They follow these sheaths on their way to the red pulp, leave the spleen and enter the circulation. Follicles and their germinal centres, which are generally assumed to play a role in the generation of B-memory cells seemed not to be an obliged station on this route.

Nevertheless, specific antibody-forming cells have been detected in the follicles of lymph nodes (2, 3) and spleen (4, 5), although in widely differing numbers.

It was the aim of the present study to investigate the conditions required for the development of specific antibody-forming cells in follicles of the rabbit spleen. It appeared that specific antibody-forming cells develop in the follicles during the late primary and secondary immune response in the spleen.

MATERIALS AND METHODS

Experimental design

Young male chinchilla rabbits were intravenously injected with human serum albumin (HSA). At different time intervals after injection, a spleen biopsy was taken from the rabbits. One till two months after the primary injection a booster injection was given. At different time intervals hereafter, another spleen biopsy was taken from the rabbits. Specific antibody-forming cells against the antigen HSA were detected using an antigen-enzyme conjugate (HSA-HRP) followed by enzymehistochemistry.

HSA-HRP conjugate

Four mg HRP in 1 ml distilled water was oxidized by 0.2 ml 0.1 M sodium periodate (6) for 20 min. Five drops ethylene glycol were added and after 5 min the reaction mixture was filtered on Ultrogel AcA 202 (7). The activated HRP was concentrated to 1 ml and 4 mg HSA or HGG in 1 ml 0.15 M NaCl was added. Immediately the pH was raised to 9.0 - 9.5 by the addition of sodium carbonate buffer pH 9.5. After 2 h at room temperature the reaction was stabilized by the addition of 0.1 ml freshly prepared sodium borohydride (4 mg/ml distilled water). After an additional 2 h at 4°C the conjugate was purified on Ultrogel AcA 44 (8).

Detection of specific antibody producing cells
The method has been described in earlier studies (4, 5). In short, pieces of spleen were frozen in liquid nitrogen and afterwards cryostat sections of 8 µm thickness were cut. Sections were air-dried overnight. They were fixed for 10 min at 20°C with acetone (analytical grade) to which H_2O_2 was added (0.02% final concentration). They were rinsed 3x in PBS (pH 7.4). The sections were incubated overnight horizontally at 4°C with the HSA-HRP conjugate in PBS + 0.05% BSA. Afterwards they were washed with PBS and treated for 8 min with 3-amino-9-ethylcarbazole (AEC) + H_2O_2 (9). Sections were briefly counterstained with haematoxylin and mounted in glycerine-gelatine.

Antigen and chemicals
HSA was obtained from Miles Laboratories Inc. Bridgend, Glamorgan, South Wales, U.K. Horseradish peroxidase (HRP, Grad I) was obtained from Boehringer, Mannheim, Germany. 3-amino-9-ethylcarbazole (AEC) was obtained from K + K Labs, Plainville, N.Y., U.S.A.

RESULTS

The first anti-HSA antibody forming cells were found in the spleen of rabbits at about 4 days after a single intravenous injection with the antigen and at 2 days after a booster injection. Peak numbers were detected at about 1 week after a single injection and at 4 days after the booster. The bulk of these antibody forming cells, developing during the early primary and secondary anti-HSA response in the spleen localized in the coaxial sheaths of lymphoid tissue around the terminal arteriolar branches and in the outer part of the PALS, as we described in earlier studies (4, 5). After reaching their peak, numbers of antibody forming cells decreased rapidly. Relatively few antibody forming cells were found in the follicles of the spleen. However, at about 10-12 days after a single HSA injection, numbers of cells in the coaxial sheaths of lymphoid tissue around the terminal arteriolar branches and in the PALS had decreased so far, that most anti-HSA antibody forming cells in the spleen were found in the follicles. At 1 week after a single injection with HSA, more than 90% of the anti-HSA antibody forming cells found in a spleen biopsy of rabbits appeared to be localized in the coaxial sheaths of lymphoid tissue around the terminal arteriolar branches and in the PALS, whereas in a second spleen biopsy of the same rabbits, taken at two weeks after injection, most antibody forming cells (more than 50%) were detected in the follicles. Many more anti-HSA antibody forming cells localized in the follicles during a secondary immune response as compared to a primary response. At 2 and 4 days after a booster injection with HSA, the bulk of anti-HSA antibody forming cells were found in the coaxial sheaths of lymphoid tissue around the terminal arterioles and in the PALS, whereas the follicles contained the majority of antibody forming cells in most rabbits at 8 days after booster.

In general anti-HSA antibody forming cells were found in the follicles of rabbits mainly during the late primary and secondary immune responses. In such animals, however, follicles contained widely different numbers of antibody forming cells. Whereas some follicles contained hundreds of antibody forming cells (as counted in serial sections, Fig. 1) most follicles contained only few cells.

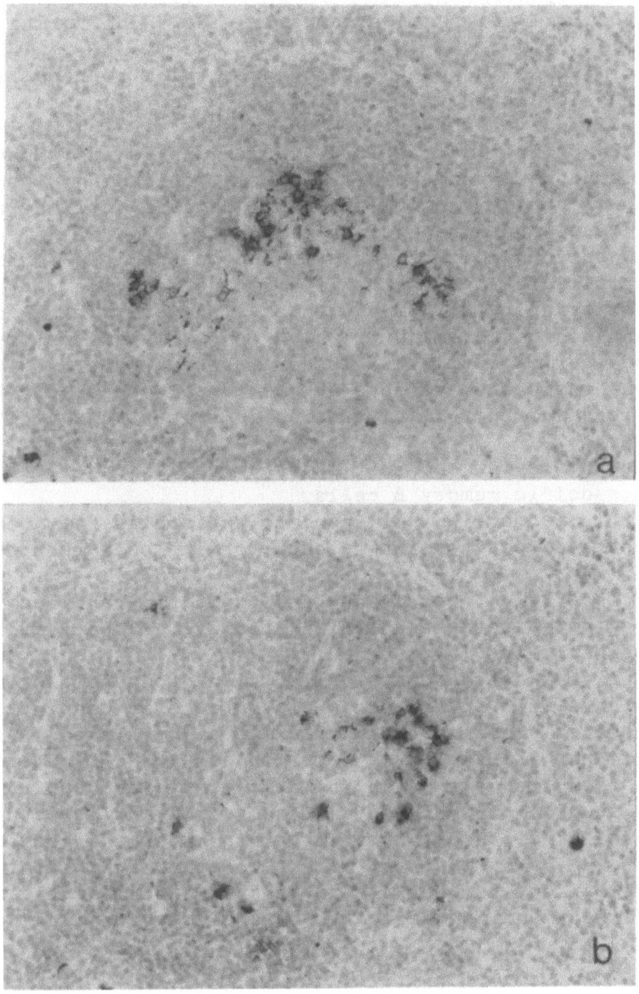

Fig. 1. Anti-HSA antibody producing cells in the spleen of a rabbit at 2 weeks after a single injection of 10 mg HSA. Antibody producing cells are shown in a follicle centre; 1a and 1b: serial sections of the same follicle.

DISCUSSION

It is concluded from the present results that specific antibody forming cells develop in the follicles during the late primary and secondary immune response in the spleen. Until now this phenomenon has escaped to our attention for three reasons:

1. We were primarily interested in the early phase of the development of specific antibody forming cells 'in situ' (1).
2. Antibody forming cells were found in widely differing numbers per follicle.
3. Antibody forming cells against the hapten trinitrophenyl (TNP) were not found in the follicles of mice (10) and rats (11) after injection of TNP-keyhole limpet haemocyanin (KLH).

The presence of specific antibody forming cells in lymphoid follicles in the late immune response was particularly intriguing, when one or two follicles showed a heavy accumulation of specific antibody forming cells in a, for the rest, unstained spleen section.

It may be speculated that the presence of specific antibody forming cells in the follicles during the late immune response corresponds with the follicular trapping of HSA-anti-HSA complexes. These complexes may have free antigenic determinants which can stimulate memory B cells to become antibody forming cells. Three important events in the later part of the immune response have been suggested to depend on this antibody complexed antigen in the follicles which is protected against degradation by macrophages:

1. The generation (review see ref. 12) and regulation (review see ref. 13) of antigen specific memory B cells.
2. The cyclical production of specific antibodies as demonstrated by clearly separated peaks in the plaque forming cell response (review see ref. 14).
3. The induction of a spontaneous and specific antibody response in animals long after antigen administration (e.g. one year) by removal of serum immunoglobulin (15).

Since other explanations have failed, antibody complexed antigens in the follicles have been suggested to play a role in the initiation of each of these phenomena, which are probably mutually related. Although it is known that neither all antigenic determinants nor all antibody combining sites of the immune complexes in the follicles are occupied (see review 13), detailed studies on the free and accessible antigenic determinants of these complexes at different time intervals after antigen administration have never been performed. At moment we are planning studies on the possible correlation between free antigenic determinants of follicular immune complexes, and specificity of antibody forming cells in the follicles.

REFERENCES

1. N. van Rooijen, E. Claassen, and P. Eikelenboom. Immunol. Today 7:193 (1986).
2. B. Sordat, M. Sordat, M.W. Hess, R.D. Stoner, and H. Cottier. J. Exp. Med. 131:77 (1970).
3. N. van Rooijen, R. van Nieuwmegen, and N. Kors. Cell Tissue Res. 239:657 (1985).
4. N. van Rooijen, N. Kors, R. van Nieuwmegen, and P. Eikelenboom. Anat. Rec. 206:189 (1983).

5. N. van Rooijen, R. van Nieuwmegen, N. Kors, and P. Eikelenboom. Anat. Rec. 208:579 (1984).
6. M.B. Wilson, and P.K. Nakane, in: 'Immunofluorescence and related staining techniques', p. 215, W. Knapp, K. Holubar, and G. Wick, eds., Elsevier, Amsterdam (1978).
7. D.M. Boorsma, and J.G. Streefkerk. J. Immunol. Meth. 30:245 (1979).
8. D.M. Boorsma, and J.G. Streefkerk. J. Histochem. Cytochem. 24:481 (1976).
9. R.C. Graham, U. Lundholm, and N.J. Karnovsky. J. Histochem. Cytochem. 13:150 (1965).
10. E. Claassen, N. Kors, and N. van Rooijen. Eur. J. Immunol. 16:271 (1986).
11. E. Claassen, N. Kors, C.D. Dijkstra, and N. van Rooijen. Immunology 57:399 (1986).
12. G.G.B. Klaus, J.H. Humphrey, A. Kunkl and D.W. Dongworth. Immunol. Rev. 53:3 (1980).
13. N. van Rooijen, in: 'Phylogeny of Immunological Memory', p. 281, M.J. Manning, ed., Elsevier, Amsterdam (1980).
14. W.O. Weigle. Adv. Immunol. 21:87 (1975).
15. S.L. Donaldson, M.H. Kosco, A.K. Szakal, and J.G. Tew. J. Leuk. Biol. 40:147 (1986).

ENRICHMENT OF GERMINAL CENTRE CELLS

A.N. Brown and H.N.A. Willcox

Department of Neurological Science
Royal Free Hospital School of Medicine
London NW3 2PF

INTRODUCTION

Culture and immunohistological results from the thymus of human myasthenia gravis (MG) patients[1] suggest that the small 'clumps' of cells in conventional suspensions (CONV) prepared by mechanical disruption may be very rich in germinal centres (GC). If these 'clumps' are rescued and dispersed with the protease dispase (referred to as D/C suspensions), they spontaneously synthesise about 4 times more autoantibody to the acetylcholine receptor (AChR) than CONV cultures. This was also sometimes the case with total IgG. It was therefore proposed that the D/C technique may be a convenient method for the isolation and enrichment of GC cells.

Studies have been conducted to confirm the above inferences, and to test whether they apply in more typical immune responses, mainly using spleens from mice immunised with a variety of antigens. In particular, we have tested for antigen-specific B memory cells.

METHODS

Thymus studies and Culture techniques

These were performed as previously described[1,2], except that dispase was used at 0.5-0.8 mg/ml.

Immunisation of mice

The majority of work was conducted with male mice (2-3 months old) of BALB/C, CBA, C57BL/6 and (CBA x C57BL/6)F_1 strains which were immunised at fortnightly intervals with 2.10^8 sheep red blood cells (SRBC) intravenously (i/v), or 1µg purified AChR intrasplenically (i/s) or intraperitoneally (i/p; occasionally with 2.10^9 B. pertussis for priming).

Adoptive transfers

Spleen cells (5.10^5 - 20.10^6) from donors last boosted 7 - 76 days in advance were transferred i/v or i/s to 500 - 600r irradiated syngeneic recipients on day 0; these were then boosted on day 1. Either their spleens were assayed for anti-SRBC plaque-forming cells (PFC) on day 6, or they were bled commencing on day 7 or 14 at weekly intervals for 3 weeks, and assayed for serum anti-AChR antibodies using a sensitive radioimmunoassay[3].

RESULTS AND DISCUSSION

Studies on human MG thymus cells and 'normal' tonsil

Results of thymus cell cultures from 30 further MG patients have been combined with the previous ones[1] (Table 1) and confirm the reported increase in spontaneously produced autoantibody and total IgG by D/C cultures compared with CONV.

Results of analogous cultures of 'normal' human tonsil cells (measuring total IgG production) have proved inconsistent within the 7 cases studied.

Table 1 : Spontaneous production in culture of Anti-AChR and total IgG by D/C as compared with CONV cell suspensions

	ANTI-AChR		TOTAL IgG	
	no. of cases / total	mean ratio (s.e.)	no. of cases / total	mean ratio (s.e.)
D/C / CONV ratio $\geqslant$ 1.5	39/52	4.2 (0.6)	20/28	5.5 (1.0)
Clear exceptions	13/52	0.9 (0.1)	8/28	0.9 (0.1)

Murine cell culture and PFC studies

Spontaneous production of total IgG in culture by D/C separated spleens has regularly been at least double that of CONV suspensions from mice immunised with such antigens as SRBC, AChR, KLH, and LPS by several different regimes.

Considering specific antibody, in 7 of 8 experiments, spleen cells from mice multiply immunised with AChR showed a strikingly similar pattern to those given by autoimmune patients' thymus cells above. A typical example is shown in Figure 1, where at optimal cell concentrations, D/C cells produced 8.5 times more anti-AChR (per 10^5 cells) than CONV.

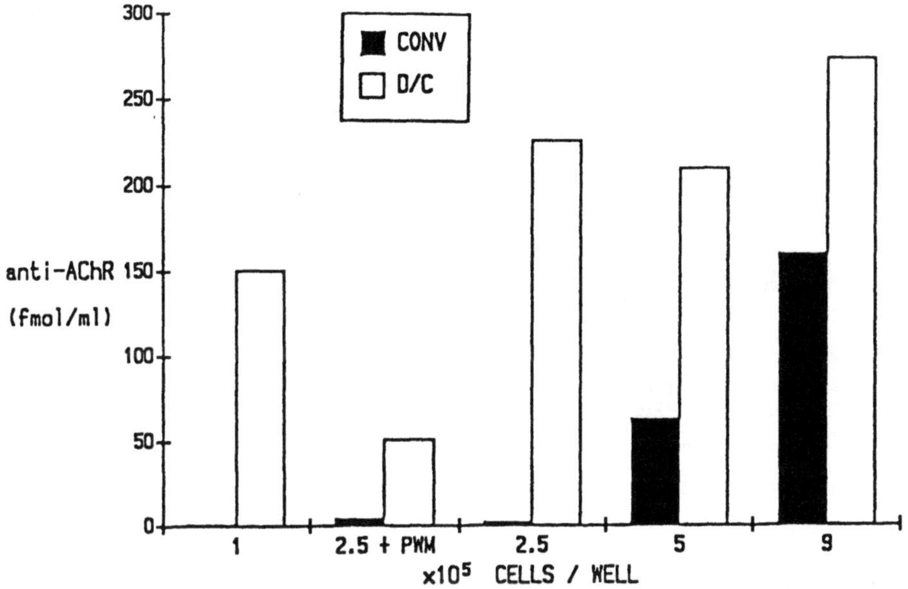

FIGURE 1. Spontaneous production of anti-AChR in culture by mouse splenocytes. Here donors received 6 immunisations, and culturing was on day 7 post final boost; in the other experiments immunisations ranged from 5-10, and culturing was on day 7-12 post final boost. The one exception to the above trend occurred when the interval between the penultimate and final boost was 14 days; in the other experiments this ranged from 20-254 days. (PWM = Pokeweed mitogen).

In the original thymus studies[1,2] much of the spontaneous anti-AChR antibody production survived 1,250 rads of irradiation; preliminary experiments have indicated that the same may be true with murine cells (not shown), suggesting an enrichment of plasma cells which are known to be radioresistant[4] but are generally rare in GC[5]. More adherent cells have consistently been observed in D/C than CONV cultures, which may have 'feeder' effects[2], but the possibility of increased numbers of follicular dendritic cells (FDC), whose isolation is helped by dispase[6], has not yet been investigated.

Anti-SRBC PFC assays conducted immediately following splenic dispersion have consistently failed to show a difference between D/C and CONV separated cells. Perhaps plasma cells are not as important here as in the anti-AChR culture responses, eg. because of feedback suppression by specific antibody.

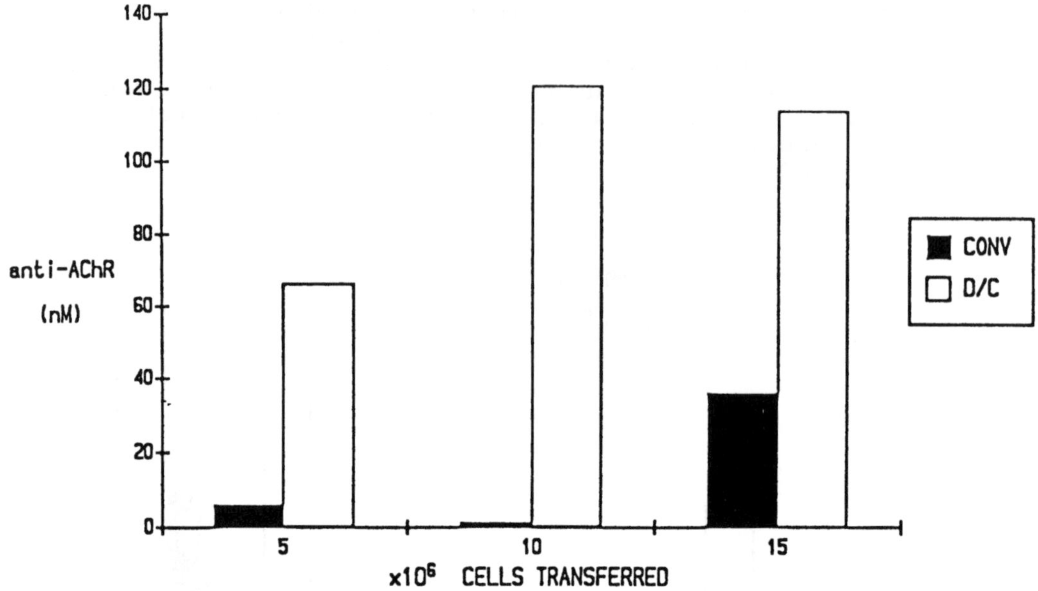

FIGURE 2(a). Adoptive transfer of AChR-primed cells (Exceptional response). (i/v transfer). Here donors received 6 immunisations, and transfer was 71 days after the final one. Comparing optimal cell transfer doses, D/C produced 5.5 times more serum anti-AChR (per 10^6 cells transferred) than CONV. This was observed in 1 of 6 transfers.

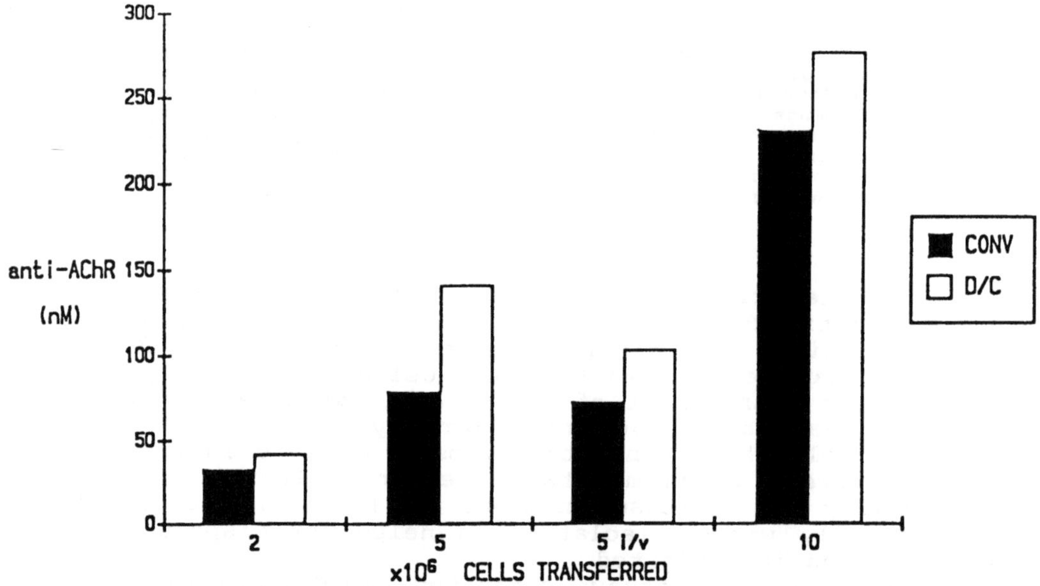

FIGURE 2(b). Adoptive transfer of AChR-primed cells (Typical response). (i/s and i/v transfer). Here donors received 5 immunisations, and transfer was 13 days after the final one; in the other typical experiments immunisations ranged from 4-5, and transfer was on day 7-76 post final boost. Comparing optimal cell transfer doses for this experiment, D/C produced only 1.2 times more serum anti-AChR (per 10^6 cells transferred) than CONV. This trend was observed in 5 of 6 transfers.

Adoptive transfer experiments

To test whether D/C cells are especially rich in antigen-specific B memory cells (which are thought to be generated in the GC[7,8]) AChR- or SRBC-immune D/C and CONV spleen cells were assayed for memory by adoptive transfer to syngeneic irradiated recipients. Intrasplenic transfer was preferred in view of the evidence for lack of homing receptors on GC B cells, and their preferential uptake by the liver[9].

In only 1 of 6 AChR transfer experiments (Figure 2(a)) did there appear to be even a two-fold increase in serum anti-AChR of D/C recipients compared with CONV recipients; in the other 5, differences were absent or minimal as in Figure 2(b).

Similar trends were shown in the transfers from SRBC-immune mice, with 7 experiments indicating little difference between D/C and CONV. In only 1 unreproducible preliminary experiment was even a two-fold increase achieved by D/C cells.

CONCLUSIONS

1. The increased antibody levels found in D/C over CONV cultures are not just a peculiarity of thymus cell cultures from human myasthenics, but occur with spleens of immunised mice.

2. This increase was striking after multiple immunisation, a possible analogy with autoimmune diseases where continuous boosting may also occur. Evidently antibody forming cells then become less accessible, and their full potential may be underestimated if no enzymes are used to disperse them.

3. The adoptive transfer experiments have not shown a consistent enrichment of B memory in the D/C cell preparations, which may imply that there is no GC enrichment. Even after intrasplenic injection, however, the cells might not settle in a favourable microenvironment (eg. if they lack homing receptors[9]), and analogous experiments to assay for memory in culture are in progress as a further test.

4. In the original thymus studies[1,2] much of the spontaneous anti-AChR production survived 1,250 rads of irradiation, and the same may be true with murine cells. This could indicate enrichment of plasma cells which are known to be radioresistant[4] but are generally rare in GC[5]. Enrichment of feeder cells and possibly FDC in D/C cultures might also contribute to raised antibody levels and a better approximation of *in vivo* conditions.

ACKNOWLEDGEMENTS

This work was supported by an SERC-CASE grant (jointly with Lilly Research Centre LTD., Windlesham) , an MRC programme grant , and the Welton Trust. We thank Drs A. Jermy and A.Vincent for help and advice, and Prof J. Newsom-Davis for access to MG material.

REFERENCES

1. N. Willcox, M. Schluep, M. Bofill and J. Newsom-Davis. Adv. Exp. Med. Biol. 186:979 (1985).

2. H.N.A. Willcox, J. Newsom-Davis and L.R. Calder. Clin. Exp. Immunol. 54:378 (1983).

3. G.K. Scadding, A. Vincent, J. Newsom-Davis and K. Henry. Neurology. 31:935 (1981).

4. T. Makinodan, P. Nettesheim, T. Morita and C.J. Chadwick. J. Cell.Physiol. 69:355 (1967).

5. H. Stein, J. Gerdes and D.Y. Mason. Clinics in Haematology 11 (3):531 (1982).

6. J.H. Humphrey and D. Grennan. In In vivo immunology (Ed by P. Nieuwenhuis, A.A. van den Broek and M.G. Hanna) pp 823-837, Plenum Publishing (1982).

7. G.G.B. Klaus, J.H. Humphrey, A. Kunkl and J. Dongworth. Immunol. Rev. 53:3 (1981).

8. G. Kraal, I.L. Weissman and E.C.Butcher. Nature 298:377 (1982).

9. R.A. Reichert, W.M. Gallatin, I.L. Weissman and E.C. Butcher. J. Exp. Med. 157(3):813 (1983).

THE GERMINAL CENTER PRECURSOR CELL IS SURFACE μ AND δ POSITIVE

H.G.Seijen, J.C.A.M.Bun, A.S.Wubbena, and
K.G. Löhlefink

Department of Histology, University of Groningen
Groningen, The Netherlands

INTRODUCTION

Primary follicles in peripheral lymphoid organs consist of large aggregates of B cells, which express both IgM and IgD on their surface (1). Upon antigenic stimulation germinal centers (GC) arise within the follicles; these GC are thought to be instrumental in the production of B memory cells (2). The cells constituting the germinal centers express only small amounts of surface IgM and sometimes of other Ig classes (3), but not of IgD (1). The question then arises whether the $sIgM^{+}$-$sIgD^{+}$ cells of primary follicles indeed are the precursor cells of germinal centers.

Treatment of rats from birth with anti-μ antibodies prevents the development of GC (4); however, mice treated from birth with anti-δ antibodies still present GC, although $sIgM^{+}$-$sIgD^{+}$ cells are virtually absent (5,6,7). These experiments, thus, strongly suggest a role for sIgM, but not for sIgD on germinal center precursor cells (GCPC) in becoming involved in GC reactions.

Treating animals with anti-μ or anti-δ antibodies interferes with the normal development of the cells populating the lymphoid tissues. We, therefore, chose to investigate the nature of the Ig receptors of GCPC as they are present in normal rats.

Thoracic duct lymphocytes (TDL) have been found to give rise to GC when transferred to lethally X-irradiated rats and challenged with SRBC (8). The B cells in these GC are $His14^{+}$, $His22^{-}$, and $sIgD^{-}$ as in GC of normal, non-irradiated rats (9). Thus, GCPC among TDL seem to react in this assay in a normal way. Separation of TDL by panning into $sIgM^{+}$ and $sIgM^{-}$ cells and into $sIgD^{+}$ and $sIgD^{-}$ cells provided a system for finding out which of these subsets the GCPC of normal rats belong to.

MATERIALS AND METHODS

Rats. Female AO/G rats were used when about 13 weeks old for providing TDL and, after 8.1 Gy total body X-irradiation, as recipients for cell suspensions.

Experimental system. 24 h after X-irradiation rats were i.v. injected with 6 x 10^7 first overnight collected TDL, or with subpopulations of TDL in numbers as present in 6 x 10^7 TDL. Simultaneously the rats were immunized with 10^9 SRBC. Spleens were taken 7 d later to assess the GC forming capacity of the transferred cells. Controls did not receive cells.

Germinal center forming capacity. Spleens were cut into 10 transverse slices, fixed, and embedded in paraffin. After sectioning and staining with methylgreen and pyronin the number of GC profiles of one section from each of the 10 slices was counted and their average (N) calculated.

Cell suspensions. TDL were fractionated into subpopulations by standard panning procedures: sIg^+-rich and sIg^--rich cell fractions were obtained with anti-rat IgG, $s\mu^+$-rich and $s\mu^-$-rich cell fractions with anti- rat µ (MARM4), and $s\delta^+$-rich and $s\delta^-$-rich cell fractions with anti- rat δ (MARD3).

Characterization of cell suspensions. Cell suspensions were incubated with anti-IgG, MARM4, or MARD3 and subsequently with appropriate fluorescent second antibodies.

Statistical methods. Two factor analysis of variance with interaction was performed to test for cooperation between cell suspensions (10).

RESULTS

The germinal center formation assay

Rats, treated with 8.1 Gy total body X-irradiation, responded hardly or not at all with the formation of GC in their spleens when stimulated with SRBC 1 day later (Fig.1). Thus, treatment with X-rays is quite appropriate to destroy the animal's own capacity to react to SRBC with the formation of GC. In contrast, rats, reconstituted 1 day after X-irradiation with 6 x 10^7 TDL and simultaneously challenged with SRBC, reacted vividly. The average number of GC profiles per spleen section of these animals was N = 3.1 + .5 (n = 3) 7 days after challenge.

Reconstitution experiments using sIg^+ and sIg^- cells

TDL were separated into sIg^+-rich and sIg^--rich cell fractions by panning with anti-rat IgG antibodies. Both fractions were given to irradiated rats separately and after recombination, in numbers as present in 6 x 10^7 TDL. Single fractions did not give rise to GC; however, in combination they reacted as well as TDL (Fig.1a).

Reconstitution experiments using sµ⁺- and sδ⁺- cells instead of sIg⁺-cells

sµ⁺-rich cell fractions and sδ⁺-rich cell fractions were prepared by panning with anti-rat µ and anti-rat δ antibodies, respectively. Replacing the sIg⁺-rich cell fraction in the previous experiments by these sµ⁺- and sδ⁺-rich cell fractions, respectively, did not change the outcome. Again, the sµ⁺- as well as the sδ⁺-rich cell fraction were virtually without effect, whereas combination of both of these fractions with sIg⁻ cells resulted in numbers of GC profiles as given by TDL (Fig.1b,c).

a

$\bar{N}$	sIg⁺: 0	sIg⁺: equiv $6x10^7$
sIg⁻: 0	.0 .1	.1 .5
sIg⁻: equiv $6x10^7$	.3 .6	2.8 3.4

b

$\bar{N}$	sµ⁺: 0	sµ⁺: equiv $6x10^7$
sIg⁻: 0	.0 .0	.1 .2
sIg⁻: equiv $6x10^7$	.1 .4	2.3 3.1

c

$\bar{N}$	sδ⁺: 0	sδ⁺: equiv $6x10^7$
sIg⁻: 0	.0 .0 .0	.0 .1 .2
sIg⁻: equiv $6x10^7$	.0 .3 .5	2.1 3.1 3.7

Fig.1. GC forming capacity of sIg⁺, sµ⁺, and sδ⁺ cells in cooperation with sIg⁻ cells.

Cell fractions of TDL, obtained by panning, together with SRBC were given to lethally irradiated rats in numbers as indicated (equiv 6 x 10^7: number of cells as present in TDL) to determine their GC forming capacities. N, the mean number of GC profiles per spleen section refers to the GC forming capacity of individual rats 7 d later. a) sIg⁺: sIg⁺-rich fraction (94 %sIg⁺); sIg⁻: sIg⁻-rich fraction (1 %sIg⁺) b) sµ⁺: sµ⁺-rich fraction (87 %su⁺); sIg⁻: sIg⁻- rich fraction (1 %sIg⁺) c) sδ⁺: sδ⁺-rich fraction (94 %sδ⁺); sIg⁻: sIg⁻-rich fraction (2 %sIg⁺). Tests for cooperation gave $p<.005$, $p<.01$, and $p<.01$, respectively.

			%	$\bar{N}$
equiv $6x10^7$	$s\mu^-$	$s\mu^+$	3	.0
				.0
		sIg^+	9	.2
	$s\delta^-$	$s\delta^+$	.6	.0
				.0
		sIg^+	5	.1

Fig.2. GC forming capacity of $s\mu^-$ and $s\delta^-$ cells. See legend to Fig.1. %: remaining immunoglobulin positive cells in depleted fractions.

		$\%sIg^+$	$\bar{N}$
equiv $6x10^7$	sIg^-	1	.2
			.4
			.9
	sIg^-	5^x	2.1
			2.3
			2.4

Fig.3. GC forming capacity of sIg^- cells, and sIg^- cells suppleted with sIg^+ cells. See legend to Fig.1. x: sIg^- cell fraction suppleted with sIg^+ cells.

Reconstitution experiments using $s\mu^-$ and $s\delta^-$ cells

Along with the $s\mu^+$- and $s\delta^+$-rich cell fractions $s\mu^-$- and $s\delta^-$-rich cell fractions were obtained. Both these fractions, although nearly devoid of $s\mu^+$ and $s\delta^+$ cells, respectively, contained appreciable numbers of sIg^+ cells. However, neither of these cell fractions had GC forming capacity (Fig.2). In contrast, sIg^- cells, deliberately contaminated with sIg^+ cells in numbers approximately as present in the $s\mu^-$- and $s\delta^-$-rich cell fractions, reacted almost like TDL (Fig.3).

DISCUSSION

Histological examination of lymphoid organs suggests that GCPC bear on their surfaces IgM as well as IgD: amidst clusters of such cells, i.e. primary follicles, GC arise upon antigenic stimulation. Our experiments clearly indicate that GCPC present in TDL indeed are sIgM as well as sIgD positive. Moreover, we found that cells, which do not bear both these immunoglobulins do not respond to antigenic stimulation with the formation of GC.

The cell fractions depleted of $sIgM^+$ or $sIgD^+$ cells may not be inherently unresponsive. Firstly, too few Ig bearing cells may be left. However, sIg^+ cells, bearing both IgM and

IgD, in comparable numbers react well (Fig.3). Secondly, cells may be damaged during isolation by panning. This seems unlikely, as the depleted fractions contain nonadherent cells. Moreover, when these fractions were combined with their respective adherent fractions their GC forming capacities were restored (results not shown). Thirdly, the sIg$^+$ cells of the depleted cell fractions may not reach the spleen. However, staining spleen sections with HIS14, an anti-rat pan-B cell monoclonal antibody (11), revealed many positive cells scattered in the white pulp (results not shown).

Our results pertain to GCPC of normal, untreated rats. Rats, treated from birth with anti-μ antibodies do not present GC, but rats treated from birth with anti-δ antibodies, although lacking sIgD$^+$ cells, nevertheless do have GC. It is puzzling that in normal rats cells lacking sIgD do not behave as GCPC, whereas such cells in anti-δ treated rats do. As yet, the relationship between sIgD$^+$ and GCPC therefore remains unsettled.

ACKNOWLEDGEMENTS

The mouse monoclonal antibodies MARM4 and MARD3 were kind gifts of Dr.H.Bazin.

REFERENCES

1. H. Bazin, D. Gray, B. Platteau, and I. C. M. MacLennan, Ann. N. Y. Acad. Sci. 399:157 (1982).
2. G. J. Thorbecke, T. J. Romano, and S. P. Lerman, Progr. Immunol. II:3 (1974).
3. G. Kraal, I. L. Weissman, and E. C. Butcher, Nature 298:377 (1982).
4. D. S. Kumararatne, H. Bazin, and I. C. M. MacLennan, Eur. J. Immunol. 11:858 (1981).
5. E. B. Jacobson, Y. Baine, Y.-W Chen, T. Flotte, M. J. O'Neill, B. Pernis, G. W. Siskind, G. J.Thorbecke, and P. Tonda, J. Exp. Med. 154:318 (1981).
6. G. J. Thorbecke, T. J. Flotte, and Y. Baine, Adv. Exp. Med. Biol. 149:845 (1982).
7. F. D. Finkelman, J. J. Mond, and E. S. Metcalf, J. Immunol. 131:593 (1983).
8. N. A. Gastkemper, A. S. Wubbena, F. J. H. Gimbrère, A. de Graaff, and P.Nieuwenhuis, Cell Tissue Res. 219:281 (1981).
9. F. G. M. Kroese, A. S. Wubbena, and P. Nieuwenhuis, Immunology 57:99 (1986).
10. J. Neter, W. Wasserman, and M. H. Kutner. Applied Linear Statistical Models. Richard D.Irwin, Inc. Sec. Ed., Chapter 20 (1985).
11. F. G. M. Kroese, A. S. Wubbena, D. Opstelten, G. J. Deenen, E. H. Schwander, L. de Leij, H. Vos, S. Poppema, J. Volberda, and P. Nieuwenhuis. Eur.J.Immunol. 17:921 (1987).

SURFACE IgD PHENOTYPE OF RAT GERMINAL CENTRE PRECURSOR CELLS

R.H. Vonderheide and S.V. Hunt

Sir William Dunn School of Pathology
University of Oxford, England

INTRODUCTION

Germinal centre B cells in lymph nodes differ from most other lymph node B cells by their lack of surface IgD (1). The question arises whether germinal centre precursor cells (GCPC) are also sIgD-negative or whether they express sIgD but lose it as they initiate or otherwise contribute to GC formation. We investigated this question by employing the monoclonal antibody HIS22, which labels all B cells in rat lymph nodes except those in GCs (2). HIS22 and anti-kappa antibodies were detected in cell suspensions by separate fluorochromes so that germinal centre B cells ($HIS22^-$ $kappa^+$) were distinguished from non-GC B cells ($HIS22^+$ $kappa^+$). Acute, non-irradiation chimeras were established by injecting B cells from TDL of one kappa allotype into host rats whose B cells expressed the opposite allotype. The relative tendency for donor cells to localise in GCs as opposed to non-GC B cell areas was measured by two-colour flow cytometry of lymph nodes sampled during a primary immune response. By using donor TDL depleted of $sIgD^+$ cells, the delta heavy chain surface phenotype of GCPC was investigated.

METHODS AND MATERIALS

Animals used in this study were SPF PVG-Ig-kappa-1a and congenic PVG-Ig-kappa-1b rats, bred and maintained in the MRC Cellular Immunology Unit, Oxford, England. Young adult males were immunised to elicit a GC reaction by injecting 2.0mg of soluble DNP-HSA s.c. in the region drained by the cervical lymph nodes.

Monoclonal antibodies used are shown in Table 1, all at saturating concentrations. Fitc-conjugated $F(ab')_2$ anti-kappa-1b was prepared by one of us (3) from immunoabsorbed alloantiserum. Blocked anti-kappa-1b was prepared by incubating the antibody with a five-fold excess of PVG-Ig-kappa-1b Fab. HIS22 was conjugated to biotin by incubating the antibody with biotin ester in a sodium bicarbonate buffer.

Rats were cannulated by standard procedures (4), and TDL were collected overnight. For rosette-depletion of $sIgD^+$ cells, donor TDL were incubated with MARD-3 (anti-delta mAb) and then mixed with SRBC coated with sheep-anti-mouse Ig antibody; rosettes were separated by indirect immunogravitation.

Table 1. Monoclonal antibodies

Antibody	Specificity	Concentration	Source
HIS22	anti-B cell subpopulation	30ug/ml	F.G.M. Kroese
HIS40	anti-μ	1/100 ascites	D. Opstelten
MRC OX11	anti-kappa-1a	1/10 TCS	
MRC OX14+16	anti-γ_2	OX14 20ug/ml OX16 10ug/ml	
R2/15S	anti-non-PVG MHC (control)	1/10 TCS	B. Roser
MARD-3	anti-δ	1/200 ascites	Serotec, U.K.

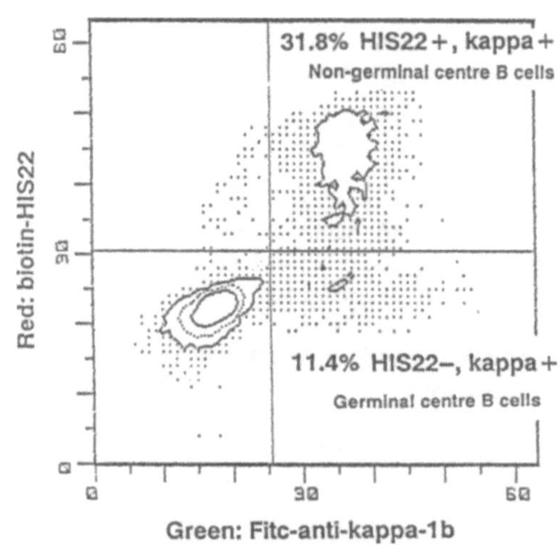

Fig. 1: FACS-generated dot plot of a lymph node cell suspension stained with HIS22 and anti-kappa-1b. Germinal centre B cells are distinguished as HIS22$^-$ kappa$^+$.

Table 2. Immunolabelling protocol for two-colour flow cytometry

Sample	1st layer	2nd layer
1	Fitc-anti-kappa-1b biotin-HIS22	PE-AV
2	MRC OX11 biotin-HIS22	Fitc-MRC OX14+16 PE-AV
3	Blocked Fitc-anti-kappa-1b biotin-HIS22	PE-AV
4	R21/5S biotin-HIS22	Fitc-MRC OX14+16 PE-AV

Cervical lymph nodes were removed from rats seven days after immunisation; cell suspensions were stained for flow cytometry by usual procedures (5), and first-layer antibodies were detected by Fitc-conjugated MRC OX14+16 and/or phycoerythrin-avidin (a gift of Dr. B. Roser, AFRC Institue for Animal Physiology). Cells were then analysed on an upgraded FACS II.

RESULTS

Cell suspensions of stimulated cervical lymph nodes from PVG-Ig-kappa-1b rats were stained for two-colour flow cytometry with biotin-HIS22 and Fitc-anti-kappa-1b. FACS-generated dot plots showed two clearly defined subpopulations of B cells: $HIS22^+$ $kappa^+$ cells and $HIS22^-$ $kappa^+$ cells (Figure 1). Flow cytometric analyses of the isotype distribution, size, and Ig expression, as well as calculations of the relative number of $HIS22^-$ $kappa^+$ cells during the time course of an immune response, identified this population as germinal centre B cells. We observed that (a) $HIS22^+$ $kappa^+$ cells were largely $sIgM^+$ $sIgD^+$ while most $HIS22^-$ $kappa^+$ cells were $sIgM^+$ and nearly all were $sIgD^-$, (b) scatter profiles for $HIS22^-$ $kappa^+$ cells were characteristic of relatively large lymphocytes, (c) $HIS22^-$ $kappa^+$ cells stained 2-3 times less brightly than $HIS22^+$ $kappa^+$ cells for the kappa light chain, and (d) the percentage of $HIS22^-$ $kappa^+$ cells among all lymph node cells increased steadily following antigenic stimulation to a peak of about 10% by day 7; by day 13 the percentage dropped to initial levels (about 3%).

Non-irradiation chimeras were established by injecting i.v. TDL from rats expressing one kappa allotype into a rat of the opposite allotype. In one set of experiments, about 10^8 TDL collected overnight were washed and injected on day 0 directly. In other experiments, TDL collected overnight were first depleted of sIgD+ cells by rosetting and then injected. On day 3, rats were immunised s.c. with DNP-HSA and seven days later, cervical lymph nodes cells were stained (Table 2). FACS analysis was based on HIS22 histogram profiles gated for the donor or host kappa allotype; among the two B cell populations -- HIS22$^+$ kappa$^+$ cells and HIS22$^-$ kappa$^+$ cells -- the percent donor among each population was calculated.

For fresh, non-depleted donor TDL, the same percentage of donor cells was found among GC B cells as among non-GC B cells. For five experiments, the mean relative tendency for GC localisation (i.e. the ratio of the donor percentage among HIS22$^-$ B cells to the donor percentage among HIS22$^+$ cells) was 0.9 ± 0.3, indicating little or no bias among donor B cells to localise in germinal centres.

We studied whether the GC localisation of donor TDL was affected by the percentage of donor B cells expressing sIgD. Among fresh TDL from SPF-raised donor rats, more than 97% of the B cells expressed detectable sIgD. When donor TDL were first depleted of this major population of sIgD$^+$ cells by rosetting, GC localisation of donor cells was enhanced. In one experiment, in which about 30% of the donor B cells injected were sIgD$^+$ (but more than 90% were sIgM$^+$), GC localisation was nine times greater than localisation of donor B cells from fresh, non-depleted TDL. The extent of this enhanced GC localisation varied according to the degree of IgD depletion achieved in different experiments; as the percentage of sIgD$^-$ B cells among donor B cells increased, there was a greater tendency for donor cells to localise in germinal centres.

DISCUSSION

From the analyses of non-irradiation chimeras, we found that germinal centre precursor cells are among TDL, and that the GCPC entering lymph node GCs are among the minor population of B cells that lack sIgD. Other studies in our laboratory concerning the recruitment of GCPC indicated that most recruitment occurs soon after immunisation (within two days) and that very few GCPC can be incorporated thereafter into developing GCs. This observation implies that sIgD$^-$ B cells that act as GCPC play a direct role in the initiation of GCs and do not simply function as late-comers to GCs that are well established.

We do not contend from these results that lymph node GCPC have never expressed sIgD, only that immediately before joining and/or initiating GCs, these precursor cells do not express the delta heavy chain. It is possible that upon activation (in non-GC areas), some sIgD$^+$ cells lose delta expression (6,7) and become able to act as GCPC upon encountering antigen associated with follicular dendritic cells. This hypothesis may reconcile our findings with observations that B cells positive for sIgD when injected i.v. with antigen can form GCs in the spleen of an irradiated host (8). Several reports have suggested that activated and/or antigen-experienced B cells may preferentially localise in GCs (9,10,11), but more work is needed to confirm the correlation of B cell activation, loss of sIgD *in vivo*, and GC formation.

Finally, the HIS22 marker employed here (negative for B cells among GCs) may be used not only analytically in acute, non-irradiation chimera studies of GCPC, but also preparatively in studies to trace the fate of GC derived cells.

ACKNOWLEDGEMENTS

We thank Dr. Frans Kroese, Dr. Davine Opstelten, and Dr. Bruce Roser for generously supplying antibodies, and Mrs. Margaret Hughes for her expert technical assistance.

REFERENCES

1. Kraal, G., I.L. Weissman, and E.C. Butcher. 1982. Germinal centre B cells: antigen specificity and changes in heavy chain class expression. *Nature*. **298**:377.

2. Kroese, F.G.M., A.S. Wubbena, D. Opstelten, G.J. Deenen, E.H. Schwander, L. De Leij, H. Vos, S. Poppema, J. Volberda, and P. Nieuwenhuis. 1987. B lymphocyte differentiation in the rat: production and characterization of monoclonal antibodies to B lineage-associated antigens. *Eur.-J. Immunol.* **17**:921.

3. Hunt, S.V. and A.F. Williams. 1974. The origin of cell surface immunoglobulin of marrow-derived and thymus-derived lymphocytes in the rat. *J. Exp. Med.* **139**:479.

4. Ford, W.L. 1977. Preparation and labelling of lymphocytes. In *Handbook of Experimental Immunology*. D.M. Weir, editor. Blackwell Scientific Publications, Oxford, England. 3rd edition. Chapter 23.

5. Parks, D.R., L.L. Lanier and L.A. Herzenberg. 1986. Flow cytometry and fluorescence activated cell sorting (FACS). In *Handbook of Experimental Immunology*. D.M. Weir, editor. Blackwell Scientific Publications, Oxford, England. 4th edition. Chapter 29.

6. Finkleman, F.D., I. Scher, J.J. Mond, J.T. Kung, and E.S. Metcalf. 1982. Polyclonal activation of the murine immune system by an antibody to IgD I. Increase in cell size and DNA synthesis. *J. Immunol.* **129**:629.

7. Monroe, J.G., W.L. Havran, and J.C. Cambier. 1983. B lymphocyte activation: entry into cell cycle is accompanied by decreased expression of IgD but not IgM. *Eur. J. Immunol.* **13**:208.

8. Seijen, H.G., J.C.A.M. Bun, A.S. Wubbena, and K.G.L. Löhlefink. The germinal centre precursor cell is surface μ and δ positive, this symposium.

9. deKruyff, R.H., S.R. Onikul, and G.J. Thorbecke. 1976. Migratory patterns of B lymphocytes. V. Surface Ig and migration properties of density gradient-separated bursa cells. *Eur. J. Immunol.* **6**:462.

10. Coico, R.F. and G.J. Thorbecke. 1985. Role of GC in the generation of B cell memory. *Folia Microbiol.* **30**:196.

11. Gray, D. Memory B cells but not virgin B cells are activated in germinal centres, this symposium.

THE *DE NOVO* GENERATION OF GERMINAL CENTERS IS AN OLIGOCLONAL PROCESS

Frans G.M. Kroese, Auk S. Wubbena, Hendrik G. Seijen and Paul Nieuwenhuis

Department of Histology, University of Groningen
Oostersingel 69/1, 9713 EZ Groningen
The Netherlands

INTRODUCTION

Germinal centers (GC) may develop in follicles of peripheral lymphoid organs as response to stimulation with T cell dependent antigens (for review see Nieuwenhuis and Opstelten, 1986). Essentially, GC are clusters of actively dividing B lymphoblastoid cells which are thought to give rise to memory-B cells. In addition to B cells, GC are composed of some T (helper) cells and a specific non-lymphoid cell, the follicular dendritic cell.

In a previous study we have demonstrated that GC may develop oligoclonally, i.e. each *de novo* generated GC on average, seems to be derived from only very few (1-3) germinal center precursor cells (Kroese et al., 1987a). In these experiments lethally X-irradiated (AO) rats were reconstituted with thoracic duct lymphocytes (TDL) from AO and (AOxBN)F1 donors in different ratios, and subsequently stimulated with sheep red blood cells (SRBC). GC in recipient spleens were analysed for the expression of major histocompatibility complex (MHC) Class II antigens of either donor. Three types of GC could be observed: GC composed entirely of lymphocytes from only one of the donors as well as GC containing a mixture of both, indicative for an oligoclonal character of the *de novo* generated GC.

Obviously, a major drawback of this system was the difference in MHC antigens between the rat strains, which may lead to rejection phenomena. This was partially circumvented by mixing cells from F1 hybrid rats with parental cells, and by using TDL derived from tolerant donors. The recent development of a mAb, HIS41, to an allotype (RT7.2) of the leucocyte common antigen (LCA) (Kampinga et al., 1987), which also reacts with germinal center cells, in combination with histocompatible rats (PVG and PVG.RT7.2) congenic in LCA allotypes, made it however now possible to perform similar experiments in a presumably less biased system. In addition, in the experiments described herein we also used the rat kappa chain congenic system to further confirm the oligoclonal development of the *de novo* generated GC in reconstituted rats.

MATERIALS AND METHODS

Animals

LOU rats and LOU rats congenic for Ig kappa light chains (LOU.Ig-k^{1b}) (Bazin et al., 1984) were originally obtained from Dr. H. Bazin (University of Louvain, Brussels, Belgium). PVG rats congenic for the RT7 allotype (PVG.RT7.2) were originally obtained from Dr. S.V. Hunt (Sir William Dunn School of Pathology, Oxford, U.K.). These rats as well as normal PVG rats were bred in our animal facility under conventional conditions.

Antibodies

The following mouse anti-rat monoclonal antibodies (mAb) were used: HIS41, directed to an allotype of the LCA (RT-7.2) (Kampinga et al., 1987), HIS22 to corona type B cells and HIS14 to all B cells (Kroese et al., 1987b) were generated in our own laboratory. RG11/15, directed to Ig kappa light chains of b but not of a allotype (Springer et al., 1982) was a kind gift of Dr. B. Roser (ARFC, Institute of Animal Fysiology, Babraham, Cambridge, U.K.). Polyclonal rabbit anti-mouse Ig, conjugated to peroxidase was purchased from Dako (Dakopatts, Glostrup, Denmark).

Reconstitution experiments

Reconstitution of lethally (9Gy) X-irradiated rats with 10^8 TDL and subsequent stimulation with 10^9 SRBC were as described (Gastkemper et al., 1981). TDL were collected using the method of Ford and Hunt (1973). X-irradiated PVG rats or LOU rats were reconstituted with TDL from PVG and PVG.RT7.2 (1:1 ratio) or LOU and LOU.Ig-k^{1b} (1:1 ratio), respectively. Reconstituted rats were killed 7-10 days after reconstitution and antigenic stimulation (see Results). Spleens were then taken out for immunoperoxidase staining of frozen sections as has been described elsewhere (Kroese et al., 1987a).

RESULTS AND DISCUSSION

Reconstitution of lethally X-irradiated PVG or LOU rats with a 1:1 mixture of TDL derived from PVG and PVG.RT7.2 congenic rats or TDL from LOU and LOU.Ig-k^{1b} congenic rats, respectively, lead to the generation of germinal centers in recipient spleens upon antigenic (SRBC) stimulation. GC could be easily detected after immunoperoxidase staining of frozen sections with mAb HIS22, directed against corona B cells and HIS14 staining all follicular elements including GC (Fig. 2a,b). When alt

Table 1. Distribution of germinal center types in the spleen of lethally X-irradiated rats reconstituted with mixtures of TDL and stimulation with SRBC.

Experiment	Days after stimulation	Proportion of GC(a)			Mean number GC/section	Number of rats
		negative	mixed	positive		
1) TDL_{PVG}+$TDL_{PVG.RT7.2}$ (1:1) --> X-irr. PVG	7	19±4(b) [16-24]	36±3 [31-38]	45±6 [38-52]	6.8±1.4	4
	10	19±2 [17,20]	47±10 [54,40]	35±8 [29,40]	5.7±2.1	2
2) TDL_{LOU}+$TDL_{LOU.Ig\text{-}k^{1b}}$ (1:1) --> X-irr. LOU	7	34±8 [39,28]	36±11 [28,43]	26±4 [23,29]	3.6±0.5	2

a) At least 18 well-separated frozen spleen sections were stained with either HIS41 (anti-RT7.2) (experiment 1) or with RG 11/15 (anti-Ig kappa b) (experiment 2). The GC were analysed for their staining pattern with the respective mAb; GC entirely stained were scored as "positive", GC entirely unstained as "negative' and GC containing a mixture of negative and positive cells as "mixed".

b) Values are given as mean ± s.d.. Between brackets the range is given for the experiment in which 4 animals were used, or the actual proportions when 2 animals per experiment were used.

frozen sections were stained with mAb directed against the allotype of one of the donors, i.e. HIS41 against RT7.2 (Fig. 1) or RG11/15 against Ig-κ^{1b} (Fig. 2c), respectively, three types of GC could be observed in both reconstitution systems: i) completely stained GC, ii) completely unstained GC and iii) GC composed of a mixture of positive and negative cells. By contrast, the lymphocyte corona always contained a mixture of positive and negative cells.

Staining of spleen sections from normal LOU.Ig-κ^{1b} rats with anti-kappa b mAb (i.e. RG11/15), showed intense dendritic staining in follicles (and GC) due to immune complexes retained by follicular dendritic cells (FDC) (Klaus et al., 1980). This dendritic staining obscured possible staining of GC cells. In GC (or follicles) of spleens from irradiated and reconstituted LOU rats, however, only little intercellular staining with anti-kappa b mAb could be detected, allowing now identification of kappa chain expression by GC cells. Apparently, Ig (with b-allotype light chains), complexed with antigen, has not yet been trapped by FDC in large amounts. This can be due to the fact that only low amounts of immune complexes have been formed and/or an X-irradiated induced failure of the immune complex trapping mechanism itself (Kroese et al., 1986).

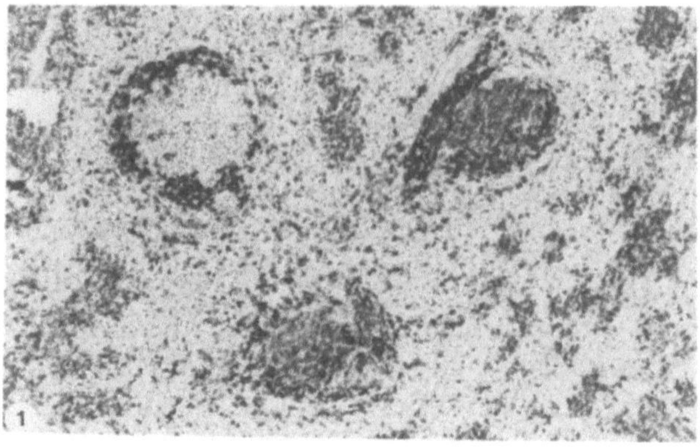

Figure 1. Spleen of PVG rat 10 days after reconstitution with a 1:1 mixture of PVG and PVG.RT7.2 TDL and challenge with SRBC, stained with HIS41 (anti-RT7.2). Note one negative (upper left) and two positive (upper right and bottom) germinal centers . (x10 obj.).

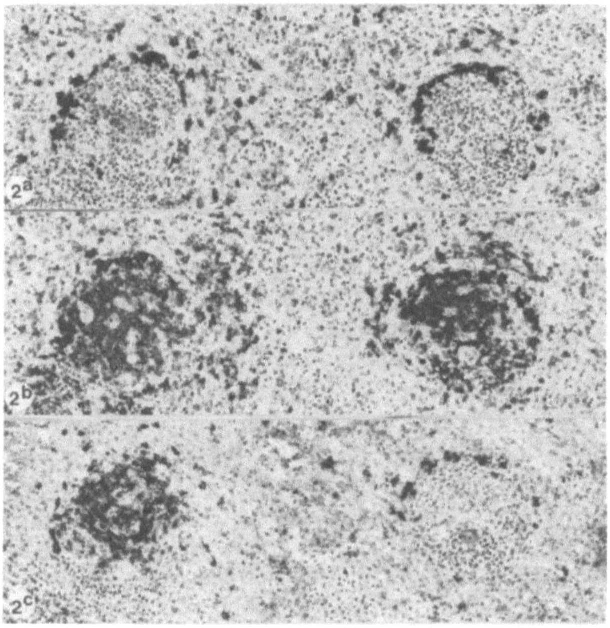

Figure 2. Spleen of LOU rat 7 days after reconstitution with a mixture of LOU and LOU.Ig-k[1b] TDL and challenge with SRBC, stained with HIS22 (anti-corona B cells; Fig. 2a), HIS14 (anti-B cells; Fig. 2b) and RG11/15 (anti-Ig kappa b; Fig 2c) respectively. Note positive (left) and negative (right) germinal center with RG11/15. (x10 obj.).

Our findings that when rats were reconstituted with mixtures of TDL, differing in allotype (Ig or RT7), upon antigenic stimulation led to three different types of GC (i.e. GC entirely composed of cells from either one of the donors and GC composed of a mixture of cells from both donors), point to an oligoclonal development of GC in these experiments. If GC had developed clonally, then only GC of either one of the donors should be found, whereas in the case of a polyclonal development all GC should be of a mixed-type. To assess from how many germinal center precursor cells (GCPC) each *de novo* GC had developed, on average, the relative frequencies of the different GC types was determined (Table 1). The number of GCPC involved in a *de novo* generated GC could be estimated using the equation I^N=S, in which I is the fraction of cells of one of the allotypes in the inoculum used for reconstitution, S the fraction of GC in the spleen that are entirely composed of cells of that allotype, and N the number of GCPC involved in the generation of GC. Since in the experiments described here I was always 0.5, this equation can be converted into : N = (log S)/-0.3. From Table 1 it follows that S varied between 0.19 and 0.45. Thus, the number of GCPC in the generation of a GC seemed to range, on average, from about 1-3. No difference was observed in the relative frequencies of GC taken 7 or 10 days after antigenic stimulation (Table 1).

These results described here confirm our previous findings in which in a similar approach, differences in MHC Class II (Ia) antigens were used to distinguish between B cells derived from two different donors (Kroese et al., 1987a). By using the RT7 and Ig kappa light chain congenic system, we expected, however, to have systems with presumably less bias, compared to the Ia-antigenic system. In a system completely free of any bias, the proportions of GC composed of either one of the donors should be exactly the same after reconstitution with equal proportions of TDL from both allotypes. Indeed, it seems that this is the case in the kappa chain congenic system (Table 1). However, in the RT7 transfer system, the proportion of GC of the RT7.2 allotype was about two times higher than the proportion of GC from the RT7.1 allotype (Table 1). The reason of this bias is unknown. It might be that TDL derived from PVG.RT7.2 rats contained relatively more GCPC and/or immunological reactions provoked by differences in RT7 allotype (or other unknown molecules that may be different in both rat strains).

ACKNOWLEDGEMENTS

The authors would like to thank Drs. H. Bazin and S.V. Hunt for their kind gift of LOU and PVG.RT7.2 rats and Dr. B. Roser for the gift of monoclonal antibodies.

REFERENCES

Bazin, H., Cormant, F., and de Clerq, L., 1984, Rat monoclonal antibodies. II. A rapid and efficient method of purification from ascitic fluid or serum. J. Immunol. Meth.71: 9.

Ford, W.L. and Hunt, S.V., 1973, The preparation and labeling of lymphocytes, in: "Handbook of Experimental Immunology", D.M. Weir, ed., p. 231., Blackwell Scientific Publications, Oxford.

Gastkemper, N.A., Wubbena, A.S., Gimbrere, F.J.H., de Graaf, A., and Nieuwenhuis, P., 1981, Germinal centres and the B cell system. V. Presence of germinal centre precursor cells among lymphocytes of the thoracic duct in the rat. Cell Tiss. Res. 219: 281.

Kampinga, J., Kroese, F.G.M., Pol, G.H., and Nieuwenhuis, P., 1987, HIS41: A new Mab detecting the RT7.2 but not the RT7.1 haplotype on lymphoid and myeloid cells. Adv. Exp. Med. Biol. (This volume).

Klaus, G.G.B., Humphrey, J.H., Kunkl, A., and Dongworth, D.W., 1980, The follicular dendritic cell: its role in antigen presentation in the generation of immunological memory. Immunol. Rev. 53: 3.

Kroese, F.G.M., Wubbena, A.S., and Nieuwenhuis, P., 1986, Germinal centre formation and follicular antigen trapping in the spleen of lethally X-irradiated and reconstituted rats. Immunology 57: 99.

Kroese, F.G.M., Wubbena, A.S., Seijen, H.G., and Nieuwenhuis., P., 1987a, Germinal centers develop oligoclonally. Eur. J. Immunol. 17: 1069.

Kroese, F.G.M., Wubbena, A.S., Opstelten, D., Deenen, G.J., Schwander, E.H., de Leij, L., Vos, H., Poppema, S., Volberda, J., and Nieuwenhuis, P., 1987b, B lymphocyte differentiation in the rat: production and characterization of monoclonal antibodies to B lineage associated antigens. Eur. J. Immunol. 17: 921.

Nieuwenhuis, P., and Opstelten, D., 1984, Functional anatomy of germinal centers. Am. J. Anat. 170: 421.

Springer, T.A., Bhattachara, A., Cardoza, J.T., and Sanches-Madrid., F., 1982, Monoclonal antibodies specific for rat IgG1, IgG2a, IgG2b subclasses, and kappa chain monocytic and allotypic determinants: reagents for use with rat monoclonal antibodies. Hybridoma 1: 257.

CHANGES IN THE GERMINAL CENTRE AND LYMPHOCYTIC CORONA

AFTER PERITONEAL INFLAMMATION

Howard W. Steer

Department of Surgery, Southampton General Hospital

Southampton, England

The inflammatory reaction following an intestinal anastomosis results in the migration of inflammatory cells from the peritoneal cavity to the parathymic lymph nodes (1). Changes occur in the plasma cells of the medullary cords of the parathymic lymph nodes (2) resulting in an initial signifcant decrease in IgA containing plasma cells at two days but a subsequent increase in IgG and IgM containing plasma cells. The increase in IgM containing plasma cells preceeded that of IgG.

The influence of the inflammatory reaction on the germinal centres and lymphocytic corona in the parathymic lymph nodes of rats has been studied at 2, 5, and 10 days after a small intestinal anastomosis and compared with control rats having no anastomosis. Immunohistological studies were carried out as previously described (2) with the quantitative analyses performed using an IBAS 2, Image Analysis System (Kontron Bildanalyse, Munich, W. Germany).

The mean volume of the germinal centres was significantly increased ($p < 0.01$) from 0.26×10^6cu µm in controls to 1.52×10^6cu µm at ten days after anastomosis. This was accompanied by a significant increase in the number of cells in the lymphocytic corona ($p = 0.001$) which increased from 145 cells in 1.7 μm^2 in the controls to 168 cells in 1.7 um^2 at ten days after the anastomosis. Of these cells in the lymphocytic corona there was a significant increase in those cells whose nuclear size was above 44 μm^2. These were large pyronophilic cells whose cytoplasm contained an abundance of polyribosomes but a sparcity of rough endoplasmic reticulum. These large cells contained internal immunoglobin and Ia antigens.

When the immunoglobulin type of these large cells was analysed there was a significant increase in these cells containing IgM at two days after the anastomosis and a significant increase in IgG containing cells at two and ten days. There were no changes in those cells containing IgA. The changes in these immunoglobulin containing large cells in the lymphocytic corona correlated with those changes which have been observed in the plasma cells of the medullary cords (2).

REFERENCES

1. H. W. Steer and D. A. Lewis, J. Pathol. 140:237 (1983).
2. H. W. Steer and R. A. Foot, J. Anat. 152:23 (1987).

THYMUS, T-CELL DIFFERENTIATION AND THE T-CELL SYSTEM

THE CHOLINERGIC AND ADRENERGIC INNERVATION OF THE RAT THYMUS

Marion D. Kendall, A. Al-Shawaf and S. A. A. Zaidi

Division of Anatomy, United Medical and Dental Schools of Guy's and St. Thomas's Hospitals London, U.K.

The first major studies on the innervation of the thymus were made by Hammar in 1935 (1) using silver stained preparations. He describes the innervation in man as reaching the gland from the vagus before the gland descends to the thorax (middle of the second month). Nerves enter medially, laterally and dorsally accompanying the blood vessels and ramify in the medulla to form a network of fibres which appears complete 16-18 weeks in utero. He and other authors (2) describe the presence of sympathetic fibres, acquired after the gland has descended into the superior and anterior mediastinum, that meet the vagal supply at the cortico-medullary junction. These are probably derived from the cervicothoracic (stellate) ganglion or ansa cervicalis and ansa subclavia with contributions from the phrenic nerves. Sergeeva (3) and Felten et al. (4) have described adrenergic fibres in the cortex, accompanying blood vessels, and running to Hassall's corpuscles (3).

Bulloch and Moore (5) retrogradely injected horse radish peroxidase into the rodent thymus. Their results implicated the vagus in thymic innervation but only underlined the complexities of the system and left the interpretation of the functional contributions from the vagus unclear. The largest population of labelled neurones were seen in the retrofacial nucleus (generally considered (6) to be the source of general visceral efferents) and throughout the rostrocaudal length of the nucleus ambiguus. This nucleus, in addition to probably giving rise to efferent parasympathetic fibres that are largely secretomotor in the gut, contains fibres that are reported (6) to go to striated muscle of branchial origin - but the thymus has none, only the myoid cells are striated. The retrofacial nucleus was also labelled. This is probably another source of general visceral efferents.

Thus the results presented here on the distribution of cholinergic cells and fibres, and adrenergic fibres in the rat requires far more work before firm conclusions can be drawn on functional interpretations. The speculations however present exciting possibilities on how the thymus actually works.

MATERIALS AND METHODS

For the acetylcholinesterase studies 8 male and 8 female adult CSE rats weighing between 240 and 260 gms were killed with chloroform, the thymus glands were dissected out and immediately frozen in isopentane cooled in liquid nitrogen. Thick cryostat sections (20-40 um) were stained with the method used by Sheheta (7) which is based on Couteaux's modification (8) of an earlier method (9). Some sections were treated with ethopropazide HCl (3.12mg/100ml) as a control for the staining reactions. Myoid cells were identified on thin sections of thymus prepared as above using a polyclonal sheep anti-troponin I antibody conjugated to fluorescein (kindly supplied by Dr Tej Dhoot).

For the simultaneous visualisation of the nerves and vasculature, 5 adult male rats (250-300 gms) were anaesthetised with 60mg/kg Sagatal injected intraperitoneally. The femoral artery was cannulated, 1 ml heparin was perfused, then 10% sodium nitrite was introduced as a vasodilator before introducing Berlin blue dye for 5 minutes. Tissue was preserved and treated as described above for the demonstration of acetylcholinesterase activity. All measurements on cells were made at a magnification of 1260 and the results expressed as equivalent diameters (square root of length x breadth). Paraformaldehyde induced fluorescence was used to examine the aminergic nerves on 20 um sections.

Electron microscope observations were made on routinely fixed and embedded material prepared for other studies (10-17) and on material kindly supplied by Dr H-J Schuurman from studies on allogeneic cultured thymic fragments (see paper in this volume) and manuscript in press (Kendall _et al._).

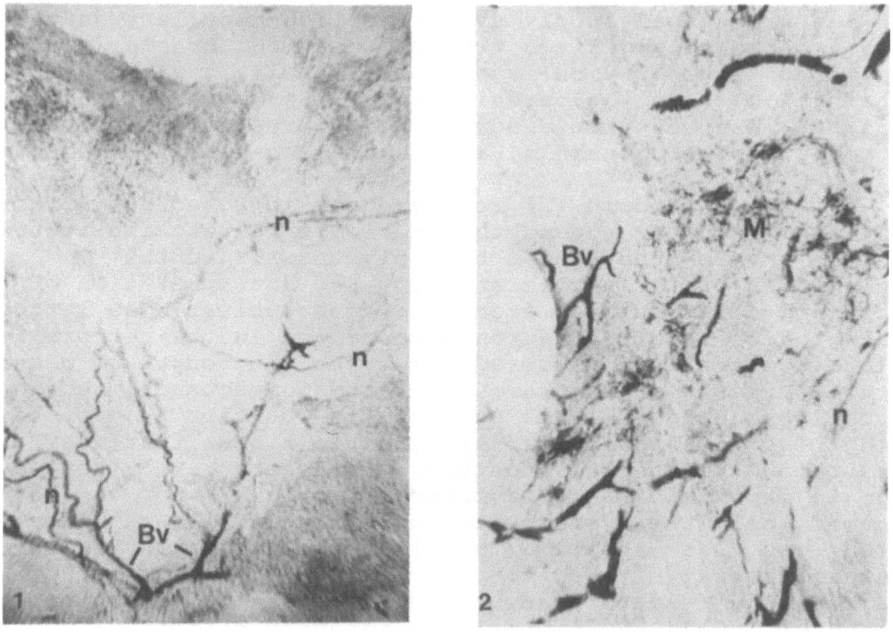

Figs. 1 and 2 show acetylcholinesterase positive nerves (n) and cells in the rat thymus. Bv = blood vessels; M = medulla x 90.

RESULTS

Acetylcholinesterase positive fibres accompanied the major blood vessels in the septa to reach the medullary region, and to remain with the major vessels in the medulla. Fine, fairly straight branches, unaccompanied by blood vessels (Fig. 1), sometimes penetrated the cortex and medulla. A similar but more sparse picture was obtained using induced fluorescence for adrenergic fibres.

In the medulla, fine fibres branched from the main nerves and formed a plexus interconnecting with acetylcholinesterase positive cells (Fig. 2). These cells (Fig. 3) were frequently stellate in shape, and although a wide spectrum of cell size was observed, they fell into 2 main size categories. The most numerous was a smaller sized group of more rounded cells that measured on average 6-7 um in diameter. The larger cells (13-17 um in diameter) were usually ganglion shaped with processes connecting them to other cells and to fine fibres. The larger sized cells were often grouped in the medulla, admixed with some smaller cells, whereas the smaller cells were often scattered in the cortex as well as occurring in the medulla.

The use of the anti-troponin I polyclonal antibody allowed the visualisation of myoid cells in the medulla. These cells were normally widely scattered (did not form groups), and although their size range was similar to the acetylcholinesterase positive cells described above, Student's t-tests applied to the data showed highly significant differences ($P<0.001$, $n= 374$) in size between the two cell types (Al-Shawaf and Kendall, in prep).

Nerve fibres around the periphery of the gland and close to blood vessels can usually be identified on sections of thymus examined under the electron microscope. Many of the bundles are unmyelinated, and may be composed of extremely fine fibres (Fig. 4). Small (100nm diameter) electron lucent vesicles are present in one nerve close to the blood vessel in Fig. 4, and other nerves sometimes also contain vesicles with electron dense cores that are approximately the same size or slightly larger. It is unusual to observe fibres in the cortex but larger bundles may be seen at the cortico-medullary junction. Sometimes these are myelinated.

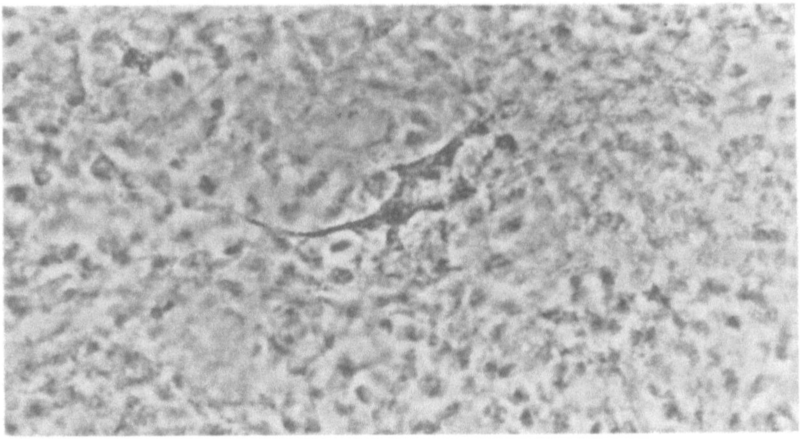

Fig. 3 is a high power view of an acetylcholine positive cell in the medulla of a rat thymus. x 560.

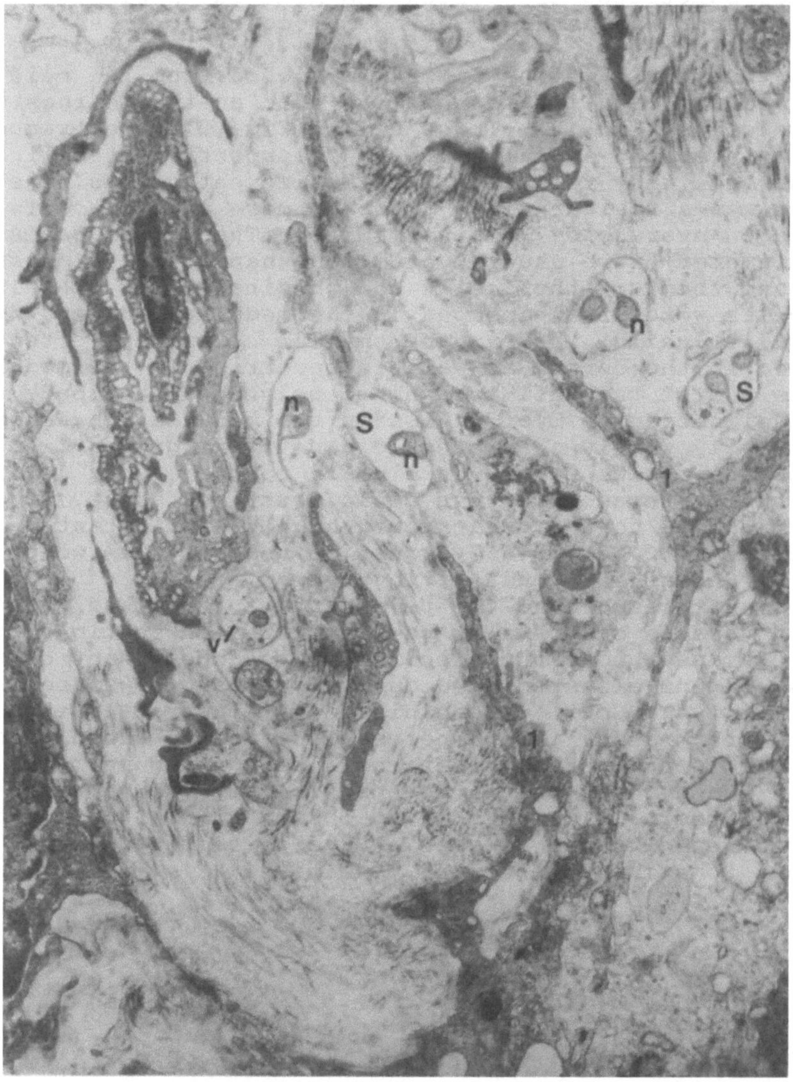

Fig. 4 is an electron micrograph from a rat with diabetes induced by a single injection of streprozotocin. Many small unmyelinated nerves (n) are present in the perivascular space which is bounded by type 1 epithelial cells (1). V = electron lucent vesicles in the nerve; S = Schwann cell. x 14,000.

The medulla, when examined under the electron microscope, contained large, fairly electron lucent cells that often occurred in small groups scattered between the other cells. These were found in all material examined: wild birds (Quelea quelea), wild rodents (Clethrionomys glareolus), laboratory rodents and man.

The cells have a centrally placed, fairly leptochromatic nucleus containing a single large nucleolus. Mitochondria are present but in general organelles are relatively sparse with only short lengths of rough endoplasmic reticulum, and occasional small electron dense granules (Fig. 5). Small circular profiles of cytoplasm are usually found around the cells, and some of these (Fig. 6) contain electron dense inclusions (100-200nm in diameter) similar to those found in the cells and in nerve fibres.

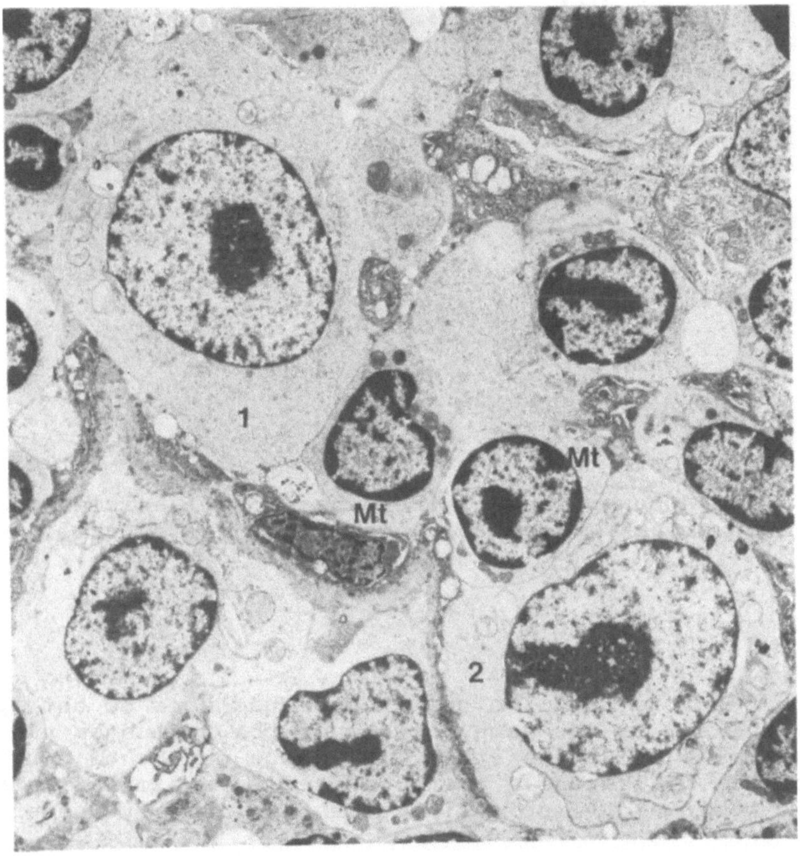

Fig. 5. Two cells (1 and 2) in the medulla of a cultured thymic fragment showing the shape and appearancce of the cells previously called 'undifferentated mesenchymal cells'. They have no tonofilaments or desmosomes. Material courtesy of Dr H-J Schuurman. Mt = medullary thymocyte. x 4,000.

DISCUSSION

Some acetylcholinesterase positive fibres have recently been described in the thymus (17), but no reference was made to any positively stained cells. For this study, we have used very thick sections and this has allowed us to clearly visualise, not only the course of the nerves and to relate this to the vasculature, but also to observe the two populations of positively staining cells. These cells do not appear to be myoid cells, and they have the form of ganglion cells. Further studies are in hand to try to confirm that these cells are indeed ganglion cells.

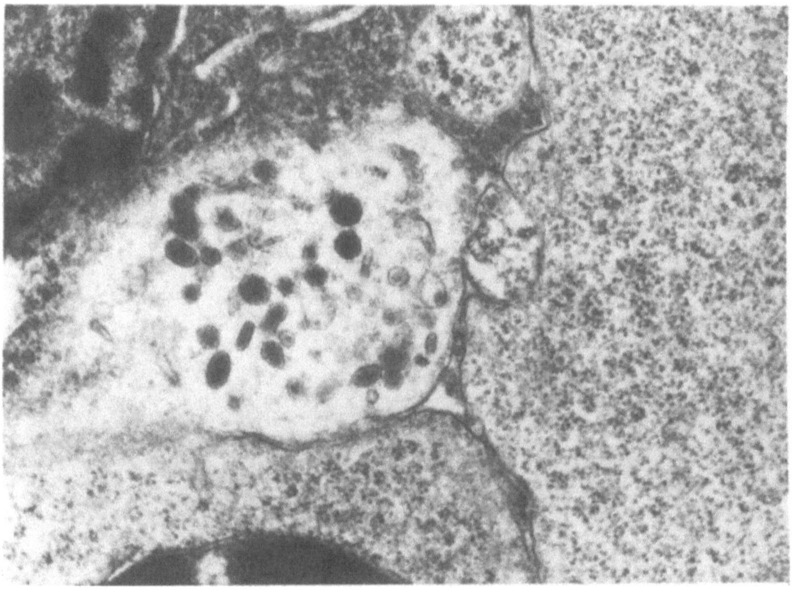

Fig. 6. A high power view of the dense-cored vesicles in a cytoplasmic process close to a possible ganglion cell. x30,000.

The possible ganglion cells described here do not appear to be localised in the same regions of the thymus as the A2B5 positive cells of Haynes et al (18) which may be of neural crest origin. Those cells are localised in a similar manner to the oxycytocin and vasopressin cells described by Geenen et al (19).

Because of the different techniques involved, we cannot be sure yet that the large electron lucent cells observed in the medulla of all species examined are the same cell types as those observed in the light microscope. These large electron lucent cells were first described in birds (20, 11) where we recognised that they were not epithelial cells as they had neither tonofilaments nor desmosomes. They were termed undifferentiated mesenchymal cells and only brief mention has been made of them subsequently. However, they are fairly common although care must be taken to distinguish them from type 5 epithelial cells (with a few tonofilaments and some desmosomes) that are present in the medulla (21).

Under the EM, most of these cells have small profiles of cytoplasm around them, and it is usually considered that these arise from epithelial cells in the vicinity, but some could be profiles of small nerves. This seems most likely when the profiles contain small electron dense vesicles as they closely resemble dense-cored neurotransmitter vesicles. Further studies are in hand to explore the possible presence of neurotransmitters in the thymus, and to try to confirm that the cells described above are neuronal cell bodies.

The authors are pleased to acknowledge the support of the Iraqi Government for a PhD studentship for Dr Ali Al-Shawaf and from the Pakistan Government for Dr S.A.A. Zaidi. The technical help from the division of Anatomy at St. Thomas's Hospital Medical School is gratefully acknowledged.

REFERENCES

1. J. A. Hammar. Z. mikr. Anat. Forsch. 8:253 (1935).
2. J.M. Williams and D.L. Felten. Anat. Rec. 199: 531 (1981).
3. V.E. Sergeeva. Byull. Eksp. Biol. Med. 77:11(1974).
4. D.L.Felten, S.Y. Felten, S.L. Carlson, J.A.Olschowka, and S.Livnat. J. Immunol. 135:755s (1985).
5. K. Bulloch and R.Y. Moore. Am. J. Anat. 162:157 (1981).
6. P.L. Williams and R. Warwick. Gray's Anatomy. 36th edn London: Longmans.
7. S.H. Sheheta. The innervation of avian muscles. PhD Thesis, University of London (1961).
8. R. Couteaux. Arch. Int. Physiol.59:52 (1951).
9. G.B. Koelle and J.S. Friedenwald. Proc. Soc. Exp. Biol. Med.70:614 (1949).
10. M.D. Kendall and P. Ward. Nature 249:366 (1974).
11. M.D. Kendall and J.A. Frazier. Cell Tiss. Res. 199, 37 (1979).
12. M.D. Kendall and J.A. Singh J. Anat. 130:183 (1980).
13. J.A. Singh. Studies on the human thymus with particular reference to age changes. PhD Thesis, University of London.
14. M.D. Kendall, P. Jaffe and Y.M. Yoffey. J. Anat. 142:85 (1985).
15. M.D. Kendall and N. Blackett. Cell Tiss Res. 232:201 (1983).
16. K. Chatamra, P. M. Daniel, M.D. Kendall and K.D.C. Lam. J. Anat. 137:413 (1983).
17. J. A. Fatani, M. A. Qayyum, L. Mehta and U. Singh. J. Anat. 147: 115 (1986).
18. B. F. Haynes, K. Shimizu and G. S. Eisenbarth. J. Clin Invest. 71:9 (1983).
19. V. Geenen, J-J Legros, P. Franchimont, M. Baudrihaye, M-P. Defresne and J. Bonivar. Science, 232:508 (1986).
20. J.A. Frazier. Z. Zellforsch. 136:191 (1973).
21. F. P. van de Wijngaert, M. D. Kendall, H-J. Schuurman, L. H. M. P. Rademakers and L. Kater. Cell Tiss. Res. 237:227 (1984).

COMPLEX HETEROGENEITY OF THE THYMIC STROMA

R.L. Boyd, D.J. Izon, D.I. Godfrey, T.J. Wilson,
H.A. Ward, and C.L. Tuček

Department of Pathology and Immunology
Monash University, Melbourne, Australia 3181

INTRODUCTION

Despite the simple structural features of the thymus which standard histological examination reveals, the very intricate pathways of T lymphocyte differentiation occuring in this organ (1) would predict that the stromal elements which regulate this process must be equally complex. Tissue culture of dispersed thymic stromal cells (2), ultrastructural studies (3) and the distribution of MHC antigens (4) provided the first indications that this was so. More recently, use has been made of M.Abs. raised against enriched preparations of thymic stromal cells to identify a limited number of specific regions of the thymus which may reflect different roles in T cell maturation (5-7). Using the chicken thymus as a model, we have been able to markedly extend these observations by finding numerous antigenically distinct regions within which were multiple populations of stromal cells and their products (8; Boyd et al. submitted). This extensive panel of M.Abs to chicken stromal antigens probably reflects the phylogenetic distance between chickens and the mice used for immunization. We were interested to see, however, whether modified immunogen preparations and immunization protocols would reveal if similar complexities exist in the mouse and human thymus. This paper summarizes the major findings of these studies which clearly indicate that this is so.

MATERIALS AND METHODS

Human thymuses were obtained from children aged between 4 weeks and 4 years, undergoing corrective cardiac surgery at the Royal Children's Hospital, Parkville, Victoria. All chickens used were closed colony Australorp x White Leghorn hybrids obtained from a local commercial source. Mice were 4-8 week old female BALB/c or CBA. Female Lou/M rats (3-12 months old) were used for immunization with murine thymic stromal cells.

Preparation of Stromal Cells

Enriched preparations of chicken, human and mouse thymic stromal cells were obtained using our standard procedures (8,9,10). Briefly, thymuses were finely minced in serum-free RPMI 1640 (SF-RPMI), the freed lymphocytes discarded and the remaining tissue fragments digested to a single cell suspension by repeated 15 min. incubation at 37°C in an enzyme solution of collagenase (Type IV, 0.15%) and DNase (0.01%). The stromal cells were enriched either by sedimentation at 1g twice through heat inactivated newborn calf serum (NBS) (1ml cells : 3ml NBS) for 30 min. at 4°C, or elutriation (11). The stromal cells so obtained consisted predominantly of thymic nurse cells (TNC), large macrophages, epithelial cells, endothelial cells and dendritic-like cells and approximately 10% thymocytes.

Production of Monoclonal Antibodies

For M.Abs to human and chicken thymus, mice were immunized i.p. three times at weekly intervals with approximately 100ul of packed stromal cells (approximately 10^7 cells). Three days after the final injection, spleen cells were fused with P3/NS-1/1-Ag/4-1 (NS-1) cells and plated into 8, 96 well flat bottom tissue culture plates. Hybridoma supernatants were initially screened by indirect immunofluorescence (12) on 4u cryostat cut tissue sections of composite blocks of thymus, bursa, spleen (for the chicken) or thymus, kidney, liver (for the human). Hybridomas showing broad tissue reactivity were rejected and those of interest cloned twice by limiting dilution.

For anti-mouse thymus M.Abs, a modified immunization protocol was developed (Izon et al. in preparation). Lou/M rats were injected subcutaneously into each hind foot pad at day 0 and day 2 with 0.1ml stromal cell suspension (approximately 10^7 cells). After 4 days the popliteal lymph node cells were fused with NS-1 cells. In some experiments the rats were immunized i.p. as for the mouse anti-chicken or human M.Abs. Hybridomas were initially screened by indirect immunofluorescence on composite blocks of thymus, kidney and small intestine. Again broad reactive clones were rejected and those of interest cloned twice by limiting dilution. All M.Abs were thoroughly characterized by immunofluorescence on cryostat sections of a wide variety of tissues, lymphoid organ cell suspensions and cultures of thymic stromal cells. Reactivity with thymic epithelial cells was investigated by double labelling with anti-keratin.

RESULTS

Many similarities in thymic stromal antigens were found between the three species, particularly the antigenically distinct nature of the cortex and medulla and the similarities between the outer cortex and medulla. Each species, however, also revealed distinct specificities presumably due to differences in the content of the stromal cell suspension and the phylogenetic distance to the immunized animal.

Anti-chicken Thymic Stroma

The staining patterns on the chicken thymus were grouped into three categories: cortex, medulla or cortex and medulla (Table 1). Collectively they identified 8 antigenically distinct regions: trabeculae, sub-capsule/sub-trabeculae, sub-capsule-medulla, outer cortex, cortex, cortico-medullary junction, medulla, vasculature and peri-vasculature. None of the antibodies was thymus-specific but many reacted only with one other organ. The major characteristics of the chicken thymus were the sub-regions within the cortex (sub-capsule, sub-trabeculae, outer cortex, entire cortex and isolated stellate macrophage-like cells) and the multiple epithelial cells within the medulla (stellate networks and isolated clusters). Several M.Abs. demonstrated shared antigens in the cortex and medulla. An interesting finding was the antigenic similarity between the thymic medulla and

Table 1 Chicken Thymic Stromal Antigens

Cortex	Medulla	M.Ab.
trabeculae	-	MUI 56,75
sub-capsule, sub-trabec.	-	MUI 53,60
outer cortex	-	MUI 59,63
ep. network	-	MUI 51,54,55,65,73
isolated stellate cells	-	MUI 52
-	epithelial clusters	MUI 61,62,68,69,72
-	epithelial stellate	MUI 58,80
-	macrophages	MUI 67
-	vasculature	MUI 82
macrophages	macrophages, clusters	MUI 66
sub-capsule epithelium	epithelial network	MUI 70
sub-trabec, vasculature	peri-vasculature	MUI 74
stellate epithelium	confluent epithelium stellate cells	MUI 78
macrophages	macrophages	MUI 79

bursal cortex (e.g. MUI 72 stains thymic medullary clusters and bursal cortical stellate cells). At least 3 M.Abs (MUI 61, 62, 69) showed granular staining patterns which were both intra- and extra-cellular, suggesting secretion of soluble products; this was most characteristic in the medulla. Sub-populations of macrophage-like cells were identified by M.Abs MUI 52 (cortical), MUI 67 (medullary), MUI 66 (cortical and medullary) and MUI 79 (pan-macrophage). MUI 78 showed staining patterns similar to that observed for anti-MHC class II (4) reagents, reacting mainly with fine stellate networks or isolated cells in the cortex and a confluent cell meshwork in the medulla.

Anti-Mouse Thymic Stroma

The anti-mouse thymic stromal M.Abs were grouped into four main catagories (Godfrey *et al.* this volume): extensive stellate network, vascular-associated stromal cells, isolated stromal cells and determinants shared between thymocytes and stromal elements. Of the first group, 5 M.Abs identified antigens on fine epithelial stellate networks throughout the cortex and medulla. Differences in the extent of staining and high magnification examination of serial sections indicated the M.Abs were reacting with distinct antigens on the same or similar networks. Four M.Abs showed a similar distribution on the adult thymus as that described for MHC-II antigens (4), but reactivity on other tissues and an ontogenic study demonstrated that only one of these (MTS6) appeared to be against a mono-morphic MHC II determinant. MTS 10 demonstrated the antigenic similarity between the sub-capsule and medulla while MTS 11 was specific for stellate medullary epithelial cells. Five M.Abs reacted with vascular components involving both the lining endothelium and connective tissue cells immediately surrounding the vessels, particularly in the medulla, sub-capsule and trabeculae.

Two other major findings also emerged from the anti-mouse thymic stroma M.Abs. These were firstly the detection by 13 M.Abs of relatively rare (3-5 per section) isolated stromal cells with extensive stellate processes which were closely associated with surrounding T cells. These cells, also found as small clusters of 2-4 cells, were present in the subcapsule, cortex and medulla. At least one M.Ab, MTS 18, detected a granular, apparently secreted, product. The second was the vast number (8 M.Abs) of antigens common to thymic stromal cells and T lymphocytes. Most cortical thymocytes shared determinants expressed on medullary epithelial clusters (MTS 33 and 34) and/or cortical epithelial cells (MTS 31 and 32) whereas a medullary thymocyte antigen (MTS 36) was also found on isolated areas of the sub-capsular stroma. This antigenic similarity between thymic stromal cells and T lymphocytes was also demonstrated for Thy-1 and L3T4.

Anti-Human Thymic Stroma

Eight M.Abs have been raised to the human thymus, 3 of which react with keratin positive cells (Izon and Boyd, in preparation). These M.Abs. identify both isolated macrophage-like stromal cells (HTS-1), stellate networks (HTS-2,3), vascular associated stromal elements (HTS-4), subcapsule/subtrabeculae and medulla (HTS-6), predominantly trabeculae (HTS-7), subcapsule and medulla (HTS-8) and stromal cells within or closely linked to Hassall's corpuscles (HTS-5).

DISCUSSION

M.Abs raised against stromal cell-rich preparations of the chicken, mouse and human thymus have collectively demonstrated that the non-lymphocytic elements of this organ are equally as complex and diverse as the T cell maturation processes they regulate. Despite their phylogenetic divergence, many equivalent stromal cell antigens were detected in each of the species reflecting the very conserved nature of this organ in evolution. Several antigenic determinants clearly demonstrated that the cortical stroma is distinct from the medulla, confirming earlier findings (5,7). Since T lymphocytes are phenotypically and functionally very different in the cortex and medulla, it is reasonable to presume that the stromal components would also have different functional effects in directing T cell maturation.

Several antigenic determinants clearly demonstrated that the cortical stroma is distinct from the medulla. In addition to the distinct stellate epithelial networks characteristic of each of these regions, the cortex was found to have stromal elements specific for sub-capsule, sub-trabeculae, trabeculae and outer cortex, as well as isolated stellate cells interspersed throughout the epithelial networks. The medullary stroma was also heterogeneous having isolated epithelial cell clusters, dendritic macrophages, granular clusters, reticulin fibres around vasculature and Hassall's corpuscles. The difference in major stellate networks between the cortex and medulla no doubt reflects distinct embryological origins of these regions, with the sub-specificites possibly being due to differential penetration by bone marrow-derived haemopoietic cells, generation of hybrid lineage cells through interactions between endoderm, ectoderm and mesoderm components early in ontogeny and divergent differentiation pathways of the stromal components. In the human, HTS-5 reacted with granules in a cell in the centre of many, but not all, Hassall's corpuscles. One proposal is that the apparently secreted product of this cell induces the formation of these swirling, epithelial cell complexes and once the completed corpuscle is formed there is a down regulation of the antigen production.

The variety of stromal components associated with the sub-capsule, sub-trabeculae extending to the cortico-medullary junction and outer cortex is of particular interest because of the proposal that these regions are the site of stem cell immigration into the thymus and subsequent rapid cellular proliferation (13). That the sub-capsule may also be involved in the final differentiation of T cell subsets was suggested by the M.Abs in all 3 species which demonstrated antigenic identity between this region and the medulla - a well documented site of functionally mature T cells (1). In this regard, thymic nurse cells which are found predominantly in the subcapsular or outer cortical regions, contain T cells with limited functional potential (10). The close association between intra-thymic T cell differentiation and the circulation was demonstrated by M.Abs reacting with capillary endothelium throughout the cortex, medulla, trabeculae and to a lesser extent the subcapsule. A number of stromal cell-types were found to be in close juxtaposition to the vasculature and it may be these cells which initiate modifications in T cell maturation patterns in response to environmental changes such as stress, reflected by altered levels of e.g. hormones in the blood. They may also regulate the immigration of precursors or the exit of mature T cells from the thymus.

The shared antigenicity between thymic lymphocytes and stromal cells is a novel finding and raises several questions. Since we have also found this to be true of Thy-1 and PgP.1 (in preparation), care must be taken in interpreting the mechanisms of inhibition of T cell differentiation by antibodies reactive to T cell antigens. Although sole reliance on phenotype as a means of determining cell lineages may not be absolutely correct, this study indicates that at least some stromal cells and T cells may derive from a common blood-borne precursor. Alternatively, the same molecule may simply have a similar function on multiple cell types.

Perhaps the most intriguing finding in this study was the isolated infrequent stromal cells in either the cortex, medulla or both, the stellate plasma membrane processes of which appear to be in close contact with surrounding T cells. These relatively rare clusters demonstrate the complex microspecificities which exist within the thymus.

In summary, the M.Abs produced in this study clearly demonstrate the extreme heterogeneity of the thymic stroma and emphasise the need for reclassification of these diverse cell-types. An essential part of this will be the isolation of the molecules identified. This will facilitate both a better understanding of their functional contribution to the specific thymic microenvironment and the relationships between M.Abs produced by different groups.

REFERENCES

1. Scollay, R., Shortman, K. Thymus, 5 : 245, (1983).

2. Loor, F. Immunology, 37 : 157, (1979).

3. Frazier, J., Z. Zellforsch, 136 : 191, (1973).

4. Van Ewijk, W., Rouse, R.V., Weissman, I.L. J. Histochem. Cytochem., 28 : 1089, (1980).

5. Haynes, B.F. Adv. Immunol., 36 : 87, (1984).

6. Van Vliet, E., Melis, M., Van Ewijk, W. Eur. J. Immunol., 14 : 524, (1984).

7. De Maagd, R.A., Mackenzie, W.A., Schuurman, H.J., Ritter, M.A. Immunology, 54 : 745, (1985).

8. Boyd, R.L., Wilson, T.J., Mitrangas, K., Ward, H.A. Avian Immunology II. Eds. W.T. Weber and D.L. Ewert, (Alan R. Liss Inc. N.Y.), p.29, (1987).

9. Boyd, R.L., Oberhuber, G., Hala, K., Wick, G. J. Immunol., 132 : 718, (1984).

10. Andrews, P., Boyd, R.L. Eur. J. Immunol., 15 : 1043, (1985).

11. Andrews, P., Shortman, K. Cell Biophysics, 7 : 251, (1985).

12. Boyd, R.L., Ward, H.A. Immunology, 34 : 9, (1978).

13. Mandel, T. Aust. J. Exp. Biol. Med. Sci., 47 : 153, (1969).

ACKNOWLEDGEMENTS

This work was supported by grants from the National Health and Medical Research Council, Australia (grant No. 860396) and the Anti-Cancer Council of Victoria (grant No. 131685).

THYMIC STROMAL ELEMENTS DEFINED BY M.ABS:

ONTOGENY, AND MODULATION IN VIVO BY IMMUNOSUPPRESSION

D.I. Godfrey, D.J. Izon, T. J. Wilson
C.L. Tuček and R.L. Boyd

Department of Pathology and Immunology
Monash University, Melbourne, Australia

INTRODUCTION

The complex intra-thymic processes culminating in the generation of fully functional, self tolerant, self MHC restricted T lymphocytes have proven difficult to decipher. They appear to involve extensive interactions between T cell progenitors and thymic stromal cells or their soluble products. A better understanding of these mechanisms, however, has been hampered through the inability to identify and isolate individual subpopulations from the extremely heterogeneous thymic stroma.

A successful approach to this has been the production of monoclonal antibodies (M.Abs.) against specific thymic stromal subpopulations (1,2,3). In line with these studies, we have produced an extensive panel of M.Abs. against a wide variety of thymic stromal components. They have identified a number of previously unrecognized, distinct regions of the thymus and further, a complex array of microspecificities interspersed within these areas.

In addition to defining the thymic stromal architecture, these M.Abs. have been extremely useful for: i) monitoring the developmental relationship of individual thymic stromal cells with T lymphocytes during ontogeny; and ii) analysing modulations in thymic stromal elements in in vivo immunosuppressed states. Two models have been used in the latter: a) natural immunosuppression associated with pregnancy; and b) severe depletion of rapidly dividing thymic lymphocytes by the administration of 5-Fluorouracil (5-Fu).

MATERIALS AND METHODS

Animals. In all experiments, 5-7 week old female Balb/c or CBA mice (both obtained from Monash Central Animal House) were used. Female Leu/M rats, approximately 1 year old (Monash Central Animal House) were used for immunization with thymic stromal cells.

Preparation of thymic stromal cells. Thymic stromal cells were prepared using standard methods (4,5). Briefly, thymuses were minced with fine scissors into small pieces, thymocytes discarded and the residual fragments subjected to enzymic digestion with 2.5 ml collagenase (3 mg/ml;

from Clostridium histolyticum, type IV) and 1 ml DNase (1 mg/ml from bovine pancreas) (All from Boehringer Mannheim, West Germany). Stromal cells were then enriched for by settling through foetal calf serum (FCS, Flow Laboratories, Sydney, Australia) at unit gravity for 30 mins.

Immunization schedule. Two immunization protocols were used for M.Ab. production. Leu/M rats were immunized i) 3 times I.P. at day 0, week 3, week 13, followed by one I.V. immunization 4 days later with enriched thymic stromal cell suspensions (1 ml of approximately 10^7 cells/ml/immunization); ii) subcutaneously into each hind footpad at day 0 and day 2 with enriched murine thymic stromal cell suspensions (0.1 ml, approximately 10^8 cells/ml). After 4 days, spleen cells or popliteal lymph node cells were fused with P3/NS-1/1-Ag/4-1 (NS-1) cells as described by Galfre (6) with modifications by Goding (7). Characterization of M.Abs. Hybridoma supernatants were screened by indirect immunofluorescence, using FITC conjugated sheep-anti-rat Ig (pre-absorbed to mouse Ig). (QMF, Silenus Laboratories, Dandenong, Australia) on 4 um frozen sections from composite blocks of murine thymus, small intestine, and kidney. Hybridomas of interest were cloned twice by limiting dilution and their specificities thoroughly characterized by screening on tissue sections of brain, liver, heart, lung, spleen and lymp node.

Double labelling with anti-keratin. Reactivity of M.Abs. with thymic epithelial cell antigens were determined via the simultaneous testing of rabbit anti-keratin (DAKO, Santa Barbara, U.S.A.), with the M.Abs. on frozen murine thymus tissue sections. Conjugates used for detection of keratin were biotinylated donkey-anti-rabbit (Amersham, U.K.), followed by phycoerythrin conjugated avidin (Serotec, Oxford, U.K.).

Ontogenic development study. M.Abs. were screened on frozen sections of thymuses from day 13, daily through to neonatal.

Analysis from mice in mid-pregnancy. M.Abs. were screened on frozen sections from thymuses from 15 day, plug determined, pregnant CBA mice. Analysis of thymuses from mice treated with 5-Fu. Five week old CBA mice were injected with 5-Fu (DBL Laboratories, Melbourne, Australia). Thymuses were taken at 12, 24, 36, 48 and 72 hours and snap frozen for screeining with M.Abs.

RESULTS

On the basis of their distribution pattern on murine thymus, small intestine and kidney tissue sections, 37 M.Abs, were selected from over 4000 screened. These M.Abs. fell into 4 broad categories of thymic reactivity: extensive stellate networks (table 1); vascular associated stromal elements (table 2); isolated stromal elements (table 3); and thymocytes and stromal elements (table 4).

i) Extensive stellate networks

This category includes 11 different M.Abs. which stain predominantly stellate networks. Double labelling experiments with anti-keratin showed these to be primarily epithelial in nature. MTS6 stained a fine cortical and dense medullary epithelial network, identical to that described for MHC class II antigens (8,9). MTS10 and 11 identified a reticular epithelial network in the medulla. The former also detected the squamous epithelium of the subcapsule, whilst the latter was completely negative on the cortex.

Abbreviations: Stell. - Stellate; n/w - network; SC - Subcapsule; gran.- granular

ii) Vascular associated stromal elements

The capillary endothelium of both the cortex and medulla was detected by MTS12. Four other M.Abs. (MTS13-16) detected stromal elements in direct contact with these capillaries as well as the capsule and trabeculae, possibly reflecting the presence of vascular channels in these connective tissue components.

iii) Isolated stromal elements

An intriguing observation in this study was the detection of brightly stained, infrequent stromal cells. These were usually observed either completely isolated or in small clusters of 2 to 4 cells and were divided into 3 groups: Cortical and medullary; cortical only; and medullary only.

iv) Thymocytes and stromal cells

The final category of M.Abs. clearly illustrated antigenic similarities between subpopulations of thymocytes and the thymic stroma. The majority of cortical thymocytes shared antigenicity with either medullary epithelial clusters (MTS33 & 34), or cortical epithelium (MTS31 & 32). Distinct subpopulations of cortical thymocytes were also shown to cross-react with isolated medullary epithelial cells, or small clusters thereof (MTS35 & 37). Similarly, a medullary thymocyte antigen detected by MTS36 was also found in isolated areas of the subcapsular stroma.

Table 1. Extensive stellate networks.

Cortex	Medulla	M.Ab.	Ontogenic Appearance
Fine stell. n/w.	Fine stell. n/w.	MTS1	13 (days)
		MTS2	13
		MTS3	13
		MTS4	13
		MTS5	13
Fine stell. n/w.	Dense stell. n/w.	MTS6	13/14
		MTS7	13/14
		MTS8	13
		MTS9	17
SC. epithelium	Stell. n/w.	MTS10	13
-	Stell. n/w.	MTS11	15

Table 2. Vascular associated stromal elements.

Cortex	Medulla	M.Ab.	Ontogenic Appearance
Capillaries	Capillaries	MTS12	13 (days)
SC / Vascular assoc'd stroma	Clusters around vasculature	MTS13	17
		MTS14	17
		MTS15	17
		MTS16	13

Ontogenic development of the antigens

The developmental relationship between the various stromal determinants and their association with different phases of T cell maturation, were investigated in an ontogenic study. First appearances of each antigen are listed in tables 1 to 4.

At day 13, the embryonic thymus is mostly cortex, with a small central medulla. The appearance of trabeculae and medullary pockets at days 15/16 represent the beginnings of lobulation, giving rise around day 18 to the characteristic cortical and medullary pattern observed in the adult. Evidence supporting the proposed embryonic "engulfment" of the cortex by

Table 3. Isolated stromal elements. * - transient appearance

Cortex	Medulla	M.Ab.	Ontogenic Appearance
Small clusters / isolated cells (SC)	Small clusters / isolated cells	MTS17	16 (days)
		MTS19	20
Isolated cells (SC., gran.)	Isolated cells	MTS18	*13/20
-	Small clusters / isolated cells	MTS20	*13/20
		MTS21	17
		MTS22	*13/20
		MTS23	20
		MTS24	13
		MTS25	20
-	Infrequent isolated cells	MTS26	20
Isolated cells	-	MTS27	*14/20
		MTS28	17
		MTS29	17

Table 4. Thymocytes and stromal cells.

Cortex	Medulla	M.Ab.	Ontogenic Appearance
Thymocytes & stromal cells	Thymocytes & stromal cells	MTS30	13 (days)
Thymocytes & stromal cells	-	MTS31	14
		MTS32	14
Thymocytes	Isolated epithelial cell clusters (gran)	MTS33	15
		MTS34	15
SC thymocytes	Infrequent isolated epithelial cells	MTS35	13
Infrequent stromal cells (SC).	Thymocyte subpopulations	MTS36	16
SC & thymocyte subpopulation	Isolated epithelial cell clusters (gran).	MTS37	13

medullary processes which extend along the trabeculae to eventually form the subcapsule (10) was provided by MTS10. Staining was initially restricted to the medulla during embryogenesis, with subcapsular reactivity not being detected until the post-neonatal stage.

Vascularization, as detected by MTS12, had already commenced by day 13. This was to be expected, as stem cells have already begun seeding the thymus at this stage during ontogeny (11). Maturation of the vascular associated stromal elements, with the exception of MTS16, did not occur until at least day 17.

The isolated stromal elements showed variable appearance, most not occuring until at least day 17. Of particular interest, MTS18, 20, 22 & 28 transiently stained early in embryogenesis (the latter two near confluence), reappearing toward the neonatal stage.

The thymocyte/stromal cell M.Abs. displayed distinctly different patterns of reactivity during ontogeny, when compared to that of the adult thymus. MTS31 and 32, normally restricted to cortical thymocytes and stroma, showed no distinction between cortex and medulla until day 16. Similarly, MTS35 and 37, normally detecting subpopulations of cortical thymocytes and medullary epithelium, also stained the entire cortex and medulla until day 16 of embryogenesis (Figs. 1 & 2). MTS33 and 34 demonstrated that the same antigen appears on different lineages at different stages of development, i.e. medullary epithelial clusters stained as of day 15, whereas the cortical thymocytes remained negative until the post-neonatal stage.

Pregnant mouse thymus

In comparison to the normal thymus, the thymuses from mice in mid-pregnancy were markedly decreased in size, with relatively disorganized cortical and medullary regions.

Major alterations in the stromal cells/molecules as detected by the M.Abs. are listed in table 5. The most interesting observation involved MTS36. This M.Ab. revealed an increase in both frequency and intensity of a medullary thymocyte subpopulation (Figs. 3 & 4).

Table 5. Pregnant mouse thymus.

M.Ab.	Normal thymus	Pregnant mouse thymus
MTS35	SC thymocytes	SC thymocytes absent
MTS36	Medullary thymocyte SP	↑ intensity and frequency
MTS14	Subcapsule/ subtrabeculae	Broader, more diffuse
MTS12	Capillaries	Capillaries dilated
MTS10	SC & medullary epithelium	~90% thymus positive

Table 6. 5-Fu treated mouse thymus.

M.Ab.	Normal thymus	5-Fu treated thymus
MTS35	SC thymocytes	SC thymocytes absent
MTS14	Subcapsule/ subtrabeculae	Broader, more diffuse
MTS10	SC & medullary epithelium	~90% thymus positive
MTS28	Isolated cortical stromal cells	24hrs: ↑ freq. and intensity 36hrs: positive pockets 48hrs: extensive staining 72hrs: predominantly cortical

5-Fu treated mouse thymus

The 5-Fu treated mouse thymus showed great similarity to the thymus during mid-pregnancy, having a collapsed cortical framework and a relatively enriched medulla. The most significant finding from this study involved MTS28. This M.Ab. normally detected isolated, infrequent cortical stromal cells (Fig. 5). Following 5-Fu treatment, however, an increase in both frequency and intensity of this antigen was observed, leading to extensive, confluent staining primarily of the cortical region by 48 hours (Fig. 6).

DISCUSSION

The main purpose of this study has been to produce and characterize M.Abs. against thymic stromal components with the ultimate aim of utilizing their unique resolving power for the identification, charcterization, and isolation of specific thymic stromal subpopulations.

Thirty seven M.Abs. were grouped into 4 broad categories, based upon their reactivity on frozen thymus sections: i) extensive stellate networks; ii) vascular associated stromal elements; iii) isolated stromal elements; and iv) thymocytes and stromal cells.

The antigenically distinct nature of thymic cortical and medullary epithelium (3), was confirmed by MTS10, even as early as day 13 in ontogeny. This M.Ab. also labelled the subcapsular cortical epithelium, an identical pattern to that detected by M.Abs. anti p-19, TE4 (2,12), and A2B5 (13) proposedly representing the neuroendocrine thymic epithelial component. This staining is of particular interest, as these regions have been proposed to be the sites of stem cell entry and initial expansion within the thymus (14).

Of the vascular related stromal elements (table 2), MTS12 enabled mapping of the thymic capillaries. Association between these capillaries and other specific stromal elements, e.g. as detected by MTS13 to 16, may implicate a role in the attraction of proteins, hormones, and cells across the blood-thymus barrier.

The isolated, sometimes very infrequent stromal cells (table 3) demonstrated the complex microspecificities which exist within the thymic cortical and medullary microenvironments, The transient, enhanced appearance of some of these antigens during ontogeny (MTS18, 20, 22 & 27), suggests they may have a role in initial events in T lymphocyte differentiation. It would be interesting to examine these antigens during reconstitution of X-irradiated thymus.

M.Abs. of the fourth category demonstrated an extensive association between thymocytes and thymic stromal cells. This association may be due to secretion of soluble factors by the stromal cells which are concordantly taken up by the thymocytes, or vice versa.

Alternatively, these M.Abs. may simply be detecting antigenic determinants shared between both cell types. This could reflect a common ontogenic lineage, or molecules having a similar function in multiple cell types. The enhanced expression of 4 of these antigens early in embryogenesis (MTS31, 32, 35 & 37), suggests that these may also play an important role in stem cell entry and/or proliferation within the thymus. Conversely, MTS33 & 34, reactive with cortical thymocytes and medullary epithelial clusters in the adult, stained stromal cells but not thymocytes during ontogeny, perhaps supporting the above proposal of the stromal cells secreting soluble factors which are concordantly taken up by the thymocytes.

In vivo modulation of the thymic microenvironment following immunosuppression was analysed using 2 models: i) Natural immunosuppression associated with pregnancy; and ii) Severe depletion of rapidly dividing thymic lymphocytes by administration of 5-Fu (a drug used in the treatment of some forms of cancer). These thymuses were commonly severely depleted of cortical thymocytes, as detected by MTS35, resulting in a collapsed cortical network and expanded subcapsular/subtrabecular regions. Mid-pregnancy thymuses displayed markedly intensified staining of a medullary thymocyte subpopulation (MTS36). This model may represent an ideal means for the enhancement of this otherwise infrequent, weakly staining subpopulation. Similarly 5-Fu treated thymuses displayed a vastly increased expression of an antigen normally restricted to isolated, infrequent cortical stromal cells. Both models thus demonstrate modulation of the thymic stroma *in vivo*, and represent excellent means for the enrichment of otherwise infrequent stromal antigens for further characterization and functiotnal analysis.

ACKNOWLEDGEMENTS

This work was supported by grants from the National Health and Medical Research Council, Australia (grant No. 860396 and the Anti-Cancer Council of Victoria (grant No. 131685).

REFERENCES

1. R. A. De Maagd, W. A. MacKenzie, H. J. Schuurman, M. A. Ritter, K. M. Price, R. Broekhuizen, and L. Kater, *Immunology* 54:745 (1984).
2. B. F. Haynes, R. M. Scearce, D. F. Lobach, and L. L. Hensley, *J. Exp. Med.* 159:1149 (1984).
3. W. Van Vliet, E. J. Jenkinson, R. Kingston, J. T. T. Owen, and W. Van Ewijk, *Eur. J. Immunol.* 15:675 (1985).
4. H. Wekerle, U. P. Ketelson, and M. Ernst, *J. Exp. Med.* 151:925 (1980).
5. P. Andrews and R. L. Boyd, *Eur. J. Immunol.* 15:36 (1985).
6. G. Galfre, S. C. Howe, C. Milstein, G. W. Butcher, and J. C. Howard, *Nature* 266:550 (1977).
7. J. W. Goding, *in*: "Monoclonal Antibodies: Principles and Practice", J. W. Goding, ed., Academic Press, London (1983).
8. R. V. Rouse, W. Van Ewijk, P. P. Jones, and I. L. Weissman, *J. Immunol.* 122:2508 (1979).
9. W. Van Ewijk, R. V. Rouse, and I. L. Weissman, *J. Histochem. Cytochem.* 28:1089 (1980).
10. D. A. Crouse, J. B. Turpen, and J. G. Sharp, *Surv. Immunol. Res.* 4:120 (1985).
11. J. F. Bach and M. Papiernik, *in*: "Microenvironments in Haemopoietic and Lymphoid Differentiation". Ciba Foundation Symposium 84, Pitman, London (1981).
12. B. F. Haynes, M. Robert-Guroff, R. S. Metzgas, G. Franchini, V. S. Kalyanaraman, T. J. Palker, and R. C. Gallo, *J. Exp. Med.* 157:907 (1983).
13. R. Kingston, *Eur. J. Immunol.* 14:1052 (1982).
14. R. Scollay, *Immunol. Today* 4:282 (1983).

KINETICS OF RAT THYMIC DENDRITIC CELLS IN BONE MARROW-RECONSTITUTED RADIATION CHIMERAS

Adrian M. Duijvestijn, Malgorzata B. Murawska, Boelo Meedendorp
Flip A. Klatter, Jaap Kampinga and Paul Nieuwenhuis

Department of Histology, Medical Faculty, University of Groningen, The Netherlands

INTRODUCTION

The thymus meets unique microenvironmental conditions essential for pre-T cell maturation and generation of the complete T cell repertoire. During intrathymic development thymocytes acquire qualities essential for adequate immune functioning of T cells, such as tolerance for self, antigen receptors, and restriction to class I or II MHC antigens. With respects to restriction of T helper cells to class II (Ia) antigens of the MHC, it is more or less generally accepted that class II positive cells in the thymus control this aspect in T cell 'education'. Potential candidates for regulating restriction of T helper cells to class II molecules are thymic reticular epithelial cells and bone marrow-derived thymic dendritic cells (DC), both known to be Ia^+. Attending the question which cell type indeed is controlling class II restriction in the thymus, one can study T helper cell restriction to Ia antigens in allogeneic or semi-allogeneic bone marrow-reconstituted radiation chimeras (BMRRC) or thymus-grafted animals. Results obtained from these studies are conflicting, suggesting on one hand, that class II restriction of T cells is dictated by thymic epithelial cells (1), or on the other hand, that DC (antigen presenting cells) in the thymus are involved in this (2,3). Of crucial importance in such restriction studies is to know at what experimental stage the thymus can be considered class II-chimeric with respects to epithelial cell versus DC, viz. how long does it take in BMRRC for the recipient thymic DC to be replaced by allogeneic or semi-allogeneic donor type DC?

We are interested in involvement of thymic stromal cells in restriction and tolerance induction in rats, and therefore developed a model of allogeneic and semi-allogeneic BMRRC. In this model, we have studied kinetics of thymic DC, and also of DC in T-dependent areas of peripheral lymphoid organs. By immuno-histology and cytology, using monoclonal antibodies discriminating between donor and recipient class II antigens, influx of donor type and disappearance of recipient type DC has been followed. The data (although not absolute conclusive, as will be discussed) indicate that DC apparently turn over rapidly and that replacement by donor thymic DC, and thus establishing thymic chimerism of epithelial cells versus DC, occurs in about 4 weeks.

MATERIALS AND METHODS

<u>Rats</u>. Used strains were inbred AO/G ($RT\text{-}1^u$), and BN/G ($RT\text{-}1^n$), and

their (AOxBN)F1 hybrids, bred in our own animal facility. Animals were male, and 8-10 weeks of age when used for preparation of chimeric rats. Animals had free access to food and water.

Bone Marrow-Reconstituted Radiation Chimeras (BMRRC). Semi-allogeneic and allogeneic chimeras were prepared by lethal total body 950 cGy (gamma) irradiation of recipients, followed 1 day later by i.v. injection of 5-10 x 10^7 freshly isolated donor bone marrow cells (bm) obtained from femur and tibia. The following BMRRC were prepared:

- AO bm ---->irradiated (AOxBN)F1
- (AOxBN)F1 bm ---->irradiated BN
- AO bm ---->irradiated BN

Controls were syngeneic bone marrow-reconstituted rats:

- (AOxBN)F1 bm ---->irradiated (AOxBN)F1
- BN bm ---->irradiated BN

Chimeric animals were sacrificed at 1,2,3,4,6,8,12, and 20 weeks.

Experimental outline. We used 2 anti-class II monoclonal antibodies (MAB) to detect donor and recipient type DC. MAB HIS-19, against a polymorphic class II MHC antigen, detects AO (RT-I^u) class II but not BN (RT-1^n) class II. MAB F17-23-2, also seeing a polymorphic class II antigen, detects BN (RT-1^n) but not AO (RT-1^u) class II.

Thus, in both AO ----> (AOxBN)F1, and AO ----> BN chimeras, F17-23-2 can be used to follow disappearance of recipient type DC. In both (AOxBN)F1 ----> BN, and AO ----> BN chimeras, HIS-19 detects incoming donor DC. AO ----> BN chimeras offer the advantage that influx of donor and disappearance of recipient DC can be measured in the same animal.

Tissue. The following tissues were taken from BMRRC and controls: thymus, spleen, cervical lymph node, Peyer's patch, tongue, salivary gland, intestine, liver, heart, and skin. Tissues were frozen in liquid freon and stored at -70°C. Freshly excised thymus material for suspension work was put in cold medium containing 10% NCS.

Monoclonal antibodies (MAB) and enzyme-conjugates. The following mouse anti-rat MABs were used: HIS-19 (anti-polymorphic class II), HIS-14 (anti-B cell), HIS-17 (anti-T cell), all from the dept. of Histology, University of Groningen; ED-2 (anti-macrophage subset), gift from dr. C. Dijkstra, dept. of Histology, Free University Amsterdam; F17-23-2 (anti-polymorphic class II) gift from dr. A. Williams, MRC Cellular Immunology Unit, University of Oxford; 1F-119 (anti-DC), gift from dr. L. Nagelkerken, dept. of Immunology, University of Limburg. Second stage antibody was peroxidase-conjugated rabbit anti-mouse Ig (from DAKO).

For double staining with HIS-19 and F17-23-2 we used biotinylated HIS-19 and peroxidase-conjugated F17-23-2. Second step for HIS-19/Biotine was Alkaline Phosphatase-conjugated Streptavidine (SBA, California).

Immunohistology. 8 um thick frozen sections were cut, and stored dry. Upon use, sections were fixed in acetone for 10 min., dried and stained in a 2 step immunoperoxidase technique as described before (4). In addition, double-immunoenzyme staining was performed with monoclonal antibodies HIS-19 and F17-23-2 on thymus and spleen sections, according to the method described by Van Rooijen (5).

Immunocytology. Pooled thymus material of 2-3 rats was used for cytocentrifuge preparations of suspensions enriched for DC. Thymus suspensions were made as described before (6). To enrich DC, thymus suspensions were centrifuged over a 1.45 gr/10 ml Nycodenz (Nygaard, Denmark) gradient, and the collected interphase cells used for preparations. Cytocentrifuge preparations were stored dry, and fixed in acetone before single or double immunoenzyme-staining.

DC quantification. Cytocentrifuge preparations of DC-enriched thymus cell suspensions from allogeneic BMRRC were used. Suspensions consisted of pooled half-thymuses of 2-3 rats per stage. Numbers of either HIS-19 or F17-23-2 positive cells with DC morphology were determined from 1000 counted cells, and percentages of recipient DC of the total DC population were calculated.

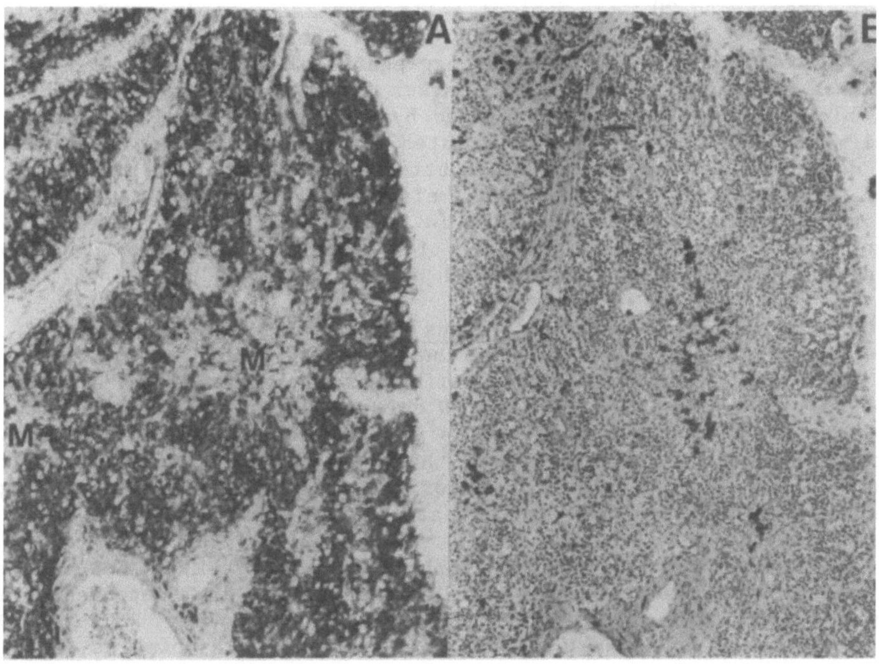

Fig. 1. Thymus of 1 wk allogeneic chimera; consecutive sections; 10x obj A) F17-23-2 staining recipient class II positive reticular epithelial cells in the cortex and scattered DC in medullary-like area (M). B) HIS-19 stains donor bone marrow-derived class II positive (dendritic) cells in medullary area and in interlobular connective tissue (arrows).

RESULTS

Overall aspects of allogeneic and semi-allogeneic BMRRC. In the 3 chimeric models no obvious signs of graft-versus-host-disease (GVHD) were observed. There was correct lymphoid-tissue compartmentalization, and no persistent increased levels of class II expression or presence of inflammatory infiltrates could be observed in the tested tissues. In immunohistology performed with the various MABs, the lymphoid organs of the 3 types chimeric rats were mutually comparable and did not differ from control or untreated rats.

Kinetics of thymic dendritic cells in BMRRC. I wk after bm reconstitution, the thymus was tiny, but in a stage of regeneration.

Thymocytes were already numerous in the cortex but almost absent in medullary-like areas in which few class II (HIS-19) positive cells of donor origin were observed. In addition, HIS-19 positive cells could be noticed in thymic capsule and interlobular septa. Recipient type class II (detected by F17-23-2) was present on cortical reticular epithelial cells and on DC in medullary areas (Fig.1). From 2 wks on, donor type class II positive DC were numerous in medullary areas. Also in the cortex scattered donor-derived class II positive cells were present, whereas frequently in thymus capsule and septa some HIS-19 positive cells could be seen. At 2 weeks, F17-23-2 staining clearly outlined the cortical epithelial reticulum and showed scattered recipient type DC in the medulla. In later stages only few to almost none scattered recipient DC were observed. However, even up to 20 wks, foci of cells expressing recipient type class II were observed in medullary areas (Fig.2A). These foci were intensely stained with F17-23-2 and resembled aggregates of DC.

Since class II is not a unique marker for DC, we concluded MAB 1F-119, which has been recently developed by Nagelkerken et al. (7), to identify DC. 1F-119 staining demonstrated the presence of DC in medullary areas during chimeric development. Controles and later-stage chimeras showed, when stained with HIS-19 anti-class II, a confluent medullary staining pattern and scattered cells in the cortex, whereas 1F-119 demonstrated a scattered distribution of DC in the medulla, and an almost absence of positive cells in cortical areas (Fig.2B & C).

Quantification of recipient versus donor thymic DC. In chimeras 2,4,6,8,12 wks of age we calculated the percentage recipient DC of the total thymic DC population. Cytocentrifuge preparations of DC-enriched thymus cell suspensions of AO ----> BN chimeric rats were stained with HIS-19 or F17-23-2, and positive cells with DC morphology were counted. At 2 weeks, recipient DC made up 28% of the total thymic DC population. At 4 wks only 3% of the DC population was of recipient origin. In later stages percentages declined to less than 1% at 12 wks (Fig.3).

Kinetics of DC in peripheral lymphoid tissues of BMRRC. Entrance of donor and disappearance of recipient DC was examined in T-dependent areas of spleen, lymph node and Peyer's patch by immunohistology. Recipient DC were still observed in small numbers in 1 and 2 wks old chimeras, but were nearly absent in T-dependent areas at later stages (Fig.4). However, up to 12 wks recipient class II positive non-lymphoid cells were observed in decreasing numbers in marginal sinuses of splenic white pulp and subcapsular sinuses of cervical lymph nodes. Incoming donor type DC were clearly present in T cell areas from 4 wks on. In earlier stages, donor type DC could not be identified, due to a rather confluent donor-class II positivity in T cell areas.

CONCLUSIONS

We have studied kinetics of thymic DC in BMRRC by using haplotype-specific anti-class II antibodies in immuno-histology and cytology. In allogeneic and semi-allogeneic rat chimeras influx of donor and disappearance of recipient type DC were examined. Section studies demonstrated class II positive donor type DC to appear in the thymus already in the first wk after bm reconstitution. Subsequently there is a rapid increase of donor type thymic DC (see also 8,9) by which at 4 wks medullary areas predominantly contain donor type DC. Our quantitative studies, in which we calculated the contribution of recipient type DC to the total DC population in chimeras showed that after 4 wks the recipient thymus DC population contributes for only 3%. Thus, our data suggest that in BMRRC replacement of recipient by donor thymic DC, and thus complete thymic chimerism of epithelial versus dendritic cells is realized in about 4 weeks.

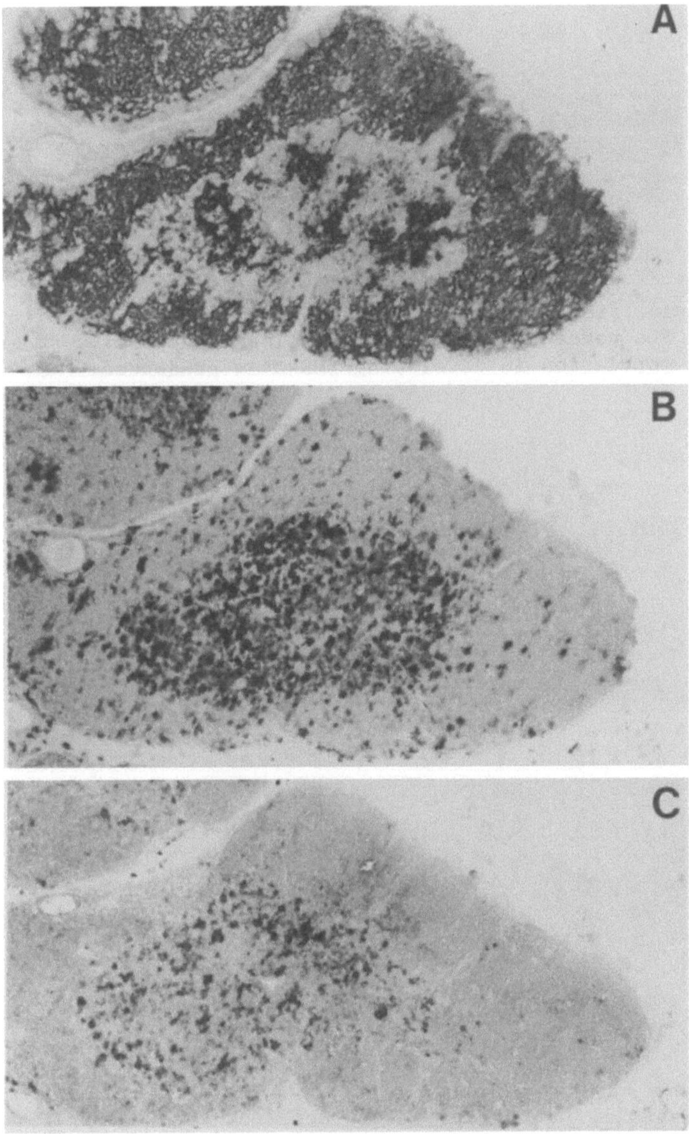

Fig. 2. Thymus of 12 wk allogeneic chimera; consecutive sections; 10x obj. A) F17-23-2 staining. A reticular pattern of recipient class II positive epithelial cells is observed in the cortex. In the medulla, foci of aggregated class II positive cells of recipient origin are present. Note the clear cut border of cortex and medulla. B) HIS-19 staining shows a rather confluent medullary class II staining pattern as observed in normal thymus. Scattered positive cells are present in the cortex. C) 1F-119 staining of DC. Positive cells are predominantly located in the medulla. Staining is much more scattered compared to class II staining.

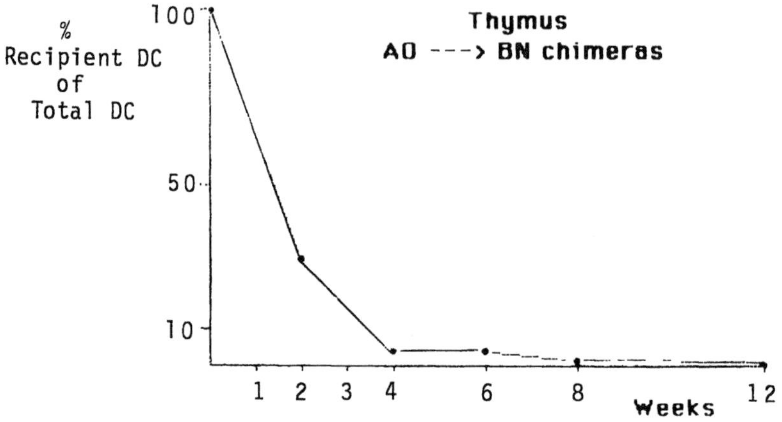

Fig. 3. Quantification of DC in thymus cell suspensions of allogeneic chimeras. The percentage of recipient class II positive DC in the total thymic DC population in developing BMRRC is presented.

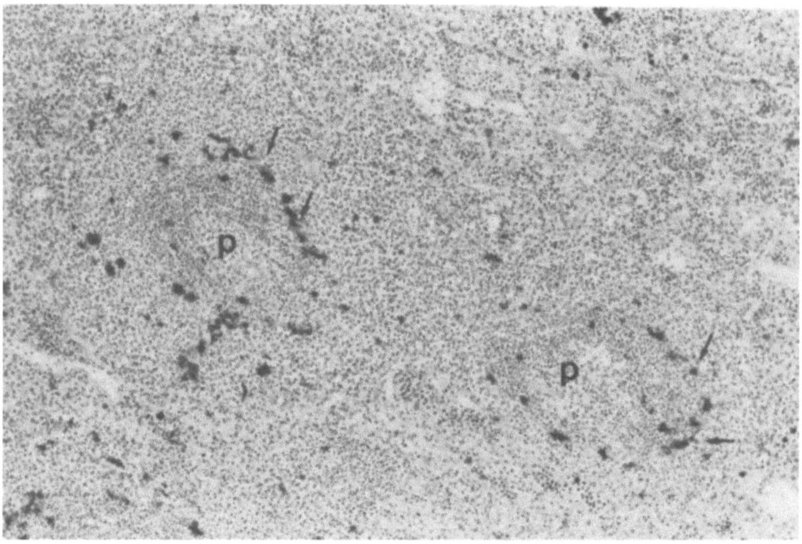

Fig. 4. Spleen of 6 wk allogeneic chimera; 10x obj. Peri-arteriolar T cell area (P) is small, but negative for recipient-type class II positive DC. Recipient class II positive cells are predominantly observed in the marginal sinus (arrows). Note the absence of recipient class II positive cells in the red pulp.

However, we can not be fully conclusive about kinetics of DC replacement, since sections of chimeric thymuses up to 20 wks showed persistent foci of aggregated cells expressing recipient class II in medullary areas. Morphologically the cells most resembled DC. Attempts to identify these cell aggregates by immunostaining with anti-keratin and 1F-119, markers for epithelial cells and DC respectively were not very succesful, demonstrating both cell types to be present in the aggregates.

Most likely, cells from the foci, that apparently are lost in suspensions, consist of recipient type DC that have been aggregated and thus immobilized as a result of irradiation and lymphocyte depletion.

Also in peripheral lymphoid tissue T-dependent areas are rapidly depleted from recipient type DC in about 4 wks. It is not clear whether the recipient class II positive cells in sinuses of spleen and lymph node observed till about 12 wks reflect entrance or exit of DC. Likely, cells in the subcapsular sinuses of cervical lymph nodes represent drainage of veiled cells/Langerhans cells from the skin (10).

Acknowledgements: We thank Jan Feringa & Dorus Zadel for animal facilities, Auk Wubbena for technical assistance, Peter van der Syde for photographic assistance, and Doete Niermeyer for secretarial support. The research of Dr. A.M. Duijvestijn has been made possible by a fellowship of the Royal Netherlands Academy of Art and Sciences.

REFERENCES

1. D. Lo and J. Sprent. Nature 319:672 (1986).
2. D.L. Longo and R.H. Schwartz. Nature 287:44 (1980).
3. D.L. Longo, A.M. Kruisbeek, M.L. Davis and L.A. Matis. Proc.Natl.Acad.Sci.USA 82:5900 (1985).
4. F.G.M. Kroese, A.S. Wubbena, D. Opstelten, G.J. Deenen, E.H. Schwander, L. de Leij, H. Vos, S. Poppema, J. Volberda and P. Nieuwenhuis. Eur.J.Immunol. 17:921 (1987).
5. N. van Rooijen, N. Kors and R. van Nieuwmegen. J.Histochem.Cytochem. 32:677 (1984).
6. A.M. Duijvestijn, R. Schutte, Y.G. Kohler, C. Korn and E.C.M. Hoefsmit. Cell Tissue Res. 231:313 (1983).
7. L. Nagelkerken, B. Schutte, R.J.M. Stet and P.J.C. van Breda Vriesman. Scand.J.Immunol. (in press).
8. A.N. Barclay and G. Mayrhofer. J.Exp.Med. 153:1666 (1981).
9. A.M. Duijvestijn and A.N. Barclay. J.Leukocyte Biol. 36:561 (1984).
10. E.C.M. Hoefsmit, A.M. Duijvestijn and E.W.A. Kamperdijk. Immunobiology 161:255 (1982).

ALLOGENEIC CULTURED THYMIC FRAGMENTS IN CONGENITALLY ATHYMIC (NUDE) RATS

Henk-Jan Schuurman, Frans J. Tielen,
Lennard M.B. Vaessen, Joseph G. Vos and Jan Rozing

Division of Immunopathology, Departments of Internal Medicine and Pathology, University Hospital, Utrecht; Institute for Experimental Gerontology, Division for Health Research TNO, Rijswijk; Department of Internal Medicine I, Erasmus University, Rotterdam; and Laboratory for Pathology, National Institute of Public Health and Environmental Hygiene, Bilthoven, The Netherlands

INTRODUCTION

Culture of small thymic explants yields fragments depleted of lymphocytes which consist of epithelial-like cells and connective tissue components (1). Such cultured thymic fragments (CTF) induce the development of thymus-dependent immunity when transplanted under the kidney capsule of congenitally athymic animals (1-3). After implantation, CTF become populated with lymphoid elements and regain their original thymus architecture. We (3) and others (2,4) have documented that both syngeneic and allogeneic CTF are effective in this respect. This phenomenon enables to study the impact of the MHC haplotype of CTF on T-cell immunity generated after implantation, especially aspects as allo-antigen recognition repertoire, self-MHC restriction specificity, and self-MHC tolerance. In this respect, the role of distinct stromal components in the thymus is subject of continuing discussion, especially the discrimination between the epithelial reticulum and medullary dendritic cells (5-9).

We here report our studies on allo-antigen recognition specificity in the model of allogeneic CTF implantation in nude rats. The data presented summarize our studies described in detail elsewhere (10,11, Kendall et al., manuscript in preparation).

MATERIALS AND METHODS

Rats. Congenitally athymic (rnu/rnu) rats and heterozygous +/rnu littermates were backcrossed in the WAG/CPB ($RT1^u$) strain. CTF was prepared by a 6-8 day culture of 1-3mm thymic fragments from PVG/CPB ($RT1^c$), DA/Ola ($RT1^a$) or BN/Ola ($RT1^n$) rats. Other rat strains used in mixed leukocyte reactions (MLR) and cell-mediated lympholysis (CML) were LEW/Ola (($RT1^l$) and RP/EUR ($RT1^{p}$(u,l)).

Antibodies. To identify T-lymphocyte subpopulations in tissue sections (12), we used the following monoclonal antibodies: MRC OX-19, binding to thymocytes and peripheral T cells (provided by Dr A.F. Williams, Oxford,

UK); ER-1, to thymocytes, with faint intensity for cells in the cortex, and peripheral T cells; ER-2, to all thymocytes in the cortex and about two-thirds of T cells in the thymus medulla and peripheral lymphoid organs, being associated with helper-inducer T-cell function; MRC OX-8, to all thymocytes in the cortex and about one-third of T cells in the thymus medulla and peripheral lymphoid organs, being associated with suppressor-cytotoxic T-cell function (Mas 041b; Sera-Lab, Crawley Down, UK); and ER-4, to cortical thymocytes in strong intensity and medullary thymocytes in weak intensity. To detect RT1 class I antigen we used MRC OX-18 to a common class I determinant (provided by Dr A.F. Williams); and U9-F4-B1, to an $RT1^{u}$ class I antigen (provided by Dr A. Santema, Leiden, The Netherlands). Anti-RT1 class II antibodies applied were: MRC OX-4 HL, binding to a common class II determinant (Mas 029b; Sera-Lab); MRC OX-3 HL, to a $RT1^{u}$ class II epitope (Mas028b; Sera-Lab); HIS 19, to a polymorphic class II determinant not present on $RT1^{n}$ molecules (provided by Dr R.J.M. Stet, Groningen, The Netherlands) (13); and F-17-23-2, which binds to a polymorphic determinant on $RT1^{n}$, $RT1^{a}$, and $RT1^{l}$ molecules. For dendritic cells, antibody 1F119 was provided by Dr L.M. Nagelkerken, Rijswijk, The Netherlands (14). This reagent preferentially labels dendritic cells in various lymphoid organs including thymus medulla, but only for rat strains of $RT1^{n}$ and $RT1^{u}$ haplotypes (strong intensity) and $RT1^{c}$ and $RT1^{l}$ haplotypes (weak intensity). To identify epithelium, a rabbit anti-human keratin antibody (Dakopatts, Copenhagen, Denmark) was applied. Second-layer reagents used in immunoperoxidase and (two-colour) immunofluorescence were rabbit anti-mouse immunoglobulin conjugated to horse-radish peroxidase (Dakopatts); fluorescein isothiocyanate (FITC)-labelled goat anti-mouse antibody (Nordic Immunological Laboratories, Tilburg, The Netherlands); and tetramethyl rhodamin isothiocyanate (TRITC)-labeled swine anti-rabbit antibody (Nordic).

Other analyses. We previously described (1,3,10-12) the procedure of CTF preparation and implantation; (immuno)histology; immunization with Ovalbumin (OA) and subsequent assessment of circulating anti-OA antibody and delayed-type hypersensitivity; skin transplantation and subsequent macroscopic reading of rejection; stimulation of spleen cells with Concanavalin A (Con A) and irradiated allogeneic leukocytes with subsequent assessment of proliferation by H-thymidine incorporation; and cell-mediated lympholysis of spleen cells stimulated by allogeneic cells, using Con A-activated and Cr-labelled spleen leukocytes as targets.

RESULTS

Induction of thymus-dependent immune competence with time after CTF implantation (Table 1)

The various parameters of thymus-dependent immune competence (Table 1) illustrate the absent T-cell immunity in nude animals. After CTF implantation, responses became detectable, first at six weeks after CTF implantation. At this moment, 3 out of 9 CTF-implanted rats manifested circulating anti-OA antibody (being immunized three weeks before). In addition to the parameters in Table 1, (immuno)histology of lymph nodes and delayed-type hypersensitivity to OA (ear challenge assay) gave similar results. At 18 weeks after CTF implantation, the thymus-dependent immunity reached values of euthymic +/rnu littermates. This concerned (immuno)histology and anti-OA responses, but mitogen responses were still lower than in euthymic littermates. The generation of anti-OA antibody after immunization was delayed, with no antibody at 10 days after immunization (in euthymic animals titers between 6 and 13), whereas antibody titers at 3 weeks after immunization were of similar value as those in euthymic rats (data not illustrated).

Table 1. Thymus-dependent immune competence in nude rats after syngeneic and allogeneic CTF implantation[a]

	Controls		CTF-implanted rnu/rnu							
	+/rnu	rnu/rnu	Syngeneic				Allogeneic			
(Immuno)histology[b]			Weeks after implantation							
Spleen			2	4	6	18	2	4	6	18
+	14	0	0	0	0	4	0	0	0	9
±	0	3	0	0	1	6	0	0	3	11
-	0	11	3	4	2	0	6	6	3	0
Total	14	14	3	3	3	10	6	6	6	20
OA immunization[c]			Weeks after implantation							
IgG-class antibody (^{2}log titer, ± SEM)					6	18			6	18
n	12	12			3	10			6	20
Mean, ± SEM	12±1	0			6±2	9±1			1±1	12±2
Con A stimulation			Weeks after implantation							
Spleen cells (stimulation index)			2-4		6	18	2-4		6	18
n	10	10	4		3	10	8		6	20
Geometric mean	95	2.6	0.9		7	30	1.4		5	44
Range: lowest value	22	1.0	0.7		5	10	0.6		2	17
highest value	320	8.5	1.6		9	65	3.8		9	84

a Data presented are combined from two separate studies (10, 11). Allogeneic CTF was from DA and BN origin.

b The cellularity of thymus-dependent areas was judged from both histopathology and immunohistology using various anti-T cell reagents. +, well-developed periarteriolar lymphocyte sheaths completely populated by T cells; ±, moderately depleted; -, depleted to near absence of cells with T markers.

c Response at 3 weeks after immunization.

In this development of immune competence, CTF of either syngeneic or allogeneic origin was effective. We here present data on CTF of DA and BN origin: similar results were obtained for CTF of PVG and RP origin. conclude that the haplotype of CTF does not influence the induction of thymus-dependent immune competence after implantation in nude recipients, and also does not influence the rate of this induction.

Responses to allogeneic cells (Table 2)

Allo-antigen responses were analysed only in animals at 14-18 weeks after CTF implantation. Nude animals revealed the absence of any response, and CTF-treated animals gave responses of about half the values of euthymic littermates. This was not only observed in MLR, but also in CML (data not illustrated). Animals implanted with CTF of either syngeneic or allogeneic origin exhibited an identical reactivity pattern towards stimulator or target cells of various RT1 haplotypes. This pattern was identical to that of +/rnu littermates. Thus, the specificity of the allo-antigen recognition repertoire in vitro assays is not influenced by the haplotype of the CTF donor, but is an intrinsic property of the nude recipient. This implies that

Table 2. Mixed leukocyte reaction of spleen cells from control +/rnu and rnu/rnu rats, and rnu/rnu rats implanted 14- 18 weeks before with syngeneic or allogeneic CTF

Stimulator cells		Controls		CTF-implanted rnu/rnu		
		+/rnu	rnu/rnu	+/rnu-CTF	DA-CTF	PVG-CTF
+/rnu ($RT1^u$)						
	Geometric mean	1.2	0.9	1.1	1.3	1.1
	Range, ± SEM	1.0-1.3	0.7-1.2	1.0-1.2	1.2-1.4	1.0-1.2
DA ($RT1^a$)						
	Geometric mean	31	0.8	19	10	23
	Range, ± SEM	24-40	0.7-1.1	15-24	7-14	20-26
PVG ($RT1^c$)						
	Geometric mean	42	1.1	17	13	13
	Range, ± SEM	40-47	1.0-1.4	12-23	10-16	11-18

Data presented are stimulation indexes (H-thymidine incorporation). Each group of animals that received the same treatment consisted of 6 rats. CTF-implanted animals were analysed individually: for control animals, spleen cells of two rats were combined before analysis.

the allogeneic response in vitro can even be directed to the haplotype of the CFT-donor. For instance, DA-CTF implanted nude rats manifested T cells in the periphery which respond to DA stimulator cells and kill DA target cells after such stimulation.

In contrast to the in vitro observations, the grafted CTF did not exhibit any sign of rejection, whereas it manifested RT1 class I and class II antigen expression of the donor haplotype (see below). Further, CTF-implanted rats did not reject skin transplants of CTF donor or recipient haplotype. Thus, DA-CTF implanted nude animals did not reject DA skin grafts and rejected "third-party" grafts, whereas DA skin grafts were rejected by +/rnu controls. This discrepancy between in vitro alloreactivity and in vivo acceptance of allografts was found in a pilot experiment on six CTF-implanted rats (2 nudes with +/rnu-CTF, 2 with DA-CTF, and 2 with BN-CTF). It confirms similar results by Hong et al. (4,15). It indicates the induction of tolerance by components in the CTF graft, presumably epithelial cells, with subsequent prevention of in vivo rejection. This tolerance apparently cannot be ascribed to the deletion of alloreactive T cells, as these cells could easily be documented in in vitro assays.

Immunohistology of CTF (Figures 1-3)

Before implantation, CTF consisted of epithelial-like cells and connective tissue components (Fig. 1a). The fragments were almost devoid of lymphocytes, with only some remnants of lymphoid elements with pycnotic nuclei. The presence of epithelium was confirmed by immunohistochemistry with anti-keratin antibody. In addition, CTF manifested RT1 class I and class II antigen expression. Dendritic cells, detectable with the antibody 1F119, were not present.

Two weeks after implantation under the kidney capsule of nude rats, CTF manifested strands of undifferentiated epithelial cells (keratin-positive) (Fig. 1b). These strands were RT1 class I and class II positive, the haplotype being that of the donor strain. RT1 antigen expression of recipient haplotype was present in the connective tissue around the epithelial strands. This observation was made in indirect immunoperoxidase and also in two-colour immuno-fluorescence with combinations of anti-keratin and anti-RT1 haplotype specific antibodies. Lymphocytes were not detectable, apart from some lymphoid foci in the connective tissue. Using the antibody 1F119, dendritic cells were not demonstrable.

CTF harvested at three weeks or later after implantation manifested a strikingly different picture. Now, the epithelium was populated by lymphoid elements. A thymus-like architecture, with cortex-medulla distinction and well-differentiated epithelial cells, was clearly identified in some cases at 4 weeks, and almost all cases at later moments after implantation (Fig. 1c). Electron microscopy, performed by Prof. Dr M.D. Kendall, London, UK, showed ultrastructural characteristics of epithelial cells similar to those in the normal thymus (Fig. 2) (16, manuscript in preparation). Histology and electron microscopy revealed two remarkable phenomena: first, the presence of mast cells in relatively high density in grafted CTF; and second, the presence of cells resembling multinucleated giant cells. The giant cells were observed both in syngeneic and allogeneic CTF implants. Hence, its presence cannot be ascribed to an immunological reaction associated with rejection. We previously documented similar cells in human CTF implants (17). At the ultrastructural level, the giant cells manifested tonofilaments and desmosomes. This indicates the epithelial origin of the cells, and not a macrophage origin.

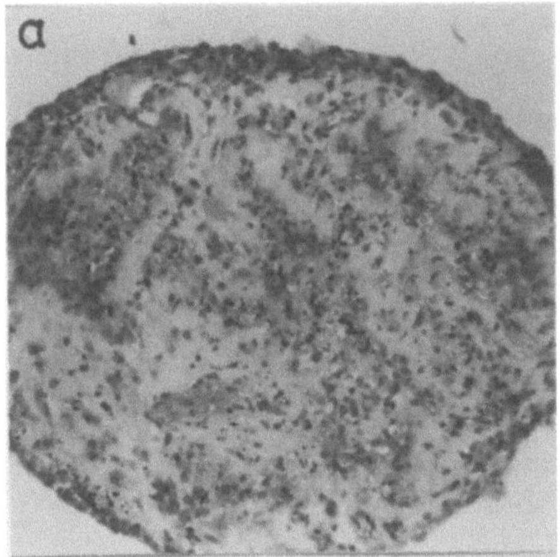

Fig. 1. Histology of CTF before implantation (a) and at two (b) and four weeks (c) after implantation under the kidney capsule of a nude rat. x110.

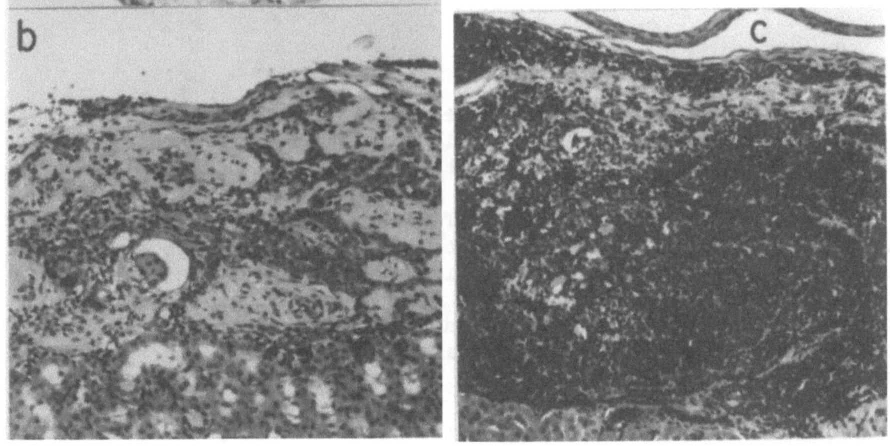

The restoration of the original thymus architecture was similar for CTF of syngeneic and of allogeneic origin. It was further illustrated in immunohistochemistry with anti-T cell reagents, the phenotypic expression being that of the normal thymus. Lymphocytes in CTF exhibited RT1 class I antigen expression of recipient haplotype. With anti-keratin antibody, the dendritic morphology of epithelial cells in cortex-like area and the more confluent staining pattern in the medulla-like area was identical to that in the normal thymus. Scattered in the medulla-like area there were cells labelled by 1F119, indicating the presence of dendritic cells. The presence

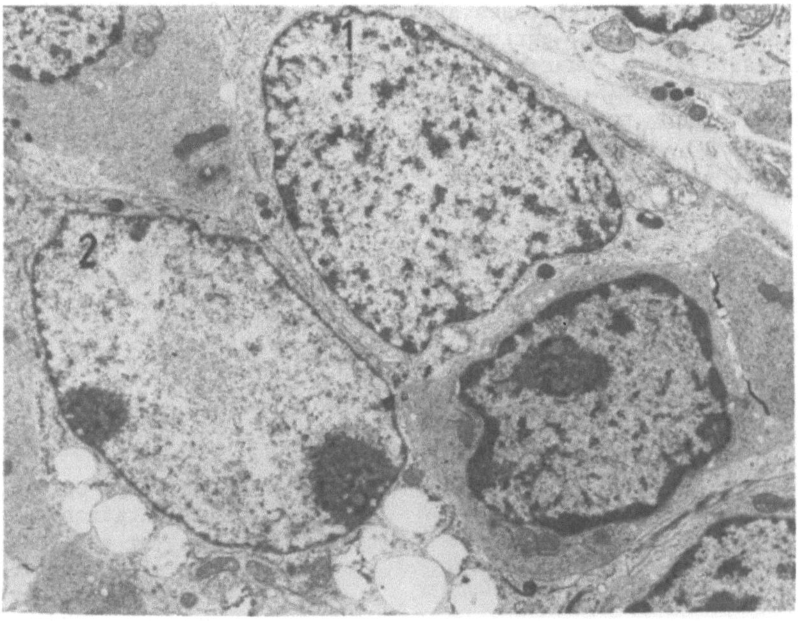

Fig. 2. Electron micrograph of epithelial cells in CTF, 4 weeks after implantation in a nude recipient. The ultrastructural morphology of a so-called type 1 and type 2 cell (16) is identical to that in the normal thymus. x6,800.

of the latter cells was further assessed in two-colour immunofluorescence using combinations of anti-RT1 class II and anti-keratin antibodies (Fig. 3). In the normal thymus, keratin-positive epithelial cells in the cortex were RT1 class II positive, but in the medulla these cells are almost RT1 class II negative. At this location, RT1 class II positive cells are supposed to be dendritic cells (of bone marrow origin). This two-colour picture in the normal thymus was identically found in CTF with restored thymic architecture after implantation (cf., Figs. 3a and 3b). Using haplotype-specific anti-RT1 class II antibodies it was documented that keratin-positive cells in the cortex-like areas were of CTF donor haplotype, whereas RT1 class II positive cells in the medulla-like area were of recipient haplotype (Figs. 3c and 3d). This observation indicates the settlement of recipient-derived dendritic cells in the grafted CTF.

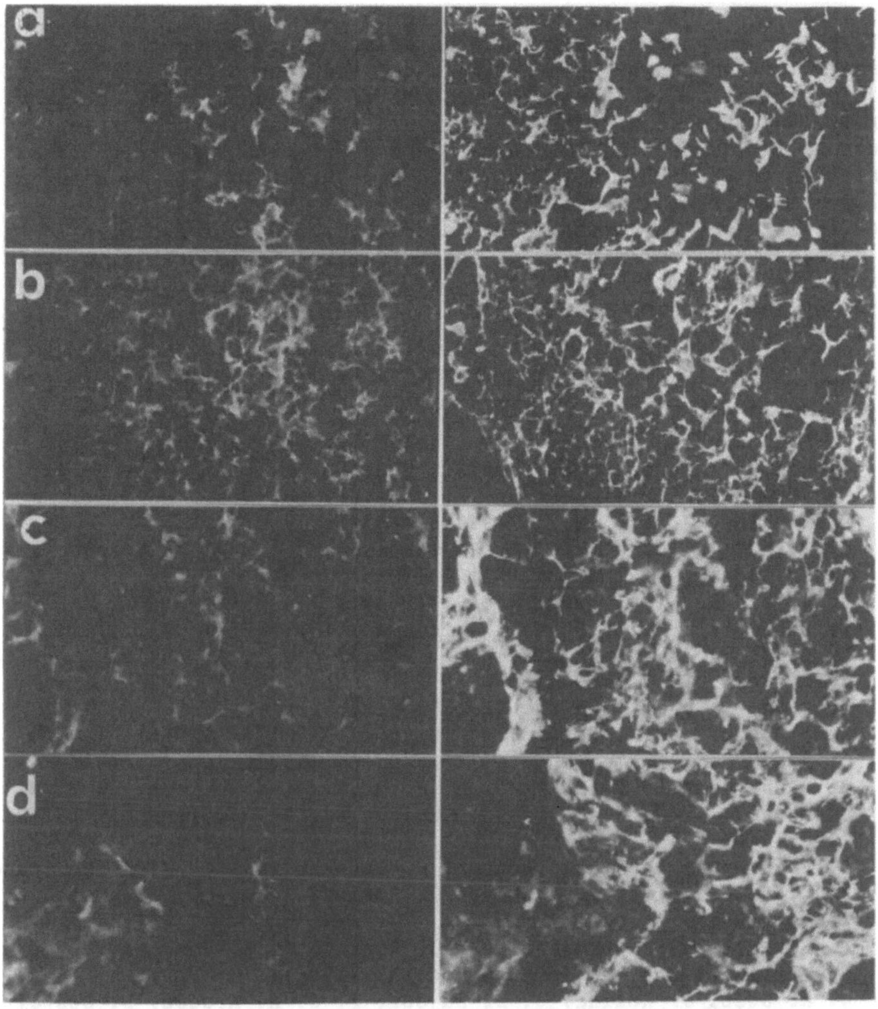

Fig. 3. Two-colour immunofluorescence with combinations of anti-RT1 class II (left panel) and and anti-keratin (right panel). a, normal thymus, common RT1 class II epitope. Epithelium in the cortex is class II-positive, and in the medulla class II-negative. Class II-positive cells at this location are sup posed to include dendritic cells. b, syngeneic CTF 4 weeks after implantation in a nude rat manifests a similar picture. c and d, CTF of DA ($RT1^a$) origin, 4 weeks after implantation in a nude ($RT1^u$) rat, with anti-RTn,a,l class II (c) or anti-$RT1^u$ class II antibody (d). Donor CTF class II positive cells are restricted to keratin-positive epithelium (c), and recipient class II-positive cells, being keratin negative, have settled in the graft (d). x250.

CONCLUSIONS

After implantation in nude recipients, CTF become populated with lymphocytes and dendritic cells from recipient origin. This influx occurs simultaneously for both cell populations, and contributes to the restoration of the original thymic architecture within about 4 weeks. First thereafter starts the periphery to become populated with immunocompetent T lymphocytes. If educated in direct contact with the stromal components in the CTF graft, polymorphic MHC antigens of donor origin on the epithelial components are ignored by the recipient-derived lymphocytes. These latter cells can exhibit an in vitro alloresponse to polymorphic MHC antigens of the graft, but in vivo a state of tolerance is generated. To explain this phenomenon, it is tempting to speculate on a difference in allo-RT1 antigen presentation between the in vitro and in vivo situation, or to a tolerizing signal emanating from the grafted CTF resulting in suppression of allo-reactive cells in vivo. The presence of donor-derived dendritic cells in the graft enables these cells to exert their function in generation of self-MHC restriction specificity, which has been claimed by Longo et al. (5,6), but disputed by the group of Sprent (7,8).

REFERENCES

1. H. J. Schuurman, J.G. Vos, E. Van de Brink, and L. Kater, Clin. Immunol. Immunopathol. 24:418 (1982).
2. R. Hong, H. Schulte-Wissermann, E. Jarett-Toth, S. D. Horowitz, and D. D. Manning, J. Exp. Med. 149:398 (1979).
3. H. J. Schuurman, J. G. Vos, R. Broekhuizen, C. J. W. M. Brandt, and L. Kater, Scand. J. Immunol. 21:21 (1985).
4. R. Hong, and R. Klopp, Thymus 4:91 (1982).
5. D. L. Longo, and M. L. Davis, J. Immunol. 130:2525 (1983).
6. D. L. Longo, A. M. Kruisbeek, M. L. Davis, and L. A. Matis, Proc. Natl. Acad. Sci. USA 82:5900 (1985).
7. D. Lo, and J. Sprent, Nature 319:672 (1986).
8. Y. Ron, D. Lo, and J. Sprent, J. Immunol. 137:1764 (1986).
9. H.-J. Schuurman, Thymus Update 1: in press (1987).
10. H.-J. Schuurman, L. M. B. Vaessen, J. G. Vos, A. Hertogh, J. G. N. Geertzema, C. J. W. M. Brandt, and J. Rozing, J. Immunol. 137:2440 (1986).
11. H.-J. Schuurman, L. M. B. Vaessen, R. Broekhuizen, C. J. W. M. Brandt, M. C. Holewijn, J. G. Vos, and J. Rozing, Scand. J. Immunol. 26: in press (1987).
12. L. M. B. Vaessen, R. Broekhuizen, J. Rozing, J. G. Vos, and H.-J. Schuurman, Scand. J. Immunol. 24:233 (1986).
13. R. J. M. Stet, J. Rozing, G. D. Majoor, F. G. M. Kroese, D. Opstelten, and P. Nieuwenhuis, Transpl. Proc. 17:1829 (1985).
14. L. M. Nagelkerken, B. Schutte, M. Henfling, and P. J. C. van Breda Vriesman, Transpl. Proc. 19:3022 (1987).
15. L. J. Jenski, and R. Hong, J. Immunol. 135:947 (1985).
16. F. P. van de Wijngaert, M. D. Kendall, H.-J. Schuurman, L. H. P. M. Rademakers, and L. Kater, Cell Tissue Res. 237:227 (1984).
17. S. A. Danner, H.-J. Schuurman, J. M. A. Lange, F. H. J. Gmelig Meyling, P. Th. A. Schellekens, J. Huber, and L. Kater, Arch. Intern. Med. 146:1133 (1986).

STUDIES ON THYMIC INVOLUTION: GROWTH POTENTIAL AND DNA REPAIR OF THE STROMAL CELLS

Myra Small and Moshe Aronson

Dept. of Cell Biology and Histology
Sackler School of Medicine
Tel-Aviv University, Ramat-Aviv, Israel

Involution of the thymus is reflected in primary cultures of mouse thymic stromal cells which we have described at the previous Germinal Center Conference (1,2). When these cultures are grown from preinvolution mice, the majority of the cells arrange themselves into networks with long cytoplasmic interconnections between the cells. In cultures from older mice, cellularity is considerably reduced especially of the cells with processes. Correspondingly, in the post involution cultures an increased proportion of cells grow as confluent carpets of paving stones (1,2). The present study was carried out a) to identify which cells each of these are and b) to use these cultures to test the hypothesis that involution, as an accelerated form of aging may result from failure to repair damaged DNA. For cell identification we used fluorescent antibody staining, and for studies of DNA repair, autoradiography of unscheduled DNA synthesis after irradiation with UV. Cortical epithelial cells turned out to be those with decreased growth in cultures from aging mice, but the cells which remain continue to repair UV-induced damage to their DNA.

Materials and Methods

Cultures. Cultures on cover slips were set up from random-bred ICR mice essentially as reported previously (1,2). The culture medium in this study was slightly changed and was prepared as follows. Valine-free DMEM was supplied by Biological Industries, Kibbutz Beit Haemek. To each 100 ml we added 9.4 mg D-valine, 10 mg glutathione, 5 mg ascorbic acid, 0.04 mg insulin, 58.4 mg glutamine, antibiotics (10 mg kanamycin and

2.5 mg gentamycin) and 10% heat inactivated fetal calf serum (from Biological Industries, Kibbutz Beit Haemek).

Antibodies. Monoclonal anti-cytokeratin 35βH11 detects cytokeratin of 54 K daltons (3). This mouse anti-human antibody was used at 1:10 dilution (culture supernatants). Binding of this IgM anti-keratin was detected with biotinylated horse anti-mouse IgG (from Vector) used at 1:100 followed by FITC-conjugated avidin (at 1:100) also from Vector. Monoclonal ER-TR5, also an IgM, combines with medullary epithelial cells of the mouse thymus (4). Culture supernatants were used undiluted. This antibody and the antibodies listed below were detected by FITC-conjugated rabbit anti-rat IgG (from Bio-Yeda, Rehovot used at 1:15 supplemented with 1% normal mouse serum to eliminate cross-reactivity. Monoclonal CDR1, a rat IgG2a was used as an ammonium sulfate precipitate and diluted 1:5. This antibody appears to stain most cortical epithelial cells of the mouse thymus (5). Rat monoclonal MD2 was used as an ammonium sulfate precipate and diluted 1:5. This antibody detects medullary epithelial cells near the cortical - medullary junction (5). Two irrelevent monoclonals against human markers were used as negative controls. One made in mice was the gift of Ruthie Lupu, and the other a rat monoclonal was the gift of Eugene Butcher.

Antibody staining of cultures. a) Fixation: for anti-keratin 5 minute fixation in cold methanol was used, for ER-TR5 fixation was for 2 min. in cold methanol, and for CDR1 and MD2 fixation was for 5 min in cold acetone. b) Primary antibody: 35 β H11 was applied for 30 min at 37°C, ER-TR5 for 30 min at room temperature, CDR1 and MD2 for 60 min at room temperature. c) Secondary antibody: FITC-conjugated rabbit anti-rat IgG was applied for 30 min. Biotinylated anti-mouse IgG was applied for 15 min and followed by d) Tertiary antibody: FITC-avidin which was applied for 15 min. Rinsing and 5 min washings between each step were in PBS. e) Counter-staining: all preparations were stained briefly with 0.1% Evans blue in PBS. The coverslips were overturned in a drop of mounting medium consisting of 50% glycerol and 50% FACS fix (0.37% formaldehyde in 1.25xPBS) onto slides. The slides were observed with a Zeiss fluorescence microscope fitted with epi-illumination. Excitation light was 450-490nm and emitted light 520nm. In all cases when the primary antibody was substituted by PBS or an irrelevant antibody, no staining was observed.

Cultures of Con A stimulated spleen cells. Double stimulation with Con A (Miles Yeda Rehovot) was performed according to the procedure of Marcucci et al (6).

Measurment of DNA repair. Since change of medium triggered a wave of DNA synthesis, for measurement of unscheduled DNA synthesis the cultures were left in 1 ml of spent medium. Tritiated thymidine of specific activity 25Ci/mMole (obtained from Amersham) was added to each culture at 10μCi/ml. Half of the dishes were irradiated with UV from a Philips germicidal UV light with a fluance of 4ergs/second. Unless otherwise indicated, a dose of $36J/m^2$ was given. After 2 hours in the 37^oCO_2 incubator, all cultures were washed four times with warm PBS and given 1ml fresh medium containing $10^{-4}M$ unlabelled thymidine (Sigma). After an additional hour in the incubator, the cultures were again washed four times with warm PBS and fixed in methanol/acetic acid (3/1). The coverslips were then glued to slides and after drying they were dipped in diluted Ilford K_5 emulsion. After one week exposure at 4° the autoradiographs were developed and stained briefly with 0.025% crystal violet. Since the background was very low and cells engaged in DNA replication were heavily labelled, cells performing excision repair could be distinguished by the presence of 5-25 grains over the nucleus. The proportion of such labelled cells in each culture was determined by counting 100 epithelial cells/coded slide.

Results and Discussion

In previous work (1,2) we had identified the cells of the confluent carpets and an unknown proportion of those connected by cytoplasmic processes as epithelial, and we considered it important to determine what percentage of the latter as well as the cells that grow individually are epithelial in origin. This was achieved by fluorescent staining with antibody 35β H11. This monoclonal antibody identifies keratin 8 in intermediate filaments of epithelia (3,7) and was the gift of Allen M Gown. Apart from macrophages and a very small number of other dispersed cells (with distinctive shape), all of the cells stained with this antibody. This confirms the epithelial nature of nearly all the cells found in these cultures (fig 1).

When antibody ER-TR5 (specific for medullary stromal cells (4) and contributed by Willem van Ewijk) was tested by fluorescent staining, all areas containing confluent carpets of cells were clearly stained (fig 2). Thus we can now identify these large cells without processes which are connected to their neighbors all along their length as medullary epithelium.

Identification of the majority of the cultured cells was achieved only after addition of Con A supernatant (6) to the medium. After 3 days with this supplement, the cultures were returned to their original medium for 4 days more and then stained with antibody CDR1. This antibody, produced by Bob Rouse (5), recognizes most of the cortical stromal cells, many of which are seen in frozen sections as long processes between the

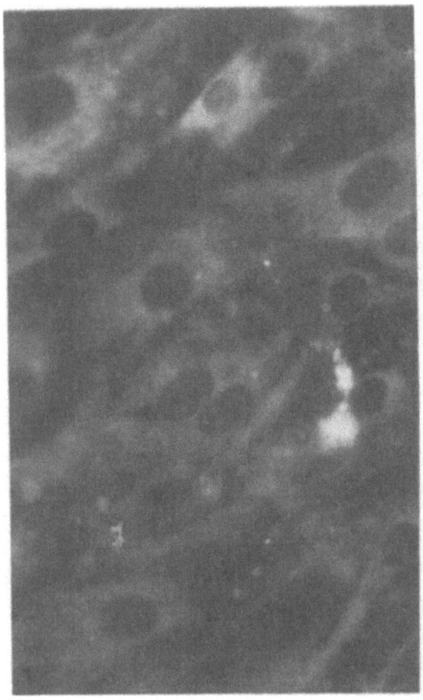

Fig. 1. Confirmation of epithelial cells. Culture stained with 35 β HII. Cells shown in network.

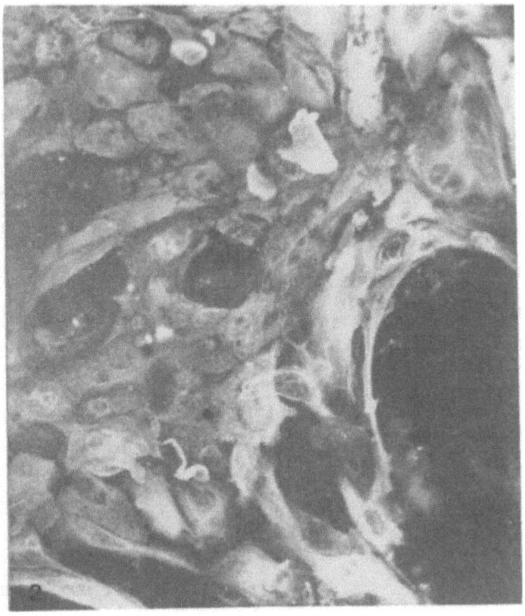

Fig. 2. Identification of medullary epithelia. Cells of confluent carpet stained with ER-TR5.

lymphocytes. As can be seen in fig. 3, such processes also stained in culture and the majority of cells in the networks and those growing as single cells were stained. However when we tested antibody MD2, also produced by Bob Rouse (5),which stains cells on the medullary side of the cortical-medullary junction, a small number of the dispersed cells and a few cells in the networks were fluorescent(8). Thus it seems that the cells organized into networks and the epithelial cells growing individually are mainly, but not exclusively, cortical in origin. Still their proportion is high enough to allow us to conclude that the number of cortical cells in culture decreases with age of the donor mouse while the cells that become more prominent with age are medullary. Interestingly, when frozen sections of thymus from mice at very advanced age were stained with the same antibodies staining of cortical antigens was diminished (M Small unpublished observations). This decreased cellularity of the cultures suggests a decline in growth potential of the cortical epithelial cells.

One possible explanation for this finding is the suggestion that DNA repair activity is impaired in the thymic stroma of postinvolution mice leading to a gradual decrease in cells. We tested this hypothesis by irradiating cultures from postinvolution mice with UV and detected

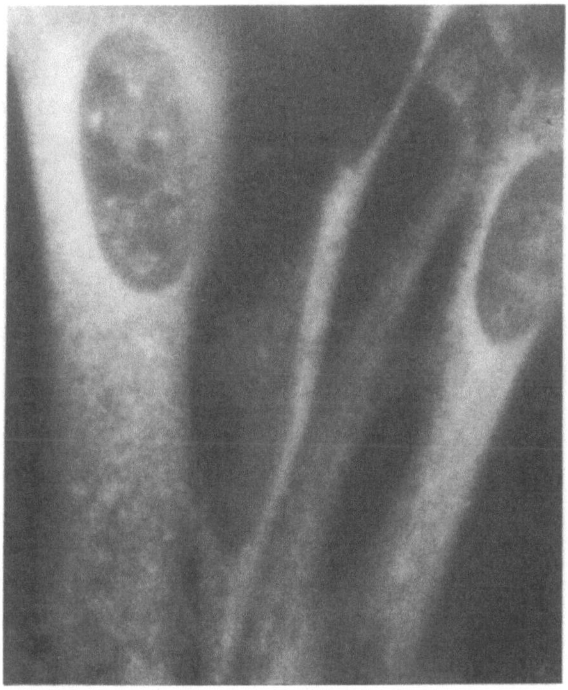

Fig. 3. Delicately labelled cells engaged in DNA repair. Culture from 9 months mouse.

the unscheduled DNA synthesis as delicate labelling of the cells (fig 4). In the cultures from older mice 70-90% of the cells still appeared to carry out DNA repair. Cultures obtained from preinvolution mice also repaired UV - induced damage to their DNA after a short time of culture but this capacity was lost with prolonged culture. This is an interesting

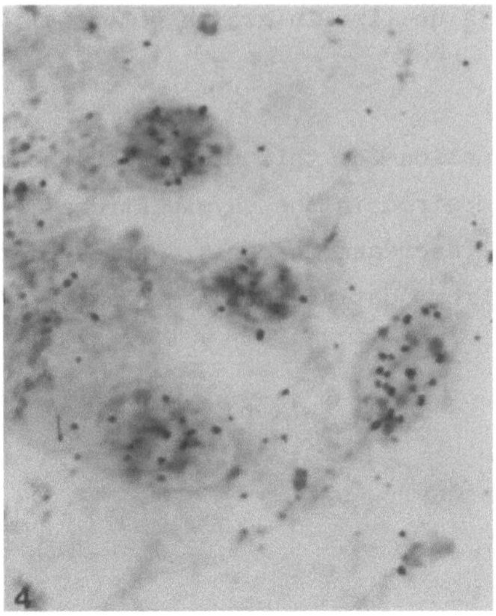

Fig. 4. Identification of cortical epithelia. Cells in network stained with CDR1.

complication especially since a similar pattern has not been observed in cultures from postinvolution mice. Perhaps the stromal cells remaining in the older mice are those which have withstood selection based on DNA repair ability ?

References

1. Small M, Barr-Nea L, and Aronson M. Eur. J. Immunol. 14:936, 1984.
2. Small M, Barr-Nea L, and Aronson M. in Microenvironments in the Lymphoid System GGB Klaus ed. Plenum Press N.Y. 1985, p. 273.
3. Gown AM, and Vogel AM. J. Cell Biol. 95:414, 1982.
4. van Vliet E, Melis M and van Ewijk W. Eur. J. Immunol. 14:524, 1984.
5. Rouse RV. Manuscript submitted.
6. Marcucci F, Klein B, Kirchner H, and Zawatky R. Eur. J. Immunol. 12:787, 1982.
7. Laster AJ, Itoh T, Palker TJ, and Haynes BF, Differentiation 31:67, 1986.
8. Small M, van Ewijk W, Gown AM and Rouse RV, manuscript submitted.

PRODUCTION OF INTERLEUKIN 1 (IL1) BY HUMAN THYMIC EPITHELIAL CELLS (TEC)

Cohen-Kaminsky S. and Berrih-Aknin S.

Centre Chirurgical Marie-Lannelongue, CNRS UA 04-1159
133, avenue de la Résistance
92350 Le Plessis-Robinson (France)

INTRODUCTION

The thymic microenvironment plays a central role in T cell proliferation and maturation, both through the secretion of thymic hormones (1), and through direct interactions between T cell precursors and their stromal partners (2, 3). Thymic epithelial cells (TEC) are ideally suited, both anatomically and functionally, as microenvironmental cells able to differentiate and educate T cells. Our investigations regarding human TEC have shown that cultures of thymic epithelium are an interesting model for phenotypic and functional studies. We have shown that cultured TEC express keratin, produce extracellular matrix components (4), and secrete the thymic hormone, thymulin (5). As a part of our effort to understand the relationship between developing T lymphocytes and the epithelial microenvironment of the thymus, we have shown that TEC class II antigen expression is lost in culture and regulated by recombinant interferon-γ(6). Interleukin 1 (IL1) represents a family of polypeptide cytokines which mediate many aspects of the immune and inflamatory responses (reviewed in ref. 7). Although they were originally considered to be strictly the product of activated monocytes or macrophages, a number of recent reports have indicated that other cell types may produce IL1-like molecules (8-11).

IL1 appears to play an important role in the early steps of T cell activation. It is required for induction of interleukin-2 receptors, secretion of IL2 and activation of T cells via the T3/Ti complex (12). Since interleukin-mediated amplification of immature T lymphocytes may be important in expanding various lineages of thymocytes selected by MHC determinants (13) expressed on stromal thymic cells (i.e., TEC in particular), we examined whether TEC could produce IL1 activity either spontaneously or after stimulation.

We demonstrate that human TEC in short term culture are effective producers of an IL1-like molecule. We also describe the characteristics of TEC-IL1 release after stimulation with various agents. We propose that TEC-IL1 production may be involved in the intra-thymic interleukin amplification circuits, and probably in the regulation of T cell selection, proliferation and differentiation.

MATERIAL AND METHODS

Culture procedure and production of TEC-IL1 activity

Fresh thymic tissue was obtained under sterile conditions from normal children (1 month to 15 years old ; 20 cases) undergoing heart surgery at Marie-Lannelongue Hospital (Le Plessis-Robinson, France). Primary cultures of human TEC were established as previously described (5). At confluence, between day 8 and 12, the cells were detached by trypsin treatment and subcultured in 24-well plates for IL1-production. In previous experiments, the optimal TEC concentration was determined as 500,000 cells/ml/well. Twenty-four hour supernatants were collected to be assayed for IL1 activity. TEC purity was assessed by the binding of an anti-keratin antibody on cytocentrifuge preparations. The contaminating macrophages or dendritic-like cells were detected by immunofluorescence and flow-cytometry analysis with anti-DR antibodies (gifts of D. Charron) which recognize monomorphic determinants and LeuM3 antibody (Becton-Dickinson) which recognizes monocytes and macrophages. Several stimulatory agents were used : Lypopolysacharides (LPS, from Escherichia Coli 0127B8 strain) tested from 0.1 to 200 μg/ml, Staphylococcal Enterotoxin A (SEA) tested from 1 to 40 ng/ml and Formalin-fixed Staphylococcus Aureus cells (Pansorbin) tested from 0.1 to 100 μg/ml.

IL1 assay

TEC-IL1 activity was measured by the classic C3H-HEJ mouse thymocyte comitogenic assay (14). All IL1 assays incorporated several relevant controls, such as response to stimulant alone with PHA. We also used as standard IL1, a purified monocyte IL1 preparation (Ultrapur interleukin 1, Genzyme), which did not contain IL2, IFN or endotoxins.

For blocking studies, rabbit polyclonal anti-IL1 purified immunoglobulin (Genzyme) which recognizes IL1α , IL1β and the precursor form of IL1, but not IL2, IL3, CSF, IFN, TNF was used to inhibit either TEC-IL1 or monocyte purified IL1 activities in the C3H-HEJ mouse thymocyte assay. Test samples or purified IL1 (10 U/ml) were pre-incubated with anti-IL1 antibodies for 2 hours at 37°C before adding thymocytes.

Immunofluorescence studies

For cytoplasmic IL1 detection, secondary cultures and stimulation were performed in 6-well plates (Falcon). Before staining, the cells were fixed by absolute methanol for 5 min and then washed. Anti-IL1 antibody was applied at 1/20 dilution for 30 min at 4°C. Cultures were then washed twice in PBS and incubated with SAR/Ig/FITC (1/40). After a second set of washings, we performed double labeling with monoclonal anti-keratin antibody (10 μg/ml) to ascertain the epithelial nature of IL1-positive cells. Staining was revealed by GAM/Ig/TRITC (1/40). In control experiments, TEC were incubated with SAR/Ig/TRITC or GAM/Ig/TRITC alone or with irrelevant antibodies (rabbit anti-collagen type III or normal rabbit serum) revealed by their specific fluorescent conjugate.

RESULTS AND DISCUSSION

In a well characterized short-term secondary culture system using marked TEC purity (> 90 % of keratin+ cells, 0.1 to 2.5 % of DR+ macrophages or dendritic-like cells, < 10 % fibroblasts), we tested the ability of TEC supernatants and lysats to induce proliferation of C3H/HEJ thymocytes. The secondary cultures were performed in serum-containing medium (complete medium) or in serum-free medium and in the presence or absence of LPS. As shown in a representative experiment (Fig. 1), thymocyte stimulating activity (TSA) could be detected spontaneously in complete medium. Moreover, the TSA increased with lower supernatant concentrations, suggesting the presence of inhibitory activities. In serum-free medium, TSA decreased with supernatant dilution. After LPS stimulation in complete medium, TSA was enhanced and the curve relating TSA to the concentration of supernatant confirmed the presence of inhibitors in these conditions. In serum-free medium, TSA was highly enhanced and decreased with supernatant dilution. Thus, inhibitory activity was detected both in stimulated and nonstimulated cultures performed in complete medium. The presence of inhibitory activity appears to be linked to the complete medium culture condition. This observation can be related to a previous work using endothelial cell cultures (8) and to the failure of Haynes et al. to detect IL1 activity in crude supernatants (15). Indeed, concentration, precipitation and dialysing steps were critical for demonstrating IL1-activity in their supernatants. Our study confirms that the inhibitory effect is reversed by dialysis. In addition, the effect of indomethacin increased TSA suggesting an inhibitory effect at the level of IL1 production. Whether an IL1-specific inhibitor is present deserves further investigation.

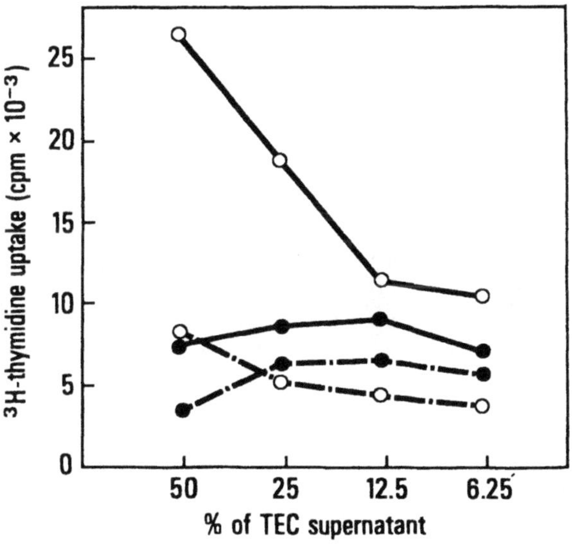

Fig.1 : Thymocyte-stimulating activity in crude supernatants of a representative thymic epithelial cell culture. (— - — - — non stimulated ; ———— LPS stimulated). Supernatants were produced with serum (●) or without serum (o). TSA is greatly increased after LPS stimulation in serum-free medium.

No interleukin 2 activity in TEC supernatants was detected in an interleukin 2-dependent proliferation assay (not shown). In addition, polyclonal antiserum to monocyte-IL1 showed dose-dependent inhibition of both purified monocyte-IL1 and TSA of TEC supernatants suggesting common antigenic determinants between the two-IL1 activity sources. Thus, TEC supernatants contained a cytokine that had the thymocyte comitogenic activity characteristic of IL1 ; hereafter, this cytokine is referred to as TEC-derived IL1 (TEC-IL1).

TEC-IL1 activity varies with cell concentration (optimum at 500,000 cells/ml/well) and duration of stimulation (detectable at 4 h and optimum within 24-48 h). TEC-IL1 production is dependent upon the nature and the concentration of the stimulating agents, with optimum at 100 µg/ml for LPS, 20 ng/ml for SEA and 10 µg/ml for Pansorbin, with a higher efficiency with SEA (Fig. 2). These three factors have no significant mitogenic effect on C3H/HEJ thymocytes in the presence of a suboptimal dose of PHA. Therefore IL1-activity found in stimulated TEC supernatants is the result of production of an active factor.

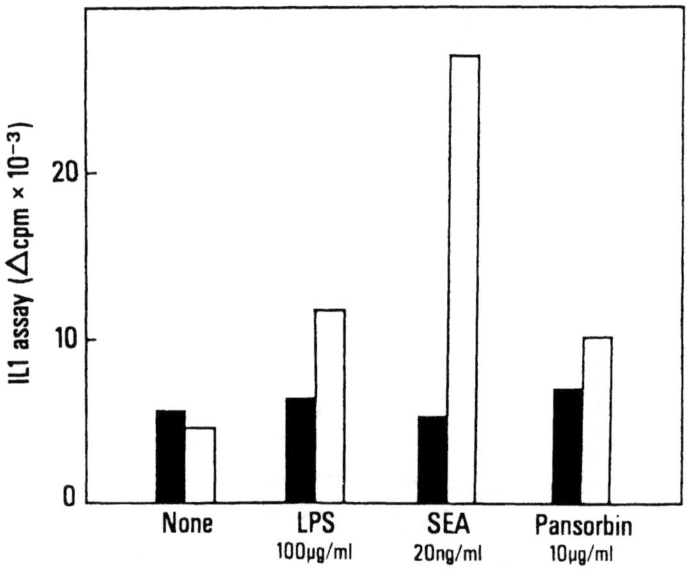

Fig. 2 : Increased TEC-IL1 production after stimulation with various agent used at optimal doses. Note the low TEC-IL1 activity in presence of serum (■) and the high activity in serum-free cultures (□), whatever the stimulation agent.

Immunofluorescence staining on stimulated TEC cultures in complete medium demonstrates the presence of IL1 in TEC cytoplasm and shows several typical perinuclear labeling patterns as granular patches or diffuse staining (Fig. 3a). The epithelial nature of IL1-positive cells was assessed by double labeling with anti-keratin antiserum (Fig. 3b). In nonstimulated cultures in complete medium, no significant differences were observed. In both stimulated and non stimulated serum-free cultures, the fluorescence intensity as well as the number of positive cells are decreased. These results were correlated with those obtained with cell lysats, showing a marked IL1 activity in complete medium and a marginal TEC-IL1 activity in serum free conditions (not shown).

Because IL1 is generated by thymic macrophages (16) and interdigitating cells of the thymic reticulum (17), consideration was given to the possible role of contamination of such cells in the IL1 activity found in TEC supernatant. This was unlikely for the following reasons. First, TEC-IL1 was detectable in unstimulated TEC lysats and supernatants, indicating that IL1 could be produced constitutively by these cells. In contrast, monocytes or macrophages are known to need previous stimulation for IL1 production

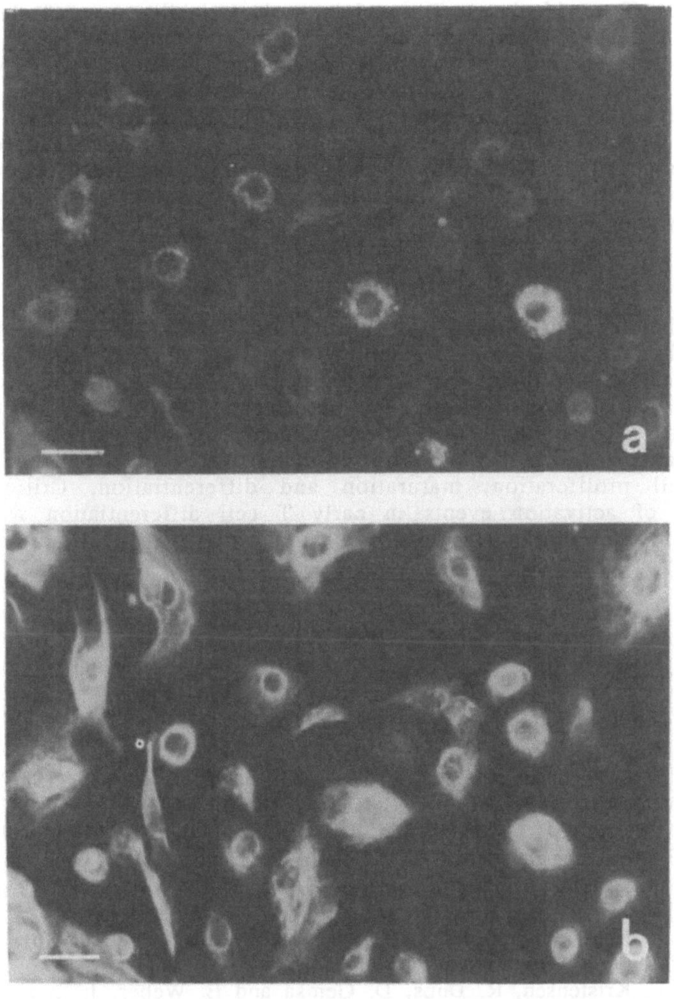

Fig. 3 : Detection of TEC-IL1 by immunofluorescence in TEC cultures after LPS stimulation. (a) Using anti-IL1 antibody, perinuclear labeling is clearly visible in some cells. (b) Anti-keratin fluorescence of the same cells. Note the bright staining of most cells. Scale bar : 15 µm.

(18). Second, TEC-IL1 production was dependent upon the dose of various stimulatory agents. However, the optimal doses found for TEC were different from those already described for monocytes and macrophages (18). For example, TEC-IL1 was maximal with 10 to 100 µg/ml of LPS stimulation, whereas the optimal dose for macrophage stimulation was .1 µg/ml. Third, no correlation was found between the percentage of DR+ cells (0.12 to 2.5 %) and IL1-activity (1000 to 30,000 cpm) tested in 20 different cultures after LPS stimulation in serum free medium. In addition, our studies were consistently performed after day 8 and it is known that cells of monocyte/macrophage lineage have a reduced IL1-producing ability after several days in culture (19, 20). Thus, it would appear that contaminating macrophages and dendritic-like cells had no significant role in the production of IL1 activity found in TEC supernatants.

Finally, we observed that the amounts of TEC-IL1 produced by 20 different cultures was very heterogeneous. No correlation could be established with the age of thymic donors (1 month to 15 years) nor with the age of the cultures (8 to 16 days). To explain such variability several theories could be suggested : 1) TEC-IL1 production could be restricted to a TEC cell subset present at various amounts in different cultures. Indeed, phenotypical and functional subpopulatons of TEC have been described by several groups (21, 22). 2) TEC-IL1 may be released at a particular TEC cell maturational stage which cannot be detected with available methods. 3) All TEC could produce IL1, but the heterogeneity of IL1 detected in the supernatants could be due to variable amounts of inhibitory activity. 4) Variable amounts of contaminating thymic dendritic-like cells or macrophages could participate in the production of TEC-IL1 activity. This hypothesis is unlikely in view of several arguments developed through this study.

Data in the literature show the ability of human TEC to interact physically with thymocytes (rosette formation) (23), to function as accessory cells for human autologous thymocyte activation in vitro (24) and to express class II antigens which are regulated by interferon- (6). Our additional finding that human TEC produce the potent immunoregulatory cytokine TEC-IL1 affirms an important role of thymic epithelial cells in the process of intrathymic T cell proliferation, maturation and differentiation. Critical to our understanding of activation events in early T cell differentiation will be the determination of the physiological stimulus for intrathymic IL1 production.

REFERENCES

1. J.F. Bach, and C. Cardaux C. Prog. Allergy, 21 : 342 (1976).
2. O. Stutman. Immunol. Rev., 42 : 138 (1978).
3. H. Wekerle, V.P. Ketelsen and M. Ernst. J. Exp. Med., 151 : 925 (1980).
4. S. Berrih, W. Savino and S. Cohen. J. Histoch. Cytoch., 33 : 655 (1985).
5. S. Cohen, S. Berrih, M. Dardenne and J.F. Bach. Thymus, 8 : 109, 1986.
6. S. Berrih, F. Arenzana-Seisdedos, S. Cohen, R. Devos, D. Charron and J.L. Virelizier. J. Immunol., 135 : 1165 (1985).
7. C.A. Dinarello. Infect. Dis. Rev., 6 : 51, 1984.
8. P. Miossec, D. Cavender and M. Ziff. J. Immunol., 136 : 2486 (1986).
9. D.N. Sauder. Lymph. Res., 3 : 145 (1984).
10. A. Fontana, F. Kristensen, R. Dubs, D. Gemsa and E. Weber. J. Immunol., 29 : 2413 (1982).
11. K. Matsushima, G. Tosato, D. Benjamin and J.J. Oppenheim. Cell. Immunol., 94 : 418 (1985).
12. E.L. Reinherz. Immunol. Today, 6 : 75 (1985).
13. K.L. Rock and F. Benacerraf. J. Immunol., 132 : 1654 (1984).
14. S.B. Mizel. Mol. Immunol., 17 : 571 (1981).
15. P.T. Le, D.T. Tuck, C.A. Dinarello, B.F. Haynes and K.H. Singer. J. Immunol., 138 : 2520 (1987).
16. R. Gallily, M. Zeira and I. Stain. Immunol., 55 : 165 (1985).
17. D.I. Beller and E.R. Unanue. J. Immunol., 124 : 1433 (1980).
18. R.C. Newton. J. Leuko. Biol., 39 : 299
19. M. Papiernik and F. Homo-delarche. Eur. J. Immunol., 13 : 689 (1983).
20. J.L. Lepe-Zuniga and I. Gery. Clin. Immunol. Immunopathol., 31 : 222 (1984).

21. B.F. Haynes; R.M. Scearle, D.F. Lobach and L.L. Hensley. J. Exp. Med., 159 : 1149 (1984).
22. R.A. De Maagd, W.A. Mackenzie, H.J. Schuurman, M.A. Ritter, K.M. Price, R. Broekhuizen and L. Kater. Immunology, 54 : 745 (1985).
23. K.H. Singer, L.S. Wolf, D.F. Lobach, S.M. Denning, D.T. Tuck, A.L. Robertson and B.F. Haynes. Proc. Natl. Acad. Sci. (USA), 83 : 6588 (1986).
24. S.T. Denning, D.T. Tuck, K.H. Singer and B.F. Haynes. Clin. Res., 34 : 669A (1986).

A LOW DENSITY FRACTION OF BONE MARROW CELLS ENRICHED IN PROTHYMOCYTE ACTIVITY

S. Ezine and B. Nabarra

INSERM U25 - Hôpital Necker

161 rue de Sèvres 75743 Paris Cedex 15, France

INTRODUCTION

The very early steps of T cell differentiation are not yet well understood. Many difficulties can be reported : 1/ the lack of surface markers on very immature population (prothymocytes and their direct progeny) and 2/ the absence of informations about the inductive signals involved in triggering the precursor subset.

In the adult animal, the bone marrow is the major source of prothymocytes. They are characterized by a functional assay described by Kadish and Basch (1) : bone marrow cells injected intraveinously are able to repopulate the thymus of a lethally irradiated animal.

We are interested in isolating a bone marrow fraction enriched in prothymocyte activity. Precursor enrichment from the bone marrow has been widely used and is mainly based on separation according to physical parameters (2). For this purpose, we used a discontinuous Ficoll gradient and looked for the prothymocyte activity in each fraction. In this study, we present the separation method and the different biological and morphological characteristics of the recovered fractions. Finally, we could determine a bone marrow fraction enriched at least 20 times in prothymocyte activity.

MATERIALS AND METHODS

Animals: Thy-1 congenic C57Bl mice (Ka : Thy 1.2 and BA : Thy 1.1) bred in our colony, were used in all the study. Host mice were 8-10 weeks, donors were 5-7 weeks. To evaluate the prothymocyte activity in the different layers of the bone marrow gradient, mice were irradiated (850 rad) and reconstituted with the appropriate preparation by intraveinous injection.

Cells: Bone marrow cells were removed from the femurs of mice by flushing with Hanks' balanced salt solution (Lab. Eurobio, Paris, France).

Reagents: The following monoclonal antibodies were used : anti-Thy 1.2 antibody (clone 30H12, Becton Dickinson), anti-Thy 1.1 (clone 19XE5, gift from R. Nowinsky), anti-Mac1 (clone M170), anti-B220 (clone 6B2) (3), clone 8C5 (gift from R. Coffman, DNAX, Palo Alto, USA) anti-Lyt2 (clone 53-6.7, Becton Dickinson) and anti-L3T4 (clone GK 1.5) (4).

Discontinuous Density-Gradient Separation: A ficoll gradient was prepared according to the method of Tak Yan Yu et al (5), modified by Papiernik and Bach (6) and consisted of Ficoll 400 (Pharmacia Fine Chemicals, Uppsala, Sweden) at a concentration of 10% (wt : wt) 14.6%, 16.1%, 17.7%, 19.2%, and 23% in 0.1 M sodium phosphate buffer, pH 7.4. The density of the gradient was controlled by measuring the refractive index with a Zeiss-Abbe refractometer (Carl Zeiss, Inc. New York). Refractive index of each layer (1.2 ml/layer) from the top of the gradient (10%) was : 1.352, 1.360, 1.363, 1.364 and 1.374.

Bone marrow cells were isolated as described above and resuspended in culture medium. Bone marrow cells from 10 to 15 mice (around 200 x 10^6 cells) were layered on the top of the Ficoll gradient in 2 ml of culture medium. The gradient was then centrifuged for 30 min in an RC2 Sorvall centrifuge (Dupont Instruments - Sorvall, Dupont Co, Newtown, Conn.) at 23,500 g at 40°C. Layer 0 was defined as the interphase between the culture medium and 10% Ficoll ; layer 1 was defined as between 10 and 14.6% Ficoll with subsequent interphases numbered sequentially. Cells from each of the six layers (the last one is the pellet) were washed twice in RPMI (Flow lab. Inc) and counted. 60-80% of the cells were recovered.

Thymus analysis. At given time after the transfert of total or fractionnated bone marrow cells, mice were killed and the thymus removed. They were homogeneized and washed in Hanks' balanced salt solution containing 5% fetal calf serum and 0.1% sodium azide. 10^6 thymocytes were incubated with a directly coupled fluoresceine isothiocyanate-anti-Thy 1.2 antibody, 30 min at 4°C. Cells were then washed twice and the positive cells (Thy 1.2+) scored under a Leitz Fluorescence microscope.

Electron microscopy preparation. Cell suspension from each layer of gradient were fixed in 1% glutaraldehyde in a 0,1 M phosphate buffer, pH 7.4 for 1 h. After washing in buffer, the cells were post-fixed in 1% osmium tetroxyde, 1 h, and then deshydrated and embedded in Epon. Between each step, cells were centrifuged at 1 200 rm. The sections are impregnated with uranyl acetate and lead citrate and observed in EM 200 Philips electron microscope.

Layers of five different gradient were examined. In each layer approximatively 300 cells were observed and counted. Identification of each cell was made on micrograph with magnifications of 3 600, 5 900 or 9 900 respectively. To avoid subjective selection of cells, micrographs were made routinely in the upperleft quarter of each grid.

RESULTS

Distribution of bone marrow cells in the discontinuous Ficoll gradient

The fractionation of bone marrow cells on a Ficoll gradient is represented in Table 1. Most of the cells are recovered in the dense fractions (layers 5 and 6). Phenotypic analysis of defined populations of T cells, macrophages, B and pre B cells, granulocytes by respectively monoclonal antibodies, anti-Thy 1.2, Mac-1, 6B2 and 8C5 shows that all these cell types are represented in the different fractions of the gradient.

Electron microscopy observation

Several cell types were found in the different layers of the Ficoll gradient. For identification and counting of cells we have mainly considered morphologic criteria, based on the aspect of the nucleus (shape, euchromatine clumping, heterochromatine aspect), of the nucleolus, of the cytoplasme and, the repartition of different cellular organite. The cell shape is rarely a good information because it is possible that centrifugation deforms the cells. In each layer, the cells are often rounded with few short pseudopods on the membrane surface.

In the immature cells, moreover the clear nucleus without clumps of chromatine and large cytoplasm scattered by many aggregated ribosomes, we sometimes observed few "primary" granules in the cell submembrane area. This fact permit to distinguish in our Table 2 , blast cells (without pre-granules) and myeloid cells and mast like cells precursors which contains "pre-granules".

Table 1. Bone marrow cells phenotypic characterization in the different layers of the discontinuous Ficoll gradient.

Layer number	Percentage of recovered cells	Percentage of cells labeled with			
		Anti-Thy 1.2	Anti-Mac1	6B2	8C5
1	0.1 - 0.4	< 1	9	13	12
2	0.9 - 2.0	1	8	51	14
3	4 - 6	2	11	30	26
4	7 - 10	1	19	33	31
5	32 - 34	< 1	16	41	30
6	46 - 48	8	4	21	11
Unfractionated		3	2	30	24

Table 2. Percentage of different type cells in layers of the Ficoll gradient

Type of cells		Layers of gradient					
		1	2	3	4	5	6
Immature cells	Blast cells Myeloid cells Mast cells like Precursors		80%	65%	35%	7%	
Mature cells	Lymphocytes		8%	12%	53%	80%	
	Granulocytes Mastocytes						+++
	Red blood cells		12%	23%	12%	13%	
Dead cells and degenerated, fragmented cells	+++	±	±	+	++	+++	

As described in Table 2 , we noted that more immature cells are present in layer 2 and 3, with a higher number in layer 2. The more mature cells are observed, mainly, in layer 5, and a few in layer 4. According to the nuclear and cytoplasm aspect, we consider different intermediary stage between immature and mature cells, but we have not separated in our counting these different forms and we have considered the first step of maturation in immature cells, and the last steps with a morphological early defined lineage in mature cells.

Macrophages are present in all layers of the gradient. It exists, also, in all layer some cellular fragments, degenerated cells and picnotic nucleus, and layer 1 and layer 6 are composed completely with dead degenerated cells and membrane fragment. In layer 6 we also observe a great number of red blood cells. Finally, a few number of mitosis are observed in less dense layer.

The layer 2, which is used in our experiment contains a high number of blast type cells, different of myeloid cells (Fig. 1). They are larger cells with a few short pseudopods on the surface. The nucleus contains clear chromatine scattered in the nucleoplasm, with only a few marginal clumps of euchromatine near the nuclear membrane. This appearance is that of an immature nucleus with one or two large nucleoli, and a round or lightly idented shape. The cytoplasm is large and contains numerous aggregated ribosomes, Golgi apparatus and a few lamellae of endoplasmic reticulum (Fig. 2). All the morphological characteristics of these cells indicate a significant metabolic activity and they are certainly pluripotent. These cells seem to be in a very early stage of differentiation and are probably precursors of various cell lines (lymphoblastes, erythroblastes, monocytes precursors). At this stage it is impossible to distinguish what is the cell line precisely.

Table 3. Study of the pre-T potential in the different layers of the bone marrow gradient 15 days after the i.v. transfert of 10^6 cells

Layer number	Total cells (per thymus) (x 10^{-6})	% Thy 1.2+ cells	Absolute number of Thy 1.2% cells
1	51	16	8 x 10^6
2	124	68	84 x 10^6
3	125	31	38 x 10^6
4	12	30	3.6 x 10^6
5	53	20	10.6 x 10^6
6	51	1	0.5 x 10^6
Unfractionated	60	6	3.0 x 10^6

Analysis of pre-T activity in the different fractions of the gradient

In order to evaluate the pre T activity in the fractionated bone marrow cells, definite numbers of each layer were i.v. injected to lethally irradiated mice. Host thymus were removed 2 weeks later and the repopulation analyzed by scoring the percentage of donor type Thy 1.2+ cells. Table 3 shows that for 10^6 cells transferred all layers are able to repopulate the thymus with a variable efficiency. The highest repopulation is obtained by cells in the lower density fractions, layer 2. The size of the thymus varies according to the cells transferred.

Comparison of the pre T potential in the low density fractions, L2 (layer 2) with total bone marrow cells

Total bone marrow cells or cells from the low density fraction were tested extensively for their pre-T potential. Irradiated mice received 10^6, or 5×10^5 or 2.5×10^5 cells and the thymus repopulation was observed 8 to 32 days later. Table 4 indicates that when examined as early as 8 days after the transfert, thymic repopulation is better with L2 selected cells than with total bone marrow cells. When fewer cells were transferred and mice observed later (2.5×10^5, 32 days), the absolute number of thymic cells recovered of donor type phenotype reveals an enrichment in pre T activity in L2 of a 20 times compare to unfractionated cells (2.6×10^6).

Table 4. Differential study of thymic repopulation after i.v. transfert of total (BM) or gradient selected bone marrow cells (L2).

Cells transferred	Days after transfert	Type of cells	Cells per thymus ($\times 10^{-6}$)	Positive mice Total	Absolute ≠ Thy1.2+ cells
10^6	8	BM	1.1	1/6	3×10^3
		L2	1.8	8/8	5.4×10^4
5×10^5	8	BM	4.5	0/2	-
		L2	1.3	8/8	2.6×10^4
2.5×10^5	32	BM	80.6	4/4	2.6×10^6
		L2	75.1	8/8	42.8×10^6

Table 5. I.V. Transfert of a limiting number of bone marrow cells selected in the low density of the gradient.

Cells transferred	Days after transfert	Positive mice Total	% Thy 1.2+ cells Total	Absolute ≠ Thy1.2+ cells
10^5	20-22	8/8	15.8	10.9×10^6
	41-45	6/6	40.6	25.8×10^6
	49-52	4/5	31.5	18×10^6
10^4	20-22	7/8	4.2	1.4×10^6
	41-45	6/8	5.6	5×10^6
	49-52	3/3	9	6.3×10^6
10^3	20-22	0/8	0	0
	41-45	4/8	2	8×10^5
	49-52	0/5	0	0
10^2	20-22	0/8	0	0
	41-45	3/6	0.6	8×10^5
	49-52	0/8	0	0

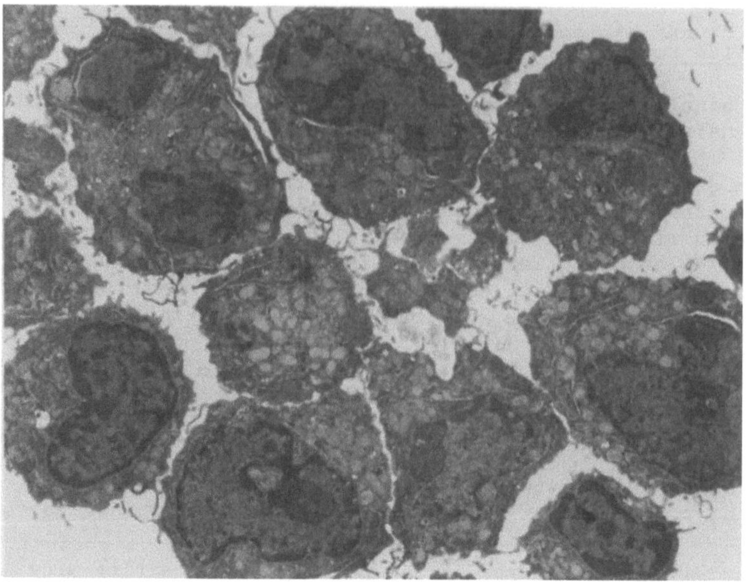

Figure 1
Morphological aspect of majority of cells in layer 2 of Ficoll gradient. Impregnated with uranyl acetate and lead citrate. X 9900.

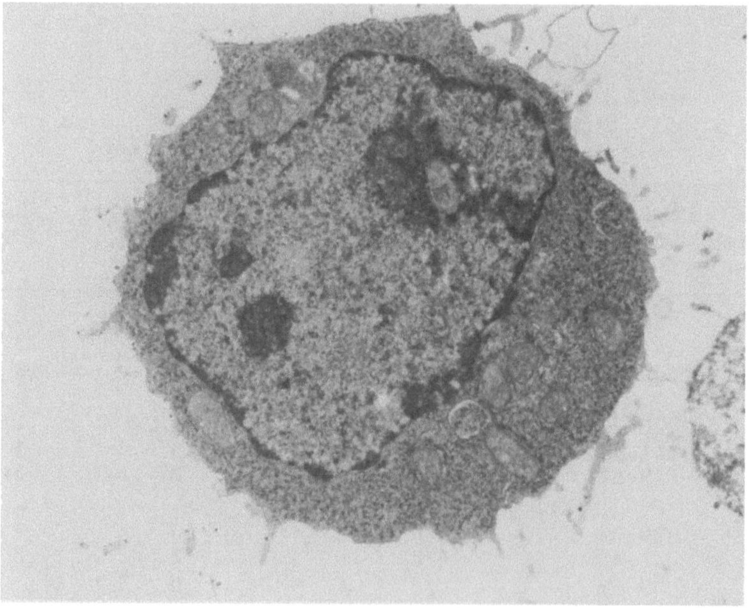

Figure 2
High magnification of blast cells without pregranules in layer 2. Impregnated with uranyl acetate and lead citrate. X 22500.

We were then interested at looking at the repopulation of the thymus after transfert of a limiting number of ractionated cells to irradiated mice. Animals received from 10^5 from up to 10^2 bone marrow cells recovered from the layer 2 and their thymus tested at 3 different times point. To fully reconstitute the hematopoietic system of the mice, 10^6 host type bone marrow cells (Thy 1.1) was injected.

Donor type cells (Thy 1.2+) can be detected in the host thymus even if as low as 10^2 fractionated cells are injected (Table 5). Mean while this dose is not large enough to maintain a permanent population.

CONCLUSION

The data presented indicate the isolation of a prothymocyte enriched in the bone marrow. Other groups have shown that this low density fraction is also enriched in CFU-s and GM-CFC activities (F. Lepault, personal communication).

From other side, electron microscopy studies could determine very immature cells in layer 2. Indeed in the present series, we have counted a high percentage of immature blast cells in this layer, which are certainly pluripotent cells. These immature cells according to the influence of different factors in vitro, or influenced by a microenvironment functional in situ, are implicated in a processus of maturation and differentiation.

In our experiment, we have selected a stem cell that under thymic environment signals could differentiate in T cells. The next experiments will analyse if a concomitant enrichment in the B lymphocyte compartment do exist.

ACKNOWLEDGMENTS

The authors thank I. Andrianarison and T. Sacksick for expert technical and photographic assistance.

REFERENCES

1. Kadish, J.L. and R.S. Basch (1976) J. Exp. Med.143, 1082.
2. El Arini, M.O. and Osoba, D. (1973). J. Exp. Med. 137, 821.
3. Coffman, R. (1982). Immunol. Rev. 69, 5.
4. Dialynas, D.P. et al. (1983). Immunol. Rev. 74, 29.
5. Tak Yan DU, D. et al. (1973). J. Immunol. 110, 151.
6. Papiernk, M. and Bach J.F. (1977). Eur. J. Immunol. 7, 800.

KINETICS OF THYMOCYTE REGENERATION IN ADULT ADRIAMYCIN TREATED RATS: EVIDENCE FOR A BONE MARROW DERIVED PROTHYMOCYTE PRECURSOR

Malgorzata B. Murawska, Adrian M. Duijvestijn,
Jaap Kampinga, Auk S. Wubbena and Paul Nieuwenhuis

Department of Histology
University of Groningen
The Netherlands

INTRODUCTION

The thymus is the central organ for T lymphocyte maturation in which prothymocytes undergo rearrangement of T cell receptor genes, acquire function associated antigens and become "educated" to respond to foreign antigens in a self MHC class I or class II restricted manner.
Although much is already known about T cell development in ontogeny, it remains controversial how the adult thymus maintains its function, among others, the continuous production of a population of functional T lymphocytes. Whether the thymocyte pool originates by expansion of intrathymic stem cells that had colonized the thymus during ontogeny or whether prothymocytes seed in continously from the bone marrow (BM) is still one of the issues (Reviewed in 1). Most studies addressing this question have been done using irradiated and reconstituted with genetically marked BM animals.

Here we present a new, more physiological, approach to study the adult thymus function using adriamycin as a depleting agent sparing BM stem cell activity in conjunction with vascular thymus transplantation.
Our results indicate that BM derived precursors are continuously recruited from the circulation to maintain the thymocyte population in adult life.

MATERIALS AND METHODS

Animals: (AO/GxBN/G)F1 hybrid rats as well as (RT1 $^{u/n}$), PVG RT7.1 and PVG RT7.2 rats (kindly provided by dr S.V. Hunt - Sir William Dunn School of Pathology, University of Oxford) were bred under conventional conditions in our own animal facility. In all experiments young adult (7 - 10 wk old) male rats were used.

Preparation of tissues and cells: For cryostat sections, small fragments of tissue were snap frozen in liquid freon and kept frozen until used. Single cell thymocyte suspensions were prepared by teasing small thymus fragments with 22 G needles in Dulbecco's A+B salt solution (DAB) (Oxoid , Basingstoke, UK) pH 7.2 , supplemented with 5% newborn calf serum (NCS) (Flow Laboratories Ltd.,Irvine, Scotland). Tissue debris was removed by 1 x g sedimentation over pure NCS. Non-sedimented cells were washed twice with DAB/NCS at 4^{0} C and resuspended in the same medium for counting. Bone marrow cell suspensions were prepared by flushing the marrow from femora and tibias with ice cold DAB/10% NCS and resuspended in DAB (serum free) to a proper concentration.

Adriamycin treatment: Rats were i.v.injected with 10 mg/kg of Adriamycin (ADR) (Doxorubicin hydrochlorid, Farmitalia Carlo Erba, Brussels, Belgium). Depletion and regeneration stages of the thymus were examined 24hr, 48hr and 3,5,7,9,12, 14,18,21 and 28 days after day 0 of ADR treatment.

Syngeneic transfer of prothymocytes: ADR treated recipients were injected with BM cell suspension 3 days after ADR treatment either systemically (i.v.) ($1x10^8$ cells) or intrathymically (i.t.) as described previously (2). Briefly 1×10^6 BM cells in a volume of 10 ul were injected into each lobe using the Hamilton automatic microdispenser equipped with a 32-gauge needle via a flexible capillary connection. Cells were injected in 5 portions of 2 ul each along the a longitudal axis of the lobe. The influence of precursor cell transfer upon the regeneration process was checked on days 14, 18 and 28 after ADR treatment (11,15,25 days after BM reconstitution).

Vascular thymus transplantation: In all instances PVG RT7.2 thymus lobes were vascularly transplanted into congenic PVG RT7.1 rats as described elsewhere (3). In one group , depleted thymus from ADR treated animals was transplanted 3 days after ADR treatment into a normal recipient . In the second group normal thymus was transplanted into congenic recipients, 3 days after ADR treatment. Immunohistology of the grafts was checked on days 14 and 18 with respect to ADR treatment.

Immunohistochemistry: Frozen sections of the thymus were air-dried, fixed in acetone for 10 min. at room temperature and stained using an indirect immunoperoxidase method. Sections were incubated for one hour with a proper dilution of Mab, washed, incubated for another 30min. with peroxidase conjugated rabbit anti-mouse antiserum (Dakopatts, Glostrup, Denmark) and after rinsing the peroxidase activity was developed with 3,3'-diaminobenzidine tetra HCl (Sigma, St. Louis,MO,USA).

The following Mab were used: HIS17, anti-pan-T lymphocytes (4); HIS19, anti-class II (5); HIS37, HIS38 and HIS39, recognizing different thymic epithelial cells (6); ER2, anti-T helper subset(7); MRC Ox-8, anti-T suppressor-/cytotoxic subset, (8); MRC Ox-19, T cell maturation marker (8); ED1, ED2, recognizing different subsets of macrophages (8), MRC Ox-39, anti-IL2 receptor/thymic IDC (dr A.F. Williams, Sir William Dunn School of Pathology, University of Oxford, personal communication); NDS 58, recognizing rat RT7.1 congenic determinant (kindly provided by dr M. Newton, Department of Zoology, University of Oxford) and HIS41 recognizing RT7.2 congenic determinant (J. Kampinga et al., this volume).

RESULTS

Cell depletion after ADR

Thymus: As seen in Figure 1,a single injection of ADR caused a gradual depopulation of thymocytes, first noticable 3 days after ADR treatment and reaching a maximum by day 7. The residual number of thymocytes did not exceed 1% of the initial population during the period of maximal depopulation (day 7 - day 14). Regeneration of thymocytes was first detected on day 18 rapidly resulting in restoration of thymus cellularity to the lower limits of control levels, but not the mean, by day 22 after treatment.

Bone marrow: ADR treatment also caused profound but incomplete BM depletion within one week after treatment. From day 7 on BM spontaneously restored its haemopoietic activity as judged by increased numbers of erythropoietic and myelopoietic foci as well megakaryocytes. Thus the regeneration of BM significantly preceded that of the thymus. Differences in kinetics of depletion and subsequent regeneration between thymus and BM are illustrated in Figure 2.

BM supplementation

To test a possible contribution of BM to thymocyte regeneration, syngeneic BM was transferred 3 days after ADR either intravenously or directly into the thymus.(See Figure 1)

Intravenous (i.v.) infusion of $1x10^8$ BM cells apparently advanced the regeneration of thymocytes, already detectable on day 14 resulting in total restoration of cell numbers comparable to control by day 28.

Intrathymic (i.t.) infusion of $1x10^6$ BM cells resulted in an even more advanced repopulation of the thymus, observed already between days 10 and 14 after ADR. By day 18 the number of thymocytes was restored to the normal level. The thymocyte number continued to increase thereafter and significantly exceeded control values by day 28.

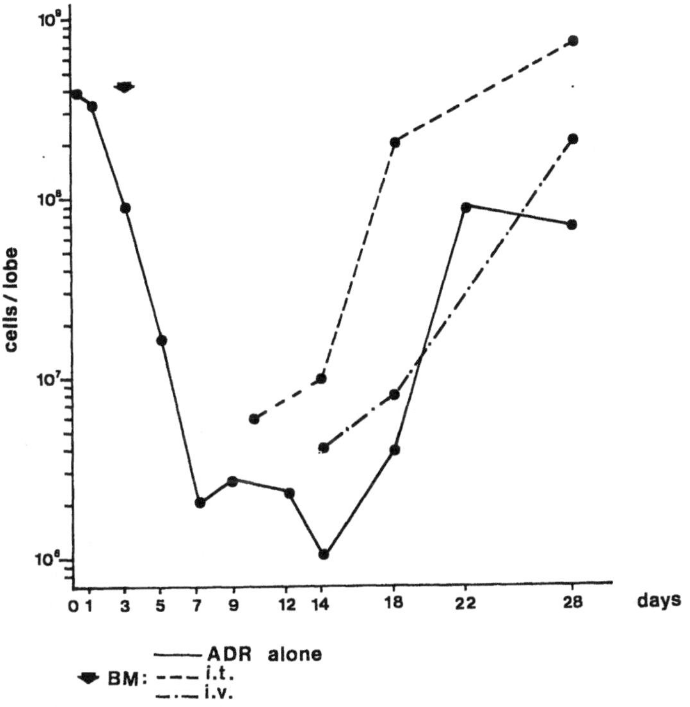

Figure 1. Kinetics of thymocyte depletion and regeneration after ADR treatment.

Immunohistological analysis of thymus sections

Lymphoid markers: Using Mab against T cell markers on cryostat thymus sections it was shown that the gradual depletion of thymocytes which started 3 days after ADR treatment was due to a disappearance of cells from the outer cortex continuing towards the medulla during consecutive days. Between days 7 and 14 after ADR virtually no cells were detected in the cortex using HIS17 (pan-T), ER2 (anti-T help) or Ox-8 (anti-T suppr/cytotox) Mab. In staining with MRC Ox-19, the normally present negative or weakly positive cells were absent from outer cortex on day 3. Subsequently positive cells also disappeared from the cortex. Medullary thymocytes when analysed with HIS17 and MRC Ox-19 decreased in number. Their depletion seemed to be random since the proportion of cells positive for ER2 (70%) to MRC Ox-8 (30%) revealed normal proportion of medullary thymocytes with respectively helper and suppressor phenotype. On day 18 the first cells, positive for HIS17, ER2, MRC Ox-8 but negative for MRC Ox-19, reappeared scattered in the cortex. The regeneration process was quite advanced on day 22 when compartmentalization of the thymus was seen again.

Nonlymphoid markers: Over the observation period there was no change in the expression of class II ($HIS19^{+}$) on either cortical epithelium or medullary IDC, despite the collapse of thymic architecture due to the depletion of thymocytes. Presence and localization of different types of epithelial cells detected with HIS37, HIS38 and HIS39 Mab as well as of medullary IDC positive for Ox-39, did not differ from normal. The macrophage population as detected by ED1 and ED2 also seemed intact, the only exception being a slight and transient increase in ED2 positive macrophages in the interlobular septa during the regeneration phase.

DAYS	0	1	3	5	7	9	12	14	18	22	28
BM	=	↓	↓↓	↑	↑↑	↑↑	ND	↑↑	=	=	=
THYMUS	=	=	↓	↓↓	↓↓	↓↓	↓↓	↓↓	↑	↑↑	=

Figure 2. Kinetics of depletion and regeneration of BM versus thymus after single ADR injection. ↓: depletion; ↑: regeneration; = : equal to normal control values

Transplantation experiments

To exclude a possible direct influence of ADR on the thymocyte regeneration process, thymus lobes were transplanted in two combinations.

ADR treated thymus into normal recipient: When thymi from 3 days previously ADR treated PVG RT7.2 rats were vascularly transplanted into normal, uncompromised PVG RT7.1 recipients, the regeneration of the transplant was found to be entirely dependent on the immigration of NDS58 positive (recipient type cells. Immunohistology of the graft after 11 days (day 14 after ADR) showed NDS58 positive cells, scattered in both cortex and medulla. By day 15 (18 days after ADR) the transplant was fully repopulated with thymocytes derived from recipient precursors. Only few donor lymphoid cells as detected with HIS41 were found scattered in the medulla on these same days (Figure 3).

Normal thymus into ADR treated recipient: When tested on days 14 and 18 after ADR treatment (11 and 15 days after grafting respecivelly) the transplants were small, with virtually empty medullary compartments and hypocellular but still present cortex. Staining with Mab recognizing the congenic determinants showed that thymocytes present were still of donor type ($HIS41^{+}$) with only scattered immigrants of recipient origin ($NDS58^{+}$). this Of this, there were few on day 14 and somewhat more on day 18 (Figure 3).

DISCUSSION

The results from this study indicate a total dependance of the adult thymus on BM derived precursors as the only means of maintaining its thymocyte population.

Using ADR we were able to totally deplete the thymus of thymocytes without damaging its microenvironment as witnessed by intact antigen profiles of stromal cells.

ADR is an antimitotic antibiotic, strongly cytotoxic for cells in G_1 and early S phase of cell cycle, by inhibition of DNA and RNA polymerases. Cells in G_0 phase of cell cycle are highly resistant to ADR killing (10). ADR treatment initially resulted in elimination of MRC Ox-19 negative (immature?) thymocytes, subsequently leading to total depletion of the cortex and to a large extent also of the medulla.

Although ADR treatment also affected the BM, after some time complete

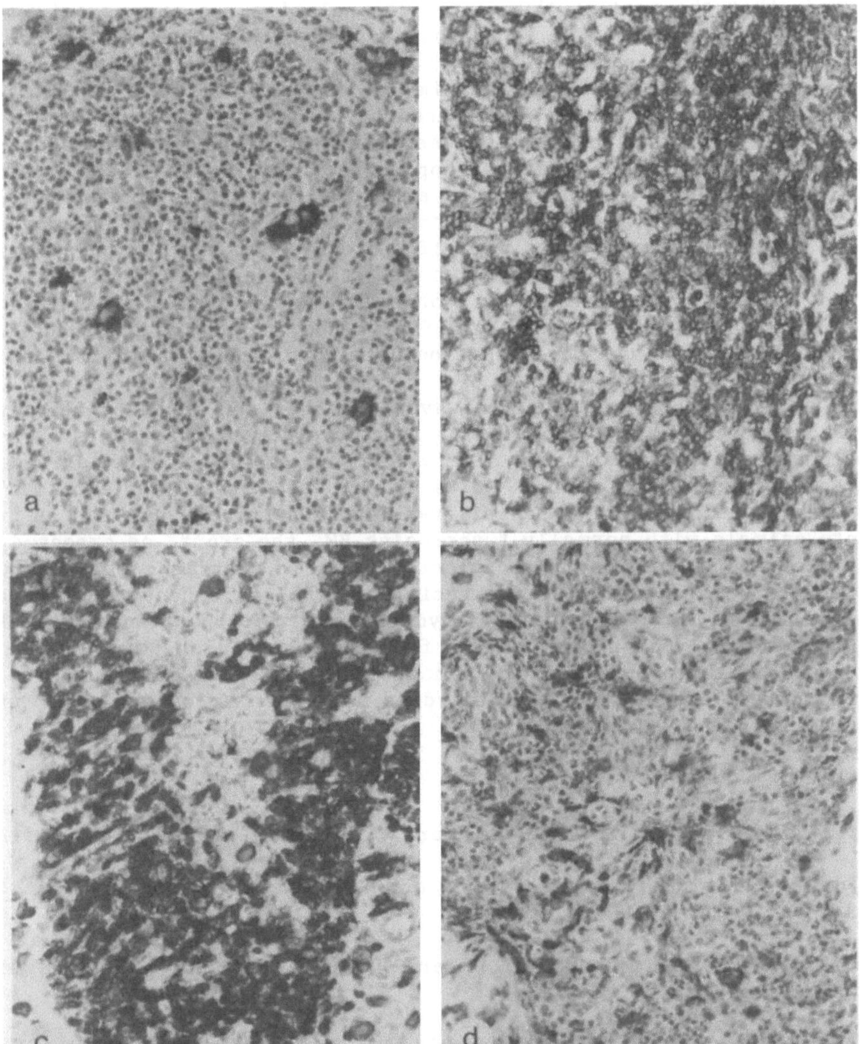

Figure 3. Identification of donor and recipient type thymocytes on frozen sections of vascularly transplanted thymi on day 18 after ADR treatment. Upper panel: ADR treated donor; lower panel: ADR treated recipient.
a and c : donor (PVG RT7.2) specific HIS41 Mab.
b and d : recipient (PVG RT7.1) specific NDS58 Mab.

thymus regeneration occurred, which process was clearly preceded by regeneration of BM haemopoietic activity.
Thus this method of thymocyte depletion seems to be advantageous over irradiation or steroids which have been described as damaging thymic stroma and thus possibly compromising the microenvironment necessary for undisturbed thymocyte development (11).

As thymus regeneration was preceded by regeneration of the BM this data suggested that the pool of thymocytes repopulating the thymus originated from newly produced prothymocytes in regenerated BM rather than from an intrathymic precursor. To further substantiate this concept syngeneic BM was given intravenously or intrathymically at the time of maximal depletion of autologous BM. By i.v. infusion of syngeneic BM apparently the time necessary for BM to restore its lymphopoietic activity was shortened, resulting in accelerated thymocyte regeneration now already apparent by day 14 after ADR treatment. Intrathymic injection of BM provided a source of prothymocytes directly to the thymus. This advanced the thymocyte regeneration for another days by overcoming the time necessery for colonization of the recipient BM prior to immigration to the thymus. Surprisingly even after $1x10^6$ cells i.t. it took 11 days before a burst of proliferation was observed, indicated by rapidly increasing cell numbers.This interval exactly reflects the lag period of 10-12 days between colonization and proliferation after i.t. injection of BM cells described earlier by others (1).

It still can be questioned, however, whether the observed regeneration of the ADR depleted thymocyte population from BM derived precursors is not exceptional resulting from selectively destroying intrathymic selfrenewing precursors by this treatment. Results of our vascular transplantation experiments seem to exclude this possibility. When ADR treated thymus was transplanted into animals with normally functioning BM, the graft, as expected, was repopulated by thymocytes of recipient origin. In contrast, when normal thymus was transplanted into ADR treated recipient, the graft became degenerated due to emigration of mature medullary thymocytes to periphery and the lack of simultaneous supply of prothymocytes from nonfunctioning BM of the recipient. Thus, although the population of thymocytes transferred in the graft possessed limited proliferating potential, does not seem to have the permanent self-renewing capacity of stem cell.

REFERENCES

1. Scollay,R., Smith, J., and Stauffer, V., (1986) Immunol. Rev. 91:129

2. Goldschneider, I., Komschlies, K. and Greiner, D. (1986) J. Exp. Med. 163:1

3. Kampinga, J., Schuil, A., Pol, G. and Nieuwenhuis, P. (1986) Transpl. Proc. 18:1429

4. Kroese, F.G.M., Wubbena, A.S., Joling, P. and Nieuwenhuis, P. (1985) Adv. Exp. Med. Biol. 186:443

5. Stet, R.J.M., Rozing, J., Majoor, G.D., Kroese, F.G.M., Opstelten, D. and Nieuwenhuis, P. (1985) Transplant. Proc. 17:1829

6. Kampinga, J., Kroese, F.G.M., Duijvestijn, A.M., Murawska, M.B., Pol, G. and Nieuwenhuis, P. (1987) Transplantation Proc. 19:3171.

7. Joling, P., Vaessen, L.M.B., Tielen, F.J. and Rozing, J. (1983) In: Modern trends in clinical immunosuppression (eds. W. Weimar, R.L. Marquet, A.B. Bijnen and R.J. Ploeg) p.201, Rotterdam, The Netherlands

8. Mason, D.W., Arthur, R.P., Dallman, M.J., Green, J.R., Spickett, G.F. and Thomas, M.L. (1983) Immunol. Rev. 74:57

9. Dijkstra, C.D., Dopp,E.A., Joling, P. and Kraal, G. (1985) Immunology 54:589

10.Sonneveld P. (1980) Pharmacokinetics of adriamycin in the rat. (Thesis for doctoral degree at the University of Leiden,The Netherlands.)

11.Huiskamp, R., van Vliet,E. and van Ewijk, W. (1985) J. Immunol. 134:2170

A STUDY OF EARLY EVENTS DURING T CELL DIFFERENTIATION IN THYMIC GRAFTED NUDE MICE

Caroline C. Chapman, Timothy L. Williams
John H. Robinson and Reg K. Jordan

Anatomy Department
Medical School
University of Newcastle upon Tyne
England. NE2 4HH

INTRODUCTION

Major progress has been made in characterizing the genes coding for the T cell antigen receptor and their products. The T cell receptor is a heterodimeric transmembrane glycoprotein consisting of disulphide linked α and β subunits and the genes encoding these proteins are arranged in a similar manner to those coding the immunoglobulins[1]. The genes are in a germ-line configuration in all tissues except for the T cell lineage where DNA rearrangements contribute to the generation of diversity of receptors. Recent evidence indicates that diversification of the T cell receptor repertoire by gene rearrangement occurs within the thymus, long considered the primary site of T cell production[2,3]. Furthermore, processes occuring within the thymus are seen as essential in moulding the repertoire of the emergent T cell pool by imprinting MHC restriction specificity and inducing self tolerance[4].

Normal thymic lymphopoiesis in the mouse commences when precursor cells from the fetal liver colonize the thymus on day 11-12 of gestation. This first cohort of lymphocyte progenitors establishes an intrathymic pool of differentiating lymphocytes which dominates the thymus throughout fetal life and into the post natal period[5]. It is assumed that the peripheral T cell compartment is initially established from this first cohort of thymocytes, and only replaced gradually by successive waves of T cell differentiation.

Before seeding by haematopoietic cells, the thymus consists of epithelial and mesenchymal cells. The epithelial cells express the surface class I and class II products of the MHC gene complex[6]. Haematopoietic colonisation of this sessile epithelial matrix establishes not only an intrathymic pool of T cell progenitors but also populations of immunologically relevant macrophages and dendritic cells or their precursors[7]. Thus the cellular components of the thymic microenvironment comprise at least four potentially separable populations (epithelium, macrophages, dendritic cells and the thymocytes themselves) which combine interactively to provide the milieu necessary for the intrathymic differentiation and maturation of T cell precursors[8]. The first cohort of thymocytes can be deleted from 14 day fetal thymus by culture at 24° or in the presence of deoxyguanosine (dGuo)[9,10]. Upon transplantation to intact recipients, thymic grafts so treated demonstrate prolonged allograft survival, suggesting that their dendritic cell population is also depleted[11,12].

We have established a model for studying the influence of the fetal thymus in shaping the emergent T cell repertoire in thymic grafted nude mice. Here we present data from animals receiving either transient or permanent grafts of syngeneic or allogeneic 14 day fetal thymus. Grafts were either untreated or enriched for thymic epithelium by low temperature culture or dGuo treatment.

These studies address two aspects of T cell ontogeny: the impact of the initial cohort of thymic emigrants on establishing a functional peripheral T cell compartment and the influence of the thymus on tolerance induction. Data from this experimental system are being published elsewhere.

METHODS

Animals: CBA-nu/nu and Balb/c-nu/nu mice were used as graft recipients. Embryonic material was obtained within the inbred congenic strains CBA-Thy-1*a, Balb/c, C57Bl/10 and Balb.K.

Thymic Grafts: Fetal thymus lobes were isolated from mouse embryos on day 14 of gestation, calculated counting the day of appearance of the vaginal plug as day 0. Grafts were either untreated or enriched for thymic epithelium by organ culture at 24° (8 days) or in the presence of 1.35 mM deoxyguanosine (5 days). The types of thymic graft, their treatment and the consequences of treatment are summarized in Table 1.

Table 1 Summary of thymic grafts and treatments

Graft	Treatment	Components Present
14 day fetal thymus	None	Thymic epithelium Macrophages/dendritic accessory cells 1st cohort lymphoid precursors
14 day fetal thymus	24° culture	Thymic epithelium Macrophages (Dendritic accessory cells??)
14 day fetal thymus	Deoxyguanosine (dGuo)	Thymic epithelium (Macrophages/ dendritic accessory cells ??)

Six lobes of 14 day fetal thymus, either treated or untreated, were grafted beneath the renal capsule of 6 week old syngeneic or allogeneic athymic nude recipients. Thy 1 allotype analysis and assessment of peripheral T cell function were performed 45-90 days post transplantation. In some experiments grafts were left in place throughout (permanent engraftment), in others grafts were recovered 14 days after transplantation (transient engraftment). In the transient engraftment protocol, the peripheral T cell compartment was assessed 30 or 60 days after the grafts had been removed.

Thy 1 Allotype Analysis: Thymocytes and ammonium chloride treated spleen cells were stained with fluoresceinated anti-Thy 1 (MCA 47F, Serotec) or anti-Thy 1.2 (1333, Becton Dickinson) monoclonal antibody for 30 mins on ice, washed and analysed by flow cytometry (FACS 420). In assessing the relative expression of each Thy 1 allotype in responding populations, mitogen or MLC stimulated blasts were separated by centrifugation over Lymphoprep and stained as above.

CTL Assay: The CTL responses of spleen cells from graft recipients were assayed in a 3 hour chromium release assay after 5 days in vitro stimulation in MLC using 5 x 10^6 responders and 5 x 10^6 irradiated (2000 rad from an x-ray source) stimulators in 2 ml RF10. The effector population was serially diluted in triplicate and incubated for three hours with 10^4 ^{51}Cr-labelled target cells. Target cells were 5 day Con A stimulated spleen cells labelled with 50 Ci ^{51}Cr-sodium chromate/10^6 cells. Results are expressed as cytotoxic index and K:T ratios use killer cell numbers after 5 day MLC.

Immunisation Against Minors: Mixed priming to minor histocompatibility antigens was achieved by giving a single i.p. dose of 7 x 10^6 B10.BR plus 7 x 10^6 Balb.K nucleated spleen cells to CBA nu/nu recipients bearing either syngeneic or Balb.K thymic grafts. After 14 days, mice were assayed for CTL responses against third party and thymic type minors, after 5 days culture in vitro with B10.BR or Balb.K irradiated stimulators (see CTL assay above).

RESULTS

Reconstitution of the immune response in nude mice

CBA-nu/nu mice (Thy 1.2) were engrafted either permanently or transiently with congenic CBA-Thy-1*a (Thy 1.1) 14 day fetal thymus. The results of an allotype analysis of Thy 1 expression by splenocytes and thymic graft lymphocytes 70 days post transplantation are presented in Table 2.

Table 2 Thy 1.1/1.2 expression by lymphocytes recovered from CBA-nu/nu mice receiving congenic CBA-Thy-1*a thymus grafts.

		% cells expressing: Thy 1.2	Thy 1.1
Permanent engraftment	thymocytes	94.9 ± 2.1	2.2 ± 0.9
	splenocytes	13.4 ± 0.9	1.6 ± 1.6
Transient engraftment	thymocytes[a]	0.5 ± 0.3	88.0 ± 5.3
	splenocytes	2.9 ± 1.3	1.2 ± 0.6

a analysis performed on cells recovered from transient grafts removed 14 days post transplantation.

In permanently engrafted animals both graft thymocytes and Thy 1 positive splenocytes were essentially of host origin. However minority populations of donor type lymphocytes (~2%) were found in both the thymic grafts and the spleens of host animals. Cell mediated immune function was reconstituted in all the permanently engrafted mice as judged by the proliferative response of spleen cells to a T cell mitogen, the DTH response to the contact sensitizing antigen fluorescein isothiocyanate and the CTL response to allogeneic major histocompatibility antigens. Thymocytes recovered from transient grafts recovered 14 days post transplantation expressed the Thy 1.1 allotype exclusively, demonstrating that the 14 day fetal thymus contained lymphoid precursor cells in sufficient number to generate all the thymic lymphocytes present. There appeared to be little or no contribution to the intragraft thymocyte population from host derived precursors. Analysis of splenocytes at 30 days or 60 days after the grafts had been removed showed that a small, but significant number expressed Thy 1.1 indicating the persistance of a donor thymus derived peripheral lymphocyte population. The percentage of splenocytes expressing Thy 1.2 was no greater than that observed in ungrafted nude mice. Transient engraftment of nude mice resulted in little reconstitution of T cell function. At 70 days post transplantation splenocytes from 5 out of 6 mice gave proliferative responses to concanavalin A, but only three gave anti-major CTL responses in vitro (Fig. 1). Thy 1 analysis of 3 day con A blasts indicated that the limited function observed could be attributed to the small number of Thy 1.1^{+} cells which had emigrated prior to graft removal and was not due to maturation of the nude recipients endogenous T cells.

It appears that a permanent thymic graft is required if full functional reconstitution is to be achieved. Lasting T cell reconstitution can be achieved not only with 14 day fetal thymus but also with 24° cultured or dGuo treated thymic grafts. Fig. 2 shows the lack of CTL in ingrafted Balb/c nude mice, as well as reconstitution of CTL responses in the spleen of Balb/c nude mice 90 days after permanent engraftment with syngeneic 14 day fetal thymus, 24° cultured thymus and dGuo treated thymus to the level of enthymic Balb/c mouse. CBA nude mice are similarly reconstituted by syngeneic foetal thymic grafts both treated and untreated.

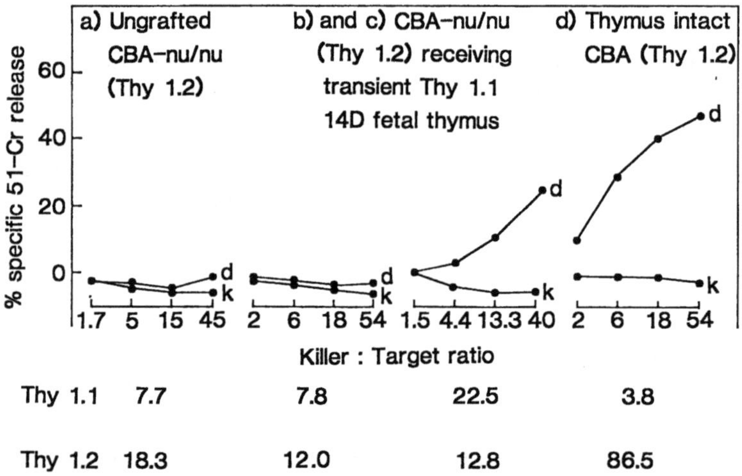

Fig. 1. Anti H-2d CTL responses and Thy 1.1/1.2 staining of 3 day con A blasts of splenocytes from CBA-nu/nu (Thy 1.2) mice receiving congenic, CBA-Thy-1*a(Thy 1.1), 14 day fetal thymus grafts.

Tolerance induction to thymic MHC

Nude mice were permanently engrafted with MHC incompatible thymus to test for the induction of tolerance to thymic MHC type.

Untreated 14 day fetal thymus: Eight nude mice were grafted with 14 day allogeneic fetal thymus. All the mice were reconstituted for third party

CTL responses and five out of the eight were tolerant of thymic MHC type by CTL (Fig. 3a) and MLC responses. The three mice which were not tolerant of thymic MHC type (Fig. 3b) all received grafts from the progeny of a single litter suggesting that the initial composition of the thymus at source may have an effect on its ability to induce tolerance.

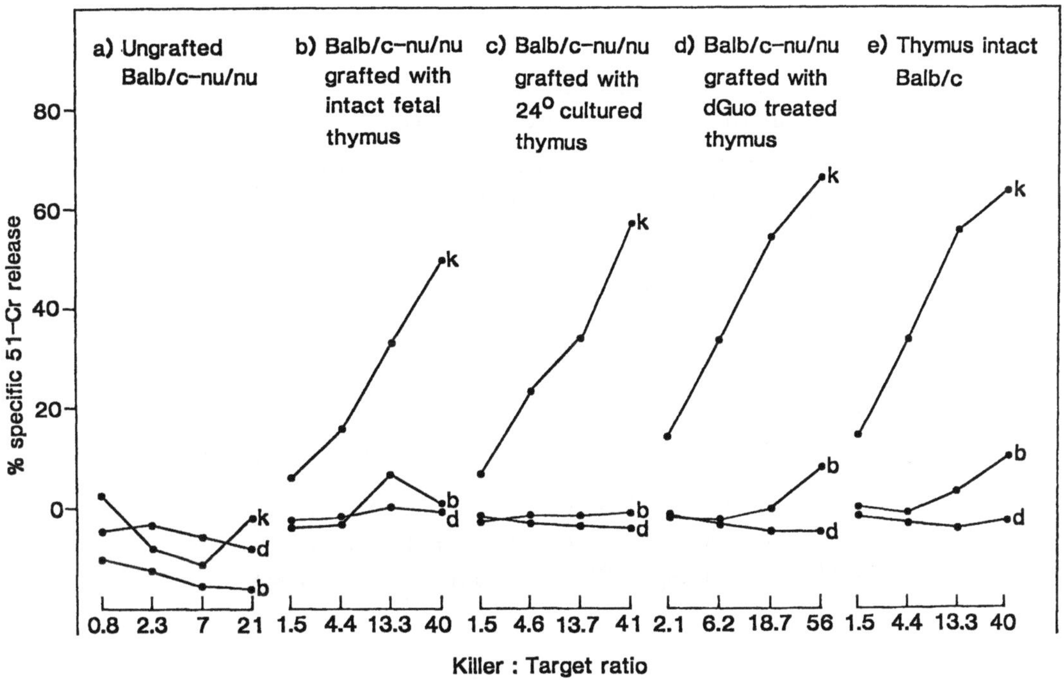

Fig. 2. Anti H-2k CTL responses of Balb/c-nu/nu mice with and without syngeneic thymic grafts. Targets: H-2k (k), H-2d (d) and H-2b (b).

24° cultured thymus: Twelve nude mice were grafted with allogeneic 24° cultured thymus. Nine out of the twelve reconstituted for CTL and MLC responses to third party stimulators and eight out of the nine mice were tolerant of thymic MHC type by CTL and MLC response whereas one of the nine was clearly not tolerant in either CTL or MLC assay. In the tolerant mice, CTL responses against MHC type were not demonstrable following in vitro stimulation in the presence of rat Con A soup suggesting the CTL precursors were tolerant in addition to the putative helper T cell measured by proliferation in MLC.

dGuo treated thymus: Ten nude mice were grafted with allogeneic dGuo treated thymus. Only four of the ten were reconstituted for third party CTL responses, suggesting that 14 day fetal thymus enriched for thymic epithelium by dGuo treatment reconstitutes nude mice poorly. However the four reconstituted mice were all not tolerant of thymic MHC type by CTL response.

Therefore 14 day fetal thymus and 24° cultured thymus can induce tolerance to thymic MHC whereas dGuo treated thymus grafts do not.

Tolerance induction to thymic minors

CBA-nu/nu (H-2k) mice receiving syngeneic 14 day fetal thymus grafts as well as ungrafted CBA-nu/nu controls were mixed primed with B10.BR and Balb.K spleen cells in vivo. B10.BR and Balb.K mice are MHC compatible with CBA-nu/nu mice but differ at multiple minor loci. Anti-minor CTL responses were seen against B10 and Balb minors in grafted but not in ungrafted mice (Fig. 4a,b).

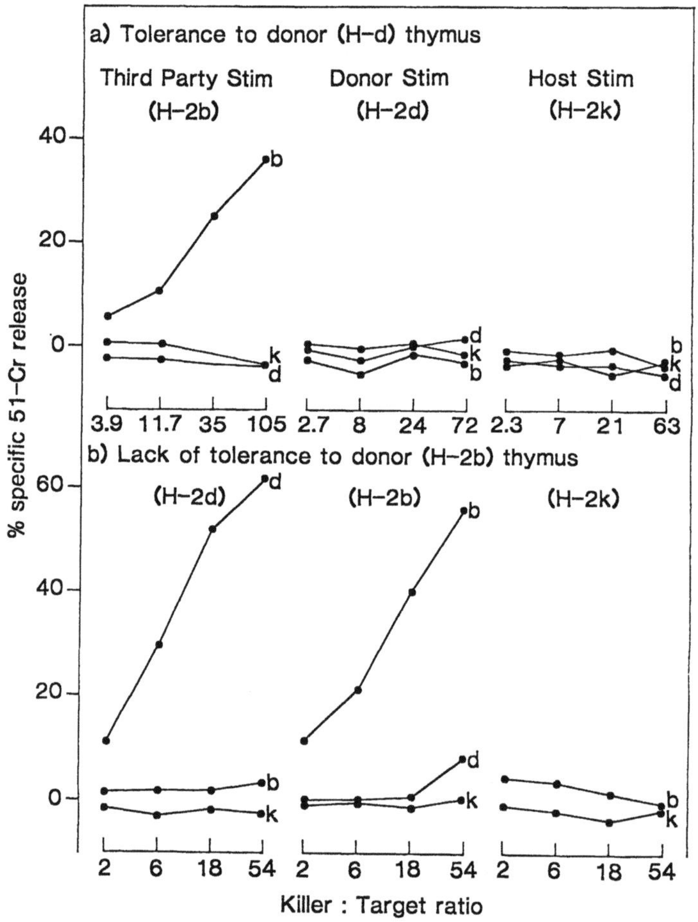

Fig. 3. CTL responses of CBA-nu/nu mice grafted with allogeneic 14 day fetal thymus. Targets: H-2k (k), H-2d (d) and H-2b (b).

14 day fetal thymus: Three CBA nude mice were grafted with Balb.k 14 day fetal thymus to test for tolerance induction by MHC compatible thymus differing from the host at multiple minor loci. All three were reconstituted for CTL responses against B10 (third party) minors following in vivo priming and were tolerant of thymic (Balb) minors by CTL response (Fig. 4c).

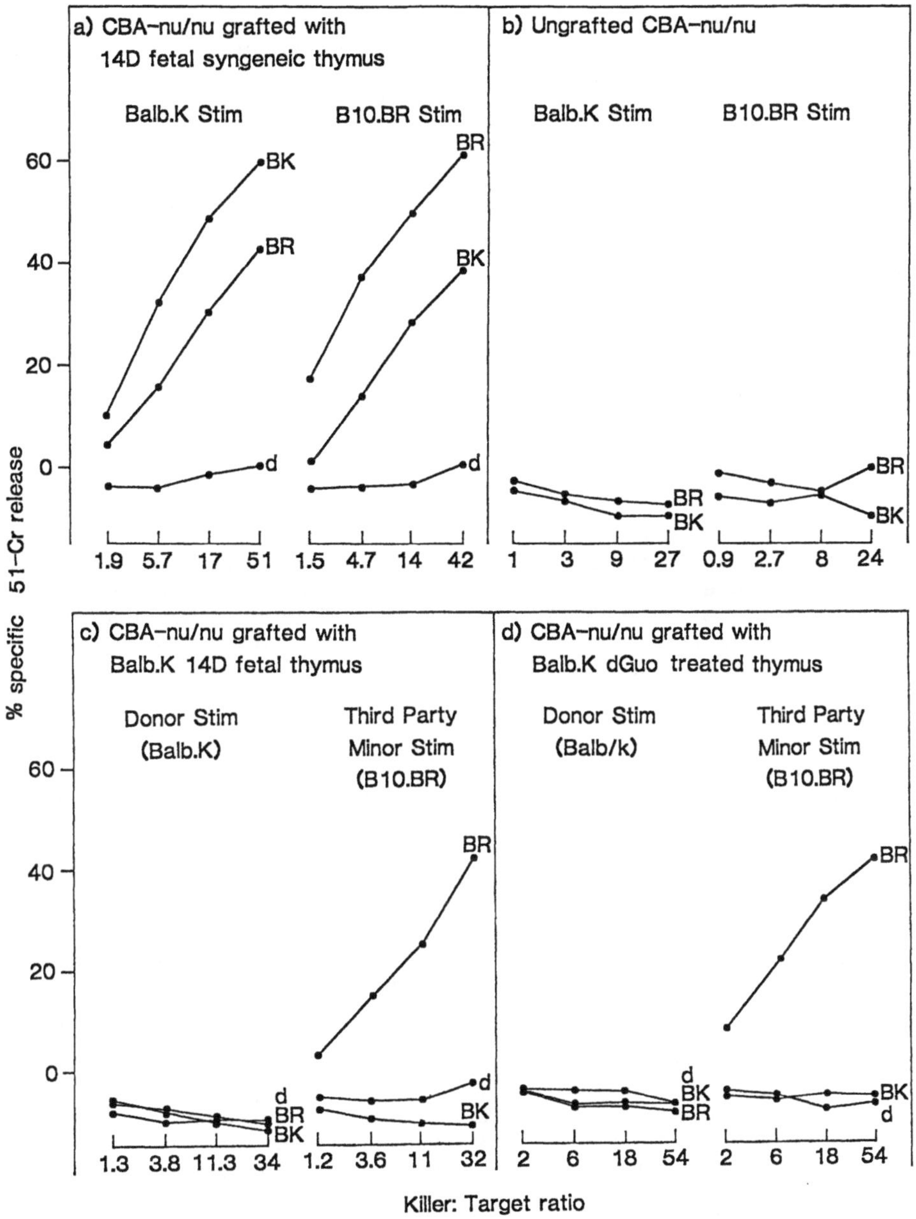

Fig. 4. CTL responses of grafted CBA-nu/nu mice 14 days after priming with 7 x 10^6 Balb.K and B10.BR spleen cells. Targets: Balb.K (BK), B10.BR (BR) and Balb/c (d).

dGuo treated thymus: Four CBA nude mice were grafted with Balb.k thymus enriched for thymic epithelium by dGuo treatment. Following in vivo priming to B10.BR and Balb.K spleen cells all four were tolerant of thymic (Balb) minors by CTL response but were reconstituted for CTL responses against the third part B10 minors (Fig. 4d).

From these data we conclude that 14 day fetal thymus and dGuo treated thymus both induce tolerance to their minor histocompatibility antigens. This includes the minors shared between B10 and Balb since no cross reactivity on Balb.k targets is seen in B10.BR stimulated spleen cells in these tolerant mice.

DISCUSSION

T cell deficient, athymic nude mice grafted with fetal thymus provide an in vivo model of T cell ontogeny. Transplantation of 14 day fetal thymus, untreated, depleted of lymphoid precursors by 24° culture, or enriched for thymic epithelium by dGuo treatment, leads to a substantial rise in the number of Thy 1 positive cells within the spleens of graft receipients and reconstitution of cell mediated immune function. Permanent thymic engraftment is essential for the reconstitution of the nude mouse from host progenitors.

Transient engraftment of 14 day fetal thymus leads to the establishment of a small, peripheral population of graft-derived functional T lymphocytes. These graft emigrants are derived from the first cohort of lymphoid precursors already resident in the donor thymus at transplantation. The population persists long term, does not appear to undergo peripheral expansion, exerts no detectable extrathymic influence upon the endogenous Thy 1 positive cells of the host, and alone is incapable of restoring cell mediated function to the nude recipients at a whole animal level. The lack of a more marked effect of the transient engraftment protocol is unexpected in view of the results reported by Jotereau et al.[5]. They suggested that there was significant emigration of first cohort cells from the fetal thymus within 21 days of transplantation to euthymic recipients. In our studies the thymic grafts were in place for only 14 days and the recipient animals were athymic.

Permanent engraftment of fully allogeneic 14 day fetal thymus, either treated or untreated, also results in the reconstitution of T cell function in nude mice, providing an experimental model for the study of tolerance induction to both thymic major and minor histocompatibility antigens in developing T cells. The major conclusions from the data presented here are that (i) 14 day fetal thymus and 24° cultured thymus usually induce MHC tolerance in reconstituted nude mouse T cells whereas, in agreement with the work of von Boehmer[13,14], dGuo treated thymus does not, and (ii) MHC compatible but minor disparate 14 day fetal thymus, either intact or enriched for thymic epithelium by dGuo treatment, induces tolerance to minor histocompatibility antigens. Deoxyguanosine treated thymus has also been shown to induce tolerance to minors when grafted to MHC incompatible new mice[15].

The simplest explanation for the pattern of MHC tolerance induction observed in these experiments and in those reported by others[2,13-15] is that the cellular composition of the initial thymic graft is critical in determining the outcome. 14 day fetal thymus coimprises thymic epithelium, the precursors for the first cohort of thymocytes and accessory cells of haemotopoietic origin. Following 24° culture the lymphoid precursors are deleted leaving an epithelial matrix together with residual macrophages.

Deoxyguanosine treatment enriches the epithelium and deletes both lymphocytes and macrophages. Whether dendritic cells persist following 24° culture or dGuo is unclear. However, the prolonged allograft survival in euthymic recipients observed following both treatments has been interpreted as suggesting that dendritic cells, if not deleted, are severely depleted. it is possible that 24° cultured thymus contains larger numbers of residual dendritic cells than dGuo treated thymus. Likewise, the lack of tolerance to thymic MHC found in three of the nude mice grafted with allogeneic 14 day fetal thymus could be explained by insufficient dendritic cells being present in the original grafts. Notwithstanding the chronological age post-conception, at 14 days gestation the thymus shows marked variation in maturation at a morphological level and there have been no reports suggesting that dendritic cells are present before 14 days. Dendritic cells can be invoked again to explain why dGuo treated thymus induces tolerance to minors but not majors[14]. One explanation is that thymic donor type minors are presented in a tolerogenic form by host-derived dendritic cells. However the explanation begs the question as to why graft majors are not similarly reprocessed by host antigen presenting cells as suggested in a different system[15]. Indeed there is no direct evidence that dendritic accessory cells can induce T cell tolerance in any experimental situation.

The demonstration that peripheral T cells from nude mice bearing allogeneic dGuo thymic grafts still respond to graft majors in vitro contrasts with the incapacity of such mice to reject test allografts of thymic donor type[2]. Thus in spite of in vitro reactivity thymic epithelial grafts appear to induce tolerance at the whole animal level.

ACKNOWLEDGEMENTS

These studies were supported by NIH grant AI 15819 and by funds from the Research Committee, University of Newcastle upon Tyne.

REFERENCES

1. Kronenberg, M.G. et al. 1986. Ann. Rev. Immunol. 4: 526.
2. Owen, J.J.T. et al. 1986. Curr. Topics. Microbiol. Immunol. 126: 35.
3. Pardoll, D.M. et al. 1987. J. Exp. Med. 165: 1624.
4. Lo, D. & Sprent, J. 1986. Nature 319: 672.
5. Jotereau, F. et al. 1987. J. Immunol. 138: 1026.
6. Owen, J.J.T. & Jenkinson, E.J. 1981. Prog. Allerg. 29: 1.
7. Robinson, J.H. 1983. J. Immunol. 130: 1592.
8. Robinson, J.H. & Jordan, R.K. 1983. Immunol. Today 4: 41.
9. Jordan, R.K. et al. 1985. J. Immunol. 134: 2155.
10. Jenkinson, E.J. et al. 1982. Eur. J. Immunol. 12: 583.
11. Jordan, R.K. et al. 1985. Nature 314: 454.
12. Ready, A.R. et al. 1984. Nature 310: 231.
13. von Boehmer, H. and Schubiger, K. 1984. Eur. J. Immunol. 14: 1048.
14. von Boehmer, H. and Hafen, K. 1986. Nature 320: 626.
15. Sherwood, R.A. et al. 1986. Eur. J. Immunol. 16: 569.

VASCULAR THYMUSTRANSPLANTATION IN RATS:

A NEW METHOD TO STUDY THYMOCYTE KINETICS

J. Kampinga, G.H. Pol, H. Bartels*, A.M. Duijvestijn, M.B. Murawska, J.M.M.M. van de Berk and P. Nieuwenhuis

Department of Histology and Central Animal House*, University of Groningen, Oostersingel 69/1, 9713 Ez Groningen The Netherlands

This work was supported by Medigon grant no. 900-505-201

INTRODUCTION

Although many studies have been performed to unravel the cellular kinetics of thymic immigration, intrathymic differentation and thymic emigration, there is still not a generally accepted view on this issue (for review see Scollay (1,2)). An important reason for the contradictory results is the lack of an experimental model mimicking the normal physiological situation of the thymus.

Recently we developed a vascular thymus transplantation technique (3). With this procedure the vessels of the right donor thymus lobe are directly connected to the blood circulation of the recipient. Immunohistology and functional studies have shown that such vascular thymus transplants (vTt's) are hardly affected after transplantation. Therefore this model offers an opportunity to study and- en out-put kinetics of thymocytes under better controlled physiological conditions. Here we want to report the use of this model in RT7 congenic rats (4) in an attempt to answer some questions on:

- the dependence on bone-marrow derived precorsors of the adult thymus
- the possible existence of long-lived intrathymic stemcells
- the turnover rate of thymocytes
- the rate of thymocyte emigration
- and the RT6 expression on early or late thymic emigrants.

MATERIALS AND METHODS

Thymus lobes of 6 months old PVG RT7.2 rats (kindly provided orginally by Dr. S.V. Hunt, Sir William Dunn School of Pathology, University of Oxford, U.K., and now bred in our own animal facilities) were transplanted in sex- and age-matched PVG (RT7.1) rats according to the method as described previously (3).

To study thymic T cell immigration vascular thymus transplants (vTt's) were removed at 1, 3, 7, 11, 19, 27 and 50 days after transplantation and single cell suspensions were double labelled using biotinylated NDS58 (5) (rat anti-RT7.1, kindly provided by Dr. M.R. Newton, Department of Zoology, University of Oxford, U.K.) and ER4 (6) (mouse anti-Thy-1, kindly

provided by Dr. J. Rozing, T.N.O., Rijswijk, the Netherlands).
To study thymic emigration recipient spleens were removed at 7, 11, 19, 27 and 50 days after transplantation and single cell suspensions of recipient spleens were double labelled with HIS41 (7) (mouse anti-RT7.2) and biotinylated DS4.23 (8) (rat anti-RT6.1, kindly provided by Dr. D.L. Greiner, Department of Pathology, University of Connecticut Health Center, Farmington, U.S.A.)
FITC conjugated Streptavidin (Amersham International, Little Chalfont, U.K.) and Goat anti-mouse IgG1 or IgG2a conjugated to TRITC (Southern Biotechnology Associates, Birmingham, Alabama, U.S.A.) were used as second stage antibodies.Labelling of single cell suspensions was performed as described by Opstelten (9).
The number of thymic immigrants (defined as both NDS58$^+$ and ER4$^+$ cells) and thymic emigrants (defined as HIS41$^+$ cells) was calculated on cytospots as follows:
The total number of lymphocytes on a cytospot was calculated by counting a representative part of it. This number was multiplied by a factor obtained after dividing the total surface of the cytospot with the surface of the counted part. Subsequently all positive cells in the cytospot were counted, and the ratio of positive cells to the total number of lymphocytes in the cytospot was converted to absolute numbers of positive cells in vascular transplants and recipient spleens, respectively.

RESULTS

Immigration

Unpublished previous experiments did show that not all recipent derived cells that had entered the vTt were bone marrow derived prothymocytes: also peripheral T and B cells as well as cells of the myeloïd lineage gained entrance into (mainly medullary part of) the vTt. This entrance may be an expression of a low degree of recirculation of leucocytes through the thymus, but it cannot be excluded that some of them entered the graft as a reaction on the slight tissue damage provoked by the transplantation procedure.
Because the Thy-1 marker in rats is only expressed on part of bone-marrow cells (a.o. stemcells and prothymocytes) as well as thymocytes but not on peripheral T cells (personal communication J. Rozing and ref. 13), recipient derived prothymocytes and their intrathymic progeny can be detected using a double label of NDS58 (recipient marker: RT7.1$^+$) and ER4 ((pro-)thymocyte marker) in single cell suspensions prepared of vTt's. Using this criterium for immigration of recipient derived prothymocytes NDS58$^+$/ER 4$^+$ cells could not be detected before day 7 after transplantation. Their number slightly increased till day 11, to rapidly increase between 11 and 27 days. After 27 days virtually all (96-100%) thymocytes were of recipient origin (fig. 1).

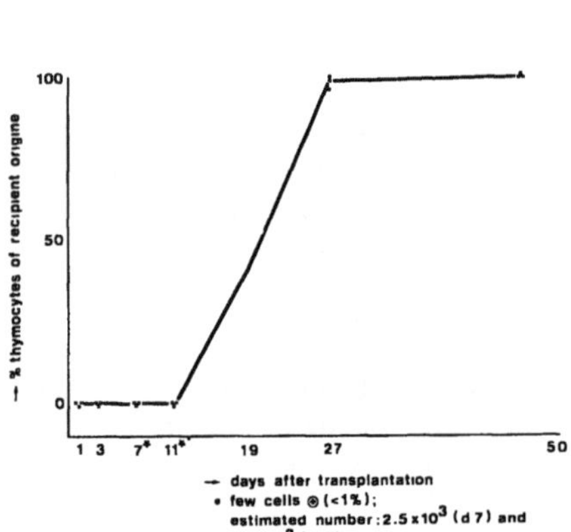

Fig.1 Immigration and intrathymic proliferation of recipient derived thymocytes.

Emigration

Single cell suspensions of spleens from thymus grafted rats were double labelled with HIS41 (donor marker: RT7.2+) and DS4.23 (RT6.1+) in order (i) to calculate the number of thymocytes leaving the vTt and (ii) to study the expression of RT6 on these thymic emigrants.
As shown in fig. 2 the number of RT7.2+ thymic emigrants in the spleen increased till day 27 and slowly decreased after this time. Suprisingly only a few of the early thymic emigrants in the spleen expressed the RT6 determinant, whereas 50 days after transplantation virtually all (94-100%) emigrated cells were RT6 positive.

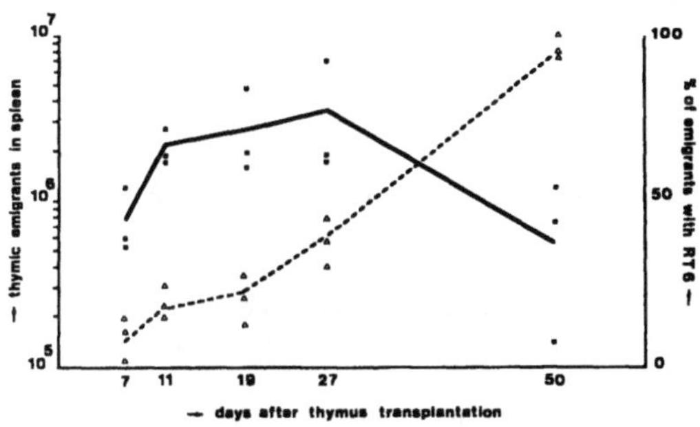

Fig.2 Emigration of donor thymic emigrants to the recipient spleen.
—— :Number of HIS41+cells in recipient spleen.
- -:Percentage of DS4.23+cells within the population of HIS41+thymic emigrants in the spleen.

DISCUSSION

The technique of vascular thymus transplantation was used to study thymocyte cell kinetics and RT6 expression on thymic emigrants, using RT7 congenic rats. The mab's NDS58 (RT7.1+) and HIS 41 (RT7.2+) were used to discriminate between T cells from recipient (RT7.1) and donor (RT7.2) origin.

Cell kinetics

Immigration: Data of this study revealed that the first thymocytes of recipient origin could not be recognized until 7 days after transplantation. A slow increase of their number was found 11 days after transplantation, followed by a rapid increase of absolute and relative cell numbers in the two weeks there after. At day 27 the orginal population of donor thymocytes had almost completely been replaced by thymocytes of recipient origin (fig. 1). These data are in agreement with the data found in thymic repopulation assays published by others (2). In these models of thymic repopulation after bone marrow transplantation in lethally irradiated mice it was found that regeneration of the depleted thymus started 7 days after bone marrow transplantation with the proliferation of radioresistant intrathymic recipient thymocytes. After a lag-time of 10-12 days the first bone-marrow derived thymocytes could be detected giving rise to a second wave of rapid dividing (donor derived) cells resulting in complete take over by donorcells between 20 and 30 days.
However, as discussed by Scollay (1,2), it is hard to draw firm conclusions regarding thymocyte kinetics under physiological conditions from this model, because of possible irradation induced artefacts (stress, thymic depletion, bone marrow damage etc.)
So, although this model had proven to be extremely useful for studying intra-thymic events, still the question remains whether the thymus in adult life is continuously seeded with bone marrow derived precursors or

whether thymocytes in adult life are the progeny of long-lived intrathymic stemcells.
In contrast to these thymic repopulation assays our model using vascular thymus transplantation highly mimics the normal physiological situation. When, with regards to thymocyte kinetics questions, our experimental data are transfered to the normal thymus situation, they strongly suggest that precursor cell immigration into the thymus is a contineous process even in adult animals (see also M. Murawska (10). Further more our data show that the progeny of intrathymic precursors ,if any, have left the thymus within 30 days.
However, our model is not completely physiological ass stress from the transplantation procedure can not be avoided, whereas the vTt is deprived of neurogeneic regulatory stimuli for at least two weeks after grafting (personal communication H.J. Schuurman and M. Kendall) .
Preliminary data obtained from adrenalectomized animals reveal that in the absence of the adrenals comparable cell entry and thymocyte turnover take place in vTt's. This was also found by Scollay (2) using the intra-thymic injection technique of Goldschneider (11).

Emigration: Our data showed that the number of donor (i.e.vTt) derived emigrants in the spleen increased till 11 days after transplantation. Between 11 and 27 days this increase was less, whereas after 27 days donor derived emigrants slowly disappeared (fig. 2). A similar picture of increase and decrease was found on cryostat sections of Peijers's patches and lymphnodes (unpublished observations).
These emigration data give some indications about the fate and daily output of thymic emigrants: As after day 27 no potential emigrants (of donor type) are left in the graft and in this period the number of thymic emigrants decreases we conclude that most emigrated thymocytes die within 50 days after emigration. However, if we assume that peripheral proliferation does not contribute significantly to the peripheral presence of graft emigrants and the rate of emigration of donor derived thymocytes is decreased because of the take over by recipient derived thymocytes, the fact that there is still an increase in the number of emigrants between 11 and 27 days suggest that most thymic emigrants do not die in the first two or three weeks after emigration.
Assuming that there is no proliferation and that recent emigrants have some weeks to live one can make an estimation of the daily number of emigrating thymocytes.
Because stress of the operation procedure can have influenced the rate of emigration directly after transplantation the daily output is calculated using the data of the period between 7 and 11 days instead of using the period of the first 7 days after transplantation.
In this 4 day period about 1.4×10^6 thymocytes have emigrated to the spleen. Scollay (12) has shown that approximately one third of the total thymic emigrants is found in the spleen. In our situation this means that the total output of the vTt in this period is 4.6×10^6 thymocytes, which gives a daily output af around 1.15×10^6 thymocytes.
Converting these data to the number of thymocytes in the vTt at the day of transplantation (about 5×10^7) gives the picture that about 2.3% of the total thymocyte population leaves the thymus daily. A figure much higher than the 0.1% found by Scollay in 6 months old mice (12).
However, when the emigration data of the first week after transplantation are used, the calculations give rise to a daily output of 0.7% .
So, more experiments are needed to estimate the daily output and to establish the role of stress and peripheral proliferation, which can be responsible for the found differences in daily output numbers.

RT6 expression on thymic emigrants

As shown in fig. 2 our data show that at 7 days only 10% of the thymic emigrants are RT6$^+$ whereas by day 50 virtually all thymic emigrants express this determinant.
These data best seem to fit the following picture:

1. The RT6 determinant is gained by thymic emigrants during their sojourn inperipheral lymphoïd organs, and
2. thymic emigrants who do not express this marker within a few weeks after emigration will die.

This hypothesis also explains the results of a study by Thiele (13) who found that in the first weeks after birth most peripheral T cells, which should be considered as recent thymic emigrants, do not express the RT6 determinant.
This notion is further supported by our data on thymectomized rats, showing that with increase in time after thymectomy the percentage of RT6$^+$ peripheral T cells increases. (J. Kampinga and J.M.M.M. van de Berg, manuscript in preparation).
All these experiments suggest that thymic emigrants are subjected to a post thymic differentiation step at which stage the RT6 marker is developed.
As to the meaning of this phenonenon we can only speculate. One posibility can be that thymic emigrants are triggered to express RT6 during an immunological reaction. These RT6$^+$ antigen experienced T cells will enter the pool of long-lived memory T cells, whereas thymic emigrants, which were not confronted with an antigen within a certain amount of time (2-3 weeks?, see discussion on emigration) after leaving the thymus, die.
This picture fits nicely to the studies of MacLennan and Gray on B cell kinetics. They found that only B cells, that have met antigen survive for a longer period in the body (14).
Further more the fact that in adult rats a high percentage of peripheral T cells (in our hands around 60%) is RT6$^-$ suggests that within the pool of peripheral T cells a high number of recent thymic emigrants circulate.
This also favours the above mentioned notion that thymic emigration is much higher than suggested by others.

In summary we showed using the model of vascular thymus transplantation performed in RT7 congenic rats, that this new technique provides an useful model in studying thymocyte kinetics.
The data observed in this model strongly suggest that in the adult thymus:

1. the postulated long-lived intra-thymic stemcell does not exist and that the thymus is continuously dependent on extra-thymic (bone-marrow derived) precursors
2. the orginal thymocyte population is totally renewed by the progeny of these precursor cells within 4 weeks
3. the daily output of thymocytes is between 0.7 and 2.3% of the total population of thymocytes
4. recent thymic emigrants do not express the RT6 determinant whereas within 50 days after emigration this marker is found on virtually all thymic emigrants

REFERENCES

1. R. Scollay and K. Shortman. In: Recognition and Regulation in Cell Mediated Immunity, eds. Watson, J.D. and Marbrook, J., p.3. Marcel Dekker, New York and Basel (1984).
2. R. Scollay, J. Smith and V. Stauffer. Immunol. Rev. 91:129 (1986).
3. J. Kampinga, A. Schuil, G.H. Pol and P. Nieuwenhuis. Transpl. Proc. 18:1429 (1986).
4. Anonymus. Transpl. Proc. 15:1683 (1983).
5. M.R. Newton, K.J. Wood and J.W. Fabre. J. Immunogenet. 13:41 (1986)

6. P. Joling, F.J. Tielen, L.M.B. Vaessen, J.C.Hesse and J.Rozing. Adv. Exp. Med. Biol. 186:235 (1985).
7. J. Kampinga, F.G.M. Kroese, G.H. Pol, D. Opstelten and P. Nieuwenhuis. This volume.
8. J.M. Ely, D.L. Greiner, D.M. Lubaroff and F.W. Fitch. J. Immunol. 130:2798 (1983)
9. D. Opstelten, G.J. Deenen, J. Rozing and S.V. Hunt. J. Immunol 137:76 (1986)
10. M.B. Murawska, A.M. Duijvestijn, A.S. Wubbena and P. Nieuwenhuis. This volume.
11. J. Goldschneider, K. Komschlies and D.L. Greiner. J. Exp. Med. 163:1 (1986)
12. R. Scollay, E. Butcher and J. Weisman. Eur. J. Immunol 10:210 (1980)
13. H.G. Thiele, F. Koch and A. Kashan. Transpl. Proc. 19:3157 (1987)
14. J.C.M. MacLennan and D. Gray. Immunol. Rev. 91:61 (1986)

M RECEPTOR AND DIFFERENTIATION OF T LYMPHOCYTES IN THE DEVELOPING HUMAN THYMUS*

Katarina Isaković, Tatjana Matejić and B.D.Janković

Immunology Research Center, Vojvode Stepe 458
11221 Belgrade, Yugoslavia

INTRODUCTION

In earlier studies we demonstrated that prothymocytes in the 14-22-week-old fetal liver contained receptors for mouse erythrocytes (1,2), and that M receptor is a differentiation marker of fetal and infant thymocytes (3). This report deals with the study of thymocyte subsets and T cell differentiation in human fetal and infant thymus, in which M receptor was used as a marker of T cells. Further, to provide a more composite picture of thymocyte differentiation and compesate the limitations of experimental methods in detecting the cells coexpressing more than two markers, a mathematical probability method defining various membrane phenotypes was used. On the basis of experimentally obtained data and results of regression analysis, a hypothetical pathway of T cell differentiation within the thymus was proposed.

MATERIALS AND METHODS

Thymus donors

Thymuses were obtained from 8.5-22-week-old thymuses after prostaglandin-induced abortion or Ceasarian sections, and from 5-month to 4-year-old children who underwent heart surgery. Thymocyte suspensions were obtained by mincing the thymus tissue over an 80 μm stainless mesh. After several washings, thymocytes were resuspended in RPMI 1640 pH 7.2, at a cell density of $5x10^6$ viable cells/ml. Such prepared cell suspensions were used in all tests.

*Supported by the Republic of Serbia Research Fund, Belgrade.

Membrane marker tests

Membrane immunoglobulins (mIg) were detected by direct immunofluorescence method using FITC-conjugated F(ab)2 fragments of sheep anti-human Ig serum (Behringwerke AG). T3, T4, T6, T8 and Ia antigens were searched with appropiate monoclonal antibodies by indirect immunofluorescent staining. For detection of pan-T and pan-B antigens a heterologous highly specific anti-T (ATS) and anti-B (ABS) sera preabsorbed with tissue antigens (4) were used. ATS stained all cells forming E rosettes, but did not stain cells lacking E receptors in E^- population. ABS stained 95% of cells in E^- population, but did not react with E rosette forming cells. Immunofluorescent staining of cytoplasmic u chain ($cIgM^+$ cells) was performed according to the method described by Parks et al. (5). Microrosette tests were applied to detect E and M rosette forming cells. For simultaneous detection of E and M receptors and antigens, rosette test was combined with immunofluorescent staining. Coexpression of two antigens on a single cell was proved by immune rosette formation (6).

Statistical mathods

The simultaneous expression of more than two markers on the same cell was calculated by a mathematical probability method, multiplying the frequencies of a single markers obtained in the experiment. The ontogenic relationship between cells at different stages of thymocyte differentiation, was tested by polynomal functions of n degree (n= 1, 2, 3.....,k). Regression analysis was done using the statistical program for the computer DECLAB PDP (7,8). Coefficients of regression were obtained by the least square method.

RESULTS AND DISCUSSION

Thymocytes obtained from human fetuses between 8.5 and 22 weks of gestation, as well as from children between 5 months and 4 years of age, were investigated for the presence of E and M receptors, cIgM, mIg, T, T3, T4, T6, T8, B and Ia antigens. 65-95% of thymocytes derived either from fetal or infant thymuses expressed T, T6, T8 antigens, and E receptors. Cells bearing T4 and T3 antigens ranged between 10 to

Table 1. Number of M^+ thymocytes present in fetal and infant thymus

group	Source of thymocytes	Age of thymus donor	Number in group	Mean of M^+ cells (%)	Statistical analysis P
A	Fetal thymus	13.5-16.5*	6	16.1	A v B >0.1
B	Fetal thymus	17 -22**	4	24.5	B v C <0.02
C	Infant thymus	5 -48	4	40.2	A v C <0.ol

*Age in weeks; **Age in months.

25%. cIgM$^+$, mIg$^+$ and Ia$^+$ cells were present in a very low number. There was no significant difference in the percentages of cells expressing T- and B-cell specific markers between fetal and infant thymuses. However, in both thymuses there was a remarkable number of cells rosetting with MRBC. The number of those cells increased during development (Table 1.).

Our further assesment was focused on phenotypic analysis of M receptor-bearing thymocytes using double marker tests, and the results are shown in Table 2.

Table 2. Membrane phenotype of M$^+$ thymocytes

Source of thymocytes	% of M$^+$ thymocytes expressing different antigens (range)							
	T	T3	T4	T6	T8	B	mIg	Ia
Fetal* thymus	78-96 (10-31)	46-59 (10-17)	o-38 (0-6)	82-88 (30-45)	72-89 (12-17)	2-10 (o.4-2)	0	0
Infant** thymus	78-88 (20-37)	12-25 (,5-11)	0-9 (0-4)	50-91 (24-37)	68-81 (29-33)	0	0	0

*10 fetuses between 13.5-22 weeks of gestation were included in the group. **Thymocytes were obtained from 4 children between 5 to 48 months of age. In parenthesis: percentage of M$^+$ cells expressing corresponding antigen within the thymus.

Almost 90% of M$^+$ cells within the thymus expressed T cell markers. Only 1-10% of thymocytes were B$^+$, while mIg$^+$ and Ia$^+$ cells were not found.

In order to overcome the experimental restrictions, a mathematical probability method was used to detect phenotypes expressing more than two markers. Phenotypes were determined in thymocyte compartments previously defined by double marker tests. In addition, it was demonstrated that there were no significant differences between experimental and inspected values concerning the expression of two markers on a single cell. The results of phenotypic analysis of thymocytes obtained by probability method are shown in Table 3. Most of predicted phenotypes expressed M receptors. However, M receptors which were expressed predominantly on immature thymocytes, were also present on T lymphocyres expressing T3 antigen, a marker of T cell maturity.

The hypothetical pathway of thymocyte differentiation is illustrated in the figure. It was proposed that, whithin the developing human thymus three main experimentally defined thymocyte compartments exist: M$^+$T$^+$, M$^+$T$^+$E$^+$, and T$^+$E$^+$. Each of these compartments contains several phenotypes defined by antigens detectable by monoclonal antibodies. In addition, M$^+$ cells lacking T and B cell markers (M$^+$T$^-$B$^-$) were also detected in a small number (about 2%), and were considered as precursors of M$^+$T$^+$ cells (1,2). It should be underlined that, a functional

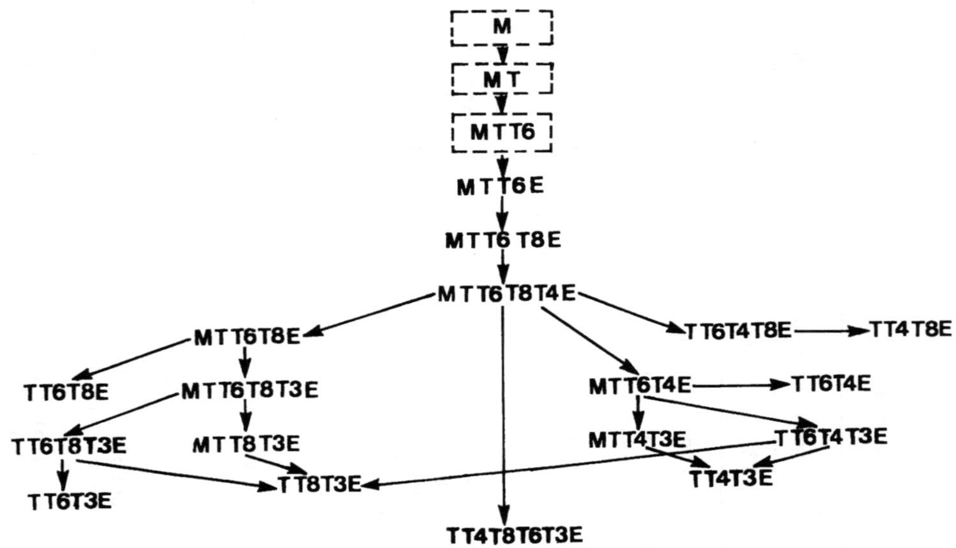

Fig. 3. Scheme of T lymphocyte differentiation in the developing human thymus.

dependancy between $M^+T^-B^-$ and M^+T^+, as well as between M^+T^+ and $M^+T^+E^+$, and $M^+T^+E^+$ and T^+E^+ compartments exist and could be approximated by linear function (data not shown). Therefore, $M^+T^-B^-$ cell differentiate into prothymocyte (M^+T^+), which than acquire T6 and E markers. The next

Table 3. Frequency of thymocytes with different membrane phenotypes present in infant human thymus

Phenotype	%	Phenotype	%
M	2.0	T T8 T3 E	19.7
M T6	4.6	T T6 T3 E	0.4
M T6 E	11.1	T T6 T4 T8 E	0.6
M T6 T8 E	20.0	T T6 T8 E	6.3
M T T6 T8 T4 E	0.3	T T6 T4 E	1.1
M T T6 T8 E	2.8	T T6 T8 T3 E	7.5
M T T6 T4 E	0.6	T T6 T4 T3 E	0.3
M T T6 T8 T3 E	3.6	T T4 T3 E	2.7
M T T4 T3 e	0.2	T T4 T8 E	0.7
M T T8 T3 E	2.4		

stage of differentiation is defined by expression of T8 and than T4 antigen. Coexpression of these two antigens, predicted in this study by mathematical method, was shown by other investigators, who also demonstrated that expression of T8 preceeds the expression of T4 molecules (9). According to our scheme, $M^+T^+T6^+T8^+T4^+E^+$ cell is a precursor cell which differentiate allong two pathways giving rise to $T8^+$ and $T4^+$ population of lymphocytes. Maturation of T cells at this stage is characterized by the loss of M receptors followed by acquisition of T3 antigen. As it is shown in the figure, during maturational processes within the human thymus several end-cells with distinct phenotypes appear. Whether those cells are involved in the differentiation of T lymphocytes is unknown. It is certain, however, that most of the phenotypes defined in this study were never found in the circulating pool of lymphocytes. Nevertheless, we assume that mathematical method applied in this study could be successfully used in phenotypic study of lymphocyte subsets and their maturational events.

REFERENCES

1. B.D.Janković, K.Isaković, and T.Matejić, in: "Microenvironments in the Lymphoid System", p.901, G.G.B.Klaus,ed.,Plenum Publishing Co., (1985).
2. K.Isaković, T.Matejić, and B.D.Janković, in: "Microenvironments in the Lymphoid System", p.909, G.G.B.Klaus, ed., Plenum Publishing Co., (1985).
3. K.Isaković, T.Matejić, G.Terzić, and B.D.Janković. Fed.Proc. 42:521 (1983).
4. K.Isaković, K.Mitrović, B.M.Marković, M.Rajčević, and B.D. Janković. J.Immunol.Meth. 7:359 (1975).
5. D.E.Parks, M.D. Doyle, and W.O.Weigle. J.Immunol. 119:1923 (1977).
6. J.D. Griffin, R.P. Beveridge, and S.F. Schlossman. Blood,60:30 (1982).
7. W.W.Cooley, and P.R.Lohnes. Multivalent Procedures for the Behavioral Sciences, John Wiley and Sons (1962).
8. B.Ostle, Statistics in Research, The Iowa State College Press (1954).
9. S.Berrish, C.Boussuges, J.P.Binet, and J.F.Bach. Cell.Immunol. 74:260 (1982).

HOMING RECEPTOR EXPRESSION IN GRAFTS OF PURIFIED THYMIC EPITHELIUM

Greg A. Perry, David A. Crouse* and Morris O. Dailey

Department of Pathology
University of Iowa
Iowa City, IA 52242 USA

*Department of Anatomy
University of Nebraska Medical Center
Omaha, NE 68105 USA

INTRODUCTION

Grafts of purified thymic epithelium have been used to study the role of the thymic microenvironment in T cell differentiation[1-6]. The ability of such grafts to immunologically reconstitute either nude mice or thymectomized, irradiated and hematopoietically reconstituted hosts has demonstrated that they are only partially able to reconstitute the host animal, as neither lymphocyte phenotype frequencies nor functional responsiveness of the graft derived peripheral (splenic) T cell population reaches normal (intact) levels[2,3,6]. More recently it has been suggested that the expression of lymphocyte homing receptors may be important in the selection of thymocytes destined to emigrate from the thymus, as well as their localization in the periphery[7-9]. In the studies reported here, we have examined homing receptor expression in grafts of purified thymic epithelium and in the peripheral lymphoid organs of grafted nude mice. Results showed that, although incomplete, T cell reconstitution did take place in these animals. The presence of MEL-14 positive cells in the graft suggests that the presence or absence of homing receptor expression in these grafts was not solely responsible for the incomplete T cell emigration and peripheralization seen in these animals.

METHODS AND MATERIALS

Animals: Female Balb/c and Balb/c nu/nu nude mice were obtained from Life Sciences, Inc. (St. Petersberg, FL). Mice were housed on laminar flow stations in sterile cages, and fed sterilized food and water *ad libitum*.

Purified Thymic Epithelial Grafts: Purified thymic epithelium was prepared by the low temperature organ culture of 14 day fetal thymic lobes as previously described[1-3,10]. Briefly, thymus lobes from 14 day fetal Balb/c mice were removed and placed in culture on 0.4 μm Nuclepore filters floating on RPMI-1640 medium containing 10% fetal calf serum, 2 mM l-glutamine, 50 U/ml penicillin and 50 µg/ml streptomycin. Thymus lobes were cultured for 7 days at 24°C, and a further 7 days at 37°C. Such cultures have been shown to be depleted of lymphocytes[1,11], however representative cultures were examined for histological confirmation of their lymphoid free nature. Treated thymic epithelial lobes were grafted under the kidney capsule of recipient nude mice, and allowed to mature for 60 days prior to sacrifice. Age matched nude and intact Balb/c controls were kept for each experimental replication.

Immunocytochemistry: The MEL-14 monoclonal antibody used in these studies was produced in the laboratory of Dr. Dailey. All antibodies were used at previously determined optimal concentrations.

Grafts and peripheral lymphoid tissues were excised and immediately frozen in OCT embedding media (Scientific Products) using isopentane and dry ice. For analysis, 6 µm sections were cut, acetone fixed and washed in phosphate buffered saline (PBS) for 30 minutes. Sections were incubated for 1 hour with the primary antibody, rinsed twice in fresh PBS, and incubated for another hour with a biotin-conjugated secondary antibody (goat anti-rat IgG, mouse serum adsorbed; Kirkegaard & Perry Laboratories). After two more washes, all sections were incubated with horesradish peroxidase-conjugated avidin (Vector) for a further hour. Sections were again washed and incubated with 0.5 mg/ml diaminobenzadine in PBS for 10 minutes. Finally, sections were post-fixed in 1% gluteraldehyde in PBS, dehydrated in alcohol and mounted with Permount. Photographs were taken using a Zeiss photomicroscope at the magnifications indicated.

RESULTS

In these studies nude mice received purified thymic epithelial grafts and examined 60 days post grafting. Animals were sacrificed and immunocytochemical analysis performed on peripheral lymph nodes and the graft itself. Results of the immunocytochemical analysis are shown in Figure 1. Sixty days after grafting the graft contains a large number of lymphocytes, a subpopulation of which can be seen to express the MEL-14 homing receptor (Figure 1B). The frequency and distribution of MEL-14 positive cells in the graft was similar to that previously reported for normal thymus[7-9]. Lymph nodes from these grafted animals were found to contain Thy-1, GK1.5 and Lyt-2 positive cells in an appropriate anatomical distribution (data not shown),in a frequency intermediate between normal (intact) and nude animals. Lymph nodes also contained cells which stained positively with the MEL-14 monoclonal antibody (Figure 1D). These cells were found throughout the node with the exception of germinal centers.

Taken together, these results demonstrate that grafts of purified thymic epithelium are capable of sustaining the differentiation of lymphocytes expressing lymphocyte homing receptors. In addition, these results suggest that homing receptor expression in these grafts is not solely responsible for the incomplete T cell emigration and peripheralization observed in these mice.

DISCUSSION

It has been shown previously that immunodeficient mice receiving grafts of purified thymic epithelium are only partially reconstituted in their ability to respond to T cell mitogens, and in the frequency of cells

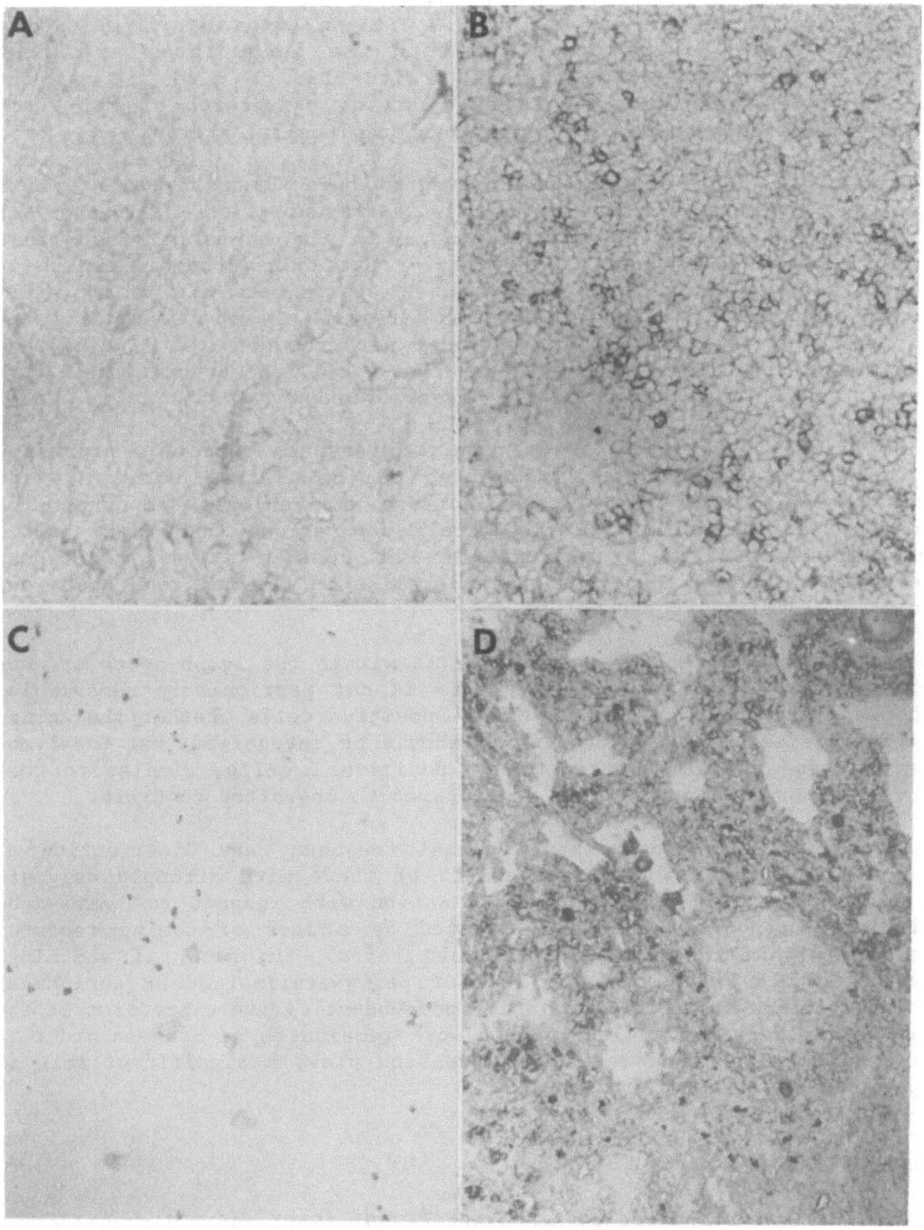

FIGURE 1. LTOC thymic epithelium 60 days after grafting (A, Negative control, /2x;B, MEL-14, 72x). Peripheral lymph node of LTOC grafted nude mouse 60 days after grafting (C, Negative control, 40x; D, MEL-14, 40x).

bearing T cell surface phenotypes[2,3,6]. Since the frequency of T cell subsets are decreased in the spleens of these animals, the incomplete functional reconstitution observed could potentially be due to abnormal or inefficient T cell emigration from the thymus graft itself. It has been suggested that expression of the MEL-14 homing receptor may play a role in the emigration and peripheral localization of T cells[7-9]. Thus inappropriate expression of the MEL-14 antigen within purified thymic epithelial grafts might correlate with the incomplete peripheral reconstitution seen in these animals. With this in mind, it was the intent of these experiments to examine the expression of MEL-14 in immunodeficient mice bearing purified thymic epithelial grafts.

Results of the experiments presented here demonstrate that such grafts are capable of sustaining the differentiation of lymphocytes expressing the MEL-14 homing receptor. In the normal thymus, cells expressing the high amounts of the MEL-14 antigen have been shown to be present by day 14 of gestation, and make up approximately 2% of the adult thymocyte population[9]. The observation that MEL-14 positive cells are found in LTOC grafts is also in agreement with the previous finding that lymphocytes within LTOC grafts express other T cell phenotypes (Thy-1, Ly-1, Lyt-2, and L3T4) in approximately normal frequency[1-3,6].

It should be noted that the results presented here only address T cell populations within the lymph nodes of the grafted animals. Flow cytometric analysis of the spleens of these same animals demonstrated a partial reconstitution of splenic T cell populations (data not shown). However such results do not exculde the possibility of functional differences between lymphocytes which localize in the in lymph nodes and those which localize in the spleen.

The presence of MEL-14 positive cells within the lymph nodes of nude mice receiving thymic epithelial grafts is not particularly remarkable. Lymph nodes of nude mice contain MEL-14 positive cells whether the animal is thymus grafted or not. However it should be recognized that the lymph nodes of these animals contain MEL-14 positive T cells, similar to that observed in normal (intact) mice as compared to ungrafted controls.

The finding of approximately normal frequency and distribution of MEL-14 positive cells within the grafts of these mice strongly suggests that the incomplete T cell peripheralization with respect to lymph node populations is not absolutely restricted by a lack of homing receptor expression within the thymic epithelial grafts. In fact, if anything, these data would argue that the lack of peripheralization or peripheral function observed in these animals is independent of the expression of the MEL-14 homing receptor. Clearly much work remains to be done in order to determine whether homing receptor expression plays a significant role in the peripheralization of T cells.

ACKNOWLEDGEMENTS

This work was supported by ACS grant PF2697 (GAP) and NIH grants AI 15819 (DAC) and AI 22730 (MOD). Morris Dailey is a fellow of the John A Hartford Foundation.

LITERATURE CITED

1) Jordan, R.K., A.L. Bentley, G.A. Perry and D.A. Crouse, Thymic epithelium I. Lymphoid-free organ cultures grafted in syngeneic intact mice. J. Immunol. 134:2155-2160 (1985).

2) Crouse, D.A., G.A. Perry and R.K. Jordan, Transplantation of lymphoid free thymic epithelium in syngeneic and allogeneic immunodeficient mice. In: Regulation of the Immune System (E. Sercarz, H. Cantor and L. Chess, eds), New York: Alan R. Liss, pp. 231-242 (1984).

3) Crouse, D.A. and R.K. Jordan, Studies of adoptive transfer of thymic rudiments., In: Leukocytes and Host Defense (J.J. Oppenheim and D.M. Jacobs, eds), New York: Alan R. Liss, pp. 11-20 (1986).

4) Jenkinson, E.J., L.L Franchi, R. Kingston and J.J.T. Owen, Effect of deoxyguanosine treatment on lymphopoiesis in the developing thymus in vitro: Application in the production of chimeric thymic rudiments. Eur. J. Immunol. 12:583-587 (1982).

5) Schulte-Wissermann, H., D. Manning and R. Hong, Transplantation of cultured thymic fragments. I. Morphologic and technical considerations. Scand. J. Immunol. 8:387-396 (1978).

6) Perry, G.A., R.K. Jordan and D.A. Crouse, Immune reconstitution in thymus grafted hematopoietic radiation chimeras. Radiat. Res. (submitted).

7) Reichert, R.A., W.M. Gallatin, E.C. Butcher and I.L. Weissman, A homing receptor-bearing cortical thymocyte subset: Implications for thymus cell migration and the nature of cortisone-resistant thymocytes. Cell 38:89-99 (1984).

8) Reichert, R.A., I.L. Weissman and E.C. Butcher, Phenotypic analysis of thymocytes that express homing receptors for peripheral lymph nodes. J. Immunol. 136:3521-3528 (1986).

9) Reichert, R.A., L. Jerabek, W.M. Gallatin, E.C. Butcher and I.L. Weissman, Ontogeny of lymphocyte homing receptor expression in the mouse. J. Immunol. 136:3535-3542 (1986).

10) Jordan, R.K. and D.A. Crouse, Non-lymphoid cells and intrathymic T-cell differentiation. In: Development and Differentiation of Vertebrate Lymphocytes (J.D. Horton, ed), Amsterdam: Elsevier-North Holland, pp. 47-61 (1980).

IN VITRO BINDING OF L3T4$^-$Lyt-2$^-$ Pgp-1$^+$ THYMOCYTES TO THYMIC ACCESSORY CELLS

Soumaya El Rouby and Martine Papiernik

INSERM U 25, Hôpital Necker,

161 Rue de Sèvres, 75015 Paris, France

INTRODUCTION

Thymic macrophages and dendritic cells (MØ-DC) have been found to play an important role in the regulation of thymocyte maturation. It has for instance been suggested that thymic MØ-DC are involved in the induction of tolerance (1,2), regulation of Ia expression on cortical epithelial cells (3) and in the stimulation of thymocyte and prothymocyte proliferation (4,5). The events regulating these maturational effects may require cell-cell contact. Indeed multicellular complexes of thymocytes and MØ-DC have been described (6). We recently described an in vitro model to study the direct interaction between thymocytes and the phagocytic cells of the thymic reticulum (P-TR) (7), which leads to complement receptor type 3-mediated rosette formation (8). Rosetting thymocytes were found to show higher responsiveness and IL-2 receptor expression during T-cell mitogen (Con A) stimulation than did non rosetting thymocytes (9). It was also found that the rosetting thymocytes have a greater capacity for migrating back to the thymus after transfer into irradiated congeneic mice (El Rouby, Ezine and Papiernik, 1978, submitted). We analysed the phenotype of the rosetting thymocytes and in the present communication we show that the rosetting cells contain a small subset of L3T4$^-$Lyt-2$^-$ {double negative (DN)} cells. These binding DN thymocytes are enriched in cells expressing Pgp-1 antigen.

RESULTS

1) Rosette formation, phenotype of the bound cells

Rosettes were performed by mixing phagocytic cells of the thymic reticulum (P-TR), isolated in a long-term culture of thymic stroma (7), and thymocytes as described in detail elsewhere (8). Rosettes formed after 2 hr incubation at + 4°C were separated from the free thymocytes by simple four-step 1-g sedimentation over ice cold undiluted fetal calf serum (FCS). Bound thymocytes were dissociated from P-TR by incubating the isolated rosettes in Petri dishes for 90 min at 37°C. This latter procedure allowed the release of the bound thymocytes which can be collected in the medium, and the adherence of P-TR to the plastic surface. Both rosetting and non rosetting thymocytes were stained with the following monoclonal antibodies (mAb), GK 1.5 anti L3T4 (10), 5763 anti Lyt-2 (11), a mixture of both, and PC-61 anti IL-2 receptors (12). Fluorescein isothiocyanate (FITC) conjugated goat anti-rat antibody (CAPPEL) was used for the second step staining. Cells were scored on a Leitz Orthoplan microscope and the percentage of positive cells for each phenotype was calculated by subtraction. Data presented in table 1 show that the majority of the rosetting thymocytes express the double positive L3T4+Lyt-2+ phenotype, that very few express the single positive mature phenotype but that they include a small subset of L3T4-Lyt-2- DN phenotype. This surface phenotype does not seem to differ from that of the non rosetting population except that the rosetting thymocytes are relatively reduced in the mature type.

Table 1. Phenotype of rosetting and non rosetting thymocytes

Phenotype	[a]ROS-thy	Non-ROS-thy
	% Positive cells	
L3T4+Lyt-2-	2.4 ± 0.2[b]	4.0 ± 0.6
L3T4-Lyt-2+	0.3 ± 0.1	3.0 ± 0.4
L3T4+Lyt-2+	91.0 ± 0.4	89.0 ± 1.0
L3T4-Lyt-2-	6.3 ± 0.2	4.0 ± 1.0
IL-2R	6.0 ± 1.8	3.5 ± 1.0

a ROS-thy = Rosetting thymocytes
b Mean value ± SEM for four experiments.

2) Binding of L3T4-Lyt-2- thymocytes to P-TR

The next step was to analyse the DN subset which binds to P-TR. For this purpose, DN thymocytes were isolated from adult thymuses by a negative selection method using anti-L3T4 (clone LICR-LAU-RL 172.4) and anti Lyt-2 clone (3-155) rat IgM mAb and low tox rabbit complement as

One representative experiment.

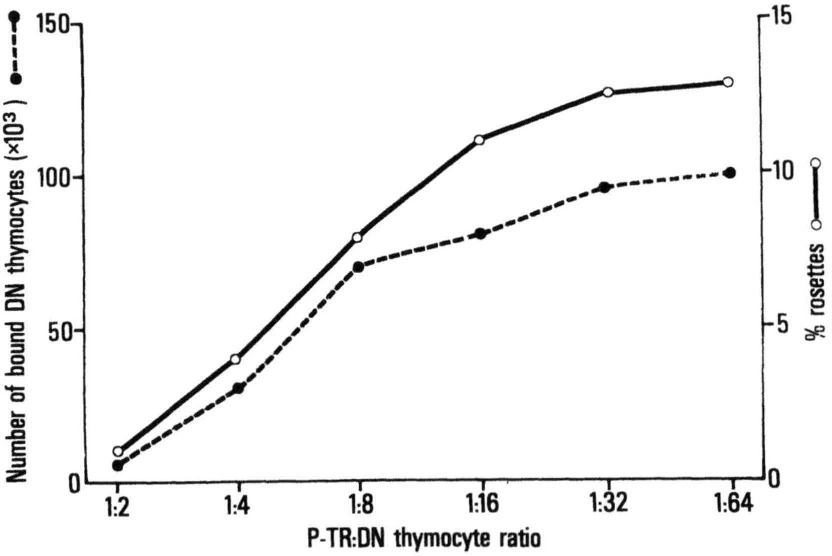

Figure 1. Binding of L3T4-Lyt-2- to P-TR

described in detail elsewhere (5). P-TR and purified DN cells were then mixed at different proportion (1:4 to 1:64) and after 2 hr and/or overnight incubation at 4°C the percentage of P-TR with bound DN cells (maximum, 3 DN/one P-TR) and the number of bound DN thymocytes were determined. Figure 1 shows that both the percentage of rosettes and the number of bound DN cells increase gradually by augmenting the DN cell number, attaining the plateau level when P-TR and DN thymocytes were mixed at a 1:32 ratio. At the plateau level, 1-1.2% DN cells bind to 10-12% P-TR.

3) Binding capacity of adult and fetal DN thymocytes to P-TR

Having established the optimal ratio of P-TR/DN thymocyte association (1:32) we compared the binding capacity of 14 day fetal DN thymocytes isolated from embryos that were obtained from timed pregnancies and purified adult DN thymocytes to adult P-TR. Results presented in table 2 {A} show that adult DN thymocytes and 14 day fetal DN thymocytes bind similarly to adult P-TR and that only a small proportion (1%) of DN cells are able to bind to P-TR. The absence of difference between adult and fetal DN thymocytes in view of the binding capacity to P-TR can be attributed to the nature of adult P-TR, which may have a certain affinity to DN cells independently of the source of DN cells. To explore this possibility we prepared P-TR from 14 day fetal thymuses by culturing fetal thymuses (a pool of 3-4 litters) after physical disruption (by teasing) in 6-well flat-bottomed tissue culture plates (Flow) in RPMI medium containing 20% FCS. As described for adult P-TR, fetal P-TR are released into the culture medium as non adherent cells and could be recovered by days 12-15 of culture. Fetal P-TR were then tested for rosette formation with fetal and adult DN thymocytes. Results presented in table 2 {B} clearly show that fetal P-TR bind similarly to either fetal or adult DN thymocytes in the same way as adult P-TR. These results suggest that independently of the age of mice, there is only a small subset of DN thymocytes which is able to bind to P-TR.

Table 2. Both adult and 14 day fetal DN cells bind similarly to P-TR

Origin of DN cells	Source of P-TR {A} Adult Thymus		Source of P-TR {B} Fetal Thymus*	
	% ROS°	N°of bound cells ($x10^3$)	% ROS	N° of bound cells ($x10^3$)
Adult Thymus	10 ± 0.9	83 ± 7 (9)	11 ±	84 (1)
Fetal Thymus*	11 ± 1	84 ± 8 (5)	12 ± 0.6	94 ± 5 (2)

( ) = Number of experiment
* = 14 day of gestation
° = Percentage of rosettes.

4) Characterization of the binding DN thymocytes

DN thymocyte binding to adult P-TR were compared to those which do not bind on the basis of the expression of Pgp-1 antigen, IL-2R and of the incorporation of Bromodeoxyuridine (BUdR) which labels cells in S-phase. Rosettes formed with purified DN cells were first separated as described above and then processed for staining with anti-IL-2R (PC-61) (12) and anti Pgp-1 (IM7) (13). For the determination of cells in S phase, purified DN cells were pulsed with BUdR (10 μM final concentration) for 30 min at 37°C before mixing with P-TR for rosette formation. The formed rosettes were then smeared onto slides, fixed and finally stained with anti-BUdR as described by C. Penit (14). Data presented in table 3 indicate that the rosetting DN thymocytes contain the same proportion of $BUdR^+$ cells as total and non rosetting populations, they are relatively reduced in cells expressing IL-2R but they are enriched in cells expressing Pgp-1 antigen (2-3 times more than total and non rosetting cells). Pgp-1 is a glycoprotein which was found to be expressed on thymus homing progenitors (15).

Table 3. Characteristics of the bound DN thymocytes

	Source of DN thymocytes					
	Adult thymus			14 day fetal thymus		
	% positive cells					
Marker	Total	Non-ROS°	ROS	Total	Non-ROS	ROS
Pgp-1	23 ± 0*	19 ± 1	56 ± 2	37 ± 3	21 ± 4	59 ± 0.3
IL-2R	50 ± 1	45 ± 4	33 ± 2		N-D	
BUdR	27 ± 2	30 ± 3	25 ± 2		N-D	

* Mean value ± SEM for 2-5 experiments
N-D = not done
° ROS = rosetting

5) <u>Proliferative response of DN rosettes; role of Pgp-1 antigen</u>

To explore the consequence of DN thymocyte binding to P-TR we attempted to study the proliferative response of the bound DN cells and the effect of anti-Pgp-1 on this proliferation after culturing the isolated DN rosettes (2×10^3/well) in the presence of r-IL-2 (Biogen, 12.5 ng/ml). Anti-Pgp-1 mAb {(hybridoma culture supernatant, a gift of Dr Hyman)} was added to some culture wells at 50% V/V (i.e., 50 µg/ml of total rat Ig) at the beginning of culture. A similar concentration of rat IgG was used as a control for specificity. Cell growth was evaluated by ^{3}H incorporation after 3 day culture. As shown in table 4 {A}, no cell proliferation was detected in the presence of medium alone. Vigorous cell proliferation was obtained when r-IL-2 was added, and this latter response was inhibited (70% of control) when anti-Pgp-1 mAb was added to the culture medium (rat IgG control for Ig class had no effect). Anti-Pgp-1 similarly inhibited the proliferation of the whole rosetting thymocytes (table 2 {B}, see table 1 for the phenotype) and this may suggest that the $L3T4^-$ $Lyt\text{-}2^-$ DN thymocytes are the only responding cells in the whole rosette complex.

DISCUSSION

This report shows that purified adult DN thymocytes and 14 day fetal thymocytes can bind <u>in vitro</u> to the phagocytic cells of the thymic reti-

Table 4. Inhibition of DN and whole rosette proliferation by anti-Pgp-1

Stimulus	Type of rosettes {A} DN Rosettes* cpm	{B} Whole Rosettes** cpm
None (medium alone)	650 ± 127°	605 ± 50
r-IL-2	17 065 ± 2162	13 432 ± 210
r-IL-2 + IM-7 (anti-Pgp-1)	4 250 ± 721	3 005 ± 307
r-IL-2 + rat IgG	15 942 ± 1196	18 086 ±1287

* rosette formed between P-TR and purified DN thymocyte
** rosette formed between P-TR and the whole thymocytes
° Mean value ± SEM for 2-4 experiments.

culum (P-TR), non epithelial stromal cells isolated from long term thymic stroma culture (7). Independently of the age of mice, a constant number (1%) of DN cells binds to P-TR. This type of interaction can be detected as early as the 14th day of gestation, and P-TR isolated from 14 day fetal thymic stroma culture are similar to adult P-TR. Rosetting DN thymocytes are distinguishable from non rosetting DN thymocytes by the higher percentage of Pgp-1$^+$ cells. DN thymocytes are heterogenous (16,17), they contain Pgp-1$^+$ and IL-2R$^+$ cells which represent non overlapping subsets (18,19). Most thymus homing progenitor cells in the adult mouse thymus are found in the Pgp-1$^+$ population (15,18). Our results showing that P-TR select the DN subset which expresses Pgp-1 antigen suggest a role of non epithelial stroma in the development of these cells while entering the thymus. In other words thymic accessory cells may provide the necessary signal for the maturation of these cells. Indeed, culture of DN rosettes with r-IL-2 led to DN cell proliferation and this proliferative response was inhibited when anti-Pgp-1 mAb was added to the culture medium. This shows that P-TR can induce the rosetting DN cell responsiveness to IL-2 as we have previously shown for the bulk of DN cells (5). It was shown that accessory cells can activate Protein Kinase C (PKC) in T lymphocytes either by arachidonate or by lipid components (20,21). The binding of DN Pgp-1$^+$ thymocytes to P-TR may represent the process of transfer of these fatty acids (20). Thus, P-TR can replace phorbol esters (activators of PKC) which were shown to induce the growth of DN cells (22,23).

Inhibition of DN and whole rosetting thymocyte proliferation by anti-Pgp-1 mAb suggests a functional role of Pgp-1 antigen and that only DN cells proliferate among the whole rosetting thymocytes. It should also be mentioned that anti-Pgp-1 inhibits the proliferative response of the bulk of DN cells associated with P-TR (not shown). However, the mechanism(s) by which anti-Pgp-1 inhibits DN cell proliferation are not known. The Pgp-1 molecule does not seem to be involved in the DN binding to P-TR (not shown) but it may play an important role in Pgp-1$^+$ DN cell interaction and activation. It may also control P-TR's function since it is also expressed on 70% of P-TR. Further studies are required to determine the role of the Pgp-1 molecule on different thymic components.

REFERENCES

1. A.R. Ready, E.J. Jenkinson, R. Kingston, and J.J.T. Owen. Nature 310:231 (1984).
2. H. Von Beohmer, and K. Schubiger. Eur. J. Immunol. 14:1048 (1984).
3. D. Lo, and J. Sprent. J. Immunol. 137:1772 (1986).
4. D.I. Beller, and E.R. Unanue. J. Immunol. 124:1243 (1980).
5. M. Papiernik, C. Penit, and S. El Rouby. Eur. J. Immunol. in press (1987).
6. B.A. Kyewski, R.B. Rouse, and H.S. Kaplan. Proc. Natl. Acad. Sci. USA 79:5646 (1982).
7. M. Papiernik, B. Nabarra, W. Savino, C. Pontoux, and S. Barbey. Eur. J. Immunol. 13:147 (1983).
8. S. El Rouby, F. Praz, L. Halbwachs-Mecarelli, and M. Papiernik. J. Immunol. 134:3625 (1985).
9. El Rouby, and M. Papiernik. Thymus, in press (1987).
10. D.P. Dialynas, D.R. Wilde, P. Marrack, A. Pierres, K.A. Wall, W. Havran, G. Otten, R.R. Loken, M. Pierres, J. Kappler, and F.W. Fitch. Immunol. Rev. 74:29 (1985).
11. J. Ledbetter, and L. Herzenberg. Immunol. Rev. 47:63 (1979).
12. R. Ceredig, J.W. Lowenthal, M. Nabholz, and H.R. MacDonald. Nature 314:98 (1985).
13. I.S. Trowbridge, J. Lesley, R. Schulte, R. Hyman, and J. Trotter. Immunogenetics 15:299 (1982).
14. C. Penit. J. Immunol. 137:2115 (1986).
15. J. Lesley, R. Hyman, and R. Schulte. Cell. Immunol. 91:397 (1985).
16. B.J. Fowlkes, L. Edison, B.J. Mathieson, and T.M. Chused. J. Exp. Med. 162:802 (1985).
17. R. Scollay, and K. Shortman. J. Immunol. 134:3632 (1985).
18. R. Hyman, J. Lesley, R. Schulte, and V. Trotter. Cell. Immunol. 101:320 (1986).
19. F. Lynch, G. Chaudhri, J.E. Allan, P.C. Doherty, and R. Ceredig. Eur. J. Immunol. 17:137 (1987).
20. W. Dröge. Immunol. Today 7:340 (1986).
21. L. McPhail, C.C. Clayton, and R. Snyderman. Science 224:622 (1984).
22. R. Palacios, and H. Von Boehmer. Eur. J. Immunol. 16:12 (1986).
23. R. Ceredig. J. Immunol. 137:2260 (1986).

INTERACTIONS BETWEEN LYMPHOID CELLS AND A THYMIC STROMAL CELL LINE *IN VITRO*

P. Hugo[1], R. M. Gorczynski[2], D. Oth[1] and E. F. Potworowski[1]

[1]Institute Armand-Frappier, Laval, QC, Canada

[2]Mount Sinai Hospital, Toronto, Ont., Canada

INTRODUCTION

Intrathymic T-cell differentiation is based on lympho-stromal interactions, resulting in the transmission of humoral and membrane-associated signals to thymocytes (Adkins et al., 1987). Particular lympho-stromal cell complexes have been identified, such as those between thymocytes and thymic nurse cells (Wekerle and Ketelson, 1980), dendritic cells (Kyewski et al., 1982) and macrophages (El Rouby et al., 1985). Using an assay developed by Stamper and Woodruff (1976) and modified by Butcher et al. (1979), we have shown specific adherence of thymocytes to a medullary thymic epithelial cell line, E-5, derived from C57BL/6 mice (Potworowski et al., 1986a,b).

In this work, we describe the conditions under which lympho-epithelial contact occurs, we present data on the phenotype of adhering thymocytes, and evidence that contact between lymphoid cells and E-5 cells induces a discrete step of T-cell differentiation.

MATERIAL AND METHODS

Adherence assay

This assay was described elsewhere (Potworowski et al., 1986b). Essentially, a 250 ul thymocyte suspension was deposited in a slide-chamber, in which E-5 cells have been cultured to 50% confluency. Co-cultures were incubated at 4°C on a rotating platform at 60 rpm for 30 min and then left to rest for an additional 60 min. The preparations were washed and fixed in glutaraldehyde, stained and mounted. Experiments were performed in quadruplicates. The level of adherence was expressed as the number of thymocytes adhered to 100 epithelial cells.

Phenotypic characterization

Adhered thymocytes were stained by immunofluorescence using antibodies to Thy-1.2, Lyt-1, Lyt-2, L3T4 and MEL 14 or with FITC-conjugated PNA. Antigenic integrity of glutaraldehyde-fixed thymocytes was first determined in suspension.

In vitro differentiation assay

Differentiation of thymocytes *in vitro* was described elsewhere (Gorczynski and MacRae, 1979). Briefly, 7.5 x 10^5 E-5 cells were cultured for 24 hr. Thymocytes or bone marrow prethymocytes, (2 x 10^7) were co-cultured with E-5 cells or incubated with 22 um filtered E-5 supernatant for another 24 hr. Lymphoid cells were then harvested and reactivity to Con-A or PHA was determined using a standard mitogen-induced proliferation assay.

RESULTS AND DISCUSSION

Optimalization of adherence assay

The optimal number of thymocytes per assay was determined in the concentration range of 2 x 10^4 to 2 x 10^7 cells/ml (Fig. 1). The adherence level increased in a linear fashion from 2 x 10^4 to 8 x 10^6 cells/ml. An exponential increase was then observed reaching a plateau at 10^7 cells/ml. Adherence to glass was noticed from 1.5 x 10^7 cells/ml. It was concluded that adherence assays should be performed using concentrations of 4 to 8 x 10^6 cells/ml, in order to observe any modulation of the phenomenon. Adherence level was also measured in a kinetic manner to determine the optimal incubation time for maximal adherence to occur (Fig. 2). For that purpose, 8 x 10^6 thymocytes were co-incubated with E-5 cells for various periods of time, from 30 min to 10 hr. Optimal adherence was reached at 1.5 hr.

Genetic influence on adherence

Since thymic epithelium was shown to be involved in thymocyte education to self (Lo and Sprent, 1986), it was relevant to determine whether adherence could occur only if MHC phenotypes of thymocytes and E-5 cells were matched. We have also monitored the influence of thymocyte genetic background on the level of adherence. Congenic mice with different H2 phenotypes (C57BL/10, $H\text{-}2^b$; B10.Q, $H\text{-}2^q$; B10.D2, $H\text{-}2^d$; B10.Br, $H\text{-}2^k$; B10.SM, $H\text{-}2^v$) or different genetic backgrounds (C57BL/10, C57BL/6, C57BL/Ka, B6lpr/lpr, NZB, Balb/c. AKR/J) were used. All these strains except B6lpr/lpr, showed levels of adherence between 40.8 and 52.9 thymocytes/100 E-5, using a thymocyte concentration of 4 x 10^6/ml. No significant difference was noticed between these groups ($p>0.05$). The immunodeficient B6lpr/lpr mice (Davignon et al., 1985) demonstrated a markedly diminished adherence of 13.7, significantly different from all the other values obtained ($p\leq 0.01$). Thus, it is unlikely that thymocyte adherence to E-5 is involved in self education.

Characterization of adhering thymocyte phenotype

Immunofluorescence staining of adhering thymocytes (Table 1) showed that they are committed to the T-cell lineage since they are all bearing the Thy-1 antigen, while neither receptor for PNA, nor MEL-14 antigen were detected at their surface. Adhering thymocytes appear to be heterogeneous based on the expression of $Lyt\text{-}1^h$ and Lyt-2. L3T4, modified by fixation, could not be assayed directly. Nonetheless, populations of enriched $L3T4^-$, $Lyt\text{-}2^-$ cells did not adhere to E-5 cells (data not shown), suggesting that some adhering thymocytes bear the L3T4 antigen, since only ≈ 20% were $Lyt\text{-}2^+$.

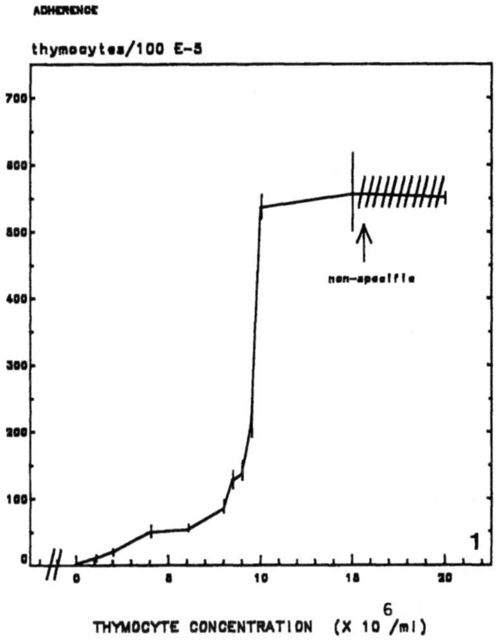

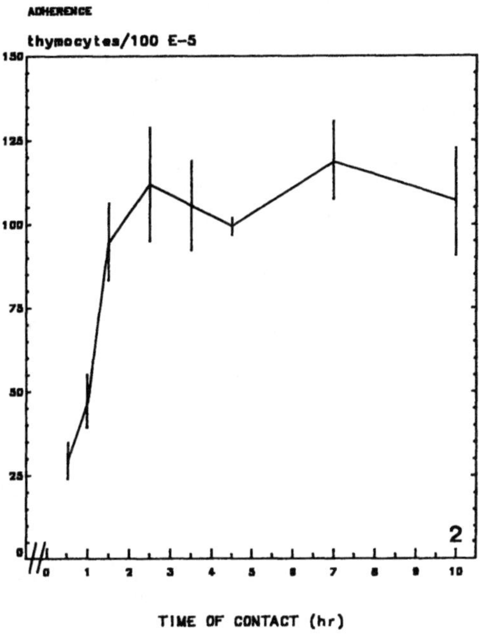

Fig. 1. (left) Effect of thymocyte concentration on the level of adherence on E-5 cells.

Fig. 2. (right) Kinetics of thymocyte adherence on E-5 cells.

Ontogeny of thymocyte adherence to E-5 cells

The above experiment indicates that adhering thymocytes are relatively mature. We therefore investigated adherence during fetal and post-natal life (Table 2). At day 14 of fetal life, adherence was very faint and noticeable levels were observed at day 16. The maximum was reached soon after birth and remained stable throughout life, unlike thymocytes reported to adhere to thymic dendritic cells (Kyewski et al., 1982).

Determination of contact duration

In order to determine the duration of contact, standard adherence and washing were performed, following which preparations were further incubated for various periods of time. After 1.5 hr, 45% of adhered thymocytes were spontaneously released (Fig. 3). These thymocytes proved to be refractive to another adherence (data not shown). Thus, two adhering populations are likely to exist.

Table 1. Phenotype of adhering thymocytes.

Ag	thymocyte suspension		adhering thymocytes
	unfixed	fixed*	fixed*
Thy-1	98.6%	98.8%	100.0%
PNA	77.5%	83.3%	0%
$Lyt\text{-}1^h$	50.0%	48.8%	50.0%
Lyt-2	80.8%	73.1%	19.7%
L3T4	82.7%	0%	?
$MEL\text{-}14^h$	5.2%	7.2%	0%

* cells labeled after fixation with glutaraldehyde 1%.

Differentiation in vitro

Mitogen reactivity of lymphoid cells was measured after co-culture with E-5 cells (Fig. 4). Thymocytes incubated with E-5 or E-5 supernatant showed an increased proliferative response to Con A, as compared to controls. When the reactivity to PHA was measured, thymocytes incubated with E-5 also showed in increased response: however, no such increase was induced with E-5 supernatant. Using bone marrow cells enriched in prethymocytes, a similar pattern of reactivity to PHA was noted, while no increase was seen in response to Con A (data not shown). Contact with E-5 is therefore essential to enhance PHA rectivity. Clonal expansion of responding cells is unlikely, since no IL-1 or IL-2 was detected in co-cultures; exogeneous IL-2 had no effect either (data not shown).

Table 2. Ontogeny of adherence

Age (days)	Adherence level* (± S.D.)
f14	6.5 + 0.6
f16	25.8 ± 5.9
f18	27.6 ± 3.2
f20	43.0 ± 4.9
p2	48.0 ± 7.7
p14	51.0 ± 11.6
p30	51.7 ± 2.6
p60	50.9 ± 6.9
p90	53.0 ± 2.1
p180	55.9 ± 2.0
p240	56.4 ± 2.3
p330	53.6 ± 3.3
p420	53.9 ± 3.6
p590	53.2 ± 3.1

*:thymocytes/100 E-5 cells.
f:fetal life p:adult life

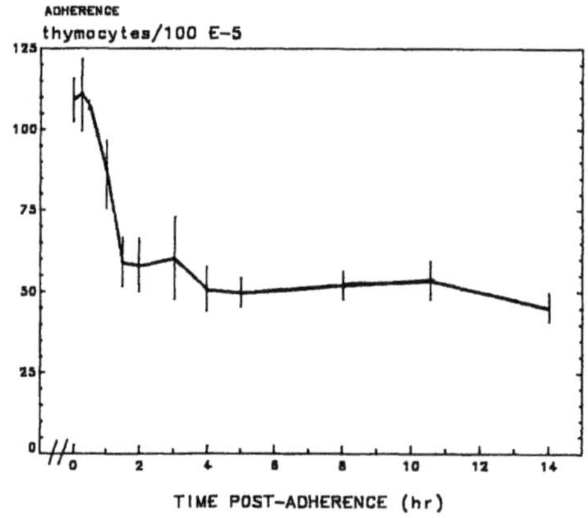

Fig. 3. Contact time. After a standard assay, thymocytes were incubated for various times.

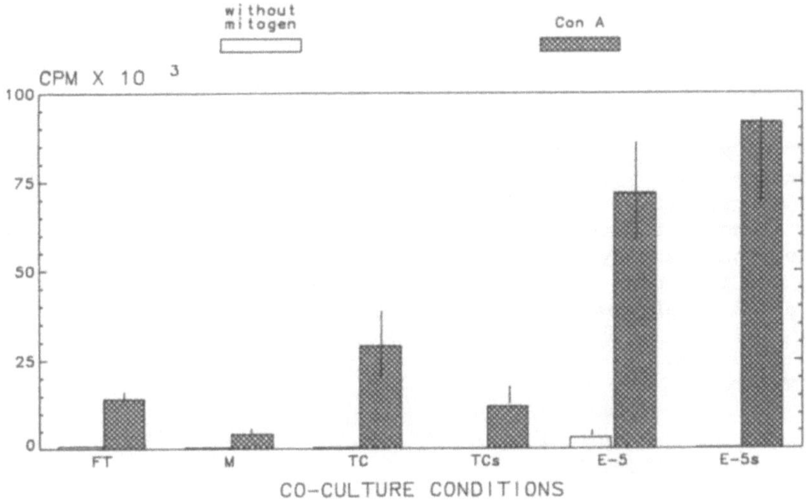

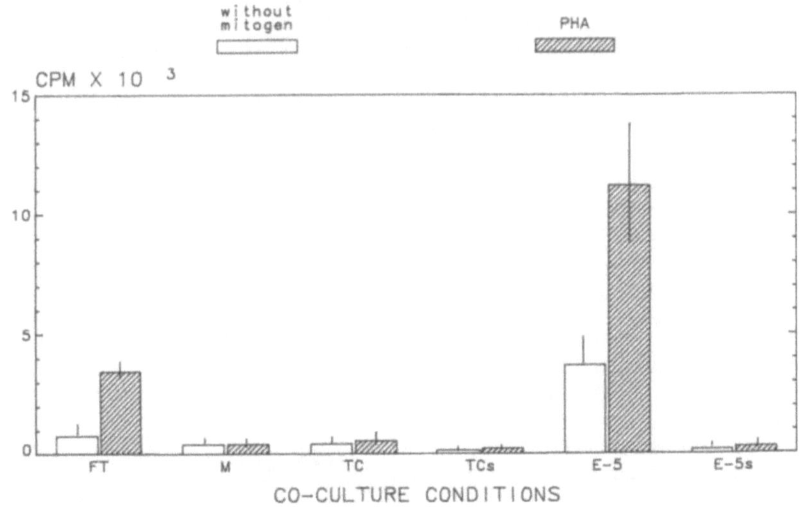

Fig. 4. Thymocyte reactivity to Con A (upper) and PHA (lower) was assessed with fresh thymocytes (FT) or after 24 hr. incubation with medium (M), primary thymic stromal culture (TC), TC supernatant (TCs), E-5 and E-5 supernatant (E-5s).

SUMMARY AND CONCLUSION

In the present study, we have optimalized the adherence assay to allow monitoring of the level of contact between thymocytes and a thymic epithelial cell line, E-5, in vitro. This type of interaction is not MHC-restricted, thus is unlikely to participate in the education of thymocytes to self. It was also shown that adherence does not vary from strain to strain, except for B6lpr/lpr immunodeficient mice which showed a markedly decreased adherence. This might be caused by the high level of $L3T4^-$, $Lyt\text{-}2^-$ thymocytes in these mice (Davignon et al., 1985), since enriched double negative cells were shown not to adhere to E-5 cells.

Preliminary characterization of adhering thymocytes suggests an heterogeneous mature phenotype. These cells appear around day 16 of fetal life and increase gradually until birth to remian constant throughout life. On the basis of contact duration, two populations of adhering thymocytes exist: one spontaneously detached after 1.5 hr, refractive to further adherence and the other which adheres for up to 14 hr. Contact between lymphoid and E-5 cells was shown to induce PHA responsiveness.

ACKNOWLEDGEMENT

Supported by the Medical Research Council of Canada and the National Cancer Institute of Canada.

REFERENCES

Adkins, B., Mueller, C., Okada, C. Y., Reichert, R. A., Weissman, I. L., and Spangrude, G. J., 1987, Ann. Rev. Immunol., 5:325.

Butcher, E. C., Scollay, R. G., and Weissman, I. L., 1979, J. Immunol., 123:1996.

Davignon, J. L., Budd, R. C., Ceredig, R., Piguet, P. F., MacDonald, H. R., Cerottini, J. C., Vassali, P., and Izui, S., 1985, J. Immunol., 135:2423.

El Rouby, S., Praz, F., Halbwachs-Mecarelli, L., and Papiernik, M., 1985, J. Immunol., 134:3625.

Gorczynski, R. M. and MacRae, S., 1979, Immunology, 38:1.

Kyewski, B. A., Rouse, R. V., and Kaplan, H. S., 1982, Proc. Natl. Acad. Sci. USA, 79:5646.

Lo, D. and Sprent, J., 1986, Nature, 319:672.

Potworowski, E. F., Turcotte, F., Beauchemin, C., Hugo, P., and Zelechowska, M. G., 1986a, In Vitro, 22:557.

Potworowski, E. F., Thibodeau, L., and Zelechowska, M.G., 1986b, Immunol. Lett., 13:89.

Stamper, H.B. and Woodruff, J.J., 1976, J. Exp. Med., 144:828.

Wekerle, H. and Ketelson, U.P., 1980, Nature, 283:402.

STRUCTURAL AND CELLULAR CHANGES IN THE MOUSE THYMUS DURING PREGNANCY

Ann G. Clarke

Department of Obstetrics and Gynaecology
University of Nottingham
Queen's Medical Centre
Nottingham, NG7 2UH
U.K.

INTRODUCTION

During pregnancy in all mammals, including man, there is a massive loss of lymphocytes from the thymus. This is reflected in an excessive involution, with decreases in weight and in the numbers of thymocytes. At the end of pregnancy in the mouse, the thymus has lost 75% of its weight and cells. Involution continues during lactation, after which the thymus regenerates, becomes repopulated with lymphocytes, and regains its original size.

The high levels of hormones associated with pregnancy have until recently, been thought to be solely responsible for these changes. In pseudopregnant mice without embryos, but with the normal conditions of pregnancy, involution occurs to the same extent as it does in normal pregnancy (Chambers and Clarke, 1979).

Although the proximate cause of pregnancy induced involution is clearly hormonal, our previous work has suggested that the behaviour of the thymus is not only a response to hormones, but also a result of immune activity. C57BL mice immunized to CBA antigens before mating to CBA males do not show the normal involution of the thymus (Clarke, 1979). As earlier pregnancies also inhibit involution, it appears that the presence of paternal immunity, whether gained naturally or induced experimentally may be involved. Furthermore, during a normal first pregnancy, a transient increase in the weight of the thymus was found around the time of implantation, before the onset of involution. This enlargement appeared to be greater in an allogeneic pregnancy than in a syngeneic one. As it also occured in mice in pseudopregnancy, the enlargement was probably not due to the presence of an embryo. Some component of the ejaculate was more likely to be responsible (Clarke, 1984).

In an attempt to extend these findings the present work has examined the changes that occur during the course of pregnancy in the total numbers of cells, and in the numbers of viable cells, in the proportion of different T cells subsets, and in the histology of one area of lymphocyte trafficking-the perivascular spaces.

MATERIALS AND METHODS

Animals

C57BL/10ScSn and CBA/Ca mice from OLAC Ltd. were used. Mature C57BL females aged 25 weeks were chosen because their thymus weights are known to decline rapidly in early adult life but later to become relatively stable (Nandi, 1959). The mice were killed as virgins, or at three stages during pregnancy (day 4-5, day 11-12 and day 17-18), or at three weeks post-partum. Mice which had been mated to vasectomized males were also studied in some of the experiments. These pseudopregnant mice lacked embryos, had been exposed to all the components of the ejaculate except the sperm and retained the hormonal conditions of pregnancy for about 10 days after mating.

The numbers of live and dead thymocytes and thymocyte cell size

Individual thymuses were dissected out, cleared of fat and lymph nodes and dissociated into suspensions of single cells. The cells were counted in a haemocytometer using trypan blue exclusion to determine total numbers of cells and the numbers of dead cells. These parameters, as well as cell size, were also determined by measuring the low angle light scatter in a Becton and Dickinson Fluorescent Activated Cell Sorter (FACS IV). After counting, the dead cells were gated out at <70 scatter units. Gates were then set to separate viable cells into 'small plus medium' (scatter units 70-120) and 'large' (scatter units >120) cells.

Immunofluorescence

The rat 1gG2b monoclonal anti mouse Thy 1, Ly 1, Ly 2, and L3T4 monoclonal antibodies (described by Gafre et al., 1979 and Cobbold et al., 1984) were used. The antibodies YTS 121.5 (Ly 1), YTS 169.4 (Ly 2), YTS 154.2 (Thy 1) and YTS 191.2 (L3T4), were kindly given by Dr. H. Waldmann and Dr. S. P. Cobbold. The thymocytes were stained by indirect immunofluorescence using fluorescein conjugated $F(ab)_2$ rabbit anti rat (DAKO) as a 2nd stage antibody. Immunofluorescence staining and washings were performed as described by Cobbold et al., (1984) using aliquots of 2×10^6 cells/tube. Cells were preserved for FACS analysis by fixing in paraformaldehyde (Lancier and Warner, 1981) and analysed on the FACS IV. Fluorescein fluorescence was excited at 488 nm and collected via a 30nm band width pass filter centered at 530 nm. Labelled latex beads were used for standardization. Dead cells were gated out and the total linear fluorescence was calculated on the viable cell populations. The results were expressed as a percentage of that found in an aged matched virgin animal included in each assay.

Perivascular Spaces (PVS)

Thymic tissue was fixed in formol saline, embedded in paraffin and sectioned at 5μ. Sections were stained with silver (Gordon & Sweets, 1936) to demonstrate the reticular fibres and to determine the size and cell content of the PVS.

RESULTS AND DISCUSSION

The numbers of lymphoctyes increased early in primiparity and also after one mating to a vasectomized male (data not shown). The effect was enhanced in multiparity, and also after several matings to a vasectomized male (Fig.1). It is greater after mating to an allogeneic strain male than to a syngeneic male. By day 11-12 the numbers of cells had fallen equally in all groups except in the allogeneically pregnant mice, in which the cell number was still elevated. By day 17-18 the numbers were beginning to recover in pseudopregnant mice, but were still falling in the pregnant mice.

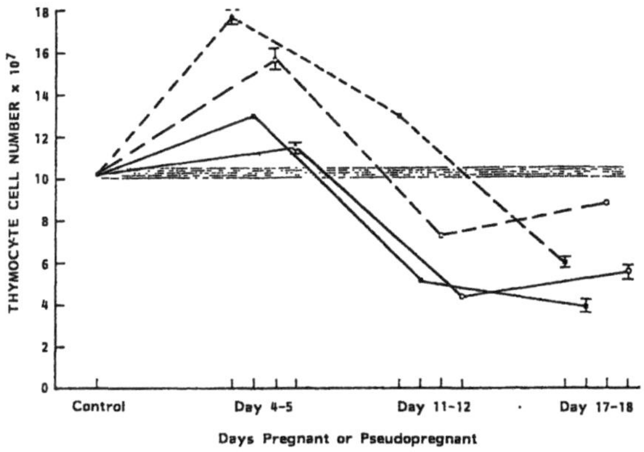

Fig. 1. The total number of thymocytes (mean ± s.e.) during normal multiparity and after several sterile matings to an allogeneic or syngeneic strain male. FACS analysis and cell counts gave similar results. o---o Allogeneic sterile mating, o——o Syngeneic sterile mating, ●---● Allogeneic pregnancy, ●——● Syngeneic pregnancy. Mean ± s.e. of virgin animals.

There has been no agreement whether during thymic involution, thymocytes die in larger numbers within the thymus, whether they are exported at a faster rate, whether they fail to arrive from the bone marrow, or whether they develop a longer cell cycle or a slower rate of cell division. However, it seems that an altered rate of cell division

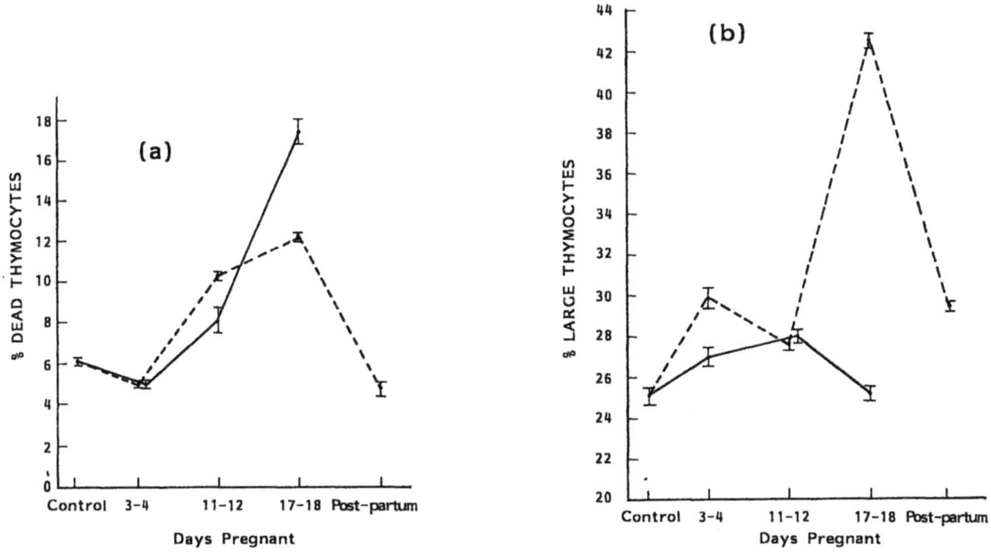

Fig. 2a. The percentage of dead thymocytes (mean ± s.e.) during a first allogeneic (---) and syngeneic (——) pregnancy, and at three weeks post-partum in allogeneic pregnancy (syngeneic not studied). Results from FACS analysis and trypan blue exclusion were similar. Fig. 2b. The percentage of large thymocytes (mean ± s.e.) during a first allogeneic (---) and syngeneic (——) pregnancy, and at three weeks post-partum after allogeneic pregnancy (syngeneic not studied). Forward light scatter >120 units.

Time	Group	PREGNANCY (a) Fluorescence (% of control)				PSEUDOPREGNANCY (b) Fluorescence (% of Control)			
		Thy 1	Ly 1	Ly 2	L3T4	Thy 1	Ly 1	Ly 2	L3T4
Day 4-5	Syngeneic	↑	N.C.	N.C.	↑	↓	↓	↑	N.C.
	Allogeneic	↑	N.C.	↑	N.C.	↓	↓	↑	↑
Day 11-12	Syngeneic	↓	↑	N.C.	↑	N.C.	N.C.	N.C.	N.C.
	Allogeneic	↓	N.C	N.C.	N.C.	N.C.	N.C.	N.C.	N.C.
Day 17-18	Syngeneic	↓	N.C.	↓	↓	↑	N.C.	N.C.	↑
	Allogeneic	↓	N.C.	↓	↓	↑	N.C.	N.C.	↑

Fig. 3. Changes in the numbers of cells staining with monoclonal antibodies to the T cell surface markers in (a) pregnancy and (b) pseudopregnancy. N.C.=no consistent significant change.

within the thymus does not play a major role during pregnancy (Clarke, 1984). The percentage of dead cells in the thymus during pregnancy are shown in Fig. 2a. After an initial slight fall, the levels increase in both syngeneic and allogeneic pregnancies. The numbers are greater in syngeneic than in allogeneic pregnancies. The levels return to normal by three weeks post-partum. This suggests that thymocyte death probably plays some part in the cellular changes that occur.

Large cells normally make up about 15% of the total thymus and are thought to be the generative cells (Scollay and Shortman, 1983). An increase in the percentage of these large cells is found during a first pregnancy (Fig. 2b). The increase is very much greater in allogeneic than in syngeneic pregnancies and seems to occur in two phases, an early small rise followed by a later and larger one. In syngeneic pregnancies there is no second peak, levels fall to normal after day 12. Whether this difference between the two kinds of pregnancy reflects a response to paternally inherited alloantigens is being further investigated. The percentage of large cells returns to normal by three weeks post-partum. Some of the early increase in the numbers of large cells is probably a reflection of the preferential loss of small thymocytes during involution. However, histological staining of tissue sections has shown that in early pregnancy large blast cells do accumulate in the subcapsular region of the

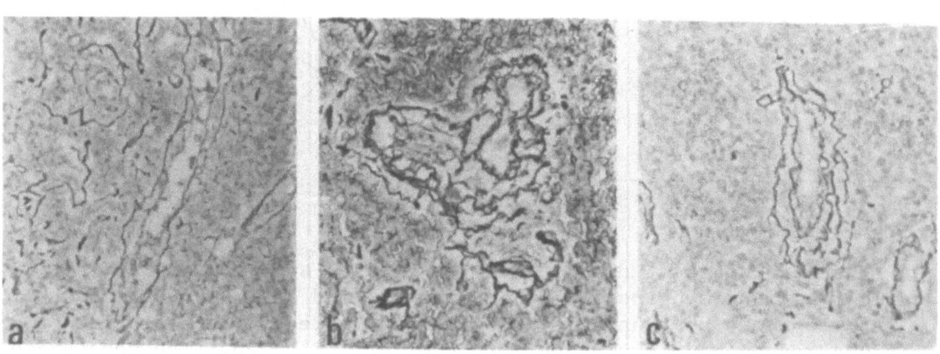

Fig. 4. A perivascular space in the thymus of (a) a virgin (b) a day 4 syngeneic pregnant mouse and (c) a day 17 syngeneic pregnant mouse. Comparable results were obtained in allogeneic pregnancy.

thymus. They are negative for the Ly, L3T4 and Mel-14 markers (data not shown). Since similar kinds of cells in the normal thymus are thought to be recent bone marrow immigrants, this suggests that a temporary increase in immigrating cells contributes to the increased cellularity associated with early pregnancy.

A summary of the changes that occur in the expression of some of the T cell subpopulations of the thymus during pregnancy and pseudopregnancy are summarized in Fig. 3. The percentage of Thy 1 positive cells increases early in both allogeneic and syngeneic pregnancy (5-10%) and then falls by as much as 25% during middle and late pregnancy. There is also some increase soon after mating in the percentage of Ly 2 positive cells in allogeneic pregnancy and of L3T4 positives in syngeneic pregnancy. A large loss occurs late in pregnancy of Ly 2, 25-27%; and L3T4, 15-30%. These results suggest that Thy 1, Ly 2, and L3T4 positive cells contribute both to the early cellular gain and to the later loss during involution. Little consistent change was found with anti Ly 1 antibody. The numbers of cells staining with these antibodies have all returned to normal by three weeks post-partum.

The changes in pseudopregnant mice show some differences and some similarities to those changes seen during normal pregnancy. There is an early loss of Thy 1 positive cells (increase in normal pregnancy) and Ly 1 positives (no change in pregnancy). A later increase in Thy 1 positives is probably associated with the return of cells, and of weight to the thymus. Some increase in Ly 2 and L3T4 positive cells also occurs soon after a sterile mating and is similar but not identical to the pattern in normal pregnancy. L3T4 positive cells would seem to repopulate the thymus in large numbers on day 17-18 after a sterile mating.

Thus, there are several alterations in the major subpopulations of thymocytes during pregnancy. Some changes seem to occur early when the thymus is undergoing an expansion. Others are seen in late pregnancy when the thymus is involuting. Some of the differences are restricted to normal pregnancy and do not occur in pseudopregnancy, others only occur after a sterile mating, and some are present in both. Further experiments are under way to determine the relative contributions of the alloantigens of paternal origin, of the embryo specific antigens, and of the hormones and proteins specific to pregnancy.

In studies carried out in collaboration with Dr. K. A. MacLennan, changes were found to occur in the areas of the perivascular spaces which are involved in the trafficking of T cells in and out of the thymus. These specialized areas are enclosed by fibres and are located at the junction of the cortex and medulla. They consist of lymphatic and the blood vessels, and an array of lymphocytes, macrophages and granulocytes which are all bounded by reticulin (argyrophilic) fibres that stain with silver. The results (Fig. 4) show that maximum size, fibre content and enclosed cells of the PVS are not associated with the time of maximum involution but occur very early when thymus cell numbers are increasing. By day 17 when the thymus involution is maximal the amount of reticulin staining and the size of the PVS is smaller than that found in the virgin animal. There are no obvious differences between allogenic and syngeneic pregnancies in this respect.

Traditionally, these complex 'highways' of the thymus have been described in association with age-related thymic involution. In the mouse, PVS first appear during puberty, grow larger and more numerous with increasing age, and reach their maximum size and fibre content in old age when the thymus has involuted. The large differences in the appearance of these areas in the involuted thymus of pregnancy and in old age supports the view that involution may be not just one but many distinct different processes (Clarke and MacLennan, 1986). Furthermore, there are now ample data to suggest that the involuted thymus of the adult is not atrophic or functionless (Kendall, 1984). Several immune activities actually increase during involution (Baroni et al., 1983). There is accumulating evidence that, at least for some of its functions, the involuted thymus may be the 'active' organ and the non-involuted thymus its 'resting' counterpart.

The significance of the changes found in the thymus, and in its lymphocyte populations, on the maternal immune response remain to be resolved. It seems unlikely that such radical alterations will not have equally large effects on the mother's immune system. Some may well be involved in her recognition and subsequent immune tolerance of foreign embryonic antigens. In this context the changes that occur early in pregnancy are the most interesting ones. One possibility is that a component of the ejaculate is sending a signal to the mother, the receipt of which is reflected in the thymus before implantation and which prepares her for the foreign antigens of the embryo to follow.

ACKNOWLEDGEMENTS

Thanks are due to Dr. R. A. Robins for FACS facilities and help with the analysis, and to O.F. Roberts and N. Bullimore for their technical assistance. The work was supported by a grant from the Journals of Reproduction and Fertility Ltd. and from the Nuffield Foundation.

REFERENCES

1. Baroni, C.D., Valtieri, M., Stoppacciaro, A., Ruco, L.P., Uccini, S. and Ricci, C., 1983, The human thymus in ageing: histologic involution paralleled by increased mitogen response and by enrichment of OKT3+ lymphocytes, Immunol. 50:519.
2. Chambers, S.P. and Clarke, A.G., 1979, Measurements of thymus weight, lumbar node weight and progesterone levels in syngeneically pregnant, allogeneically pregnant and pseudopregnant mice, J. Reprod. Fertil., 55:309.
3. Clarke, A.G., 1979, Pregnancy-induced involution of the thymus can be prevented by immunizing with paternal skin grafts: a stain-dependent effect, Clin. Exp. Immunol., 35:421.
4. Clarke, A.G., 1984, Immunological studies on pregnancy in the mouse, in: "Immunological Aspects of Reproduction in Mammals", D.B. Crighton, ed., Butterworths, London.
5. Clarke, A.G. and MacLennan, K.A., 1986, The many facets of thymic involution, Immunol. Today, 7:204.
6. Cobbold, S.P., Jayasuriya, A., Nash, A., Prospero, T.D. and Waldman, H., 1984, Therapy with monoclonal antibodies by elimination of T-cell subsets in vivo. Nature, 312:549.
7. Galfre, G., Milstein, C. and Wright, B., 1979, Rat x rat hybrid myelomas and a monoclonal anti-Fd portion of mouse lgG, Nature, 277: 131.
8. Gordon, H. and Sweets, H.H., 1936, A simple method for the silver impregnation of reticulum, Am. J. Pathol., 12:545.
9. Kendall, M., 1984, Have we underestimated the importance of the thymus in man? Experientia, 40:1181.
10. Lancier, L.L. and Warner, N.L., 1981, Paraformaldehyde fixation of haematopoietic cells for quantitative flow cytometry (FACS) analysis. J. Immunol. Methods, 47:25.
11. Nandi, S., 1959, Hormonal control of mamogenesis and lactogenesis in the C_3H/He Crgl mouse. Univ. Calif. Publ. Zool., 65:1.
12. Scollay, R. and Shortman, K., 1983, Thymocyte subpopulations: an experimental review, including flow cytometric cross-correlations between the major murine thymocyte markers, Thymus, 5:245.

ANALYSIS OF LYMPHOCYTE PHENOTYPE IN HUMAN THYMOMAS

Berrih-Aknin S., Safar D. and Cohen-Kaminsky S.

Centre Chirurgical Marie-Lannelongue, CNRS UA 04-1159
133, avenue de la Résistance
92350 Le Plessis-Robinson (France)

Thymomas are tumors originated from the epithelial component of the thymus. This definition according to Rosai and Levine (1) excludes germ cell tumors, malignant lymphomas and neuroendocrine tumors (carcinoids and small cell carninoma). Thymomas are heterogeneous with regard to size, epithelial cell type and the ratio of lymphocyte to epithelial cells (1-3).

Stromal and nonstromal cells from thymomas have been identified by different groups, using a large variety of monoclonal antibodies (4-7). Several authors found that the predominant lymphocyte type in thymomas is cortical (4, 8). Evidence of medullary type thymocyte differentiation has also been reported in some cases (5). Thymoma epithelial cells were shown to contain thymic hormones but to be defective in expression of HLA class II molecules (6). Other findings suggested an association, although not absolute, between the lack of HLA class II antigens on neoplastic epithelial cells and decreased numbers of associated lymphocytes (9).

A recent study attempted to analyse the origin or subtype of epithelial and other stromal cells in relation to the surface antigens expressed by tumor lymphocytes. This study of 13 thymomas showed that the cortical type T cell-inducer capacity of thymomas is independent of the epithelial type predominant in the tumor (10).

The aim of our report was to further define the phenotype of associated lymphocytes in 22 thymomas and to seek correlations with clinical data. Several parameters have been investigated :
1) malignancy or benignancy of the tumor
3) association with Myasthenia Gravis
3) predominant cell type (epithelial or lymphoid cells).

PATIENTS AND METHODS

Patients

Thymomas were diagnosed and classified according to Verley and Hollman (2). Twenty-two patients, 12 females and 10 males, aged from 33 to 78 years were studied. Among them, 14 presented Myasthenia Gravis associated to thymomas. Eleven were diagnosed as benign thymomas (without any gross or microscopic signs of invasiveness) and the others were malignant. Sixteen contained a high proportion of lymphocytes while 6 were essentially constituted of epithelial cells with few lymphocytes.

Immunofluorescence on isolated thymus cells

Thymoma samples were minced in Hank's solution. The cell suspensions were washed three times before counting. Fluorescence staining was performed with several monoclonal antibodies provided by Immunotech (France) : anti-CD3 (immunocompetent T cells), anti-CD4 (T helper and thymic cortical cells), anti-CD8 (T cytotoxic and thymic cortical cells), anti-CD1 (cortical thymocytes), IOM (anti-monocytes) and IOB1 (anti-B cells). OKT10 (Ortho Diagnostics) recognizes all thymocytes and NK peripheral lymphocytes. Anti-DR monoclonal antibody was a kind gift from D. Charron.

Cell suspensions (10^6 cells in 45 µl) were incubated with 5 µl monoclonal antibody at its optimal dilution for 30 min at 4°C in Hank's solution supplemented with 5 % fetal calf serum and 0.1 % sodium azide. The cells were then washed twice, resuspended in 45 µl of medium, incubated with 5 µl of fluorescein-conjugated goat anti-mouse Ig antiserum for 30 min at 4°C, and finally washed twice. Peanut-agglutinin coupled to fluorescein-isothiocyanate (PNA-FITC) was also used to evaluate PNA+ and PNA- subsets, as previously described (11). The percentage of labeled cells was evaluated with a Leitz Orthoplan microscope.

RESULTS AND DISCUSSION

The distribution of T cell subsets in thymomas compared to normal adult thymuses shows clear modification of several subsets (Table 1). The CD4+ CD8+ subset is drastically diminished (33.3 % versus 74.3 %, $p < 0.001$). The percentage of the CD1+ subset is also reduced (65.2 % versus 84.4 %, $p < 0.05$). Interestingly, the percentages of CD4+ CD8+ cells and of CD1+ cells are different in thymomas while they are similar in normal thymuses. These results demonstrate a subset CD1+ CD4 CD8 which is not clearly represented in the normal thymus. In addition, the decreased number of CD4+ CD8+ is compensated by an increase in CD4+ (36.3 % versus 12.2 %, $p < 0.001$) and in CD8+ (14.1 % versus 6.9 %) subsets. In contrast, the percentages of CD3+ and B cells are similar in thymomas and adult thymuses. These data indicate that both cortical and medullary cell types are represented in thymomas but that the cortical cell type subset is reduced in favor of medullary cell type population.

No correlations could be established with malignancy of the tumor, association with Myasthenia Gravis disease or epithelial cell type (spindle, ovoid or round cells). On the other hand, interesting findings were observed in correlations with the cell type most represented. The 22 thymomas were thus divided into 2 subgroups according to the predominant cell type. Results concerning 16 thymomas with a high number of lymphocytes and 6 thymomas containing mainly epithelial cells are reported in Table 2. Clear differences were observed between the 2 categories of thymomas : 1) the number of CD4+ CD8+ cells is lower in thymomas containing moderate numbers of lymphocytes compared to lymphocyte-rich thymomas. 2) Similarly, the amount of CD1+ is decreased, as well as the number of OKT10+ cells. As underlined above, the number of CD1+ cells is significantly larger than that of CD4+ CD8+ cells. The proportion of CD1+ cells which do not bear simultaneously CD4 and CD8 antigens is 20.3 % and 36.4 % for epithelial cell-predominant thymomas and lymphocyte-rich thymomas, respectively. 3) The

Table 1 : Distribution of T cell subsets in thymomas compared to controls

Surface marker	Thymoma	Normal adult thymus	
CD3	48.0 %	48.9 %	
CD4 CD8	33.3 %	74.3 %	$p < 0.001$
CD4	36.3 %	12.2 %	$p < 0.001$
CD8	14.1 %	6.9 %	
CD1	65.2 %	84.4 %	$p < 0.05$
OKT10	74.7 %	93.0 %	$p < 0.05$
IOB	1.4 %	2.9 %	
n	22	13	
Age (yrs)	32-78	17-60	

Table 2 : Distribution of T cell subsets in two subgroups of thymoma

	Predominance		
Surface marker	Epithelial cells	Lymphoid cells	
CD3	56.2 %	44.9 %	
CD4 CD8	14.6 %	40.1 %	$p < 0.03$
CD4	39.9 %	34.9 %	
CD8	18.2 %	12.6 %	
CD1	34.9 %	76.5 %	$p < 0.002$
OKT10	41.8 %	88.9 %	$p < 0.001$
IOM	8.0 %	0.2 %	$p < 0.04$
IOB	3.0 %	0.8 %	
DR	11.4 %	1.7 %	$p < 0.03$
PNA	19.6 %	54.7 %	$p < 0.002$
n	6	16	
Age (Yrs)	33-72	32-78	

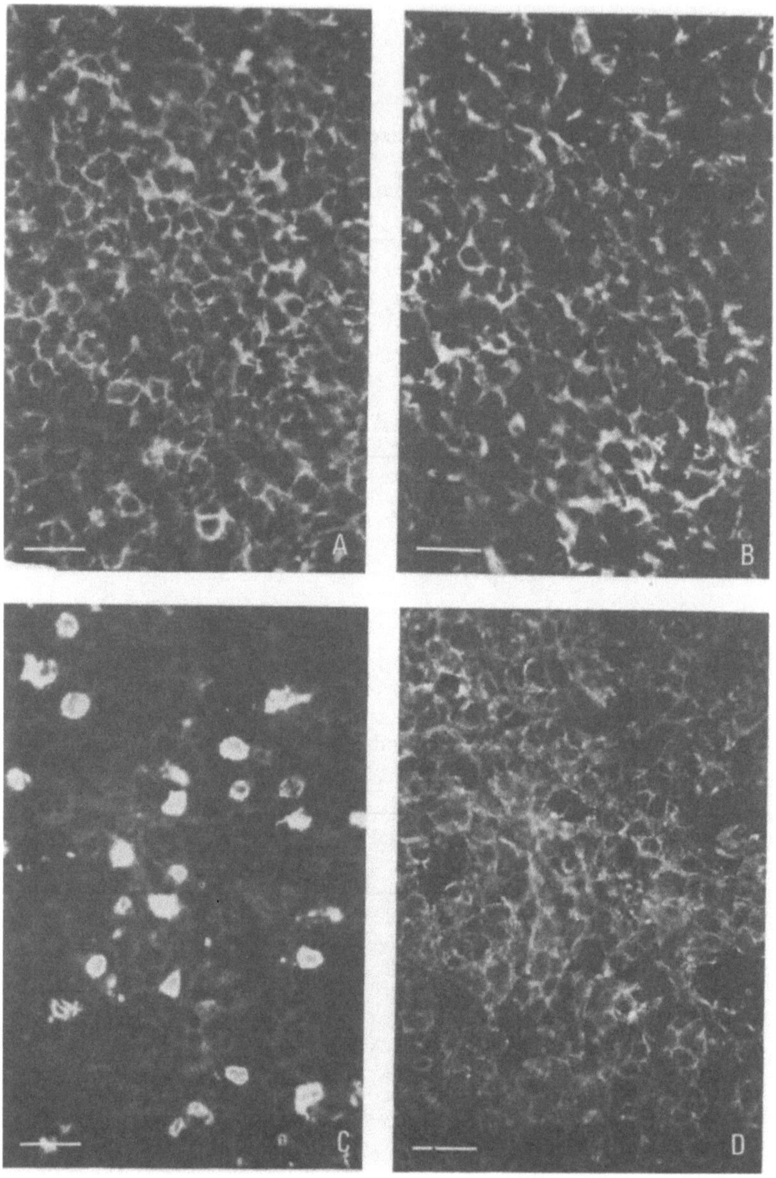

Fig.1: Double immunofluorescence staining with anti-CD8 mouse monoclonal antibody revealed by GAM/Ig/TRITC (a, c) and anti-keratin rabbit polyclonal antibody revealed by GAR/Ig/FITC (b, d). In a lymphocyte-rich thymoma (a, b), the keratin network is clearly delineated and the number of CD8+ cells is large. Scale bar = 10 µm. In thymoma with few lymphocytes (c, d), the keratin network is dense and the number of CD8+ cells is reduced. Scale bar = 15 µm. Similar patterns were observed using anti-CD4 and anti-CD1 antibodies.

percentage of PNA+ cells is also decreased in thymoma with few lymphocytes, which is correlated to decreased CD1+ cell percentage. However, in both types of thymoma, the number of PNA+ cells is slightly lower than that of CD1+ cells, in accordance with a previous report (11). 4) Conversely, numbers of monocytes and DR+ cells are increased in lymphocyte-poor thymoma and the number of DR+ cells corresponds to the sum of monocytes and B lymphocytes.

These results indicate that in thymomas with few lymphocytes, the phenotype of lymphoid cells is mostly of the medullary cell type, similar to the peripheral cell type. One cannot exclude the contamination with blood vessel peripheral lymphocytes. However, such lymphocytes cannot account for the majority of cells. Indeed, several reasons could be suggested : 1) more than 40 % of cells are OKT10+, strongly suggesting a thymic origin of these cells. 2) In frozen sections (Figures 1 and 2), T cells were clearly observed throughout the parenchyma and not only around the vessels. 3) A significant number of CD4+ CD8+ cells, which are specifically thymic cells, were consistently observed in these thymomas. This result is in accordance with that of Takacs et al. (10) showing that whatever the epithelial cell type, thymomas contain the cortical T cell type.

In both types of thymoma, the number of cells which do not display CD4 or CD8 antigens (double negative) is increased compared to normal thymus (27 %, 12 % and 6 % in thymomas containing few lymphocytes, numerous lymphocytes and normal thymus, respectively). Although these numbers include some B cells, it is clear that double negative cells represent a significant subset in thymomas and particularly in thymomas containing few lymphocytes.

Altogether our results indicate that although the lymphocyte cortical phenotype is consistently observed in thymomas, its percentage depends upon the proportion of epithelial cells in the thymoma. In addition, these results show the presence of the CD1+ CD4- CD8- subset which could represent a particular stage in intrathymic differentiation. The increased numbers of double-negative cells indicate that thymomas could be an interesting tool for studying these cells in humans.

REFERENCES

1. J. Rosai and G.D. Levine, Tumors of the thymus, in : Fascile 13, Atlas of Tumor Pathology. Washington : Armed Forces Institute of Pathology (1976).
2. J.M. Verley and K.H. Hollmann, Thymoma. A comparative study of clinical stages and histology of 200 cases, Cancer, 55 : 1074 (1985).
3. T. Eimoto, K. Teshima, T. Shirakusa, M. Takeshita, H. Okamuro, H. Waito, T. Mitsui and M. Kikuchi, Heterogeneity of epithelial cells and reactive components : an ultrastructural and immunohistological study, Ulstrastruc. Pathol., 10 : 157 (1986).
4. N. Mokhtar, S.M. Hsu, R.P. Lad, B.F. Haynes and E.S. Jaffe, Thymoma : lymphoid and epithelial components mirror the phenotype of normal thymus, Hum. Pathol., 15 : 378 (1984).
5. H.K. Muller-Hermelink, M. Marino, G. Palestro, V. Schumacher and T. Kirchner, Immunohistological evidences of cortical and medullary differentiation in thymoma, Virchows Arch. (A) 408 : 143(1985).
6. W. Savino, G. Manganella, J.M. Verley, A. Wolff, S. Berrih, P. Levasseur, J.P.Binet, M. Dardenne and J.F. Bach, Thymoma epithelial cells secreting thymic hormone but not expressing class II antigens of the major histocompatibility complex, J. Clin. Invest., 76 : 1140 (1985).

7. W. Savino, S. Berrih and M. Dardenne, Thymic epithelial antigen acquired during ontogeny and defined by the anti-p19 monoclonal antibody is lost in thymomas, Lab. Invest., 51 : 292 (1984).
8. A.C. Aisenberg, B. Wilkes, N.L. Harris and W.H. First, The predominant lymphocyte in most thymomas and in non-neoplastic thymus from patients with myasthenia gravis is the cortical thymocyte, Clin.Immunol. Immunopathol., 35 : 130 (1985).
9. R.V. Rouse, Is Ia antigen-expression by thymic epithelial cells constitutive ? In "Microenvironments in the lymphoid system". Adv. Exp. Med. Biol., 186 : 915-919, 1985. G.G.B. Klaus ed., Plenum Press, New York.
10. L. Takacs, W. Savino, E. Monostori, I. Ando, J.F. Bach and M. Dardenne, Cortical thymocyte differentiation in thymomas : an immunohistologic analysis of the pathologic micro-environment, J. Immunol., 138 : 687-698 (1987).
11. S. Berrih, C. Boussuges, J.P. Binet and J.F. Bach, Studies on human thymocytes using anti-T cell monoclonal antibodies and peanut agglutinin, Cell. Immunol., 74 : 260 (1982).

THYMIC HYPERPLASIA IN MYASTHENIA GRAVIS : IMMUNO-HISTOLOGICAL ABNORMALITIES AND RESPONSIVENESS TO RECOMBINANT IL2

Cohen-Kaminsky S., Berrih-Aknin S., and Safar D.

Centre Chirurgical Marie-Lannelongue, CNRS UA 04-1159
133, av. de la Résistance
92350 Le Plessis-Robinson (France)

Myasthenia Gravis (MG) is an autoimmune disorder of neuromuscular transmission characterized by the presence of autoantibodies directed against acetylcholine receptor (AChR), which are detected in more than 85 % of patients' sera (1). The role of the thymus in MG has been strongly suggested by the favorable clinical effects of thymectomy and by the high incidence of thymic abnormalities (15-20 % thymomas and 50 % hyperplastic thymuses) (2).

In recent studies, we have shown that MG patients presenting thymic hyperplasia are mainly young women (< 30 years) with high anti-AChR antibody titer (3). Several other arguments suggest the existence of a close relationship between disease pathogenesis and thymus function : first, thymic lymphocytes are able to produce anti-AChR Ab in vitro without polyclonal activation (4), second, an AChR-like antigen has been extracted from human thymus (5), and finally thymic epithelial cells synthesize increased amounts of thymic hormones (6), one of which (thymopoietin) has been reported to depress neuromuscular conduction and to bind specifically to AChR with a high affinity (7, 8).

The aim of this study was to further characterize thymic hyperplasia and to investigate the activation status of thymic lymphocytes in MG. We tested the ability of freshly isolated lymphocytes from normal and from MG thymuses to respond to recombinant IL2 (r-IL2). In MG a high proportion of T lymphocytes showed functional signs of preactivation as evidenced by hypereactivity to r-IL2 with no previous stimulation.

PATIENTS AND METHODS

Patients

MG thymocytes were mechanically extracted from thymic fragments obtained from patients (n = 24, aged from 11 to 57 years) thymectomized at Marie-Lannelongue Hospital. Disease severity was graded according to Osserman's classification (9). Two major groups were represented : IIA, mild disease (9 cases) and IIB, generalized disease (8 cases). Normal human thymic fragments were obtained from subjects (n = 17, aged from 2 month to 25 years) undergoing surgery for cardiac lesions.

Immunofluorescence studies on frozen sections

Thymic fragments were frozen in liquid nitrogen just after surgery. Several antibodies were used : goat anti-human IgM antiserum coupled to FITC was provided by Dako ; mouse monoclonal anti-fibronectin antibody (Immunotech) revealed by SAM/Ig/TRITC (Cappel) was used in double labeling experiments ; monoclonal anti-CD21 antibody (IOB1a) which recognizes non-circulating B lymphocytes was provided by Immunotech and was revealed by SAM/Ig/TRITC (Cappel). Double labeling was performed with rabbit anti-keratin antibody (Dako) revealed by SAR/Ig/FITC (Cappel).
Unfixed frozen sections (2 µm thick) were processed as described below. For single labeling experiments, sections were incubated with the first layer specific antibody for 15 min and washed twice in PBS for 5 min. If necessary, they were then subjected to the appropriate fluorescent conjugate for 15 min, washed again in PBS and mounted in glycerol/PBS. When double-labeling experiments were performed, the specimens were sequentially incubated with the first specific antibody and its proper fluorescent conjugate, followed by the second specific antibody and its corresponding conjugate.

Measurement of responsiveness to r-IL2

Freshly isolated thymocytes were cultured at concentrations of 5×10^6 cells/ml in RPMI 1640 medium supplemented with 2 % AB human serum. Recombinant-IL2 (r-IL2, produced after genetic recombination in Escherichia Coli ; Boehringer) was added at serial doses from 3 to 50 U/ml for 2 to 12 days. Proliferation was measured by ^{3}H-thymidine (CEA, France) incorporation. To test the specificity of the response, a monoclonal anti-IL2 receptor blocking antibody (IOT14a, Immunotech) was added to the cultures at serial doses from 0.1 to 100 nM.

Detection of the IL2 receptor (TAC) antigen

For some patients, TAC antigen expression was studied in parallel with proliferation. Thymic lymphocytes were cultured in 24-well plates (1 ml/well) with an optimal dose of 20 UI/ml r-IL2 for 3 to 6 days. The percentage of TAC+ cells was evaluated by indirect immunofluorescence using the rat monoclonal IOT14a. Double labeling experiments with anti-CD1 (IOT6), anti-CD3 (IOT3), anti-CD2 (IOT11), anti-CD37 (IOB1) (Immunotech), and anti-DR 206 (a gift of D. Charron) mouse monoclonal antibodies were performed to determine the phenotype of TAC+ cells. The percentages of positive cells were evaluated by fluorescence microscopy and cytofluorography analysis.

RESULTS AND DISCUSSION

It is generally accepted that T lymphocyte proliferation subsequent to the interaction between IL2 and its receptor depends upon activation of lymphocytes by mitogen or antigen (10). However, recent reports have clearly demonstrated that recombinant IL2 (r-IL2) induces lymphocyte proliferation without any previous stimulation and regulates IL2 receptor expression (11, 12). In other respects, thymic hyperplasia, which is the main abnormality of the MG thymus, is characterized by several immunohistological features. Numerous germinal centers (stained by anti-B cell monoclonal antibody) are located in keratin-negative areas (Fig. 1a and 1b), which include an abnormally extended connective framework (Fig. 2a

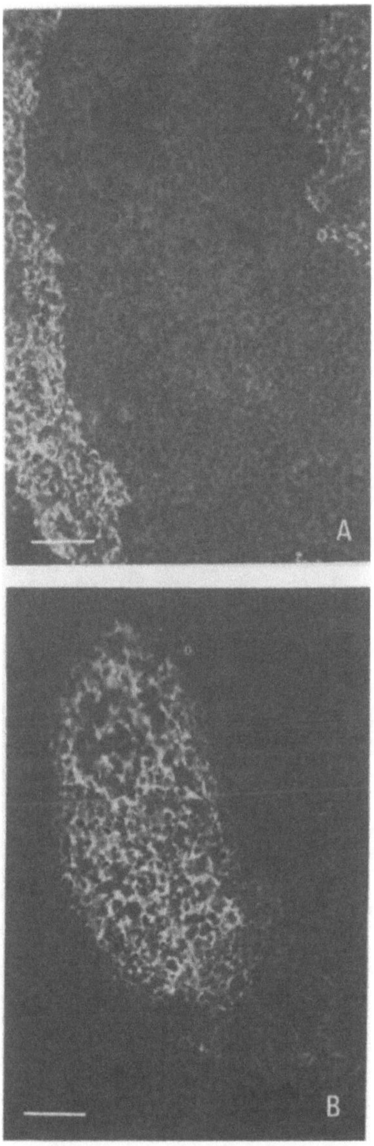

Fig. 1 : Frozen section of hyperplastic MG thymus : the epithelial network stained by a polyclonal anti-keratin antibody (a) is highly compressed. Lymphoid follicles are found in the large keratin-negative areas (b). Here, it is revealed by labeling with monoclonal anti-CD21 antibody which recognizes non-circulating B lymphocytes. Scale bar : 25 μ.

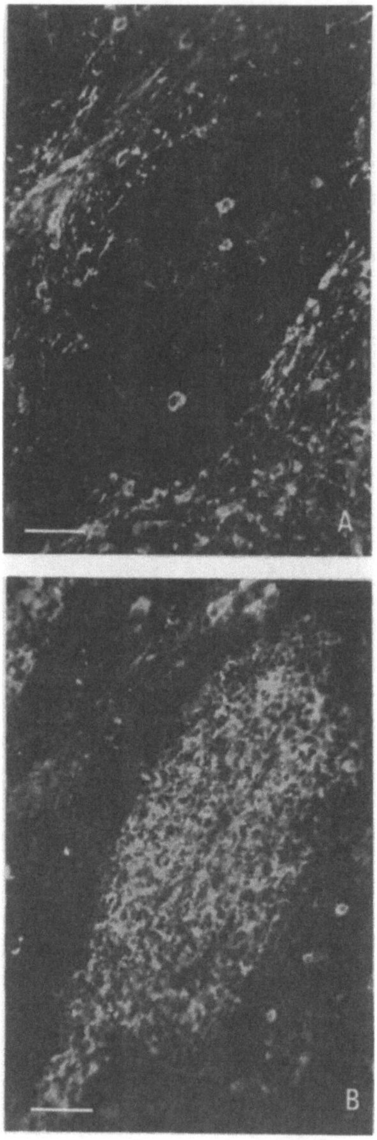

Fig. 2 : Frozen section of hyperplastic MG thymus : the connective fibers visualized by monoclonal anti-fibronectin antibody are abnormally extended (a) around the B cell area of the germinal center stained by a Goat anti-human IgM antibody (b). Scale bar : 25 µm.

and 2b). As in lymph nodes, the germinal centers are formed by a core of B cells (IgM+ cells with a ring of IgM+IgD+ cells) surrounded by a T cell area (mostly CD4+, partially CD8+).

The observation of these thymic abnormalities led us to investigate the activation status of thymic lymphocytes. We evaluated the response to r-IL2 of thymic lymphocytes freshly extracted from MG or control thymuses. Proliferative response measured by ^{3}H-thymidine incorporation was genrally higher in MG patients compared to controls (Fig. 3) ($p < 0.02$ for the 12 U/ml IL2 concentration and $p < 0.001$ for the 25 and 50 U/ml IL2 doses). However, differences in the magnitude of the response among experiments were observed in MG, as well as in controls, probably due to the influence of individual immune status at the time of the study.

The response to r-IL2 was dose-dependent and could be inhibited by a blocking anti-IL2 receptor antibody (anti-TAC), which supports the specificity of the response to r-IL2 and proves that the proliferation observed is mediated by the binding of IL2 to its receptor. Dose effect study of r-IL2 and kinetic study on ^{3}H-thymidine uptake showed that the optimal conditions were 6 day treatment with 20 to 50 U/ml of r-IL2 for all patients tested.

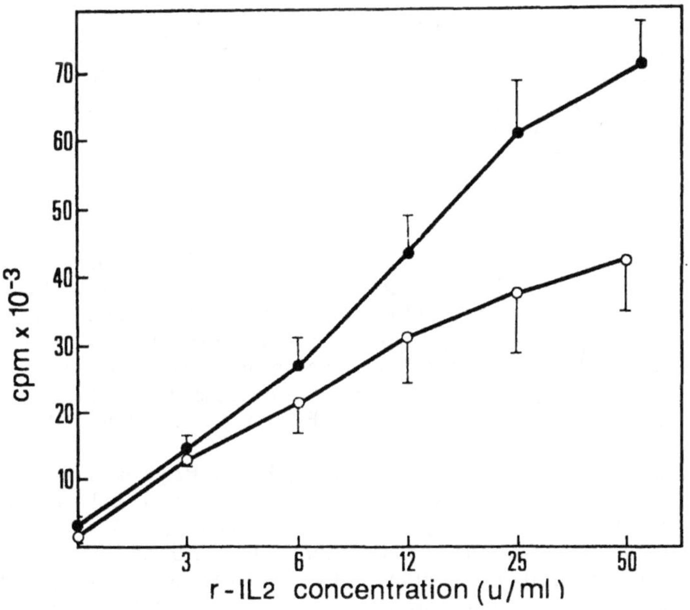

Fig. 3 : Proliferative response to 6 day treatment with r-IL2 of thymocytes from normal (O) and MG (●) subjects. Results are expressed as the mean of cpm. The bar represents SEM. Significant differences were observed between MG and controls from12 to 50 U/ml IL2 concentration.

In order to evaluate the number of activated cells, monoclonal anti-TAC antibody was used to quantify the cells expressing IL2 receptors. The high proliferative response of MG thymic lymphocytes was well correlated to an increased TAC+ cell number evaluated by immunofluorescence and flow cytometry analysis. The maxima (nearly 50 % of TAC+ cells) was reached after 5-6 day treatment with the optimal dose of r-IL2. Double labeling experiments revealed that TAC+ cells are T cells, totally CD2+ (IOT11+) and CD37-(IOB1-), CD1- (IOT6), mostly CD3+ (IOT3+) and partially DR+.

Altogether, these data suggest that in the MG thymus a high proportion of T lymphocytes shows functional signs of pre-activation which is reflected by hypereactivity to r-IL2 of MG lymphocytes without previous stimulation. These lymphocytes could represent autoreactive cells involved in the pathogenesis of MG. However, it should be pointed out that phenotypic signs of activated T cells (namely IL2 receptor expression) could not be detected on freshly extracted cells. Indeed, the percentage of TAC-positive cells is similar in MG and normal subjects before culture with r-IL2 (5 % of TAC+ cells at maxima in MG or normal subjects).

To explain the apparent contradiction between a high proliferative response to r-IL2 and the low number of TAC-positive cells before r-IL2 treatment, several non-mutually exclusive theories can be suggested :
1) pre-activated cells from MG patients may bear IL2 receptors in low density (therefore undetectable by anti-TAC binding).
2) MG lymphocytes may have a higher ability to express high affinity IL2 receptors which are known to be involved in the interaction with IL2 and in proliferative response. Therefore a different regulation of IL2 receptor expression could occur in MG, as has been described in multiple sclerosis (13).
3) MG lymphocytes could have receptor dysfunction related to the expression of an abnormal IL2-receptor molecule, as has been shown in cell leukemia (14).

These particular autoreactive T cells are likely to be located in the abnormal zones of hyperplastic thymuses, namely in germinal centers developed in the extended connective framework, and could result from a primary autosensitisation against an acetylcholine receptor-like antigen previously demonstrated in the thymus (5).

REFERENCES

1. J.M. Lindstrom, M.L. Seybold, V.A. Lennon, S. Whittingham and D.D. Duane. Antibody to acetylcholine receptor in Myasthenia Gravis : prevalence, clinical correlates, and diagnostic value. Neurology (Minneap.) 26 : 1054-1059 (1976).
2. G.D. Levine. Pathology of the thymus in myasthenia gravis : current concepts. In : Dau P.C., ed. Plasmapheresis and the immunobiology of myasthenia gravis. Boston : Houghton Mifflin : 113-123 (1979).
3. S. Berrih, E. Morel, C. Gaud, F. Raimond, H. Le Brigand and J.F. Bach. Anti-AChR antibodies, thymic histology and T cell subsets in myasthenia gravis. Neurology, 34 : 66-71 (1984).
4. D. Safar, S. Berrih-Aknin and E. Morel. In vitro anti-acetylcholine receptor antibody synthesis by Myasthenia Gravis patient lymphocytes : correlations with thymic histology and thymic epithelial cell interactions. J. Clin. Immunol., 7 : 225-234 (1987).
5. F. Raimond, E. Morel and J.F. Bach. Evidence for the presence of immunoreactive acetylcholine receptors on human thymus cells. J. Neuroimmunol., 6 : 31-34 (1984).

6. J.J. Twomey J.J., U.M. Lewis, B.M. Patten, G. Goldstein and R.A. Good. Mysthenia gravis, thymectomy and serum thymic hormone activity. Am. J. Med., 66 : 639-643 (1979).
7. K. Venkatasubramanian, T. Audhya and G. Goldstein. Binding of thymopoietin to the acetylcholine receptor. Proc. Natl. Acad. Sci. USA, 83 : 3171-3174 (1986).
8. S. Berrih-Aknin, E. Morel, F. Raimond, S. Safar, C. Gaud, J.P. Binet, P. Levasseur and J.F. Bach. The role of the thymus in myasthenia gravis. Immunohistological and immunological studies in 115 cases. Ann. N.Y. Acad. Sci. (in press).
9. V.P. Perlo, D.C. Poskanzer, R.S. Schwab, H.R. Viets, K.E. Osserman and G. Genkins. Myasthenia gravis : evaluation of treatment in 1 355 patients. Neurology (Minneap.), 16 : 431-439 (1966).
10. E.L. Larsson, N.N. Iscove and A. Coutinho.. Two distinct factors are required for induction of T-cell growth. Nature, 283 : 664-666 (1980).
11. F.H. Reem, N.H. Yeh, D.L. Urdal, P.L. Kilian and J.J. Farrar. Induction and upregulation by interleukin 2 of high-affinity interleukin 2 receptors on thymocytes and T cells. Proc. Natl. Acad. Sci. USA, 82 : 8663-8666 (1985).
12. S.M. Hammer and J.M. Gillis. Effects of recombinant inter-leukin-2 on resting human T lymphocytes. J. Biol. Resp. Modifers, 5 : 36-44 (1986).
13. E.C. De Freitas, M. Sandberg-Wollheim, K. Schonely, M. Boufal and H. Koprowsky. Regulation of interleukin 2 receptors on T cells from multiple sclerosis patients. Proc. Natl. Acad. Sci. (USA), 83 : 2637-2641 (1986).
14. J. Yodoi and T. Uchiyama. IL2 receptor dysfunction and adult T-cell leukemia. Immunol. Reviews, 92 : 135-156 (1986).

THYMUS HOMING AND RECOSTITUTION IN SPLIT DOSE IRRADIATED PRELEUKEMIC MICE (1)

M. P. Defresne (2), C. Humblet, R. Greimers
and J. Boniver

Laboratory of Pathological Anatomy
B.23, Université de Liège
Sart Tilman, B. 4000 Liège (Belgique)

The incidence of thymic lymphomas in C57BL/Ka mice treated with four weekly irradiations of 1.75 Gy, reaches more than 90% (1). The tumors appear after a long latent period (4 to 9 months), during which the thymuses contain preneoplastic cells (called preleukemic cells) requiring the thymic microenvironment for their progression into frank neoplasia (2).

During this latency, alterations of the marrow-thymus interactions are observed: the pool of bone marrow prothymocytes is impaired (3, 4), the function and the phenotype of thymic environment are altered (5) and the pattern of thymus repopulation is modified.

Bone marrow grafting immediately after the last irradiation is followed by the restoration of the pool of prothymocytes (3, 4) and the thymic microenvironment (6), facilitates thymus regeneration and prevents the development of lymphomas (7). Under these conditions, the appearance of preleukemic cells is not inhibited but they disappear one month later (6). The mechanisms inducing this disappearance and thus the protection against lymphomas have not been demonstrated. It has been proposed that marrow cells actively repopulating the irradiated thymuses are responsible for the protection against tumors, either by "competition" with preleukemic cells, either by restoring some components of the thymic microenvironment.

Such bone marrow precursors grafted later on after the last irradiation, for example one month later are unable to repopulate the irradiated thymuses and do not inhibit the lymphoma development (8). The present work was undertaken to test whether the injury to thymic stromal cells is the major factor involved in the lack of reconstitution in preleukemic mice. For this purpose, the homing of normal marrow precursors to thymus and their capacity to repopulate the thymus of sublethally irradiated mice were compared in normal and preleukemic thymuses.

---

(1) Supported in part by the Medical Scientific Research Foundation and the Anticancer Centre by the University of Liège.
(2) Senior Research Associate of the Belgian National Scientific Research Foundation.

MATERIALS AND METHODS

Mice: One to two-months old C57BL/Ka mice of both sexes were used. Four week old congenic Thy-1.1 C57BL/Ka mice (called BL/1.1) were used as donors of bone marrow cells.

Irradiations: For the induction of lymphomas, mice were given four doses of 1.75 Gy applied at weekly intervals. For the thymus homing and for the repopulation assay, mice were whole-body irradiated with a single dose of 4 Gy. The irradiation was delivered by an X-ray apparatus (Stabilivolt Siemens, 190 Kv, 18 mA, HVL: 0,5 mm Cu) at a dose rate of 1,60 Gy/min.

In vivo thymus homing assay was done as described (9). Briefly, 10.10^6 FITC labelled bone marrow cells (10) were injected i.v. into 4 Gy irradiated animals within 2 hours following X-ray exposure. Twenty-four to 72 hours later, recipients were sacrificed and the number of FITC labelled cells was determined in cell suspensions treated with propidium iodure to exclude death cells.

The thymus repopulation assay was done as described previously (11). Aliquots of 5.10^6 bone marrow cells collected in BL/1.1 mice (Thy-1.1) were injected i.v. into C57BL/Ka mice (Thy-1.2) within 2 hours following a single 4 Gy exposure. Fifty, 20 and 30 days later, recipients were sacrificed and thymus cell suspensions were treated with monoclonal anti-Thy-1.1 and anti-Thy-1.2 antibodies followed by a FITC labelled second stage antibody.

Fluorescence analysis: Cell suspensions were analyzed on a fluorescence activated cell sorter (FACS IV, Becton-Dickinson, Sunnyvale, California). Only the fluorescence of viable cells (discriminated on the basis of labelling with propidium iodure) was considered.

RESULTS

The first experiment was undertaken to test whether radiation-induced injury to thymus inhibits the homing of normal bone marrow precursors. For this purpose, FITC labelled bone marrow suspensions were injected in the four Gy treated C57BL/Ka mice, either normal, either 30 days after the end of the leukemogenic treatment. The percentage of viable fluorescent cells was estimated in the whole thymus 24 and 72 hours after the graft. As seen in Table 1, the proportion of fluorescent cells at each time interval is similar in both conditions. The data indicate that homing of normal precursors is similar in normal and i preleukemic thymuses.

The following experiment was designed to compare the capacity of normal marrow precursors to repopulate 4 Gy irradiated normal and preleukemic thymuses. Bone marrow from BL/1.1 mice (Thy-1.1) were injected in 4 Gy irradiated C57BL/Ka mice (Thy-1.2) either normal, either preleukemic. Fifty, 20 and 30 days later, the percentage of donor (Thy-1.1) thymocytes was scored. As seen in fig. 1, the percentage of donor cells in normal thymuses increases gradually from the 15th to the 30th day after the irradiation to reach about 90%. This evolution is quite different in preleukemic mice: the percentage of donor cells decreases from day 15 to day 20, before to increase and to reach normal values at day 30. The data indicate that reconstitution of preleukemic thymuses is delayed as compared with that of normal thymuses.

Table 1. Migration of FITC labelled bone marroe cells in 4 Gy treated thymuses

Delay after bone marrow grafting (hours)	Frequency of FITC labelled cells	
	In normal thymuses	In preleukemic thymuses (30 days after the leukemogenic irradiation)
24	8.5/10.000	10/10.000
72	10/10.000	12/10.000

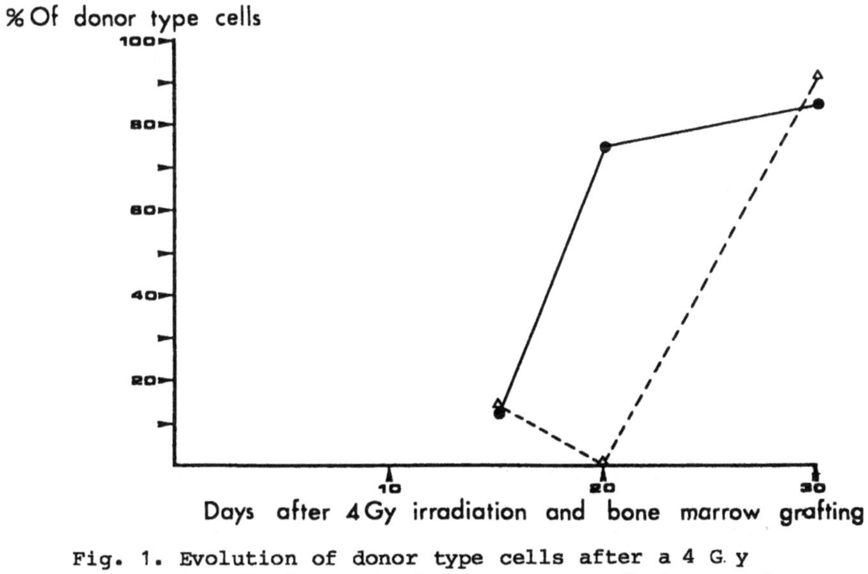

Fig. 1. Evolution of donor type cells after a 4 G.y irradiation and a bone marrow graft in normal (●——●) and preleukemic (Δ----Δ) mice treated 30 days after the leukemogenic irradiation.

DISCUSSION

Previous studies have shown that marrow thymus interactions are altered in preleukemic C57BL/Ka mice (3, 4, 6, 7). Therefore we have designed experiments to test whether the homing or normal marrow precursors to preleukemic thymuses and/or their repopulation capacities were modified.

It has been demonstrated that in vitro fluorochrome labelling of lymphocytes does not alter their migration properties when they are injected into animals (10). With this technique, we studied the migration of normal bone marrow precursors to thymuses of normal and preleukemic mice (taken 30 days after the end of the leukemogenic irradiation), and we demonstrated that it was similar in both conditions.

By contrast, using radiation chimeras, we showed that reconstitution of preleukemic thymuses by normal bone marrow cells is delayed as compared with that of normal thymuses.

There is no evidence that short term thymus homing bone marrow cells in fact are the precursors that allow long term thymic reconstitution. Because fluorescence fades rapidly, it is not possible to follow the evolution of immigrants for a long period of time. However, their phenotype (12) and their localization in the outer region of the thymus (data not shown) which is known to contain precursors of thymocytes suggest that they might be responsible for thymus repopulation. If it is true, one may conclude that the altered thymic microenvironment during the preleukemic period is responsible for the impaired proliferation of precursors within the thymus. In agreement with this hypothesis, we have shown that the phenotype and the function of thymic nurse cells which are responsible for some early steps of T cell proliferation and differentiation (13) are altered (5). During the preleukemic period, they are unable to establish in vitro interactions with immature thymocytes.

It remains to explain why the thymus reconstitution is only delayed and not totally abolished during the preleukemic period and how this phenomenon contributes to the leukemogenic process.

REFERENCES

1. H. S. Kaplan, Cancer Res. 27:1325 (1967).
2. G. Goffinet, M. P. Houben-Defresne, and J. Boniver, Cancer Res. 43:5416 (1983).
3. J. Boniver, A. Decleve, M. Lieberman, C. Honsik, M. Travis, and H. S. Kaplan, 41:390 (1981).
4. N. H. Pazimo, R. McEwan, and J. N. Ihle, J. Exp. Med. 148:1338 (1978).
5. M. P. Defresne, A. M. Rongy, R. Greimers, and J. Boniver, Leukemia Res. 10:783 (1986).
6. M. P. Defresne, R. Greimers, P. Lenaerts, and J. Boniver, J. Nat. Cancer Inst. 77:1079 (1986).
7. H. S. Kaplan, M. B. Brown, and J. Paull, J. Nat. Cancer Inst. 14:303 (1953).
8. D. W. Van Bekkum, W. J. A. Boersma, J. F. Eliason, and S. Knaan, Leukemia Res. 8:461 (1984).
9. F. Lepault and I. L. Weissman, Nature 293:151 (1981).
10. E. C. Butcher and I. L. Weissman, J. Immunol. Methods 37:97 (1980)
11. M. P. Houben-Defresne, A. Varlet, and J. Boniver, J. Thymus 6:324 (1984).
12. F. Lapault, R. L. Coffman, and I. L. Weissman, J. Immunol. 131:64 (1983).
13. B. A. Keyewski, Immunology Today 7:374 (1986).

SYNOVIAL T CELLS DETECT POLYMORPHISM OF CLASS II

HLA ANTIGENS

J.S.H. Gaston, P. Life, and P.A. Bacon

University of Birmingham
Department of Rheumatology
The Medical School
Birmingham, B15 2TJ
United Kingdom

INTRODUCTION

Rheumatoid arthritis (RA) is associated with the class II MHC antigen HLA-DR4[1]. Since MHC antigens control recognition of antigens by T cells[2], this association with HLA-DR4 strongly implies that a T cell-mediated immune response is important in the pathogenesis of RA. This response may occur within the joint since extensive T cell infiltration is characteristic of RA synovium[3,4].

We have isolated RA synovial T cells and expanded them in vitro as continuously growing T cell lines in order to examine their properties. They have been shown to recognise autologous class II MHC antigens, either as restricting elements for the response to an undefined nominal antigen, or in an "autoreactive" fashion. Because of this property it may be possible to utilise the T cell lines as "cellular typing reagents" in order to define the particular Class II MHC determinants recognised by RA synovial T cells. This approach compliments serological and molecular studies[5,6,7] and has the eventual aim of investigating the frequency of these T cell-defined MHC determinants amongst populations of normal individuals, RA patients, and patients with other arthropathies.

MATERIALS and METHODS

1. Derivation of T cell lines. Synovial specimens were obtained by closed needle biopsy or at arthroscopy from patients with RA or other inflammatory arthropathies. Synovial biopsies were subdivided and cultured for 4-6 days in the presence of 10-20U/ml recombinant IL-2 (Cetus, Emeryville, CA, ref.8) in RPMI 1640 supplemented with 2mM glutamine, 100U/ml penicillin, 100mg/ml streptomycin, 10% foetal bovine serum (all from Gibco U.K.) and 1% human serum. T cell blasts were harvested from these cultures and expanded using IL-2 and irradiated autologous feeder cells, either peripheral blood mononuclear cells (PBMC)(15Gy)or Epstein-Barr virus-transformed lymphoblastoid cells (EBLCL)(50Gy). T cell lines were also derived from PBMC in parallel with the synovial T cell cultures.

2. Testing the specificity of T cell lines. The specificity of the T cell lines was tested in proliferation assays as follows:- 2×10^4 T cells were co-cultured with 5×10^4 irradiated autologous or allogeneic stimulator cells for 72 hours in U-bottomed wells. ^{3}H-thymidine incorporation was measured during the last 18 hours of culture. Results are expressed as d.p.m. after subtracting the proliferation due to irradiated EBLCL alone, invariably <2000 d.p.m.

In some experiments, various concentrations of monoclonal antibodies (mabs) were added to the cultures and their effect on proliferation measured.

TABLE 1.

Proliferative responses (d.p.m.) of synovial T cell lines

T cell line		Stimulator cells			
	DR	auto	DR-	DR+	
MA syn T	1	13717	328	-	
GL syn T	2,5	14120	883	3789	
				DR4+	DR4+
WC syn T	4	19002	1941	6704	2899
LH syn T	4	12968	1165	6145	1245
GOP syn T	4	16177	3794	19477	2284

RESULTS

1. Synovial T Cell Lines Recognise MHC Antigens

Table 1 displays results obtained for 5 synovial T cell lines, 4 from RA patients and 1 (WC syn T) from a patient with reactive arthritis. The HLA-DR type of each of these patients is also shown. For each T cell line the proliferative responses to autologous EBLCL (auto) and to HLA-DR mismatched EBLCL (DR-) appear in columns 1 and 2 respectively. For T cell line GL the response to HLA-DR2 or -DR5-matched stimulators (DR+) is shown in the third column. Each of the T cell lines from DR4+ patients (WC, LH, GOP) distinguished amongst DR4-matched stimulators; the mean response to stimulatory DR4+ cells is shown in column 3, whereas the response to non-stimulatory DR4+ cells is shown in column 4. The results represent the mean proliferative responses obtained when each of the T cell lines was tested against 8-24 different EBLCL of known HLA-DR type.

It may be seen that each of the synovial T cell lines tested gave a preferential proliferative response to autologous, compared with allogeneic, EBLCL. When the allogeneic stimulator cells were grouped according to their degree of sharing of HLA-DR antigens with the autologous stimulators, it was clear that DR-mismatched stimulators did not elicit a response whereas those which were matched for one or two DR antigens were stimulatory, usually to a lesser extent than the autologous EBLCL. An additional complication was the heterogeneity found within DR4+ stimulators, some of which were highly stimulatory whilst others gave responses equivalent to those observed using DR-mismatched EBLCL.

TABLE 2. Inhibition of GOP synovial T cell proliferation by monoclonal antibodies

Antibody		Proliferation (d.p.m.)
None		7741
Anti-DR (A21BC1)	1:500	2837
	1:100	1048
	1:50	649
Control (10DF5)	1:500	8473
	1:100	8707
	1:50	6036

To provide additional evidence that the response to stimulator cells reflected recognition of class II MHC antigens, blocking studies were carried out using monoclonal antibodies. A representative experiment is shown in Table 2. Mab A21B1, which has a broad specificity for human class II antigens, blocked proliferation effectively, whereas the control mab, which binds to other antigens on EBLCL, did not do so. When antibodies specific for HLA-DR (L243) and -DQ (Leu 10) were used only L243 blocked proliferation efficiently (data not shown).

2. Synovial T Cell Lines Show Different Patterns of Recogniton of DR4+ Cells to PBMC-derived T Cell Lines.

The suggestion from Table 1 that the synovial T cells distinguished between different HLA-DR4+ stimulator cells was investigated further. As a control for the results obtained from synovial T cell lines, T cell lines were also established from PBMC of RA patient GOP using an identical protocol. Table 3 shows results of two experiments in which the responses of synovium- and PBMC-derived T cell lines were compared.

TABLE 3. Proliferative responses of synovial- and PBMC-derived T cell lines.

Stimulator	HLA-DR	Proliferation (d.p.m.)	
EXPT. 1		GOP syn T	GOP PBMC T
None		117	153
GOP-EB	4	20626	15946
WC-EB	4	19832	14814
MS-EB	4	1793	12103
DG-EB	4	2072	7614
SC-EB	3	3832	1118
SG-EB	3	2072	1116
EXPT. 2			
None		380	94
GOP-EB	4	17230	27911
WC-EB	4	25216	29602
LH-EB	4	-1510	13362
NB-EB	4	737	11945
GP-EB	4	419	7846

The GOP synovial T cell line proliferated in response to the autologous EBLCL, and also to the DR4+ EBLCL derived from patient WC (who has reactive arthritis, rather than RA). In contrast, DR4+ EBLCL from either normal individuals (MS and DG) or RA patients (LH, NB, GP) were not recognised. Patients LH, NB, and GP all have classical rheumatoid factor-positive RA, whereas GOP has a rheumatoid factor-negative, somewhat atypical, form of RA. When the same stimulator cells were tested using two independently generated PBMC-derived T cell lines, all of the DR4+ stimulator cells were recognised. Table 4 indicates that, like the synovial T cell line, the proliferation of the PBMC-derived T cells could be inhibited most effectively by HLA-DR specific mabs. Since it was possible that prolonged stimulation with autologous EBLCL had selected T cell lines with the properties observed, the times of culture of the synovial- and PBMC-derived T cell lines were compared. In expt. 1 the GOP synovial T cell line was tested after 89 days in culture, but only 43 days after first exposure to autologous EBLCL. The GOP PBMC T cell line was assayed 49 days after first stimulation with autologous EBLCL. In expt. 2 the GOP synovial T cell line was tested 41 days, and the GOP PBMC T cell line 29 days, after first stimulation with EBLCL. Thus the T cells from the different sources were exposed to autologous EBLCL for broadly similar periods of time.

TABLE 4. Inhibition of GOP PBMC-derived T cell line proliferation by monoclonal antibodies.

Antibody		Proliferation (d.p.m.)
None		40807
Anti-DR (L243)	0.05ug/ml	33421
	0.5ug/ml	15910
	5.0ug/ml	9408
Anti-DQ (Leu 10)	0.05ug/ml	29637
	0.5ug/ml	31691
	5.0ug/ml	32772
Anti-ABC (W6/32)	0.1ug/ml	34285
	0.5ug/ml	26589
	1.0ug/ml	23701
	5.0ug/ml	20494

DISCUSSION

The association between autoimmune diseases and class II MHC antigens implies that T cell-mediated responses are important in the pathogenesis of those diseases. In this respect the properties of the T cells at the site of disease may be especially relevant. This approach has been utilised in the case of thyroiditis, where there is some evidence of T cells which recognise thyroid-specific antigens within the thyroid[9]. In RA the critical antigens which are recognised within the synovium are not known, and it will be important to define these. However, in this study we have used T cell lines, derived from the synovium, which respond to polymorphic (MHC) antigens on autologous stimulator cells. Whether these MHC antigens are restricting elements for the recognition of an unknown nominal antigen, or are recognised directly by the T cells ("autoreative" recognition) is not clear. Antigens derived from foetal bovine serum were not responsible for the proliferation, and although in some cases the proliferative response was seen with autologous EBLCL rather than other autologous stimulator cells, this observation is more likely to reflect the high level of expression of MHC antigens on EBLCL rather than the recognition of virus-specific antigens.

T cells which appear to recognise autologous MHC antigens have been described in many experimental situations in both man and rodents[10,11]; some antigen-specific cells show a partial reponse to autologous MHC alone[12], and "autoreactive" cells are commonly found amongst polyclonal T cell populations responding to a known antigen[13]. These T cells may have regulatory functions.

Irrespective of their true antigen specificity, the present study has shown that synovial T cells may be used to investigate and define MHC antigens. This approach may indeed be especially relevant, since the determinants which are recognised by autologous T cells are those which function physiologically, whereas the significance of determinants defined by serology or by allo-reactive T cells is less clear. In addition, the differences observed between T cells from the synovium and those from the peripheral blood may indicate that the particular MHC determinants are selected from amongst all of the autologous class II HLA antigens by T cells responding to a particular antigenic stimulus, or carrying out a particular immunoregulatory role. A combination of cellular and molecular techniques may thus be required to define such MHC determinants, and to investigate their prevalence amongst RA patients compared with normal individuals, or patients with arthropathies which are clinically distinct from RA.

CONCLUSIONS

1. T cell lines derived from synovium specifically recognise autologous class II MHC antigens, either as restricting antigens or directly.

2. Synovial T cells can be used to distinguish between serologically identical antigens, particularly within HLA-DR4. The distribution of such T cell-defined MHC determinants in normal and arthritis patient populations may help to define disease-susceptibility MHC antigens more exactly.

3. The difference between the MHC determinants recognised by synovial and PBMC T cells cultured under identical conditions might reflect prior in vivo selection of T cells which recognise particular MHC determinants. These may be the restricting elements for recognition of antigen by T cells within the synovium, or determinants which can be recognised by "auto-reactive" T cells.

REFERENCES

1. P. Stasny, Association of the B cell alloantigen DRw4 with rheumatoid arthritis, N. Engl. J. Med. 298:869 (1978).

2. R.H. Schwartz, Immune response (IR) genes of the murine major histocompatibility complex, Adv. Immunol. 38:31 (1986).

3. C.L. Young, T.C. Adamson, J.H. Vaughan, and R.I. Fox, Immunologic characterization of synovial membrane lymphocytes in rheumatoid arthritis, Arthritis Rheum. 27:32 (1984).

4. D.G. Malone, S.M. Wahl, M. Tsokos, H. Catell, J.L. Decker, and R.L. Wilder, Immune function in severe, active rheumatoid arthritis, J. Clin. Invest. 74:1173 (1984).

5. B.S. Nepom, G.T. Nepom, E. Mickelson, J.G. Schaller, P. Antonelli, and J.A. Hansen, Specific HLA-DR4-associated histocompatibility molecules characterize patients with seropositive juvenile rheumatoid arthritis, J. Clin. Invest. 74:287 (1984).

6. S.C. Holbeck, S.J. Kim, J. Silver, J.A. Hansen, G.T. Nepom, HLA-DR4-associated haplotypes are genotypically diverse within HLA, J. Immunol. 135:637 (1985).

7. P.K. Gregersen, M Shen, Q.L. Song, Molecular diversity of HLA-DR4 haplotypes, Proc. Natl. Acad. Sci. (USA) 83:2642 (1986).

8. S.A. Rosenberg, E.A. Frimary, M. McGrogan, M. Doyle, E. Kawaski, K. Koths, and D.F. Mark, Biological activity of recombinant interleukin-2 produced in Escherichia coli, Science 223:1412 (1984).

9. M. Londei, G.F. Bottazzo, and M. Feldmann, T cell clones from autoimmune thyroid glands: specific recognition of autologous thyroid cells, Science 228:85 (1985).

10. P.B. Hausmann and J.D. Stobo, Specificity and function of a human autologous reactive cell, J. Exp. Med. 149:1537 (1979).

11. L.H. Glimcher, D.L. Longo, I. Green, and R.H. Schwartz, Murine syngeneic mixed lymphocyte response. I. Target antigens are self Ia molelcules, J. Exp. Med. 154:1632 (1981).

12. G.A. Dos Reis and E.M. Shevach, Analysis of autoreactive I region-restricted T cell colonies isolated from guinea pig syngeneic mixed leukocyte reaction and from immune responses to conventional foreign antigens, Eur. J. Immunol. 15:466 (1985).

13. H. Kotani, H. Mitsuya, R.F. Jarrett, G.G. Yenokida, S.P. James, and W. Strober, An autoreactive T cell clone that can be activated to provide both help and suppressor function, J. Immunol. 136:1951 (1986).

ACTIVATED INFLAMMATORY T LYMPHOCYTES IN RHEUMATOID SYNOVITIS - EFFECT OF CYCLOSPORINE A AND ANTI-TAC

Ø. Førre, K. Waalen, and J. B. Natvig

Oslo Sanitetsforenings Rheumatism Hospital, Akersbakken 27, and Institute of Immunology and Rheumatology, University of Oslo, Fr. Qvamsgt. 1, 0172 Oslo 1, Norway

One of the characteristic features in rheumatoid arthritis (RA) is a chronic inflammation with an extensive infiltration of lymphocytes in the synovitis in the joints of these patients. Immunohistological studies of rheumatoid synovitis have revealed that the infiltrating cells are organized in follicles-like structures with inflammatory T cells of the CD4 phenotype localized in close contact with strongly HLA-DR positive, dendritic-like cells (2,3). A similar infiltration of plasma cells have also been described (4). In recent studies some investigators have reported that synovial tissue (ST) and synovial fluid (SF) T lymphocytes are activated as shown by high incorporation of ^{3}H-labelled thymidine, and by expression of various activation markers (5,6). In contrast, other investigators have reported a defective T cell system in RA patients as shown by low or deficient production of interleukin 2 (IL-2) (7). Very recent studies have shown that lymphoid dendritic cells (DC) isolated from inflammatory sites like the ST and SF are very potent stimulators of T cells in autologous and allogeneic mixed leucocyte reaction (MLR) and are very effective accessory cells for antigenic and mitogenic T cell responses (8,9,10). To assess whether the rheumatoid patients have a defective or an activated T cell system, inflammatory T cells from rheumatoid synovitis were compared with normal T cells activated by blood DC in autologous MLR with respect to activation markers, and to production and consumption of IL-2. The effect of anti-Tac (anti-CD25) and cyclosporine A was also investigated. T cells (more than 98% pure) isolated from ST, SF and peripheral blood (PB) of RA patients and from normal PB were analyzed for expression of HLA-DR antigens, receptors for transferrin (TfR) and interleukin 2 receptors

(CD25). The following mean percentages of T cells positive for HLA-DR, TfR and CD25 were in ST: 32.3, 17.5 and 17.1%; in SF: 25.6, 5.3 and 10.4%; in RA PB: 3.8, 1.1 and 1.9%; and in normal PB: 2.1, <1 and 1.0%, respectively (6). Activation of resting T cells by DC in MLR gave a similar expression of these markers as in ST and SF with 31.8, 11.2 and 18.8%, respectively. Activation by DC were significantly higher than activation by monocytes with 6.4, 1.9 and 1.7%, respectively (6). In addition, we found that a significantly increased number of synovial inflammatory T cells expressed both early (TLiSA1) and late (VLA1) T cell activation antigens with 7.1 and 10.5%, respectively, compared to less than 1% in control PB T cells. Isolated ST and SF T cells also had high spontaneous incorporation of ^{3}H-labelled thymidine as compared to RA and normal PB T cells with 3,958, 3,448, 164 and 117 cpm, respectvely (6). The spontaneous poliferation could be further increased by addition of exogeneous recombinant interleukin 2 to 11,812 and 19,856 cpm compared to 2,630 and 820 cpm by RA and normal PB T cells, respectively. The spontaneous poliferation of ST and SF T cells could be inhibited by 47 to 53% by the anti-Tac antibody, and by 65% using cyclosporine A (200 ng/ml). When analyzing ST mononuclear cells (MNC) for production of IL-2 after stimulation with PHA (10 µg/ml) a content of 12.0 U/ml of IL-2 was detected in the culture supernatant compared to 37.5 U/ml in supernatants from normal MNC. However, the IL-2 production by ST MNC could be divided into two groups, one with very low (3.8 U/ml) and one with high or normal IL-2 content (39.5 U/ml). Also when analyzing the IL-2 content in supernatants from PHA stimulated T cells preactivated in MLR by DC a very low amount of IL-2 (<2 U/ml) was found compared to the pared normal control (20.3 U/ml) (6). To account for the observed low content of IL-2 in the supernatants above the ability of T cells purified from ST and SF and of T cells activated by DC in MLR to absorb IL-2 from a standard IL-2 supernatant (= 100%) was tested. Significant absorption of IL-2 was obtained both by ST and SF T cells and by T cells activated by DC. The remaining IL-2 activity in the supernatants was 31.2, 81.8 and 78.4%, respectively, compared to 98.3% after absorption with the CD25 negative CEM T cell line (6). These studies show that synovial inflammatory T cells express activation markers and produce and consume IL-2 to a similar extent as T cells activated by DC in MLR. Thus we conclude that the RA patients have an activated T cell system.

REFERENCES

1. G. Janossy, O. Duke, L. W. Poulter, G. Panayi, M. Bofill, and G. Goldstein. Lancet 2:839 (1981).
2. Ø. Førre, J. Thoen, J. H. Dobloug, O. J. Mellbye, J. B. Natvig, J. Pahle, and B. G. Solheim. Scand. J. Immunol. 16:315 (1982).
3. T. G. Abrahamsen, S. S. Frøland, J. B. Natvig, and J. Pahle. Scand. J. Immunol. 5:1057 (1976).
4. E. Munthe, and J. B. Natvig. Clin. Exp. Immunol. 12:55 (1972).
5. G. R. Burmester, B. Jahn, M. Gramatzki, J. Zacker, and J. R. Kalden. J. Immunol. 133:1230 (1984).
6. K. Waalen, Ø. Førre, M. Linker-Israeli, and J. Thoen. Scand. J. Immunol. 25:367 (1987).
7. B. Baron, and N. Talal. Clin. Exp. Immunol. 59:520 (1985).
8. K. Waalen, J. Thoen, Ø. Førre, T. Hovig, J. Teigland, and J. B. Natvig. Scand. J. Immunol. 23:233 (1986).
9. K. Waalen, Ø. Førre, J. Pahle, and J. B. Natvig. Scand. J. Immunol. 23:273 (1986).
10. K. Waalen, Ø. Førre, J. Teigland, and J. B. Natvig. Clin. Exp. Immunol. 69:1 (1987).

INHIBITION OF SPLENIC T LYMPHOCYTES AND ACCESSORY CELL INTERACTION BY ANTI-MAC-1 (MONOCLONAL ANTIBODY TO COMPLEMENT RECEPTOR TYPE 3)

S. El Rouby and M. Papiernik

INSERM U 25, Hôpital Necker

161 Rue de Sèvres, 75015 Paris, France

INTRODUCTION

Macrophages express a family of structurally and functionally related glycoproteins: LFA-1, Mac-1 and P150,95, which are involved in cell adhesion (1). Each of these molecules contains a different α subunit but they share an identical β subunit (1-2). LFA-1 mediates the attachment of macrophages to tumor cells and the binding of cytotoxic cells to targets (3,1,4). Mac-1 antigen is identical to complement receptor type 3 (CR3) which mediates adhesion of monocytes and granulocytes to C3bi-sensitized particles (5,6). In addition recent findings suggest the role of Mac-1 in the activation of mouse peritoneal macrophages (7).

We have recently described a direct cell-cell contact between thymocytes and thymic macrophages generated in a long-term thymic stroma culture (termed phagocytic cells of the thymic reticulum (8)) that was prevented by anti-Mac-1 monoclonal antibody (mAb) or C3bi (CR3 ligand) treatment (9). Splenic macrophages can similarly be isolated from splenic stroma culture (called phagocytic cells of the splenic reticulum or P-SR which are Mac-1^+, FcR^+ and Ia^-). These P-SR are able to bind to splenocytes (T and B cells) and form rosettes. We characterized the T cell subset which binds spontaneously to P-SR and found that their binding to P-SR was also abolished by anti-Mac-1 mAb. Contrary to what we observed in the thymus, the rosetting spleen T cells are almost undetectable in newborn mice and their number increases gradually with the age of the mice, reaching a maximal level (2% of the total population) at 8-12 weeks of age. We also showed that both $L3T4^+$ and Lyt-2^+ splenocytes bound to P-SR and that their binding was not restricted to the H-2 haplotype (10). The possibility was considered that T cell/P-SR interaction may reflect a role of Mac-1 antigen in T/accessory cell cooperation and consequently T cell activation. To test this hypothesis, we evaluated the effect of anti-Mac-1 mAb in the immunological assays which require direct cell-cell contact between T cells and accessory cells (e.g mixed lymphocyte culture and mitogeneic response). The present report shows the result of such an experiment.

MATERIALS AND METHODS

1. Animals

Different mouse strains (indicated in Table 1) were bred and maintained in our animal facilities. Female mice were used between 8-10 weeks of age.

2. Mixed lymphocyte reaction (MLR) and mitogeneic response

For MLR, 10^5 responder splenocytes (see Table 1 for mouse strain) were cultured with 5 x 10^5 irradiated stimulator cells (2000 rad) in a final volume of 200 µl/well culture medium {RPMI 1640 (GIBCO) containing 1% L glutamine, 1% sodium pyruvate, 100 U/ml penicillin, streptomycin and 2.5% human serum (groupe AB)}. For mitogeneic response, 5 x 10^4 splenocytes prepared from DBA/2 mice were stimulated with Con A (2.5 µg/ml) in 100 µl/well of the culture medium. Anti Mac-1 mAb rat IgG (clone M1/70) Seralab, Crawley-Down, UK) was added to the culture medium at the beginning of culture at a total concentration of 20 µg/ml. A similar concentration of purified rat IgG (Nordic Immunology, Tilburg, the Netherlands) was used as a control for specificity for rat IgG. Cell growth was evaluated by ^{3}H thymidine incorporation on day 4 (allogeneic response) and day 3 (mitogeneic response) of culture. Thymidine (2 µCi, Amersham, UK) was added during the last 16 hr of culture.

RESULTS

Blocking effect of anti-Mac-1 mAb on the mixed lymphocyte reaction and mitogeneic response :

Results presented in Table 1 show that anti-Mac-1 inhibited T cell proliferation in the MLR. 20 µg/ml (a dose which was found to inhibit T cell binding to P-SR) were sufficient to inhibit 83% of the response. The same dose of rat IgG did not modify the response. Anti-Mac-1 also blocked T cell proliferation during Con A stimulation (70% of the response, Table 2).

Table 1. Effect of anti-Mac-1 on T-cell proliferation in the MLR cpm

Exp.	Stimulators (cell origin)	Responders (cell origin)	MLR	MLR + rat IgG	MLR + anti-Mac-1	% inhibition
1	445 ± 118* (DBA/2 H-2^d)	641 ± 28 (C57 H-2^b)	17,499 ± 464	16,545 ± 103	1,947 ± 208	88,8
2	367 ± 91 (C3H H-2^k)	2197 ± 321 (DBA/2 H-2^d)	41,713 ± 2133	37,700 ± 1343	7,239 ± 1342	83
3	612 ± 131 (C57 H-2^b)	800 ± 80 (DBA/2 H-2^d)	123,224 ± 2215	N.D	24,699 ± 4544	80

* Mean value ± SEM for triplicate culture

Table 2. Effect of anti-Mac-1 on the mitogeneic response cpm

	Cells	Cells + Con A	Con A + rat IgG	Cells + Con A + anti-Mac-1	% inhibition
Exp.1	1033	75 267	72 711	17 841	76
	± 109*	± 1110	± 1736	± 327	
Exp.2	103	96 246	N.D	34 083	65
	± 85	± 5049		± 2587	

* Mean value ± SEM for triplicate culture

DISCUSSION

The aim of this study was to determine the role of Mac-1 antigen in T-cell activation during lectin or specific antigen stimulation, reactions which require the participation of accessory cells. Our results show that monoclonal antibody to Mac-1 antigen can block T cell proliferation during allogeneic and mitogeneic responses. Mac-1 antigen and the LFA-1 molecule are similar in having an identical β chain. However, contrary to LFA-1, Mac-1 antigen is not expressed on the lymphocytes, and its expression is restricted to monocytes, granulocytes, macrophages and dendritic cells (11,12,13). Thus the mAb used in these experiments appeared to act directly on the accessory cells. These results are too preliminary to analyse the mechanism(s) of the blocking effect induced by anti-Mac-1. Whether anti-Mac-1 binds directly to CR3 on the accessory cells and affects their function or whether it prevents the attachment of T lymphocytes to accessory cells (through CR3), which subsequently may lead to the activation of the accessory cells, is not known. Recent data of Ding et al (14) have shown that suitable ligation of CR3 induces various aspects of macrophage activation. These findings support the possibility that T lymphocyte/CR3 interaction may induce the activation of accessory cells (e.g lymphokine secretion) and that the prevention of this interaction abolishes accessory cell function and consequently T cell activation. Studies are now underway to explore this hypothesis.

REFERENCES

1. T.A. Springer, D. Davignon, M.K. Ho, K. Kurzinger, E. Martz, and F. Sanchez-Madrid. Immunol. Rev. 68:171 (1982).
2. F. Sanchez-Madrid, J.A. Nagy, E. Robbins, P. Simon, and T.A. Springer. J. Exp. Med. 158:1785 (1983).
3. G.S. Strassmann, T.A. Springer, S.D. Somers, and D.O. Adams. J. Immunol. 136:4328 (1986).
4. D. Davignon, E. Martz, T. Reynolds, K. Kurzinger, and T.A. Springer. J. Immunol. 127:590 (1981).
5. D.I. Beller, T.A. Springer, and R.D. Schreiber. J. Exp. Med. 156:1000 (1982).
6. S.D. Wright, P.E. Rao, W.C. Van Voorhis, L.S. Craigmyle, K. Iida, M.A. Talle, G. Goldstein, and S.C. Silverstein. Proc. Natl. Acad. Sci. USA 80:5699 (1983).
7. A. Ding, S.D. Wright, and C. Nathan. J. Exp. Med. 165:733 (1987).

8. M. Papiernik, B. Nabarra, W. Savino, C. Pontoux, and S. Barbey. Eur. J. Immunol. 13:147 (1983).
9. S. El Rouby, F. Praz, L. Halbwachs-Mecarelli, and M. Papiernik. J. Immunol. 134:3625 (1985).
10. S. El Rouby. Doctorat d'Etat Thesis, Faculty of Science, University of Pierre et Marie Curie, Paris 6ème, France (1987).
11. T.A. Springer, G. Galfre, D.S. Secher, and C. Milstein. Eur. J. Immunol. 9:301 (1979).
12. M. Bekoff, T. Kakiuchi, and H.M. Grey. J. Immunol. 134:1337 (1985).
13. G. Schuler, and R.M. Steinman. J. Exp. Med. 161:526 (1985).

MYCOBACTERIUM TUBERCULOSIS ANTIGEN SPECIFIC HUMAN T-CELL LINES ARE CYTOLYTIC TO AUTOLOGOUS ANTIGEN PULSED MACROPHAGES

D.S Kumararatne[1], P. Drysdale[1], J.-S. Gaston[2], P. Stacey[3], P. Richardson[1], and R. Wise[3]

Department of Immunology[1] and Rheumatology[2], The Medical School, Birmingham B15 2TJ, and the Department of Microbiology[3], Dudley Road Hospital, Birmingham B18 7QH, U.K.

INTRODUCTION

Mycobacterium tuberculosis is a bacterial pathogen capable of survival and replication within human macrophages. Current ideas about antimycobacterial immunity grew out of the pioneering work of Lurie[1] and Mackaness[2]. In summary, it is believed that acquired resistance to tuberculosis depends on T-cell mediated macrophage activation, resulting in an enhancement of the bacteriostatic or bacteriocidal capacity of the latter cells[3]. However, experimental confirmation of the above using human T-cells and macrophages has been very difficult (reviewed by Rook, 1987)[4]. Recent evidence indicates that murine T-cells possess cytolytic capacity against mycobacterial antigen bearing macrophages[5]. Adoptive transfer of such L3T4 negative Lyt-2 positive cells induces protection against intracellular bacterial infections including tuberculosis[5,6]. Therefore, we have investigated the generation of human T-cell lines with cytotoxic capacity against mycobacterial-antigen-pulsed macrophages.

MATERIALS AND METHODS

Subjects: Six healthy adult human volunteers were studied. Four were Caucasians and two were of Asian origin.

Generation of T-cell lines: Peripheral blood mononuclear cells (PBMC) were separated from defibrinated blood by Ficoll-Hypaque centrifugation. These were adjusted to 10^6 cells per ml in RPMI 1640 (Gibco), supplemented with 20% autologous serum and incubated with M. Tuberculosis sonicate antigen[7] (kindly provided by Dr Anne Rees, MRC Tuberculosis Research Centre, U.K.) at an optimum concentration (5 g protein/ml) at 37° in 5% CO_2 in air for 7 days. Some T-cell lines were expanded for up to 4 weeks by weekly feeding with irradiated antigen presenting cells (PBM or autologous EBV transformed B-cell line), antigen and recombinant IL2 (Cetus corp) at optimum concentrations, as described in detail elsewhere[8]. These T-lines were cryopreserved until ready for use.

Antigen specificity of T-cell lines was tested in a standard ^3H Thymidine incorporation assay[8].

Preparation of macrophages: Plastic adherent cells obtained from PBMC[9] were cultured in 10% unheated autologous serum supplemented RPMI 1640 for 7 days in flat bottom microculture plates (Nunc) at a concentration of 10^5 cells/well.

Mycobacterial antigen dependent cytotoxicity assay: Graded numbers of T-cell blasts were added to 10^5 macrophages to give effector: target cell ratios of 0.2, 1 or 5 respectively. Replicate wells were set up with varying concentrations of MTSA (see results section) and IL2 (20 μ/ml). Control macrophage monolayers were set up omitting (a) T blasts or (b) antigen and IL2, respectively. After 48 hours' incubation target cell lysis was estimated by a photometric method[10] of measuring binding of dye (0.5% Napthol black) to the macrophage monolayer. The principle of this method is the removal of T-cell blasts and dead macrophages by rinsing with tissue culture medium. The binding of dye to the residual adherent cells is a measure of the surviving number of macrophages.

BCG infection of target macrophages: In some experiments the macrophages used as targets in the cytotoxicity assay were infected 48-72 hours earlier with small numbers (bacteria to macrophage ratio 1:1) of M. bovis (BCG strain) (Glaxo Ltd) as described in detail elsewhere[3,9].

RESULTS

Table 1 summarises the surface phenotypic data of 5 of the T-cell lines used for the cytotoxicity experiments.

Table 1. Surface phenotyping data for polyclonal MTSA induced T-cell lines used in cytotoxicity experiments

Surface marker analysis was done by a sensitive Rosette assay[11] using monoclonal antibody[12] coated indicator red cells.

Antibody[12]	Specificity[12]	% positive cells Donor number				
		1	2	3	4	5
OKT3	CD3	74	68	98	82	74
OKT11	CD2	90	92	97	92	91
OKT4	CD4	36	52	80*	19	0
B9.4.1	CD8	46	40	75*	52	50
Bu26	(MHC Class 11)	60	50	60	74	90
8EB1	CD19	4	12	2	4	0
OKT10	CD38	60	75	94	86	84
AC2	CD39	83	54	69	86	60
N2C1	CD9	12	42	4	26	16
HNK1	(?)	10	28	35	27	56

*Note: Double expression of CD4 and CD8. These assays were done with a rosette assay[11] which is more sensitive than immunofluorescence and hence may detect very low levels of antigen expression.

MTSA responsive T-cell lines are cytotoxic to antigen pulsed autologous macrophages

Results summarised in figure 1 and table 2 show that human T-cell lines generated against M. tuberculosis sonicate antigen are cytolytic to autologous macrophages in the presence of antigen and IL2. This cytolysis did not occur in the absence of antigen or IL2; antigen and IL2 on their own did not cause the death of the macrophages, either (data not shown).

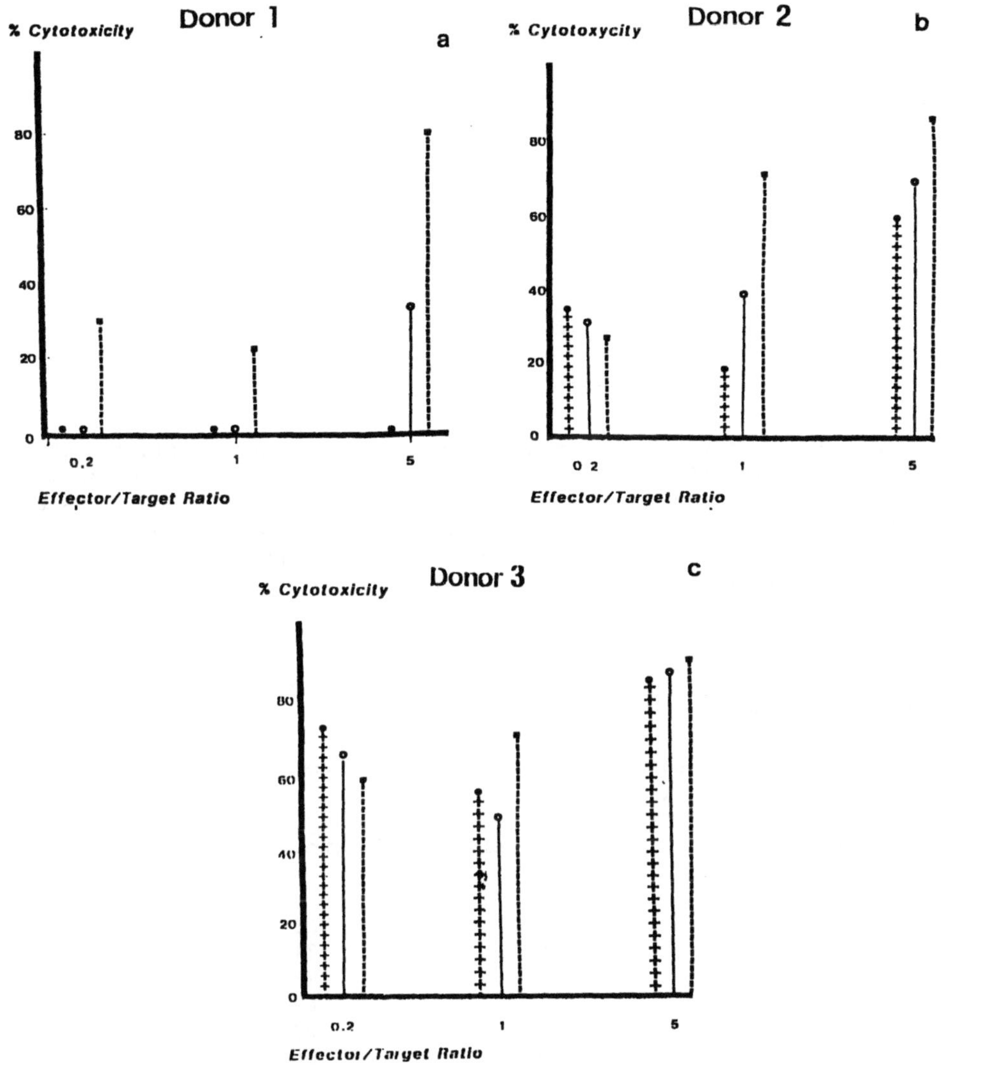

Figure 1. Cytolytic activity of M. Tuberculosis antigen reactive T-cell lines against autologous macrophages. Fig 1 a-c shows the results of three representative experiments done with cells from 3 separate healthy adult Caucasian donors. Cytotoxicity has been tested at three effector target ratios of 0.2, 1 and 5 respectively. For each E/T ratio the three bars represent the degree of cytolysis at antigen concentrations of 0.05μg/ml (‡) 0.5μg/ml (♀) and 5μg/ml (▮) respectively.

The results of the above cytotoxicity experiments were confirmed by direct morphological assessment of the macrophage monolayers. Figure 2 shows a photograph of macrophage monolayers at the end of a cytotoxicity experiment.

Note that the control macrophages treated with antigen and IL2 are intact (fig 2a), while these cells have been in killed cultures containing T-cell lines (fig 2b). Fig 2c shows that most macrophages have survived but have undergone morphological alteration in the presence of an allogeneic T-cell line (macrophage donor 4, effector cell donor 1 as in table 2).

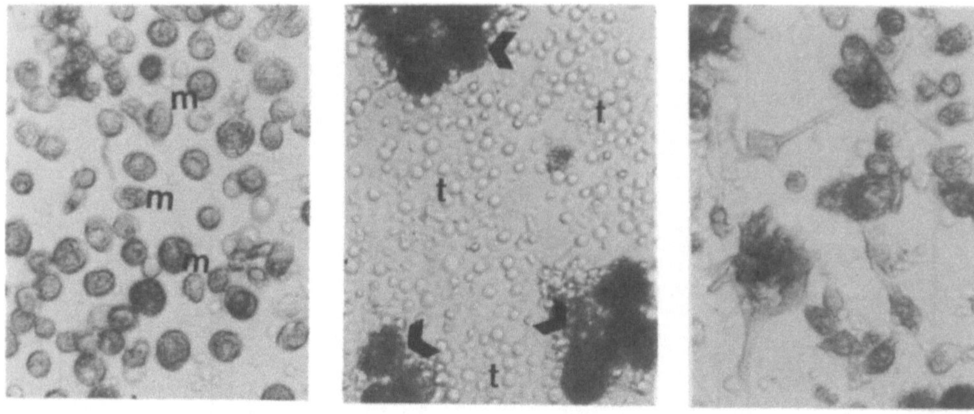

A B C

Figure 2. Phase contrast photomicrographs of macrophage monolayers at end of cytotoxicity experiment detailed in table 2. 2a: Macrophages from donor 5, no T-cells added. x 400. 2b: Macrophages from donor 5, T-cell lines from same donor added. x 400. 2c: Macrophages from donor 4, T-cell lines from donor 1 added. x 400. Key: macrophage - m, t lymphoblast - t, clump of dead macrophages - arrow.

Table 2 The cytotoxicity of MTSA-reactive Polyclonal T-cell lines against MTSA antigen pulsed macrophages.

(All cell donors were healthy adults and were Caucasian except for donors 4 and 5 who were Asian).

Macrophage donor	Effector cell donor	% Cytotoxicity	
1	1	22	
2	2	71	
3	3	72	
4	4	69	
5	5	69	
6	6*	88	
4	1*	12	) Allogeneic
4	3*	62	) combination

In all experiments initial effector/target ratio was 1:1 and the antigen concentration used 5 μg/ml, and the IL2 concentrations was 20 μ/ml.

NB: Antigen and IL2 alone or T cells without antigen and IL2 did not produce cytotoxicity (data not shown).

* These were cryopreserved T cell lines obtained by pulsing with antigen and IL2 for 3-4 weeks.

<u>Fate of BCG infected macrophages when treated with MTSA reactive T-cell lines</u>

BCG infected macrophages are lysed by MTSA reactive T-cell lines when co-cultured for 5-7 days in the presence of antigen (5 μg/ml) and IL2 (20 μ/ml). When these cell preparations were stained by the Ziehl Nielsen method at the end of the co-culture period very few surviving

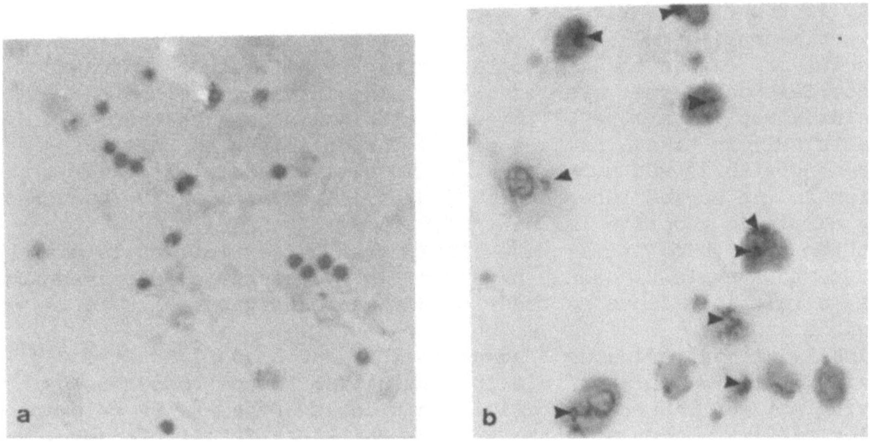

Figure 3. Ziehl Nielsen stained preparatons of cells to show intracellular BCG bacilli (arrowed) (See text for explanation).

macrophages were seen in the T-cell treated cultures (fig 3a) while the control cultures contained numerous macrophages full of intracellular BCG bacilli (fig 3b). These cell preparations were stained with fluorescein diacetate (FDA) and ethidium bromide (EB). FDA stains viable mycobacteria green-yellow and EB stains dead bacilli red when viewed by epifluorescent illumination[13]. In such preparations most of the intracellular BCG (60-70%) in control macrophage cultures appeared to be viable (i.e. stained green). T-cell treated cultures showed sparse extracellular bacilli, many of which appeared to be viable (i.e. stained green).

DISCUSSION

The results presented here show that M. tuberculosis antigen reactive T-cell lines can kill autologous macrophages in the presence of M. tuberculosis sonicate antigen and IL2. A significant amount of cytotoxicity can be seen at relatively low effector/target ratios (0.2 to 5) and varies with antigen concentration in a dose dependent manner.

Preliminary results show that the number of viable (FDA staining[13]) intracellular M. bovis (BCG) are decreased in T-cell treated macrophage cultures, in the presence of MTSA and IL2. However, since extracellular bacilli might be more easily washed off in making stained preparations from these cultures this observation needs to be confirmed by direct estimation of colony forming units. Such experiments are in progress.

Murine T-cells which are cytolytic against mycobacterial antigen pulsed macrophages are of CD4 or CD8 phenotype 5. Cytolysis by the former is MHC class II restricted while the latter are restricted by the class I locus[5]. Human cytotoxic T-lines showing tuberculin (PPD) specific cytotoxicity against monocyte targets have been described previously by Hansen et al (1986)[14]. Cytolysis by these T-lines appear to be MHC class II restricted. It would be relevant to examine the ability of cloned mycobacterial antigens to induce such cytolytic T-cells and to repeat the current work using defined T-cell clones.

The in vivo significance of cytolytic T-cells to antimycobacterial immunity is uncertain. Virulent mycobacteria possess the ability to resist antimicrobial mechanisms of mononuclear phagocytes[15]. Lysis of these cells may expose the intracellular bacteria to other (unknown) antibacterial mechanisms. For example during in vitro culture mycobacteria replicate more efficiently within macrophages than extracellularly in tissue culture medium, especially when this is supplemented with 10-20% human serum[9]. In experimental listerial infection and in murine tuberculosis CD8 positive T-cells have been shown to play a role in adoptive immunity[5,6].

On the other hand it is quite likely that cytolytic T-cells may contribute to the tissue damage occurring in mycobacterial diseases and may play a role in caseous necrosis or damage to Schwann cells in tuberculoid leprosy.

Finally, recent evidence[16] suggests that MHC class II restricted cytotoxic T cells may play a role in terminating immune responses by destroying antigen presenting cells. This could of course be relevant for physiological immunoregulation as well as M. leprae specific anaergy which occurs in Lepromatous leprosy[17].

ACKNOWLEDGEMENTS

This work was supported by generous grants from the Endowment Fund of the United Birmingham Hospitals and the West Birmingham Health District Research Fund.

REFERENCES

(1) Lurie M B (1942). J. Exp. Med. 75, 247.
(2) Mackaness G B (1969). J. Exp. Med. 129, 973.
(3) Hahn H, Kaufmann S H (1985). REv. Inf. Dis. 3, 1221.
(4) Rook G A W (1987). Clin. Exp. Imm. 69, 1.
(5) Kaufmann S H, Chiplunkar S, Fleisch I, De Libero G (1986). Lep. Rev. 57 Suppl 2, 101.
(6) Orme I (1987). J. Immunol. 138, 293.
(7) Matthews R, Scoging A, Rees A D M (1985). Immunology 54, 17.
(8) Gaston J S H, Rickinson A B, Epstein M A (1983). J. Exp. Med. 158, 1804.
(9) Hassan N F, Campbell D E, Douglas S D (1986). J. Imm. Methods 95, 273.
(10) Leu R W, Herriott M J (1984). J. Imm. Methods 67, 63.
(11) Ling N R, Richardson P (1981). J. Imm. Methods 47, 265.
(12) Leucocyte Typing III. (1987). Ed. A.J. McMichael et al, Oxford University Press.
(13) Odinsen O, Nilson T, Hunber D P (1986). Int. J. Leprosy 54, 403.
(14) Hansen P and Kristensen T (1986). Tissue Antigens 27, 217.
(15) Moulder J W (1985). Microbiol. Rev. 49, 298.
(16) Braakman G, Rotteveil F T M, Bleek van G, et al (1987). Immunology Today 8, 265.
(17) Mustafa A S, Godal T (1985). Clin. exp. Immunol. 62, 474.

CORRELATION OF IL2 PRODUCTION AND LYMPHOID TISSUE MAKE-UP IN TWO INBRED STRAINS OF RATS

M.Mostarica-Stojković, S.Vukmanović, and M.L.Lukić

Institute of Microbiology and Immunology, School of Medicine, University of Belgrade and Institute for Biological Research, Belgrade, Yugoslavia

INTRODUCTION

Strain differences in susceptibility to the experimentally induced autoimmunity provide valuable tools for analysing and understanding cellular and genetic deffects leading to autoimmunity in humans. We have shown that AO and DA inbred strains of rats differ in susceptibility to the induction of several experimental T cell dependent autoimmune diseases, the former being resistant and latter susceptible to these disorders. In search for the cellular basis of the observed differences and in view of the central role played by Interleukin 2 (IL2) in regulating T cell response,we have analysed the pattern of IL2 production in DA and AO strain and found striking difference between these two strains (1).Supernatants obtained from ConA stimulated AO lymphocytes revealed lower IL2 activity when compared with preparations derived from DA cells (1). It appeared that a gene not linked to the MHC exerts the major control on the level of IL2 production in this strain combination (2).

It remains to elucidate whether the differences in susceptibility to autoimmunity and in the level of IL2 production are related to different control mechanisms and cellular make-up of lymphoid tissue or to the inherent capacity of AO and DA T cells subsets. Additionally, we followed the effect of elimination of Ts precursors by low dose of cyclophosphamide (Cy) on these parameters.

MATERIALS AND METHODS

Animals. Inbred AO and DA rats, when 12-16 weeks old, were used in the experiments.

Treatment with Cy. Freshly prepared solution of Cy was given i.v. in the dose of 20 mg/kg body weight.

IL2 production and assay. IL2 containing supernatants were obtained and assayed as previously described (1).

Determination of T cell subsets. The proportion of phenotypically distinct T cell subsets was determined in lymph node, spleen and peripheral blood using indirect immunofluorescence. Spleen and peripheral blood mononuclear cells were obtained using density gradient centrifugation on Lymphoprep (Pharmacia). For T cell analysis the following monoclonal antibodies (kindly donated by Dr D.W.Mason, Sir William Dunn School of Pathology, University of Oxford, UK) were used: MRC OX19, recognizing all peripheral nonactivated T cells and thymocytes; MRC OX8, binding to suppressor/cytotoxic subsets of T cells, and W3/25 directed against helper/inducer subset of T cells (3). 1×10^6 cells were incubated with appropriate dilution (1:100) of the above monoclonal antibodies for 60 min at 4°C. A FITC-conjugated sheep anti-mouse Ig (INEP) was used as a second antibody. The cells were analysed either visually using fluorescent microscope or by cytofluorometric analysis on the automated cell sorter EPICS-C. For each specimen analysed, the percentage of T ($OX19^+$), helper/inducer ($W3/25^+$) and suppressor/cytotoxic ($OX8^+$) cells were determined and the ratio of W3/25 to OX8 positive cells was calculated.

RESULTS AND COMMENTS

In order to find out whether differences in susceptibility to T cell dependent autoimmunity and capacity to produce IL2 stemed from the differences in T cell population of AO and DA rats, we analysed T cell subsets and $W3/25^+$:$OX8^+$ (Th/Ts) ratio in lymphoid tissues of these strains. As shown in Table I, there was no significant difference in the percentage of $OX19^+$ cells in the peripheral lymphoid tissue indicating that the relative number of T cells do not differ in AO and DA rats. However, AO and DA rats exhibited significant differences in the percentage of $OX8^+$ cells in the analysed lymphoid tissues. Consequently, W3/25:OX8 ratios were always lower in AO rats when compared with DA strain.

Table I

T cell subsets in peripheral lymphoid tissues of AO and DA rats

Tissue	Strain	OX19	W3/25	OX8	Th/Ts
Spleen	AO	74.2±3.2[a]	39.2±1.5	26.8±2.2	1.46±0.10
	DA	64.1±1.2	43.1±1.1	16.8±1.1	2.58±0.23
Lymph node	AO	69.5±2.1	34.4±3.2	17.0±3.9	1.95±0.68
	DA	62.4±3.1	34.0±2.4	11.9±2.1	2.98±0.67
Blood	AO	74.5±1.8	46.6±6.4	24.9±3.1	1.92±0.46
	DA	76.9±7.6	48.0±4.1	17.1±1.3	2.83±0.40

[a] Percentage of positive cells (M±SD)

Further, the simultaneous analysis of IL2 production and Th/Ts ratio in these strains indicated that AO rats, showing lower level of IL2 production have also much lower Th/Ts ratio when compared with DA rats (Table II).

We have previously shown that low dose of cyclophosphamide abolished the resistance of AO rats to the induction of EAE (4). Therefore, we tested the effect of the same regimen on the IL2 production and T cell subsets in peripheral lymphoid tissues of AO rats. As shown in Table III, the procedure that facilitates the induction of T cell dependent autoimmunity also induces striking increase in IL2 production and elevates Th/Ts ratio.

Thus, it appears that the lower level of IL2 production in EAE resistant rats is related to the presence of relativly high proportion of $OX8^+$ cells in their lymphoid tissues. It could be assumed that these $OX8^+$ cells suppress the expansion of EAE effector cells. It is clearly shown that IL2 is required for the expansion of autoreactive cells (5) and that

Table II

IL2 production and Th/Ts ratio in peripheral lymphoid tissues of AO and DA rats

	Spleen		Lymph node	
Strain	IL2	Th/Ts	IL2	Th/Ts
AO	69.9[a]	1.5[b]	19	1.8
DA	151.4	2.6	65.1	2.9

[a] U/ml

[b] Percentage of $W3/25^+$:percentage of $OX8^+$ cells as determined by cytofluorometric analysis

Table III

Effect of cyclophosphamide on IL2 production and Th/Ts ratio in AO rats

	Spleen		Lymph node	
Cy[a]	IL2	Th/Ts	IL2	Th/Ts
-	96.5[b]	0.8[c]	19	1.2
+	168	1.8	40	3.2

[a] Cy was given 10 days before testing [b] U/ml

[c] As determined by fluorescent microscopy

EAE producing T cell lines are of W3/25 phenotype (6).It is also well established that W3/25$^+$ cells or a subset of these cells are major producers of IL2 in rats (7). The putative mechanism by which OX8$^+$ cells in rats and Lyt2$^+$ cells in mice affect IL2 production is not understood. Data obtained by Hardt et al. (8) suggested that Lyt 2$^+$ cyclophosphamide-sensitive T cells produced IL2 inhibitor. More recently, Chiba et al. (9) have shown in rats that T cells responsible for the production of a lymphokine inhibitory for cellular DNA synthesis are OX8$^+$ and cyclophosphamide-sensitive. However, it is also possible that OX8$^+$ cells prevent the expansion of autoreactive T cells in AO rats simply by consuming IL2. Further studies are needed in order to elucidate the molecular mechanism by which OX8$^+$ cells exert their suppressive effects in low responder AO rats.

REFERENCES

1. M. Mostarica Stojković, L. Ejdus-Konstantinović, M. Kostić, and M.L. Lukić. Adv. Exp. Med. Biol. 186:713 (1985).
2. M.L. Lukić, M. Mostarica Stojković, M. Kostić, N. Tucić, and S. Vukmanović. Transpl. Proc. 19:3137 (1987).
3. D.W. Mason, R.P. Arthur, M.J. Dallman, J.R. Green, G.P. Spickett, and M.L. Thomas. Immunol. Rev. 74:57 (1983).
4. M. Mostarica Stojković, M. Petrović, and M.L. Lukić. Clin. Exp. Immunol. 50:311 (1982).
5. L. Ortiz-Ortiz, and W.O. Weigle. J. Immunol. 128:1545 (1982).
6. H.J. Schluesener, and H. Wekerle. J. Immunol. 135:3128 (1985).
7. R.P. Arthur, and D.W. Mason. J. Exp. Med. 163:774 (1986).
8. C. Hardt, M. Röllinghoff, K. Pfizenmaier, H. Mosmann, and H. Wagner. J. Exp. Med. 154:262 (1981).
9. K. Chiba, T. Nishimura, and Y. Hashimoto. J. Immunol. 134:3172 (1985).

p145-MR6: A NOVEL T CELL SURFACE MOLECULE WITH A ROLE IN TOLERANCE TO SELF

Mark Larche°, Jonathan R. Lamb* and Mary A. Ritter°

°Department of Immunology, Royal Postgraduate Medical School
*MRC Tuberculosis and Related Infections Unit
Hammersmith Hospital
Ducane Road
London W12 OHS
England

INTRODUCTION

The role of the thymus in the generation of the T cell repertoire is well established (Gelfand et al., 1980; Cantor and Weissman, 1976), but remains poorly understood. In order to study the molecular influences within the microenvironment of the thymus we raised monoclonal antibodies to the stromal components of the thymic microenvironment as described previously (DeMaagd et al., 1985; Ritter et al., 1985). This study focuses on one such antibody, MR6, which reacts strongly with the cortical epithelium and medullary macrophage/dendritic cells (MO/DC), and less strongly with thymocytes and T cells (Larche et al., 1987). Functional studies on this molecule (p145-MR6) lead us to believe that it may play a central role in the clonal "deletion" of autoreactive T cells within the thymus.

MATERIALS AND METHODS

The production and screening of MR6 has been described in detail elsewhere (DeMaagd et al., 1985). In Western blotting analysis, 10^8 thymocytes or $1cm^3$ of normal thymus were lysed in 300μl or 1ml respectively of 10mM Tris-HCl, 150mM NaCl pH 7.9 containing 0.5% Nonidet P-40 (NP40; SIGMA, U.K.) and 10μg/ml Aprotonin. After centrifugation, lysates were subjected to SDS-PAGE separation in a 10% gel and electroblotted onto nitro-cellulose membrane (Bio-Rad, Watford, U.K.). Unoccupied charged sites on the membrane were blocked and strips of membrane were then incubated with either MR6, an anti-class II antibody as a positive control, or an irrelevant antibody (anti-beta 2 microglobulin). After washing, strips were incubated with peroxidase-conjugated rabbit anti-mouse Ig (Dakopatts, Copenhagen, Denmark) and staining was developed using a combination of 3,3-diaminobenzidine and 4-chloro-1-napthol with imidazole (SIGMA, U.K.). For functional studies, T cells of clone HA 1.7 (Lamb et al., 1982) (10^4/well) were cultured with peptide 14 (p14; residues 306-319 of influenza HA-1 molecule; 3μg/μl) in the presence of inactivated autologous EBV-transformed B cells (▲; 10^4/well). MR6 (▲—▲) or anti-beta 2 microglobulin (△—△) (Becton Dickinson, CA, U.S.A.) were added at the initiation of these cultures in the concentrations indicated (Fig.3). Additionally, T cells of HA 1.7 were cultured alone with IL-2 (●) or in the presence of MR6 (●—●) or anti-beta 2 microglobulin (○—○). The response of T cells alone to MR6 (■—■) was also determined. After 60 hours, the cultures were pulsed with 1μCi of tritiated methyl thymidine (^{3}H-TdR; Amersham International, Bucks, U.K.) for 8-16 hours and then harvested onto glassfibre filters. Proliferation as correlated with ^{3}H-TdR incorporation was measured by liquid scintillation spectroscopy.

Within the same functional system, cells were removed and stained with antibodies to several cell surface molecules involved in T cell activation, in order to assess changes in the levels of these molecules. 10^6 cells were incubated on ice for 1 hour with primary antibodies, washed in PBS and further incubated with FITC-conjugated rabbit anti-mouse Ig (Dakopatts, Copenhagen, Denmark). Cells were then washed and analysed by flow cytofluorimetry (EPICS C; Coulter, U.S.A.).

RESULTS AND DISCUSSION

In frozen tissue sections of normal human thymus, MR6 stains the cortical epithelium and MO/DC in the medulla (DeMaagd et al., 1985; Ritter et al., 1985) (Fig.1). By flow cytofluorimetry MR6 also stains cells of the T lineage including thymocytes, T cell clones and long term T cell acute lymphoblastic leukaemias (T-ALL) (Larche et al., 1987) proving that the molecule is synthesised by T cells and not merely passively acquired from the cortical epithelium. Biochemical analysis by Western blotting shows that the molecule detected by this antibody has a molecular weight of 145,000; it has thus been provisionally termed p145-MR6 (Fig.2). In functional studies with T cell clone HA 1.7 and p14 (Fig.3), peripheral blood mononuclear cells (PBMC) with the soluble extract of Mycobacterium tuberculosis (MTSE), and several alloreactive T cell clones with their allo-MHC targets, we have shown that MR6 causes inhibition of cellular proliferation in response to both specific antigen and to IL-2. This inhibition is dose dependent and >50% inhibition is typically seen at 1-5μg/ml of MR6. In modulation experiments the levels of Tac (beta chain of

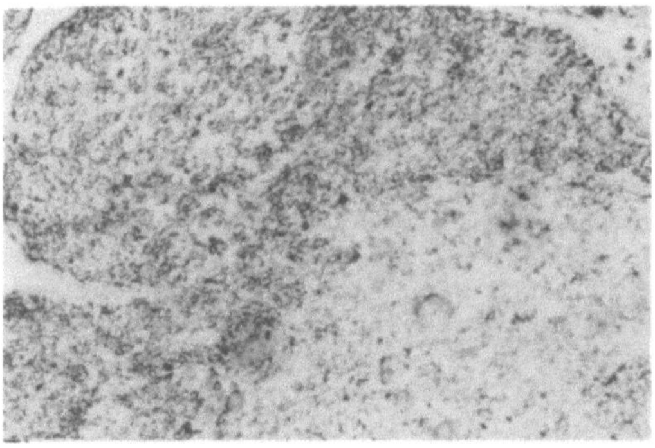

Fig. 1 Immunolabelling of frozen section of paediatric human thymus using an indirect immunoperoxidase technique. Monoclonal antibody MR6 binds to cortical epithelial cells and medullary macrophage/dendritic cells.

the IL-2 receptor) were found to be consistantly reduced after incubation of the cells with antigen in the presence of IL-2. This reduction was small (typically the proportion of stained cells was reduced by 12-17%); however, the inability of the anti-Tac antibody to distinguish between the low and high forms of the receptor may mask a more functionally significant reduction in the levels of the high affinity receptor whose expression appears to be some 50 fold lower than the low affinity form (Kondo et al., 1987).

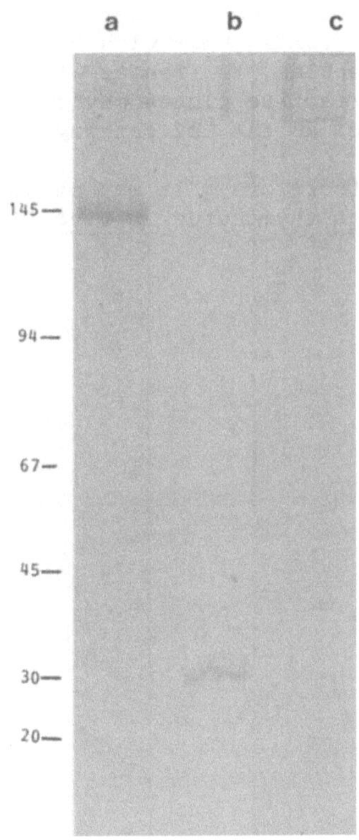

Fig. 2 Western Blotting analysis of the MR6 antigen: lane 1, thymocyte lysate with monoclonal antibody MR6; lane 2, thymocyte lysate with monoclonal antibody CR3/43 (binds to the beta chain of the HLA-DR family); lane 3, control liver lysate with monoclonal antibody MR6.

From these studies it appears that MR6 defines a previously unidentified molecule of relative molcular mass 145K which is present on the surface of T cells, and the cortical epithelium and medullary MO/DC of the thymus. When this molecule is bound by an artificial ligand (MR6) in culture, it inhibits cellular proliferation in response to antigen and to IL-2. This inhibition may be mediated by a change in the levels of the IL-2 receptor (IL-2R).

The physiological role of p145-MR6 in the thymus and on the surface of T-cells can only be speculated upon. It is, however, possible that autoreactive T cells that recognise and respond to self-MHC products, expressed either on thymic epithelial cells or on MO/DC, may be tolerized by the additional interaction of p145-MR6 with its ligand - as simulated with T cells in vitro. This clonal unresponsiveness may result from a relative lack of IL-2 during the cells' first contact with antigen (Malkovsky and Medawar, 1984), since the IL-2 receptor appears to be modulated during this interaction. It may also be possible within this system that thymocytes whose receptors for antigen react only relatively weakly with self-MHC products (and thus perhaps strongly with self + 'X') do not interact with the stromal cells with high enough affinity to engage sufficient p145-MR6 molecules to render them tolerant, these MHC-restricted but not autoreactive clones may then be expanded by antigen independant mechanisms such as the CD2 pathway (Reinherz, 1985).

From this study, therefore, it may be possible to explain clonal "deletion" of autoreactive thymocytes in vivo in terms of a lymphocyte-

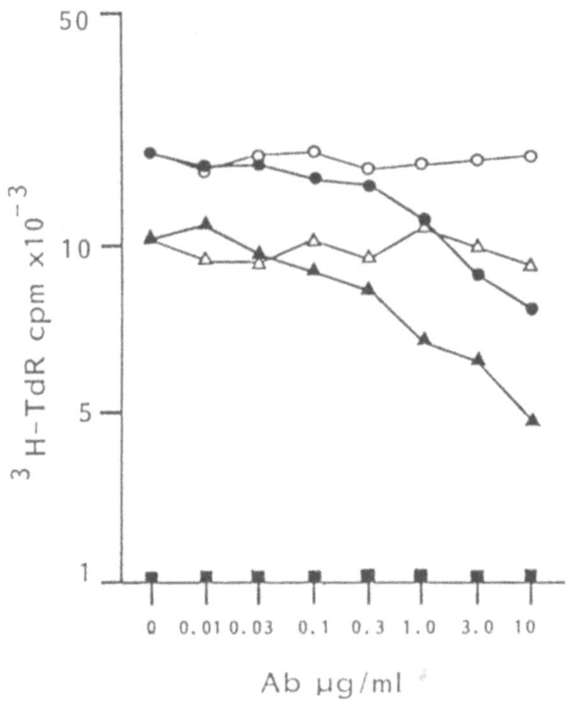

Fig. 3 Inhibition of T cell proliferation by monoclonal antibody MR6. MR6 inhibits both antigen (▲—▲) and IL2 (●—●) induced proliferation of cloned helper T cells (clone HA1.7). Addition of a control antibody, anti-beta 2 microglobulin does not inhibit either antigen (△—△) or IL2 (○—○) induced proliferation. MR6 in the absence of either antigen or IL2 does not induce proliferation (■—■).

microenvironmental interaction involving p145-MR6 on the T lymphocyte surface and either the same molecule (homophilic interaction) or a different ligand molecule (heterophilic interaction) on the surface of thymic cortical epithelium and medullary MO/DC. Finally the tissue and cellular distribution of p145-MR6 coupled with its molecular weight indicate that this is a previously unidentified molecule. Studies are currently underway to further characterise this molecule.

ACKNOWLEDGEMENTS

We thank Claire Green, Phil Mason, Nikki Dobson, Wilma Mackenzie and Mary Bowe for their help with various aspects of this manuscript. This work was supported by a grant from the Medical Research Council (no.98502997CA).

REFERENCES

Cantor, H. and Weissman, I., 1976, Development and function of subpopulations of thymocytes and T lymphocytes, Prog. Allergy, 20:1.

DeMaagd, R.A., Mackenzie, W.A., Schuurman, H-J., Ritter, M.A., Price, K.M., Broekhuizen, R. and Kater, L., 1985, The human thymus microenvironment: heterogeneity detected by monoclonal anti-epithelial cell antibodies, Immunology, 54:745.

Gelfand, E.W., Dosch, H-M., Shore, A., Limbatibul, S., and Lee, J.W.W., 1980, The role of the thymus in human T cell differentiation, in: "Biological Basis of Immunodeficiency", E.W. Gelfand and H-M Dosch, eds., New York Raven Press, p. 39.

Kondo, S., Kinoshita, M., Shimuzu, A., Saito, Y., Konishi, M., Sabe, H. and Honjo, T, 1987, Expression and functional characterisation of artificial mutants of interleukin-2 receptor, Nature, 327:64.

Lamb, J.R., Eckels, D.D., Lake, P., Woody, J.N. and Green, N.,1982, Human T cell clones recognise synthesised peptides of influenza haemagglutinin, Nature, 300:66.

Larche, M., Ladyman, H., and Ritter, M.A., 1987, Human thymic epithelial cells and lymphocytes may share common epitopes, in: "Leukocyte Typing III", A. McMichael et al., eds., Oxford University Press, Oxford, p.257.

Malkovsky, M. and Medawar, P.B., 1984, Is immunological tolerance (non-responsiveness) a consequence of interleukin 2 deficit during the recognition of antigen? Immunology Today, 5: 340.

Reinherz, E.L., 1985, A molecular basis for thymic selection: regulation of T11 induced thymocyte expansion by the T3-Ti antigen/MHC receptor pathway, Immunology Today, 6:75.

Ritter, M.A., Schuurman, H.J., MacKenzie, W.A., DeMaagd, R.A., Price, K.M., Broekhuizen, R. and Kater, L., 1985, Heterogeneity of human thymus epithelial cells revealed by monoclonal anti-epithelial antibodies, Advances in Experimental Mwedicine and Biology, 186: 283.

CELL TYPES INVOLVED IN NON-ADAPTIVE OR NATURAL IMMUNITY

SUPPRESSION OF IMMUNOGLOBULIN PRODUCTION BY GERMINAL CENTRE HNK-1+ CD3+ CELLS

D. Banerjee, J. Baril, D.A. Bell, D. McFarlane, and R. Karim

Department of Pathology and Medicine
University of Western Ontario, London, Canada

This study was funded by the Medical Research Council of Canada.

ABSTRACT

Germinal centre T cells co-expressing HNK-1 antigen have little lytic activity against NK targets (K562 cells). In order to determine whether these cells regulate B cell function, they were purified from human tonsils by panning on anti-HNK-1 antibody coated Petri dishes and co-cultured with autologous and allogeneic tonsillar T and B cells in the presence of pokeweed mitogen. At the end of 7 days of culture, supernatants were assayed for immunoglobulin concentration by ELISA. A dose- dependent suppression of both IgG and IgM production was demonstrated at ratios from 1:125 to 1:16 of HNK-1+ cells to B cells, but enhancement was observed at very low ratios (less than 1:500) or ratios exceeding 1:16 in some tonsil preparations. Similar results were obtained with peripheral blood HNK-1+ cells but without enhancement in some cases at the extremes of HNK-1+ cells to B cell ratios. The suppression was not MHC-restricted. These preliminary experiments indicate that germinal centre HNK-1+ cells may be intrafollicular suppressor cells.

INTRODUCTION

A subset of T lymphocytes within follicular centre of lymph nodes, tonsils and spleen expresses the antigen HNK-1 and such cells express the phenotype CD3+CD11+HNK-1+ (1-6). These are not related functionally to natural killer (NK) cells despite the expression of HNK-1 antigen as they lack significant lytic activity against NK-sensitive targets (7). We (1) and others (8) have postulated that such cells may represent regulatory cells strategically located in B cell compartments. Since CD4+ T helper cells are found in both intra- and inter- follicular areas while CD8+ T suppressor cells are normally found only in interfollicular areas and not within follicular centres (2,3,5,6), we were interested in determining whether the intrafollicular CD3+HNK-1+ cells are suppressor cells as such cells may be necessary to counteract the influence of CD4+ T helper cells present within follicles.

MATERIALS AND METHODS

Tonsillar and peripheral blood mononuclear cells were isolated by centrifugation on Ficoll Hypaque gradients. Plastic adherent cells were removed by incubating the cells in fetal calf serum (FCS) pre-coated polystyrene Petri dishes. T and B enriched fractions were isolated by differential adherence to nylon wool columns.

Using the panning technique described by Wysocki and Sato (9), HNK-1+ and HNK-1- T cells were isolated by panning on anti-HNK-1 antibody coated Petri dishes.

Cell purities were monitored by staining aliquots of the various fractions for CD3, CD20 and HNK-1 antigens by an indirect immunofluorescence technique and counting fluorescent cells by fluorescence microscopy or flow cytometry.

Equal numbers of B cells and HNK-1 depleted T cells (10^5 each) were cultured with various numbers of purified HNK-1+ cells per well in 96 well round bottomed plates (Nunc). Each well contained a total volume of 200 ul of complete medium with 0.5% PWM (Gibco Laboratories, Grand Island, New York). The plates were incubated at 37^oC in an atmosphere of 5% carbon dioxide in air for 7 days, following which supernatants were collected for immunoglobulin quantitation by ELISA.

RESULTS

The effect of adding enriched HNK-1+ cells on the immunoglobulin production by B cells in the presence of T cells and PWM is summarized in figure 1. Suppression of IgM production could be detected at HNK-1+ cell:B cell ratios as low as 1:500. IgG production could be suppressed at ratios as low as 1:250. However, at ratios less than 1:500 upto 40% enhancement of IgG production was observed on some tonsil preparations. At ratios higher than 1:16 marked enhancement of both IgG and IgM production was noted.

The suppression by HNK-1+ cells was not MHC restricted as both autologous and allogeneic peripheral blood HNK-1+ cells suppressed IgG production (figure 2) and IgM production (data not shown). As shown in figure 2, enhancement is not usually observed at ratios exceeding 1:16 with blood derived HNK-1+ cells unlike tonsillar HNK-1+ cells. The lack of MHC restriction is also true for tonsillar HNK-1+ cells (data not shown).

DISCUSSION

The presence of HNK-1+ T cells in germinal centres has been reported from this laboratory and others (1-6). It is clear from previous reports that these cells are not NK cells as they do not lyse K562 cells effectively and lack the CD16 antigen (7). We have also noted that these cells do not express an additional NK-related antigen NKH-1 (11 and Banerjee, unpublished observations). A recent report indicates that these cells can be cloned and that cloned tonsillar follicular centre HNK-1+ cells co-express the antigen CD4 but lose the expression of HNK-1 while acquiring HLADR antigens (12). Despite the expression of T helper cell related antigen CD4, these cells have been shown not to produce B cell growth factors (12). The suppresive effect of the cloned cells was not investigated in that study. We report here that freshly isolated tonsillar and blood HNK-1+ cells are potent suppressors of immunoglobulin production by autologous and allogeneic

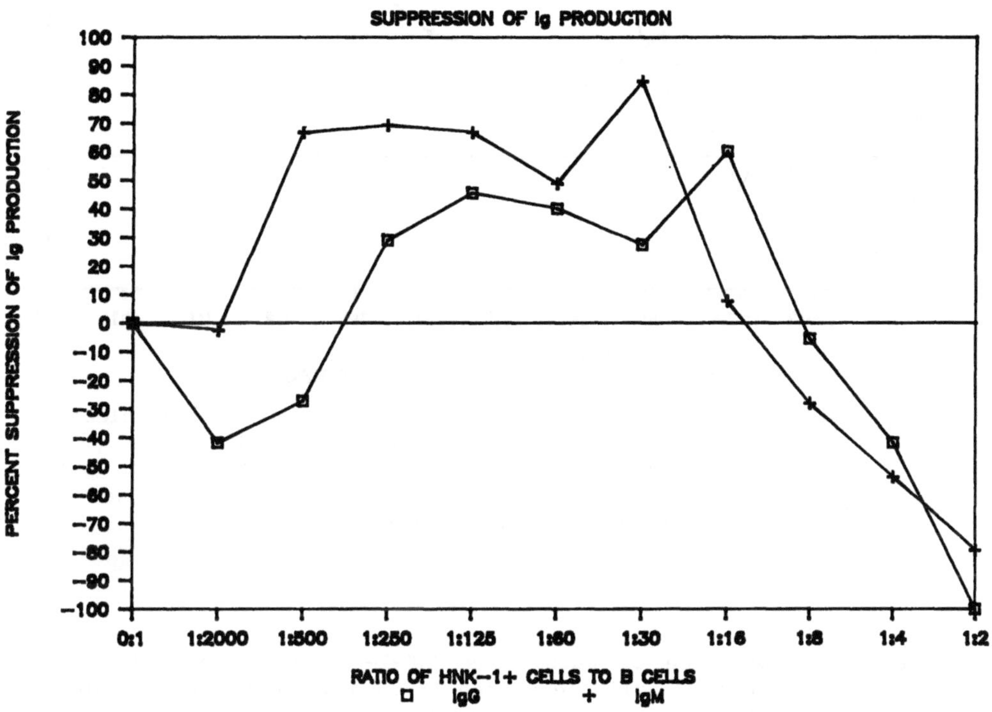

Figure 1. Autologous combination of tonsillar B cells (10^5 cells per well), tonsillar HNK-1+ cell-depleted T cells (10^5 cells per well) and varying numbers of tonsillar HNK-1+ cells added to each well. Supernatants were analysed 7 days after initiaton of the culture. in the presence of PWM. Graph shows the percent suppression at each ratio. Negative values indicate enhancement.

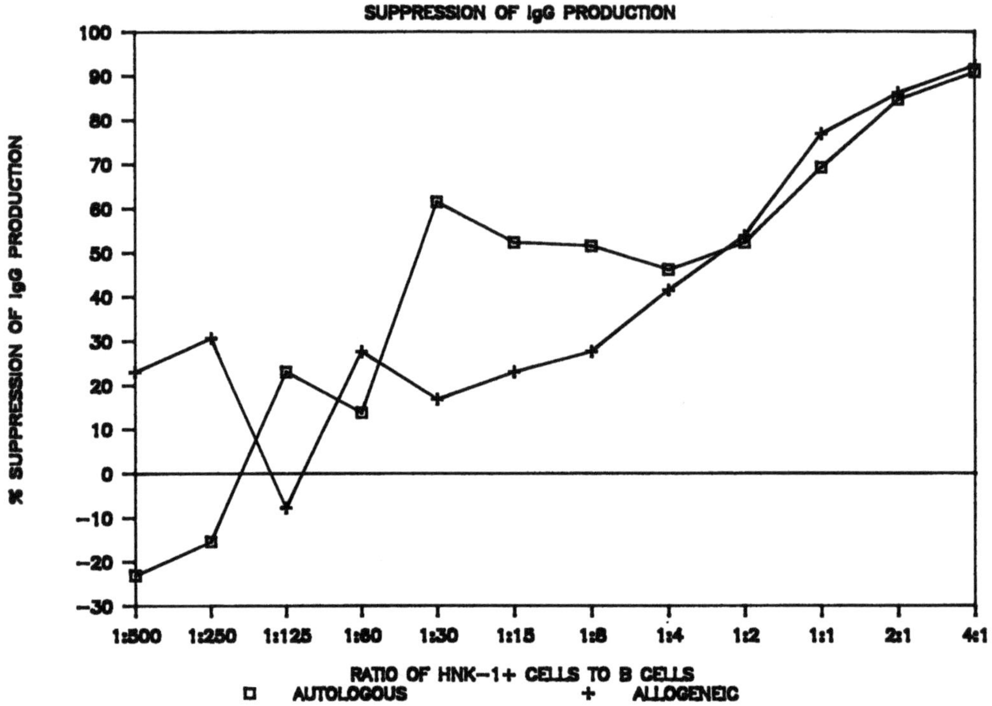

Figure 2. Autologous and allogeneic combinations of peripheral blood B cells (10^5 per well), HNK-1+ cell-depleted T cells (10^5 cells per well) and varying numbers of HNK-1+ cells per well. Graph shows percent suppression at each ratio. Negative values indicate enhancement.

B cells in the presence of autologous HNK-1- T cells and PWM. At ratios of HNK-1+ cells:B cells less than 1:500 and greater than 1:16, however, enhancement of immunoglobulin production may be observed. The mechanism of suppression is not yet clear and may be due to direct effects on B cells, effects on T helper cells or the production of soluble suppresor factors. These possibilities will be explored in future experiments. The bi-phasic effects of suppression and enhancement at certain ratios cannot be explained at this time but one can speculate that more than one target or mechanism may be involved or that there may be functional heterogeneity amongst follicular centre HNK-1+ cells. Now that cloning of such cells is possible, it may be feasible to determine the basis for variability of the effects observed in-vitro.

REFERENCES

1. Banerjee D, and Thibert RF. 1983. Natural killer-like cells found in B-cell compartments of human lymhoid tissues. Nature **304**:270.
2. Poppema S, Visser L, and De Leij. 1983. Reactivity of presumed anti-natural cell antibody Leu 7 with intrafollicular T lymphocytes. Clin Exp Immunol **54**:834.
3. Si L, and Whiteside TL. 1983. Tissue distribution of human NK cells studied with anti-Leu 7 monoclonal antibody. J Immunol **130**:2149.

4. Porwit-Ksiazek A, Ksiazek T and Biberfeld P. 1983. Leu 7+ (HNK-1+) cells: I. Selective compartmentalization of Leu 7+ cells with different immunophenotypes in lymphatic tissues and blood. Scand J Immunol **18**:485.
5. Von Gaudecker B, Pfingsten U, and Muller-Hermelink H-K. 1984. Localization and characterization of T-cell subpopulations and natural killer cells (HNK-1+ cells) in the human tonsilla palatina: An ultrastructural-immunocytochemical study. Cell Tissue Res **238**:135.
6. Ritchie AWS, James K, and Micklem HS. 1983. The distribution and possible significance of cells identified in human lymphoid tissue by the monoclonal antibody HNK-1. Clin Exp Immunol **51**:439.
7. Abo T, Miller CA, and Balch M. 1984. Characterization of human granular lymphocyte subpopulations expressing HNK-1 (Leu 7) and Leu 11 antigens in the blood and lymphoid tissues from fetuses, neonates and adults. Eur J Immunol **14**:273.
8. James K, and Ritchie AWS. 1984. Do natural killer cells regulate B cell activity? Immunol Today **3**: 193.
9. Wysocki LJ, and Sato VL. 1978. "Panning" for lymphocytes: A method for cell selection. Proc Natl Acad Sci USA **75(6)**:2844.
10. Abo T, and Balch CM. 1981. A differentiation antigen of human NK and K cells identified by a monoclonal antibody (HNK-1) J Immunol **127**:1024.
11. Hercend T, Griffin JD, Bensussan A et al. 1985. Generation of monoclonal antibodies to a human natural killer clone: Characterization of two natural killer-associated antigens, NKH1A and NKH2, expressed on subsets of large granular lymphocytes. J Clin Invest **75**:932.
12. Velardi A, Mingari MC, Moretta L, and Grossi CE. 1986. Functional analysis of cloned germinal center CD4+ cells with natural killer cell-related features. Divergence from typical T helper cells. J Immunol **137**:2808.

MAST CELL RESPONSE TO TUMOR GRAFT

Robert Courtoy, Dominique Degraef-Lovens, Karin Dupont, Nicole Schaaf-Lafontaine, and Jacques Boniver

University of Liege, Laboratory of Pathology, (B33) Sart-Tilman, B-4000 Liège1 (Belgium)

INTRODUCTION

The association of mast cells (MC) with tumor tissue has been frequently reported since the first description of these cells by Ehrlich (1). Mast cells were found in abundance at the periphery of carcinogen-induced tumor as well as of transplanted fibrosarcoma (2-4).

Substantial evidence exists in the literature to suggest that mast cells may exert considerable influence on the inflammatory or on the host-immune responses against tumor growth (5-6). Tumor draining lymph nodes also have been reported to exhibit a pronounced mast cell reaction. Higher number of MC have been observed in lymph nodes of syngeneic and allogeneic skin-grafted mice (7).

It is known that mast cell autopharmacological mediators can influence the T cell host system via H1 and H2 receptors and thus either favor or inhibit the progression of the tumor (8). However there is no demonstration that, in the vicinity of the tumor, mast cells are secreting their granules components. On the contrary, in tumor bearing mice an increase in histamine levels in numerous tissues has been associated with a decrease in anaphylactic response (5).

The Alcian blue-Safranin (AB-S) two steps reaction has been frequently used to stain MC, resulting in blue or red granules. Safraninophilic granules were considered as mature type MC granules and the alcianophilic ones as young type MC granules (4) or as granules having already released their basic protein components while conserving their polyanion matrix (9). The two types of granules were observed in distinct mast cells as well as within the same cell (10).

In this paper, we present the follow-up of MC- number in tissue sections of graft and of sarcoma-draining lymph nodes stained with AB-S reaction. We have used a mouse MCA-fibrosarcoma in both allogeneic and syngeneic situation. The evolution of the MC response to the tumor was compared to the response induced by a s.c. graft of normal tissue. Although the sarcoma cells induce an immune response in the syngeneic mice (11), they are not rejected in unimmunized hosts. However, by repeated injections of inactivated tumor cells, we can establish a process of tumor rejection in the syngeneic mice (12). During the rejection, histological studies show lymphocytic and macrophagic effector cells infiltrating the tumor implant (13). We have also studied the MC reaction in the lymp nodes during the tumor regression period.

MATERIALS AND METHODS

MICE: C57BL/6 and CBA male mice from our own breeding colony were used at 3 months of age.
TUMOR: a transplantable fibrosarcoma T2 (originally induced in C57BL/6 mice by MCA) maintained by serials s.c. transplantation of solid tumor fragments. In these experiments, tumor pieces of about 10.6 living tumor cells were implanted s.c. with a trocar.
CONTROLS: normal spleen fragments implanted in the same conditions in syngeneic mice.
HISTOLOGY: draining lymph nodes to be studied for MC content were removed from freshly killed mice (4 mice, in each condition every 2 days following the graft). They were fixed for 1 hr in Carnoy's fluid. After embedding in paraffin, 3 tissue sections were performed.. The sections were stained for 1 hr in Alcian Blue (3% solution in 2N sulfuric acid), rinced in water, stained for 3 min. in Safranin (0.1% solution in water) and mounted in Depex medium.
IMMUNIZATION PROCEDURE: solid tumor fragments were implanted s.c. and 7 days later, the growing fibrosarcoma was surgically excised. Mitomycin-C treated tumor cells (10.6) were s.c. injected on day 14 and on day 21. Living tumor cells (10.6) were then s.c. injected on day 28 after the primary graft (12). Three mice were sacrified on day 5, 8, 12 and 15 following the second living tumor cells challenge. Control mice received a single injection of 10.6 living sarcoma cells.
NUMERATION: 3 sections/lymp node were analyzed. Red and blue MC were counted on the total lymph node area for each section.

RESULTS

By use of the AB-S staining, the analysis of MC content in lymph node (LN) sections from normal mice demonstrated that two types of MC were to be distinguished: the red mast cells (rMC) and the blue mast cells (bMC). The mean values have been established as 10rMC/16bMC and 9rM/2bMC per section in C57BL/6 and in CBA LN sections respectively.

a) Tumor graft in sygeneic mice (C57BL/6):
After s.c. trocar transplantation of living T2 sarcoma fragment, the tumor grew progressively and killed the syngeneic host within 30-40 days. Its mean diameter increased from about 1 mm (day 1) and reached a size of 7.5 and 16 mm on day 10 and 22 respectively. In the draining LN, the rMC number drastically increased on the second day after the graft; it remained higher than normal values during the observation period (Fig. 1a). The bMC number fluctuated around normal values but remained below the rMC number.

b) non-tumor graft in syngeneic mice (C57BL/6):
The s.c. graft of normal spleen fragment resulted in a drastic increase of rMC in LN on day 2 after the implantation. From day 4 to day 10, the rMC number was about the normal values (Fig. 1b). The bMC number falled down below normal values on day 4 to day 10. During the experiment, the bMC appeared fewer than the rMC.

c) Tumor graft in allogeneic mice (CBA):
The tumor grew slowly to a maximum diameter of 5 mm, then it was rejected between day 14 and 18. The number of bMC increased in the draining LN2 days after the tumor injection and remained higher than normal values till day 8. On day 10, it falled down to normal values (Fig. 1c). On the contrary, the rMC number was near normal values to the 8th day, remaining below the bMC number. On day 10, rMC number suddenly increased.

d) Non-tumor graft in CBA mice:
Normal spleen graft induced the increase of rMC number in LN on day 2; this level remained about constant during the period of observation. On the contrary, the number of bMC was about the normal values except on day 4 when it increased to the value of the rMC. Otherwise, the bMC appeared fewer than rMC (Fig. 1d).

e) Tumor graft in immunized syngeneic mice:
According to our immunizing procedure allowing the tumor rejection (12), 96% of the immunized mice were still alive three months after the injection of the second living tumor cells challenge; no control mice survived at this period. In

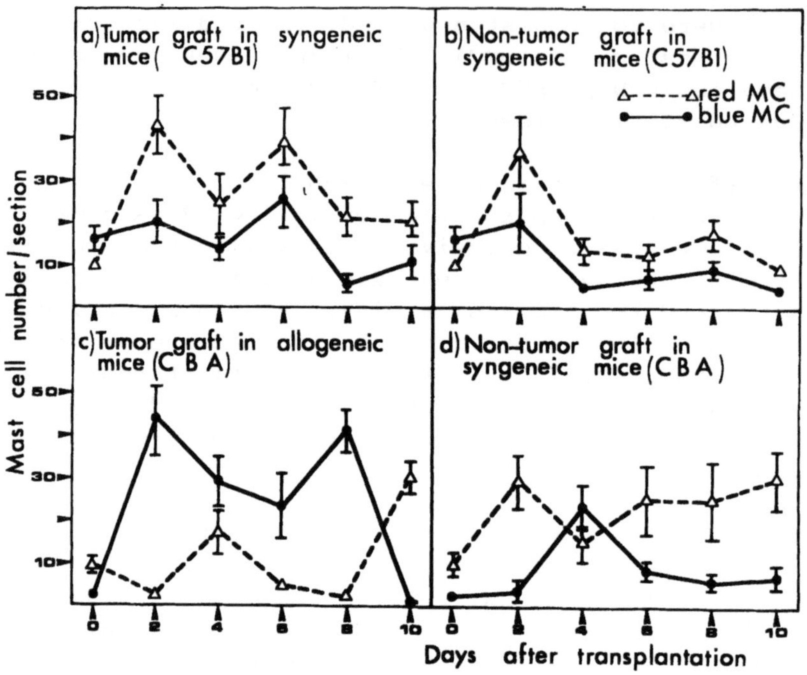

Figure 1. Lymph Node Mast Cells in Grafted Mice.

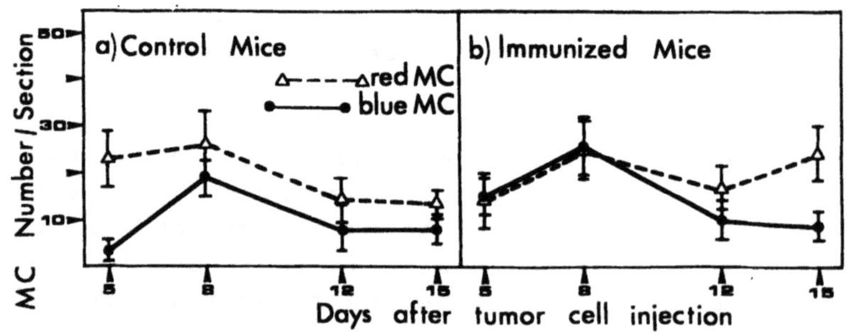

Figure 2. Lymph Node MC in Immunized Mice.

immunized mice, the content of rMC and bMC is similar on day 5, 8 and 12 but the number of rMC increased on day 15 (Fig. 2).

DISCUSSION AND CONCLUSION

The s.c. transplantation of spleen or tumor tissue induced an important increase of cell number in the draining LN. These MC were stained either blue or red by use of the AB-S reaction. When the grafts were tolerated (syngeneic tumor or spleen), red MC were generally more numerous than blue MC during 10 days after the tranplantation. On the contrary, allogenic tumor graft induced an important increase in blue MC number during 8 days, after what red MC increased and blue MC falled down.

It is admitted that rMC are mature cells, containing polyanion--basic protein complexes in the granules. Blue MC are either immature cells or cells that have released the basic component of their granules (9,10). Preliminary study in electron microscopy shows that bMC in LN shared the morphological and cytochemical characteristics of "degranulated" mast cells (unshown results). However, we do not know whether bMC have released their mediators in the LN or if they are coming back to the LN after secretion elsewhere (in the vicinity of the tumor?). Moreover, though it has been reported that MC precursors are enhanced in the LN after injection of antigen (14), there is no evidence that the MC increase observed after the graft resulted of an "in situ" proliferation or if it reflected the accumulation of MC coming from elsewhere.

The rejection of living tumor cells consequently to an immunization treatment was preceded by a relative but less evident increase of bMC on day 5 and 8, and by enhancement of rMC on day 15. Experiments are in progress to accurately compare the MC response in LN of tumor grafted allogeneic mice and in the immunized tumor grafted mice. The MC response in the graft and the spleen will be also studied.

REFERENCES

1. P.Ehrlich, Beiträge zur Kerntniss der granulierten Bindegewalzellen und der eosinophilen Leucocyten. Arch. Anat. Physiol.U, 3: 166 (1879).
2. R.G.Herne, M.Gregoire and G.Rosa. Mast cells and 5-hydraxytryptamine in precancerous mouse skin. Nature 187:1128 (1960).
3. B.C.Broom and P.Alexander. Mast cell and anaphylactic antibody responses in inbred rats to syngeneic fibrosarcomas. Int. Arch. Allergy appl. Immunol., 49:627 (1975).
4. W.R.Roche. The nature and significance of tumour associated mast cells. J. Path. 148:175 (1986).

5. R.Lynch and J.C.Salomon. Tumour-associated inhibition of immediate hypersensitivity reactions in mice. Immunology, 32:645 (1977).
6. M.K.Dabbous, R.Walker, L.Haney, L.M.Carter, G.L.Nicholson and D.E.Wooley. Mast cells and matrix degranulation at sites of tumour invasion in rat mammary adenocarcinoma. Brit. J. Cancer, 54: 459 (1986).
7. L.J.Simar. L'ultrastructure des ganglions lymphatiques au cours des réactions immunitaires. Université de Liège Ed., Liège (1973).
8. J.Bartholeyns and M.Boucher. Involvment of histamine in growth of mouse and rat tumors: antitumoral properties of monofluoromethylhistidine, an enzyme-activated irreversible inhibitor of histidine decarboxylase. Cancer Res., 44:639 (1984).
9. R.Courtoy, J.Boniver and L.J.Simar. Cytochemistry of mouse mast cell reaction to polylysine. Histochemistry 66:49 (1980).
10. M.Miyata and K.Takaya. Effect of strong electrolytes on the iron alun-alcian blue-safranin staining of mast cells in vitro during the growth of a syngeneic mouse tumour. Int. J. Cancer 21: 329 (1978).
12. N.Schaaf-Lafontaine, P.Maquet, D.De Graef-Lovens, R.Courtoy and H.Firket. Mise au point d'une procédure d'immunisation efficace contre un sarcome murin syngénéique. C.R.Soc Biol. 179: 122 (1985).
13. D.De Graef-Lovens, N.Schaaf-Lafontaine, R.Courtoy, J. Boniver. Characterization of immunocompetent cells at the tumor site following an immunizing procedure against a murine syngeneic sarcoma. Cancer Letters 36: 47 (1987).
14. R.M.Crapper and J.M.Schrader. Frequency of mast cell precursors in normal tissues determined by an in vitro assay: antigen induces parallel increases in the frequency of P cell precursors and mast cells. J.Immunol. 131: 923 (1983).

INTERLEUKIN-2 INDUCED CHANGES IN MONONUCLEAR CELL SURFACE CHARGE

R. Jenkins, A.C. Williams, G.A. Limb, K.A. Brown and D.C. Dumonde

Department of Immunology, United Medical and Dental Schools of Guy's and St. Thomas' Hospitals, St. Thomas' Campus, London SE1 7EH, U.K.

INTRODUCTION

Interleukin-2 (IL-2) is not only the hormone responsible for progression of the T cell cycle, but it is also a potent activator of cytolytic (lymphokine-activated killer, LAK) activity against a wide variety of autologous and allogeneic tumours[1]. These effects are mediated through specific target cell membrane receptors.

Surface electric charge has been shown to be a sensitive and widely applicable parameter for the study of plasma membrane features and ligand-receptor interactions[2]. Mammalian cells, like all others studied to date, have a net negative surface charge at physiological pH, due mainly to the carboxyl groups of sialic acids (particularly N-acetyl neuraminic acid) on the plasma membrane[3]. Microelectrophoresis provides an indirect measurement of this surface charge in vitro, and is one of the few physicochemical assessments which can be made on living cells without producing damage. This technique has already been used in the characterisation of granulocyte and macrophage receptors for molecules with "migration-inhibition"-like properties[4]; in these experiments ligand binding was accompanied by a decrease in the net negative cell surface charge. However, in recent preliminary work recombinant IL-2 (rIL-2) was found to increase the electrophoretic mobility (EPM) of murine thymocytes, but to exert no effect on guinea-pig macrophages[5].

The aim of the present investigation is to determine whether human peripheral blood mononuclear cells exhibit an in-vitro electrophoretic response to rIL-2. Thus mononuclear cell surface charge has been studied during preparation of LAK cells in culture.

MATERIALS AND METHODS

Preparation of mononuclear cells

Samples of heparinised venous blood (10 u/ml preservative-free heparin, Leo Labs.) were obtained from healthy volunteers and cancer patients. Mononuclear cells were isolated by buoyant density centrifugation on Lymphoprep™ (Nycomed, Norway)[6] and subsequently washed three times in serum-free RPMI 1640 medium (Gibco Biocult) containing 2 mM L-Glutamine (Gibco Biocult), 40 μg/ml Gentamycin (Cidomycin, Roussel), 100 u/ml Penicillin (Crystapen, Glaxo) and 2 μg/ml Fungizone (Gibco Biocult). At this stage the cell yield was determined with an Improved Neubauer haemocytometer and the viability estimated by the Trypan Blue dye exclusion test. Cell viability was always greater than 95%. Following these procedures the cells were washed once more in the same medium prior to use in cell culture/cytolytic assays, or once in RPMI 1640 medium without either serum or antibiotics for EPM analysis.

Cell cultures

Mononuclear cells were activated in 72 hour incubations at 37°C, using 25 ml plastic culture flasks and a humidified atmosphere of 5% CO_2 in air. Cultures were prepared with RPMI 1640 medium containing the antibiotics described above and 2% heat-inactivated human AB serum. The cells were incubated, at a concentration of 2×10^6/ml, either alone or with 1000 u/ml rIL-2 (Amgen). In certain experiments additional cultures were prepared with 10 μg/ml PHA (Sigma). After incubation the cells were always washed twice with the serum-free RPMI Medium before further use. Phenotypic analysis by flow cytometry has confirmed that the inevitable loss of cells during culture was not associated with any major depletion of particular cell subsets.

Cytotoxicity assays

Mononuclear cell preparations were tested, before and after culture, in 4 hour ^{51}Cr-release assays[7] with NK-sensitive (K562) and NK-resistant (Daudi) target cells. These target cells were first labelled with 100 μCi sodium chromate (Amersham, UK) in 45 minute incubations at 37°C, then washed twice with RPMI medium containing antibiotics and 10% heat-inactivated foetal calf serum and once more with the serum-free RPMI medium.

The cytotoxicity tests were carried out in triplicate using 96-well microtitre plates (Nunc, Denmark) and target/effector cells suspended in RPMI medium with antibiotics and 5% human AB serum. Samples of the labelled targets (100 μl containing 1.0×10^4 cells) were admixed with preparations of mononuclear cells (100 μl) such that the final effector: target ratios ranged from 50:1 to 1:1. After incubation the supernatants were harvested and the radioactivity measured in a computerised gamma counter (LKB - model 1282). Each assay was controlled for maximum isotope release - by incubation of target cells with 5% SDS, and spontaneous release - incubation of targets with medium alone. Cell killing was expressed as follows:

$$\% \text{ specific lysis} = \frac{\text{cpm test culture} - \text{cpm spontaneous release}}{\text{cpm maximum release} - \text{cpm spontaneous release}} \times 100$$

EPM analysis

The electrophoretic mobility of randomly selected, individual cells was measured in a cylindrical microelectrophoresis apparatus[8]. All measurements were performed in RPMI 1640 medium without antibiotics or serum, and with a constant voltage at 25°C. Paired timings were recorded on 100-150 cells from each mononuclear cell preparation by means of an electronic system operating at 1/40th second intervals. Mean values were subsequently converted to electrophoretic mobilities by reference to the migration distance (25 μm) and field strength (8.00 V/cm) used in the cytopherometer. Freshly aspirated human erythrocytes (blood group A_2, Rh^+) were always examined prior to the measurement of test samples, in order to check the apparatus and provide a reproducible reference mobility for each experiment.

Statistical analysis

Tests of significance were applied to the differences between the mean electrophoretic mobilities of mononuclear cell samples before and after culture with rIL-2 (Student's t test).

RESULTS

In these experiments the mean EPM recorded for human erythrocytes was 1.08 ± 0.03 $\mu m\ s^{-1} V^{-1} cm$. This value was in good agreement with previous findings.[8]

Fig. 1 illustrates the distribution of electrophoretic mobilities obtained with mononuclear cells from a healthy individual. The mean EPM of such "normal" cells was found to be significantly reduced ($P<0.001$) after incubation with either rIL-2 or PHA (Table 1). Similar electrophoretic responses were observed with cells from six cancer patients tested (Fig. 2), even though more slower-moving cells contributed to their EPM distributions at the preculture stage. The cultures derived from healthy volunteers and cancer patients showed comparable rIL-2-induced NK and LAK activity (Fig. 3), although native levels detected during the preculture analyses were rather more variable. No significant enhancement of cytolysis or reduction in EPM were detected with cells from control cultures. More experimental data must be collected with both assays before any statistical comparison can be made between the subject groups.

DISCUSSION

This work clearly indicates that IL-2-induced LAK activity in cultures of human blood mononuclear cells is accompanied by a significant decrease in the net negative cell surface charge. However, the demonstration of similar EPM changes in PHA-stimulated cultures lends further support to the concept of surface charge involvement in a common pathway of cell response

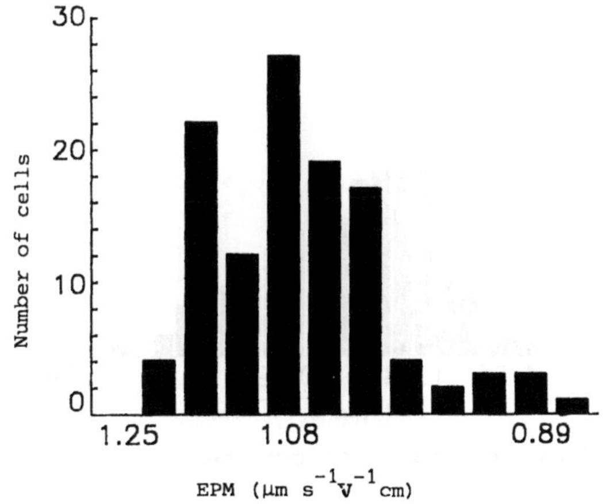

Fig. 1 The EPM distribution of peripheral blood mononuclear cells from a healthy donor.

Table 1. Electrophoretic response of mononuclear cells to rIL-2 and PHA

Cell Sample	Electrophoretic Mobility ($\mu m\ s^{-1}V^{-1}cm$)	
	rIL-2	PHA
	(n=4)	
Cells alone	1.06 ± 0.02	1.01 ± 0.01
Cells + ligand	0.97 ± 0.02	0.88 ± 0.02
	(10% reduction, P<0.001)	(14% reduction, P<0.001)

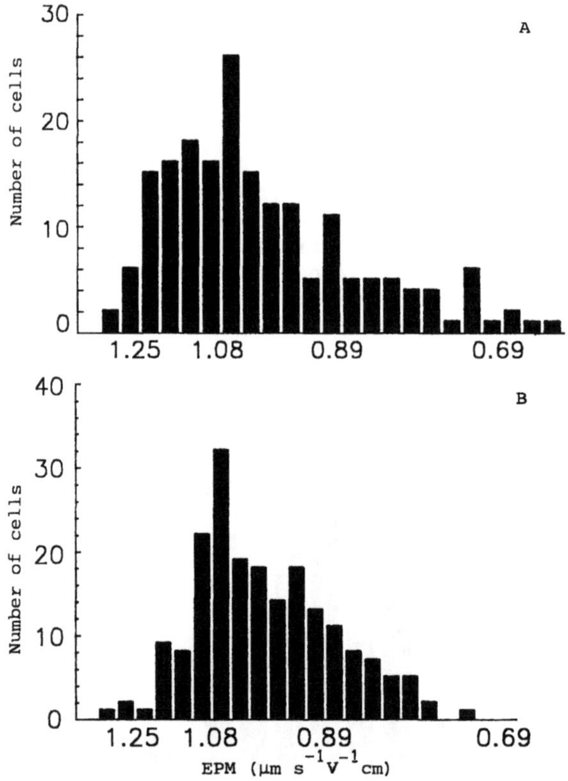

Fig. 2. EPM distributions of mononuclear cells from a cancer patient.
A. Preculture B. After culture with rIL-2 (P<0.001)

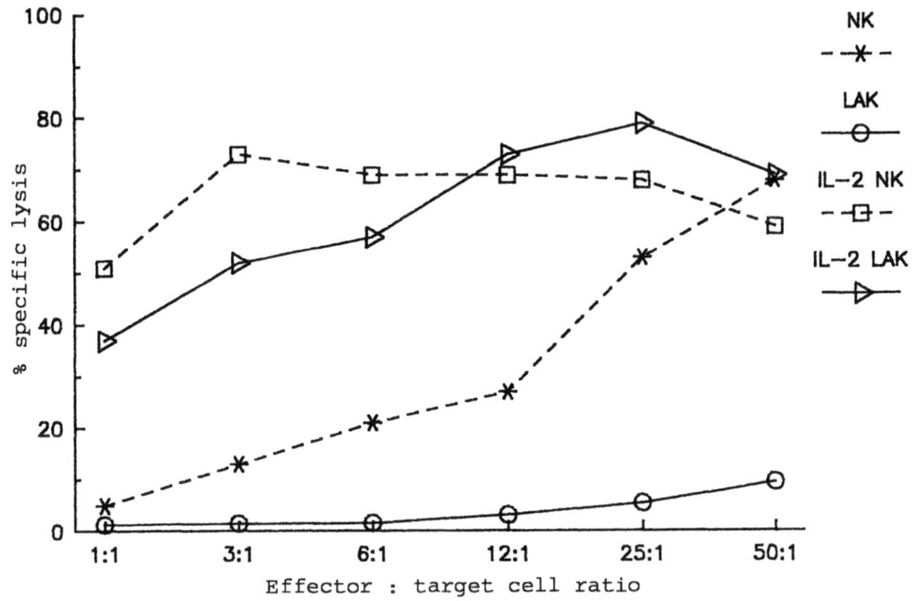

Fig. 3. The effect of rIL-2 on NK and LAK activity of mononuclear cells from a cancer patient.

following specific ligand-receptor interactions.

Preparations of natural and rIL-2 have already been shown to inhibit the in-vitro capillary migration of human lymphocytes[9] and monocytes[10] in a dose-dependent manner. As these effects were manifest by direct exposure of the cells to the ligand, the IL-2 activity may have been analogous to classic "migration inhibitory" lymphokines, such as MIF, which have also been characterised by their reduction of net negative target cell surface charge in various microelectrophoresis systems.

In conclusion surface charge is considered to be a useful parameter for studies on IL-2-triggered cell activation. Further investigations are now being carried out in this laboratory to establish the precise relationship between the surface charge response and the sequence of events initiated by the binding of IL-2 to its receptor.

REFERENCES

1. E. A. Grimm, A. Mazunder, H. Z. Zlang, and S. A. Rosenberg, Lymphokine-activated killer cell phenomenon, J. Exp. Med. 155:1823 (1982).
2. A. W. Preece, and D. Sabolovic, eds., "Cell Electrophoresis: Clinical Application and Methodology", North Holland, Amsterdam (1979).
3. G. M. W. Cook, D. H. Heard and G. V. F. Seaman, Sialic acids and the electrokinetic charge of the human erythrocyte, Nature 191,44 (1961).
4. R. Jenkins, P. Balding, P. Ann Light and A. W. Preece, The role of α-L-fucose in the lymphokine receptor sites of macrophages and granulocytes, in: "Cell Electrophoresis: Clinical Application and Methodology", ibid.
5. K. A. Brown, A. W. Preece, R. A. Wolstencroft and D. C. Dumonde, Use of laser-Doppler spectroscopy for the evaluation of IL-1 and IL-2 activities, 5th International Lymphokine Workshop, Clearwater, USA, abs (1987).

6. R. Harris and E. O. Ukaejiofo, Tissue typing using a routine one-step lymphocyte separation procedure, Brit. J. Haemat., 18:229 (1970).
7. J. R. Ortaldo, G. D. Bonnard and R. B. Herberman, Cytotoxic reactivity of human lymphocytes cultured in-vitro. J. Immunol. 119:1351 (1977).
8. A. W. Preece and P. Ann Light, The macrophage electrophoretic mobility (MEM) test and studies on macrophage slowing factors. Clin. Exp. Immunol. 18:543 (1974).
9. D. S. B. Hoon, D. R. Wen, M. Stone, R. K. Gupta and A. J. Cochran, Inhibition of lymphocyte mobility by IL-2. Clin. Exp. Immunol., 66:566 (1986).
10. G. B. Thurman, A. E. Maluish, J. L. Rossio, E. Schlick, K. Onozaki, J. E. Talmadge, A. D. G. Procopio, J. R. Ortaldo, F. W. Ruscetti, H. C. Stevenson, G. B. Cannon, S. Iyar and R. B. Herberman, Comparative evaluation of multiple lymphoid and recombinant human IL-2 preparations, J. Biol. Response Modifiers, 5:85 (1986).

THE TARGET CELLS FOR RAT NK CELLS AMONG ALLOGENEIC BONE MARROW CELLS

B. Rolstad[1], S. Fossum[1] and H.B. Benestad[2]

Institutes of [1]Anatomy and [2]Physiology

University of Oslo, N-0162 Oslo 1, Norway

INTRODUCTION

The mechanisms governing the rejection of bone marrow allografts in mammals are complex, and may in addition to adaptive immune responses mediated by T and B cells also involve non-adaptive, NK cell mediated rejection mechanisms (Cudkowicz and Bennet, 1971; Kiessling et al., 1977; Lotzová, 1977). Thus, cloned NK cells can adoptively transfer the resistance to bone marrow allografts to NK-deficient recipients (Warner and Dennert, 1982), and NK cells are able to inhibit the proliferation of committed haematopoietic precursors after coculture in vitro (Bordignon et al., 1985).

We have recently developed an in vitro assay for measuring rat NK cell mediated destruction of allogeneic lymphocytes (ALC) (Rolstad and Fossum, 1987) or allogeneic bone marrow cells (ABC) (Rolstad and Benestad, 1984). The killing takes place within 4 h of coculture of the effector and target cells and discriminates between allogeneic and syngeneic cells, as has been observed for ALC and ABC in vivo (Rolstad and Ford, 1983; Rolstad and Benestad, 1984). However, previously we used unfractionated bone marrow cells as targets, and these experiments could not decide whether the killing was indiscriminate, affecting any bone marrow cells, or whether the NK cells selected certain cell types among the bone marrow cells, as has been suggested by results from syngeneic systems (Barlozzari et al., 1987; Matera et al., 1986). Here we have approached this problem by using subpopulations of bone marrow cells as targets in ABC, and we have compared the cytotoxicity against cell fractions rich in blast cells and progenitor cells with that against the more differentiated cells of the marrow.

MATERIALS AND METHODS

Animals

Rats of the inbred strains PVG(RT1^{c}), AO(RT1^{u}) and PVG nude (rnu) (RT1^{c}) were bred in conventional animal houses. The rats were used for experiments between 8 and 12 weeks of age.

Preparation of effector cells

Spleen cells (SC) were isolated from PVG rnu spleens that had been gently disrupted in ice-cold phosphate-buffered saline (PBS). The fragments were squeezed with a rocker and the suspension filtered through loose cotton-wool. Erythrocytes were removed by centrifugation on Lymphoprep (Nycomed A/S, Oslo, Norway) (density 1.077 g/cm^3), and B cells and monocyte-macrophages were removed by incubation on columns of nylon wool (NW) (Leukopak, Fenwal Laboratories, Illinois, USA) for 1 h at 37°C, as detailed elsewhere (Timonen et al., 1982; Reynolds et al., 1981b, 1982). The medium used was RPMI 1640 with HEPES buffer (25 mM/l), 5% foetal calf serum (FCS), and a standard antibiotic-antimycotic solution added (all reagents purchased from Gibco, Paisley, Scotland). This will later be referred to as complete medium (CM). This procedure yielded 10-20x10^6 cells per nude spleen, with 20-30% large granular lymphocytes (LGL). In some experiments LGL were further enriched by separation on a 4-step discontinuous Percoll density gradient (Percoll purchased from Pharmacia Fine Chemicals AB, Uppsala, Sweden). For details of the procedure, see Timonen et al., 1982, Reynolds et al., 1981, and Rolstad et al., 1986. The Percoll solution was adjusted to 285 mOsm/l with a PBS solution with salt concentration 10 times the physiological value (Gibco), and CM to 290 mOsm/l with sterile water. Usually, the gradients were composed so that densities relative to water were 1.053 for the upper fraction, 1.065, 1.075, and 1.082 for the subsequent three fractions. The separated effector cells were stored overnight at 4°C in CM before being used in the assay.

Target cells

Both femora and tibiae were flushed with CM, the BM cells were washed once in PBS + 2% FCS and separated on a 4 step Percoll density gradient, with the same osmolarities and densities as for separation of the effector cells. 30-40x10^6 bone marrow cells were layered onto each gradient. The bone-marrow fractions F_1, F_2, F_3 and F_4 are the cell layers just above the Percoll fractions with densities 1.053, 1.065, 1.075, and 1.082, respectively.

Measurement of ^{3}H-thymidine incorporation into the target cells

Rats got 1 μ Ci ($\sim$40 kBq) of ^{3}H-thymidine (Code TRK 120, Amersham, England) per gram of body weight i.v. 1 h before sacrifice. The cells of the different fractions were counted in a Packard Tricarb scintillation spectrometer (mod. 460c) and the radioactivity per 10^5 cells was expressed as the % of the F_2 value (=100%).

In vitro ABC assay

The target cells (the human erythroleukemic cell line K562, or the different bone marrow cell fractions) were labelled with $Na_2{}^{51}CrO_4$ (Amersham, England), at 200 μ Ci($\sim$8 MBq)/ml cells at the concentration of 5-10x10^6 cells/ml for 1 h at 37°C in CM. The cells were washed three times in PBS-FCS, and between the second and third wash incubated in ca 40 ml PBS-FCS for approx. 1/2 h at room temperature, to elute loosely bound ^{51}Cr.

Effector and target cells were incubated in U shaped Costar microtiter plates in a volume of 200 ul (Costar, cat. no 3799) at 10^4 targets/well, in CM for 4 h at 37°C in 5% CO_2 in air. Phytohaemagglutinin (PHA) (Wellcome) was added to the wells to the final dilution of 1/100 of the standard preparation. The supernatants were harvested with a Titertek harvesting system (Skatron, Lierbyen, Norway) removing approx. 85% of the supernatant. The filters containing the supernatants were counted in a Packard Minaxi auto-gamma 5500 counter. The % experimental ^{51}Cr release (Exp) was calculated according to the formula:

$$\text{Exp} = \frac{x - SR}{T - SR}$$

where x was the radioactivity in the test sample, SR (spontaneous release), was the amount of ^{51}Cr released from the target cells alone, and T was the total counts per minute (cpm) in the target cells. For bone marrow cells as targets the SR ranged from 15 to 35% of the total cpm (T) in the targets, and the total cpm in 10^4 cells ranged from 1500-4000 cpm. The T value for K562 was 5000 - 10000 cpm. The results are presented as median values from triplicate assays for each effector-target cell ratio.

RESULTS

The separation of dispersed bone-marrow cells on a 4-step discontinuous Percoll density gradient yielded markedly different compositions of bone marrow cells in the various fractions. The more immature precursor cells were preferentially localized in the lighter density fractions, especially among F_2 cells (Table 1). Among the higher density cells in F_4 we found the more mature members of the granulocyte series, with only minimal contamination of immature blasts (Table 1). This finding was further substantiated by measuring cell proliferation in the various cell fractions, estimated by ^3H-thymidine incorporation (Table 1). The ^3H-thymidine incorporation of F_4 cells was less than 10% of that found for F_2 cells, indicating that the proliferative (i.e. the more immature) precursors resided mainly in F_2.

Despite the large variations in cell composition between the bone marrow cell fractions only small differences were observed in the susceptibility to attack by allogeneic NK cells in vitro (Fig. 1). Although allogeneic F_2-BM cells were most susceptible to lysis, the more mature cells in F_4 were also efficiently killed (Fig. 1). Whether this somewhat lower degree of cytotoxicity against F_4 cells reflected their composition of different cell types, or whether mere physical characteristics, like the target cell size (F_2 cells being on an average larger than F_4 cells) influenced the degree of killing, could not be determined from the present experiments. However, the killing must have included some of the more mature members of the granulocyte series, since these cells altogether constituted about 85% of the F_4 target cells (Table 1), and the cytotoxicity indicated that the majority of these targets had been killed at the highest effector to target cell ratio (Fig. 1).

A low degree of killing of syngeneic bone marrow cell fractions was also observed (Fig. 1). However, the cytotoxicity against all the allogeneic cell fractions (F_2-F_4) on the one hand, as opposed to the feeble or non-existent killing of the syngeneic bone marrow cell fractions on the other, indicated that in this system the NK cells recognized a histocompatibility difference between AO and PVG target cells, and not just differentiation antigens common to all rat strains.

DISCUSSION

The target specificity of NK cells, concerning both cell population and cell surface molecules, has remained an intriguing puzzle, despite several efforts to characterize effector-target interactions in NK phenomena. One school of thought holds NK cells to be entirely unspecific, so that any foreign cell within the species, which does not present on its cell surface the right inhibitory, or "pass-word" molecules, will be susceptible to NK cell attack (Kärre et al., 1986). If MHC molecules were such important inhibitory molecules (Grönberg et al., 1985), any cell with either poor expression of "self" MHC (immature cells within the haemopoietic system or

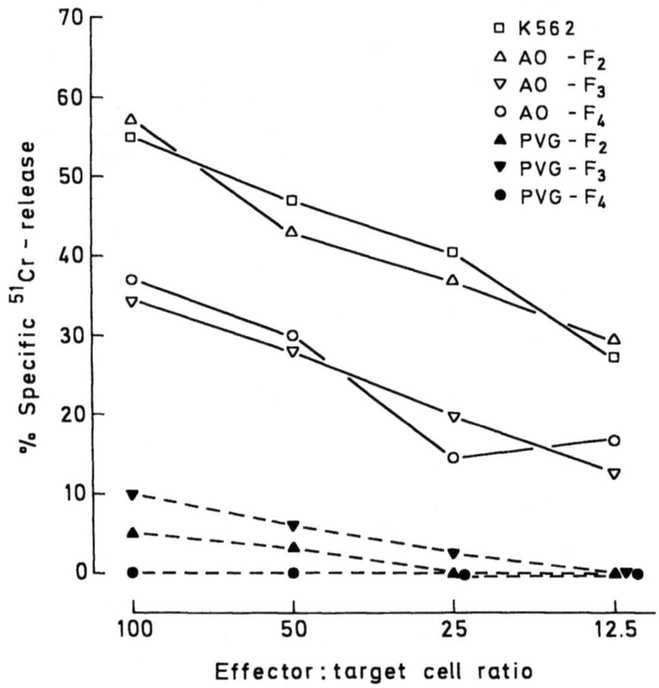

Fig. 1. Cytotoxic activity of NK cells from PVG rnu spleens against allogeneic AO, or syngeneic PVG, bone marrow cell fractions, and against the human erythroleukemic cell line K562.

Table 1. Cell number, morphology and proliferative rate in various fractions (F_1 - F_4) of bone marrow cells, separated with Percoll density gradient centrifugation

BM fractions:	F_1	F_2	F_3	F_4
Cell yield (% of input)	1.1	5.5	35	24
Mitoses	0.2	1.0	0.2	0
Blasts [1]	9	33	8	3
Proliferative granulocytes [2]	18	30	34	11
Non-proliferative granulocytes [3]	10	3	11	73
Non-proliferative erythroblasts [4]	37	6	25	9
Lymphocytes	18	19	26	5
^{3}H-thymidine uptake [5]	60%	100%	65%	8%

Median values from 7 observations in each fraction are presented. The fraction of destroyed cells was negligible (except in F_1) and these cells have been excluded from the counts.

1 Includes all immature, proliferative cells that have not yet developed characteristics of the granulocyte series (cytoplasmic granules) or erythrocyte series (staining of hemoglobin in the cytoplasm).
2 Promyelocytes and myelocytes.
3 Metamyelocytes and more differentiated stages of the neutrophilic and eosinophilic series. The eosinophils constituted only 1-15% of the total number of cells in the granulocyte series, with the highest concentration in F_4.
4 Polychromatophilic erythroblasts and more mature nucleated cells within the erythrocyte series.
5 Dpm per 10^5 cell, relative to the F_2 value.

certain tumor cells) or with "wrong" password (MHC histoincompatible cells), should be susceptible to NK cell attack. Such mechanisms are, however, not compatible with our findings from the NK cell mediated destruction of allogeneic lymphocytes (ALC) (Fossum et al., 1987). In this system we have found exquisite allospecificity of the NK cells.

Another school of thought likewise believes that NK cells have no inherent allospecificity. The discrepancy between the allospecificity of NK cells against haematopoietic grafts in vivo and the apparent non-MHC-directed killing of tumour cells in vitro could be explained by naturally occurring antibodies directed against allo-MHC-determinants. The NK cells, also known to possess F_c receptors for immunoglobulin, would then act as effector cells in antibody-dependent cellular cytotoxicity (Warner and Dennert, 1985). This hypothesis cannot satisfactorily explain our findings, however, since the selective destruction of allogeneic lymphocytes in vitro (Rolstad and Fossum, 1987) and of allogeneic bone marrow cells, as shown here, can occur in the absence of rat serum or serum factors (Rolstad and Benestad, 1984).

A third school of thought claims that NK cells recognize and kill allogeneic hematopoietic or lymphocyte grafts through positive allorecognition mechanisms. In contrast to cytotoxic T cells, however, the NK cells recognize other antigens than the codominantly expressed MHC antigens. Although this haematopoietic-histocompatibility system (Hh) is poorly defined, the postulation of an alloantigen system with non-codominant expression of target structures, having intimate linkage to the MHC, fits central features of both ALC and ABC (Snell, 1976; Lotzová, 1977; Rolstad and Ford, 1983; Daley and Nakamura, 1984; Bordignon et al., 1985; Rolstad et al., 1985).

Since, according to the latter hypothesis, the Hh antigens are present on all cells within the haematopoietic system (Lotzovà, 1977), one prediction would be that all allogeneic bone marrow cells should be susceptible to NK cell attack, irrespective of cell lineage and cell maturation, a prediction that was strongly supported by the present experiments. Another prediction would be that the genetics and strain distribution of the cytotoxicity against the various categories of bone marrow cells and against lymphocytes should be identical. A third prediction is that different categories of target cells from a given strain should compete for the same NK cell receptor, while targets from unrelated, NK susceptible strains would not necessarily do so in cold target inhibition studies. At least for lymphocytes as targets, a high degree of NK cell specificity against different alloantigen complexes has been found (Fossum et al., 1987), and experiments are in progress to determine whether NK cells recognize the alloantigens present on bone marrow cells vs. lymphocytes through the same receptor structures.

ACKNOWLEDGEMENT

The financial support of Norsk Forening til Kreftens Bekjempelse, The Norwegian Cancer Society, The Norwegian Research Council for Science and the Humanities, Anders Jahre's foundation, and the Nansen and Langfeldt's Funds is greatly appreciated. Ms Inger Strøm-Gundersen, Anna-Bjørg Bore and Wendi Jensen gave expert technical assistance.

REFERENCES

Barlozzari, T., Herberman, R., and Reynolds, C. W., 1987, Inhibition of pluripotent hematopoietic stem cells of bone marrow by large granular lymphocytes, Proc. Natl. Acad. Sci., 84:1.

Bordignon, C., Daley, J. P., and Nakamura, I., 1985, Hematopoietic histoincompatibility reactions by NK cells in vitro: model for genetic resistance to marrow grafts, Science, 230:1398.

Cudkowicz, G., and Bennett, M., 1971, Peculiar immunobiology of bone marrow allografts. I. Graft rejection by irradiated responder mice, J. Exp. Med., 134:83.
Daley, J. P., and Nakamura, I., 1984, Natural resistance of lethally irradiated F_1 hybrid mice to parental marrow grafts is a function of H-2/Hh-restricted effectors, J. Exp. Med., 159:1132.
Fossum, S., Ager, A., and Rolstad, B., 1987, Specific inhibition of natural killer (NK) activity against different alloantigens, Immunogenetics, 26:329
Grönberg, A., Kiessling, R., and Fiers, W., 1985, Interferon- is a strong modulator of NK susceptibility and expression of β 2-microglobulin but not transferrin receptors of K562 cells, Cell. Immunol. 95:195.
Kiessling, R., Hochman, P. S., Haller, O., Shearer, G. M., Wigzell, H., and Cudkowicz, G., 1977, Evidence for a similar or common mechanism for natural killer cell activity and resistance to hemopoietic grafts, Eur. J. Immunol., 7:655.
Kärre, K., Ljunggren, H. G., Piontek, G., and Kiessling, R., 1986, Selective rejection of H-2-deficient lymphoma variants suggests alternative immune defence strategy, Nature, 319:675.
Lotzová, E., 1977, Resistance to parental, allogeneic and xenogeneic hemopoietic grafts in irradiated mice,. Exp. Hemat., 5:215.
Matera, L., Santoli, D., Garbarino, G., Pegoraro, L., Bellone, G., and Pagliardi, G., 1986, Modulation of in vitro myelopoiesis by LGL: different effects on early and late progenitor cells, J. Immunol., 136:1260.
Reynolds, C. W., Timonen, T., and Herberman, R. B., 1981b, Natural killer (NK) cell activity in the rat: I Isolation and characterization of the effector cells, J. Immunol., 127:282.
Reynolds, C. W., Timonen, T., Holden, H. T., Hansen, C., and Herberman, R. B., 1982, Natural killer cell activity in the rat. Analysis of effector cell morphology and effets of interferon on natural killer cell function in the athymic nude rat, Eur. J. Immunol., 12:577.
Rolstad, B., and Benestad, H. B., 1984, The "natural resistance" to bone marrow allografts in normal and athymic nude rats. Rapid cytotoxic reactions both in vivo and in vitro, Eur. J. Immunol. 14:793.
Rolstad, B., and Ford W.L. (1983) The rapid elimination of allogeneic lymphocytes: relationship to established mechanisms of immunity and to lymphocyte traffic, Immunological Rev., 73:87.
Rolstad, B., and Fossum, S., 1987, Allogeneic lymphocyte cytotoxicity (ALC) in rats: Establishment of an in vitro assay, and direct evidence that cells with natural killer (NK) activity are involved in ALC, Immunology, 60:151.
Rolstad, B., Fossum, S., Bazin, H., Kimber, I., Marshall, J., Sparshott, S. M., and Ford, W. L., 1985, The rapid rejection of allogeneic lymphocytes by a non-adaptive, cell-mediated mechanism ("NK" activity), Immunology 54:127.
Rolstad, B., Herberman, R. B., and Reynolds, C. W., 1986, Natural killer cell activity in the rat. V. The circulation patterns and tissue localization of peripheral blood large granular lymphocytes (LGL), J. Immunol., 136:2800.
Snell, G. D., 1976, Recognition structures determined by the H-2 complex, Transplant. Proc., 8:147.
Timonen, T., Reynolds, C. W., Ortaldo, J. R., and Herberman, R. B., 1982, Isolation of human and rat natural killer cells, J Immunol. Methods, 51:269.
Warner, J. F., and Dennert, G., 1982, Effects of a cloned cell line with NK activity on bone marrow transplants, tumor development and metastasis in vivo, Nature 30:31.
Warner, J. F., and Dennert, G., 1985, Bone marrow graft rejection as a function of antibody-directed natural killer cells, J. Exp. Med., 161:563.

ARE LYMPHOKINE ACTIVATED KILLER CELLS A POSSIBLE FORM OF LEUKAEMIA IMMUNOTHERAPY?

J.A. Tratkiewicz, J. Szer and R.L. Boyd

Alfred Hospital and Monash University
Commercial Road, Prahran 3181, Australia

INTRODUCTION

The culture of freshly isolated lymphocytes for three to five days *in vitro* with a lymphokine preparation rich in interleukin 2 (IL-2), results in the generation of effector cells capable of lysing a variety of tumour cells[1]. This phenomenon has been described as lymphokine activated killing (LAK) and represents a lytic system distinct from that of natural killer (NK) cells[2] and cytotoxic T lymphocytes[3]. Subsequent studies have revealed that LAK cells mainly arise from non-B, non-T lymphoid cell precursors and are capable of expanding for short times both *in vivo* and *in vitro* when stimulated by IL-2[4].

LAK cells are lytic for a wide range of tumour cell lines as well as freshly isolated autologous tumours which are generally poorly attacked by unstimulated cytotoxic effector cells. The specificity of generated LAK cells raised the possibility of immunotherapy by infusion of LAK cells and IL-2 *in vivo*. The potential therapeutic use of LAK cells *in vivo* has been studied in murine[5] and human[6] systems and has proved useful in the regression of many established cancers.

It has been reported that levels of natural cytotoxicity against fresh allogeneic leukaemias are generally low[7]. In this study, we analysed the ability of non-stimulated (NK) and activated (LAK) cytotoxic cells to lyse fresh, non-cultured human leukaemias.

MATERIALS AND METHODS

Patients: NK and LAK activity was tested in forty-one controls using blast cells removed from the peripheral blood of twenty-three patients with acute leukaemia, as a target cell source. Diagnosis of leukaemia was made according to standard Cancer and Leukaemia Group B criteria[8].

Effector Cells: Effector cells from forty-one healthy volunteers were prepared from heparinized diluted blood samples separated over lymphocyte separation medium (Flow Laboratories, Australia). Peripheral blood mononuclear cells (PBMC) were collected from the interface, washed and resuspended in supplemented RPMI 1640 (10% foetal bovine serum, glutamine, penicillin, streptomycin).

PBMC were used untreated or following incubation with recombinant IL-2 (Boehringer Mannheim) at a concentration of 1500U/ml for three to five days at 37° in a CO_2 atmosphere. Control cultures were incubated in supplemented RPMI 1640 alone. Following incubation, effector cells were washed and resuspended in supplemented medium.

Target Cells: The K562 cell line[9] was used as the source of target cells to estimate NK cell activity (NKA). The U937 cell line[10] was used as the source of target cells to estimate LAK activity. While U937 is sensitive to LAK cytotoxicity, it is resistant to NK cytotoxicity. Leukaemic blast cells (LBC) were obtained from twenty-three patients with acute leukaemia. Heparinized peripheral blood was obtained prior to treatment and blast cells were collected after density gradient centrifugation. The blast cell counts ranged between 50-90%. Some of these target cells were then labelled with isotope for immediate use, set up in short term culture in supplemented RPMI 1640 and also cryopreserved for future use.

Cytotoxicity Assay: Cytotoxicity tests were performed using a standard ^{51}Cr release assay[11]. All target cells were labelled with 100-200 uCi $Na_2{}^{51}CrO_4$ for one hour at 37°C. Labelled target cells were washed and resuspended in supplemented medium. Generally, LBCs showed lower labelling compared to the cell lines used. Spontaneous release varied between 10-30% on all occasions. Different concentrations of effector and target cells were added in equal volumes in triplicate, to give final effector:target ratios ranging from 50:1 to 3:1. After a four hour incubation period at 37°C, supernatants were collected and percent cytotoxicity assessed by the following formula; % cytotoxicity = 100 x (median experimental release-median spontaneous release)/(median total release-median spontaneous release). For simplicity of presentation, data are given at an effector:target ratio of 50:1.

RESULTS

A total of twenty-three fresh tumour targets and two cell lines were used in combination with effector cells taken from forty-one different donors. NK cell cytotoxicity was evaluated in eighteen healthy volunteers. The median NKA against K562 was 25.3% while NKA against LBCs was not detected (Table 1).

TABLE 1. NKA AGAINST K562 CELLS AND LEUKAEMIC BLAST CELLS (LBC)

TARGET	%CYTOTOXICITY AT 50:1	RANGE
K562	25.3	11.2-48.5%
LBC	0	0-1.2%

Incubation of control PBMC with IL-2 produced increased cytotoxicity against K562 and U937 cells. In 23 healthy controls, the median LAK activity against K562 was 52.8%. The median LAK activity against U937 was 41.3% (Table 2) which differs from NKA against U937 where cytotoxicity is minimal. In contrast to the cell lines tested, fresh leukaemia targets were rarely killed by LAK cells where a median cytotoxicity of 0% was observed.

TABLE 2: LAK CYTOTOXICITY AGAINST LEUKEMIC BLAST CELLS (LBC) AND CELL LINES K562 AND U937

TARGET	%CYTOTOXICITY AT 50:1	RANGE
K562	52.8	29.6-77.3%
U937	41.3	15.4-64.8%
LBC	0	0-16.8%

Although LAK cytoxicity of LBCs was not evident in 21/23 cases, two controls (Table 3) displayed cytotoxicity against allogeneic LBCs.

TABLE 3: CYTOTOXICITY OF LAK CELLS FROM 2 DONORS WITH ACTIVITY AGAINST LEUKAEMIC BLAST CELLS (LBC)

	% CYTOTOXICITY AT 50:1	
TARGET	CONTROL 1	CONTROL 2
K562	62.9	33.5
U937	64.8	23.4
LBC	16.8	14.6

DISCUSSION

Many studies have now demonstrated that the adoptive transfer of LAK cells, in conjunction with recombinant IL-2, can mediate the regression of various established metastases[6,12]. This is dependent upon the ability of LAK cells to recognize and mediate killing which was demonstrated *in vitro* by measuring cytotoxicity. In the case of freshly isolated leukaemias, however, such LAK recognition does not seem to occur as *in vitro* tests showed minimal cytotoxicity against LBCs. Our results showed that although the formation of LAK cells increases the cytotoxic capacity of healthy volunteers against a range of cell lines, these cells do not seem to alter cytotoxic activity against LBCs. The killing of fresh LBC does not appreciably alter with IL-2 activated or untreated PBMCs. The susceptibility of leukaemic blasts to LAK cytotoxicity rarely approaches that demonstrated in solid tumours thus it would be safe to assume that in leukaemia, simple LAK activation alone would not be an effective form of therapy.

Two healthy volunteers displayed LAK cytotoxicity against allogeneic LBCs. We are uncertain as to the explanation for this killing, however, it was interesting to note that one of the controls displayed LAK cytotoxicity against LBCs taken from his father. In this case, LAK cytotoxicity against LBC was 16.4% however, the same LAK cells from this control were unable to lyse a non-related freshly isolated leukaemia.

Our future studies will be aimed at the autologous system. LAK cells will be produced from PBMC of patients who achieve complete remission and LAK cytotoxicity assessed against previously stored autologous LBC. We will also test the cytotoxicity of relative's LAK cells against LBCs. By endeavouring to amplify LAK eradication of acute leukaemia.

REFERENCES

1. Lotze M.T., Grimm E.A., Mazumder A., Strausser J.L., Rosenberg S.A., Lysis of fresh and cultured autologous tumor by human lymphocytes cultured in T cell growth factor. Cancer Res. 41:4420, (1981).

2. Grimm E.A., Maxumder A., Zhang H.Z., Rosenberg S.A., Lymphokine activated killer cell phenomenon. Lysis of natural killer resistant fresh solid tumor cells by interkeukin 2 activated autologous human peripheral blood lymphocytes. J. Exp. Med. 155:1823, (1982).

3. Grimm E.A., Ramsey K.M. Mazumder A., Wilson D.J., Djeu J.Y., Rosenberg S.A., Lymphokine activated killer cell phenomenon II. Precursor phenotype is serologically distinct from peripheral T lymphocytes, memory cytotoxic thymus-derived lymphocytes and natural killer cells. J. Exp. Med. 157:884, (1983).

4. Ettinghausen S.E., Lipford E.H., Mule J.J., Rosenberg S.A., Recombinant interleukin 2 stimulates in vivo proliferation of adoptively transferred lymphokine activated killer cells. J. Immunol. 135:3623, (1985).

5. Ettinghausen S.E., Rosenberg S.A., Immunotherapy of murine sarcomas using lymphokine activated killer cells: Optimization of the schedule and route of administration of recombinant interleukin 2. Cancer Res. 46:2784, (1986).

6. Jacobs S.K., Wilson D.J., Kornblith P.L., Grimm E.A., Interleukin 2 and autologous lymphokine activated killer cells in the treatment of malignant glioma. J. Neurosurg. 64:743, (1986).

7. Zarling J.M., Eskra L., Borden E.C., Horoszewicz J., Carter W.A., Activation of human natural killer cells cytotoxic for human leukemia cells by purified interferon. J. Immunol. 123:63, (1979).

8. Yates J., Glidewell O., Wiernik P., Cooper M.R., Steinberg D., Dosik H., Levy R., Hoagland C., Henry P., Gottlieb A., Cornell C., Berenberg J., Hutchinson J.L., Raich P., Nissen N., Ellison R.R., Frelick R., James G.W., Falkson G., Silver R.T, Haurani F., Green M., Henderson E., Leone L., Holland J.F., Cytosine arabinoside with daunorubicin or adriamycin for therapy of acute myelocytic leukemia: a CALGB study. Blood 60:454, (1982).

9. Andersson C.C., Nilsson K., Gahmberg C.G., K562- a human erythroleukaemic cell line. Int. J. Cancer 23:143, (1979).

10. Sundstron C., Nilsson K., Establishment and characterization of a human histiocytic lymphoma cell line (U937). Int. J. Cancer 17:565,(1976).

11. Brunner K., Engers H., Cerottini J.C., The ^{51} Cr release assay as used for the quantitative measurement of cell-mediated cytolysis in vitro, in: "In vitro methods in cell mediated and tumor immunity," Bloom B, David J eds., Academic Press 1976, p 423.

12. Mazumder A., Rosenberg S.A., Successful immunotherapy of natural killer-resistant established pulmonary metastases by the intravenous adoptive transfer of syngeneic lymphocytes activated in vitro by interleukin 2. J. Exp. Med. 159:495, (1984).

BLOOD LEUKOCYTE NATURAL KILLER ACTIVITY AS AN EARLY DIAGNOSTIC INDICATOR OF LEUKAEMIC RELAPSE

J.A. Tratkiewicz, J. Szer and R.L. Boyd

Alfred Hospital & Monash University
Prahran, Victoria, 3181, Australia

INTRODUCTION

Natural cell mediated immunity is a well-defined system that may play a role in cancer defence. In contrast to specific T-cell mediated immunity, the mediators of natural immunity, Natural Killer (NK) cells, appear to be non-specific and their lytic activity is not restricted by major histocompatibility (MHC) determinants[1].

NK cells are a distinct population of non-T, non-B lymphoid cells[2] which bear monocytic[3], lymphocytic[4] and unique markers[5]. They are characterized morphologically as large granular lymphocytes[6] and are present in the circulating mononuclear cell pool and lymphoid tissues.

In this study, we confirm reports of depressed NK activity (NKA) in patients with untreated leukaemia[7] and demonstrate the return of NKA in patients who achieve complete remission. We also demonstrate that in some circumstances it is possible to predict leukaemic relapse by a sustained fall in NKA, even where the patient otherwise still has the morphologic and cytogenetic criteria of remission.

MATERIALS & METHODS

Patients: NKA was tested in patients with acute leukaemia. In all, seventy-six patients with active disease (i.e. at presentation or relapse) and fifty-six patients in complete remission were assessed. A normal range of NKA was determined in sixty healthy volunteers. Diagnosis of leukaemia and remission status was made according to standard Cancer and Leukaemia Group B criteria[8].

Effector Cells: Heparinized peripheral blood samples were collected, diluted one to one with Phosphate buffered saline and separated by density gradient centrifugation[9] using lymphocyte separation medium (Flow Laboratories, Australia). Peripheral blood mononuclear cells (PBMC) were washed and resuspended in RPMI-1640 (Flow laboratories, Australia) containing 10% heat-inactivated foetal bovine serum, 2mM glutamine, 100units/ml penicillin and 100mg/ml streptomycin (RPMI-FBS).

Target Cells: The K562 cell line, derived from a patient with chronic myeloid leukaemia in blast crisis[10], was used as the target source. This cell line was maintained in RPMI-FBS at 37° in a 5% CO_2 humidified atmosphere.

Cytotoxicity Assay: Cytotoxicity tests were performed by a standard ^{51}Cr release assay[11] in a total volume of 200ul. Effector cells and ^{51}Cr labelled target cells were added, in varying ratios ranging from 50:1 to 3:1, in triplicate. After a four hour incubation, supernatant was collected and percent cytotoxicity was assessed according to the following formula: % cytotoxicity = 100 x (median experimental release - median spontaneous release)/(median total release-median spontaneous release).

Monoclonal Antibody Analysis: Isolated PBMC from patients with active disease, patients in remission and healthy volunteers were analysed by indirect immunofluorescence with anti-leu 7 and anti-leu 11 antibodies in a flow cytometer (FACS II, Becton-Dickinson, Sunnyvale CA, USA).

Statistics: The Wilcoxon two sample test was used to compare NK cell activity between the patient groups. A simple regression analysis was used to correlate NKA with phenotypic markers.

RESULTS

NK Cell Mediated Cytotoxicity: The data for all groups are shown in Table 1. Results were expressed as percentage cytotoxicity at an effector to target ratio of 50:1. The median NKA in healthy volunteer controls was 33.6% with a range of 15.4 - 71%. Patients who were in complete remission had a median NKA of 22.7% with a range of 9.5 - 64.4%. The median NK cell cytotoxicity in patients with active disease was 2.4% with a range of 0-13.4%, significantly less than the cytotoxicity in the control group ($p < .001$) and the remission group ($p < .0.001$).

Prediction of Leukaemic Relapse: NKA was tested at regular intervals (4-8 weekly) in patients who achieved a complete remission. In sixteen patients, we noticed a sudden, significant and maintained drop in NKA or failure to normalize NK activity at the time of apparent complete remission. Within one to ten weeks, cytogenetic and morphologic relapse was documented (Table 2).

No correlation was observed between NKA at diagnosis and prognosis as all patients had extremely low initial NKA.

Monoclonal Antibody Analysis: An analysis of PBMC for the relative proportions of leu 7 and leu 11 bearing cells was performed on 16 and 9 subjects, respectively, using flow cytometry. No statistical correlation could be drawn between NKA and the percentages of leu 7 and leu 11 positive cells in PBMC populations.

DISCUSSION

This study demonstrated that adult patients with acute leukaemia have significantly reduced peripheral blood NKA compared to healthy volunteers and patients who have achieved complete remission. Patients in remission displayed NK cell cytoxicity lower than that of healthy controls ($p < 0.001$). Furthermore, there was a large variation in the range of NKA between individuals in the healthy volunteer group and patients in complete remission. Individual subjects, however, generally showed little variability in sequential testing. These findings support earlier reports of a correlation between NKA and the course of the disease in patients with leukaemia[12].

Various explanations for the depressed NKA have been proposed. The presence of suppressor factors[13], defective function of NK cells[14] and reduced numbers of circulating effector cells[12], have all been described to explain this phenomenon.

A drop in NKA preceeded or coincided with leukaemia relapse in twenty-two of the patients tested. A sudden and maintained drop in NKA was not seen other than in patients who subsequently relapsed.

TABLE 1. NK ACTIVITY AND DISEASE STATUS

	n*	% Cytotoxicity @ 50.1 Median	Range
Controls	60	33.6	15.4-71
Active Disease	76	2.4	0-13.4
Remission	56	22.7	9.5-64.4

*n denotes number of patients in each group
Controls vs active disease $p<0.001$
Remission vs active disease $p<0.001$

TABLE 2. SEQUENTIAL ASSAYS IN PATIENTS WHO SUBSEQUENTLY RELAPSE

PATIENT	DISEASE	CONSECTIVE NK ASSAYS % CYTOTOXICITY AT 50:1			INTERVAL TO RELAPSE
1	ANLL	20.7	5.2	5.5	5
2	ANLL	27.6	12.9	7.6	6
3	ANLL	39.7	15.5	1.8	1
4	ANLL	27.5	6.2	1.1	8
5	ANLL	21.0	12.7	4.6	8
6	ANLL	14.2	5.0	2.7	4
7	ANLL	16.0	8.6	1.0	5
8	ANLL	21.8	4.5	2.3	3
9	ANLL	29.5	9.0	4.3	1
10	ANLL	19.0	0.9		10
11	ANLL	37.0	9.0	0	5
12	ALL	17.7	5.7	1.8	2
13	ALL	38.0	9.0	0	8
14	ALL	25.6	8.0	5.0	7
15	ALL	15.5	4.8	3.9	9
16	ALL	23.4	2.3	0	5

* Abbreviations: ANLL - Acute Myelogeneous Leukaemia
ALL - Acute Lymphoblastic Leukaemia

The monoclonal antibodies anti-leu 7 and anti-leu 11 have been used in recent studies to characterize subpopulations of NK cells[5]. Quantitative analysis with these antibodies by indirect immunofluorescence revealed no correlation between NKA and relative proportions of leu 7 and leu 11 bearing cells. These results indicate that while these markers may give an estimate of NK cell numbers, they do not allow for the functional evaluation of NKA.

In conclusion, we suggest that there exists a correlation between NKA and leukaemia cell load, with a sensitivity greater than that of light microscopy of bone marrow. The routine use of this assay, when combined with morphologic analysis may permit earlier diagnosis of leukaemic relapse and firmer diagnoses of complete remission.

REFERENCES

1. Ortaldo J.R. & Herberman R.B., Heterogeneity of natural killer cells. Annu. Rev. Immunol. 2:359 (1984).

2. Lanier L.L., Cwirla S., Federspiel N., & Phillips J.H., Human natural killer cells isolated from peripheral blood do not rearrange T cell antigen receptor B chain genes. J. Exp. Med. 163:209 (1986).

3. Barada F.A. Jr., Kay H.D., Emmons R., Davis J.S., Horwitz D.A., Natural cytotoxicity of adherent cells from human blood: role of the Fc receptor. J. Immunol.125:865, (1980).

4. Bakacs T., Gergely P., Cornain S., Klein E., Characterization of human lymphocyte subpopulations for cytotoxicity against tumor-derived monolayer cultures. Int. J. Cancer 19:441, (1977).

5. Lanier L.L., Le A.M., Phillips J.H., Warner N.L., Babcock G.F., Subpopulations of human natural killer cells defined by expression of the Leu-7 (HNK-1) and Leu 11 (NK-15) antigens. J. Immunol. 131:1789, (1983).

6. Timonen T., Ranki A., Saksela E., Hayry P., Fractionization, morphological and functional characterization of effector cells responsible for human natural killer activity against cell-line targets. Cell. Immunol. 48:121 (1979).

7. Sorskaar D., Lie S.O., Forre O., Natural killer cell activity of peripheral blood and bone marrow mononuclear cells from patients with childhood acute lymphoblastic leukemia. Acta. Paediatr. Scand. 74:433 (1985).

8. Yates J., Glidewell O., Wiernik P., Cooper M.R., Steinberg D., Dosik H., Levy R., Hoagland C., Henry P., Gottlieb A., Cornell C., Berenberg J., Hutchinson J.L., Raich P., Nissen N., Ellison R.R., Frelick R., James G.W., Falkson G., Silver R.T, Haurani F., Green M., Henderson E., Leone L., Holland J.F., Cytosine arabinoside with daunorubicin or adriamycin for therapy of acute myelocytic leukemia: a CALGB study. Blood 60:454, (1982).

9. Boyum A., A one-stage procedure for isolation of granulocytes and lymphocytes from human blood. General sedimentation properties of white blood cells in a 1g gravity field. Scand. J. Clin. Lab. Invest. 21 (Suppl 97): 51, (1968).

10. Andersson L.C., Nilsson K., & Gahmberg C.G., K562 - A human erythroleukemic cell line. Int. J. Cancer 23:143 (1979).

11. Brunner K.T., Engers H.D., Cerottini J.C., The ^{51}Cr release assay as used for the quantitative measurement of cell-mediated cytolysis in vitro, in: "In vitro methods in cell-mediated and tumor immunity," Bloom B.R, David J.R, eds., Academic Press, New York (1976).

12. Dickinson A., Proctor S.J., Jacobs E., Reed M.M., Walker W., Craft A.W., Kernahan J., Natural killer cell activity in childhood acute lymphoblastic leukaemia in remission. Br. J. Haematol. 59:45 (1985).

13. Kay N.E, Zarling J.M., Impaired natural killer activity in patients with chronic lymphocytic leukemia is associated with a deficiency of azurophilic cytoplasmic granules in putative NK cells. Blood 63:305, (1984).

14. Ruco L.P, Procopio A., Maccallini V., Calogero A., Uccini S., Annino C., Mandelli F., Baroni C.D.: Severe deficiency of natural killer activity in the peripheral blood of patients with hairy cell leukemia. Blood 61:1132 (1983).

PROLIFERATION OF LARGE GRANULAR LYMPHOCYTES IN THE RAT

J. T. Vaage, S. Fossum, and B. Rolstad

Anatomical Institute, University of Oslo, Karl Johansgt. 47

N-0162 Oslo 1, Norway

INTRODUCTION

In rats, most of the natural killer cell activity against neoplastic cells is associated with a class of leukocytes with an easily recognizable morphology. In Giemsa-stained cytocentrifuge preparations they appear as mononuclear cells with a kidney-shaped nucleus and with azurophilic granules in their cytoplasm, usually located in the cytocentre (Fig. 1). Cells with this morphology are usually referred to as large granular lymphocytes (LGL). These cells are found in many different organs of the body, but are particularly numerous in blood, spleen, liver and lungs, from where they are easily isolated. Although these cells have received interest for several years as putative effector cells in the surveillance against neoplasia, little is as yet known about their physiological turnover and migration patterns. A bone marrow origin of LGL is well documented (1), and the majority of these cells apparently mature in the absence of a thymus since they are found in increased numbers in athymic nude rats (2). However, we do not yet know if LGL can proliferate under physiological conditions <u>in vivo</u>, and if they can, to what extent their proliferation rate can be modulated, e.g. by cytokines. The intentions of the present study were:

a) to compare the rate of proliferation of LGL in blood, spleen and liver of normal, unstimulated rats with that of cell populations with a known long life span (small recirculating lymphocytes) or short life span (an LGL leukemic cell line), and

b) to look at the effect of the interferon inducer Poly I:C on the rate of proliferation of LGL.

MATERIALS AND METHODS

Animals

Adult, inbred AO male rats 2-4 months old from our own animal facilities were used.

Poly I:C treatment

Rats were injected iv. with 1 mg of Poly I:C (Sigma, St. Louis, USA) dissolved in 1 ml of PBS approximately 20 h before killing.

Preparation of cells

Peripheral blood leukocytes were obtained by cardiac puncture, single cell suspensions from the spleen were prepared by mincing spleen fragments in ice-cold phosphate buffered saline (PBS), and liver cells were obtained by high pressure perfusion of the liver sinusoids, as described (3). The cell suspensions were centrifuged on Lymhoprep (Nycomed A/S, Oslo, Norway - density 1.077) and LGL were enriched by removing adherent B cells and monocyte-macrophages by incubation on columns of nylon wool (NW) (Leukopak, Fenwal Laboratories, Ilinois, USA) for 1 h at 37°C. The medium used was RPMI 1640 with HEPES buffer (25 mM/l) and 5% foetal calf serum (FCS) and a standard antibiotic-antimycotic solution added (all reagents purchased from Gibco, Paisley, Scotland), hereafter referred to as complete medium (CM). In some experiments, LGL were further purified by density gradient centrifugation on Percoll (Pharmacia A/B, Uppsala, Sweden) (4,5). Thoracic duct cells were isolated by a standard technique (6). The IL-2 independent LGL tumor cell line CRC- was a kind gift from dr. C.W. Reynolds, NIH, USA (7).

Labelling of cells with ^{3}H-thymidine

Cell suspensions were incubated for 1 h in CM in vitro at 37°C with 0.25 uCi ^{3}H-thymidine (9.3 kBq) (Code TRK. 120, Amersham, England) added per ml and thereafter washed three times before making cytocentrifuge preparations of the cells.

Autoradiography

Cytospins fixed in methanol and stained with Giemsa (4) were covered with a layer of PVC (a copolymer of vinylchloride and vinylidenechloride) (8) (Ixan, S.G.A. Solvay S.A., Brussels) (12% PVC in butylacetate, trichloroethylene and dibutylphtalate) before dipping in the photographic emulsion (Kodak NTB-2 diluted 1:1 in dest. water) (8,9). Exposure time was usually 4-5 weeks, and cells that had more than 10 grains over their nucleus, were assessed as S-phase cells.

RESULTS

Cytocentrifuge preparations of LGL were stained with Giemsa and covered with a layer of PVC before the photographic emulsion was applied. The stained cells were thereby protected against the photographic processing fluids. Furthermore, positive chemography generated by the Giemsa stain was prevented (8,9). In this way the LGL morphology was well preserved, and S-phase LGL were easily identified (Fig. 1).

The pulse labelling of LGL in vitro showed that even in unstimulated normal rats about 5% of blood LGL were in the S-phase (Table 1). These results were confirmed by pulse labelling the cells in vivo with the isotope before separation of the LGL (data not shown). Also the liver and spleen LGL populations contained S-phase LGL. In all three compartments the labelling index was significantly higher than that of thoracic duct lymphocytes, which contained virtually no labelled cells (Table 1).

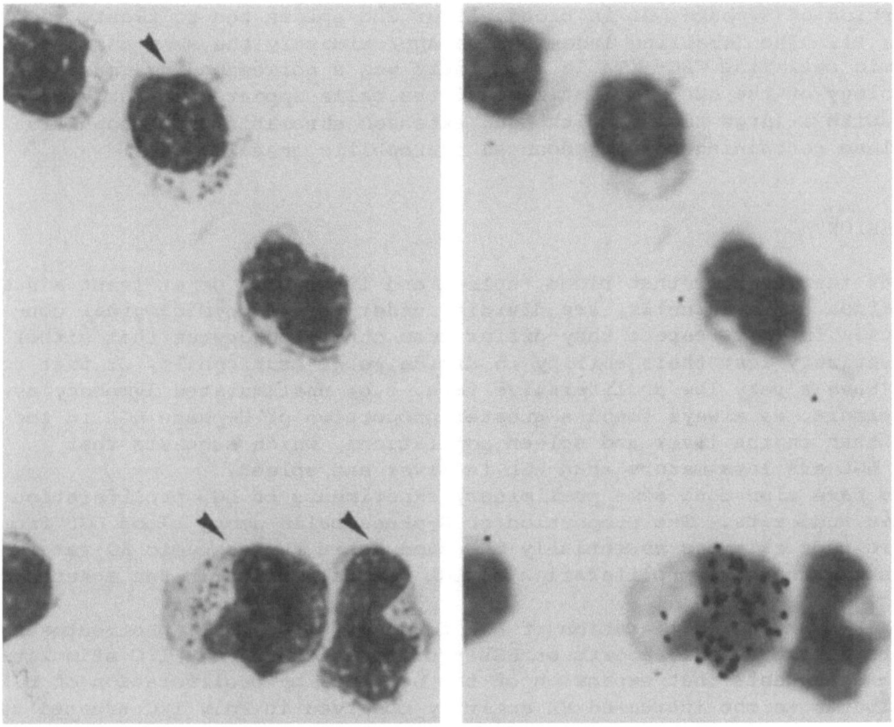

Fig. 1. S-phase LGL from blood shown by autoradiography.
3 LGLs with azurophilic cytoplasmic granules are shown (arrowheads, left picture), one of which has incorporated ^{3}H-thymidine (numerous silver grains above the nucleus, right picture)

Table 1. The Proportion of S-Phase LGL Determined by ^{3}H-thymidine Incorporation and Autoradiography

	Labelling index, LGL	
	Normal rats	Poly I:C stimulated rats
Blood	5.1%	34.3%
Spleen	2.4%	32.4%
Liver	1.3%	27%
	Labelling index, control cells	
Small thoracic duct lymphocytes		<< 1%
LGL leukemic cell line (RNK)		27.1%

Pretreatment of rats with the interferon inducer Poly I:C increased the proportion of S-phase LGL in blood, liver and spleen ten to twenty fold (Table 1). The labelling index became approximately the same as in the LGL leukemic cell line CRC- (Table 1). There was a corresponding change in the morphology of the LGL, in that most of the cells appeared as activated cells with a large nucleus with much extended chromatin and basophilic cytoplasm containing more pronounced azurophilic granules.

DISCUSSION

Our results show that blood, spleen and liver LGL, or at least a subpopulation of these cells, are dividing under normal physiological conditions. In this respect they differ from other leukocytes that either have entirely lost their ability to divide, e.g. neutrophils, or that normally have a very low proliferative rate, e.g. unstimulated lymphocytes. Furthermore, we always found a greater proportion of S-phase LGL in the blood than in the liver and spleen populations, which suggests that blood LGL are less mature than LGL in liver and spleen.

We have also done some preliminary experiments on LGL proliferation in athymic nude rats. The proportion of S-phase cells among blood LGL from athymic nude rats was essentially the same as in the euthymic AO-rats. This shows that the proliferation of LGL can take place in the absence of a thymus.

The enhanced proliferation of LGL found after Poly I:C pretreatment corroborates previous reports on LGL proliferation in Poly I:C stimulated rats and suggests that expansion of the LGL pool by proliferation of LGL contributes to the increased NK activity observed in Poly I:C treated rats (10). However, we do not yet know whether the Poly I:C treatment expands the already existing cell populations in the three compartments, or mobilizes immature LGL into the tissues or both.

REFERENCES

1. O. Haller, R. Kiessling, A. Örn, and H. Wigzell, Generation of natural killer cells: an autonomous function of the bone marrow, J. Exp. Med., 145:1411, (1977).
2. C. W. Reynolds, T. Timonen, H. T. Holden, C. T. Hansen, and R. B. Herberman, Natural killer cell activity in the rat. Analysis of effector cell morphology and effects of interferon on natural killer cell function in the athymic (nude) rat, Eur. J. Immunol., 12:577 (1982).
3. L. Bouwens, L. Remels, M. Baekeland, H. V. Bossuyt, and E. Wisse, Large granular lymphocytes or "Pit cells" from rat liver: isolation, ultrastructural characterization and natural killer activity, Eur. J. Immunol., 17:37 (1987)
4. T. Timonen, C. W. Reynolds, J. R. Ortaldo, and R. B. Herberman, Isolation of human and rat natural killer cells, J. Immunol. Meth., 51:269 (1982).
5. C. W. Reynolds, T. Timonen, and R. B. Herberman, Natural killer (NK) cell activity in the rat. I. Isolation and characterization of the effector cells, J. Immunol., 127:282 (1981).
6. W. L. Ford, The preparation and labelling of lymphocytes, in: "Handbook of Experimental Immunology", D. M. Weir, ed., Blackwell Scientific Publications, London (1978)

7. C. W. Reynolds, E. W. Bere, Jr., and J. M. Ward, Natural killer activity in the rat. III. Characterization of transplantable large granular lymphocyte (LGL) leukemias in the F344 rat, J. Immunol., 132:534 (1984).
8. A. Keyser, and C. Wijffels, The preparation and use of polyvinylidenechloride protective films in autoradiography, Acta Histochem., Suppl. 8:359 (1968).
9. H. B. Benestad, A modified technique for combination of cytochemistry and radioautography of haemic cells, Scand. J. Haematol., 18:54 (1977).
10. C. A. Biron, K. F. Pedersen, and R. M. Welsh, Purification and target cell range of in vivo elicited blast natural killer cells, J. Immunol., 137:463 (1986).

CELL TRAFFIC IN THE IMMUNE SYSTEM

ESTABLISHMENT OF LYMPH NODE DERIVED ENDOTHELIAL CELL LINES WHICH SHOW LYMPHOID CELL BINDING

Reiner Harder, Birgit Schütt, Adrian Duijvestijn,
Heinz-Günter Thiele and Alf Hamann

Abt. f. klin. Immunologie, I. Med. Klinik
Univ.krankenhaus, 2 Hamburg 20, F.R.G.

INTRODUCTION

The recirculation of lymphocytes through distinct parts of the body is regulated at least partially by sets of homing receptors on lymphocytes recognizing organ specific determinants on high endothelial venules (HEV). In the mouse, interaction between lymphocytes and peripheral node HEV can be inhibited by the monoclonal antibody MEL-14, reacting with a ubiquitinated 90 kD glycoprotein on the lymphocyte surface (Ref.1,2). Recently monoclonal antibodies against putative ligands for lymphatic receptors on endothelial cells have been obtained (Ref.3), and an HEV-specific marker, defined by the monoclonal antibody MECA 325, has been described (Ref.4). However, molecular analysis of such molecules is complicated by the limited availability of endothelial cells carrying homing receptor ligands. Attempts to bring these specialised endothelial cells into culture succeeded eventually in the rat system (Ref.5), but not in the mouse. Prerequisite for a molecular characterisation is a source with larger quantities of endothelial cells than those from the lymphatic organs of the mouse. To achieve this we tried to transform high endothelial cells to establish permanently growing endothelial cell lines of mouse origin.

METHODS

1) Isolation and transformation of lymph node endothelial cells

Peripheral lymph nodes from CBA mice were disrupted mechanically and after washing out the majority of the lymphocytes the remainder was dissociated enzymatically (collagenase/dispase). Clusters of endothelial cells were enriched by sedimentation through a cushion of 30% FCS in RPMI 1640 and transferred to plastic tissue culture dishes. Non-adherent cells were washed off after a 16 hours incubation period (37°C, 6% CO_2).
The adherent cell fraction was incubated over night with SV40 virus. After cultivation for some weeks the obtained transformed, permanent growing cells were cloned by limiting dilution and screened with the HEV-specific marker MECA 325 (Ref.4). An initial analysis revealed that

binding of the MECA 325 antibody to HEV-cells of peripheral lymph nodes was not affected by trypsin treatment. Therefore subclones of the transformed cells could be screened with antibody after trypsinisation.

2) Binding assay

For the determination of lymphocyte binding to the transformed cell lines a modified binding assay (Ref.6) was used: Microtiterplates were coated over night with 50 µl/per well of a collagen solution (0,5mg/ml; type III from calf skin, Sigma). Transformed endothelial cells were trypsinized and plated at 1.5×10^4 /cup . The next day the monolayer cells were washed and incubated in the presence or absence of MEL-14 and/or anti-LFA-1 (H35.89.9; Ref.7) at 37°C with 1.5×10^5/cup lymphocytes or cells of the T-cell hybridoma 16F3 (Ref.8), previously biotinylated by reaction with biotinyl N-hydroxysuccinimide ester. Cells were sedimented by a short centrifugation at 25 x g. After 30 min the cups were filled with buffer, sealed with an adhesive plastic sheet and centrifuged upside down for 2 minutes at 750 x g. The supernatant was decanted, and the number of remaining bound lymphocytes was determined in an ELISA-type procedure after incubation with avidin-peroxidase.
The proper adherence of the transformed cells to the culture vessel (microtiter plate) was essentiell for the centrifugation step in that assay. This adherence was greatly increased by coating the plastic with collagen, such that centrifugation at 750 x g the endothelial cells remained attached to the substrate, while unspecifically bound lymphocytes or 16F3 cells were centrifuged off.

3) Frozen section assay

The assay was essentially done as described (Ref.9).

4) Immunofluorescence screening procedure:
Cells were trypsinised, washed and incubated with MECA 325 (30yg/ml) or an isotype-matched control antibody. Second antibody was a FITC-mouse anti rat IgG (Jackson/Dianova, Hamburg), diluted 1:20 in 10% FCS. Clones showing at least partial reactivity with MECA 325 were selected and further subcloned.

RESULTS

1) Expression of the HEV-marker MECA 325 on transformed cells

Cloned lines of the transformed, permanently growing cells were screened for the expression of the HEV-specific marker MECA 325 (Ref.4). The different MECA 325 positive clones expressed the antigen in different amounts. Only 1 clone (3H3B1) showed a 100% expression of that antigen while in the other cases only 1% to 50% of the cells of a given clone were MECA 325 positive. Attemps to increase the proportion of MECA 325 positive cells by panning with that antibody were not successful, the following generations of these selected cells expressed the antigen to the same amount as the original clone.
It appears that the expresssion of the MECA 325 antigen decreases upon time of culture (about 10 month up to now) and its density on the transformed cells is weaker than on native HEV-cells.

2) Reactivity of monoclonal antibodies raised against the transformed cell lines

A rat was immunized with the transformed cell line 3H3B1. Hybridoma supernatants obtained from a fusion using this rat were tested on lymph node sections. 8 out of 130 supernatants tested in a first series stained different kinds of endothelial cells in the lymph node. Similar results were reported by George Kraal (personal communication).

3) Binding of lymphocytes

The most important functional criterium for HEV-type endothelial cells of lymph nodes is the homing receptor mediated binding of lymphocytes. To examine this question a modified binding assay (Ref.6) was applied.

Table 1: A representative experiment for inhibition of lymphocyte binding to monolayers of transformed cells. Shown are the O.D.495 values of at least 6 wells (for details see `Method` section).

Cell line	PMA	Antibody Control	MEL-14	MEL-14 + αLFA-1	% Inhibition MEL-14	MEL-14 + αLFA-1
1E7	-	1,913	1,780	1,688	7,0	11,5
±SD		±0,040	±0,061	±0,059	±3,2	±3,1
3H3B1	-	1,871	1,653	1,532	11,65	18,1
±SD		±0,057	±0,048	±0,077	±2,6	±4,1
1E7	+	1,876	1,503	1,293	19,5	26,7
±SD		±0,062	±0,149	±0,145	±7,9	±7,7
3H3B1	+	1,609	1,421	1,159	11,7	30,0
±SD		±0,078	±0,088	±0,143	±5,5	±8,9

The results show that the 2 transformed MECA 325 positive endothelial cell lines under study are capable of binding lymphoid cells. This binding is weakly, but significantly reduced by the MEL-14 antibody. Additional anti-LFA-1 antibody increased the inhibitory effect on binding. Preincubation of the endothelial cells with Phorbol-12-myristate-13-acetate (PMA; 10 ng/ml, 4 hours at 37°C) reduced the basal binding (3H3B1) and increased the inhibitory effect of MEL-14 and anti-LFA-1 on binding (3H3B1 and 1E7).

4) Functional relation to lymphocyte-HEV-binding

Since it is known that cells shed surface molecules into the culture medium, we wanted to examine whether material produced by the cell lines is capable of inhibiting lymphocyte-HEV interaction. Lymphocytes were preincubated with culture supernatant of transformed MECA 325 positive cell lines and then overlaid on frozen sections of peripheral lymph nodes from CBA mice. The data (table 2) show a marked inhibition in the mean number of lymphocytes bound to an HEV with both of the supernatants in comparison with control supernatants.
Similarly, a polyclonal rat antiserum raised against a transformed cell line inhibited lymphocyte binding to HEV by 49%.

Table 2: Inhibition of lymphocyte binding on frozen sections of peripheral lymph nodes

Preincubation lymphocytes	Average bound lyphocytes per HEV	Number of counted HEV`s	% Inhibition
Control	9,5	48	0
1E7 Supern.	5,1	34	46
3H3B1 Supern.	5,8	65	39
MEL-14	0	62	100

DISCUSSION

It has been shown, that at least 2 different receptor systems for binding of recirculating lymphocytes to HEV of either peripheral lymph nodes or Peyers patches do exist (Ref.1,10). Molecules on the lymphocyte surface acting as putative receptors for organ-specific determinants on HEV have been identified by monoclonal antibodies in the mouse, rat and human (Ref.1,11,12,13,14). An additional system has been described for lymphocyte binding to HEV of rheumatoid arthritis patients (Ref.15). However, until recently little was known about the ligands for lymphocyte homing receptors on HEV-cells.
To establish the prerequesites for a molecular characterization of homing receptor ligands and other molecules relevant for lymphocyte binding to HEV, we tried to establish permanent lines of HEV derived cells by transformation with the tumor virus SV40. Data reported here demonstrate the endothelial nature of the transformed cells:
1.) They express, at least transiently, the HEV-specific marker MECA 325 and
2.) supernatants of hybridomas obtained from a fusion using a rat that was immunised with a transformed cell line stain in a high percentage endothelial cells in sections of lymph nodes.
3.) Functional properties: The binding of the T cell hybridoma 16F3 or mouse lymphocytes to the transformed cell lines is weakly but significantly inhibited by the homing receptor specific MEL-14 antibody. However this receptor system apparently makes only a small contribution to total binding. The same applies for the LFA-1 adhesion mechanism. LFA-1 has been reported to have an accessory role in lymphocyte binding (Ref.16) and also in binding of human lymphocytes to human endothelial cells (Ref.17). It can be suggested, that hitherto not identified mechanisms contribute to the adherence of lymphoid cells to the endothelial cell lines described here.
Preliminary experiments show that antibodies against putative ligands for homing receptors on HEV, raised recently (Ref.3) gave no detectable staining. Eventually transformed cells represent a differentiation state of lymph node cells, in which the ligands for the homing receptor are not constitutively or only in low amounts expressed. This may be concluded from experiments (Ref.18) where the influx of lymphocytes from the afferent lymph is interrupted and subsequently the binding of lymphocytes to HEV disappears. Also the inducibility of HEV-specific morphology and -markers in inflamed tissue and upon induction with y-interferon supports this view (Ref.19,20). It is conceivable that distinct factors may induce lymph node endothelial cells (possibly other endothelial cells as well) to express the ligands for homing receptors.
Inhibition of lymphocyte binding in the frozen section assay by supernatants from cultured transformed cells or by polyclonal antiserum raised against these cells shows that at least part of the binding structures, that are expressed on the transformed endothelial cell lines,

play a role for physiological binding of lymphocytes to HEV. Nevertheless, it is unclear, whether these structures are related to the homing receptor itself or rather to accesssory molecules relevant for lymphocyte-HEV adhesion.
In conclusion, transformed cell lines of mouse lymph node origin could be a tool for investigating the role of endothelium in the recirculation and homing of lymphocytes and in the study of mechanisms regulating lymphocyte extravasation.

ACKKNOWLEDGEMENTS: We thank Dr. M. Breindl for SV40 virus, P. Streeter and E.C. Butcher for supplying monoclonal antibodies, reacting with putative HEV-ligands, M. Pierres for anti-LFA-1 producing hybridomas, and S. Kaufmann for the T-cell hybridoma 16F3. This work was supported by the DFG.

REFERENCES

1. Gallatin, W.M., I.L. Weissman and E.C. Butcher. 1983. A cell-surface molecule involved in organ-specific homing of lymphocytes. Nature 303:30.
2. St. John, T. W.M. Gallatin, M. Siegelman, H.T. Smith, V.A. Fried and I.L.Weissman. 1986. Expression cloning of a homing receptor cDNA: Ubiquitin is the reactive species. Science 231:845.
3. Streeter, P., E. Lakey, B.T. Rouse, R. Bargatze and E.C. Butcher. A tissue-specific endothelial cell molecule involved in lymphocyte homing. Submitted.
4. Duijvestijn, A.M., M. Kerkhove, R.F. Bargatze and E.C. Butcher. 1987. Lymphoid tissue- and inflammation-specific endothelial cell differentiation defined by monoclonal antibodies. J. Immunol. 138:713.
5. Ager, A. 1987. Isolation and culture of high endothelial cells from rat lymph nodes. J. Cell Science 87:133.
6. Somers, S.D., C.C. Whisnant and D.O. Adams. 1986. Quantification of the strength of cell-cell adhesion: The capture of tumor cells by activated murine macrophages proceeds through two distinct stages. J. Immunol. 136:1490.
7. Pierres, M., C. Goridis, and P. Golstein. 1982. Inhibition of murine T cell-mediated cytolysis and T cell proliferation by a rat monoclonal antibody immunoprecipitating two lymphoid cell surface polypeptides of 94000 and 180000 molecular weight. Europ. J. Immunol. 12:60.
8. Kaufmann, S.H.E., K. Eichmann, I. Müller and L. Wrazel. 1985. Vaccination against the intracellular bacterium Listeria monocygotes with a clonotypic antiserum. J. Immunol. 134:4123.
9. Woodruff, J.J., I.M. Katz, L.E. Lucas and H.B. Stamper, Jr. 1977. An in vitro model of lymphocyte homing. II.:Membrane and cytoplasmic events involved in lymphocyte adherence to specialized high-endothelial venules of lymph nodes. J. Immunol. 119:1603.
10. Chin, Y.-H., R.A. Rasmussen, A.K. Cakirhoglu and J.J. Woodruff. 1984. Lymphocyte recognition of lymph node high endothelium. VI. Evidence of distinct structures mediating binding to high endothelial cells of lymph nodes and Peyer`s patches. J. Immunol. 133:2961.
11. Haskard, D., D. Cavender, P. Beatty, T. Springer and M. Ziff. 1986. T lymphocyte adhesion to endothelial cells: Mechanisms demonstrated by anti-LFA-1 monoclonal antibodies. J. Immunol. 137:2901.
12. Chin, Y.-H., R.A. Rasmussen, J.J. Woodruff and T.G. Easton. 1986. A monoclonal anti-HEBFpp antibody with specificity for lymphocyte surface molecules mediating adhesion to Peyer`s patch high endothelium of the rat. J. Immunol. 136:2556.
13. Rasmussen, R.A., Y.H. Chin, J.J. Woodruff and T.G. Easton. 1985. Lymphocyte recognition of lymph node high endothelium. VII. Cell surface proteins involved in adhesion defined by monoclonal

anti-HEBFln (A.11) antibody. J. Immunol. 135:19.
14. Jalkanen, S.T., R.F. Bargatze, L.R. Herron and E.C. Butcher: 1986. A lymphoid cell surface glycoprotein involved in endothelial cell recognition and lymphocyte homing in man. Eur. J. Immunol. 16:1195.
15. Jalkanen, S., A.C. Steere. R.I. Fox and E.C. Butcher. 1986. A distinct endothelial cell recognition system that controls lymphocyte traffic into inflamed synovium. Science 233:556.
16. Hamann, A., Jablonski-Westrich, D., Duijvestijn, A., E.C. Butcher, H. Baisch, R. Harder and H.-G. Thiele. Evidence for a accessory role of LFA-1 in lymphocyte-high endothelium interaction during homing. J. Immunol., in press.
17. Mentzner, S.J., S.J. Burakoff and D.V. Faller. 1986. Adhesion of T lymphocytes to human endothelial cells is regulated by the LFA-1 membrane molecule. J. Cell. Physiol. 126:285.
18. Hendriks, H.R., C. Korn, A.M. Duijvestijn and G. Kraal. 1987. Changes in lymphocyte binding and expression of His 22 of high endothelial venules in rat lymph nodes after occlusion of afferent lymph flow. 9th international conference on lymphatic tissues and germinal centres in immune reactions. Oslo, M-15.
19. Duijvestijn, A.M., A.B. Schreiber and E.C. Butcher. 1986. Interferon-regulates an antigen specific for endothelial cells involved in lymphocyte traffic. Proc. Natl. Acad. Sci. 83:9114.
20. Horst, E., S.T. Pals, A.M. Duijvestijn, J.v. Maurik, G. Kraal, E.C. Butcher andC.J.L.M. Maijer. 1987. Expression and regulation of an antigen specific for andothelium involved in human lymphocyte homing. 9th conference on lymphativ tissues and germinal center in immune reactions. Oslo, M16.

THE EXPRESSION OF LYMPHOCYTE ADHESION MOLECULES ON CULTURED HIGH ENDOTHELIAL CELLS

Ann Ager and Shilaben Mistry

Immunology Group, Cell and Structural Biology
University of Manchester
Manchester M13 9PT

INTRODUCTION

Lymphocytes enter lymph nodes (LN) from the blood by penetrating the walls of high endothelial venules (HEV). The initial step in this process involves the selective adhesion of lymphocytes to the luminal surface of high endothelial cells. Detailed studies of this adhesion process using the frozen section assay developed by Stamper and Woodruff (1) have identified receptors on the surfaces of rat (2) and mouse (3) lymphocytes which mediate adhesion to HEV in LN cryostat sections. These receptors also mediate lymphocyte migration into LN of intact recipients. A natural conclusion from these studies is that HEV express surface adhesion molecules (ligands or complementary receptors) for lymphocytes and the expression of adhesion molecules by HEV regulates lymphocyte migration from the blood. At least two sets of lymphocyte receptor-HEV adhesion molecules exist in rodents. One directs lymphocyte migration into peripheral LN and the second directs lymphocyte migration into Peyer's patches (3-5). The expression of lymphocyte adhesion molecules by HEV is poorly understood. The role of endothelium in this intriguing process could be studied using isolated high endothelial cells (HEC). The recent successes achieved with the isolation and long-term culture of aortic, venous and capillary endothelium prompted an attempt at the isolation of HEC from rat lymph nodes.

CULTURE OF HIGH ENDOTHELIAL CELLS

The success of this project depended on the ability to identify HEC after isolation from enzyme digested lymphoid tissue. There are no monoclonal antibodies available which are absolutely specific for HEV. However a unique metabolic pathway for sulphate has been identified in HEV of rat LN. Autoradiographic localisation of (^{35}S)-sulphate following short incubations of up to 60 min readily identified HEV from all other vascular and stromal cells in which this metabolic pathway was absent (6). Using this marker collagenase digestion of (^{35}S)-sulphate labelled LN yielded primary cultures which were enriched for HEC to at least 95% purity. Isolated HEC proliferated readily in RPMI 1640 growth medium containing 20% foetal calf serum with a doubling time of 1.6 days. The identity of cultured HEC was substantiated using a polyclonal antiserum raised against these cultures which stained HEV in cryostat sections of LN. Lymphocyte adhesion to monolayers of cultured HEC was 40-fold higher than lymphocyte adhesion to either aortic endothelial cells or aortic fibroblasts demonstrating that cultured HEC continued to express specific adhesion molecules for lymphocytes (7).

Lymphocyte adhesion to cultured HEC has been studied in detail and the results demonstrate that the lymphocyte-HEC interaction shares many properties with lymphocyte adhesion in the frozen section assay and lymphocyte migration in vivo. Some of these results will be presented in this paper. Using this model we have found that peripheral node HEC in long-term culture express lymphocyte adhesion molecules which bind B lymphocytes preferentially.

METHODS

The methods used have been described in detail elsewhere (7). Briefly, high endothelial cells (HEC) were cultured from cervical LN of AO rats and used between 8 and 15 cell doublings. Thoracic duct lymphocytes (TDL) were collected from AO rats for 2-4 hr at room temperature. A suspension of lymph node cells (LNC) was prepared by teasing apart cervical lymph nodes and collecting all cells released into the supernatant. Lymphocytes were labelled with 25µCi/ml 4,5 (^{3}H) leucine (8) in Nunc tissue culture flasks to remove tissue culture plastic adherent cells. Lymphocytes were suspended in RPMI 1640 plus 1% foetal calf serum at 3 x 10^6 cells/ml and incubated for 60 min with monolayers of HEC plated in 16mm diameter wells of tissue culture dishes.

After rigorous washing lymphocyte adhesion was measured by scintillation counting after dissolving the cell layer in 0.1M NaOH. The number of HEC per well was measured by electronic particle counting after detachment using trypsin : EDTA (7). The results of the radioassay were expressed as % lymphocytes bound per 3 x 10^4 HEC.

Immunocytochemistry

In order to classify lymphocytes bound to cultured HEC the microscope adhesion assay was performed as described (7) except that the HEC layer was fixed in methanol at -20°C for 10 min. Smears of TDL or LNC were prepared from 10µl of cell suspension at 30 x 10^6/ml in RPMI 1640 plus 10% foetal calf serum and air dried before methanol fixation. Immunocytochemical staining of T and B lymphocytes was carried out using W3/13 and MRC OX-12 mouse monoclonals respectively and peroxidase conjugated second antibody technique as described previously (9). Slides were examined by high power light microscopy and 200 lymphocytes were counted per slide.

RESULTS AND DISCUSSION

Trypsin sensitivity of lymphocyte adhesion

The capacity of lymphocytes to migrate into LN is reduced by brief exposure of lymphocytes to trypsin prior to intravenous administration. This enzyme treatment is thought to remove lymphocyte surface receptors involved in adhesion to HEV. The effect of trypsin on lymphocyte adhesion to cultured HEC was therefore determined.

Pre-treatment of LNC with increasing concentrations of trypsin from 2 to 200 µg/ml for 10 min at 37°C caused a dose-dependent inhibition of lymphocyte adhesion to cultured HEC (Table 1a). Lymphocyte adhesion at 20 µg/ml trypsin was 51% of control. This dose of trypsin reduced lymphocyte migration into isolated mesenteric LN to 60% of control (10). Lymphocyte adhesion in the frozen section assay was also reduced by 20 µg/ml trypsin to 10% of control adhesion (11). When a direct comparison is made between lymphocyte adhesion to cultured HEC and the frozen section assay the latter reflects the in vivo situation more closely. The effect of trypsin was much less marked when the frozen section assay was conducted at 23°C (12). We feel that the migration of lymphocytes into isolated mesenteric LN is most comparable with lymphocyte-adhesion to cultured HEC but results with this model are relatively few in number. However, it is noteworthy that prior exposure of lymphocytes to trypsin significantly inhibits lymphocyte adhesion to cultured HEC as it does in all other assays of lymphocyte-HEC adhesion.

Divalent cation requirements for lymphocyte adhesion

The role of divalent cations was studied by comparing the adhesion of LNC to cultured HEC in Ca^{2+} - and Mg^{2+} - containing phosphate buffered saline (Ca-Mg-PBS) with that in phosphate buffered saline (PBS) alone. In the absence of divalent cations very few lymphocytes adhered to HEC. Adhesion was fully restored by the addition of 1.1mM $CaCl_2$. Adhesion was partially restored by 0.49mM $MgCl_2$ to 55% of adhesion in Ca-Mg PBS (the concentrations of these salts in Ca-Mg-PBS) (Table 1b). Lymphocyte adhesion in Ca-Mg-PBS was lower than in RPMI 1640. The increased adhesion in RPMI 1640 is unlikely to be due to the concentration of divalent cations since both Ca^{2+} and Mg^{2+} levels are lower than in Ca-Mg-PBS at 0.43mM and 0.40mM respectively. It could, however, be due to an additional metabolic requirement of either the lymphocyte or the HEC or both for the adhesion reaction (see next section).

Lymphocyte adhesion in the frozen section assay is dependent on Ca^{2+} ions. Magnesium partially restores adhesion in this assay to 15% of the level with calcium ions alone (15).

Effects of fixation on lymphocyte adhesion

LNC adhesion to cultured HEC was dependent on viable lymphocytes. Glutaraldehyde, neutral buffered formalin or methanol fixation of lymphocytes completely abolished adhesion. However viable HEC was not an absolute requirement for lymphocyte adhesion. Glutaraldehyde or formalin fixation of HEC only partially reduced lymphocyte adhesion to 40% and 50% respectively of binding to unfixed HEC. Adhesion to methanol fixed HEC was lower at 20% of control (Table 1c). The frozen section assay demonstrates an absolute requirement for viable lymphocytes whereas HEV is not viable in this assay. This lack of requirement for HEV viability suggests that HEV is the passive partner in the lymphocyte adhesion reaction. Lymphocyte adhesion to cultured HEC supports this conclusion but it should be emphasised that 40-50% of lymphocyte adhesion is dependent on viable HEC. This component could be a metabolic requirement on thepart of HEC as discussed in the previous section.

Lymphocyte adhesion molecules on cultured high endothelial cells preferentially bind B lymphocytes

TDL from AO x DA rats contained 64 ± 5% $W3/13^+$ cells and 39 ± 1% $OX12^+$ cells. LNC prepared from cervical LN of AO x DA rats contained 59 ±7% $W3/13^+$ cells and 38 ± 2% $OX12^+$ cells. Since there was no difference between these two populations and HEC used in this study were cultured from cervical LN it was expected that $W3/13^+$ and $OX12^+$ cells would adhere to cultured HEC in a similar ratio. This was not observed.

All of the cells which adhered to HEC from the TDL suspension were accounted for by these T and B cell markers. When LNC were used in the adhesion assay 90% of the cells which adhered to cultured HEC were accounted for by T and B cells. Lymphocytes bound to HEC were enriched for B cells

Table 1. The effects of trypsin, divalent cations and cell fixation on lymphocyte adhesion to cultured high endothelial cells

Treatment	Lymphocytes bound +	% of control#
a. 0 μg/ml trypsin, 10 min, 37°C*	12.1±0.4 (7)	100
2 " " " "	11.1±0.4 (3)	91
20 " " " "	6.2±1.3 (6)	51
200 " " " "	3.9±0.7 (3)	32
b. PBS-Ca-Mg*	12.3±2.4 (7)	100
PBS	1.7±0.7 (4)	13
PBS-Ca (1.19 mM)	10.1±2.1 (4)	77
PBS-Mg (0.49 mM)	7.3±0.9 (4)	55
c. RPMI 1640*	19.1±0.1 (4)	100
Pretreatment, 30 min, 22°C of:		
Lymphocytes:		
3% glutaraldehyde pH 7.4	0.0	0
10% formalin, pH 7.4	0.0	0
methanol	0.0	0
High endothelial cells:		
3% glutaraldehyde pH 7.4	7.5±0.6 (4)	39
10% formalin pH 7.4	9.2±0.1 (4)	48
methanol	3.8±0.8 (4)	20

+ values represent mean percentage of lymphocytes bound to $3x10^4$ high endothelial cells ± S.D. (no. of observations)

values represent lymphocytes bound for each treatment as a percentage of control adhesion

* control adhesion for each group

over T cells in comparison with the cell suspension plated. The ratio of T:B cells bound to cultured HEC was reduced from 1.6:1 in the starting population to 0.6:1 for TDL and from 1.5:1 to 0.4:1 for LNC (Fig 1).

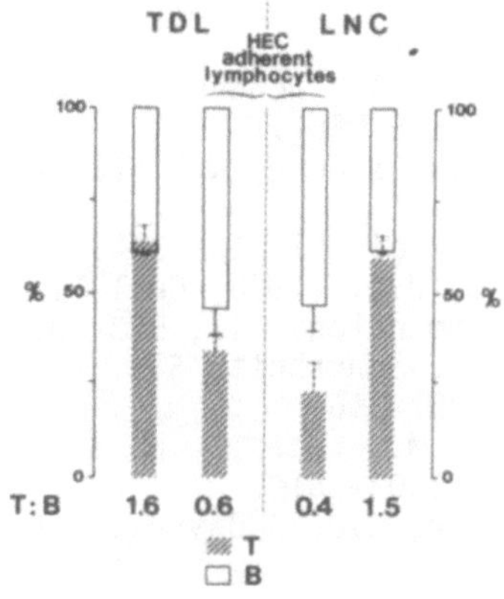

Fig. 1. The adhesion of T lymphocytes (▨ ; W3/13$^+$) and B lymphocytes (□ ; OX12$^+$) to cultured high endothelial cells from suspensions of thoracic duct lymphocytes (TDL) and lymph node cells (LNC).

Different affinities of HEV for T and B cells have been documented in peripheral LN and Peyer's patches both in vivo (13) and in vitro (14). Cultured HEC (from peripheral LN) tends to resemble Peyer's patch HEV in which T:B cell binding is 1:1 rather than peripheral LN in which the ratio is higher at 2:1. The significance of these observations with cultured HEC is not yet clear. It needs to be determined whether T and B cells are binding to cultured HEC via the same adhesion molecules and whether these adhesion molecules are related to those on peripheral LN or Peyer's patches. The use of monoclonal antibodies to rat lymphocyte surface receptors (4,5) should answer these points. As yet there is no evidence to support the alternative hypothesis that T and B cells adhere to HEV via different receptors.

CONCLUSION

Lymphocyte adhesion to cultured high endothelial cells is regulated in a similar manner to lymphocyte adhesion in the frozen section assay and lymphocyte migration into lymph nodes in vivo . Therefore for the first time it is possible to study the expression of lymphocyte adhesion molecules by HEV in vitro using cultured HEC. We have found that peripheral node HEV in long term culture express adhesion molecules which preferentially bind B lymphocytes.

Acknowledgements

We thank Pamela Walker for expert technical assistance and Jackie Jolley for typing the manuscript. This work was supported by the Medical Research Council (UK).

REFERENCES

1. Stamper, H.B. & Woodruff, J.J. (1986) J. Exp. Med. 144, 828-833.
2. Chin, Y.-H. Carey, G.D. & Woodruff, J.J. (1982) J. Immunol. 129, 1911-1915.
3. Gallatin, W.M., Weissman, I.L. & Butcher, E.C. (1983) Nature, 304, 30-34.
4. Rasmussen, R.A., Chin, Y.-H., Woodruff, J.J. & Easton, T.G. (1985) J. Immunol. 135, 19-24.
5. Chin, Y.-C., Rasmussen, R.A., Woodruff, J.J. & Easton, T.G. (1986) J. Immunol. 136, 2556-2561.
6. Andrews, P., Milsom, D.W. & Ford, W.L. (1982) J. Cell Sci. 57, 277-292.
7. Ager, A. (1987) J. Cell Sci. 87, 133-144.
8. Nieuwenhuis, P. & Ford, W.L. (1976) Cell Immunol. 23, 254-267.
9. Barclay, A.N. (1981) Immunology, 42, 593-600.
10. Ford, W.L., Sedgley, M., Sparshott, S.M. & Smith, M.E. (1976) Cell Tissue Kinet. 9, 351-356.
11. Woodruff, J.J., Katz, I.M., Lucas, L.E. & Stamper, H.B. (1977) J. Immunol. 119, 1603-1610.
12. Braaten, B.A., Spangrude, G.J. & Daynes, R.A. (1984) J. Immunol. 133, 117-122.
13. Fossum, S., Smith, M.E. & Ford, W.L. (1983) Scand. J. Immunol. 17, 539-549.
14. Stevens, S.K., Weissman, I.L. & Butcher, E.C. (1982) J. Immunol. 128, 844-851.
15. Woodruff, J.J. & Kuttner, B.J. (1980) Ciba Foundation Symposium, 71, 243-263.

EXPRESSION AND REGULATION OF AN ANTIGEN SPECIFIC FOR ENDOTHELIUM INVOLVED IN HUMAN LYMPHOCYTE HOMING

Eveliene Horst[1], Steven T. Pals[1], Ad M. Duijvestijn[3], Jan v. Mourik[4], Georg Kraal[2], Eugene C. Butcher[5], and Chris J.L.M. Meijer[1]

[1]Dept. of Pathology and
[2]Histology, Free University
Amsterdam, The Netherlands

[3]Dept. of Histology, State University
Groningen, The Netherlands

[4]Dept. of Blood Coagulation, C.L.B.
Amsterdam, The Netherlands

[5]Dept. of Pathology, Stanford School of Medicine
Stanford, U.S.A.

INTRODUCTION

Lymphocyte migration from blood to lymph is mediated by high endothelial venules (HEV), a specialized class of postcapillary venules characterized by plump or "high" endothelial morphology (1,2,3). HEV are normally present in the paracortex of lymph nodes and mucosa associated lymphoid tissues, moreover vessels with HEV-like morphology can arise at sites of inflammation (4,5).

Recently we have generated a rat Monoclonal Antibody (MoAb) HECA-452 that specificially recognizes human high endothelium but does not identify common "flat" vascular endothelium. HECA-452 is also reactive with a subset of cells belonging to the macrophage/Dendritic Cell population (6).

The specificity of HECA-452 for HEV prompted us to study whether the antibody would also identify HEV-like vessels in a variety of pathological tissues showing lymphocytic infiltrates.Moreover,since HECA-452 offers an unique probe to analyse the regulation of high endothelial differentiation, we studied the in vitro requirements for HECA-452 antigen induction by stimulating cultured Human Umbilical Vein Endothelium (HUVE) with a number of cytokines (U937sup, r-γIFN, r-IL-1, r-IL-2 and r-TNF) and Biological Response Modifiers (BRM)(PMA and calciumionophore A23187)

MATERIALS AND METHODS

Indirect immunoperoxidase technique

Aceton fixed 5µm cryosections and 5µm paraffin sections (sublimated formaldehyde fixed) were stained immunohistochemically as described (6).

Human Umbilical Vein Endothelium (HUVE) culture conditions

Human vascular endothelial cells were isolated from umbilical veins according to the method described by Jaffe et al. with some modifications (7). The cells were suspended in culture medium containing 50% Medium 199 and 50% RPMI 1640 supplemented with 20% human serum and cultured on human fibronectin precoated culture flasks (Corning) at 37°C in 5% CO_2. Endothelial cells were identified by their typical characteristics. After confluency was reached, they were serially passed by brief exposure to trypsin. Suspensions of second and third passage endothelial cells ($1,5.10^6$/ml) were subcultured on 1% gelatin coated coverslips in a 24 well tissue plate (Costar). The plates were incubated until monolayers had been formed. At confluence the cells were cultured for one day in culture medium without 20% human serum. As control we used cultures of Smooth Muscle Cells (SMC).

Induction of HUVE with cytokines and BRM

HUVE and SMC cultures were incubated with the following mediators: Supernatant of U937 cells, purified human IL-1 and r-IL1 (10U/ml-100U/ml) from Dr. L. Aarden, Dept of Autoimmune Disease, C.L.B, Amsterdam, The Netherlands, r-TNF (1ng/ml-1µg/ml) from Dr. R. Roozemond, Dept. of Histology and Cellbiology, A.M.C.,Amsterdam, The Netherlands, r-IL-2 (10U/ml-100U/ml) and r-γIFN (10U/ml-100U/ml)(Roche Nutley),PMA (Sigma)(1ng/ml-1µg/ml) and calciumionophore A23187 (Sigma)(1µM-100µM). Monolayers of HUVE and SMC were cultured in serum free culture medium as described, in the presence of the indicated mediators for 1-48 houres. The induction was stopped by washing the cells with culture medium without serum. The cells were fixed in methanol for 10 minutes, washed twice with PBS and the HECA-452 antigen expression was subsequently studied by indirect immunofluorescence using FITC-conjugated Rabbit anti Rat Ig (DAKO) in the second step.

RESULTS AND DISCUSION

Our previous observations that the MoAb HECA-452 recognizes a HEV specific determinant is illustrated in Figure 1; whereas the high endothelial cells in lymph nodes and mucosa associated lymphoid tissue showed strong membrane and cytoplasmic expression of the HECA-452 determinant, expression was not observed on "flat" vascular endothelium of either lymphoid tissues (Figure 1) or other normal tissues (data not shown)(6).

The specificity of HECA-452 for HEV prompted us to study whether the antibody would also identify HEV-like vessels in a variety of pathological tissues showing lymphocytic infiltrates.

The results of our immunohistochemical studies indicate that the HEV specific HECA-452 antigen is indeed expressed on vascular endothelium in a number of conditions characterized by dense lymphoid infiltrates including Autoimmune Thyroiditis, Crohn's Disease and Lymphoma benigna cutis (Table 1, Figure 2).

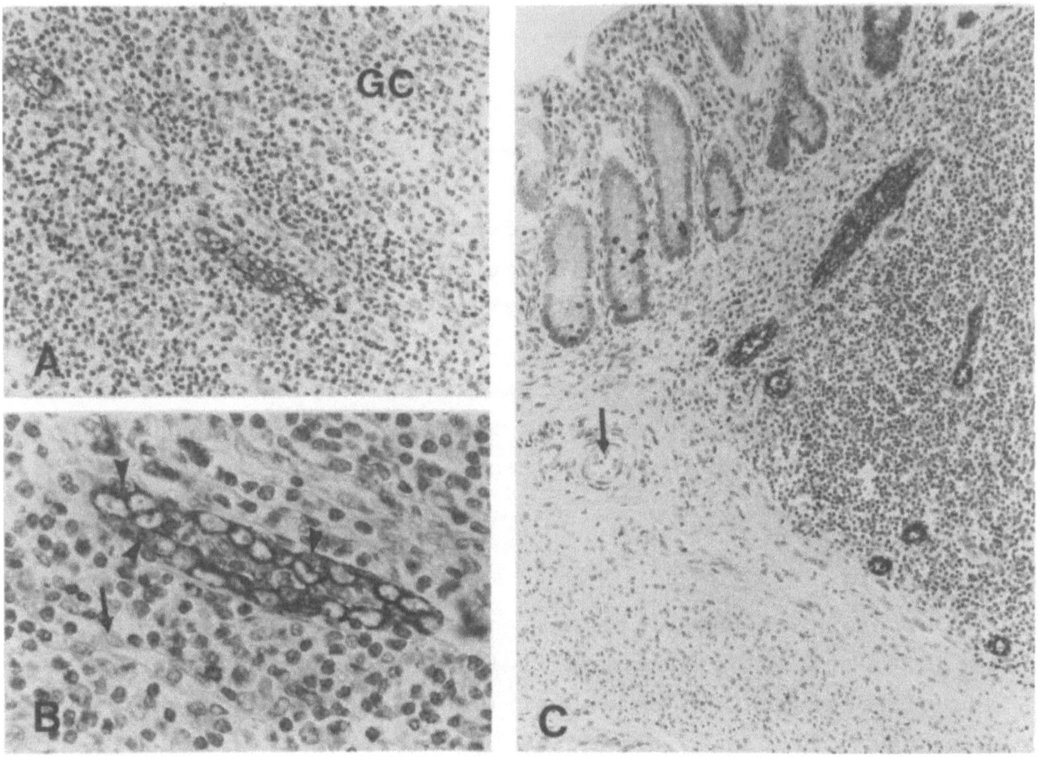

Figure 1

Expression of HECA-452 on HEV in normal lymphoid tissue

A) HECA-452 expression on HEV in paracortical area of a tonsil. GC=Germinal Center (Immunoperoxidase on paraffin section,40x)

B) Detail of A showing predominance of HECA-452 expression on the luminal endothelial border. Lymphocytes are present in the vesselwall (arrowheads). Flat endothelium is negative (arrow)(250x)

C) HECA-452 expression on HEV in the wall of normal appendix Flat endothelium of arteriole is negative (arrow)(40x).

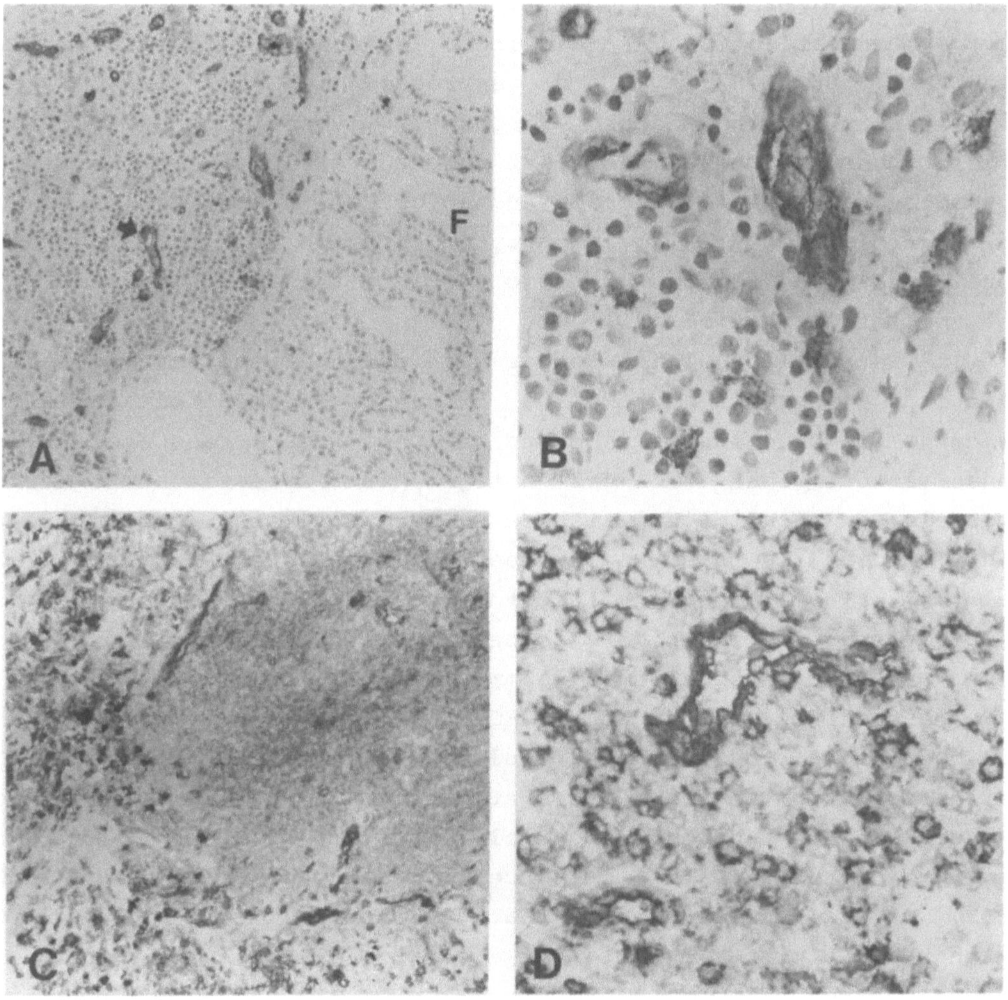

Figure 2

Expression of Heca-452 on HEV-like vessels and Dendritic Cells at sites of inflammation

A) HECA-452 positive vessels in an infiltrate in the thyroid of a patient with Graves Disease. F=Follicles of the thyroid
(Immunoperoxidase on cryosection,40x)

B) Detail of A . Note HECA-452 positive Dendritic Cells surrounding the vessels (250x)

C) HECA-452 positive vessels at the border of a GC in a case of Lymphoma benigna cutis (40x)

D) Detail of C showing many HECA-452 expressing Dendritic Cells surrounding the vessels (400x)

In addition to their lymphoid nature , these infiltrates containing HECA-452 positive vessels generally showed compartimentatization in distinct T- and B-cell areas. By contrast HECA-452 was not expressed in conditions with infiltrates lacking a dense "organoid" lymphoid organization. These observations suggest that the HECA-452 positive HEV-like vessels in the above described pathological situations not only play a role in mediating lymphocyte migration from the blood to the surrounding tissues but also regulate the specific structure of the infiltrates.

It is still unclear whether high endothelium represents a distinct lineage of endothelium or can be induced from any flat endothelium by products released in local immune response. Our observations that HECA-452 antigen is expressed on HEV-like vessels in pathological tissues favour the latter possibility. This concept is also supported by studies of Duijvestijn et al.(8). They showed that in the mouse the expression of an antigen specific for high endothelium can be induced in cultured lung vascular endothelium by gamma IFN.

The specifity of the MoAb HECA-452 for human HEV makes it an unique probe to study the regulation of high endothelial cell differentiation in the human system. To gain insight into this regulation we studied the possible induction of HECA-452 antigen in HUVE cultured in the presence of a number of cytokines and BRM. Our results demonstrate that 6-48 houres incubation of HUVE with U937sup or r-γIFN but not with r-IL-1, r-IL-2, r-TNF or BRM, induces HECA-452 expression in over 80% of the endothelial cells (Table 2). By contrast, no expression is induced when SMC are incubated with U937sup or r-γ IFN. These data strongly suggest the existance of a high endothelial differentiation mechanism regulated by local products of the immune response. Moreover the results indicate a cell-type specific induction of the antigen recognized by HECA-452. In contrast to the membrane and cytoplasmic HECA-452 antigen expression seen on HEV and Dendritic Cells, we observed only cytoplasmic HECA-452 positivity in r-γIFN and U937sup induced HUVE. This lack of membrane expression on HUVE in vitro might be due to the absence of an appropriate micro-environment for HEV development.

Table 1

Reactivity of HECA-452 with endothelium in sites of inflammation

Disease	Pos. cases/Tested cases	Vesseltype
Autoimmune thyroiditis		
- Graves Disease	3/3	HEV (1)
- Hashimoto's Disease	2/3 (2)	HEV
Crohn's Disease	2/2	HEV
Lupus erythomatodes	0/8	-
Lymphoma benigna cutis	2/2	-
Lichen planus	0/10	-
Acute and subacute dermatitis	0/7	-
Peptic ulcers	0/2	-

* Reactivity with Dendritic Cells could be observed in most cases.

HEV (1): High Endothelial Venule

(2): negative case was less inflamed

Table 2

Induction of HECA-452 in HUVE by cytokines and BRM

	time(h)			
Mediators	2	6	24	48
U937sup	-	+	++	++
IL-1	-	-	-	-
r-IL-1	-	-	-	-
r-IL-2	-	-	-	-
r-TNF	-	-	-	-
r-γ IFN	-	+	++	++
PMA	-	-	-	-
A23187	-	-	-	-

Alternatively it might be the result of an intrinsic inability of HUVE to express the molecule on the cellmembrane.

Preliminary immunoprecipitationresults indicate that the antigen identified by HECA-452 is a highly glycosilated molecule with a molecular weight of ± 90 kD (Horst et al., unpublished observations). The function of this antigen remains to be elucidated. Although the expression of HECA-452 determinant on the membrane of both HEV and Dendritic Cells suggests that the molecule might play a role in cell-cell interactions, HECA-452 did not inhibit the binding of lymphocytes to HEV in an in vitro binding assay (9), nor did it block the clustering of PMA stimulated U937cells in an aggregation assay (10). These results indicate that the epitope recognized by HECA-452 is not functionally involved in adhesion mechanisms. Alternatively HECA-452 might identify a differentiation antigen reflecting a stage of cell activation.

In conclusion the data presented in this paper suggest that the high endothelial phenotype can be induced in vascular endothelium by products such as r-γ IFN released in local immune response. Furthermore the expression of high endothelial phenotype in inflammatory tissues is accompanied by organoid organization of the lymphocytic infiltrate.

REFERENCES

1) Gowans JL, Knight EJ. Proc.R.Soc.Lond.(Biol.)(1964) 157:257
2) Stamper HB, Woodruff JJ. J.Exp.Med.(1976)144:828
3) Butcher EC. In:Curr.Top.Microbiol.Immunol.(1986)Vol.128
4) Smith JB, Mcintosh GH, Morris B. J.Pathol.(1970)100:2
5) Freemont AJ. J.Clin.Pathol.(1980)36:161
6) Duijvestijn AM, Horst E, Pals ST, Rouse BT, Steere A, Picker L, Meijer CJLM, Butcher EC. Am.J.Pathol.,in press
7) Jaffe EA, Nachman RL, Becker CG, Minick CR. J.Clin.Invest.(1973)52:2745
8) Duijvestijn AM, Schreiber AB, Butcher EC. Proc.Nat.Acad. Sci USA (1986) 83:9114
9) Pals ST, Kraal G, Horst E, Groot A de, Scheper RJ, Meijer CJLM. J.Immunol.(1986)137:760
10) Rothlein R, Springer TA. J.Exp.Med.(1986)163:1132

CHANGES IN LYMPHOCYTE BINDING AND EXPRESSION OF HIS 22 OF HIGH ENDOTHELIAL VENULES IN RAT LYMPH NODES AFTER OCCLUSION OF AFFERENT LYMPH FLOW

H.R. Hendriks, C. Korn, A.M. Duijvestijn* and G. Kraal

Dept. Cell Biology, Medical Faculty, Free University
Amsterdam and *Dept. Histology, State University Groingen
The Netherlands

INTRODUCTION

Recirculating lymphocytes migrate from the blood into lymph nodes through the walls of the high endothelial venules (HEV). HEV are lined by large plump so-called high endothelial cells (HE cells) to which lymphocytes specifically adhere. This specific interaction can be studied _in vitro_ by overlaying lymphocyte suspensions on cryostat sections of lymph nodes (1-3).

Previously we found that occlusion of the afferent lymph flow to the lymph node by interruption of its afferent lymphatic vessels results in both flattening of the HEV-endothelium and a severe decrease in numbers of lymphocytes in transit across the walls of the flattened HEV (4,5). In this study we have investigated with the _in vitro_ lymphocyte binding assay the capacity of HEV to bind lymphocytes on various timepoints after occlusion of the afferent lymph flow. In addition the expression of two antigens normally found on HEV, recognized by the monoclonal antibodies (mabs) His 22 and OX 2, was studied in such operated lymph nodes.

MATERIAL AND METHODS

Animals

Inbred male WAG/Rij (REPGO-TNO, Rijswijk, The Netherlands) and outbred Wistar rats (CPB-TNO, Zeist, The Netherlands), age 8 to 9 weeks, were used.

Surgical procedure

Deafferentization of the popliteal lymph node was done as described elsewhere (5,6). Briefly, the popliteal lymph node was exposed whereafter the afferent lymphatic vessels were interrupted. The efferent lymphatic vessels and the blood vessels both emerging from the hilus were left intact. Prior to excision 0.05 ml 10% Indian ink solution was injected into the ipsilateral footpad to determine the completeness of the operation. Lymph nodes that would take up ink were discarded. The contralateral lymph node served as unoperated control.

Rats were killed at various timepoints after deafferentization. Operated and unoperated lymph nodes were dissected, snap-frozen in liquid nitrogen and stored at -70°C untill use.

In vitro assay of lymphocyte binding

The technique for the in vitro binding of lymphocytes to rat HEV, originally developed by Stamper and Woodruff (7), has been described before (8). Freshly cut 8 μm thick cryostat sections of lymph nodes were overlaid with mesenteric lymph node cell suspension and agitated for 30 min at 4°C. Next the sections were rinsed, fixed with 1% glutaraldehyde and processed for methylgreen-pyronin staining.

In order to establish the pattern of adherence of lymphocytes to HEV of lymph nodes at various timepoints after operation morphometric analysis was used. Therefore the number of bound lymphocytes per HEV was counted on methylgreen-pyronin stained sections. At the same time the area of the HEV-endothelium and the contourlenght of the HEV were measured using a microscope (40x objective) coupled to an interactive measuring program running on an IBAS image analysis sytem of Kontron (Munich, Fed. Rep. Germany) developed by Mr. H.C. den Harink (Dept. Medical Informatics, Free University, Amsterdam, The Netherlands).

Immunoperoxidase staining

Acetone-fixed cryostat sections of both operated and unoperated popliteal lymph nodes were incubated with His 22 or OX 2 mab for 60 min at room temperature. After rinsing in phosphate-buffered saline (PBS) the slides were incubated with peroxidase-conjugated rabbit anti mouse IG (Dako, Copenhagen, Denmark). Peroxidase activity was demonstrated by 3',3'-diaminobenzidine tetrahydrochloride (Sigma, St. Louis, USA) in PBS containing 0.01% H_2O_2 for 10 min. After rinsing in PBS and tap water the sections were lightly counterstained with hematoxylin, rinsed, dehydrated and mounted in malinol.

RESULTS

In normal lymph nodes lymphocytes adhere to virtually all HEV using the in vitro lymphocyte binding assay. However, 1 and 2 weeks after deafferentization lymphocytes bound to only 50-60% of the HEV and by 3-6 weeks 20-30%. The total numbers of lymphocytes bound to these HEV had also diminished to 10% of the control value by 3-6 weeks. The reduction in numbers of lymphocytes adhering to HEV in operated lymph nodes could be due to either the flattening of the HEV-endothelium or a loss of the capacity of the HE cells to bind lymphocytes. In order to discriminate between these possibilities morphometric analysis was used. From the measured parameters the height of the HEV-endothelium and the lymphocyte density was calculated as follows: Height HEV = area HEV/contourlength HEV, and Lymphocyte density = lymphocytes bound to HEV/area HEV. Fig. 1 showed that the height of the HEV-endothelium gradually decreased from 80% of the control value one week after operation to 45% at 6 weeks. The lymphocyte density, which is the number of lymphocytes bound per unit area HEV and thus independent of the available HEV area in the lymph node sections, decreased from approximately 50% at one week to 20-25% at 3 and 6 weeks. The final reduction in the number of adherent lymphocytes is the result of both the decrease in HEV height and a strongly decreased capacity of the flattened HE cells to bind lymphocytes.

His 22 stains the majority of B lymphocytes, and HEV-endothelium, but does not cross-react with any other endothelium (9). The B cells stained more intensely than HEV in normal lymph nodes. Whereas the intensity of the staining of B cells remained unchanged, the expression of His 22 on HEV decreased after operation and had disappeared by 3 weeks (Fig.2). OX 2 recognizes a surface glycoprotein on a variety of cells such as B

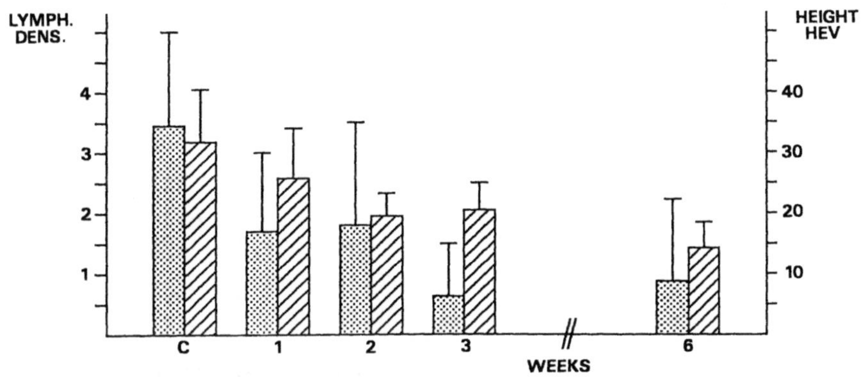

Fig. 1. Average and standard deviation of HEV height (hatched bars) and lymphocyte density (dotted bars) in arbitrary units of normal lymph nodes (c) and lymph nodes 1-6 weeks after operation.

cells and endothelium of all blood vessels (10). The expression of this antigen did not change in operated lymph nodes (Fig.3). The results showed that His 22 expression of flattened HEV disappeared concurrently with the decline in lymphocyte adherence to such HEV. Therefore we studied whether His 22 was involved in the binding of lymphocytes to HEV. Table 1 showed that neither His 22 nor OX 2 blocked the binding of lymphocytes to HEV.

DISCUSSION

Deafferentization of lymph nodes results in flattening of HEV-endothelium. We investigated with the in vitro lymphocyte binding assay the ability of HEV in cryostat sections of normal and deafferentized lymph nodes to bind lymphocytes. Both the number of HEV to which lymphocytes adhered and the numbers of lymphocytes bound to HEV in operated lymph nodes decreased rapidly. Morphometric analysis showed that this was not only due to the flattening of HEV-endothelium, which results in a decrease of HEV area available for lymphocyte binding, but also to a strongly decreasing capacity of HE cells to bind lymphocytes. The diminished binding of lymphocytes to flattened HEV is in agreement with the reduced lymphocyte influx into operated lymph nodes (11). Concurrently the expression of His 22 on flattened HEV disappeared. His 22, however, was not able to block the adherence of lymphocytes to HEV in normal lymph nodes.

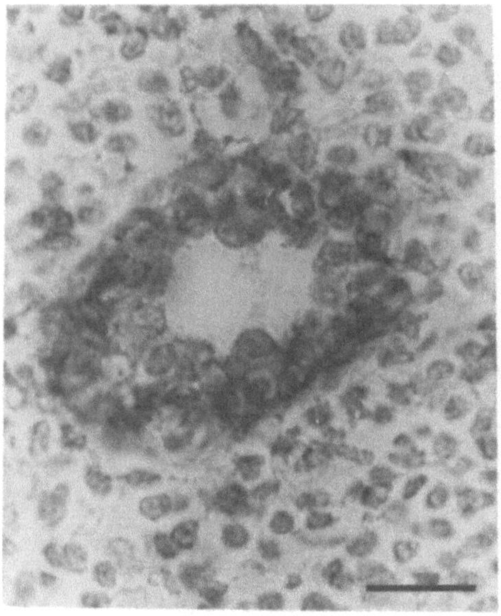

Fig. 2a. His 22, HEV in normal lymph node. Bar = 20 µm.

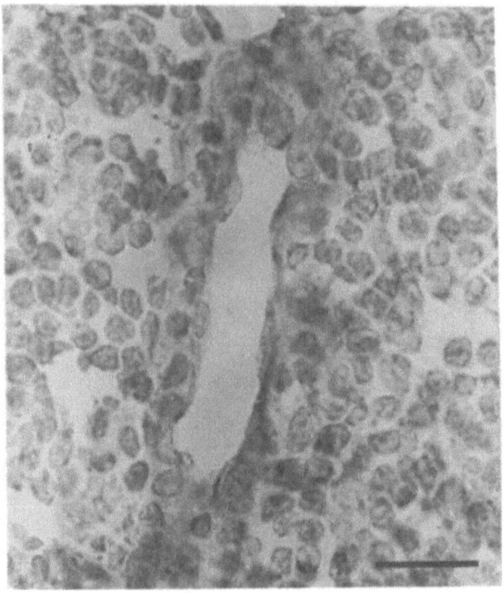

Fig. 2b. His 22 expression absent in flattened HEV 3 weeks after operation. Bar = 20 µm.

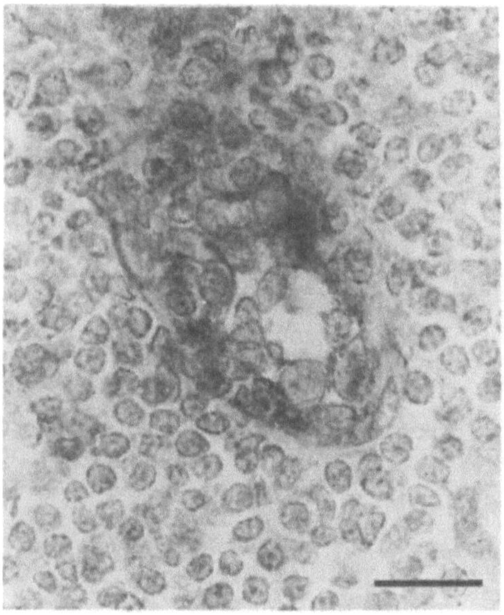

Fig. 3a. OX 2, HEV in normal lymph node. Bar = 20 µm.

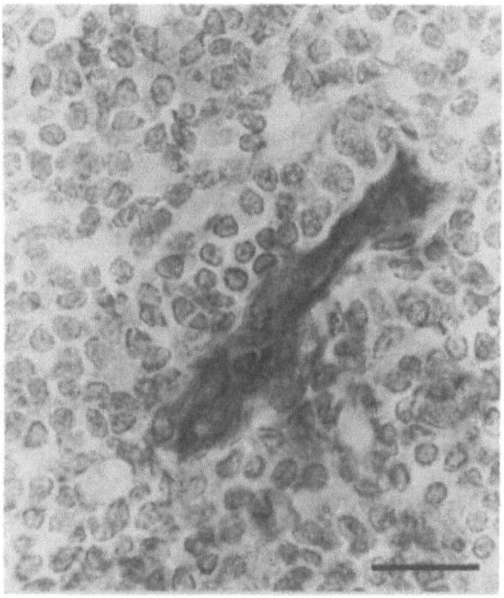

Fig. 3b. Intensity of OX 2 in flattened HEV 3 weeks after operation similar to that in HEV of normal lymph node. Bar = 20 µm.

The present model of lymphocyte-HEV interaction assumes the existance of at least two complementary sets of lymphocyte receptor-HE cell surface determinants. HEV in peripheral lymph nodes differ from HEV in Peyer's patches (2). HE cells in deafferentized lymph nodes are metabolically inactive (5). In addition we now showed that such flattened HE cells had also lost their ability to bind lymphocytes. These results suggest that HE cell determinants involved in lymphocyte binding had rapidly disappeared from the surface of the flattened HE cells. Since the decline in the capacity of HE cells to bind lymphocytes preceded the reduction in height, the height of HEV seems to be of minor importance than the ability to bind lymphocytes.

Table 1. In vitro HEV binding of lymphocytes to HEV in normal lymph nodes after preincubation with mab.

	number of HEV	averaged number of bound lymphocytes/HEV
control	123	12 ± 2.6
OX 2	61	11.5 ± 0.4
His 22	59	12.7 ± 1.0
irrelevant mab	27	12.1 ± 1.2

The results are from a representative experiment.

His 22 recognizes an antigen on B lymphocytes, which seems to be an epitope of the leucocyte common antigen (9). The finding that HEV-endothelium stained less intensely with the mab suggest that the binding of His 22 to HEV is a cross-reaction. The disappearance of His 22 from the metabolicly, and functionally inactive flattened HE cells showed that the expression of His 22 is related with the activity of these cells. Preliminary results indicate that His 22 reappeared on HE cells from flattened HEV, which had increased in height after antigenic stimulation of operated lymph nodes (Hendriks, Korn and Kroese, unpublished results). So His 22 seems to be an activation antigen on HEV, which is not directly involved in lymphocyte binding.

It has been shown that the absence or presence of antigen and antigen-stimulated macrophages clearly affects the HEV morphology and the number transmigrating lymphocytes in HEV of operated and unoperated lymph nodes (5,12). So in our view local factors in the lymph node produced by antigen-stimulated cells may modulate the ability of HE cells to bind lymphocytes and thus regulate indirectly the lymphocyte traffic into lymph nodes and the expression of His 22.

REFERENCES

1) J.L. Gowans, and E.J. Knight. Proc. R. Soc. B 159:257 (1964).
2) E.C. Butcher. Curr. Topics Microbiol. Immunol. 128L85 (1986).
3) G. Kraal, A.M. Duijvestijn, and H.R. Hendriks. Exp. Cell Biol. 55:1 (1987).
4) H.R. Hendriks, I.L. Eestermans, and E.C.M. Hoefsmit. Cell Tissue Res. 211:375 (1980).
5) H.R. Hendriks, and I.L. Eestermans. Eur. J. Immunol. 13:663 (1983).

6) H.R. Hendriks. Z. Versuchstierk. 28:105 (1978).
7) H.B. Stamper, and J.J. Woodruff. J. Exp. Med. 144:828 (1976).
8) G.J. van der Brugge-Gamelkoorn, and G. Kraal. J. Immunol. 134:3746 (1985).
9) F.G.M. Kroese. Acad. Thesis, State University Groningen, Groningen, The Netherlands (1987).
10) A.N. Barclay. Immunology 44:727 (1981).
11) M.T. Drayson, and W.L. Ford. Immunobiol. 168:362 (1984).
12) H.R. Hendriks, H.A.B. van Hemert, and M. van der Heijden. in: "In Vivo Immunology", p. 207, P. Nieuwenhuis, A.A. van den Broek, and M.G. Hanna, eds., Plenum Publishing Corporation, New York (1982).

REGULATION OF FUNCTIONAL AND MORPHOLOGICAL ASPECTS OF HIGH ENDOTHELIUM IN MOUSE

Adrian M. Duijvestijn[1], Martin Rep[2], Eugene C. Butcher[3], Hans R. Hendriks[2], and George Kraal[2]

Depts. of Histology, University of Groningen[1], and Cell Biology, Free University, Amsterdam[2], The Netherlands, and Dept. of Pathology, Stanford School of Medicine[3], California, USA

INTRODUCTION

Lymphocyte migration and recirculation between lymphoid and also non-lymphoid tissues is essential for effective immunological surveillance. The entrance of blood-borne lymphocytes into peripheral lymphoid organs, such as lymph nodes and Peyer's patches, occurs at particular vascular sites, so called post-capillary high endothelial venules (HEV;1,2). The endothelium of these HEV is unique in it's capacity to mediate lymphocyte extravasation, and its characteristic cuboidal or 'high' cellular appearance. From recent investigations, it has become clear that subtile mechanisms are controlling a balanced and selective immigration of lymphocytes and lymphocyte subsets via HEV into mucosal and non-mucosal lymphoid tissues (3,4). Major contributions to studying lymphocyte extravasation and homing via interaction with high endothelium have been the production of monoclonal antibodies (MABs) against lymphocyte 'homing-receptors' for HEV (5,6,7), and the development of an elegant in vitro assays for lymphocyte-binding to HEV in frozen sections by Stamper and Woodruff (8). With these tools the nature and many aspects of the regulation of homing receptors on lymphocytes and lymphocyte subsets have been unraveled. However, little is known of the nature and regulation mechanisms of the specific sets of ligands on high endothelium involved in lymphocyte binding, nor about why this endothelium develops it's characteristic high appearance and what mechanisms regulate this process. In this paper we will address mechanisms controlling specific functional and morphological qualities of high endothelium. We summarize some previous results obtained with HEV-specific MAB MECA-325, e.g. expression of the MECA-325 antigen in chronic inflammation and induction of the antigen in cultured endothelial cells by lymphokines (9,10). In addition, an in vivo model is described in which severe depletion of recirculating lymphocytes by irradiation affects endothelial height but not its lymphocyte-binding capacity or MECA-325 antigen expression. By injecting recirculating lymphocytes the characteristic high endothelial cell appearance can be quickly restored.

MATERIALS AND METHODS

Mice. BALB/c mice were from the Frederick Cancer Research Center

(Frederick, MD, USA) and C3D2 mice from Bomholtgard Ltd. (Rye, Denmark). All animals had access to food and water ad libitum.

Monoclonal antibody MECA-325. The MECA-325 hybridoma is derived from a fusion of Sp 2/0 myeloma cells with spleen cells of a rat immunized with stromal preparations of multiple mouse lymph nodes. MECA-325 has been tested on lymphoid organs and a variety of non-lymphoid tissues, and appears to be remarkable specific for HEV present in lymph nodes and Peyer's patches. The antibody is of the IgG1 isotype, and in immunostaining studies we used isotype matched J4.B anti-NIP as control antibody.

Immunoperoxidase staining. Acetone-fixed frozen sections were stained with MAB as previous described (11). MAB were used as hybridoma supernatant or purified antibody at 100 ug/ml. The second stage antibody, horse radish peroxidase conjugated goat anti-rat (DAKO; 1:40 dilution) contained 5 % normal mouse serum.

Immunofluorescence staining. Cell cultures were grown on coverslips. MECA-325 and control antibody J4.B anti-NIP were used as hybridoma supernatant or purified antibodies in the first incubation step. After 30 min. the coverslips were washed, and incubated with a fluorescein-conjugated rabbit anti-rat antibody (Sigma; used at 100 ug/ml) containing 5 % normal mouse serum for 30 min. After staining the cultures were washed, fixed with 3.7 % paraformaldehyde in PBS and examined in a fluorescence microscope under epi-illumination.

Granulomas. These were induced by injecting 200-300 ul of a 1:1 mixture of packed SRBC and complete Freund's adjuvant subcutanously in the thigh or flank of BALB/c mice. At 2,6,12,21, and 28 days, inflammatory sites were excised and processed for immunohistology.

Cell cultures and induction studies. LE-2 (gift of A. Curtis, Glasgow) and MCC (gift of D. Gospodarowicz, San Francisco) endothelial cell and L-929 fibroblast (ATCC) were of mouse origin. They were cultured on coverslips for 2-4 days in Dulbecco's modified Eagle's medium with 10 % fetal calf serum at $37^{o}C$, 5 % CO2. Stimulating agents were added to the cells from the start of culturing. IL-1 (from S. Gilles, Immunex Co., Seattle), IFN-beta (Lee Biomolecular), and recombinant murine IFN-gamma (gift from Genentech, South San Francisco) were used at 100 units/ml. Murine EGF (from A. Schreiber, Syntex Corp., Palo Alto) FGF (gift from D. Gospodarowitcz) and ECGF (gift from T. Maciag, Revlon Rockville) were used at 5-10 ng/ml.

Irradiation and reconstitution studies. Mice were 950 cGy total-body (gamma) irradiated. 4 days later animals were sacrificed and lymph nodes excised and frozen or fixed in glutaraldehyde and Epon embedded. In reconstitution experiments, 4 days after irradiations mice were injected i.v. with 0.2 ml of either fresh mouse serum or ConA supernatant, or with thymocytes, 1500 cGy irradiated lymph node cells, or viable lymph node cells in cell numbers of $1.5\text{-}2.0 \times 10^{7}$. From 0.5 to 24 hr after injection mice were sacrificed and lymph nodes frozen, or fixed and Epon embedded.

In vitro lymphocyte-binding to HEV. This assay, originally developed by Stamper and Woodruff (8), was applied as described before (12). The average numbers of bound lymphocytes per HEV in treated or control lymph nodes were calculated from counts of 20-30 HEV. T/B binding ratios for lymph node HEV were roughly determined from binding assays in which mixtures of labeled peripheral node cells (for the majority T cells) and Peyer's patch cells (for the majority B cells) were used for incubation.

TABLE 1

Appearance of MECA-325$^+$ HEV in granulomas

Antibody	Staining of granulomas in function of age in days				
	2	6	12	21	28
MECA-325	-	+/-	+	++	++
Anti Thy-1 *	+/-	+	++	++	+++

* T cell marker

- , absent ; +/-, few vessels/cells ; etc.

TABLE 2

Induction of MECA-325 antigen in cultured endothelial cells by IFN-gamma

Cell line	Staining * of cultured cells with MECA-325 after treatment with :					
	IFN beta	IFN gamma	IL-1	EGF	FGF	ECGF
LE-2	-	+	-	-	-	-
MCC	-	+	-	-	-	-
L-929	-	-	-	-	-	-

* Immunofluorescence
LE-2 : lung capillary endothelial cells
MCC : bone marrow endothelial cells
L-929 : fibroblast cells

- , negative
+ , positive

Morphometric analysis of HEV. Frozen sections immunostained with MECA-325 were used for morphometric analysis. The height of the endothelium of HEV in lymph node of treated and control mice was calculated using a microscope coupled to an interactive measuring program (developed by H. den Harink, Amsterdam) running on an IBAS image analysis system (Kontron, Munich) Briefly: both the exposed surface of HEV in cross-section and it's contour length were measured; endothelium-height in arbitrary units was calculated as HEV-surface-area divided by HEV-contour-length.

RESULTS

Expression of MECA-325 positive HEV in granulomas (Table 1). MECA-325 as a unique marker for high endothelium offered the possibility to define if and when MECA-325 positive HEV appear in sites of chronic inflammation, known to develop mononuclear cell infiltrates. For this purpose we induced granulomas by injecting SRBC/CFA s.c. In these inflammatory sites blood vessels were MECA-325 negative after 2 days, a stage of acute inflammation. At day 6 some HEV-like MECA-325 positive vessels were observed, whereas later stages, in which granulomatous tissue developed, showed increasing numbers of vessels expressing MECA-325 antigen. During inflammation development we examined the presence of lymphocytes. Few Thy-1 positive T lymphocytes (B cells are hardly present in these inflammatory sites) were observed already at day 2, when MECA-325 positive vessels were still absent. In later stages T cells rapidly increased in numbers and formed a considerable part of the mononuclear cell infiltrate.

Induction of MECA-325 antigen in cultured endothelial cells (Table 2). An important issue in high endothelial cell research is whether this type of endothelium is ontogenetically defined or can differentiate from flat endothelium under proper conditions. To address this, we took cultures of mouse endothelial cells know to be derived from normal flat endothelium, and attempted to induce the MECA-325 antigen (as a phenotypic marker for high endothelium) in these cell cultures by stimulating the cells with a variety of agents including immunological ones such as lymphokines IFN-beta and gamma, and monokine IL-1, and a couple of growth factors reacting with endothelium. When we incubated LE-2 and MCC endothelial cell cultures with these various agents, it was IFN-gamma, and non of the other tested agents, that was capable to induce MECA-325 antigen in LE-2 as well as in MCC endothelial cells. MECA-325 antigen induction was endothelium specific, since L-929 fibroblasts could not be induced to express the antigen by incubation with IFN-gamma.

The influence of recirculating lymphocytes on high endothelium in lymph nodes. Apparently lymphokines, such as IFN-gamma, are factors involved in the differentiation of high endothelium. To get more information about the role of lymphocytes in the regulation of functional and morphological aspects of high endothelium we depleted mice of circulating lymphocytes by lethal total body irradiation. 4 days later virtually no recirculating lymphocytes could be observed as demonstrated by the absence of lymphocytes in the wall of HEV and of MEL-14 (homing receptor) positive cells in lymph node sections. When the height of endothelium of (still recognizable) HEV was measured by morphometry, it appeared reduced with about 30 % as compared to HEV in sections of control lymph nodes (Fig.2). The decrease in height of the endothelium was mainly due to reduction of cytoplasm (Fig.1A). The endothelial cells still had rather oval nuclei and numerous (pyroninophilic) ribosomal RNA. Endothelium was MECA-325 positive (Fig.1C), and moreover, it's capacity to bind lymphocytes, as measured in the frozen section assay, was comparable with that of high endothelium in control lymph nodes (Fig.3). Also the specificity of peripheral node HEV to bind more T cells than B cells (T/B ratio about 3) was still measured in this flattened endothelium (Fig.4).

Irradiated mice offered the opportunity to define what mechanisms are controlling endothelial cell height. We approached this question by injecting mice 4 days after irradiation with: 1) material known to contain immunological or other regulatory facors, and 2) lymphoid cells. At several stages up to 24 hrs endothelial cell height was measured. Neither injection of ConA supernatant nor of freshly isolated mouse serum could restore endothelial cell height. Also when thymocytes or lethally irradiated lymph node cells were injected, the endothelium remained flattened. However upon injection with viable lymph node cells the endothelium regained it's normal cell height. The recovery of height occurred rapidly and was already obvious at 3 hrs after injection (Fig.1B & 2). Mel-14 staining of sections after lymph node cell injection

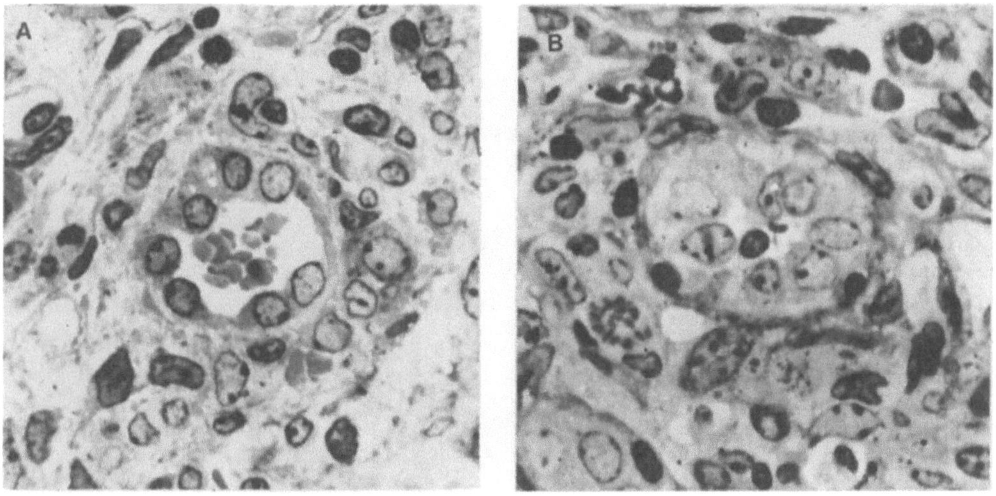

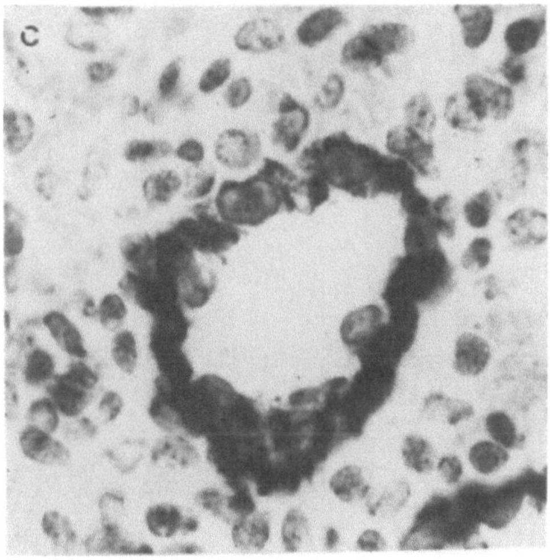

FIG. 1

A) 1 µm Epon section; flattened HEV in lymph node 4 days after irradiation. B) Epon section. When lymph node cells are injected 4 days after irradiation, characteristic 'high' appearance of HEV is restored within several hours. C) Frozen section of lymph node immunoperoxidase stained with MECA-325. Flattened HEV in lymph node are still MECA-325 positive.

demonstrated that the injected cells extravasated at sites of HEV. Positive lymphocytes were observed in the high endothelial vessel wall, and frequently cuffs of immigrated cells were present around HEV.

DISCUSSION

We have tried to define mechanisms regulating functional and morphological aspects of high endothelium in mouse. An important question is whether high endothelium is an ontogenetically defined cell lineage or a particular endothelial differentiation induced by microenvironmental conditions. We have approached this issue *in vivo* and *in vitro* respectively. *In vivo*, we studied granulomas, as sites of chronic inflammation, to check if MECA-325 positive HEV developed and at what stage of inflammation. Indeed, HEV-like vessels bearing the MECA-325 antigen were demonstrated in the granulomas from about 1 week, in parallel with development of a persistent mononuclear cell infiltrate containing numerous T lymphocytes (see also 9). Interestingly, during the acute phase of inflammation no MECA-325 positive vessels were observed. Appearantly, high endothelium is not involved in the acute phase of

FIG. 2

Average height of HEV in arbitrary units (with standard error), measured by morphometry in sections of lymph node.
Columns from left to right:
- untreated
- 4 days after 950 cGy irradiation
- 3 hrs after ConA sup injection, given 4 days after irradiation
- 3 hrs after injection of lymph node cells, given 4 days after irradiation

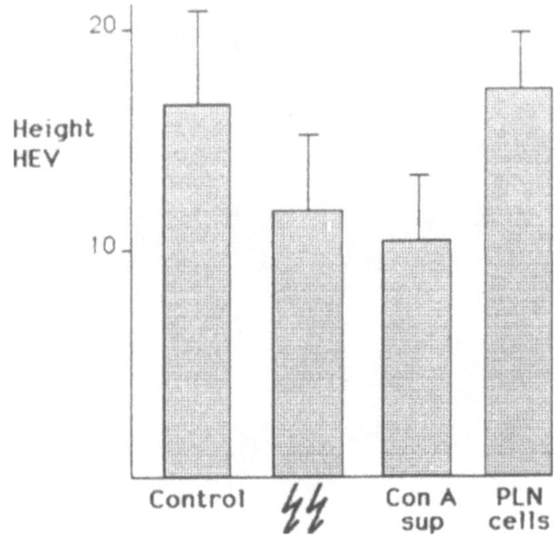

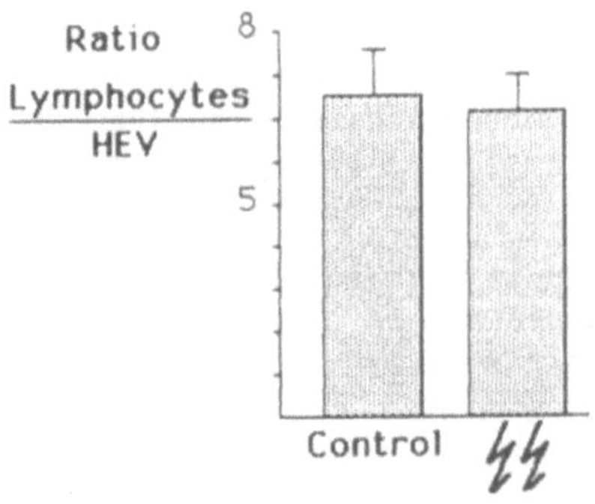

FIG. 3. Average numbers (and standard error) of bound lymphocyes per HEV. Right column is 4 days after irradiation.

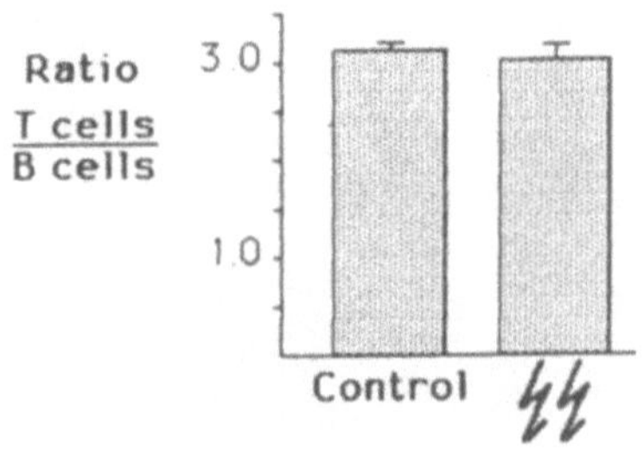

FIG. 4. Average ratio of T cells over B cells bound to HEV. T/B = 1 in incubation suspensions. Right column is after radiation.

inflammation, which is likely, since infiltrating lymphocytes are few in numbers and don't play a major role in that stage. The fact that few T cells were already observed before HEV-like vessels were present, may indicate that high endothelial differentiation only occurs whan inflammation gets a chronic character and lymphocytes majorly contribute to the population of extravasating and infiltrating cells.

The granuloma studies strongly suggest that high endothelial differentiation can be induced by local immune regulatory conditions. Our in vitro studies with endothelial cell cultures may support this. The MECA-325 antigen was induced in LE-2 and MCC endothelial cells exclusively by lymphokine IFN-gamma, and not by IL-1 or IFN-beta (see also 10). This suggest that lymphocytes and their immune factors are essential for differentiation of high endothelium. Interestingly, there has been recently described an endothelial activation antigen, recognized by a monoclonal antibody, that is induced in human endothelial cells in vitro by IL-1. Antigen expression is related to increased neutrophil/monocyte adhesion in vitro and in vivo (13). Appearantly, monocytic immune factors are involved in regulating endothelial activation to support infiltration of acute inflammatory cells. From our studies in mice severely depleted of recirculating lymphocytes it can be concluded that at least part of the characteristic high endothelial cell morphology, viz., the voluminous cytoplasm and the height of HEV cells, is dependent on direct interaction with transmigrating lymphocytes. Height of endothelium of HEV declined with about 30 % in lymphocyte-depleted mice, and could be restored by injecting recirculating MEL-14 positive lymphocytes. However, lymphocyte binding-capacity was not affected after lymphocyte depletion. This suggest that endothelial cell height and expression of high endothelial cell determinants involved in lymphocyte binding are regulated by different mechanisms respectively. Expression of endothelial determinants involved in lymphocyte adherence seems controlled by locally produced immune-regulatory factors. Endothelial cell height appearantly is regulated by cell-cell interaction of high endothelial cells and transmigrating lymphocytes. This is supported by previous studies in lymphopenic animals, such as thymectomized mice (14), nude rats (15) and thoracic-duct-drained rats (16).

Our data suggest that lymphocytes, by means of released lymphokines and direct endothelial cell interaction, play an essential regulatory role in the special function and morphology of high endothelium.

Acknowledgements: We thank Doete Niermeyer for secretarial assistance.

REFERENCES

1. J.L. Gowans and E.J. Knight. Proc.R.Soc.Lond. 159:257 (1964).
2. K. Yamguchi and G.I. Schoefl. Anat.Rec. 206:419 (1983).
3. S.K. Stevens, I.L. Weissman and E.C. Butcher. J.Immunol. 128:844 (1982).
4. G. Kraal, I.L. Weissman and E.C. Butcher. J.Immunol. 130:1097 (1983).
5. W.M. Gallatin, E.C. Butcher and I.L. Weissman. Nature 304:30 (1983).
6. R.A. Rasmussen, Y.H. Chin, J.J. Woodruff and T.G. Easton. J.Immunol. 135:19 (1982).
7. S.T. Jalkanen, R.F. Bargatze, L.R. Herron and E.C. Butcher. Eur.J.Immunol. 16:1195 (1986).
8. H.B. Stamper, Jr. and J.J. Woodruff. J.Exp.Med. 144:828 (1976).
9. A.M. Duijvestijn, M. Kerkhove, R.F. Bargatze and E.C. Butcher. J.Immunol. 138:713 (1987).
10. A.M. Duijvestijn, A.B. Schreiber and E.C. Butcher. Proc.Natl.Acad.Sci. USA 83:9114 (1986).
11. E.C. Butcher, R.V. Rouse, R.L. Coffman, C. Nottenburh, R.R. Hardy and I.L. Weissman. J.Immunol. 129:2698 (1982).
12. E.C. Butcher, R.G. Scollay and I.L. Weissman. J.Immunol. 123:1996 (1979).
13. R.S. Cotran, M.A. Gimbrone, Jr., M.P. Bevilacqua, D.L. Mendrick and J.S. Pober. J.Exp.Med. 164:661 (1986).
14. O. Jørgensen and M.H. Claësson. Z.Zellforsch. 132:347 (1972).
15. S. Fossum, M.E. Smith, E.B. Bell and W.F. Ford. Scand.J.Immunol. 12:421 (1980).
16. I. Goldschneider and D.D. McGregor. J.Exp.Med. 127:155 (1968).

SUBSTANCE P-MEDIATED MODULATION OF THE PRIMARY ANTIBODY RESPONSE

A. Eglezos*, R.D. Helme**, G.W. Dandie*, P.V. Andrews** and R.L. Boyd*

* Monash University Department of Pathology and Immunology, Alfred Hospital, ** Monash University Department of Medicine, Prince Henry's Hospital, Melbourne, Australia

The undecapeptide Substance P (SP) which is contained in primary afferent nerves, has been shown to affect certain components of the neurogenic inflammatory response including vasodilatation and plasma extravasation. Substance P also modulates the *in vitro* responses of lymphocytes and accessory cells. Rats treated neonatally with the sensory neurotoxin capsaicin, show a diminished neurogenic inflammatory response, but the effects of capsaicin treatment on cells of the immune system have not been previously documented. Using Sprague-Dawley rats at 6-12 weeks of age that had been pretreated neonatally with capsaicin, we examined the primary antibody response of popliteal lymph-node lymphocytes to antigenic stimulation *in vivo* with sheep red blood cells. It was observed that the number of cells secreting antigen-specific immunoglobulins was decreased in the capsaicin pretreated rat by more than 85%. This reduced response was reversed by a sub-cutaneous infusion of SP at the injection site immediately following antigenic stimulation. The results suggest that the decreased response in the capsaicin treated rats is secondary to the neurotoxic effects of capsaicin, and is most probably mediated by SP.

INTRODUCTION

Antidromic stimulation of sensory nerves has been shown to cause vascular permeability changes characteristic of inflammation, namely plasma extravasation (1) and flare (2). These effects are thought to be mediated by neurotransmitter substances contained in the peripheral sensory nerves, especially Substance P (SP). The neural component in this response can be shown by using the capsaicin treated rat. Neonatal treatment of rats with the sensory neurotoxin capsaicin, at a dose of 50 mg/kg, permanently destroys the unmyelinated primary afferent nerves (3) and depletes them of their neurotransmitters (4).

In previous studies the plasma extravasation component of the inflammatory response has been studied. The cellular component of this response has rarely been studied. It is known, however, that there are specific SP receptors on lymphoid cells (5) and that capsaicin treated rats have a reduced accumulation of phagocytic cells at the site of thermal injury (6). These primary studies suggest a possible novel route of cellular modulation of the immune response by neuropepties released from the primary afferent nerves.

In the present study we examined the effects of capsaicin treatment and SP on the in vivo antibody response using a modified method of Cunninghams Plague forming cell (PFC) assay.

METHODS

Sprague-Dawley rats (Monash University Central Animal House) were given a sub-cutaneous injection of capsaicin (50 mg/kg, in alcohol (10%), Tween 80 (10%) and physiological saline) between day 2-6 og life. The effect of the capsaicin treatment was tested by the topical chemical stimulation test described by Jancso et al. 1961 (7). Capsaicin pretreated rats had an 80% decrease in this response (data not shown). Antibody-secreting cells capable of forming haemolytic plaques in a monolayer of target cells and complement, were assayed using a modified Cunningham PFC assay (8). Antigenic stimulation was initiated by a sub-cutaneous injection of 10^7 fresh sheep red blood cells (SRBC) in phosphate buffered saline in both hind foot-pads of ether anaesthetised rats. After six days, the popliteal lymph nodes (PLN) were removed and washed in supplemented M199 (40 ml x10 concentrate M199 in Hanks salts (CSL), 340 ml sterile distilled water, 22 ml 2.8% w/v sterile Sodium Bicarbonate (Ajax), 2.8 ml Hepes Buffer (CSL). 0.25 ml Steptomycin Sulphate (Glaxo), 0.4 ml Amphotericin B (Sqibb), 0.5 ml Penicillin (Crystapen; Glaxo) and 0.4 ml DGP amino-acid supplement (CSL). The PLN cell suspensions were adjusted to 10^6 and 10^7 cells/ml using a Coulter Counter (Coulter Electronics). Five hundred microlitres of the cell suspension was added to 50 microlitres neat guinea-pig serum and 100 microlitres M199 washed SRBC (20% v/v). Fifty microlitres of this reaction mixture (direct assay) was then added to a Cunningham chamber and incubated at 37°C in a humidified incubator. The indirect assay included 10 microlitres of rabbit anti-rat IgG (Cappel) into the reaction mixture. Plaques were counted at x20 magnification using a light microscope. Infusions of SP were performed using a Microperpex S (LKB) peristaltic pump at a rate of 100 microlitres/hour over 4 hours. Throughout the infusion procedure rats were anaesthetised with Nembutal (Ceva). Statistical significance was determined using one way analysis of variance.

RESULTS

Neonatal capsaicin treatment resulted in a 50% decrease in the number of cells in the regional PLN (figure 1) and an 85% decrease in the number of antigen-specific IgM and IgG secreting cells (figure 2) when compared to the normal response. An infusion of SP (10^{-5} M in normal saline) restored both the total and responding cell number in the PLN from capsaicin pretreated rats (figures 1 & 2). This reversal effect was absent when either saline or normal rat plasma were infused (data not shown).

DISCUSSION

The immunoregulatory potential of SP and other neuropeptides has recently received much attention. Substance P, Somatostatin (9) and Vasoactive Intestinal Polypeptide (10) have been shown to bind and/or modulate the actions of lymphoid and accessory cells. However, little is known of the potential role of these peptides as modulators of the immune response in vivo.

In the present study, we explored the effects of capsaicin treatment and the neuropeptide SP on the primary antibody response of PLN lymphocytes to the T-dependant antigen SRBC, using a modified PFC assay technique. The results suggest that neonatal capsaicin treatment of rats has a significant effect on the capacity of the regional lymph node to respond to a sub-cutaneously administered particulate antigen. This effect could be due to a long lasting effect of neonatally administered capsaicin on the immune system, or to its neurotoxic effect on small diameter primary afferent neurons. A 50 mg/kg sub-cutaneous injection of capsaicin in the neonatal rat results in a complete loss of unmyelinated fibres in the range of 0-3

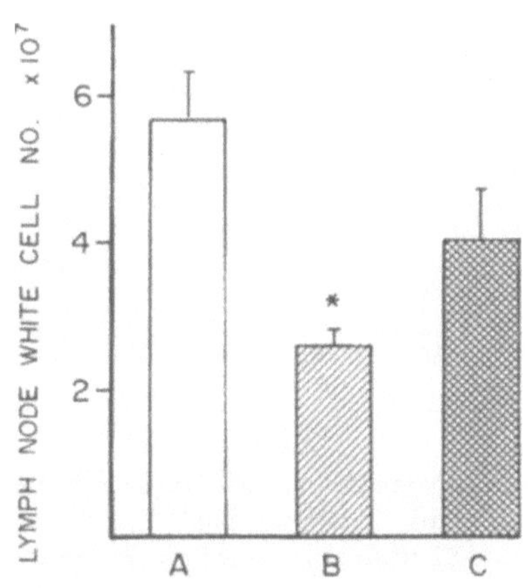

Fig. 1. Lymph-node white blood cell numbers from (A) normal, (B) capsaicin pretreated and (C) capsaicin pretreated Substance P infused rats. Values represent mean ± SEM. n=16 (A and B); 6 (C). Tests of significance by one-way analysis of variance (*$p<0.001$).

micrometers (3). Its direct effects on the immune response are not known although limited studies suggest it does not have such an effect (11, 12). Substance P has been previously shown to effect the *in vitro* antibody response (13). In our model, SP appears to effect early processing of the antigen, as the number of total and responding cells in the regional PLN were reversed by a sub-cutaneous infusion of SP immediately following antigenic challenge. It is possible that the increased plasma extravasation due to the infused SP may have been the necessary element to induce the observed reversal. However, this is unlikely as an infusion of either saline or normal rat plasma did not produce the same effect. It is also possible that neurogenically mediated neutrophil accumulation is important, as capsaicin treated rats accumulate less neutrophils to a site of inflammation (6) and we have shown that SP is a chemotactic agent for rat neutrophils (12). Other effects of SP may be mediated through activation of T_H and T_S cells (5).

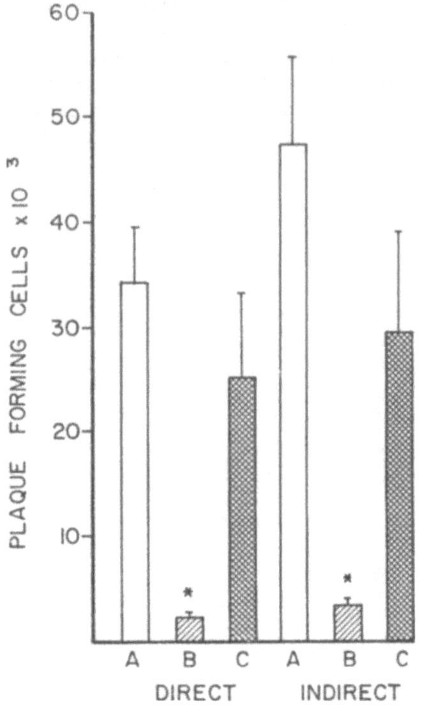

Fig. 2. Plaque forming cell number using both direct and indirect assyas for the same three groups of rats as shown in figure 1. Values represent mean ± SEM. n=16 (A and B); 6 (C). Tests of significance by one-way analysis of variance (*p<0.001).

We propose that the neurogenic inflammatory response is an important component of the modulation of the antibody response in PLN. Whether this effect is due to a primary neural modulation of lymph-node function has yet to be determined. Although it is not clear how this effect is mediated it is evident that SP may act as an *in vivo* immunomodulator.

REFERENCES

1. N. Jansco, A. Jansco-Gabor, and J. Szolcsanyi, *Br. J. Pharmacol. Chemoter.* 312:138 (1967).
2. W. M. Bayliss, *J. Physiol. (Lond.)* 26:173 (1901).
3. J. I. Nagy, L. L. Iversen, M. Goedert, D. Chapman, and S. D. Hunt, *J. Neurosci.* 3:399 (1983).
4. M. Fitzgerald, *Pain* 15:109 (1983).
5. D. G. Payan, D. R. Brewster, A. Missirian-Bastian, and E. J. Goetzl, *J. Clin. Invest.* 74:1532 (1984).
6. R. D. Helme and P. V. Andrews, *J. Neurosci. Res.* 13:453 (1985).
7. N. Jansco, A. Jansco-Gabor, and I. Takats, *Acta Physiol. Acad. Sci. Hung.* 19:113 (1961).
8. A. J. Cunningham and A. Szenberg, *Immunology* 14:599 (1968).
9. D. G. Payan, J. D. Levine, and E. J. Goetzl, *J. Immunol.* 132:1601 (1984).
10. M. S. O'Dorisio, *Fed. Proc.* 46:192 (1987).
11. R. D. Helme, A. Eglezos, and P. V. Andrews, *Clin. Exp. Neurol.* In press.

12. R. D. Helme, A. Eglezos, and C.S. Hosking, Immunol. Cell Biol. 65 (Pt.3):267 (1987).
13. A. M. Stanisz, D. Befus, and J. Bienenstock, J. Immunol. 136:152 (1986).

MEL-14 AND LFA-1, ORGAN SPECIFIC AND NON-SPECIFIC ADHESION MOLECULES INVOLVED IN HOMING OF HUMAN LYMPHOCYTES

Steven T. Pals[1], Annelies den Otter[1], Frank Miedema[2], Peter Kabel[1], Gerrit D. Keizer[3], Rik J. Scheper[1] and Chris J.L.M. Meijer[1]

1) Department of Pathology, Free University, Amsterdam, The Netherlands
2) Central Laboratory of The Netherlands Red Cross Blood Transfusion Service, incorporating the laboratory for Exp. and Clin. Immunology, University of Amsterdam, The Netherlands
3) Department of Immunology, The Netherlands Cancer Institute, Amsterdam, The Netherlands

INTRODUCTION

In the process of recirculation lymphocytes migrate from the blood into the surrounding tissues by adhering to and crossing the wall of high endothelial venules (1-4). Recently we and others have provided evidence that in man, like in rodents, a system of organ-specific receptors mediates lymphocyte-HEV recognition and thus presumably controls lymphocyte homing to lymphoid tissues and sites of inflammation (4-7).

The fact that a family of three leucocyte adhesion molecules, LFA-1, CR3 (Mac-1) and P150,95 plays an important role mediating cell adhesion in a wide variety of granulocyte, monocyte and lymphocyte cell-substrate and cell-cell-interactions (8-9), prompted us to test whether these molecules could also be involved in the binding of lymphocytes to HEV. Here we describe the results of monoclonal antibody (Mab) inhibition studies of in vitro-lymphocyte-HEV interaction. These studies in which T cells from both normal donors and from a LFA-1-deficient patient were used demonstrate 1) that LFA-1 is involved in human lymphocyte-HEV adhesion and 2) that LFA-1, in contrast to MEL-14, functions as a non-organic specific adhesion molecule.

Together with previous data our results indicate that both organ-specific (MEL-14 like) and non-organic specific (LFA-1) adhesion molecules are involved in lymphocyte recirculation and homing.

MATERIALS AND METHODS

The anti-LFA-1α antibody was SPV-L7 specific for a determinant on the α-chain of LFA-1 (8). Similar results were obtained with the anti-LFA-1α antibodies SPV-L12 and NKI-L5 (8). The anti-LFA-1β antibody was CLB-LFA-1/1, which is specific for a determinant shared by LFA-1, CR3 and p150,95 (9). The anti-CR3α and p150/95α antibodies were Bear-1 (8) and S-HCL3 respectively (11). MEL-14, a rat Mab, recognizes a determinant on the receptor molecule on mouse and human lymphocytes that mediates lymphocyte binding to PN HEV (12). Negative control antibodies were W6/32 which is specific for a determinant of HLA-ABC and HECA-452, a rat Mab specific for a determinant on HEV (5). Antibody concentrations were based on results of titrations to give optimal inhibitory effect and were 5-10μg/ml.

T-lymphocytes were prepared from the peripheral blood of normal donors by Ficoll-isopaque density gradient centrifugation and subsequent depletion of B-cells and monocytes by panning as described previously (3). The T-cell enriched population contained more than 95% CD3+ cells as determined by immunoflorescence.

The in vitro binding of human lymphocytes to HEV was assayed as described previously (3). In brief freshly cut 11 μm crystostat sections of normal peripheral lymph nodes (PN), tonsils and GALT were incubated for 30 min. at 7°C with 1.10^6 lymphocytes in 100 μl Hepes-buffered RPMI 1640 containing 5% newborn calf serum on microprint stock slides in the presence or absence of different Mab (see above). The sections were agitated during incubation. Subsequently, they were fixed in 1% glutaraldehyde and were examined under a phase contrast microscope to identify HEV-adherent cells. HEV-adherent cells were quantified per standard unit HEV length by means of a graphic tablet fitted to a micro computer. All samples were assayed in triplicate. Per spot approximately 20-30 HEV with 50-200 adherent cells were scored. The adherence of Mab treated lymphocytes was expressed as a percentage of that of the control (no antibody added). From this the percentage of inhibition of lymphocyte binding was calculated.

RESULTS AND DISCUSSION

Blocking studies with Mab to inhibit LFA-1 dependent adhesion in a number of systems demonstrated that anti-LFA-1 consistently inhibited the in vitro adhesion of human-peripheral blood T lymphocytes to HEV (table 1). In contrast to Mab against the α-chain of LFA-1, Mab against the CR3α and p150,95α cell surface molecules which are structurally related to LFA-1 since they share the same β-chain (8) did not interfere with T lymphocyte-HEV interaction (table 1).
This finding is in agreement with the observation that CR3 and p150,95 are in contrast to LFA-1 generally absent on lymphocytes (6,7). In addition we found that these molecules are, by immunohistohemical techniques, undetectable on HEV (data not shown). Blocking of T-lymphocyte adhesion to PN HEV by optimal concentrations of the three different anti-LFA-1α and the anti-LFA-1β antibodies used, ranged from 45-70% ($x \pm SD = 58 \pm 8$) (table 1) and there was no significant additive effect by mixing of the Mab (data not shown). In this context

Table 1. Inhibition of adhesion of human peripheral blood T-lymphocytes to periheral lymph node HEV.

Mab	Determinant	% inhibition
W6/32	HLA-ABC	10 ± 7
HECA-452	HECA	2 ± 3
CLB-LFA-1/1	LFA-1β	62 ± 6
SPV-L7	LFA-1α	63 ± 10
BEAR-1	CR3α	6 ± 8
SHCL-3	p150,95α	11 ± 10
MEL-14	MEL-14	61 ± 4
MEL-14 + CLB-LFA-1/1		68 ± 4

Table 2. Organ-specific and non-specific inhibition of adhesion of normal peripheral blood T-lymphocytes to peripheral lymph node (PN), tonsil (T) and GALT HEV by MEL-14 and anti-LFA-1.

	% inhibition (± sd)		
Mab	PN	T	GALT
anti-MEL-14	58 ± 8	0 ± 2	0 ± 6
anti-LFA-1β	53 ± 3	40 ± 8	55 ± 7
anti-LFA-1α	58 ± 7	52 ± 5	60 ± 8

it is of interest that ICAM-1, a recently described putative counter structure of LFA-1, is strongly expressed on HEV (13).

How does LFA-1 relate to the forementioned organ-specific lymphocyte-HEV recognition system? The existence of distinct lymphocyte-surface-recognition elements for PN versus GALT HEV is strongly supported by experiments using MEL-14 (12). This Mab specifically blocks adhesion of mouse lymphocytes to PN HEV in vitro and interferes with their homing to these lymph nodes in vivo (2,12). It precipitates a highly glycosylated protein of about 90 kD molecular weight and has recently been reported to crossreact with a putative human lymphocyte receptor for HEV (2,4). Our studies using MEL-14 confirm the observation by Jalkanen et al. who found that this antibody also selectively blocks adhesion of human lymphocytes to HEV (2,4).

Whereas the adhesion of lymphocytes to PN HEV by optimal concentrations of MEL-14 was blocked by 50-70% ($x \pm SD = 59 \pm 7$) adhesion to GALT and also to tonsil HEV was not influenced (table 2). In sharp contrast to the selective effect of MEL-14, blocking by anti LFA-1 was entirely non-organic specific (table 2) Mixes of MEL-14 and LFA-1 assayed for inhibitory capacity failed to show significant additive inhibitory effects (Table 1). Taken together the above findings indicate that in addition to organ-specific (MEL-14 like) recognition elements, LFA-1 also plays an important role in the adhesion of human lymphocytes to HEV. Moreover absence of significant additive effects with a combination of anti-LFA-1 and MEL-14 suggests that the molecules recognized by these antibodies function in conjunction in the same adhesion pathway rather than in two distinct pathways. It is conceivable that the role of LFA-1 in this pathway is supportive, increasing the strength of adhesive bonds.

To further corroborate our finding that LFA-1 is involved in lymphocyte-HEV interaction we examined the HEV-binding of T-lymphocytes from an LFA-1/CR3 deficient patient. A detailed account of this patient, whose lymphocytes express low but detectable amounts (<5% of control) of LFA-1 has been published previously (9). Patient's lymphocytes showed normal adhesion to HEV (absolute binding 14.8 ± 2.2 and 16.4 ± 2.3 cells/mmHEV length for patient and controls respectively) which was like that of the controls, sensitive to blocking by both anti-LFA-1 and MEL-14. However, the adhesion of the patient's T-lymphocytes was 10-100 fold more sensitive to the inhibiting effects of LFA-1 than that of the T-lymphocytes of the normal controls, while no differences in sensitivity to the blocking effect of MEL-14 were found. A similar, increased sensitivity of other, otherwise normal or near normal lymphocyte functions to the blocking effect of anti-LFA-1 has previously been reported in this and other LFA-1 deficient patients (6,9,10). These observations indicate that despite the minimal amount of LFA-1 on the patient's cells, this scant expression was still required and sufficient for lymphocyte functional activities and adhesion to HEV. The relative independence of lymphocyte, as opposed to granulocyte and monocyte functions of LFA-1 is consistent with the occurrence of recurrent bacterial and not of viral or fungal infections in most LFA-1 deficient patients (6). It should be pointed out, however, that clinical evidence of lymphocyte dysfunction including overwhelming viral infection has been reported in patients with a total lack of LFA-1 expression on their leucocytes (6). Moreover, the recent finding by Fisher et al, that treatment of recipients of HLA-mismatched bone morrow grafts with Mab to LFA-1 leads to successful engraftment (14), likewise suggests a dramatic impact of LFA-1 on in vivo lymphocyte function. On the basis of our present results we believe that in addition to defective lymphocyte-target cell interaction, defective lymphocyte recirculation and homing may contribute importantly to abnormal in vivo lymphocyte function in the above mentioned clinical situations. We thank Dr. W.van Ewijk for the hybridoma MEL-14 and S.A.A.B. Haverkamp and M.J.J. Strooper for typing the manuscript. This work was supported by grant IKA 87-17 from the Netherlands Cancer Organization.

REFERENCES

1. Gowans JL, Knight EJ. Proc. R.Soc. Lon. (Biol.) 159: 257 (1964).
2. Jalkanen S, Reichert RA, Gallatin WM, Baegatze RF, Weissman IL, Butcher EC. Immunol. Reviews 91: 39 (1986).
3. Pals ST, Kraal G, Horst E, Groot A, Scheper RJ, Meijer CJLM. J.Immunol. 137: 760 (1986)
4. Jalkanen S, Steere AC, Fox RI, Butcher EC. Science 233:556 (1986)
5. Duijevestein AM, Horst E, Pals ST, Rouse BT, Steere A, Picker L, Meijer CJLM, Butcher EC, Am J. Pathol., In press.
6. Anderson DC, Springer TA, Ann. Rev. Med. In press.
7. Martz E. Human Immunol. 18: 3 (1987).
8. Keizer GD, Borst J, Figdor CG, Spiets H, Miedema F, Terhorst C, De Vries JE. Eur. J. Immunol. 15:1142 (1985).
9. Miedema F, Tetteroo PAT, Terpstra FG, Keizer G, Roos M, Weening RS, Weemaes CMR, Roos D, Melief CJM. J. Immunol. 134: 3075 (1985).
10. Krensky AM, Mentzler SJ, Clayberger C, Anderson DC, Schmalstieg FC, Burakoff SJ, Springer TA. J. Immunol. 135: 3102 (1985)
11. Schwarting R, Stein H, Wang CY. Blood 65: 974 (1985)
12. Gallatin WM, Weissman IL, Butcher EC. Nature 304: 30 (1983).
13. Dustin ML, Rothlein R, Bhan AR, Dinarello CA, Springer TA. J. Immunol. 137: 245 (1986).
14. Fisher A, Blanche S, Verber F, LeDeist F, Genota I, Lopez M, Durandy A, Griscelli C. In: recent advances in bone marrow transplantation (RP Gale ed.) AR Liss, New York, in press.

HOMING RECEPTOR EXPRESSION AND MIGRATION OF ACTIVATED LYMPHOCYTES

Alf Hamann, Dorothee Jablonski-Westrich, Reiner Harder and Heinz-Günter Thiele

Abt. f. klin. Immunologie, I. Med. Klinik, Univ.krankenhaus 2 Hamburg 20, F.R.G.

SUMMARY

The experiments show that homing receptors are regulated in a complex fashion during initial cellular activation:

Signals leading to blast formation induced either:
-- a decrease of homing receptor expression in the majority of blasts;
-- an increase of the Mel-14 expression in 20 - 40% of the blasts, and
-- a selective down-regulation in the capacity to bind to Peyer's patch HEV even under conditions, where binding to peripheral HEV is high.

Submitogenic stimuli in partially activated cultures induce a rise in Mel-14 antigen expression and binding to peripheral node HEV, whereas Peyer's patch binding is unchanged or lowered.

Thus, a selective and differential regulation of organ-specific homing receptors takes place under distinct activation conditions.
The mucosal system-related receptor is more easily down-regulated upon activation.

The in vivo homing experiments indicate that mitogen activation induces one dominant migratory phenotype. Alterations in homing receptor expression seem to be associated with changes in further cellular functions leading to reduced entry into lymphatic tissues and increased localization of these cells in lung or liver. The mechanisms regulating the differential expression of organ-specific homing receptors and additional homing-relevant properties of the cells are still unknown.

INTRODUCTION

The migration and recirculation of lymphocytes through the body is a carefully regulated process, governed at least partially by the existence of organspecific homing receptors on lymphocytes which recognize determinants on high endothelial venules (HEV).

While differences in the homing pattern of virgin lymphocytes are not very pronounced, distinct migration routes are found for different populations of activated and perhaps also of memory lymphocytes. Examination of mouse lymphocytes activated in vitro indicated a loss of the putative homing receptor defined by the Mel-14 antibody (1), but this correlates not with in vivo studies showing recirculatory capabilities of several kinds of blasts. We therefore asked the following questions:

1. What are the effects of different in vitro activating procedures on the expression of the (peripheral node-specific) homing receptor Mel-14?
2. What is the influence of activation on the in vitro binding of lymphocytes to HEV of peripheral nodes and Peyer's patches ?
3. Is in vivo migration behavior correlated with homing receptor expression or determine further factors the migration pattern of activated lymphocytes ?

METHODS

Lymphocytes of mesenteric + peripheral lymph nodes of CBA mice were used. The cells were stimulated by 0.5 mM periodate or a 16 h pulse of Con A or LPS, or cultivated without mitogen (autologous MLR). Adherent cells were removed after 16 h. Immunofluorescence and FACS analysis was done as described (2). In vitro assessment of HEV-binding was done by the frozen section assay as described (3), using FITC-labeled lymphocytes as internal standard. For in vivo homing experiments lymphocytes were labeled by 20 µCi 51-chromium/ml and freed of dead cells by centrifugation on 18% isotonic metrizamide. $5*10^6$ cells were injected into the tail vein and after the time indicated 0.5 ml blood was withdrawn and the mice killed by ether. Peripheral lymph nodes removed consisted of the superficial cervical, brachial and superficial inguinal nodes. All Peyer's patches were collected. The radioactivity of the organs and the remaining body was determined in a y-counter.

RESULTS and DISCUSSION

We analysed the expression of the Mel-14 antigen, the putative receptor for peripheral lymph node HEV (4), on lymphocytes stimulated in vitro by different mitogens (Fig. 1, 3) and from autologous cultures (Fig. 2). The results of the FACS analyses indicate that full blast transformation is associated with a decrease or complete loss of Mel 14 antigen expression on

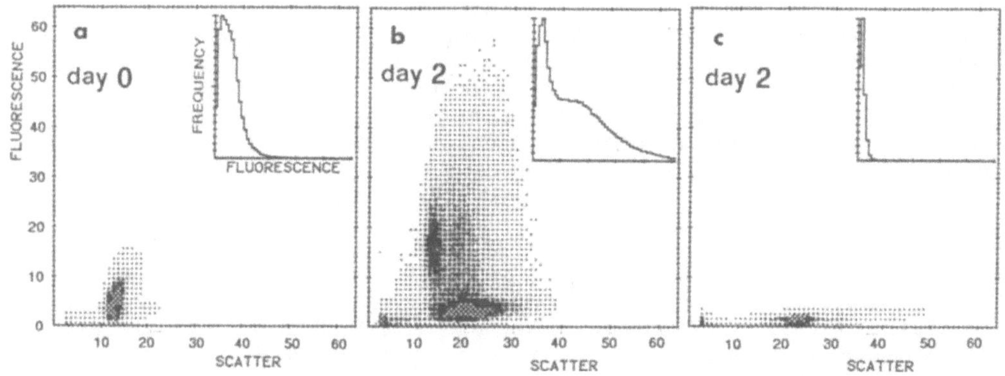

Fig. 1. Expression of MEL-14 antigen on periodate-activated lymph node lymphocytes as analyzed by the FACS. Distribution of cells according to fluorescence intensity and light scatter (y- and x-axis respectively, arbitrary units, linear scale). Inset: one-dimensional plot of cellular fluorescence. a) fresh, unstimulated lymphocytes. b) 2 days after stimulation. c) control (2nd antibody only).

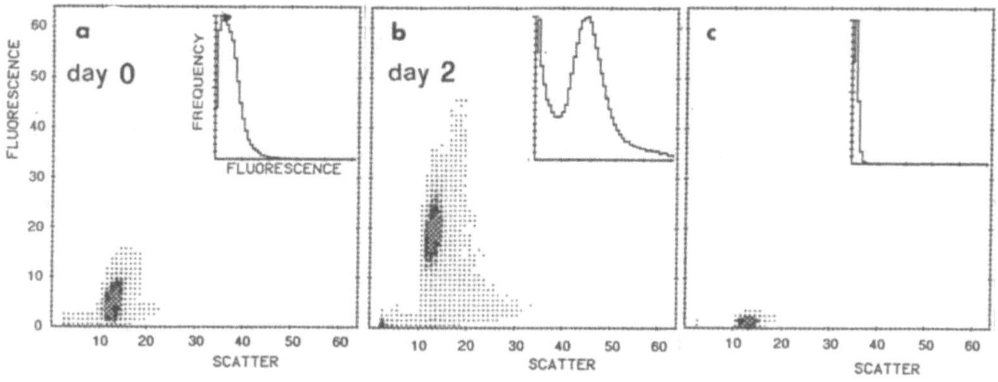

Fig. 2. Expression of MEL-14 antigen on lymphocytes in an autologous MLR. a) fresh lymphocytes. b) after 2 days of culture. c) Control (2nd antibody only).

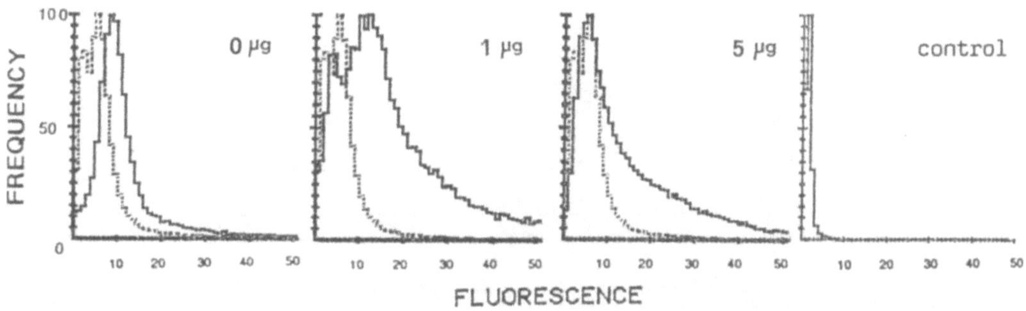

Fig. 3. Expression of Mel-14 antigen on lymphocytes stimulated with different concentrations of Con A at day 2 of culture. Overlaid curve (broken line): Fluorescence pattern of fresh lymphocytes. Control: 2nd antibody only.

the majority of blasts in cultures stimulated by periodate, LPS (not shown), optimal doses of Con A or in autologous cultures (for further data see ref.2).

However, a subset of 20-40% of fully activated cells may nonetheless express very high levels of the Mel 14 antigen. The failure to detect this Mel-14-positive blasts in previous studies (1,5) may be due to the use of prolonged stimulation schedules in contrast to the pulse stimulation used here.

In contrast, submitogenic stimuli induce a general increase in the expression of the Mel-14 antigen on lymphocytes. This is most clearly seen in autologous cultures of lymphocytes, where only a small portion of the cells becomes fully activated, and in cultures stimulated by different doses of Con A. Low, submitogenic doses of Con A let rise the Mel-14 expression, whereas the majority of blasts stimulated by high doses has rather low levels of Mel-14. Similarly, the non-transformed part of lymphocytes in periodate- (or LPS-, not shown) -stimulated cultures also expresses high levels of the Mel-14 antigen.

Use of the frozen section assay reveals that periodate- (or Con A-, not shown) stimulation leads to a selective and nearly complete suppression of the capacity to bind to HEV of Peyer's patches, even under conditions, where overall binding to peripheral nodes is increased (Fig. 4). This indicates a differential regulation of the two respective receptors, with the mucosa-system specific homing receptor being more prone to down-regulation in these activation systems.

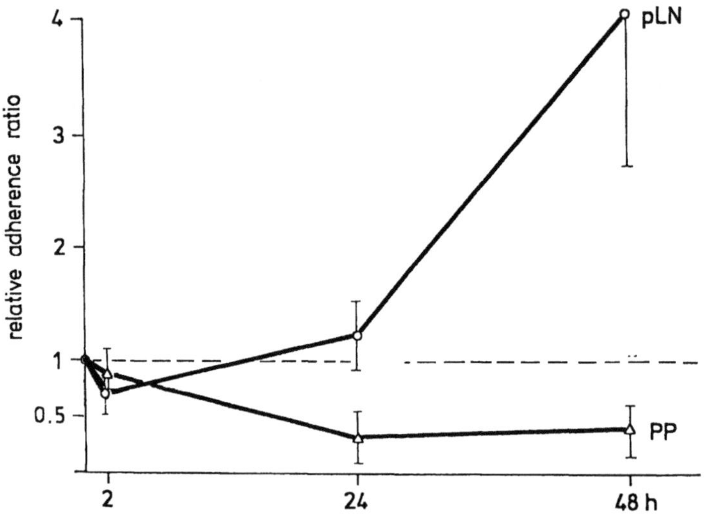

Fig. 4. Binding of periodate-activated lymph node lymphocytes to HEV on frozen sections of peripheral nodes and Peyer's patches, at different times after stimulation. Given is the relative adherence ratio, indicating the binding of activated cells in relation to that of control cells (fresh lymphocytes, = unity), each standardized by internal reference cells (FITC-labeled fresh lymphocytes). Bars: Standard deviation.

In short term in vivo homing experiments, periodate-, LPS- (or Con A-, not shown) activated lymphocytes enter lymph nodes and Peyer's patches only in small numbers (Fig. 5, 7), in spite of their ability to bind to HEV in vitro. On the other hand, localization in the lung (and in the liver at later time) is enhanced.

This homing pattern is not due to non-specific alterations related to in vitro cultivation, since unstimulated lymphocytes of autologous cultures migrate as well to lymphoid organs as do fresh lymphocytes (Fig. 6).

The gross migration pattern displaid by mitogen-activated lymphocytes is not dependent on the organ source of lymphocytes or source of accessory cells present during activation nor of type of mitogen (Fig. 7). Its distribution pattern is very similar to that found by Smith, Martin and Ford (6; Fig. 5 d) for in vivo alloantigen-activated peripheral node ("ABA"-) blasts in the rat, whereas a pattern ressembling that of thoracic duct blasts in that study could not be elicited by mitogen activation. Thus, mitogen activation seems to induce only one dominant migratory phenotype.

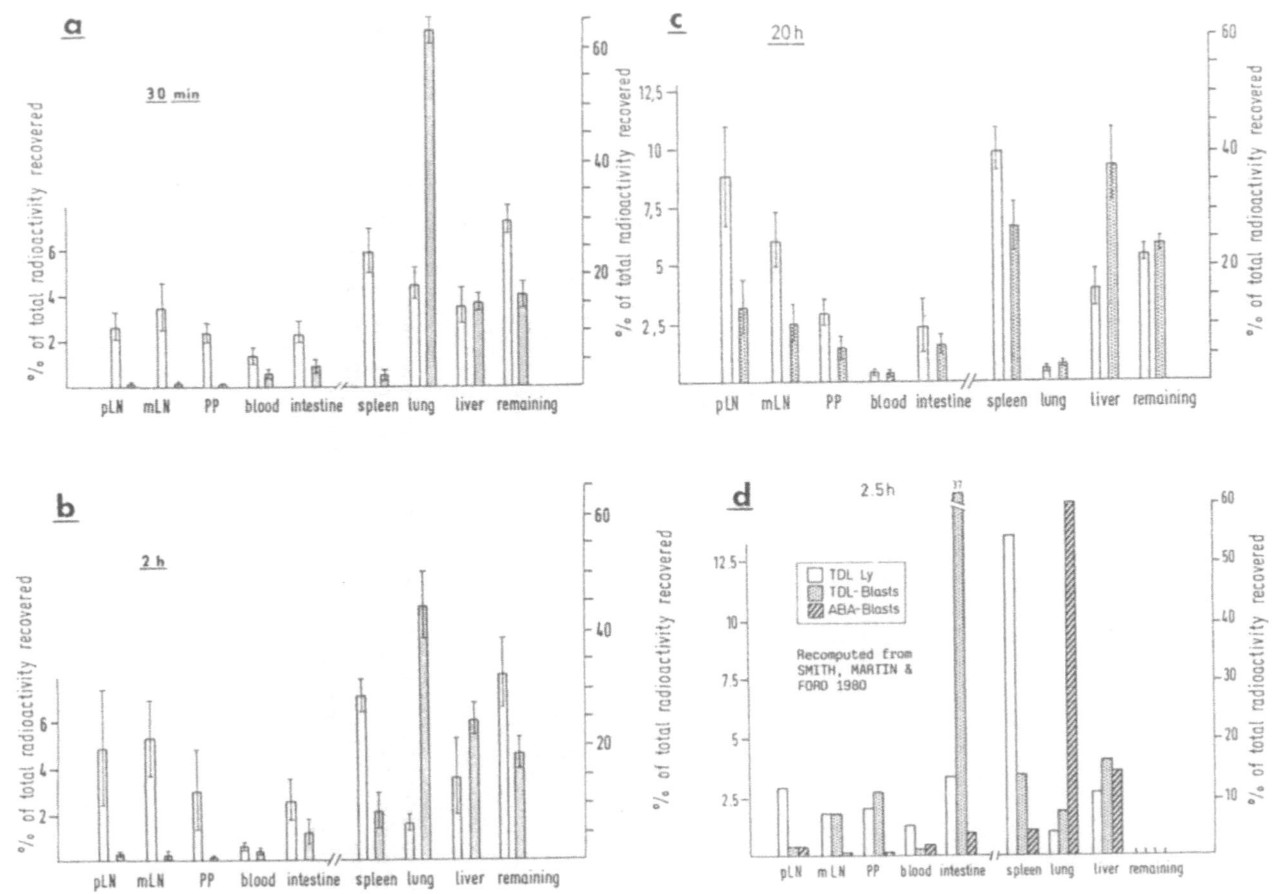

Fig. 5. In vivo localization of ^{51}chromium-labeled, periodate-activated lymph node lymphocytes (day 2). a) 30 min after injection into the tail vein. b) 2 h after injection. c) 20 h after injection. Open columns: control (fresh) lymphocytes; dotted columns: activated cells. Mean of 4-6 animals and its 95% confidence interval. d) For comparision: In vivo localization of rat lymphocytes and blasts, data from SMITH, MARTIN and FORD 1980, recomputed.

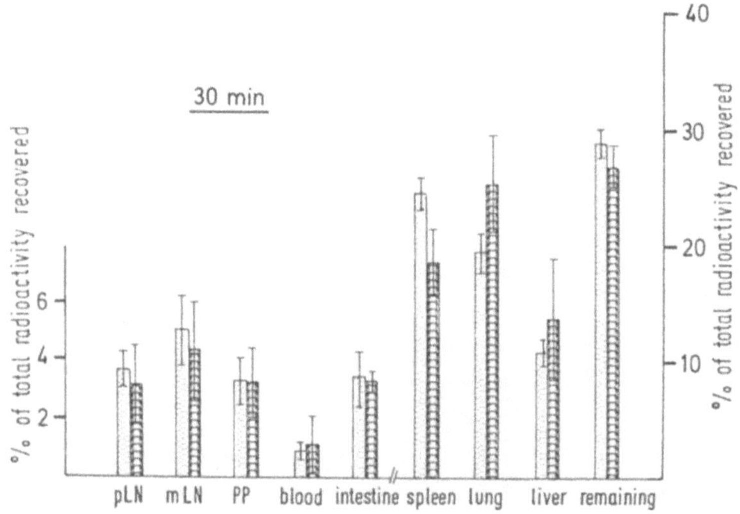

Fig. 6. In vivo localization of lymph node lymphocytes after 2 days autologous MLR. Dotted columns: control (fresh) lymphocytes; hatched columns: cultured cells. 30 min after injection into the tail vein. Bar: 95% confidence limit (4 animals).

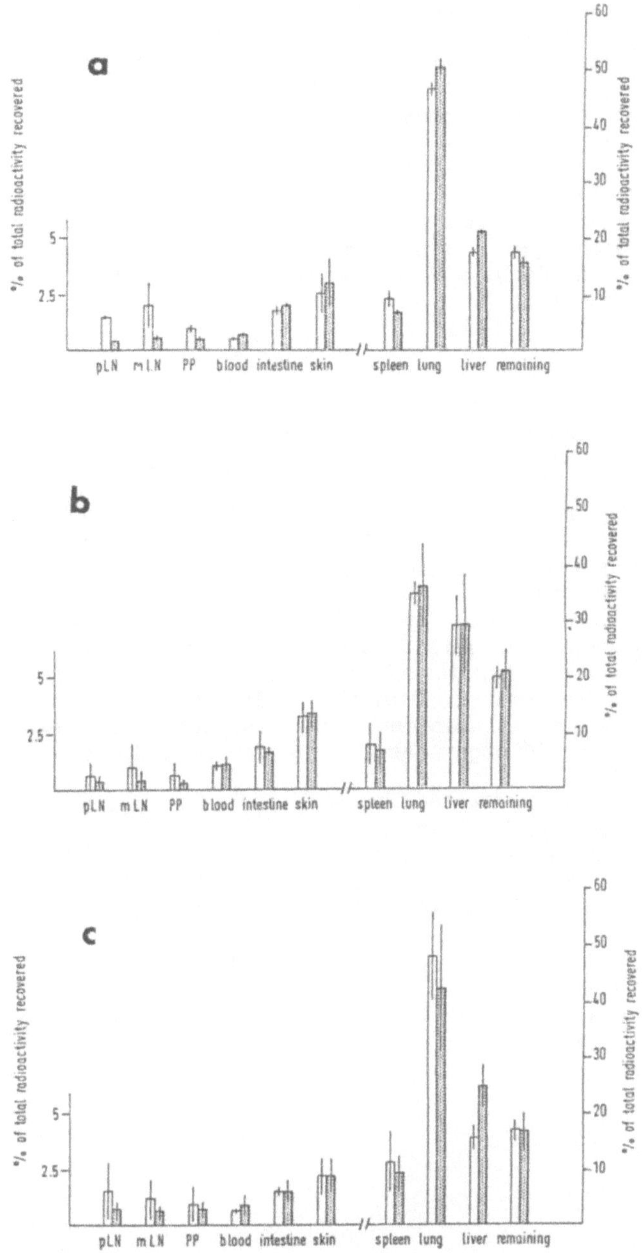

Fig. 7. In vivo localization of either lymph node (open column) or Peyer's patch (dotted column) lymphocytes. a) Periodate activated. b) LPS-activated. c) Periodate-activated spleen lymphocytes, stimulated in the presence of accessory cells of either lymph nodes (open column) or Peyer's patches (dotted column). 1 h after injection into the tail vein. Bar: standard deviation (3-6 animals).

The cellular mechanisms influencing the recirculatory properties of activated lymphocytes in addition to homing receptors are still obscure. The LFA-1 related accessory interaction mechanisms was found to contribute not only to HEV-binding but also to the homing of activated lymphocytes into the lung (7). It may be speculated, that the induction of additional adhesive interaction systems and receptors could play a role for the pronounced lung- and liver-seeking properties of the blast populations formed under the activation conditions used here.

ACKNOWLEDGEMENTS
We thank H. Bollmann, K.-U. Scholz, H. Baisch (Department of Biophysics, Hamburg), for carrying out FACS-analyses. E.C.Butcher, Stanford, and A. Duijvestijn, Groningen, supported the work by practical advice and stimulating discussions. This work was supported by the Deutsche Forschungsgemeinschaft.

REFERENCES

1. Dailey, M. O., W. M. Gallatin, I. L. Weissman, and E. C. Butcher. 1983. Surface phenotype and migration properties of activated lymphocytes and T cell clones. In: Intercellular Communication in Leucocyte Function. J. W. Parker, and R. L. O'Brien (Eds.), Wiley, p. 641.
2. Hamann, A., D. Jablonski-Westrich, K.-U. Scholz, A. Duijvestijn, E. C. Butcher, and H.-G. Thiele. Regulation of lymphocyte homing (I): Alterations in homing receptor expression and organ-specific HEV-binding of lymphocytes upon activation. J. Immunol., submitted.
3. Butcher, E. C., R. G. Scollay, and I. L. Weissman. 1979. Lymphocyte adherence to high endothelial venules: characterization of a modified in vitro assay and examination of the binding of syngeneic and allogeneic lymphocyte populations. J. Immunol. 123:1996.
4. Gallatin, W. M., I. L. Weissman, and E. C. Butcher. 1983. A cell-surface molecule involved in organ-specific homing of lymphocytes. Nature 303:30.
5. Jung, T. M. and M. O. Dailey. Reversibility of loss of homing receptor following lymphocyte activation. This volume.
6. Smith, M. E., A. F. Martin, and W. L. Ford. 1980. Migration of lymphocytes in the rat. Preferential localization of DNA-synthesizing lymphocytes in particular lymph nodes and other sites. Monogr. Allergy 16:203.
7. Hamann, A., D. Jablonski-Westrich, A. Duijvestijn, E. C. Butcher, R. Harder, H. Baisch and H.-G. Thiele. Evidence for an accessory role of LFA-1 in lymphocyte - high endothelium interactions during homing. J. Immunol., in press.

REVERSIBILITY OF LOSS OF HOMING RECEPTOR EXPRESSION FOLLOWING ACTIVATION

Thomas M. Jung and Morris O. Dailey

Departments of Pathology and Microbiology
University of Iowa, College of Medicine
Iowa City, Iowa 52242 USA

The ability of lymphocytes to recirculate from peripheral blood to lymphoid tissue is necessary in order to maintain an effective immune response, irrespective of an antigen's port of entry. Recirculation and homing to lymph nodes (LN) and Peyer's patches (PP) is initiated by the binding of cells to specialized post-capillary high endothelial venules (HEV's), followed by their passage through the vessel walls into the surrounding lymphoid tissue.

The binding of lymphocytes to the HEV's is mediated by specific cell surface receptors. Cells which lack this receptor, such as most thymocytes, do not home to LN or PP. The lymphocyte homing receptor specific for lymph node HEV is recognized by the monoclonal antibody MEL-14 and is designated $gp90^{MEL-14}$ (gp90) (1).

Previous studies with germinal center B cells and T cell clones, both of which are activated populations, demonstrated that they do not express the cell surface marker gp90 (2,3). This present report examines the relationship between cellular activation and lymphocyte homing receptor expression. The results reviewed here demonstrate that activation of T cells by Con A results in a down regulation of the homing receptor and that homing receptor re-expression occurs on these cells after cessation of antigenic stimulation. Thus, there appears to be a reciprocal relationship between homing receptor expression and cellular activation. Activation leads to a decrease in homing receptor expression and this loss is reversible, with the lymphocytes re-expressing normal levels of homing receptor as they become small, resting cells.

MATERIALS AND METHODS

Reagents

The following FITC or biotin conjugates of monoclonal anti-bodies were prepared in this laboratory: GK1.5, 53-6.7, C2 and 7D4, which are directed against L3T4, Lyt-2, the transferrin receptor and the IL-2 receptor, respectively. MEL-14 (Rat IgG2a) and normal rat immunoglobulin were used unconjugated with 2nd stage rabbit anti-rat-allophycocyanin (APC) or FITC conjugates, also prepared in this laboratory. Phycoerythrin-streptavidin was obtained from Biomeda, (Foster City, CA.). Propidium iodide was obtained from Sigma Chemical Co.,(St. Louis, MO).

Cell Cultures

Three x 10^7 spleen cells from 7-10 week old female Balb/c mice (Jackson Labs., Bar Harbor, ME.) were stimulated with 2 ug/ml of Con A (Sigma) in 25 ml of RPMI-1640 supplemented with 1 mM non-essential amino acids, 1 mM HEPES, 1 mM glutamine, $5x10^{-5}$ mM 2-ME and 10% FCS. Cell cultures were split on day 2 and 10% rat Con A supernatant was added on day 3. Cells were centrifuged, washed and analyzed on day 4. For the resting protocol, two x 10^6 washed 4 day blasts were resuspended in 6 ml of RPMI-1640 supplemented media with 20% FCS and 20 mM alpha-methyl-mannoside and were added to T25 flasks containing underlayers of freshly isolated, irradiated (2000 rads), adherent spleen cells. These cultures were then analyzed 3 days later (day 7 of total culture).

Cell staining

Five x 10^5 cells were incubated with 25 ul of unconjugated rat monoclonal antibody, washed and centrifuged through a FCS underlayer, and then stained with APC- or FITC-conjugated rabbit anti-rat Ig. Twenty min. later, 5 ul of normal rat serum was added to block residual anti-rat Ig binding sites. After 1 wash, the cells were then incubated with 25 ul of FITC- and biotin-conjugated antibodies and washed as before. Finally, the cells were incubated with 25 ul of PE-streptavidin and then washed and resuspended in HBSS and 5% FCS. Just prior to analysis 0.5 ug of propidium iodide was added for dead cell gating. Cells were then analyzed on a fluorescence activated cell sorter (FACS 440) using logarithmic amplification for fluorescence.

RESULTS

Loss of Homing Receptor with T Cell Activation. In order to directly examine the relationship between activation and homing receptor expression, cells were stained for two activation markers, the IL-2 receptor (IL-2R) and the transferrin receptor (TfR) (4,5). Figure 1 shows that most normal resting lymphocytes express high levels of gp90 and are TfR and IL-2R negative. However, after 36 hr of activation, three-color staining with anti-IL-2R (7D4), anti-TfR (C2) and anti-homing receptor (MEL-14) antibodies, demonstrates the presence of two populations of cells. One population expresses high levels of IL-2R and TfR and the other is negative for the activation markers (Fig. 2). In addition, these two populations of cells express different levels of gp90. Those cells that express high levels of homing receptor are negative for IL-2 and transferrin receptors, and those cells that express low levels of homing receptor stain brightly for IL-2R and TfR. Thus, the loss of gp90 occurs almost exclusively on activated cells in the cultures. This correlates well with forward scatter and microscopic identification of both large blast cells and residual unactivated cells in these cultures at 36 hr.

By day 4, the entire population is composed of activated cells which express high levels of IL-2R and TfR (data not shown). FACS analysis of gp90 expression shows that the surface density of homing receptor is much lower on the 4 day blasts than on freshly isolated spleen cells (fig. 3). (Compare modal fluorescence of positive cells to that of the negative control cells. Note that these activated cells are highly autofluorescent, shifting the histograms to the right.) Thus, by day 4 these large activated cells are essentially all MEL-$14^{\pm}$ (neg. or dim).

In order to determine whether the loss of homing receptor expression is merely the result of overgrowth of a MEL-14^- population, sorted MEL-14^+ cells were cultured with Con A and an irradiated spleen cell feeder layer. After 6 days of activation, MEL-14^+ cells became entirely MEL-14 neg. In addition, single cell cloning of sorted MEL-14^+ cells cultured with Con A and an irradiated spleen cell underlayer resulted in 45 of 45 clones being MEL-14$^-$. These experiments show that MEL-14$^-$ lymphocytes can arise from MEL-14$^+$ precursors.

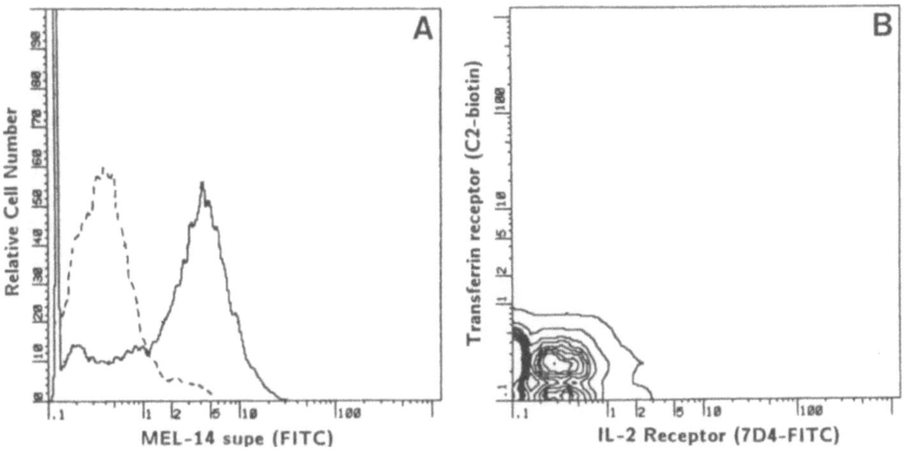

Fig. 1 Immunofluorescence histograms of freshly isolated Balb/c spleen cells. A) Approximately 65% of normal spleen cells stain brightly for the LN-specific homing receptor recognized by the monoclonal Ab MEL-14. (Dashed curve represents negative control staining with normal rat Ig.) B) Contour plot demonstrating lack of expression of transferrin (TfR) and IL-2 (IL-2R) receptors by these normal lymphocytes.

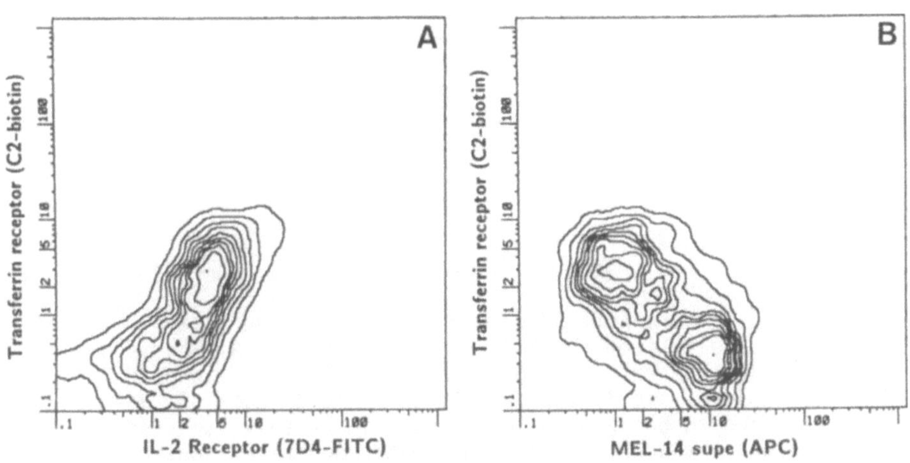

Fig. 2 Immunofluorescence two-color plots of 36 hr Con A blasts. A) Contour plot demonstrating coordinate expression of IL-2R and TfR. Staining with anti-IL-2R and anti-TfR reveals two populations of cells, one population is positive for both IL-2R and TfR, and one is negative for both IL-2R and TfR. B) Staining with MEL-14 and anti-TfR reveals two populations of cells. One population that is MEL-14$^+$ (high) and negative for TfR and the other population that is MEL-14$^\pm$ (neg. or dim) and TfR high. Thus, those cells that express both of these activation markers are MEL-14$^\pm$ and those that do not express IL-2R and TfR are MEL-14$^+$.

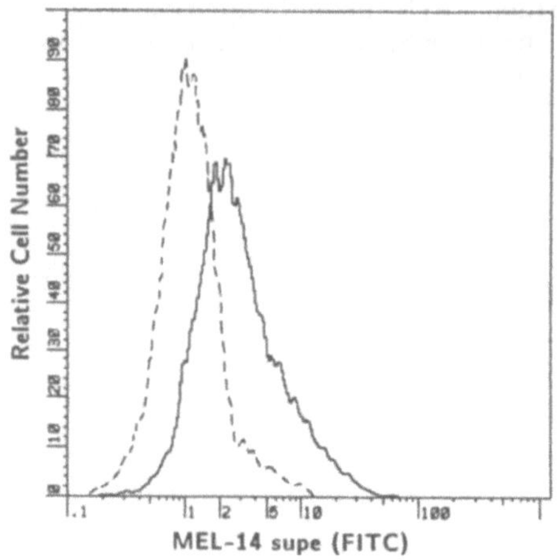

Fig. 3 Immunofluorescence histograms of day 4 Con A blasts. The entire population has become MEL-14$^{\pm}$ (neg. or dim), as compared to negative control staining. (The dashed curve represents negative control staining with normal rat immunoglobulin.) The high background of the control cells is due to inherent autofluorescence of these blast cells and not to increased nonspecific binding of antibodies (data not shown).

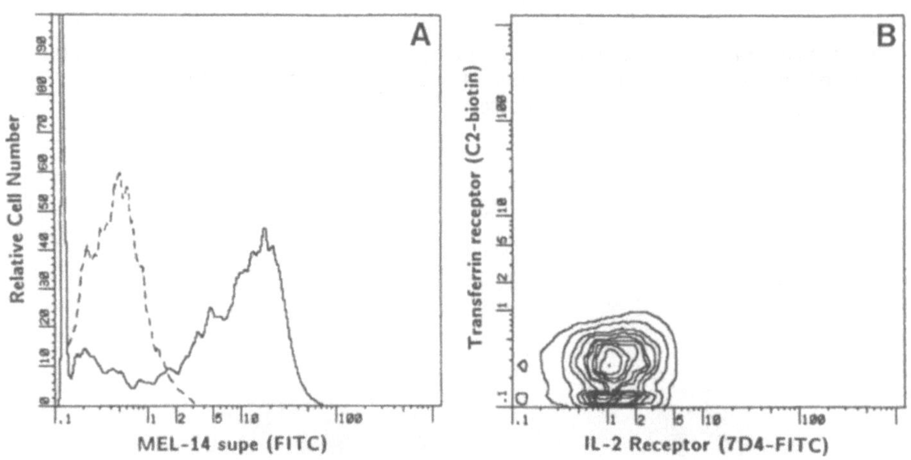

Fig. 4 Immunofluorescence analysis of day 7 rested cells from the 4 day Con A blasts. A) MEL-14 staining has returned to the high levels seen in normal spleen, and B) the expression of IL-2R and TfR has returned to background levels seen in freshly isolated spleen. (Dashed line represents negative control staining.)

<u>Return of Homing Receptor Expression on Resting Cells</u>. By day 4 of activation with Con A, the entire culture is composed of large activated cells that express little or no homing receptor and express high levels of TfR and IL-2R. When these activated blast cells are placed in resting cultures without mitogenic stimulation, the expression of IL-2R and TfR

returns to their normal low levels. In addition, the homing receptor becomes re-expressed on these rested cells to a level ~20 fold higher than the level seen on the 4 day activated cells (again, compared to negative control staining). Interestingly, the intensity of re-expression is somewhat higher than the level seen in freshly isolated spleen cells. This re-expression of the homing receptor is not due to selective death of MEL-14$^-$ cells as the viability of the rested cells was over 70%. Thus as the cells become quiescent and small in size, they lose their receptors for IL-2 and transferrin and re-express high levels of surface homing receptors.

Furthermore, subset studies have shown that the re-expression of homing receptor occurs on both the helper and cytotoxic T cell subsets (Data not shown). Therefore, this re-expression of the homing receptor after activation seems to be inversely related to cellular activation alone and does not appear to be selectively up regulated on different T cell subsets.

DISCUSSION

The studies summarized in this report demonstrate that activation of murine T cells by concanavalin A leads to a decrease in the expression of the lymph node-specific homing receptor recognized by MEL-14. This down regulation is time-dependent, associated with the expression of activation antigens (IL-2 receptor and transferrin receptor) and associated with blastogensis. Furthermore, the down regulation of the homing receptor is reversible. That is, as MEL-14$^-$ cells are placed in cultures without exogenous stimulation, as the cells become smaller and enter the resting state, homing receptor expression returns to levels at or above those seen in freshly isolated spleen cells. Preliminary studies have shown a functional correlation between homing receptor expression and the binding of cultured cells to HEV in the frozen section binding assay. Those blast cells that express low levels of homing receptor bind poorly to frozen section HEV, and those cells that express high levels of homing receptor bind as well as normal spleen cells.

These results correlate well with previous studies that have shown that germinal center B cells, T cell clones, and T cells stimulated by the mixed lymphocyte reaction express little or no LN-specific homing receptor (2,3,6). The present report has extended these findings to show that the loss of expression of the homing receptor is reversible and that homing receptor expression is inversely correlated with cellular activation. This reciprocal relationship between cellular activation and homing receptor expression may be physiologically relevant in the maintenance of an immune response. Specifically, the loss of homing receptors would result in dramatic alterations in the pattern of lymphocyte localization in vivo. Re-expression of homing receptors after cessation of antigenic stimulation would again permit trafficking to lymph nodes and allow for the dissemination of immunologic memory to distant lymphoid tissues.

This work was supported by NIH grant AI-22730 and grant IM56 from The American Cancer Society. Morris Dailey is a fellow of the John A. Hartford Foundation.

REFERENCES

1. Gallatin, W.M., I.L. Weissman and E.C. Butcher. A cell sur face molecule involved in organ-specific homing of lymphocytes. Nature 303:30 (1983).

2. Reichert, R.A., W.M. Gallatin, I.L. Weissman and E.C. Butcher. Germinal center B cells lack homing receptors. J. Exp. Med. 157:813 (1983).

3. Dailey, M.O., C.G. Fathman, E.C. Butcher, E. Pillemer, and I.L. Weissman. Abnormal migration of T lymphocyte clones. J. Immunol. 128:2134 (1982).

4. Haynes, B.F., M. Hemler, T. Cotner, D.L. Mann, G.S. Eisenbarth, J.L. Strominger, and A.S. Fauci. Characterization of a monoclonal antibody (5E9) that defines a human cell surface antigen of cell activation. J. Immunol. 127:347 (1981).

5. Helmes, M.E., M.B. Brenner, J.M. McLean, and J.L. Strominger. Antigenic stimulation regulates the level of expression of interleukin 2 receptor on human T cells. Proc. Natl. Acad. Sci. 81:2172 (1984).

6. Dailey, M.O., W.M. Gallatin, I.L. Weissman, and E.C. Butcher. Surface phenotype and migration properties of activated lymphocytes and T cell clones. In Intercellular Communications in Leukocyte Function, J.W. Parker and R.L. O'Brien, eds., Wiley, pp.641-644 (1983).

ENDOCYTIC STRIPPING OF LIGANDS FROM MIGRANT LYMPHOCYTES IN HIGH ENDOTHELIAL VENULES (HEV): IMPLICATIONS FOR IMMUNOMODULATION vs VIRAL PATHOGENESIS

Arthur O. Anderson and Jerrold M. Ward

USAMRIID, Fort Detrick, Frederick, MD 21701-5011
and NCI/FCRF, Frederick, MD 21701-1013

INTRODUCTION

"Grooming" of membrane detritis from lymphocyte surfaces by mononuclear phagocytes during migration through lymphatic tissues was proposed as a cause of differences in the potency of comparable populations of lymphocytes to produce the normal lymphocyte transfer reaction (LTR) when equivalent numbers of T cells from blood, afferent lymph, and efferent lymph were injected SC[1]. Three- to five-fold greater LTR activity was found with efferent lymph cells compared to blood or afferent lymph cells. It is possible that these differences in reactivity were due to anti-idiotypic antibodies or other "blocking factors" occupying lymphocyte receptors. Isotypic, allotypic and idiotypic immunoregulation occurs in mice, rats, rabbits, and man[2-5]. The present studies employ light and electron microscopy, cytochemistry and immunochemistry to analyze the endocytic potential of Lewis Rat HEV endothelium toward both particulates and toward autologous lymphocytes whose surfaces have been modified by liganding. Results indicate that HEV endothelial cells nonspecifically remove receptor-blocking proteins during blood to tissue transit.

MATERIALS AND METHODS

Male, Lewis Rats of 180-200g body weight were used in these studies. Outbred Rabbits were used for antibody production. Colloidal carbon, inactivated Pneumococci type 25, zymosan, glutaraldehyde-fixed sheep erythrocytes (SRBC), rabbit anti-thymocyte serum (ATS)-treated thoracic duct lymphocytes and chicken erythrocytes were suspended in tissue culture media and injected retrograde into the surgically-exposed brachial arteries in order to perfuse the axillary lymph node vasculature. The axillary

---

In conducting the research described in this report, the investigator(s) adhered to the "Guide for the Care and Use of Laboratory Animals," as promulgated by the committee on Care and Use of Laboratory Animals of the Institute of Laboratory Animals Resources, National Research Council. The facilities are fully accredited by the American Association for Accreditation of Laboratory Animal Care.

The views of the author(s) do not purport to reflect the positions of the Department of the Army or the Department of Defense.

Table 1. RETICULOENDOTHELIAL FUNCTION OF HEV FOLLOWING I.V. INFUSION

	Phagocytosis by cells in Tissue		
Agent	Lymph node HEV	Blood Monocytes	Liver and Spleen Phagocytes
Colloidal Carbon	+	+	++++
Type 25 Pneumococci	-	++	+++
Zymosan Particles	-	++	+++
Saccharomyces Cerevisiae	±	++	+++
Sheep Red Blood Cells	+	±	ND
Chicken Red Blood Cells	+	+	ND
Autologous Lymphocytes + Anti-Immunoglobulin	-	-	±
Autologous Lymphocytes + Anti Thymocyte Serum	++	+	+++

lymph nodes were removed at 30 min, 3 hours and 18 hours after perfusion and either snap frozen for histochemistry or fixed in phosphate buffered glutaraldehyde for transmission electron microscopy. Lymph nodes from HIV virus-infected patients (from ref. 6) were also examined by immunochemistry for viral antigen and lymphocyte surface receptors. Thoracic duct lymph (TDL) was obtained via surgically prepared Bollman fistulae as described elsewhere[7]. Aliquots of 3 x 10^8 Fresh TDL were washed and suspended in media containing ^{3}H-Uridine 10 mCi/ml and incubated for 30 minutes at 37°C, chased with cold Uridine, washed and resuspended in aliquots of Alcian blue dye 2-200 µg/ml, Concanavalin A 1-5 µg/ml, Rabbit anti-rat Immunoglobulins (IgG fraction) FAB_1[8,9], or Rabbit anti-rat thymocyte serum (IgG fraction) FAB_1[8,9], and then infused intravenously (i.v.). Lymphatic tissues were removed, dissolved in scintillant and the percent of injected CPM calculated. TDL treated with Alcian blue dye or ferritin or horseradish peroxidase (HRP)[10] conjugated FAB_1 antibody fragments of anti-immunoglobulins and anti-thymocyte serum. Autologous Lewis rat peripheral blood lymphocytes were collected on Percoll gradients after lysing erythrocytes and surface ^{125}I iodinated using lactoperoxidase (11). Axillary lymph nodes of these rats were excised, fixed and processed for EM peroxidase cytochemistry or autoradiography.

RESULTS

Endothelial cell "coated pits" were frequently found associated with sites of membrane contact with untreated lymphocytes emigrating along interendothelial spaces in HEV (Fig. 1a). Lymphocyte contacts were in the form of microvillus-pit interactions or adhesion plaques[15]. The endothelial cells of HEV exhibited modest endocytic activity for colloidal carbon, CRBC and SRBC, whereas intra-arterial infusion of opsonized microorganisms did not reveal significant reticuloendothelial function (Table 1). Carbon was found in vacuoles within most of HEV endothelial cells after intra-arterial injection. Much less carbon and none of the other particles were found in HEV following intravenous infusion of equivalent doses. Carbon particles, Pneumococci, zymosan granules and fixed SRBC were sequestered in Kupffer cells in the liver and macrophages in the red pulp and marginal zone of the spleen after intravenous infusion. Autologous ^{3}H-Uridine labeled TDL treated with FAB_1 fragments of anti-immunoglobulin, ATS and alloantisera, Con A and Alcian blue exhibited reduced homing to lymph nodes and spleen without concommitant sequestration in the liver (Table 2). Intact Rabbit anti-rat thymocyte serum resulted in significant reticuloendothelial sequestration in the liver, spleen and in the endothelium of lymph node HEV. Infusions of lymphocytes coated with

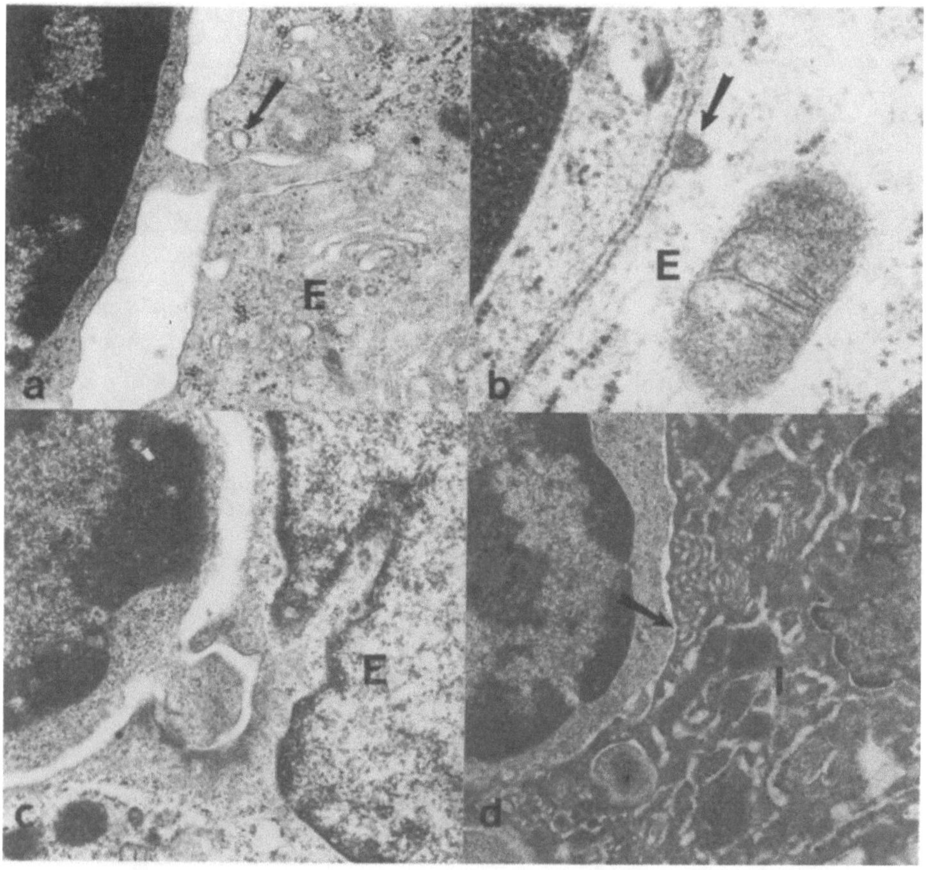

Fig. 1(a-d). a. Lymphocytes home to HEV endothelium via microvilli that interact with pits on endothelial (E) surfaces, presumably through receptors aggregated at these sites. "Coated pits" (arrow) are often found associated with these points of intimate contact. b. Peroxidase tagged FAB_1 fragments of anti-thymocyte IgG (arrow) are endocytosed off the surface of migrating lymphocytes by HEV endothelium (E). c. ATS-treated TDL are "manipulated" by endothelial cells of HEV, and d. a tubular reticulum can be seen emptying on the surface of a migrating lymphocyte (arrow).

monovalent ligands resulted in endocytic but not phagocytic activity by HEV endothelial cells. In vitro treatment of TDL with doses of HRP-labeled FAB fragments prepared from anti-lymphocyte serum and selected ligands, (that permitted reversible suppression of entry into lymph nodes) resulted in stripping of the labeled ligand from the surface of lymphocytes as they emigrated from the blood across HEV (Fig. 1b). The HRP complexes were "zippered" from the lymphocyte surface by coated pits in the endothelial cell membrane while sparing the lymphocyte. In vitro treatment of TDL with higher doses of ATS resulted in endocytosis and destruction of the labeled cells by HEV after intra-arterial infusion (Fig. 2). Fragments of migrating lymphocytes appeared to be in the process of being pinched off by the adjacent endothelial cells (Fig. 1c). In other preparations, active pinocytosis and membrane ruffling was associated with endothelial membranes that were in contact with anti-immunoglobulin-treated lymphocytes. Smooth membrane arrays of tubules were rarely seen emptying into the space between

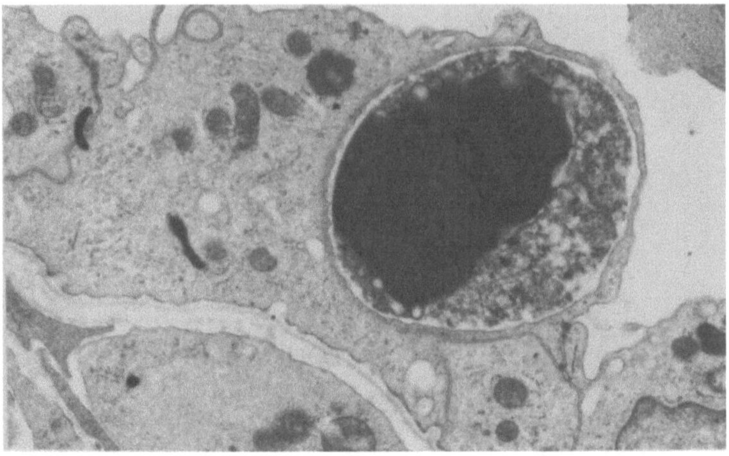

Fig. 2. Cellular debris is found in phagocytic vacuoles of HEV after infusion of ATS-treated TDL.

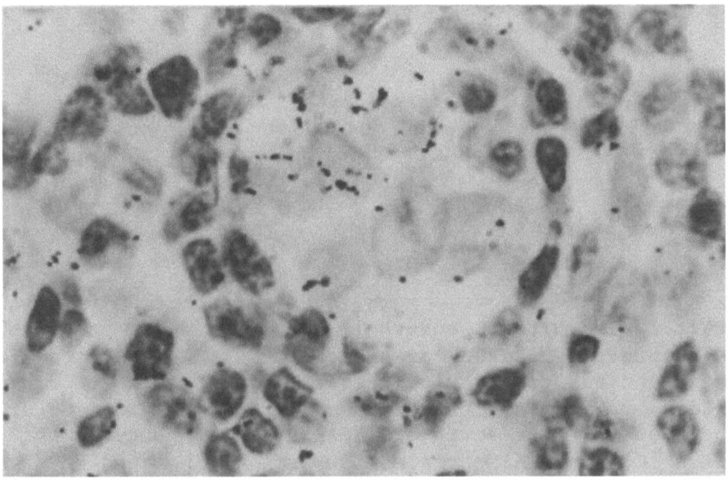

Fig. 3. Silver grains associated with radioactivity left behind by surface-radioiodinated TDL can be seen lined up over HEV endothelial cells.

endothelial cell and lymphocytes that had been treated with ATS (Fig 1d). Lymphocyte grooming was further studied using non-specific lactoperoxidase ^{125}I iodination of peripheral blood lymphocyte surface proteins followed by autoradiography of lymph node sections from rats who received i.v. infusions of surface-labeled lymphocytes. Lactoperoxidase iodination labels endogenous anti-lymphocyte antibodies as well as membrane glycoproteins because freshly isolated periperal blood lymphocytes label 80-85% with FITC-anti-immunoglobulins before pronase treatment and 15-20% after treatment and reexpression in culture (unpublished data, Anderson and Anderson). Silver grains were found in the autoradiographs above inter-endothelial clefts and lysosomal granules of HEV and perivascular lymphocytes (Fig. 3). There were fewer labeled cells by this method than was seen for peroxidase labeled antibodies. This can be attributed to the

Table 2. EFFECT OF LIGAND SURFACE BINDING ON LYMPHOCYTE FUNCTION

Dose (μg/ml)	Percent Viability by Trypan Blue	Lymphocyte Immigration[a] Lymph Node	Spleen
	Alcian Blue Treatment		
200	10.4	3.72	2.96
20	93.8	4.75	1.52
2	96.6	13.41	4.40
Control	96.8	17.90	17.15
	CON A Treatment		
5.0	82.0	0.30	0.35
2.5	86.0	0.12	14.90
1.0	96.0	17.70	6.37
Control	96.4	17.82	5.44
	Antibody Treatment		
FAB-α-IgG; 50	98.2	17.72	7.15
FAB-α-IgG; 5	97.3	18.41	--
Control Rat Serum	96.8	17.00	6.80
FAB-ATS; 50	0	0.31	0.42
FAB-ATS; 5	34.0	3.28	2.28

[a] =% injected CPM at 24 hr after 3×10^8 ^{3}H-U lymphocytes.
ATS = anti-thymocyte serum.

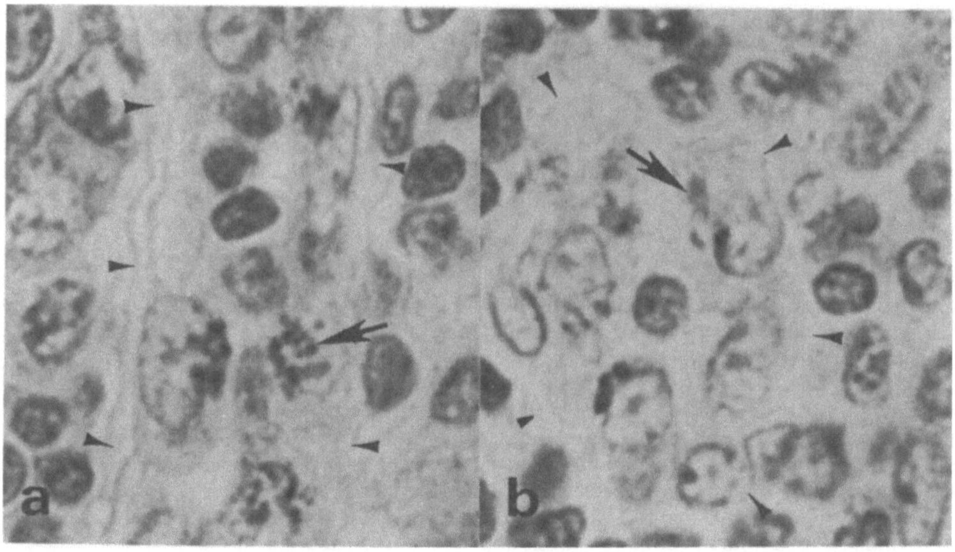

Fig. 4(a & b). a. HEV (outlined by small arrows) in a human lymph node from a patient with AIDS[6] contain immunoperoxidase labeled P24 antigen from HIV. b. Analogous sites in HEV from a deeper section also label with peroxidase tagged OKT-4 monoclonal antibody.

relative instability of the iodine label over the time course of the study. The ferritin labeled Con-A was endocytosed by the lymphocytes. Few of the Con-A treated TDL could be found to be migrating at HEV in EM preparations.

In lymph nodes of selected AIDS cases[6] (Fig. 4), the P24 antigen of HIV virus was found in endocytic vacuoles near Golgi regions of HEV endothelium. Antibodies to the CD-4 molecule, expressed on helper T-lymphocytes, labeled identical vacuoles only in HEV from HIV (+) lymph nodes. This suggests that endocytic stripping of lymphocyte CD4 molecules which are receptors for HIV may be a pathogenic mechanism for penetration of virus into HEV endothelium.

DISCUSSION

These data indicate that HEV endothelial cells exhibit active endocytic interactions with recirculating lymphocytes enroute from blood to lymphatic tissue which results in removal of ligands from the lymphocyte surface. In the extreme case, xenogeneic and heavily opsonized cells are destroyed as has been described in earlier studies of the effects of ATS in vivo[12]. Griffin et al. have shown that mononuclear cells participate in removing capped immunoglobulin from B cells incubated with anti-immunoglobulin in vitro[13]. Studies described here indicate that lymphocytes might also be "groomed" during migration into lymph nodes. FAB_1 fragments were used to eliminate Fc receptor- or complement receptor-mediated endocytosis as an explanation for this phenomenon. Endocytic grooming of lymphocyte surfaces by HEV endothelium may be an essential mechanism for restoring responsiveness to idiotype-suppressed circulating cells as they enter lymphatic tissues. Idiotype suppression reverses in time and continuous exposure to anti-idiotype is necessary to maintain suppression[5]. If grooming is occurring, then the intranodal lymphocytes should be more responsive than peripheral blood or spleen cells. Indeed, Charpentier et al.[14], described spontaneous recovery of lymphocytes from idiotype suppression of responsiveness to major histocompatibility antigens quantitated by mixed lymphocyte responses, as might have occurred if lymphocytes were groomed during transit from blood to lymph as proposed by Scollay[1]. In the case of endocytosis of HIV virus, pathogenesis of Acquired Immunodefficiency may follow if HIV destroys HEV's or if antigen-receptor bearing T-cells were denuded of CD-4 molecules which are necessary for proper interaction with antigen presenting cells. There are no data presently available about whether HIV replicates in HEV or not.

Endocytosis of receptor complexes at sites of adhesion between migrating lymphocytes and HEV endothelial cells (Fig. 1a) may also represent a mechanism by which the HEV endothelium breaks the chemical bond between the "homing" receptor[15-18] and its ligand so the lymphocyte can leave the interendothelial space. It is plausible that cleavage of the homing receptor is what we are observing because heterologous antibodies and lactoperoxidase ^{125}I iodination were used to nonspecifically label glycoproteins on lymphocyte surfaces. Liganding receptors may alter or denature the surface glycoprotein so that accessory proteins that preserve native receptor structure are eliminated[19]. The lymphocyte "homing" receptor[18] recognized by MEL-14 may, in fact, be one of the molecular chaperones[19] that have been associated with ubiquitous heat shock proteins[20]. Finally, it is intriguing that the HEV presumably endocytosed HIV virus and its receptor[21] from transmigrating helper T-cells. Could part of the loss of immunity in AIDS patients be due to loss of lymphocyte recirculation?

REFERENCES

1. R. Scollay, J. Hall, and E. Orlans, Eur. J. Immunol. 6:121, (1976).
2. S. Burdette and R. S. Schwartz, N. Engl. J. Med. 317:219 (1987).
3. A. F. Schrater, E. A. Goidl, G. L. Thorbecke, and G. S. Siskind, J. Exp. Med. 150:138, (1979).
4. M. S. Sy, J. W. Moorehead, and H. N. Claman, J. Immunol. 123:2593 (1979).
5. M. S. Sy, B. A. Bach, Y. Dohi, A. Nisonoff, B. Benacerraf, and M. I. Greene, J. Exp. Med. 150:1216, (1979).
6. J. Ward, T. J. O'Leary, G. B. Baskin, R. Benveniste, C. A. Harris, P. L. Nara, and R. H. Rhodes, Amer. J. Pathol. 127:199 (1987).
7. J. L. Gowans and J. Knight, Proc. Roy. Soc. Lond. (B) 159:257, (1964).
8. Chapter 7, "Handbook of Experimental Immunology 3rd (Ed), D.M. Weir, Blackwell Scientific, London Publ. (1978).
9. Chapter 6, "Handbook of Experimental Immunology 3rd (Ed), D.M. Weir, Blackwell Scientific, London Publ. (1978).
10. Chapter 37, G. A. Andres, K. C. Hsu, and B. C. Seegal, "Handbook of Experimental Immunology 3rd (Ed), D.M. Weir, Blackwell Scientific, London Publ. (1978).
11. A. L. Hubbard and Z. A. Cohn, Specific labels for cells surfaces in E. H. Maddy (Ed), Biochemical Analysis of Membranes Wiley, N.Y.
12. W. D. Perkins, M. R. Schwarz, and J. K. Koehler, Clin. Immunol. and Immunolpathol. 1:27 (1972).
13. F. M. Griffin, J. A. Griffin, and S. C. Silverstein, J. Exp. Med. 144:788 (1976).
14. B. Charpentier, D. Housin, P. Long, and D. Fries, 17.6.07 in: Immunology IV, Fourth internation Congress of Immunology (ed. M. Fougereau and J. Dausset). London and New York: Academic Press, p 1213 (1980).
15. A. O. Anderson and N. D. Anderson, Immunol. 31:731 (1976).
16. Y. H. Chin, R. A. Rasmussen, J. J. Woodruff, and T. G. Easton, J.Immunol. 136:2556 (1986).
17. R. A. Rasmussen, Y. H. Chin, J. J. Woodruff, and T. G. Easton, J. Immunol. 135:19 (1985).
18. W. M. Gallatin, I. L. Weissman, and E. C. Butcher, Nature 304:30 (1983).
19. J. Ellis, Nature 328:378 (1987).
20. T. St. John, M. Gallatin, M. Siegelman, H. T. Smith, V. A. Fried, and I. L. Weissman, Science 231:845 (1986).
21. D. D. Ho, R. J. Pomerantz, and J. C. Kaplan, N. Engl. J. Med. 317:278 (1987).

XENOGENEIC LYMPHOCYTE MIGRATION IN THE FOETAL LAMB AND UNSUCKLED NEONATAL PIGLET: EVIDENCE FOR THE EVOLUTIONARY CONSERVATION AND HETEROGENEITY OF MIGRATION DETERMINANTS

R.M. Binns and S.T. Licence

Immunology Department, AFRC Institute of Animal Physiology and Genetic Research, Babraham, Cambridge CB2 4AT, UK

INTRODUCTION

Lymphocyte recirculation provides the means of rapidly bringing the pool of antigen-reactive cells to specialized sites of antigen presentation in organised lymphoid tissues. Entry into these is via postcapillary venules (PCV) with high endothelium (HEV)(1) and evidence is emerging of differences in the HEV determinants and lymphocyte receptors involved in this traffic, for instance into Peyer's patches (PP), surface and mesenteric lymph nodes (LN) and chronic inflammatory tissue(1,2,3). This 'active' entry is reversibly inhibited by trypsin-pretreatment of the lymphocytes, distinguishing it from the passive arrival of lymphocytes in spleen, bone marrow, liver, lung and other non lymphoid tissues(1). Recent detailed studies have charted the migration behaviour of lymphocytes in the young pig and revealed evidence of active entry and heterogeneity of lymphocyte migration into the surface and visceral LN, jejunal, ileal and colonic PP, tonsils and tongue papillae and thymus(4,5,6). The unusual structural inversion and the lack of lymphocytes in the draining lymph of LNs(7) led to the finding that lymphocyte traffic both in and out of LNs is via PCVs(5,6,7). This communication uses the transfer of labelled xenogeneic lymphocytes into foetal lambs and unsuckled neonatal piglets, which are immunologically virgin and lack natural antibodies and NK cell activity(8), and other evidence to elucidate the HEV heterogeneity and unusual two-way lymphocyte traffic through PCVs in the pig.

MATERIALS AND METHODS

Lymphocytes(Ly), labelled with ^{51}Cr or FITC as described before(4), were injected i.v. into histocompatible young Large White pigs (trypsin experiments), unsuckled newborn piglets and Clun Forest lambs, and ~135 day foetal lambs immediately after thoracic duct cannulation (methods modification of Pearson(9)). The lymphocytes were isolated from i) citrated blood of donor female pigs and horses, as before(4) and of young adult ewes using osmotic RBC lysis with glycerol(10) or 15 sec. water treatment; ii) cervical/mandibular and mesenteric LNs of ~3 month old Lou/c rats and Large White pigs and iii) spleens of ~18 month old crossbred hens and young pigs. Trypsin treatment of pig lymphocytes was at 37° for 5 minutes using 10^8 ^{51}Cr-labelled cells and 200μg Trypsin (Sigma) per ml. RPMI 1640 without FCS, halted by addition of an equal volume of cold RPMI 20% FCS, passage through nylon mesh and centrifugation. Weighed recipients of labelled cells were exsanguinated under general anaesthesia at day 1, except using trypsinised cells (1 hour). In foetal lambs, this followed continuous fractionated collection of thoracic duct lymph (TDL) in utero which was stored at 4° for cell counting. Some 30 standard tissues(4) were isolated, cleaned, weighed and estimated for ^{51}Cr content (counts per minute-CPM) to calculate their labelled "cell concentration factors"(4) (CCF=tissue sample CPM over weight, divided by total injected CPM over body weight). The labelled cell output in foetal TDL was estimated in pooled washed concentrated 1 hour collections of cells both by ^{51}Cr gamma counting and by direct enumeration of living $FITC^+$ and unlabelled lymphocytes, checked by phase contrast microscopy. Newborn piglet mesenteric LNs, with the gut tied off, were normothermically perfused with RPMI-dextran as described previously(11), performing similarly.

RESULTS AND DISCUSSION

1. Differential trypsin-sensitivity of the 'active' entry of labelled lymphocytes into lymphoid tissues.

Figure 1 shows that pretreatment of ^{51}Cr-PBL with high levels of trypsin (200μg for 5 min) selectively inhibits localization in all LN, all PP, thymus, tonsils and tongue papillae at 1 hour (~5-30% of control cells) in contrast to the other sites of localization (here termed 'passive') in spleen, lung, liver and possibly bone marrow (BM). By 1 day, much of the inhibition is reversed. This supports in the pig the observations made in other mammals that lymphocyte localization in PP and LN involves an active process initiated by interaction of renewable Ly receptors with PCV determinants(1) and indicates that entry into the thymus, tonsil and tongue papillae involves similar determinants, confirming observations of PCV-like traffic in these sites(see 6).

Fig. 1 also shows that trypsin treatment both at high and low levels (10μg for 5 minutes) tends to inhibit this active localization differentially, affecting entry into surface (~26 and 60% of controls) less

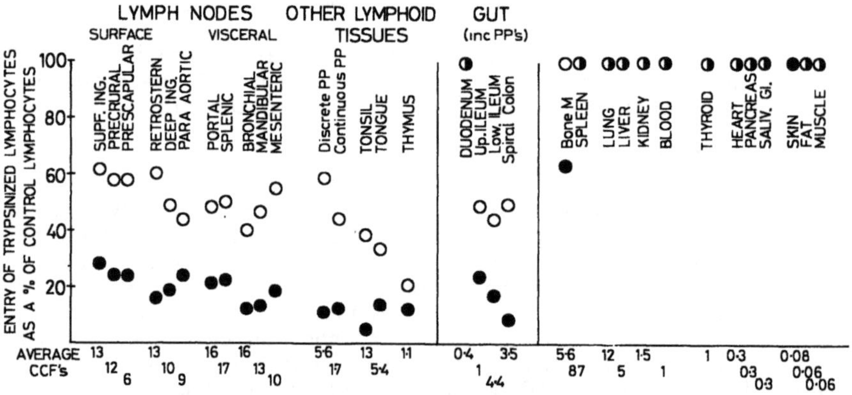

Figure 1 The selective inhibition of entry of lightly (10μg/ml - ○) and heavily (200μg/ml - ●) trypsinized lymphocytes in pig lymphoid tissues and the CCFs of control lymphocytes. Entry into tissues containing organised lymphoid tissue is inhibited to different degrees while that into other tissues is not significantly different for trypsinized and untreated lymphocytes.

than visceral LNs (18 and 49%) and PP (12 and 52%), and all these less than tonsils (9 and 36%) and thymus (13 and 22%). These findings are consistent with a heterogeneity of these Ly receptor:PCV interactions, previously noted in the localization of lymphocytes from thymectomized donor pigs(4) (entry into LN and thymus inhibited, into PP not and into tonsil partially). It is noteworthy that, as in other species(1), the passive localization in lung, liver and spleen, though considerable and probably important both immunologically (spleen)(12) and as lymphocyte depots (lung and liver)(13), apparently does not involve trypsin-sensitive protein receptors on the lymphocytes. The localization in pig BM is more ambiguous (~30% inhibition) as was noted in rats(14).

2. Interspecies lymphocyte recognition of 'active' migration determinants

It is striking that, in spite of their ~50 million years of evolutionary divergence from a common ancestor(15) and of their marked differences in lymphoid tissue structure and function(6), pig and sheep lymphocytes localize in these sites of active immigration remarkably well across species in the foetal lamb and unsuckled neonatal piglet (Fig.2) as shown by ^{51}Cr-labelled cell CCFs in the isolated tissues at 1 day and confirmed by observation of living motile normal FITC^{+} lymphocytes.

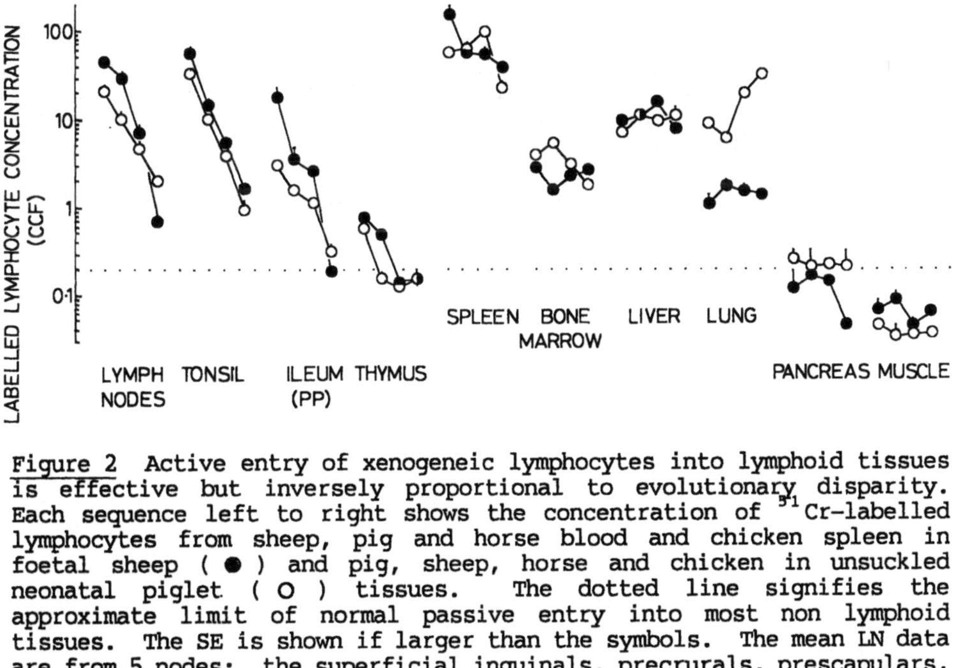

Figure 2 Active entry of xenogeneic lymphocytes into lymphoid tissues is effective but inversely proportional to evolutionary disparity. Each sequence left to right shows the concentration of ^{51}Cr-labelled lymphocytes from sheep, pig and horse blood and chicken spleen in foetal sheep (●) and pig, sheep, horse and chicken in unsuckled neonatal piglet (O) tissues. The dotted line signifies the approximate limit of normal passive entry into most non lymphoid tissues. The SE is shown if larger than the symbols. The mean LN data are from 5 nodes: the superficial inguinals, precrurals, prescapulars, and mesenterics in both species plus the popliteals in sheep and mandibulars in pigs.

Moreover for many hours the kinetics of arrival of ^{51}Cr-sheep and ^{111}Ind-pig PBL in the main sites of active entry (LN, PP, tonsil and thymus) as well as in the spleen in newborn unsuckled piglets is similar (data not shown). This active entry of xenogeneic lymphocytes in LN, PP, tonsil and thymus apparently also involves their specific receptors since trypsin treatment of labelled pig lymphocytes selectively inhibited their localisation in these tissues in the newborn unsuckled lamb in a preliminary experiment. Apart from demonstrating remarkable conservation of the migration determinants, these experiments show that during the first 2 days after birth, the natural cytotoxicity seen later in life for allogeneic and xenogeneic cells in several species(16) including the pig(17), is of a very low order, paralleling a lack of natural antibodies and NK cell activity(8). This suggests that natural allogeneic cell cytotoxicity(16) may arise from cross-reactive stimulation with irrelevant environmental antigens.

However, the active localization of xenogeneic lymphocytes in LN, ileum (PP), tonsil and thymus, unlike the entry into other sites, is not uniform but inversely related to their evolutionary disparity (Fig.2), as was noted for various species' lymphocytes binding to HEV in vitro (18). Horse PBL (~60 million years divergent from pig and sheep (15)) and even chicken spleen lymphocytes (~ 200 million years divergent (15)) show entry rates lower than those of pig and sheep lymphocytes but still significantly above those into similarly vascularized non-lymphoid tissues, in spite of the fact that chickens lack classical lymph nodes (19). Nevertheless the entry of chicken lymphocytes into ileum and thymus in both species and of horse and sheep into xenogeneic thymus is of such a low order that it is not clear that any active arrival is present.

3. PCV emigration from pig LNs does not involve new receptors

Figure 3A shows that, after entry into the xenogeneic lymphoid tissues, FITC-labelled pig lymphocytes were able to emigrate from lymph nodes into thoracic duct lymph (TDL) in the foetal lamb as well or better than sheep lymphocytes, even though almost none follow this route in the pig(6). Similar data were obtained using gamma-counting for ^{51}Cr-labelled lymphocytes (not shown here).

The kinetics of this emigration for pig and sheep lymphocytes were

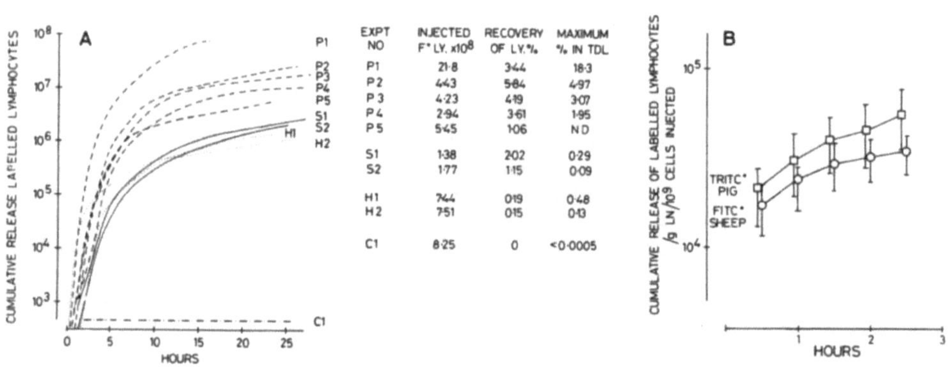

EXPT NO	INJECTED F* LY. x10⁸	RECOVERY OF LY.%	MAXIMUM % IN TDL
P1	21·8	3·44	18·3
P2	4·43	5·84	4·97
P3	4·23	4·19	3·07
P4	2·94	3·61	1·95
P5	5·45	1·06	ND
S1	1·38	2·02	0·29
S2	1·77	1·15	0·09
H1	7·44	0·19	0·48
H2	7·51	0·15	0·13
C1	8·25	0	<0·0005

Figure 3 A. The cumulative release of fluorochrome-labelled lymphocytes from lymph nodes. A. Sheep (S1,S2 ——), pig (P1–5 – – –), horse (H1,H2 ······) and chicken (C1–·–·) lymphocytes in ~135 day foetal sheep TDL. B. Sheep and pig lymphocytes in mesenteric vein draining isolated perfused mesenteric nodes of unsuckled neonatal piglets 1 day after i.v. injection of labelled cells.

similar but those of the more xenogeneic horse and chicken lymphocytes showed differences. Horse lymphocytes showed a phase of rapid emigration into TDL, very similar to that of pig lymphocytes but then slowed after ~4 hours and while the mean % recoveries for sheep and pig were ~2% and ~4% that for horse was ~0.2%. Chicken lymphocytes did not emigrate into TDL.

Sheep lymphocytes were also able to follow the peculiar direct venous route of emigration from pig lymph nodes. Fig. 3B shows that FITC-labelled sheep and TRITC-labelled pig lymphocytes which had been injected i.v. into unsuckled piglets shortly after birth and had localized in lymph nodes, both emerged continuously from the mesenteric LNs isolated and perfused in vitro 1 day later, joining a direct venous emigration of unlabelled autologous lymphocytes similar (cells/g LN) to that found in older pigs, apparently occurring via PCVs(11). So xenogeneic pig and sheep lymphocytes both follow the entry and exit pathways of the recipient species, using routes hardly employed at all in their own species. This shows that the direct vascular emigration of T and B, but not Null, lymphocytes(5,6) from the LN in the pig involves determinants which, though probably different from those mediating entry(6), are not "new" specificities since they are recognised by sheep lymphocytes but are new to the basement membrane side of the vascular endothelium since traffic by this route in sheep and most other species is negligible. The functional role of the receptors on sheep lymphocytes which allow this atypical exit is unknown, unless they are involved in emigration into lymph sinuses.

The differences in the migration behaviour of the horse and chicken lymphocytes are more complex. Detailed comparison of the migration capacity of lymphocytes from donors of different ages and species will be subject to some variation arising from their differing ratios of functional lymphocyte subsets. But the reduced efficiency of these species' lymphocytes appear to involve both inefficiency of entry and an increased loss of cells, apparently shown by the increased localisation of xenogeneic lymphocytes in neonatal pig lung (Fig.2). At present it is not clear what mechanisms may be involved in the slower rate of continued emigration of

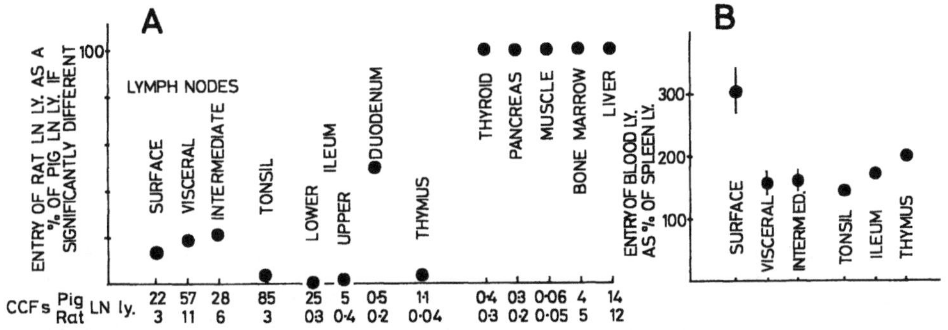

Figure 4 The selectivity of the active entry of rat LN and pig spleen lymphocytes in newborn piglet lymphoid tissues. A. Entry of rat lymph node as a % of pig LN lymphocytes and their mean concentrations. Note that entry of the rat LN cells into thymus and gut is not above non lymphoid tissues and in gut does not differ in tissues with (ileum) and without (duodenum) PP. B. Entry of pig blood lymphocytes as a proportion of spleen lymphocytes. Note that this reveals a difference in surface LN from visceral LN and other tissues.

horse lymphocytes after 4 hours and in the failure of emigration of chicken lymphocytes in sheep TDL but we cannot rule out the existence of metabolic or immunological factors interfering with the normal physiological behaviour of these lymphocytes from a different Phylum! Nevertheless, except for the localization of the xenogeneic lymphocytes in thymus, each of these cells showed some traffic into each of the lymphoid tissues (Fig.2). However comparison of the localization of labelled rat and pig LN cells in unsuckled piglet tissues (Fig.4A) revealed both quantitative and qualitative differences. Rat lymph node cells localized relatively well in LNs (R/P = 10-30% in different LNs), significantly in tonsils (R/P = ~ 5%) but showed no increment above non lymphoid tissues in thymus and PP, though pig LN cells localized efficiently in all these tissues. Thus rat LN apparently lack the capacity to recognise pig PP or thymus migration determinants but do recognise LN and tonsil.

Further evidence of determinant heterogeneity was evident in the comparison of spleen and blood lymphocyte localization. Thus Fig. 4B shows that pig spleen lymphocytes are relatively poor at entry into surface LN in contrast to visceral and intermediate type LNs and other lymphoid tissues. This is consistent with the notion that surface and visceral LNs bear different migration determinants (1) and with previously observed differences in the localization preferences of precrural and mesenteric LN cells in the pig(5). So several complementary lines of evidence point to the probable difference between migration determinants involved in entry into surface and visceral LN, tonsils, PP and thymus (summarised in Table 1) including the selective localization of thymectomized (Tx) donor (4,5,6), trypsinized (Fig.1), rat LN (Fig.4A) and pig spleen (Fig.4B) lymphocytes compared with normal pig blood lymphocytes.

Table 1. Evidence of heterogeneity of active lymphocyte localization.

Lymphocytes	Surface LN	Visceral LN	Tonsil	PP	Thymus
Tx donor entry	±*	+	++	+++	±
Trypsinized entry	++	(++)	+	+	±
Rat cf Pig LN entry	+	++	(+)	–	–
Spleen vs PBL entry	+	(+++)	+++	(+++)	++

* Scale from Normal = +++ down to – = no significant entry above non-lymphoid tissues.

CONCLUSIONS

These experiments have been revealing in many ways. 1) The immunologically-virgin unsuckled Artiodactyl foetus/neonate provides a valuable model for short-term studies of migration showing little sign of interference with the normal behaviour of transferred closely xenogeneic lymphocytes. 2) The recognition of migration determinants between sheep and pigs is remarkably efficient but for horses is less efficient and for chickens detectable for entry but not exit from tissues. 3) Several experimental criteria indicate that there are differences in the migration determinants involved in entry into surface and visceral LN, PP, tonsil and thymus, including the differential localization of partially trypsinized, thymectomized-donor and spleen lymphocytes of pigs and of rat LN lymphocytes supplementing the known heterogeneity(1) and making particularly pressing now the development of monoclonal reagents which descriminate these specificities in the pig. 4) The unusual direct venous exit from the pig LN appears due to the expression of unknown but conventional determinants in a new site, which also require analysis at a molecular level.

ACKNOWLEDGEMENTS

We thank Drs. W.R. Allen and J. Kidd for the horse blood, Messrs. R. Fulton and P. Byrd and their staff for provision and care of the animals. Mr. N. Miller for help with the figures and Mrs. Jenny Richards for typing the manuscript.

REFERENCES

1. Woodruff J.J., Clarke L.M. & Chin Y.H. Ann. Rev. Immunol. 5, 201 (1987).
2. Jalkanen E., Steere A.C., Fox R.I. & Butcher E.C. Science 233, 556 (1986).
3. Chin W. & Hay J.B. Gastroenterology 79, 1231 (1980).
4. Binns R.M. & Licence S.T. Microenvironments in the lymphoid system p.661, Plenum NY (1985).
5. Binns R.M., Pabst R. & Licence S.T. Swine in Biomedical Research, Vol. III p.1837 Plenum NY (1986).
6. Binns R.M. & Pabst R. Migration and homing of lymphoid cells. Vol. II Ch. 16, CRC Press. (In press)
7. Binns R.M. Vet. Immunol. Immunopath. 3, 95 (1982).
8. Charley B., Petit E. & Bonnardiere C.La. Ann. Recherche Vet. 16, 399 (1985).
9. Pearson, L.D., Simpson-Morgan M.W. & Morris B. J. exp. Med. 143, 167 (1976).
10. Binns R.M. J. Immunol. Meth. 21, 197 (1978).
11. Pabst R., Binns R.M. & Licence S.T. Immunology 56, 301 (1985).
12. Pabst R. Migration and homing of lymphoid cells. Vol. I. CRC Press (in press).
13. Pabst R., Binns R.M., Peter M. and Licence, S.T. This proceedings.
14. Rannie G.H., Smith M.E. & Ford W.L. Nature 267, 520 (1977).
15. Young J.Z. The life of vertebrates. Oxford Univ. Press (1962)
16. Rolstad B. & Ford W.L. Immunol. Rev. 73, 87 (1986).
17. Binns R.M. & Licence S.T. Immunology, 49, 727 (1983).
18. Butcher E.C. Scollay R. & Weissman I. Nature, 280, 496 (1979).
19. Biggs P.M. Acta Anat. 29, 36 (1957).

HEV DEVELOPMENT AND TRYPSIN-INHIBITABLE LOCALIZATION OF LYMPHOCYTES IN THE REACTION TO INTRADERMAL PHA: A RECEPTOR-DEPENDENT IMMIGRATION AND POSSIBLE MODEL FOR PERIPHERAL CMI REACTIONS?

R.M Binns, J.F.T. Griffin S.T. Licence and F.B.P.Wooding

Immunology Department, AFRC Institute of Animal Physiology and Genetics Research, Babraham, Cambridge CB2 4AT.

Lymphocyte recirculation is essential for the efficient recognition of antigenic challenge in the organised lymphoid tissues and the periphery. Lymphocyte entry into most organised lymphoid tissues such as the lymph nodes (LN), Peyer's patches (PP)(1) tonsils, tongue papillae and thymus(2) is inhibitable by trypsin(1,3), apparently involving the interaction of protein receptors on the circulating lymphocytes with determinants on specialized post capillary venular (PCV) endothelial cells (1). This interaction is also involved in the lymphocyte localization in granulomata and rheumatoid joints (chronic inflammatory tissue)(4) but not in spleen, lung and liver(1,3) or the minor traffic in other tissues, none of which is significantly reduced by trypsinisation. It is probable that the lymphocyte receptors and endothelial cell (EC) determinants involved in active entry into different LN types, into PP and into chronic inflammatory tissue are different(1,4).

Recent intensive research in the young pig, a species with interesting differences in its lymphocyte migration patterns(2), has documented the detailed migration behaviour of lymphocytes(2,5,6), confirming and extending observations from other species(1). A complex of data in this species suggest that the active typsin-sensitive lymphocyte localization in somatic and visceral LN, PP, tonsils and thymus may each involve different lymphocyte:PCV interactions(3). In addition some of the migration determinants in lymphoid tissues appear to be inherent and independent of antigenic stimulation (developing in utero, some during early foetal life e.g. thymus, tonsils, LNs, jejunal and ileal PP), while traffic to others develops after birth (tongue papillae, colonic PP) suggesting that antigenic stimuli may play a part in the development of their migration determinants(2), as is evident in chronic inflammatory tissue(4). This report shows that a feature of the vigorous DTH-like reaction in young pig skin(7), which follows intradermal (id) injection of the T cell mitogen(8) Phytohaemagglutinin (PHA-P), is the rapid development both of structures reminiscent of high endothelium blood vessels (HEV) and of a major short term trypsin-sensitive traffic of lymphocytes, pointing to the existence of a mechanism for acute acquisition of endothelial determinants which might play an important role in defensive lymphocyte infiltrations.

MATERIALS AND METHODS

The donors were female and the recipients females and males of our MHC homozygous (SLAb/b) line of Large White pigs which accept each other's lymphocytes for at least 10 days (2,5,6), revealing migration behaviour similar to autologous lymphocytes (5,6). Blood lymphocytes were prepared and labelled with ^{51}Cr using 10-20μCi and 10^8 cells per ml as described previously(6). Entry into most tissues is expressed as the labelled cell concentration factor - CCF (Tissue label concentration divided by computed mean body label concentration, were cell distribution uniform) (3,6)and the entry into skin sites as the CCF or absolute content of labelled cells (CPM) in an excised skin disc (normally 2.5 or 3cm in diameter). Tissue radiolabel was counted in an LKB γ spectrometer for 3-30 minutes, depending on the label content and corrected for background. Red cell entry was studied using RBC's extensively washed, removing the buffy coat and superficial light RBCs at each step, and labelled with 100μCi ^{51}Cr per 10^9 RBC/ml. Skin reactions against the PHA-P (Sigma) were usually assessed by measurement of the diameter of visible reactions and the double skin thickness at 1 day. In histological studies, thin slices of skin were fixed in glutaraldehyde and embedded in acrylic or epoxy resins. Light micrographs were taken of H & E or toluidine blue stained sections. Electron micrographs were of osmicated, uranyl and lead stained sections (9).

RESULTS AND DISCUSSION

A. The macroscopic reaction and lymphocyte entry following intradermal PHA

The reaction one day after intradermal injection of PHA-P in RPMI in the mid abdominal skin of young pigs (~3-12 weeks old) proved to be remarkably reproducible and dose-dependent in most pigs (Fig. 1). The swelling (diameter and skin thickness), erythema and induration (expressed

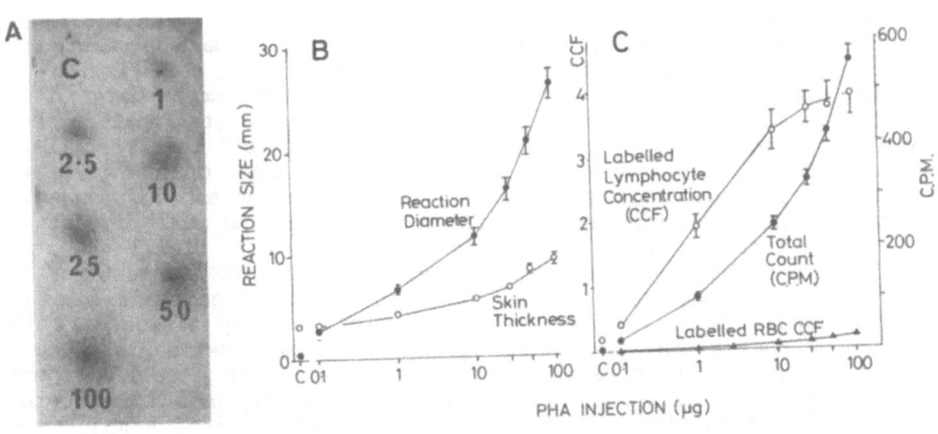

Figure 1 Skin reactions on the ventral abdomen 1 day after intradermal injection of varying doses (μg) of PHA-P in 100 μl medium.
A. The reactions. B. Increasing reaction diameters and skin thickness (n = 9); data is M $\pm$ SE. C. Increasing localization of labelled lymphocytes, measuring total label content - CPM and concentration - CCF of lymphocytes and of autologous red cells.

± - ++++) were proportional to PHA concentration and related to the log PHA dose injected (~0.1 - 100µg PHA in 100µl RPMI) (Figure 1B). Labelled lymphocytes localized in these sites in large numbers and this entry was also highly dose-dependent and related to the macroscopic reactions (Fig. 1C).

In practice, using larger doses of PHA and of labelled cells in young piglets, the maximum labelled cell entry could be >100-fold that in appropriate control normal skin at ~1 day with CCFs about half of those in normal surface lymph nodes (e.g. precrural LN). In the course of these studies, the effects of varying both the volume of the intradermal injection (see below) and the weight of PHA were examined. The combination chosen as standard to analyse these PHA reactions was 25 µg PHA in 100 µl RPMI. It is noteworthy that the PHA injections also induced a marked increase in lymphocyte localization and swelling in the local draining superficial inguinal lymph nodes in one day.

At this dose and above, the reactions show a central red indurated area of ~10-12 mm diameter (sometimes with a focal necrotic spot centrally if volumes of 10-20 µl were used) surrounded by a less erythematous area and oedema, related to the PHA dose, up to ~30 mm in diameter. Preliminary histology at 1 day after the standard dose showed a central spot (~1-2mm; perhaps the needle track) containing mononuclear cells, neutrophils and erythrocytes. Outside and below this was a zone containing large numbers of perivascular lymphoid and mononuclear cells with many vessels showing changes in their endothelium (especially in the fat immediately below the dermis) (see below).

The red cell extravasation in these reactions was also measurable by an escape of iv injected ^{51}Cr-labelled autologous RBC (Fig.1C) and related to the tissue reddening,though red cell CCFs were very low (similar to those in mildly bruised natural skin lesions which show almost no increased lymphocyte entry). Within the PHA reactions, labelled lymphocyte entry appeared to be predominantly in the reddened region of PHA deposition (outside the necrotic spot, if present), rather than in the surrounding paler area which showed degrees of pinkness, swelling and oedema dependent on the site and dose. The relationship to the site of PHA deposition and spread was shown both by a tendency for labelled cell entry to increase when 25 or 50 µg PHA was injected in increasing volumes (10-500 µl) and by showing that following such a dose in 10 µl medium, half the labelled cells are in the central ~15% (by weight or area) of the reaction and the highest concentration in the central ~5mm disc. So the zone of active labelled cell localization related more to the spread of PHA than of the inflammatory reaction. Moreover it was evident that inflammation and induration alone are not sufficient to cause major lymphocyte localization. Injection of PMA (100 ng) caused inflammation and induration equivalent to >50 µg PHA but lymphocyte arrival was much less than for 1 µg PHA (Mean CCF = 0.27 for PMA versus 1.3 for PHA) and, injected with the latter, did not potentiate the lymphocyte entry it induced.

The relatively hairless skin of the ventral abdomen (which is similar to that of man) was chosen for almost all experiments, but in exploratory experiments sites on the medial thighs and base of the ears proved similarly reactive to PHA (reaction and lymphocyte entry). Other sites such as on the neck, shoulder, hip and lower leg were significantly less reactive. On the abdomen, the pubic area was most reactive both macroscopically and in lymphocyte entry making the matched distribution of experimental and control sites important. Incidentally we also found that sick or stressed pigs were relatively unresponsive, that newborns respond and also that occasional pigs were significantly more reactive than normal for reasons that are unclear.

B. The rapid onset and termination of labelled-lymphocyte entry to PHA sites

Figure 2 shows that both the macroscopic reactions and lymphocyte entry were maximal by 18 hours and yet were detectable within 2 hours. Further experiments will establish the time of peak entry between 8 and 18 hours. By 25 hours, the lymphocyte localization and macroscopic reactions were significantly lower and in experiments allowing 1 day localization in sites 1,2,3,4 and 7 day old, it was evident that the largest entry was at 1 day and that entry was negligible after 2 days (i.e. in reactions receiving labelled cells from 48 to 72 hours after PHA injection), inspite of the fact that macroscopic reactions were still readily visible at day 7.

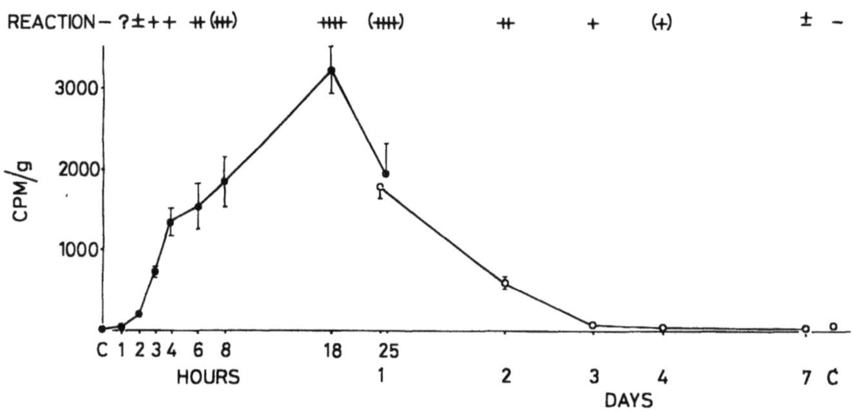

Figure 2 The rapid onset and end of lymphocyte entry into skin reactions to 25 μg PHA-P. Expt. 1 (●) - 3 pigs injected id with PHA 1,2,3,4,6,8,18 and 25 hrs and iv with ^{51}Cr lymphocytes 2 hrs before exsanguination. Data is M ± SE CPM/g in 12 sites. Expt. 2 (○) - 2 pigs injected id with PHA 1,2,3,4, and 7 days and iv with ^{51}Cr lymphocytes 1 day before exsanguination. Data is M ± SE in 8 sites, corrected for size of pigs and injected cell number to give similar conditions as Expt. 1.

So apparently the mechanisms involved in localization of labelled lymphocytes in the PHA site can be switched on and off rapidly, at peak resulting in a >200 fold increase in traffic of lymphocytes. Moreover the increased localization appears to be largely due to this increased lymphocyte arrival, since no evidence was found of reduced dispersion of lymphocytes once within the site, in experiments in which labelled lymphocytes were injected intradermally with or 1 hr after varying doses of PHA and the residual counts studied at day 1.

C. Induced HEVs and trypsin-dependent traffic in PHA sites

The major induced lymphocyte traffic into PHA sites is reminiscent of that into lymph nodes, which involves the efficient interaction of trypsin-sensitive receptors on lymphocytes with endothelial cell determinants on the specialized high endothelium of post capillary venules. Fig. 3 shows that, as for the other lymphoid tissue sites of active lymphocyte migration (LN, PP, tonsils, tongue papillae and thymus)(1,2,3), localization of labelled lymphocytes in PHA sites at 1 hour is markedly inhibited by trypsin pretreatment (in contrast to the entry to normal skin, lung, liver, spleen and other non lymphoid tissues which is not significantly reduced). By 1 day in vivo the distribution has shown a marked recovery. Moreover Figure 4 shows that 1 day PHA sites contain many blood vessels (particularly just below the dermis) with enlarged endothelium, often including large numbers of lymphoid cells in the lumen and on and within the walls, giving a picture similar to classical HEVs. Both of these observations warrant further extensive study but are consistent with the rapid development of endothelial function and determinants, analogous to those in lymph nodes, responsible for initiation of this high rate traffic through the tissues.

Figure 3 Inhibitory effect of trypsin pretreatment of labelled lymphocytes on their entry at 1 hour into PHA sites and lymphoid tissues in contrast to control normal skin, medium-injected sites and other tissues. Plot of CCFs of trypsinized over control lymphocytes in tissues in experiments using 50 μg and 25 μg PHA-P. Other tissue data from both experiments pooled - mesenteric (Mes) and precrural (PC) LNs, jejunal discrete Peyer's patches (PP), tonsils (Ton), tongue papillae (Tng) and thymus (Thy); spleen, bone marrow (BM), liver, lung, pancreas and submandibular salivary gland. The concentration of the normal lymphocytes (CCF) is also shown.

The nature of the inductive signal to the endothelium provided by PHA remains obscure, though its similarity to the location of the PHA itself is striking (above). However it is probably not caused indirectly by activated T cell products induced by this T cell mitogen(8); firstly because of the rapid onset of traffic and secondly because whole-body irradiation of pigs (500 rads 250 KEV Gamma Cell 40 on day -3), which suppresses the macroscopic reaction compared with mock-irradiated controls, did not significantly alter localization of labelled lymphocytes into standard 25μg PHA sites, (both injected at day -1) even though circulating lymphocytes are reduced >90% by such a treatment(10). However we still cannot decide whether PHA acts directly on the endothelium or indirectly via mediators from local radio-resistant cells since other cells show much less change after radiation. An examination of the possible roles of a variety of mediators which are released during delayed and acute inflammation in this inductive change of the endothelium remains a pressing priority.

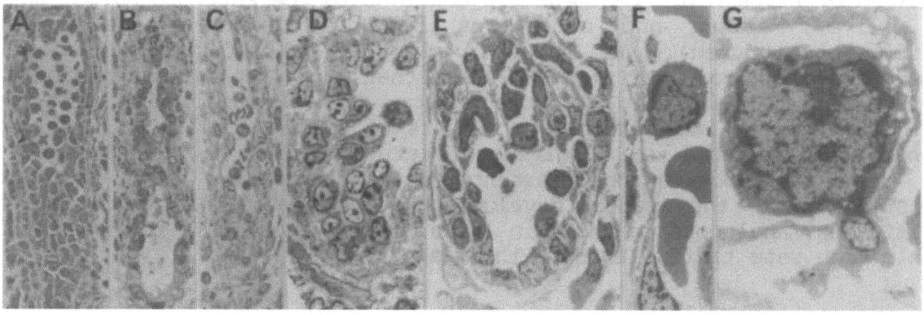

Figure 4 HEV-like blood vessels in skin reactions, with large numbers of lymphocytes in the walls and lumen 1 day after 25μg PHA. Left to right A, B & C light micrographs LS mag. ~ x300. Varying levels of lymphocyte migration. D & E Electron micrographs mag. ~ x1000. Large numbers of lymphocytes within endothelium. F & G. Electron micrographs mags. ~ x2500 and ~ x4000. Lymphocytes apparently in transit through endothelial pores; E mainly in lumen and F mainly in the wall.

So these experiments indicate the following conclusions: 1) Intradermal injection of PHA results in a rapid development of a DTH-like skin reaction which includes HEV-like venular differentiation and arrival of large numbers of blood lymphocytes. 2) This localization, which is largely due to increased entry rather than decreased dispersion of lymphocytes, is mediated by the rapid development of induced endothelial determinants recognised by trypsin-sensitive receptors on the blood lymphocytes, apparently like those involved in entry across PVCs into LNs, PP and other sites of active immigration, which are resynthesised. 3) The "switching on" of such EC determinants apparently within 1-3 hours, accompanied by a "switching off" within 48 hours, provides a mechanism which might operate in the development of lymphocyte infiltrations in peripheral immunological reactions and supports the findings of Bill Ford's group(11) that 1 and 2 hour entry of labelled lymphocytes in some DNCB contact sensitivity reactions in the rat are significantly inhibited (~70% of control) by trypsin pretreatment.

ACKNOWLEDGEMENTS

We thank N. Miller for help with the Figures, G. Morgan for preparing the micrographs and Mrs. Jenny Richards for typing the manuscript.

REFERENCES

1. Woodruff J.J., Clarke L.M. & Chin Y.H. Ann. Rev. Immunol. 5, 201 (1987).
2. Binns R.M. & Pabst R. Migration and homing of lymphoid cells, Vol. II Ch.16, CRC Press (in press).
3. Binns R.M. & Licence S.T. These proceedings.
4. Jalkanen E., Steere A.C., Fox R.I. & Butcher E.C. Science 233, 556 (1986).
5. Binns R.M., Pabst R. & Licence S.T. Swine in Biomedical Research, Vol 3, 1837, Plenum N.Y. (1986).
6. Binns R.M. & Licence S.T. Microenvironments in the lymphoid system. p.661, Plenum N.Y. (1985).
7. Blecha F., Pollman D. & Nichols D. J. Anim Sci. 56, 396 (1983).
8. Binns R.M. Vet. Immunol. Immunopath 3, 95, (1982).
9. Wooding F.B.P. Biol. Reprod. 22, 357, (1980).
10. Binns R.M. Mongr. in Allergy 16, 37 (1980).
11. Rannie G.H., Smith M.E. & Ford W.L. Nature 267, 520, (1977).

THE MIGRATION OF LYMPHOCYTE SUBSETS FROM BLOOD TO LYMPH IN THE NORMAL RAT

J. Westermann, Z. Puskas and R. Pabst

Centre of Anatomy, Medical School of Hannover

D-3000 Hannover 61, West Germany

INTRODUCTION

Individual lymphocyte subsets have different functions in the immune response, but only very few data exist on their blood transit time and their migration from blood to lymph. These data were obtained using thoracic duct lymphocytes (with up to 40% contaminating B-lymphocytes) as a source of T-lymphocytes (1) and either thymectomized, irradiated and bone marrow reconstituted rats (2) or nude rats (3) as a source of B-lymphocytes. No data have been recorded for the T-helper or T-suppressor lymphocytes.

The aims of the present study were to investigate the migration of the different lymphocyte subpopulations from blood to lymph under physiologic conditions using monoclonal antibodies and to estimate both their blood transit time and their mean transit time from blood to lymph.

MATERIALS AND METHODS

Thoracic duct lymphocytes were labeled with fluorescein isothiocyanate (FITC). The cells ($50x10^6$/ml) were incubated at a final concentration of 50fg/ml FITC for 15 min at 37°C (4). They were centrifuged through an FCS-cushion, resuspended in 2ml PBS with 4% FCS and injected i.v. over 2 min into restrained rats (Lewis, male). Samples were collected using either a central venous catheter (n=6) or a thoracic duct cannula (n=4). Within the first hour after injection 9 blood samples were taken. The lymph was collected in 2 h samples up to 48 h after injection.

The labeling index and, in the case of the lymph samples, the absolute numbers of cells were determined. Then all samples were incubated with the monoclonal antibodies Ox12, Ox19, W3/25, Ox8. These antibodies are used to label the lympocyte subpopulations in the rat (5). The second stage antibody was rhodamine labeled.

In the following text the positive cells are refered to as B-, T-, T-helper and T-suppressor lymphocytes, respectively. The percentage of each subset in the labeled cells was determined using a UV-microscope and counting at least 100 FITC+ cells. Mean and S.E.M. are indicated unless stated otherwise. Probability (p) values were determined by using Student's t-test.

RESULTS

Blood. By 30 min the labeling index in the blood had decreased to about 50% of its initial value (Fig.1). As soon as 5 min after injection, the percentage of labeled B-lymphocytes was greatly reduced while that of T-lymphocytes was greatly increased. This increase continued over the whole observation period of 60 min. The percentage of T- and B-lymphocytes in the injected lymph was 66.3 ± 2.5 % and 32.7 ± 1.7 %, respectively. In the blood the percentage of T-lymphocytes in the labeled cells was never below 83.7 ± 2.0 %, and never higher than 15.7 ± 2.8 % for the B-lymphocytes. This increase was caused by the T-helper lymphocytes, their percentage being 52.0 ± 2.3 % in the injected lymph and never less than 65.7 ± 2.4 % in the blood. There was no change in the T-suppressor lymphocytes over 60 min. This is also reflected in the T/B and the T-helper/T-suppressor ratios (Fig.2). The T/B ratio of the labeled lymphocytes in the blood was increased from 5 min after injection onwards whereas the T-helper/T-suppressor ratio increased from 20 min onwards compared to that of the injected cells, due to the increased T-helper lymphocytes.

Thoracic duct lymph. For T lymphocytes the peak recovery occurred between 16 and 20 h after injection with a maximum of 5.7 ± 0.9 % of the injected cells in the sample by 19 and 20 h. B lymphocytes showed a much broader, lower and later peak between 20 and 30 h with a maximum of 3.0 ± 0.4 % by 27 and 28 h. There is a tendency for T-suppressor lymphocytes to be initially slower than T-helper lymphocytes which is reflected by the slight increase in the T-helper/T-suppressor ratio in the first 8 hours after injection. Thereafter no differences were

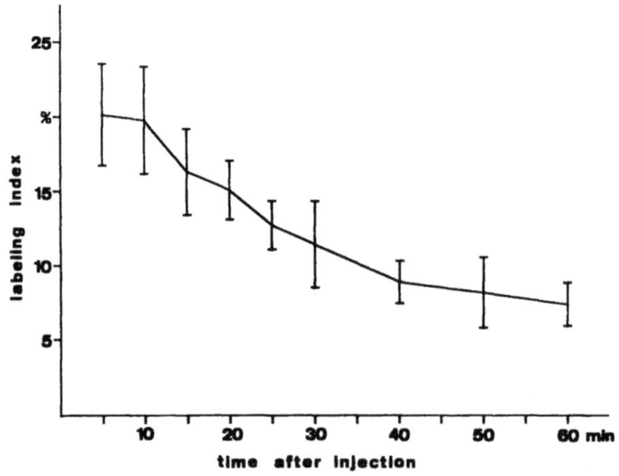

Fig. 1: Disappearance of labeled cells from the blood afte: injection of thoracic duct lymphocytes. The labeling index ha: been corrected for $100x10^6$ injected lymphocytes.

seen between T-helper and T-suppressor lymphocytes (Fig.3). In contrast only after 28 h did the T/B ratio of the recovered cells reach the ratio of the cells before injection (Fig.3).
The recovery of B lymphocytes (31.5 ± 4.2% of the injected cells) was significantly lower than that of T-lymphocytes (66.3 ± 6.5, Fig.4). No such difference could be observed between T-helper and T-suppressor lymphocytes.

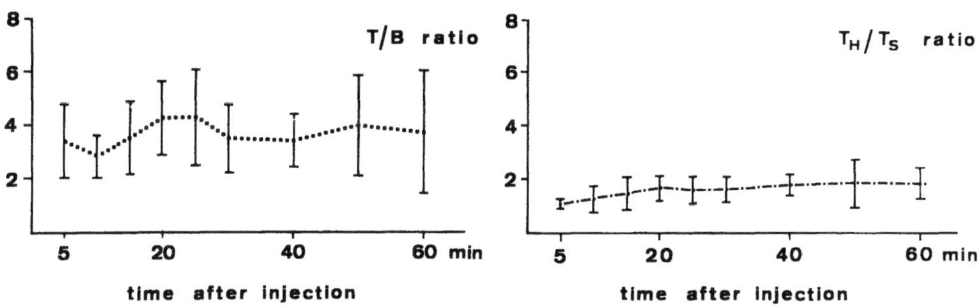

Fig.2: T/B and T-helper/T-suppressor ratios of labeled cells in the blood during the first hour after injection of thoracic duct lymphocytes. The T/B and T-helper/T-suppressor ratios have been divided by the corresponding ratios of the injected lymph. Therefore the value one indicates that the two ratios are identical (Mean + SD).

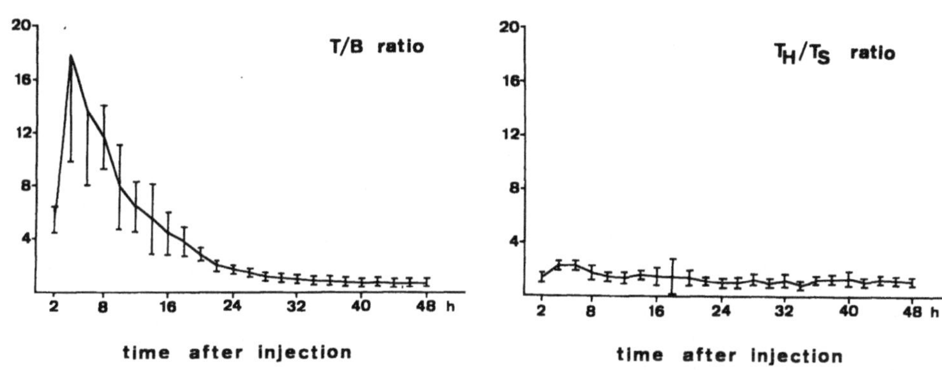

Fig.3: T/B and T-helper/T-suppressor ratios of labeled cells in the thoracic duct over 48 hours. The T/B and T-helper/T-suppressor ratios have been divided by the corresponding ratios of the injected lymph. The value one therefore indicates that the two ratios are identical (Mean ± SD).

DISCUSSION

From 5 to 60 min after injection the T/B ratio was increased, which means that either a proportion of B-lymphocytes initially leaves the blood more rapidly than T-lymphocytes or that T-lymphocytes initially stay longer in the blood (Fig.2). The observation that the T-helper lymphocytes increased whereas the T-suppressor lymphocytes stayed constant indicates that B-lymphocytes initially leave the blood very rapidly. This is in agreement with other studies (3).

T-helper lymphocytes seemed to stay longer in the blood than T-suppressor lymphocytes, an effect which was obvious from 20 min after injection onwards (Fig.2). The lung and/or the liver might be the organs responsible for this effect. In addition a different margination of each subset could be the reason (3).

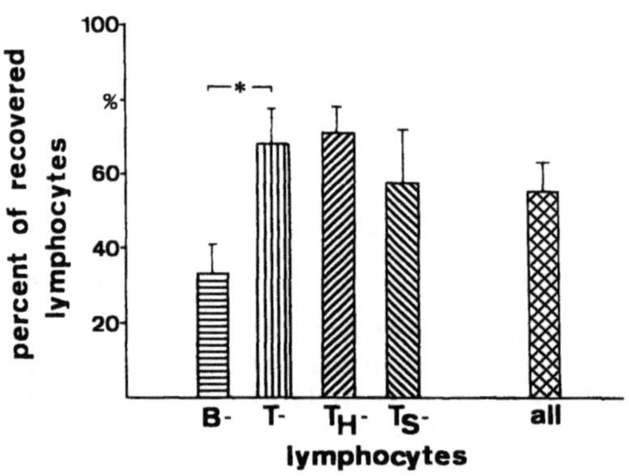

Fig.4: Total recovery of the injected cells in the thoracic duct lymph over 48 hours. The recovery for each subset is indicated. Note that the recovery of the B-lymphocytes is significantly lower than that of the other subsets (* :$p<0.05$).

Because the percentage of each subset did not change in the first hour the blood transit times for B-, T-, T-helper and T-suppressor lymphocytes were similar. Thirty minutes is probably an overestimation of the blood transit time because the first value was not obtained immediately, but only 5 min after injection. Taking this into account the blood transit time of the present study fits in well with other studies (6,7,8).

B-lymphocytes were recovered in much smaller quantities over 48 h than T-lymphocytes (Fig.4), which is due to the reduced appearance of B-lymphocytes in the first 28 h after injection (Fig.3). Thereafter, the T/B ratio was almost the same as before injection, with a value of 0.76 ± 0.01 at 48 h after injection.

The data obtained with this method are in accordance with published data (9,10) concerning B-and T-lymphocytes. However,

the recovery of injected cells is at least 5 times higher than with 51Cr-labeled cells collected under normal conditions (10) and about two times higher than with 51Cr-labeled cells collected under optimal conditions (11). This indicates that FITC-labeling does not interfere with lymphocyte migration.
Although B-cell deficient rats and T-cell deficient rats are available, rats deficient in T-helper cells and in T-suppressor cells are not. The present method, however, allows the migration of each cell type for which antibodies are available to be traced without the problems of sorting and panning.

ACKNOWLEDGEMENTS

The authors wish to thank Ms I.Dressendörfer and P.Peschel for their technical assistance and Ms S.Fryk for correcting the english of the manuscript.

REFERENCES

1. W.L.Ford, and S.J.Simmonds. Cell Tissue Kinet. 5:175 (1972).

2. J.C.Howard. J.Exp.Med. 135:185 (1972)

3. S.Fossum, M.E.Smith, and W.L.Ford. Scand.J.Immunol. 17:539(1983).

4. R.M.Binns, D.Blakely, and S.T.Licence. Int.Arch.Allergy Appl.Immunol. 66:341 (1981).

5. J.Westermann, and R.Pabst. Clin exp.Immunol. 64:188 (1986).

6. D.W.Whitelaw and H.F.Batho. Cell tissue Kinet. 5:21 (1972).

7. R.Pabst, and F.Trepel..Cell Tissue Kinet. 9:179 (1976).

8. D.Schick, F.Trepel, M.Eder, M.Matzner, S.Benedek, H.Themel, W.Kaboth, H.Begemann and T.M.Fliedner Acta haemat. 53:206(1975)

9. J.C.Howard, S.V.Hunt, and J.L.Gowans. J.Exp.Med. 135:200(1972).

10. S.Fossum, M.E.Smith, and W.L.Ford. Scand.J.Immunol. 17:551 (1983).

11. M.E.Smith, and W.L.Ford. Immunology 49:83 (1983).

THE PHYSIOLOGICAL ROLE OF THE LUNG IN LYMPHOCYTE MIGRATION

R. Pabst, R.M. Binns[1], M. Peter and S.T. Licence[1]

Centre of Anatomy, Medical School of Hannover,
D-3000 Hannover 61, West Germany, and
[1]AFRC, Institute of Animal Physiology, Babraham,
Cambridge, England

INTRODUCTION

The high numbers of lymphocytes found in the lungs early after i.v. injection of labelled cells have often been interpreted as a passive filter function of the first capillary bed lymphocytes have to pass. When normal human lung tissue had been digested, large numbers of lymphocytes were recovered, which approximately amount to the same pool size as the peripheral blood pool (1). Intravascular lymphocytes have been demonstrated histologically in perfusion-fixed lungs (2). One aim of the present study was therefore to find out whether the early homing, at 30 min, of up to 40% of the i.v. injected lymphocytes (3,4,5) is a physiological phenomenon or an artifact. The homing to the lung after i.v. and after intraaortal injection of labelled lymphocytes was compared. The pool size of the pulmonary vascular system was quantitated by extracorporeal perfusion and collection of lymphocytes in the venous effluent. Similar perfusions were carried out with segments of the liver and kidney for comparison. Finally, lymphocytes homing to the lung and being released from it were characterized by surface markers. Normal young pigs were used because the migration kinetics of lymphocytes to lymphoid and non-lymphoid organs are well described in this species (5,6), and the lungs are large enough for a controlled ex vivo perfusion with a system previously used for spleen perfusion (7).

MATERIALS AND METHODS

Twenty-eight normal young pigs were used. In the first series of experiments ~4 x 10^8 autologous peripheral blood lymphocytes labelled with FITC in vitro were retransfused into unanaesthetized animals (8). In 4 pigs the lymphocytes were injected into the external jugular vein and in 4 into the ascending aorta. Thirty minutes later the pigs were exsanguinated and samples of both lungs, 3 different lymph nodes, spleen and blood were taken and lymphocyte cell suspensions prepared. The incidence of FITC positive lymphocytes was counted using a UV microscope.

Lung perfusions

In 12 pigs the left pulmonary artery and the left atrium were cannulated and after ligating all connections to the right lung and heart the left lung was perfused with a cell free perfusate under controlled conditions as described (7,9). In 3 experiments a closed circuit system was used. In all other experiments an open system was started, that means for up to 4 hours fresh perfusate without any cells was pumped into the pulmonary artery and the cell number in the collected venous return determined. Red cells and leucocytes were counted in each sample and differentials carried out on Giemsa stained smears. In some experiments lymphocytes were classified as B and T cells using the established rosetting techniques for pig lymphocytes (10). For comparison of the vascular pool of lymphocytes in other non-lymphoid organs, segments of 4 kidneys (30.5 ± 5.0% of the organ) and of 4 livers (6.1 ± 1.1% of the organ) were perfused using the same system.

RESULTS AND DISCUSSION

There were no differences in the numbers of lymphocytes recovered from the lung, spleen, blood and different lymph nodes 30 min after injection of labelled cells, whether i.v. or intraaortally (Fig. 1).

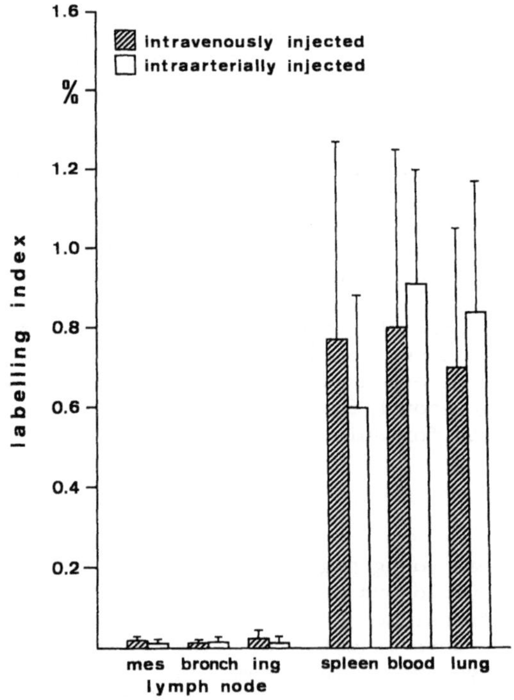

Fig. 1 Labelling index of lymphocytes 30 min after i.v. or intraaortal injection of autologous lymphocytes labelled in vitro with FITC.

Thus, lymphocytes were not trapped in other capillary beds after intraaortal injection but traversed these to arrive in the pulmonary vascular bed in similar numbers as follows i.v. injection. The early homing of lymphocytes to the lung as described in other experiments (3,4,5) seems to be a specific function of the lung.

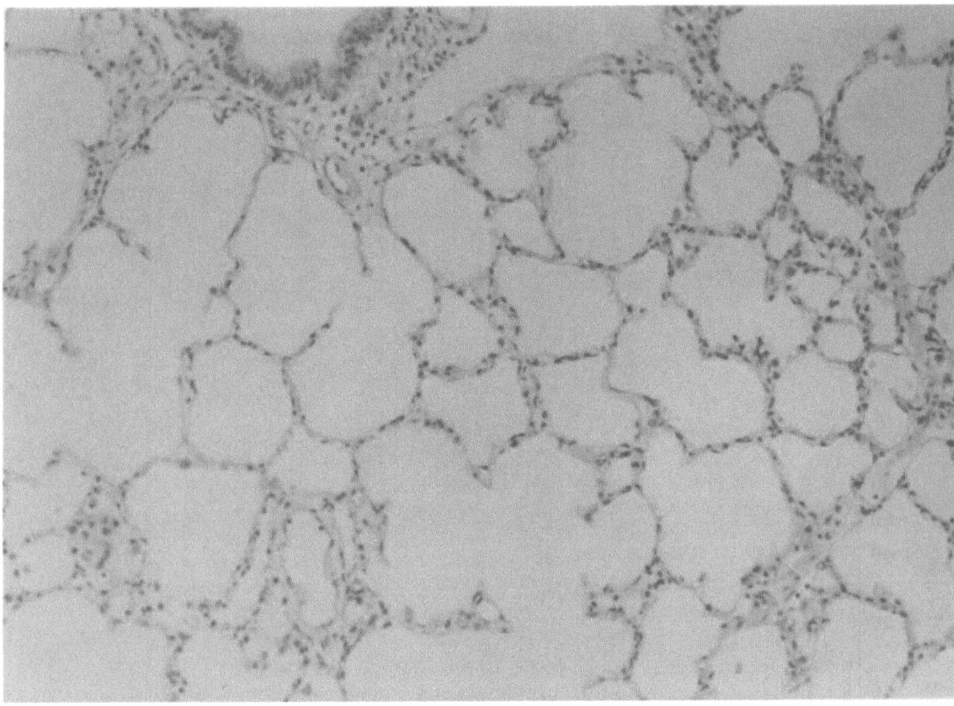

Fig. 2 Histological section of a pig lung after 2.5 hours of isolated perfusion in vitro. The normal alveolar structure can be identified. (Glycolmethacrylate embedded, 2 μm, Giemsa stain)

Isolated perfused lungs

The left lungs were perfused at an almost constant vascular resistance. There were no differences in the number of lymphocytes released, whether the lung was ventilated continuously or intermittently. In histological sections, the structure of the lung parenchyma was shown to be normal (Fig. 2). In closed circuit perfusions an equilibrium of about 500-800 leucocytes/mm^3 was reached at about 30 min. In two experiments lymphocytes not homing to the lung, i.e. remaining in the perfusate, and those released from the lung were classified as T, B or NULL lymphocytes (Table 1). In comparison to the blood, NULL cells were enriched among lymphocytes retained in the perfusate. Thus, lymphocyte subsets differed in their capacity to home to the lung.

Table 1 The incidence (%) of lymphocyte subpopulations in the blood and in the perfusate one hour after adding FITC labelled lymphocytes to a closed circuit perfusion of a pig lung.

Marker	Pig	Blood	Lung perfusate lymphocytes Unlabelled	FITC+
DARR (B cells)	1	26.4	17.9	31.2
	2	26.0	31.3	10.2
DS (T cells)	1	54.5	46.4	32.1
	2	54.9	41.4	26.8
NULL	1	19.1	35.7	36.7
	2	19.1	27.3	63.0

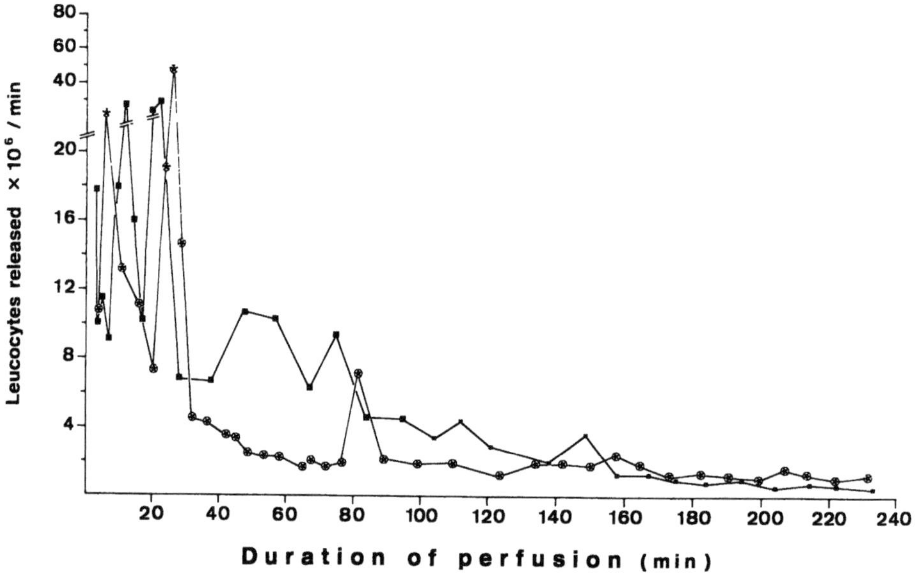

Fig. 3 Release of leucocytes during about 4 hours of isolated perfusion of the left lung in young pigs.

Although a cell-free perfusate was pumped into the pulmonary artery for up to 4 hours, leucocytes were released into the venous effluent continuously (Fig. 3). The marginal vascular pool and probably the interstitial lymphocyte pool had not been emptied during that period. In the first 4 samples about 75% but later >95% of the leucocytes were lymphocytes. To exclude that the lymphocytes released during the perfusion are the result of remaining blood, the ratio of red cells to leucocytes was calculated. In the blood the ratio is 1000:1 but in the venous effluent the ratio dropped to less than 10:1 in the second hour of perfusion.

In two experiments the venous effluent was pooled and the incidence of T, B and NULL lymphocytes determined in early, middle and late periods. The B lymphocytes showed no major differences while T lymphocytes were enriched in the perfusate and increased with time (Table 2).

Table 2 The incidence (%) of lymphocyte subpopulations in the blood and released from isolated perfused lungs.

Marker	Pig	Blood	5-30 min	60-90 min	110-150 min
DARR	1	29.8	22.6	20.9	22.3
(B cells)	2	23.5	18.8	17.0	9.3
DS	1	32.9	43.2	56.6	59.3
(T cells)	2	28.4	51.0	62.3	63.8
NULL	1	37.3	34.2	22.5	18.4
	2	48.1	30.2	20.7	26.9

The total number of leucocytes and lymphocytes released from the lung and for comparison from the liver and kidney are summarized in Table 3. A total release of ~1.5×10^9 lymphocytes was extrapolated for the whole lung. The vascular bed of the liver and kidney also contained considerable numbers of intravascular lymphocytes.

Table 3 Ratio of lymphocytes/erythrocytes and lymphocyte numbers released during normothermic perfusion of the pig lung, liver and kidney.

	Lymphocytes / erythrocytes	Lymphocytes released/g/h $\times 10^3$	Lymphocytes released/h and organ $\times 10^6$
Lung (n=9)	1:11.9 ($\pm$ 3.4)	3140 ($\pm$1670)	616 ($\pm$201)
Liver (n=4)	1:14 ($\pm$11)	281 ($\pm$ 225)	189 ($\pm$158)
Kidney (n=4)	1:32 ($\pm$25)	293 ($\pm$ 286)	17 ($\pm$ 16)

CONCLUSION

The initial high uptake of i.v. injected in vitro labelled lymphocytes by the lung (discussed in detail in 9) is not a graveyard for lymphocytes damaged by the in vitro procedure, but is a physiological role of the lung and might be important for immune reactions of the respiratory tract (11,12). The experiments characterizing subsets on lymphocytes homing to the lung or being released from the organ indicate obvious differences, as shown previously for the spleen (13) and mesenteric lymph nodes. All these findings document the specific homing and release of lymphocytes to the lung vascular bed. In previous studies (9) ^{51}Cr-labelled lymphocytes homed to the lung in vivo and their release was studied by selective perfusion at 15 min or 1 d. These experiments indicated that the ^{51}Cr-labelled lymphocytes trace labelled a large pulmonary physiological pool and later emigrated typically with them.

ACKNOWLEDGEMENTS

Supported by the Deutsche Forschungsgemeinschaft (Pa 240/3) and the European Economic Community Agricultural Division (0172).

REFERENCES

1. P.G. Holt, B.W.S. Robinson, M. Reid, U.R. Kees, A. Warton, V.H. Dawson, A. Rose, M. Schon-Hegrad, and J. M. Papadimitriou. Clin. Exp. Immunol. 66:188 (1986).
2. D.W. Scheuermann. Acta Anat. 113:264 (1982).
3. G.H. Rannie, and K.J. Donald. Cell Tissue Kinet. 10:523 (1977).
4. M.E. Smith, and W.L. Ford. Immunology 49:83 (1983).
5. R.M. Binns, R. Pabst, and S.T. Licence, in: "Swine in Biomedical Research," Vol. 3, p. 1837, M.E. Tumblesom, ed., Plenum Press, New York (1986).
6. R.M. Binns, and R. Pabst, in: "Migration and Homing of Lymphoid Cells," A.J. Husband, ed., CRC Press, Boca Raton (in press 1987).
7. R. Pabst, and F. Trepel. Cell Tissue Kinet. 8:529 (1975).
8. R.M. Binns, D. Blakely, and S.T. Licence. Int. Arch. Allergy Appl. Immunol. 66:341 (1981).
9. R. Pabst, R.M. Binns, S.T. Licence, and M. Peter. Am. Rev. Res. Dis. (in press 1987).
10. R.M. Binns. Vet. Immunol. Immunopathol. 3:95 (1982).
11. J. Bienenstock. "Immunology of the Lung and Upper Respiratory Tract," McGraw Hill, New York (1984).
12. J. Hall. Immunol. Today 6:149 (1985).

LYMPHOCYTE EMIGRATION FROM THE OVINE THYMUS: CHARACTERIZATION OF LYMPHATIC EMIGRANTS AND VENOUS EMIGRANTS

M. Miyasaka[1,5], R. Pabst[2], K. Yamaguchi[3,5] and V. Colombo[4]

1) Basel Institute for Immunology, Basel, Switzerland, 2) Center Anat., Hannover Univ. Sch. Med., Hannover, West Germany, 3) Institute for Laboratory Animals, Yamaguchi University Sch. Med., Ube, Japan, 4) Central Research Unit, Hoffmann La-Roche, Co. Ltd., Basel, Switzerland, 5) Tokyo Metropolitan Institute of Medical Science, Tokyo, Japan

INTRODUCTION

Although it has been generally agreed that mammalian thymus produces lymphocytes of the T lineage and releases them as T cells into the general circulation, there have been differences in opinion expressed concerning the rate of thymocyte emigration compared to that of production and also concerning the maturity of emigrant cells. Cell traffic studies by the use of an intrathymic injection technique by Scollay and his associates indicated that, in the mouse thymus, only a small fraction of the cells produced in the thymus emigrate and that those emigrant cells are functionally mature (Scollay, Chen and Shortman, 1984). In contrast, experiments in other animal species often yielded results which indicated a high rate of thymocyte emigration with the emigrants being at least phenotypically immature (Williams et al., 1971).

In order to examine the export of cells from the thymus, samples of emigrant cells were taken directly from thymus-associated lymphatics and also from the thymic vein. In this paper we report a preliminary characterization of these thymic emigrant cells.

MATERIALS AND METHODS

White Alpine and Black Jura lambs of both sexes (1 to 4 weeks old) were obtained from Versuchsbetrieb Sennweid, Olsberg, Switzerland. Lymphatic emigrants were obtained by cannulating thymic lymphatics as previously described (Yamashita et al., 1985, Miyasaka and Trnka, 1986). Venous emigrants were identified as fluorescent cells from amongst cells collected from the thymic vein blood after extracorporeal perfusion of the cervical thymus with fluorescein isothiocyanate (FITC) (Pabst et al., 1986). The phenotype of the emigrants was determined by either an indirect immunofluorescence technique using a fluorescence-activated cell sorter and a fluorescence microscope (Miyasaka and Trnka, 1985) or by an immunoalkaline phosphatase technique (Cordell et al., 1984). T cell separation was performed by the use of the Ig panning method (Miyasaka and Trnka, 1986). Scanning electron microscopic (SEM) studies were carried out as described in this volume (Yamaguchi and Miyasaka, 1987).

RESULTS AND DISCUSSION

Lymphatic vessels and veins leaving the cervical thymus of the sheep were cannulated and the thymocyte emigration studied directly. In the sheep the thymus is located not only in the thoracic cavity but also extrathoracically in the cervical area. Both lymphatic and blood vessels associated with the cervical thymus could be readily found. Lymphatic emigrants were collected from the thymic lymphatics and their morphology studied with SEM. Unlike peripheral T cells a considerable proportion of the lymphatic emigrants possessed a motile morphology as shown in Figure 1 We have previously shown that lymphatic emigrants could be distinguished from peripheral T cells by light microscopy in that; i) the nuclear chromatin has a fine appearance of euchromatin, and ii) there is almost no peripheral basophilia in the cytoplasm of lymphatic emigrants (Yamashita, Miyasaka and Trnka, 1985).

Phenotypically also, the lymphatic emigrants differed from peripheral T cells, particularly, in terms of MHC expression (Table 1); i) whereas peripheral T cells expressed MHC I antigens uniformly, lymphatic emigrants expressed variable levels of class I antigens as assessed by the staining of collected cells with an immunoalkaline phosphatase technique, and ii) while the majority of resting small T cells found in efferent lymphatics from peripheral lymph nodes did not have any detectable levels of MHC class II antigens on the cell surface, a substantial proportion of lymphatic emigrants had easily demonstrable levels of membrane-bound class II antigens. Whether the detected class II antigens were the product of the emigrants themselves or passively adsorbed material remains to be explored.

Table 1. Surface phenotype of thymic lymph lymphocytes as determined by immunoalkaline phosphatase technique

mAb	Ag or cell population recognized	Thymic lymph lymphocytes	Peripheral T cells	Thymocytes
ST-1	CD5	94[1]	98	97
T-80	T_H containing cell pop.	78	85	82
ST-8	CD8	12	17	85
E-79	Subcapsular blast	0	0	0.1
197	220 Kd gp (T subset)	28	12	2
175	Mono, granulocytes	0.4	0	0
SO64.6	MHC class I antigens	85[2]	100	5
SW73.2	MHC class II antigens	42	5	3

1) % positive cells are indicated. These figures are obtained from one representative experiment. Three experiments were carried out and similar results were obtained. Antibodies used have been described elsewhere in detail (Miyasaka et al., 1987).

2) Staining intensity was heterogenous.

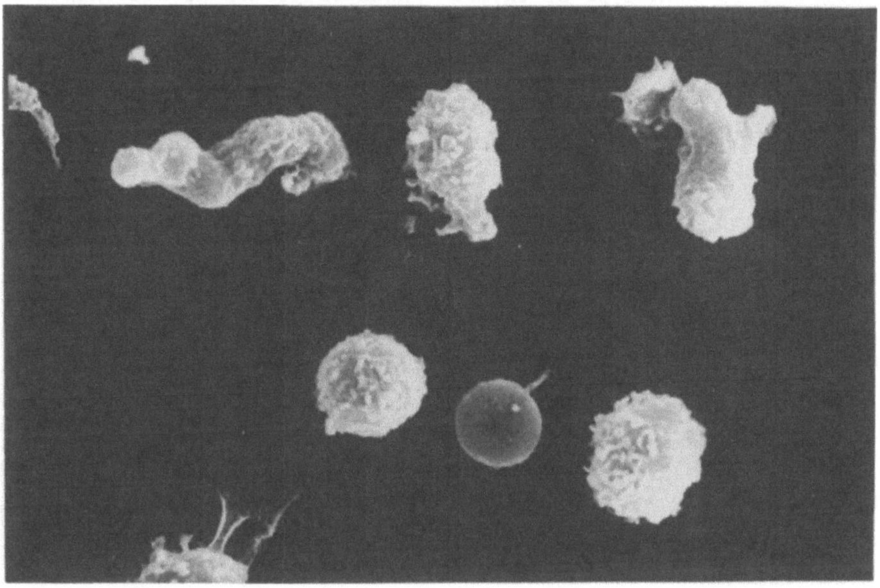

Fig. 1 SEM of thmic lymph lymphocytes. (x 2800)

Venous emigrants were identified among cells in the thymic vein blood as fluorescent cells after in situ labelling of cells in the cervical thymus with a fluorescent dye, FITC, by means of an extracorporeal perfusion system. This system had been successfully used to label bone marrow cells or Peyer's patch cells in vivo (Pabst, Miyasaka and Dudler, 1986; Reynolds and Pabst, 1984). Since the perfusion was carried out through the regional blood vessels at a constant temperature, flow rate, O_2 tension and pH, the procedure left no recognizable damage to the tissue architecture in contrast to the transcapsular injection of the material into the thymus. As was observed in the case of lymphatic emigrants, a majority of the venous emigrants were phenotypically T cells with very few B cells and monocytes being recognized, thereby suggesting that the labelled cells were not those 'in transit' or recirculating cells but were those in the process of emigration after having been produced within the thymus (Table 2). These venous emigrants also demonstrated variable levels of class I antigens with a small, but significant, proportion of the emigrants bearing detectable levels of class II antigens on the cell surface. These cells are phenotypically unique, since T cells in this species can express class II antigens upon activation in vivo (Hopkins, Dutia and McConnell, 1986) and in vitro (Miyasaka; unpublished observations) but the activated T cells invariably express easily detectable levels of the class I antigens, irrespective of whether they are activated in vivo or in vitro (Miyasaka; unpublished). Thus, these cells bearing low levels of the MHC class I antigens and high levels of the class II antigens appear to be different from ordinary activated T cells. Further characterization will be required to verify this point.

Functional characterization of these lymphatic and venous emigrant cells remains to be performed. This is mainly due to the difficulty in obtaining sufficient number of cells required for the functional assays. An attempt to examine the functional capacity of thymic emigrants has been carried out in the mouse (Scollay, Chen and Shortman, 1984), but this was done after expanding emigrant cells in vitro. Hence, the function of

Table 2. Surface phenotype of thymic venous emigrants as determined by indirect immunofluorescence technique

Ab	Ag or cell population recognized	Thymic venous emigrants	Peripheral blood T cells
ST-1	CD5	71 (1	96
T-80	T_H containing population	46	35
ST-8	CD8	15	18
anti-Ig	B cells	2	0
175	Mono, granulocytes	2	0
S064.6	MHC class I antigens	97 (2	100
SW73.2	MHC class II antigens	11	5

1) % positive cells determined by visual immunofluorescence are shown. Six experiments were performed and a representative result is given here.

2) Staining intensity was heterogenous.

emigrant cells at the time of emigration is as yet unexplored in any species of experimental animals. Improvement of techniques to collect emigrant cells in large animals may enable the extent of maturity and a rate of the emigration to be directly examined.

ACKNOWLEDGEMENT

We thank Dr. W. Hein and Ms. V. Stauffer who kindly performed immunoalkaline phosphatase staining for us.

The Basel Institute for Immunology was founded and is supported by F. Hoffmann-La Roche and Co. Ltd, Basel, Switzerland.

REFERENCES

Cordell, J.L. et al., 1984, Immunoenzymatic labelling of monoclonal antibodies using immune complexes of alkaline phosphatase and monoclonal anti-alkaline phosphatase (APAAP) complexes. J. Histochem. Cytochem. 32:219.

Hopkins, J., Dutia, B.M. and McConnell, I., 1986, Monoclonal antibodies to sheep lymphocytes. I. Identification of MHC class II molecules in lymphoid cells in resting lymph and after in vivo antigen inoculation. Immunology 59:433.

Miyasaka, M. and Trnka, Z., 1985, Sheep as an experimental model for immunology: Immunological techniques in vitro and in vivo. in: Immunological Methods, Vol. III, Lefkovits, I. and Pernis, B. eds. p.403, Academic Press, New York.

Miyasaka, M. and Trnka, Z., 1986, Lymphocyte migration and differentiation in a large-animal model: The sheep. Immunol. Rev. 91:87.

Miyasaka, M. et al., 1987, Differentiation antigens in lymphohemopoietic system of the sheep. in: Differentiation Antigens in Lymphohemopoietic Tissues, Miyasaka, M. and Trnka, Z. eds., Marcel Dekker, Inc., New York.

Pabst, R., Miyasaka, M. and Dudler, L., 1986, Numbers and phenotype of lymphocytes emigrating from sheep bone marrow after in situ labelling with fluorescein isothiocyanate. Immunology 59:217.

Reynolds, J. and Pabst, R., 1984, The emigration of lymphocytes from Peyer's patches in sheep. Eur. J. Immunol. 13:627.

Scollay, R., Chen, W.F. and Shortman, K., 1984, The functional capabilities of cells leaving the thymus. J. Immunol. 132:25.

Scollay, R. and Shortman, K., 1985, Cell traffic in the adult thymus, in: Recognition and Regulation in Cell-Mediated Immunity, Watson, J.D. and Marbrook, J., eds. p.3, Marcel Dekker, Inc., New York.

Williams, R.M. et al., 1971, Antigenic markers on cells leaving calf thymus by way of the efferent lymph and venous blood. J. Immunol. 106:1143.

Yamaguchi, K. and Miyasaka, M., 1987, The microvasculature of the sheep thymus, In this volume.

Yamashita, A., Miyasaka, M. and Trnka, Z., 1985, Early post-thymic T cells: Studies on lymphocytes in the lymph coming from the thymus of the sheep. in: Immunology of the Sheep, Morris, B. and Miyasaka, M., eds., p.162, Editiones Roche, Basel.

T LYMPHOCYTE PRECURSORS MIGRATE TOWARDS CHEMOTACTIC PEPTIDES SECRETED BY EMBRYONIC OR JUVENILE THYMUS[1]

B. A. Imhof, M. A. Deugnier, T. Itoh*, and J. P. Thiery

C.N.R.S., Ecole Normale Supérieure, Physiopathologie du Développement, 46 rue d'Ulm, 75230 Paris Cedex 05, France,
*Tohoku University School of Medicine, Seiryo-cho, Sendai 980, Japan

During embryonic life the thymic rudiment is extensively seeded by invading hemopoietic precursors that give rise to mature T cells (1). In birds thymus, colonization by precursor cells is limited to two defined time periods of embryonic development (2). The first hemopoietic precursor cells appear at day 5 in the thymus of quail embryos. At this stage of development the precursors were enriched in the perithymic region. These precursors are often in contact with endothelial cells of the jugular vein, a prefered site of extravasation; they are also dispersed within the mesenchymal tissue around the thymus, and adhere to the basal lamina that separates the thymus from the mesenchyme (3).

A chemotactic mechanism is involved in colonization of embryonic thymus

Recently our group described that the embryonic thymus secretes a peptide of 4kd which is chemotactic for hemopoietic precursor cells (4,5). The intact avian thymus secretes this chemoattractant only during embryonic stages coinciding with the two colonization periods. The chemotactic mechanism has been demonstrated *in vitro* both by using Zigmond chambers where migrating cells can be observed simultaneously by a video microscopy system and, by using Boyden blind-well chambers where cells can be counted after migration through nuclepore filters (5µm pore size). Primary cultures of quail thymus taken during the second colonization phase of embryogenesis contain fibroblasts, macrophages and epithelial cells. Our laboratory developed a technique to separate these cells and to establish pure thymic epithelial cell cultures. The secretion of the chemotactic peptide is unique to these thymic epithelial cells. Other control cells derived from the same embryo such as kidney

[1]Supported by grants from CNRS (950914), INSERM CRE 86-4015, the ligue Française contre le Cancer, and the Association pour la Recherche sur le Cancer.

epithelial cells, heart fibroblasts or liver cells do not secrete such peptides. In contrast to the *in vivo* situation, thymic epithelial cells in culture constitutively produce the chemotactic peptide. This suggests an endocrine type of regulation *in vivo* to permit the discontinous release of the chemoattractant. Such hormonal control may activate the cAMP pathway since drugs that increase intracellular c-AMP concomitantly increase the secretion of chemotactic peptides (5).

Origin and properties of hemopoietic precursor cells during embryogenesis

Hemopoietic precursor cells responding to the chemotactic peptides can be found in the bone marrow. During embryogenesis the percentage of responders can be rather high (1-2% in 11 1/2 day-old quail embryos). In older embryos or in the adult animal this percentage drops drastically by a factor of 10 to 100. It is noteworthy that the second colonization period of the embryonic thymus (at day 11 in quail embryos) coincided with a maximal number of responding bone marrow cells . This colonization wave may be the most important period with respect to the number of precursors colonizing the thymus. Hemopoietic precursors from bone marrow selected by *in vitro* migration

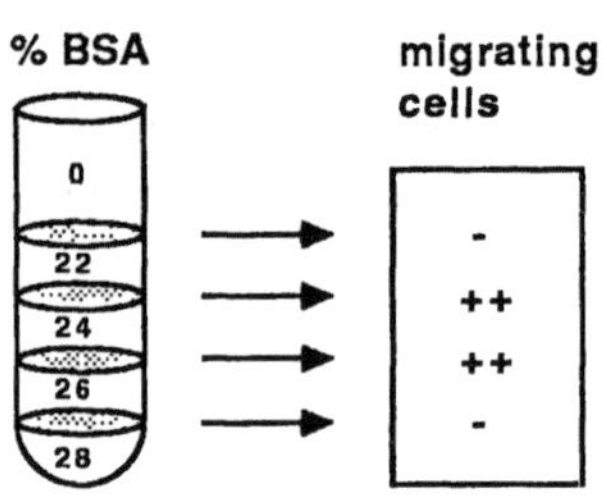

Figure 1. Hemopoietic precursors enriched on a BSA gradient contain migating cells
Bone marrow cells from three week old rats were loaded onto a discontinuous BSA gradient (0-28%) and centrifuged at 1000g for 30min. Cells from interphases were tested for chemotactic migration towards medium that was conditioned by thymic epithelial cells or fresh control medium.Boyden chambers of 100µl volume contained $3x10^5$ bone marrow cells and were incubated for three hours at 37°C. (++) 0.5% of incubated cells migrated specifically towards thymic products. (-) no specific migration.

towards thymic peptides in Boyden Chambers can still colonize a thymus *in vitro* . When this colonized thymus is grafted back to an embryo the precursors mature into T cells suggesting that migrating precursors in embryos are T-cell precursors. Intrathymic T-cell precursors might persist as self-renewing cells, although with a limited cycle number (6). For continuous seeding of this pool of precursors the same homing mechanism may operate in the juvenile animal, but at a low level (7,8)

Hemopoietic precursors from juvenil mammals migrate towards thymic peptides

In juvenile rats hemopoietic precursors are found as a minor subpopulation in the bone marrow. These cells can be efficiently enriched on a discontinuous BSA (bovine serum albumin) gradient ranging from 0 to 28% at 1000 g for 30 min. (9). Four cell fractions from such a gradient were tested for chemotactic migration towards thymic epithelial cell-conditioned medium.Two gradient fractions contained hemopoietic cells that migrated chemotactically when tested by the Boyden chamber assay (Fig. 1). Interestingly these two fractions found in the interphases of 22/24% and of 24/26% BSA have been shown to contain T-cell precursors (10). All erythrocytes, polymorphonuclear leukocytes, granulocytes and mastocytes were found in the pellet fraction below 28% BSA which contained usually between 80 to 90% of the total cells. Therefore cells centrifuged onto a cushion of 28% BSA were sufficiently enriched for hemopoietic precursors and were routinely used for experiments.

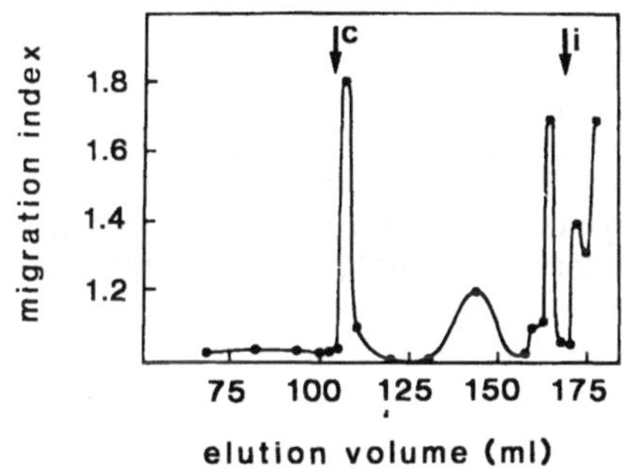

Figure 2. Purification of chemotactic peptides produced by thymic epithelial cells
The conditioned medium was concentrated and desalted on Sep-Pak cartridges. This material was then fractionated on a Sephadex G-50 column and the fractions were subsequently tested for chemotactic migration of hemopoietic precursors in Boyden chambers. Molecular weight markers are cytochrome c (**c**), and insulin B chain (**i**). Migration index was calculated as the ratio between the number of cells attracted by peptides and the number of cells found in the presence of control medium (DMEM).

Isolation of thymic peptides that are chemotactic for hemopoietic precursors in mammals

The epithelial cell line IT-45 R1 derives from a juvenile rat thymus and has been shown to express several thymus specific properties. The cells secrete functional thymic hormones such as thymopoietin and thymulin (11). Furthermore, IT-45 cell conditioned medium directed hemopoietic precursors to undergo first events in T-cell differentiation (10). For production of chemotactic substances IT-45 cells were grown in large scale cultures and the conditioned medium was produced under serum-free conditions for 3 days. Rapid concentration was achieved when conditioned medium was passed through Sep-Pak C_{18} cartridges and the chemotactic activity was subsequently eluted by 80% acetonitrile. This concentrate was gel filtered on a Sephadex G-50 column and the fractions were tested using the migration assay of bone marrow precursors in Boyden chambers. Chemotactic activity was found in two distinct zones of molecular weight of 11kDa and 4kDa (Fig. 2).

Table I. Morphological phenotype of migrated cells

	lymph.	immat. blasts	erythr. prog.	myel. prog.	pmn	monoc. macro.
total cells before migration	**29**±1	12±2	46±3	9±3	2±1	0.0
total IT-45 medium	**29**±8	8±5	27±18	26±20	7±6	2±1
11kDa peptides	**57**±12	9±4	31±11	2±2	0.0	1±1

Bone marrow derived precursor cells were taken before and after chemotactic migration towards IT-45 conditioned medium and 11kDa peptides. These cells were cytocentrifuged and stained in May-Grünwald-Giemsa solution. Numbers represent the percentage of at least 250 counted cells. (lymph.) Lymphoid cells, (immat.blasts) immature blast cells that did not contain any granules in the cytoplasm and were probably not commited cells, (erythr. prog.) erythroid progenitor cells, (myel. prog.) myeloid progenitor cells, (pmn) polymorphonuclear leukocytes, (monoc. macro.) monocytes and macrophages.

Table II. Characterization of surface antigens on migrated cells

	[a]thy-1 [b]Ox-7	CD-8 Ox-8	CD-5 Ox-19	CD-4 W3-25	IgM Marm 4	Ia Ox-6	TdT
total cells before migration	**50**	1	1	1	2.5	1	31
total IT-45 medium	**16**	0.4	0.0	0.3	0.8	0.7	0.5
11kDa peptides	**71**	0.8	0.0	0.4	0.4	0	n.d.
4kDa peptides	**34**	2.0	0.0	0.4	1.5	0.4	n.d.

Bone marrow derived precursor cells were taken before and after chemotactic migration towards IT-45 conditioned medium or towards the purified peptide fractions of molecular weights 11kDa and 4kDa.Numbers represent the percentage of differentiation antigen positive cells stained by immunofluorescence. (a) antigen, (b) corresponding antibody.

Phenotype of hemopoietic precursor cells that migrate towards thymic peptides

May-Grünwald-Giemsa staining was used for morphological characterization of bone marrow cells after chemotactic migration in Boyden chambers towards thymic peptides. The purified 11kDa material attracted preferentially cells of the lymphoid type and was much more selective than crude conditioned medium (57 versus 29% of total migrating cells, Table I). However these lymphoid cells were often contaminated by erythroid precursors. It remains to be shown whether this may be due to other chemotactic peptides in our preparations, or whether very immature cells of the red blood lines also bear receptors for the chemotactic material. Analysis of cell surface antigens showed that a significantly higher percentage of Thy-1-positive cells migrated towards the purified thymic peptides of 11kDa than towards the total IT-45 conditioned medium (Table 2).

Concluding remark

Thy-1 positive cell fractions from bone marrow that do not bear differentiation antigens of T cells, B cells, granulocytes or macrophages are thought to contain pre-T cells (12). We hypothetize that hemopoietic precursors colonizing the thymus respond to chemotactic peptides, have a lymphoid morphology, are Thy-1 positive, and are negative for differentiation markers (Fig. 3). Some of these migrated precursors can be differentiated into T cells in an *in vitro* assay system that uses rat embryonic thymus (B.A. Imhof et al.; M.A. Deugnier et al.; both manuscripts submitted for publication). Whether these migrating cells are pluripotent and can also reconstitute the B cell pool, the myeloid lineages or the erythroid cells lineage waits further analysis.

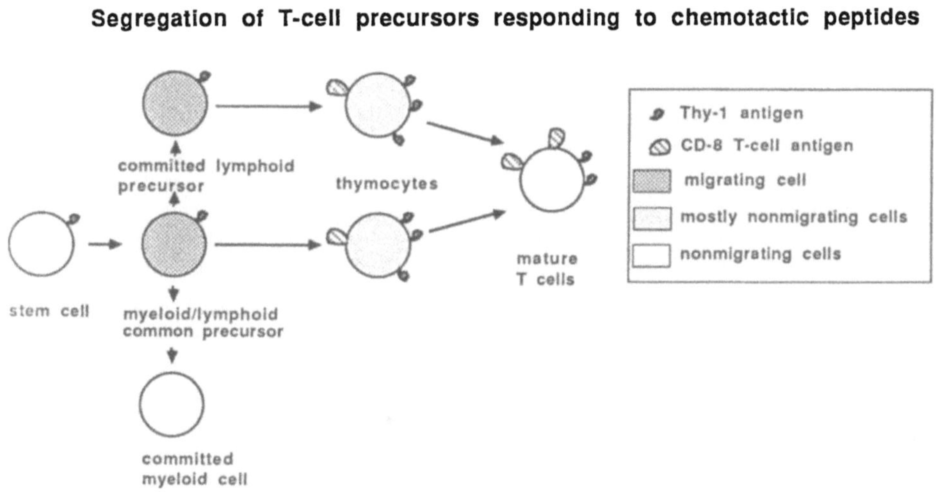

Figure 3. Segregation of T-cell precursors responding to chemotactic peptides

REFERENCES

1. N.M. Le Douarin, and F.V. Jotereau. In Immunology 80. M. Fougereau, and J. Dausset, eds. (London: Academic Press, pp 285 (1980).
2. F.V. Jotereau, and N.M. Le Douarin. J. Immunol. 129: 1869 (1982).
3. P. Savagner, B.A. Imhof, K.M. Yamada, and J.P. Thiery. J. Cell Biol. 103: 828 (1986).
4. S. Ben Slimane, F. Houllier, G.C. Tucker, and J.P. Thiery. Cell Diff. 13: 1 (1983).
5. S. Champion, B.A. Imhof, P. Savagner, and J.P. Thiery. Cell. 44: 781 (1986).
6. I. Goldschneider, K.L. Komschlies, and D.L. Greiner. J. Exp. Med. 163: 1 (1986).
7. F. Lepault, and I.L. Weissman. Nature. 293: 151 (1981).
8. S. Ezine, I.L. Weissman, and R. Rouse. Nature. 309: 629 (1984).
9. T. Itoh, S. Aizu, S. Kasahara, and T. Mori. Biomed. Res. 2: 11 (1981).
10. T. Itoh, S. Kasahara, and T. Mori. Thymus 4: 69 (1982).
11. W. Savino, T. Itoh, B.A. Imhof, and M. Dardenne. Thymus, 8: 245 (1986).
12. B. Adkins, C. Müller, C.Y. Okada, R.A. Reichert, I.L. Weissman, and G.J. Spangrude. Ann. Rew. Immunol. 5: 325 (1987).

MICROVASCULATURE OF THE SHEEP THYMUS

K. Yamaguchi[1,3] and M. Miyasaka[2,3]

1) Institute of Laboratory Animals, Yamaguchi University School of Medicine, Ube, Japan, 2) Basel Institute for Immunology, Basel, Switzerland, 3) Tokyo Metropolitan Institute of Medical Science, Tokyo, Japan

INTRODUCTION

The thymus plays a crucial role in T lymphocyte production and differentiation. It has been suggested that lymphocytes become mature while passing through the cortex to the medulla of the thymus (1,2) and then emigrate to the peripheral lymphoid tissues via either blood (3,4) or lymphatic pathways from the medulla (5-7).

In this study the microvasculature of the sheep thymus was examined in relation to thymocyte emigration by light microscopy (LM), scanning (SEM) and transmission electron microscopy (TEM).

MATERIALS AND METHODS

Cervical thymi from five 6-week-old lambs, White Alpine and Black Jura, were used in this study. The hilar region was exposed surgically under halothane anesthesia. For LM studies, 5% gelatin-carbon solution was infused into the thymus via the thymic artery. Slices of the thymus, 0.5 - 2mm in thickness, were prepared, dehydrated and mounted in neutral mounting medium. To prepare vascular casts, thymi were infused with Mercox resin via the thymic artery and corroded with 10-20% KOH solutions at 37°C. The resin cast was freeze-cracked and dried, coated with platinum and examined with a JEOL T-300 microscope. For TEM and SEM studies, thymi were infused with 2% glutaraldehyde/paraformaldehyde mixture in 0.1M cacodylate, pH 7.4, and cut into small pieces. The tissue was postfixed with 2% osmic acid for 2 hrs, dehydrated with graded acetone and embedded in Epon 812 for TEM studies. For SEM studies, thymus slices were also postfixed, dehydrated and critical point-dried using CO_2.

RESULTS

The vascular architecture of the sheep thymus was investigated in detail by the use of either gelatin-carbon infused specimens or resin casts. Individual thymic lobules were supplied by two different sets of arteries and veins. With respect to the arterial system, main vessels were the medullary (Med) arteries which stemmed from the interlobular artery and entered the medulla together with the accompanying Med veins.

The Med arteries, with ramifying arterioles, ran along the cortico-medullary junction (CMJ) and supplied capillaries to the cortex and medulla (Fig. 1). The other set of arteries was the cortical (Cor) arteries which entered the thymus from the capsular side and supplied capillaries to the cortex. Arterial anastomoses were frequently observed. Hair-pin-shaped arterioles, arising from the Med artery and supplying capillaries to the cortex, were often found to form thoroughfare channels between the arterial and venous vessels (Fig. 2).

Capillaries were observed to be more densely distributed in the cortex than in the medulla (Fig. 1 & 2). Cor capillaries displayed a radial distribution pattern extending from the CMJ to the outer cortex and then converged into venules at the cortical edge. In the outer cortex many of the Cor capillaries demonstrated a spiral or tortuous shape (Fig. 2) emptying also into venules at the cortical surface. In contrast, the Med capillaries formed a rough meshwork and emptied into venules (Fig. 3).

In the venous system, the distribution of venules was denser in the medulla than in the cortex (Fig. 1 & 2). In the medulla (Fig. 3), tortuous or winding Med venules bearing a wide lumen were frequently observed at the CMJ (Fig. 1-3). These venules are derived from either the medulla or cortex and, in turn, become the Med veins which are connected with the interlobular vein. On the luminal surface of these Med venules, lymphocytes penetrating the endothelium were frequently observed (Fig. 4). Furthermore, many depressions or pits, possibly representing lymphocyte penetration through the venular wall, were abundant on the outer surface of these venules (Fig. 3). As shown in Fig. 5 & 6, most of the Med venules possessed a perivascular space (PVS) located between the wall of Med venules and the thymic parenchyma, although the outline of the PVS was not distinct around the Med venules located in the deep medulla. The PVS frequently contained one to several layers of lymphocytes. Granulocytes, macrophages and mast cells were also occasionally found (Fig. 6). The PVS appeared to extend to areas located between the cortical septa and capsule. These septal spaces also contained one or two layers of lymphocytes (Fig. 5 & 6).

On the capsular surface of the thymus, there were many Cor venules which collected Cor capillaries and then converged into the interlobular vein (Fig. 1 & 2). Many of these Cor venules drained into, or anastomosed with, the Med venules through the cortical septa or parenchyma. The Cor venules did not possess any PVS.

DISCUSSION

In this study, the microvasculature of the sheep thymus was investigated. The vascular architecture of the sheep thymus is similar to that of rats (8), guinea-pigs and pigs (9). The sheep thymus possesses a dual vascular drainage system consisting mainly of Med and Cor venules. This differes from the mouse thymus which has only medullary drainage (10). On the inner wall of the Med venules, lymphocytes penetrating the vascular endothelium were frequently observed, while on the outer surface of these venules abundant depressions or pits, possibly representing the lymphocyte penetration through the venular wall, were observed. These Med venules possess PVS which is absent around the Cor venules. The PVS contained one to several layers of lymphocytes and appeared to extend to the lymphatic space located in the lobular septum near the capsule.

From these results we consider that in the mammalian thymus, two different routes of thymocyte emigration are present; one is the blood vascular route by which lymphocytes enter the lumen of postcapillary Med venules by crossing the endothelial lining and the other is the lymphatic

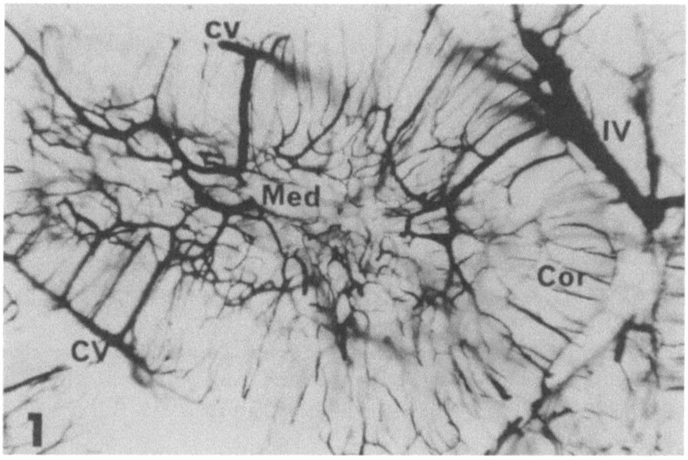

Fig. 1. A thymic lobule in a gelatin-carbon infused specimen. The cortex (Cor) showing a radial distribution pattern of capillaries is distinguished from the medulla (Med). CV: Cor venule, IV: Interlobular vein. (x 25)

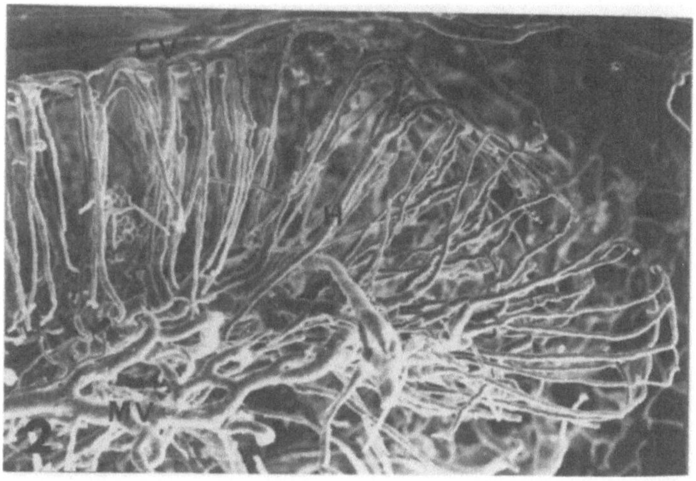

Fig. 2. SEM of the vascular cast. The Cor capillaries extend in a radial distribution pattern running from the CMJ towards the cortical margin. A hair-pin-shaped arteiole (H) arises from the medulla. A number of tortuous venules (MV) are distributed in the medulla. CV: Cor venule. (x 90)

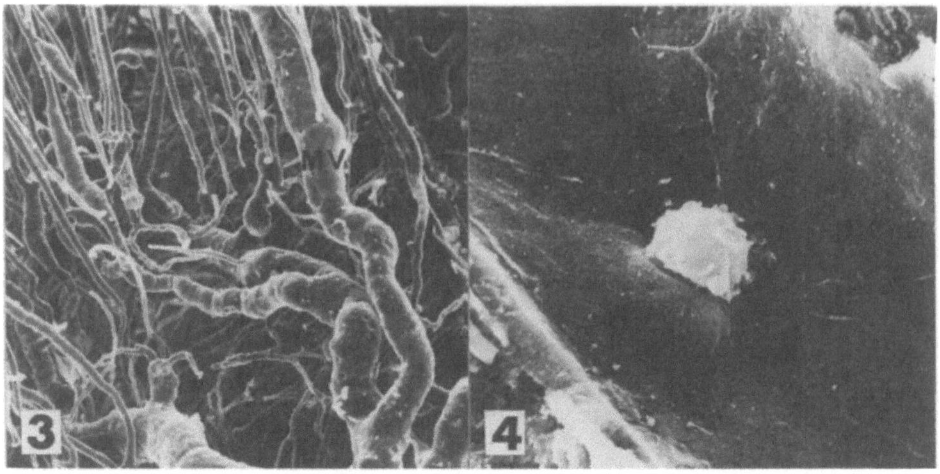

Fig. 3. Higher magnification of the Med venules (MV). Many depressions or pits are evident on the venule casts and may reflect the site where lymphocytes penetrate the vascular encothelium. (x 180)

Fig. 4. A lymphocyte penetrating the endothelial wall of a Med venule. The endothelium of this venule is flat. (x 3000)

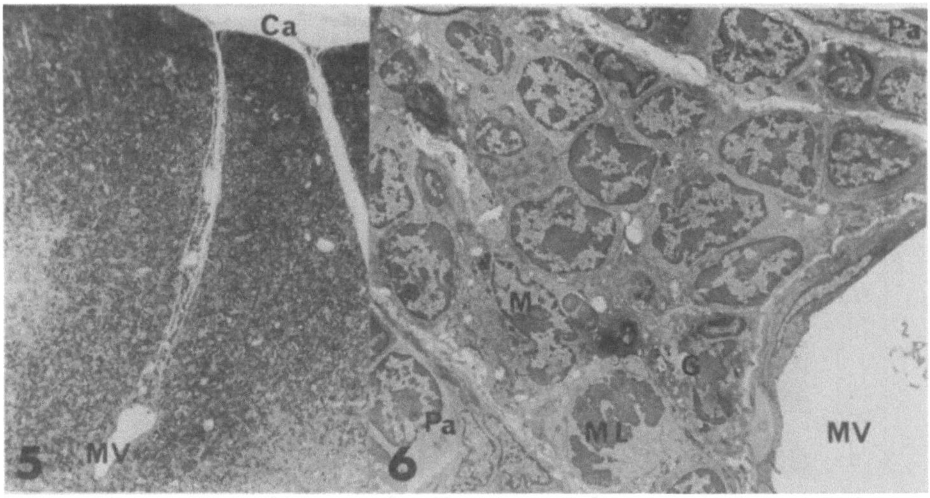

Fig. 5. Semithin section of the thymus. The perivascular space (PVS) around a Med venule (MV) extends into the space formed by the cortical septum and capsule (Ca). Several layers of lymphocytes are present. (x 100)

Fig. 6. TEM of the PVS around a Med venule (MV). Lymphocytes, a granulocytes (G) and a macrophage (M) are also prsent in the PVS. A mitotic lymphocyte (ML) is also present. Pa: Parenchyma. (x 4000)

route by which lymphocytes emigrate into the PVS and leave the thymus via lymphatics. We are currently conducting further studies concerning the structure of the PVS.

REFERENCES

1. I.L. Weissman, Thymus cell migration. Studies on the origin of cortisone-resistant thymic lymphocytes. J. Exp. Med. 137:504 (1973).

2. W. van Ewijk, Immunohistology of lymphoid and non-lymphoid cells in the thymus in relation to T lymphocyte differentiation. Am. J. Anat. 170:311 (1984).

3. I. Törö and I. Oláh, Penetration of thymocytes into the blood circulation. J. Ultrastruct. Res. 17:439 (1967).

4. D. K. Bhalla and M. J. Karnovsky, Surface morphology of mouse and rat thymic lymphocytes: An in situ scanning electron microscopic study. Anat. Rec. 191:203 (1978).

5. G. Sainte-Marie and C. P. Leblond, Cytologic features and cellular migration in the cortex and medulla of thymus in the young rat. Blood 23:275 (1964).

6. T. Ito and T. Hoshino, Light and electron microscopic observations on the vascular pattern of the thymus of the mouse. Arch. histol. Jap. 27:351 (1966).

7. A. Yamashita, M. Miyasaka and Z. Trnka. Early post-thymic T cells: Studies on lymphocytes in the lymph coming from the thymus of the sheep. in: Immunology of the Sheep, Morris, B. and Miyasaka, M. eds., p.162, Editiones Roche, Basel (1985).

8. G. Sainte-Marie, F.S. Peng and D. Marcoux, The stroma of the thymus of the rat: Morphology and antigen diffusion, a reconsideration. Am. J. Anat. 177:333 (1986).

9. I. A. Olson and M. E. Poste, The vascular supply of the thymus in the guinea-pig and pig. Immunol. 24:253 (1973).

10. C. Smith et al., Studies on the thymus of the mammal. VI. The vascular pattern of the thymus of the mouse and its changes during aging. J. Morph. 91:199 (1952).

TISSUE LOCALIZATION AND MIGRATION OF MURINE SPLEEN MACROPHAGES CARRYING THE FORSSMAN GLYCOSPHINGOLIPID ANTIGEN

Peter Conradt, Rainer v. Kleist and Peter F. Mühlradt

GBF, Gesellschaft für Biotechnologische Forschung
Immunobiology Group, D 3300 Braunschweig, FRG

INTRODUCTION

Glycosphingolipids (GSLs) are receptors for virus (1), bacteria (2) and toxins (3). Their functions, if any, for the cells on whose surface they are expressed are still unclear. Recent data suggest the involvement of a sulfated GSL (4) in the cell-cell interactions of developing nervous tissue (5).

Our interest in GSL antigens on cells of the immune system lies in their hypothetical role in the tissue localization of these cells. One possible candidate for a GSL with such a role is globoside on murine T cells. This GSL is detected on precursors of helper and cytolytic T cells of the spleen, but not on thymocytes nor on mature cytolytic effector cells (6). Another, biosynthetically closely related GSL which has the characteristics of a differentiation antigen on murine macrophages (MPh) is the Forssman (F) GSL of the following structure (7):

GalNAcα(1-3)GalNAcß(1-3)Galα(1-4)Galß(1-4)Glcß(1-1)Ceramide

We have shown in a recent study with monoclonal antibodies (MAB) against this GSL that it is an antigenic marker for a major (7o%) subpopulation of murine spleen and peripheral lymph node MPh. This antigen is absent from monocytes and polymorphonuclear cells of the blood, MPh in mesenterial lymph nodes, alveolar and liver MPh and those found in the peritoneal cavity (8). The present work was done to study the localization of F^+ cells in the spleen by immunofluorescence, to compare it with the staining for nonspecific esterase as a general marker for MPh, and to study the migration of F^+ MPh from the spleen to the peritoneum under the influence of inflammatory stimuli.

MATERIALS AND METHODS

Preparation of glass adherent cells

CBA/J mice were killed by asphyxiation in CO_2. Single cells from spleen tissue were prepared by gently teasing the minced tissue through a nylon sieve. The yield of F^+ MPh could be enhanced by 60% when the remaining tissue was treated with the following enzymes at 37 C° for 1h: o.o8% trypsin, 0.1 U/ml collagenase from *Clostr. histolyticum*, 4o µg/ml DNAse in PBS containing 1o mM glucose (all enzymes from Boehringer Mannheim, FRG). Peritoneal cells were obtained by injecting 3ml ice-cold RPMI 164o medium with 2% FCS i.p. and withdrawing the liquid after gently massaging. Cells were left for 2 h at 37 C° to adhere to microscope slides, nonadherent cells were washed off with warm medium. The remaining cells were allowed to spread overnight in the CO_2 incubator and were then washed once again.

Immunostaining of tissue sections and glass adherent cells

Spleens from 2 to 3 months old CBA/J mice (Bomholtgaard, Ry, Denmark) were removed and snap-frozen in liquid nitrogen. Four µm sections were cut on a cryostat, air dried, dipped for 5 sec in cold acetone, and kept at 4 C° until further processed. Sections or glass adherent cells were fixed for 10 min in 3.7% formalin in PBS (immunostaining) or for 1 min in cold 4o% acetone, 2o% formalin in PBS, pH 6.4 (for immunostaining followed by staining for nonspecific esterase). Rat F-specific MAB IgG_{2c} from clone III E2 (8) was purified over protein A sepharose and used at 4 ug/ml, FITC-labeled goat anti rat second antibody was purchased from Medac (Hamburg, FRG) and used at 1/1o dilution. Antibodies were diluted in PBS, containing 2% FCS and o.o2% azide. Incubations were done for 3o min in the cold, followed by 3 washes with cold PBS.

Inflammatory stimuli

Mice were injected i.p. with the following stimulating agents: 170 µg live *bacilli of Calmette and Guerin* (BCG), 1.4 mg *Corynebacterium parvum* (CP) (both from Calbiochem), or 2o µg phenol extracted smooth-type lipopolysaccharide (LPS, endotoxin) from *Salmonella typhimurium*. Endotoxin from *S.minnesota* is not suitable for these studies as it cross-reacts with F^+ specific antibodies. Peritoneal lavage followed after 4 d.

RESULTS AND DISCUSSION

Localization of F^+ cells in the spleen

F^+ cells were predominantly seen in the red pulp (Fig.1). Staining with anti Mac-1 antibody (9) was found in the same areas, but was not as sharply confined to the red pulp (not shown). Distinctly fewer cells were Mac-1^+ than F^+. This seemed a dis-crepancy at first, but staining of glass adherent spleen cells with MPh morphology with MAB against F and Mac-1 showed that, whereas 65% of these cells were F^+, only 18% were Mac-1^+. We think that Mac-1^+ cells are included in the F^+ MPh population.

When spleen sections were stained for nonspecific esterase oflocalize MPh irrespective of their antigenic markers, most of he esterase$^+$ cells were concentrated in the same areas as F$^+$ hells. However, some esterase$^+$ cells were also seen in the center of the white pulp (Fig.2). These were always F$^-$. In T cell areas compare neighboring section in Fig.3 stained with anti Thy-1) another type of F$^+$ structure appeared with a striated appearance. he think it possible that these may be interdigitating dendritic eells. In all areas F$^+$ reactivity disappeared after extraction with organic solvents, in support of the lipid nature of this entigen.

We have sofar not been able to isolate from spleen any F$^+$ cells other than MPh. Even mild protease treatment only increased she yield of F$^+$ MPh, as was to be expected from the data of Humphrey et al. (10) who reported that e.g. marginal zone MPh were difficult to obtain unless collagenase were used during isolation. Our findings do not exclude that, besides MPh and possibly interdigitating den- dritic cells, there are other F$^+$ cells in spleen tissue which we have neither identified nor isolated in viable form. Such cells could be stromal or reticular cells.

In a report by Ghandour et al. (11) a MESA-1 (mouse endothelial cell surface antigen) specific MAB was used to identify endothelial cells lining blood vessels in murine brain, heart, lung, kidney and liver. The MESA-1 MAB did not stain endothelial cells lining liver sinus (11). Dr. Ghandour was kind enough to supply us with a sample of his MAB which we now know to react with F GSL when tested in an immunostaining assay (12). As expected, the staining patterns of spleen sections with our anti F MAB and MESA-1 MAB were identical (not shown). We were unable to detect any staining of cells lining vessels in the spleen with either MAB. This would indicate that, as in liver sinus (11), endothelial cells of the spleen are F$^-$.

Studies similar to ours have been reported by others employing their MAB F4/80 (13) or ERTR9 (14). Obviously it is difficult to exactly correlate our findings with theirs without comparing the different MAB in the same experiments. It is clear, though, from the published cell specificities (F4/80 stains blood monocytes and Kupffer cells (13), which are F$^-$; ERTR9 stains Ia$^-$ cells (14), whereas we found all F$^+$ cells to be Ia$^+$ (8)) that different cell populations are detected by these 3 MAB. We think that F$^+$ cells may be included in the F4/80$^+$ spleen cell population, but that ERTR9$^+$ cells, since they are Ia$^-$, may not be.

Migration of F$^+$ MPh to the peritoneum

In a previous paper we had shown that i.p. thioglycollate administration led to the appearance of a few % F$^+$ MPh in the peritoneal cavity which is otherwise devoid of F$^+$ MPh (8). In continuation of this work we have now tested the influence of other inflammation stimulating compounds on this effect. Age-matched groups of 3 mice from the same cage were used for each experiment. Only glass-adherent, esterase$^+$ peritoneal cells were counted when evaluating F specific immunostaining.

Fig.1

Immunofluorescence staining of Forssman specific areas in a cryosection of a mouse spleen. Note intense fluorescene in the red pulp, and weak, striped fluorescence in the PALS.

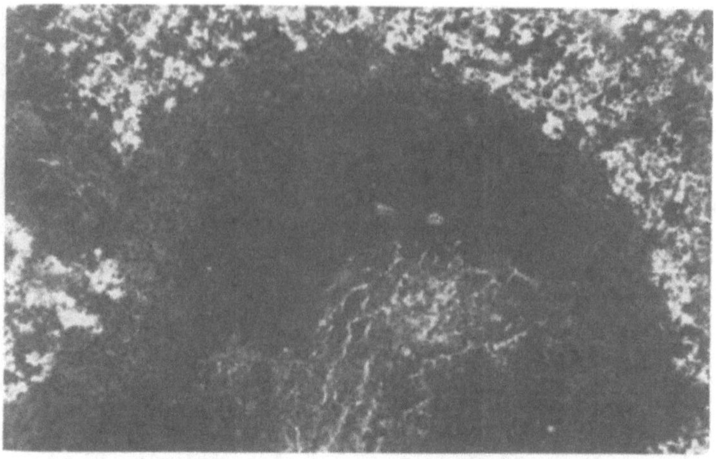

Fig. 2

Staining for nonspecific esterase in adjacent cryosection.

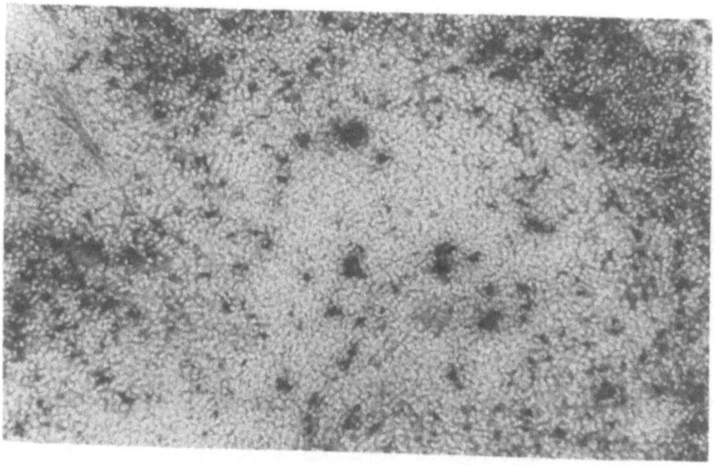

Fig. 3

Immunofluorescence staining of neighboring section with anti-thy-1 to visualize the PALS.

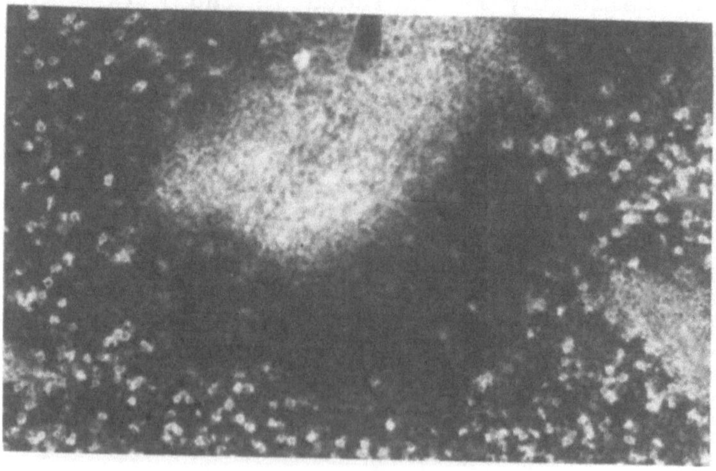

BCG, which is known to activate MPh in infected mice to increased superoxide production (15) or lytic activity towards tumor cells (16), had no significant effect on the accumulation of F^+ MPh in the peritoneum. When peritoneal cells were harvested 2o d post infection, there was still no effect. In contrast, CP, which also activates MPh to increased phagocytosis (17) or increased H_2O_2 production (18), led to an overall increased leukocyte count and an increase in the percentage of F^+ MPh in the peritoneum (Table 1, exp. 2). In exp. 3, only the increase in the percentage of F^+ MPh was seen, probably because these particular animals had an abnormally high number of peritoneal leukocytes to start with. Endotoxin, which reportedly gives rise to a fast increase in neutrophil blood count in man (19), and activates murine MPh to become cytotoxic (20), had no effect on the yield of peritoneal leukocytes at the doses tried, although an increase was reported by others (21). However, the percentage of F^+ MPh was significantly increased in LPS treated animals and could be further enhanced to 25% by harvesting the cells 24 h after LPS application. As expected, the LPS resistent C3H/HeJ mice (22) showed no response in this system (Table 1, exp. 6).

Table 1. INFLUENCE OF INFLAMMATORY STIMULI ON APPEARANCE OF F^+ MACROPHAGES IN THE PERITONEAL CAVITY AFTER 4 DAYS

Exp.No.	Inflamm. stimulus	Yield of perit. leukocytes	F^+ glass adherent MPh
		(x $1o^{-6}$)	(%)
1	BCG	2.o ± o.5	o.5
	-	1.5	o.5
2	CP	6.1 ± 1.3	2.6 ± o.5
	-	1.7	o.6
3	CP	11.o ± 3.o	9.o ± 4.4
	-	12	2.9
4	LPS	5.7 ± 1.1	11.2 ± 3.2
	-	5.7	o.7
5	LPS	5.9 ± 1.8	3.3 ± o.9
	-	3.9	1.3
6[a)]	LPS	3.3	o.6
	-	2.8	<o.3

[a)] Exp. 6 was carried out in the LPS resistent C3H/HeJ mice, all others in LPS sensitive CBA/J mice.

Of the stimulants tried, only BCG had no effect on the appearance of F^+ MPh in the peritoneal cavity. This could mean that activation of MPh is not the primary cause for this effect, but that a local focus of inflammation giving rise to a gradient of chemoattractants, rather than a systemic infection, as would happen with live BCG, is required to cause accumulation of the F^+ MPh population in the peritoneum where they are not normally observed. Since monocytes newly recruited from the bone marrow are F^-, we believe that the F^+ MPh have migrated to the peritoneum from the spleen where F^+ MPh reside under normal circumstances. This hypothesis should be reconfirmed in splenectomized mice. Such experiments are underway.

Acknowledgements: We thank Mrs. Toni Mühlradt for excellent technical assistance, and the Fond der Chemischen Industrie for financial help.

REFERENCES

1. M.A.K. Markwell, L. Svennerholm, & J.C. Poulson. Proc.Natl. Acad. Sci. USA 78: 5406 (1981).
2. K. Bock, M.E. Breimer, A. Brignole, G.C. Hansson, K-A. Karlsson, G.Larson, H.Leffler, B.E. Samuelsson, N.Strömberg, C.S. Eden, & J.Thurin. J.Biol.Chem. 260: 8545 (1985).
3. W.E. van Heyningen. Nature 249: 415 (1974).
4. D.K.H. Chou, A.A. Ilyas, J.E. Evans, C. Costello, R.H. Quarles & F.B. Jungalwala. J.Biol.Chem. 261: 11717 (1986).
5. G. Keilhauer, A. Faissner, & M. Schachner. Nature 316: 728 (1985).
6. P.F. Mühlradt, U.Bethke, D.A. Monner, & K. Petzold. Eur.J.Immunol. 14: 852 (1984).
7. B. Siddiqui, & S.I. Hakomori J.Biol.Chem. 346: 5766 (1971).
8. U. Bethke, B. Kniep, & P.F.Mühlradt. J.Immunol. 138: 4329 (1987).
9. T. Springer, G. Galfre', D.S. Secher, & C. Milstein. Eur. J. Immunol. 9: 301 (1979).
10. J.H. Humphrey, & D. Grennan. Eur.J.Immunol. 11: 221 (1981).
11. M.S. Ghandour, K.Langley, G. Gombos, M. Hirn, M-R. Hirsch, & Ch. Gordis. J. Histochem. Cytochem.30: 165 (1982).
12. U. Bethke, J. Müthing, B. Schauder, P. Conradt, & P.F. Mühlradt. J. Immunol. Methods 89: 111 (1986).
13. D. A. Hume, A. P. Robinson, G. C. Macpherson, & S. Gordon. J. Exp. Medicine 158: 1522 (1983).
14. C.D. Dijkstra, E. van Vliet, E.A. Döpp, A.A. van der Lelij, & G. Kraal. Immunology 55: 22 (1985).
15. R.B. Johnston Jr., C. A. Godzik, & Z. A. Cohn. J. Exp. Medicine 143: 1551 (1978).
16. R.P. Cleveland, M.S. Meltzer, & B. Zbar. J.Natl. Cancer Inst. 52: 1887 (1974).
17. B. Halpern, A.R. Prevot, G. Biozzi, C. Stiffel, D. Mouton, J. C. Morard, Y. Bouthilier, & C. Decreusefond. J. Reticuloendothel. Soc. 1: 77 (1964).
18. C.F. Nathan, & R.K. Root. J. Exp. Medicine 146: 1648 (1977).
19. S.M. Wolff, M. Rubenstein, J. H. Mulholland, & D.W. Alling. Blood J. Hematol. 26: 190 (1965).
2o. P. Alexander & R. Evans. Nature (New Biol.) 232: 76 (1971).
21. G.R. Moeller, L. Terry, & R. Snyderman. J. Immunol. 120: 116 (1978).
22. B.M. Sultzer. Infect. Immun. 5: 107 (1972).

MIGRATION PATTERNS OF DENDRITIC CELLS IN THE MOUSE

J.M. Austyn, J.W. Kupiec-Weglinski* and P.J. Morris

Nuffield Department of Surgery, University of Oxford
John Radcliffe Hospital, Oxford OX3 9DU

INTRODUCTION

Dendritic cells (DC) in lymphoid organs and related cells in non-lymphoid tissues play a pivotal role in the induction of T lymphocyte responses (reviewed in Austyn, 1987). DC stimulate primary immune responses in culture and in vivo, and trigger allograft rejection. However, their mode of entry to the tissues and subsequent fate, and the significance of apparently mature DC in blood is unclear.

Here we summarize our findings on the migration patterns of dendritic cells in the mouse. These were made using three different techniques: a) use of radiolabelled DC to trace the traffic of DC after intravenous or subcutaneous administration, b) use of fluorochrome-labelled DC to localize administered cells within the tissues, c) development of a frozen section assay to define the interaction of DC with particular lymphoid elements. These results are discussed in detail elsewhere (Kupiec-Weglinski et al., 1988; Austyn et al., 1988).

METHODS

Cells

Dendritic cells were prepared from the spleens of mice by density fractionation and adherence to tissue culture dishes. The resultant low-density (<1.08) adherent cells were used to prepare populations of a) "unsorted DC" (70-85% pure) from which many macrophages (MΦ) were removed by re-adherence; b) "sorted DC" (generally >>95% pure) after negative selection on an Ortho Cytofluorograf to deplete cells with markers of T cells, B cells and macrophages (MΦ); and c) "EA-negative DC" (generally >95% pure) after depleting cells that formed rosettes with antibody-coated sheep erythrocytes (EA).

T lymphocytes (>95% pure) were prepared by passage of higher density (>1.08) spleen cells through nylon wool. MΦ were from resident peritoneal cells.

Radiolabelling

Cells were labelled with ^{111}Indium (^{111}In) chloride in water-soluble carrier tropolone. They were administered intravenously (IV) via the penile vein or subcutaneously (SC) into the footpad of anaesthetized recipients.

Fluorescent labelling

Cells were labelled with the (blue fluorescing) DNA-binding fluorochrome Hoescht 33342 (H33342) and administered as above. Frozen sections of spleen or node were fixed in acetone and stained with rat monoclonal antibodies to Ly1 plus Ly2 (for T cells), or Ia (for interdigitating cells and B cells) and (green-fluorescing) fluorescein-isothiocyanate (FITC) rabbit anti-rat Ig.

Frozen section assays

H33342-labelled cells were placed on freshly-prepared, unfixed, frozen sections in serum-containing medium and incubated at 4°C or at 37°C. After 30 min, 1% formalin was added for 10 min, and the sections were then rinsed gently until only firmly-adherent cells remained, and examined by ultraviolet microscopy.

Control experiments

These established that: a) the labelling techniques did not alter DC function, either for stimulation of the allogeneic mixed leukocyte reaction in culture, or for triggering rejection of vascularized cardiac allografts; b) cells that were double-labelled with 111In and H33342 migrated similarly to those labelled with either alone; c) the results were not due to release and reutilization of the radiolabel or fluorochrome (shown by culture of unlabelled cells in isotope released from dying cells, and injection of cell-free ^{111}In or H33342); and d) glutaraldehyde-fixed DC did not migrate in the manner described.

RESULTS

Quantitative analysis of traffic of radiolabelled DC

Unsorted splenic DC (70-85% pure) or T cells for comparison were labelled with ^{111}Indium (^{111}In) and adoptively transferred (IV) to syngeneic mice. DC were sequestered primarily in the lungs from as early as 1min until 1h after transfer, but the radioactivity declined exponentially to minimal levels (<1%) by 24h. At the same time there was a progressive increase in the numbers of DC within the liver and spleen. These reached plateau levels by 3h and they were maintained until at least 24h. Data for the localization of unsorted DC at 3h and 24h are in Table 1, which also compares their traffic in syngeneic and allogeneic recipients. The results are expressed both as the radioactivity in each organ as a percentage of the total dose injected, and as a percentage of the total label recovered, per 0.1g tissue sample (i.e. specific activity).

Table 1. Tissue Localization of unsorted dendritic cells

	Radioactivity per organ (% total dose injected)				Radioactivity per 0.1g (% total label recovered)			
	Syngeneic		Allogeneic		Syngeneic		Allogeneic	
	3h	24h	3h	24h	3h	24h	3h	24h
	A*	B*	C*	D*				
Liver	43.31	41.58	47.17	48.15	5.85	5.90	6.62	6.56
Spleen	12.22^{x}	13.42	9.81	10.76^{x}	17.55	20.14	14.55	15.51
Lungs	7.32^{z}	0.96^{z}	2.88^{z}	0.73	8.04	1.11	3.31	0.81
MLN	0.05	0.06	0.11	0.09	0.18	0.23	0.41	0.33
Peyer's	0.03	0.03	0.05	0.04	0.14	0.14	0.23	0.18
Gut	1.50	2.70^{z}	1.38^{Y}	2.85	0.28	0.53	0.27	0.53
Skin	2.96	4.90^{Y}	2.94	3.89	0.19	0.32	0.19	0.24
Kidney	4.30	5.88	5.92	5.56	1.83	2.61	2.60	2.38
PBL	1.12	0.66^{z}	1.06^{z}	0.49	N/A	N/A	N/A	N/A
TOTAL	72.74	70.19	71.32	72.56				
(n)	(7)	(6)	(4)	(4)				
(p)*	A:C	A:B	C:D	B:D				

* x = $p<0.05$; y = $p<0.01$; z = $p<0.001$; blank = not significant
MLN = mesenteric lymph node; PBL = peripheral blood

Several points are evident from this analysis. Firstly, more DC entered the liver than the spleen, but the spleen was the principal site of localization in terms of specific activity. Secondly, the entry of DC was not substantially affected by histocompatibility barriers as no major differences were observed between syngeneic and allogeneic recipients. Nor did it represent the preferential return of splenic DC to their site of origin since in other experiments we found that DC enriched from mesenteric lymph nodes migrated similarly. Thirdly, the migration of DC to lymph nodes and Peyer's patches was negligible. It should be noted that the activity in other sites was not significantly different to that found after injection of cell-free isotope or ^{111}In-labelled glutaraldehyde-fixed cells.

In contrast to the results with unsorted DC, T lymphocytes also migrated to spleen and liver, after initial retention in the lungs, but their plateau levels at 3h and 24h were opposite to those of DC: some 15% of T cells migrated to liver but 36% to spleen. In addition, T cells homed readily to other lymphoid tissues and some 4-8% and 3-5% entered the mesenteric nodes and Peyer's patches respectively.

Splenectomy of the prospective recipients (n=4) did not influence the inability of unsorted DC to enter lymph nodes, but the otherwise spleen-seeking DC were quantitatively diverted to the liver (41.6% and 53.0% in normal and splenectomized mice respectively: $p<0.01$). In contrast, the migration of T cells to the mesenteric lymph nodes rose sharply (6.6% to 19.8%: $p<0.001$), while the increase in the liver was less dramatic (14.7% to 20.5%: $p<0.01$). Thus, one of the distinctive features of DC is their inability to migrate from peripheral blood into lymph nodes and hence to recirculate.

Qualitatively similar results were obtained when the traffic of radiolabelled sorted DC, or EA-negative DC (>95% pure), was examined. The principal differences were increased accumulation of cells in the liver but decreased levels in the spleen (to about 57% and 7% respectively in syngeneic recipients at 24h). Again the spleen was the principal site of localization of DC in terms of specific activity, followed by the liver (about 9/0.1g and 7/0.1g respectively) while nodes and Peyer's patches were devoid of radiolabelled cells; these differences could be explained by the removal of cells with the migratory phenotype of T cells.

When ^{111}In-labelled (EA-negative) DC were inoculated SC into the footpad, a large amount of activity remained at the site of injection. However, some DC entered the draining popliteal lymph node by 3h, and increased numbers were present by 24h, showing that the physiological route of entry of DC to the lymph nodes is perhaps solely via the afferent lymphatics, in agreement with data of others. DC were not found in the next (inguinal) lymph node of the chain, again in keeping with the known scarcity of DC in efferent lymphatics.

DC home to T-dependent areas of spleen

It has long been suspected that the in situ counterparts of lymphoid DC are the interdigitating cells (IDC) of the T cell areas, despite the lack of direct evidence. Sorted DC were labelled with H33342, so they could be visualized in tissue sections in relation to particular areas defined by monoclonal antibodies and FITC-conjugated anti-Ig.

Three hours after IV injection, about three-quarters of the DC were located in the red pulp of the spleen. In contrast, by 24h at least two-thirds of the DC were within T-dependent areas, and were clearly excluded from the B cell areas. Thus, there is active migration of DC from the blood into splenic T-dependent areas where they may become IDC. DC thus have characteristic migration patterns that differ from those of T or B lymphocytes (which migrate to T or B areas), or large granular lymphocytes and unfractionated bone marrow cells which remain in the red pulp.

Preliminary results indicate that H33342-labelled DC home similarly within popliteal lymph nodes after SC injection into the footpad; at 3h they tend to be outside of T-dependent areas, but they are within them by 24h.

T-cell-dependent and -independent migration into lymphoid tissues

When ^{111}In-labelled unsorted DC were administered IV to nude mice, the number of cells that entered the spleen was drastically reduced compared to euthymic recipients (about 3% and 13% respectively at 24h) and insignificant compared to controls (cell-free isotope or fixed DC). But when nude mice were reconstituted with splenocytes or T cells, 24h before transfer of DC, the number of DC entering the spleen (about 9-11%) approached that of euthymic mice. We conclude that entry of DC from the blood to the spleen is dependent on the presence of T cells.

In nude mice the DC that would have gone to the spleen were quantitatively diverted to the liver, and were correspondingly reduced in numbers in the livers of T-cell-reconstituted nude mice; this situation appears similar to that in splenectomized hosts. Although we have not excluded the possibility that some DC in liver are damaged or dying cells, the equilibrium between liver- and spleen-seeking DC may be analogous to that of spleen- and lymph node-seeking T lymphocytes. Fossum (pers. comm.) has reported that rat DC (derived from central lymph after mesenteric lymphadenectomy) migrate from the blood into the liver and thence to the coeliac nodes.

A similar analysis of the migration of DC after SC transfer was carried out in nude and euthymic recipients. In both cases DC entered the popliteal nodes, showing that migration to the nodes via afferent lymphatics does not require T cells, unlike entry to the spleen from the blood. Therefore DC exhibit T-cell-dependent and -independent migration to lymphoid tissues.

DC bind to the marginal zone in frozen sections of spleen

The T-dependency of DC migration into the spleen does not appear to be a result of direct DC-T cell clustering, like that studied in vitro, because DC are initially localized in the red pulp and outside of the T areas. To examine whether DC could interact specifically with particular regions of lymphoid tissues, we developed a frozen section assay. This was based on that used to study the binding of T cells to high endothelial venules (HEV) of lymph nodes.

When DC were incubated on frozen sections of spleen at 37°C they attached specifically to the marginal zone; no binding occurred at 4°C. In contrast, T cells did not bind specifically and Mϕ attached to the red pulp at 37°C and to a lesser extent at 4°C. However, DC did not bind to sections of mesenteric lymph node, unlike T cells which attached at 4°C presumably to HEV. Therefore, the ability of DC to adhere to frozen sections of spleen or node seems to correlate with their ability to enter these tissues from the bloodstream.

CONCLUSIONS

The foregoing results have lead us to several conclusions and hypotheses.

Firstly, lymphoid DC can migrate from the blood into the spleen and they probably become interdigitating cells in the T-dependent areas. However, as isolated, these cells may be a more motile and migratory form that could be localized in situ within the marginal zone and/or red pulp; they may also be present in the blood. Why they should migrate first to the red pulp and thence to white pulp (apparently unlike T cells) is open to conjecture.

Secondly, migration of DC into the spleen may depend on their ability to first attach to the marginal zone; we assume they adhere to the endothelium rather than to marginal zone Mϕ in sections but this has not been proven. Thus DC may express homing receptors specific for certain endothelia that are similar to those of lymphocytes. Conceivably the expression of these may differ on different stages of the lineage: DC in blood may have receptors for splenic marginal zone endothelium but veiled cells of afferent lymph might have no need of them and one might predict that these cells would not localize in the spleen to the same extent after IV injection. A lack of receptors specific for the endothelium in lymph nodes might account for the inability of DC from spleen, and perhaps blood, to enter them.

Thirdly, it is possible that the presence of T cells in certain areas might modify the endothelium and thereby control recruitment of DC from the circulation. This would account for the T-dependency of DC entry from the blood into the spleen. Since the DC that are obtained from human peripheral blood seem to be mature, and unlike Langerhans' cells which are DC precursors, it is conceivable that they could enter non-lymphoid sites and contribute to the generation of extra-lymphoid immune responses. We are currently evaluating whether DC can enter sites of inflammation and allografts, and if so, whether this is a T-dependent phenomenon.

REFERENCES

Austyn JM. (1983) Lymphoid dendritic cells. Immunology. 62:161-170.

Austyn JM, Kupiec-Weglinski JW, Hankins DF and Morris PJ. (1988) Migration patterns of dendritic cells in the mouse. Homing to T-cell-dependent areas of spleen, and binding within marginal zone. J. Exp. Med. In press.

Kupiec-Weglinski JW, Austyn JM and Morris PJ. (1988) Migration patterns of dendritic cells in the mouse. Traffic from the blood, and T-cell-dependent and - independent entry to lymphoid tissues. J. Exp. Med. In press.

---

* Present address for JK-W: Surgical Research Laboratories, Harvard Medical School, Boston, Massachusetts, USA.

A MONOCLONAL ANTIBODY WHICH REACTS SPECIFICALLY WITH A POPULATION OF BOVINE LYMPHOCYTES LACKING B CELL AND T CELL MARKERS

W.I. Morrison, N.D. MacHugh, A. Bensaid, B.M. Goddeeris, A.J. Teale and D.J. McKeever

International Laboratory for Research on Animal Diseases
P.O. Box 30709, Nairobi, Kenya

INTRODUCTION

In the last few years a number of monoclonal antibodies (MAb) specific for bovine T lymphocytes has been produced. These include antibodies which react with the sheep erythrocyte receptor (designated BoT2) (Davis et al., in press) and thus can be used as pan-T cell markers, and antibodies to the two major subset markers BoT4 and BoT8 (Baldwin et al., 1986; Ellis et al., 1986). When these antibodies are used, in conjunction with reagents that identify B lymphocytes and monocytes, to phenotype bovine leukocytes, there is a residual population of lymphocytes comprising 10-20% of peripheral blood mononuclear cells (PBM) that is unreactive with these reagents. Recently, we have produced a MAb that reacts with a proportion of these "null" lymphocytes. Herein, we will summarise our findings with this antibody.

MATERIALS AND METHODS

Animals

Samples of PBM and lymphoid tissues were obtained from female or castrated male Boran (Bos indicus) cattle aged 1-36 months.

Production of the monoclonal antibody

Balb/c mice were inoculated intraperitoneally with 10^7 cells of a bovine lymphoblastoid cell line infected with the protozoan parasite Theileria parva and 4 weeks later they received 2×10^7 bovine PBM by intravenous injection. Spleen cells harvested 3 days later were fused with x63-Ag myeloma cells and hybridomas selected and cultured as described elsewhere (Lalor et al., 1986). The hybridoma producing MAb IL-A29 was selected following screening of supernatants by immunofluoresence staining of PBM, using a fluorescence-activated cell sorter (FACS II, Mountain View, California). The MAb is of the IgG1 isotype.

Samples of bovine lymphoid cells

Peripheral blood mononuclear cells, monocyte-depleted PBM and populations of cells enriched for monocytes were obtained as previously described (Goddeeris et al., 1986a). The monocyte-depleted PBM contained less than 0.2% monocytes compared to between 5 and 20% in intact PBM. Suspensions of cells were prepared from samples of lymph node, spleen and thymus collected immediately following slaughter of healthy animals, as described elsewhere (Ellis et al., 1986).

Staining of cell suspensions and tissues

Suspensions of lymphoid cells were stained by indirect immunofluorescence and analysed on a FACS. Two colour analyses were carried out using reagents conjugated with fluorescein isothiocyanate or phycoerythrin (Ellis et al., 1986). Cryostat sections of bovine lymphoid tissues were stained by the indirect immunoperoxidase technique.

Functional assays

All cultures were conducted in RPMI 1640 medium supplemented with 10% foetal bovine serum, 5×10^{-5} M 2-mercaptoethanol, 2mM glutamine and 50 ug/ml gentamycin. For stimulation with mitogens, PBM were added at 1.25×10^6 per ml to 96 well or 24 well plates in the presence of 2.5 ug/ml concanavalin A (ConA) or 10 ug/ml pokeweed mitogen (PWM). Allogeneic and autologous mixed leukocyte cultures (MLC) were established in 96 well or 24 well plates as described previously (Goddeeris et al., in press). Intact PBM were used both as stimulator cells and responder cells in the allogeneic MLC, whereas the autologous MLC was established with monocyte-depleted PBM as responder cells and intact PBM as stimulator cells. Cultures of T lymphoblasts specific for cells infected with *T parva* were established by stimulation of intact PBM from immune cattle, in autologous *Theileria* MLCs (Goddeeris et al., 1986b). Stimulator cells consisted of parasitised cells from cell lines maintained *in vitro*. Proliferation of cultures was evaluated by measuring the incorporation of ^{125}I -iododeoxyuridine. This was done after 3 days of culture with mitogens and after 5 days in the case of MLCs.

Immunoprecipitation

Peripheral blood mononuclear cells were surface-labelled with ^{125}I by the lactoperoxidase technique. The cells were solubilised, IL-A29-reactive molecules immunoprecipitated with protein A-Sepharose and precipitates

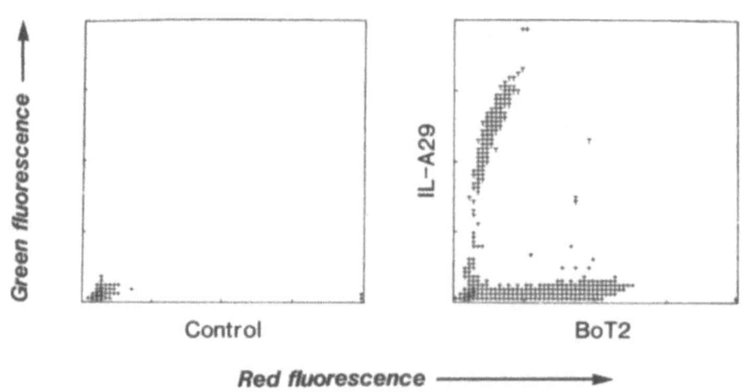

Fig. 1. Two colour FACS profile of PBM stained with IL-A29 and a MAb specific for the pan-T cell marker BoT2. The two positive populations are distinct. Control cells were exposed only to second step reagents.

subjected to sodium dodecyl sulphate/polyacrylamide gel electrophoresis (SDS-PAGE) (Naessens et al., 1985). Precipitated molecules in the gels were visualised by autoradiography.

RESULTS AND DISCUSSION

The MAb IL-A29 reacts strongly with a small discrete population of lymphocytes in PBM. In animals of 6 months of age or older, the population represents about 15% of PBM, ranging from 6% to 25%. The percentage of positive cells is somewhat higher in younger animals. Two colour immunofluorescence staining of PBM has shown that these cells do not express surface immunoglobulin (sIg) and are negative for the T cell-specific markers BoT2, BoT4 and BoT8 (Fig.1). The antibody also does not react with monocytes. Less than 3% of cells in suspensions of lymph node cells react with IL-A29, and in tissue sections these positive cells are found in the peripheral cortex adjacent to the sinuses (Fig.2a). A similar small population of positive cells is found in the spleen where they are located in the red pulp and marginal zones. Thus, the tissue distribution of these cells in secondary lymphoid tissues is distinct from that described for bovine B and T lymphocytes (Morrison et al., 1987). In the bovine thymus, approximately 80% of thymocytes express the sheep erythrocyte receptor BoT2.

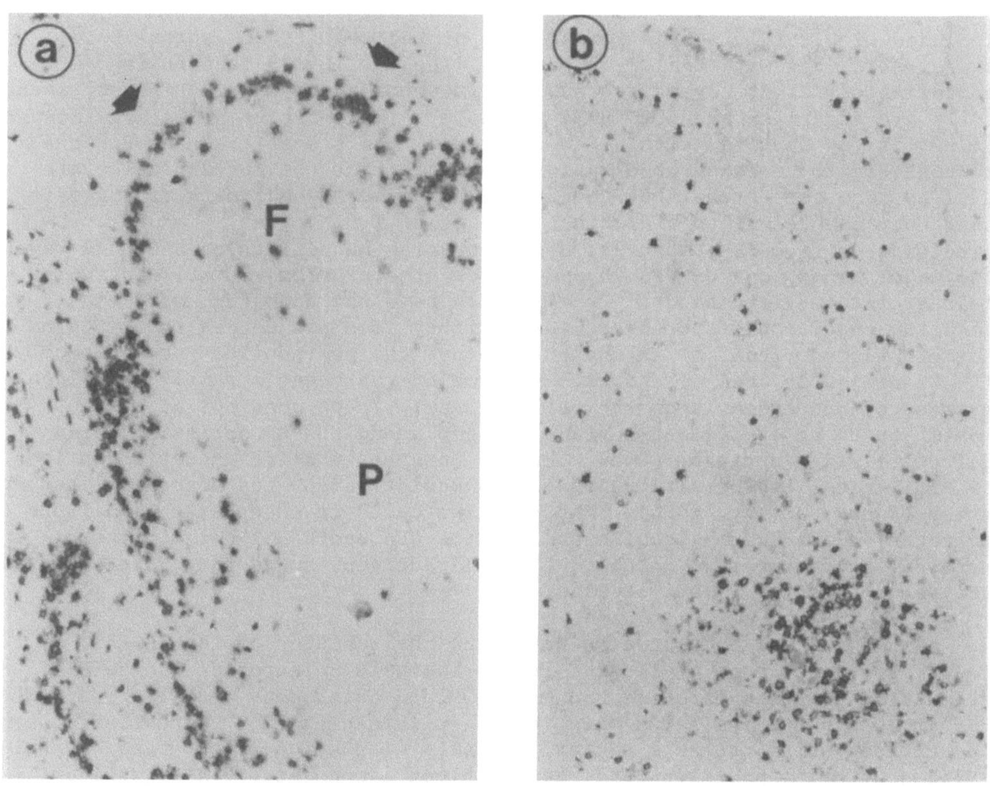

Fig. 2. Sections of bovine lymphoid tissues stained by immunoperoxidase with MAb IL-A29. (a). Lymph node: Positive cells are found adjacent to the subcapsular sinus (arrows). F - follicular region; P - paracortex. (x 125). (b). Thymus: Positive cells are present mainly in the medulla, shown in the lower part of the figure. (x 120).

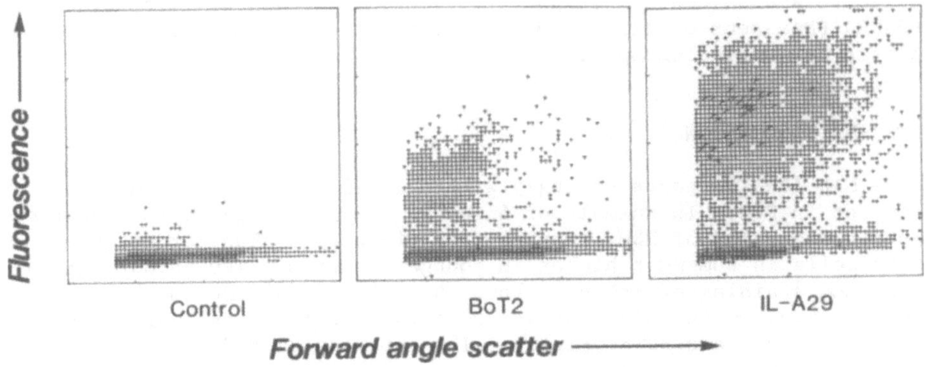

Fig. 3. FACS profiles of cells from day 5 of an autologous MLC, stained with a MAb specific for BoT2 and with IL-A29. Fluorescence intensity is displayed on the vertical axis and forward angle light scatter (as an estimate of cell size) on the horizontal axis. A large proportion of the large cells are positive for the IL-A29 determinant and negative for BoT2.

Included within this population are all of the cells that express BoT4 and/or BoT8 (Baldwin et al., in press). The MAb IL-A29 reacts with approximately 2% of thymocytes and, as in secondary lymphoid tissues, these cells are negative for BoT2, BoT4 and BoT8. The positive cells are located mainly in the medulla, with a few individual cells scattered through the cortex (Fig.2b). These observations indicate that the IL-A29$^+$ cells are derived from the thymus, but the absence of other T cell markers suggests that they may belong to a distinct lineage of lymphocytes. Immunoprecipitates from PBM, when run under reducing conditions on SDS-PAGE reveal two major bands, one of 215 kD and one of greater than 250 kD, and two fainter intermediate bands. Precipitates have not been run under non-reducing conditions. Moreover, the antibody does not detect the proteins in Western blots. Thus, it is unclear whether the polypeptides recognised by IL-A29 are associated on the cell surface or represent a family of molecules expressed independently on the cell surface. There does not at present appear to be an equivalent surface antigen among the CD series of human differentiation antigens (Shaw, 1987). However, a differentiation antigen of similar cellular distribution and molecular weight has been described in sheep (Mackay et al., 1986). The positive cells in sheep were also found to be negative for immunoglobulin and for the CD4 and CD8 T cells markers. However, they expressed a CD5-like T cell antigen, suggesting that the cells are related to the T cell lineage.

As an initial step towards examining the function of the IL-A29$^+$ cells, the appearance of IL-A29$^+$ lymphoblasts in cultures of bovine PBM stimulated with the mitogens ConA and PWM, in autologous and allogeneic MLCs and in autologous Theileria MLCs, has been investigated. Lymphoblasts were identified, using the cell sorter, as cells displaying larger forward angle light scattering profiles than resting blood lymphocytes. A significant number of IL-A29$^+$ blast cells (greater than 5%) was observed in both allogeneic and autologous MLCs, in autologous Theileria MLCs and in ConA-stimulated cultures, but not in cultures stimulated with PWM. The number of IL-A29$^+$ lymphoblasts was particularly high in the autologous MLC, in some instances representing 50% or more of the cultured cells (Fig. 3). Staining of these cells by two colour immunofluorescence again showed that they were negative for the available B and T cell markers.

From these data, it would appear that, with the exception of PWM, stimuli which induce proliferation of bovine T lymphocytes also result in proliferation of the IL-A29$^+$ cells. However, among a large number antigen-specific clones of bovine lymphocytes that have been derived from allogeneic MLCs (Teale et al., 1986) or autologous Theileria MLCs (Goddeeris et al., 1988c; Baldwin et al., 1987; Morrison et al., 1987) all express T cell markers, and a small number of these which have been tested with IL-A29 are negative for the IL-A29 determinant. These observations, together with the finding that the numbers of IL-A29$^+$ lymphoblasts were particularly high in autologous MLCs, suggest that, rather than having specificity for foreign antigens, this population of cells may react to signals provided by other populations of autologous leukocytes and/or growth factors. Further conclusions regarding the function of these cells await studies with purified cell populations.

REFERENCES

Baldwin, C.L., Goddeeris, B.M. and Morrison, W.I., 1987, Bovine helper T-cell clones specific for lymphocytes infected with Theileria parva (Muguga), Parasite Immunol., 9 : 499

Baldwin, C.L., MacHugh, N.D., Ellis, J.A., Naessens, J., Newson, J. and Morrison, W.I., Monoclonal antibodies which react with bovine T-lymphocyte antigens and induce blastogenesis : Tissue distribution and functional characteristics of the target antigens, Immunology, in press.

Baldwin, C.L., Teale, A.J., Naessens, J.G., Goddeeris, B.M., MacHugh, N.D. and Morrison, W.I., 1986, Characterization of a subset of bovine T lymphocytes that express BoT4 by monoclonal antibodies and function : Similarity to lymphocytes defined by human T4 and murine L3T4, J. Immunol., 136 : 4385.

Davis, W.C., Ellis, J.A., MacHugh, N.D. and Baldwin, C.L., Bovine pan T-cell monoclonal antibodies reactive with a molecule similar to CD2, Immunology, in press.

Ellis, J.A., Baldwin, C.L., MacHugh, N.D., Bensaid, A., Teale, A.J., Goddeeris, B.M. and Morrison, W.I., 1986, Characterization by a monoclonal antibody and functional analysis of a subset of bovine T lymphocytes that express BoT8 a molecule analogous to human CD8, Immunology, 58 : 351

Goddeeris, B.M., Baldwin, C.L., ole MoiYoi, O. and Morrison W.I., 1986a, Improved methods for purification and depletion of monocytes from bovine peripheral blood mononuclear cells. Functional evaluation of monocytes in responses to lectins J. Immunol. Methods, 89 : 165.

Goddeeris, B.M., Morrison, W.I., Naessens, J. and Magondu, J., The bovine autologous mixed leukocyte reaction : A proliferative response of non-T cells under the control of monocytes, Immunobiology, in press.

Goddeeris, B.M., Morrison, W.I. and Teale, A.J., 1986b, Generation of cytotoxic cell lines specific for cells infected with the protozoan parasite Theileria parva and restricted by products of the major histocompatibility complex, Eur. J. Immunol., 16 : 1243.

Goddeeris, B.M., Morrison, W.I., Teale, A.J., Bensaid, A. and Baldwin, C.L., 1986c, Bovine cytotoxic T-cell clones specific for cells infected with the protozoan parasite Theileria parva : Parasite strain specificity and class I major histocompatibility complex restriction, Proc. Natl. Acad. Sci. USA, 83 : 5238

Lalor, P.A., Morrison, W.I., Goddeeris, B.M., Jack, R.J. and Black, S.J., 1986, Monoclonal antibodies identify phenotypically and functionally distinct cell types in the bovine lymphoid system, Vet, Immunol. Immunopathol., 13 : 121.

Mackay, C.R., Maddox, J.F. and Brandon, M.R., 1986, Three distinct subpopulations of sheep T lymphocytes, Eur. J. Immunol., 16 : 19.

Morrison, W.I., Baldwin, C.L., MacHugh, N.D., Teale, A.J., Goddeeris, B.M. and Ellis, J., 1987, Phenotypic and functional characterisation of bovine lymphocytes, in, "Moving Frontiers in Veterinary Immunology", R. Pandey, ed., Progress in Veterinary Microbiology and Immunology, Vol.4, S.Karger, Basel, p.134.

Naessens, J., Newson, J., Bensaid, A., Teale, A.J., Magondu, J. and Black S.J., 1985, De novo expression of T cell markers on *Theileria parva*-transformed lymphoblasts in cattle, *J. Immunol.*, 135 : 4183.

Shaw, S., 1987, Characterization of human leukocyte differentiation antigens, *Immunol. Today*, 8 : 1.

Teale, A.J., Baldwin, C.L., Ellis, J.A., Newson, J., Goddeeris, B.M. and Morrison, W.I., 1986, Alloreactive bovine T lymphocyte clones: An analysis of function, phenotype and specificity, *J. Immunol.*, 136: 4392.

SECRETORY IMMUNITY

MECKEL'S DIVERTICLE: A GUT ASSOCIATED LYMPHOID ORGAN IN CHICKENS

Suzan H.M. Jeurissen, E. Marga Janse
and Guus Koch

Central Veterinary Institute, Dept. Virology

P.O. Box 365, 8200 AJ Lelystad, The Netherlands

INTRODUCTION

Halfway along the jejunum, closer to the ileum than to the duodenum, a blind ending tube, the Meckel's diverticle, can be seen in more than 60% of all adult chickens (1). This diverticulum caecum vitelli is the remnant of the omphalomesenteric or vitelline duct that forms the intermediate between the yolk sac and the primitive gut of the embryos. In the first days after hatching, the Meckel's diverticle is unusually large and filled with yolk, which still functions as food supply for the young chicken (2).

Although this diverticle was described by Meckel long ago, little attention has been paid since then to this organ. Our attention was drawn by the Meckel's diverticle during our studies of lymphoid organs and tissues associated with the mucosae. This report deals with the development, structure, and function of the Meckel's diverticle, using routine and immunohistochemical methods.

MATERIALS AND METHODS

Animals

White leghorn strain A (WLA) chickens, 1 and 5 days old as well as 2, 6, 12, and 15 weeks old, were bred and kept under routine s.p.f. conditions. Chicken embryos were used at 15, 19, 20, and 21 days of incubation. After exsanguination, part of the jejunum with the Meckel's diverticle was removed and frozen in liquid nitrogen or treated for routine histology.

Accessibility of the Meckel's diverticle to gut antigens

In order to investigate whether antigens from the gut lumen can enter the Meckel's diverticle, a large quantity of India ink was injected into the jejunum of exsanguinated animals. The Meckel's diverticles were macroscopically observed for 30 min; then the diverticles along with the adjacent part of the jejunum were removed and treated for routine histology.

Routine histology

Tissues were fixed in sublimate formaline for at least 2 hours at RT. After dehydration in acetone, tissues were embedded in Paramat (Gurr, BDH Chemicals Ltd, Poole, U.K.) with a congealing point of 56-57°C. Sections of 8 um thickness were stained with Azan, PAS, andf methyl-green pyronin.

Immuno- and enzymehistochemistry

Cryostat sections of 8 μm thickness were picked up on slides, air-dried, and fixed in pure acetone for 10 min. Sections were then incubated for 60 min with one of the following monoclonal antibodies: HISC 1 (anti-B lymphocytes; Ekino et al., submitted) HISC 7 (anti-leucocytes; Jeurissen et al., submitted), HISC 12 (anti-IgM; Ekino et al., submitted), CVI-Chγ-47.3 (anti-IgG; 3), CVI-Chα-46.5 (anti-IgA; 3), CVI-ChNL-68.1 (anti-monocytes/macrohages; Jeurissen and Janse, submitted). After rinsing in phosphate-buffered saline (PBS, 0.01 M, pH 7.4), slides were covered with rabbit anti-mouse Ig peroxidase (Dakopatts, Glostrup, Denmark) diluted in PBS containing 0.2% bovine serum albumin for 60 min. After rinsing, the slides were treated for peroxidase activity woth 0.5 mg 3,3'-diaminobenzidine-tetrahydrochloride (DAB; Sigma Chemicals Co. St. Louis, M.O.) per ml Tris-HCl buffer (0.05 M, pH 7.6) containing 0.01% H_2O_2. Control slides were incubated as described above, omitting only the first step.

RESULTS

Macroscopical observations

Chicken embryos incubated for 15 and 20 days appeared to internalize the yolk. At that time, the yolk can be observed in the abdomen as well as on the skin. After careful removal of the skin, a preliminary Meckel's diverticle could be seen as a 10 to 15 mm long stalk with a diameter of 1-2 mm, connecting the yolk sac and the gut (Fig. 1a).

One day after hatching, the yolk was completely internalized in to the abdomen, leaving a superficial navel-like mark on the skin. The Meckel's diverticle was still the same size as those of 20 day-old embryos.

In contrast, the Meckel's diverticles of 5- and 14-day-old chickens contained only small amounts of yolk (Fig. 3a) and had begun to acquire and adult appearance. In adult birds, the Meckel's diverticle was 4 to 8 mm long with a diameter of 2 to 5 mm and, though attached to the gut, was clearly recognizable as a separate organ (Fig. 4a).

Accessibility to gut antigens

In order to gain insight into the accessibility of Meckel's diverticles to antigens present in the gut lumen, a large amount of India ink was injected into the jejunum. Macroscopical observations revealed that India ink streamed freely through the embryonic Meckel's diverticle and mixed with the yellow yolk (Fig. 1a). In contrast to chick embryos, Meckel's diverticles of 1-day-old and adult chicken seemed impermeable to India ink from the gut lumen (Fig. 4a). After 30 min, Meckel's diverticles were removed to be examined by light microscopy.

The macroscopical observations were only partially confirmed on microscopical level. India ink was present in the lumen of the jejunum of all chickens examined. The ink was usually attached to the mucus of the villi. In the Meckel's diverticle of a 15-day-old embryo, small amounts of ink could be detected (Fig. 1b). Unlike macroscopical observations, microscopical examinations revlealed that Meckel's diverticles of newly hatched chickens contained only insignificant amounts of ink, when compared to the large amount of ink injected. Meckel's diverticles of adult chickens were completely devoid of ink.

Routine histology

Routine staining of Meckel's diverticles showed a resemblance to other gut-associated lymphoid tissues, such as Peyer's patches or coecal tonsils. The inside of the Meckel's diverticle was lined by epithelium, identical to intestinal epithelium, though goblet cells were less frequently present, and by basal membrane. Serosa and muscle layers were observed on the outside. Connective tissue with an age-dependent number of leucocytes was situated between epithelium and muscle layers. In 15- and 20-day-old embryonic Meckel's diverticles, small infiltrates had developed in the connective tissue beneath the epithelium (Fig. 1b). In older chicken embryos, the flat inner surface of the diverticle had transformed into smaller villi-like folds.

In Meckel's diverticles of 5-day-old chickens, a ring of connective tissue developed around the connection of the diverticle and the gut; in only 2 of the 3 birds could an opening from the gut to the diverticle be detected. The diverticles consisted of an increased number of larger infiltrates situated in connective tissue.

In chickens 2 and 6 weeks old, the resemblance between Meckel's diverticles and coecal tonsiles or Peyer's patches had increased, since the individual infiltrates had connected to form one large mass of lymphoid cells. In contrast to coecal tonsils, however, germinal centers were not detected in Meckel's diverticles at this stage. Examination of random

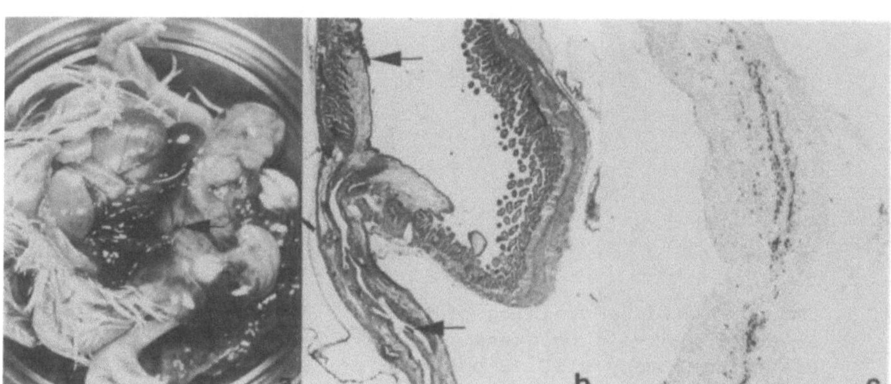

Fig. 1. Meckel's diverticles of chicken embryos. a) After intra-intestinal injection, India ink was detected in the yolk sac (arrow). b) On microscopical level, small amounts of India ink were seen in the jejunum as well as in the Meckel's diverticle (arrows). c) Immunoperoxidase staining with HISC 7 showed small infiltrates of leucocytes in the connective tissue beneath the epithelium.

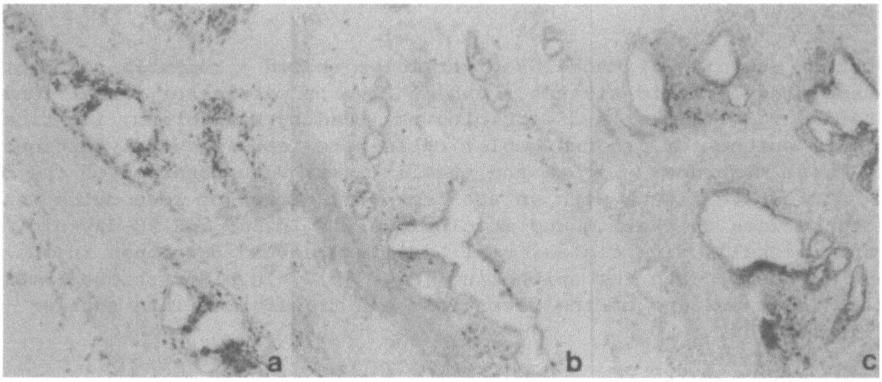

Fig. 2. Meckel's diverticles of 5-day-old chickens. Immunoperoxidase staining with a) HISC 7, b) HISC 1, and c) HISC 12, showed that the infiltrates of leucocytes (a), present in the epithelium and the underlying connective tissue, contained B lymphocytes (b) and IgM-positive cells (c).

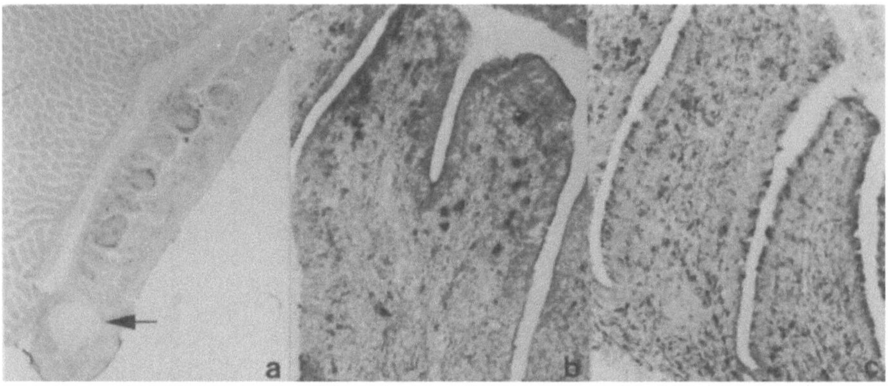

Fig. 3. Immunoperoxidase staining of Meckel's diverticles of 2- and 6-week-old chickens. a) In a 2-week-old Meckel's diverticle, many HISC 12-positive cells were present. The remainder of the yolk was seen at the distal end (arrow). b) CVI-Igα-46.5 stained IgA-positive cells with cytoplasmic and membrane-bound molecules. c) Using CVI-ChNL-68.1, non-lymphoid cells were detected throughout the organ.

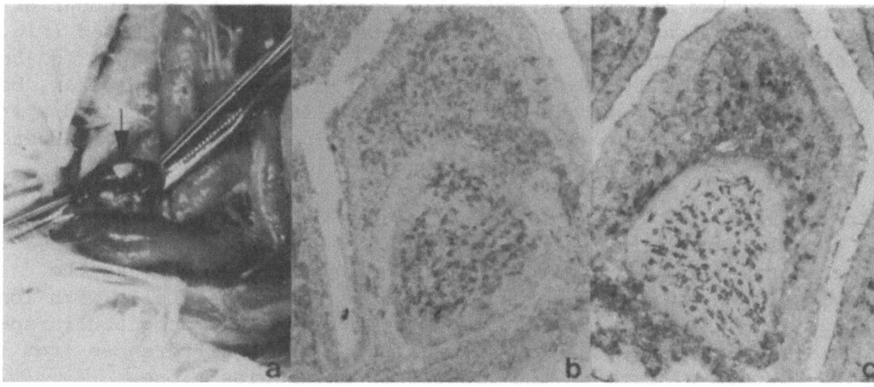

Fig. 4. Meckel's diverticles of 12-week-old chickens. A) After intra-intestinal injection, India ink could not be observed in the diverticle (arrow). b and c) Immunoperoxidase staining with HISC 12 (b) and CVI-Igα-46.5 (c) revealed that dendritic cells bearing IgM (b) and/or IgA (c) were present in germinal centers.

sections of Meckel's diverticles showed no opening from the diverticle to the gut; the connections consisted of a thick layer of connective tissue and muscularis externa. Serial sections revealed, however, that a small opening with a diameter up to 100 μm was present.

Germinal centers in Meckel's diverticles were detected only in 12- and 15-week-old birds. An opening from the diverticle to the gut was sporadically detected.

Immuno- and enzymhistochemistry

Monoclonal antibodies against chicken leucocytes (HISC 7), B lymphocytes (HISC 1), IgM (HISC 12), IgC (CVI- Chγ-47.3), IgA (CVI-Chα-46.5), and monocytes/macrophages (CVI-ChNL-68.1) were used to characterize the cells in Meckel's diverticles of different ages.

In embryonic Meckel's diverticles, small cell infiltrates were recognizable by HISC 7 (Fig. 1c). Within these infiltrates HISC 1-positive and CVI-ChNL-68.1-positive cells were seen sporadically, whereas no cells were positive for immunoglobulin markers.

In 5-day-old chickens, HISC 7 stained not only all of the infiltrates but also single cells in the epithelium and connective tissue (Fig. 2a). In contrast, HISC 1 stained single cells in the infiltrates and the epithelium (Fig. 2b). Staining with an IgM-specific monoclonal antibody showed a cell distribution of membrane-positive cells identical to HISC1, though with fewer cells (Fig. 2c). In contrast, CVI-ChIgγ47.3 and CVI-ChNL-46.5 recognize no cells.

In chickens aged 2 and 6 weeks, Meckel's diverticles had increased in size, but cell distribution was similar to that in 5-day-old chickens. The only difference observed was in the number of immunoglobulin-containing cells. The number of IgM-bearing cells seemed equal to the number of HISC 1-positive cells, whereas few IgM plasma cells could be detected (Fig. 3a). In chickens 2 weeks old or more, Meckel's diverticles contained IgG-bearing cells and, sporadically, IgG plasma cells, In 6-week-old chickens, both IgA plasma cells and membrane-positive Iga cells were detected, dispersed over the diverticle (Fig. 3b). CVI-ChNL-68.1-positive cells were detected as single cells dispersed over the lymphoid tissue (Fig. 3c).

In chickels aged 12 and 15 weeks, both B cell and non-B cell areas could be distinguished in Meckel's diverticles. The newly developed germinal centers stained positive with HISC 7 and HISC 1, whereas the immunoglobulin-specific monoclonals selectively stained cells with long processes (Fig. 4b and c). Staining with all three immunoglobulin-specific monoclonals on the same section showed little or no differences from the staining pattern of each of the monoclonal antibodies alone.

DISCUSSION

In this study, we investigated the development, histology, and function of a gut-associated lymphoid organ, the so-called Meckel's diverticle. During embryogenesis, the ductus vitellinus, which forms the connection between the yolk and the embryonic gut, becomes surrounded by leucocyte infiltrates. After birth, these infiltrates increase in size until they resemble other mucosal lymphoid organs, such as Peyer's patches and coecal tonsils.

Experiments concerning the uptake of antigens from the gut lumen and the histology of newly hatched chickens demonstrated clearly that the connection between the diverticle and the gut lumen acquires a different appearance after hatching. In 1-day-old chickens, Meckel's diverticles appear to be less permeable by India ink injected into the lumen of the jejunum than was expected, since histological observations had demonstrated the presence of an opening; the explanation may lie in the continuation of the mucus layer, which is present in the intestinal villi, over the opening. In 2-week-old chickens the interface between gut and diverticle is partly fortified with connective tissue and jejunum-derived muscularis externa. In adult chickens, Meckel's diverticles were impermeable by India ink though serial sections revealed that a small opening did exist. If India ink is representative of gut antigens, such antigens essentially cannot enter the lumen of Meckel's diverticles, except possibly in tracer amounts.

The results show that Meckel's diverticles continue to mature immunologically and develop germinal centers at older ages. Studies in rodents have demonstrated that the induction of germinal centers is an antigen-dependent process (4,5). This discrepancy may be explained by the assumption that yolk-derived antigens induce lymphoid infiltration. The furhter development of these infiltrates may be an endogenous process or be dependent upon very small amounts of intestinal antigens. These findings may explain the late appearance of germinal centers compared ot other mucosa-associated lymphoid organs that encounter antigens in large amounts, e.g. coecal tonsils that contain germinal centers from 2 weeks onward.

The histology of Meckel's diverticles is characterized by an abundance of B cells. These B cells are present throughout the organ but are generally underneath the epithelium. In adult chickens, distinct areas can be detected that are devoid of B cells and these areas stain with a rabbit anti-T cell serum (unpublished results). It is remarkable that these T-cell areas are always situated adjacent to a germinal center.

After staining with antibodies against chicken immunoglobulins, germinal centers exhibit strongly positive cells with a dendritic appearance. Combined staining using all three monoclonals showed no marked difference in number of cells or staining itensity when compared with staining of a single monoclonal antibody. This finding may indicate that these dendritic cells from germinal centers bear immunoglobulins of different classes on their surfaces and represent the follicular dendritic cells.

In conclusion, this study indicates that Meckel's diverticles serve an immunological function. The production of secretory immunoglobulins to protect the gut mucosa against invading micro-organisms is unlikely to be the main function of Meckel's diverticles, since plasma cells that are present in Meckel's diverticles, unlike those in the gut lamina propria, are not situated directly beneath the epithelium. Possibly, Meckel's diverticles, functionally analogous to Peyer's patches (6), play a role in memory-formation during immune responses against intestinal antigens. The function of Meckel's diverticles can be further studied by _in ovo_ immunizations and long-term feeding experiments.

REFERENCES

1. V. Cittero, A. Istvanffy, and B. Trosarelli, _Ref. Anat. Ber._ 30 (1935).
2. W. Kruger, _Berl. tierarztl. Wschr._ (1931).
3. G. Koch and I. C. M. van Jongelen. This volume.
4. I. L. Weissman, G. A. Gutman, S. H. Friedberg, and L. Jerabek, _Adv. Exp. Med. Biol._ 66:229 (1976).
5. P. Nieuwenhuis, and K. Lennert, _in_: "Malignant lymphoproliferative diseases", J. G. van den Tweel, C. R. Taylor, and F. T. Bosman, eds., Martinus Nijhof Publishers, The Hague (1980).
6. S. H. M. Jeurissen, G. Kraal, and T. Sminia, _in_: " Recent Developments in Mucosal Immunology", in press.

SUBPOPULATIONS OF NON-LYMPHOID CELLS IN BRONCHUS ASSOCIATED LYMPHOID TISSUE AND LUNG OF THE MOUSE

M. Breel, M.B. van der Ende, T. Sminia and G. Kraal

Department of Histology, Medical Faculty, Free University, P.O. Box 7161, 1007 MC Amsterdam, The Netherlands

Bronchus associated lymphoid tissue has been observed in lungs of a number of mammalian species and of chickens (1). It is characterized by a dense accumulation of lymphoid cells which are in contact with the bronchus epithelium. It occurs mainly along the main bronchi and is situated between a bronchus and an artery. The lymphocytes are located in discrete T and B cell areas and may migrate into the BALT by high endothelial venules (HEV), which are clearly present in the T cell areas (2-6). These data together indicate that the general composition of BALT resembles that of a predetermined lymphoid organ and that BALT is not a random accumulation of lymphocytes. The precise role BALT plays in the immunological defense mechanism of the lung, however, has yet to be determined. Studies on BALT have mainly been performed in larger rodents like rats and guinea pigs where BALT is relatively well developed. In mouse only circumstantial evidence for the presence of BALT has been given (7) and no systematic study on the morphology and function has been performed. This may be due to the fact that in mice BALT is rather inconspicious, probably also based on strain differences. Considering the existence of many markers for lymphoid and non-lymphoid cells in the mouse and the feasability of many experimental approaches in this species we studied murine BALT using immunohistochemistry with monoclonal antibodies against lymphocytes and non-lymphoid cells, in combination with enzyme histochemistry. Special attention was paid to the presence of macrophage subpopulations in BALT and surrounding lung tissue.

MATERIALS AND METHODS

Animals and antibodies

C57Bl/6, $(C_3D_2)F_1$ and BALB/c mice were obtained from Bomholtgard, Denmark. The following rat anti mouse monoclonal antibodies were used as culture supernatant: anti Thy-1, Lyt-2 (8); L3T4 (9) and anti kappa (clone 187/1) directed against lymphoid subpopulations. F4/80 (10); Mac-1 (11); Moma-2 (12) directed against macrophages. NLDC-145 (13), and MIDC-8 (14) directed against interdigitating cells. MOMA-1 (15) and ERTR9 (16) directed against subpopulations of macrophages. MECA-325 (17) directed against high endothelial venules. Ia.m5 (anti $Ia^{b,d,k}$; 18), an anti-class II monoclonal antibody was used as biotin conjugate in combination with peroxidase conjugated avidin (19). As control antibody 2H10 was used, a rat anti-adenovirus antibody with no known reactivity against mouse tissues (20).

Tissue sections and cell preparations of alveolar macrophages

Mice were killed by carbon dioxide inhalation. Pieces of lung comprising BALT were frozen in liquid nitrogen. Alveolar macrophages were retrieved from mice which were anaesthetized with 0.5 ml Ketalar (Parke-Davis, Canada Inc., Scarborough, Ontario) injected intraperitoneally. The trachea was cannulated with a nylon tube, and the lungs were lavaged six times with 0.3 ml PBS containing 0.6 mM EDTA. The lavage fluid was pooled, washed and cytocentrifuge preparations were made of the retrieved cells.

Immuno- and enzyme-histochemistry

Cryostat sections (8-12μm) were airdried and fixed in acetone for 10 min. Cytospin preparations were airdried and fixed in a mixture of aceton and formol (40:1). Preparations were incubated for 60 minutes with culture supernatant or biotinylated monoclonal antibody in proper dilution in PBS with 0.2% BSA (PBS/BSA). After washing in 0.01 M phosphate-buffered saline (PBS) slides were covered with a 1:200 dilution of rabbit anti-rat Ig peroxidase (DAKO,Copenhagen,Denmark) in PBS/BSA and 1% normal mouse serum for 30 min or with avidine-peroxidase conjugate (in a proper dilution). After thoroughly washing in PBS, the preparations were stained for peroxidase activity with 3,3'-diaminobenzidine-tetrahydrochloride (DAB,Sigma,St Louis,MO) in 0.5 mg/ml Tris-HCl buffer pH 7.6 containing 0.01% H_2O_2. The staining was intensified by incubating the preparations for 15 minutes in 0.5% $CuSO_4.5H_2O$ in 0.9% NaCl after washing the preparations with 0.9% NaCl. Control slides were incubated in the same way, omitting the first step or with an irrelevant monoclonal antibody 2H10. Non-specific esterase (NSE) activity was demonstrated with α-naphthyl acetate as a substrate. Acid phosphatase (AP) activity was demonstrated using naphtol-AS-BI-phosphate as substrate (21) with hexazotized pararosaniline as diazonium salt (22). Enzyme substrate was obtained from Sigma. When immunohistochemistry and enzyme-histochemistry were combined on the same section, immuno-histochemistry preceded the enzyme histochemistry.

RESULTS

As in other mammalian species in the lung of the mouse Bronchus Associated Lymphoid Tissue (BALT) is located between an artery and a bronchus as an accumulation of lymphoid cells (Fig. 1). BALT was detected in different strains of mice: Balb/c, (C3D2)F1 and C57Bl/6. From the latter the results are shown in this paper.

Distribution of lymphoid cells in BALT and lung

Immunohistochemical staining was performed with T and B cell markers on cryostat sections of non-stimulated lungs. In BALT, distinct T and B cell areas were detected, although the localization of T cell areas with regard to the bronchus was not fixed from one mouse to an other. Investigation of T lymphocyte subpopulations showed the presence of predominantly the T helper phenotype L3T4 (Fig. 1). In the lung many Thy-1+ cells could be detected outside the blood vessels in the interstitium. In contrast only a very few B lymphocytes were found.

Distribution of non-lymphoid cells in BALT

To investigate the macrophage and dendritic cell populations several monoclonal antibodies were used in combination with enzymehistochemistry.

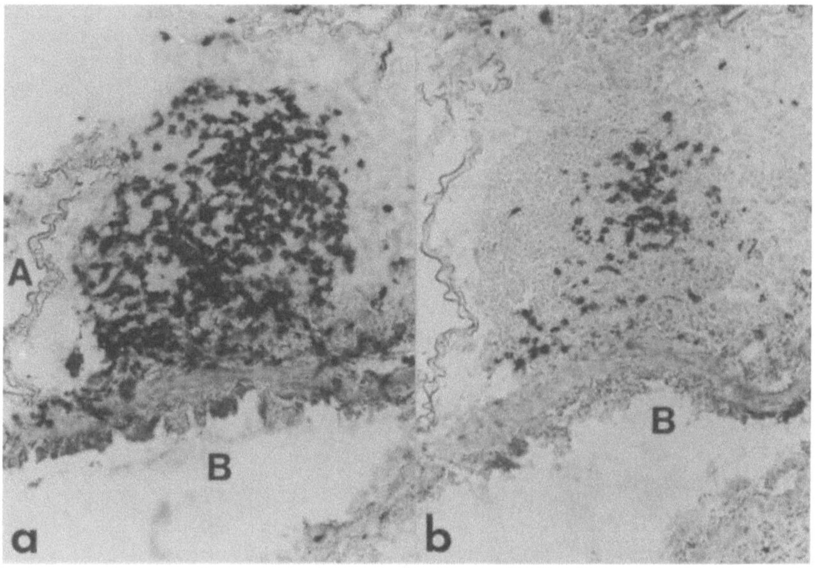

Fig. 1. T cell subpopulations in BALT

BALT is positioned between a bronchus (B) and an artery (A). The majority of T cells belongs to the L3T4+ phenotype. 1a: L3T4+ T cell (T helper) subpopulation (20x objective); 1b: Lyt2+ T cell (T suppressor/cytotoxic) subpopulation (20x objective).

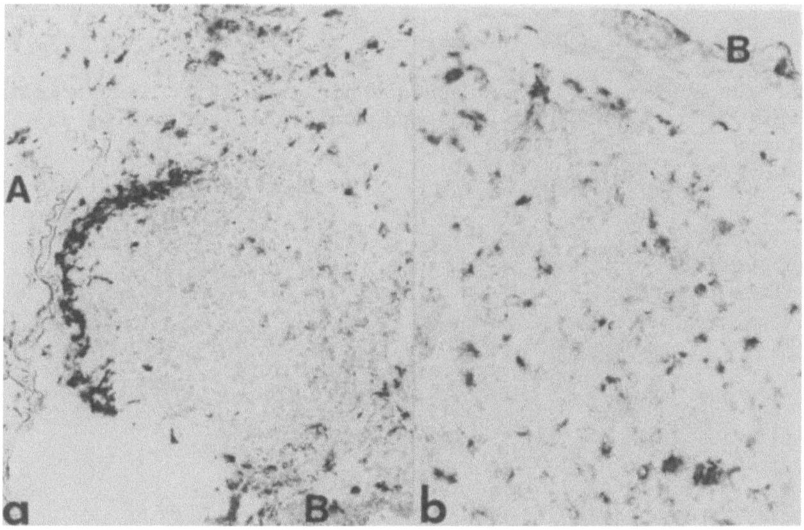

Fig. 2. Heterogeneity of macrophages in the BALT

2a: Macrophages at the rim of the BALT are recognized by MOMA-1 (40x objective); 2b: Macrophages dispersed in BALT are recognized by MOMA-2 (40x objective). A = artery; B = bronchus.

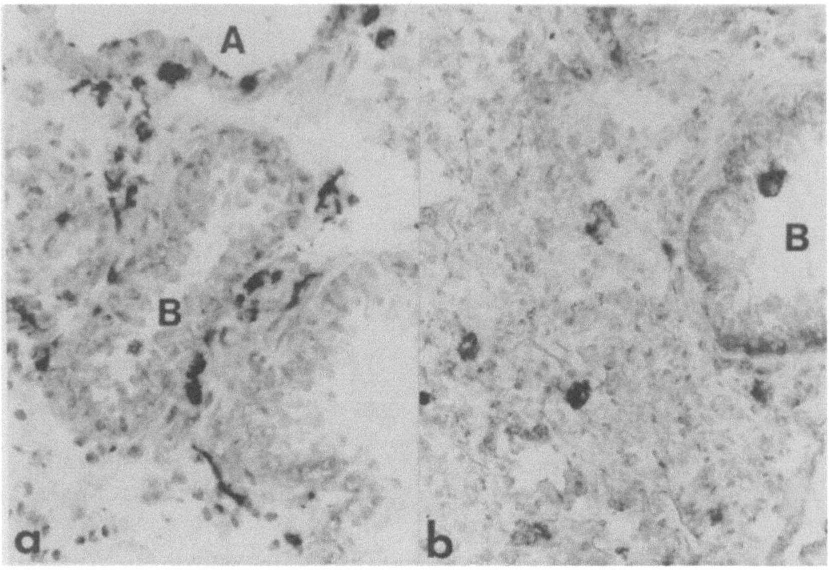

Fig. 3. NLDC-145 staining of BALT and interstitium of the lung

3a: The majority of macrophages in the interstitial tissue of the lung are stained with MOMA-1 (40x objective); **3b:** In the alveoli as well as in the interstitium several macrophages are detected by the NLDC-145 monoclonal antibody (40x objective). A = artery; B = bronchus.

Table 1

Staining of subpopulations of macrophages and dendritic (interdigitating) cells in various parts of the lung using monoclonal antibodies and enzyme histochemistry

	BALT	Lung tissue	Macrophages in alveoli
- Acid phosphatase	++	+	+
- Non specific Esterase	-	+	+
- NLDC 145	+	+	+
- MIDC-8	+	-	-
- Ia.m5	+	+	-
- F4/80	-	-	-
- Mac-1	-	+	-
- MOMA-1	+	+	+
- MOMA-2	+	+	+
- ERTR9	+	+	-

Strong acid phosphatase activity was found in cells dispersed through BALT. In contrast no non-specific esterase positive cells were demonstrated. Based on localization, morphology and immunohistochemistry, at least two populations could be discriminated. One population, which cells were situated dispersed through the BALT, showed a more rounded morphology and were often found to contain cellular remnants. This population comprised tingible body macrophages and was only recognized by the MOMA-2 monoclonal antibody (Fig. 1). The other macrophage subpopulations, located at the border of BALT, had the characteristic appearance of tissue macrophages and reacted with the MOMA-1 macrophage marker and with the ERTR9 antibody. Staining with F4/80 and Mac-1 revealed very little activity in the BALT; only an occasional cell at the border of the BALT was stained with Mac-1.

Next to the macrophage subpopulations in BALT, interdigitating cells (IDC) could be found in the T cell areas as detected by immunohistochemical staining with anti class II antibody as well as with two monoclonal antibodies which reacted with IDC: NLDC-145 and MIDC-8. In T cell areas also high endothelial venules were determined by the MECA-325 antibody.

Non-lymphoid cells in lung sections and lung lavages

In the lung different populations of macrophages could be distinguished by location, morphology, and immunohistochemistry.

Typical tissue macrophages could be found in the connective tissue around blood vessels and bronchi. These cells with slender protrusions showed high activity for acid phosphatase and non specific esterase as determined by enzyme-histochemistry. They stained positive with the macrophage markers Mac-1, MOMA-1 (Fig.3A) and ERTR9, but failed to react with the marker F4/80. Remarkable was the finding that a small proportion of interstitial cells also stained with the interdigitating cell-specific antibody NLDC-145 (Fig 3B), but not with MIDC-8.

Pulmonary alveolar macrophages, discriminated by their position inside the bronchioli and their strong activity for acid phosphatase and non-specific esterase, could not be recognized by Mac-1 and ERTR9. In contrast to this was the finding that alveolar macrophages reacted strongly with MOMA-1, MOMA-2 and also, as in the case of the interstitial macrophages, strongly with NLDC-145 (Fig 3B).

The findings on macrophages in tissue sections were confirmed using bronchal lavages in which the majority of cells were NLDC-145 positive. The alveolar macrophages did not stain with MIDC-8.

DISCUSSION

In the lung of the mouse the position of the BALT and the appearance of discrete B and T cell areas in BALT correspond with the results obtained in rats and guinea pigs (2, 3, 5). The localization of T and B cell areas seems not to be fixed in relation with the bronchus. The same phenomenon was also observd in rats but to a much lesser extent in guinea pigs (6). This is in contrast to the determined T and B cell areas in other lymphoid organs in relation to the site from which antigens are transported.

A marked heterogeneity in macrophage subpopulation was determined in BALT and lung as detected by acid phosphatase and non-specific esterase staining. This heterogeneity was further confirmed with monoclonal antibodies against macrophage subpopulations (Table 1).

The F4/80, and Mac-1 monoclonal antibodies did not recognize any macrophages in the BALT, which is not surprising because, also in T and B cell areas of other lymphoid organs virtually no reaction of these antibodies with macrophages has been demonstrated (11, 23, 24).

The MOMA-1 and ERTR9 monoclonal antibody showed differential staining in the lung. In spleen they also recognize different subpopulations, although in lymph node medulla they detect identical populations. In BALT both monoclonal antibodies recognize macrophage populations, situated at the border of the BALT. In the interstitium of the lung many MOMA-1 positive macrophages and only a few ERTR9 positive macrophages are detected, whereas the ERTR9 determinant is absent from alveolar macrophages. In this respect MOMA-1 staining resembles the staining patterns found in the rat lung with the monoclonal anti-macrophage antibody ED2 (5). The presence of MOMA-1 positive cells under the epithelium of the bronchus as found in this study and a comparable position of MOMA-1 macrophages under the epithelium of Peyer's patches as found previously (15) is suggestive for a role of these cells in antigen handling.

Concerning the NLDC-145 antibody, in BALT interdigitating cells were detected in T cell areas as was established by staining with anti-class II and MIDC-8 antibodies. Remarkably is that part of the interstitial tissue macropages and the majority of the alveolar macrophages are recognized by the NLDC-145 antibody. Alveolar macrophages are derived from (NLDC-145 negative) monocytes, which have been reported to reside in the interstitial tissue for two days at most before migrating into the alveoli (25). This suggests an inductive role of the lung environment for the expression of the NLDC-145 determinant. It would be interesting to compare this situation in the lung with that in the skin where monocytes are suggested to differentiate in NLDC-145 positive Langerhans cells. The role of Langerhans cells in antigen handling has been reported extensively (26, 27) and under certain conditions they can become potent antigen presenting cells (28). Alveolar macrophages have a variety of functions (29) and the presence of the NLDC-145 marker is suggestive for a role as antigen presenting cell, whereby in analogy to the Langerhans cell, they may be involved in the transport of attached or processed antigen into draining lymph nodes and BALT. Studies on the antigen presenting capacity of isolated alveolar macrophages are, however, controversial. Suppressive as well as stimulatory activities of alveolar macrophages have been found employing in vitro models of both polyclonal and antigen-specific T-cell activation (30-33). This may be caused by the heterogeneity in the population of alveolar macrophages as demonstrated in functional assays (29, 34, 35) and in this paper. Another possibility is the necessity for a differentiation step, as found in Langerhans cells, which they may undergo during their travelling into the lymph nodes. The large population of NLDC-145 positive cells in the lung interstitial tissue may therefore represent 'veiled cells'.

The complex structure of the mouse BALT, in which the various cell types and structures necessary for normal immunological functioning are present, certainly suggests a direct role of these structures in the physiology of the lung. The presice relationship of BALT and the macrophages in alveoli and interstitial tissues on the one hand and the draining lymph nodes of the lungs on the other remains an area of further investigation.

ACKNOWLEDGEMENTS

We are grateful to Karin Hoeben-Schornagel for her practical assistance and Martine Roest for her professional secretarial assistance.

REFERENCES

1. J. Bienenstock, N. Johnston, and D.Y.E. Perey. Lab. Invest. 28:686 (1973a).

2. B.E.C. Plesch, in: 'In vivo immunology', p. 491, Plenum Publishing Corp. (1982).
3. G.J. van der Brugge-Gamelkoorn, M.B. van der Ende, and T. Sminia. Cell Tissue Res. 239:177 (1985a).
4. G.J. van der Brugge-Gamelkoorn, C.D. Dijkstra, and T. Sminia. Immunobiol. 169:553 (1985b).
5. G.J. van der Brugge-Gamelkoorn, and T. Sminia, in: 'Microenvironments in the lymphoid system', p. 323, G.G.B. Klaus, ed., Plenum Publishing Corp. (1985c).
6. G.J. van der Brugge-Gamelkoorn, and G. Kraal. J. Immunol. 134:3746 (1985d).
7. P.A. Crabbe, D.R. Nash, H. Bazin, H. Eyssen, and J.F. Heremans. Lab. Invest. 22:448 (1970).
8. J.A. Ledbetter, and L.A. Herzenberg. Immunol. Rev. 47:63 (1979).
9. A. Pierres, P. Nagnet, A. van Agthoven, F. Bekkhoncha, F. Demiot, Z. Mishal, A.M. Smit-Verhulst, and M. Pierres. J. Immunol. 132:2775 (1984).
10. D.A. Hume, A.P. Robinson, G.G. MacPherson, and S. Gordon. J. Exp. Med. 158:1522 (1983).
11. T. Springer, D. Galfre, S. Scher, and C. Milstein. Eur. J. Immunol. 9:301 (1979).
12. G. Kraal, M. Rep, and M. Janse. Scand. J. Immunol. (1987) in press.
13. G. Kraal, M. Breel, M. Janse, and G. Bruin. J. Exp. Med. 163:981 (1986).
14. M. Breel, R. Mebius, and G. Kraal. Eur. J. Immunol. (1987) in press.
15. G. Kraal, and M. Janse. Immunology 58:665 (1986).
16. C.D. Dijkstra, E. van Vliet, E.A. Döpp, A.A. van der Lely, and G. Kraal. Immunology 55:23 (1985).
17. A.M. Duijvestijn, M. Kerkhove, R. Bargatze, and E.C. Butcher. J. Immunol. 138:137 (1987).
18. H. Lemke, G.J. Hämmerling, and U. Hämmerling. Immunol. Rev. 47:175 (1979).
19. D.M. Boorsma, J. van Bommel, and J. van den Heuvel. Histochemistry 83:333 (1986).
20. A. Zantema, J.A.M. Fransen, A. Davis-Olivier, F.C.S. Ramaekers, G.P. Vooys, B. Deleys, and A.J. van der Eb. Virology 142:44 (1985).
21. A.G.E. Pearse. Histochemistry, Theoretical and Applied, 3rd ed., vol. 1, Churchill Livingstone (1968).
22. B.J. Davis, and L. Ornstein. J. Histochem. Cytochem. 7:297 (1959).
23. S.H. Lee, P.M. Starkey, and S. Gordon. J. Exp. Med. 161:475 (1985).
24. M.D. Witmer, and R.M. Steinman. Am. J. Anat. 170:465 (1984).
25. D.H. Bowden, and I.Y.R. Adamson. Lab. Invest. 42:511 (1980).
26. G. Stingl, L.A. Gazze-Stingl, W. Aberer, and K. Wolff. Immunology 127:1707 (1981).
27. G. Stingl, L.A. Stingl-Gazze, and K. Wolff. J. Invet. Dermatology 79:129 (1982).
28. G. Schuler, and R.M. Steinman. J. Exp. Med. 161:526 (1985).
29. D.B. Chandler, W.C. Fuller, R.M. Jackson, and J.D. Fulmer. J. Leukocyte Biol. 39:371 (1986).
30. S.W. Ullrich, and H.N. Herscowitz. J. Reticuloendothel. Soc. 28:111 (1980).
31. P.G. Holt, L.A. Warner, and G. Mayrhofer. Cell. Immunol. 63:57 (1981).
32. C.R. Lyons, J.W. Uhr, and G.B. Toews. J. Immunol. 126:286 (1981).
33. M.R. Schuyler, and L.W. Todd. Am. Rev. Respir. Dis. 123:53 (1981).
34. J. Shellito, and H.B. Kaltreider. Am. Rev. Respir. Dis. 131:678 (1985).
35. D. Sandron, H.Y. Reynolds, A. Venet, A. Laval, D. Israel-Biet, and J. Chretien. Eur. J. Respir. Dis. 69:226 (1986).

2. R.H.C. [illegible], in: 'In vivo Immunology', p. 131. Plenum [illegible] (198[illegible]).
3. [illegible] van der Bruggen-[illegible], M.H. van der [illegible] and [illegible] Cell [illegible] (198[illegible]).
4. [illegible] van der Bruggen-Gameelkoorn, C.D. Dijkstra, and [illegible] (198[illegible]).
5. [illegible] van der Bruggen-Gameelkoorn, and [illegible] Sminia, in: [illegible] 'p. 113, [illegible] Plenum Publ. [illegible] (198[illegible]).
6. [illegible] van der Bruggen-Gameelkoorn, and [illegible] (198[illegible]).
7. [illegible]
[illegible]

MIGRATION AND LODGING OF PEYER'S PATCH IgA PRECURSOR CELLS IN THE GUT LAMINA PROPRIA

Jeenan Tseng

Department of Experimental Pathology, Division of Pathology, Walter Reed Army Institute of Research, Washington, D.C. 20307-5100

INTRODUCTION

A striking feature of the intestine is the presence of a huge number of IgA plasma cells in the lamina propria. The source of these IgA plasma cells is the Peyer's patches (PP), in which the precursor cells are committed by the antigens from the lumen (1, 2). The committed precursor cells then migrate to mesenteric lymph nodes (MLN), enter the blood circulation via the thoracic duct, and eventually arrive in the gut lamina propria (GLP) where they differentiate into resident plasma cells. During the migration, the precursor cells differentiate and expand (2). Two types of IgA precursor cells are present in Peyer's patches: the IgA bearing cells in the germinal center and the IgM-IgD double bearing cells outside the germinal center (3). The former are sessile cells reactive with peanut agglutinin (PNA) but not MEL-14 monoclonal antibody while the latter are migratory or recirculating cells bearing MEL-14 but not PNA reactive determinants; they are designated as MEL-14(-) PNA(+) and MEL-14 (+) PNA(-) cells, respectively (4). The IgA precursor cells differentiating into IgA containing blasts in the MLN and thoracic duct migrate directly to and lodge permanently in the GLP (5, 6). The small IgA precursor cells migrating indirectly to the GLP expand and differentiate enroute (2). How the differentiation accompanying with migratory-sessile changes occurs in the journey is still unknown. The present study was undertaken to investigate the possible relationship between the reactivities of MEL-14 and PNA and the differentiation-migration of IgA precursor cells. Passive transfers using Ig allotype congeneic mice were conducted; MEL-14 and PNA reactivity were examined for the lymphoid cells. The results suggest that when IgA precursor cells differentiate into blasts and plasma cells they lose MEL-14 and PNA reactivity. Lodging of IgA cells in the gut may require factor other than MEL-14 and PNA reactive determinants.

RESULTS AND DISCUSSION

1. *Most IgA precursor cells divide and differentiate into IgA containing cells in the GLP*

To study if IgA precursor cells divide and differentiate into IgA containing (cIgA) cells in the GLP, passive transfers of CB-20 PP cells to BALB/c recipients were conducted. At days 7-14, the recipients were sacrificed; gut sections were doubly stained with anti-CB-20 IgA allotype (anti-Ig-2b) and anti-α. As shown in Table 1, CB-20 IgA containing cells started to repopulate BALB/c IgA cells at day 7 and completely repopulated them at days 12-14. When

TABLE 1. Donor IgA Containing Cells in the Gut Lamina Propria of the Recipients

Day	% repropulation	% dividing Ig-2b cells in total IgA cells
7	5	3
8	16	7
9	34	15
10	44	0
12	98	0
14	99	0

Day 0: CB-20 PP cells to BALB/c (400 R). Days 5-6: ^{3}H-thymidine, ip. into BALB/c recipien
Days 7-14: Double immunofluorescence and radioautography.

radioautography was performed, dividing CB-20 IgA cells were seen. In addition, clusters of 2 CB-20 cIgA cells were seen at day 10 and essentially no dividing CB-20 cIgA cells were seen during days 7-14 in the spleen which is the intermediate tissue for the prescursor cells (2, 7). Thus, cIgA cells can indeed divide when they arrive at the GLP.

2. *Repopulating donor cIgA cells do not express MEL-14 and PNA reactive determinants*

Previous reports indicated that IgA precursor cells in PP are mainly small resting lymphocytes bearing both IgM and IgD or only IgA on the membrane (3). While most of the IgM-IgD bearing cells are outside the germinal center and are MEL-14(+) PNA(-) cells, the IgA bearing cells are in the germinal center and are MEL-14(-)PNA(+) cells (4). Attempts were made to se if a correlation exists between the migration-differentiation of cIgA cells, and the expression of MEL-14 and PNA reactive determinants. The results would shed some insight into the mechanism by which cIgA cells migrate to and lodge in the GLP. Thus passive transfers of CB-20 PP cells to BALB/c were performed and, at days 6-20, lymphoid cells from the spleen and the GLP were isolated and doubly stained with MEL-14 (or PNA) and anti-α (or anti-CB-20 IgA) fluorescent reagents. The results are shown in Table 2. Although a large portion of spleen cells from the BALB/c recipients and control mice (HBSS injected) were MEL-14(+) PNA(-) cells,

TABLE 2. MEL-14(+) IgA Containing Cells in the Spleen and Gut Lamina Propria of BALB/c Mice Transferred with CB- 20 PP Cells

	BALB/c	Spleen			Gut		
Day	Recipient	M-14	M-14,cIgA	cIgA	M-14	M-14,cIgA	cIgA
6	PP cells	65	0	4	0	0	2
	HBSS	90	0	9	0	0	2
8	PP cells	65	0	1	0	0	1.5
	HBSS	65	0	1.5	0.2	0	0.2
11	PP cells	35	0	4	1.2	0	2.5
	HBSS	40	0	0.2	0.2	0	0.2
15	PP cells	60	0	1.5	4	0	7
	HBSS	56	0	4	5	0	6
20	PP cells	62	0	0.1	42	3.1	12
	HBSS	52	0	0.2	40	2.5	10

Day 0: CB-20 PP cells to BALB/c (400R). Days 6-20: membrane and cytoplasmic staining.
HBSS = Hanks balanced salt solution injected.
M-14 =membrane MEL-14(+) cells; cIgA = cytoplasmic IgA(+) cells.
M-14,cIgA = cIgA cells with membrane MEL-14 positivity.

essentially no cIgA cells bore either MEL-14 or PNA reactive determinants. In the gut, very few MEL-14(+) small lymphocytes were seen; blasts were MEL-14(-)PNA(-) cells. Some cIgA cells at day 20 were MEL-14(+) PNA(-) cells ; however, they were of host origin. Thus, it appears that most donor cells migrating to the GLP lose their MEL-14 and PNA reactivity. These results, coupled with the results mentioned above, suggest that differentiation of PP IgA precursor cells while enroute to the GLP may result in a loss of MEL-14 and PNA reactivity, and that lodging in the gut may require factors other than MEL-14 and PNA reactive determinants.

3. *Most cIgA cells in the GLP of normal mice are MEL-4(-) cells*

To study the possibility that PP IgA precursor cells lose MEL-14 and PNA reactive determinants while differentiating and dividing enroute to the GLP, cells from PP, MLN and the GLP of normal mice were isolated and examined for MEL-14 and PNA reactivity. The results are shown in Table 3. The majority (57-58%) of lymphoid cells in PP and MLN are MEL-14(+) PNA(-) small lymphocytes; however, these cells are present in the GLP in substantially smaller numbers (14%). Essentially no cIgA cells in the PP and MLN react with MEL-14 or PNA.

TABLE 3. MEL-14(+) and IgA Containing cells in the Peyer's Patches, Mesenteric Lymph Nodes, and Gut Lamina Propria of CB-20 Mice.

	% positive cells		
Cells with	PP	MLN	GLP
mMEL-14	57(55-60)	58(57-60)	14(12-16)
cIgA	0.3(02-0.5)	1.2(0.6-1.5)	21(18-25)
mMEL-14, cIgA/cIgA	0	0	21(16-25)

PP = Peyer's patches; MLN = mesenteric lymph nodes; GLP =gut lamina propria. m = membrane; c = cytoplasmic. Numbers are averages; numbers in parenthesis are ranges. mMEL-14, cIgA/cIgA = % MEL-14(+) cIgA cells relative to total cIgA cells.

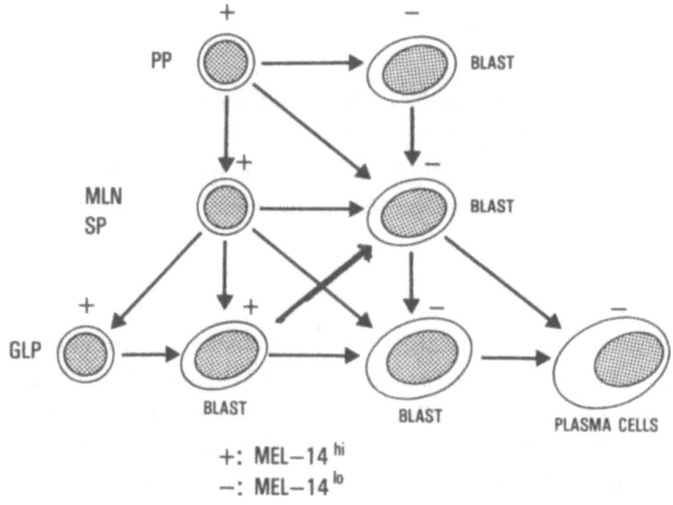

Figure 1. Loss of MEL-14 reactive determinants on the surface of PP IgA precursor cells while dividing and differentiating enroute to the gut lamina propria.

In contrast, approximately 20% of cIgA cells in the GLP react only with MEL-14. Obviously, these cIgA cells are derived from small resting MEL-14(+) PNA(-) IgA precursor cells in PP and MLN which migrate to the GLP, differentiate into cIgA cells and lodge there. This is supported by the facts that cIgA cells in PP and MLN are MEL-14(-) cells (Table 3) and are capable of migrating directly to and lodging in the GLP (5, 6). Thus, MEL-14 and PNA reactive determinants appear not to be the factors required in the migration and lodging of IgA cells in the GLP.

CONCLUSIONS

Most of the PP IgA precursor cells divide and differentiate into plasma cells in the gut lamina propria, although some of them may divide and differentiate before arriving there. When the IgA precursor cells differentiate and divide, they lose MEL-14 and PNA reactive determinants. A factor(s) other than MEL-14 and PNA reactive molecules may be required for the lodging of IgA cells in the gut lamina propria. The loss of MEL-14 reactive determinants on the IgA cells in the migration can be depicted as the following (Figure 1):

REFERENCES

1. S. W. Craig and J. J. Cebra. J. Exp. Med. 134: 188 (1971).
2. J. Tseng. J. Immunol. 127: 2039 (1981).
3. J. Tseng. J. Immunol. 132: 2730 (1984).
4. E. C. Butcher, R. J. Rouse, R. L. Coffmans, N. C. Nottenburg, R. Hardy, and I. L. Weissman. J. Immunol. 129: 2698 (1982).
5. D. C. Guy-Grand, C. Griscelli, and P. Vassali. Eur. J. Immunol. 4: 435 (1974).
6. M. McWilliams, J. Phillips-Quagliata, and M.E. Lamm. J. Immunol. 115: 54 (1975).
7. J. J. Cebra, P. J. Gearhart, R. Kamat, S. M. Robertson, and J. Tseng. Cold Spring Harbor Symp. Quant. Biol. 41: 201 (1976).

LOCALIZATION, ISOLATION AND HISTOCHEMICAL CHARACTERIZATION OF MACROPHAGES IN THE GUT AND MESENTERIC LYMPH NODES OF THE RAT

Taede Sminia and Marja B. van der Ende

Department of Histology, Medical Faculty, Free University, P.O. Box 7161, 1007 MC Amsterdam, The Netherlands

Cells of the mononuclear phagocytic system occur widespread in the body (1). Among mononuclear phagocytes there consists a marked heterogeneity in morphology, enzyme contents, cell surface markers and function. This heterogeneity is related to the developmental stage of the individual cells, the state of activity of the cells and the microenvironment of the diverse tissues and organs (2, 3). Also in lymphoid organs a heterogeneous population of mononuclear phagocytes occurs (4, 5, 6). From numerous studies on these cells in bone marrow, blood and lymphoid organs it has become clear that in addition to the 'classical' macrophages - cells with the typical features of scavenger cells - a second population of cells (dendritic or interdigitating cells) exists (7). This latter population is highly efficient in presenting antigen to T cells but show a low phagocytic activity (8). Besides the large population of these mobile mononuclear phagocytes, which are present in all compartments of lymphoid organs, the germinal centre of lymphoid follicles is characterized by a population of 'fixed' non-lymphoid cells, the follicular dendritic cells (6). These cells bind and retain antigen-antibody complexes on their extensive cell processes (9).

Since 1980 we study the localization, structure and function of non-lymphoid cells in mucosa-associated lymphoid tissue (GALT and BALT; 10-13). We have obtained indications for the presence of an antigen-presenting cell system in Peyer's patches and BALT, which shows similarities with the antigen-presenting cell system in the skin (11, 14). Furthermore, we found that the non-lymphoid cells in mucosa-associated lymphoid tissues differs in several aspects from those in the spleen and peripheral lymph nodes (15). In the present study, which is an extension of the previous ones, we focus on the macrophages in the gut and mesenteric lymph nodes. The macrophage population has characterized both in situ and in cell suspensions using ED-monoclonal antibodies (16), anti-Ia antibodies, enzyme-cytochemical tests and lectins.

MATERIALS AND METHODS

Immunocytochemistry

Cryostat sections of mesenteric and facial lymph nodes (MLN and FLN), and of the small intestine including Peyer's patches of adult rats were air-dried for 1 h and fixed for 10 min in pure acetone. Macrophages were detected in these sections using monoclonal antibodies and lectins. For staining with monoclonal antibodies the two-step immuno-peroxidase technique was used. The lectins, which were all directly conjugated with peroxidase, were stained with a one-step peroxidase technique. For the combined staining with monoclonal antibodies and lectins the following procedure was used:

1. Incubation with monoclonal antibody for 1 h. Carefully rinsing in phosphate-buffered saline (PBS). Incubation with rabbit-anti-mouse IgG peroxidase (30 min). Staining with 3-amino-9-ethylcarbazole (17).
2. Incubation with lectin (45 min), rinsing in PBS. Staining with 4-chloro-1-naphthol (18).
3. Counterstaining with haematoxylin, rinsing in tap water and mounting in aquamount.

The following monoclonal antibodies against macrophage subpopulations were used: ED1, ED2 and ED3 (16). The used lectins were Griffonia simplicifolia 1, GS1, and Dolichos biflorus, DbA (Sigma, USA). Some sections were also stained for acid phosphatase activity to get an overall view of the macrophage and monocyte population.

Suspension of lymph nodes

Suspensions of lymph nodes were made by digesting the MLN and FLN for 2 h in Earle's Balanced Medium (EBM), which contained 1 mg collagenase/ml (I and III , Cooper Biomed. USA) and 0.1 mg DNAse/ml (BDH Chem., England), at 37°C on a magnetic stirrer. The now obtained suspension was poured over nylon gauze and then washed throughly with EBM, containing 0.1% BSA. Cytocentrifuge preparations were made, dried, fixed in 60 ml acetone with 1.5 ml formaldehyde during 10 sec and stored in PBS. After fixation staining was carried out as described for cryostat sections.

Suspension of the lamina propria of the gut

Animals were kept without food, during 15 to 20 h, to get the gut as clean as possible. The next morning the empty gut was collected, and the last remaines of food were removed. After removal of the Peyer's patches the gut was cut in pieces of approximately 0.5 mm. These pieces were incubated in Ca-Mg free PBS with EDTA (0.2 g/100 ml) during 4 h at 37°C on a magnetic stirrer. The solution of PBS with EDTA was four times refreshed, two times after half an hour and two after one hour. Before removing the 'old' solution the pieces were shaken using a Vortex, this enabled us to remove the epithelium together with the 'old' solution. Then the pieces of gut were incubated for 1 h in EBM, which contained 1 mg collagenase III/ml and 0.1 mg DNAse/ml, at 37°C on a magnetic stirrer. The remaining pieces were minced through nylon gauze. After that the cell suspension was washed thoroughly with EBM, which contained 0.1% BSA. Cytocentrifuge preparations were fixed and stained using the same procedure as used for the cytocentrifuge preparations of lymph nodes.

RESULTS AND DISCUSSION

Peyer's patches (PP). Previous observations on non-lymphoid cells were confirmed and extended (cf. 12, 13). Macrophages were present in all compartments of PP (T cell-, B cell- and subepithelial area). The macrophages, except the tingible body macrophages (TBM) in the germinal centre, stained with ED1 and ED2. Only a few ED3 positive macrophages which are characteristic of lymphoid organs (16) were found. The majority of the macrophages in the B cell area and in the subepithelial region displayed a high acid phosphatase activity, in contrast to those in the T cell (interfollicular) area. Ia-staining was mainly found on macrophages in the subepithelial region and on interdigitating cells (IDC) in the T cell area. With respect to the lectins used we found that DbA reacted strongly with TBM (Fig. 1), whereas GS1 stained almost all macrophages including IDC weakly.

The results obtained on cell suspensions were comparable to those made on tissue sections. It was, however, remarkable that IDC in the cell suspensions often formed rosettes with T cells, whereas the majority of the classical macrophages do not show such a reaction. The function of rosette formation might be facilitation of antigen presentation by physical contact.

Villi. The lamina propria of the villi comprises numerous macrophages which constitute morphologically and phenotypically a heterogeneous population of cells. The villi macrophages are ED1 and ED2 positive, but do not express ED3 antigens; they show acid phosphatase activity and are Ia-positive (Fig. 2; cf. 15).

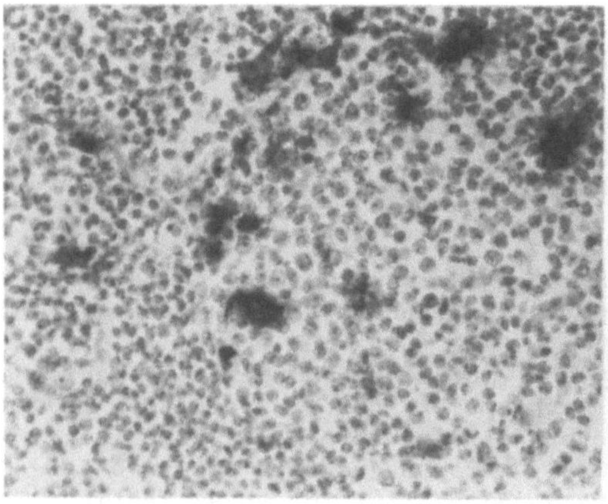

Fig. 1. DbA reaction in GC. Note the positive TBM (dark cells).

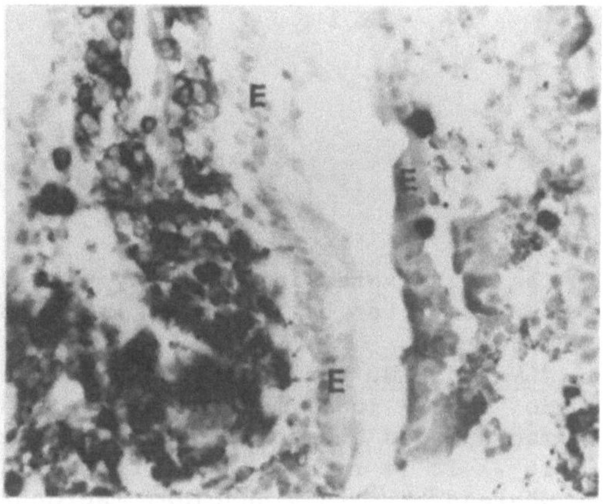

Fig. 2. Ia staining on villi showing numerous positive macrophages. E, epithelium.

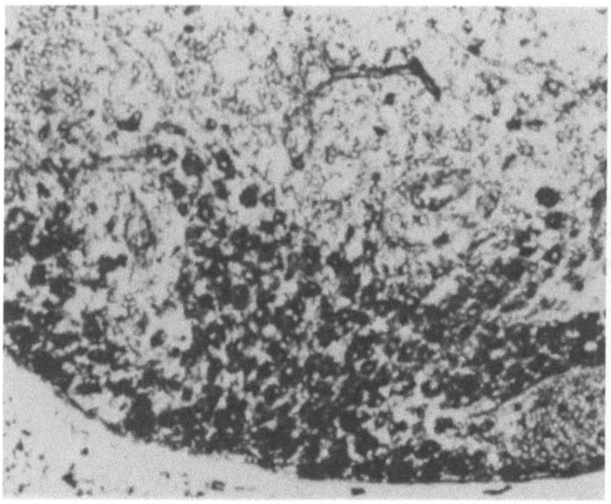

Fig. 3. GS1 staining of ED3 positive medullary macrophages in MLN.

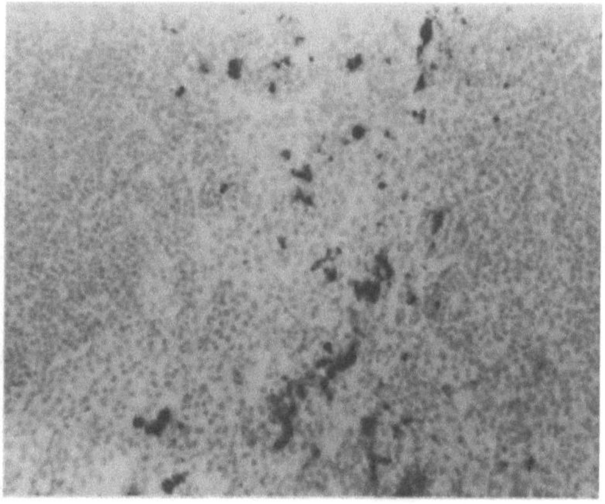

Fig. 4. ED2 positive macrophages in medulla of FLN.

Quantitative observations on cell suspensions of the villi showed that about 5.5% of the total cell population of the lamina propria consisted of ED1 and ED2 positive macrophages (Fig. 5). All macrophages were Ia positive (Table 1); they were relatively large with a round to oval nucleus.

Table 1 Percentages of macrophage populations in the villi of the gut

	villi
ED1	3.5%
ED2	3.4%
Ia	5.5%

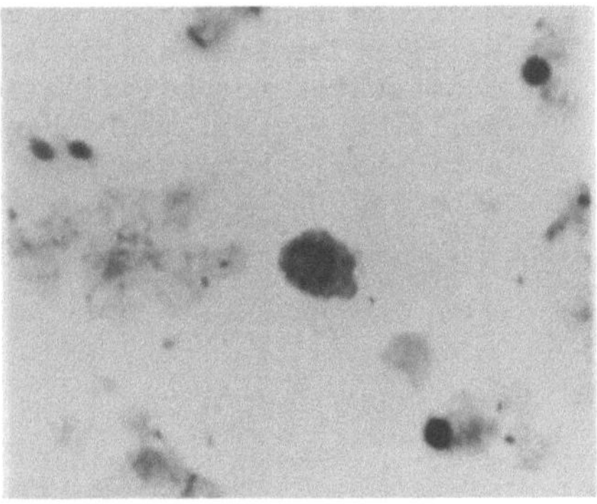

Fig. 5. ED2 positive macrophage in cell suspension of villi

The high expression of Ia by villi macrophages is surprising as the villi are supposed to be sites of immune suppression rather than sites of immune induction.

MLN and FLN. As described previously ED1-, ED2- and ED3 positive macrophages do occur in lymph nodes (Fig. 6); the major portion of the medullary macrophages and the subepithelial macrophages are strongly ED3 positive (cf. 16). It is remarkable that the medullary macrophages were positive for GS1, whereas the subsinusoidal macrophages were negative for this lectin. These data indicate that the ED3 positive subpopulation of lymph node macrophages is not only morphologically heterogeneous (16), but also phenotypically. Whether this implies that these two subpopulations of ED3 positive macrophages display other functions is a matter for further research. Observations on tissue sections showed that the total number of ED1 positive cells in FLN is higher than in MLN. This could be confirmed by quantitative observations on cell suspensions.

Table 2 Percentages of macrophage populations in Facial (FLN) and Mesenteric lymph node (MLN)

	FLN	MLN
ED1	6.3%	3.3%
ED2	1.5%	0.3%
ED3	3.0%	2.5%

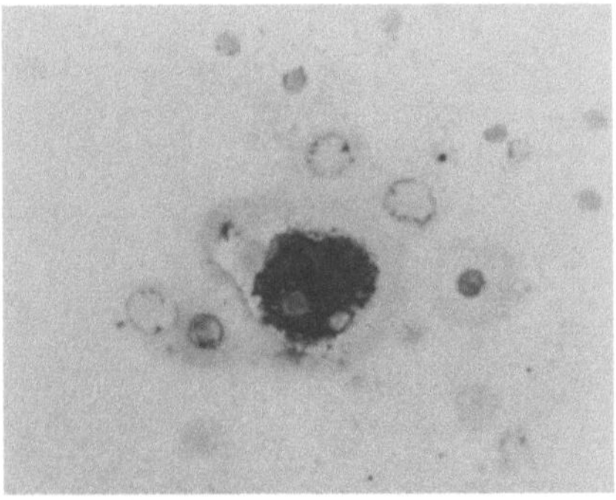

Fig. 6. Cluster of 2-3 ED1 positive macrophages in cell suspension of MLN.

Another difference between these two types of lymph nodes is the percentage of ED2 positive cells, which is relatively high in the FLN and low in the MLN. Comparable observations have previously been done on ED2 positive cells in popliteal lymph nodes (PLN), paratracheal lymph nodes (PTLN) and MLN; the percentage of ED2 positive cells in PLN was about twice as high as that in MLN and PTLN, which are both mucosa-associated lymph nodes (19). These data point to functional differences between mucosa-associated and peripheral lymph nodes. With respect to the results of the lectin reactions it was found that TBM in MLN strongly stained with DbA, in contrast to the TBM in FLN. As was already mentioned TBM in PP follicles do also stain with DbA. Using this lectin it might be possible to isolate these cells and to characterize them functionally. No differences were found between both types of lymph nodes after staining with GS1.

CONCLUSIONS

- Peyer's patches contain strongly Ia-positive interdigitating cells and diverse types of scavenger macrophages, part of them are also Ia positive. The lamina propria of the villi comprises a heterogeneous population of macrophages, which display a high acid phosphatase activity and express Ia (observations on tissue sections).
- All macrophages in cell suspensions of the villi are Ia-positive, which is surprising since the villi are supposed to be sites of immune suppression rather than sites of immune induction.
- The number of ED2-positive macrophages in the mesenteric lymph node is very low as compared to other (non-mucosal) lymph nodes, e.g. facial lymph nodes or popliteal lymph nodes. The functional significance of this difference is still unknown.
- Lectins are a usefull tool to further characterize macrophages (of GALT and BALT), especially when used in combination with other markers for macrophages.

ACKNOWLEDGEMENTS

The authors wish to thank Martine Roest for secretarial assistance.

REFERENCES

1. R. van Furth, in: Mononuclear phagocytes: functional aspects, p. 1, R. van Furth, ed., Martinus Nijhoff, The Hague, The Netherlands (1980).
2. Z.A. Cohn. J. Immunol. 121:813 (1978).
3. P.S. Morahan. J. Reticuloendothel. Soc. 27:223 (1980).
4. E.C.M. Hoefsmit, E.W.A. Kamperdijk, and B.M. Balfour, in: Mononuclear phagocytes: functional aspects, p. 1809, R. van Furth, ed., Martinus Nijhoff, The Hague, The Netherlands (1980).
5. E.C.M. Hoefsmit, A.M. Duijvestijn, and E.W.A. Kamperdijk. Immunobiol. 161:225 (1982).
6. S. Fossum, and W.L. Ford. Histopathology 9:469 (1985).
7. L.W. Poulter, D.A. Campbell, C. Munro, and G. Janossy. Scand. J. Immunol. 24:351 (1986).
8. R.M. Steinmann, and M.C. Nussenzweig. Immunol. Rev. 53:127 (1980).
9. G.J.V. Nossal, A. Abbot, J. Mitchell, and Z. Lummus. J. Exp. Med. 127:277 (1968).
10. B.E.C. Plesch, in: In vivo immunology, p. 491, P. Nieuwenhuis, A.A. van den Broek, and M.G. Hanna, eds., Plenum Publ. Corp., New York/London (1982).

11. G.J. van der Brugge-Gamelkoorn, M.B. van der Ende, and T. Sminia. Cell Tissue Res. 239:177 (1985).
12. T. Sminia, M.M. Wilders, E.M. Janse, and E.C.M. Hoefsmit. Immunobiol. 164:136 (1983).
13. T. Sminia, and S.H.M. Jeurissen. Immunobiol. 172:72 (1986).
14. M.M. Wilders, H.A. Drexhage, T. Sminia, H. Mullink, E.F. Weltevreden, B.E.C. Plesch, and S.G.M. Meuwissen. Acta Chirurgica Scand. Suppl. 525:93 (1985).
15. S.H.M. Jeurissen, and C.D. Dijkstra. Eur. J. Immunol. 16:562 (1986).
16. C.D. Dijkstra, E.A. Döpp, P. Joling, and G. Kraal. Immunology 54:589 (1985).
17. R.C. Graham, M.C. Karnovsky, and U. Lundholm. J. Histochem. Cytochem. 13:150 (1965).
18. P.N. Nakane. J. Histochem. Cytochem. 16:557 (1968).
19. T. Sminia, M.B. van de Ende, and C.D. Dijkstra. Eur. J. Cell Biol. Vol. 39, Suppl. 11:35 (1985).

DIRECT TRANSDIAPHRAGMATIC TRAFFIC OF PERITONEAL MACROPHAGES TO THE LUNG

M. L. M. Pitt and A. O. Anderson

U. S. Army Med. Res. Inst. of Inf. Dis., Fort Detrick
Frederick, MD 21701-5011

INTRODUCTION

Intraperitoneal (i.p.) inoculation is a parenteral route of immunization which results in priming of the mucosal immune response. We hypothesized that this may be related to the processing of antigen either in the lymphoid tissue associated with the peritoneal cavity or in the sites that drain it. Particulates injected into the peritoneal cavity are removed principally by way of transdiaphragmatic lymphatic channels situated under the diaphragmatic mesothelium[1,2]. These channels drain into larger intrathoracic lymphatics which in turn enter anterior mammary and parathymic lymph nodes before emptying into the bloodstream.

We have identified a novel pathway for the clearance of particulates from the peritoneal cavity which could be a significant source of lung interstitial macrophages as well as a pathway for immune priming of the lung and mucosa-associated lymphoid tissue (MALT). Some of the injected colloidal carbon is phagocytosed by peritoneal macrophages. These macrophages migrate across the diaphragm through the pleural space entering the lung at specific sites along the visceral pleural surface and then penetrate into the interstitium through short channels that serve as potential lymphatics.

MATERIALS AND METHODS

Adult female Swiss Webster mice were given a single i.p. injection of 0.1 ml of a 1:2 dilution of Pelikan special black ink in phosphate-buffered saline (PBS). This dose was empirically selected to prevent rapid overflow into the blood. Four animals per group were killed 5 minutes, 10 minutes, 1, 2, 3, 24 and 72 hours post i.p. injection. Three additional groups of

---

In conducting the research described in this report, the investigator(s) adhered to the "Guide for the Care and Use of Laboratory Animals," as promulgated by the Committee on Care and Use of Laboratory Animals of the Institute of Laboratory Animals Resources, National Research Council. The facilities are fully accredited by the American Association for Accreditation of Laboratory Animal Care.

The views of the author(s) do not purport to reflect the positions of the Department of the Army or the Department of Defense.

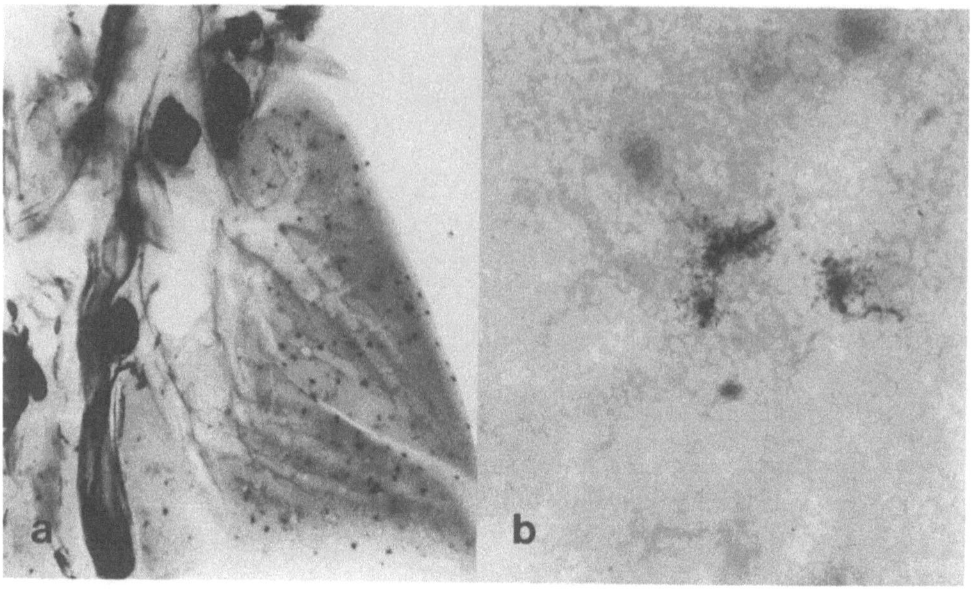

Figure 1. (a) Slice of a whole lung 18 hrs post i.p. inoculation of carbon showing accumulation in mediastinal and parathymic lymph nodes and in the short lymphatic channels leading from the visceral pleura. (b) Macroscopic view of the lymphatic channels showing dissemination into the alveolar septae.

mice were given an equivalent amount of carbon either intravenously, intranasally or intrapleurally and killed 30 minutes and 24 hours post injection.

Blood was collected in EDTA and the leucocytes isolated. Cells present in both the pleural cavity and lungs were harvested by washing with Ca^{++}- and Mg^{++}-free Jokliks medium. The resulting cell suspensions were centrifuged at 200 x g for 10 minutes at room temperature and resuspended in RPMI 1640 plus 10% foetal calf serum. The number of macrophages/monocytes present in the suspensions of blood leucocytes, pleural cavity cells and bronchial lavage was estimated using a haemocytometer. Cytocentrifuge preparations were made and the percentage of macrophages/monocytes containing carbon was determined using light microscopy.

Following bronchial lavage the lungs were inflated with either 10% formalin or 2% glutaraldehyde in PBS and fixed overnight at $4^{o}C$. Formalin-fixed lungs were processed for histology. Glutaraldehyde-fixed lungs were dehydrated in ethanol and cleared in methyl salicylate to study the carbon deposition in situ. The cleared lungs were cut into thick slices and embedded in plastic. The plastic blocks were oriented along the planes of carbon-laden channels and 1 μm sections of these channels were cut, mounted and studied by light microscopy. Also, the pathways of carbon movement across fixed and cleared diaphragms were studied. The diaphragms of untreated mice were examined by transmission electron microscopy for identification of the cell types present in diaphragmatic lymphatics.

RESULTS

Translocation of intraperitoneal colloidal carbon: Carbon was observed in the diaphragmatic, mediastinal and internal mammary lymphatics

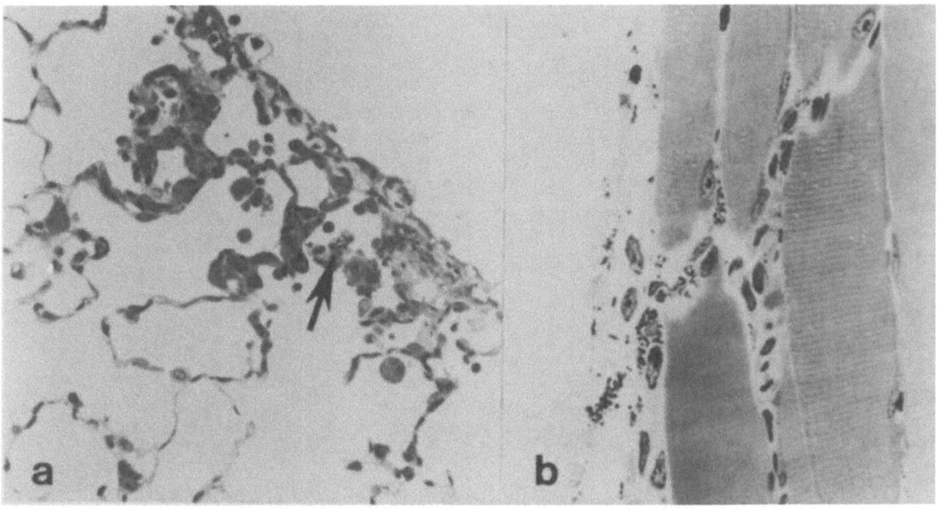

Figure 2. 1 μm sections through (a) a lymphatic channel in the lung showing carbon in the macrophages of the potential lymphatic. A labelled macrophage (⟵) can be observed in an alveolus, (b) a diaphragm showing macrophages containing carbon entering a channel on the peritoneal cavity side and moving through a channel. These were obtained from a mouse 18 hrs post i.p. inoculation.

and had reached the mediastinal and parathymic lymph nodes within 5 minutes of i.p. introduction. Eighteen hours post i.p. inoculation the carbon was also present in discrete areas of the lung pleura (Figure 1a & 1b).

Light microscopic examination of 1 μm sections of the cleared lungs showed that the carbon was in fact present in macrophages which were delineating unique lymphatic channels leading from the visceral surface of the lung into interalveolar septae deep within the lobules (Figure 2a). Examination of cleared diaphragms showed macrophages containing carbon making their way in channels across the diaphragm (Figure 2b). Lymphocytes, small monocytes, macrophages and Langerhans-like cells were present in diaphragmatic lymphatics of both normal mice and mice inoculated i.p. with carbon (Figure 3).

Carbon introduced intravenously did not result in carbon-laden macrophages being present in discrete areas associated with the visceral pleura as with i.p. introduced carbon although free carbon was spread diffusely in the perivascular spaces throughout the lung. Carbon given intranasally resulted in macrophages containing carbon in the alveolar spaces while carbon injected into the pleural space resulted in carbon being situated in discrete areas similar to i.p. inoculation.

Transport of colloidal carbon in macrophages from the peritoneal cavity to the lung: Macrophages containing carbon were first observed in the pleural cavity 5 minutes post i.p. inoculation. The number of carbon-laden pleural macrophages increased with time to reach a maximum at 24 hours post injection (Figure 4a) while there was no significant change in the total number of pleural macrophages obtained throughout the study period (Figure 4b).

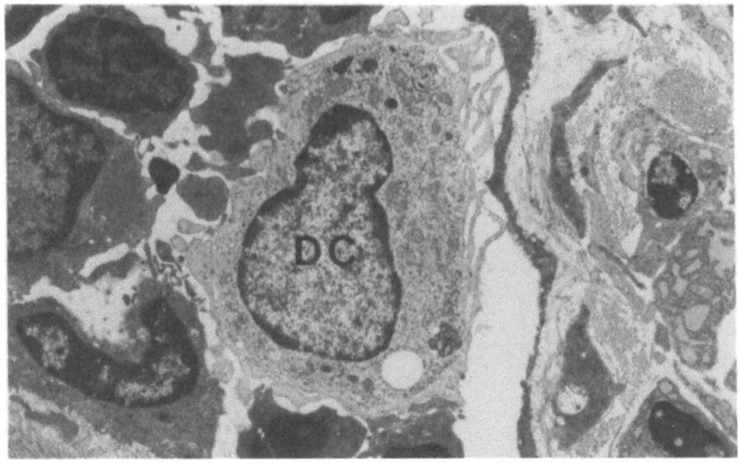

Figure 3. An electron micrograph of a diaphragmatic lymphatic showing a lymphoid cell (L) and a dendritic cell (DC).

Macrophages containing carbon were first encountered in bronchial lavage 1 hour post i.p. inoculation and in blood after 3 hours. While the number of carbon-laden macrophages present in bronchial lavage remained very low, less than 2%, throughout the time studied, the numbers in blood increased with time to 10% at 24 hours post i.p. inoculation (Figure 4a). No free carbon was found in the washings from the pleural cavity.

Intravenous injection of carbon did not lead to the presence of labelled macrophages in the pleural cavity and intranasally introduced carbon did not produce labelled macrophages in either the pleural or peritoneal cavities. Carbon introduced intrapleurally did not result in carbon-laden peritoneal macrophages although they were present in the bronchial lavage after 24 hours.

DISCUSSION

We have described in this study the unidirectional traffic of macrophages from the peritoneal cavity through the diaphragmatic lymphatics, across the pleural space to the lung. Introduction of carbon i.p. did not elicit this macrophage traffic because the total number of macrophages isolated from the pleural cavity did not increase during the time period studied. This suggests that the transition of peritoneal macrophages across the diaphragm to the pleural cavity is a normal ongoing movement. The dissemination of carbon through the bloodstream could not be responsible for the presence of macrophages containing carbon in channels leading from the visceral pleura into interalveolar septae as intravenous carbon introduction did not produce the same carbon deposition patterns in the lung as the i.p. introduced carbon. Also, carbon was not detected in the blood following i.p. inoculation until well after the appearance of labelled cells in the pleural cavity and lung. The appearance of carbon-labelled macrophages in the blood at 3 hours in the present study may have been due to passage up the mediastinal lymphatics or via thoracic duct as described by Drayson et al.[3]

The absence of carbon-labelled macrophages in the peritoneal cavity after intrapleural introduction and in both the pleural and peritoneal

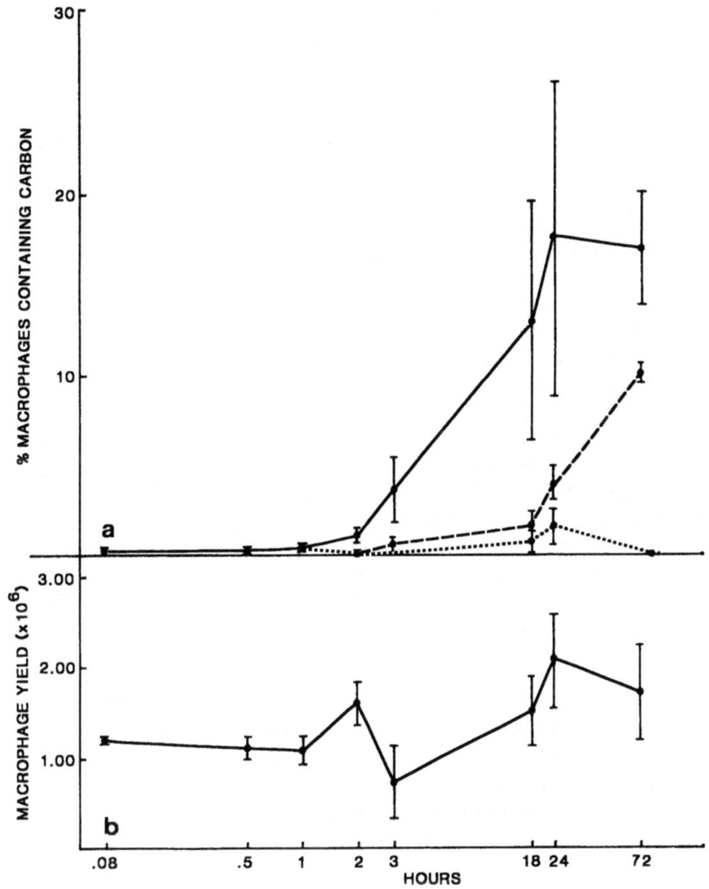

Figure 4. (a) Dissemination of i.p. inoculated carbon with time (hrs in log scale). The percentage of carbon laden macrophages in pleural cavity cells (———), blood (-----) and bronchial lavage (............) were estimated. (b) The total number of pleural cavity macrophages/monocytes isolated throughout the time period studied.

cavities after intranasal inoculation emphasizes the unidirectional movement of peritoneal macrophages to the lung. The finding of carbon-labelled macrophages in bronchial lavage following intrapleural instillation supports this.

The lack or very low numbers of carbon-laden macrophages found in the bronchial lavage suggests that the peritoneal macrophages that migrate to the lung are not a significant source of alveolar macrophages. Instead, they enter the interstitium. We propose that these macrophages are a significant source of long-surviving interstitial macrophages. This is supported by the observation that mice which were killed 5 months after receiving carbon i.p. still had distinct areas of carbon-labelled macrophages in channels associated with the periphery of the lungs.

Intraperitoneal introduction of antigen primes for a mucosal immune response. Pierce and Koster[4] found the i.p. route primed for a mucosal IgA response to cholera toxin and in their opinion appeared to reflect enhanced encounter of the i.p. introduced antigen with IgA-committed lymphocytes in extra-intestinal MALT. It is probable that the peritoneal macrophages which migrate to the lung carrying antigen could be antigen-presenting cells which prime T cells either in the lung or associated lymphoid tissue. The observations of Davies and Parrott[5] support this. They detected increased cytotoxic T cell activity in the lungs after an i.p. injection of tumour cells compared to subcutaneous or i.v. injection. The presence of lymphocytes and Langerhans-like cells in diaphragmatic lymphatics as well as labelled and non-labelled macrophages, which appear to be moving together, provide evidence that these mononuclear cells could be antigen presenting cells which lodge and reside in the interstitium.

Macrophages laden with carbon in potential lymphatics dissect along interalveolar septae presumably within an extravascular compartment. Aggregates of lymphocytes and monocytes were found along the pathway intermingled with the labelled cells. The labelled macrophages may slowly migrate through the interstitium until they reach lymphoid aggregates including the bronchial-associated lymphoid tissue (BALT). Support for this theory comes from unpublished data by Potter et al.[6], in their study they observed oil droplet-containing macrophages resulting from i.p. inoculation of pristane in the BALT. Green[7] observed that pigment-laden macrophages, labelled by airborne exposure, move through neighbouring alveoli and interstitial locations through lymphoid nodules out into the bronchiolar mucosa at the level of broncho-alveolar junctions. The intraperitoneally labelled macrophages appear to take a similar route.

Carbon deposition in the macrophages of lung interstitium is much less than that in the mediastinal internal mammary and parathymic lymphatics and nodes indicating that passage across the diaphragm to the pleural cavity and lung is not the predominant route of clearance of peritoneal lymph. However, the longevity of carbon-laden peritoneal macrophages in the lung interstitium points to a potentially important role of these cells in presenting antigens to lung-homing lymphoid cells.

REFERENCES

1. N. L. Tilney, J. Anat., 109:369 (1971).
2. E. C. Tsilibary and S. L. Wissig, Microvasc. Res., 25:22 (1983).
3. M. T. Drayson, D. R. Canning, and E. B. Bell, Adv. Exp. Med. Biol. Res., 186:401 (1985).
4. N. F. Pierce and F. T. Koster, J. Immunol., 124:307 (1980).
5. M. D. J. Davies and D. V. M. Parrott, Immunol., 44:367 (1981).
6. M. Potter, J. S. Wax, A. O. Anderson, and R. P. Nordan, J. Exp. Med., 161:996 (1985).
7. G. M. Green, Arch. Intern. Med., 131:109 (1973).

QUANTIFICATION AND CLASS DISTRIBUTION OF IMMUNOGLOBULIN-SECRETING CELLS IN MUCOSAL TISSUES OF THE CHICKEN

G. Koch and I.M.C.A. Jongenelen

Central Veterinary Institute
Dept. of Virology
P.O. Box 365, 8200 AJ Lelystad
the Netherlands

INTRODUCTION

Immunoglobulin-secreting cells (Ig-SC) can be observed in animals that are not deliberately immunized. This phenomenon has been referred to as spontaneous "background" Ig production (1). The background Ig production is induced both by endogenous and exogenous antigenic stimuli (2,3,4). Exogenous stimuli mainly depend on antigens that cross the mucosae. Therefore, studying the development of the spontanous Ig-SC response at mucosal sites is an appropriate way to acquire more insight into the functioning of the mucosal immune system in chickens.

In this study, Ig-SC were detected by employing the reverse ELISA spot test. In this test, monoclonal antibodies (McAbs) to the three major chicken Ig classes were used as developing antisera. The number and class distribution of Ig-SC in various lymphoid organs and at different mucosal sites were studied in SPF chickens of different ages.

MATERIALS AND METHODS

Chickens

White Leghorn type chickens (WLA) were reared under SPF conditions and kept in isolator facilities.

Cell suspensions

Chickens were deadened by an electrical shock and killed by exsanguination. Spleen, bursa, caecal tonsils (CT), Harderian glands (HGs), and small intestines (SI) were removed. Single-cell suspensions of these organs were prepared, as described for mice (5). For the preparation of intestinal cell suspensions, the SI was cut into three segments, i.e., duodenum, jejunum, and ileum, which were distinguished by the vascularization of the segments. Single-cell suspensions of the intestinal tissues were prepared, according to the method described by Arnaud-Battandier et al. (6) with some modification. The precise procedure will be published in greater detail elsewhere (Koch et al., manuscript in preparation).

Detection of Ig-SC by ELISA spot assay

The IgM-, IgG-, and IgA-SC in chicken cell suspensions were counted, using the reverse ELISA spot test as described by Czerkinsky et al.(7). Microtiter plates (Greiner, FRG) pre-incubated with 0.2% glutaraldehyde were coated with 70 μg/ml of the Ig-fraction of a goat-anti-chicken Ig serum, and subsequently, remaining sites were blocked by the addition of 5% fetal calf serum in phosphate-buffered saline (PBS). One hunderd microliters of cell suspension were then added to each well and incubated at 37 °C for 2-4 hours. Cells were removed by careful washing with running tap water. After removal of the cells, the test proceeded as for a normal ELISA. Successively, the plates were incubated with optimal dilutions of ascites fluid of McAbs CVI-Chμ-59.7 specific to chicken IgM, CVI-Chγ-47.3 to IgG, CVI-Chα-46.5 to IgA, or CVI-Chλ-47.5 to the L-chain (Koch et al., manuscript in preparation); with affinity-purified rabbit anti-mouse IgG alkaline phosphatase conjugate (Dakopatts, Denmark) 1:300 diluted in PBS containing 0.2% tween and 1 % bovine serum albumin; and finally with substrate solution consisting of 4 volumes of 5-bromo-4-chloro-3-indolyl phosphate (Sigma, USA) in 2-amino-2-methyl-1-propanol working buffer (Sigma) and 1 volume of low-gelling agarose. Between incubation steps, the plates were washed with running tap water. Blue spots appeared after 1 hr and were counted after an incubation period of at least 24 hrs.

RESULTS

The Ig-class specificity of the reverse ELISA spot test

The specificity of the McAbs to Ig classes employed were extensively tested in indirect ELISAs and with the immuno-electroblot technique (Koch et al., submitted for publication).

Table 1. Comparison of the total number of Ig-seceteting cells (SC) determined directly with L-chain specific McAb and calculated as the sum of the number of IgM-, IgG-, and IgA-SC.

Organ	Ig-SC/10^6 cells				
	Ig	IgM	IgG	IgA	IgM+IgG+IgA
exp. 1					
Spl	1,675[a]	150	1,250	565	1,965
	(260)	(64)	(188)	(136)	(241)
Duod.	33	36	2	15	53
	(7)	(5)	(1)	(2)	(5)
exp. 2					
Spl	375	60	130	17	207
	(49)	(19)	(30)	(5)	(36)
Duod.	12,484	158	1,044	11,720	12,922
	(739)	(17)	(222)	(547)	(591)

a) Mean of 6 determinations with standard deviation between brackets. One chicken was used in each experiment.

Two experiments (one with a one-week-old and another with an adult chicken) were performed to test whether the class-specific McAbs detected all Ig-SC. Two methods were employed to detect the total number of Ig-SC. First, Ig-SC were calculated as the sum of the number of IgM-, IgG-, and IgA-SC (Table 1, 4th column); then the Ig-SC were compared to the number of Ig-SC in the same cell suspension detected by only McAb Chλ specific for the L-chain (Table 1, 1st column). Both methods yielded very similar numbers of total Ig-SC, even when one Ig class was more dominantly secreted by the Ig-SC than others in the cell suspension.

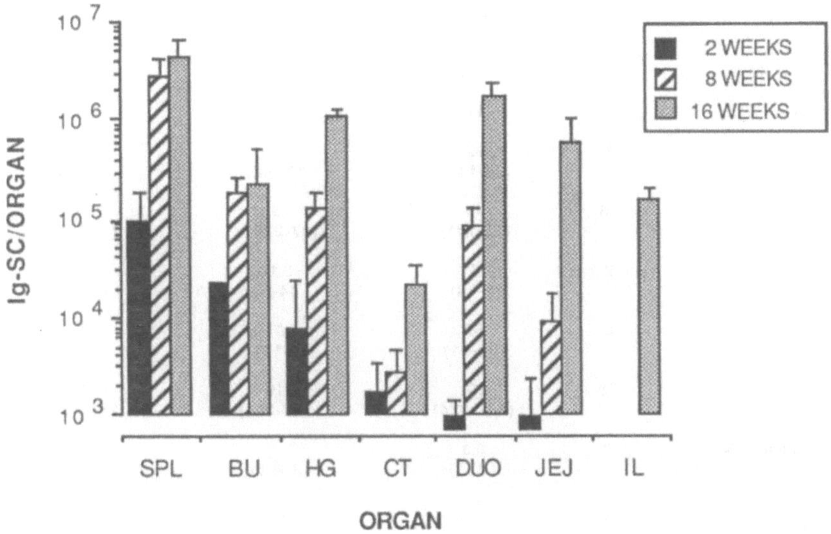

Figure 1. Number of Ig-SC in lymphoid tissues of the chicken. Columns represent the geom. mean (± s.e.m.) of the number of Ig-SC in cell suspensions of spleen (SPL), bursa of Fabricius (BU), Harderian gland (HG), caecal tonsils (CT), duodenum (DUO), jejunum (JEJ), ileum (IL) obtained from chickens at the age of 2 (closed), 8 (hatched) and 16 weeks (dotted columns). n=5., except for the bursa at 2 weeks of age, n=1.

Total number of Ig-secreting cells in chickens of different age

The total number of Ig-SC detected by the ELISA spot test with an McAb specific for the L-chain is shown in Fig. 1.

With increase of age, a number of Ig-SC were observed in all organs tested. In 2- and 8-week-old chickens, the majority (75-90%) of the total number of Ig-SC were located in the spleen. The contribution of the spleen to the total number of Ig-SC decreased from 75-90% to 54 % at 16 weeks of age. This decrease was caused mainly by the increased contribution from HGs and SI. Especially in the SI, the number of Ig-SC present started to increase substantially between 2 and 8 weeks of age. The majority of the intestinal Ig-SC were located in the duodenum. The number of Ig-SC detected in the caecal tonsils was low and never exceeded 2% of the total number of Ig-SC in all organs tested. The HGs contained a substantial number of Ig-SC already at 2 weeks of age. At this age, 10% of the Ig-SC were located in this organ.

Table 2. Class Distribution of Ig-SC in Chicken Lymphoid Tissues.

organ	Age (weeks)	Ig-SCs x 10-3/organ IgM	IgG	IgA
spleen:	2	73[a] (31-171)	0.7 (.3-2)	0.5 (.3-.7)
	8	1753 (1162-2647)	144 (63-329)	5 (3-8)
	16	2334 (15119-3587)	1481 (1105-1986)	76 (56-104)
bursa:	2	11	5	0.4
	8	92 (60-141)	41 (19-89)	0.7 (.3-1.5)
	16	40 (16-101)	79 (27-229)	19 (10-35)
Harderian glands:	2	5 (2-14)	0.1 (.04-.4)	0.4 (.1-1)
	8	57 (35-91)	6 (4-11)	6 (2-18)
	16	261 (182-374)	210 (145-280)	368 (286-476)
caecal tonsils:	2	0.3 (.1-.7)	0.1 (.04-.5)	0.1 (.06-.4)
	8	0.3 (.1-.6)	0.2 (.1-.5)	0.3 (.1-.7)
	16	1 (.6-2)	5 (3-8)	9 (5-18)
duodenum:	2	0.4 (.2-.8)	0.1 (0.09-1.8)	0.1 (.1-.2)
	8	9 (5-17)	8 (4-19)	38 (25-60)
	16	45 (22-94)	202 (83-492)	853 (429-1697)
jejunum:	2	0.4 (.1-1)	0.2 (.1-.6)	0.2 (.1-.4)
	8	1 (.7-2)	1 (1-2)	3 (2-6)
	16	8 (4-15)	56 (30-102)	500 (280-835)
ileum:	2	nd	nd	nd
	8	nd	nd	nd
	16	4 (3-6)	39 (20-80)	11.1 (87-141)

a) Numbers represent geometric mean of 5 individually tested chickens. Numbers between brackets represent the mean ± s.e.m.

Class distribution of Ig-SC in different organs

Until 2 weeks of age, most of the Ig-SC secreted IgM. In older chickens, increasing numbers of IgG- and IgA-SC were detected (Table 2).

The Ig class distribution of the Ig-SC also depended on whether the tissue belonged to the systemic or mucosal lymphoid tissues. Only around 10% of the Ig-SC in the spleen secreted IgA, whereas more than 70% of the Ig-SC in the SI secreted IgA (Table 2). The class distribution pattern of the HGs fell between both extreme patterns, since the SC were evenly distributed over all three Ig classes.

DISCUSSION

Comparison of the "background" Ig-SC response in germfree and conventional animals indicated that this response is the result of both endogenous and exogenous antigen stimulation (2). Experiments with germ-free mice fed on a chemically defined, synthetic diet showed that the development of the Ig-SC response and the class distribution are greatly influenced by exposure to exogenous antigens (3,4). We compared the development and class distribution of Ig-SC response in mucosal tissues and systemic lymphoid organs of SPF chickens.

The Ig-SC were counted by means of a reverse ELISA spot assay. The main advantage of this assay over the reverse haemolytic plaque assay (RHPA) is that McAbs can be used as developing antibodies. An important finding is that the number of Ig-SC detected by a McAb specific for the L-chain is about equal to the sum of IgM-, IgG-, and IgA-SC; this shows that by using the reverse ELISA spot assay, no major subpopulations of Ig-SC remain undetected.

By means of immunofluorescence, cytoplasmic IgM-containing cells were already observed in the bursa of embryo's from day 14 of incubation onward and in spleen from day 17. IgG-containing cells appeared later in bursa on day 21 of incubation and in spleen on day 4 after hatch (8). Lawrence et al (9), using the RHPA, showed that in the spleen, cells that actually secreted IgM were not apparent before day 3 after hatch, and those secreting IgG- and IgA-SC not before day 6.

In this study, we showed that the development of Ig-SC in the SI lags behind the response in the bursa and spleen. Only in chickens 2 weeks old or older did we find substantial numbers of Ig-SC in the SI. The number of Ig-SC decreased from duodenum to ileo-caecal junction, which was also concluded from previous immunofluorescence studies performed on tissue sections of SI (10).

In 2- and 8-week-old chickens, the majority of the Ig-SC in all organs tested secreted IgM (Table 2). Although this result is in agreement with results in germ-free mice (2), it is in contrast to the results obtained in chickens (8). Different antigenic exposure of the chickens may explain this discrepancy, since the chickens we used were kept in isolation and were fed with irradiated food and water. Antigen deprivation may have retarded the development of the background Ig-SC response.

In older chickens, the majority (70%) of the Ig-SC in the SI produced IgA. Although no quantitative data were presented, a similar class distribution was also reported in several previous studies (8,10,11,12). On the other hand, Lawrence et al. (9) reported nearly equal numbers of IgG- and IgA-SC in the SI. They asserted that the prominence of IgG-SC in chicken as compared to mammalian intestine was due to species differences (13). Comparison of their results and ours shows that we observed a much higher number of IgA-SC, yet a comparable number of IgG-SC. Therefore, we may conclude that the difference in findings is rather a result of strain than of species differences. It should be noted, however, that methodological difference could explain the different results, since we used highly specific McAbs, whereas they used polyclonal rabbit sera to identify the Ig classes (13).

The HG functions primarily as a secretor of lubricatant for the third eyelid of chickens. The gland is populated with Ig-SC, predominantly IgM-SC, around the age of two weeks (14, Table 2). In 16-week-old chickens, the three Ig classes are equally represented among the Ig-SC. This class distribution differs from observations of Bienenstock et al. (10), who reported that the HG produced IgM and IgG, but hardly any IgA, and observations of Albini et al.(14), who reported a predominance of IgA and almost no IgM immunofluorence-positive cells.

Of the lymphoid cells in the HG, 60% are Ig-SC. This percentage is strikingly high, especially in comparison to the spleen with only 0.5% Ig-SC (data not shown). The high frequency of Ig-SC is compatible with the high content of plasma cells (14,15,16). This dominance of Ig-SC and the low T-cell percentage of the gland (16) suggest that the Ig-SC immigrated from lymphoid tissues located elsewhere (17).

The relative contribution of different organs to the total number of Ig-SC is difficult to establish, since lymphoid tissue in chickens is distributed much more diffusely over the whole body than in mammals. Lawrence et al. (9) concluded that bone marrow (BM) is the major site of immunoglobulin production, since only a small sample of BM contained as much Ig-SC as the spleen.

We measured a much higher number of Ig-SC in the SI than was observed by Lawrence's group. The higher number can be explained by the fact that the cell yield we obtained from the SI was 10-fold higher than the yield they reported. Possibly the gradient purification step, which was omitted in our isolation procedure, caused substantial cell loss. Nevertheless, at the age of 16 weeks the number of Ig-SC in the SI was still smaller than in the spleen. Therefore, although we did not investigate the BM ourselves, the latter tissue seems indeed to be the major reservoir of Ig-SC. These results are at variance with recent data from mice studies at our institute that show the SI is by far the major reservoir of Ig-SC (18). Antigenic exposure and the age of the chickens may explain this difference. Nevertheless, taking into account the results of Lawrence et al. (9) regarding the relative number of IgA-SC in the BM we may conclude that in chickens, the intestine is the major reservoir of IgA-SC.

ACKNOWLEDGEMENTS

We thank Drs. A.T.J. Bianchi, G.F. de Boer and P.J. van der Heijden for their critical reading and mrs. V. Wensvoort-Thatcher for editing of the manuscript, Mr. L. Hartog for his excellent technical assistence during the production of the McAbs.

REFERENCES

1. R. Benner, A. van Oudenaren, M. Björkland, F. Ivars and D. Holmberg, <u>Immunology Today</u> 3:243 (1982).

2. R. Benner, A. van Oudenaren, J.J. Haaijman, J. Slingerland-Teunissen, B.S. Wostmann, W. Hijmans, Int. Arch. Allergy Appl. Immunol. 66:404 (1981).
3. H. Hooijkaas, R. Benner, J.R. Pleasants, B.S. Wostmann, Eur. J. Immunol. 14:1127 (1984)
4. N.A. Bos, C.G. Meeuwsen, H. Hooijkaas, R. Benner, B.S. Wostmann, and J.R. Pleasants, Cell. Immunol. 105:235 (1984).
5. R. Benner, A. van Oudenaren, G. Koch, in "Immunological Methods", Vol II, p.247, I. Lefkovits and B. Pernis eds., Academic Press, New York (1981).
6. F. Arnaud-Battandier, E.C. Lawrence and R.M. Blaese, R. Digestive Dis. Sci., 25:252 (1980).
7. C.C. Czerkinsky, A. Tarkowski, L.A. Nilsson, O. Ouchterlony and C. Gretzer, J. Immunol. Methods 72:489 (1984).
8. P.W. Kincade and M.D. Cooper, J. Immunol. 106:371 (1971).
9. E.C. Lawrence, F. Arnaud-Battandier, J. Grayson, I.R. Koski, N.J. Dooley, A.V. Muchmore and R.M Blaese, Clin. exp. Immunol. 43:450 (1981).
10. J. Bienenstock, J. Gauldie and D.Y.E. Perey, J. Immunol. 111:1112 (1973).
11. A-M. Lebacq-Verheyden, J.P. Vaerman and J.F. Heremans, J. Immunol. 109:652 (1972).
12. Kincade P.W. and M.D. Cooper, Science 179:398 (1973).
13. E.C. Lawrence, F. Arnaud-Battandier, I.R. Koski, A.V. Muchmore and R.M Blaese, J. Immunol. 123:1767 (1979).
14. B. Albini, G.Wick, E. Rose and E. Orlans, Int. Arch. Allergy 47:23 (1974).
15. B.G. Bang and F.B. Bang, Amer. J. Pathol. 53:735 (1968).
16. A.P. Mueller, K. Sato and B. Glick, Immunology 2:140 (1971).
17. B.D. Survashe and I.D. Aitken, Res. Vet. Sci. 24:182 (1978).
18. Ph.J. Heijden, W. Stok and A.T.J. Bianchi, Immunology, in press (1987).

THE IDENTIFICATION OF MURINE PEYER'S PATCH T CELL-DERIVED FACTORS WHICH ENHANCE IgA RESPONSES

K.W. Beagley, J.H. Eldrigde, H. Kiyono and J.R, McGhee

Department of Microbiology, Oral and Preventive Dentistry
University of Alabama at Birmingham, Birmingham
Alabama 35294, U.S.A.

INTRODUCTION

The humoral immune response to ingested antigens is dominated by antibody of the IgA isotype. In the mouse as in other higher mammals the major inductive site of secretory IgA responses is the Peyer's patch (1). Work in this and other laboratories has indicated that regulatory T cells (2) and accessory cells (3) within the PP microenvironment play an important role in controlling the isotype specificity of mucosal immune responses. We have previously shown that treatment of whole PP cell populations, derived by enzymatic disaggregation, with the oxidative mitogen sodium periodate results in the formation of cell clusters enriched in L3T4$^+$ T cells and dendritic cells (DC) (4). Co-culture of these cluster cells with either PP B cells or bone marrow mononuclear cells enhances polyclonal IgA production (5). Furthermore, the IgA inductive effect of the cluster cells is also found in 7 day culture supernatants of isolated clusters, indicating that the effect is mediated by soluble factors. T cell hybridomas have been produced from cluster T cells and the culture supernatants screened for IgA enhancing activity on whole PP cultures. Culture supernatants which showed IgA enhancing activity were tested for well characterized lymphokines including IL-2, IL-4 and IL-5. Attempts were also made to generate DC hybridomas by fusion of DC enriched cluster populations with a HAT sensitive subclone of the macrophage line P388D1. These initial fusions however, produced further T cell hybridomas as demonstrated by expression of surface Thy 1.2 antigen and the production of multiple T cell lymphokines. Culture supernatants from several of these hybridomas also showed potent IgA enhancing activity when added to whole PP cell cultures or LPS stimulated spleen B cell cultures. In all cases, hybridoma supernatants exhibiting IgA enhancing activity were positive for IL-4 and IL-5. To further investigate the role of these lymphokines in IgA synthesis, LPS stimulated spleen B cells and nonstimulated PP B cells were cultured with recombinant IL-4 and/or IL-5 to see if the IgA enhancing activity was associated with these molecules.

RESULTS

Sodium periodate induced clusters constitute 12-15% of the starting PP cell population and are enriched 4 fold in L3T4 positive T cells and contain numerous Ia positive DC (5). Cluster cells, pretreated with anti-immunoglobulin (Ig) plus complement (C) to remove contaminating B cells, enhanced polyclonal antibody synthesis when added to PP B cell cultures and bone marrow mononuclear cells. Maximum enhancement was seen in the IgA isotye. This isotype specific enhancement was also seen when 7 day culture supernatants from PP cluster cells were added to whole PP cultures.

Culture supernatants from T cell hybridomas produced by fusion of the T cell lymphoma R1.1 with cluster T cells were added to PP cell cultures and after 7 days antibody levels in culture fluids were assayed. Hybridomas that caused a greater than 3 fold increase in polyclonal IgA production were selected for further study. All hybridomas expressed Thy1.2 and L3T4 molecules. Hybridoma supernatants were assayed for several interleukins and all were found to contain both IL-4 and IL5. Levels of both interleukins were increased by overnight Con A stimulation of hybridoma cells.

Attempts to produce DC hybridomas by fusing DC enriched cell populations with the P388D1 cell line also resulted in several hybridomas which stained for surface Thy 1.2 and produced multiple T cell lymphokines. Addition of supernatants from these hybridomas to PP cell cultures also resulted in enhancement of polyclonal antibody production with some hybridomas producing enhancement mainly in the IgA isotype (Fig. 1).

Since the PP are the major inductive site for IgA immune responses it could be argued that PP B cells are already committed to the IgA isotype.

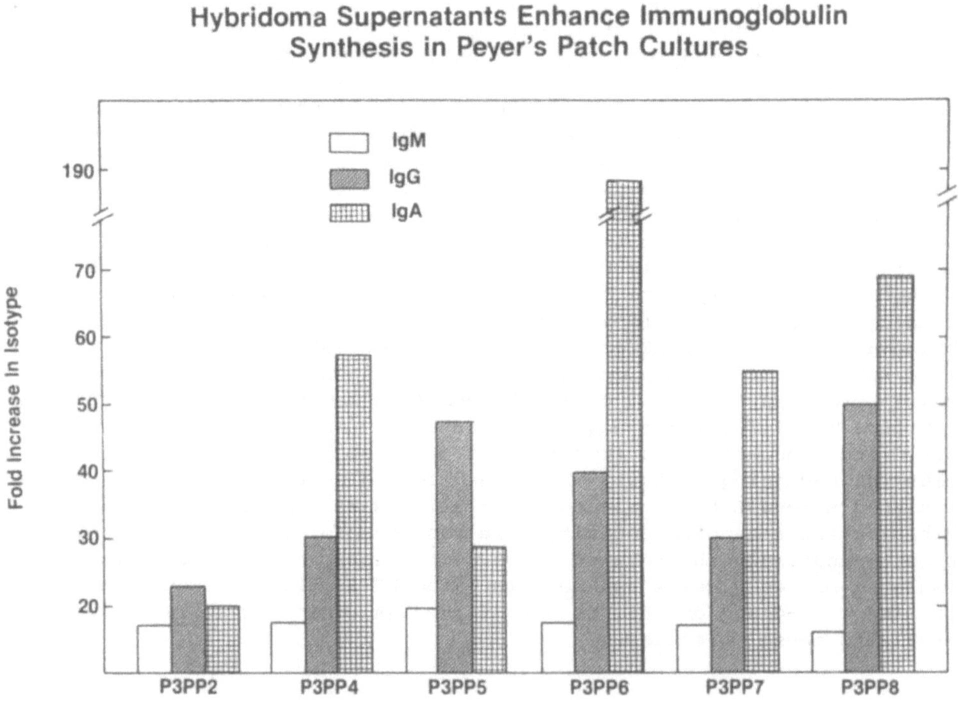

Fig. 1. Whole PP cells (2.5 x 10^5) were cultured for 7 days in RPMI 1640 plus 10% FCS. Hybridoma supernatants were added at a final concentration of 33%. Background immunoglobulin levels were IgM 132 ng/ml, IgG 107 ng/ml and IgA 454 ng/ml.

Supernatants from IgA enhancing hybridomas were therefore added to cultures of LPS prestimulated spleen B cells and after 7 days antibody levels in the culture supernatants were determined. The hybridoma supernatants caused significant enhancement of IgA responses, with no enhancement of the polyclonal IgG response and some enhancement of the polyclonal IgM response (Fig. 2). The IgM augmentation may be due to the LPS prestimulation.

Since the majority of the IgA enhancing hybridomas produced IL-4 and IL-5, recombinant interleukins were obtained from transfected HeLa (6) cells and added alone and in combination to LPS prestimulated spleen B cells to test the possibility that IL-4 and/or IL-5 are involved in augmentation of IgA synthesis. IL-4 alone caused no enhancement of polyclonal IgA responses, however some increase in the IgG response was seen as has been reported by others (7). When recombinant IL-5 was added to murine spleen B cells which had been preactivated with four different LPS preparations the polyclonal IgA response was enhanced in all cases (Fig. 3). In some experiments the addition of IL-4 in combination with IL-5 further enhanced the polyclonal IgA response although IL-4 alone was without effect.

As mentioned earlier the PP are thought to be the major inductive site for IgA responses. We therefore asked the question are PP B cells more responsive to the effects of IL-5 than B cells isolated from another

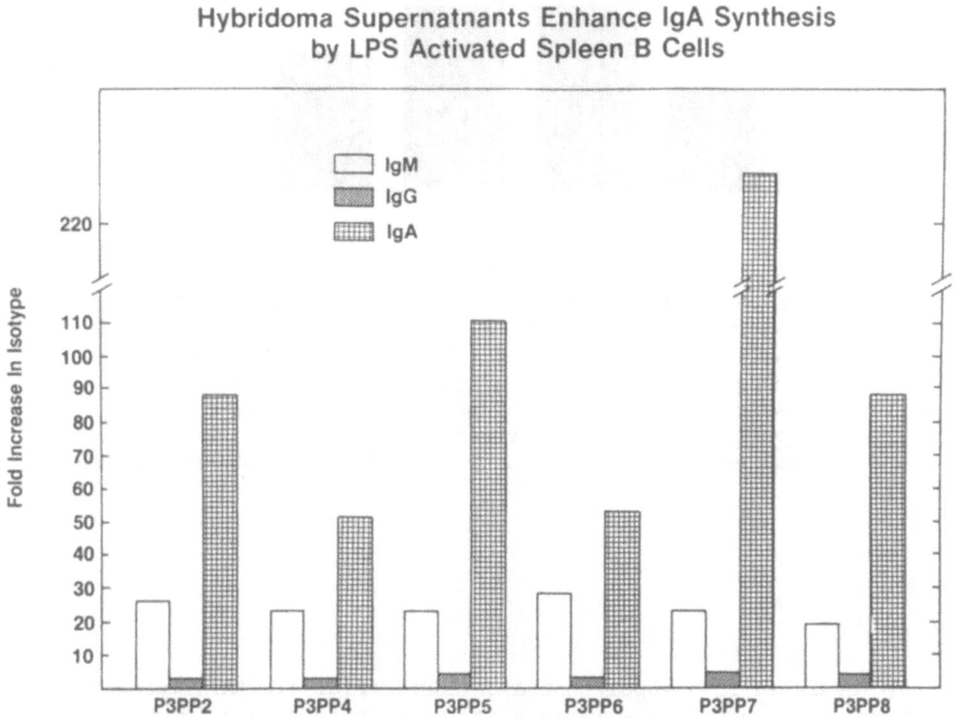

Fig. 2. Spleen B cells were cultured for 24 hours with LPS (5 µg/ml), washed 3 times, then cultured for 7 days at 2 x 10^5 cells per well in RPMI 1640 plus 10% FCS. Hybridoma supernatants were added at a final comcentration of 33%. Background immunoglobulin levels were IgM 1476 ng/ml, IgG 200 ng/ml and IgA less than 160 ng/ml.

microenvironment? B cells from PP and spleen were cultured without LPS prestimulation for seven days in the presence of IL-5 and/or IL-4 and the levels of each isotype were assayed by RIA. The IL-5 caused no significant enhancement of polyclonal IgA production by unstimulated spleen B cells (Fig. 4). Slight enhancement of IgM production was seen in IL-4 supplemented cultures and both IgM and IgG in IL-5 supplemented cultures. In contrast IgA production by PP B cells was increased by more than 2 µg above

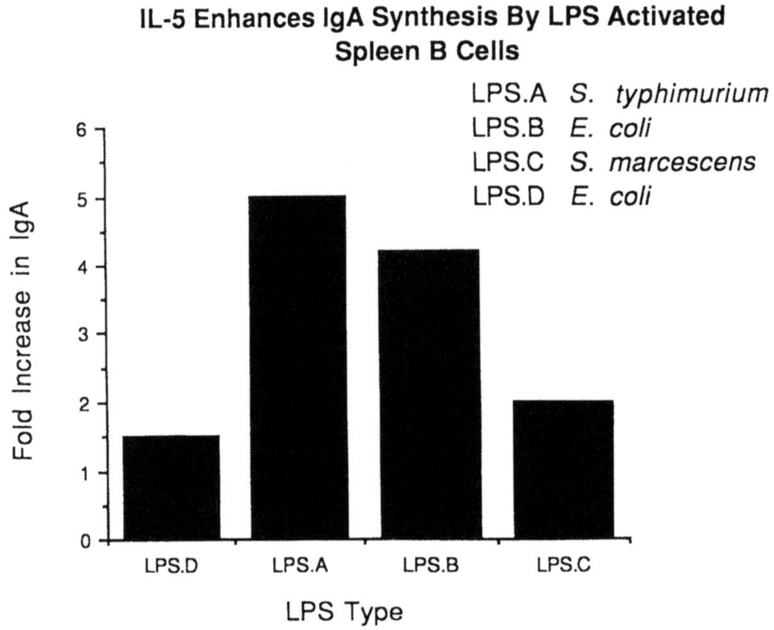

Fig. 3. Spleen B cells were cultured for 24 hours with LPS (5 µg/ml), washed 3 times then cultures for 7 days at 2 x 10^5 cells per well in RPMI 1640 plus 10% FCS. IL-5 (HeLa supernatant) was added at a final concentration of 33% (8U/culture). LPS-A from S. typhimurium, phenol extract; LPS-B from E . coli K235, butanol extract: LPS-C from S. marcescens, phenol extract: LPS-D from E. coli 0128B12, phenol extract.

control levels in IL-5 supplemented cultures (Fig. 4). Minimal increases in IgM and IgG production were seen in response to either lymphokine in PP cultures. We have preliminary evidence showing that a greater proportion of PP B cells are in cell cycle than spleen B cells. The PP also contains many more surface IgA positive B cells than the spleen. These observations suggest that IL-5 may therefore act as a differantiation factor on B cells that are already in cell cycle, and possibly already committed to IgA production.

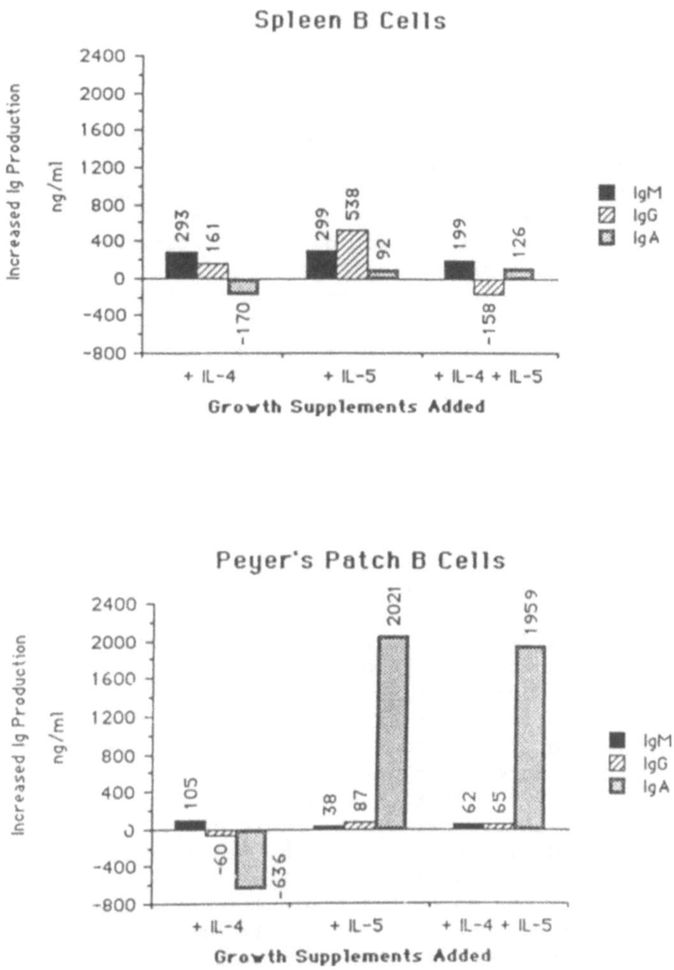

Fig. 4. PP ans spleen B cells were cultured for 7 days at 2×10^5 cells per well in RPMI 1640 plus 10% FCS. IL-4 and IL5 were added at a final concentration of 25% (IL-4 10U/culture, IL-5 8U/culture)

DISCUSSION

Immune responses at mucosal surfaces are dominated by antibodies of the IgA isotype. Many antigens which elicit immune responses at mucosal sites are first encountered by the gut-associated lymphoreticular tissue (GALT). The object of these studies has been to investigate interactions between T cells and accessory cells in GALT which regulate the priming of B cells in the PP microenvironment. Sodium periodate induced clusters from PP are enriched for L3T4 positive T cells and Ia positive DC and these cells selectively induce B cells from several tissues to differentiate to IgA synthesis. Similarly induced clusters from mouse spleen enhance only IgM and IgG responses (3). The IgA enhancing activity of PP cluster cells is found in cell free supernatants indicating that these effects are mediated by soluble factors.

T cell hybridomas produced from PP cluster cells produced factors which selectively enhanced polyclonal IgA responses. All these hybridoma supernatants contained IL-4 and IL-5 (4). Both of these interleukins have been cloned and cDNA transfected into stable cell lines. IL-4, formerly BSF-1 is known to have a variety of effects on both B and T cells. IL-4 enhances isotype specific IgG1 and IgE production in LPS stimulated spleen B cell cultures (7). IL-5, formerly termed BCDF-2 or TRF also has multiple effects on B cells and other cell types. One commonly used source of IL-5 is the T cell hybridoma B151K12. The product secreted by this cell exhibits two well defined B cell activities; induction of IgM secretion by the BCL-1 leukemic B cell line and induction of secondary IgG responses in vitro by DNP-primed B cells (8). Using LPS prestimulated spleen B cells as targets we have shown that IL-5 can selectively enhance IgA production by these cells. IL-4 alone had little effect on IgA synthesis but in some experiments did augment the effect of IL-5.

Two mechanisms for this isotype specific enhancement have been proposed. The first mechanism involves T cell directed switching of the B cell heavy chain genes by T cell derived factors (9). The second mechanism involves the activation of isotype committed B cells by T cell derived B cell differentiation factors (10). The two mechanisms are certainly not mutually exclusive and could be expected to operate in combination in the PP microenvironment. With regard to the IgA enhancing activity of IL-5 shown in this study we favor the latter mechanism. Studies with non-LPS activated B cells from spleen and PP show that only the latter can be stimulated by IL-5 to produce polyclonal IgA. The PP contain greater numbers of surface IgA^+ B cells and it is likely that these cells are already activated by LPS from the intestinal flora. This suggests that IL-5 may be acting on cells already in cell cycle, i.e., as a differentiation factor. Cell cycle analysis of PP and spleen B cells by flow cytometry using propidium iodide shows that the PP contains significantly greater numbers of B cells in G2 and S phase. Many of these cells may already be committed to IgA production. Several of the hybridoma supernatants are now being used for purification of factor(s), separate from IL-5, which enhance IgA synthesis and which may act via a switching mechanism.

ACKNOWLEDGEMENTS

HeLa cells transfected with the genes for murine IL-4 and IL-5 were a generous gift from Dr T. Honjo (Kyoto, Japan) to Dr Michael P. Everson in our laboratory. We also thank Ms Charlotte Jones and Mrs Mary-Lou McGhee for expert technical assistance.

REFERENCES

1. J. R. McGhee, S. M., Michalek, H. Kiyono, S. I. Williamson, T. A. Brown, and J. Mestecky, Mucosal immunoregulation: IgA inductive sites, isotype-specific helper T cells, gut influence, and subclass distribution of IgA antibodies. Bayer Symposium VIII, Springer Verlag, pp. 85-98 (1985).
2. H. Kiyono and J. R. McGhee, Mucosal T cell networks: Role of Fc R T cells and IgBF in regulation of the IgA response. Intern. Rev. Immunol. 2:157 (1987).
3. D. M. Spalding, S. I. Williamson, W. J. Koopman, and J. R. McGhee, Preferential induction of polyclonal IgA secretion by murine Peyer's patch dendritic cell-T cell mixtures, J. Exp. Med. 160:941 (1984).

4. K. W. Beagley, J. H. Eldridge, H. Kiyono, and J. R. McGhee, Analysis of Peyer's patch T cell and accessory cell populations involved in IgA isotype-specific immunoregulation, Federation of American Societies for Experimental Biology Proceedings 46:783 Abst 2725 (1987).
5. K. W. Beagley, J. H. Eldridge, C. D. Alley, H. Kiyono, and J. R. McGhee, Enhancement of IgA isotype-specific responses by Peyer's patch dendritic cell activated T cells. Proc. Am. Soc. Microbiol. Abst E-53 (1987).
6. T. Kinashi, N. Harada, E. Severinson, T. Tanabe, P. Sideras, M. Konishi, C. Azuma, A. Tominaga, S. Bergstedt-Lindqvist, M. Takahashi, F. Matsuda, Y. Yaoita, K. Takatsu, and T. Honjo, Cloning of complementary DNA encoding T-cell replacing factor and identity with B-cell growth factor II. Nature 324:70 (1986).
7. P. C. Isakson, E. Pure, E. S. Vitetta, and P. H. Krammer, T cell-derived B cell differentiation factors. J. Exp. Med. 155:734 (1982).
8. K. Takatsu, N. Harada, Y. Hara, Y. Takahama, G. Yamada, K. Dobashi, and T. Hamaoka, Purification and physicochemical characterization of murine T cell replacing factor (TRF). J. Immunol. 134:382 (1985).
9. H. Kawanishi, L. E. Saltzman, and W. Strober, Mechanisms regulating IgA class-specific immunoglobulin production in murine gut-associated lymphoid tissues. I. T cells derived from Peyer's pathces that switch sIgM B cells to sIgA B cells in vitro. J. Exp. Med. 157:433 (1983).
10. H. Kawanishi, L. E. Saltzman, and W. Strober, Mechanisms regulating IgA class-specific immunoglobulin production in murine gut-associated lymphoid tissues. II. Terminal differentiation of postswitch sIgA-bearing Peyer's patch B cells. J. Exp. Med. 158:649 (1983).

ROLE OF THE GALT CONTRASUPPRESSOR T CELL CIRCUIT IN ISOTYPE-SPECIFIC IMMUNOREGULATION

Kohtaro Fujihashi, Hiroshi Kiyono, Kenneth W. Beagley,
John H. Eldridge and Jerry R. McGhee

Departments of Microbiology and Oral Biology
University of Alabama at Birmingham
University Station, Birmingham, Alabama 35294

1. Introduction

Contrasuppression has been shown to be an important regulatory mechanism where mediation of T helper (Th) and T suppressor (Ts) cell activities are balanced in order to maintain appropriate immune responses in the host. In this regard, the functions of T contrasuppressor (Tcs) cells have been described in a number of experimental systems including the enhancement of *in vitro* anti-sheep erythrocyte (SRBC) responses, the adoptive transfer of contact sensitivity to hapten, the regulation of B cell responses to type 3 pneumococcal polysaccharide and in the abrogation of systemic unresponsiveness (1-5). Effector Tcs cells in these systems share the phenotypic characteristics of being Lyt-1$^+$, -2$^-$ and I-J$^+$, and binding to *Vicia villosa* lectin (6).

Human mucosal surfaces cover an enormous body area (over 200 times larger than the area covered by skin), and the major class of immunoglobulin present in external secretions protecting these surfaces is secretory IgA (S-IgA). In this regard, S-IgA production in the average human is 6-7 grams which is much higher than serum IgG subclasses and IgM. Ingested antigens are selectively taken up into the Peyer's patch (PP) which are a major component of the gut associated lymphoreticular tissue (GALT). The PP contain isotype-specific Th cells and a high proportion of IgA committed B cells which play major roles in the induction and regulation of subsequent IgA responses (7). Under certain conditions, systemic unresponsiveness (oral tolerance) to ingested antigens is simultaneously induced and this tolerance is largely mediated by Ts cells which are initially activated in GALT and which subsequently migrate to systemic tissue, i.e., spleen (8). The simultaneous occurrence of mucosal immune responses and systemic-unresponsiveness is important for host protection, first from penetration of antigen across the gut barrier (e.g., S-IgA antibodies) and from untoward systemic responses to antigen which escapes the gut S-IgA barrier. In order to maintain this dual response mediated by opposing regulatory T cells, clearly requires another level of control. In this brief review, we will describe the evidence for isotype-specific Tcs cell involvement in the protection of antigen-specific IgA responses in the face of oral tolerance.

2. GALT Derived Isotype-specific Tcs Cells Protect IgA Responses

Peyer's patches (PP) are a major IgA inductive site in higher mammals, and the PP possess all major immunocompetent cells, including helper, suppressor and contrasuppressor T cells, IgA committed B cells and accessory cells (7). In general, the oral administration of a T cell dependent (TD) antigen stimulates regulatory T cells in GALT, and results in the induction of isotype specific Th cells for IgA responses and Ts cells which mediate oral tolerance. The C3H mouse system has been a useful model to study how regulatory T cells are involved in the induction of mucosal responses and systemic unresponsiveness. Continuous gastric intubation of C3H/HeN mice (normal) with heterologous erythrocytes (e.g., SRBC) results in the induction of antigen-specific Ts cells and systemic unresponsiveness (8). The same immunization procedure in C3H/HeJ mice (high IgA responder mice) results in active Th cell pathways and good secondary type responses including IgM, IgG and IgA isotypes (8). Our recent studies have shown that C3H/HeJ mice given SRBC by the oral route exhibit enhanced Tcs cell activity, and account for the enhanced Th cell pathway and IgM, IgG and IgA responses to orally administered antigen (4,5). Adoptive transfer of Tcs cells from the spleens of C3H/HeJ mice orally immunized with SRBC, reversed oral tolerance in C3H/HeN mice and allowed IgM, IgG and IgA anti-SRBC responses (4). This response pattern was similar to that seen in C3H/HeJ mice given SRBC by the oral route. Tcs cells also share the common property of effector contrasuppressor cells, including binding to V. villosa and expression of Lyt-1 and I-J determinants (4,5).

Since the Lyt-1 phenotype is also expressed on L3T4 positive Th cells and a subpopulation of B cells, it was essential to examine whether Tcs cells are a distinct cell subset or whether contrasuppressor cells occur in the Th or B cell pools. Treatment of Tcs cells with anti-L3T4 and complement (C) did not effect their contrasuppressor activity (5,9). Furthermore, flow cytometer (FACS)-separated Lyt-1^+ and L3T4^- T cell subpopulations possessed contrasuppressor activity upon adoptive transfer to orally tolerant C3H/HeN mice (9). On the other hand, Lyt-1^+ and L3T4^+ T cells did not contain Tcs cell activity, but showed a classical helper T cell function in B cell cultures. In addition, anti-Lyt-2 and C did not remove contrasuppressor activity (4,5,9). Adoptive transfer of purified splenic B cells from orally immunized C3H/HeJ mice to tolerant C3H/HeN mice did not abrogate oral tolerance. Similar results were also observed when enriched Lyt-1 positive B cells were transfered to orally tolerant mice (9). These results clearly indicated that Tcs cells are a distinct subset of regulatory T cells which does not belong to helper ($CD4^+$) and suppressor ($CD8^+$) T cell or B cell subpopulations (Table 1).

When Tcs cells were obtained from the spleens or PP of C3H/HeJ mice previously fed SRBC for 28 days, and adoptively transfered to tolerized C3H/HeN mice, two types of immune response were noted. Mice that received splenic Tcs cells supported almost equal numbers of IgM, IgG and IgA plaqueforming cells (PFC). However, PP-derived Tcs cells converted systemic unresponsiveness to a response predominated by IgA (with some IgM and IgG responses) (Table 1) (5). When the numbers of IgA to IgM and IgA to IgG response were compared between mice that received either splenic Tcs cells or PP Tcs cells, the ratios of IgA/IgM and IgA/IgG in mice that received PP Tcs cells were 3.4 and 4.3, respectively. In contrast, ratios of 1.4 and 1.5 were noted in mice that received splenic Tcs cells (5). These results led to an additional experiment where PP Tcs cells were separated according to their ability to bind to FcαR$^+$ T cells. Interestingly, adoptive transfer of the nonadherent Tcs cell fraction abrogated tolerance but supported only IgM and IgG responses, but not IgA. On the other hand, Tcs cells which bound FcαR$^+$ T cells possessed the

Table 1

Characterization of Contrasuppressor T Cells Which Convert Oral Tolerance to Isotype-Specific Responses

Route of Immunization	Source of Tcs cells	Phenotype	*V. villosa*	Isotype-specific Responses
Oral	PP	Lyt-1$^+$, -2$^-$, L3T4$^-$	Adherent	IgA >>> IgG = IgM
Oral	Spleen	Lyt-1$^+$, -2$^-$, L3T4$^-$	Adherent	IgA > IgG = IgM
IV (Low dose of Ag)	Spleen	Lyt-1$^+$, -2$^-$, L3T4$^-$	Adherent	IgM & IgG only
IV (High dose of Ag)	Spleen	Lyt-1$^+$, -2$^-$, L3T4$^+$	Adherent	IgM & IgG only

ability to support a predominant IgA response (5). Based on these findings, we have postulated that at least two subsets of effector Tcs cells occur in GALT, e.g., one subset of Tcs cell is IgA isotype-specific and the other subset of Tcs cell supports responses in a pan isotype manner. Further, IgA specific Tcs cells may control the balance between Th cell-induced IgA responses and Ts cell-evoked systemic-unresponsiveness to maintain homeostasis against foreign antigens which enter the host via the gut.

3. Systemic Versus Local Priming Results In Antigen-Specific Tcs Cells Which Correspond To The Isotype Of Antibody Produced

Since oral administration of antigen induces effector Tcs cells in the PP which mainly support IgA responses, it was of interest to determine if the route of immunization affects the type of Tcs cell induced, and thus the isotype pattern of antibody response seen when oral tolerance was ablated. In this study, C3H/HeJ or C3H/HeN mice were systemically primed with SRBC (intravenously; IV) at doses of 0.01 to 10%, and purified splenic Tcs cells were assessed for contrasuppressor activity by adoptive transfer to orally tolerant mice. Purified Tcs cells from IV-primed C3H mice converted oral tolerance to responses predominated by IgM and IgG isotypes but not IgA (10). This is logical, since systemic immunization generates typical secondary type responses (IgM and IgG) in secondary lymphoid tissues (e.g., spleen and peripheral lymphoid tissue) where a classical Th cell pathway is induced. These Tcs cells also share the common characteristics of contrasuppressor cells including *V. villosa*-adherence, and expression of Lyt-1 and I-J but not Lyt-2 (Table 1).

Interestingly, systemically induced Tcs cells can be divided into two groups based on their expression of L3T4. Low doses of antigen induced effector Tcs cells only in the L3T4-negative subset, while both the L3T4-positive and -negative subsets possess contrasuppressor activity when a high dose of antigen was used for priming (Table 1) (10). One possible explanation for these results is that the low dose of antigen induces Th cell pathways and L3T4-negative Tcs cells may protect Th cells from suppressor effects. On the other hand, higher antigen exposure generally induces Ts cell-mediated suppression and both the L3T4$^+$ and L3T4$^-$ Tcs cells may be necessary to down regulate this predominant Ts cell pathway.

Taken together, these results showed that systemic-priming induces Lyt-1$^+$/L3T4$^-$ Tcs cells and/or Lyt-1$^+$/L3T4$^+$ Tcs cells which convert oral tolerance to only IgM and IgG responses. However, oral immunization induces at least one subset of Tcs cells which specifically protect IgA responses in the presence of systemic unresponsiveness. We postulate that the route of immunization is a major determinant for the type of Tcs cell induced, and this influences the isotype of immune response which occurs at this site.

Both orally and systemically induced contrasuppressor cells are antigen-specific (9,10). For example, Tcs cells from SRBC primed mice converted SRBC-specific oral tolerance to secondary responses, but were without effect when transferred to mice orally tolerized with horse erythrocytes (HRBC). The reciprocal experiment gave similar results, in that HRBC induced Tcs cell abrogate HRBC-specific tolerance, but not SRBC-specific systemic unresponsiveness. Our results also indicated that Tcs cells may possess a functional T cell receptor (TCR) which accounts for the antigen-specificity of contrasuppression. In order to address this point, we examined whether Tcs cells express T3 molecules presumably associated with a functional TCR. In the initial experiment, Lyt-1 positive T cells were fluorescently stained for T3 positive cells using monoclonal anti-T3 (145-2C11) and isolated by flow cytometry. These Lyt-1$^+$ and T3$^+$ T cells were further separated into V. villosa-adherent and -nonadherent fractions, and tested for contrasuppressor activity. Orally tolerized mice which received T3$^+$ and V. villosa-adherent T cells supported secondary responses, while V. villosa-nonadherent and T3$^+$ cells did not affect the unresponsive state (9). In further experiments, Lyt-2$^-$ and L3T4$^-$ T cells were isolated by FACS sorting and then were stained with 145-2C11. The T3$^+$ T cells were obtained by additional sorting and tested for contrasuppressor function. Adoptive transfer of Lyt-1$^+$, Lyt-2$^-$, L3T4$^-$ and T3$^+$ Tcs cells to tolerized mice resulted in conversion of systemic-unresponsiveness to secondary responses. These results support our hypothesis that L3T4 (CD4) and Lyt-2 (CD8) negative effector Tcs cells possess a functional TCR-T3 complex. This is intriguing in light of the fact that a minor population of thymus- and peripheral blood-derived CD4-negative and CD8-negative T cells express T3 protein in association with a rearranged gamma form of TCR for antigen recognition (11,12).

4. Summary

GALT contains Tcs cells which protect IgA responses under conditions of oral tolerance. Oral immunization of erythrocyte antigen induces antigen-specific Tcs cells in Peyer's patches which predominantly support IgA responses with minor IgM and IgG responses upon adoptive transfer to tolerized mice. On the other hand, systemically-primed mice possess Tcs cells which convert oral tolerance to only IgM and IgG responses. In general, effector Tcs cells which abrogate antigen-specific oral tolerance are Lyt-1$^+$, Lyt-2$^-$, L3T4$^-$, I-J$^+$ and V. villosa-adherent. Interestingly, these Tcs cells express T3 molecules which suggest the expression of a TCR which may account for their antigen specificity.

Acknowledgement

Authors are grateful to Ms. Sandra Roberts for preparation of the paper. This work was supported by U.S.P.H.S. grants AI 19674, AI 21032, AI 18958, DE 04217, P60 AM 20614, DE 02670, AI 21774 and a grant from the Procter and Gamble Company.

References

1. R.K. Gershon, D.D. Eardley, S. Durum, D.R. Green, F.W. Shen, K. Yamauchi, H. Cantor and D.B. Murphy. Contrasuppression. A novel immunoregulatory activity. J. Exp. Med. 153:1533 (1981).
2. M. Iverson, W. Ptak, D.R. Green and R.K. Gershon. Role of contrasuppression in the adoptive transfer of immunity. J. Exp. Med. 158:982 (1983).
3. H. Braley-Mullen. Regulation of the antibody response to type III pneumococcal polysaccharide by contrasuppressor T cells. J. Exp. Med. 160:42 (1984).
4. I. Suzuki, H. Kiyono, K. Kitamura, D.R. Green and J.R. McGhee. Abrogation of oral tolerance by contrasuppressor T cells suggests the presence of regulatory T cell networks in the mucosal immune system. Nature 320:451 (1986).
5. I. Suzuki, K. Kitamura, H. Kiyono, T. Kurita, D.R. Green and J.R. McGhee. Isotype-specific immunoregulation: Evidence for a distinct subset of T contrasuppressor cells for IgA responses in murine Peyer's patches. J. Exp. Med. 164:501 (1986).
6. D.R. Green and W. Ptak. Contrasuppression in the mouse. Immunol. Today 7:81 (1986).
7. H. Kiyono and J.R. McGhee. Mucosal T cell networks: role of $Fc\alpha R^+$ T cells and $IgBF\alpha$ in regulation of the IgA response. Intern. Rev. Immunol. 2:157 (1987).
8. H. Kiyono, J.R. McGhee, M.J. Wannemuehler and S.M. Michalek. Lack of oral tolerance in C3H/HeJ mice. J. Exp. Med. 155:605 (1982).
9. K. Kitamura, H. Kiyono, K. Fujihashi, J.H. Eldridge, D.R. Green and J.R. McGhee. Contrasuppressor cells which break oral tolerance are antigen-specific T cells distinct from T helper ($L3T4^+$), T suppressor ($Lyt\text{-}2^+$) and B cells. (Submitted for publication).
10. K. Kitamura, H. Kiyono, K. Fujihashi, J.H. Eldridge, K.W. Beagley and J.R. McGhee. Isotype-specific immunoregulation: systemic antigen induces splenic T contrasuppressor cells which support IgM and IgG subclass but not IgA responses. (Submitted for publication).
11. L.L. Lanier and A. Weiss. Presence of Ti (WT31) negative T lymphocytes in normal blood and thymus. Nature 324:268 (1986).
12. A.M. Lew, D.M. Pardoll, W.L. Maloy, B.J. Fowlkes, A. Kruisbeek, S.-F. Chens, R.N. Germain, J.A. Bluestone, R.H. Schwartz and J.E. Coligan. Characterization of T cell receptor gamma chain expression in a subset of murine thymocytes. Science 234:1401 (1986).

EVIDENCE FOR CONTRASUPPRESSION IN PATIENTS WITH CROHN'S DISEASE

A. Raedler, S. Schreiber, A. de Weerth, B. Brinkmann, K. Sandgren H.G. Thiele, and H. Greten

Med. Dept. Univ. D. 2000, Hamburg 20, FRG

INTRODUCTION

Vicia villosa agglutinin (VVA) is a lectin that binds specifically to glycoconjugates with terminal N-acetyl-galactose. Originally murine T cells binding to VVA are reported to be involved in cytotoxicity (1) however this has been contradicted by others (2,3). Moreover murine and human T cells reacting with VVA were suggested to be involved in contrasuppression (4-7). T cells displaying contrasuppressor activity inhibit the action of suppressor cells on helper cells, thus enhancing immune response. Beside binding sites for VVA, contrasuppressor T cells in mice express Lyt-2 (corresponding to CD8, contrasuppressor inducer), Lyt-1 (corresponding to CD4, contrasuppressor effector) or both determinants (contrasuppressor acceptor or transducer cells). Additionally I-J determinants are exhibited that are however different from those found on feedback suppressor cells (8). It has been suggested that contrasuppression is responsible for regulation of local immune response in particular in the gut mucosa (5). Alterations in contrasuppression have further been suspected to be involved in autoimmunity (9), although experimental data to support this thesis are sparse (1o).

In a previous study we have shown that peripheral VVA binding T cells are increased in number in patients with CD and represent a major subset of activated peripheral T cells (11). These cells do not exhibit helper activity. In a subfraction of patients removal of these cells causes reduction of Ig and IgA secretion in autologous B and T cell cultures. Thus it has been suggested that these cells may function as contrasuppressor cells, although the possibility remains that these cells are helper inducer cells. Moreover from these experiments there were no data obtained concerning the IgA subtype. The following experiments were undertaken to analyse the phenotype of VVA binding in regard to the Leu8 determinant, which is known to be not expressed on helper inducer cells and to determine IgA subtypes in the supernatant of B and T cell cocultures.

MATERIAL AND METHODS

Patients

For analysis of activated peripheral lymphocytes, patients with active CD (n=12, CDAI > 15o, activated peripheral T cells > 15%, age: 19-48 ys) ulcerative colitis (n=10), autoimmune diseases (n=8) including systemic lupus erythematodes (n=4), immunovasculitis (n=2), Wegener's granulomatosis (n=2), viral and bacterial enteritis (n=5) and normal healthy controls (NC, n=12) were examined.

Preparation of T lymphocytes

Samples of 10 ml blood were drawn and added to 10 ml 0.9% saline supplemented with 2000 units heparine. 2.5 ml of 'Lymphoprep' (Nyegaard) were overlayered with 5.0 ml of diluted blood and centrifuged for 30 min (20°C, 560 g). The lymphocytes were recovered from the interphase and washed twice with phosphate-buffered saline (PBS, GIBCO, Mass.). The pellet was resuspended in 15 ml of cell culture medium (RPMI, GIBCO, Mass.) supplemented with 10% fetal calf serum (FCS, MEDAC, FRG) and cultured for 2 h (37°C, 5% CO_2) for removal of monocytes. Non-adherent cells were then collected and layered on Petri-dishes (5.5 cm, 5 x 10^6 cells per plate coated with Fab_2 goat anti-human immunoglobulin (TAGO, Cal., 4 ml, 10 /u/ml). After incubation for 60 min at room temperature non-adherent (non-surface Ig bearing) cells were recovered.

Panning procedures

Panning via Vicia-villosa agglutinin was achieved giving 5 x 10^6 T cells on Petri-dishes previously coated with VVA (3 ml, 10 /ul/ml diluted to 0.001%, 12 h, 4°C). Non-adherent cells were carefully decanted after 1 h. Harvesting of adherent cells were carefully decanted after 1 h. Harvesting of adherent cells was facilitate by supplementation of a 0.2 mmol solution of N-acetyl-galactosamine. Controls of panning were performed by labeling of non-adherent cells and adherent cells with Texas-red coupled VVA. Impurities of less than 5% were tolerated. If this percentage was exceeded the same panning procedure was repeated.

Detection of lectin binding sites

Texas-red or FITC-coupled Vicia villosa agglutinin (VVA, MEDAC, FRG) were used incubating 1 x 10^6 non-glas-adherent, surface-Ig-negative cells with 20 /ul lectin (1 mg/ml,diluted 1 : 10 with PBS 30 min, 4°C). Controls were performed using a solution of inhibitory sugars (N-acetyl-galactosamine, D-galactose respectively, 0.2 mmol, 30 min, 4°C, MEDAC, FRG).

Detection of cell surface determinants using monoclonal antibodies

The following monoclonal antibodies were used: OKT3 (Ortho-mune, peripheral T cells) for efficiency-control of B cell panning, Leu8 (determination of suppressor inducer cells), OKT9 (determination of activated T cells), Leu12 (Becton Dickinson, Utah, peripheral B cells for efficiency of B cell panning), Leu M1 (Becton Dickinson, Utah, peripheral monocytes for efficiency control of monocyte panning). All monoclonal antibodies were used incubating 1 x 10^5 - 1 x 10^6 non-glas-adherent, surface-Ig-negative cells in 5 /ul (30 min, 4°C) of the antibody under study and 45 /ul medium supplemented with FCS. Control experiments were performed omitting the specific antibody or using normal mouse serum instead. Additionally cell-lines were used where the percentage of positive cells for the antibodies was previously standardized.

As second antibodies TRITC- or FITC-coupled antibodies (Fab_2) goat-anti-mouse Ig of the subclass appropriate for the monoclonal antibody were used (5 /ul diluted 1 : 60 in medium/FCS, 30 min at 4°C). After each labeling step, cells were thoroughly washed twice in medium supplemented with FCS 1 : 10.
In double labeling experiments T cells under study were incubated in each monoclonal antibody/VVA alone and in a mixture of antibodies and Texas red coupled VVA respectively. The subsets expressing both determinants were assessed by comparing the numbers of labeled cells in both single incubations with labeled cells in the incubation of a mixture of both antibodies. Additionally, direct double labeling with FITC-coupled monoclonal antibodies and Texas red coupled VVA were performed.

Assessment of fluorescence labeled cells

In double labeling experiments cells were analysed by a blinded examiner counting 1 x 100 cells under a fluorescence microscope (ZEISS, FRG). Single labeling analysis was performed by a FACS IV cell sorter (Becton Dickinson, Utah).

Fluorescence cell-sorter analysis

After removal of glass-adherent and surface-Ig-bearing cells, the remaining lymphocytes were labeled with VVA-FITC as already described. Cells were measured for fluorescence and small-angle scatter (cell size) with a FACS IV cell sorter (Becton, Dickinson). The blue laser line (488 nm,0.2W) was used for excitation. The green fluorescence was measured using the filter combination KP 650 and LP 530 and stored correlated with the scatter signal in a 4 x 64 channel histogram. Results were assessed by a mathematical evaluation program for two-parameter flow cytometric histograms.

Autologous B cells assays

1 x 10^6 separated B cells were cultured for 7 days in RPMI 1640 supplemented with 10% FCS and 50 mg/ml pokeweed mitogen (PWM, GIBCO, Mass.) and antibiotics. 1 x 10^6 peripheral autologous T cells or the same amount of VVA^+ Tc or Tc minus VVA^+ Tc separated by panning procedures as described above were added. Supernatants were analysed for IgA1 and IgA2 using an ELISA-technique as described priviously (9).

Statistics

Differences of group means were tested using the Students's t-test, the Wilcoxon-test or the Mann-Whitney U-test resp.. Multiple regression analysis was performed using the stepwise selection algorhythm.

RESULTS

1. Determination of VVA binding T cells

As shown in Fig. 1 patients with CD (n=12) exhibit elevated numbers of VVA binding T cells. In ulcerative colitis (UC, n=10) comparable numbers of VVA positive T cells were assessed. In contrast their number is low in patients with infectious (bacterial or viral) enteritis (IE, n=5) in autoimmune diseases (AD, n=8) and in healthy volunteers (NC, n=12, p= .001 (CD and UC versus controls).

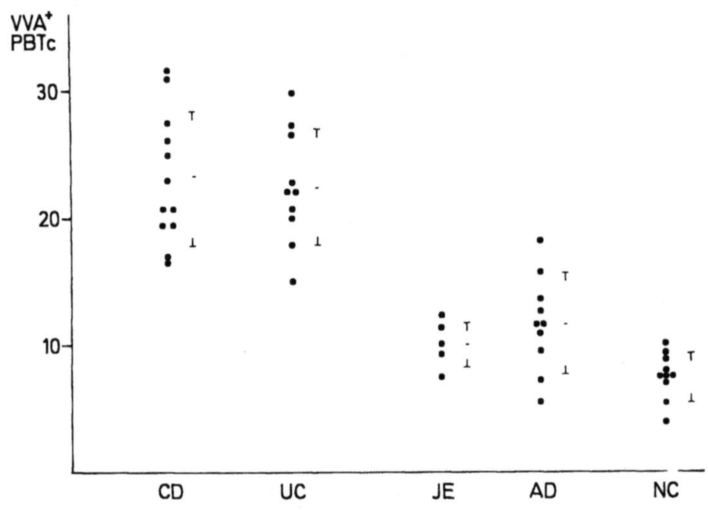

Fig. 1. Percentages of Vicia villosa agglutinin (VVA) binding peripheral T cells (PBT) in patients with Crohn's disease (CD), ulcerative colitis (UC), infectious enteritis (IE), autoimmune disease (AD), normal controls (NC).

2. Determination of VVA binding T cells coexpressing Leu8

Double labeling experiments using a monoclonal antibody reacting with Leu8 and Texas red coupled VVA and Goat-anti-Mouse-Ig-FITC showed that 78,6% of VVA positive cells coexpress the Leu8 determinant in cells derived from patients with CD (n=12). In normal controls (n=10) almost all VVA positive lymphocytes display binding sites for anti-Leu8 antibodies (Fig.2).

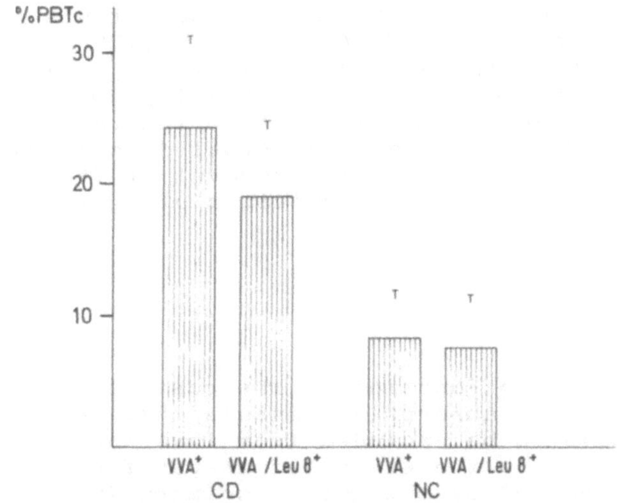

Fig. 2. Percentage of peripheral blood T cells (PBT) expressing binding sites for Vicia villosa agglutinin (VVA) or coexpressing Leu8 in patients with Crohn's disease (CD, n=12) and normal controls (NC, n=10).

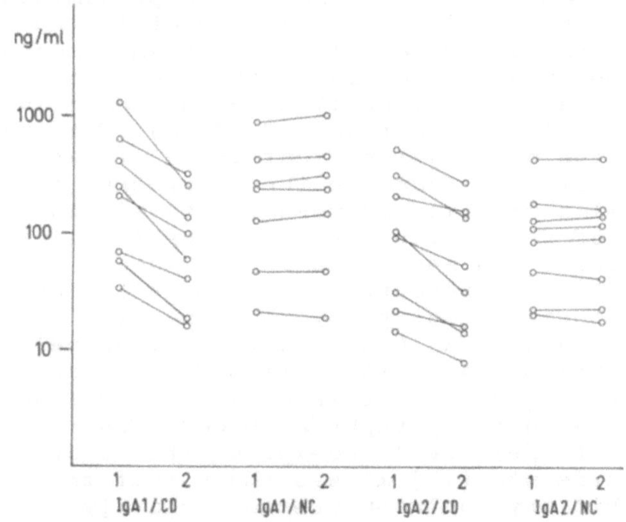

Fig. 3. Assessment of IgA1 and IgA2 in the supernatants of T and B cells cultures (ratio 1:1) in patients with Crohn's disease (CD) and normal controls (NC). 1= total T cells, 2= T cells depleted of VVA^{+}T cells.

3. Assessment of Ig, IgA, IgA1 and IgA2 in the supernatant of cell cultures containing autologous B and T cells

Concentration of Ig and IgA in supernatants of B cell cultures after 7d without T cells is low (12 ng/ml+-8, 4 ng/ml+-4). Supplementation with VVA positive T cells does not enhance Ig or IgA production significantly. Determination of IgA subtypes in the supernatants of peripheral B and total T cells (B and T cell ratio 1:1) revealed that IgA1 and IgA2 were assessed in a ratio of 2:1. Removal of VVA positive T cells (ratio 1:1) leads to a reduction of IgA1 and IgA2 as well (Fig.3).

DISCUSSION

In spite of the function of the gastrointestinal mucosa as the main site of absorption, agents that potentially act as antigens have to be excluded from invasion. It is generally accepted that beside the so-called unspecific defense mechanism this is achived by the gut associated lymphoid tissue (Galt, 12). Bearing in mind the great variety of antigens within the intestinal content, exclusion of antigens by the immune system without provoking a general response needs special regulation of local mucosa associated immunity (13). The major Ig isotype of Galt is IgA and to a much lesser degree IgM (14). The advantage of IgA antibody mediated immune response is the resistance of this isotype against proteolysis, transport to the intestinal lumen by epithelial secretory pieces and biological properties that favour protection instead of inflammation (14). In CD the predominance of IgA is restricted by the relative increase of IgG and IgM secreting B cells that accumulate in the inflamed mucosa (15,16). Moreover properties of B cells in the mucosa of patients with CD resemble those of peripheral B cells suggesting a breakdown of the barrier of Galt in CD (17,18). This hypothesis is supported by the observation, that in CD elevated numbers of activated T cells are found in the peripheral blood (19) and that these cells exhibit suppressor activity for IgA (20). Another kind of regulation that is thought to influence Ig and in particular IgA secretion in the mucosa is contra-suppression (5). Contrasuppressor T cells are characterized in regard to their phenotype by their ability to bind to Vicia villosa agglutinin (VVA, 4,5). They act upon helper cells via ligands that are in competition with factors secreted by suppressor cells (10). There is only a minor fraction of VVA binding T cells in the peripheral blood of normal controls. The majority of these cells coexpress the T8 determinant. In previous experiments (11) it was shown that these cells neither exhibit any contrasuppressor activity, nor show helper activity. In patients with CD elevated numbers of VVA binding cells are detectable in the peripheral blood, whereas their number is decreased in the inflamed mucosa. VVA binding cells in CD express either the T8 or the T4 determinant. In analogy to the murine system (4,8) this could be explained as an increase of contrasuppressor effector cells.

Functional analysis of these cells in autologous B cell assays shows that VVA binding T cells in CD as in normal controls, does not possess any helper activity. However in half of the patients with CD a reduction of Ig and IgA secretion after removal of VVA binding T cells is demonstrable. In the present experiments it was shown that IgA1 and IgA2 were produced in a ratio of 2:1. Both IgA subtypes were reduced in the supernatants of 7d cell cultures by the removal of VVA binding. Thus VVA binding cells seem to be able to influence IgA1 and IgA2 producing B cells as well. The high amount of IgA2 however can be explained by the presence of pokeweed mitogen which is known to stimulate B cells producing IgA2 preferentially (21).

From T and B coculture experiments it has been deduced that VVA binding T cells in CD represent contrasuppressor cells. Another explanation of this regulatory effect - that this cell subset acts as helper inducer cells - is not completely ruled out. It is however improbable, for less than one third of these cells do not express the Leu8 determinant which is known to be correlated with helper inducer function.

It seemed interesting to look if patients with CD showing contrasuppression are characterised in terms of clinical, laboratory of anamestic data. In previous experiments it was shown by multiple step analysis that of all data considered patients exhibiting contrasuppression lacked extraintestinal manifestations, abscesses and fistulas. Thus these experiments do not give evidence that contrasuppression is involved in the pathogenesis of CD. However contrasuppression might be involved in the manifestation of extramucosal symptoms.

SUMMARY

Patients with Crohn's disease (CD) have elevated numbers of Vicia villosa agglutinin (VVA) binding cells in the peripheral blood. These cells represent a major subset of activated peripheral T cells. VVA binding T lymphocytes express either the T8 or the T4 determinant on their cell surface. In contrast in normal controls only a minor subset of peripheral T cell expresses binding sites for VVA. The majority of these cells coexpress T8. VVA binding T cells display no helper activity. Only in a subfraction of patients with CD and not in normal controls these cells mediate contrasuppressor activity for Ig and in particular for IgA. This subgroup of patients is characterized by the lack of extramucosal manifestations. It has now been shown that VVA binding T cells in their majority do not possess phenotypic features of helper inducer cells. This further supports the hypothesis of their involvement in contrasuppression. Moreover it was shown that IgA produced in the presence of VVA binding T cells is IgA1 and IgA2 (ratio 2:1) which are both modulated by VVA binding T cells.

REFERENCES

1. A. Kimura, H. Wigzell, G. Holmquist, B. Ersson and P. Carlson, Selective affinity fractionation of murine cytotoxic lymphocytes (CTL), J.exp.Med.149: 473 (1979).
2. V. I. Braciale, H. P. Friedman and T. J. Braciale, A method for the preparation of Vicia villosa lectin

and rosette procedure for fractionation of lectin binding lymphocytes, J.Immunol.Meth. 43:241 (1981.
3. L. Lang, J. P. Banga, A. M. Varey, H. Gunn, A. Cooke and I. M. Roitt, Direct staining of mouse T lymphoblasts with fluoresceinated Vicia villosa lectin, Immunol. 46:769 (1982).
4. R. K. Gershon, D. D. Eardley, S. Durum, D. R. Green F. W. Shen, K. Yamauchi, H. Cantor and D. B. Murphy, Contrasuppression, a novel immunoregulatory activity, J.exp.Med. 153:1533-46 (1981).
5. D. R. Green, J. Gold, S. St.Martin and R. Gershon, Microenvironmental immunoregulation: possible role of contrasuppressor cells in maintaining immune responses in gut-associated lymphoid tissue, Proc.Natl.Acad., Sci. USA, 79:889-92 (1982).
6. T. Lehner, J. Avery and T. Jones, Separation and Characterisation of a subset of human $T8^+$ cells which function as antigen presenting and contrasuppressor cells, Immunol. 54:713-22 (1985).
7. T. Lehner, J. Avery and T. Jones, The role of a human antigen specific $T8^+$ cell subset in antigen presentation, helper function and contrasuppression, Clin. exp.Immunol. 61:2o3-213 (1985).
8. K. Yamauchi, M. Taniguchi, D. Green and R. K. Gershon, The use of a monoclonal I-J-specific antibody to distinguish cells in the feedback suppression circuit from those in the contrasuppressor circuit, Immunogenetics 16:551-58 (1982).
9. T. Lehner, Antigenpresenting contrasuppressor human T cells, Immunol. Today 7:87-92 (1986).
10. D. R. Green, Immunoregulation by contrasuppressor cells, Eos 4, 3:155-61 (1983).
11. A. Raedler, H. J. Lenz, K. Sandgren, A. de Weerth, K. H. Schulz, H. G. Thiele and H. Greten, Phenotype and functional properties of Vicia villosa agglutinin (VVA) binding T cells in patients with Crohn's disease, Submitted.
12. C. D. Elson, M. F. Kagnoff, C. Fiocchi, A. D. Befus and S. Targan, Intestinal immunity and inflammation: Recent progress, Gastroenterology 91: 746-68 (1986).
13. C. D. Elson, E. Machelski, D. B. Weiserbs, T cell B cell regulation in the intestinal lamina propria in Crohn's disease, Gastroenterology 89:321-27 (1985).
14. B. J. Underdown and J. M. Schiff, Immunoglobulin A: Strategic defense initiative at the mucosal surface, Ann.Rev.Immunol. 4:389-417 (1986).
15. K. Baklien and P. Brandzaeg, Comparative mapping of the local distribution of immunoglobulin-containing cells in ulcerative colitis and Crohn's disease of the colon, Clin.exp.Immunol. 22:197 (1975).
16. D. F. Keren, H. Appelman, W.D. Dobbins, J. J. Wells, B. Whisenant, J. Foley, R. Dieterle and K. Geisengra, Correlation of histopathologic evidence of the disease activity with the presence of immunoglobulin-containing cells in the colon of patients with inflammatory bowel disease, Hum.Pathol. 15:757 (1984).
17. R. P. MacDermott, G. S. Nash, M. J. Bertovich, M. V. Seiden, M. J. Bragdon and M. G. Beale, Activation of IgM, IgG and IgA synthesis and secretion by peripheral blood and intestinal mononuclear cells from

patients with ulcerative colitis and Crohn's disease, Gastroenterology 81:844 (1981).
18. R. P. MacDermott, M. G. Scott, G. S. Nash, K. Macke, M. J. Bertovich and M. H. Nahm, Increased spontaneous synthesis and secretion of immunoglobulin G subclass 1 (IgG1) by ulcerative colitis intestinal mononuclear cells, Gastroenterology 88:1483 (1985).
19. A. Raedler, S. Fraenkel, G. Klose, K. Seyfarth and H. G. Thiele, Involvement of the immune system in the pathogenesis of Crohn's disease. Expression of the T9 antigen on the peripheral immunocytes correlates with the severity of the disease, Gastroenterology 88:978-83 (1985).
2o. A. Raedler, S. Fraenkel, G. Klose and H. G. Thiele, Elevated numbers of peripheral T cells in inflammatory bowel diseases displaying T9 antigen and Fc-alpha receptors, Clin.exp.Immunol. 6o:518-24 (1985).
21. W. H. Kutteh, W. J. Koopman, M. E. Conley, M. L. Egnan and J. Mestecky, Production of predominately IgA by human peripheral blood lymphocytes stimulated in vitro with mitogens, J.exp.Med. 152:1424-29 (198o).

ACTIVATED Fc ALPHA T CELLS IN CROHN'S DISEASE ARE INVOLVED IN REGULATION OF IgA

A. Raedler, S. Schreiber, K.H. Schulz, S. Peters, H. Greten, H.G. Thiele

Med. Dept. Univ. D 2000 Hamburg 2o, FRG

INTRODUCTION

In patients with Crohn's disease (CD) elevated number of peripheral T cells exhibit activation associated determinants on their cells surface (1-3). The number of activated peripheral T cells (APT) is correlated to the clinical activity (2), to the number of activated T cells in the inflamed mucosa and to the pathohistological features of ileal biopsies (4). One possible explanation of the close relationship between inflammation in the mucosa and peripheral activation of lymphocytes could be an origin of APT from inflamed areas of the gut mucosa. Since IgA is the major isotype of the humoral immune response of the gut associated lymphoid tissue (Galt), APT were analysed in regard to their ability to modulate IgA secretion of autologous B cells.

MATERIAL AND METHODS

Patients

For analysis peripheral lymphocytes of patients with active CD (n=41, CDAI > 150, T9 > 15%) and controls (n=37) including other autoimmune diseases (n=12), systemic lupus erythematodes (n=6), immunovasculitis (n=4), Wegener's granulomatosis (n=2), infectiousenteritis (n=5) and normal controls (n=20) were examined. Immunohistological analysis of ileal mucosa was performed on biopsies taken from patients with active CD (n=11) and normal controls (n=7).

Preparation of T lymphocytes

Samples of 10 ml blood were drawn and added to 10 ml 0.9% saline supplemented with 200 units heparine. 2.5 ml of 'Lymphoprep' (Nyegaard) were overlayered with 5.0 ml of diluted blood and centrifuged for 30 min (20°C, 560 g). The lymphocytes were recovered from the interphase and washed twice with phosphate-buffered saline (PBS, GIBCO, Mass.). The pellet was resuspended in 15 ml of cell culture medium

(RPMI, GIBCO, Mass.) supplemented with 10% fetal calf serum (FCS, MEDAC, FRG) and cultured for 2 h (37°C, 5% CO_2) for removal of monocytes. Non-adherent cells were then collected and layered on Petri-dishes (5.5 cm, 5 x 10^6 cells per plate coated with Fab_2 goat anti-human immunoglobulin (TAGO, Cal., 4 ml, 10 µl/ml). After incubation for 60 min at room temperature non-adherent (non-surface Ig-bearing) cells were recovered.

Phenotype analysis using monoclonal antibodies

The following monoclonal antibodies were used: OKT3 (Ortho-mune, peripheral T cells) for efficiency-control of T cell panning and for immunohistologic experiments, Leu 12 (Becton Dickinson, Utah, peripheral B cells for efficiency of B cell panning), Leu M1 (Becton Dickinson, Utah, peripheral monocytes for efficiency of monocyte panning). All monoclonal antibodies were used incubating 1 x 10^5 - 1 x 10^6 cells in 5 µl (30 min, 4°C) of the antibody under study and 45 µl medium supplemented with FCS. Control experiments were performed omitting the specific antibody or using normal mouse serum instead. Additionally, cell-lines were used to standardized the percentage of positive cells for the antibodies used.
As second antibodies FITC-coupled (Fab_2) goat-anti-mouse Ig of the subclass appropriate for the monoclonal antibody were used (5 µl diluted 1 : 60 in medium/FCS, 30 min at 4°C). After each labeling step cells were thoroughly washed twice in medium supplemented with FCS 1 : 10.

Detection of IgA-receptors on T cells (Fcα-R^+ T cells)

5 x 10^6 glas adherent, non-surface-Ig-bearing cells were incubated in 25 µl heataggregated human IgA (10 µ/ml, 30 min, 4°C, BEHRING, FRG) and then reacted with TRITC-coupled (Fab_2) goat anti-human IgA (1 : 5, 30 min, 4°C, MEDAC, FRG). Additionally biotin-coupled aggregated human IgA was used as controls (10 mg/ml, 30 min, 4°C) followed by incubation in avidin-FITC (1 : 10, 30 min, 4°C, MEDAC, FRG). Further controls were performed using heat-aggregated IgM and IgG-preparations, rosetting with IgA-coupled ox-erythrocytes or omitting incubation with aggregated IgA.

In double labeling experiments the T cells under study were incubated in monoclonal antibodies or aggregated human IgA alone and in a mixture of both antibodies and IgA, respectively. The subsets positive for both ligands were assessed by adjusting the numbers of labeled cells in single incubation with labeled cells and in the incubation of a mixture of both ligands.

Assessment of fluorescence labeled cells

Samples were analysed counting 1 x 100 cells under a fluorescence microscope (ZEISS, FRG) and/or measured by a FACS IV cell sorter (Becton Dickinson, Utah) as described previously (2).

Panning procedures

Panning via Fcα-positive T cells was performed by incubation of 5 x 10^6 cells in 25 µl of heat-aggregated human IgA (10 µl, 30 min, 4°C, BEHRING, FRG) and layering of these cells on Petri-dishes previously coated with (Fab_2) anti-human IgA (4 ml, 10 µl/ml, MEDAC, FRG) or (Fab_2) goat-anti-human Ig- light chains (4 ml, 10 µl/ml, 12 h, 4°C, MEDAC, FRG). Panning via PNA was achieved giving 5 x 10^6 T cells on Petri-dishes previously coated with PNA (3 ml, 10 µl/ml, diluted to 0.001%, 12 h, 4°C). Non-adherent cells were carefully decanted after 1 h and adherent cells were harvested after a further culture interval of 12 h in RPMI 1640 supplemented with 10% FCS at 37°C.

Controls of panning were performed by labeling of non-adherent cells and adherent cells with aggregated human IgA.

Autologous B cells assays

1 x 10^6 separated B cells were cultured for 7 days in RPMI 1640 supplemented with 10% FCS and 50 mg/ml pokeweed mitogen (PWM, GIBCO, Mass.) and antibiotics. 1 x 10^6 autologous T cells or Fcα-R^+ T cells or Fc -R^- T cells separated by panning procedures as described above, were added. Supernatants were analysed for Ig and IgA using an ELISA-technique.

Immunhistological analysis of biopsies

Biopsies of the ileum won by endoscopy were immediately deep-frozen and then cut into slices of 6 µm at a temperature of -18°C. Slices were mounted on glass slides, airdried and fixed in acetone. After washing in TRIS-buffered saline the specimens were incubated either in a mixture of OKT3 (Orthomune, 5 µl/45 µl AB Rh^+ serum 30 min at 20°C) and twice washed in TRIS-buffered saline. The preparations were then incubated with aggregated human IgA (10 µl/ml, 30 min at 20°C) FITC-coupled goat-anti-mouse (Fab_2) IgG (TAGO, 2.5 µl/45 µl AB Rh^+ serum 30 min at 20°C) and TRITC-coupled (Fab_2) goat anti-human IgA (1 : 5, 30 min at 20°C) was used as second ligands. The specimens were then washed as described above and embedded in glycerine-jelly. The number of Fc alpha receptor positive T lymphocytes was determined as double-labeled cells by means of a fluorescence microscope (ZEISS, FRG).

Statistics

Differences of group means were tested using the Student's t-test. Correlation analyses were performed using Pearson's product-moment correlation-coefficience and linear-regression analysis.

RESULTS

1. Determination of activated ($T9^+$), Fc-alpha binding sites bearing peripheral T cells

Peripheral T cells have been separated by panning procedures and double labeled by OKT 9 (recognizing the activation associated antigen T9 = transferrin receptor) and aggregated human IgA. It is shown that a significantly higher proportion of peripheral T cells coexpresses Fc-alpha receptors an T9 in patients with CD (n=41) compared to controls (n=37), the latter including healthy volunteers and patients with autoimmune diseases as well (Fig.1).

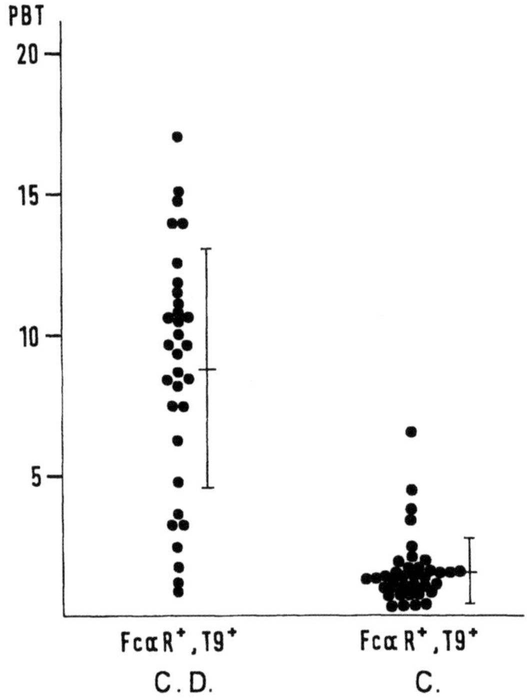

Fig. 1. Percentages of peripheral blood T cells (PBT) that coexpress T9 (=activation associated determinants) and Fc alpha receptors (Fc alphaR) in patients with Crohn's disease (CD) and controls (C) including normal volunteers (n=20), patiens with infectious enteritis (n=5) and patients with autoimmune diseases (n=12)

2. Determination of Fc-alpha receptor bearing T cells in the mucosa

Biopsies of the ileum of patients with active CD (n=11) and normal controls (NC, n=7) have been analysed by means of immunohistological methods. It is shown that the number of cells with in the lamina propria that coexpress Fc-alpha binding sites and T3 (as indicator of their T cell origin) is slightly elevated in CD compared to normal controls (p= .05) (Fig.2).

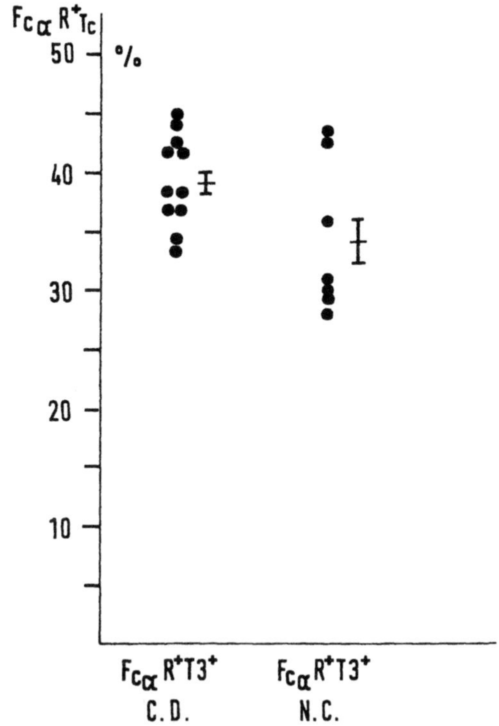

Fig. 2. Percentages and standard error of $T3^+$ T cells in the lamina propria of biopsies of the ileum coexpressing Fc alpha receptors (Fc alpha R) in patients with Crohn's disease (CD) and normal controls (NC).

3. Assessment of Ig and IgA in the supernatants of autologous B and T cell cultures

Autologous peripheral B and T cells (ratio 1 : 1) were cultured for 7d in the presence of pokeweed mitogen. Depletion of Fc-alpha receptor bearing T cells leads to an increase of IgA in patients with CD but not in NC (Fig. 3). Total Ig secretion was not significantly influenced in both experimental groups. Addition of Fc-alpha receptor expressing T cells to autologous B cells does not enhance Ig or IgA in the supernatants significantly.

DISCUSSION

Due to its biological properties IgA is the predominant isotype of the gut associated lymphoid tissue (Galt, 5,6). Intestinal secretory IgA is thought to be produced by plasma cells of the lamina propria of the gut mucosa (7). Precursor

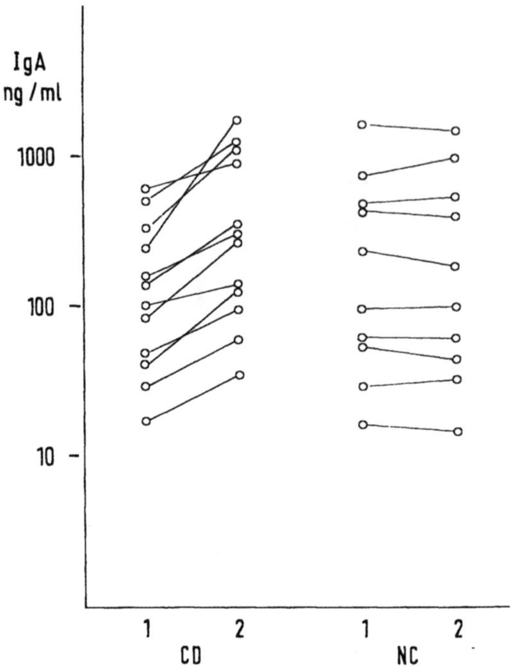

Fig. 3. Assessment of IgA in the supernatants of autologous pokeweed mitogen driven B and T cell cultures in patients with Crohn's disease (CD) and normal controls (NC). 1= B cells and total T cells (ratio 1:1), 2= B cells and T cells depleted from Fc alpha receptor bearing cells (ratio 1:1).

of IgA secreting plasma cells of the intestinal mucosa are surface IgM and IgD bearing B cells that are demonstrable in the Peyer plaques (8). Differentiation and maturation of B cells, in particular switch from IgM/IgD to IgA is suggested to be controlled by regulatory T cells (9). These T cells are thought to express already partly receptors for the Fc fragment of IgA on their cell surface (Fc alpha receptor bearing T cells) (10,11).

In patients with CD an accumulation of B cells and plasma cells in the inflamed mucosa has been reported (12). This results in a relative increase of IgG and IgM compared to IgA producing B cells and plasma cells (12,13). Moreover it has been shown by MacDermott and coworkers that IgA positive B cells in the mucosa of patients with CD exhibit properties of peripheral B cells compared to mucosa associated B cells of normal controls (14,15). This observation points to a less effective demarcation of the Galt compartment at least in functional terms.

The experiments presented here showed, that within the population of APT a subset is demonstrable that coexpresses Fc alpha binding sites on their cell surface. A comparable subset of Fc alpha receptor positive APT could not be detected in other disorders that are accompanied with increase of APT. Fc alpha receptor positive APT exhibit suppressor activity to autologous B cells in regard to IgA whereas other Ig isotypes were not significantly influenced. T cells bearing Fc alpha receptors that exhibit suppressive effects on the production of the IgA response (16). These results must be interpreted however cautiously since the B and T lymphocytes under study were cultured in the presence of pokeweed mitogen which is known to modulate IgA secretion of B cells (17,18).

The observation that activated peripheral T cells in CD are competent for IgA regulation supports the hypothesis that these cells may be involved in the process of inflammation in the gut mucosa. Indeed an increased number of Fc alpha receptors bearing T cells have been detected in the mucosa of patients with CD. However whether a relative downregulation of IgA take place also in the involved mucosa could not be deduced from the data presented.

Further experiments using mucosa standing regulatory T cells in the presence of a more appropriate mitogenic stimulus are under way.

SUMMARY

Activated peripheral T cells (APT) of patients with Crohn's disease (CD) have been analysed for the expression of IgA-Fc receptors and for competence of IgA regulation. It was found that within the subset of APT an increased number of cells express binding sites for IgA (IgA-Fc), that was not found in other diseases with elevated numbers of APT. Moreover the number of IgA-Fc receptor expressing T cells was found to be increased in the inflamed mucosa too. Cocultures with autologous B cells revealed that isolated IgA-Fc receptor bearing T cells of patients with CD suppress IgA secretion. These data support the hypothesis that APT are involved in the immunopathogenesis of CD.

REFERENCES

1. F. Pallone, S. Montano, S. Fais, M. Boirvant, A. Signore and P. Pozzilli, Studies of peripheral blood lymphocytes in Crohn's disease: Circulating activated T cells, Scand.J.Gastroenterol. 18:1oo3-8 (1983).
2. A. Raedler, S. Fraenkel, G. Klose, K. Seyfarth and H. G. Thiele, Involvement of the immune system in the pathogenesis of Crohn's disease: Expression of the T9 antigen on peripheral immunocytes correlates with the severity of the disease, Gastroenterol. 88:978-83 (1985).
3. A. Raedler, S. Fraenkel, G. Klose and H. G. Thiele, Elevated numbers of peripheral T cells in inflammatory bowel diseases displaying T9 antigen and Fc-alpha receptors, Clin.exp.Immunol. 6o:518-24 (1985).
4. A. Raedler, S. Schreiber, B. Brinkmann, A. de Weerth, H. G. Thiele and H. Greten, Assessment of activated T alpha cells in Crohn's disease, Submitted for publication.

5. P. A. Crabbé, A. D. Carbonara and J. J. Heremans, The normal human intestinal mucosa as a major source of plasma cells containing gamma A-immunoglobulin, Lab. Invest. 14:235 (1965).
6. B. J. Underdown and J. M. Schiff, Immunoglobulin A: Strategic defense initiative at the mucosal surface, Ann. Rev.Immunol. 4:389-417 (1986).
7. A. J. Husband and J. L. Gowans, The origin and antigen-dependent distribution of IgA-containing cells in the intestine, J.exp.Med. 148:1146-53 (1978).
8. J. Tseng, A population of resting IgM-IgD double bearing lymphocytes in Peyer's plaques: The major precursor cells for IgA plasma cells in the gut lamina propria, J.Immunol. 132, 6:273o-35 (1984)
9. H. Kawanishi, L. E. Saltzman and W. Strober, Mechanisms regulating IgA class-specific immunoglobulin production in murine gut-associated lymphoid tissues, J.exp. Med. 157:433-5o (1983).
10. H. Kiyono, J. R. McGhee, L. M. Mostellerr, J. H. Eldridge, W. J. Koopman, J. F. Kerney and S. M. Michalek, Murine Peyer's patch T cell clones: Characterisation of antigen specific helper T cells for immunoglobulin A responses, J.exp.Med. 156:1115-30 (1982).
11. M. Endoh, H. Sakai, Y. Nomoto, Y. Tomino and H. Kaneshige, IgA-specific helper activity of T cells in human peripheral blood, J.Immunol. 127:2612-13 (1981).
12. K. Baklien.and P. Brandzaeg, Comparative mapping of the local distribution of immunoglobulin-containing cells in ulcerative colitis and Crohn's disease of the colon, Clin.exp.Immunol. 22:197 (1975).
13. D. F. Keren, H. Appelman, W. D. Dobbins, J. J. Wells, B. Whisenant, J. Foley, R. Dieterle and K. Geisengra, Correlation of histopathologic evidence of the disease activity with the presence of immunoglobulin-containing cells in the colon of patients with inflammatory bowel disease, Hum.Pathol. 15:757 (1984).
14. R. P. MacDermott, G. S. Nash, M. J. Bertovich, M. V. Seiden, M. J. Bragdon and M. G. Beale, Activation of IgM, IgG and IgA synthesis and secretion by peripheral blood and intestinal mononuclear cells from patients with ulcerative colitis and Crohn's disease, Gastroenterology 81:844 (1981).
15. R. P. MacDermott, M. G. Scott, G. S. Nash, K. Macke, M. J. Bertovich and M. H. Nahm, Increased spontaneous synthesis and secretion of immunoglobulin G subclass 1 (IgG1) by ulcerative colitis intestinal mononuclear cells, Gastroenterology 88:1483 (1985).
16. R. G. Hoover and R. G. Lynch, Isotype-specific suppression of IgA: Suppression of IgA responses in Balb/c mice by T alpha cells, J.Immunol. 1320, 2:521-23 (19 (1983).
17. W. H. Kutteh, W. J. Koopman, M. E. Conley, M. L. Egan and J. Mestecky, Production of predominantely poly-

meric IgA by human peripheral blood lymphocytes stimulated in vitro with mitogens, J.exp.Med. 152:1424-29 (1980).
18. J. Pryjma, H. D. Flad, M. Ernst and A. Pituch-Noworolska, Pokeweed mitogen activated suppressor T cells preferentially reduce immunoglobulin secretion by differentiated B lymphocytes, Immunolog.letters 13:273-79 (1986).

INDUCTION OF A COMBINED MUCOSAL AND SYSTEMIC ANTI-OVALBUMIN RESPONSE

A.T.J. Bianchi, R.J. Zwart and P.J. van der Heijden

Dept. Immunology, Central Veterinary Institute, P.O.Box 65
8200 AB, Lelystad, The Netherlands

INTRODUCTION

It is generally accepted that a relatively independent mucosal immune system exists next to a systemic immune system (1). Antigen presentation by the mucosal route can evoke mucosal immune responses (2), while the systemic immune response is suppressed. Contrary to the latter oral tolerance phenomenon (2) parenteral immunization for induction of efficient systemic immunity, sometimes has a suppressive effect on the mucosal immune response (3,4).

After vaccination these opposite effects can be a handicap since mutual activation of both limbs of the immune system will be more effective. Up till now most vaccination regimes mainly induce efficient systemic immunity and leave the mucosa, which is the porte d'entree of most infectious agents, relatively unprotected.

The introduction of new vaccines, like subunit or peptide vaccines, introduces an additional problem since these vaccines are less effective than live vaccines, especially in relation to induction of mucosal immunity (5). Therefore there is a great need for the development of more effective vaccination regimes.

Using ovalbumin (OA) as a passive protein antigen, we investigated in this study the feasibility to induce a combined systemic and local immune response in mice.

MATERIALS AND METHODS

Mice: Female C3H/He mice were purchased from HARLAN-OLAC Ltd., U.K. and used at an age of 10-14 weeks.

Immunization: Mice were injected intraperitoneally (ip) with 0.1 ml of a W/O emulsion (6), containing 0.1 mg glutaraldehyde-polymerised-OA (Grade V, Sigma, U.S.A.). Four weeks later mice were boosted either ip with 0.1 mg polymerised OA in 0.1 ml saline or orally (or) with 80 mg OA and 10 μg of the B-subunit of cholera toxin (Sigma) in 0.5 ml $NaHCO_3$.

In situ detection of anti-OA containing cells: For detection of OA-specific antibody containing cells (OACC) essentially the method described by Sminia et al. (9) was followed. Cryostat sections of lymphoid tissue and microscopical preparations of BM cell suspensions were incubated respectively with OA, Rabbit-anti-OA (Pel-Freeze, U.S.A.) and peroxidase-coupled Swine-anti-rabbit Ig (DAKO, Denmark). From each individual mouse the number of OACC was counted in an average section of the various organs. The mean OACC response per PP was calculated after counting the OACC in all visible PP in one average Swiss roll section of the SI.

Assay for OA specific plaque forming cells: Cell suspensions from spleen, Peyer's patches (PP), mesenteric lymph nodes (MLN), bone marrow (BM) and the small intestine (SI) were prepared and assayed for IgM-, IgG- and IgA OA-specific plaque forming cells (PFC) as described previously (7,8).

RESULTS

Semi-quantitative analysis by in situ detection of OACC after secondary immunization: Mice were ip primed with OA in a W/O emulsion and boosted by ip or oral OA application 4 weeks later. On day 2, 3, 4, 5, 6, 7 and 9 after booster, organs of five mice were prepared for immunohistological study. The number of OACC was enumerated in sections of PP, MLN, spleen and SI and in microscopical preparations of BM suspensions. The localization was studied in the sections of these organs. To evaluate the contribution of a primary response, the same study was done on day 3, 5, 7 and 32 after ip priming. The quantitative results are shown in fig. 1 A-E.

Only a few OACC are observed in the white pulp of the spleen during the first week after priming. After 4 weeks a small persisting primary response can be found in the spleen only.

Both ip and oral boosters give rise to significant OACC responses in all organs. The kinetics of the OACC responses in spleen, MLN and BM are almost identical after both booster regimes (Fig. 1 B, C, E). OACC responses in MLN and spleen are observed already on day 2 after booster and reach their maximum around day 4, while in BM the OACC responses reach their maximum around day 5.

Striking differences between the ip and oral boosted mice are found in PP and SI (Fig. 1 A, D). In the oral boosted mice, the response in PP reaches its maximum on day 3, while the maximal response in the ip boosted mice is found 3 days later. The response in the SI of the oral boosted mice is about 10 times as high as in the ip boosted mice. The intestinal response is persisting at its maximal level during a longer period than in the other organs.

There is no difference in the localization of the OACC in ip or orally boosted mice. In the PP, most OACC are localized in the interfollicular area and in the MLN mainly in the paracortical zone. The OACC in the spleen are at first localized in the T-cell area of the PALS, but are also observed in the red pulp later during the response. In the SI most OACC are observed in the lamina propria at the base of the villi.

In general most OACC are located in the T-cell areas. Some secondary B-cell follicles with OA-positive B-cells are observed in MLN later during the secondary response.

Quantification of OA-specific PFC after secondary immunization: PFC tests were performed with cell suspensions of PP, MLN, spleen, SI and BM five days after ip or oral booster immunization (Table 1). The results show the number of OA-specific antibody secreting cells, and the isotype distribution of these cells.

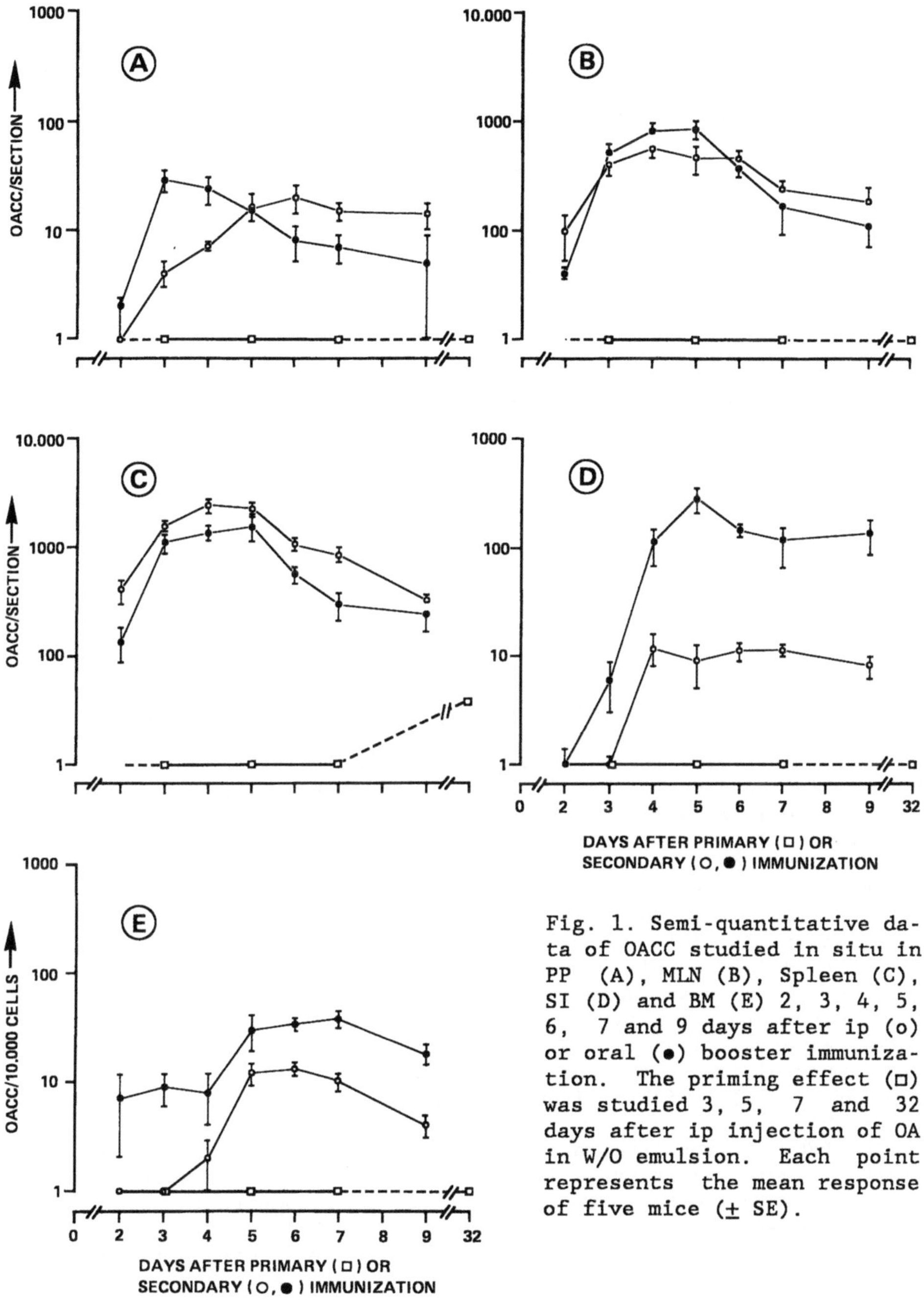

Fig. 1. Semi-quantitative data of OACC studied in situ in PP (A), MLN (B), Spleen (C), SI (D) and BM (E) 2, 3, 4, 5, 6, 7 and 9 days after ip (o) or oral (•) booster immunization. The priming effect (□) was studied 3, 5, 7 and 32 days after ip injection of OA in W/O emulsion. Each point represents the mean response of five mice (± SE).

After oral booster the OA-specific response has shifted from a dominant IgG-response to an IgG- and IgA response. The most striking effect was found in the SI: After oral booster the Ig-PFC response to OA is about 50-fold higher than in ip boosted mice.

Table 1. The isotype distribution of OA-specific antibody secreting cells in different organs.

	IgM		IgG		IgA	
	ip*	or	ip	or	ip	or
PP	0**	20	285	570	60	750
MLN	0	0	6.000	10.000	50	2.600
Spleen	2.300	750	250.000	134.500	5.050	34.000
SI	0	0	620	1.000	520	50.000
BM	232	7	23.600	16.300	168	1.600

* C3H/He mice, ip primed with OA, were boosted ip or orally 4 weeks later. Five days after booster PFC responses were tested in different organs.

** The PFC numbers present the mean IgM-, IgG- or IgA PFC response of 5 mice.

DISCUSSION

To evaluate the feasibility of a mutual activation of the mucosal and the systemic immune system, we studied a combination of ip priming in adjuvant followed by an oral booster. The anti-OA response in orally boosted mice was compared with the response in ip boosted mice. Several studies have shown that ip priming in adjuvant may prime for a mucosal immune response (10,12) and in general, this route is appropriate for systemic immunization. Moreover, Hanson et al. (13) demonstrated with OA as antigen that such a priming prevents induction of oral tolerance after a subsequent oral booster.

The results described in this paper clearly show (Fig. 1 & Table 1) that ip priming followed by oral booster induces both a mucosal and a systemic anti-OA B-cell response. The induction of an efficient mucosal immune response against a passive protein like OA is rarely observed, since protein antigens are normally bad immunogens for mucosal immune responses (19). For that reason, studies describing the induction of mucosal immune responses are often using cholera toxin as antigen (3-5,10,11), which by itself has special "mucosal" adjuvant properties (5).

The PFC test results (Table 1) show that, in spite of a relatively high systemic IgA response in the spleen after oral booster, a substantial splenic IgG response is found, which is comparable with the IgG response after ip booster immunization.

Several studies (15,16) showed that oral or mucosal priming with protein antigens induces IgA memory which could be expressed at mucosal as well as at systemic sites by local or systemic booster. Our results cannot be explained by the induction of similar IgA memory cells after ip priming, because after ip booster hardly any OA-specific IgA PFC response is found (Table 1). It is more likely that ip injection of OA in adjuvant mainly primes for an efficient IgG memory. The adjuvant seems to be essential (17), as it has been shown that ip priming without adjuvant or priming by other parenteral routes induces a state of mucosal suppression (3).

The induction in our experiment of OA-specific IgA secreting (B)--cells after booster can be explained as follows:

1. After booster at least part of the or presented OA will be processed by the PP where IgG memory cells may undergo heavy chain class switching by IgA specific T helper cells (18) to IgA producing (B)cells. Early activation of the PP after oral booster is also suggested by the early appearance of OACC in the interfollicular areas, which are absent in ip boosted mice (Fig. 1 A).

2. Subsequently, most OA specific IgA-positive B-cells migrate from the PP to the MLN and ultimately to the intestinal mucosa, but a substantial part also migrate to the spleen. The same explanation for IgA memory expression in the intestine has been suggested in other studies (11,19).

The existence of sessile IgA memory cells e.g. in the mucosa of the SI (11), directly induced after ip priming cannot be ruled out, but these memory cells cannot explain the discrepancy between the high splenic IgA response after oral booster and the low splenic IgA response after ip booster (Table 1).

Although several aspects of the mechanisms involved remain unexplained this study clearly indicates that ip priming with OA in adjuvant followed by oral booster is successful for mutual activation of both limbs of the immune system.

ACKNOWLEDGEMENTS

We thank Drs. B.A. Bokhout and G. Koch for their critical reading of the manuscript and Bernie Moonen-Leusen for excellent technical assistance.

REFERENCES

1. Bienenstock, J. and A.D. Befus. Immunology 41 : 249 (1980)
2. Challacombe, S.J. and T.B. Tomasi. J. Exp. Med. 152 : 1459 (1980)
3. Pierce, N.F. and F.T. Koster. J. Immunol. 124 : 307 (1980)
4. Koster, F.T. and N.F. Pierce. J. Immunol. 131 : 115 (1983)
5. Lycke, N. and J. Holmgren. Immunology 59 : 301 (1986)
6. Bokhout, B.A., C. van Gaalen and P.J. van der Heijden. Vet. Immunol. Immunopathol. 2 : 491 (1981).
7. Benner, R., F. Meima, G. van der Meulen and W.B. van Muiswinkel. Immunology 26 : 247 (1974).
8. Van der Heijden, P.J. and W. Stok. J. Immunol. Meth. in press (1987)
9. Sminia, T., F. Delamarre and E.M. Janse. Immunology 50 : 53 (1983)
10. Pierce, N.F. and J.L. Gowans. J. Exp. Med. 142 : 1550 (1975)
11. Husband, A.J. and J.L. Gowans. J. Exp. Med. 148 : 1146 (1978)
12. Scicchitano, R., A.J. Husband and R.L. Clancy. Immunology 53 : 375 (1984)
13. Hanson, D.G., N.M. Vaz, L.C.S.Maia and J.M. Lynch. J. Immunol. 123 : 2337 (1979)
14. Elson, C.O. and W. Ealding. J. Immunol. 133 : 2892 (1984)
15. Fuhrman, J.A. and J.J. Cebra. J. Exp. Med. 153 : 534 (1981)

16. Jeurissen, S.H.M., E. Claassen, N. van Rooijen and G. Kraal. Immunology 56 : 417 (1985)
17. Keren, D.F., P.J. Scott, R.A. McDonald and M. Wiatrak. Inf. Immun. 42 : 202 (1983)
18. Kawanishi, H., L.E. Saltzmann and W. Strober. J. Exp. Med. 157 : 433 (1983)
19. Lycke, N. and J. Holmgren. Scand. J. Immunol. 23 : 611 (1986)
20. Pierce, N.F. and W.C. Cray. J. Immunol. 128 : 1311 (1982)

DIFFERENTIAL EXPRESSION OF EPITHELIAL MHC CLASS II DETERMINANTS (HLA-DR, -DP, AND -DQ) AND INCREASED CLASS I EXPRESSION IN INFLAMED SALIVARY GLANDS

Per S. Thrane[1,2] and Per Brandtzaeg[1]

[1]Laboratory for Immunohistochemistry and Immunopathology (LIIPAT), Institute of Pathology, Medical Faculty, The National Hospital, Rikshospitalet, and [2]Department of Oral Surgery and Oral Medicine, Dental Faculty, University of Oslo, N-0027 Oslo 1, Norway

INTRODUCTION

Sjögren's syndrome (SS) affects salivary and lacrimal glands, causing xerostomia and keratoconjunivitis. An increased number of infiltrating leukocytes are present and the glandular tissue is gradually destroyed. Increased epithelial HLA-DR expression in salivary glands from SS patients has been reported (1,2,3), but the other MHC class II subregion products (HLA-DP and -DQ) and MHC class I determinants have not been studied.

Expression of the various class II subregion products is of interest in light of the recently published association between SS and HLA-DQ (6). A potential role of epithelial class II expression in the initiation of autoimmune disease (4,5) is indeed intriguing. However, evaluation of this hypothesis requires inclusion of inflamed glands unrelated to autoimmunity; such controls have been lacking in previous studies. There is, furthermore, disagreement with regard to the number of activated T cells in SS glands (7,8,9).

Several lymphokines derived from activated mononuclear cells [interferon-γ (IFN-γ), lymphotoxin (LT) and probably also tumor necrosis factor-α (TNF-α)] have been shown to modulate MHC class I and II expression (10,11). We therefore wanted to study whether a spatial relationship exists between T-cell infiltration and epithelial expression of MHC class I and II determinants in SS and various control glands.

MATERIAL AND METHODS

Patients

The tissue sites of the material studied will be apparent from Table 3.

Normal control subjects. Tissue specimens from five normal salivary glands were included. The subjects (three women and two men: median age, 48 years; range, 37-55 years) were operated because of benign tumors or preprotetic adjustment, and the specimens were taken well away from the lesions.

Control patients. Four women and one man (median age, 56,2 years;

range, 34-76 years) with obstructive adenitis were included. Special care was taken to exclude that these patients had a history or clinical signs of any autoimmune disease. Four of them were operated because of obstruction of the salivary gland duct, with subsequent inflammation of the glands, and one because of pleomorphic adenoma.

Sjögren patients. Salivary gland specimens from eight patients with SS (six women and two men: median age, 58,6 years; range, 41-70 years) were examined. They had all reduced salivation rate and fulfilled a focus score of ≥1 (12) on histological examination. Six of the patients had positive Rose-Bengal and/or Schirmer's test, indicating reduced tear flow and conjuntitis. Furthermore, six had elevated erythrocyte sedimentation rate (ESR), five had positive Waaler's and/or Latex fixation test, and four had joint problems. None of them had any verified rheumatoid disease, and they had not been treated with corticosteroids (two used Naprosyn, 250 mg x 4). Four of the patients had elevated serum IgG levels. The clinicopathological data on the Sjögren patients are reported elsewhere (13).

Immunohistochemistry

Tissue processing and immunostaining. The salivary gland specimens were immediately fixed in cold 96% ethanol and embedded in paraffin (14). Serial sections were cut at 6 µm and incubated for 20 h at room temperature with various monoclonal antibodies (Table 1). Paired staining was performed according to Brandtzaeg and Rognum (15) and alkaline phosphatase anti-alkaline phosphatase (APAAP) staining (16) according to Brandtzaeg et al. (17).

Antibodies. The specificities and sources of the primary monoclonal antiboies are listed in Table 1. A polyclonal antibody reagent reacting with keratin (R505) (23) was used to demonstrate epithelium. Secondary antibodies included a biotinylated horse anti-mouse IgG (Vector laboratories, Burlingame, USA), used at 1:10 and incubated for 3 h at room temperature, and fluorescein-conjugated avidin (Vector) applied at 1:100 for 30 min. The combinations of markers used for paired staining appear in Table 2.

Microscopy and photography. A Leitz Orthoplan microscope equipped with x10 and x25 immersion objectives, a x10 ocular and a Ploem-type vertical illuminator with interference filters for selective filtration of red (rhodamine) and green (fluorescein) emission was used. Findings were recorded on Agfa 1000 ASA film, and double exposures were taken to document results of paired staining.

Staining evaluation. Epithelial fluorescence was scored semiquantitatively on an arbitrary scale from (-) to (+++), which reflected the overall distribution of the epithelial staining: virtually no epithelial staining, (-); <5% overall epithelial staining, (+); 5-50% epithelial staining, (++); >50% epithelial staining, (+++). Furthermore, the intensity was scored as intense staining (I), faint (F) or heterogeneous, that is, both intense and faint (IF).

RESULTS

Epithelial expression of MHC class I. MHC class I (HLA-A, -B, -C) determinants were expressed by most epithelial cells (>50%) in normal as well as in inflamed salivary glands. Both acini and minor and major ducts were stained, although rather heterogeneously (Fig. 1f). Faint staining was observed in normal as compared with inflamed glands in which the staining was intensified adjacent to larger T-cell infiltrates, including numerous activated (DR-positive) T lymphocytes. The intensity of the staining thus parallelled the severity of inflammation in the glands of both control patients and SS patients.

Table 1. Murine monoclonal antibodies.

Mo Ab	Specificity	Isotype	Preparation	Working dilution	References
L 243[a]	HLA-DR	IgG2a	Purified Ig	1:20	16
HKB1[b]	HLA-DR, DP (except DP2)	IgM	Ascites	1:4000	
Anti Leu 10[a]	HLA-DQ (except DQ_w2)	IgG1	Purified Ig	1:20	17
B7/21[a]	HLA-DP	IgG1	Purified Ig	1:20	18
W6/32[c]	HLA-A,B,C	IgG2[a]	Ascites	1:80	19
Anti-Leu 4[a]	T cells, CD3	IgG1	Purified Ig	1:20	20

[a]Purchased from Becton-Dickinson, Sunnyvale, California, U.S.A.
[b]Kindly provided by Dr. S. Funderud, Oslo, Norway
[c]Purchased from Sera-lab, Sussex, U.K.

Table 2. Combination of markers demonstrated by paired immunofluorescence staining.

Rhodamine label	Fluorescein label
1. Broad keratin	HLA-DR
2. Broad keratin	HLA-DP
3. Broad keratin	HLA-DQ
4. Broad keratin	MHC class I

Table 3. Epithelial staining of MHC class I (HLA-A, -B, -C) and II (HLA-DR, -DP, -DQ) determinants.

Patients	Age/sex	Gland	Diagnosis	HLA-DR	HLA-DP	HLA-DQ	HLA-A-B-C
1	47/♀	P[1]		++[2]	+	-	+++ IF[3]
2	37/♂	P		+	+	-	+++ IF
3	49/♀	P		+	+	+	Nd[4]
4	52/♀	L[5]		-	-	-	++ F
5	55/♂	L		-	-	-	+++ F
6	34/♀	P		+++	+	+	+++ IF
7	55/♀	L		-	-	-	+++ F
8	58/♀	L		+	-	-	+++ IF
9	58/♂	SM[6]		+++	+	+	+++ IF
10	76/♀	SM		+++	+	+	+++ IF
11	64/♀	P		+++	++	+	+++ IF
12	65/♂	P		+++	+	-	+++ IF
13	66/♂	P		+++	++	+	+++ IF
14	41/♀	P		+++	++	++	+++ IF
15	57/♀	L		+++	Nd	Nd	++ F
16	70/♀	L		++	+	-	+++ IF
17	47/♀	L		++	-	-	++ IF
18	59/♀	L		++	+	-	+++ IF

[1]Parotid gland
[2]-, virtually no staining; +, 0-5% of the epithelial cells stained; ++, 5-50% epithelial staining; +++, >50% epithelial staining
[3]I, intense; F, faint; IF, both intense and faint epithelial cells (heterogeneous staining)
[4]Nd, not determined
[5]Inferior labial gland
[6]Submandibular gland

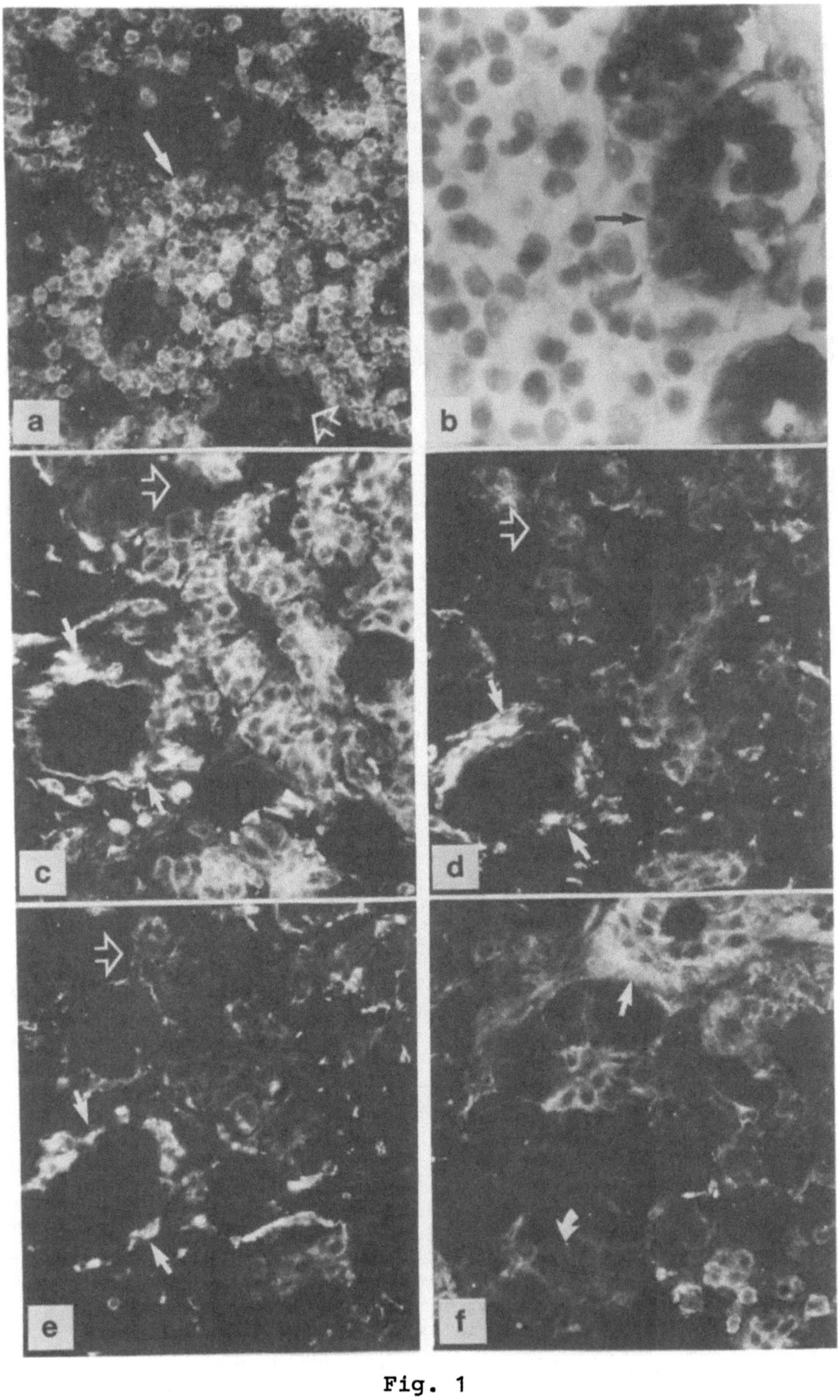

Fig. 1

Fig. 1a. Immunofluorescence staining for Leu 4 (CD3) in ethanol-fixed salivary gland tissue from patient with Sjögren's syndrome. Note numerous infiltrating T cells (closed arrow) surrounding remaining minor ducts (open arrow). Magnification: x420.

Fig. 1b. APAAP staining for HLA-DR determinants in ethanol-fixed salivary gland tissue from patient with Sjögren's syndrome. Remaining ducts (arrow) express HLA-DR determinants. Magnification: x625.

Fig. 1c. Immunofluorescence staining for HLA-DR determinants in ethanol-fixed salivary gland tissue from patient with Sjögren's syndrome. Note extensive epithelial DR positivity (area between open arrows). The DR-positive vessel (closed arrow) allows orientation in serial sections, cf. Figs. 1d and 1e. Magnification: x420.

Fig. 1d. Immunofluorescence staining for HLA-DP in section adjacent to Fig. 1c. Note faint DP positivity in some of the epithelial cells that expressed DR determinants (area between open arrows) in Fig. 1c. The DR-positive vessel (closed arrow) in Fig. 1c co-expresses HLA-DP determinants. Magnification: x420.

Fig. 1e. Immunofluorescence staining for HLA-DQ in section adjacent to Fig. 1d. Only faint epithelial DQ staining is seen in a few epithelial cells (area between open arrows). DQ positivity is also seen in vessel wall (closed arrow) and scattered leukocytes. Magnification: x420.

Fig. 1f. Immunofluorescence staining for MHC class I determinants in ethanol-fixed salivary gland tissue from patient with Sjögren's syndrome. MHC class I molecules are most extensively expressed by various duct cells (straight arrow). The class I staining is rather faint and sometimes absent from acinar cells (curved arrow). Magnification: x420.

Differential expression MHC class II. HLA-DR, -DP, and -DQ determinants were differentially expressed in both normal and inflamed glands of both disease categories (Table 3). Glandular epithelium (both acinar and minor duct cells) stained with various intensities following the pattern DR>DP>DQ. This differential expression was furthermore related to the severity of inflammation and the proximity to accumulations of T lymphocytes. SS glands generally contained larger numbers of leukocytes (and had higher focus scores) than both normal and inflamed control specimens (Fig. 1a).

In normal parotid glands, only some intercalated duct cells (ICD) and scattered acinar cells (AC) expressed class II determinants (<5% of total number of epithelial cells); the relative proportion of positive epithelial cells (HLA-DR/-DP/-DQ ~6/2/1) indicated differential expression. Major salivary glands from patients with obstructive adenitis expressed HLA-DR determinants extensively but only little HLA-DP and -DQ (DR/DP/DQ ~ 10/1/1). Epithelial staining for HLA-DR determinants was widely distributed in the SS (Fig. 1b,c). Significant epithelial staining for HLA-DP (Fig. 1d) and -DQ (Fig. 1e) was also seen in these specimens (DR/DP/DQ ~ 10/5/4). This contrasted with the more restricted distribution of HLA-DP and -DQ in obstructive adenitis. Staining in serial sections indicated that HLA-DP and DQ were expressed concomitantly with HLA-DR.

DISCUSSION

The most striking observation in this study was the virtual absence of epithelial HLA-DP and DQ expression in chronic obstructive salivary glands in contrast to SS glands. This might merely reflect the fact that a smaller number of activated T cells is present in obstructive adenitis than in SS specimens, and suggests that these subregion products apparently need a stronger stimulus than HLA-DR for their expression. The close spatial relationship between activated T cells and class II-positive epithelium in salivary glands in both disease categories, furthermore suggested an influence of immunoregulatory cytokines (IFN-γ, LT and TNF-α). Epithelial class II expression might hence be a secondary phenomen rather than an initiating event in autoimmune disease as proposed by Bottazzo et al. (4). A recent report on the effect of cyclosporin A (CyA) on the immunopathological lesion of the labial minor salivary glands from patients with SS favours this explanation (24). All the CyA-treated patients showed a decrease in the number of T "helper" (CD4 positive) lymphocytes. Concomitantly epithelial HLA-DR expression was abrogated in these patients.

Another explanation is equally conceivable. If the primary disease gene assosiation in SS is HLA-DQ (or perhaps HLA-DP) rather than HLA-DR, this would agree with a more frequent expression of this subregion product in SS as compared with obstructive adenitis. A recently published association between SS and HLA-DQ (DQ1 and 2) supports this possibility (6). DQ rather than DR molecules might thus provide the salivary gland epithelium with the putative disease-associated restriction element and be a prerequisite for the development of SS. It is also possible that DP or some still unidentified class II molecule or hybrid molecule is the disease-associated restriction element. Perhaps aberrant epithelial HLA-DQ-expression in the course of an ordinary inflammatory disease or as a result of a particular virus infection, may result in SS in genetically susceptible individuals. Epstein-Barr virus has been assosiated with SS (2) and might represent one potentially initiating agent.

In addition, our findings raise the possibility that HLA class I-associated (cytotoxic) immune reactions may occur in SS on the basis of enhanced epithelial expression of such determinants.

CONCLUSIONS

In patients with SS, the total number of HLA-DR-positive (activated) T cells was increased compared with controls. Glandular epithelium adjacent to such lymphocyte infiltrates expressed MHC class II subregion products extensively (>50% of epithelial cells positive: HLA-DR/-DP/-DQ ~ 10/5/4). A similarly increased epithelial expression of HLA-DR but with surprisingly little concomitant DP and DQ, was observed in chronically inflamed control glands (5-50% of epithelial cells positive: HLA-DR/-DP/-DQ ~ 10/1/1). MHC class I molecules were widely distributed (>50%) but the staining was heterogeneous; the most intense staining was observed adjacent to dense leukocyte infiltrates. HLA-DP and -DQ expression was not observed without concomitant HLA-DR. It seems justified to speculate that leukocyte-derived immunoregulatory factors (e.g., interferon-γ, lymphotoxin and tumor necrosis factor-α) are responsible for the observed modulation of epithelial MHC class I and II expression. It is unknown if such enhanced expression is related to the autoimmune process in SS, but it might be a prerequisite for the disease process to develop in genetically susceptible individuals.

ACKNOWLEDGEMENTS

This study was supported by the Norwegian Cancer Society, The Norwegian Research Council for Science and the Humanities, and Anders Jahre's Fund. We thank Dr. S. Funderud (Oslo, Norway) for the kind gift of reagent and Dr. L. M. Sollid and Dr. H. Scott for valuable advice.

REFERENCES

1. G. Lindahl, E. Hedfors, L. Klareskog and U. Forsum, Epithelial HLA-DR expression and T lymphocyte subsets in salivary glands in Sjögren's syndrome, Clin. Exp. Immunol. 61:475 (1985).
2. R. I. Fox, T. Bumol, R. Fantozzi, R. Bone and R. Schreiber, Expression of histocompatibility antigen HLA-DR by salivary gland epithelial cells in Sjögren's syndrome, Arthr. Rheum. 29:1105 (1986).
3. H. M. Moutsopoulos, J. J. Hooks, C. C. Chan, Y. A. Dalavanga, F. N. Skopouli and B. Detrick, HLA-DR expression by labial minor salivary gland tissues in Sjögren's syndrome. Ann. Rheum. Dis. 45:677 (1986).
4. G. F. Bottazzo, R. Pujol-Borrell and T. Hanafusa, Role of aberrant HLA-DR expression and antigen presentation in induction of endocrine autoimmunity, Lancet ii:1115 (1983).
5. I. Todd, M. Londel, R. Pujol-Borrell, R. Mirakian, M. Feldmann and G. F. Bottazzo, HLA-D/DR expression on epithelial cells: the finger on the trigger?, Ann. N.Y. Acad. Sci. 475:241 (1986).
6. J. B. Harley, M. Reichlin, F. C. Arnett, E. L. Alexander, W. B. Bias and T. T. Provost, Gene interaction at HLA-DQ enhances autoantibody production in primary Sjögren's syndrome, Science 232:1145 (1986).
7. R. I. Fox, A. N. Theofilopoulos and A. Altman, Production of interleukin 2 (IL 2) by salivary gland lymphocytes in Sjögren's syndrome. Detection of reactive cells by using antibody directed to synthetic peptides of IL 2, J. Immunol. 135:3109 (1985).
8. M. Segerberg-Konttinen, V. Bergroth, P. Jungell, M. Malmström, D. Norström, J. Sane, I. Immonen and Y. T. Konttinen, Activated lymphocytes and lymphoblasts in the minor salivary glands of patients with Sjögren's syndrome, Scand. J. Immunol. 24:79 (1986).
9. D. A. Isenberg, D. Rowe, A. Tookman, A. Hopp, Griffiths, E. Paice, J. Stewart and P. C. L. Beverley, An immunohistological study of secondary Sjögren's syndrome, Ann. Rheum. Dis. 43:470 (1984).
10. R. Pujol-Borell, I. Todd, M. Doshi, G.F. Bottazzo, R. Sutton, D. Gray, G.R. Adolf and M. Feldmann, HLA class II induction in human islet cells by interferon- plus tumour necrosis factor or lymphotoxin, Nature 326:304 (1987).
11. K. Pfizenmaier, P. Scheurich, C. Scluter and M. Krönke, Tumor necrosis factor enhances HLA-A,B,C and HLA-DR gene expression in human tumor cells, J. Immunol. 138:975 (1987).
12. T. E. Daniels, Labial salivary gland biopsy in Sjögren's syndrome. Assessment as a diagnostic criterion in 362 suspected cases, Arthr. Rheum. 27:147 (1984).
13. P. S. Thrane, L. M. Sollid, P. Brandtzaeg and H. R. Haanes, Increased expression of secretory component (SC) and HLA-DR in salivary gland epithelium from patients with Sjögren's syndrome, submitted (1987).
14. P. Brandtzaeg, Mucosal and glandular distribution of immunoglobulin components. Immunohistochemistry with cold ethanol-fixation technique, Immunology 26:1101 (1974).

15. P. Brandtzaeg and T. O. Rognum, Evaluation of tissue preparation methods and paired immunofluorescence staining for immunocytochemistry of lymphomas, Histochem. J. 15:655 (1983).
16. D. Y. Mason, Techniques in Immunohistochemistry, vol. 3, 25-43 (1985).
17. P. Brandtzaeg, I. Dale and M. K. Fagerhol, Distribution of a formalin-resistant myelimonocytic antigen (L1) in human tissues. I. Comparison with other leukocyte markers by paired immunofluorescence and immuinoenzyme staining, Am J. Clin. Pathol. 87:681 (1987).
18. L. A. Lampson and R. Levy, Two populations of Ia-like molecules on a human B cell line, J. Immunol. 125:293 (1980).
19. Y. X. Chen, R. L. Evans, M. S. Pollack, L. L. Lanier, J. H. Phillips, C. Rousso, N. L. Warner and F. M. Brodsky, Characterization and expression of the HLA-DC antigens defined by anti-Leu 10, Hum. Immunol. 10:221 (1984).
20. A. J. Watson, R. DeMars, I.S. Trowbridge and F.H. Bach, Detection of a novel class II HLA antigen, Nature 304:358 (1983).
21. C. J. Barnstable, W. F. Bodmer, G. Brown, G. Galfre, C. Milstein, A. F. Williams and A. Ziegler, Production of monoclonal antibodies to group A erythrocytes, HLA and other human cell surface antigens - new tools for genetic analysis, Cell 14:9 (1978).
22. J. A. Ledbetter, R. E. Evans, M. Lipinski, C. Cunningham Rundles, R. A. Good and L. A. Hertzenberg, Evolutionary conservation of surface molecules that distinguish T lymphocyte helper/inducer and T cytotoxic/supressor subpopulations in mouse and man, J. Exp. Med. 153:310 (1981).
23. H. Huitfeldt and P. Brandtzaeg, Various keratin antibodies produce immunohistochemical staining of human myoepithelium and myometrium, Histochemistry 83:381 (1985).
24. Y. A. Dalavanga, B. Detrick, J. J. Hooks, A. A. Droosos and H. M. Moutsopoulos, Effect of cyclosporin A (CyA) on the immunopathological lesion of the labial minor salivary glands from patients with Sjögren's syndrome, Ann. Rheum. Dis. 47:89 (1987).

JEJUNAL EPITHELIUM OF PATIENTS WITH COELIAC DISEASE SHOWS ENHANCED EXPRESSION OF MHC CLASS II SUBREGION PRODUCTS

H. Scott, L.M. Sollid, P. Brandtzaeg and E. Thorsby

Laboratory for Immunohistochemistry and Immunopathology (LIIPAT), Institute of Pathology, and Institute of Transplantation Immunology, University of Oslo, The National Hospital, Rikshospitalet, Oslo, Norway

INTRODUCTION

We have previously shown that HLA class II molecules are expressed by small intestinal epithelial cells, both in control subjects with a normal mucosa (1) and in coeliac disease (CD) patients with total villous atrophy (2). CD is HLA-associated, and particularly related to DR3, DR7 (3) and DQ2 (4,5). It develops in genetically susceptible individuals because of gluten ingestion. Antigens of gluten may be presented by enterocytes to intraepithelial T cells as an initial step in the pathogenesis of CD. Class II-positive rat enterocytes have in fact been reported to perform antigen presentation (6). In this study we used a panel of monoclonal antibodies to better define the epithelial expression of class II subregion products in CD.

MATERIALS AND METHODS

Immunohistochemical methods

Biopsy specimens of proximal jejunal mucosa from 11 untreated CD patients, 11 CD patients on a gluten-free diet and 11 controls, were immediately fixed in cold 96% ethanol and embedded in paraffin. Tissue sections were cut serially at 6 µm. One section was stained with haematoxylin and eosin for histological evaluation and the others were subjected to immunofluorescence staining. After dewaxing, they were first incubated with murine monoclonal antibodies to different HLA class II determinants or T lymphocytes (L243 and P4.1 to HLA-DR; anti-Leu 10 and Tü22 to HLA-DQ; B7/21 to HLA-DP; and UCHL1 and anti-Leu 4 to T cells). The incubation took place at room temperature for 20 h. Biotinylated horse anti-mouse IgG (Vector Laboratories, Burlingame, CA, USA), extensively absorbed with insolubilized human serum, was used as secondary reagent (0.05 g/l; room temperature; 3 h) followed by fluorescein isothiocyanate-conjugated avidin (Vector; 0.025 g/l; room temperature; 30 min). The sections were finally mounted in buffered (pH 8) polyvinyl alcohol.

Enumeration of T cells was performed in a defined "mucosal unit" consisting of a 500-µm wide tissue block of full mucosal height. The mucosal unit was photographed, projected on a paper screen and T cells were counted in the surface epithelium, in the crypt epithelium and in the lamina propria. For every specimen cell enumerations were based on two tissue units located as far apart as possible in the section.

RESULTS

HLA class II antigens

Epithelial expression of HLA-DR appeared mainly in a granular distribution apically; fainter staining was seen basolaterally and in the cytoplasm of the columnar surface cells, both in the controls and in the untreated and treated CD patients. In the untreated CD patients HLA-DR was, in addition, generally expressed by the columnar crypt cells in a similar pattern with decreasing intensity towards the bottom of the crypts. Faint staining was noted in the upper part of the crypts in five of the untreated CD patients and in two of the controls (Table 1).

HLA-DP was present apically in the surface epithelium of all untreated CD patients (Table 1); the distribution was granular as for DR. In the treated CD patients, a similar appearance of DP was seen in three patients and two others showed small surface areas positive for DP. Two control subjects showed generalized surface staining for DP and two others showed limited patchy positivity, whereas the remainders were negative despite positive lamina propria elements. DP was noted apically in the columnar cells of the upper parts of the crypts in five of the untreated CD patients but was absent from glandular epithelium in treated patients and controls.

HLA-DQ was only seen apically in surface patches in three untreated patients (Table 1). All treated patients and controls were negative despite positive lamina propria elements.

In the lamina propria, class II determinants were expressed by lymphocytes, histiocytes and capillary endothelium. However, whereas such elements were always strongly positive for HLA-DR, the staining intensity and also the number of cells positive for DP and DQ varied strikingly within the three subject categories.

Intraepithelial and lamina propria T cells

There was no difference in the number of intraepithelial T cells per mm of muscularis mucosae when untreated CD patients were compared with treated patients or controls (Table 2). Conversely, the number of T cells per mm of surface epithelium was significantly ($p<0.01$, Wilcoxon) increased in untreated CD patients (Table 2). While the intraepithelial T cells normally were located mainly along the basement membrane, they were distributed throughout the coeliac surface epithelium. The number of intraepithelial T cells in the crypts was also higher ($p<0.05$) in untreated than in treated CD patients or controls.

Relation between HLA class II expression and intraepithelial T cells

In sections with epithelial staining for HLA-DP, the number of T cells per mm of surface epithelium was significantly higher ($p<0.01$) than in sections without such staining. HLA-DQ expression was only seen in surface areas with very large numbers of intraepithelial T cells; the number of such cells per mm surface epithelium was above the median for the untreated CD group in all of the three specimens showing DQ expression.

TABLE 1. Proportion of subjects showing expression of HLA class II subregion products (DR, DP and DQ) in jejunal surface and crypt epithelium.

Subject category	DR		DP			DQ	
	Total surface	Crypts	Total surface	Surface patches	Crypts	Surface patches	Crypts
Controls	11/11	2/11	2/11	2/11	0/11	0/11	0/11
CD-GFD*	11/11	5/11	3/11	2/11	0/11	0/11	0/11
CD untreated	11/11	11/11	11/11	N.A.**	5/11	3/11	0/11

*Patients with coeliac disease on a gluten-free diet.
**N.A. = not applicable

TABLE 2. T-cell distribution (median and observed range) in various small intestinal mucosal compartments of patients with coeliac disease (CD) and controls.

Mucosal compartment	Controls	CD treated	CD untreated
Surface epithelium[a]	20 (9- 50)	32 (19- 98)	80 (32-120)*
Surface epithelium[b]	84 (44-204)	106 (42-156)	80 (32-114)
Crypt epithelium[b]	18 (6- 34)	22 (6- 32)	40 (22-112)*
All epithelium [b]	102 (46-155)	128 (48-156)	132 (88-285)
Lamina propria[b]	556 (378-840)	466 (388-758)	802 (251-1048)
Total mucosa[b]	668 (480-988)	600 (484-896)	890 (350-1216)*

[a]Number per mm epithelium
[b]Number per mm muscularis mucosae

*Significant differences between untreated patients and controls ($p<0.05$)

DISCUSSION

In patients with active CD, the jejunal enterocytes were intensely stained for HLA-DR both in the surface and most of the crypt epithelium. This contrasted with the virtual absence of crypt staining in control patients as reported previously by us (1,7) and others (8,9). In the treated CD group, the DR staining of the crypt epithelium was variable and completely absent in six of the eleven patients.

In histologically normal small intestinal mucosa, we found strong DR expression with apical granular intensification in the surface epithelium of all biopsy specimens but DP was seen in only four.

Like Natali et al. (10) we did not detect HLA-DQ determinants in normal small intestinal epithelium. DQ was only present in small areas of the surface epithelium in some of the untreated CD patients with very high numbers of intraepithelial T cells. Like Marsh (11) and Jenkins et al. (12) we found that the total number of T cells in the surface epithelium was not increased in untreated CD, but there was an increased epithelial T-cell density in agreement with previous observations (13,14). Moreover, the number of T cells in the crypts was significantly raised in untreated CD.

Epithelial MHC class II expression may be modulated during immune responses by interferon-γ(15,16) and lymphotoxin produced by T cells and by tumor necrosis factor-α released from macrophages (17,18). Intraepithelial lymphocytes have been shown to be involved in the modulation of intestinal class II expression (19). Expression of DP, DQ and DR may follow a hierarchical pattern: DR>DP>DQ (16,20). Our findings of a relation between epithelial DP expression and local T-cell number, and the fact that DQ was seen only in three untreated CD patients with high epithelial density of T cells, are in accordance with a T-cell-mediated differential class II modulation.

Epithelial expression of the HLA class II gene product DQ that is primarily associated with CD, may be important in the pathogenesis of this disease; and the limited sensitivity of our staining method does not exclude that epithelial DQ is expressed in every untreated CD case. It has been reported that CD is primarily related to a DQ allelic variant (21,22). However, the 4.0 kilobase class II β-chain fragment found to be most strongly associated with CD (22) was recently identified as a fragment of DP rather than DQ as initially believed (23). It is therefore possible that DP, or some still unidentified class II molecule or hybride HLA molecule, is the disease-associated restriction element in CD.

The observed differences in the staining intensity of the three epithelial class II subregion products may be a matter of antibody avidity. Moreover, the monoclonal antibodies used may not bind to the actual disease-associated gene product. Nevertheless, we found increased expression and an altered distribution of HLA class II molecules in jejunal epithelium of untreated CD, and there was a relationship between both DP and DQ and the epithelial T-cell infiltration. These observations suggested that activated T cells modulate epithelial class II expression. Since the distribution of the class II molecules was fairly similar in treated CD patients and controls, and since a gluten-like lectin was unable to induce class II expression in a colonic carcinoma cell line (16), it is possible that an event (e.g. virus infection) leading to local release of interferon-γ or other lymphokines, may be the initial step in the pathogenesis of CD. On the other hand, it cannot be excluded the epithelial class II expression is induced by gluten, perhaps secondarily to gluten-dependent epithelial lymphocyte infiltration.

REFERENCES

1. H. Scott, B. G. Solheim, P. Brandtzaeg, and E. Thorsby, Scand. J. Immunol. 12:77 (1980).
2. H. Scott, P. Brandtzaeg, B. G. Solheim, and E. Thorsby. Clin. Exp. Immunol. 44:233 (1981).
3. W. Strober, pp. 169-182, in: "Perspectives in Coeliac Disease," B. Mc Nicholl, C. F. Mc Carthy, and P. F. Fottrell, eds. MTP Press Limited, Lancaster, England (1978).
4. R. Tosi, N. Tanigaki, I. Polanco, M. De Marchi, J. C. Woodrow, and P. A. Hetzel, Clin. Immunol. Immunopathol. 39:168 (1986).
5. R. Tosi, D. Vismara, N. Tanigaki, G. B. Ferrara, F. Cicimarra, W. Buffolano, D. Follo, and S. Auricchio. Clin. Immunol. Immunopathol. 28:395 (1983).
6. P. W. Bland, and L. G. Warren, Immunology 58:1 (1986).
7. P. Brandtzaeg, K. Valnes, H. Scott, T. O. Rognum, K. Bjerke, and K. Baklien, Scand. J. Gastroenterol. 20 (suppl. 114): 17 (1985).
8. F. Arnaud-Battandier, N. Cerf-Bensussan, R. Amsellem, and J. Schmitz, Gastroenterology 91:1205 (1986).
9. W. S. Selby, G. Janossy, G. Goldstein, and D. P. Jewell, Clin. Exp. Immunol. 44:453 (1981).
10. P. Natali, A. Bigotti, M. Cavalieri, R. Nicotra, R. Tecce, D. Manfredi, Y. X. Chen, L. M. Nadler, and S. Ferrone, Human Immunol. 15:220 (1986).
11. M. N. Marsh, Scand. J. Gastroenterol. 20 (suppl. 114):55 (1985).
12. D. Jenkins, A. Goodall, and B. B. Scott, Gut 27:1330 (1986).
13. A. Ferguson, J. P. Mc Clure, and R. R. W. Townley, Acta Paediatr. Scand. 65:541 (1976).
14. U. B. Schaad, H. Gaze, and B. Hadorn, Am. J. Dis. Child. 135:272 (1981).
15. N. Barclay, and D. W. Mason, J. Exp. Med. 156:1665 (1982).
16. L. M. Sollid, G. Gaudernack, G. Markussen, D. Kvale, P. Brandtzaeg, and E. Thorsby, Scand. J. Immunol. 25:175 (1987).
17. K. Pfizenmaier, P. Scheurich, C. Schlüter, and M. Krönke, J. Immunol. 138:975 (1987).
18. R. Pujol-Borrell, I. Todd, M. Doshi, G. F. Bottazzo, R. Sutton, D. Gray, G. R. Adolf, and M. Feldmann, Nature 326:304 (1987).
19. N. Cerf-Bensussan, A. Quaroni, J. T. Kurnick, and A. K. Bhan, J. Immunol. 132:2244 (1984).
20. I. Todd, R. Pujol-Borrell, and G. F. Bottazzo, Immunology Today 7:6 (1987) (letter).
21. G. A. Hitman, M. J. Niven, H. Festenstein, P. G. Cassell, J. Awad, J. Walker-Smith, J. N. Leonard, L. Fry, P. Ciclitira, P. Kumar, and J. S. Sachs, J. Clin. Invest. 79:609 (1987).
22. M. D. Howell, R. K. Austin, D. Kelleher, G. T. Nepom, and M. F. Kagnoff, J. Exp. Med. 164:333 (1986).
23. M. F. Kagnoff, and M. D. Howell. Abstract L52. International symposium on the immunology of the gastrointestinal tract and the liver. Jerusalem, Israel, March (1987).

T CELLS AND EPITHELIAL EXPRESSION OF HLA CLASS II DETERMINANTS IN RELATION TO PUTATIVE M CELLS OF FOLLICLE-ASSOCIATED EPITHELIUM IN HUMAN PEYER'S PATCHES

K. Bjerke and P. Brandtzaeg

Laboratory for Immunohistochemistry and Immunopathology (LIIPAT), Institute of Pathology, The National Hospital Rikshospitalet, N-0027 Oslo 1, Norway

INTRODUCTION

M cells of follicle-associated epithelium (FAE) transport luminal antigens into Peyer's patches (PP), but the complete role of these cells in local immunity has not been defined. To this end it would be of interest to evaluate their spatial relationship to T cells and their possible capacity to present antigens to T cells.

MATERIAL, METHODS AND RESULTS

In an immunohistochemical study we quantified intraepithelial T cells, mapped their distribution in relation to M cells, and examined if the M cells were positive for MHC class II (HLA-DR) determinants or not. DR expression might be a prerequisite for an antigen-presenting function of the M cells. Due to the limited resolution of the light microscope, we tried to identify M cells indirectly by their reduced expression of brush border alkaline phosphatase as described by Owen and Bhalla (1). Tissue sections cut at 1-2 μm were costained for alkaline phosphatase and $CD3^+$ T cells or HLA-DR determinants. Enumeration of T cells was performed with a grid moved systematically along the FAE with one side of the frame touching the epithelial surface. The epithelium was thereby divided into 0.055 mm long segments.

The material consisted of normal distal ileum obtained from kidney donors with maintained peripheral circulation. We found a significantly higher number of T cells in the FAE of PP than in the epithelium of distant villi (Table I). The T cells tended to be clustered in all layers of FAE and were more numerous adjacent to interruptions of the brush border as revealed by lack of staining for alkaline phosphatase (Figure 1, Table II). This indicated a spatial relationship between M cells and accumulation of T cells. However, in contrast to other epithelial cells of the FAE, the putative M cells did generally not express HLA-DR determinants (Figure 2).

Table 1. T cells/grid (median and 95% confidence interval) enumerated in the ileum of 10 subjects*

FAE	Villous epithelium
3.0 (2.1-5.3)	1.2 (1.0-1.6)

*The median of examined grid segments per subject in PP and villi was 14 (range, 5-30) and 15.5 (range, 14-19), respectively.

Table II. T cells/grid (median and 95% confidence interval)enumerated in FAE of 10 subjects.

Adjacent to putative M cells*	Remaining FAE
3.6 (3.3-4.9)	2.6 (1.7-4.0)

*The median of examined grid segments was 4.5 per subject (range, 3-14).

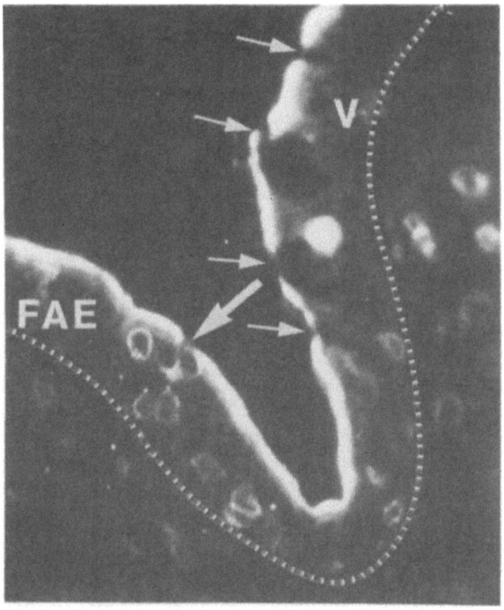

Figure 1. Fluorescence co-staining for CD3 and alkaline phosphatase. Note difference of intraepithelial $CD3^+$ T-lymphocyte distribution in follicle-associated epithelium (FAE) and in villous epithelium (V). The brush border is stained for alkaline phosphatase. Lymphocytes cluster near an interruption (bold arrow) in FAE, whereas they appear as single lymphocytes adjacent to the basement membrane in villous epithelium. Broken line indicates basement membrane. Thin arrows indicate goblet cells. (Magnification: X610).

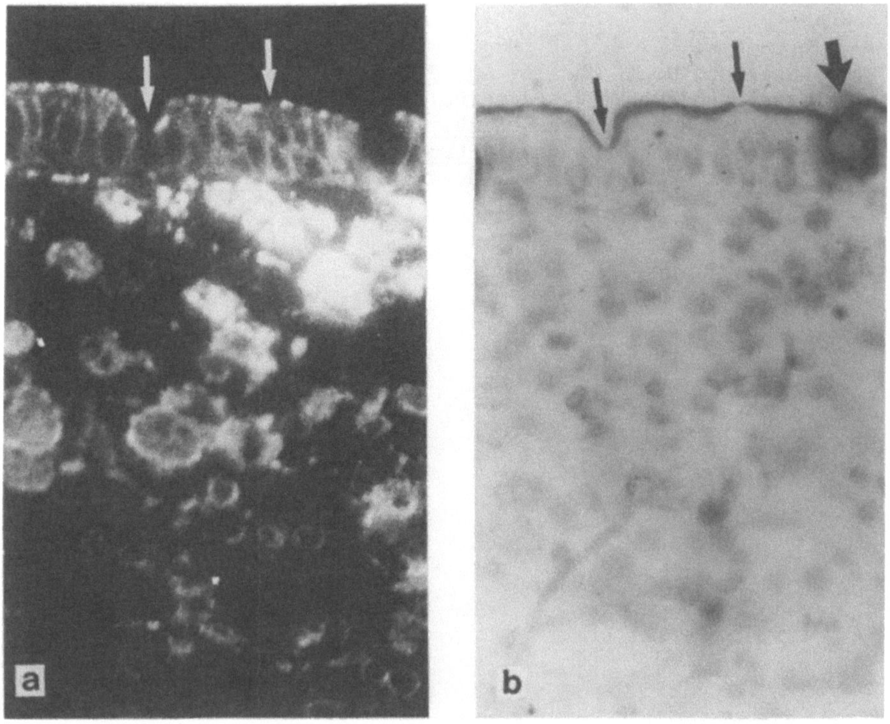

Figure 2. Co-staining for (a) HLA-DR (immunofluorescence) and (b) alkaline phosphatase/mucin (alcian blue/PAS) in same field from Peyer's patch. Interruptions in the brush border staining for alkaline phosphatase indicate M cells (thin black arrows). Goblet cell is positive for mucin (bold black arrow). Note that there is little or no epithelial expression of HLA-DR corresponding to putative M cells (white arrows). There are numerous HLA-DR-positive histiocytic cells in the subepithelial dome area. (Magnification: X610).

DISCUSSION

Ferguson and Parrott first reported that antigens in the gut lumen were required for infiltration of the FAE by lymphocytes (2). Our study showed that significantly more T cells appeared in the FAE than in the villous epithelium. This observation accorded with the notion that the FAE is a main intestinal entrance of foreign antigens.

The presence of both $CD8^+$ and $CD4^+$ T cells in FAE in contrast to virtually only $CD8^+$ cells in villous epithelium (3), makes a particular functional relationship between the FAE and T cells likely. As previously indicated by Owen for FAE (4), our study suggested that there is a preferential spatial relationship between lymphocytes and M cells. However, it is still uncertain whether M cells differentiate from absorbing cells under the influence of T lymphocytes or from primitive crypt cells (5). Our findings indicated lack of general HLA-DR expression on M cells which suggested that these cells are probably not involved in presentation of antigens to T cells. Such presentation might rather be performed by DR-positive subepithelial dendritic cells or macrophages to which luminal antigens gain access by transport through M cells.

REFERENCES

1. R. L. Owen and D. K. Bhalla, Am. J. Anat. 168:199 (1983).
2. A. Ferguson and D. M. V. Parrott, Clin. Exp. Immunol. 12:477 (1972)
3. T. H. Ermak and R. L. Owen, Anat. Rec. 215:144 (1986).
4. R. L. Owen and A. I. Jones, Gastroenterology 66:189 (1974).
5. W. A. Bye, C. A. Allan and J. S. Trier, Gastroenterology 86:789 (1984).

MODULATION OF HISTAMINE SECRETION FROM RAT AND HUMAN MAST CELL SUBPOPULATIONS

M. Swieter, C. Rimmer, B. Webb, W. Ghali, F. Pearce* and D. Befus

Immunological Sciences Research Group, Department of Microbiology and Infectious Diseases, University of Calgary, Calgary, Alberta, Canada and University College London, London, England*

INTRODUCTION

Mast cells are distributed throughout the connective tissues of the body where they interact with numerous environmental agents including other cells and their products, blood-bourne materials, and foreign substances that penetrate the overlying epithelia and enter subjacent connective tissue compartments. Many of these agents are capable of stimulating mast cells to release their cache of multipotent mediators. For example, antigens, macrophage (19) and T cell factors (11), vagus nerve stimulation (13) and neuropeptides (20) have all been shown to activate mast cells. Furthermore, although they are best known for their responses to allergens in immediate hypersensitivity reactions, mast cells are involved in such diverse immune and inflammatory phenomena as cell and organismal toxicity, immunoregulation, promotion of neovascularization and fibroblast proliferation (1).

The importance of environmental influences on phenotypic expression of genetic potential in mast cells is a major area of research (2). For example, although in rats, mice and humans two histochemically distinct mast cell populations have been described, and in rats these subtypes have been shown to be biochemically and pharmacologically disparate as well, the mechanisms underlying these differences are unknown. Because these differences may relate to environmental factors that modulate cell responsivity and function, we studied the effects of adenosine, complement and interferon on mast cells and basophils.

SOURCES OF CELLS

Animals

Male Sprague-Dawley rats were purchased from Charles River, Montreal, Canada. Male Hooded Lister rats were obtained from a breeding colony maintained at the University of Calgary. Animals were infected with 3,000 third-stage larvae of the intestinal nematode parasite *Nippostrongylus brasiliensis* by subcutaneous injection 4-5 weeks prior to mast cell isolation. Procedures and materials used in the isolation and purifica-

Table 1. Effects of adenosine on anti-IgE (1/50)-induced histamine secretion by human intestinal mast cells

Adenosine dose	Net percent release +/- S.E.M. (n)*	
	20 min preincubation	concurrent administration
none	22.6 $\pm$ 6.5 (6)	22.6 $\pm$ 6.5 (6)
10^{-3} M	6.9 $\pm$ 2.4 (6)	N.D.
10^{-4} M	8.7 $\pm$ 4.8 (2)	15.2 $\pm$ 5.8 (6)
10^{-5} M	27.6 (1)	17.7 $\pm$ 13.4 (3)

* in excess of spontaneous release (3.9 $\pm$ 0.9%, n=6)

tion of rat mast cell subtypes were as previously reported (12).

Human Peripheral Blood Leukocytes and Intestinal Mast Cells

Peripheral blood samples were drawn from healthy adult volunteers. Pieces of small and large intestine were obtained from uninvolved areas of bowel segments removed during surgical resection for carcinoma, Crohn's disease, diverticular disease or ulcerative colitis. Materials and procedures used to isolate peripheral blood leukocytes and human intestinal mast cells from these specimens have been reported previously (3).

RESULTS AND DISCUSSION

Studies with adenosine

The data in Table 1 show that preincubation of dispersed human intestinal mast cells with 10^{-3} to 10^{-4} M adenosine brings about marked inhibition of anti-IgE-induced histamine release. Similar results have been found for dispersed human lung mast cells (8) and human basophils (4,16). In contrast, adenosine appears to have cell type-specific effects when added concurrently with anti-IgE. In human intestinal mast cells (Table 1) adenosine is either without effect or only slightly inhibitory; with human lung mast cells (8) it is either ineffective or slightly potentiating; and with human basophils (4) it potentiates IgE-mediated secretion in a dose-dependent manner. Furthermore, because adenosine apparently exerts its influence on human basophils and lung mast cells through cell surface receptors (A2 or Ra) that activate adenylate cyclase and give rise to elevated intracellular levels of cyclic AMP (4,8), it is likely that the same is true of human intestinal mast cells. This hypothesis must be tested.

Adenosine's potentiating effect, seemingly independent of time of administration, on antigen-induced histamine release from rat peritoneal mast cells is well documented (7,14,15,17). In addition, it has recently been found (unpublished observations) that immunologic secretion of histamine from rat intestinal mucosal mast cells is also enhanced by adenosine. When taken together with the human cell data, these results suggest species-specific effects of adenosine on mast cell mediator release.

Studies with Complement

Human peripheral blood leukocytes, human intestinal mast cells or mast cells taken from the peritoneal cavity or intestinal lamina propria of Hooded Lister rats were incubated with various amounts of zymosan-

Table 2. Effect of activated serum (complement) on specific histamine release from human basophils and human and rat mast cells

ul serum	Percent histamine release +/- S.E.M.			
	HUMAN (n=5-7)		RAT (n=3)	
	basophils	IMC	IMMC	PMC
50	12.4 ± 3.9	2.5 ± 2.2	5.5 ± 3.0	12.4 ± 5.4
100	42.6 ± 8.8	11.9 ± 1.7	14.9 ± 5.2	35.4 ± 11.3
150	58.8 ± 8.2	13.3 ± 2.1	N.D.	N.D.
200	62.4 ± 5.2	20.8 ± 7.8	11.1 ± 4.7	47.0 ± 14.5
300	59.2 ± 8.4	23.3 ± 4.7	13.7 ± 9.2	57.2 ± 11.8

N.D.=not done, IMC=intestinal mast cells, IMMC=intestinal mucosal mast cells, PMC=peritoneal mast cells

activated human serum at 37°C for 10 min (rat cells) or 45 min (human cells). Cells and supernatants were analyzed for histamine by radioenzymatic assay. Goat anti-human complement antisera were purchased from Cappel (CooperBiomedical, West Chester, PA). Anti-C3a (36 mg total protein/ml) was usedat 1:5; anti-C5 (18 mg total protein/ml) was used at 1:10.

Zymosan-activated human serum (complement) stimulates histamine secretion by human basophils and intestinal mast cells, as well as by rat peritoneal and intestinal mucosal mast cells, although intestinal mast cells are less sensitive to complement activation than peritoneal mast cells or basophils (Table 2). Zymosan is known to activate the alternative complement pathway resulting in the production of the anaphylatoxins, C3a and C5a (5,6). Because antibodies to C5a but not C3a inhibit serum-stimulated histamine release from human basophils (5,6), it has been suggested that C5a is the major anaphylatoxin secretagogue for basophils. Interestingly, we found that anti-C3a antibodies were more potent inhibitors of human serum complement-activated rat peritoneal mast cell histamine release than were anti-C5 antibodies (data not shown). This suggests that although purified human or porcine C3a and C5a are both capable of stimulating histamine release from rat peritoneal mast cells (10), C3a in human serum may be a more potent stimulator of rat serosal mast cell degranulation than C5a. It should be noted, however, that in the present study complete inhibition of activated serum-induced rat peritoneal mast cell histamine secretion required antibodies to both C3a and C5. Given the responsiveness of mast cells and basophils to activated complement components, it is clear that in considering the role of these cells in inflammmatory diseases of, for example, the intestine, stimuli other than IgE antibodies may be of etiological significance.

Studies with Interferon

Peritoneal mast cells from Sprague-Dawley rats were incubated for 24 hr at 37°C in 1 ml RPMI 1640 culture medium containing 10% fetal calf serum, 1% penicillin/streptomycin (Gibco, Grand Island , NY), and various concentrations of rat alpha/beta interferon (10^5 International Reference Units/mg, Sigma Chemical Company, St. Louis, MO; 2.7 X 10^5 units/mg, Lee Biomolecular, San Diego, CA) in the presence or absence of neutralizing antibodies (rabbit anti-rat interferon, 10^4 neutralizing units/ml, Lee Biomolecular). The cells were resuspended in medium lacking both interferon and antibody and stimulated to release by the addition of *N. brasiliensis* antigen (expressed in worm equivalents), compound 48/80 (1 μg/ml, Sigma) or substance P (5 X 10^{-5}M, Sigma). Phosphatidylserine (30

Table 3. Effects of interferon (IFN) on specific histamine secretion by rat peritoneal mast cells

Dose IFN (U)	Percent inhibition of histamine release +/- S.E.M. Ag (3WE) alone n=3	Ag (10WE) + PS n=2	Substance P n=1-2	48/80
15	N.D.	1.9 ± 3.3	3.3 (1)	1.1 ± 2.9
150	22.3 ± 4.2	19.0 ± 0.5	7.0 ± 4.3	5.7 ± 1.7
450	28.6 ± 5.3	N.D.	N.D.	N.D.
750	36.5 ± 7.5	37.8 ± 5.1	9.7 ± 1.2	5.4 ± 2.4
500	53.4 ± 4.6	35.1 ± 4.5	18.0 (1)	8.3 ± 2.2

U=units of interferon, Ag=antigen, PS=phosphatidylserine, N.D.=not done

μg/ml, Sigma), an enhancer of antigen-induced histamine release, was used with antigen in some experiments. Cells and supernatants were separated and the relative proportion of histamine in each fraction was determined by fluorimetric assay (12).

Antigen-induced degranulation of rat peritoneal mast cells is inhibited in a dose-dependent manner following 24 hr exposure of the cells to alpha/beta (Type I) interferon (Table 3). A similar pattern is seen whether or not phosphatidylserine is present with antigen, although it appears that the maximum level of inhibition is reached at a lower concentration of interferon when the phospholipid is added with antigen- than when it is not. Why this should be so is not presently clear. Furthermore, antibodies capable of neutralizing the antiviral properties of interferon also completely block the anti-mast cell effects (25-75 neutralizing units of antibody ablated the inhibitory influence of 750 units of interferon on histamine secretion).

Human peripheral blood basophils respond to 24 hr preincubation with interferon in the opposite fashion. Rather than decrease their output following anti-IgE provocation, human basophils preincubated 24 hr with Type I interferons increase their release of histamine above that of the controls (9). However, whether this reflects a direct effect of interferon on basophils is speculative because human peripheral blood leukocyte preparations, which were used in the aforementioned study, are heterogeneous and basophils usually comprise only 0.5-1% of the total cells.

Beta interferon, in concentrations ranging from 160 to 2560 units/ml, has been shown to affect structural changes in the lipid bilayers of cell membranes in a time-dependent fashion (18). Plasmalemmal rigidity in human HeLa-S_3 cells increased by 30 min after interferon exposure, declined within 3-5 hr, but by 24 hr had again increased and remained so for at least 48 hr longer. The doses of interferon used in the present study and time course of effect suggest that similar events may occur in rat peritoneal mast cells exposed to Type I interferons. Increased membrane rigidity would hinder IgE receptor cross-linking, a requisite event in mast cell activation by antigen or anti-IgE, and inhibit histamine release. Degranulation mechanisms not so dependent on membrane fluidity should, therefore, be less affected. This too was observed in the present study where in contrast to stimulation with antigen, histamine release induced by the neuropeptide substance P and the polyamine compound 48/80 were only modestly affected by interferon (Table 3).

Taken together, these results support the concept that interferon exhibits a cell and/or species-specific effect on immunologically-mediated histamine secretion.

OVERALL CONCLUSIONS

1. Microenvironmental components: complement, adenosine and alpha/beta interferon modulate histamine secretion by mast cells and basophils.
2. Complement fragments activate histamine secretion in all mast cells studied and in basophils.
3. Adenosine inhibits histamine secretion by human mast cells and basophils, but enhances secretion by rat mast cells.
4. Interferon alpha/beta inhibits antigen-induced histamine secretion by rat peritoneal mast cells; earlier studies showed that interferon enhances secretion by human basophils.
5. It is necessary to uncover further information about microenvironmental factors that modulate mast cell activation before the role of mast cells in immune and inflammatory reactions is clarified.

REFERENCES

1. Befus, D. Intestinal mast cell polymorphism: new research directions and clinical implications. J. Pediat. Gastroenterol. Nut. 5: 517-520. 1986.
2. Befus, A.D., Denburg, J.A., and Bienenstock, J. (eds). Mast Cell Differentiation and Heterogeneity. Raven Press. N.Y. 1986.
3. Befus, A.D., Dyck, N., Goodacre, R. and Bienenstock, J. Mast cells from the human intestinal lamina propria. Isolation, histochemical subtypes, and functional characterization. J. Immunol. 138: 2604-2610. 1987.
4. Church, M.K., Holgate, S.T. and Hughes, P.J. Adenosine inhibits and potentiates IgE-dependent histamine release from human basophils by an A2-receptor mediated mechanism. Brit. J. Pharmacol. 80: 719-726. 1983.
5. Grant, J.A., Dupree, E., Goldman, A.S., Schultz, D.R. and Jackson, A.L. Complement-mediated release of histamine from human leukocytes. J. Immunol. 114: 1101-1106. 1975.
6. Hook, W.A., Siraganian, R.P. and Wahl, S.M. Complement-induced histamine release from human basophils. I. Generation of activity in human serum. J. Immunol. 114: 1185-1190. 1975.
7. Hughes, P.J. and Church, M.K. Separate purinoceptors mediate enhancement by adenosine of concanavalin A-induced mediator release and the cyclic AMP response in rat mast cells. Agent. Action. 18: 81-84. 1986.
8. Hughes, P.J., Holgate, S.T. and Church, M.K. Adenosine inhibits and potentiates IgE-dependent histamine release from human lung mast cells by an A2-purinoceptor mediated mechanism. Biochem. Pharmacol. 33: 3847-3852. 1984.
9. Ida, S., Hooks, J.J., Siraganian, R.P. and Notkins, A.L. Enhancement of IgE-mediated histamine release from human basophils by viruses: role of interferon. J. Exp. Med. 145: 892-906. 1977.
10. Johnson, A.R., Hugli, T.E. and Muller-Eberhard, H.J. Release of histamine from rat mast cells by the complement peptides C3a and C5a. Immunology 28: 1067-1080. 1975.
11. Kops, S.K., Van Loveren, H., Rosenstein, R.W., Ptak, W. and Askenase, P.W. Mast cell activation and vascular alterations in immediate hypersensitivity-like reactions induced by a T cell-derived antigen-binding factor. Lab. Invest. 50: 421-434. 1984.
12. Lee, T.D.G., Shanahan, F., Miller, H.R.P., Bienenstock, J. and Befus, A.D. Intestinal mucosal mast cells: isolation from rat lamina propria and purification using unit gravity velocity sedimentation. Immunology 55: 721-728. 1986.
13. Leff, A.R., Stimler, N.P., Munoz, N.M., Shioya, T., Tallet, J. and

Dame, C. Augmentation of respiratory mast cell secretion of histamine caused by vagus nerve stimulation during antigen challenge. J. Immunol. 136: 1066-1073. 1986.
14. Leoutsakos, A. and Pearce, F.L. The effect of adenosine and its analogues on cyclic AMP changes and histamine secretion from rat peritoneal mast cells stimulated by various ligands. Biochem. Pharmacol. 35: 1373-1379. 1986.
15. Lohse, M.J., Maurer, K., Gensheimer, H.P. and Schwabe, U. Dual actions of adenosine on rat peritoneal mast cells. Naunym-Schmied. Arch. Pharmacol. 335: 555-560. 1987.
16. Marone, G., Findlay, S.R. and Lichtenstein, L.M. Adenosine receptor on human basophils: modulation of histamine release. J. Immunol. 123: 1473-1477. 1979.
17. Marquardt, D.L., Parker, C.W. and Sullivan, T.J. Potentiation of mast cell mediator release by adenosine. J. Immunol. 120: 871-878. 1978.
18. Pfeffer, L.M., Landsberger, F.R. and Tamm, I. Beta-interferon-induced time-dependent changes in the plasma membrane lipid bilayer of cultured cells. J. Interferon Res. 1: 613-620. 1981.
19. Schulman, E.S., Liu, M.C., Proud, D., MacGlashan, D.W., Lichtenstein, L.M. and Plaut, M. Human lung macrophages induce histamine release from basophils and mast cells. Amer. Rev. Resp. Dis. 131: 230-235. 1985.
20. Shanahan, F., Denburg, J., Fox, J., Bienenstock, J. and Befus, D. Mast cell heterogeneity: effects of neurenteric peptides on histamine release. J. Immunol. 135: 1331-1337. 1985.

BACTERIAL CHARACTERISTICS AND FOLLICLE SURFACE STRUCTURE: THEIR ROLES IN PEYER'S PATCH UPTAKE AND TRANSPORT OF *Vibrio cholerae*

Robert L. Owen,[1] William C. Cray, Jr.,[2,3]
Thomas H. Ermak[1] and Nathaniel F. Pierce[2,4]

[1]Intestinal Immunology Research Center, University of California, San Francisco and Cell Biology and Aging Section (151E), Veterans Administration Medical Center, San Francisco, California 94121

[2]Department of Medicine, Johns Hopkins University School of Medicine and Francis Scott Key Medical Center, Baltimore, Maryland

[3]Current Location:Department of Veterinary Science University of Nebraska, Lincoln, Nebraska 68583-0905

[4]Current Location: Diarrheal Diseases Control Program, World Health Organization, CH-1211 Geneva 27, Switzerland

INTRODUCTION

Structural and functional specializations in the intestinal mucosal barrier over Peyer's patches facilitate contact with and uptake of adherent luminal particles and soluble molecules. These specializations include local reduction in mucus-secreting goblet cells, interruption of the muscularis mucosae by lymphoid follicles (which may alter mucosal motility), and the presence of M cells. M cells are epithelial cells with irregular, widely spaced microvilli, reduced or absent lysosomes, an active transcellular transport system, and a pliable basolateral surface usually invaginated by lymphoid cells. At their luminal surfaces M cells form coated and uncoated vesicles, which are transported to intercellular spaces in the follicle epithelium where initial contact of intestinal antigen and lymphoid cells occurs.[1] The mucus blanket which lies across the microvilli of enterocytes (intestinal absorptive cells) is broken over M cells, allowing microorganisms and particles to approach M cell luminal membranes.[2] Studies with labeled lectins have demonstrated a thin but detectable glycocalyx coat on M cell membranes.[3-5] Sugars detected by these lectins include beta-D-galactose (*Ricinus communis* agglutinin), sialic acid and N-acetyl-D-glucosamine (wheat germ agglutinin), but not alpha-D-mannose or alpha-D-glucose (Con A binds only sparsely). Cationized ferritin distributes evenly on M cell luminal membranes indicating that anionic sites are present.

Attachment sites, such as those revealed by lectin binding studies, are of potential significance for adherence of luminal microorganisms. Thus far no lectin binding sites have been identified over M cells which are not shared by enterocytes. Reovirus[6] and rabbit diarrhea Escherichia coli (RDEC)[7] adhere preferentially to M cells, indicating that the combination or configuration of glycocalyceal structures over M cells differs from surrounding enterocytes, even if comprised of the same biochemical components. Luminal constituents which bind to M cell membranes are taken up in coated pits which typically incorporate clusters of membrane-bound ligands.[5] Although fluid phase molecules such as horseradish peroxidase[1] and native ferritin[5] are also taken up from the unstirred layer immediately above M cell luminal membranes, they are taken up at least 50 times less efficiently than are adherent proteins.[5] Material endocytosed by either mechanism enters flattened cisternae and tubules and passes through to M cell basolateral surfaces where it is released into pockets in the epithelium occupied by lymphoid cells. Adherent and fluid phase proteins appear to travel across M cells at approximately the same rate (about 15-20 minutes).[5]

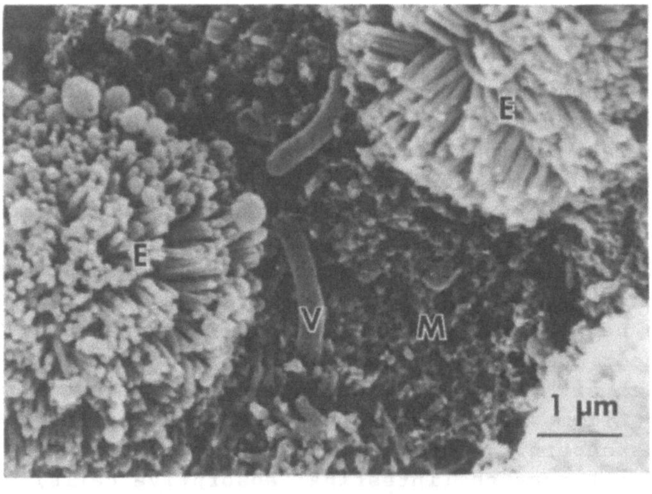

Figure 1. In a rabbit Peyer's patch, M cells (M) with short and branched microvilli are recessed below the level of surrounding enterocytes (E) with their long, straight microvilli. Curved cholera vibrios (V) adhere to the M cell surface. Scanning electron micrograph.

Bacteria, protozoa and large particles are taken up from the lumen by phagocytosis by the M cell and are carried in large vacuoles.[2,8,9] M cells have a greatly reduced or absent lysosomal system so that phagocytosed as well as endocytosed material is transported across M cells rather than diverted into the lysosomal degradation system, as occurs in enterocytes.[10] *Vibrio cholerae* 01, the cause of cholera, do not penetrate the mucosal barrier but induce disease by colonizing the intestinal surface and releasing a toxin which initiates diarrhea. In Peyer's patches, M cells in the epithelium of lymphoid follicles take up vibrios from the intestinal lumen and carry them into follicle domes.[2,11] Although both *Vibrio cholerae*[2] and *Salmonella typhi*[12] are taken up by M cells, the consequences are quite different. *Vibrio cholerae*, after transport through M cells are taken up by macrophages in the epithelium and in the dome, where immune responses are initiated. *Salmonella typhi* destroy M cells and produce ulcerations over Peyer's patch follicle surfaces.[12]

Microbial characteristics and M cell surface factors which affect adherence and uptake of bacteria are thus not only of consequence for antigenic processing and initiation of mucosal immune responses but also for pathogenesis of local and systemic infection by luminal microorganisms. We have carried out a series of studies with mutant strains of *Vibrio cholerae* to examine the effects of bacterial characteristics on adherence to and uptake by M cells and studies with ruthenium red staining to further characterize the luminal surface of M cells.

METHODS

To determine the effects of bacterial characteristics on uptake into Peyer's patches, we inoculated ligated intestinal segments of non-immune adult rabbits with 10^7 colony forming units (CFU) of [2-^{3}H]-adenine labeled vibrios of each of the following strains and characteristics:

Virulent Ogawa 395	A^+B^+ (toxin producing)	motile	colonizing
Ogawa 395 NT	A^-B^- (nontoxin producing)	motile	moderately colonizing
Ogawa RV 79	A^+B^+ (toxin producing)	motile	poorly colonizing
Virulent Inaba N16961	A^+B^+ (toxin producing)	motile	colonizing
CVD 49 (mutant of N16961)	A^+B^+ (toxin producing)	nonmotile	moderately colonizing

(For further discussion of the origin, derivation and characteristics of these strains, see reference 13.) After 90 or 180 minutes Peyer's patches from inoculated segments were removed, fixed and prepared for light microscopy, transmission electron microscopy and scanning electron microscopy. Well-oriented sections through Peyer's patch follicles were examined by autoradiography as previously described[2] and compared for adherence and uptake of labeled vibrios.

Ruthenium red was dissolved in cacodylate buffer which was used for preparation of fixatives and buffers as previously described.[2] Rabbit intestinal segments containing Peyer's patches were fixed three hours after inoculation with sterile medium or Ogawa 395 *Vibrio cholerae* cultures. Peyer's patches were post-fixed in osmium with ruthenium red and all rinses contained ruthenium red. Transverse and coronal sections of follicle epithelium surfaces were examined by transmission electron microscopy.

RESULTS

Studies with radiolabeled vibrios demonstrated adherence and uptake of virulent and motile Ogawa 395 and Inaba N16961 strains as well as the non-toxigenic Ogawa 395 NT strain and the poorly colonizing RV 79 strain. Adherence was patchy and follicle epithelial uptake intensity was directly related to the concentration of vibrios at the surface. Attachment and uptake were greatest with the fully

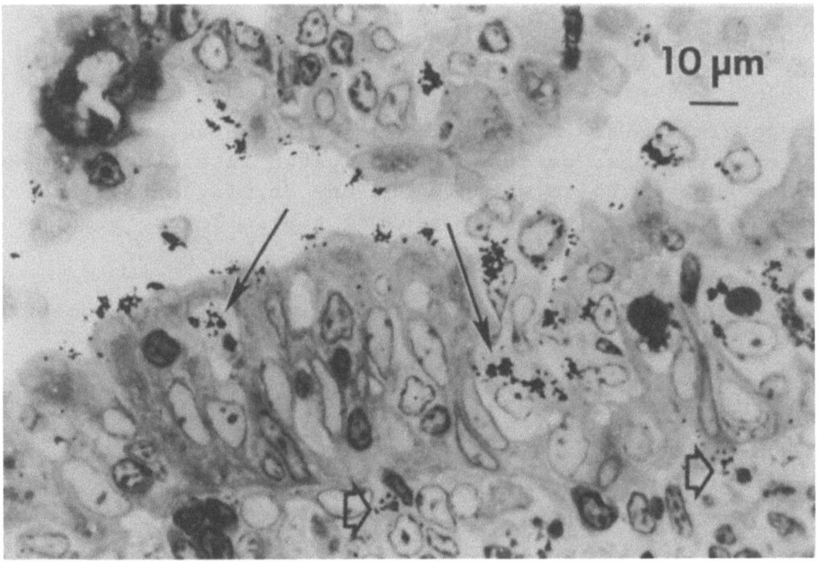

Figure 2. In a furrow over the surface of a rabbit Peyer's patch lymphoid follicle, ^{3}H-labeled Ogawa 395 cholera vibrios, revealed by black autoradiographic grains, adhere to the surface of M cells and enterocytes 90 minutes after inoculation into an intestinal loop. They are also seen in association with lymphoid cells in pockets invaginating M cells (arrows) and have already reached macrophages at the basal lamina (open arrows). Light micrograph.

virulent Ogawa 395 strain but were almost as great with the virulent Inaba N16961 strain. With both these virulent strains attachment over follicle domes occurred in sheets. By scanning electron microscopy vibrios were recognized as curved rods, adhering to the surfaces of M cells, which were recessed below surrounding enterocytes [Figure 1]. At the point of entry, vibrios were ensheathed by fused M cell microvilli. Vibrios adhered to surfaces of villi and Peyer's patches after 90 minutes and were taken up into follicle epithelium [Figure 2]. By transmission electron microscopy vibrios were seen in phagocytic vacuoles where they remained segregated from M cell cytoplasmic contents during transport [Figure 3]. After 180 minutes, vibrios were also observed in domes of lymphoid follicles. Adherence and uptake of the mutant Ogawa NT (non-toxigenic) and RV 79 (poorly colonizing) strains were detected less frequently than were adherence and uptake of the virulent Ogawa 395 and Inaba N16961 strains. When these mutant strains were found over follicle surfaces, they occurred singly or in small clusters rather than in sheets. When uptake was

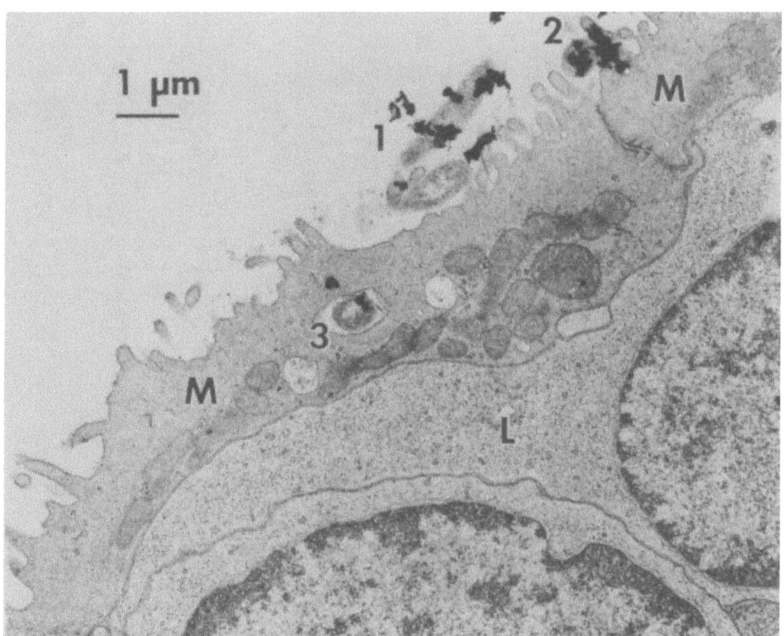

Figure 3. After 90 minutes in the intestinal loop of an adult rabbit, ^{3}H-labeled Ogawa 395 vibrios (1) lie in the lumen over Peyer's patch follicle M cells (M), are adherent to microvilli (2), and lie in a vacuole (3) within an M cell separating lymphocytes (L) from the lumen. Transmission electron micrograph.

detected with these mutant strains, however, it was below or near vibrios attached to the surface. We were unable to detect either adherence or uptake into Peyer's patch lymphoid follicles of the non-motile but toxin producing strain, CVD 49, even though these vibrios displayed normal ultrastructure and were as heavily labeled with ^3H as the other strains.

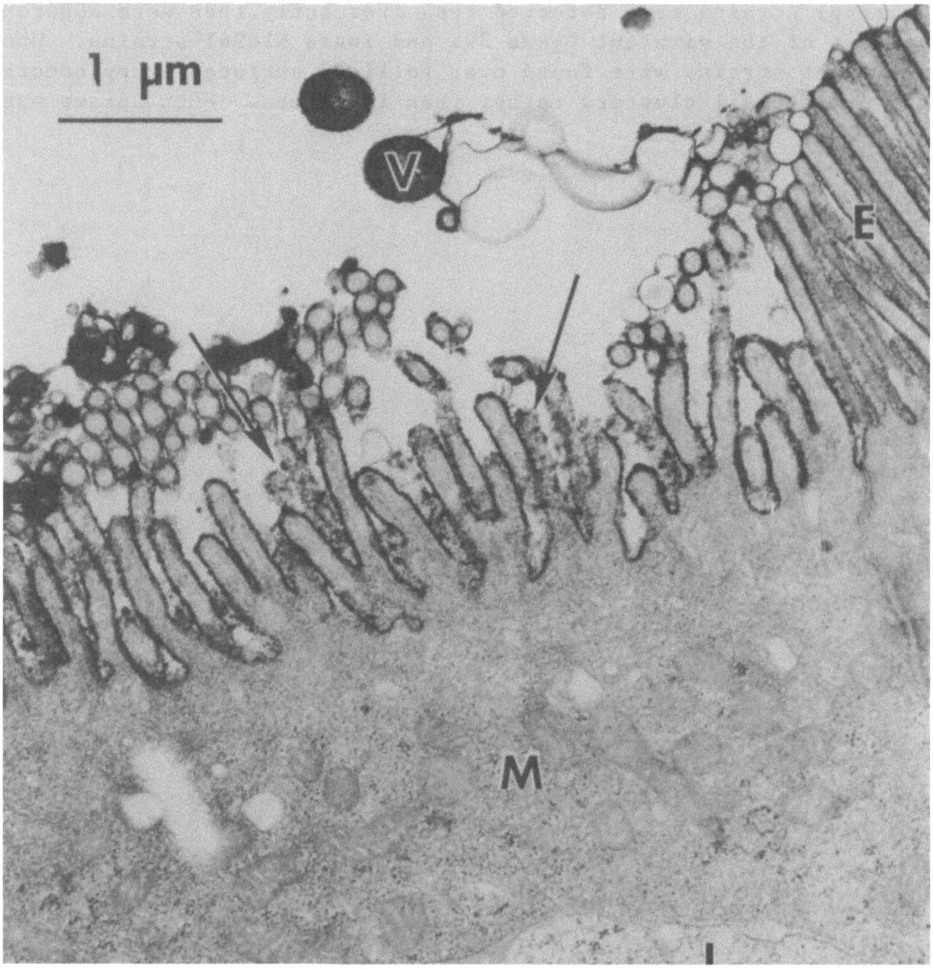

Figure 4. Ruthenium red evenly coats glycocalyx along surfaces of straight microvilli of an enterocyte (E), but is aggregated in clumps, forming knobs on surfaces of irregular and branching microvilli of an M cell (M). Aggregates continue down the margins to pits forming pinocytotic vesicles between microvilli, and stud surfaces of microvilli cut in grazing sections (arrows). Vibrios (V) adhere to shed membrane in the lumen. Lymphocyte (L) in M cell invagination.

Ruthenium red revealed a mucus blanket across the tops of enterocyte microvilli approximately 0.1 micron in thickness, which was interrupted over the surface of M cells. Ruthenium red also revealed a distinct glycocalyx coat lining the surfaces of M cells, which was aggregated in knob-like protuberances along the surface of the stubby branching microvilli and microfolds of the M cells. This pattern continued into pits at the bases of M cell microvilli and contrasted with the even glycocalyx layer on the surfaces of enterocyte microvilli [Figure 4]. Sections through the bases of M cell microvilli showed that this knob-like pattern formed spokes around the circumference of microvilli but was somewhat irregular in spacing [Figure 5]. Glycocalyx was also seen on vibrios adherent to M cells, bridging the gap between microvilli and vibrio surfaces [Figure 6].

DISCUSSION

In ligated gut segments, poorly colonizing RV 79 vibrios and non-toxigenic Ogawa 395 NT vibrios reach the intestinal surface and are then taken up into Peyer's patch lymphoid follicles, whereas nonmotile CVD 49 vibrios fail to reach the surfaces of Peyer's patches, do not

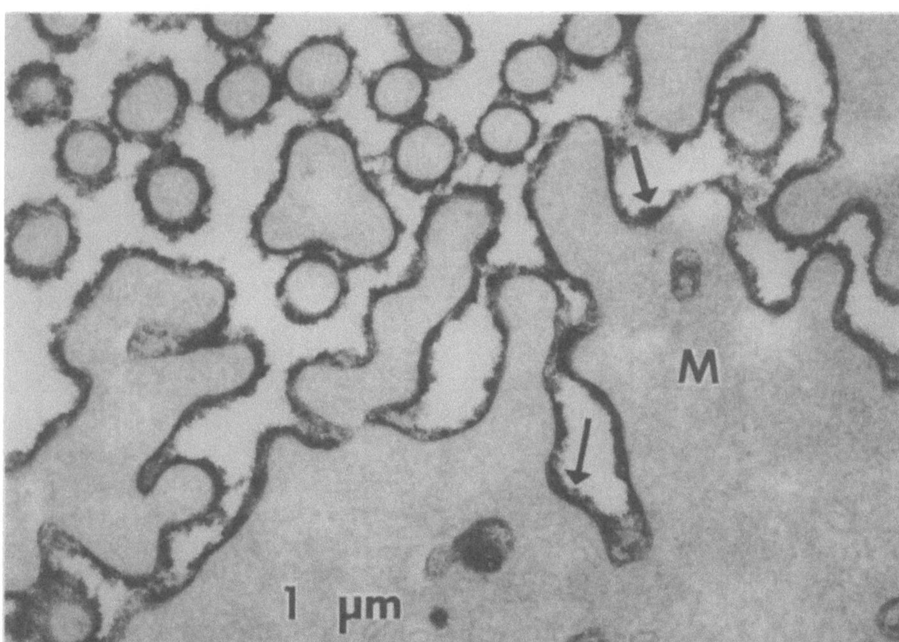

Figure 5. In a tangential section near the surface of a Peyer's patch follicle M cell (M), ruthenium red-stained glycocalyx is distributed irregularly around the circumference of single and branched microvilli and also froms aggregates in pits in the M cell surface (arrrows). Transmission electron micrograph.

adhere, and are not taken up. This difference may be attributable to the absence of peristalsis and the lack of its stirring of gut contents within ligated gut segments. Vibrios must reach the surfaces of Peyer's patches before uptake can occur. In the intact animal, the dynamic situation produced by intestinal peristalsis and continual shedding of enterocytes and enterocyte membranes can have a very different total effect than in ligated gut segments. In the latter, bacterial motility may be necessary for vibrios to reach the epithelial surface so that attachment and subsequent phagocytosis can occur. In the intact animal, peristalsis sweeps poorly colonizing strains such as RV79 down and out of the intestine even though they are motile and produce toxin.[13] Conversely, peristaltic movement in the intact animal may provide exogenous mobility for the non-motile CVD49 mutant, which is moderately colonizing and able to elicit a mucosal IgA antitoxin response after oral feeding, unlike the poorly colonizing RV79 strain.[13] In the intact animal, non-motile vibrios may be passively brought into contact with the mucosal surface, and reach M cell surfaces by proliferating and spreading laterally across the mucosal surface. Results of the present studies on segments of ligated bowel, when considered in conjunction with investigations of the same strains in intact animals, emphasize the care that must be exercised in extrapolation from ligated gut segment experiments.

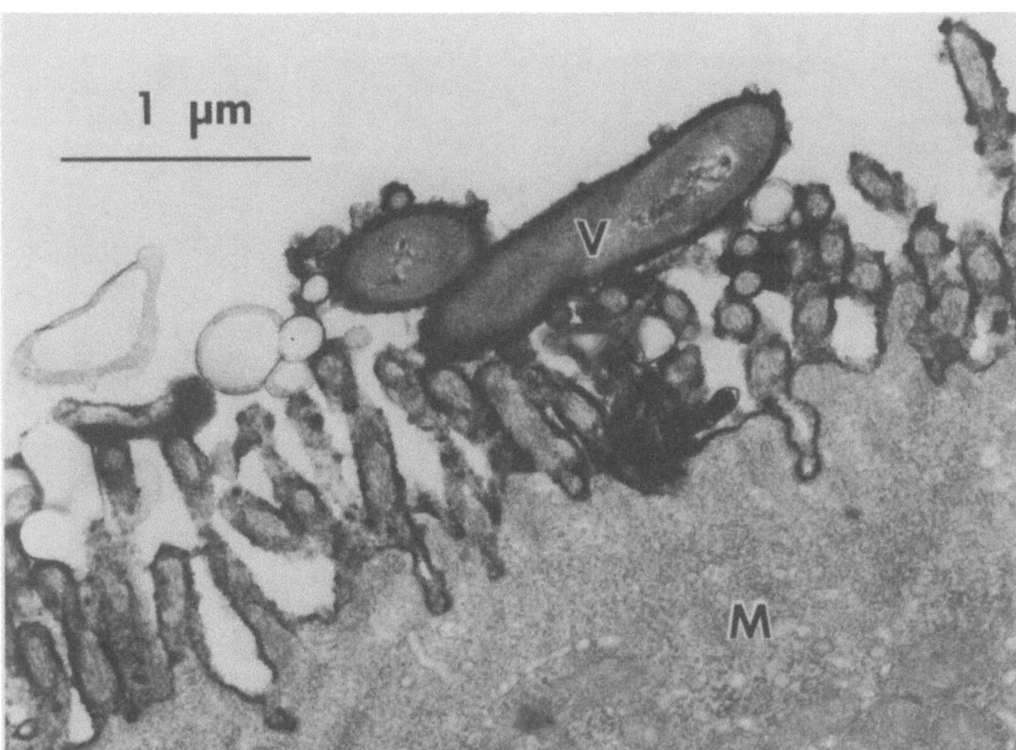

Figure 6. A cholera vibrio (V) is attached to microvilli over an M cell (M). Ruthenium red stain demonstrates glycocalyx bridging the space between the vibrio and the microvillus surface. Transmission electron micrograph.

A combination of evolved structural and functional characteristics of the intestine and of the mucosal lymphoid structures must be considered in evaluating the ability of microorganisms to attach and colonize the intestine. Secretion of mucus, rapid turnover of glycocalyx, and desquamation of enterocytes provide constant cleansing of the intestinal surface[14] and may inhibit colonization. Enterocytes and enterocyte products within the intestinal lumen provide diversionary targets for microbial attachment [Figure 4]. Even though constant cell renewal has a great metabolic cost, this mechanism along with the mechanical mucosal barrier and mucosal immune responses, are generally effective in restricting microorganisms to the intestinal lumen. In contrast, certain characteristics of Peyer's patches may enhance attachment. Interruptions in the muscularis mucosae and lymphoid infiltration may alter local motility in Peyer's patches. Reduced or absent goblet cells over follicle surfaces reduce the mucus blanket, revealing glycocalyceal structures which are accessible to microorganisms especially over the splayed and rounded microvilli of M cells. Absence of secretory component (SC) over lymphoid follicles in gut associated lymphoid tissues suggests that local SC-mediated transport of IgA may be diminished, reducing the immunologic barrier over M cells as well.[15] There is some evidence that M cell life span is longer than that of other enterocytes, prolonging opportunities for contact with luminal microorganisms, particles and molecules. After endocytosis and transport of attached luminal material, turnover of M cell luminal membranes provides a fresh surface for further uptake.[5] Invasive organisms, such as Salmonella typhi, which are not sequestered in transport vacuoles, can lead to M cell destruction and follicle ulceration.

Further study of receptors on M cell surfaces will be of great interest for understanding the process by which bacteria, such as Vibrio cholerae,[2,11] Salmonella typhi[12] and Mycobacteria,[8] and viruses, such as reovirus,[6] parvovirus-2,[16] Bredavirus,[17] and astrovirus,[17] adhere to and enter M cells. Of particular interest is whether M cells have attachment sites which may play a role in uptake of human immunodeficiency virus from the intestinal lumen, although we have not found the CD4 receptor of helper T-lymphocytes on follicle epithelial cells.[18] Identification of bacterial characteristics which affect uptake into Peyer's patches may also provide a rational basis for development of strains for oral immunization.

Acknowledgments: This work was supported by NIH grants DK-38550 and AI-19547 and by the United States Veterans Administration. Research facilities were provided by the Gerontology Research Center of the National Institute of Aging under its Guest Scientist program. Expert technical assistance was provided by R. Ted Apple, Elizabeth Juhasz and Harold Steger.

REFERENCES

1. R. L. Owen, Sequential uptake of horseradish peroxidase by lymphoid follicle epithelium of Peyer's patches in the normal unobstructed mouse intestine: an ultrastructural study, Gastroenterology 72:440 (1977).
2. R. L. Owen, N. F. Pierce, R. T. Apple, and W. C. Cray Jr., M cell transport of Vibrio cholerae from the intestinal lumen into Peyer's patches: a mechanism for antigen sampling and for microbial transepithelial migration, J Infect Dis 153:1108 (1986).
3. R. L. Owen and D. K. Bhalla, Cytochemical analysis of alkaline phosphatase and esterase activities and of lectin-binding and anionic sites in rat and mouse Peyer's patch M cells, Amer J Anat 168:199 (1983).
4. W. A. Bye, C. H. Allan, and J. S. Trier, Structure, distribution, and origin of M cells in Peyer's patches of mouse ileum, Gastroenterology 86:789 (1984).
5. M. R. Neutra, T. L. Phillips, E. L. Mayer, and D. J. Fishkind, Transport of membrane-bound macromolecules by M cells in follicle-associated epithelium of rabbit Peyer's patch, Cell Tissue Res 247:537 (1987).
6. J. L. Wolf, R. S. Kauffman, R. Finberg, R. Dambrauskas, B. N. Fields, and J. S. Trier, Determinants of reovirus interaction with the intestinal M cells and absorptive cells of murine intestine, Gastroenterology 85:291 (1983).
7. L. R. Inman and J. R. Cantey, Specific adherence of Escherichia coli (strain RDEC-1) to membranous (M) cells of the Peyer's patch in Escherichia coli diarrhea in the rabbit, J Clin Invest 71:1 (1983).
8. Y. Fujimura, Functional morphology of microfold cells (M cells) in Peyer's patches - phagocytosis and transport of BCG by M cells into rabbit Peyer's patches, Gastroenterologia Japonica 21:325 (1986).
9. M. A. Marcial and J. L. Madara, Cryptosporidium: cellular localization, structural analysis of absorptive cell-parasite membrane-membrane interactions in guinea pigs, and suggestion of protozoan transport by M cells, Gastroenterology 90:583 (1986).
10. R. L. Owen, R. T. Apple, and D. K. Bhalla, Morphometric and cytochemical analysis of lysosomes in rat Peyer's patch follicle epithelium: their reduction in volume fraction and acid phosphatase content in M cells compared to adjacent enterocytes, Anat Rec 216:521 (1986).
11. R. L. Owen, N. F. Pierce, and W. C. Cray Jr., Effects of bacterial inactivation methods, toxin production and oral immunization on uptake of Vibrio cholerae by Peyer's patch lymphoid follicles, in: "Advances in Research on Cholera and Related Diarrheas," Vol. 4, S. Kuwahara and N. F. Pierce, eds., KTK Scientific Publishers, Tokyo (1987).
12. S. Kohbata, H. Yokoyama, and E. Yabuuchi, Cytopathogenic effect of Salmonella typhi GIFU 10007 on M cells of murine ileal Peyer's patches in ligated ileal loops: an ultrastructural study, Microbiol Immunol 30:1225 (1986).

13. N. F. Pierce, J. B. Kaper, J. J. Mekalanos, W. C. Cray Jr., and K. Richardson, Determinants of the immunogenicity of live virulent and mutant Vibrio cholerae 01 in rabbit intestine, Infect Immun 55:477 (1987).
14. H. J. A. Egberts, J. F. J. G. Koninkx, J. E. van Dijk, and J. M. V. M. Mouwen, Biological and pathobiological aspects of the glycocalyx of the small intestinal epithelium, a review, Veterinary Quarterly 6:186 (1984).
15. J. Pappo, J. P. Kraehenbuhl, and R. L. Owen, Absence of expression of secretory component by epithelial cells overlying gut-associated lymphoid tissue, Clin Res 35:190A (1987).
16. P. S. Carman and R. C. Povey, Pathogenesis of canine parvovirus-2 in dogs: histopathology and antigen identification in tissues, Res Vet Sci 38:141 (1985).
17. G. N. Woode, J. F. Pohlenz, N. E. Kelso Gourley, and J. A. Fagerland, Astrovirus and Breda virus infections of dome cell epithelium of bovine ileum, J Clin Micro 19:623 (1984).
18. T. H. Ermak and R. L. Owen, Differential distribution of lymphocytes and accessory cells in mouse Peyer's patches, Anat Rec 215:144 (1986).

MUCOSAL PRIMING ALTERS PATHOGENESIS OF RIFT VALLEY FEVER

Arthur O. Anderson, Lisa F. Snyder, M. Louise Pitt and Owen L. Wood.

Laboratory of Respiratory and Mucosal Immunity
USAMRIID, Ft. Detrick, Frederick MD 21701-5011

INTRODUCTION

The concept of a common Mucosal Immune System implies that enteric priming would be effective in initiating secretory immunity in local and non-intestinal mucosal sites. If this[1,2] tenet also applies to protective immunity, enteric immunization could be utilized to advantage in vaccination against viruses that may impinge upon the conjunctiva, nasopharynx, respiratory, and gastrointestinal tracts. Recent studies showed that parenteral immunization with non-replicating viral antigens was ineffective in protecting rodents from aerosol challenge with virulent virus, although similarly vaccinated animals were completely protected from parenteral challenge with the same dose of virus[3]. Live attenuated viral vaccines elicited protective immunity for either route of challenge regardless of the route of immunization. The present studies were conducted in order to ascertain the optimum route of immunization for respiratory immunity and the effects of mucosal VS peripheral immunization on Rift Valley fever Virus (RVFV) pathogenesis in mice.

MATERIALS AND METHODS

Adult female C3H/HeJ mice were immunized with 0.2 ml formalin-inactivated RVFV vaccine (NDBR-103)[10] by direct intraduodenal (ID) injection via laparotomy on anesthetized mice[4], intraperitoneally (IP) or subcutaneously (SC). Intranasal and SC boosters were given to the appropriate groups at 14 and 21 days following priming. On day 28 the 3 vaccinated groups and a control group were halved and distributed either into SC or aerosol challenge groups. They were challenged with a 10 x LD_{50} dose[7,9] of virus strain ZH-501[8] (SC dose of 600 plaque forming units (PFU)

In conducting the research described in this report, the investigator(s) adhered to the "Guide for the Care and Use of Laboratory Animals," as promulgated by the Committee on Care and Use of Laboratory Animals of the Institute of Laboratory Animal Resources, National Research Council. The facilities are fully accredited by the American Association for Accreditation of Laboratory Animal Care.

The views of the author(s) do not purport to reflect the positions of the Department of the Army or the Department of Defense.

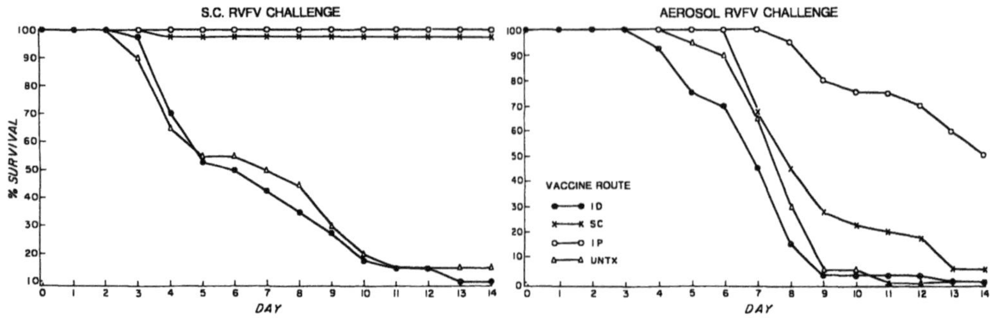

Fig. 1. Above are cumulative survival curves of untreated controls (UNTX) or mice vaccinated with RVFV by intraduodenal (ID), subcutaneous (SC) or intraperitoneal (IP) routes and then challenged with 10 x LD_{50} doses of RVFV (ZH-501) as indicated above.

Table 1. HISTOPATHOLOGY OF RVFV NON-SURVIVORS

Vaccine Route	SC Challenge			Aerosol Challenge		
	%HEP	%OBE	%ENC	%HEP	%OBE	%ENC
S.C.	3	0	0	68	61	29
I.D.	100	0	15	100	18	5
IP	0	0	0	0	67	100
Untx	81	19	19	74	58	26

Data represents percent of mice who died with the following lesions: HEP = hepatitis, OBE = olfactory bulb encephalitis (perineural), ENC = multifocal encephalitis (angiocentric). S.C. = subcutaneous, ID = intraduodenal, IP = intraperitoneal, UNTX = untreated.

and an aerosol dose of 4700 PFU of virulent RVFV per mouse). All procedures were conducted by immunized personnel in a biohazard containment laboratory. After challenge, mice were housed in filter top cages located in ventilated P-3 Biohazard glove boxes.

All mice were necropsied as they died. Fourteen days following challenge the surviving mice were euthanized and necropsied. Serum and Bile samples were taken for plaque reduction/neutralization titers against RVFV (ZH-501 strain) using standard techniques[8]. The liver, brain, nasal turbinates, and lungs were prepared for histopathology and examined for characteristic lesions of hepatitis, encephalitis, and mucosal damage. Selected specimens were examined by ABC-peroxidase immuno-histochemistry with biotinylated RVFV-specific monoclonal antibodies[11].

RESULTS

The survival for the SC and IP vaccinated groups challenged SC (Fig. 1) was 97.5% and 100% respectively. The ID vaccinated mice exhibited mortality to SC challenge identical to that of untreated controls. In the aerosol-challenged groups, survival at 10 days was 75% for i.p. vaccinated,

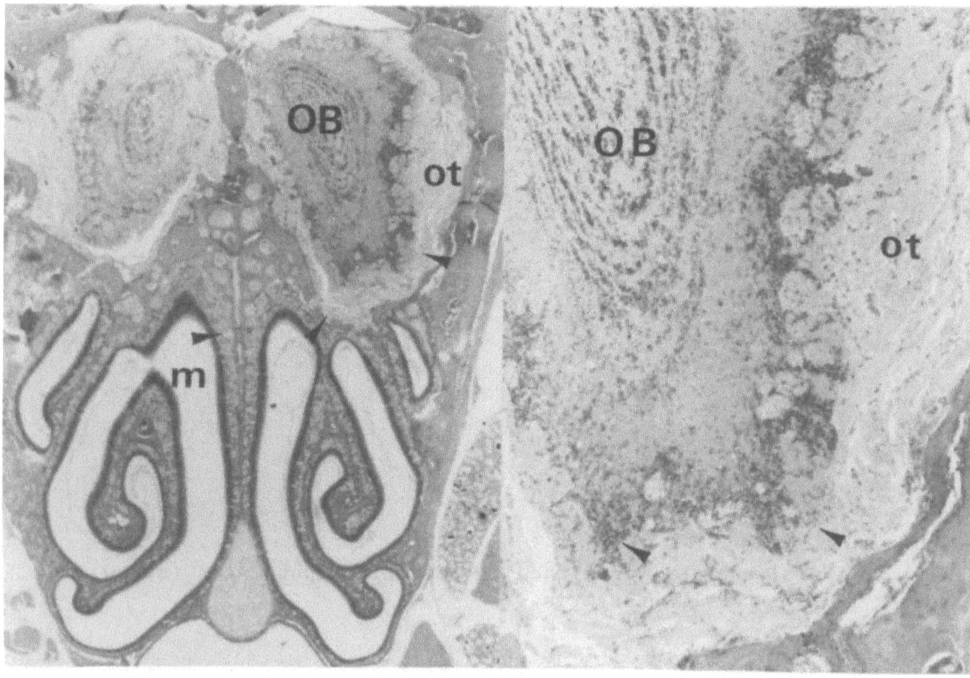

Fig. 2. Aerosol exposure to RVFV often results in olfactory bulb encephalitis. On the left is frontal section through a mouse head showing the anatomical relationships between nasal mucosa (m) olfactory tracts (arrows) and olfactory bulb (OB). Note that necrosis and leukocyte infiltration is greatest in the lateral olfactory tracts (ot).

Table 2. CROSS REGULATION IN ANTIBODY TO RVFV

Vaccine Route	Serum Titer			Bile Titer: PRN80 at 1/200	
	PRN80(x10-1)	IgA	IgM	% Pos.	# Pos./Total
S.C.	306	67	147	36	4/11
I.D.	40	190	140	60	6/10
UNTX	0	0	0	0	0/5
I.P.	N	N	N	N	N

N = Not done

22% for SC vaccinated, 1% for ID vaccinated and 5% for the untreated controls. Based on the survival curves, there did not appear to be any benefit to intraduodenal vaccination in producing protection against either SC or aerosol exposure.

Histopathology of the mice revealed interesting alterations of RVF viral pathogenesis which correlated with the immune effects of mucosal versus peripheral priming (Table 1).

Subcutaneous vaccination with RVFV vaccine prevented both hepatitis and encephalitis after SC challenge but not after aerosol challenge. In addition to increased incidence of hepatitis in the aerosol challenge group there was also an appreciable occurrence of olfactory bulb encephalitis (Fig. 2) that was more than double that of multifocal encephalitis which was primarily angiocentric in pattern. In contrast, ID vaccination

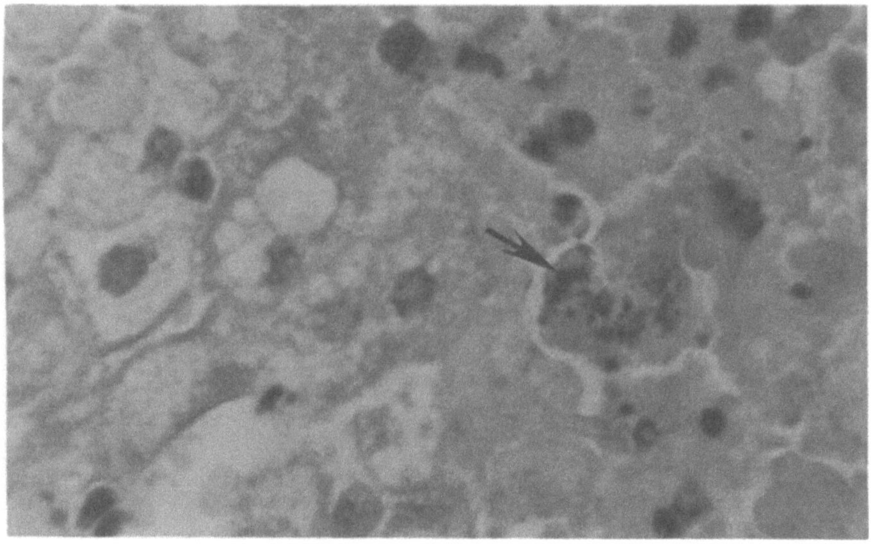

Fig. 3. Viral antigens localized in necrotic hepatocytes (arrow) during fulminant RVFV hepatitis.

increased the proportion of mice dying with hepatitis in both challenge groups compared to the untreated controls. Fulminant hepatitis occurred on days 4-12 in the ID vaccinated group compared to days 2-6 in the untreated controls. The frequency of olfactory bulb and multifocal encephalitis was significantly reduced in both challenge groups compared to the untreated controls and the SC challenged mice that had been vaccinated SC. There were no mortalities in the IP vaccinated SC challenged group, but of the 25% mortalities that occurred in the aerosol challenged group none had fulminant hepatitis. All those that died had extensive multifocal encephalitis which also involved the olfactory bulbs. Although it was difficult to distinguish between the angiocentric and perineural patterns of encephalitis in these mice, there appeared to be fewer of the latter. Immunohistochemistry with monoclonal antibodies specific for RVFV confirmed that the hepatitis and encephalitis were both associated with significant deposition of viral antigen. In the liver of RVFV infected mice, viral antigen is frequently found extracellularly because of the extensive liver cell necrosis (Fig. 3).

There were two patterns of viral antigen deposition in the brain lesions. Aerosol exposure with RVFV resulted in olfactory bulb encephalitis with neuronal cell necrosis occurring in nuclei supplied by myelinated nerve tracts, principally the lateral olfactory tracts. RVFV antigen was found in neuronal cells and in the axons of myelinated tracts of these sites (Fig. 4a). The multifocal encephalitis which occurred after either SC or aerosol exposure to RVFV was centered around blood vessels whose endothelium was frequently positive for RVFV antigen (Fig. 4b). The geometric mean 80% plaque neutralizing titers in serum samples were 1/40 (n=10) for the ID vaccinated group versus 1/306 (n=11) for the SC vaccinated group. Six of ten (60%) and four of eleven (36%) bile samples (approximate dilution 1/200) were positive for neutralizing antibody in the ID and SC vaccinated groups, respectively (Table 2).

DISCUSSION

The initial observations by Chase in the 1940's that mucosal priming with skin sensitizing agents diminishes peripheral cell-mediated responses

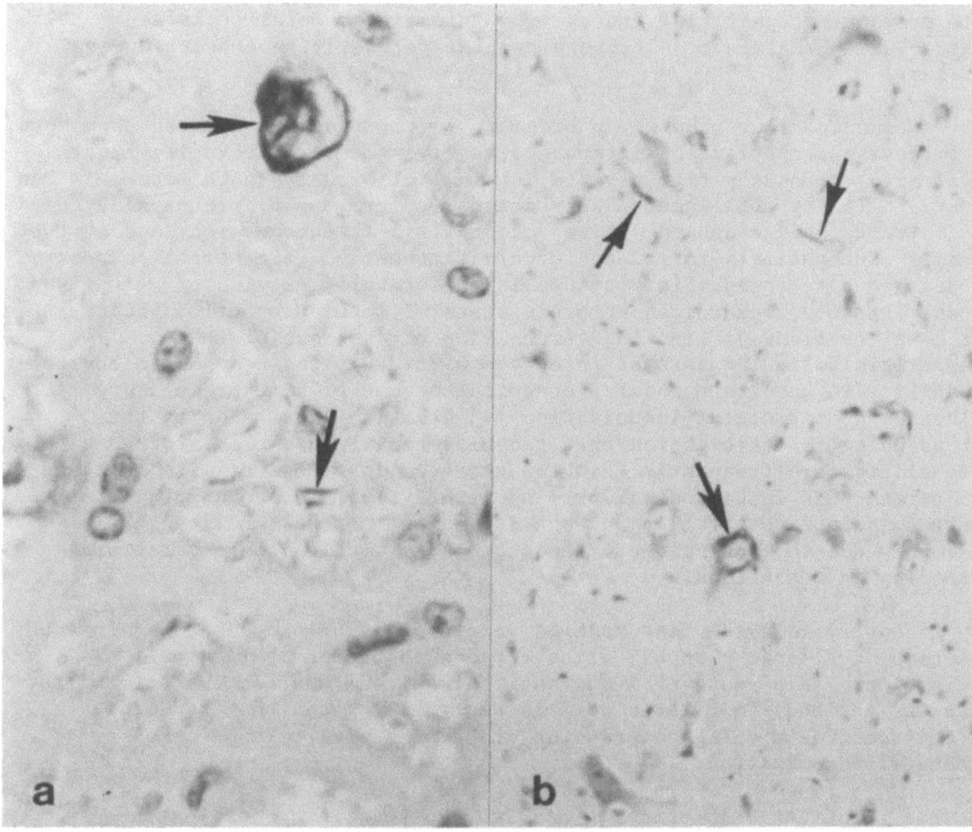

Fig. (4 a & b) a. RVFV antigen in neurons and axons of intracerebral myelinated nerve tracts (arrows) after aerosol exposure. b. Late encephalitis in RVFV infected mice often shows an angiocentric pattern. Visual antigen is located in neurons (arrows) and some endothelial cells (arrows).

in guinea pigs[12] and recent studies by Pierce and Koster where mucosal priming resulted in suppression of peripheral immunity and vice versa[13,14] has led to a better appreciation of cross regulation between the mucosal and peripheral immune systems. Furthermore, this cross regulation seems to be linked to phenomena that control the committment of specific B-Cells to IgA or IgG isotypes[15], where antigen presentation in mucosal-associated lymphatic tissues[16] and mucosally committed, immunoregulatory T-cells play obligatory roles in the selection of IgA B-cell precursors[17].

It is customary to test vaccine efficacy by IP or SC vaccination followed by IP or SC challenge. When aerosol challenge was introduced as a routine test of vaccine efficacy[3], (E. H. Stephenson, unpublished), surprising results were noted in animal models. Not only did parenteral vaccination with non-replicating viral antigens fail to protect mice, rats, and hamsters from aerosol challenge, but SC-primed animals appeared less resistant to lethal infection than untreated controls. The cause of death was usually encephalitis. Fortunately this phenomenon only occurred when the antigen was a non-replicating protein or peptide. Live, attenuated vaccines exhibited complete protection against both parenteral and aerosol challenge[3]. These phenomena suggested that the mucosa of the conjunctiva and the nasopharynx had been "tolerized" against induction of secretory immunity by parenteral priming, which left no immunological barrier that would prevent encephalitis viruses from infecting mucosal cells and penetrating ophthalmic or olfactory cranial nerves lying beneath mucosal surfaces.

We conducted our studies in order to better understand these phenomena and to devise prophylactic regimens to resolve them. Rift Valley fever virus rapidly produces high viremia in mice followed by death between 4 and 10 days following challenge. Early deaths (before day 6) are usually due to massive hepatic necrosis, while encephalitis is encountered late in the disease. Encephalitis in rodents occurs frequently after aerosol exposure, and death due to encephalitis occurs in parenterally vaccinated mice given aerosol exposure, regardless of their state of total protection from parenteral challenge[9]. The results confirm our impression that encephalitis following aerosol infection with RVF virus was due to loss of the ability to acquire mucosal protection because of immunoregulatory influences of parenteral immunization[14]. This is illustrated by the histopathological observation that the olfactory bulbs were involved with encephalitis more frequently than the cerebrum in groups of mice that had been primed SC. Intraduodenal priming with RVF viral vaccine prevented the occurrence of olfactory bulb encephalitis after aerosol challenge, presumably due to increased commitment to specific IgA and neutralizing antibodies in secretions.

In our system, the best mode of vaccination for protection against SC or aerosol challenge with RVF virus appears to be intraperitoneal vaccination. This route of immunization resulted in effective priming for secretory antibody in bile as well as neutralizing antibody in serum whereas commitment to IgA expression might otherwise be suppressed by parenteral immunization.

The relative roles of s-IgA and IgG isotypes in protecting the respiratory tract from infectious diseases is a topic of interest in many laboratories. The consensus is that s-IgA is important for protection of upper respiratory tract structures and IgG isotypes become more important deeper into the lungs[18]. Therefore, it would be undesirable to develop a vaccine that inhibited the production of either isotype, except for agents that exhibit very limited tropisms, for instance, organisms that have never been shown to penetrate beyond the mucosa. In the present study, we have uncovered complications, e.g. altered pathogenesis rather than protection, when using a strict mucosal or parenteral approach to protection against an organism that exhibits divergent tropisms. The observation that i.p. vaccination primes both mucosal and peripheral systems may be of value in respiratory vaccine development.

ACKNOWLEDGEMENTS

We are most grateful to John Cooley for technical assistance and Susan Wilders and Rita May for typing.

REFERENCES

1. S. W. Craig, and J. J. Cebra, J. Exp. Med. 134:188, (1971).
2. M. R. McDermott, and J. Bienenstock, J. Immunol. 122:1892, (1979).
3. P. B. Jahrling, and E. H. Stephenson, J. Clin. Microbiol. 19:429, (1984).
4. A. O. Anderson, T. T. MacDonald, and D. H. Rubin, Int. J. Immunother. 1:107, (1985).
5. D. H. Rubin, A. O. Anderson, and D. Lucis, Ann. N.Y. Acad. Sci. 409:866, (1983).
6. N. Pierce, and J. B. J. Sacci, Infect. Immun. 44:469, (1984).

7. W. S. Miller, P. Demchak, C. R. Rosenberger, J. W. Dominik, and J. L. Bradshaw, Amer. J. Hygiene. 77:114, (1963).
8. J.M. Meegan, Trans. Roy. Soc. Trop. Med. and Hygiene, 73:618, (1979).
9. J. L. Brown, J. W. Dominik, and R. L. Morrissey, Infect. Immun. 33:848, (1981).
10. R. Randall, L. N. Binn, and V. R. Harrison, J. Immunol. 93:923, (1964).
11. H. I. Tappert, J. M. Meegan, J. M. Dalrymple, and C. J. Peters, (Submitted).
12. M. W. Chase, Proc. Soc. Exp. Biol. Med. 61:257, (1946).
13. N. F. Pierce, J. Exp. Med. 148:195, (1978).
14. F. T. Koster, and N. F. Pierce, J. Immunol. 131:115, (1983).
15. J. A. Fuhrman, and J. J. Cebra, J. Exp. Med. 153:534, (1984).
16. M. L. Dunkley, and A. J. Husband, Immunol. 57:379, (1986).
17. S. C. Gautom, and J. R. Battisto, J. Immunol. 135:2975, (1985).
18. J. Bienenstock, (ed.), IMMUNOLOGY OF THE LUNG AND UPPER RESPIRATORY TRACTS, McGraw-Hill Book Co., (1984).

J CHAIN IS EXPRESSED BY MORE IgA2- THAN IgA1-PRODUCING CELLS IN COLONIC MUCOSA, BUT IT IS REDUCED IN BOTH SUBCLASSES IN INFLAMMATORY BOWEL DISEASE

K. Kett and P. Brandtzaeg

Laboratory for Immunohistochemistry and Immunopathology (LIIPAT), Institute of Pathology, University of Oslo Rikshospitalet, N-0027 Oslo 1, Norway

INTRODUCTION

We have recently observed a shift from the normal predominance of IgA2- (1) to IgA1- producing cells in colonic mucosa from patients with ulcerative colitis (UC) and Crohn's colitis (CC) (2). This disease-associated alteration may be of great biological importance in terms of secretory immunity. Only J-chain-containing polymeric immunoglobulins (pIg), such as dimeric IgA (pIgA) and pentameric IgM will bind to secretory component (SC) which functions as a pIg transport receptor on secretory epithelial cells (3). In this study the capacity of mucosal IgA1- and IgA2-producing cells for pIgA formation was examined by paired immunofluorescence staining for subclass and J chain, both in normal and inflamed colonic tissue specimens.

MATERIALS AND METHODS

Colonic tissue specimens from six normal controls, ten UC patients and eight CC patiens, were included in the study. Thin tissue slices were extracted in cold phosphate-buffered isotonic saline to remove diffusible, extracellular proteins prior to ethanol fixation and paraffin embedding (4). From each tissue block, three serial sections were cut at 6 μm; one section was stained with haematoxylin and eosin for histological examination, and two were subjected to paired immunofluorescence staining for J chain and IgA subclass. To unmask antigenic determinants of the J chain, the latter sections were pre-treated with 6 M acid urea (pH 3.3) for 1 h at 4°C (5). Further incubation took place at room temperature for 20 hours with a mixture of unlabelled murine monoclonal antibody to IgA1 or IgA2 and polyclonal (rabbit) IgG (0.04 g/l) anti-J-chain conjugated with tetramethylrhodamine isothiocyanate (TRITC). A second overnight incubation included a mixture of TRITC-labelled anti-J chain (0.04 g/l) and fluorescein isothiocyanate (FITC)-labeled rabbit anti-mouse IgG. The characteristics and working concentrations of the various reagents have been described previously (1,5,6). Only fluorescent cells with distinct diffuse cytoplasmic red staining were considered as J-chain positive. For each tissue specimen, an average of 200 green (subclass-specific) cells were examined for concomitant red (J-chain) staining.

RESULTS

Table 1 summarizes the percentages of mucosal IgA1- and IgA2-producing cells positive for J chain in normal colon and in colonic mucosa from patients with UC or CC. Compared with normal, the J-chain positivity of both immunocyte subclasses was significantly reduced in the two disorders ($p<0.002$). A significantly higher proportion of IgA2- than IgA1-producing cells showed co-staining for J chain both in normal ($p<0.005$) and diseased colonic mucosa ($p<0.004$).

Table 1. J-chain positivity of IgA subclass-producing cells

	Median percentage (range)	
	IgA1	IgA2
Normal	88 (79-98)	100 (94-100)
Ulcerative colitis	46 (33-86)	69 (53-94)
Crohn's colitis	46 (26-66)	74 (63-85)

DISCUSSION

We observed that IgA2-producing cells expressed more J chain than the IgA1 counterpart in normal colonic mucosa. This was in agreement with a previous report (7). However, we found an overall higher percentage of J-chain positivity, probably because we unmasked J-chain determinants with acid urea (5).

The J-chain expression of the mucosal IgA cells was strikingly down-regulated in inflammatory bowel disease. We have seen previously a shift from the normal predominance of IgA2- to IgA1-producing cells in these disorders (1,2). Both in terms of the changed subclass proportions and the reduced J-chain expression, the local IgA-population in inflammatory bowel disease seems to become more similiar to that found in peripherial lymphoid organs (5,7,8).

The observed immunopathological alterations may reflect an influx of monomer-expressing IgA1 cells representing relatively more mature B-cell clones (with down-regulated J chain) of secondary systemic immunity. These cells may originate directly from the circulation and migrate from the basal part of the mucosa into the lamina propria and the submucosa. This possibility is supported by our finding of an even higher proportion of IgA1 cells in the submucosa than in the mucosa (2).

Another possibility is that IgA1 cells may develop within the inflamed tissue as a result of heavy-chain switching from precursors of the IgG1 isotype, which we have shown to be abundantly present in diseased colonic mucosa (9). According to the arrangement of human heavy chain genes (10), an IgG1 to IgA1 switch pattern may be expected, as recently suggested by studies of human myeloma cells (11,12).

In the normal state the follicle-associated B cells of the gut (e.g. in Peyers s patches) are considered to migrate through lymphe and peripherial blood and extravasate in distant lamina propria where they end up as J-chain positive IgA cells (13). A third alternative explanation for the disease-associated alteration observed for the IgA-cell population, is that more local retention of mature B-cell clones with repressed J-chain expression may take place in overstimulated mucosal lymphoid follicles exposed to abnormally large amounts of luminal antigens and mitogens.

CONCLUSIONS

J chain is a key factor in secretory immunity. Only J-chain-expressing IgA cells can contribute to the secretory IgA system. The decrease of J chain in mucosal IgA-producing cells seen in inflammatory bowel disease therefore reflects an important immunoregulatory alteration. However, down-regulation of J chain may in terms of secretory immunity be compensated for by a two-fold numerical increase of IgA-producing cells occuring in inflamed colonic mucosa (14).

REFERENCES

1. K. Kett, P. Brandtzaeg, J. Radl, J. J. Haaijman, Different subclass distribution of IgA-producing cells in human lymphoid organs and various secretory tissues, J. Immunol. 136:3651 (1986).
2. K. Kett, P. Brandtzaeg, Local IgA subclass alterations in ulcerative colitis and Crohn's disease of the colon, Gut 28:1013 (1987).
3. P. Brandtzaeg, H. Prydz, Direct evidence for an integrated function of J chain and secretory component in epithelial transport of immunoglobulins, Nature 311:71 (1984).
4. P. Brandtzaeg, Mucosal and glandular distribution of immunoglobulin components. Immunohistochemistry with a cold ethanol-fixation technique, Immunology 26:1101 (1974).
5. P. Brandtzaeg, Immunohistochemical characterization of intracellular J-chain and binding site for secretory component (SC) in human immunoglobulin (Ig)-producing cells, Molec. Immunol. 20:941 (1983).
6. P. Brandtzaeg, K. Kett, T. O. Rognum, et al, Distribution of mucosal IgA and IgG subclass-producing immunocytes and alterations in various disorders, Monogr. Allergy 20:179 (1986).
7. S. S. Crago, W. H. Kutteh, I. Moro, et al, Distribution of IgA1-, IgA2-, and J chain-containing cells in human tissues, J. Immunol. 132:16 (1984).
8. P. Brandtzaeg, F. R. Korsrud, Significance of different J chain profiles in human tissues: generation of IgA and IgM with binding site for secretory component is related to the J chain expressing capacity of the total local immunocyte population, including IgG and IgD producing cells, and depends on the clinical state of the tissue, Clin. exp. Immunol. 58:709 (1984).
9. K. Kett, T. O. Rognum, P. Brandtzaeg, Mucosal subclass distribution of Immunoglobulin G-producing cells is different in ulcerative colitis and Crohn's disease of the colon, Gastroenterology (in press).
10. J. G. Flanagan, T. H. Rabbits, Arrangement of human immunoglobulin heavy chain constant region genes implies evolutionary duplication of a segment containing γ, ε, and α genes, Nature 300:709 (1982).
11. L. Hammarström, H. Mellstedt, M. A. A. Persson, C. I. E. Smith, A. Ahre, IgA subclass distribution in paraproteinemias: suggestion of an IgG-IgA subclass switch pattern, Acta Pathol. Microbiol. Immunol. Scand. (C) 92:207 (1984).
12. M. Takahashi, T. Tsukada, M. Kojima, et al, Immunoglobulin class switch from IgG to IgA in a patient with smoldering multiple myeloma, Blood 67:1710 (1986).

13. P. Brandtzaeg, Research in gastrointestinal immunology - state of the art, Scand. J. Gastroenterol. 20(Suppl. 114):137 (1985).
14. P. Brandtzaeg, K. Valnes, H. Scott, T. O. Rognum, K. Bjerke, K. Baklien, The human gastrointestinal secretory immune system in health and disease, Scand. J. Gastroenterol. 20(Suppl. 114):17 (1985).

ANTIGEN PRESENTATION AND CELLULAR INTERACTIONS IN INITIATION AND REGULATION OF IMMUNE RESPONSES

ONTOGENY OF MACROPHAGES AND DENDRITIC CELLS OF THE RAT

Christine D. Dijkstra, E.P. Van Rees, and E.A. Döpp

Department of Histology, Medical Faculty, Free University Amsterdam, The Netherlands

INTRODUCTION

Earlier studies on the ontogeny of macrophages have demostrated that the immunological immaturity of newborn rodents is at least in part due to the immaturity of the macrophage system (1,2). In particular the capacity of macrophages to present antigen and thus induce an adequate immune respons was impaired in neonates as compared to adults (13,19,21), which appeared to be due to a low number of Ia-positive antigen presenting cells (15,20). A considerable number of studies are now available on the ontogenetic development of antigen presenting cells in the different lymphoid organs as thymus (10,12), gut and gut-associated tissue (17,25,26), spleen (7,11) and lymph node (24) or in cell suspensions of these organs (14,16). In these studies, the antigen presenting cells, in particular the interdigitating cells (IDC, in tissues) respectively dendritic cells (DC, in cell suspensions) were studied using either their expression of Ia-antigen or their antigen presenting capacity as characteristic.
The ontogenetic development of the "classical macrophages" in the lymphoid organs is less well established than that of the Ia-positive antigen presenting cells. This lack of information is partly due to the lack of markers for classical macrophages. The development of a panel of monoclonal antibodies against rat macrophage subpopulations (8) now enables us to study the phenotypical properties of macrophages of fetal and neonatal rats. Aim of the present study was to obtain more information on the lines of differentiation of macrophages as compared to dendritic cells. Furthermore study of the ontogeny of macrophage subpopulations provides information on whether these subpopulations migrate to certain sites or express their phenotypical properties after arrival at such a site. The latter possibility would indicate micro-environmental influences on the expression of their membrane antigens.

MATERIALS AND METHODS

Whole rat fetuses (from Wistar rats, TNO, Zeist, The Netherlands) of 14 till 21 days after conception were frozen in liquid nitrogen. Neonatal rats were killed by decapitation at day 0, 2, 4, 6, 8, 10, 12 and 14 after birth, relevant tissues (containing lymphoid organs) were disected frozen in liquid nitrogen and stored at -20°C. Cryostat sections were cut, air-dried, fixed in pure acetone (pro analysi) and air-dried again. A two step immunoperoxidase method was used to demonstrate the different subpopulations of macrophages (8) using the following monoclonal antibodies in the first step: ED1, ED2 and ED3 against rat macrophage subpopulations (8) and OX4 against Ia-antigen (Serotec, UK). For the second step peroxidase conjugated rabbit-anti-mouse immunoglobulin was purchased from DAKO (Denmark). Peroxidase activity was demonstrated with 3,3'-diaminobenzidine-tetra-hydrochloride (Sigma, USA). Acid phosphatase activity was demonstrated according to Burstone (5), using naphtol AS-BI phosphate as substrate and pararosaniline as diazonium salt (22).

RESULTS

Yolk sac

None of the used monoclonal antibodies recognized any cells in the yolk sac.

Liver

At day E15 after conception the first ED1-positive cells were present in the liver; at no other sites ED1-positive cells were present at that stage. The number of positive cells increased rapidly during the following days, showing two types of cells: small round cells and larger cells with cell processes. From day E16 on, a gradually increasing number of ED2-positive cells was present in the liver, all of them large cells with cell processes. The number of ED2-positive cells was always lower than that of ED1-positive cells. Similar cells were stained with ED3 just before birth, but only very weakly. Occasionally isolated Ia-positive cells were found in liver from day E16 on.

Mesenteric lymph node

The ontogenetic development of macrophages in gut and gut-associated tissue has been described extensively before (25). Briefly, the results for the mesenteric lymphe node were as follows: very early in ontogeny, at day E16 after conception, Ia-positive cells could be found in the loose mesenchymal tissue between the primitive gut loops. In the following days these Ia-positive cells accumulated around the larger blood vessels between the kidney and the primitive gut. At a later stage, these accumulations were surrounded by endothelial tissue forming a sinus, thus representing the anlages of the mesenteric lymph nodes (MLN). These primitive MLN's were relatively very large organs within the tiny fetus. ED1-positive cells appeared later in these strings of cells along the abdominal vessels and showed a different tissue distribution as compared to the Ia-positive cells. Only very few ED2-positive cells were present in the MLN before birth as well as in the early postnatal period. From day E19 on ED3 recognized almost all macrophages in the developing MLN. The MLN was the first site of expression of ED3.

Other lymph nodes

At other sites of the body, in the inguinal and axillary region and in the thoracal mesenchymal tissue close to the thymus accumulations of Ia-positive cells developed to form primitive nodes in a similar way as described for the mesenteric lymph node. However, these nodes developed later and were of smaller size. After birth the popliteal lymph node showed a similar distribution of ED1- and ED3-positive cells as in adults beit in lower numbers; the ED2-positive cells showed a sudden increase in number in the third week of life (24).

Gut and gut-associated tissue

Very early in ontogeny Ia-positive cells were found in the gut wall preceeding the appearance of ED1-positive cells. At a later stage at certain sites in the gut wall primitive Peyer's patches were found. The tissue distribution of cells positive for each of the monoclonals was similar to that of adults. One exception was the large number of strongly ED3-positive cells in the developing Peyer's patches, whereas adult Peyer's patches contain only very few ED3-positive cells (22).

Thymus

Ia-positive cells were found at day E15 after conception (cf.10) in the developing thymus but could at that stage be recognized as epithelial cells. At a later stage (E16/E17), Ia-positive non-epithelial cells entered the thymic region, although it was hard to discriminate them, due to the abundant Ia expression of the epithelial cells. The first ED1-positive cells were found in the thymus from E19 on. The ED2-postive cells appeared at the day of birth, dispersed throughout the thymus. At day 7 numerous ED2-positive cells were present in the thymus, now confined to the cortical area, displaying extensive cytoplamic processes. The tissue distribution of the macrophage subpopulations from now on resembled that of adult thymus (23).

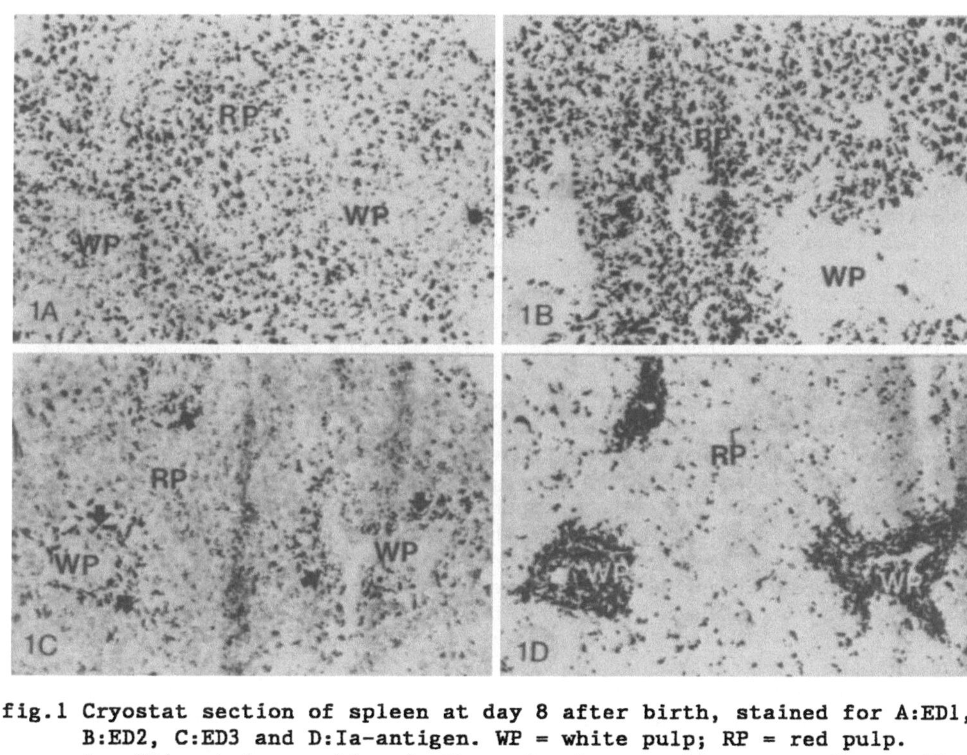

fig.1 Cryostat section of spleen at day 8 after birth, stained for A:ED1, B:ED2, C:ED3 and D:Ia-antigen. WP = white pulp; RP = red pulp.
A: ED1 stains cells in the red pulp as well as a few cells scattered in the white pulp, B: ED2-positive cells are confined to the red pulp, C: strongly ED3-positive cells are surrounding the small white pulp areas, D: Ia-positive cells are present in the white pulp only.

Spleen

At birth no compartments can be recognized in the splenic pulp. Soon after birth the white pulp compartment could be recognized as accumulations of cells around the arterioles (7). ED1 and ED2 stained from day 0 on all macrophages in the red pulp. In addition ED1 stained a few cells in the developing white pulp. ED3 also stained cells in the red pulp, but fewer than ED1 and ED2 and only weakly. At day 2 few strongly ED3-positive cells were present at the periphery of the small white pulp compartments; at day 8 a rim of such cells surrounded the white pulp. With the development of the marginal zone, ED3 also stained the marginal zone macrophages very strongly. Ia-positive cells with dendritic features were observed as soon as white pulp was recognizable (7).

Skin

Ia-positive cells were found in the subcutaneous tissue from day day E17 on, only a few large cells with blunt cell processes. At the same stage also occasionally ED1 positive cells were found in the skin. At day E19 still few Ia-positive cells were present in the subcutaneous tissue; the number of ED1-positive cell had increased considerably. In the cutis Ia-positive cells with dendritic processes (Langerhans cells) were observed on the day of birth at first, in the basal layer of the epithelium. Staining with ED1 revealed no positive cells in the cutis, in the subcutis the number of ED1-positive cell was considerably higher than that of Ia-positive cells.

DISCUSSION

The blood islands of the yolk sac are known to be the first sites of haematopoiesis in mammals (18). Also progenitors of macrophages are present there, beit not recognizable as macrophage precursors by morphological and functional characteristics (6). The present study shows that the monoclonals against macrophage subpopulations also failed to detect these precursors in the yolk sac. ED1 recognizes monocytes as well as almost all macrophages but the recognized antigen is apparently not expressed by the progenitor in yolk sac. Janossy et al. (14) described that during ontogeny two seperate subpopulations could by distinguished in the human yolk sac, type i, the majority, expressing typical macrophage markers and type ii, the minority, expressing Ia-antigen. Our study, however, revealed that no Ia-positive cells were found in the yolk sac. It is possible that this is a matter of species difference, since bone marrow precursors of the Ia-positive DC in rat have also been described to be Ia-negative (4).
The progenitor cells migrate from the yolk sac to proliferate in the hemopoietic organs (18), that is at that stage of development the liver. The first ED1-positive cells were detected in liver at day 15, they rapidly increased in number and thereafter seeded out to all connective tissues of the body. Part of the ED1-positive cells in liver became ED2-positive cells. These cells were large cells with well-developed cell processes and recognizable as Kupffer cells. These dates indicate that ED1 recognizes the macrophage precursor as well as mature macrophages, whereas ED2 rather recognizes a differentiation antigen on mature macrophages. This is in line with our finding that ED2-positive peritoneal macrophages show the endogenous peroxidase pattern of resident macrophages (2).
The most conspicuous finding with ED3 was that the strongly positive cells were, exactly like in adults, confined to lymph node and spleen, with a similar distribution pattern as in adults (8). Obviously the

microenvironment of these lymphoid organs influences the expression of the antigens recognized by ED3. Strongly ED3-positive cells occur outside lymphoid organs only in pathological conditions, such as antigen-induced arthritis (9). Many questions remain to be answered on this antigen, such as which factors play a role in its induction and what the functional significance is of such an antigen.
The development of Ia-positive cells differed from that of ED1, ED2 and ED3-positive cells. Very early in ontogeny, Ia-positive cells with dendritic features were present in the gut and gut-associated tissue (cf.25,26). Although in adult rats IDC and DC are positive for ED1 (8), the results in fetal and neonatal rats showed a dicrepancy between the expression of Ia-antigen and ED1. At several sites (mesenteric lymph node, gut wall, skin), Ia-positive cells appeared earlier than ED1-positive cells. Since ED1 recognizes monocytes, these results indicate that the Ia-positive cells originate from a "premonocytic precursor. Furthermore the differences in the early localization of ED1, and later ED2 and ED3-positive cells, on the one hand and that of Ia-positive cells on the other, is indicative for seperate different pathways of the classical macrophages and the dendritic cells. Janossy et al. (14) came to the same conclusion in their study of the ontogenetic development of macrophages and dendritic cells in human.

Acknowledgements

Thanks are due to Marja van de Ende and Marga Janse for their skillful technical assistance.

REFERENCES

1 Argyris BF. J Exp Med 128: 459 (1968).
2 Beelen RHJ, Eestermens IL, Döpp EA, Dijkstra CD. Transpl Proc (1987) in press
3 Blaese RM in: The phagocyte cell in host resistance, ed. JA Bellanti & DH Dayton, Raven press, New York, (1975).
4 Bowers WE, Berkowitz ME. Adv Exp Med Biol 186: 377 (1985).
5 Burstone MS. J Natl Cancer Inst 20: 601 (1958).
6 Cline MJ & Moore MAS. Blood 39: 842 (1972).
7 Dijkstra CD & Döpp EA. Cell Tissue Res 229: 351 (1983)
8 Dijkstra CD, Döpp EA, Joling P, Kraal G. Immunol 54: 589 (1985).
9 Dijkstra CD, Döpp EA, Vogels IMC, Van Noorden CJF. Scand J Immunol (1987) in press.
10 Duijvestijn AM, Sminia T, Köhler TG, Janse EM, Hoefsmit ECM. Dev Comp Immunol 8: 451 (1984).
11 Edwards JA, Durant BM, Jones DB, Evans PR, Smith JL. J Immunol 137:490 (1985)
12 Higley HR, O'Morchoe CCC. Dev Comp Immunol 8: 711 (1984).
13 Inaba K, Masuda T, Miyama-Inaba M, Aotsuka Y, Kura F, Komatsu-Bara S, Ido M, Maramatsu S. Immunol 47: 449 (1982).
14 Janossy G, Bofill M, Poulter LW, Rawlings E, Burford GD, Navarette C, Ziegler A, Kelemen E. J Immunol 136: 4354 (1986).
15 Lu CY, Calamai EG, Unanue ER. Nature 282: 327 (1979).
16 Lu CY, Beller DI, Unanue ER. Proc Natl Acad Sci USA 77: 1597 (1980).
17 Mayrhofer G, Pugh CW, Barclay AN. Eur J Immunol 13: 112 (1983).
18 Moore MA and Owen JJT. Lancet 2: 658 (1967).
19 Murgita RA & Wigzell H. Ontogeny of the immune system Progress in Allergy 29: 54 (1981).
20 Nadler PI, Klingenstein RJ, Hodes RJ. J Immunol 125: 914 (1980).
21 Nakano K, Hosokawa T, Muramatsu S. Developm Comp Immunol 2: 505 (1978).
22 Pearse AGE. Histochemistry. Theoretical and applied Vol. I, 3rd ed., Churchill Livingstone, Edinburgh (1968)

23 Sminia T, Van Asselt AA, Van de Ende MB, Dijkstra CD. Thymus 8: 141 (1986).
24 Van Rees EP, Döpp EA, Dijkstra CD, Sminia T. Cell Tissue Res 242: 391 (1985).
25 Van Rees EP, Dijkstra CD, Van der Ende MB, Janse EM, Sminia T. Immunol (1987) in press.
26 Wilders MM, Sminia T, Janse EM. Immunol 50: 303 (1983).

POSTNATAL DEVELOPMENT OF NON-LYMPHOID AND LYMPHOID CELL POPULATIONS IN SITU IN DIABETES-PRONE BB RATS

E.P. van Rees and C.D. Dijkstra

Department of Histology, Medical Faculty, Free University
P.O. Box 7161, 1007 MC Amsterdam, The Netherlands

INTRODUCTION

The BB rat serves as an animal model for human insulin-dependent diabetes mellitus (1). Independent of the development of diabetes mellitus, BB rats show a profound T-cell lymphocytopenia (2-7,18). Both the T-helper lymphocyte and the T-suppressor/cytotoxic subsets are depressed (3,5-7) and T-cell mediated immuneresponses in vitro are impaired (3,4,6,7). There are indications, that this T-cell immunoincompetence in adult rats is a defect in T-cell maturation or differentiation which manifests either in the thymus (6) or has a postthymic origin (3,5). However, few data are available on the microenvironment in the BB rat where T-cell maturation takes place, namely the thymus and the other lymphoid organs. In the T-cell areas of these organs interdigitating cells are thought to figure in the homing process of T lymphocytes (8-10,22) or in the creation of a suitable environment for T-cell proliferation and differentiation (11-13). Furthermore, the epithelial component of the thymic stroma (29) and thymic macrophages (24,25) probably play a prominent role in this process.

The reduction of T-lymphocytes increases with age (5,6). As far as we know, the youngest BB rats examined with mab's against T-cell subsets were 3-4 weeks old (3-5). Hence, hardly any information is available either on the onset and progress of T-cell lymphocytopenia or on the development of the non-lymphoid cells in newborn and sibling BB rats. Therefore, the neonatal development of non-lymphoid cells in thymus, spleen and lymph nodes of BB rats was studied, in correlation with that of the T lymphocyte subsets. Non-lymphoid cells were detected using a panel of monoclonal antibodies (mab's) aimed specifically at rat macrophages (14) and Ia-antigen.

MATERIALS AND METHODS

BB rats were obtained from Organon (Oss, the Netherlands). The Oss colony of BB rats was derived from BB rats originally coming from the BB colony in Worcester (U.S.A.). The Oss colony has been inbred selectively for over 10 generations, to acquire a high incidence of diabetes mellitus. The incidence of diabetes is now 80%. The age of onset is between 10 and 15 weeks after birth and is determined by positive urinary glucosis. Diabetic rats are daily treated with 2 IU protamine zinc insulin. BB rats were kept at Organon under routine laboratory conditions, with free access to commercial rat food and water.

Wistar rats, that served as controls, were obtained from TNO (Zeist, The Netherlands) and were kept in our own laboratory under similar conditions.

At least 2 BB rats were killed at the following ages: day of birth (day 0), day 1, 4, 7, 10, 12, 14, 18, 21, 23, 26, 28, 35 and 3 months after birth. Rats younger than 3 weeks were sacrificed by decapitation, older rats by exposure to CO_2.

Thymus, mesenteric and popliteal lymph nodes and spleen were removed, frozen in liquid nitrogen and stored at -70° C.The frozen tissue was cut on a cryostat to 8 µm sections that were picked up on glass slides. The sections were air-dried, fixed in pure acetone for 10 min and subsequently air-dried again. A two-step immunoperoxidase method was carried out (15), applying various mab's (Table 1). Slides were incubated with the first-step mab's during 60 min at room temperature. After being rinsed in 0,01 M phosphate-buffere saline (PBS) for 3 x 10 min slides were subsequently incubated with the second-step antiserum conjugated with peroxidase (Miles, USA, dilution 1:200) during 30 min. After being washed 3 x 10 min in PBS the sections were stained for peroxidase activity with 3,3'-diaminobenzidine-tetra-hydrochloride (Sigma, St. Louis, USA) in 0,5 mg/ml Tris HCl (pH 7,6), containing freshly made 0,01 % H_2O_2 (14). The slides were washed in PBS, slightly counterstained with haematoxylin, rinsed, dehydrated and mounted in Entellan.

To quantify the T-cell subsets, counts were carried out on at least 5 sections of each examined organ. The number of ER2 (T-helper) and Ox8-positive cells (T-suppressor/cytotoxic) was determined in a given area of the T-cell region.

Table 1. Monoclonal antibodies being applied in the present study.

mab	from	against	reference
ER2	Erasmus University Rotterdam,the Netherlands	T-helper cells	26
OX8	Seralab,England	T-suppressor cells	27
ER13	Erasmus University	Ia-antigen	28
ED1,2,3	Free University Amsterdam,the Netherlands	macrophage subpopulations	14
ED8	Free University	dendritic cells, macrophages,granulocytes	

RESULTS

1. Thymus

On the day of birth, no differences could be observed with reference to cells recognized by mab's ER2 and Ox8, between sections of thymus obtained from Wistar rats and BB rats. The cortical areas were filled with cells, recognized by ER2 as well as by Ox8. In the medullar areas cells with a T-helper phenotype (ER2-positive) outnumbered those with a T-suppressor phenotype (Ox8-positive). The T-cell populations gradually increased with age, no dissimilarities could be seen between BB and Wistar rats until day 10 after birth. From day 10 on, the medullar areas of the BB rats contained less T-suppressor cells than Wistar rats of the same age. Whereas the number of the T-suppressor cells in the thymus of Wistar rats continued to increase, this subpopulation did not enlarge in BB rats older than 10 days. Notably, from day 18 on the number of suppressor cells even showed a decrease compared to younger BB rats, especially in the medullary areas. The few cells that were recognized by Ox8 in the thymic medulla were weakly stained. These differences between BB and Wistar rats were not observed in the T-helper cells.

With reference to the macrophages, on the day of birth ED1 recognized in the Wistar rats many large cells with dendritic cellprocesses. Although most of the ED1-positive macrophages were located in the medulla, they also were present in the cortex. The number of these cells, counted per given area, gradually increased with age. In the thymic medulla of BB rats ED1 stained slightly less cells than in the Wistars on day 0, but in the cortex of BB rat thymus only few cells were found. Whereas the number of ED1 positive cells gradually increased in Wistars, especially in the medulla, this population hardly increased in BB rats. In this way the difference between Wistars and BB rats became larger with age (Fig.1a,b). Notably, the cell processes of macrophages recognized by ED1 in BB rats were less extensive than in Wistars.

Even more remarkable points of difference between Wistars and BB rats could be observed in the ED2-positive macrophages. In newborn Wistars, ED2-positive macrophages could be subdivided into two groups. Darkly stained spindle-shaped cells were lined up in capsules and connective tissue septa, and larger round-oval cells with cytoplasmatic processes scattered in the medulla. During the development, the ED2-positive cells in the medulla tended to concentrate in the cortico-medullar area. This group slowly enlarged. In adult Wistar rats, the large ED2-positive cells were only found in the cortex. Along the border between cortex and medulla, ED2-positive macrophages were lined up in a not completely continuous rim. (Fig.1c). In BB rats, in the first place, only the spindle-shaped cells in the connective tissue were demonstrable on day 0. The other kind of ED2-positive cell was found on day 7 for the first time in the cortico-medullar regio, albeit in small numbers. This population did not change much either in number or in distribution pattern until day 21. From day 21, ED2-positive macrophages started to increase in number, although always to a lesser extent than in Wistars. On day 35, aforementioned cells were mostly evenly spread in the cortex. Furthermore, a special zone consisting of macrophages recognized by ED2 around the medulla like the Wistars showed, could never be detected (Fig.1d).

In Wistar rats ED3-positive macrophages were only found in the medulla of animals younger than 14 days. These cells always were very weakly stained and after day 14 they could only occasionally be demonstrated. In all BB rats ED3 recognized only on occasion a single cell.

The subpopulations stained by ED8 hardly differed between Wistars and BB rats. Lightly stained cells with extensive cell processes were mostly located in the medulla. The amount of these cells increased slowly, in Wistars and BB rats likewise.

The mab ER13 was being applied to recognize Ia-antigen. The staining patterns of the two rat strains were indistinguishable: the thymic cortical epithelium was Ia-positive in a similar reticular pattern, the medulla showed the more confluent staining of dendritic cells. The number of Ia-positive dendritic cells slowly increased in Wistars and BB rats with the same rate.

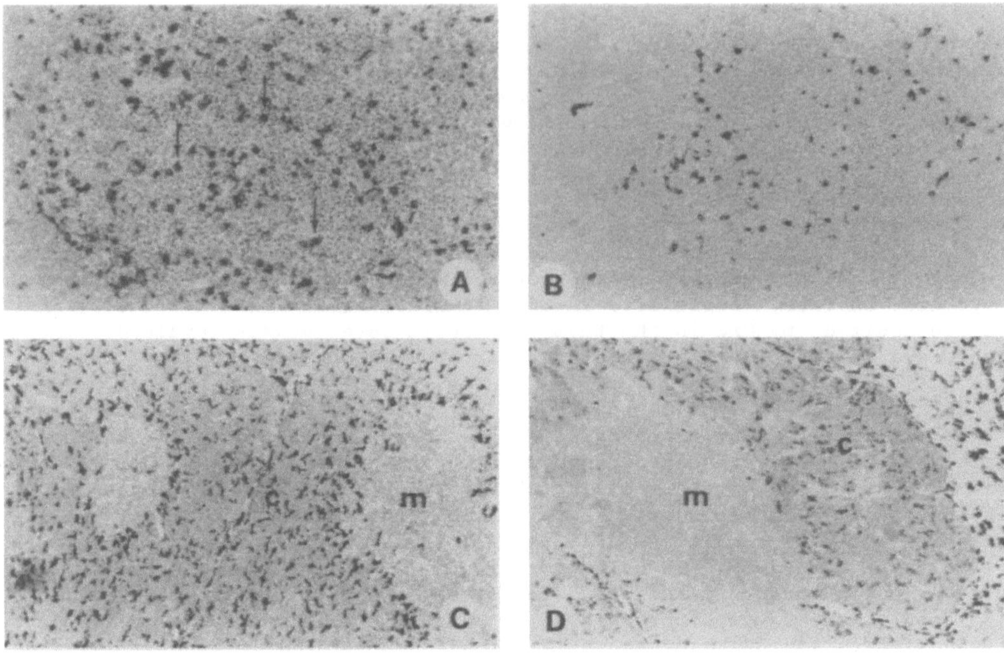

Fig.1 ED1 and ED2 recognize less macrophages in the thymus of BB rats than in that of Wistars. A: ED1-positive cells in the thymus of a 5-weeks-old Wistar rat (arrows). B: Thymus of a 5-weeks-old BB rat, stained with ED1. C: ED2 recognizes many cells in 5-weeks-old Wistars thymus. D: Age-matched BB rat shows a pronounced reduction in ED2-positive cells. c = cortex, m = medulla.

2. Spleen

Until day 14 after birth, no discrepancy could be established between the developing spleen in Wistars and BB rats. In newborn rats, the spleen mainly consisted of red pulp. On the day of birth, lymphocytes with the T-helper phenotype were found gathered around the arterioles in the developing white pulp. Cells with the T-suppressor phenotype were concentrated in a smaller ring around the arterioles. Several T-suppressor cells were scattered outside this zone and intermingled with the T-helper cells. In the Wistar rats as well as in the BB rats of comparable ages, the T-cell subsets gradually enlarged. Concentric rings developed in the periarteriolar areas forming periarteriolar lymphocyte sheaths (PALS).

On day 14 a difference was noticed between spleens of BB rats and Wistars regarding the T-suppressor population. At this stage 2-4 concentric rings of Ox8-positive cells had developed in the small PALS in Wistars besides many cells spread in the white pulp outside of the rings. In contrast with this distribution pattern, no concentric rings of T-suppressor cells were present in BB rats whereas 1-2 of these rings had been found in BB rats younger than 14 days. Now, BB rat spleen showed only few T-suppressor cells scattered over the developing PALS. Moreover, these were very weakly stained cells. Because in younger BB rat spleen T-suppressor cells had been present in a larger amount this alteration in both distribution pattern and number is an obvious reduction. During further development the T-suppressor subset in the BB rats increased only slightly in number, but the weak staining pattern persisted.

Concerning the macrophages, no disparity was visible between BB rats and Wistars. On day 0, ED1 and ED2-positive cells were largely demonstrated in the red pulp, ED1 also recognized a few small cells in the white pulp. The cells, stained by ED3 could be divided into two groups: many weakly stained cells were detectable in the red pulp while a few darkly stained macrophages were seen around the arterioles. Gradually these populations developed, showing the adult distribution pattern (14) by the end of the third week after birth.

3. Lymph nodes

Examination of lymph nodes of young BB and Wistar rats yielded information similar to that of the spleen. Until day 14 after birth, no conspicuous differences could be reported. On the day of birth, both T-helper and T-suppressor cells were present, spread over the cortex of the lymph nodes in comparable quantities. After day 14, the amount of T-suppressor cells gradually decreased. As the T-helper subset did not seem to diminish this resulted in a larger ratio of T-helper to T-suppressor cells. By the end of the fourth week, sections of lymph nodes did not show more than 10-20 weakly stained Ox8-positive cells.

Parallel to the situation in the spleen, no differences could be detected in the non-lymphoid cell populations as recognized by the ED mab's and ER13. These populations developed in a similar way in BB rats and Wistars and conform the results of an earlier study (17).

DISCUSSION

During postnatal development several dissimilarities were demonstrable between BB rats and Wistars of comparable ages. In the thymus, the T-cell subset recognized by Ox8 did not increase anymore after day 10 and even started to diminish and express less antigen after day 14. In peripheral lymphoid organs like spleen and lymph nodes an analogous reduction could be observed from day 14 on.

T-cell lymphocytopenia has been reported before (2-7,16), which was due to a decrease of T-helper cells as well as of T-suppressor cells. However, we only could determine an obvious reduction in Ox8-positive suppressor cells. As in the present study no precise countings were carried out in vitro, no definite conclusion can be made regarding this point. Furthermore, it is possible that BB rats in different colonies have different characteristics (19,20).

As far as we know, the youngest BB rats examined with mab's against T-cell subsets were 3-4 weeks old (3-5). The present study gives evidence for an onset of the T-cell abnormalities during the second week of life. During normal maturation, thymocytes migrate from the cortex to the medulla. Most thymocytes in the cortex are double-positive for mab's against T-helper and T-suppressor cells, and in the medulla for either of these. Single positive cells do emerge from double positive cells (21). The sequence of appearance of the T-cell decrease, viz. firstly in the medulla of the thymus and later in spleen and lymph nodes, combined with the information that the cortex of the thymus seemed to be normal, is suggestive for an intrathymic maturational defect.

Probably, non-lymphoid cells play a major role in maturation and differentiation of lymphoid cells (11-13). In the thymic medulla of BB rats less ED1-positive cells were visible than in Wistars thymus, whereas no differences were found either in the ED8-positive population or in the expression of Ia. In normal adult rats, interdigitating cells in the medulla of the thymus are recognized by ED1 and ED8(14) and they express Ia. Therefore, BB rats' interdigitating cells may not be recognized by mab ED1. The first Ia-positive cells with dendritic features that appeared during fetal life in Wistars were not feing recognized by ED1 (23). Hence, ED1-negative Ia-positive dendritic cells may represent immature dendritic cells.

However, the most conspicuous findings in the present study were the dissimilarities between BB rats and Wistars in the cortical thymus macrophages. To macrophages in the cortex also a role in the thymocyte maturation and proliferation has been ascribed (24,25), these cells are ED2-positive and show a distinct pattern of Ia-expression (25). BB rats have a remarkably diminished ED2-positive population in the cortex of the thymus, and a different distribution pattern as well, as they fail to show the conspicuous rim of ED2-positive cells in the corticomedullar region that Wistars show (14,25). Thymus sections, stained with routine hematoxylin and eosin, showed a less distinct demarcation of the medulla in BB rats (3), probably due to the reduction of ED2-positive cells in the corticomedullar area.

As both the amount of ED1 and ED2-positive cells had decreased and the distribution pattern of ED2 was abnormal, this may be accounted for the development of the T-cell lymphocytopenia.

REFERENCES

1. C.I. Chappel, W.R. Chappel. Metabolism 32, 8 (suppl.1)(1983)
2. R. Jackson, N. Rossi, T. Crump, N. Haynes, G.S. Eisenbarth. Diabetes 30, 887 (1981)
3. M.E. Elder, N.K. MacLaren. J. Immunol. 130, 1723 (1983)
4. D. Bellgrau, A. Naji, W.K. Silvers, J.F. Markmann, C.F. Barker. Diabetologia 23, 359 (1982)
5. D.L. Greiner, E.S. Handler, K. Nakano, J.P. Mordes, A.A. Rossini. J. Immunol. 136, 148 (1986)
6. Y.F. Yale, E.B. Marliss. Clin. Exp. Immunol. 57, 1 (1984)
7. R. Jackson, P. Kadison, J. Buse, N. Rassi, B. Jegasothy, G.S. Eisenbarth Metabolism 32, 83 (suppl.1)(1983)
8. W. van Ewijk, J.H.M. Verzijden, Th.M. Van der Kwast, S.W.M. Luyckx-Meijer. Cell Tiss. Res. 149, 43 (1974)
9. A. Friess. Cell Tiss. Res. 170, 43 (1976)
10. R. Markgraf, B. von Gaudecker, H.K. Müller-Hermelink. Cell Tiss. Res. 225, 387 (1982)
11. A.J.P. Veerman, W. van Ewijk. Cell Tiss. Res. 156, 417 (1975)
12. E. Rausch, E. Kaiserling, M. Goos. Virch. Arch. B Cell Path. 25, 327 (1977)
13. A. Villena, A. Zapata, J.M. Rivera Pomar, M.G. Barrutia. Cell Tiss. Res. 229, 219 (1983)
14. C.D. Dijkstra, E.A. Döpp,P. Joling, G. Kraal. Immunol. 54, 589 (1985)
15. C.D. Dijkstra, E.A. Döpp. Cell Tiss. Res. 229,351 (1983)
16. R.C. Graham, M.C. Karnovsky. J. Histochem. Cytochem. 14, 291 (1966)

17. E.P. van Rees, E.A. Döpp, C.D. Dijkstra, T. Sminia. Cell Tiss. Res. 242, 391 (1985)
18. B.A. Woda, A.A. Like, C. Padden, M.L. McFadden. J. Immunol. 136, 856 (1986)
19. J.R. Wright, A.J. Yates, H.M. Sharma, P. Thibert. Metabolism 32, 101 (suppl.1) (1983)
20. A.M. Varey, B.M. Dean, R. Walker, A.J. Bone, J.D. Baird, A. Cooke. Immunol. 60, 131 (1987)
21. L. Smith. Nature 326, 798 (1987)
22. R. Namikawa, T. Mizuno, H. Matsuoka, H. Fukami, R. Ueda, G. Itoh, M. Matsuyama, T. Takahashi. Immunol. 57, 61 (1986)
23. E.P. van Rees, C.D. Dijkstra, M.B. van der Ende, E.M. Janse, T. Sminia. Immunol. accepted (1987)
24. D.I. Beller, E.R. Unanue. J. Immunol. 121, 1861 (1978)
25. T. Sminia, A.A. van Asselt, M.B. van der Ende, C.D. Dijkstra. Thymus 8, 141 (1986)
26. P. Joling, F.J. Tielen, L.M.B. Vaessen, C.J. Hesse, J. Rozing. Adv. Exp. Med. Biol. 186, 235 (1985)
27. R.J. Brideau, P.B. Carter, W.R. McMaster, D.W. Mason, A.F. Williams. Eur. J. Immunol. 10, 609 (1980)
28. R.J.M. Stet, J. Rozing, G.D. Majoor, F.G.M. Kroese, D. Opstelten, P. Nieuwenhuis. Transplant. Proceed. 17, 1829 (1985)
29. J.J.T. Owen, E.J. Jenkinson. Amer. J. Anat. 170, 301 (1984)

IMMUNOLOGICAL CHARACTERISTICS OF MILKY SPOTS IN THE OMENTUM OF RATS

R.H.J. Beelen*, I.L. Eestermans, E.A. Döpp, and C.D. Dijkstra

Department of Cell Biology, Medical Faculty, Free University
P.O. Box 7161, NL-1007 MC Amsterdam, The Netherlands

In the rat model we have clearly shown in the past that the diversity of cytochemically heterogeneous populations of macrophages is best explained by stage of maturation differentiation of cells belonging to one lineage (1) in the sequence monocyte (or exudate macrophage) - exudate-resident macrophage - resident macrophage, the dynamics being related to the state of inflammation induced (2,3). Our earlier studies on milky spots in the omentum indicated that they are a source of the free peritoneal macrophages as judged by the maturation pattern based on morphology (4) and cytochemistry (5). Moreover, precursor cells as promonocytes may settle and proliferate in milky spots and so locally generate as a pool of monocytes and macrophages. Interestingly in the milky spots the more immature macrophages were found inside and the mature resident macrophages outside, indicating their maturation route. These results were later confirmed by others (6,7).

Recently a panel of three monoclonal antibodies (MoAb) recognizing exclusively cells belonging to the mononuclear phagocyte system in the rat have been developed in our laboratory (8): ED1 recognizing an antigen on all types of macrophages; ED2 represents a differentiation antigen and ED3 recognizes and antigen present on a special type of macrophage in the lymphoid organs (9).

In the literature milky spots have also been mentioned as a possible omental lymphoid organ (10,11). Therefore in this study we have further investigated the immunological characteristics of milky spots in the rat - also with the use of MoAb - with special attention to the macrophages and lymphoid cell composition.

MATERIALS AND METHODS

Animals

Male Wistar rats weighing 180-200 g were used.

and correspondence
*Present adress: /Department of Haematology, University Hospital, Free University, Amsterdam, The Netherlands

The greater omentum

An acute inflammatory state was induced by i.p. administration of 10 ml sterile newborn calf serum (NBCS) according to Beelen et al. (4,12). The greater omentum was dissected free and the milky spots were dissected free and processed for electron microscopy as described by Beelen et al. (4), for cytochemistry or immunohistochemistry (cryostat sections; 8) or completely dissociated with collagenase according to Dux et al. (11).

Cytochemical procedures

The milky spots were fixed 10 minutes in 1.5% glutaraldehyde at 4°C, washed and processed for peroxidatic activity (PA) in DAB, postfixed in OsO_4 for 30 minutes at 4°C and processed further as described in detail elsewhere (5).

Immunohistochemistry

Immunohistochemistry was done exactly as described by Dijkstra et al. (8). Briefly, cryostat sections were fixed in aceton, incubated in diluted ascites fluid, washed, incubated with rabbit anti mouse IgG peroxidase in PBS with 0.2% BSA and 1% normal rat serum for 60 minutes and stained for peroxidase in DAB.

RESULTS AND DISCUSSION

In the omentum of a normal steady state rat numerous milky spots can be detected (6-30; n=8). After i.p. administration of NBCS the blood flow of the omentum increased strongly, followed by an increase in number (11) of the milky spots (30-100; n=4). Apparently the omentum was a very reactive site in the induction phase of an acute inflammatory state. The number of cells isolated after collagenase digestion increased about 5-fold.

Electron microscopy

As shown in Fig. 1 small blood vessels run invariably through the milky spots. The milky spots shown a framework of reticulum and mesothelial cells and always contain immunologically active lymphocytes and macrophages (the macrophages of the milky spots can actively phagocytose, bear Fc receptors and Ia antigen and also are active immune effector cells, data not shown). Occasionally a cluster of plasma cells was found.

Cytochemistry

Macrophages showing PA like normally seen in blood monocytes were found inside the milky spots (perivascularly) and the resident macrophages outside. The results of our studies led to the scheme (5,13) of Fig. 2. So macrophages do originate at least in part from blood borne mononuclear phagocytes of the milky spots.

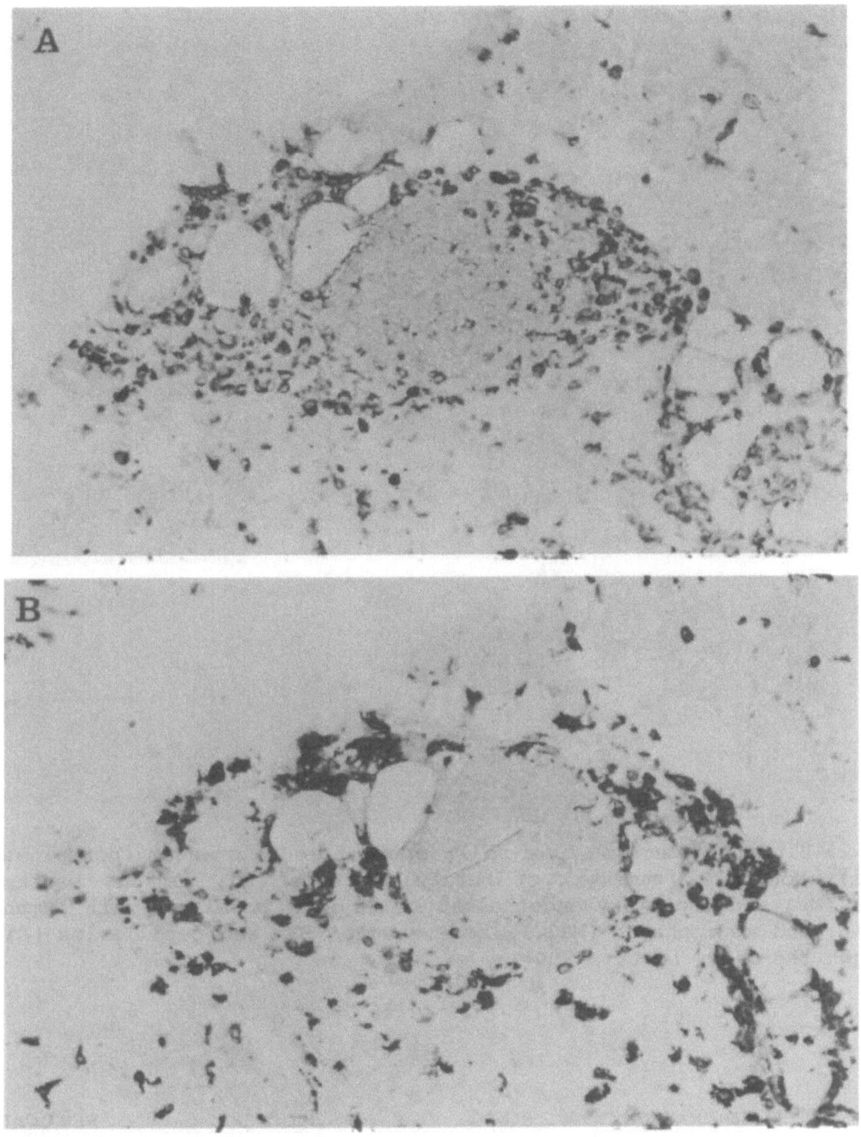

Fig. 3. Same milky spot after NBCS immunolabeled with ED1 (A) and ED2 (B).

A. Outside all macrophages are positive and the few inside macrophages are positive.

B. Outside the resident macrophages are positive and inside macrophages are negative. Note the follicle-like area.

Immunohistochemistry

Macrophages. With the panel of MoAb which recognizes rat macrophages we found that virtually all macrophages were positive for ED1, i.e. all mature macrophages outside and the few immature macrophages inside (Fig. 3A). ED2, which recognizes only mature macrophages, was only seen on the

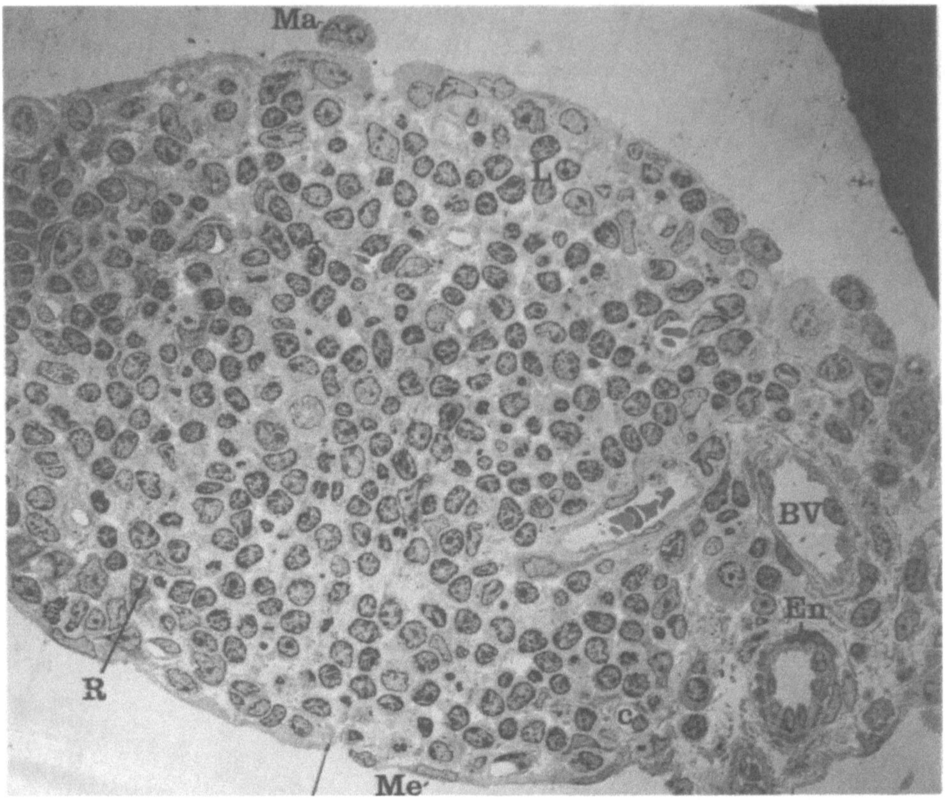

Fig. 1. Ultrathin section of a milky spot of mouse omentum (normal steady state) showing a framework of reticulum cells (R), lining mesothelial cells (Me) and capillary endothelial cells (En) populated with lymphocytes (L) and macrophages (Ma). Blood vessels (BV) and capillaries (C) run through the milky spot. x 900

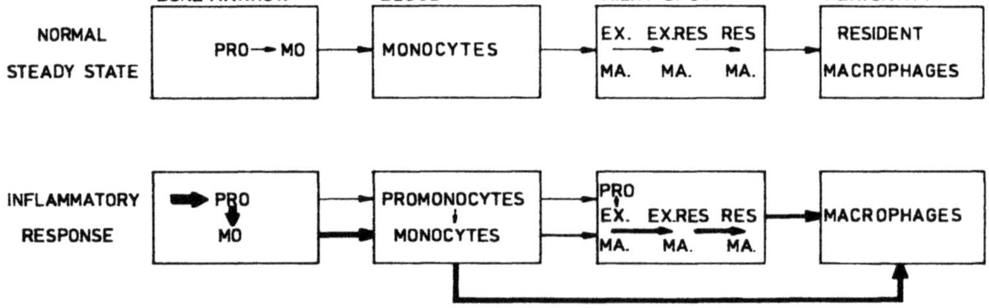

Fig. 2. Schematic diagram of the mononuclear phagocytes in 4 compartments under normal, steady state conditions and during acute inflammation. During the inflammatory response there is an increase in the formation of monocytes in the bone marrow and an accelerated influx of these cells into the milky spots and the circulation, which may contribute to the population of macrophages into the peritoneal cavity.

macrophages outside the milky spot (Fig. 3B) confirming the maturation route of macrophages. Most remarkably, as shown in Fig. 3, a follicle--like area is seen almost completely devoid of macrophages, containing mainly lymphocytes and so suggesting a lymphoid characteristic of this milky spot. Also (not shown) a large number of the milky spot macrophages were positive for ED3, which is normally only seen in lymphoid organs but not in the free peritoneal cavity (8,9). Finally, dendritic-like cells were found. From the cell suspensions obtained after collagenase digestion it appeared that about 30% of the milky spot cells are macrophages. From these macrophages ~100% are ED1 positive, ~70% ED2 positive and ~60% ED3 positive.

Lymphocytes. With a panel of MoAb which recognizes T and B cells as also T and B subsets it appeared that already after NBCS clusters of both T cells (Fig. 4A) and B cells (Fig. 4B) were found. Always cells of both the T helper and T suppressor type were found and for B cells subclasses with IgG, IgM, IgG and IgA surface immunoglobulins were found. From the cell suspensions it appeared that ~30% of the cells were lymphocytes being ~10% B cells and ~20% T cells.

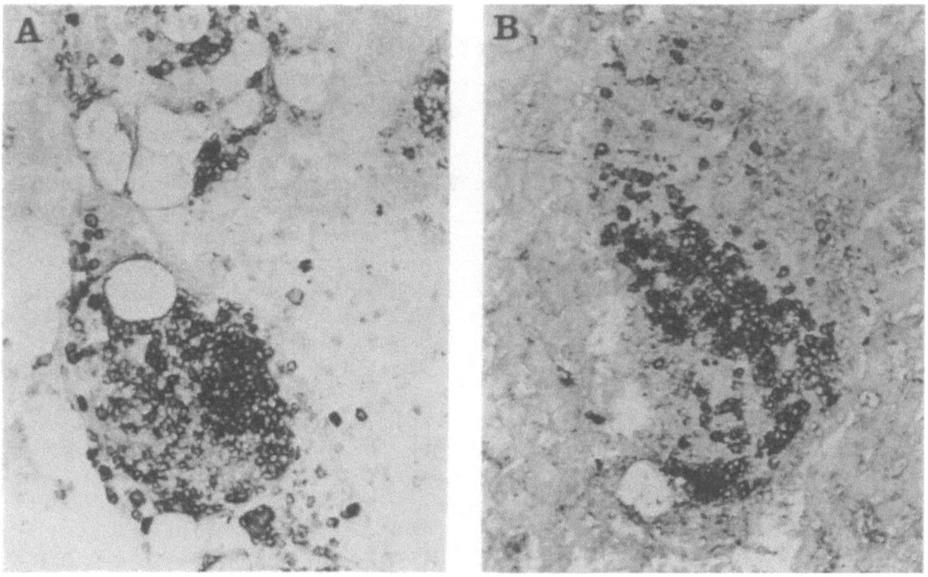

Fig. 4. Two milky spots after NBCS immunolabeled with anti T (A) and anti B (B). In A and B clusters of lymphocytes are positive.

In conclusion, our results do indicate that milky spots are already upon stimulation with NBCS remarkably changed in number and size as also in composition. On one hand the results do suggest that milky spots are very reactive inflammatory tissue, on the other hand the presence of ED3 positive macrophages, dendritic-like cells, follicle-like areas, clusters of T and B cells, do suggest more or less lymphoid characteristics. Whether milky spots are really "immune factories" or "omental lymphoid organs" (10,11) is now being investigated after specific primary and secondary immunization with TNP-KLH and TNP-LPS.

Acknowlegements

We are greatly indebted to Mrs. Anjo Steenvoorden-Bosma for typing the manuscript and Mr. Shimon Paniry for preparing the photographs.

REFERENCES

1. R. van Furth, in: "Mononuclear Phagocytes. Functional Aspects", R. van Furth, ed., Martinus Nijhoff, The Hague (1980).
2. R.H.J. Beelen, and D.M. Fluitsma. Immunobiol. 161:266 (1982).
3. R.H.J. Beelen, and W.S. Walker. Cell. Immunol. 82:246 (1983).
4. R.H.J. Beelen, D.M. Fluitsma, and E.C.M. Hoefsmit. J. Reticuloendothel. Soc. 28:585 (1980).
5. R.H.J. Beelen, D.M. Fluitsma, and E.C.M. Hoefsmit. J. Reticuloendothel. Soc. 28:601 (1980).
6. E. Mandache, E. Moldoveany, and G. Savi. Rev. Roum. Morphol. Embryol. Physiol. 23:137 (1985).
7. J.M. de Bakker. Thesis. State University Leiden, The Netherlands (1983).
8. C.D. Dijkstra, E.A. Döpp, P.P. Joling, and G. Kraal. Immunology 54: 584 (1985).
9. R.H.J. Beelen, I.L. Eestermans, E.A. Döpp, and C.D. Dijkstra. Transplant. Proc. 19:3166 (1987).
10. K. Dux. Cell. Immunol. 32:97 (1977).
11. K. Dux, R.V. Rouse, and B. Kyewski. Eur. J. Immunol. 16:1029 (1986).
12. R.H.J. Beelen, D.M. Fluitsma, and E.C.M. Hoefsmit. J. Reticuloendothel. Soc. 23:203 (1978).
13. R.H.J. Beelen. Blut 27:89 (1981).
14. H. Fischer, W. Ax, and H. Malchow. Klin. Wochenschrift 47:1014 (1969).

LYMPHOID DENDRITIC CELLS IN RHEUMATOID TISSUE AND NORMAL BLOOD - CHARACTERISTICS AND FUNCTIONS[1]

K. Waalen, Ø. Førre and J.B. Natvig

Institute of Immunology and Rheumatology, University of Oslo
The National Hospital and Oslo Sanitetsforenings Rheumatism Hospital, Oslo, Norway

Rheumatoid arthritis (RA), a disease of unknown etiology, is characterized by a chronic inflammation with an extensive infiltration of lymphocytes in the synovial membrane of these patients. Immunohistological studies of RA synovial tissue (ST) have also revealed an increased infiltration of strongly HLA-DR positive cells with dendritic morphology (1). These cells are in close contact with T helper ($CD4^+$) lymphocytes and are organized in germinal centers (1,2). In order to characterize the function of rheumatoid synovial dendritic cells (DC) in vitro, methods for isolation of these cells have been established (3). Recent studies have revealed that synovial DC have functions comparable to blood DC. Thus they are potent accessory cells for T cell activation (3,4). Furthermore, interleukin 1 (IL-1) production has been shown to be necessary for synovial and blood DC in order to exhibit accessory functions (4,5). Synovial and blood DC have been further investigated with respect to expression of cell membrane markers, to accessory function and expression of HLA-DR antigens after short term culture, and with respect to cluster formation with autologous and allogeneic T cells. DC were isolated as described previously (3). Briefly, T cells were removed from the Isopaque-Ficoll separated mononuclear cells and non adherent cells isolated by incubation on plastic for 45 minutes. Semiadherent cells (including DC) were collected from the non T adherent cells after over night culture, and Fc-receptor positive cells were removed by incubation on IgG-coated culture dishes.

---

[1] Supported by grants from Norsk Hydro and Grethe Harbitz Legacy, Oslo, Norway.

Finally, remaining T cells, and monocytes/macrophages were removed by using a mixture of cytotoxic monoclonal antibodies specific for pan-T cell, pan-B cell and monocytes/macrophages plus rabbit complement. T cells (more than 98% pure) were isolated from peripheral blood (PB) after rosetting with AET-treated SRBC and subsequent gradient centrifugation on Isopaque-Ficoll as previously described (6). In indirect immunofluorescence (IF) murine monoclonal antibodies with the following specificities (source) were used: CD45 (7) (Dr. S. Funderud); HLA-DP, HLA-DQ, HLA-DR, CD4, CD16, Leu 7 and transferrin receptor (Becton Dickinson); CD1a, CD3 and CD11a (Ortho); CD22, CD30, CD35 and dendritic reticulum cell (Dakopatts). The following monoclonal antibodies were also used: 3.9 detecting monocytes (Dr. N. Hogg) (8); RF-D1, -D2, -D7 and -D9 specific for thymic interdigitating cells, monocytes/macrophages, mature tissue macrophages and epitheloid cells, respectively (Dr. L. Poulter) (9) and 83c2, 123b1 and 124a5 specific for fibroblasts (Dr. G. Burmester) (10). Polyspecific rabbit anti-mouse immunoglobulin antibodies conjugated with fluorescein isothiocyanate (FITC) were used as the second antibody. In culture experiments the DC ($5x10^5$/ml) were incubated in medium RPMI 1640 (Gibco, UK) containing 20% human AB serum for 3 to 5 days. To some cultures were also added 10% rheumatoid synovial fluid (SF). The cells were harvested using cold HBSS (without Ca^{2+} and Mg^{2+}) containing 1 mM EDTA, and washed three times in cold HBSS before further studies. Mixed leucocyte reaction (MLR) were performed by incubating $5x10^4$ DC (irradiated 3000 rad) with 10^5 autologous or allogeneic T cells. T cell proliferation was measured at day 6 by adding ^{3}H-labelled thymidine (0.5 µCi/well) the last 18 hours of the incubation. The cells were harvested with a semiautomatic cell harvesting machine (Skatron, Norway). Cluster formation was analysed after 1 to 2 days by using an inverted microscope. To some cultures was also added mouse monoclonal anti-HLA-DR (0.5 µg/ml) or anti-CD2 (Becton Dickinson) (0.5 µg/ml) antibodies. Isolation of clusters was done by gradient centrifugation on Percoll as described elsewhere (11). Marker studies showed that both synovial and blood DC were highly positive (mean percentages of positive cells) for the CD45 antigen with 90.2% and >98%, respectively. This suggests that both types of DC are bone marrow derived. In addition, the cells were strongly positive for HLA-DP, -DQ and -DR major histocompatibility class II antigens with the following staining pattern for synovial DC: 39.2%, 65.0% and 91.5%, and for blood DC: 67.1%, 58.9% and 96.4%, respectively (12). In contrast, both synovial and blood DC lacked or were mainly negative (range) for markers characteristic for monocytes/macrophages (1.4-11.5%),

T cells (1.3-2.5%), B cells (1.8-4.7%), Langerhans cells and thymocytes (CD1a) (<1-3.2%) and NK cells (<1%). In addition, they were negative to receptors like CR1 (CD35) (1.0-4.4%), CR3 (CD11a) (2.5-5.8%) and transferrin receptors (<1-2.9%). Furthermore, the synovial DC were devoid of markers expressed by dendritic reticulum cells (<1-2.2%), Reed Sternberg cells (CD30) (<1%), epitheloid cells (<1%) and fibroblasts (1.1-9.8%) (12). Based on these results the purity of the isolated synovial and blood DC have been estimated to be 80% and 85%, respectively. Both synovial and blood DC showed a similar staining pattern with the various monoclonal antibodies used. Thus they appear to represent cells of the same lineage, different from other bone marrow derived lymphoid cells. The synovial DC were also devoid of fibroblast antigens (<10%) both before and after 3 to 5 days in culture and still expressed high numbers of HLA-DR antigens (>70%) (12). Thus the isolated synovial DC do not represent fibroblasts with dendritic morphology with passively absorbed or gamma interferon induced HLA-DR antigens. When blood and synovial DC were cultured for 5 days the number of cells expressing HLA-DR antigens were reduced from 95% to 50% and from 90% to 70%, respectively. However, when blood DC were cultured in 10% rheumatoid SF the number of cells expressing HLA-DR antigens at day 5 were 80%, indicating that the SF might contain factors like gamma interferon which have been shown to induce new HLA-DR antigens during culture. Similarly, when analysing the accessory function of synovial and blood DC after culture for 3 days, they were still potent inducers of allogeneic T cell

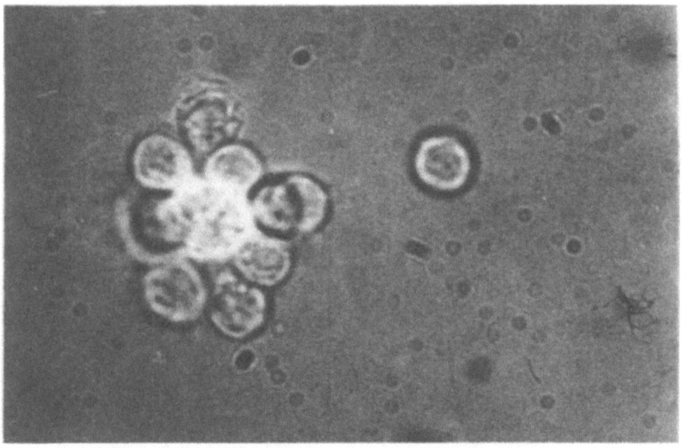

Fig. 1. A DC/T cell cluster stained with anti-HLA-DR antibody in indirect IF. In the centre is shown a HLA-DR positive DC. A combination of light and fluorescence microscopy is used. (x620).

responses (12). The accessory function was reduced by 73-76% by adding monoclonal anti-HLA-DR antibodies (12). In addition, the stimulatory ability of blood DC cultured in the presence of 10% rheumatoid SF for 3 days increased by a mean of 25% as compared to cultures without SF. These results are consistent with the observed retention of HLA-DR antigens by DC during culture. In cultures of T cells and accessory cells characteristic clusters have been observed (11). Both synovial and blood DC formed similar clusters with autologous and allogeneic T cells during the first 24 to 48 hours of culture (12). In Fig. 1 is shown a typical cluster between blood DC and autologous T cells. In the centre of the cluster is shown a HLA-DR positive DC after staining with monoclonal anti-HLA-DR antibody plus FITC-labelled rabbit anti-mouse antibody. The majority of the T cells in the clusters was of the CD4 phenotype, but also some CD8 positive cells were involved in cluster formation (not shown). The formation of clusters could be prevented by monoclonal anti-HLA-DR and anti-CD2 antibodies (12). This indicates that both these structures are important for cluster formation and cell adhesion between DC and T cells. By carefully resuspending cultures after cluster formation with a wide bore pipette it permitted us to isolate clustered cells from non-clustered cells by gradient centrifugation. As shown in Fig. 2 the proliferative capacity of the T cells in the isolated cluster fraction was 5 to 10 times higher compared to the non-cluster fraction. These observations are in agreement with the fact

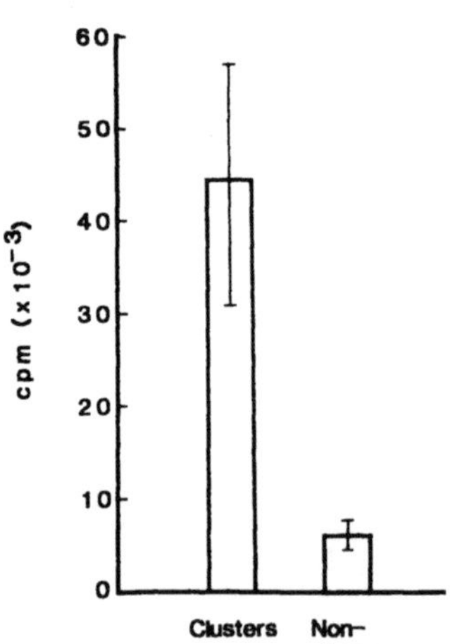

Fig. 2. Proliferation of isolated DC/T cell clusters and nonclustered cells from allogeneic MLR cultures. The mean (±SEM) results from three experiments.

that rheumatoid synovial DC are in close contact with T cells in inflammed ST (1). The synovial DC might thus be responsible for some of the observed activation process seen in rheumatoid joints.

REFERENCES

1. G. Janossy, O. Duke, L. W. Poulter, G. Panayi, M. Bofill, and G. Goldstein. Lancet 2:839 (1981).
2. Ø. Førre, J. Thoen, J. H. Dobloug, O. J. Mellbye, J. B. Natvig, J. Pahle, and B. G. Solheim. Scand. J. Immunol. 16:315 (1982).
3. K. Waalen, J. Thoen, Ø. Førre, T. Hovig, J. Teigland, and J. B. Natvig. Scand. J. Immunol. 23:233 (1986).
4. K. Waalen, Ø. Førre, J. Teigland, and J. B. Natvig. Clin. Exp. Immunol. 69:1 (1987).
5. K. Waalen, G. Duff, Ø. Førre, E. Dickens, and G. Nuki. Scand. J. Immunol. 23:365 (1986).
6. J. H. Dobloug, Ø. Førre, T. Lea, B. G. Solheim and J. B. Natvig. Clin. Immunol. Immunopathol. 19:224 (1981).
7. Leucocyte Typing III. White Cell Differentiation Antigens, p 988, A. J. McMichael, ed., Oxford University Press (1987).
8. N. Hogg, Y. Selvendran, C. Allen, and D. Palmer. Immunobiology 167:117 (1983).
9. L. W. Poulter, D. A. Campbell, C. Munro, and G. Janossy. Scand. J. Immunol. 24:351 (1986).
10. G. R. Burmester, A. Dimitriu-Bona, and S. J. Waters. Scand. J. Immunol. 17:69 (1983).
11. K. Inaba, A. Granelli-Piperno, and R. M. Steinman. J. Exp. Med. 158:2040 (1983).
12. K. Waalen, Ø. Førre, J. Pahle, J. B. Natvig, and G. R. Burmester. Scand. J. Immunol. 26:525 (1987).

INTERDIGITATING CELLS AND DENDRITIC CELLS OF THE MOUSE IDENTIFIED BY MONOCLONAL ANTIBODIES: INTERACTION WITH LYMPHOCYTE SUBSETS

Matthijs Breel, R. Mebius and G. Kraal

Department of Histology, Medical Faculty, Free University
P.O. Box 7161, 1007 MC Amsterdam, The Netherlands

Interdigitating cells (IDC) of the T cell dependent areas in lymphoid organs are considered to be the in vivo equivalent of the in vitro isolated non-adherent dendritic cells (1, 2). Dendritic cells are extremely potent accessory cells, in particular involved in the induction of T cell dependent responses (3, 4). They are thought to be derived from monocytes, although they show very little phagocytic activity (5, 6).

Interdigitating cells keep in close contact with the surrounding lymphocytes in lymphoid organs and it may well be that their presence is a basic requirement for the integrity of the T cell compartment. Characterization of IDC is mostly performed by the expression of high levels class II MHC antigen whereas in mouse IDC can also be discriminated by the NLDC-145 monoclonal antibody (7).

In this report we use the NLDC-145 antibody in combination with a new IDC specific antibody, MIDC8, to investigate the distribution of the antigens recognized by these antibodies on dendritic cell populations isolated from spleen, lymph nodes and thymus. Isolated rosettes, which reflected the interaction of IDC and lymphocytes, consist predominantly of T lymphocytes. However, after antigenn stimulation of the lymph nodes, a shift in the surrounding lymphocyte population was seen in the favor of T helper cells and B cells. The NLDC antibody does not influence the interactive bonds of IDC and lymphocytes.

MATERIALS AND METHODS

Animals

BALB/c, (C3D2)F1 and C57Bl/6 mice were obtained from Bomholtgard, Denmark. LTS/A mice were obtained from the Netherlands Cancer Institute, Amsterdam.

Antibodies

The following rat anti mouse monoclonal antibodies were used: NLDC-145, with specificity for interdigitating cells (7), MIDC8 with specificity for interdigitating cells, anti Thy-1 (8), anti-Lyt-2 (8), anti-L3T4 (9) and anti-kappa (Clone 187/1). 2H10, as control antibody with no known specificity for mouse tissues (10). All rat antibodies were used as culture supernatants, unless stated otherwise, in combination with a highly specific rabbit anti rat peroxidase conjugated second stage. Anti Ly5.1 antibody was purchased from NEN Dupont and was used in biotinylated form in combination with avidin peroxidase.

Immuno- and enzyme-histochemistry

Cryostat sections (8-12 μm) were airdried and fixed in acetone for 10 min. Cytospin preparations were air-dried and fixed in a mixture of acetone and formol (40:1). Sections were incubated for 60 min with culture supernatant. After washing in 0.01 M phosphate-buffered saline (PBS) slides were covered with a 1:200 dilution of rabbit anti rat Ig peroxidase (Dako, Copenhagen, Denmark) in PBS with 0.2% BSA and 1% normal mouse serum for 60 min. After washing in PBS, the slides were stained for peroxidase activity with 3,3'-diaminobenzidine-tetrahydrochloride (DAB, Sigma, St. Louis, MO) in 0.5 mg/ml Tris-HCl buffer, pH 7.6 containing 0.01% H_2O_2. Control slides were incubated in the same way, omitting the first step. Acid phosphatase (AP) activity was demonstrated using naphthol-AS-BI phosphate as substrate (11) with hexazotized pararosaniline as diazonium salt (12). Enzyme substrate was otained from Sigma. When immunohistochemistry and enzyme-histochemistry were combined in the same section, immunohistochemistry preceded the enzyme-histochemistry.

For two-color immunohistochemistry, sections or cytospin preparations were stained consecutively with the two antibodies. The first antibody was visualized using a peroxidase labeled second stage and 3-amino-9-ethylcarbazole (Sigma) as substrate resulting in a red color, whereas the second antibody was stained using an alkaline phosphatase conjugated second stage (anti-rat IgG alkaline phosphatase, Sigma). Alkaline phosphatase was demonstrated according to Burstone (11) using Fast Blue BB salt (no. F0250, Sigma) as substrate resulting in a blue color.

Cell suspensions

Adherent spleen dendritic cells were prepared according to the method of Steinman and Cohn (1). Briefly, spleen cells were centrifuged over discontinuous bovine serum albumin, whereafter the interphase was collected and cultured on glass for 90 min. Non-adherent cells were discarded, and after further culturing for 18 h the resulting non-adherent cells were separated by Fc receptor rosetting. FcR negative cells were used as adherent spleen dendritic cells. Dendritic cells from lymph nodes and thymus were obtained according to the method of Klinkert et al. (13) using radioresistance to separate dendritic cells from contaminating lymphocytes after discontinuous gradient centrifugation. The gradient used in these methods was a 14.5% metrizamide (Nyegaard Ltd.) gradient according to Knight et al. (14). The cells were irradiated with 1500 rad using a Cobalt source before further culturing for 18 h. Dead cells were removed from the resulting non-adherent fraction by Ficoll centrifugation and the remaining cells were separated in Fc-receptor negative and positive cells by rosetting. Again, as in the case of the spleen cells, Fc receptor negative cells were used as dendritic cells. Cytocentrifuge preparations of the cell suspensions were used for immuno- and enzyme-histochemistry.

Rosette isolation

Axillary, brachial and inguinal lymph nodes from 1-2 mice were cut into small fragments and gently agitated in 2 ml Earle's medium with 3% Hepes (flow) for 10 min at room temperature. Medium containing free lymphocytes was carefully removed and discarded. The tissue fragments were thereafter incubated in 1 ml medium with 0.5-0.8 mg collagenase IV (Sigma, St. Louis, MO) for 15 min at

25°C under gentle agitation. The supernatant was removed and replaced by fresh medium with collagenase. This was repeated four times. The collected fractions were kept on ice and carefully layered upon a column of Fetal Calf serum. The serum columns were left for 15 min, after which the top layer was removed and discarded and the serum was spun at 200 g for 12 min at 20°C. Cytocentrifuge preparations were made from the pellets after gentle resuspending.

RESULTS

Tissue distribution of MIDC8 compared to NLDC-145 in situ

The tissue distribution of the MIDC8 monoclonal antibody corresponds with the NLDC-145 antibody. It reacts with interdigitating cells, in T cell areas of lymphoid organs and dendritic cells in the thymus medulla. The MIDC8 antigen is also expressed on veiled cells and relatively weak on Langerhans cells. Two-color immunohistochemical staining with MIDC8 and NLDC-145 clearly showed that the two antibodies stain identical cell populations of interdigitating cells in situ. The MIDC8 antibody predominantly reacts with a cytoplasmic component, whereas the NLDC-145 also reacts with an antigen on the membrane (Fig. 1). No MIDC8 expression was found in other tissues, so the MIDC8 monoclonal antibody is even more specific than the NLDC-145 antibody which also reacts with the cortical for IDC epithelial cells of the thymus and the intestinal epithelium.

Distribution of MIDC8 and NLDC-145 on dendritic cells in vitro

Because of the suggested relationship of interdigitating cells and dendritic cells we wished to determine whether both the MIDC8 and NLDC-145 antibodies would react with in vitro isolated dendritic cells. Therefore dendritic cells were isolated from different organs in various ways and their phenotype was determined by immunocytochemical staining of cytospin preparation. From the spleen dendritic cells were isolated as Fc-receptor negative, non-adherent cells using a discontinuous BSA gradient. Cells were determined as dendritic cells by morphology and the expression of high levels of class II antigens. Furthermore, low numbers of these cells showed to be active as antigen presenting cells in an antigen dependent T cell proliferation assay (results not shown). As shown in Fig. 2a both NLDC-145 and MIDC8 reacted with the majority of these splenic non-adherent Fc negative dendritic cells. Staining of in vitro isolated IDC confirmed the observation that MIDC8 reacts with a cytoplasmic component, whereas the NLDC-145 is also expressed on the membrane.

Isolation of dendritic cells from lymph nodes and thymus was performed on a metrizamide density gradient. Because it was found that freshly isolated dendritic cells from lymph nodes and thymus do not adhere, lymphocytes were depleted by irradiation. The Fc negative, non-adherent cell population was analyzed for expression of anti-Ia, NLDC-145 and MIDC8 (Fig. 2b). Concerning the thymus, dendritic cell populations were always contaminated with epithelial cells, sometimes up to 50% of the cells. However, these cells could be easily identified from the dendritic cells by morphology and when staining with NLDC-145 and MIDC8 is performed only the cells with dendritic morphology were positive for these markers.

Isolation of rosettes

A large number of rosettes, consisting of a central cell surrounded by lymphocytes, were found after dissociation of lymph nodes by collagenase treatment (Fig. 1c). The central cell was characterized as interdigitating cell by staining with NLDC-145 antibody in combination with acid phosphatase staining, which was often detected in a small spot near the nucleus. Besides, IDC could be identified by long cell processes and their irregular nucleus.

Not all of the NLDC-145 positive interdigitating cells showed to be surrounded by lymphocytes. The percentage of rosettes (scored in NLDC-145 positive cells surrounded by at least 3 lymphocytes) of the total number of NLDC positive cells varied somewhat between experiments. However, in lymph node preparations of stimulated animals, always a larger number of rosettes was found in comparison with lymph nodes preparation of control animals, which were obtained under identical conditions. Allotypic mixture experiments were performed to investigate whether these rosettes were represecuted on in situ interaction from IDC with lymphocytes or whether these rosettes were merely aggregates formed during isolated (Table 1).

Table 1 Allotypic distribution of lymphocytes in mixed rosettes

type of rosette	number of rosettes	Ly 5.1 positive lymphocytes	Ly 5.1 negative lymphocytes	% contaminating lymphocytes
Ly 5.1	95	356	55	13.4
Ly 5.2	42	34	175	16.3

Lymph nodes from (C3D2)F1 (Ly5.1) and LTS/A (Ly5.2) mice were cut in very small fragments and mixed before collagenase treatment. Rosettes were isolated as described in Materials and Methods and stained for Ly 5.1. A rosette was scored Ly5.1 or Ly5.2 when more than 50 % of the surrounding cells were of the given allotype. The results are data from a representative experiment.

Table 2 The phenotype of lymphocytes in the rosettes

		Thy-1	L3T4	Lyt2	Kappa
control lymph node	A	65.1 ± 5.2	54.1 ± 5.0	26.9 ± 3.6	24.6 ± 5.2
	B	70.4 ± 3.3	51.5 ± 3.9	22.3 ± 3.1	23.7 ± 4.0
Stimulated lymph node	C	45.9 ± 3.2	45.2 ± 4.4	9.2 ± 2.3	58.7 ± 4.4
	D	47.4 ± 4.4	46.3 ± 4.1	10.5 ± 2.4	57.0 ± 4.4

After collagenase treatment of axillary and brachial lymph nodes from control (A and B) or oxazolone stimulated (C and D) animals, cytospin preparations were stained with NLDC-145 in combination with one of the T cell markers or anti Kappa together with acid phosphatase enzyme histochemistry. Per staining 30 - 50 rosettes were scored and the number of positive lymphocytes surrounding the rosettes was determined. The results are expressed as the percentage (± S.E.) of positive cells in the rosettes from an individual animal. All countings were done singleblind.

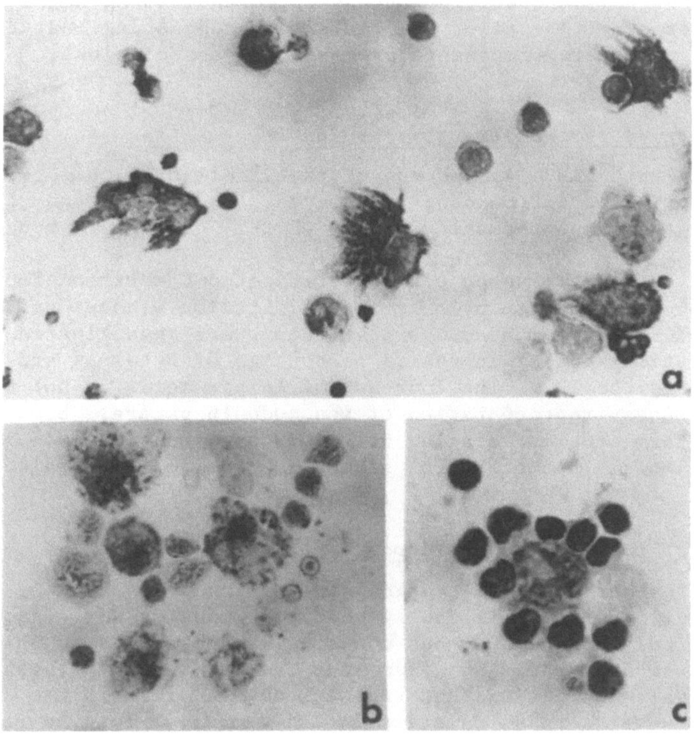

Fig. 1. Dendritic cells isolated from lymph nodes

Cytospins were obtained after metrizamide density gradient. Several lymphocytes are still present.
1a: NLDC-145: recognizes both membrane and intracellular component (40x); 1b: MIDC8: predominantly intracellular component is recognized (40x); 1c: rosette: a central interdigitating cell surrounded by lymphocytes (40x).

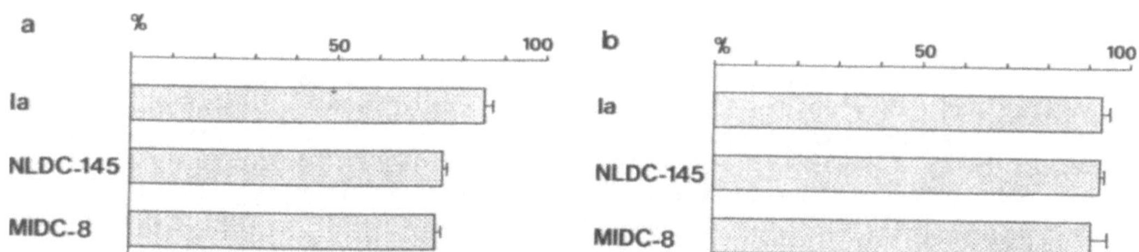

Fig. 2. Dendritic cells from spleen (a) and lymph node (b)

Dendritic cells were isolated as described in Materials and Methods. Cytospin preparations of these cells were stained with either anti-Ia, NLDC-145 and MIDC8. The results are from a representative experiment and expressed as the mean number (± SE) of positive cells for a given staining. Per sample 200-300 cells were scored.

It was clear that only a small percentage of contaminating cells were present in the rosettes of either allotype (Table 1), suggesting that the rosettes found in these procedures represent pre-existing associations.

Characterization of lymphocytes surrounding the rosettes

Double staining with NLDC-145 and one of the T cell or B cell markers showed that the majority of surrounding lymphocytes are T cells and that both T helper (LsT4) and T suppressor/cytotoxic (Lyt-2) cells were present in the rosettes.

The ratio of the two subpopulations reflected the ratio found in cell suspensions obtained through nylon screen filtration without enzyme digestion (data not shown). When lymph node preparations were investigated from animals that had been stimulated an increased percentage of T helper and B cells are found surrounding the IDC. The increase of helper cells is not a reflection of changes in subset representation of the node in general, because when this ratio was determined in single cell suspensions of stimulated nodes comparable ratios as in the control situation were found (data not shown).

The influence of the NLDC-145 antibody on the rosettes

To find out whether the NLDC-145 antigen was involved in the interaction between interdigitating cells and lymphocytes rosettes were incubated with the antibody after isolation. When staining was performed on these rosettes using only the second stage antiserum, the central interdigitating cells were clearly positive, indicating that the antibody had reached the cells during incubation. However, no differences in the number of rosettes or the number of cells present in the rosettes was found.

DISCUSSION

Both the monoclonal antibodies MIDC8 and NLDC-145 can be used to discriminate interdigitating cells in situ and dendritic cells in vitro. The staining pattern confirmed a direct relationship between the interdigitating cells in situ and the dendritic cells in vitro (1, 2). The nature of the MIDC8 antigen is not known. However, interference in a double staining method staining was interfered when NLDC-145 was preceded. This probably suggests that the two antibodies recognize determinants on the same molecule. The NLDC-145 antibody recognizes both the intercellular and the membrane component. Steric hindrance cannot be excluded although this double staining technique was successfully uwed to determine membrane and intracellular components in the same cell (15, 16).

The rosettes we describe in this paper must be regarded as in vitro correlates of in vivo pre-existing associations. This was established by allotypic mixture experiments. We therefore believe that the rosettes described in this paper are different from the clusters of antigen presenting cells and activated T cells found after co-culturing <u>in vitro</u> (17-19). The formation of these clusters depends on high cell density, the presence of antigen and the use of activated T cells. Most antigen presenting cells including macrophages and B cells, were found to form these clusters with primed T cells, but only dendritic cells were able to bind T cells in at antigen-dependent way (19). This difference was explained by assuming that dendritic cells in particular present antigen in primary activation Interdigitating cells, as the <u>in vivo</u> equivalent of the dendritic cell may therefore constitute a suitable microenvironment for T cell priming, but perhaps also for B cell activation. MacLennan and Gray (20) have recently suggested that during the primary phase of B cell activation, virgin B cells gain access to antigen which is presented on interdigitating cells, in particular in the interfollicular regions of the lymph nodes. In this area many invading veiled cells can be seen after antigenic stimulation of the lymph node with oxazolone (7) and it may well be that the rosettes isolated in our experiments are selectively derived from interactive associations in these areas. When sections of stimulated lymph nodes are examined after staining with anti-kappa an increase of the number of B cells can be found in the T cell areas, in concordance with the findings in the rosettes (results not shown). The NLDC-145 antibody did not influence the association of IDC and lymphocytes after incubation the rosettes in vitro or injection of the antibody in vivo. The NLDC-145 antigen may therefore not directly be involved in cell-cell interaction although the antibody was administered after the interactions had taken place in vivo. However, it cannot be ruled out that other epitopes on the NLDC-145 antigen, not recognized by NLDC-145 antibody, are involved in such a function.

REFERENCES

1. R.M. Steinman, and Z.A. Cohn. J. Exp. Med. 137:1142 (1973).
2. E.W.A. Kamperdijk, M.L. Kapsenberg, M. van den Berg, and E.C.M. Hoefsmit. Cell Tissue Res. 242:469 (1985).
3. R.H. Schwartz, A. Yano, and W.E. Paul. Immunol. Rev. 40:153 (1978).
4. R.M. Steinman, and M.C. Nussenzweig. Immunol. Rev. 53:127 (1980).
5. R.M. Steinman, and Z.A. Cohn. J. Exp. Med. 139:380 (1974).
6. C.W. Pugh, G.G. MacPherson, and H.W. Steer. J. Exp. Med. 157:1758 (1983).
7. G. Kraal, M. Breel, M. Janse, and G. Bruin. J. Exp. Med. 163:981 (1986).
8. J.A. Ledbetter, and L.A. Herzenberg. Immunol. Rev. 47:63 (1979).
9. A. Pierres, P. Naquet, A. van Agthoven, F. Bekkhoucha, F. Denizot, Z. Mishal, A.M. Smit-Verhulst, and N. Pierres. J. Immunol. 132:2775 (1984).
10. A. Zantema, J.A.M. Fransen, A. Davis-Olivier, F.C.S. Ramaekers, G.P. Vooys, B. Deleys and A.J. van der Eb. Virology 142:44 (1985).
11. M.S. Burstone. J. Natl. Cancer Inst. 20:601 (1958).
12. B.J. Davis, and L. Ornstein. J. Histochem. Cytochem. 7:297 (1959).
13. W.E.F. Klinkert, J.H. La Badie, and W.E. Bowers. J. Exp. Med. 156:1 (1982).
14. S.C. Knight, J. Mertin, A. Stackpoole, and J. Clark. Proc. Natl. Acad. Sci. USA 80:6032 (1983).
15. N. van Rooijen, N. Kors, and R. van Nieuwmegen. J. Histochem. Cytochem. 32:677 (1984).
16. N. van Rooijen, and N. Kors. J. Histochem. Cytochem. 33:175 (1985).
17. K. Inaba, W.D. Witmer, and R.M. Steinman. J. Exp. Med. 160:858 (1984).
18. J. Green, and R. Jotte. J. Exp. Med. 162:1546 (1985).
19. K. Inaba, and R.M. Steinman. J. Exp. Med. 136:247 (1986).
20. I.C.M. MacLennan, and D. Gray. Immunol. Rev. 91:61 (1986).

PHENOTYPIC CHANGES IN CULTURED RAT DENDRITIC CELLS: EXPRESSION OF IL-2 RECEPTOR, THY-1 AND OTHER ANTIGENS

G.G. MacPherson

Sir William Dunn School of Pathology
South Parks Road
Oxford OX1 3RE, England

INTRODUCTION

The lymphoid dendritic cell (DC) is thought, mainly as a result of 'in vitro' studies, to play a central role in the initiation of immune responses, particularly in the activation of resting T lymphocytes (1). However, despite considerable investigation, relatively little is known of the 'in vivo' physiology and life history of the DC.

A current hypothesis for DC life history is that a bone marrow derived (2, 3), blood-borne precursor (as yet unidentified), gives rise to populations of Ia positive non-lymphocytic cells in the majority of peripheral tissues (4), e.g. the Langerhans cells in the skin, and that a proportion of these cells migrate via afferent lymphatics, as veiled cells (5), or LDC (3), to regional lymph nodes where they are recognized as interdigitating cells (IDC) (6) and that IDC are extracted as DC when cell suspensions are prepared from lymphoid tissues. However, evidence that this life history explains all the relationships of peripheral DC, IDC and lymphoid DC is far from complete (Reviewed in 1).

We have shown previously that rat DC, migrating between small intestine and mesenteric lymph nodes(LDC), are heterogenous in terms of surface antigen expression and some other characteristics (7) but it was not clear if this represented different DC lineages, different stages of maturation or maturation in different micro-environments.

In the investigation of the origins of this heterogeneity an important initial step is to define the stability of the various markers. Thus, purified LDC have been cultured under different conditions and the expression of surface antigens examined by immunocytochemistry.

The results show that the surface phenotype of LDC is not a fixed characteristic and that the expression of several cell membrane antigens, including the IL-2 receptor(IL-2R), CD4, the C'3bi receptor and Thy1 can be modulated rapidly in culture.

Modulation of surface antigens has also been shown to occur when murine Langerhans cells are cultured 'in vitro' (8), but apart from the C'3bi receptor, none of the antigens described in this report have been examined on either Langerhans cells or DC in culture.

Materials and Methods.

Animals SPF PVG rats were mesenteric lymphadenectomized, irradiated and cannulated as described previously (3).

Cells Thoracic duct cells were collected overnight and enriched for LDC by centrifugation over 15% Metrizamide. LDC were 65-85% pure.

In some experiments T lymphocytes and/or Fc receptor-bearing cells were removed by rosetting techniques using antibody-coated SRBC (9) or magnetic beads (Dynal).

Culture Cells were cultured in 96-well flat or round bottomed plates or in Teflon beakers in RPMI 1640 or Hb102 (a serum-free medium) with different supplements as described below.

Culture supplements Recombinant human IL-1α was a kind gift from Dr. P.T. Lomedico, Hoffman-La Roche. Recombinant human IL-2 was a kind gift from Biogen, Geneva. Recombinant murine IL-3 was a kind gift from Dr S. Gordon.

Interleukins were added at concentrations from 1 to 100u/ml.

Conditioned media from WEHI cells and from rat spleen cells cultured with 1μg ConA/ml for four days were added at final concentrations from 1 to 20%.

Immunocytochemistry Fresh or cultured LDC were adhered to 4-spot poly-l-lysine-coated slides or cytocentrifuged onto glass slides.
After fixation with cold ethanol, cells were stained using mouse-anti-rat monoclonal antibodies as first layer and peroxidase-conjugated rabbit-anti-mouse IgG (Dako, P161) as second layer.
LDC were scored using MRC OX21 (anti-human C'3bi inactivator)(10) as the negative control. In some experiments the degree of staining was scored on + to +++ scale.

Results and Discussion

Phenotype of cultured LDC

Purified LDC were cultured in Hb102 with 10% CAS for 5h., 20h., or 48h.and cytospin or poly-L-lysine preparations stained using mouse-anti-rat monoclonals. The results shown in Table 1 represent observations from several different experiments and are based on 20h cultures. Where differences are time dependent they are mentioned in the text.

The antigens detected on fresh and cultured LDC are shown in Table 1. It is clear that the changes seen after relatively short periods of culture vary for different antigens. Those of particular interest are discussed in more detail.

Ia The expression of Ia antigen on LDC as detected by immunocytochemistry appears to remain constant in culture. This conclusion is supported by results from indirect binding assays on cytospin preparations of LDC (GGM, n in preparation) and contrasts with observations on murine Langerhans cells which significantly increase Ia expression in culture (8).

The stability of Ia expression by LDC, even in the absence of exogenous factors, is typical of DC and differs from Ia expression in macrophages, where expression decreases rapidly in the absence of lymphokines.

Table 1 Expression of surface antigens by fresh and cultured LDC

Mab	Antigen	Mr(kD)	Fresh LDC % Positive	Cult. LDC % Positive	REF
MRC OX2	OX2	43	>80 (+)	>90 (+-++)	(11)
MRC OX6	Ia	33/29	100 (+++)	100 (+++)	(11)
MRC OX7	Thy1	18	50-60 (+-++)	>90 (+-+++)	(12)
MRC OX19	CD5	67	40-45 (+-++)	30-35 (+)	(13)
MRC OX34	CD2	50	50-70 (+)	50-70 (+)	(14)
MRC OX39	IL-2R	50	<5 (+)	>90 (+-++)	(15)
MRC OX42	C'3biR	95/103	40-50 (+-++)	10-20 (+)	(16)
MRC OX48	OX48	95	0-5 (+)	>90 (++-+++)	(15)
W3/25	CD4	52	40-60 (+-++)	variable	(13)
W3/13		95	45-50 (+-++)	45-50 (+-++)	(13)

LDC isolated from thoracic duct lymph were stained immunocyto-chemically either immediately after separation or after 16-20h in culture supplemented with 10% CAS. Positive cells were scored on a + to +++ scale, corresponding to weak, intermediate, or strong staining. The table shows the proportion of positive cells and the range of staining intensities of positive cells.

Thy1 Both the proportion of positive cells and the amount of antigen per positive LDC appear to increase in culture. Quantitative assessment of cell numbers discounts the possibility that this is due to selective loss of negative cells. Unlike murine T cells, rat T cells do not express Thy1. The functional significance of Thy1 is not known but it does not seem likely that there is any relationship between LDC and the Thy1-positive Ia-negative murine epidermal cell, recently shown to possess rearranged T cell receptor gamma chain genes (17).

IL-2 Receptor(IL-2R) Very few freshly isolated LDC express the IL-2R recognized by MRC OX39. However, in culture, LDC begin to express the receptor within 5h (Fig 2). and the proportion of positive cells rises rapidly to almost 100% within 24h. By immunocytochemistry the amount of antigen per cell appears to be similar to that expressed by T lymphocytes activated in a MLR, and binding assays using iodinated recombinant human IL-2 confirm this impression (GGM, MS in preparation). It is not known if LDC express any high affinity IL-2R or if IL-2R on LDC have a functional role in the activation of T cells.

Several other non-lymphocytic cell types have been shown to express IL-2R. Herrmann et al, (18), showed that human monocytes and a macrophage-like line could be induced to express IL-2R by incubation with τ-interferon and Steiner et al, (19), showed that cultured murine Langerhans cells expressed IL-2R. However in both these cases IL-2R expression was only seen after 24h or more in culture, in contrast to the very rapid expression by LDC seen in these experiments.

It remains to be established if DC from lymphoid organs express or can be induced to express the IL-2R.

C'3bi Receptor(CR3) MRC OX42 is thought to recognize CR3 on rat macrophages,(16). About 50% of freshly isolated LDC are OX42-positive but both the proportion and intensity of staining decline in culture and the intensity is always less than that seen on peritoneal macrophages. A lower receptor density may explain the failure to show significant binding of C'3-coated SRBC by fresh LDC (3).

Freshly isolated Langerhans cells express CR3 but also lose the antigen in culture (8).

OX2 MRC OX2 recognizes a molecule found on rat thymocytes, nervous tissue and follicular dendritic cells (11). More than 80% of fresh LDC are weakly OX2-positive. The proportion of positive cells increases slightly in culture but the intensity of staining increases markedly. It is known that cells in afferent lymph can enter follicles and germinal centres (20,21) but it is not known if LDC can do so.

CD4 W3/25 recognizes the rat CD4 molecule(13). About 50% of fresh LDC are CD4 positive. In culture the expression of CD4 is variable and is influenced by the presence of rat serum. CD4 expression by rat macrophages is similarly variable (Crocker,P. and Henwood, N., pers. comm.) The function of CD4 on these cells is not understood.

Conclusions Rat LDC are a heterogeneous population of cells travelling from intestine to lymph node. Their function in the periphery is not understood although an attractive hypothesis is that they may be able to transport antigen to the node.

It is clear that LDC are not terminally differentiated and that their expression of a variety of antigens changes dramatically in culture. It remains to be determined how this antigenic modulation is regulated, whether the same modulation can be induced 'in vivo' and which of the changes have physiological significance in the life history and functions of dendritic cells.

Acknowledgements I am grateful for the excellent technical assistance of Mr. C. Jenkins, and to members of the MRC Cellular Unit for the supply of monoclonal antibodies and for helpful discussions.

References

1. R.M. Steinman, K. Inaba, G. Schuler, and M. Witmer in "Mechanisms of host resistance to infectious agents, tumors and allografts", p.71, R.M. Steinman and R.J. North, eds., The Rockefeller University Press, (1986).
2. R.M. Steinman, D.S. Lustig, and Z.A. Cohn. J. Exp. Med. 139: 1431 (1974).
3. C.W. Pugh, G.G. MacPherson, and H.W. Steer. J. Exp. Med. 157: 1758 (1983).
4. D.N.J. Hart and J.W. Fabre. J. Exp. Med. 154: 347 (1981).
5. R.H. Kelly, B.M. Balfour, J.A. Armstrong, and S. Griffiths. Anat. Rec. 190: 5 (1978).
6. I. Silberberg-Sinakin, G.J. Thorbecke, R.L. Baer, S.A. Rosenthal, and V. Berezowsky. Cell. Immunol. 25: 137 (1976).
7. G. Schuler, and R.M. Steinman. J. Exp. Med. 161: 526 (1985).
8. G.G. MacPherson, and C.W. Pugh. Immunobiol. 168: 338 (1984).
9. D.W. Mason. Transplant. 32: 222 (1981).
10. L. Hsiung, A.N. Barclay, M.R. Brandon, E. Sim, and R.R. Porter. Biochem. J. 203: 293 (1982).
11. W.R. McMaster, and A.F. Williams. Immunol. Rev. 47: 117 (1979).
12. D.W. Mason, and A.F. Williams. Biochem. J. 187: 1 (1980).

13. D.W. Mason, R.P. Arthur, M.J. Dallman, J.R. Green, G.P. Spickett, and M.L. Thomas. Immunol. Rev. 74: 57 (1983).
14. A.F. Williams, A.N. Barclay, S.J.Clark, D.J. Patterson, and A.C. Willis. J. Exp. Med. 165: 368 (1986).
15. D.J. Patterson, W.A. Jefferies, J.R. Green, M.R. Brandon, P. Cortese, M. Puklavec, and A.F. Williams. Molec. Immunol. In press.
16. A.P. Robinson, T.M. White, and D.W. Mason. Immunol. 57: 239 (1986).
17. W.A Kuziel, A. Takashima, M. Bonyhadi, P.R. Bergstresser, J.P. Allison, R.E. Tigelaar, and P.W. Tucker. Nature. 328: 263 (1987).
18. F. Herrman, S.A. Cannistra, H. Levine, and J.D. Griffin. J. Exp. Med. 162: 1111 (1985).
19. G. Steiner, E. Tschachler, M. Tani, T.R. Malek, E.M. Shevach, W. Holter, W. Knapp, K. Wolff, and G. Stingl. J. Immunol. 137: 155 (1986).
20. M. Kotani, T. Ezaki, H. Fuji, K. Matsuno, S. Ekino, and S. Harada. Acta. Anat. 104: 406 (1979).
21. G.J. Deenan, D. Opstelten, and P. Niewenhuis. Cell. Tiss. Res. 238: 183 (1984).

LYMPH NODE MACROPHAGE HETEROGENEITY : A SUBPOPULATION FAILS TO EXPRESS ACTIVATION PHENOTYPE

D. Chao and G. G. MacPherson

Sir William Dunn School of Pathology, University of Oxford, South Parks Road, Oxford OX1 3RE, England

INTRODUCTION

There is evidence to suggest that tissue-specific macrophages may have different characteristics to the peritoneal macrophage population which has been the most extensively studied [1-3]. These differences may be related to specialised functions of the macrophages in the tissues. The lymph node macrophage population was studied with the objective of shedding light on the functions of this population in lymph node function.

We report here that there is at least one macrophage subpopulation in the rat lymph node with different characteristics, and the functional significance of this population is discussed.

MATERIALS AND METHODS

All rats used were from the SPF unit, MRC Cellular Immunology Unit, Oxford. PVG rats more than 12 weeks old were used routinely. For induction of Graft-versus-Host disease (GvH), (PVG X DA) F_1 hybrids were used [4].

Cervical and popliteal lymph nodes were digested with collagenase (Boehringer) and macrophages isolated by adherence onto acid-cleaned coverslips. The Thymus-Independent type 2 (TI-2) antigen Fluorescein- conjugated hydroxyethyl starch (FOEt) was the kind gift of Professor J.H. Humphrey, Hammersmith Hospital, London U.K. To analyse TI-2 antigen uptake by lymph node cells, 20μg FOEt was injected into the afferent drainage beds of the lymph nodes [5]. For live cell immunofluorescence, OX6 anti-Ia monoclonal antibody (the kind gift of the MRC Cellular Immunology Unit, Oxford) and 1in30 dilution of TRITC-conjugated rabbit anti-mouse polyclonal (Dakopatts) were sequentially incubated on the coverslips for one hour at 4°C.

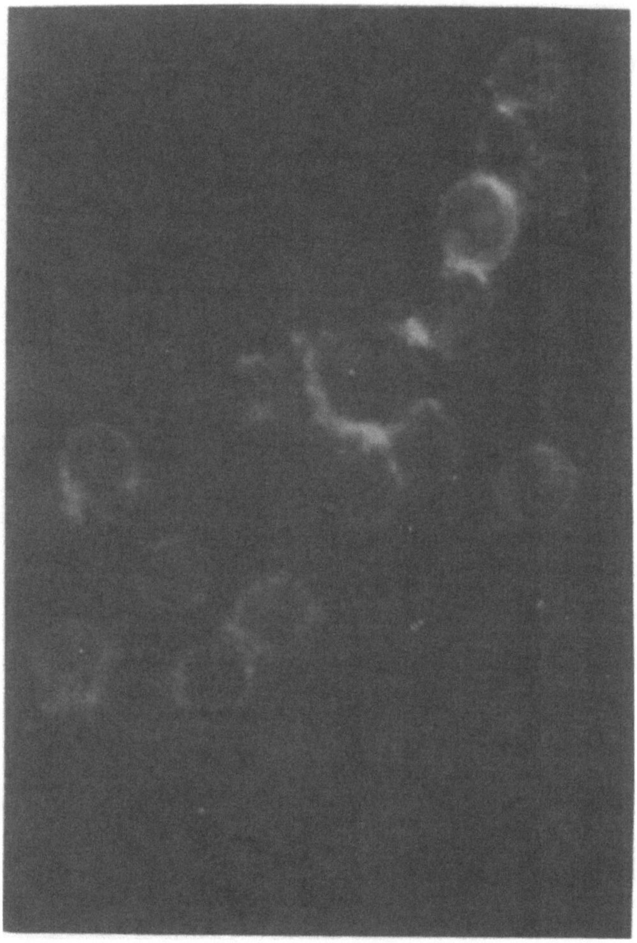

Figure 1. (X 4000) Under epifluorescence and TRITC-filter, the Ia^+ clustering cells contrast with the negative underlying stromal macrophage; compare with fig. 2 to identify the underlying stromal macrophage.

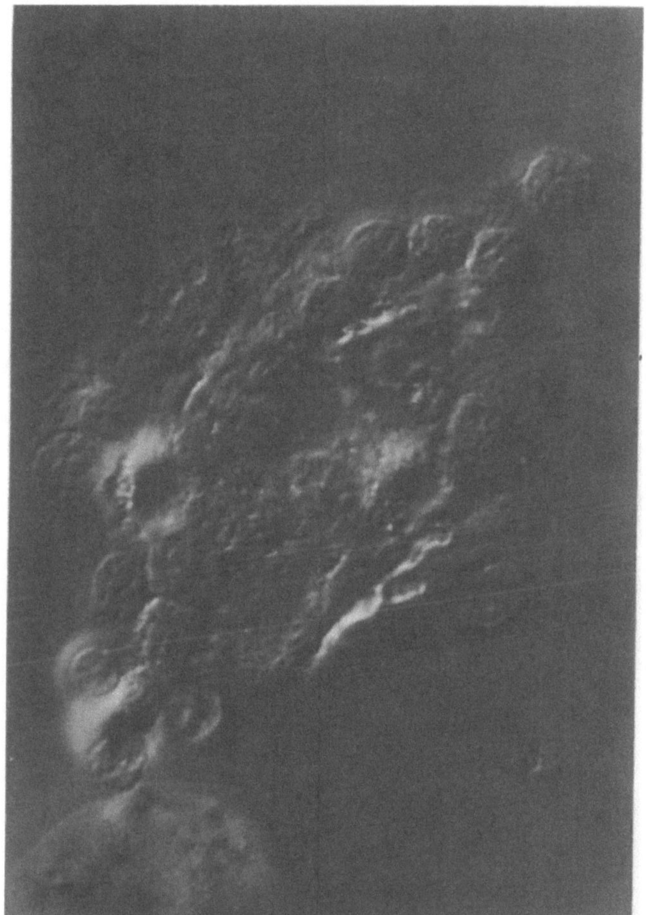

Figure 2. Viewed under Nomarski optics, the extent of the membrane spread of the stromal macrophage can be seen.

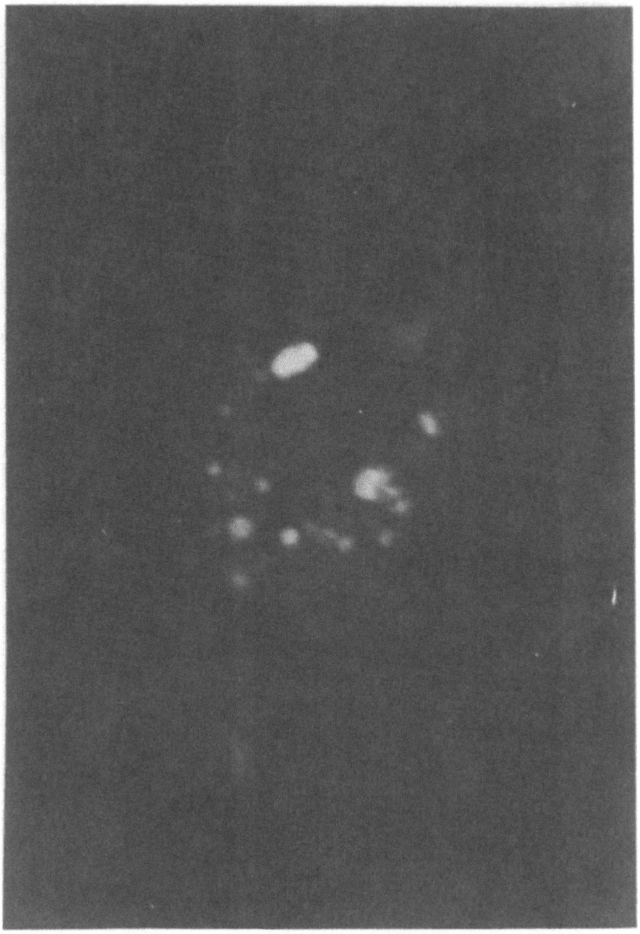

Figure 3. Under epifluorescence and the FITC-filter, the stromal macrophage would seem to have taken up the FOEt; note that the overlying cells are negative for uptake.

RESULTS

The major characteristics which distinguish the " stromal " lymph node macrophage population from the peritoneal macrophage population are summarised in the following photomicrographs. These are all of the same cells isolated from 4day Graft-versus-Host disease lymph nodes, labelled in vivo with FOEt for the last day, and then stained for Ia in vitro.

DISCUSSION

The stromal macrophage shows three characteristics not expressed by the peritoneal macrophage population;

(1) Clustering of Ia^+ cells. These cells also stain positively for immunoglobulin but not for T cell markers [6] and so it is thought that they are B cells. Peritoneal macrophages have been reported to interact with T cells [7], but not B cells. Furthur studies are under way to examine this B cell population, particularly to see if they are similar to the B cell population found in the splenic marginal zone [8] given the similarities between the lymph node stromal and splenic marginal zone macrophages.

(2) Selective uptake of TI-2 antigens. Only the splenic marginal zone macrophage population has also been reported to take up FOEt [3].

(3) Failure after activation to express detectable surface Ia. The Graft-versus-Host disease was used to maximally stimulate the lymph nodes and should generate the lymphokine gamma-interferon which activates macrophages and one change is an increase in surface Ia expression. The splenic marginal zone macrophage population has also been shown to fail to express Ia after in vivo activation [3]; unfortunately little is known about the marginal zone macrophage population and so the similarities with this population have little functional signiificance. The ability to express Ia is very widespread and not restricted to macrophages [9] which suggests that a macrophage-T cell interaction had to be actively avoided.

These characteristics are suggestive of a role in B cell rather than T cell function.

SUMMARY

We have identified and characterised a subpopulation of macrophages in the rat lymph node. The stromal macrophage shows many characteristics not expressed by peritoneal macrophages; clustering of B cells, selective uptake of TI-2 antigens, failure to express surface Ia after activation. These differences are suggestive of a regulatory role in B cell rather than T cell function

REFERENCES

[1] D.A. Lepay, C.F. Nathan, R.M. Steinman, H.W. Murray & Z.A. Cohn J. Exp. Med. 161:1079 (1985)

[2] P.R. Crocker & S. Gordon J. Exp. Med. 162:993 (1985)

[3] J.H. Humphrey & D. Grennan Eur. J. Immunol. 11:221 (1981)

[4] W.L. Ford, W.Burr & M. Simonsen Transplant. (Balt.) 10:258 (1970)

[5] N.L. Tilney J. Anat. 109:369 (1971)

[6] D. Chao & G.G. MacPherson, manuscript in preparation.

[7] Rosenwasser L.T., Werdelin O., Braendstrup O. & Rosenthal A.S. in : " Mononuclear phagocytes. Functional Aspects. ", p.1887, R. van Furth ed., Martinus Nijhoff Publishers (1980)
[8] P.J.L. Lane, D. Gray, S. Oldfield & I.C.M. MacLennan Eur. J. Immunol. 16:1569 (1986)
[9] Pober J.S., Gimbrone M.A., Cotran R.A., Reiss L.S., Burakoff S.J., Fiers W. & Ault K.A. J. Exp. Med. 157:1339 (1983)

INDUCTION OF THE EXPRESSION OF AN FcR FOR MONOMERIC RAT IgG2b ON RAT PERITONEAL MACROPHAGES

S. Denham and R.K. Barfoot

Experimental Unit, Section of Medicine
Institute of Cancer Research, Block X, Belmont
Sutton, Surrey, SM2 5PX, United Kingdom

INTRODUCTION

The existence of receptors for the Fc moiety of mouse monomeric IgGs on mouse macrophages is well documented[1] but, by contrast, far less is known about analogous Fc receptors (FcR) on rat macrophage populations. Our studies of rat macrophage FcR arose from observations made by other members of our department on the localisation of rat ^{125}I-labelled anti tumour monoclonal antibodies (McAbs) administered to rats bearing (syngeneic) sarcomata possessing the specific antigens. It was found that non-specific uptake of antibody by normal tissues occurred predominantly in the spleen, was associated with use of rat antibodies of the IgG2b isotype and did not occur if $F(ab')_2$ fragments of IgG2b antibodies were used[2]. Subsequently, we were able to show that macrophages from rat spleen bound monomeric rat ^{125}I-IgG2b *in vitro* whilst, in similar conditions, the binding of monomeric IgGs of other subclasses could not be detected[3]. Macrophages from the peritoneal cavity of rats however bound far less ^{125}I-IgG2b than macrophages from the spleen and the degree of binding varied with the batch of rats used as cell donors. The purpose of the experiments reported here was to investigate a possible relationship between the state of activation of rat macrophages and the expression of a receptor for rat IgG2b as it is reported that activation modulates the expression of FcR on mouse macrophages[4].

MATERIALS AND METHODS

Rats

Inbred, specific pathogen-free rats of the CBH/Cbi (RT1^c) strain and an athymic (rnu/rnu) variant of the same strain were used; these rats were maintained in isolators until sacrificed as cell donors. Conventionally housed inbred Wistar strain (RT1^u) AND CBH/Cbi rats were also used.

Cells

Donors of peritoneal macrophages (PMØ) were pretreated with a variety of eliciting agents i.p.. Sterile solutions of 10% thioglycollate or proteose peptone (8-10ml/rat) were given 4-7 days before cell collection; 2mg *corynebacterium parvum* (Wellcome, U.K.), 5 x 10^7 BCG organisms (Glaxo, U.K.) or 5 x 10^7 allogeneic spleen cells were given twice and cells collected 5-7 days after the final immunisation. All macrophage donors were given 400R total body irradiation 20hr before cells were taken to reduce the numbers of contaminating lymphocytes in macrophage preparations. Macrophages were washed from the peritoneal cavity with cold, serum-free medium containing 10 U/ml of heparin, washed once more in a large volume of cold medium and kept on ice until used. Spleen, liver, bone marrow and tumour cell suspensions were obtained by collagenase digestion (0.1% for 1hr); macrophages were isolated on gradients of 50% isotonic Percoll or by adherence to gelatin coated plasic. Monocytes were removed from peripheral blood mononulcear cells by adherence to plastic.

Immunoglobulins (Igs)

Affinity purified rat monoclonal antibodies of various specificities were prepared from concentrated hybridoma supernatants in the manner described in a previous publication[3]. Igs were iodinated by the chloramine T method.

^{125}I-Ig binding assay

A modified version of the assay described by Segal and Hurwitz[5] was used. ^{125}I-Igs were added to 0.5ml volumes of target cells suspended in MEM + 1% BSA in plastic tubes. The tubes were incubated for 2-3 hr at 4°C with continual rotation; the contents of the tubes were then layered over an equal volume of an oily gradient (dibutyl and dioctyl phthalates at 1:1.1) and spun for 1 min. The total acqueous supernatant was removed and radioactivities of the unbound Ig in the supernantant and bound Ig in the cell pellet measured separately. The level of non-specifically bound ^{125}I-Ig associated with the cells (which was related to the volume of the cell pellet) was about 1% of the input.

Comparative assay for superoxide production

The micro-densitometric assay for nitroblue tetrazolium (NBT) reduction was used[6]. Formazan was solubilised with dimethyl sulphoxide (DMSO) before measuring absorbance at 550nm with an automatic multichannel photometer (Multiskan, Flow, U.K.). 100μl of NBT at 1mg/ml + phorbol myristic acetate (PMA) at 100ng/ml were added to PMØ for 2hr at 37°C in phenol red-free PBS.

Ia staining and cytotoxic assay

PMØ were stained by incubation with mouse anti rat Ia antibody OX6 (Serotec) followed by FITC labelled rabbit a-mouse Ig antibody. Stained cells were counted by microscopic observation: 200 cells were counted/sample. Mouse L cells were used as targets in the cytotoxicity assay: 10^4 targets were added to each well of 96-well plates and incubated with various numbers of macrophages for 48hr. Surviving cells were counted in a cell counting chamber. Control cultures contained no macrophages.

EXPERIMENTS AND RESULTS

In the monomeric form, rat Igs of the IgG2b subclass only bound to rat macrophages in amounts that were, in our assay system, clearly in excess of background levels of binding. ^{125}I-IgG2b binding was most clearly exhibited by spleen macrophages which at concentrations of radiolabelled Ig in the range of 100-500 ng/ml, bound a large proportion of the radioactive input (Table I). Partially enriched preparations of tumour-macrophages (containing 50% macrophages) from immunogenic sarcomata also bound high levels of ^{125}I-IgG2b. Preparations of peritoneal cells, comprised of approximately 90% macrophages, from conventionally housed rats displayed some ability to bind ^{125}I-IgG2b, but PMØ from isolator bred and maintained rats did not bind significant amounts. Table I gives data obtained with two IgG2b Igs but at least two other Mc IgG2bs and one other Ig representing the other IgG subclasses have been tested with similar results. The differences between PMØ of SPF and conventional rats and the nature of the tissues in which maximum binding of ^{125}I-IgG2b took place, i.e. spleen and immunogenic tumour, suggested that binding was related to the degree of immunological stimulation of the cells.

Table 1. Tissue distribution of cells binding monomeric IgG2b

	% ^{125}I-Ig bound				
Ig isotype and code:	IgG2b (A2aC7)	IgG2b (4B1F8)	IgG2a (H6B5)	IgG2c (2C4bc)	IgG_1 (1A4B)
Tissue origin of cells					
Spleen	36.1*	28.7	1.5	1.6	1.2
Liver	7.7	9.3	2.2	-	2.0
Bone marrow	6.3	5.5	0.8	1.2	1.1
HSN tumour	25.5	25.8	3.2	-	2.7
PB, monocytes	4.6	-	0.6	-	0.5
PE, MØ	3.8	5.9	0.7	1.6	1.5
PE, MØ (Isolator-maintained rats)	0.8	1.7	0.7	1.1	0.5

Concentration of cells: 10^7/ml; concentration of ^{125}I-Ig:100ng/ml

*Figures represent mean of triplicate assays from at least two experiments

The effects of i.p administration of various inflammatory and macrophage-activating agents on the binding of monomeric IgGs to PMØ from isolator maintained rats are shown in Table II. Only the cells from rats given <u>C.parvum</u> bound markedly increased amounts of ^{125}I-IgG2b and increased binding of monomeric Igs of other IgG isotypes was not detected. The association constant (Ka) for the binding of ^{125}I-IgG2b to <u>C.parvum</u> stimulated PMØ was calculated (data not shown) and values similar to those estimated for the binding of IgG2bs to PMØ of conventionally housed rats were obtained[3]. <u>C.parvum</u> treatment of rnu/r nu rats also induced increased binding of ^{125}I-IgG2b by their PMØs comparable in extent to that induced on cells of euthymic rats Table II).

Macrophages removed from treated rats were simultaneously assayed for properties that would characterise them as being in the "activated" state, i.e. the expression Ia antigens, increased respiratory activity with production of superoxide anions and enhanced tumour cytotoxicity. The results of these assays are shown in Table III.

Table 2. Effect of agents given to SPF rats on binding of ^{125}I-labelled monomeric IgGs to peritoneal macrophages

	% ^{125}I-Ig bound			
Ig isotype and code:	IgG2b (A2aC7)	IgG2a (H6B5)	IgG2c (2C4bc)	IgG_1 (1A4b)
Agent given i.p				
None	0.7 ± 0.4	-	-	-
Proteose peptone	1.1 ± 0.4	0.5 ± 0.2	1.1 ± 0.5	0.8 ± 0.3
Thioglycollate	1.0 ± 0.3	1.0 ± 0.4	-	-
Allogeneic cells	2.0 ± 0.7	0.5 ± 0.3	-	0.6 ± 0.1
BCG	2.1 ± 0.6	0.8 ± 0.5	0.5 ± 0.1	0.7 ± 0.1
C.parvum	9.5 ± 0.7	0.6 ± 0.4	0.9 ± 0.3	0.6 ± 0.2
None (rnu/rnu donors)	1.0 ± 0.5			
C.parvum " "	8.3 ± 0.9			

Concentration of cells: 5 x 10^6/ml; concentration of Ig: 200ng/ml.

Figures represent mean of triplicate assays from 2-3 experiments ± standard deviation.

Table 3. Increase in IgG2b binding by peritoneal MØ accompanies increases in properties associated with MØ activation

Agent given ip to rats	% ^{125}I-IgG2b* bound	% cells Ia +ive	Formazan formation/ $2.5x10^5$ cells O.D at 550nm.	Surviving LC as % of control at 48 hr MØ:LC 50:1	25:1
Protease peptone	0.9	8	0.057**	98±1++	100±0
Allogeneic cells	1.8	87	0.066	100±0	95±2
BCG	2.2	17	0.082	2±1	33±7
C.parvum	9.7	69	0.217	5±1	36±12

* Concentration of cells: 5.10^6/ml; concentration of Ig (A2aC7) 200ng/ml.

** Mean absorbance by 8 wells; s.d. of the mean value was less than 10% of the mean.

++ Figures represent the mean ± s.d. of quadruplicate assays. LC: mouse L cells.

Attempts to activate peritoneal macrophages *in vitro* to express an Fc receptor for IgG2b have so far produced variable results. PMA, the ionophore A23187, bacterial lipopolysaccharide (LPS) and crude supernatants from spleen cells stimulated with concanavalin A (ConA), PMA and combinations of ConA, PMA and purified recombinant tumour necrosis factor (TNF) have been used as activating agents. Short term culture (i.e. 2hr, 5hr or 24hr) of macrophages in suspension with any of these agents did not induce increased binding of ^{125}I-IgG2bs. Longer term culture - 3 days or more - of macrophages on gelatin coated petri dishes resulted in slightly increased binding of ^{125}I-IgG2b (2-3% of the radiolabelled input) by cells from some cultures particularly those stimulated with PMA (100ng/ml). However, some untreated macrophages cultured for three or more days also displayed slightly elevated levels of IgG2b binding.

DISCUSSION

In a previous publication[3] we described the presence of an FcR for monomeric rat IgG2b on macrophages from rat spleen which was also detectable, albeit at lower levels, on resident PMØs. The results reported here, however, indicate that the binding of monomeric IgG2b to the PMØs of conventionally housed rats was probably induced by microbial fauna in the environment as unstimulated macrophages from SPF isolator-maintained rats did not bind IgG2b above the background values in our assay. The administration of *C.parvum* organisms to rats either induced or upgraded the expression of the FcR for monomeric IgG2b on rat PMØs. *C.parvum* stimulated macrophages also displayed increases in Ia antigen expression, ability to reduce NBT and cytotoxic activity, all of which characterise the "activated macrophage" phenotye[7]. BCG stimulated macrophages were as cytotoxic to mouse L cells as the *C.parvum* activated cells and it is probable that increased or repeated administration of BCG would have induced expression of the IgG2b FcR on these cells also. (It was necessary to give *C.parvum* to rats at least twice to achieve a consistent effect). The results obtained with *C.parvum* on rat macrophages are very similar to the reported effects of BCG on mouse macrophages[4]: the expression of the FcR for monomeric mouse IgG2a (FcRI) on mouse cells is upregulated by activation with BCG which is accompanied also by increased respiratory activity and tumour cell toxicity. These similarities suggest that the mouse macrophage FcRI and the rat FcR for rat IgG2b may be functionally homologous structures.

The mechanism of activation of rat macrophages by *C.parvum* is not clear. The lymphokine products of T cells are known to be important in macrophage activation but *C.parvum* induced expression of the FcR for IgG2b on macrophages occurred also in athymic rats. A similar result has been reported for the effects of BCG on the FcR for IgG2a of nude mouse macrophages[4]. Either rnu/rnu rats are not completely deprived of T-cell activity or T independent mechanisms of macrophage activation exist. The question of whether normal PMØs can be induced to express enhanced binding of IgG2b monomer by exposure to activating agents *in vitro* has not yet been answered.

REFERENCES

1. Unkeless, J.C., Fleit, H. & Mellman, I.S. Adv. Immunol. 31, 247. 1981.
2. Eccles, S.A., Purvies, H.P., Styles, J.M., Hobbs, S.M. & Dean, C.J. Cancer Immunol. and Immunother. In Press. 1987.
3. Denham, S., Barfoot, R. & Jackson, L. Immunology 62, 69. 1987.
4. Ezekowitz, R.A.B., Bampton, M. & Gordon, S. J. Exp. Med. 157, 807. 1983.
5. Segal, D.M. & Hurwitz, E. J. Immunol. 118, 1338. 1977.
6. Pick, E., Charon, J. & Mizel, D.J. Reticuloendothel. Soc. 30, 581. 1981.
7. Hamilton, T.A. and Adams, D.O. Immunology Today 8, 151. 1987.

This work was supported by the Cancer Research Campaign.

ACTIVATION OF MONONUCLEAR PHAGOCYTE FUNCTIONS BY STRUCTURALLY DIFFERENT BACTERIAL ENDOTOXINS

B.Johne, K.Bryn, G.Gaudernack, V.Myhrvold and B.Morland

Inst. of Microbiol., Dental fac., Univ. of Oslo; National Inst. of Public Health, Oslo and Institute of Transplantation Immunology, National Hospital, Oslo, Norway

INTRODUCTION

Mononuclear phagocytes(monocytes, macrophages) are multipotent cells with immunomodulatory properties, as well as important effector activities in inflammatory reactions (1). Much interest is at present focused on endotoxins (lipopolysaccharide,LPS) as modulators of the immune response, and particularly on the relations between structure and biological activity of LPS (2). The macrophage is suggested as the central effector cell in endotoxic reactions (3), and stimulatory as well as suppressive effects on mononuclear phagocytes have been reported (4,5).

B. intermedius and other black pigmented bacteroides species are associated with tissue destructive periodontal disease (6). Endotoxins from *Bacteroides intermedius* (B-LPS) and from *Escherichia coli* (E-LPS) differ structurally and in their biological activities (7-10). Effects of B-LPS on mononuclear phagocytes may have importance in the pathogenesis of tissue destructive periodontitis(6,10,11).

The aim of our investigation was to study effector functions and surface antigen expression in murine and human mononuclear phagocytes after exposure to structurally different endotoxins *in vitro*, and thus elucidate mechanisms with possible importance for the biological effects of endotoxins *in vivo*.

METHODS

B-LPS and E-LPS were isolated with hot phenol, purified by ultracentrifugation and gelfiltration and analyzed by gas chromatography (7).

Mouse peritoneal macrophages were harvested from responder (C3H/Tif and C57BL/6J) and from low responder (C3H/Hej) mouse strains by peritoneal lavage (9). Intracellular acid phosphatase was determined in the cell lysates by a colorimetric assay.

Supported by The Norwegian Research Council for Science and Humanities.

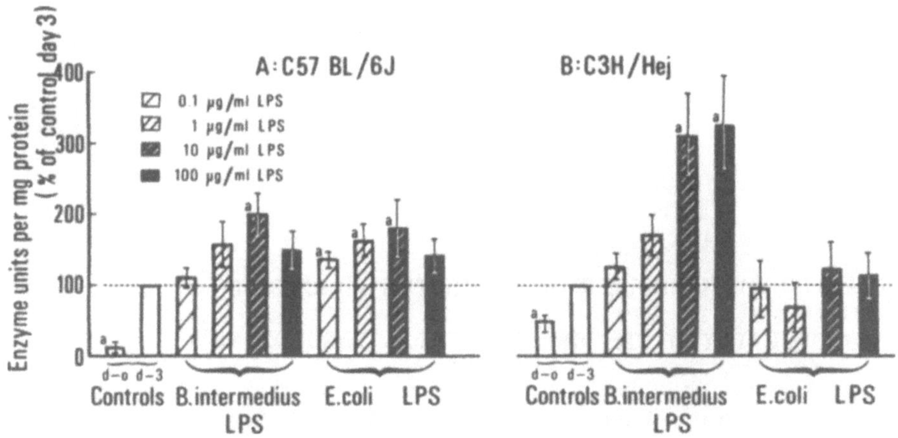

Fig. 1. Effects of B. intermedius LPS or E. coli LPS stimulation on intracellular level of acid phosphatase in peritoneal macrophages from LPS responder (A) and LPS low responder (B) mouse strains. (a) =significantly different ($p<0.01$) from controls day 3. (From reference 9)

Macrophage mediated cytotoxic activities were investigated in macrophage supernatants and in cocultures of stimulated macrophages and target cells. Release of radioactivity from 14-C-thymidine labelled L929 cells, and microscopic counting of L929 cells were used as parameters (9).

Human peripheral blood was obtained from voluntary blood donors, and monocytes isolated by density gradient centrifugation and adherence to glass (12). Acid phosphatase and complement receptor mediated phagocytosis was assayed. Binding of the activation related monocyte specific monoclonal antibody 1D5 (13) to the monocytes was measured by a direct radioimmuno assay.

RESULTS

Both endotoxins gave enhanced acid phosphatase activity in normal mouse macrophages, however the B-LPS also stimulated enzyme activity in the low responder C3H/Hej macrophages (Fig. 1). Stimulatory effects of both endotoxins was inhibited with Polymyxin B, 10 times the LPS dose (data not shown).

Macrophage mediated cytotoxic activity evaluated by target cell counting was notably more enhanced in C3H/Tif compared to C3H/Hej macrophages. E-LPS had a somewhat stronger effect than B-LPS in the responder macrophages (Table 1). Cytotoxic activity

Table 1. Cytotoxic effects of endotoxin stimulated C3H/Hej and C3H/Tif macrophages evaluated by microscopic counting of target cells. (Adapted from reference 9)

		C3H/Hej			C3H/Tif		
	Days in culture	Control	(% of control) B-LPS	E-LPS	Control	(% of control) B-LPS	E-LPS
Supernat.	1	100	79.8±13.1	101.6±33.8	100	66.7± 6.1	59.7± 5.5
+	2	100	77.5±12.7	73.3± 9.0	100	59.8±13.7	52.4± 7.6
L-929 cells	3	100	72.3± 8.9	81.0± 5.7	100	65.0±13.1	54.9±8.6

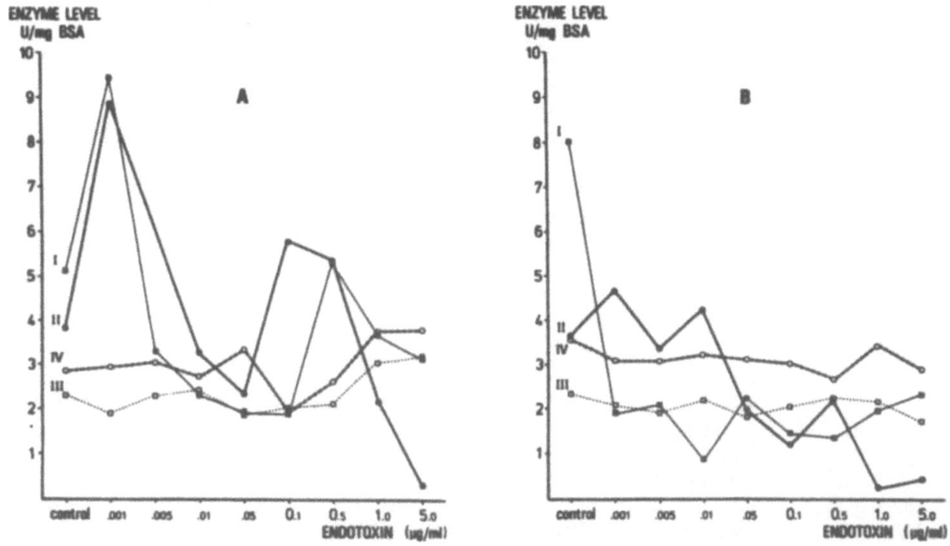

Fig. 2. LPS dose response curves. Intracellular acid phosphatase in human monocytes on day 3 in vitro. A: B-LPS. B: E-LPS. Roman numerals I-V denotes four different blood donors. (From reference 23)

against L929 cells, measured by release of radioactivity , was enhanced only in the LPS responder macrophages (data not shown).

In human blood monocytes, E-LPS caused suppression of intracellular acid phosphatase, while the B-LPS was able to increase enzyme activity after 3 days in culture (fig. 2). Variance analysis (23) demonstrated significantly different donor groups (high and low responders), significant stimulation with B-LPS in single donors, and a significant biphasic response with B-LPS.

Both endotoxins gave increased expression of the activation related cell surface antigen characterized by the monocyte specific moAb 1D5 (Fig. 3). Other moAbs were also investigated

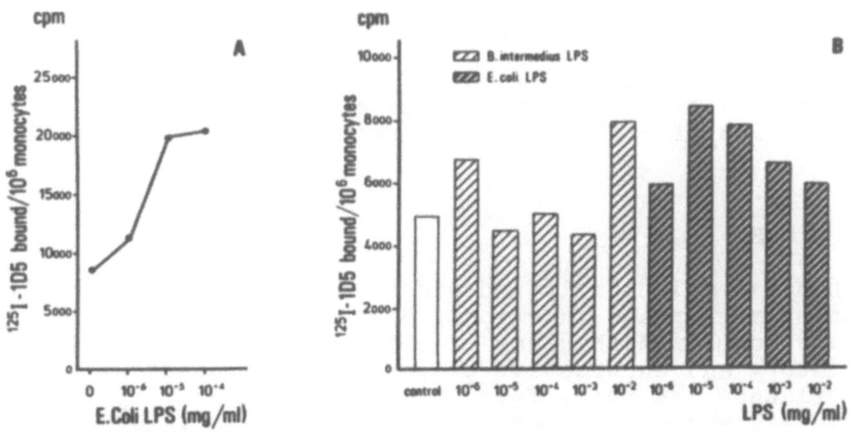

Fig. 3. Expression of an activation related cell surface antigenic determinant after LPS stimulation of human monocytes. Binding of 1D5 moAb was measired by RIA. A:n=5. B:n=1. (From reference 23)

in a similar manner, and increased as well as decreased expression of their corresponding cell surface antigens were observed.

Biphasic dose response relations were observed in assays for intracellular acid phosphatase (Fig. 2), C3b reseptor mediated phagocytosis (data not shown) and binding of 1D5 monoclonal antibody (Fig. 3).

Human monocytes were more susceptible to toxic/supressive LPS effects than murine macrophages (Figs. 1&2). The B-LPS showed a strong stimulatory potential and was significantly less toxic compared to E-LPS.

DISCUSSION

The stimulatory effects of B-LPS and E-LPS was ascribed to the lipid-A moiety, since both endotoxins were inhibited with Polymyxin B. Lipid-A carries the endotoxic principle of bacterial lipopolysaccharide (2). Furthermore the core region linking lipid-A to the polysaccharide chain has importance for endotoxic activity (5). Bacteroides LPS has atypical lipid-A as well as core structure compared to enterobacterial LPS (7,16,17).

The selective and simultaneous up and down regulation of cell functions are in accordance with other reports which have also discussed the contradiction of LPS toxic and stimulatory properties (18,19). Activation occurs via different mechanisms and pathways for the different macrophage cellular functions (20,21). It has also been suggested that different macrophage populations express different functions (22). Furthermore different macrophage functions are induced by different determinants on the LPS molecule (14). This complexity may have a bearing on the biphasic dose response relations which may concievably be a characteristic feature of LPS biological activity (23).

Both endotoxins were strong stimulators in the murine macrophage system, while the toxic properties were much more dominating in the human monocyte system. The strong enhancement of intracellular enzyme activity obtained with B-LPS compared to E-LPS in human monocytes may be due to the low toxicity of this LPS towards the cells. This may indicate that the balance between stimulatory and suppressive properties of an LPS is important for its biological effects.

The complex repertoir of LPS biological activities may be explained by selective activation of macrophage functions (20), cooperation with other immunocompetent cells (1), and by the participation of different structural parts of the LPS molecule (14).

CONCLUSIONS

LPS can selectively and simultaneously up or down regulate different macrophage functions. Biphasic dose response relations indicate complex mechanisms of action. The balance between stimulatory and suppressive properties of a particular endotoxin may influence its net effect on mononuclear phagocytes. Our results may suggest Bacteroides endotoxins as interesting in the concept of non toxic immunomodulators.

REFERENCES

1. C.A. Dinarello. Rev. Infect. Dis. 6:51 (1984).
2. E.Th. Rietschel, H. Brade, W. Kaca, K. Kawahara, B. Lindner, T. Luderitz, T. Tomita, U. Schade, U. Seydel, and U. Zahringer, in: "Bacterial endotoxins: Structure, biomedical significance and detection with the Limulus amebocyte lysate test", p. 31, J.W. ten Cate, H.R. Buller, and A. Sturk, eds., Alan R. Liss Inc. (1985).
3. D.C. Morrison, R.L. Duncan, and S.A. Goodman, in: "Bacterial endotoxins: Structure, biomedical significance and detection with the Limulus amebocyte lysate test", p. 81, J.W. ten Cate, H.R. Buller, and A. Sturk, eds., Alan R. Liss Inc. (1985).
4. W.H.D. Riches, and P.M. Henson. J. Leuk. Biol. 36:143 (1984).
5. C.C. Leslie, R.A. Musson, and P.M. Henson. J. Leuk. Biol. 36:143 (1984).
6. J. Slots, and R.J. Genco. J. Dent. Res. 63:412 (1984).
7. B. Johne, and K. Bryn. Acta Path. Microbiol. Immunol. Scand. Sect. B. 94:265 (1986).
8. R. Storeng, and B. Johne. Acta Path. Microbiol. Immunol. Scand. Sect. C. 95:135 (1987).
9. B. Johne, V. Myhrvold and B. Morland. Acta Path. Microbiol. Immunol. Sect. C. 95:85 (1987).
10. C.G. Daly, G.J. Seymour, and J.B. Kieser. J. Oral Path. 9:1 (1980).
11. S.N. Vogel, and S.E. Mergenhagen, in: "Host parasite interactions in periodontal disease.", p.160, R.J. Genco and S.E. Mergenhagen, eds, ASM (1982).
12. A. Boyum. Scand. J. Clin. Lab. Invest. 21:77 (1968).
13. G. Kaplan, and G. Gaudernack. J. Exp. Med. 156:1101 (1982).
14. M. Caroff, J-M. Cavaillon, C. Fitting, and N. Haeffner-Cavaillon. Infect. Immun. 54:465 (1986).
15. H. Brade, and C. Galanos. Infect. Immun. 42:250 (1983).
16. A. Weintraub, U. Zahringer, and A.A. Lindberg. Eur. J. Biochem. 151:657 (1985).
17. A. Weintraub, B.E. Larson, and A.A. Lindberg. Infect. Immun. 49:197 (1985).
18. K.A. Barker, and S.C. Holt. Can. J. Microbiol. 29:1552 (1983).
19. J. Hammerstrom and G. Unsgaard. Acta Path. Microbiol. Immunol. Scand. Sect. C. 87:381 (1979).
20. B. Morland, and J. Morland. J. Reticuloendothel. Soc. 23:469 (1978).
21. W.S. Futch, and L.B. Schook. Infect. Immun. 50:709 (1985).
22. P.A. Campbell, C.J. Czuprynski, and J. Cook. J. Leukocyte Biol. 36:293 (1984).
23. B. Johne, G. Gaudernack and B. Morland. Acta Path. Microbiol. Immunol. Scand. Sect. C. 95: (1987) in press.

EXPRESSION AND FUNCTION OF MHC CLASS II ANTIGENS ON MACROPHAGES AND DENDRITIC CELLS

E.W.A. Kamperdijk, M.A.M. Verdaasdonk, and R.H.J. Beelen*

Department of Cell Biology, Medical Faculty
Free University, NL-1007
MC Amsterdam, The Netherlands

Macrophages bearing MHC class II antigens (Ia antigens) play an important role in presenting antigen to specifically primed T lymphocytes (1). Since populations of macrophages are heterogeneous in the expression of detectable Ia antigen (2,3), the presence of the determinant marks the macrophage subpopulation responsible for this important immunoregulatory function. It has been shown that also dendritic cells (4) play a crucial role in antigen presentation to immunocompetent T lymphocytes. In this study we have investigated the expression and function of Ia antigen on different populations of macrophages and dendritic cells from inbred ACI/Mal rats, which are genetically defined high responders for the copolymer GT (glutamine tyrosine, 5). The expression of Ia antigen was detected by EA rosetting on both vital and paraformaldehyde fixed cells and by immunocytochemistry (immunoperoxidase).

MATERIALS AND METHODS

Resident peritoneal macrophages (RPM) were obtained by peritoneal lavage of naive and BCG immunized animals (6). The naive resident macrophages were studied before and after culture in normal medium and in the presence of lymphokines. Immunization with BCG was according to Hoefsmit et al. (6) and bone marrow derived macrophages were obtained according to Walker et al. (7). Dendritic cells were isolated from lymph nodes as described earlier (8). Immunization of lymph node cells (LNC) by GT was according to Amerding et al. (5).

Rosetting assay. Ia bearing macrophages were detected by a rosetting assay using SRBC to which a monoclonal antibody reactive to Ia (monoclonal anti Ia.17) had been coupled by the chromic chloride method of Parish and Hayward (9).

Fixation. RPM were fixed with 1% paraformaldehyde following Beller and Unanue (10). Immunocytochemistry for Ia on macrophages and dendritic cells was performed using the method described by Hoefsmit et al. (11).

---

*Present adress: Department of Haematology, University Hospital, Free University, Amsterdam, The Netherlands.

Lymphokines (LK). T cell LK was prepared as described by Calamai et al. (12). Isolation of T cell enriched lymph node cells was performed following a three step procedure (adherence, nylon wool and Sephadex G10) as described for the mouse system (7).

GT primed proliferation assay. Antigen presentation was detected by ^{3}H thymidine incorporation in co-cultures of the antigen presenting cells, GT primed T cell enriched lymph node cells (LNC) and GT. To get more information about the role of Ia antigen in antigen presentation, some blocking experiments were performed using monoclonal anti Ia.17 (Seralab).

RESULTS

Expression of Ia antigen on RPM

Using the rosetting assay on non-fixed RPM no loss of Ia bearing cells during culture was observed (Table 1). However, when we fixed the cells in 1% paraformaldehyde (10) before rosetting, we detected a significant reduction of Ia bearing cells. Moreover, if the cells were fixed after culture almost all cells had lost the ability to bind E-anti Ia.17 indicator cells. If lymphokines were added to the culture medium, the relative number of non-fixed rosetting cells was only slightly increased whereas after fixation still a relatively high number of Ia positive macrophages (32%) was detected. So the influence of lymphokines is much more pronounced when detected with the rosetting assay on fixed cells than on non-fixed cells. After immunization with BCG almost all peritoneal macrophages expressed Ia antigen (as dendritic cells do) as detected both by the rosetting assay as by immunocytochemistry (> 90%).

Table 1. Influence of lymphokines (LK) on the expression of Ia on resident peritoneal macrophages (Mφ) using the rosetting assay.

Days in culture	LK	Percentage Ia positive Mφ	
		not fixed	paraformaldehyde fixed
0	-	47	28
4	-	46	3
4	+	54	32

Expression of Ia antigen on bone marrow derived macrophages

Bone marrow derived macrophages, cultured solely with LCM showed about 10% Ia antigen positive cells by immunocytochemistry and even slightly more by rosetting on non-fixed cells. The addition of LK strongly increased the number of Ia positive cells as detected by both assays (± 90% positive). Moreover, we observed not only a visual expression of Ia antigen, but also a functional expression of Ia. Presentation of the copolymer Gt to GT primed lymph node cells occurred in an Ia dependent manner with and without the addition of lymphokines in culture (manuscript in preparation). This means that at least part of the bone marrow derived macrophages express Ia antigen, which may be enhanced by the addition of LK.

Expression of Ia antigen on dendritic cells

Using immunocytochemistry it was shown that virtually all dendritic cells expressed Ia antigen (Fig. 1).

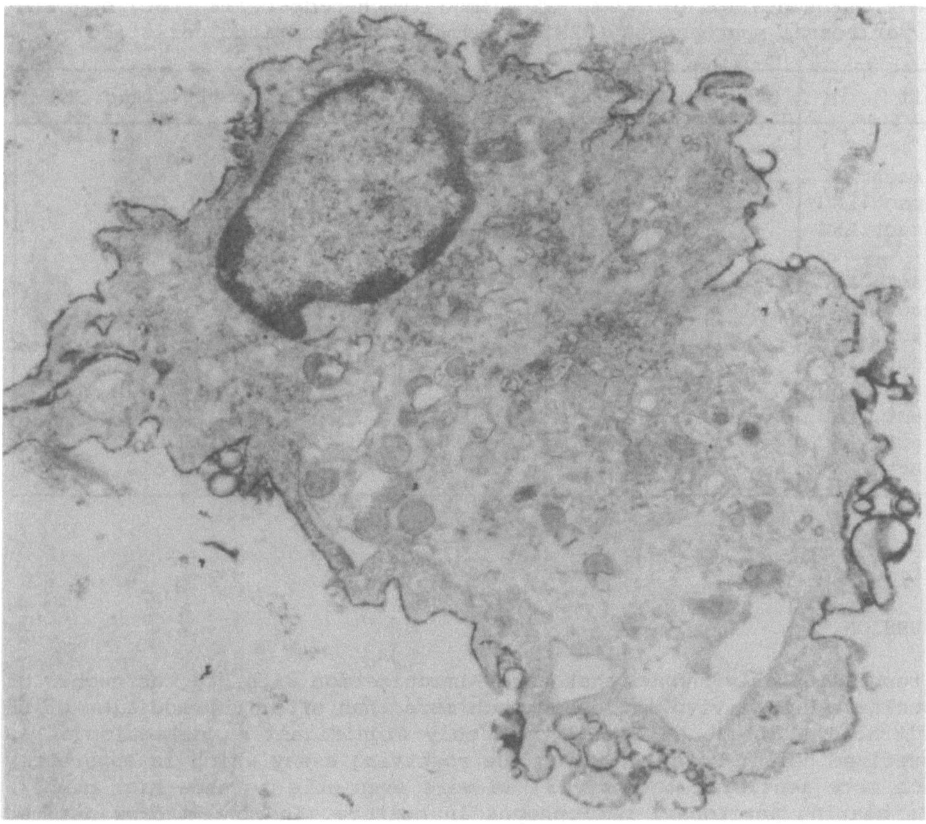

Fig. 1. Dendritic cell of lymph node. The cell is strongly positive for Ia (black reaction product).

Presentation of GT to GT primed lymph node cells (LNC) by resident peritoneal macrophages (RPM) or lymph node dendritic cells (DC)

The expression of Ia antigen on many cultured RPM and on almost all DC suggests that populations of such cells might be able to present to primed T lymphocytes an antigen which is under Ir-gene control. This possibility was tested by priming ACI/Ma1 rats with GT in complete Freund's adjuvant and measured their LNC two weeks later for responsiveness to the antigen (5). It was shown (Table 2) that unfractionated LNC responded to GT as measured by ^{3}H thymidine (^{3}H Thd) incorporation in an Ia dependent manner, whereas suspensions of fractionated LNC failed to respond to the added antigen. The addition of reshly obtained RPM or DC ($4x10^{3}$ cells/ well) restored the proliferation response of the fractionated LNC only after the addition of GT. Both responses could largely be blocked by monoclonal anti Ia. DC appeared to be much more effective in presenting GT to the primed "enriched T cells" than the RPM (up to 40 DC were already able to induce a significant response). Preliminary results showed that RPM from BCG immunized rats, which were almost all Ia positive, were not better in presenting GT to primed T cells than naive RPM.

Table 2. Presentation of GT to GT primed lymph node cells (LNC) by resident peritoneal macrophages (RPM) or lymph node dendritic cells (DC).

cell	RPM	DC	GT	anti-Ia	^{3}H Thd incorp. (cpm ± SE)
unfract.LNC	-	-	-	-	4,200 ± 400
unfract.LNC	-	-	+	-	32,000 ± 5,000
unfract.LNC	-	-	+	+	3,000 ± 500
fract. LNC	-	-	-	-	1,100 ± 600
fract. LNC	-	-	+	-	3,500 ± 700
fract. LNC	+	-	-	-	4,000 ± 500
fract. LNC	+	-	+	-	34,000 ± 5,000
fract. LNC	+	-	+	+	7,000 ± 800
fract. LNC	-	+	-	-	8,000 ± 600
fract. LNC	-	+	+	-	180,000 ± 10,000
fract. LNC	-	+	+	+	7,500 ± 600

DISCUSSION

Our results clearly showed that after immunization with BCG the number of Ia positive RPM in vivo increased much more than after the addition of LK in vitro. The effect of LK (10) was only significant by detection of Ia on prefixed cells. However, using the rosetting assay which is apparently a much more sensitive method (13) we were even able to show high numbers of Ia bearing peritoneal macrophages in culture. In bone marrow derived macrophages cultured solely with LCM a steady state expression of Ia is already present which may be enhanced upon the addition of LK, we observed the expression of Ia on bone marrow derived macrophages as demonstrated by rosetting by immunocytochemistry (immunoperoxidase) after a mild fixation with glutaraldehyde and by function (i.e. presenting GT to GT primed LNC in an Ia dependent manner). In accordance with Calamai et al. (12) we observed an enhancement of Ia on bone marrow derived macrophages after the addition of LK (± 90% Ia positive cells).

Dendritic cells isolated from lymph nodes of ACI/Mal rats appeared to be persistent Ia positive as was already shown for Wistar rats (14). These cells were superior in antigen presentation of GT, but this function can also be carried out by Ia positive macrophages.

It has been suggested that a relation exists between the presence of Ia antigen and antigen presentation. However, in our experiments an increase of Ia on RPM by immunization with BCG dit not result in an enhanced capacity of these macrophages to present the copolymer. This confirms that other factors also play an important role in antigen presentation.

REFERENCES

1. E.R. Unanue. Adv. Immunol. 31:1 (1981).
2. C. Cowing, B.D. Schwartz, and H.B. Dickler. J. Immunol. 120:378 (1978).
3. H.R. Schwartz, H.B. Dickler, D.H. Sachs, and B.D. Schwartz. Scand. J. Immunol. 5:731 (1976).
4. R.M. Steinman, and Z.A. Cohn. J. Exp. Med. 139:380 (1974).
5. D. Amerding, D.H. Katz, and B. Benacerraf. Immunogenetics 1:329 (1974).
6. E.C.M. Hoefsmit, and R.H.J. Beelen, in "Mononuclear phagocytes. Characteristics, physiology and function, p. 19, R. van Furth, ed., Martinus Nijhof, Dordrecht (1985).
7. W.S. Walker, R.B. Hester, and R.H.J. Beelen. Cell. Immunol. 79:125 (1983).
8. E.W.A. Kamperdijk, M.L. Kapsenberg, M. van den Berg, and E.C.M. Hoefsmit. Cell Tissue Res. 242:469 (1985).
9. C.R. Parish, and J.A. Hayward. Proc. R. Soc. Lond. Ser. B 197:47 (1974).
10. D.I. Beller, and E.R. Unanue. J. Immunol. 126:263 (1981).
11. E.C.M. Hoefsmit, I.L. Schadee-Eestermans, and R.H.J. Beelen. J. Histochem. Cytochem. 34:633 (1986).
12. E.G. Calamai, D.I. Beller, and E.R. Unanue. J. Immunol. 128:1692 (1982).
13. W.S. Walker, R.H.J. Beelen, P.J. Buckley, S.L. Melvin, and S. Yen. J. Immunol. Methods 67:89 (1984).
14. E.W.A. Kamperdijk, M. van den Berg, M.L. Kapsenberg, and E.C.M. Hoefsmit. Cell Biol. Int. Rep. 8:180 (1984).

INDUCTION OF CLASS II MHC ANTIGEN EXPRESSION IN RAT ORGANS AFTER SYSTEMIC APPLICATION OF RECOMBINANT GAMMA INTERFERON

B. Steiniger[1], P.H. van der Meide[2], P. Falk[1] and J. Klempnauer[3]

[1]Centre of Anatomy and [3]Centre of Surgery, Hannover Medical School, FRG; [2]TNO Primate Centre, Rijswijk, Netherlands

INTRODUCTION

Class II MHC antigens appear in a variety of previously negative cell types during immune reactions in vivo and may decisively contribute to antigen presentation and target cell destruction. In cell culture systems gamma interferon was identified as one of the most potent mediators inducing class II antigen expression (1). In vivo effects of gamma interferon on MHC antigen distribution have been described but have not been analyzed exhaustively (2,3). Detailed dose-response studies in experimental animals are lacking. We have thus investigated the class II MHC antigen inducing capacity of different amounts of recombinant rat gamma interferon in lymphatic and non-lymphatic organs after prolonged intravenous infusion.

MATERIALS AND METHODS

Recombinant rat gamma interferon was produced in CHO cells and subsequently purified by monoclonal antibody affinity chromatography (4). Its specific activity was 3-4 x 10^6 U/mg measured as cytopathic effect inhibition. Interferon was diluted in PBS pH 8.2 plus 0.5% normal rat serum. LEW ($RT1^l$) male rats two months of age, weighing 270g, were infused at 3 ml/h via a central venous catheter with increasing doses of 15(n=2), 30(n=2), 50(n=10), 100(n=2), 200(n=1) and 400(n=1) x 10^3 U/h for three days. Two LEW animals received 15 x 10^3U/h for six days. Control animals were treated with PBS and normal serum only. Class II MHC antigens were demonstrated by antibodies 0x6 (5) and 0x17 (6) directed against products of the I-A and I-E subregion equivalent of the rat MHC using an indirect immunoperoxidase technique on cryostat sections.

RESULTS

In control rats class II MHC antigens were present only in interstitial dendritic cells in the pancreas, liver and heart. The only parenchymal cells expressing class II antigens in the normal state were parts of proximal kidney tubules and villous epithelium in the small intestine. Endothelia did not react with any of the class II reagents. Lymphocytes in the T cell areas of normal lymph nodes and spleen and most cells in the splenic red pulp were also unreactive (Fig. 1a). In the brain no class II positive cells were found at all.

After continuous intravenous infusion of recombinant gamma interferon for three days class II MHC antigens were, however, induced in all organs investigated. With each interferon dose applied Ox6 and Ox17 reacted with endothelia of all big vessels, while capillaries stayed negative. The number of class II positive interstitial dendritic cells (and perhaps fibroblasts) increased substantially. In addition, some perivascular infiltration with strongly class II reactive round cells was seen. In the individual organs the following alterations of class II antigen expression occurred.

Spleen

With the exception of megakaryocytes, nearly all cells in the red pulp expressed class II antigens after all interferon dosages so that the follicles and marginal zone could hardly be distinguished any longer (Fig. 1b). However, even after the largest dose of 400×10^3 U/h a certain amount of lymphocytes in the PALS remained class II negative, although the class II negative area appeared smaller than in normal spleens. After six days of interferon infusion a marked splenomegaly developed with increased hematopoiesis in the red pulp.

Mesenteric and submandibular lymph nodes

The amount of class II antigen expression in the lymph node medulla increased and all the previously negative high endothelial venules turned class II positive (Fig. 2a,b). In the paracortex, however, clusters of unreactive lymphocytes still persisted after the maximum dose was applied.

Pancreas

At a dose of 15×10^3 U/h for three or six days strong class II reactivity appeared in duct epithelia while acinar cells and endocrine islet cells stayed negative. With increased dosages more intercalated ducts and centroacinar cells were stained. At 400×10^3 U/h acinar cells focally started to express class II antigens, especially in the vicinity of ducts and vessels. With this dose non-myelinated nerves also turned class II positive. The majority of acinar cells and all hormone producing islet cells however did not express class II antigens.

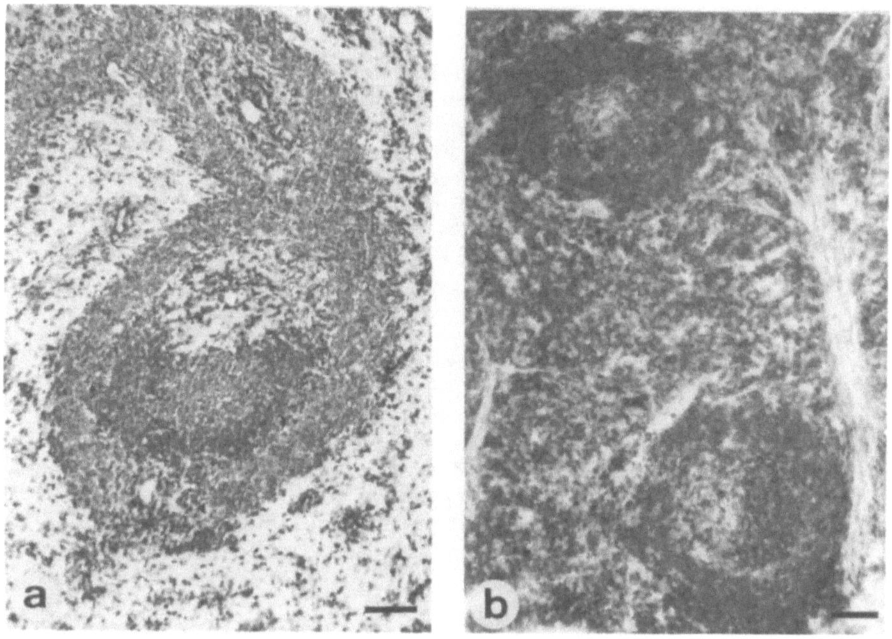

Fig. 1 (a) Normal distribution of class II reactivity in LEW spleen. Lymphocytes in the PALS and the majority of cells in the red pulp are class II negative. (b) LEW spleen after 400 x 10^3 U/h of gamma interferon for 3 days. Almost all cells in the red pulp express class II antigens. Class II negative lymphocytes are still present in the PALS. (Indirect immunoperoxidase staining with Ox6 antibody, bar = 100 μm).

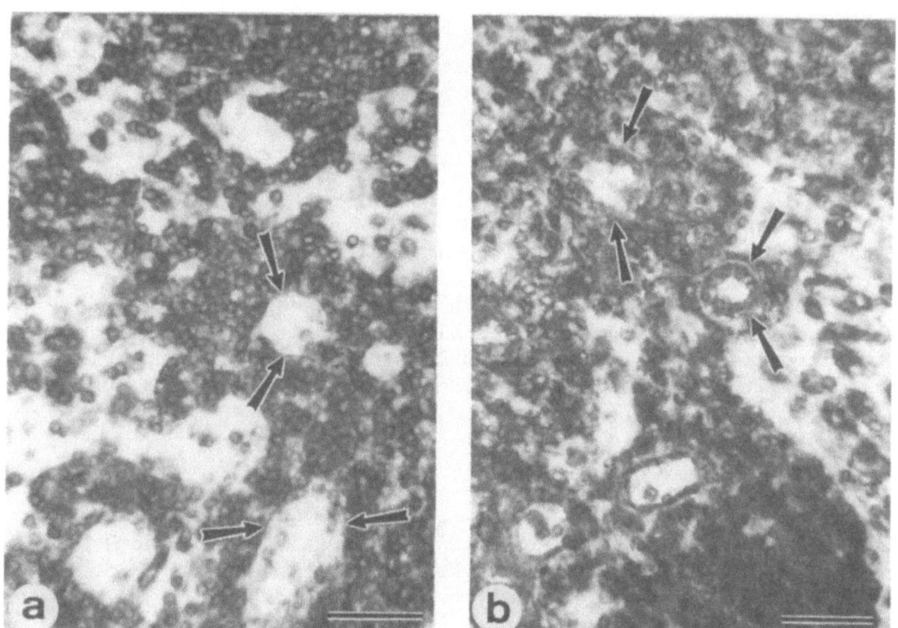

Fig. 2 (a) In normal LEW mesenteric lymph nodes high endothelial venules (HEVs) are class II negative. (b) After 15 x 10^3 U/h of gamma interferon for 3 days all HEVs express class II antigens. (Indirect immunoperoxidase staining with Ox6 antibody, bar = 50 μm).

Liver

Kupffer cells and bile duct epithelia were converted to class II positivity with all doses. Hepatocytes also expressed class II antigens on their membranes and intracellularly, although less strongly than the previous cell types. Hepatocytes adjacent to central veins were most intensely stained. Clusters of strongly class II positive infiltrating cells were present in areas of hepatocyte necrosis, especially after a six day course of interferon.

Kidney

The amount of class II positive cortical tubules increased with rising interferon doses. In addition, all parenchymal structures in the kidney medulla turned strongly class II positive. The urothelium also expressed class II antigens. Small parts of convoluted distal tubules and the straight segments of distal tubules in the outer part of the renal medulla stayed class II negative with all doses tested. In contrast, collecting tubules in the outer medulla were strongly reactive.

Heart

Intermediate class II reactivity appeared at the membranes and intracellularly in myocytes with any dose. Membrane staining was especially evident at the intercalated discs. The endocardium also turned strongly class II positive.

Gut

The previously negative crypt epithelia of the small and large intestine expressed class II antigens after all doses.

Brain

In addition to the endothelium of major vessels, evenly arranged stellate cells in the gray and white matter turned class II positive in the brain. In the cerebellum the distribution of these cells differed from the pattern stained with antibodies against glial fibrillary acidic protein. Thus the class II positive cells most likely cor pond to microglial cells. However, unless double staining experiments are performed it cannot be entirely excluded that they form a subpopulation of astrocytes.

General pathology

After a six day course of 15 x 10^3 U/h certain pathological features were more clearly apparent than after a three day application period. In the liver focal hepatocyte necrosis increased. There was a partial occlusion of veins in all organs by class II positive cells with irregular or elongated nuclei, which showed mitotic figures as well as cell necrosis. These cells tended to infiltrate the perivascular space. Their surface phenotype indicated that they represented an immature myeloid stage, most likely myelocytes. Some animals developed a generalized interstitial edema, which was most easily seen in the pancreas, combined with ascites.

SUMMARY AND DISCUSSION

Recombinant rat gamma interferon was continuously infused intravenously into LEW rats at doses ranging from 15 to 400 x 10^3 U/h for three or six days. These activities correspond to about 0.4 to 12 mg/kg/d of protein. The animals showed a rather uniform induction pattern of I-A and I-E like class II MHC antigens in all organs investigated, which was not directly dose-dependent within a ten to twentyfold increase in dosage. There was a substantial difference in the sensitivity of individual cell types to the inducing effect of gamma interferon. Even differently specialized subpopulations of one cell type did not behave identically as observed with endothelial cells. In general, endothelia in big vessels, certain epithelial cells like duct epithelia in the liver and pancreas and the urothelium were most easily induced to express strong class II reactivity. Myocytes in the heart and hepatocytes stained only moderately and needed higher interferon doses. Neurons, pancreatic endocrine cells, capillary endothelia and certain parts of distal kidney tubules did not respond to gamma interferon in the dose range tested. These cells may only express low levels of gamma interferon receptors or may require mixtures of different mediators to be stimulated for class II gene transcription. At least within an interval from three to six days different kinetics of MHC antigen expression do not seem to play a role.

Interestingly, acute toxic effects did not limit the amount of interferon applicable in this study. With prolonged administration hepatocyte damage became more prominent and immature hemopoietic cells were mobilized into the circulation. Interferon obviously crossed the blood-brain barrier leading to class II induction in glial cells. This seems to be due to direct action of interferon on endothelial cells increasing vascular permeability in all organs. Thus, the specific consequences of class II antigen induction - for example in high endothelial venules - will have to be differentiated from the sequelae of increased leakiness of vessel walls, if alterations in immune reactivity or circulatory behaviour of lymphocytes occur in interferon treated animals.

ACKNOWLEDGEMENTS

This study was supported by grant Ste 360/2-1 of the Deutsche Forschungsgemeinschaft. We thank S. Fryk for secretarial assistance.

REFERENCES

1. M. Fellous, U. Nir, D. Wallach, G. Merlin, M. Rubinstein and M. Revel. Proc. Natl. Acad. Sci. USA 79:3082 (1982).
2. F. Momburg, N. Koch, P. Möller, G. Moldenhauer and G.J. Hämmerling. J. Immunol. 136:940 (1986).
3. K.J. Skoskiewicz, R.B. Colvin, E.E. Schneeberger, and P.S. Russell. J. Exp. Med. 162:1645 (1985).
4. P.H. van der Meide, M. Dubbeld, K. Vijverberg, T. Kos and H. Schellekens. J. Gen. Virol. 67:1059 (1986).
5. W.R. McMaster and A.F. Williams. Eur. J. Immunol. 9:426 (1979).
6. T. Fukumoto, W.R. McMaster and A.F. Williams. Eur. J. Immunol. 12:237 (1982).

HUMAN PERITONEAL MACROPHAGES ENHANCE HLA-DR EXPRESSION IN CULTURE AND ARE VERY POTENT STIMULATORS OF AN ALLOGENEIC MIXED LEUCOCYTE REACTION[1]

H.J. Bos, B. Kranenburg, and R.H.J. Beelen*

Departments of Electron Microscopy and Haematology*, Medical Faculty and University Hospital*, Free University, NL-1081 BT Amsterdam, The Netherlands

The expression of Major Histocompatibility Complex (MHC) class II antigens on macrophages in the rat, mouse and in the human system has been shown to be essential for antigen presentation. In both rat and mice Ia expression of peritoneal macrophages decreases after culture (1-4), but in the human system this point has been poorly investigated. In the human system almost only the adherent monocyte subpopulations have been studied for HLA-DR expression in culture. Some investigators showed an enhancement (5,6) or a more or less stable (3,7) HLA-DR expression in culture. Others even demonstrated a decrease of HLA-DR expression (8-11).

We examined the HLA-DR expression of blood monocytes and peritoneal macrophages in culture. In addition with the HLA-DR expression in culture we studied the stimulation of an one-way allogeneic mixed leucocyte reaction by peritoneal macrophages and the influence of different MHC class II antigens (HLA-DR and HLA-DQ) in the proliferative response. Different macrophage subpopulations were isolated by adherence steps and especially the non adherent cells were potent stimulators of an allogeneic mixed leucocyte reaction.

MATERIALS AND METHODS

Human monocytes and peritoneal cells

Rat peritoneal cells (> 80% macrophages) were isolated from the unstimulated peritoneal cavity of male Wistar rats weighing 180-200 g as described in detail elsewhere (12). Human blood was drawn from healthy persons by venipuncture into EDTA syringes (Searle, Hortòn, England) and mononuclear cells (90% lymphocytes and 10% monocytes) were isolated by the Böyum technique (13,14).

Peritoneal cells (> 80% macrophages) were isolated from patients with end-stage renal failure undergoing continuous ambulatory peritoneal dialysis (CAPD). No patient was studied within one month of an episode of peritonitis. Peritoneal cells were isolated from the peritoneal effluent volume of an overnight exchange of at least 8 hours.

---

[1]Supported in part by a grant from the Kidney Foundation, Bussum, The Netherlands.

Monoclonal antibodies (MoAbs)

Mouse anti rat Ia antibody 29c (Seralab) was used for the detection of Ia antibodies on cells from rat peritoneal macrophages.

MoAb H2.5.10 (mouse anti HLA-DR + HLA-DP) was a kind gift from Dr. P.M. van Landsdorp (Central Laboratory of the Netherlands Red Cross Blood Transfusion Service, Amsterdam), MoAb 3B3 (mouse anti HLA-DR) and MoAb SPV L3 (mouse anti HLA-DQ; 15) was a kind gift from Dr. G.D. Keizer (The Netherlands Cancer Institute, Amsterdam). As determined by immunoprecipitation, sodium dodecyl sulfate, polyacrylamide gel electrophorese, autoradiography and cytofluorometric experiments (16,17), MoAb H2.5.10 binds on the same (common) determinant of the B-subunit of HLA-DR and to HLA-DP as MoAb Q5/13 (18). As determined by immunoprecipitation and gel electrophorese MoAb 3B3 binds to a common determinant of HLA-DR.

Culture of monocytes and peritoneal cells

For culture, cells were adjusted to a concentration of $1x10^6$/ml in RPMI 1640 medium containing 10% heat inactivated fetal calf serum (60 min, 56°C), 2 mM L-glutamine, 100 µg/ml penicillin and 100 µg/ml streptomycin (αRPMI). The total cell population was cultured in teflon bags (recovery > 80%). The adherent cells were cultured in plastic dishes. After 2 hr of incubation the non adherent cells were washed off the dishes and the adherent cells were incubated for further culture, processed for immunocytochemistry or, after removal with trypsine, used for the MLR.

Immunocytochemistry

Immunocytochemistry was done exactly as described earlier for F4/80 (19). Additional processing of cell suspensions or adherent cells for electron microscopy was performed as described earlier (12).

Radiolabeling of MoAb H2.5.10 and the binding assay

The radiolabeling and binding assay was performed as previously described (20,21). The dialyzed radioactive MoAb H2.5.10 had a specific activity of $0.2x10^6$ cpm/µg protein and > 90% of the label was insoluble in trichloro acetic acid. The saturation of specific binding of ^{125}I anti HLA-DR occured at a concentration of antibody in excess of 75 ng per $1x10^6$ unfixed human peritoneal macrophages. Therefore in the binding assay we used 75 ng of ^{125}I anti HLA-DR. The specific binding of MoAb H2.5.10 was determined by substracting the radioactivity bound to human peritoneal macrophages by the activity bound to rat peritoneal cells.

Allogeneic mixed leucocyte reaction (MLR)

The MLR was performed in αRPMI as described by Mottolese et al. (22). Experiments were done in triplicate in samples of 200 µl.

RESULTS AND DISCUSSION

Immunocytochemistry

It has been determined that glutaraldehyde (GA) reduces or abolishes the detectability of Ia antigen or HLA-DR on cells (21), so by mild fixation with 0.05% GA we do not detect cells with only a few HLA-DR or Ia antigens. As shown in table 1 on human monocytes and macrophages in culture the HLA-DR expression is enhanced. This is is contrast to the Ia expression on rat macrophages in culture. The enhancement of HLA-DR expression could not be due to the effect of lymphokines produced by contaminating lymphocytes, since after removing the non adherent cells the enhancement of HLA-DR expression was still seen. In addition, the increase of HLA-DR expression could not be due to the binding of MoAb H2.5.10 to Fc receptors of macrophages, because preincubation with human IgG did not alter the percentage macrophages staining positive for HLA-DR.

Table 1. The percentage of rat peritoneal macrophages and human monocytes and macrophages (Mφ) positive* for respectively Ia or HLA-DR in an immunoperoxidase assay before and after culture.

Source of material	Hours cultured**				
	0	2 (AC)	5 (AC)	20	20 (AC)
Rat peritoneal macrophages	11	-	-	0	-
Human monocytes	64	69	94	-	96
Peritoneal Mφ of CAPD patient 3	7	-	-	93	98

Three representative cultures are shown. AC: adherent cells.
*Cells were fixed with 0.05% glutaraldehyde prior to the immunolabeling. Cells were labeled with H2.5.10 for HLA-DR and 29c for Ia antigens.
**Preincubation with human IgG did not change the percentage of HLA-DR positive cells in culture.

Quantitative determination of HLA-DR expression with ^{125}I anti HLA-DR (H2.5.10) in culture

The amount of ^{125}I anti HLA-DR bound before and after culture is shown in table 2. In culture the HLA-DR expression is enhanced and the results are in agreement with the results of the immunoperoxidase staining. Culturing with cycloheximide, an inhibitor of the protein synthesis, completely blocks the enhancement of HLA-DR expression (Table 2). Therefore the increase of HLA-DR expression is a protein synthesis dependent process.

Table 2. Specific binding of ^{125}I anti HLA-DR (MoAb H2.5.10) to $1x10^6$ peritoneal cells from CAPD patients 4, 5 and 6 in culture with or without cycloheximide by means of three experiments as presented.

Patient	Hours cultured in teflon bags	^{125}I anti HLA-DR bound (cpm ± SE)	^{125}I anti HLA-DR bound after culture with 10 µg/ml cycloheximide* (cpm ± SE)
4	0	4,961 ± 134	-
4	20	5,516 ± 143	-
5 t0	0	1,487 ± 9	-
5 t0	20	2,113 ± 72	1,355 ± 12
5 t1	0	1,596 ± 25	-
5 t1	5	2,994 ± 161	1,497 ± 21
6	0	1,389 ± 7	-
6	20	4,321 ± 56	-

*An inhibiting agent of the protein synthesis.
t0 = peritoneal cells from patient 5 at the beginning of the experiment
t1 = peritoneal cells from patient 5 five days later.

Allogeneic one-way mixed leucocyte reaction (MLR) with (subpopulations) peritoneal cells (PC)

As shown in table 3, PC are about 10-fold better stimulators of a MLR when compared with peripheral blood monocytes (PBM; table 3, numbers 5 and 7). The PC population of CAPD patients can induce a MLR better than the PBM and much more better than alveolar macrophages (23). Therefore PC could, in comparison with PBM and alveolar macrophages, contain more dendritic cells (DC), because especially DC are good stimulators of a MLR (24).

The non adherent PC after 90 min of culture (NAC 90') and the PC which become non adherent after 16 hours of culture (NAC 16hr) are 2 to 5-fold better stimulators of a MLR than the starting material (table 3, numbers 13, 15 and 19). These subpopulations of PC (NAC 90' and NAC 16hr) could be enriched with DC, because DC do not adhere (25) or become non adherent after 16 hours of culture (26) and DC are good stimulators of a MLR (24).

The proliferation in the allogeneic MLR is dependent of MHC class II antigens, since the addition of MoAbs directed against HLA-DR or HLA-DQ decreased the response (table 3, numbers 6, 8, 9, 10 and 14). The MLR with NAC 90' and NAC 16 hr could not be completely blocked with a MoAb directed against HLA-DR (table 3, numbers 16 and 20). From other studies it appeared that the MLR with NAC 90' and NAC 16hr could only be completely blocked by addition of both HLA-DR and HLA-DQ MoAbs (Bos et al., unpublished results). So for the DC enriched fractions both HLA-DR and HLA-DQ are important in stimulating a MLR.

Table 3. Representative mixed leucocyte reactions with (subpopulations) peritoneal macrophages of CAPD patients.

Number	Mon donor	Mon donor	Mφ donor	MoAb	ConA	^{3}[H]dThd incorp (cpm±SE x10^3)
1	3	-	-	-	-	10,2 ± 2.2
2	3	-	-	-	+	217,7 ± 58,9
3	-	4	-	-	-	1,8 ± 0,5
4	-	4	-	-	+	138,6 ± 42,9
5	3	4	-	-	-	27,9 ± 11,0
6	3	4	-	3B3	-	0,9 ± 0,2
7	3	-	9	-	-	240,7 ± 8,1
8	3	-	9	H2.5.10	-	10,1 ± 1,1
9	3	-	9	3B3	-	4,4 ± 0,1
10	3	-	9	SPV L3	-	36,9 ± 5,9
11	5	-	-	-	-	7,1 ± 6,6
12	5	-	-	-	+	388,0 ± 94,7
13	5	-	10(TOT)	-	-	41,2 ± 10,9
14	5	-	10(TOT)	3B3	-	5,1 ± 1,9
15	5	-	10(NAC 90')	-	-	183,3 ± 91,0
16	5	-	10(NAC 90')	3B3	-	29,3 ± 0,3
17	5	-	10(AC 90')	-	-	34,9 ± 0,8
18	5	-	10(AC 90')	3B3	-	10,3 ± 0,0
19	5	-	10(NAC 16hr)	-	-	89,3 ± 1,1
20	5	-	10(NAC 16hr)	3B3	-	53,6 ± 9,0
21	5	-	10(AC 16hr)	-	-	36,8 ± 6,0
22	5	-	10(AC 16hr)	3B3	-	0,7 ± 0,7

Mon donor: mononuclear cells (90% lymphocytes and 10% monocytes) isolated from blood of healthy donors 3, 4 and 5 (concentration 1x10^6 cells/ml); Mφ donor: peritoneal macrophages from CAPD patients 9 and 10 (concentration 1x10^5 cells/ml); MoAb: monoclonal antibody (5 μg/ml); ConA: concanavalin A (concentration 12.5 μg/ml); TOT: total cell suspension; NAC 90': non adherent cells after 90 min of culture in plastic dishes; AC 90': adherent cells after 90 min culture in plastic dishes; NAC 16hr: cells which become non adherent after 16 hours of culture; AC 16hr: adherent cells after 16 hours of culture; 3B3: MoAb which binds to a determinant of HLA-DR; H2.5.10: MoAb which binds to a determinant of HLA-DR and HLA--DP; SPV L3: MoAb which binds to a determinant of HLA-DQ.

In conclusion, after culture on human PBM and peritoneal macrophages, in contrast to the animal model, the expression of HLA-DR increases. Peritoneal cells and especially the DC enriched fractions are very potent stimulators of a one-way allogeneic MLR when compared with PBM and alveolar macrophages. As already suggested by other investigators (27,28) for DC both HLA-DR and HLA-DQ could be important for antigen presentation.

Acknowledgements

We want to thank Mrs. Anjo Steenvoorden-Bosma for typing the manuscript and Mr. Adrie v.d. Muijsenberg for excellent technical assistance.

REFERENCES

1. D.I. Beller, and E.R. Unanue. J. Immunol. 126:263 (1981).
2. W.S. Walker, R.B. Hester, and R.H.J. Beelen. Cell. Immunol. 79:125 (1983).
3. S.L. Kunkel, D.A. Campbell Jr., S.W. Chensue, and G.I. Higashi. Cell. Immunol. 99:140 (1986).
4. E.W.A. Kamperdijk, M.A.M. Verdaasdonk, and R.H.J. Beelen. Transplant. Proc. 19:3 (1987).
5. S. Becker, and A. Takiguchi. Macrophage Biology, p. 455. Alan R. Liss Inc., New York (1985).
6. G.S. Firestein, and N.J. Zwaifler. Cell. Immunol. 104:343 (1986).
7. D.M. Glover, D. Brownstein, S., Barchett, A. Larsen, and C.B. Wilson. Immunol. 61:195 (1987).
8. G. Gaudernack, and S. Bjercke. Scand. J. Immunol. 21:493 (1985).
9. T.A. Gonwa, and J.D. Stobo. J. Clin. Invest. 74:859 (1985).
10. M.R. Sztein, P.S. Skeg, H.M. Johnson, and J.J. Oppenheim. J. Clin. Invest. 73:556 (1984).
11. G. Kaplan, and G. Gaudernack. J. Exp. Med. 156:1101 (1982).
12. R.H.J. Beelen, M. van 't Veer, D.M. Fluitsma, and E.C.M. Hoefsmit. J. Reticuloendothel. Soc. 24:351 (1978).
13. A. Böyum. J. Clin. Lab. Invest. (suppl.) 97:29 (1968).
14. A. Böyum. Scand. J. Clin. Lab. Invest. (suppl.) 145:264 (1977).
15. H. Spits, J. Borst, M. Giphart, J. Coligan, C. Terhorst, and J.E. de Vries. Eur. J. Immunol. 14:299 (1984).
16. G.D. Keizer, J. Borst, W. Visser, R. Schwarting, J.E. de Vries, and C.G. Figdor. J. Immunol. in press (1987).
17. A. Dimitriu-Bona, G.R. Burmester, S.J. Waters, and R.J. Winchester. J. Immunol. 130:145 (1983).
18. V. Quarantana, N. Tanigaki, and S. Ferrone. Immunogenet. 12:175 (1981).
19. E.C.M. Hoefsmit, I.L. Schadee, and R.H.J. Beelen. J. Histochem. Cytochem. 34:633 (1986).
20. M.A. Markwell, and C.F. Fox. Biochem. 17:4807 (1978).
21. W.S. Walker, R.H.J. Beelen, P.J. Buckley, S.L. Melvin, and S.E. Yen. J. Immunol. Methods. 67:89 (1984).
22. M. Mottolese, P.G. Natali, G. Atlante, A. Cavallari, F. Difilippo, and S. Ferrone. J. Immunol. 135:200 (1985).
23. M.E. Lipscomb, C.R. Lyons, G. Nunez, E.J. Ball, P. Stastny, W. Vial, V. Lenn, J. Weissler, L.M. Miller, and G.B. Toews. J. Immunol. 136: 497 (1986).
24. W.C. van Voorhis, L.S. Hair, R.M. Steinman, and G. Kaplan. J. Exp. Med. 155:1172 (1982).
25. S. Bjercke, and G. Gaudernack. Scand. J. Immunol. 21:501 (1985).
26. S. Knight, J. Farrant, A. Byant, A.J. Edwards, S. Burmen, A. Lever, J. Clarke, and D.B. Webster. Immunology 57:595 (1986).
27. G. Nunez, E.J. Ball, L.K. Meyers, and P. Stastny. In: Histocompatibility Testing (eds. E.D. Albert, M.P. Baur, and W.R. Mayr), p. 636. Springer Verlag, Berlin (1984).
28. S.C. Knight, P. Fryer, S. Griffiths, and B. Harding. Immunology 61:21 (1987).

THE EFFECTS OF BACTERIAL LIPOPOLYSACCHARIDE (LPS) ON HIGH ENDOTHELIAL VENULES, T LYMPHOCYTES AND INTERDIGITATING CELLS IN MOUSE LYMPH NODES

Agaath J.T. Twisk, Paul H.P. Groeneveld, and Georg Krall

Department of Histology, Medical Faculty, Free University
P.O. Box 7161, 1007 MC Amsterdam, The Netherlands

Lymphocytes can recirculate continuously throughout the body from blood to lymph and from lymph to blood. The lymphocyte extravasation from blood into lymphoid tissues is mediated by a specialized endothelium, morphologically characterized by high, cuboidal cells, found in venules, located in lymph nodes and mucosal lymphoid organs (1, 2). Recent studies suggest that lymphocyte migration through the wall of these so-called high endothelial venules (HEV) is accomplished by a ligand receptor system present on the endothelium of HEV and on the surface of lymphocytes, which differs between lymph nodes and Peyer's patches. This organ-specific HEV-lymphocyte interaction selectively controls the lymphocyte migration through the body (3-7). Little is known about the mechanisms that regulate the activity of the HEV and the specificity of interaction with lymphocytes in various lymphoid organs. In the lymph node HEV are located in the T cell dependent areas. These areas are characterized by the presence of non-lymphoid cells, the interdigitating cell (IDC), which may play an essential role in maintaining the integrity of the T cell area. The aim of this study was to determine whether by disturbing the integrity of the T cell area the function of the HEV could be influenced. This was studied in mice by injecting bacterial lipopolysaccharide (LPS) subcutaneously in the hind footpad. LPS was chosen because of its effects on IDC (8) and T cells (9) in the mouse spleen. Using immunohistochemistry the influence of LPS on the T cell dependent areas was examined and using short-term in vivo homing experiments special attention was paid to the HEV specificity.

MATERIALS AND METHODS

Animals: (C3D2)F1 mice were obtained from Bomholtgard Ltd., Rye, Denmark.

Reagents: Lipopolysaccharide (LPS W) from E. Coli 055:B5 was obtained from Difco Lab. (Detroit, USA) and injected subcutaneously in the hind footpads (100 µg in saline per footpad).

Immunohistochemistry: At 24, 48, 72, 96, 120 h, 7 and 9 days after LPS administration, mice were killed and their popliteal lymph nodes taken and prepared for a two-step immunoperoxidase procedure. Sections of the popliteal lymph nodes (LN) were covered in the first step with culture supernatants of the monoclonal rat anti mouse antibodies 59-AD-22 (directed to Thy-1+ cells; 10), NLDC-145 (directed to IDC; 11) or MECA-325 (directed to HEV; 12) at saturating concentrations in 0.01 M phosphate buffered saline (PBS, pH 7,0), containing 0,2% bovine serum albumin (PBS/BSA) for 60 min. After washing thoroughly in PBS, the slides were covered in the second stage with a 1:100 dilution of peroxidase conjugated rabbit anti rat immunoglobulin (Dako, Denmark) in PBS/BSA containing 1% normal mouse serum (60 min). B lymphocytes were demonstrated using peroxidase conjugated rabbit anti mouse Ig (Dako, Denmark). The sections were rinsed in PBS (three times) and stained for peroxidase activity with 3,3'-diaminobenzidine-tetra-HCl in 0.5 mg/ml Tris-HCl buffer, pH 7.6, containing 0.01% H_2O_2 for 10-15 min.

In vivo homing studies: Spleen cells from unprimed donors were labeled with tetramethyl-rhodamine isothiocyanate (TRITC) as described (13) and injected into syngeneic recipients ($5x10^7$ cells/animal). The popliteal LN of the recipients had been challenged by an injection of 100 µg LPS at 72 and 96 h before. Two h after cell transfer the recipient's popliteal LN suspensions, as well as the original injected sample population, were stained with anti-Thy-1 in combination with a highly specific fluoresceinated rabbit anti-rat Ig. Staining conditions have been described (14). Using a Leitz fluorescence microscope donor-derived cells were detected using filters for rhodamine and by switching to fluorescein filters their phenotype was determined. At least 400 cells/sample were counted. Results are expressed as the ratio of T cells to non-T cells.

RESULTS

Control mouse popliteal lymph node: The localization of T lymphocytes as demonstrated with the monoclonal antibody (mAb) Thy-1 is shown in Fig. 1a. The positive lymphocytes are predominantly found in the paracortex. The follicles are negative for this mAb although scattered positive cells can be observed.

Staining with the NLDC-145 mAb, directed to IDC, demonstrated that the distribution of these cells is confined to the paracortical areas. The finger-like processes are clearly visible (Fig. 2a).

Fig. 3a shows the large cuboidal endothelial cells of the HEV present in the paracortex, as recognized by the mAb MECA-325.

LPS-treated mouse popliteal lymph node: Fig. 1b shows the effect of LPS on the T cell localization after 72 h. In the paracortex, a marked fall in T cell concentration could be found at this time point, which had already started at earlier times (24 and 48 h). At 96 and 120 h, the area populated by the Thy-1+ cells was still very small, this in contrast to the B cell area, which increased in size. Starting from day 7 the T cell area began to restore.

Changes in the localization and staining pattern of interdigitating cells in the LN could be observed 48 and 72 h after LPS injection (Fig. 2b). From that time point their number reduced, whereas the remaining IDC were weakly NLDC-145 positive. In contrast to the staining of IDC in control LN (Fig. 2a), their dendritic cell processes were harder to discriminate. This staining pattern could be observed up until day 9, then the characteristic appearance of IDC was restored and their localization comparable with an untreated LN.

Considering the HEV it was found that after LPS treatment the HEV remained unchanged in appearance, as determined by staining with mAb MECA-325. Their size, length and localization was comparable to the normal situation at any time point examined (Fig. 3a, 3b).

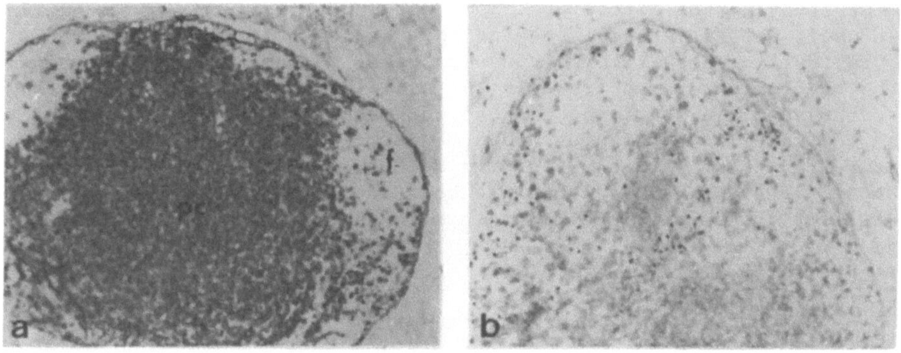

Fig. 1. Sections of mouse popliteal lymph nodes were stained with Thy-1. a: In control animals Thy-1 positive cells are seen predominantly in the paracortex. b: 72 h after LPS administration the area populated by T-lymphocytes was reduced. pc: paracortex; f: follicle (x10).

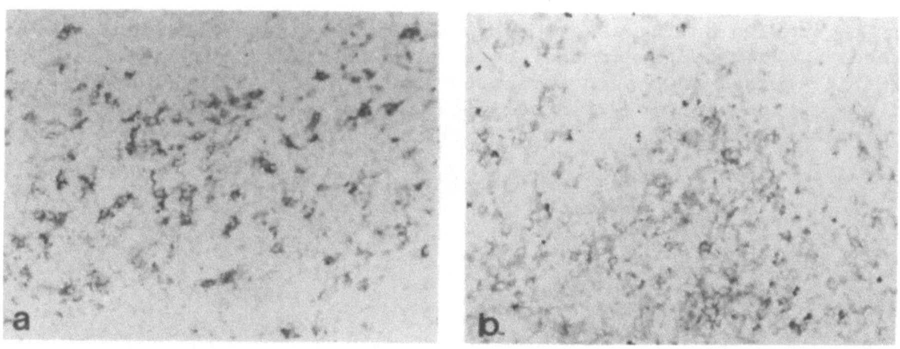

Fig. 2. Interdigitating cells in the paracortex of popliteal lymph nodes as demonstrated with NLDC-145. a: control popliteal LN. b: 72 h after LPS their number was reduced, whereas the remaining cells were weakly positive (x20).

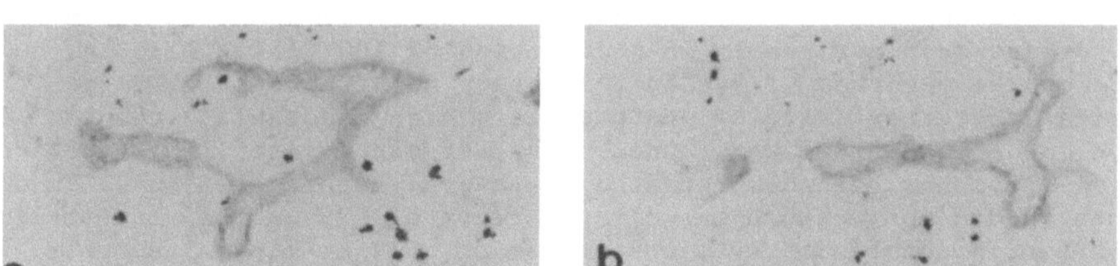

Fig. 3. High endothelial venules in the paracortex of popliteal lymph nodes as demonstrated with MECA-325. Comparing control (a) and LPS-treated animals (b) no visible changes could be observed (x20).

For studying the effect of LPS on the specificity of HEV rhodaminated spleen cells were injected into mice that had been treated with LPS 72 and 96 h previously. These time points were taken because of the maximum effect on the T cell dependent areas in the draining lymph node. At both time points the specificity of the HEV had not changed and a preference of the T cells to localize into lymph nodes, control as well as LPS treated was demonstreated shortly (2 h) after the cell transfer. The 2 h time point was chosen to exclude recirculation so that the localization of donor cells will reflect the initial localization as determined by the HEV (Fig. 4).

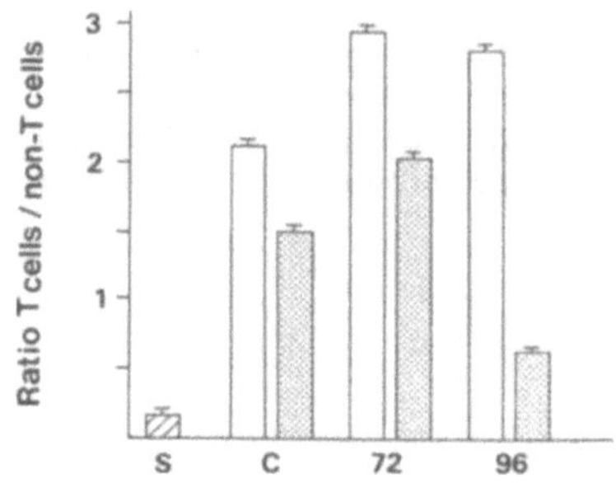

Fig. 4. The in vivo homing assays were performed as described in 'Materials and Methods'. With double immunofluorescence the ratio of T over non-T (B) cells was determined of the rhodaminated immigrant (open bars) and resident (dotted bars) population. The striped bar represents the ratio of the injected spleen cells (S). C = control animal.

DISCUSSION

After LPS administration the number of interdigitating cells and T lymphocytes, located in the paracortex of the draining lymph node decrease, whereas the B cell area increases in size. These results, in accordance with those of Groeneveld et al. (8, 9) in the spleen, show a direct effect of LPS on the localization of both IDC and T cells. Interdigitating cells, with their dendritic protrusions make close contacts with the surrounding T cells in the T cell-dependent areas of lymphoid organs. Since their association with T cells has led to the suggestion that IDC create a suitable environment for proliferation and differentiation of T cells (15), the loss of IDC may explain why the number of T cells in the paracortex is decreased. On the other hand it is not likely that loss of T cells induces the death or depletion of IDC since it has been demonstrated in the case of lethal irradiation that a complete depletion of T lymphocytes from the lymphoid organs does not result in a depletion of IDC (11). Also in nude mice IDC can be observed (11). In this study no effects of LPS on the morphological characteristics of postcapillary high endothelial venules in the lymph node paracortex was found. That LPS can exert influences on endothelial cells has been demonstrated in several reports (16, 17). In our experiments we did not measure the direct effect of LPS on the endothelium because LPS would not be in the blood stream or only at minimal amounts. It has been suggested that factors produced or present in the paracortex play an important role in maintaining the HEV function (18). IDC are one of the cells, that are thought to produce such factors. Firstly because of their proximity with HEV; secondly, the fact that IDC are able to produce different kinds of glycoproteins (19). In agreement with this idea is the observation that UV-irradiation, for which IDC are sensitive, greatly interferes with the lymphocytes influx into the exposed lymph node (20). The observation, however, that the LPS-induced loss of IDC does not influence the specificity of HEV, presented in this paper, is not a direct support for an involvement of IDC in the lymphocyte-HEV interaction.

REFERENCES

1. J.L. Gowans, and E.J. Knight. Proc. R. Soc. Lond. Biol. 159:257 (1964).
2. G. Kraal, A.M. Duijvestijn, and H.R. Hendriks. Cell Biol. 55:1 (1987).
3. E.C. Butcher, R.G. Scollay, and I.L. Weissman. Eur. J. Immunol. 10:556 (1980).
4. W.M. Gallatin, I.L. Weissman, and E.C. Butcher. Nature 304:30 (1983).
5. M.O. Dailey, W.M. Gallatin, and I.L. Weissman. J. Mol. Cell. Immunol. 2:27 (1985).
6. R.A. Rasmussen, Y.-H. Chin, J.J. Woodruff, and T.G. Easton. J. Immunol. 135:19 (1985).
7. Y.-H. Chin, R.A. Rasmussen, J.J. Woodruff, and T.G. Easton, J. Immunol. 136:2556 (1986).
8. P.H.P. Groeneveld, T. Erich, and G. Kraal. Immunology 58:285 (1986).
9. P.H.P. Groeneveld, and N. van Rooijen. Cell Tissue Res. 236:637 (1984).
10. J.A. Ledbetter, and L.A. Herzenberg. Immunol. Rev. 47:63 (1979).
11. G. Kraal, M. Breel, M. Janse and G. Bruin. J. Exp. Med. 163:981 (1986).
12. A.M. Duijvestijn, M. Kerkhove, R. Bargatze, and E.C. Butcher. J. Immunol., in press (1987).
13. E.C. Butcher, and I.L. Weissman. J. Immunol. Meth. 37:97 (1980).
14. S.K. Stevens, I.L. Weissman, and E.C. Butcher. J. Immunol. 128:844 (1982).
15. W. van Ewijk, J.H.M. Verzijden, T.M. van der Kwast, and S.W.M. Luijcx-Meijer. Cell Tissue Res. 149:43 (1974).
16. M.A. Reidy, and D.E. Bowyer. Atheroscl. 26:319 (1977).
17. E. Gaynor. Lab. Invest. 24:318 (1971).
18. H.R. Hendriks, and I.L. Eestermans. Eur. J. Immunol. 13:663 (1983).
19. A. Friess. Cell Tissue Res. 170:43 (1976).
20. M.T. Drayson, and W.L. Ford. Immunobiol. 168:362 (1984).

INTESTINAL UPTAKE OF CARRAGEENAN: DISTRIBUTION AND EFFECTS ON HUMORAL IMMUNE COMPETENCE

S. Nicklin, K. Baker and K. Miller

Dept. Immunotoxicology
The British Industrial Biological Research Assoc.
Carshalton, Surrey UK

INTRODUCTION

Carrageenan is the generic name given to high-molecular-weight (100,000 daltons) sulphated polygalactans derived from certain species of red algae. Three basic types can be isolated: k, i, and l. These have differing colloidal properties and are characterised by the type of intergalactan bonding and the degree of sulphation (Rees, 1972). Carrageenans are used extensively in the food industry as thickening, gelling and protein-suspending agents. Initially, all grades of carrageenan were considered safe by the Food and Drug Administration (FDA, 1959) and were permitted for use as regulated food additives (FDA, 1969). However, subsequent reports implicated degraded forms of carrageenan (mol wt 30,000) in the induction of ulcers and metaplastic changes in the intestinal tract of a number of species of experimental animals (Fabian, <u>et al</u>., 1973; Sharratt, <u>et al</u>., 1970; Watt & Marcus, 1970). Although some of these findings were contested by other studies (Maillet, <u>et al</u>., 1970; Sharratt, <u>et al</u>., 1971) the FDA ruled that food-grade carrageenans should have an average molecular weight exceeding 100,000 (FDA, 1972). It is known however that carrageenans may undergo some degree of hydrolysis at gastric pH (Stoloff, 1959; Glicksman, 1969) and a small proportion of the lower molecular weight polygalactan units produced can apparently gain access to the body tissues by persorption (Pittman, <u>et al</u>., 1976).

Whilst there is no evidence to suggest that the persorption of small quantities of carrageenan across the epithelial barrier poses an acute toxic hazard, carrageenan is biologically active, and several authors have reported that parenterally administered carrageenans, can markedly affect a number of normal immune responses. These studies include the suppressive effect of carrageenans on macrophage function (Di Rosa, 1972; Fowler, et al., 1980), antibody production (Rumjanek, et al., 1977; Thomson, et al., 1979) and cell-mediated immune responses (Rumjanek & Brent, 1978; Sakemi, et al., 1980). Furthermore, Bash & Vago (1980), demonstrated that the oral administration of carrageenan to rats resulted in a dose-dependent suppression of lymphocyte responsiveness in vitro.

These results are in accordance with more recent studies which demonstrated that the intestinal persorption of small quantities of orally administered food grade carrageenan was associated with a transient depression in systemic humoral immunity (Nicklin & Miller, 1984). This was attributed to the selective toxic effect of carrageenan on antigen processing macrophages. However the assumption that persorbed carrageenan necessarily depressed the systemic antibody response by a direct cytotoxic effect on macrophages is open to criticism, first because quantitative data and organ distribution of the carrageenan was not available from the persorption study and secondly, it was suggested that resting peritoneal macrophages may be more sensitive to carrageenan than the antigen processing macrophages normally found within the GALT.

In view of the potent biological properties of carrageenan and its wide application in the food industry, a more detailed study of the effects of this material on various components of the immune system was undertaken.

UPTAKE AND ORGAN DISTRIBUTION OF ORALLY ADMINISTERED FOOD GRADE CARRAGEENAN

The intestinal absorption of carrageenans is difficult to study and consequently has been examined by relatively few groups. The studies conducted by Grasso, et al., (1975) and Abraham, et al., (1972) indicated that substantial amounts of ingested degraded carrageenan were absorbed by a variety of experimental animals. Pittman, et al., (1976) also studied the absorption and excretion of various grades/molecular weights of carrageenans in guinea-pigs, monkeys and rats. This study confirmed the previous work on degraded carrageenans but also suggested that small amounts of relatively high molecular weight carrageenan may also be absorbed. However they implied that this uptake may be a consequence of high molecular weight carrageenan undergoing minor degradation within the gastrointestinal tract. This suggestion was contested by Stancoff & Renn (1975). Nevertheless our own studies have also demonstrated carrageenan within intestinal macrophages of carrageenan fed rats (Nicklin & Miller, 1984). In these studies however carrageenan was detected using Alcian blue metachromasia, which although specific for carrageenan, lacks sensitivity.

In view of the paucity of data concerning the intestinal uptake of orally administered high molecular weight carrageenan a series of experiments were performed using ^{3}H-labelled carrageenan. These studies were designed to quantitate uptake and provide information on subsequent organ distribution.

EXPERIMENTAL

PREPARATION OF ^{3}H-LABELLED CARRAGEENAN

Carrageenan was internally labelled with tritium using a modification of the galactase oxidase/tritiated borohydride method described by Malmstrom, et al., (1976). A solution of iota-carrageenan (1 mg/ml; 25 μl) (Food grade - Hercules Ltd. UK) in 0.1M phosphate buffer pH 6 was incubated with 100 units galactose oxidase for 2 hr at 37°C. The carrageenan was then precipitated from the mixture using cold ethanol, collected by filtration and redissolved in 25 ml 0.1M HCl/borate buffer pH 9. Sodium boro [^{3}H] hydride (25 mCi) was then added and the mixture incubated for 1 hr at room temperature in a fume hood. Unlabelled sodium borohydride (5 g) was then added as a cold "chase" and the mixture incubated overnight. The radiolabelled carrageenan was collected following extensive dialysis against deionised water, ultrafiltration using a 30,000 dalton cut off Diaflow membrane filter and partition chromatography using Sepharose 6B. The final preparation had a specific activity of approximately 10^{7} dpm/mg.

UPTAKE AND DISTRIBUTION STUDIES

Groups of 200-250 g male PVG strain rats were starved overnight prior to gavage with 50 μg [^{3}H] carrageenan in 1 ml deionised water. Following treatment, the rats were housed individually in all glass metabolism cages and urine and faeces collected. Groups of animals were killed by exsanguination under pentabarbitone anaesthsia at 6, 24, 48 and 72 hr and the following tissues removed for sample oxidation and scintillation counting:- liver, kidney, spleen, mesenteric lymph node, caecal lymph node, Peyer's patches and washed duodenum (free of Peyer's patches). Results are presented as mean dpm ± SEM (n = 3).

ASSESSMENT OF HUMORAL IMMUNE COMPETENCE

Groups of rats were maintained on tap water (controls) or tap water containing 0.25% carrageenan. On day 184 of treatment all rats were challenged intraperitoneally with 1 ml 5% sheep red blood cells (SRBC). Serum was collected 7, 14, 21 and 28 days later and analysed for anti-SRBC antibody activity by the haemagglutination method. Results indicate mean end point haemagglutination titre ± SEM (n = 4).

RESULTS AND CONCLUSIONS

Although the majority of the administered radioactivity was excreted in the faeces, uptake into specific organs was evident. Counts present within the washed intestinal wall presented a plateau-like profile for 48 hr followed by a rapid decline in activity at 72 hr (Fig.1). In contrast, counts in Peyer's patches peaked at 6 hr and declined rapidly during the following 3 days (Fig.2). The mesentereic lymph nodes presented a different time course profile, with very low levels of activity at 6 hr, which increased significantly to peak at 24 hr.

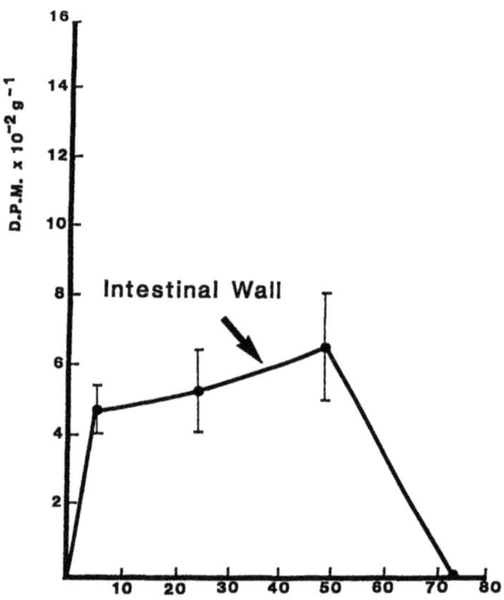

Fig. 1

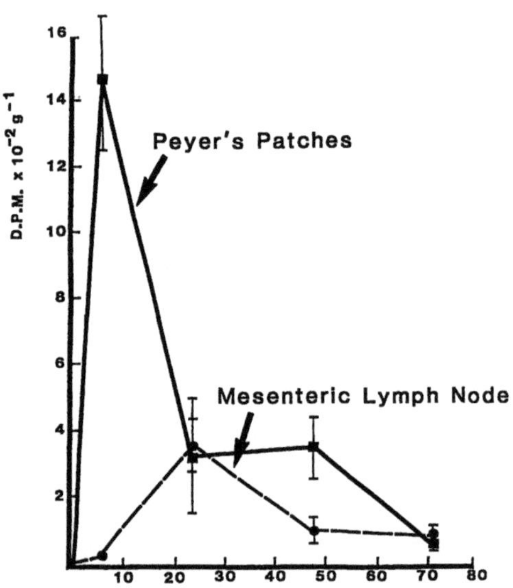

Fig.2

It is also clear from the data that the Peyer's patches are not the only route of access into the tissues. Examination of the caecal lymph node revealed substantial accumulation of radioactivity in this organ, with levels consistently higher than in the Peyer's patches (Fig.3). Whilst this suggests that caecal lymph nodes may be more efficient at antigenic uptake than Peyer's patches, it is also possible that the high level of counts are a consequence of the prolonged retention of intestinal contents in the caecum thereby extending the time of contact between the carrageenan and the lymphoid tissues. The latter possibility is supported by the finding of counts in these lymph nodes at 72 hr.

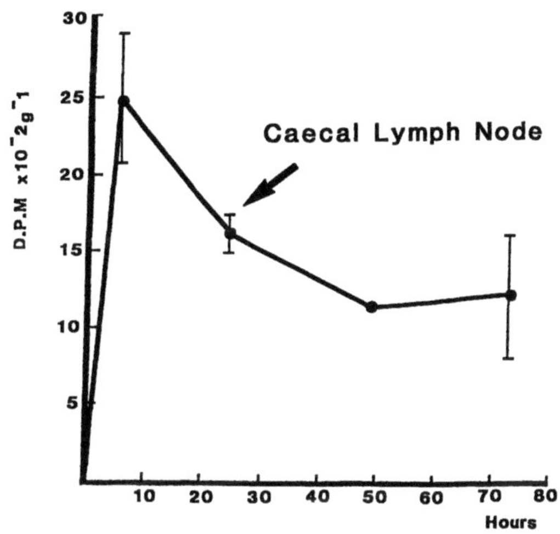

Fig.3

Low levels of activity were also detected in the liver, spleen, skeletal muscle and kidney (Figs. 4 & 5). The retention of counts within the kidney parallels the appearance of counts within the urine (Fig.6). The fact that free counts were also detected in serum (Fig.7) is indicative of tritium exchange between the ^{3}H-labelled carrageenan and body water. In this study the carrageenan was internally labelled by the reduction of -CHO groups (produced after galactose oxidase treatment) using tritiated borohydride to produce tritiated $-CH_2OH$ groups. Each -CHO radical can therefore receive two tritium atoms one bonded directly to the carbon and one to the oxygen. While the carbon carbon bound tritium is stable the oxygen bound tritium is labile and capable of exchange with hydrogen from water (Feinendegen, 1967). As a consequence of this exchange only organ counts over and above serum levels can be considered significant.

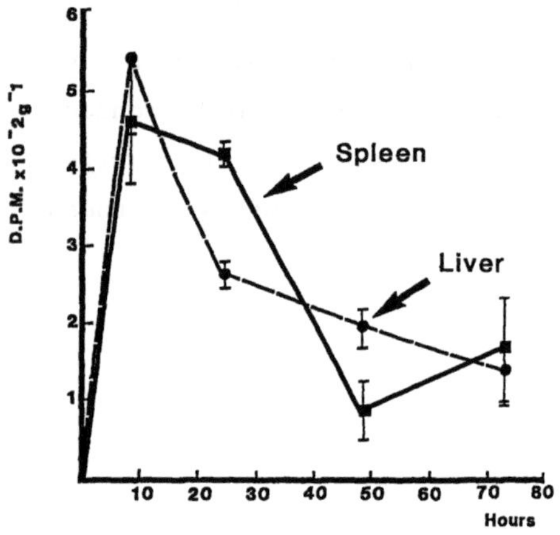

Fig.4

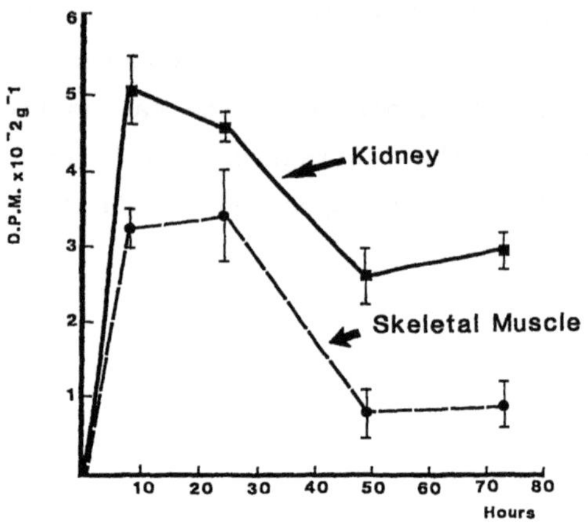

Fig.5

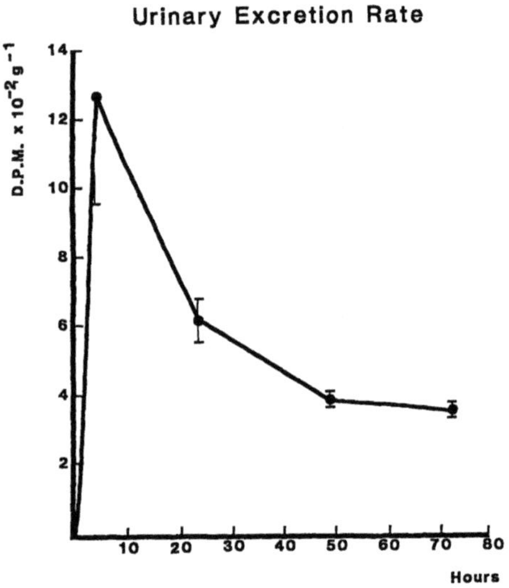

Fig.6

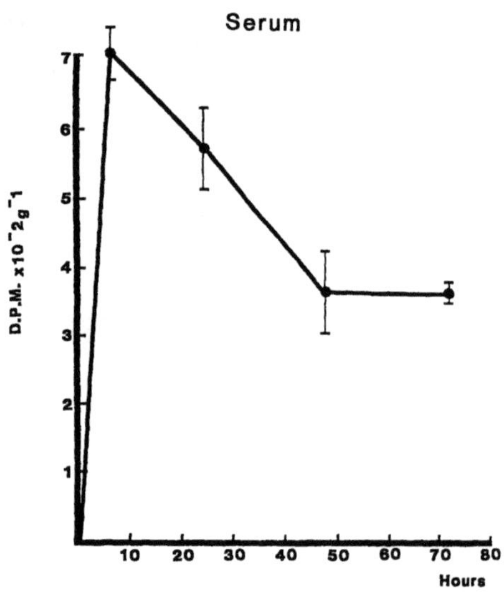

Fig.7

Examination of selected tissue samples by autoradiography revealed that the bound tritium was invariably associated with macrophages. This is consistent with our earlier studies (Nicklin & Miller, 1984) and clearly indicates that components of orally administered food-grade iota-carrageenan can gain access to the body tissues via the Peyer's patches and caecal lymph node. It is likely that carrageenan enters the body via macrophages and is transported intracellularly to the mesenteric lymph nodes. As carrageenan accumulates within macrophages, organ distribution may not necessarily be relevant to cell toxicity. Since carrageenan is known to possess immunoregulatory properties (see Thompson & Fowler, 1981 for review) its presence within these lympoid organs might be expected to influence immune competence. Evidence in support of this postulate is provided by experiments which demonstrated that animals maintained on unlabelled carrageenan prior to parenteral challenge with test antigen presented both a delayed and significantly reduced antibody response (Fig.8).

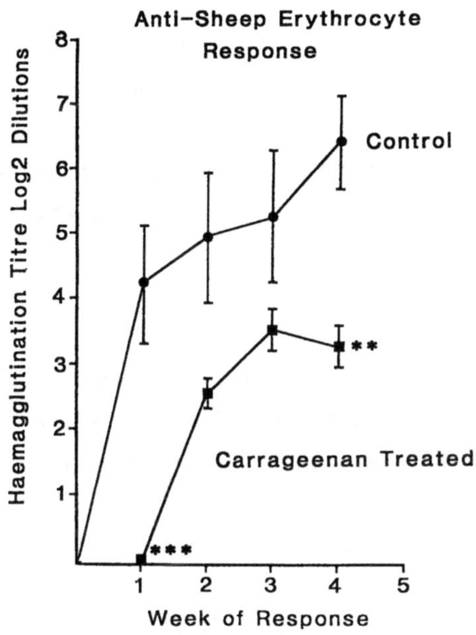

Fig.8

Further studies designed to elucidate the various interactions that can occur between carrageenan and macrophages are presented in the accompanying paper (Nicklin, Baker & Miller in this volume).

ACKNOWLEDGEMENTS

These studies form part of a research project funded by the U.K. Ministry of Agriculture, Fisheries and Food. The results of this research are Crown copyright.

REFERENCES

Abraham, R., Golberg, L. & Coulston, F. Exp. Mol. Path. 17, 177 (1972).
Bash, J.A. & Vago, J.R. Reticuloendothel. Soc. 28, 213 (1980).
Di, Rosa, M. J. Pharm. Pharmac, 24, 89 (1972).
Fabian, R.J., Abraham, R., Coulston, F. & Golberg, L. Gastroenterology, 65, 265 (1973).

FDA (1959). Federal Register, 24, 9368.

FDA (1969). Federal Register, 26, 94.

FDA (1972). Federal Register, 37, 15434.

Feinendegen, L.E. Tritium labelled molecules in medicine and biology. Academic Press, N. York, p.26 (1967).

Fowler, E.F., Simpson, J.G. & Thompson, A.W. J. Path. 132, 63 (1980).

Glicksman, M. In: Gum Technology in the Food Industry 1st Ed. p.222. Academic Press, New York (1969).

Grasso, P., Gangolli, S.D., Butterworth, K.R. & Wright, M.G. Fd. Cosmet, Toxicol. 13, 195 (1975).

Maillet, M., Bonfils, S. & Lister, R.E. Lancet II, 414 (1970).

Malmstrom, B.G., Andreasson, L.E. & Reinhammar, B. In: The Enzymes 3rd Edition. Vol XII. Ed. P.D. Boyer. pp 529, Academic Press (1976).

Nicklin, S. & Miller, K. Fd Chem. Toxic. 22, 615 (1984).

Pittman, K.A., Golberg, L. & Coulson, F. Fd. Cosmet. Toxicol. 14, 85 (1976).

Rees, D.A. Biochem. J. 126, 257 (1972).

Rumjanek, V.M. & Brent. L. Transplantation 26, 113 (1978).

Rumjanek, V.M., Watson, S.R. & Sljivic, H. Immunol. 33, 423 (1977).

Sakemi, T., Kuroiwa, A. & Nomoto, K. Immunol. 41, 297 (1980).

Sharratt, M., Grasso, P., Carpanini, F. & Gangolli, S.D. Lancet II, 932 (1970).

Sharratt, M., Grasso, P., Carpanini, F. & Gangolli, S.D. Gastroenterology 61, 410 (1971).

Smith, G. & Nicklin, S. J. Imm. Method. 25, 265 (1979).

Stancoff, D.J. & Renn, K.W. Amm. Chem. Soc. Acss. Symp. 15, 282 (1975).

Stoloff, L. (1959). Polysaccharides and Their Derivatives. Edited by R.L. Whistler & J.N. BeMiller, p.83. Academic Press, New York.

Thomson, A.W., Fowler, E.F. & Pugh-Humphreys, R.G.P. Int. J. Immunopharmac 1, 247 (1979).

Thompson, A.W. & Fowler, E.F. Agents & Actions 11, 3 (1970).

Watt, J. & Marcus, R. Gastroenterology 59, 760 (1970).

CARRAGEENAN INDUCED CHANGES IN THE IMMUNOMODULATORY CAPACITY OF MACROPHAGES: AN IN VITRO STUDY

S. Nicklin, K. Baker, and K. Miller

Dept. Immunotoxicology
The British Industrial Biological Research Association
Carshalton, Surrey, U.K.

INTRODUCTION

Experimental studies using rats have shown that small quantitites of orally administered carrageenan accumulate in the Peyer's patches and caecal lymph node prior to transport to the mesenteric lymph node and peripheral tissues. Whilst there is no evidence to suggest that the persorption of small amounts of carrageenan across the intestinal barrier in any way poses an acute toxic hazard, carrageenan is a biologically active molecule (Di Rosa, 1972) and its extended oral administration was shown to modify systemic immune competence (see Nicklin and Miller, 1984 plus Nicklin, Baker & Miller in this volume). Whereas this effect could be attributed to carrageenan having a selective toxic effect on antigen processing macrophages, additional studies suggest that macrophages may also influence immune responses by the release of immunoregulatory mediators. In view of the continuing wide application of carrageenan in the food and drug industry an investigation into the effects of this material on the immunoregulatory role of macrophages was considered particularly relevant. The present study described a series of experiments designed to determine whether macrophages treated with toxic/subtoxic doses of carrageenan are capable of affecting lymphocyte activity *in vitro*.

Effect of carrageenan on macrophage viability

PVG strain male rats were injected ip with either 10 ml thioglycollate broth or saline and killed two days later using CO_2. The resident activated macrophages were collected from the peritoneal cavity by lavage using cold (4°C) phosphate buffered saline (PBS) pH 7.2. The cells were washed using tissue culture medium (RPMI 1640), counted and seeded (1 x 10^6 cells) onto 12 mm circular glass cover slips in Costar tissue culture plates. They were cultured at 37°C in 5% CO_2/air in RPMI 1640 tissue culture medium supplemented with 20% foetal calf serum. Non-adherent cells were washed off after 1 hr in culture to yield a macrophage monolayer. Graded doses of carrageenan were then added in fresh culture medium. After 16 hr culture, the cells were washed again, and incubation continued for up to 72 hr. Cover slips for cell counting were taken at 24, 48 and 72 hr of culture, washed gently with warm PBS to remove non-viable cells, fixed in Bouins fixative for 30 min and stained with either haematoxylin and eosin or Alcian blue (to demonstrate intracellular carrageenan). The coverslips were mounted onto glass slides examined microscopically and cell counts performed on 30 high power fields. Results are presented as the number of adherant cells in test culture expressed as a percentage of those present in control cultures. Relative toxicity of the various carrageenans is indicated as the minimal cytotoxic dose (MCD) which is the lowest dose of carrageenan required to induce significant ($P<0.05$) cell death at any given time point.

Table 1

Carrageenan Treatment	Time in culture		
	24 hours	48 hours	72 hours
Resident Macrophages			
Sigma-Kappa	NS*	100	100
Sigma-Lambda	NS	10	10
Sigma-Iota	NS	10	10
Hercules-Kappa	1000	100	100
Hercules-Lambda	1000	10	10
Hercules-Iota	NS	10	1
Marine Colloids-Food Grade Iota	NS	NS	10
Marine Colloids-Degraded Iota	NS	NS	NS
Thioglycollate Elicited Macrophages			
Sigma-Kappa	NS	100	10
Sigma-Lambda	NS	10	10
Sigma-Iota	NS	100	10
Hercules-Kappe	NS	100	10
Hercules-Lambda	NS	10	10
Hercules-Iota	NS	100	10
Marine Colloids-Food Grade Iota	NS	10	10
Marine Colloids-Degraded Iota	NS	NS	NS

Macrophage cultures were exposed to carrageenan for 16 hours.
Numbers represent the minimum dose (ug/ml) inducing significant ($P< 0.05$) cell death within 24, 48 or 72 hours of commencement of culture.

* NS = No significant death compared to control cultures.

Table 1 compares the minimal cytotoxic doses of the various carrageenans for thioglycolate activated and resident macrophages. Toxicity was dose dependent and varied with the source and grade of carrageenan. In the main iota and lambda grade carrageenans proved more toxic than kappa grades. These results indicate that activated and resident peritoneal macrophages are equally sensitive to microgramme quantities of carrageenan.

Evidence in the literature however suggests that macrophages exposed to sub-toxic doses of carrageenan can modify *in vitro* antibody responses through the release of immunoregulatory mediators (Rumjanek, *et al*., 1977; Evelegh, *et al*., 1982). In order to determine whether similar factors are produced by macrophages following exposure to the food grade carrageenans used in the present study, experiments were performed to determine whether macrophages treated with toxic/sub-toxic doses of carrageenan release factors capable of affecting lymphocyte proliferation *in vitro*.

Effect of carrageenan exposure on ^{3}H-uridine incorporation by macrophages

Uridine incorporation into the RNA precursor pool in normal and carrageenan treated macrophages was determined as described previously (Smith & Nicklin, 1979). Thioglycollate-elicited macrophages were obtained and established in culture at 5 x 10^3 cells/cover slip. Non-adherent cells were washed off after 1 hr of incubation and graded doses of food grade iota-carrageenan added in fresh culture medium. After 16 hr culture, the cells were again washed and incubation continued for upto 72 hr. During the incubation period cultures were pulsed with 2 µCi ^{3}H-uridine at 6, 24, 48 and 72 hr. Two hours later the radioactive medium was removed and the cells washed with cold PBS. The intra-cellular uridine was collected following cell disruption using 100 µl cold 10% TCA and its concentration determined by scintillation counting.

Figure 1 indicates the influence of various carrageenan concentrations on uridine incorporation by peritoneal macrophages. Whilst 100 µg carrageenan treatment did not significantly influence uridine uptake, pretreatment with 500 µg carrageenan produced a significant increase in incorporation at both 48 hr ($P < 0.05$) and 72 hr ($P < 0.001$) of culture. Increasing the level of carrageenan exposure to 1000 µg/ml produced a further increase in the level of incorporation with the 48 hr counts being 8 x greater than control cultures, although the level of incorporation declined slightly at 72 and 96 hr values remained significantly different from control levels. These results indicate that toxic doses of carrageenan stimulate uridine incorporation in macrophages and thus by implication suggest that cells damaged by carrageenan may undergo a cycle of protein synthesis prior to death.

Macrophage products are numerous and much work has been published indicating that macrophages can release a variety of biologically active molecules, some of which have been shown to influence the immune system (Bash & Vago, 1980; Bash & Cochran, 1980; Evelegh et al., 1980; Cochran & Baxter, 1984; Quan, et al.,1980). The following a series of experiments were designed to examine the possibility that carrageenan treated macrophages release similar immuno-regulatory mediators prior to cell death.

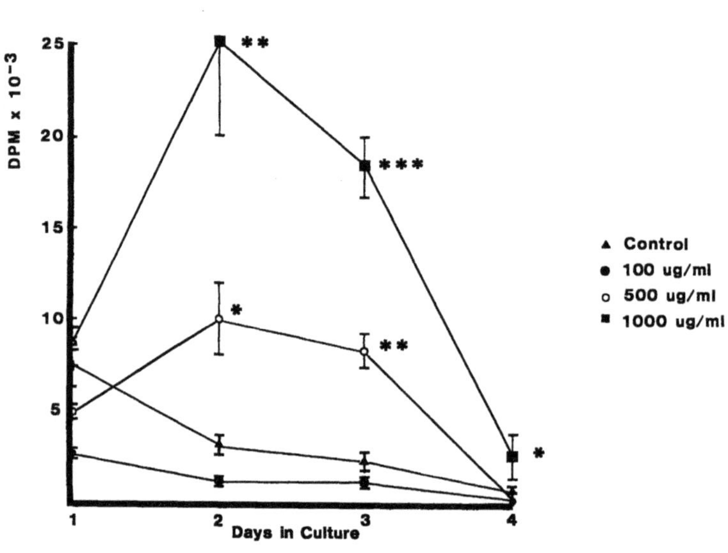

Fig. 1. Effect of Food Iota-Carrageenan on Uridine Incorporation by Rat Peritoneal Macrophages.

Effect of food grade iota-carrageenan on the lymphocytes

In order to study the effect of carrageenan treatment on the immunomodulatory function of macrophages it was first necessary to evaluate the direct effect of carrageenan on lymphocyte viability/ functions, as shown by their ability of incorporate ^{3}H-thymidine. The effect of graded doses of Marine Colloids food grade iota-carrageenan on

Table 2 Effect of food grade iota-carrageenan on thymidine incorporation by rat peripheral blood lymphocytes

Carrageenan conc. (μg/ml)	Period in culture: 24 hours	48 hours	72 hours	96 hours
0	361 ± 113	276 ± 44	384 ± 81	411 ± 45
1	504 ± 112	364 ± 50	570 ± 109	258 ± 69
10	522 ± 114	602 ± 170	421 ± 86	392 ± 16
100	633 ± 193	356 ± 68	308 ± 68	783 ± 64
PHA* (3.75 μg/ml)	13186 ± 1603	193224 ± 31805	841200 ± 13632	802102 ± 25364

Table 3 Effect of food grade iota-carrageenan on thymidine incorporation by rat thymocytes

Carrageenan conc. (μg/ml)	Period in culture: 24 hours	48 hours	72 hours	96 hours
0	321 ± 86	320 ± 58	414 ± 58	305 ± 97
1	448 ± 109	424 ± 59	614 ± 117	191 ± 52
10	490 ± 129	740 ± 199	453 ± 94	291 ± 13
100	453 ± 29	522 ± 144	297 ± 58	191 ± 49
1000	198 ± 29	357 ± 105	208 ± 26	289 ± 131
PHA* (1 μg/ml)	1314 ± 45	37783 ± 747	46125 ± 3703	54960 ± 2855

* Positive control, figures indicate mean dpm ± S.E.M. (n = 4).

^{3}H-thymidine incorporation by rat peripheral blood lymphocytes and thymocytes is shown in Tables 2 and 3 respectively. It is apparent that carrageenan at doses up to 1000 µg/ml does not significantly influence lymphocyte viability, and does not possess overt mitogenic properties.

Analysis of macrophage products

Macrophage-conditioned medium was produced according to the method described by Chu, et al., 1984. Briefly 12.5 x 10^6 thioglycollate-elicited peritoneal exudate cells were seeded into 60 mm Nunc tissue culture dishes. The cells were then cultured for 16 hr at 37°C either in normal media, with or without graded doses of carrageenan or in normal medium plus 10 µg/ml concanavalin-A (CON-A). Supernatant fluids were collected, (where appropriate CON-A activity was neutralised by the addition of β-D-methyl manoside) filtered through a 0.45 µm millipore membrane and dialysed against PBS for 7 hr and RPMI 1640 tissue culture mediumfor 16 hr. The supernatant (referred to as conditioned medium) was filtered as above and stored at -20°C prior to testing.

The effect of including 5% conditioned medium from normal or carrageenan treated macrophages in a PHA induced peripheral blood lymphocyte mitogenesis assay is shown in Table 4. Conditioned medium has no effect on unstimulated lymphocytes (no PHA) but significantly enhances (P = 0.05 - 0.001) optimally stimulated (7.5 µg PHA/ml) cultures in a dose dependent manner. This finding implies that the product is specific for antigen/mitogen driven responses rather than a non-specific proliferation factor. An obvious contender for this factor is interleukin-1 which is released by antigen-activated macrophages and initiates the proliferation of antigen or mitogen triggered T cells but not resting T cells (Oppenheim, et al., 1984). However, as T cell proliferation is associated with the release of interleukin-2, a secondary T derived agent that promotes the long term proliferation of activated T cells (Gillis & Watson, 1980) it was necessary to differentiate between the two factors.

Assessment of interleukin-1 (IL-1) production by carrageenan treated macrophages

Interleukin-1 (IL-1) production by macrophages was assessed as described by Bendtzen (1983). This method utilises the fact that IL-1 is not species specific and that C3H/HeJ strain mouse thymocytes are particularly sensitive to IL-1. Briefly, thymocytes were obtained from

Table 4 Effect of conditioned medium from carrageenan#-treated macrophages on PHA-induced lymphocyte+ mitogenesis

	PHA concentration (µg/ml)		
Carrageenan conc. (µg/ml)	0	3.75	7.50
0	535 ± 80	17320 ± 3000	54307 ± 3930
10	313 ± 91	17855 ± 2246	87256 ± 5887*
100	322 ± 53	29252 ± 2533*	120349 ± 2024**
1000	314 ± 66	24940 ± 2309	125092 ± 2116**

Figures indicate mean ± S.E.M. (n = 3). + Rat peripheral blood.

* $P < 0.05$; ** $P < 0.001$. # Food grade iota-carrageenan.

8-10 week old C3H/HeJ mice (3 x 10^5 cells) and incubated under standard tissue culture conditions in the presence of phytohaemagglutin (PHA) with or without the addition of macrophage-conditioned medium. Control cultures received media from untreated macrophage cultures to which carrageenan had been added after culture. The cells were then incubated for 48 hr prior to the addition of 2 µCi ^{3}H-thymidine. The cells were harvested in the usual fashion and the level of ^{3}H incorporation determined by scintillation counting.

Assessment of Interleukin-2 (IL-2) activity in macrophage conditioned media

The interleukin-2 (IL-2) assay utilised CTLL cells, an IL-2 dependent untransformed primary cell line derived from cloned murine cytotoxic T cells. These cells require IL-2 for proliferation and can consequently be used to monitor the levels of IL-2 in tissue culture supernatants (Ishizaka & Stutman, 1983).

Medium to be tested (100 µl) was added to the first well of a round bottomed microtitre plate and serially double-diluted across twelve wells in normal tissue culture medium supplemented with 2.5 x 10^{-5} M 2-mercaptoethanol. CTLL cells (10^4) obtained from a primary culture maintained in this laboratory using IL-2 supplemented medium were then added to each well. The plates were cultured under standard conditions for 18 hrs prior to receiving a 2 μCi pulse of ^{3}H-thymidine in fresh medium. Incubation was then continued for a further 6 hr, the cells were then harvested and the level of ^{3}H- incorporation determined by scintillation counting. An IL-2 standard (4 μnits/ml) provided by Dr Mirek Malkovsky, Clinical Research Centre, Harrow, UK was used to

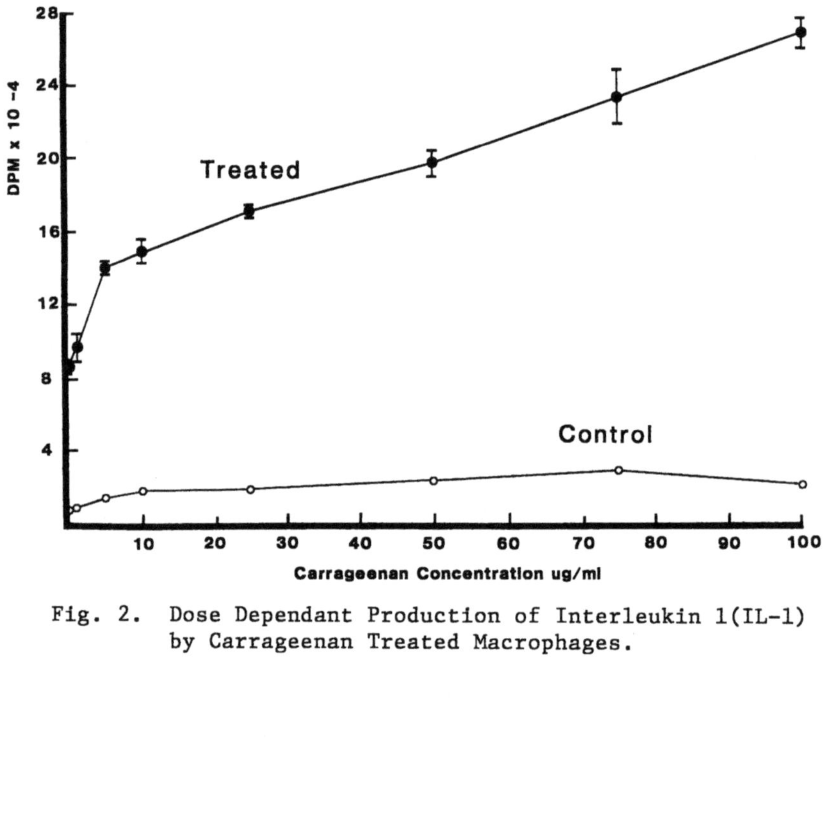

Fig. 2. Dose Dependant Production of Interleukin 1(IL-1) by Carrageenan Treated Macrophages.

calibrate the system. The production of IL-1 by macrophages exposed to graded doses of food-grade iota-carrageenan, as compared with control medium containing carrageenan alone is shown in Fig.2. It is apparent that carrageenan in control medium has a slight stimulatory effect on C3H thymocytes presumably because the macrophages associated with the thymocytes (i.e. of thymic origin) are also stimulated to release IL-1. This system is also responsive to the low level of IL-1 normally released by macrophages in culture, since supernatant obtained from macrophage cultured in the absence of carrageenan was also slightly stimulatory as compared to fresh medium. Nevertheless the addition of microgramme quantities of carrageenan to cultured macrophages caused a marked dose-dependent and highly significant increase of mediator release into the supernatant.

These results, taken together with the finding that conditioned medium was comitogenic with PHA but not mitogenic alone, are consistent with the factor being IL-1. However, they do not exclude the possibility that the mediator is, or is possibly contaminated with, IL-2. To examine this possibility the supernatants were examined for IL-2 activity using the IL-2 dependent CTLL cell line. IL-2 was not detected (data not shown).

DISCUSSION

It is clear from the literature and studies performed at BIBRA that by the correct adjustment of dosage, route and time of administration, carrageenan can enhance both cell mediated and humorally mediated immune responses (see Nicklin, et al., 1985). It is possible that this effect may be a consequence of the ability of carrageenan to initiate IL-1 release by macrophages, which if triggered at the appropriate time during an immune response, may cause preferential amplification of the T-helper population and thereby enhance immune responsiveness. At the same time it is equally clear that carrageenan given prior to antigen either parenterally or orally can lead to reduced immune responsiveness. Whilst it is tempting to speculate that this could be a consequence of IL-1 expanding a regulatory suppressor T cell population which in turn depresses immunity, we have also demonstrated during the course of these experiments that the direct addition of either food grade or degraded iota-carrageenan in fresh (unconditioned) medium to a PHA-induced rat thymocyte transformation assay can result in a potent dose dependent suppression of mitogenesis (Fig.3 & 4).

Since thymocyte cultures contain approximately 0.1 - 1.00% thymic macrophages it is reasonable to assume that carrageenan may initiate the release of an inhibitory factory from these cells. This release is not a consequence of toxicity per se since both non-toxic doses of food grade carrageenan and non-toxic degraded carrageenan were both associated with depression. It therefore appears that carrageenan can cause macrophages to release both inhibitory and stimulatory signals.

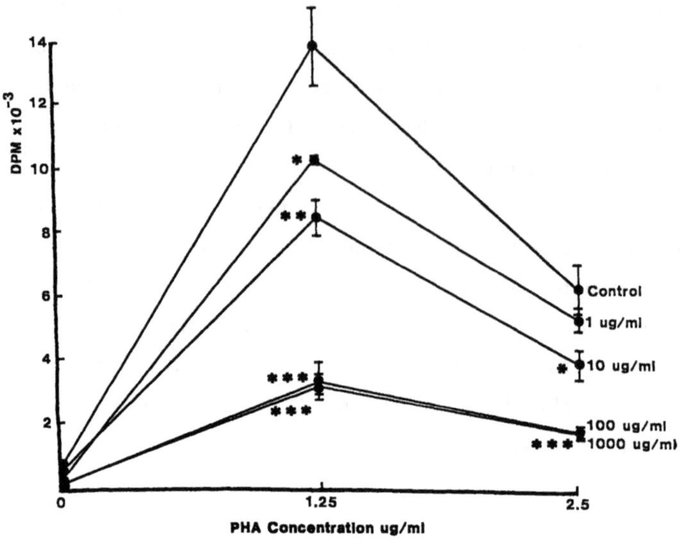

Fig. 3. Effect of Food Grade Iota-Carrageenan on PHA Induced Thymocyte Proliferation.

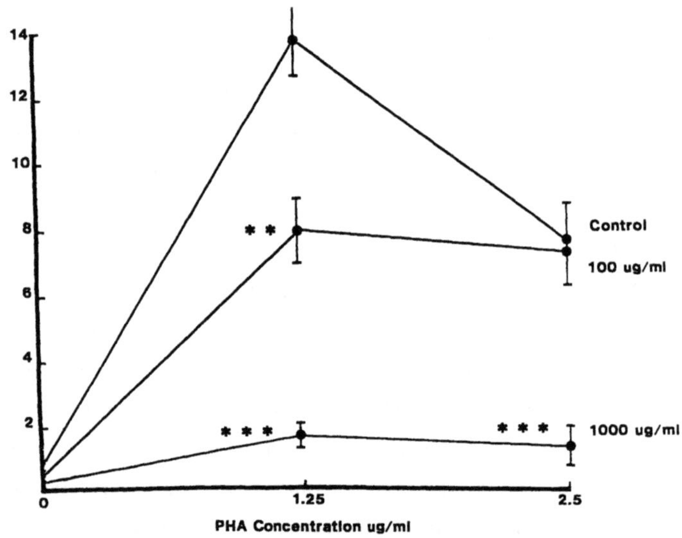

Fig. 4. Effect of Degraded Iota-Carrageenan on PHA Induced Thymocyte Proliferation.

The latter is probably IL-1, however the former, which is yet to be formally identified, is released early in reactions under non-toxic influences. Since dialysed and frozen supernatants from carrageenan treated macrophages have stimulatory properties (presumably mediated by IL-1) it is probable that the inhibitory factor is either dialysable, labile, or both. Prostaglandins particularly those of the E series are obvious contenders for this inhibitory factor.

Prostaglandins have been shown to effectively inhibit a range of *in vitro* immune responses including humoral antibody production (Webb & Nowowiejski, 1977) mitogen induced lymphocyte transformation (Rao, *et al.*, 1979) lymphocyte mediated cytolysis (Henney, *et al.*, 1972) and both the production of leukocyte (Lomniter, *et al.*, 1976) and macrophage migration inhibition factor (Gordon, *et al.*, 1976). Further evidence directly implicating prostaglandins in the role of negative immunoregulatory molecules is based on the ability of inhibitors of prostaglandin synthesis to enhance antibody production (Webb & Nowowieski, 1977) and mitogenicity responses (Vosixa & Thies, 1979). Furthermore Bash & Cochran (1980) reported that a suppressor factor produced by incubating rat spleen cells with carrageenan was inhibited by indomethacin (a prostaglandin synthetase inhibitor). These data taken as a whole strongly suggest a role for prostaglandins in the regulation of immune function. The fact that prostaglandins appear so ubiquitously but act on specific cells at specific times may be attributable to their short biological half-life which in itself makes them ideal candidates for immunoregulatory molecules.

We are now in a better position to explain certain aspects of carrageenan induced modification of normal immune responses. From the above studies it seems likely that the adjuvant properties of carrageenan (Nicklin *et al.*, 1985) stem from the ability of carrageen to function as a protein carrier and at the same time elicit enhanced IL-1 production by antigen presenting macrophages. The immune depressive effects observed when carrageenan is given prior to antigen is more difficult to explain but probably involves carrageenan-induced expansion of a T suppressor rather than T helper cell population, the release of inhibitory prostaglandin molecules and ultimately direct macrophage toxicity - clearly if these events occur prior to antigen presentation then subsequent immune responses will be markedly effected.

A more detailed knowledge of the interaction between carrageenan and the various components of the immune responses is obviously central to our understanding of the complex interactions demonstrated in the present report. However although it is clear that carrageenan can under defined conditions function both as an immunostimulatory and immunedespressive molecule, the precise inter-relationships that occur between carrageenan and the natural regulatory mechanisms within the immune response remains to be determined.

ACKNOWLEDGEMENTS

These studies form part of a research project funded by the U.K. Ministry of Agriculture, Fisheries and Food. The results of this research are Crown copyright.

REFERENCES

Bash, J.A. & Cochran, F.R. J. Reticuloendothel. Soc. 28, 203 (1980).

Bash, J.A. & Vago, J.R. J. Reticuloendothel. Soc. 28, 213 (1980).

Bendtzen, K. (1983). Allergy, 38, 219 (1983).

Chu, E., Rosenwasser, L.S., Dinarello, S.A., Lareau, M. & Geha, R.S. J. Immunol. 132, 1311 (1984).

Cochran, F.R. & Baxter, C.S. Immunology, 53, 291 (1984).

Di Rosa, M. J. Pharm. Pharmac, 24, 89 (1972).

Evelegh, M.J., Clark, D.A. & McCandless, E.L. Immunology letters, 5, 247 (1982).

Gillis, S. & Watson, J.J. Exp. Med 152, 1709 (1980).

Gordon, D., Bray, M.A. & Morley, J. Nature (Lond), 262 (1976).

Henney, C.S., Bourne, H.R. & Lichtenstein, L.M. J. Immun. 108, 1526-1534 (1972).

Ishizaka, S.T. & Stutman, O. Eur. J. Immunol 13, 936 (1983).

Lomnitzer, R., Rabson, A.R. & Koornhof, H.J. Clin. exp. Immun. 24, 42-48 (1976).

Nicklin, S. & Miller, K. Fd Chem. Toxic. 22, 615 (1984).

Nicklin, S., Atkinson, H.A.C. & Miller, K. Int. J. Immunopharmacol. 7, 677 (1985).

Oppenheim, J.J., Scala, G., Kuang, Y.D., Matsushima, K., Sztein, M.B. & Steeg, P.S. Progress in Immunology 5, 285 (1984).

Quan, P.C., Kolb, J.P. & Lespinats, G. Immunol. 40, 495-503 (1980).

Rao, K.M.K., Schwartz, S.A. & Good, R.A. Cell Immun. 48, 155-165 (1979).

Rumjanek, V.M., Watson, S.R. & Sljivic, H. Immunol. 33, 423 (1977).

Smith, G. & Nicklin, S. J. Imm. Methody. 25, 265 (1979).

Vosixa, G. & Thies, J. J. clin. Immun. Immunopath. 13, 30-38 (1979).

Webb, D.R. & Nowowiejski, I. Cell Immun. 33, 1-10 (1977).

IMMUNOREGULATION IN EPIDERMIS: GLUCOCORTICOIDS INDUCED EPIDERMAL CELL DERIVED INHIBITOR OF INTERLEUKIN 1 ACTIVITY

M. L. Lukič, S. Stošić-Grujičić, Dj. Lalošević and S. Vukmanović

Institute of Microbiology and Immunology, School of Medicine, and Institute for Biological Research University of Belgrade, Belgrade, Yugoslavia

Epidermis, in addition to being a surface barrier, actively participates in various immunologic and inflammatory reactions in the skin. It is already well established that dendritic cells within epidermis such as Langerhans cells may function as antigen presenting cells and thereby play an important role during the initiation of an immune response (1). Among immunologically active elements resident in the epidermis the major cell population is the keratinocytes. They are able to produce several nonspecific immunoregulatory mediators including an immunoenhancing cytokine - epidermal cell (EC) derived thymocyte activating factor - which is indistinguishable from macrophage - derived interleukin 1 (IL 1) (reviewed in 2). We have analyzed recently the accessory cell function of mouse and human epidermal cells (ECs) by measuring their capacity to support the proliferative response of lectin-stimulated nonadherent thymic T cells (NATs) (3). NATs exhibit low responsiveness to lectins, while the addition of ECs induced in a dose-dependent way either supportive or suppressive effects. Having in mind that functional capacity of epidermal cells can be altered by various physico-chemical agents we have analysed the effects of glucocorticoids on the accessory function of epidermal cells. We present here evidence that _in vivo_ exposure of ECs to glucocorticoids not only inhibits the production of IL 1 but also favors the production of factors which suppress NAT proliferation. Additionally we have shown the existence of an immunoinhibitory factor(s) produced by ECs which represent(s) an inhibitor of IL 1 - activity.

Glucocorticoids abrogate the immunoenhancing effects of epidermal cells. We have shown previously that macrophage-depleted nylon wool nonadherent thymocytes (NATs) are able to proliferate in response to concanavalin A (Con A) only in the presence of accessory ECs (3). Addition of various numbers of mouse ECs to a constant number of syngeneic NATs resulted in varying degrees of enhancement or inhibition of Con A-induced proliferation. Lower numbers of added ECs progressively increased proliferation until an optimum ratio of ECs to NATs in cell mixture was reached. The addition of supraoptimal number of ECs ($6x10^4$ ECs : $1x10^6$ NATs) resulted in progressive inhibition of Con A-induced proliferation of NATs (data not shown and Fig. 1). Using the same experimental system we analyzed the effects of topical steroid treatment on ECs accessory function. Synthetic preparation of glucocorticoid (tincture containing 2 mg/ml triamcinolone acetonide) or vehicle preparation only were applied topically on the mice ears once daily for 1, 3 or 5 consecutive days. One day after cessation of treatment EC suspension was prepared as descibed elsewhere (3) and cocultivated with 2.5 ug/ml Con A stimulated $1x10^6$ NATs in flat-bottomed microculture plates. In contrast to the cell viability, which remained unaltered, the function of ECs was greatly impaired by in vivo steroid treatment. As can be seen (Fig. 1) the capacity of ECs to support Con A-induced proliferation of thymic T cells was completely abolished. In contrast to control mice, none of the concentrations of ECs from steroid-treated animals induced enhancement regardless to the duration of steroid treatment.

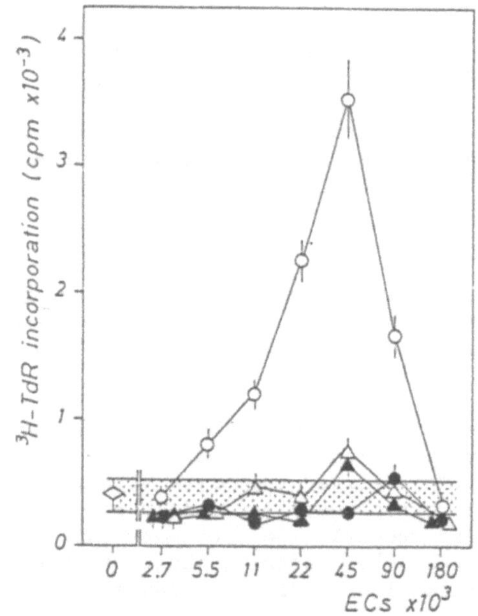

Fig. 1. Effect of in vivo glucocorticoid treatment of ECs on their accessory function. ECs were obtained from Balb/c mice to which steroid preparation was topically applied 1 (●), 3 (△) or 5 (▲) days, or from control mice treated with vehicle only (o). Indicated numbers of ECs were cocultured with Con A-stimulated Balb/c NATs. ^{3}H-TdR incorporation was measured on the third day of culture. Shaded area represents isotope incorporation by Con A-stimulated NATs cultured alone.

In vivo pretreatment with glucocorticoids facilitates the production of immunoinhibitory factors. Immunoenhancing effects of ECs are mediated by their soluble products (4, 3). We considered the possibility that corticosteroid effects observed after topical administration might be due to the impaired production and/or release of soluble factors. Therefore we prepared the supernatants of ECs suspensions obtained from untreated and glucocorticoids treated mice. ECs were cultured ($1x10^6$ cells per ml) in round-bottomed tubes containing 0.5 - 1 ml culture medium (Medium 199 with 20 mM HEPES, 5% heat-inactivated fetal calf serum, $5x10^{-5}$ M 2-mercaptoethanol and antibiotics). After 24 h culture supernatants were harvested by centrifugation, dialyzed and used in the experiments.

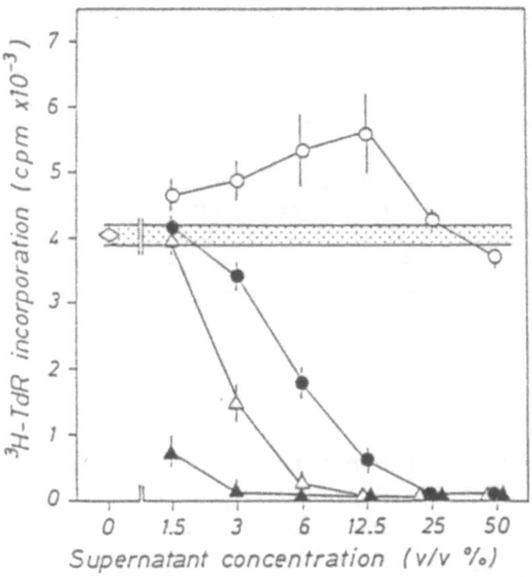

Fig. 2. Effect of in vivo glucocorticoid treatment of ECs on their soluble factor production. ECs were obtained from Balb/c mice to which steroid preparation was topically applied 1 (●), 3 (▲) or 5 (▲) days, or from control mice treated with vehicle only (o). ECs were cultured for 24 h. Their culture supernatants were dialyzed and tested at indicated concentrations on 1×10^6 Con A-stimulated Balb/c NATs supplemented with 1×10^3 irradiated syngeneic peritoneal macrophages. ^{3}H-TdR incorporation of Con A-stimulated NATs supplemented with macrophages only is shown by the shaded area.

As a test population we used NATs supplemented with suboptimal number of 3000 R X-irradiated syngeneic peritoneal macrophages and considered their proliferative response as a baseline (Fig. 2). Supernatants of EC cultures from 1, 3 or 5 day corticosteroid-treated mice induced suppression rather than enhancement of Con A-induced proliferation of NATs. The suppressive activity remained in supernatants even after extensive dialysis, and was proportional to the duration of corticosteroid pretreatment as well as the concentration of supernatant tested.

Glucocorticoid-induced epidermal cell-derived inhibitors interfere with interleukin 1-effects whereas interleukin 2-dependent proliferation is not altered. We attempted to dissect the mechanisms of action of the inhibitory factors (IFs) derived from glucocorticoid pretreated ECs. To this end we analysed the effects of IFs on IL 1- and IL 2-dependent events in T cell proliferation. Therefore we tested the effect of preincubation of NATs with supernatants derived from control ECs (EC-SUP) and from glucocorticoid-pretreated ECs (GC-EC-SUP) on proliferative response of NATs cultured in medium with Con A and 20 ng/ml recombinant (r) IL 1 (kindly donated by Dr Charles Dinarello). As shown in Table 1, GC-EC-SUP in contrast to the effect of control EC-SUP exhibited profound effects in IL 1-promoted lectin-induced proliferation. These results indicated that IFs interfere with IL 1-effects in T cell proliferation rather than on macrophage capacity to produce IL 1.

Further we tested whether IFs act on IL 1-promoted acquisition of responsiveness to IL 2. As shown in Table 2 NATs preincubated in GC-EC-SUP and then pulsed with Con A and rIL 1 for 18 h acquired markedly lower reactivity to recombinant (r) IL 2 (kindly donated by Dr Kendall Smith) when compared with the cells preincubated in control EC-SUP.

As IL 1 is required for IL 2 production we tested the effects of pretreatment with IFs on IL 2 production under optimal conditions. As shown in Table 3 preincubation with GC-EC-SUP containing IFS for 2 h diminished the production of IL 2 after appropriate stimulation with rIL 1 and lectin, while EC-SUP had an enhancing effect.

Table 1. Effects of ECs supernatant on T cell responsiveness to interleukin 1.

NATs preincubated in medium with	^{3}H-TdR incorporation (cpm ± SE)
-	7,977 ± 1,280
EC-SUP	10,510 ± 441
GC-EC-SUP	3,857 ± 267

NATs were preincubated for 2 h in medium alone, or in medium supplemented with supernatant of control ECs or glucocorticoid-treated ECs (2% v/v) as indicated. After preincubation cells were cultured for 64 h in the presence of 2.5 ug/ml Con and 20 ng/ml rIL 1. ^{3}H-TdR was added 16 h prior to cell harvesting.

Table 2. Effects of ECs supernatant on IL 1-promoted Con A-induced acquisition of reactivity to Il 2.

NATs preincubated in medium with	^{3}H-TdR incorporation (cpm ± SE)
-	5,660 ± 247
EC-SUP	6,500 ± 229
GC-EC-SUP	3,390 ± 124

NATs were preincubated for 2 h in medium alone, or in medium supplemented with supernatant of control ECs or glucocorticoid-treated ECs (2% v/v) as indicated. After preincubation cells were pulsed for another 18 h with 2.5 ug/ml Con A in the presence of rIL 1 (20 ng/ml). After that cells were washed in medium containing 0.1 M -MM. IL 2 reactivity of preinmcubated cells was estimated by their cultivation in the presence of standard preparation of 25 U/ml rIL 2. ^{3}H-TdR incorporation was measured during the last 8 h of 2 days culture period.

Table 3. Effects of ECs supernatant on IL 1-promoted Con A-induced production of IL 2

NATs preincubated in medium with	^{3}H-TdR incorporation (cpm ± SE)
-	790 ± 53
EC-SUP	1,247 ± 78
GC-EC-SUP	657 ± 59

NATs were preincubated for 2 h in medium alone, or in medium supplemented with supernatant of control ECs or glucocorticoid-treated ECs (2% v/v) as indicated. Preincubated cells were cultured for 24 h in medium containing 2.5 ug/ml Con A and 20 ng/ml rIL 1. IL 2 activity in culture supernatants was assayed on the IL 2-dependent indicator cells (2×10^4 thymus-derived T cell blasts per well) cultured 48 h in the presence of 50% v/v test supernatant. ^{3}H-TdR was added 8 h prior to cell harvesting.

Finally we tested whether the presence of IFs affects responsiveness of T cell blasts to rIL 2. It was found that neither IFs-containing nor control supernatant did affect response to IL 2 (data not shown).

Thus, we provide the evidence of the existence of an inhibitory cytokine produced by ECs which production is facilitated by glucocorticoids and which interfere with IL 1 effects. A variety of IL 1-inhibitors produced by distinct cells have been described. Most recently Schwarz et al. (5) have shown that UV irradiated epidermal cells produce a specific inhibitor of IL 1 activity. However, they did not analyse the precise mechanism of action of this ECs product. Therefore it remains to be elucidated whether glucocorticoid-induced and UV-induced inhibitors produced by ECs are the same entities. Nevertheless, their existence supports the concept that the epidermis may function as an immunoregulatory tissue.

REFERENCES

1. G. Stingl, K. Tamaki, and S. I. Katz, Immunol. Rev. 53:149 (1980).
2. J. J. Oppenheim, E. J. Kovacs, K. Matsushima, and S. K. Durum, Immun. Today 7:45 (1986).
3. S. Stošić-Grujičić, Dj. Lalošević, and M. Lukić, Int. J. Immunopharmac. (1987) In press.
4. T. A. Luger, B. M. Stadler, S. I. Katz, and J. J. Oppenheim, J. Immunol. 127:1493 (1981).
5. T. Schwarz, A. Urbanska, F. Gschnait, and T. A. Luger, J. Immunol. 138:1457 (1987).

MONOCYTE DEFECT CAUSES DECREASED AUTOMLR IN MULTIPLE SCLEROSIS PATIENTS

C.N. Baxevanis, G.J. Reclos, P. Arsenis, A. Katsiyiannis, N. Matikas* and M. Papamichail

Department of Immunology, Hellenic Anticancer Institute, Athens
*Neurology Department, Evangelismos State Hospital, Athens

SUMMARY

We have recently demonstrated that peripheral blood monocytes from patients with multiple sclerosis (MS) have a defect in stimulating autologous and allogeneic T lymphocytes. This defect was found to correlate with disease activity. We report here that T_4^+ cells from MS patients proliferate weakly in the autologous mixed lymphocyte reaction (autoMLR). Furthermore, we provide direct proof that the depressed T_4^+ cell proliferation is due to the monocyte functional (stimulatory) defect.

INTRODUCTION

Multiple sclerosis (MS) has been reported to be associated with disorders of T cell regulatory functions[1] and decreased percentage of suppressor T cells in the peripheral blood and the cerebrospinal fluid[2], although the latter has not been a consistent finding[3]. Decreased autoMLR responses have been previously reported in patients with MS[4,5]. We have recently demonstrated that monocytes from MS patients have a defect in stimulating autoMLR which correlated with disease activity[6,7]. We report in this study that the MS-monocyte defect influences T_4^+ cell proliferation in the autoMRL. The importance of this finding in understanding the aetiology of immune abnormalities in MS patiens is discussed.

MATERIALS AND METHODS

Subjects. Thirty-five patients with MS were studied. All patients were examined in the Neurology department of Evangelismos State Hospital, Athens. Most MS patients had active disease, seven patients had started a relapse less than one month before study, and 5 patients had a progressive form. Fourteen patients with active disease were also tested in remission. Staff members served as healthy controls

Cell preparation. Highly purified T cells or monocytes were isolated from peripheral blood as described previously[6,7].

Preparation of T-cell subsets. T_4^+ or T_8^+ cells were obtained through negative selection by using the monoclonal antibody (MoAb) OKT-8 (cytotoxic/suppressor) or OKT-4 (helper/inducer; Ortho Laboratories) respectively as previously described[8].

AutoMLR. This has been described in detail elsewhere[6,7]. In brief, responder T cells (or T-subsets) ($1x10^5$ cells/well) were cultured with irradiated (3300 rad) autologous stimulatory monocytes ($2x10^5$ cells/well) in flat-bottomed microtiter plates in 0.2 ml of RPMI-1640 supplement with 10% pooled, heat-inactivated human AB serum, antibiotics and L-glutamine for 6 days. Proliferation was measured by ^{3}H-TdR uptake.

Cryopreservation of PBL. PBL from MS patients were cryopreserved in liquid nitrogen[9]. After thawing, the cells were incubated for at least 1 h at room temperature in culture medium supplemented with 10% pooled human AB serum.

Statistical analysis. Data were analysed with Student's t-test.

RESULTS

T cells from MS patients demonstrated a significant proliferating defect in response to autologous stimuli in comparison to normal controls (Table 1). This defect was attributed to the T_4^+-cell subset (Table 1). T_8^+ cells as also shown by others[10] did not proliferate in the autoMLR. We further investigated whether the functional abnormality of T_4^+ cells in the autoMLR was secondary to the MS monocyt stimulatory defect. To test this we conducted the following experimentation: peripheral blood lymphocytes (PBL) from seven patients with active MS were preserved in liquid nitrogen. Eight months later when these patients were in remission (they had been stable for the last six months) PBL were again collected. PBL from both stages of MS were then used for the preparation of T_4^+ cells and monocytes. These cells were then used as responders (T_4^+) and stimulators (monocytes) in the autoMLR. All four combinations were tested (Table 2, upper part). Similarly, cryopreserved PBL from 7 MS patients with inactive disease (they had been stable for the last six months) were tested in the autoMLR with autologous PBL from an acute relapse (Table 2, intermediate part). As a control, PBL from 10 healthy subjects were cryopreserved and tested in the autoMLR with fresh autologous PBL in parallel with the MS patients (Table 2, lower part). The data clearly demonstrate that MS monocytes dictate T_4^+ cell proliferation in autoMLR. Thus, when MS monocytes (used as stimulators) were harvested from an active stage of MS, T_4^+ cell responses were significantly lower compared to that with MS monocytes from an inactive stage of the disease. These data prove that the decreased proliferative responses of MS-T_4^+ cells in the autoMLR are due to the MS-monocyte stimulatory defect which fluctuates with disease activity.

DISCUSSION

In this report we directly prove that the defective autoMLR in MS patients is caused by a functional (stimulatory) defect of MS monocytes. Thus, T_4^+ cells tested with autologous stimulatory monocytes harvested from an inactive stage of MS, exhibited considerably higher proliferative responses than when stimulated with autologous monocytes obtained from an acute relapse (active MS). A decreased autoMLR in MS patients was also demonstrated by others[4,5]. However, in those studies the stimulatory cell-population consisted of a variety of cells (non-T cells) so that the reason for low responses in the autoMLR was rather attributed to the responder T cells (purified T_4^+ cells) than to the monocytes. A decreased autoMLR response has been demonstrated in other diseases of suspected autoimmune aetiology[11,12]. A depressed autoMLR response in these diseases may have

Table 1. MS-T cell and T-cell subset responses in the autoMLR

Responders v.s. monocytes	cpm
MS patients (n=35)	
T	8,582+2,658[a] (874+150)[b]
T_4^+	12,288+3,956[a] (1,240+352)
T_8^+	2,918+1,013 (976+280)
Normal subjects (n=39)	
T	22,758+6,012 (915+302)
T_4^+	34,792+5,981 (1,503+658)
T_8^+	3,272+976 (1,456+472)

[a] $p<10^{-3}$ (compared to the respective values of normal subjects).
[b] Indicates background values.

Table 2. MS monocyte defect causes inability of T_4^+ cells to proliferate in the autoMLR

Source of responding T_4^+ cells	Stage of stimulating monocytes	cpm
MS-active[a] (n=7)	MS-active[a]	6,932+1,959
	MS-inactive	18,764+3,520
MS-inactive (n=7)	MS-inactive	17,212+3,840
	MS-inactive[a]	7,220+2,021
MS-inactive[a] (n=7)	MS-inactive[a]	14,852+3,288
	MS-active	8,202+2,520
MS-active (n=7)	MS-active	7,949+1,923
	MS-inactive[a]	15,850+4,385
Normal Subjects[a] (n=10)	Normal[a]	27,872+7,894
	Normal	24,964+6,312
Normal Subjects (n=10)	Normal	26,298+7,820
	Normal[a]	27,008+6,982

[a] Indicated T_4^+ cells or monocytes which had been cryopreserved. These cells were thawed and tested with fresh autologous cells in the autoMLR as indicated above

major significance to the pathogenesis or progression of disease. First, in the type of autoMLR studied here, in which monocytes are used as stimulators, the majority of responding cells have the T_4^+ phenotype[4,5,10]. Furthermore, different subsets of T_4^+ cells which proliferate in the autoMLR develop both helper/inducer and suppressor cell activity. Since MS patients have depressed suppressor T cell function, especially during active disease[2], it is possible that the decreased proliferative activity seen in the autoMLR represents the inability of MS monocytes to stimulate MS-T_4^+-cells to proliferate and generate suppressor cell function. Since the previously reported decreases in T_8^+ cells in active MS patients have not been confirmed in several laboratories it is possible that the depression of suppressor cell activity in active MS is associated with a population of cells other than the T_8^+ cell. The precise characterization of the MS monocyte defect might explain the aetiology of MS.

REFERENCES

1. B. G. W. Arnason and J. Antel, Suppressor cell function in multiple sclerosis, Ann. Immunol. (Paris), 12a C:159 (1978).
2. H. L. Weiner and S. L. Hauser, Neuroimmunology I: Immunoregulation in neurological disease, Ann. Neurol. 11:437 (1982).
3. R. L. Hirsch, J. Ordonez, H.S. Panitch, and K. P. Johnson, T_8 antigen density on peripheral blood and lymphocyte remains unchanged during exacerbations of multiple sclerosis, J. Neuroimmunol. 9:391 (1985).
4. R. L. Hirsch, Defective autologous mixed lymphocyte reaction in multiple sclerosis, Clin. Exp. Immunol. 64:107 (1986).
5. D. A. Hafler, M. Buchsbaum, and H. L. Weiner, Decreased autologous mixed lymphocyte reaction in multiple sclerosis, J. Neuroimmunol. 9:339 (1985).
6. C. N. Baxevanis, G. J. Reclos, C. Sfagos, E. Doufexis, C. Papageorgiou, and M. Papamichail, Multiple sclerosis: I. Monocyte stimulatory defect in mixed lymphocyte reaction associated with clinical disease activity, Clin. Exp. Immunol. 67:362 (1987).
7. G. J. Reclos, C. N. Baxevanis, C. Sfagos, C, Papageorgiou, G. C. Tsokos, and M. Papamichail, Multiple sclerosis: II. Effects of Prothymosin a on the autologous and allogeneic MLR in patients with multiple sclerosis, Clin. Exp. Immunol. 70:336 (1987).
8. C. N. Baxevanis, G. J. Reclos, S. Perez, D. Kokkinopoulos, and M. Papamichail, Immunoregulatory effects of fraction 5 thymus peptides. I. Thymosin a_1 enhances while thymosin b_4 suppresses the human autologous mixed lymphocyte reaction, Immunopharmacology 13:133 (1987).
9. A. A. Bom-Van Noorloos, A. A. Van Beek, and C. J. M. Melief, Cryopreservation cells for immunological typing of non-Hodgkin-lymphomas, Cancer Res. 40:2890 (1980).
10. J. F. Smolen, T.A. Luger, T. M. Chused, and A. D. Steinberg, The human autologous mixed lymphocyte reaction, J. Immunol. 129:1051 (1982).
11. T. Sakane, H. Kotani, S. Takada, Y. Murakawa, and Y. Ueda, A defect in the suppressor cicuits among OKT4+ cell populations in patients with systemic lupus erythematosus occurs independently of a defect in the OKT8+ suppressor T cell function, J. Immunol. 131:753 (1983).
12. K. G. Chandy, A. M. Charles, A. Kershnar, B. Buckingham, N. Waldeck, and S. Gupta, Autologous mixed lymphocyte reaction in man. X. Cellular and molecular basis of deficient autologous mixed lymphocyte response in insulin-dependent diabetes mellitus. J. Clin. Immunol. 4:424 (1984).

MORPHOLOGICAL STUDIES ON MITOGEN AND ANTIGEN PRESENTATION BY HUMAN AND RABBIT CELLS

B.M. Balfour, A. Gautam*, J. Clark, T. Goscicka**
S. Griffiths, P. Fryer, A.D.B. Webster, and
J. Farrant

Clinical Research Centre, Harrow, UK, *Institute of Nurology, London UK, *** University of Lodz, Poland

MATERIALS AND METHODS

Mitogen presentation was studied using human peripheral blood cells from normal subjects. The technique involved defibrination of the blood, separation of the mononuclear cells on Ficoll, followed by removal of adherent macrophages. The non-adherent cells were separated on a metrizamide gradient into a low density fraction enriched for dendritic cells (LDC) and a high density fraction containing lymphocytes.(1) The lymphocytes were further separated into T and non-T fractions by rosetting with neuraminidase-treated sheep red blood cells (SRBC). The LDC fraction was pulsed with pokeweed mitogen (PWM) 100 ng/ml and mixed with autologous lymphocytes in the ratio of one LDC to ten lymphocytes. The lymphocyte fraction contained T and non-T cells in the ratio of two to one.

The mixture of cells was observed on the warm stage of an inverted phase contrast microscope. Clustering of lymphocytes around LDC occurred and the number of clusters was recorded at intervals up to 24 hours after mixing the cells. In some experiments the cells were filmed by time-lapse cinemicrography, then fixed and embedded for examination in the electron microscope. In this technique the same cluster was followed throughout the procedures.(2)

RESULTS

When first isolated some of the LDC put out small processes but they did not move directionally. After being pulsed with the mitogen they put out larger processes, but still did not move directionally. Cluster formation was first observed three hours after mixing the cells when the mean number was 3.6/1000 LDC. The maximum number for any one individual was 16/1000 and this was observed at 6 hours, when the mean was 8.3/1000, thereafter the number of clusters declined and at 24 hours the mean was 3.5/1000. Fig. 1. upper graph.

The mode of formation of the clusters was also studied. Two LDC already very close were observed to fuse together and immediately afterwards, three lymphocytes already joined together and also very close, attached themselved to one or other LDC forming a tight cluster. This process took 2-3 minutes. Although the constituent cells, both LDC and lymphocytes were very active, extending their processes in all directions, the whole cluster moved sluggishly or simply rotated on its axis. When first formed the clusters were often visited by other very active

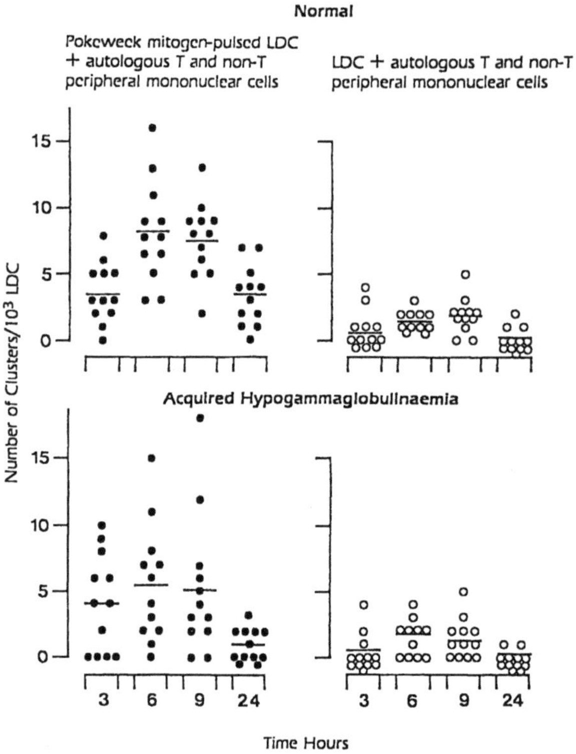

Fig.1 Cluster formation recorded at 3,6,9 and 24 hours after mixing PWM-pulsed or untreated LDC plus autologous T and non-T cells. Number of clusters/ 1000 LDC. Closed circles = LDC pulsed with PWM. Open circles = untreated LDC. Upper graph cells from normal subjects. Lower graph cells from patients with AHG.

lymphocytes which moved towards them, circled round and then went away. Later cohorts of even more active cells arrived, burrowed their way into the clusters, sometimes dislodging one of the LDC. Later some clusters began to move rather slowly towards nearby clusters and sometimes coalesce with them.

In one group of experiments the LDC fraction was prepared as described above, but the lymphocytes were not separated into T and non-T cells. In these mixtures clustering around the LDC was minimal, although a few large aggregates were formed at later times.

Mitogen-induced clusters were also examined in the electron microscope and it was observed that the ultrastructure of the two presenting cells was different. One contained a number of membrane-bound inclusions with a high lipid content. The other cell had relatively few inclusions, but a large array of rough endoplasmic reticulum, arranged in the form of concentric rings. Fig. 2A. These two cells were in intimate contact with each other and the lymphocytes were attached to them by a series of small interdigitations extending over a considerable part of the cell surface. Fig. 2B. The large aggregates which were formed in cultures containing PWM-pulsed LDC and unseparated lymphocytes were also examined. In these clusters the contacts between the cells were never very close and interdigitations between LDC and lymphocytes were not seen.

Mitogen-induced clusters differed in many respects from antigen-induced clusters already observed in cultures containing antigen-pulsed veiled cells and augologous lymphocytes from rabbits immunized with Keyhole Limpet haemocyanin (KLH) or human gammaglobulibn (HgG) (3). In antigen-induced clusters the veiled cells put out large processes, often moved directionally and sometimes turned towards lymphocytes. Clustering of lymphocytes about veiled cells was slower and first

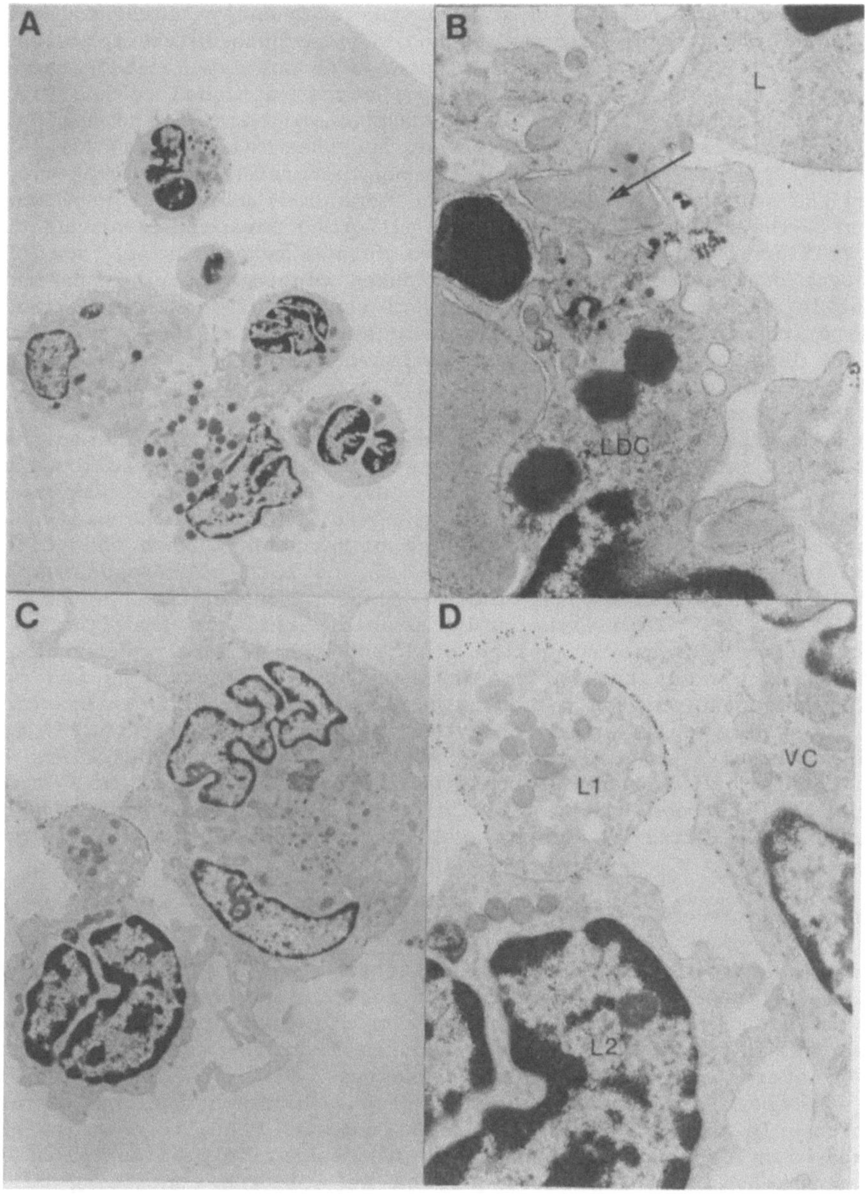

Fig.2A Electron micrograph of cell cluster from a culture containing PWM-pulsed LDC and autologous PBL from a normal individual. Two large LDC of dissimilar type are present and in intimate contact with each other. x 3,000. Fig.2B High power view of area of contact between LDC and lymphocyte (L). The lymphocyte has inserted a process into the LDC (arrow). x 19,000. Fig.2C Electron micrograph of cell cluster containing HgG-pulsed veiled cell plus autologous lymphcytes from a rabbit immunized with HgG. x 6,000. Fig.2D High power view showing monoclonal antibody staining of rabbit Ia antigen plus immuno-gold. Accumulation of gold particles along line of contact between VC and L1 and absence of gold in contact between L1 and L2. x 14,000.

observed 3-6 hours after mixing the cells, but the number of clusters was still increasing at 24 hours. In these cultures, particularly those containing KLH-immune cells, the lymphocytes frequently aggregated with each other, forming trios or quartets and these groups interacted with a single veiled cell forming a stable cluster with one lymphocyte attached to the veiled cell and the other two lymphocytes attached to the first cell by their uropods. Interactions between a veiled cell and a single lymphocyte were also frequently observed, but these did not usually last longer than half an hour. In HgG immune cultures single lymphocytes attached themselves to a veiled cell and some of these bonds were more permanent. The attached cell sometimes performed peripolesis around the veiled cell until it was joined by two more lymphocytes to form the stable type of cluster. The antigen-induced clusters were always much more mobile than the mitogen-induced clusters, often moving directionally at considerable speed. They were also visited by other active cells, but less frequently than mitogen-induced clusters. Occasionally a large cell would contact a cluster and proceed to dislodge all the lymphocytes attached to a veiled cell.

In some experiments antigen-induced clusters were first filmed then stained with a monoclonal antibody against rabbit Ia antigen followed by a gold-labelled anti-mouse globulin antibody. The veiled cell was found to be Ia^+, also the first attached lymphocyte, but the other attached lymphocytes were Ia^-. Fig. 2C. The line of contact between the veiled cell and the first lymphocyte was also Ia^+ and both cells appeared to contribute to this, whereas the line of contact between the lymphocytes was Ia^-. Fig. 2D. This distribution probably reflects the fact that the interaction between the veiled cell and the first lymphocyte was Ia-restricted, whereas the interaction between the lymphocytes was not (4). In mitogen-stimulated responses on the other hand the interaction between presenting cells and lymphocytes was not MHC-restricted (5) and preliminary results suggest that LDC expressing almost no HLA Class II antigen interact with Class II positive lymphocytes. This would imply that the bond between these cells did not depend on a reaction of identity between Class II antigen molecules expressed by both interacting cells.

Immunodeficiency

Peripheral blood cells from patients with acquired hypogammaglobulinaemia (AHG) (6,7) were examined in the same way as cells from normal individuals. Initially cells from six patients with AHG were examined and it was found that lymphocytes from five of the patients formed clusters around PWM-pulsed autologous LDC. The LDC were also tested for their ability to present PWM and induce polyclonal B-cell activation. In this test the LDC were pulsed with the mitogen as before and mixed with T and non-T lymphocytes. The secretion of immunoglobulin was measured by the ELISA technique and in cultures containing cells from normal individuals a significant increase in immunoglobulin secretion was observed. When LDC from the six AHG patients were pulsed with PWM and mixed with lymphocytes from normal individuals there was no increase in immunoglobulin secretion. Since PWM activation is not MHC-restricted this result established that the LDC themselves were functionally defective (8,9). Cells from another 6 AHG patients were examined for their ability to form clusters with PWM and again only one failed to form clusters. Taking the results from the whole group of twelve patients the number of clusters formed at each time point was not significantly different from the number formed by cells from normal subjects, except at 24 hours after mixing the cells, when the mean number was 1/1000 as against 3.5/1000 LDC for normal cells. Figure 1, lower graph.

In preliminary experiments time-lapse cinemicrographic records of cluster formation by cells from AHG patients showed that these clusters did not coalesce in the same way as clusters formed by cells from normal individuals.

The LDC fraction from a normal subject and from one of the AHG patients whose cells were able to form clusters with PWM were examined in the electron microscope and found to be very similar. Fig. 3.1 and 3.2. Both groups of cells had kidney-shaped nuclei, fairly dense cytoplasm containing very few vacuoles or membrane-bound inclusions some of the cells had long processes with irregular outlines. By contrast the LDC from the two patients whose cells failed to form clusters, were both very abnormal. In one case the LDC had convoluted nuclei and phagocytic vacuoles containing cellular debris as well as vacuoles with electron-lucent contents. Fig. 3.3. The lymphocyte fraction from this patient was also abnormal and contained lymphoblasts and multinucleate cells. The LDC from the other patient had even more convoluted nuclei and electron-lucent cytoplasm. Many cells also had large lipid inclusions, while others had enormous vacuoles, most of which appeared to be empty. These cells had extremely long veils with smooth outlines. Fig. 3.4.

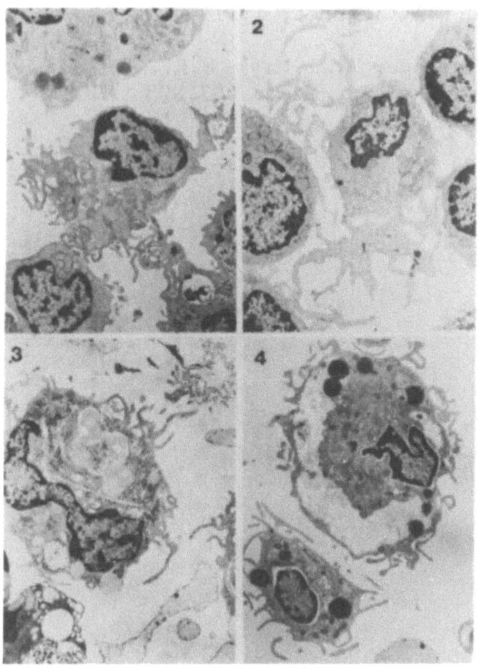

Fig. 3.1 Electron micrograph of LDC from a normal subject. 3.2 LDC from a patient with AHG whose cells formed clusters around PWM-pulsed autologous LDC. 3.3 and 3.4 LDC from two patients with AHG whose cells failed to form clusters around autologous PWM-pulsed LDC. DLC from all AHG patients were functionally defective. x 3750.

The results showed that within AHG there were at least two categories of patients, those whose LDC were morphologically normal and able to form clusters with PWM and those whose LDC were morphologically abnormal and unable to form clusters. Although the LDC from both groups of patients exhibited a functional defect, the majority of cases were of the first type, in that clusters were formed, but they were less stable than those formed by normal cells, therefore the defect in the presenting cells was operative in the later phase of the response to PWM. In the second group of AHG patients the obvious structural abnormality of the presenting cells was associated with a complete inability to form clusters.

References

1. Knight S.C., Farrant J., Bryant A., Edwards A.J., Burman S., Lever A.M.L., Clark J., and Webster A.D.B. Immunology, 57:595 (1986).

2. Balfour B.M., Clark J., Farrant J., and Gautam A. J. Physiol., 384:13P (1986).

3. Balfour B.M., Buttifant L., O'Brien J., Clarke J., and Knight S.C. Advances in Experimental Medicine and Biology, 186:395 (1985) Ed. G.G. Klaus, Plenum Press, New York.

4. Braendstrup O., Werdelin O., Shevach E.M., and Rosenthal A.S. J. Immunol., 122:1608 (1979).

5. De la Concha E.G., Oldham G., Webster A.D.B., Asherson G.L., and Platts-Mills T.A.E. Clin. Exp. Immunol., 27:208 (1977).

6. Cooper A.M., and Lawton A.R. Am. J. Path., 69:513 (1972).

7. Choi Y.S., Biggar W.D., and Good R.A. Lancet, i:1149 (1972).

8. Eibl M.M., Mannhalter J.W., Zlabinger G., Mayr W.R., Tilz G.P., Ahmed R., and Zielinski C.C. New Engl. J. Med., 307:803 (1982).

PHYSICAL INTERACTION BETWEEN DENDRITIC CELLS AND LYMPHOCYTES IS NOT LFA-1 MEDIATED

Eliane Buss, Heinz-Günter Thiele and Alf Hamann

Abt. f. klin. Immunologie, I. Med. Klinik, Univ.krankenhaus
2 Hamburg 20, F.R.G.

INTRODUCTION

Dendritic cells from spleen, lymph nodes and other organs have been found to represent the most efficient accessory cell population for the induction of a primary immune response. Interaction with lymphocytes is thought to take place via direct cell-cell contact and indeed a striking property of dendritic cells is their ability to form clusters with lymphocytes containing that T-cell fraction becoming activated subsequently (1).

Activation of T-cells not only requires presence and proper combination of T-cell receptor, antigen and class II- molecules, but in general is additionally dependent on the function of accessory molecules such as LFA-1 and L3T4, suggested to act by mediating adhesion between the cells. Especially LFA-1 was found to have an apparently universal role in different systems of lymphocyte interactions such as cytotoxic T-cell function or T-cell activation (2,3) and homing-related lymphocyte/high endothelium adherence (4).

It was tempting to suggest, that the role of LFA-1 in activation could be to establish bonds between accessory cells and lymphocytes. We tested therefore, wether binding between lymphocytes and dendritic cells (cluster formation) is prevented by antibodies against LFA-1.

METHODS

Dendritic cells (Fig. 1) were isolated from lymph nodes and thymus of mice by density separation on BSA after irradiation (5). In some experiments the isolation method of Steinman for DCs of spleen (6) using 2 h -adherence and density separation was applied. Fc-receptor positive cells were removed by rosetting with Ig-SRBC. Lymphocytes were activated by 1 mM periodate, 15 min at 4 °C, and cultivation for 2 days.

To analyze the formation of clusters between dendritic cells and lymphocytes, the latter cells were labeled with the fluorescent dye H 33342 (Bisbenzimide, (7)) or fluoresceine isothiocyanate (8). A mixture of both populations was allowed to aggregate for 3 h on a rocking platform in the presence or absence of 50µg/ml anti-LFA-1 antibody (H-35.89.9 (3), Fab). Resulting clusters were analyzed as shown in the scheme. The same approach was used for the evaluation of divalent cation dependency, except that the

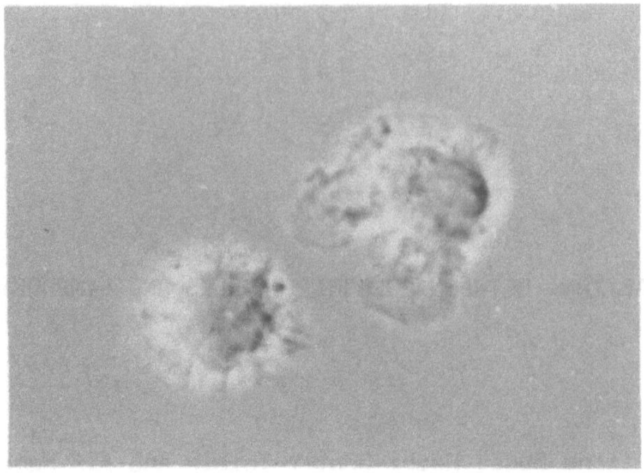

Fig. 1. Purified dendritic cells. Phase contrast (1200x), cells fixed by glutardialdehyde at 37°C.

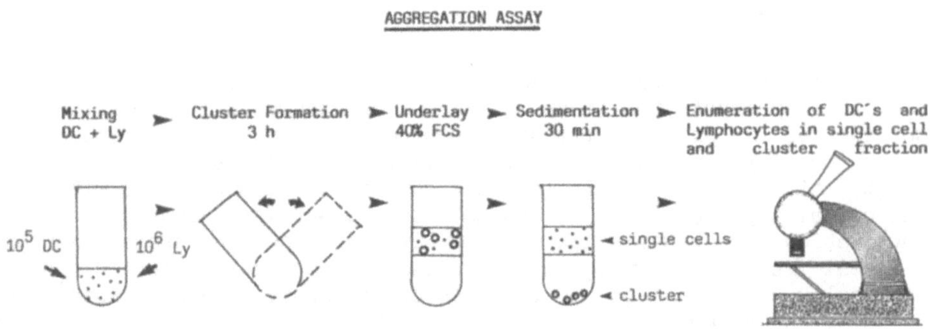

cells were washed in 0.1 mM EDTA before incubation in Hank's BSS + 1% BSA and the additions indicated.

RESULTS

As shown in Fig. 2 and 4, the formation of mixed clusters between activated lymphocytes and dendritic cells is only slightly impaired by antibodies against LFA-1. Most dendritic cells adhere to lymphocytes even in the presence of concentrations of anti LFA-1 more than fivefold higher than that preventing homotypic adhesions (cluster formation, Fig. 3) between activated lymphocytes (9). The same is found when analyzing unstimulated lymphocytes + dendritic cells (not shown). Since lymphocytes are present in a tenfold escess over dendritic cells, it is conceivable that the small inhibitory effect in the quantitative assay and the reduced size of clusters in the presence of anti LFA-1 antibody is due to this inhibition of lymphocyte-lymphocyte adhesion in the mixed clusters and not to an impairment of the adherence of lymphocytes to dendritic cells.
Contact interaction between lymphocytes and dendritic cells therefore seems not to rely primarily on a mechanism involving LFA-1..

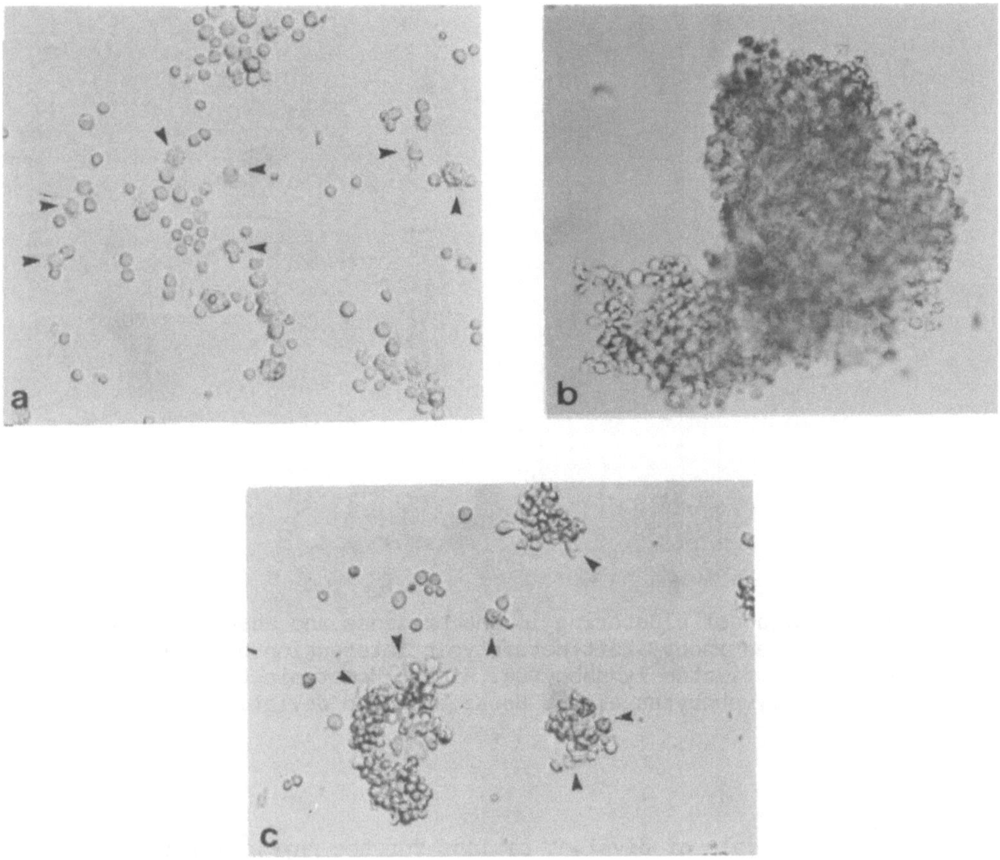

Fig. 2. Effect of anti LFA-1 antibody on the cluster formation between dendritic cells (➤) and (periodate-activated) lymphocytes (130x). a) Cells before incubation. b) After 3 h incubation; large mixed clusters have formed. c) Incubation in the presence of anti LFA-1 antibody (50 µg/ml H.35.89.9, Fab); aggregates are smaller, but the majority of dendritic cells are found within the clusters.

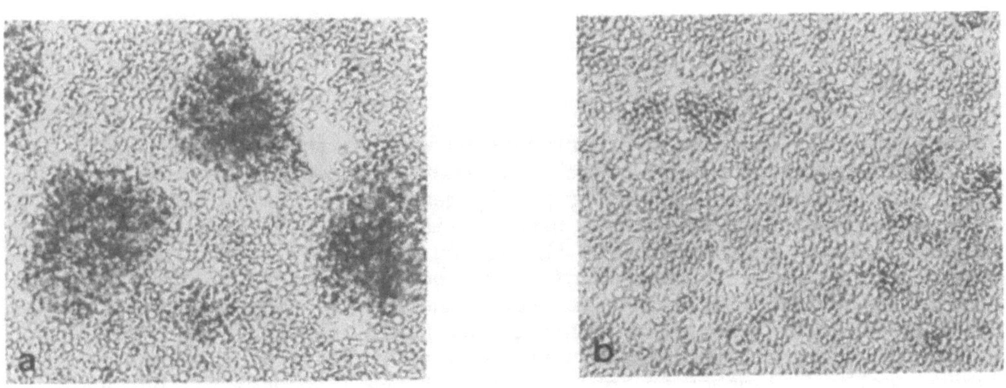

Fig. 3. Clustering of activated lymphocytes, no dendritic cells added (90x) a) Control. b) In the presence of 20 µg/ml anti LFA-1.

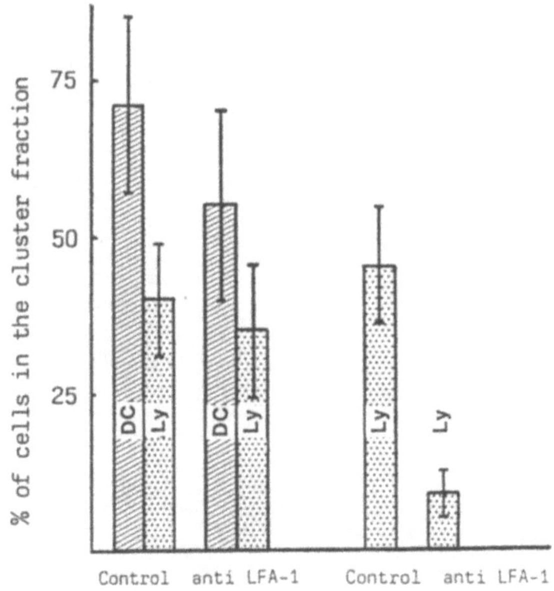

Fig. 4. Determination of clustering in the presence and absence of 50µg/ml anti LFA-1 antibody. Left:Heterotypic interaction between dendritic cells and activated lymphocytes. Right: Homotypic clustering of activated lymphocytes alone. Bars: Standard deviation (n=6).

In Fig. 5 the role of divalent cations for the physical interaction of lymphocytes with dendritic cells is analyzed. The formation of mixed clusters between dendritic cells and lymphocytes (either fresh or activated) is partially inhibited by EDTA. However, the aggregation capability is restored as well by calcium as by magnesium, in contrast to the LFA-1 mediated homotypic interaction between activated lymphocytes, which is supported by magnesium only (10). Dependency on magnesium has been shown to be a characteristic feature of LFA-1 mediated interactions as e.g. cytolytic T-cell/target cell conjugate formation (11).

CONCLUSIONS

The antibody inhibition experiments as well as the analysis of divalent cation requirements reveal that cell contact formation between dendritic cells and lymphocytes is not primarily mediated by LFA-1. Similar results using spleen dendritic cells have been published recently (12).

This finding contrasts with the strong inhibitory action of anti LFA-1 antibodies in several systems of T cell activation (2,3,13,14). Supposing that cell contact formation reflects the functionally important role of LFA-1 in cellular interaction, this result could indicate that T-cell activation in systems strongly inhibitable by anti-LFA-1 requires not only accessory cell / lymphocyte interaction (not LFA-1 dependent), but additionally lymphocyte / lymphocyte interaction (LFA-1 dependent). Alternatively, LFA-1 had to fullfil further functions in T-cell activation in addition to its well established role as an adhesion strenghtening mechanism.

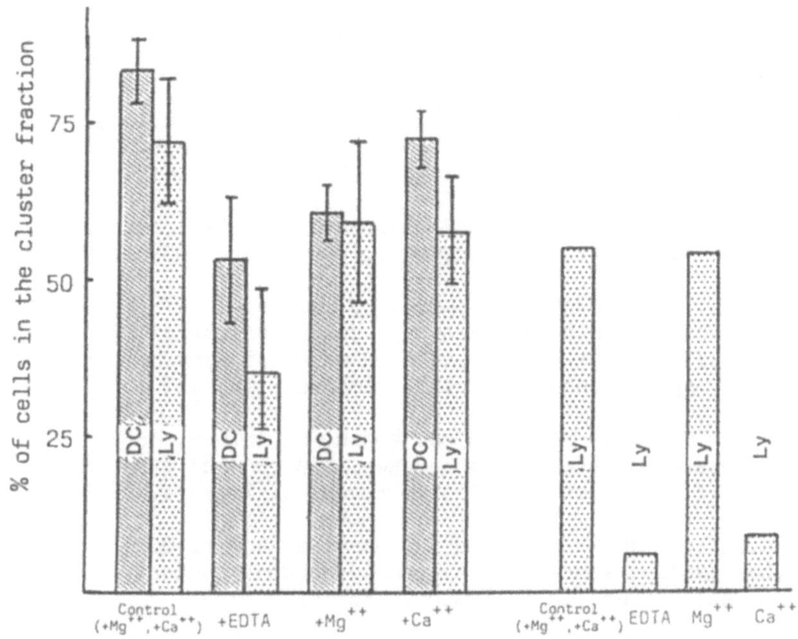

Fig. 5. Role of divalent cations for dendritic cell-lymphocyte clustering. Left: Heterotypic interaction of washed dendritic cells+activated lymphocytes in the presence of 1 mM Mg++ and Ca++, EDTA, or either Mg++ or Ca++. Right: The same for activated lymphocytes alone.

ACKNOWLEDGEMENTS

We thank W.E.F. Klinkert for valuable help in preparing dendritic cells and M. Pierres for kindly supplying anti LFA-1 antibody producing hybridomas. This work was supported by the Deutsche Forschungsgemeinschaft.

REFERENCES

1. Inaba, K., M.D. Witmer, and R.M. Steinman. 1984. Clustering of dendritic cells, helper T lymphocytes, and histocompatible B cells during primary antibody responses in vitro. J. Exp. Med. 160:858.

2. Davignon, D., E. Martz, T. Reynolds, K. Kürzinger, and T. A. Springer. 1981. Monoclonal antibody to a novel lymphocyte function-associated antigen (LFA-1): mechanism of blockade of T lymphocyte-mediated killing and effects on other T and B lymphocyte functions. J. Immunol. 127:590.

3. Pierres, M., C. Goridis, and P. Golstein. 1982. Inhibition of murine T cell-mediated cytolysis and T cell proliferation by a rat monoclonal antibody immunoprecipitating two lymphoid cell surface polypeptides of 94000 and 180000 molecular weight. Europ. J. Immunol. 12:60.

4. Hamann, A., D. Jablonski-Westrich, A. Duijvestijn, E.C. Butcher, H. Baisch, R. Harder, and H.-G. Thiele. Evidence for an accessory role of LFA-1 in lymphocyte-high endothelium interaction during homing. J. Immunol. in press.

5. Klinkert, W.E.F., J.H. LaBadie, K.P. O'Brien, C.F. Beyer, and W.E.

Bowers. 1980. Rat dendritic cells function as accessory cells and control the production of a soluble factor required for mitogenic responses of T lymphocytes. Proc. Natl. Acad. Sci. USA 77:5414.

6. Nussenzweig, M.C., R.M.Steinman, B. Gutchinov, and Z.A.Cohn. 1980. Dendritic cells are accessory cells for the development of anti-trinitrophenyl cytotoxic T lymphocytes. J. Exp. Med. 152:1070.

7. Brenan, M. and C. R. Parish. 1984. Intracellular fluorescent labeling of cells for analysis of lymphocyte migration. J. Immunol. Meth. 74:31.

8. Butcher, E. C., R. G. Scollay, and I. L. Weissman. 1979. Lymphocyte adherence to high endothelial venules: characterization of a modified in vitro assay and examination of the binding of syngeneic and allogeneic lymphocyte populations. J. Immunol. 123:1996.

9. Hamann, A. , Jablonski-Westrich, D. and H.-G. Thiele. 1986. Contact-interaction between lymphocytes is a general event following activation and is mediated by LFA-1. Europ. J. Immunol. 16:847

10. Hamann, A.,D. Jablonski-Westrich, A. Raedler, and H.-G. Thiele. 1984. Lymphocytes express specific antigen-independent contact interaction sites upon activation. Cell. Immunol. 86:14.

11. Martz, E. LFA-1 and other accessory molecules functioning in adhesions of T and B lymphocytes. 1986. Human Immunol. 18:3.

12. Inaba, K. and R. Steinman. 1987. Monoclonal antibodies to LFA-1 and to CD4 inhibit the mixed leucocyte reaction after the antigen-dependent clustering of dendritic cells and T-lymphocytes. J. Exp. Med. 165:1403.

13. Regnier-Vigouroux, A., D. Blanc, S. Pont, S. Marchetto, and M. Pierres. 1986. Accessory molecules and T cell activation. I. Antigen receptor avidity differentially influences T cell sensitivity to inhibition by monoclonal antibodies to LFA-1 and L3T4. Eur. J. Immunol. 16:1385.

14. Gougeon, M.-L., Bismuth, G., and J. Theze. 1985. Differential effects of monoclonal antibodies anti-L3T4 and anti-LFA-1 on the antigen-induced proliferation of T-helper-cell clones: correlation between their susceptibility to inhibition and their affinity for antigen. J. Cell. Immunol. 95:75.

THE INVOLVEMENT OF T-CELLS IN THE STIMULATORY EFFECT OF A WATER-IN-OIL EMULSION ON THE B-CELL RESPONSE

A.T.J. Bianchi, J.W. Scholten and B.A. Bokhout

Dept. Immunology, Central Veterinary Institute, P.O.Box 65
8200 AB Lelystad, The Netherlands

INTRODUCTION

Originally the immunostimulatory effect of water-in-oil (W/O) emulsions, used as adjuvant, has been explained by depot formation resulting in a chronic stimulation of the immune system by small amounts of released antigen (1,2). Later studies by ourselves (3,4) clearly demonstrated that other mechanisms are involved also.

In our studies (3,4) we excluded all possible stimulatory effects from antigen-adjuvant interactions by choosing an immunization regime in which antigens and adjuvant were administered with an interval and by different routes. These studies showed that intraperitoneal (ip) or subcutaneous (sc) injection of adjuvant stimulates the splenic IgM plaque forming cell (PFC) response intravenously (iv) induced against *soluble* T-cell dependent (TD) antigens, which have hardly any intrinsic adjuvanticity, like dinitrophenyl-chickengammaglobulin (DNP-CGG) (4). Stimulation was also shown of the splenic IgM-, IgG- and IgA response against *particulate* TD antigens with intrinsic adjuvanticity, like sheep red blood cells (SRBC) (3).

It is generally accepted that modulation of macrophage activity plays a crucial role in stimulation of B-cells after adjuvant application (1,5). In this study we investigated whether T-cells are involved in the stimulation by a W/O emulsion of the B-cell response against a T-cell independent (TI) antigen in normal and T-cell deficient nude mice.

MATERIALS AND METHODS

Mice: Female C57Bl/6J, C3H/He, BALB/c and BALB/c nu/nu mice were purchased from HARLAN-OLAC Ltd., U.K., and female C57Bl/6J nu/+ and nu/nu mice were purchased from Bomholtgard, Denmark. All mice were used at an age of 10-14 weeks.

Adjuvant, antigen and immunization: A stable W/O emulsion was used as adjuvant. The composition and preparation of the emulsion has been described before (6). The adjuvant was applied ip in a volume of 0.1 ml without antigen.
Dinitrobenzene sulfonic acid (Eastman Kodak, U.S.A.) was conjugated with Ficoll (Pharmacia, Sweden) as described by Koch et al. (7). Dose response experiments (data not shown) demonstrated that after iv immu-

nization with 10 μg or 0.01 μg DNP-Ficoll in 0.5 ml saline, opptimal respectively suboptimal anti-DNP responses were obtained in all mouse strains used.

Assay for PFC: Trinitrobenzene sulfonic acid (Eastman Kodak) was conjugated to SRBC to obtain TNP-SRBC, which were used for assaying DNP-specific PFC's. The splenic antigen-specific PFC assay and isotype-specific protein-A PFC assay has been described by Van der Heijden et al. (3). The Rabbit-anti-mouse IgM, IgG and IgA sera were obtained commercially (Nordic, the Netherlands).

RESULTS

The adjuvant effect on the response against DNP-Ficoll in normal and T-cel deficient nude mice: C57Bl/6J, BALB/c or C3H/He mice were ip injected with adjuvant or were left untreated. Within one hour half of both groups of mice was iv immunized with a suboptimal dose of the TI antigen DNP-Ficoll (0.01 μg) and the other half was iv immunized with an optimal dose of DNP-Ficoll (10 μg). DNP-specific PFC responses were measured on day 4 or 5 after iv immunization, when normally the highest anti-DNP-Ficoll responses are obtained for the mouse strains tested.

From the results presented in Fig. 1 it appears that adjuvant application evokes a substantial stimulation of the anti-DNP B-cell response in C57Bl/6J mice after optimal as well as after suboptimal immunization. In BALB/c and C3H/He mice a minor stimulation or no stimulation is observed.

Using an optimal dose of DNP-Ficoll, an identical experiment was performed with normal and T-cell deficient C57Bl/6J and BALB/c nude mice. The results are presented in Fig. 2. Again only a significant stimulation of the anti-DNP PFC response is found in the normal C57Bl/6J mice. In all other mouse strains tested, only minor adjuvant effects are observed. It is remarkable that in normal non-adjuvated DNP-Ficoll immunized C57Bl/6J mice the absolute number of DNP-PFC is always at least two times lower than in the other mouse strains tested (Fig. 1 & 2).This observation also applies to the nude mice.

In Fig. 1 & 2 the IgM DNP-PFC responses are shown. No IgG and IgA responses are found in adjuvant treated and control C3H/He and BALB/c mice. In normal C57Bl/6J mice, very small splenic IgG (300 PFC's) and IgA (50 PFC's) DNP-PFC responses are detected 7 days after immunization with an optimal dose of DNP-Ficoll. Moreover, in these C57Bl/6J mice, adjuvant treatment causes a strong increase of the splenic IgG (17.000 PFC) and IgA (3.000 PFC) DNP-PFC responses. Low IgG and IgA DNP-PFC responses are also detected in C57Bl/6J nude mice, but the adjuvant effect in these mice for IgG and IgA is absent (data not shown).

The adjuvant effect on the spontaneous (background) PFC response in the spleen of normal and T-cell deficient mice. It was investigated whether adjuvant application has a stimulatory effect on the total number of spontaneously Ig-secreting cells in the spleen. C57Bl/6J nu/+, C57Bl/6J nu/nu, BALB/c +/+ and BALB/c nu/nu mice were ip injected with 0.1 ml of adjuvant, and the numbers of IgM-, IgG- and IgA secreting PFC's were determined 4, 5 and 6 or 7 days later. The results in Fig. 3 are expressed as a stimulation coëfficiënt. The adjuvant causes a clear stimulation of all isotypes in C57Bl/6J nu/+ and BALB/c +/+ mice. In the BALB/c nu/nu mice we solely found stimulation of the IgM response, while in C57Bl/6J nu/nu mice a stimulation of all isotypes is observed.

MOUSE STRAIN	ADJUVANT TREATMENT 100 µl I.P. ON DAY 0	IMMUNIZATION ON DAY 0 WITH DNP–FICOLL	STIMULATION COEFFICIENT
C57BL/6J	+	0.01 µg I.V.	1.9 ± 0.3
C57BL/6J	-	0.01 µg I.V.	
BALB/c	+	0.01 µg I.V.	1.2 ± 0.1
BALB/c	-	0.01 µg I.V.	
C3H-He	+	0.01 µg I.V.	1.4 ± 0.1
C3H-He	-	0.01 µg I.V.	
C57BL/6J	+	10 µg I.V.	3.1 ± 0.8
C57BL/6J	-	10 µg I.V.	
BALB/c	+	10 µg I.V.	1.0 ± 0.1
BALB/c	-	10 µg I.V.	
C3H-He	+	10 µg I.V.	1.1 ± 0.1
C3H-He	-	10 µg I.V.	

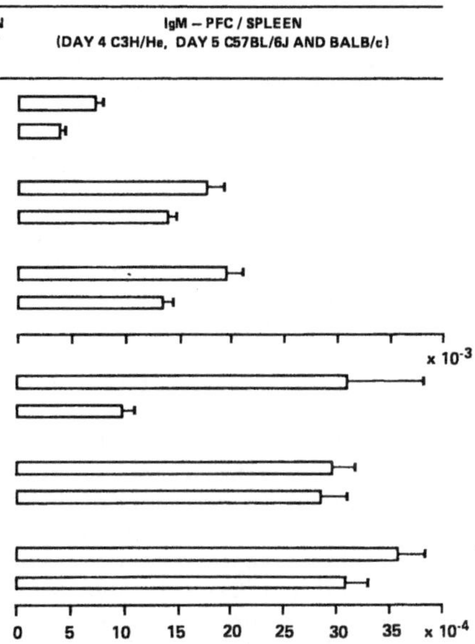

Fig. 1. The adjuvant effect on the anti-DNP PFC response in normal mice immunized with DNP-Ficoll. C57Bl/6J, BALB/c or C3H/He mice were ip injected with 0.1 ml adjuvant and iv immunized with a suboptimal dose of 0.01 µg or an *optimal* dose of 10 µg DNP-Ficoll. Control groups of iv immunized mice, which were not injected with adjuvant, were included. DNP-specific PFC responses were measured on day 4 or 5 after iv immunization. The DNP-PFC responses are expressed as the arithmetic mean of 5 mice ± SE. The stimulation coëfficiënt (± SE) is calculated as the mean number of DNP-PFC of adjuvant treated mice divided by the mean number DNP-PFC of control mice (n=5).

MOUSE STRAIN	ADJUVANT TREATMENT 100 µl I.P. ON DAY 0	IMMUNIZATION ON DAY 0 WITH DNP–FICOLL	STIMULATION COEFFICIENT
C57BL/6J nu/+	+	10 µg I.V.	2.7 ± 0.7
C57BL/6J nu/+	–	10 µg I.V.	
C57BL/6J nu/nu	+	10 µg I.V.	1.5 ± 0.5
C57BL/6J nu/nu	-	10 µg I.V.	
BALB/c +/+	+	10 µg I.V.	1.2 ± 0.1
BALB/c +/+	-	10 µg I.V.	
BALB/c nu/nu	+	10 µg I.V.	1.2 ± 0.1
BALB/c nu/nu	-	10 µg I.V.	

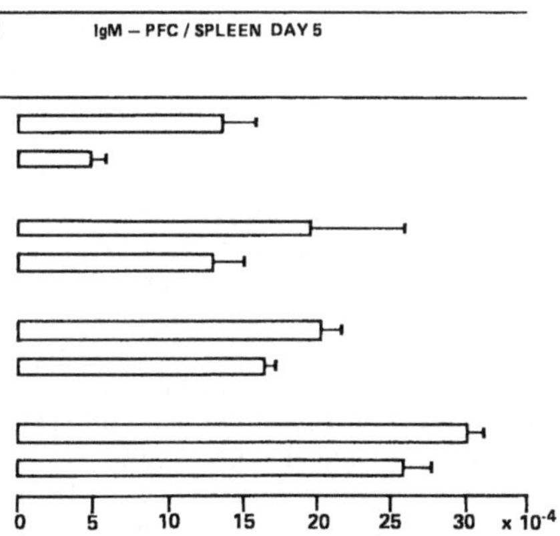

Fig. 2. The adjuvant effect on the anti-DNP PFC response of T-cell deficient nude mice. C57Bl/6J nu/nu and BALB/c nu/nu mice were ip injected with 0.1 ml adjuvant and iv immunized with an optimal dose DNP-Ficoll. For comparison also normal C57Bl/6J nu/+ and normal BALB/c +/+ mice were included in this experiment. The experimental details are the same as in Fig.1.

DISCUSSION

The *DNP-Ficoll results* obtained with the C3H/He, BALB/c and BALB/c nude mice are relatively easy to interpret. The W/O emulsion (6) used in our experiments hardly influences the PFC response against the TI-antigen DNP-Ficoll (Fig. 1 & 2). In earlier studies (3,4) a clear stimulation of the B-cell response against particulate and soluble TD antigens has been demonstrated. From these results (Fig. 1 & 2 and ref. 3,4) it can be concluded that in BALB/c and C3H/He mice the involvement of T-cells is needed for an efficient stimulatory adjuvant effect on the B-cell response.

Other investigators (8-10), using alum or aluminiumhydroxide as adjuvant, found that this type of adjuvant too stimulates the B-cell response against TD antigens, not against TI antigens. This result (8) was also obtained after separate application of adjuvant and antigen. From the combined results, it is now obvious that adjuvants, like W/O emulsions and aluminium compounds, also have direct stimulatory effects on the cells co-operating in the antibody response like macrophages and T-cells. Originally only depot formation effects were ascribed to these adjuvants when they are used as antigen vehicle (1,2).

The need for T-cell involvement, to obtain an adjuvant induced stimulation of the B-cell response, seems to be denied by the results found with the C57Bl/6J mice, since in this mouse strain the response against the TI-antigen DNP-Ficoll is stimulated by application of the W/O emulsion (Fig. 1&2). Moreover, besides a stimulated anti-DNP IgM response a stimulated IgG and IgA response is found in these mice as well. Such IgG and IgA responses against TI antigens have been described earlier for C57Bl/6J mice and appear to be mouse strain dependent (11). In contrast to the result in normal C57Bl/6J mice, only a weak stimulation of the anti-DNP IgM response is found (Fig. 2) in T-cell deficient C57Bl/6J nude mice and the IgG and IgA responses in these mice are not stimulated at all. This suggests that also for TI antigens, like DNP-Ficoll, the adjuvant effect by the W/O emulsion in C57Bl/6J mice mainly depends on a stimulated T-cell compartment. Other studies (12,13) which describe the existence of T-cell regulation of B-cell responses against the TI antigen DNP-Ficoll support this suggestion.

The possibility to stimulate the anti DNP-Ficoll IgM response in C57Bl/6J mice by adjuvant may also depend on the relatively low response obtained without adjuvant in this mouse strain compared to the response in BALB/c and C3H/He mice (Fig. 1, 2&3). Therefore, it is imaginable that in non-adjuvated C57Bl/6J mice the capacity for expansion of the number of DNP-Ficoll specific B-cells is not fully utilized, while in the other mouse strains the responses without adjuvant are already on their maximum.

The adjuvant effect investigated in relation to the *polyclonal IgM, IgG and IgA background response* again shows that the most effective stimulation of all isotypes is found in normal T-cell bearing mice (Fig. 3) as has been shown for antigen-specific responses. On the other hand, remarkable results are obtained in the T-cell deficient nude mice, since in C57Bl/6J nude as well as in BALB/c nude mice the IgM background response is slightly but significantly stimulated (Fig. 3). In the C57Bl/6J nude mice even a small stimulation of the "T.D. isotypes" IgG and IgA is found. Effective stimulation of B-cell responses in T-cell deficient mice mediated by other types of adjuvant, like LPS (8) and Quil A (9,10) was reported earlier. However, these adjuvants very efficiently stimulate the response against TI antigens in nude mice as well, while the W/O emulsion used in our experiments is not able to do so (Fig. 2).

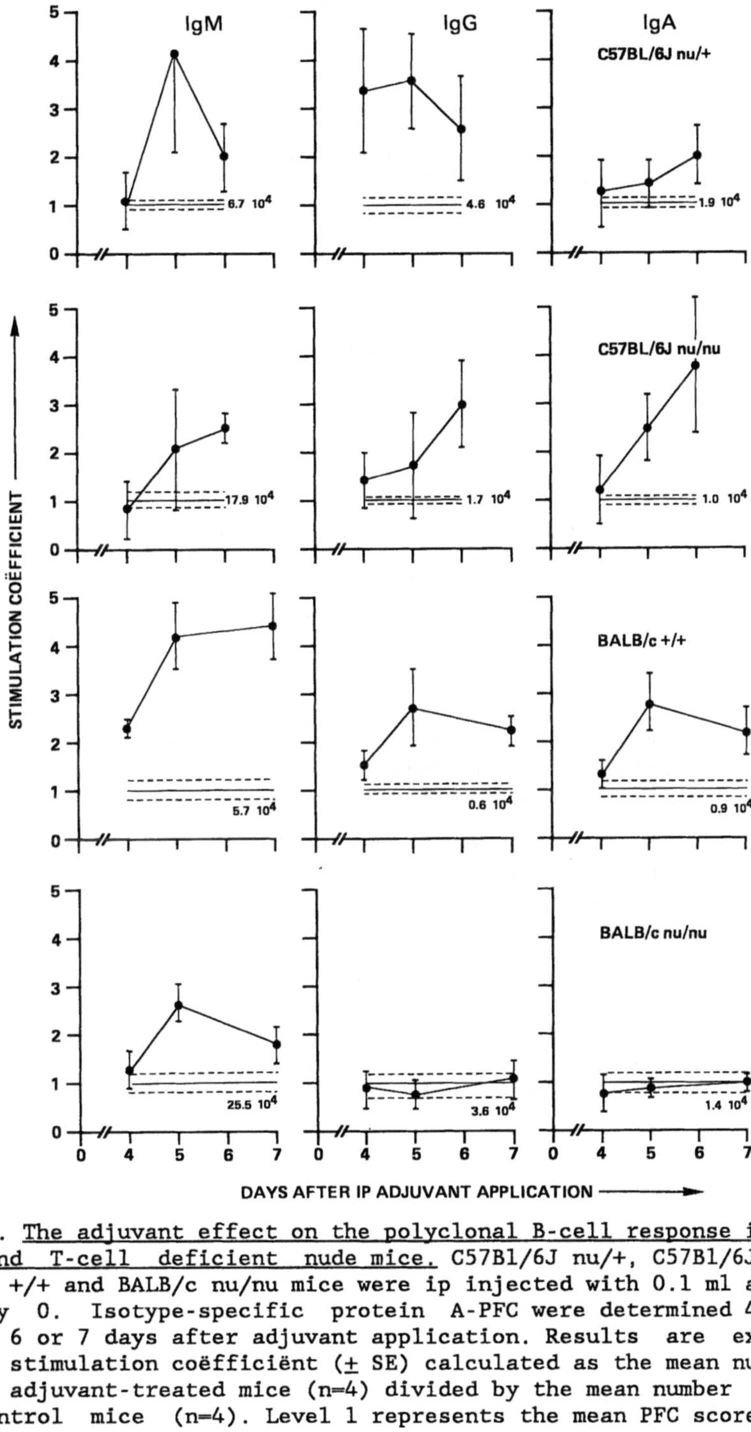

Fig. 3. The adjuvant effect on the polyclonal B-cell response in normal and T-cell deficient nude mice. C57Bl/6J nu/+, C57Bl/6J nu/nu, BALB/c +/+ and BALB/c nu/nu mice were ip injected with 0.1 ml adjuvant on day 0. Isotype-specific protein A-PFC were determined 4, 5 and either 6 or 7 days after adjuvant application. Results are expressed as a stimulation coëfficiënt (± SE) calculated as the mean number of PFC of adjuvant-treated mice (n=4) divided by the mean number of PFC of control mice (n=4). Level 1 represents the mean PFC score (± SE) for the control mice.

The adjuvant effects in T-cell deficient mice (Fig. 3 and ref. 8-10) cannot be ascribed to "T-cell replacing effects", since responses against TD antigens are not facilitated by these adjuvants in T-cell deficient mice (14).

In conclusion, it appears from our results that a most prominent stimulatory effect by the W/O emulsion on the B-cell response depends on the presence of T-cells. Moreover, direct effects by an activated macrophage compartment (1,2) on the B-cell can contribute to the observed stimulatory adjuvant effect.

ACKNOWLEDGEMENTS

We thank Drs. P.J. van der Heijden and G. Koch for their critical reading of the manuscript.

REFERENCES

1. Freund, J. Adv. Tubercul. Res. 7 : 130 (1956)
2. Allison, A.C. and N.E. Byars. J. Immunol. Meth. 95 : 157 (1986)
3. Heijden, P.J. van der, B.A. Bokhout, A.T.J. Bianchi, J.W. Scholten and W. Stok. Immunobiol. 171 : 143 (1986)
4. Bokhout, B.A., A.T.J. Bianchi, Ph.J. van der Heijden, J.W. Scholten and W. Stok. Comp. Immun. Microbiol. Infect. Dis. 9 : 161 (1986)
5. Spitznagel, J.K. and A.C. Allison. J. Immunol. 104 : 128 (1970)
6. Bokhout, B.A., C. van Gaalen, P.J. van der Heijden. Vet. Immunol. Immunopathol. 2 : 491 (1981)
7. Koch, G., B.D. Lok, A. van Oudenaren and R. Benner. J. Immunol. 128 : 1479 (1982)
8. Seppälä, I.J.T. and O. Mäkela. Immunology 53 : 827 (1983)
9. Flebbe, L.M. and H. Braley-Mullen. Cell. Immunol. 99 : 119 (1986)
10. Flebbe, L.M. and H. Braley-Mullen. Cell. Immunol. 99 : 128 (1986)
11. Ivars, F., G. Nyberg, D. Holmberg and A. Coutinho. J. Exp. Med. 158 : 1498 (1985)
12. Mongini, P.K.A., K.E. Stein and W.E. Paul. J. Exp. Med. 153 : 1 (1983)
13. Endres, R.O., E. Kushnir, J.W. Kappler, P. Marrack and S.C. Kinsley. J. Immunol. 130 : 781 (1983)
14. Allison, A.C. and A.J.S. Davies. Nature 233 : 330 (1971)

IMPACT OF IgD RECEPTOR-POSITIVE T CELLS ON THE IMMUNE RESPONSE

R. F. Coico, C. Swenson, G. W. Siskind and G. J. Thorbecke

CUNY Medical School, Cornell University Medical College and NYU School of Medicine, New York, NY, USA

INTRODUCTION

IgD is present on the surface of the majority of B lymphocytes, whereas serum IgD levels are low. This has suggested to previous investigators[1], that the major function of IgD is to serve as a surface receptor, but no definite biological role different from surface IgM has been found for this molecule. Investigation of the effect of IgD on immune responses was prompted by our observation that IgD-myeloma-bearing mice exhibit greatly enhanced immune responses[2]. We find that injections of IgD enhance antibody production and that this effect of IgD is correlated with the induction of receptors for IgD on helper T cells[3]. Such T cells are able to transfer immuno-augmentation to normal recipients[4]. It appears that "Tδ" cells can be induced by cross-linked IgD on the surface of B cells, and lead to enhanced: 1) primary and secondary responses, 2) Ig production of all isotypes studied, and 3) germinal center formation. The findings supporting these conclusions are reviewed below.

MATERIALS AND METHODS

Mice: BALB/c and $CB6F_1$ mice were obtained from Charles River Breeding Laboratories, Inc. (Wilmington, MA). Aging BALB/c mice were obtained under contract # NO1-AG-3-2130 through the courtesy of the NIA. B10.A and SJL/J mice were purchased from Jackson Labs. (Bar Harbor, ME).
Antigens: Sheep and horse red blood cells (SRBC and HRBC) were purchased from Colorado Serum Co. (Denver, CO), bovine serum albumin (BSA) from Sigma Chemical Co. (St. Louis, MO).
Cell Lines: The IgD-secreting plasmacytoma, TEPC-1017, was maintained intraperitoneally (ip) in BALB/c mice. The B cell hybridoma, B1-8.δ1, which secretes IgD specific for the hapten 4-hydroxy-3-nitrophenylacetyl (NP) was maintained intraperitoneally in $CB6F_1$ mice. The murine IgM^+, IgD^+ B cell lymphoma, CH15, was obtained from Dr. G. Haughton (U. of North Carolina, Chapel Hill, NC) and maintained *in vitro*[5]. The myoglobin-specific T cell clones and hybridomas have been previously described[6,7]. The T cell hybridoma 2.1-6#16, which constitutively expresses Fcα receptors[8], was obtained from Dr. A. Stanisz and J. Bienenstock (McMaster University, Ontario, Canada).
Purification of IgD: TEPC-1017-derived and monoclonal B1-8.δ1 IgD were

purified by affinity chromatography on rabbit anti-IgD-coupled sepharose columns[2].

Lymphokines: Samples of recombinant interleukin 2 (IL-2), were kindly provided by Genentech Inc. (So. San Francisco, CA) and Cetus Corp. (Emeryville, CA). Recombinant tumor necrosis factor (TNF) was also provided by Genentech, Inc. Recombinant interleukin 4 (IL-4) was kindly provided by Dr. S. Gillis (Immunex, Seattle, WA).

IgD Rosette-Forming Cell Assay (IgD-RFC): Lymphoid cells were examined for their expression of receptors for IgD using IgD-coated SRBC in a rosette assay as previously described[3].

Histology: Brachial, axillary and mesenteric lymph nodes from control and IgD-injected mice were removed and fixed in Carnoy's solution for 30 min., paraffin embedded, sectioned, and stained with methylgreen pyronin.

Plaque-Forming Cell (PFC) Assay: Anti-HRBC splenic PFC were assayed by the slide modification of the technique of Jerne *et al.*[9]. IgG-producing cells were developed with rabbit anti-mouse Ig in the complement and goat anti-μ in the agar.

RESULTS AND DISCUSSION

Requirements for induction of Tδ cells: It was of interest to consider the possibility that IgD may function as a regulatory molecule involved in normal humoral immune responses through induction of Tδ cells. We have therefore investigated whether Tδ cells are induced in response to normal events associated with immune responses, such as lymphokine

Table 1

INDUCTION OF IgD-RFC IN SPLENIC T CELLS BY INCUBATION WITH IgD OR LYMPHOKINES

Stimulus Used:$^{\Theta}$	% IgD-RFC* in T cells from: Young adult mice	Aged mice
TEPC-1017 ascites(10%), 18hr	26	1
TEPC-1017 IgD (100 μg/ml), 18hr	35	ND
B1-8.σ1 ascites(10%), 18hr	6	ND
B1-8.σ IgD (100 μg/ml), 18hr	1	ND
NP-BSA (500 μg/ml), 18hr	1	ND
B1-8.σ IgD + NP-BSA, 18hr**	34	ND
TEPC-IgD, 1 hr	27	0
rIL-2 (10^3 U/ml), 1hr	3	1
rIL-2, 1hr; then IgD, 1hr	34	16
rIL-2 (250 U/ml), 18 hr	37	12
rIL-2 (10^3 U/ml), 18 hr	37	11
rIL-4 (10 U/ml), 18hr	44	ND
rTNF (500 U/ml), 18hr	3	ND
rIFN-γ (10^3 U/ml), 18hr	25	0

Θ Incubation was at 37°C. T cells isolated as in ref. 4.

* The values in the table represent means from 2-4 experiments. Background IgD-RFC after incubation with medium alone varied from 5-14% and were subtracted from the values given in the table. Complete inhibition of IgD-RFC could be obtained with IgD, but none with IgG1, IgG2, IgA or IgM.

** IgG or IgM antibodies to NP with NP-BSA did not induce IgD-RFC.

production and cell surface Ig crosslinking. The results in Table 1 confirm our previous observations that IgD receptors are readily induced on T cells following their exposure to TEPC-1017-derived IgD. In contrast, the monoclonal IgD anti-NP, B1-8.δ1, only induces Tδ when it is complexed with NP-haptenated antigens such as BSA. These results are consistent with the notion that a prerequisite for IgD-mediated induction of Tδ cells is that oligomeric forms of IgD be employed. Indeed, the TEPC-1017-derived IgD is secreted as an oligomer, whereas the B1-8.δ1 has been reported to be secreted as a monomer[10]. Control monoclonal antibodies with the same specificity as B1-8.δ1, but of different isotypes, do not induce Tδ cells in native form or as immune complexes. These results do not provide direct information regarding the affinity of IgD-specific receptors, but they suggest a relatively low affinity of interaction between univalent forms of the IgD molecule and receptors on the cells and/or the need for crosslinking of IgD receptors for the upregulation of such receptors.

The results in Table 1 also show that certain lymphokines including IL-2, IL-4 and IFN-γ are capable of inducing Tδ cells. In contrast, exposure of T cells to TNF does not induce the expression of IgD receptors. Although, as also shown, T cells from aged mice are unresponsive to IgD, IL-2 does induce an increase in IgD receptors on such cells. However, the % IgD-RFC induced in T cells from aged mice is usually lower than values obtained with T cells from young mice. These

Table 2

INDUCTION OF IgD RECEPTORS ON HELPER PHENOTYPE T CELL CLONES AND HYBRIDOMAS BY OVERNIGHT INCUBATION WITH IgD

Cells Assayed*	Phenotype	% IgD-RFC Following Incubation with: Medium	IgD#
Resting myoglobin-** specific T cell clone 2	Lyt-1$^+$,2$^-$,L3T4$^+$	3	73
Myoglobin-stimulated** clone 2	"	47	86
Myoglobin-specific T cell hybridoma A3.4c6	"	46	83
Myoglobin-specific T cell hybridoma C7R14	"	38	83
BW5147 Thymoma	Lyt-1$^-$,2$^-$	3	4
FcαR$^+$ T cell hybridoma 2.1-6#16	Lyt-1$^-$,2$^-$	6	6
R.1.1 Thymoma	"	3	3
CTLL clone+	Lyt-1$^-$,2$^+$	2	ND

* BW5147 thymoma cells were used as the fusion parent for hybridomas A3.4c6 and C7R14, and R.1.1 thymoma cells for hybridoma 2.1-6#16

\# Cells were incubated at 2.5 x 10^6 cells/ml with TEPC-1017 IgD containing ascites fluid 10% in MEM at 37°C for 18 hours; final IgD concentration 340μg/ml. After incubation cells were washed twice and examined for RFC.

** Myoglobin stimulated clone 2 had been exposed to myoglobin 4 days earlier, resting clone 2 had not been stimulated for 14 days.

\+ IL-2 dependent cytotoxic T cell clone which had been continuously grown in the presence of IL-2 for more than 5 years.

results cannot be explained by a deficiency in the overall number of helper T cells in aged mice since Lyt1^{+}2^{-} and Lyt1^{-}2^{+} T cell subsets are not much lower in the spleens of aged than of young mice[11].

The observation that IL-2 but not IFN-γ is capable of inducing IgD receptors on T cells from aged mice led us to investigate the possibility that IL-2 might be capable of acting in concert with IgD in the induction of Tδ cells. As shown in Table 1, when T cells from aged mice are incubated with either IgD or IL-2 alone for 1 hour, Tδ cells are not induced. T cells from young mice show maximal induction of Tδ cells following a 1 hour exposure to IgD alone. However, when T cells from aged mice are sequentially exposed to IL-2 and IgD, each for one hour, IgD receptors are induced. It could be speculated that IL-2 may facilitate

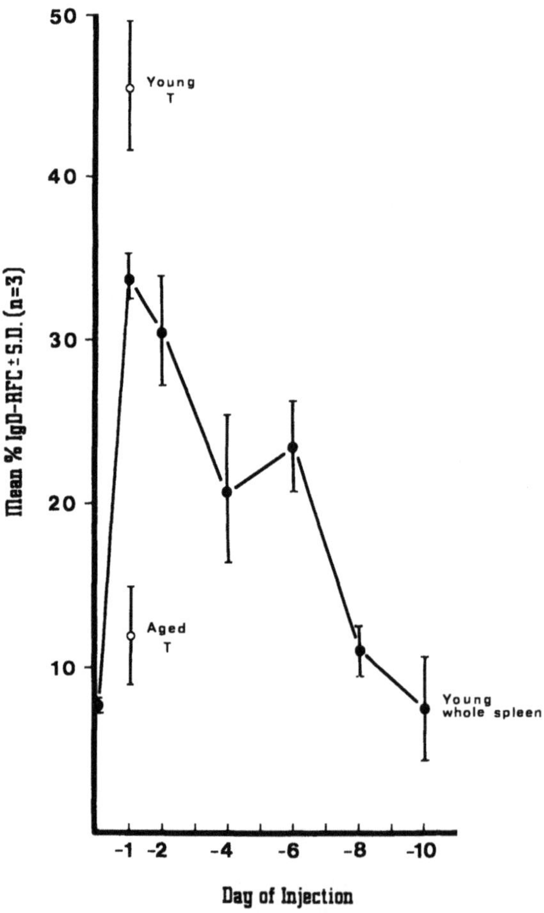

Fig. 1 Induction of IgD-RFC by injection of goat anti-IgD. Young (2-3 months) and aged (20-24 months) BALB/c mice were injected iv with 800 μg affinity-purified anti-IgD. One, 2, 4, 6, 8 and 10 days later spleen cells were examined for IgD receptors. In addition, isolated splenic T cells (T) were examined on day 1.

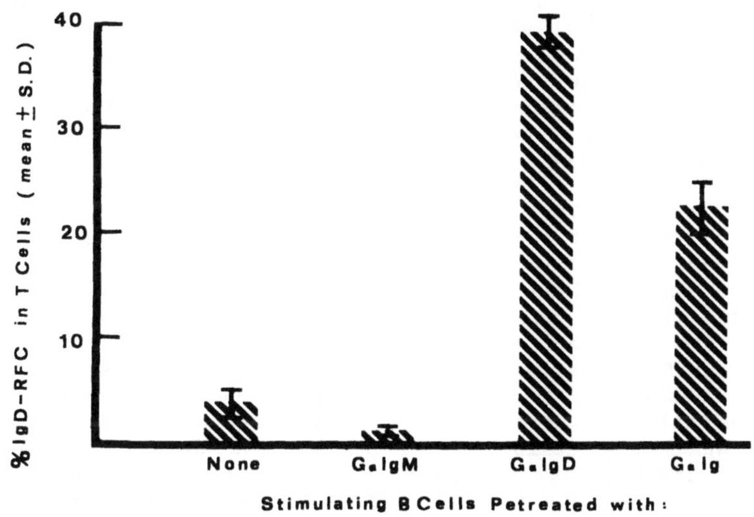

Fig. 2 Induction of IgD-RFC in T cells in vitro by B lymphoma cells bearing crosslinked IgD. Equal numbers (10^6/ml) of splenic B10.A T cells and pretreated CH15 lymphoma cells were incubated in 5 ml of RPMI-1640 with 2% fetal calf serum for 24 hrs at 37°C. Pretreatment of CH15 cells with the indicated antibody was followed by fixation with paraformaldehyde and washing.

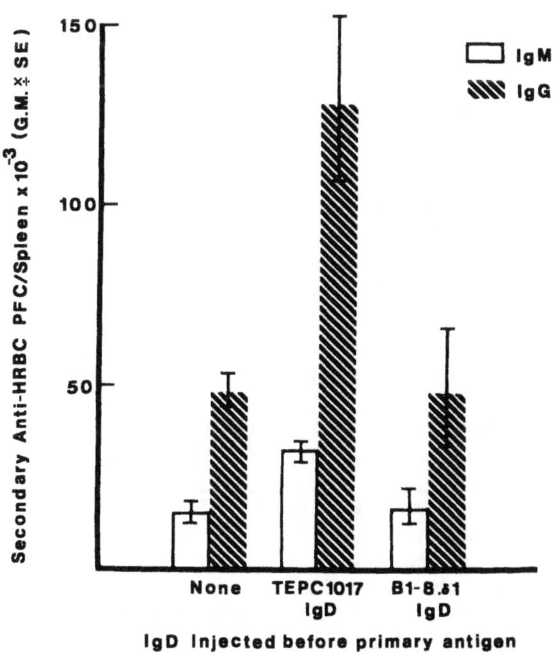

Fig. 3 Injection of IgD preparations prior to the primary antigen injection: Effect on the secondary response to HRBC. Days -1 and 0: 1 -1.5 mg IgD injected ip; Days 0 and 10: 10^7 HRBC injected iv; Day 14: anti-HRBC PFC assayed.

responsiveness to IgD in T cells by causing neosynthesis of Tδ receptors. Following such short term neosynthesis further upregulation of IgD receptors might then result from exposure of these cells to IgD. With T cells from young mice, adequate cytoplasmic stores of these receptors would allow for rapid responses to IgD as evidenced by maximal induction of IgD receptors within 1 hour. T cells from aged mice, however, may be deficient in this putative IgD receptor pool.

Earlier observations have shown that intravenous injection of antigens, such as TNP-KLH or TNP-Ficoll, results in the induction of a small, but significant increase in the frequency of splenic IgD-RFC[3]. We speculated that antigen might have this effect because it crosslinks either serum IgD or B cell surface IgD. The results shown in Fig. 1 provide additional support for this idea. We reasoned that iv injection of affinity-purified goat anti-IgD would mimic the effects of antigen and cause such crosslinking to occur at maximal levels. Indeed, within 24 hours of injection of goat anti-IgD, maximal numbers of IgD-RFC are induced in splenic T cells. In light of the deficient IgD receptor expression observed following exposure of T cells from aged mice to IgD, it is of interest to note that goat anti-IgD is just as deficient as IgD itself in the induction of Tδ cells in aged mice (Fig. 1). Since the frequency of IgD^+ B cells as well as serum IgD levels do not decrease

Table 3

FUNCTIONAL EFFECTS OF PRETREATMENT WITH IgD* IN MICE OF DIFFERENT AGES

Age**	Pretreated With IgD	Mean % of IgD-RFC(±SE)	Geometric Mean Reverse PFC/Spl(x/÷SE)+ for: IgG1	IgG2	Percentage of Follicles With GC (±SE)
7-10 day old	-	3.4(0.7)	75(1.6)	183(1.5)	0
	+	8.5(0.8)^a	119(1.5)	266(1.3)	0
17-21 day old	-	6.7(0.6)	260(1.3)	900(1.5)	5.6(3.5)
	+	26.8(1.7)^b	625(1.3)^a	5,200(1.2)^b	53.4(12.0)^b
24-28 day old	-	7.5(2.2)	1,615(1.2)	5,130(1.7)	65.5(3.1)
	+	30.9(1.7)^d	33,470(1.2)^c	35,900(1.2)^b	64.4(16.6)
2 month old	-	8.0(1.2)	275(1.4)	590(1.5)	23.0(6.0)
	+	30.0(1.7)^d	6,800(1.3)^d	46,630(1.3)^d	56.0(1.2)^b
20month old	-	5.4(2.0)	4,770(1.2)	2,875(1.4)	21.6(7.0)
	+	6.0(2.0)	3,900(1.2)	3,058(1.6)	37.0(18.0)

* Mice 7-10 days old received a single injection and the other age groups 2-3 weekly injections i.p. of TEPC-1017 IgD (1.7 mg) containing ascites or of saline. Their spleens were analyzed in the IgD-RFC and reverse PFC assay 5-7 days after the last injection. The percentage of follicles in axillary, brachial and mesenteric lymph nodes showing germinal centers (GC) was determined histologically.

\+ Reverse PFC were assayed on protein A-coated SRBC with isotype-specific antibody in the agar and represented as geom.mean x/: SE.

** SJL x BALB/c mice were used at ages 7-28 days and BALB/c mice at 2 and 20 months of age (n = 4-12).

Significantly different from control: ^{a} $p<0.05$, ^{b} $p\leq0.01$, ^{c} $p\leq0.001$, ^{d} $p\leq0.0001$. Values without footnotes were NS.

with age, these results reenforce the interpretation that this functional defect in aged mice is a property of the T cell population.

Studies reported by Finkelman _et al_. have shown that injection of goat anti-mouse IgD leads to polyclonal activation of the B cell population[12,13]. This effect is due, in part, to the production of lymphokines including IL-2[14]. The lymphokine response first appears 72 hours following injection of goat anti-IgD. Since Tδ cell frequencies are at peak levels within 24 hours of injection, we conclude that the early Tδ cell response to this stimulus is mediated by IgD crosslinking. However, maintenance of high Tδ cell levels through day 5 may be the result of lymphokine-induced receptor expression.

In order to assess whether crosslinking of B cell surface IgD is sufficient for optimal induction of Tδ cells, we performed _in vitro_ experiments with the IgM^+, IgD^+ B cell lymphoma CH15[15]. Our strategy for resolving this question was again based on the use of affinity-purified goat antibodies specific for either IgM or IgD (kindly donated by Dr. F. Finkelman, USUHS, Bethesda, MD). The CH15 lymphoma cells were first incubated with goat anti-IgM, goat anti-IgD, goat anti-Ig or control medium as previously described[15]. The cells were then washed and fixed with 0.1% paraformaldehyde as described by Allen and Unanue[16]. These treated lymphoma cells were subsequently co-cultured with syngeneic B10.A splenic T cells for 18 hours to determine their effects on T cell IgD receptor expression. As shown in Fig. 2, T cells exposed to CH15 lymphoma cells bearing crosslinked IgM + IgD (incubated with anti-Ig) or IgD alone respond with increased expression of IgD receptors, whereas T cells exposed to CH15 cells bearing crosslinked IgM alone do not. Control CH15 cells subjected to paraformaldehyde fixation without antibody treatment are also unable to induce Tδ cells. These findings provide direct support for the conclusion that interaction between T cells and IgD^+ B cells, under conditions where the surface IgD is crosslinked, results in increased expression of IgD receptors on the T cells. It is interesting to speculate that these effects of goat anti-Ig or anti-IgD may mimic those associated with physiologic B cell stimulation by antigen. Within a local microenvironment, the functional consequences of such an interaction could result in specific upregulation of B cell responses by the newly induced Tδ cells.

<u>Characterization of cells expressing IgD receptors</u>: Our previous studies showed that Tδ cells could be induced in $Lyt1^+2^-$, $L3T4^+$ T cells, but not in $Lyt2^+$ T cells or thymocytes[3]. The data shown in Table 2 provide additional support for the helper T cell-restricted expression of IgD receptors. Resting, myoglobin-specific T cell clones of helper phenotype are induced to express receptors for IgD following overnight incubation with IgD[7]. In addition, exposure of these cloned T cells to antigen (in the absence of IgD) also causes increased expression of IgD receptors, probably as the result of lymphokine stimulation, since antigen exposure induces IL-2 production in these cells. It is important to note that these T cell clones were not maintained in IL-2-containing medium. Under such growth conditions, cloned T cells which have not been stimulated with antigen for a period of 12 days show a low % of IgD-RFC. However, $Lyt1^+2^-$ T cell hybridomas produced from cloned T cells show high background IgD-RFC levels and are induced by exposure to IgD to express even higher levels of IgD receptors.

Similar to thymus cells, $Lyt1^+2^-$ thymoma cells show only a minimal % of IgD-RFC either before or after exposure to IgD. Also negative are $Lyt1^-2^-$ T hybridoma cells which constitutively express Fc receptors specific for IgA[8]. In addition, cloned $Lyt2^+$ T cells do not express IgD receptors (Table 2). In contrast, receptors for IgA, IgE and IgG appear rarely on helper T cells and are frequently observed on Lyt2+ T cells[17].

Functional effects of IgD: Since we established a direct correlation between the immunoaugmenting effects of IgD and induction of increased numbers of Tδ cells, we determined whether pretreatment of mice with the soluble IgD of B1-8.δ1 origin, which fails to induce Tδ cells, would also fail to augment antibody responses. Indeed, in experiments reported elsewhere[15], we showed that pretreatment of mice with the B1-8.δ1 IgD had no effect on primary antibody responses. The data in Fig. 3 show that pretreatment of BALB/c mice with TEPC-1017-derived IgD prior to injection of the primary dose of HRBC results in significant augmentation of the secondary response to this antigen. Thus, the modulation of a secondary immune response is evident even though IgD exposure, and the resulting Tδ cell induction, has occurred at the time of priming only. In contrast, similar pretreatment with the B1-8.δ1 IgD has no effect.

The effect of IgD pretreatment on the secondary response suggests that memory cell generation and, therefore, germinal center (GC) formation[18], might be affected by Tδ cells. The results in Table 3 show that repeated injections of TEPC-1017 IgD causes increased IgG1 and IgG2 production as well as GC formation only in mice which are also capable of responding with increased IgD-RFC in their spleens. Mice of the youngest age-group have low percentages of IgD+ B cells and of L3T4+ T cells, either of which might account for the unresponsiveness to IgD[19]. The aged mice fail to show Tδ cell induction and are also functionally unresponsive to IgD. It should be noted that GC formation at the ages of 24-28 days is unusally high in control mice and not further enhanced by injections of IgD. This possibly reflects a physiological state of development associated with the establishment of the normal bacterial flora as a consequence of weaning.

The mechanism responsible for the immunoaugmenting properties associated with Tδ cells has not been elucidated. Possible mechanisms for the humoral regulatory effects of such Tδ cells include: (1) facilitated interaction between IgD receptor-positive T cells and IgD+ B cells; (2) signal transduction which results in down-regulation of IgD transcription by the responding B cells, a phenomenon known to be associated with B cell activation[20,21]; and (3) production of biologically active soluble mediators, which might include IgD-binding factor (BF), by Tδ cells. Indeed, T cells induced to express Fc receptors for IgE are reported to mediate molecular events which cause down-regulation of IgE transcription[22]. In general, Ig-BFs produced by Fc receptor-positive lymphocytes regulate the production of the Ig heavy chain for which they are specific[23]. Recent studies reported by Adachi and Ishizaka suggest that Tδ cells are capable of producing IgD-BF[24]. Our own preliminary findings show that supernatants from hybridoma A3.4c6 also contain IgD-BF (A. Amin *et al.*, unpubl. obs.).

Further studies are needed to determine whether a direct Tδ-IgD+ B cell interaction and/or soluble factors mediate the immunoaugmenting effects of Tδ cells. Whatever the mechanism, assuming that IgD+ B cells are the only targets of Tδ cells, we have to conclude that the enhancing influence of Tδ cells on GC formation is exerted on IgD+ precursors of GC cells rather than on the IgD− GC cells themselves.

Acknowledgements: The expert technical assistance of Marilyn Shapiro, Melvin Bell and Ariel Soriano is gratefully acknowledged. Supported by USPHS grants AI-22645, AG-04980, and AI-11694.

REFERENCES

1. B. Pernis. Lymphocyte membrane IgD, Immunol. Rev. 37:210 (1977).
2. B. Xue, R. Coico, D. Wallace, G.W.Siskind, B. Pernis, and G.J. Thorbecke. Physiology of IgD. IV. Enhancement of antibody production in mice bearing IgD-secreting plasmacytomas, J. Exp. Med. 159:103 (1984).
3. R.F. Coico, B. Xue, D. Wallace, B. Pernis, G.W. Siskind, and G.J. Thorbecke. T cells with receptors for IgD, Nature 316:744 (1985).
4. R.F. Coico, B. Xue, D. Wallace, G.W. Siskind, and G.J. Thorbecke. Physiology of IgD. VI. Transfer of the immunoaugmenting effect of IgD wiht Tδ-containing helper cell populations, J. Exp. Med. 162:1852 (1985).
5. G. A. Bishop and G. Haughton. Induced differentiation of a transformed clone of Ly-1^{+} B cells by clonal T cells and antigen, Proc. Natl. Acad. Sci. USA 83:7410 (1986).
6. I. Berkower, H. Kawamura, L.A. Matis, and J.A. Berzofsky. T cell clones to two major T cell epitopes of myoglobin: effect of I-A/I-E restriction on epitope dominance, J. Immunol. 135:2628 (1985).
7. R.F. Coico, J.A. Berzofsky, J. York-Jolley, S. Ozaki, G.W. Siskind, and G.J. Thorbecke. Physiology of IgD. VII. Induction of receptors for IgD on cloned T cells by IgD and interleukin 2, J. Immunol. 138:4 (1987).
8. H. Kiyono, L.M. Mosteller-Barnum, A.M. Pitts, S.I. Williamson, S.M. Michalek, and J.R. McGhee. Isotype-specific immunoregulation. IgA-binding factors produced by Fcα receptor-positive T cell hybridomas regulate IgA responses, J. Exp. Med. 161:731 (1985).
9. N.K. Jerne, A.A. Nordin, and C. Henry. The agar plaque technique for recognizing the antibody producing cells, in "Cell Bound Antibody", Amos and Koprowsky, eds., Wistar Institute Press, Philadelphia, 109 (1963).
10. M.S. Neuberger and K. Rajewsky. Switch from hapten-specific immunoglobulin M to immunoglobulin D secretion in a hybrid mouse cell line, Proc. Natl. Acad. Sci. USA 78:1138 (1981).
11. R.F. Coico, S.R.S. Gottesman, G.W. Siskind, and G.J. Thorbecke. Physiology of IgD. VIII. Age-related decline in the capacity to generate T cells with receptors for IgD and partial reversal of the defect with IL 2, J. Immunol. 138:2776 (1987).
12. F.D. Finkelman, I. Scher, J.J. Mond, J.T. Kung, and E.S. Metcalf. Polyclonal activation of the murine immune system by an antibody to IgD. I. Increase in cell size and DNA synthesis, J. Immunol. 129:629 (1982).
13. F.D. Finkelman, J. Smith, N. Villacreses, and E.S. Metcalf. Polyclonal activation of the murine immune system by an antibody to IgD. VI. Influence of dose of goat anti-mouse δ chain and normal goat IgG on B lymphocyte proliferation and differentiation, Eur. J. Immunol. 15:315 (1985).
14. F.D. Finkelman, J. Smith, N. Villacreses, and E.S. Metcalf. Polyclonal activation of the murine immune system by an antibody to IgD. VII. Demonstration of the role of nonantigen-specific T help in in vivo B cell activation, J. Immunol. 133:550 (1984).
15. R.F. Coico, F. Finkelman, C.D. Swenson, G.W. Siskind, and G.J. Thorbecke. Exposure to crosslinked IgD induces receptors for IgD on T cells in vivo and in vitro, Proc. Natl. Acad. Sci. USA, in press (1987).
16. P.M. Allen and E.R. Unanue. Differential requirements for antigen processing by macrophages for lysozyme-specific T cell hybridomas, J. Immunol. 132:1077 (1984).
17. R.G. Hoover and R.G. Lynch. Lymphocyte surface membrane immunoglubulin in myeloma. II. T cells with IgA-Fc receptors are

markedly increased in mice with IgA plasmacytomas, J. Immunol. 125:1280 (1980).

18. R.F. Coico, B.S. Bhogal, and G.J. Thorbecke. Relationship of germinal centers in lymphoid tissue to immunologic memory. VI. Transfer of B cell memory with lymph node cells fractionated according to their receptors for peanut agglutinin, J. Immunol. 131:2254 (1983).
19. C.D. Swenson, R.F. van Vollenhoven, B. Xue, G.W. Siskind, G.J. Thorbecke, and R.F. Coico. Physiology of IgD. IX. Effect of IgD on immunoglobulin production in young and old mice (submitted for publication).
20. D. Yuan. Regulation of IgM and IgD synthesis in B lymphocytes. II. Translational and post-translational events, J. Immunol. 132:1566 (1984).
21. S. Kanowith-Klein, E.S. Vitetta, E.L. Korn, and R.F. Ashman. Antigen-induced changes in the proportion of antigen binding cells expressing IgM, IgG and IgD receptors, J. Immunol. 122:2349 (1979).
22. A. Mathur, D.M. Kamat, B.G. Vanness, and R.G. Lynch. Thymus-dependent in vivo suppression of IgE synthesis in a murine IgE-secreting hybridoma, J. Immunol., in press (1987).
23. W.H. Fridman, C. Rabourdin-Combe, C. Naeuport-Sautes, and R.H. Gisler. Characterization and function of T cell Fcγ receptors, Immunol. Rev. 56:51 (1981).
24. M. Adachi and K. Ishizaka. IgD-binding factors from mouse T lymphocytes, Proc. Natl. Acad. Sci. USA 83:7003 (1986).

SPECTROTYPE MODULATION OF BALB/C ANTI (1-3>)

DEXTRAN 1355S RESPONSES[1]

Franz Stäb, Kerstin Boss and Eckehart Kölsch

Department of Immunology
University of Münster
Domagkstr. 3, D-4400 Münster, FRG

The immune response of inbred mice against the "thymus-independent" antigen (1->3) Dextran 1355S (Dex) is genetically controlled. Congenic BALB.Igh^b mice are considered to be non-responders (1). BALB/c mice (Igh^a) are high responders (2) with expression of anti-Dex antibodies sharing idiotopes of public (MOPC 104, J 558) and private idiotypes (3). Conventionally raised euthymic BALB/c mice, due to T cell regulation, produce predominantly IgM, but only rarely IgG anti-Dex antibodies of these idiotypes (4,5). In BALB/c nu/nu and isolator raised (germfree) BALB/c mice, however, the class switch to IgG occurs readily in all animals (4).

Dex is an environmental bacterial antigen. The induction of down-regulation of the Dex response by environmental and consequently also by maternal influences can occur early in life, during pre- and perinatal development of the immune system (6,7). Our studies on age dependency in isotype and idiotype expression in strains of the haplotype Igh^a and Igh^b have revealed that previous pictures concerning the anti-Dex response have to be modified in several aspects.

MATERIALS AND METHODS

Dex 1355S has been kindly provided by Dr. M.E. Slodki, USDA, Preoria, IL, USA. BALB/c, BALB/c nu/nu and allotype congenic BALB.Igh^b mice were bred in our animal facilities. All other materials and methods have been described previously (4,5).

RESULTS AND DISCUSSION

Age dependency of the IgG anti-Dex response

Fig. 1a shows that the primary anti-Dex IgG response is more intensive in old BALB/c mice ($\geq$ 9 months) than in young mice (2-4 months). In addition a new IgG3 spectrotype (IEP: pHI 6.8) which is rarely expressed in young anti-Dex IgG3 producers, appears in most of the old animals.

[1]Supported by the Deutsche Forschungsgemeinschaft through SFB 310.

Furthermore, aged BALB.Igh^b and C57BL/6 mice immunized at the age of 9-12 months produce IgM and occasionally IgG3 anti-Dex antibodies, BALB.Igh^b mice mainly of the pH_I 6.8 spectrotype (Fig. 1b). Young BALB.Igh^b and C57BL/6 mice cannot mount any (1->3) Dex response (data not shown).

In BALB/c mice the anti-Dex response seems to be down regulated at the level of B-memory cells (5). This can be demonstrated by adoptive transfer of Dex primed spleen cells from old and young BALB/c donors (Fig. 2a,b) into X-irradiated syngeneic, congenic and (BALB/c x BALB.Igh^b)F1 recipients. Dex primed spleen cells of young BALB/c mice mount only a weak, those from old BALB/c mice, immunized at old age, show a strong secondary anti-Dex IgG

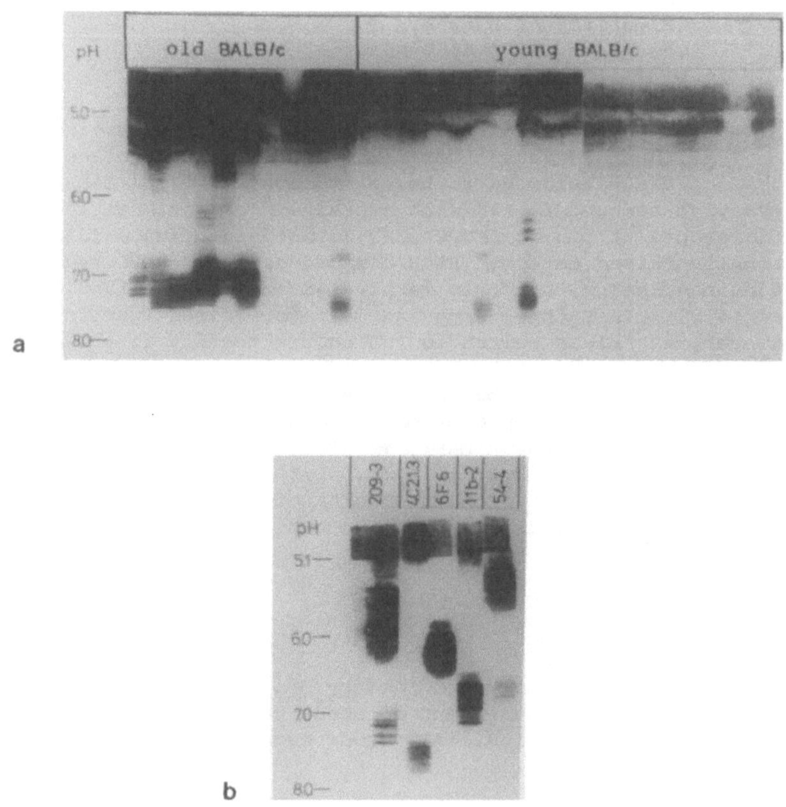

Fig. 1. IEF pattern of the anti-Dex IgG response in old and young mice

a) IEF pattern of old and young BALB/c mice. Sera of 10 old (11-16 months) and 18 young (2-3 months) BALB/c mice focussed 14 days p. inj. of 10 ug Dex i.p.

b) Sera of two old C57/BL6 (12 months) - 209.3 and two old BALB/c.Igh^b mice (9 and 12 months) - 11b.2 and 54.4 focussed 14 days p. inj. of 10 ug Dex i.p. 4C2.1.3 (IgG3) and 6F6 (IgG1) are anti-Dex mab (monoclonal antibodies).

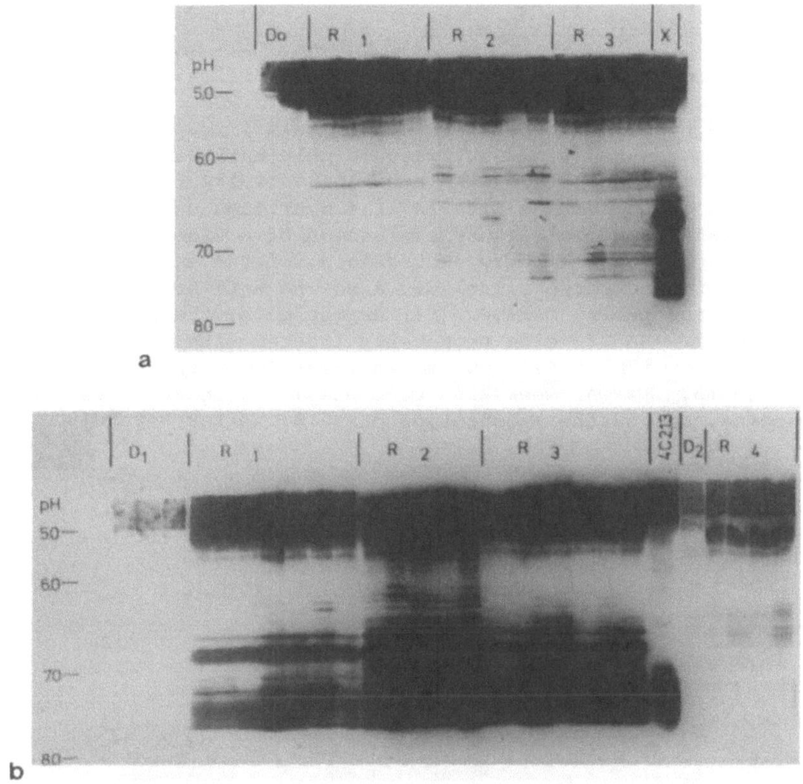

Fig. 2. Secondary anti-Dex response of primed spleen cells in adoptive transfer experiments

a) 1 x 10^7 spleen cells (Do) of two young BALB/c mice primed with 10 ug Dex i.p. were adoptively transfered (i.v.) into X-irradiated (650r) young recipients (R1 = 6 BALB/c, R2 = 5 BALB.Igh^b, R3 = 4 (BALB/c x BALB.Igh^b)F1) 4 weeks p. inj. of Dex. For comparison serum of one young recipient reconstituted with spleen cells of old, Dex primed donors (X).

b) 1 x 10^7 spleen cells (D1) of three old BALB/ mice (18 months) were adoptively transfered (i.v.) into groups of X-irradiated young recipients (R1, R2, R3) 40 weeks after priming with 10 ug Dex i.p. Control: 1 x 10^7 spleen cells (D2) of one naive BALB/c mouse (15 months) adoptively transfered into 4 X-irradiated BALB.Igh^b recipients (R4).

response in X-irradiated recipients, suggesting an accumulation of B-memory cells during life time. In BALB.Igh^b recipients the secondary anti-Dex IgG response of pooled BALB/c spleen cells is stronger as in syngeneic recipients presumably due to remaining suppressor mechanisms in X-irradiated young BALB/c mice.

Modulation of the anti-Dex IgG response

Young BALB/c mice challenged by s.c. injection of living hybridoma cells (7D7 3.4) secreting anti-Dex IgM of the public MOPC 104 idiotype (IdX) reject these hybridoma cells within 2-4 weeks (Fig. 3). These mice, when subsequently immunized with Dex, produce only anti-Dex IgG3 of the pH_I 6.8 spectrotype. Their sera have also anti-IdX activity (data not shown). Several challenges of these mice with 7D7 3.4 hybridoma cells and Dex simultaneously lead to an oscillatory expression of a wide panel of anti-Dex IgG3 spectrotypes presumably by neutralising of the anti-IdX regulation. This panel of spectrotypes includes also the anti-Dex IgG spectrotype (IEP: pH_I 7.4) of the public Id-MOPC 104, expressed by the hybridoma cell line 7D7 5.4 (IgG3). BALB7c mice neonatally treated with IdX anti-Dex IgG3 monoclonal antibodies (7D7 5.4) mount to anti-Dex IgG response up to 24 weeks post natum (p.n.). At week 26th p.n. one mouse showed a weak anti-Dex IgG3 response, but of the spectrotype pH_I 6.8. After several

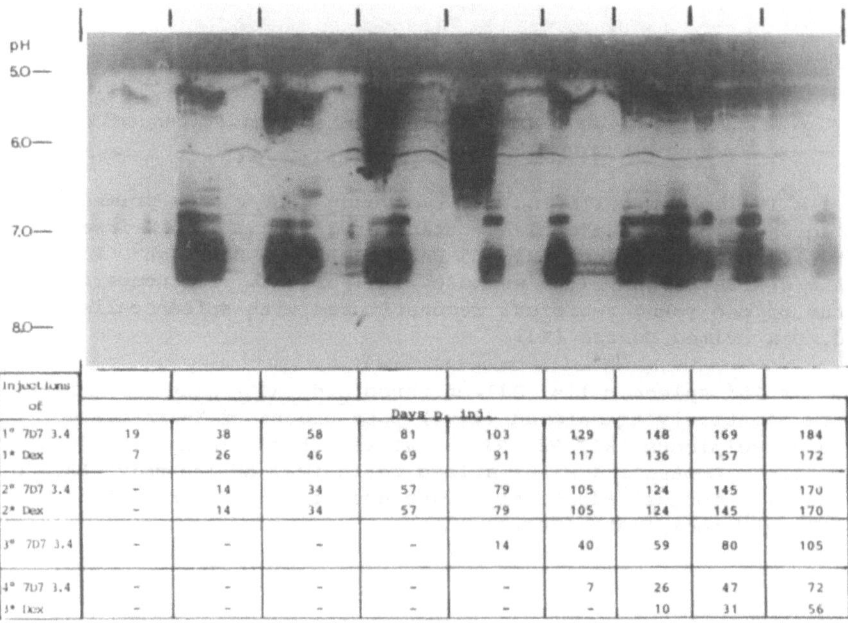

Injections of	Days p. inj.								
1° 7D7 3.4	19	38	58	81	103	129	148	169	184
1° Dex	7	26	46	69	91	117	136	157	172
2° 7D7 3.4	-	14	34	57	79	105	124	145	170
2° Dex	-	14	34	57	79	105	124	145	170
3° 7D7 3.4	-	-	-	-	14	40	59	80	105
4° 7D7 3.4	-	-	-	-	-	7	26	47	72
3° Dex	-	-	-	-	-	-	10	31	56

Fig. 3. IEF pattern of sera from 4 young BALB/c mice (2 months) challenged with living anti-Dex hybridoma 7D7 3.4 (IgM, public MOPC 104 idiotype) and subsequently immunized with 10 ug Deg i.p.

challenges with living hybridoma cells (7D7 5.4) of the pH_I 7.4 spectrotype, the hybridoma cells were rejected rapidly. Additionally the pH_I 6.8 spectrotype was oscillatorily expressed in some mice. Sera of those mice had anti-IdX activity. In controls age matched naive BALB/c mice could not reject the hybridoma cells (7D7 5.4) and died within 4 weeks (data not shown).

The data show that specific anti-Id suppressor mechanisms can be generated by idiotypic stimulation early in life. Similar regulation mechanisms have been induced by maternal influences (6,7).

Regulation of the anti-Dex IgG response in T cell reconstituted BALB/c nu/nu mice

It has been shown previously that the pronounced overall IgG anti-Dex response of BALB/c nu/nu mice is reduced in animals reconstituted with T cells of naive BALB/c mice (5). The IEF analysis of anti-Dex immune sera of BALB/c nu/nu mice, reconstituted with T cells from young BALB/c mice neonatally treated with syngeneic (mab) anti-Dex IgG3 (7D7 5.4) shows only the IEP pH_I 6.8 spectrotype. These IgG3 spectrotype with IEP pH_I 6.8 has kindly been characterized by Drs. G. Lehle and E. Weiler, University of Konstanz, FRG, as private anti-Dex IgG idiotype only weakly crossreacting with the public anti-Dex MOPC 104 idiotype (personal communications).

With regard to spectrotype analysis the data suggest a specific T cell regulation of the anti-Dex response at the level of an anti-idiotype specific T suppressor cell or a T helper cell for anti-idiotypic B cells. It seems that the anti-Dex response of the BALB/c mouse has a crucial regulatory element at the level of B-memory cells, whereas in mice of the Igh^b allotype critical steps in down regulation of the anti-dex responsiveness manifest themselves earlier in B cell ontogeny (8,9).

REFERENCES

1. W. Geckeler, B. Blomberg, C. de Preval, and M. Cohn, C.S.H.S. Quant. Biol. 41:743 (1977).
2. B. Blomberg, W. Geckeler, and M. Weigert, Science 177:178 (1972).
3. B. Clevinger, J. Schilling, L. Hood, and J.M. Davie, J. Exp. Med. 151:1059 (1980).
4. W. Schuler, G. Lehle, E. Weiler, and E. Kölsch, Eur. J. Immunol. 12:120 (1982).
5. W. Schuler, A. Schuler, and E. Kölsch, Eur. J. Immunol. 14:578 (1984).
6. I. J. Weiler, E. Weiler, R. Sprenger, and H. Cosenza, Eur. J. Immunol. 7:591 (1977).
7. M. Vakil and J. F. Kearney, Eur. J. Immunol. 16:1151 (1986).
8. J. Pene, H. Zaghouani, and M. Stanislawski, Ann. N.Y. Acad. Sci. 418:296 (1983).
9. B. G. Froscher and N. R. Klinmann, J. Exp. Med. 162:1620 (1985).

B LYMPHOMA CELLS PROCESS AND PRESENT THEIR ENDOGENOUS IG: IMPLICATIONS FOR NETWORK THEORY

Bjarne Bogen[o+] and Siegfried Weiss[+]

[o]Institute of Immunology and Rheumatology, University of Oslo, Fr. Qvamsgt. 1, 0172 Oslo 1, Norway and [+]Basel Institute of Immunology, Basle, Switzerland

We show here that B lymphoma cells not only express native immunoglobulin (Ig) on their cell surface but also processed forms of the same Ig in the context of MHC products. Secretion and uptake of Ig does not seem to be necessary for processing and presentation to take place.

The BALB/c T cell clones used in this study are $CD4^+$ and have an α, β heterodimeric T cell receptor. They react to an idiotope (Id) on the syngeneic myeloma protein M315 (α,λ2) in the context of the class II MHC molecule $I\text{-}E^d$. The Id is expressed on a synthetic peptide which contains amino acids (AA) 91-108 of $\lambda 2^{315}$; the expression is influenced by AA exchanges at positions 94, 95 and 96 (1,2).

Because of somatic mutations, the λ2 light chain of M315 differs from the germ line-encoded λ2 light chain at the crucial 94, 95, 96 residues (3). Therefore, to obtain a clonal population of B cells expressing endogenously produced idiotype, BALB/c B lymphoma cell lines were transfected with the $\lambda 2^{315}$ gene (Fig. 1). We used the plasmid pSV2λ2, which consists of a 6.6 kb EcoRI fragment containing the λ2 gene of MOPC 315 cloned into the expression vector pSV2neo (ref. 3, the construct was kindly provided by Dr. Wu, Basle). To increase the λ2 expression, the 1 kb XbaI fragment of the heavy chain enhancer (4) was isolated from the cloned Sp6 heavy chain gene (provided by A. Traunecker, Basle) and inserted into the λ2 gene with the same orientation as it is found in the heavy chain gene. This construct was transfected by protoplast fusion into 3 sublines of the A20J BALB/c lymphoma (5). As a representative example, the F10 transfectant (derived from A20/10) expresses the λ2 light chain together with the parental κ and heavy chain on the cell surface as detected by immunoprecipitation (Fig. 1). F10 and A20/10 express similar amounts of I-A^d and I-E^d molecules. Supernatants of F10 cultures contain 50-500 ng/ml of λ2 chains.

Table I demonstrates that mitomycin C-treated parental A20/46 cells only stimulate an Id-specific T cell clone when exogenous $\lambda 2^{315}$ is added. Glutaraldehyde fixation of B lymphoma cells reduces their capacity to present soluble $\lambda 2^{315}$ chains drastically. However, such fixed cells still efficiently present a synthetic peptide containing the idiotope. This suggests that Id-specific T cells recognize a processed form of the idiotope, similar to what has been found for other protein antigens (6). In contrast to A20/46, both mitomycin C-treated and glutaraldehyde-fixed F9 cells (transfectant derived from A20/46) spontaneously stimulate 4B7 T cells in an I-E^d restricted manner (Table I). This indicates that the transfected cells continuously process their endogenously produced $\lambda 2^{315}$ polypeptide chain.

It is unlikely that the processed form of $\lambda 2^{315}$ stems from secreted and endocytosed $\lambda 2^{315}$. First, the concentration of free $\lambda 2^{315}$ in supernatants of F9 is many times less than that of exogenously added $\lambda 2^{315}$ needed to induce proliferative responses (no response to 1 µg/ml of $\lambda 2^{315}$). Second, we failed to demonstrate in a Costar Transwell (cat. No. 3410) diffusion chamber system that exposure to $\lambda 2^{315}$ - containing F9 supernatant could antigen-prime A20/46 cells. After cocultivation in Transwells, F9 T.W. and A20/46 T.W. cells were tested as APCs in a growth inhibition assay (Table II). (We have previously shown that our long-term T cell clones are cytotoxic to L cells in an antigen and MHC-specific fashion, ref.1). Only F9 T.W. cells and not A20/46 T.W. cells were killed by the Id-specific T cells. Diffusion of ^{125}I-labelled M315 across the membrane (3.0 µm pores) seperating inside and outside wells reached equilibrium within 2-4 hours (data not shown). Furthermore, concentration of λ2 was about the same in supernatants from inside and outside wells in cocultivation experiments (Table II). It should also be noted that growth-inhibition experiments were set up after 4 and 5 days (or 10 and 11 days) of cocultivation to ensure that

A20/46 at day 5 were exposed to higher concentrations of $\lambda 2^{315}$ than F9 at day 4 (Table II). Thus, in disagreement with a recent hypothesis (7), the data suggest that cells may process and present endogenous self proteins not only in the context of their class I (8) but also their class II molecules.

The findings have important implications for network theory (9). On the basis of earlier experiments, we have previously suggested that B cells may process their Ig and present peptides to Id-specific, MHC-restricted T cells (1,10,11). Since then similar ideas have been advanced by others (12,13). It has also recently been shown that unfractionated spleen cells stimulate Ig allotype-specific, class II molecule-restricted T cells (14). The present data provide the first direct evidence that B lymphoma cells process their endogenous Ig and present Id-containing peptides in the context of class II molecules. The findings suggest that B cells spontaneously process their Ig without interaction with ligand (antigen or anti-Id antibodies). Thus, the clone sizes of B cells may be directly regulated by Id-specific, MHC-restricted T cells. However, as emphasized earlier (1,11,15), this idea has several limitations. First, T cells may be tolerant to germ line-encoded peptides (15). (Therefore, Id-specific, MHC-restricted T cell help may play a role in the somatic diversification of Ig.) Secondly, B cells must have proper class II molecules to present V-region peptides (16). Thirdly, some Ig may be fairly resistant to processing (1,17).

Polyclonal T cells induced by $\lambda 2^{315}$ immunization help secondary anti-Id^{315} B cells in vivo (18). This observation has since been interpreted (1,11) within current concepts of linked recognition and T-B collaboration, i.e. that anti-Id B cells are probably far more potent APCs of Id^+ Ig than B cells of other specificities. From this, together with the observations presented here, one is compelled to conclude (for discussion see ref. 11) that any B cell processes both endogenous Ig and soluble Ig which binds sIg. Therefore, Id^+ and anti-Id B cells may be recognized by both Id-specific and anti-Id specific, MHC-restricted T cells. However, on a B cell surface, the ratio of peptides originating from Id^+ and anti-Id Ig could vary according to the stage of B cell differentiation (as could class II molecule expression). Also, the limitations discussed in the preceding paragraph suggest that only certain peptides may be presented and recognized by Id-specific T cells. Finally, further V-gene mutations may change antibody or T cell-defined idiotopes so that B cells or T and B cells can no longer communicate. In conclusion, more experiments are needed to delineate the functional outcome of Id-related T-B interactions _in vivo_.

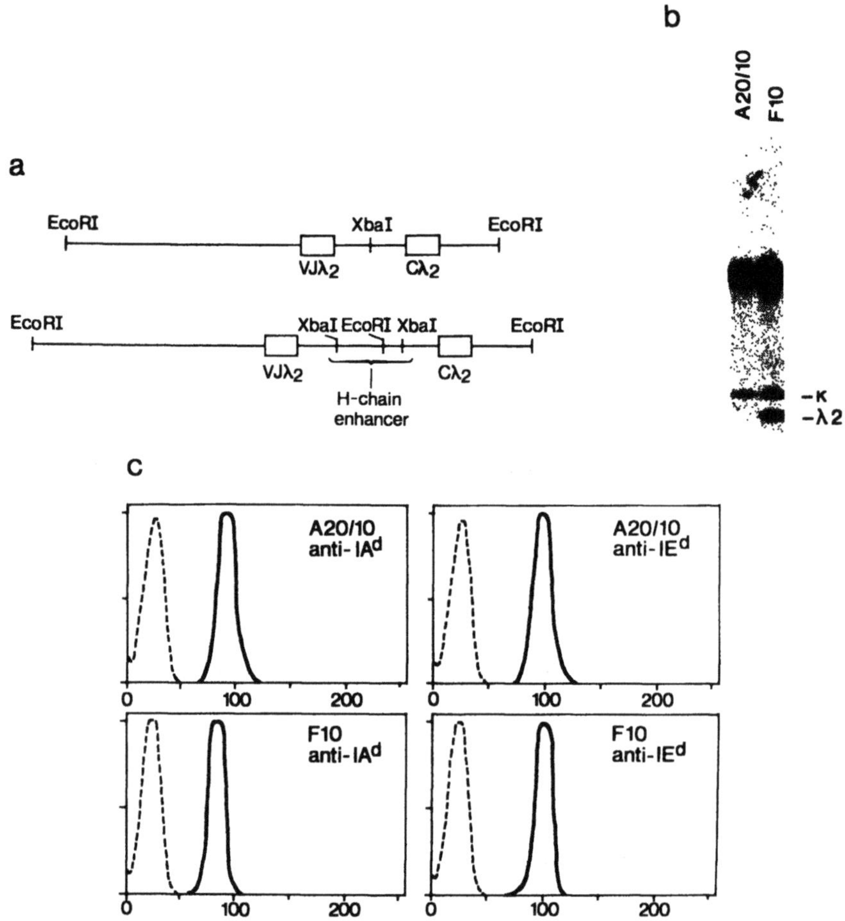

Fig. 1

a) Construct of the λ2 gene of MOPC315 containing the 1 kb XbaI heavy chain gene enhancer fragment. The modified λ2 construct is integrated in the EcoRI site of the expression vector pSV2neo.

b) Electrophoretic analysis of ^{125}I-labelled surface proteins of A20/10 and the corresponding transfectant F10 precipitated with anti-mouse Ig and protein-A Sepharose.

c) FACS analysis of A20/10 and F10 using fluorescein conjugated monoclonal anti-I-A^d and anti-I-E^d antibodies.

Table I Proliferation of cloned 4B7 Id-specific T cells ($cpm \times 10^{-3} \pm SD$)[a)]

Antigen	Mitomycin C- treated APC		Glutaraldehyde-fixed APC	
	A20/46	F9	A20/46	F9
None	3 ± 1	14 ± 2	1 ± 1	36 ± 2
$\lambda 2^{315}$ 10 μg/ml	27 ± 5	n.d.	1 ± 1	n.d.
FRN 5 μg/ml	n.d.	n.d.	14 ± 1	n.d.

a) Mitomycin C treatment (1), gluteraldehyde fixation (6) and proliferative assays (1) were essentially performed as described. FRN is a synthetic peptide (kindly provided by Dr. Hannestad, Tromsø) stretching from residues 91-108 of the $\lambda 2^{315}$ sequence. The proliferative T cell response of F9 cells was inhibited by anti-I-E^d but not anti-I-A^d antibodies. Similar results were obtained with other A20J sublines and transfectants (not shown).

Table II. F9 supernatant does not $\lambda 2^{315}$ prime A20/46 cells[a)]

	A20/46 T.W.		F9 T.W.	
	day 4	day 5	day 4	day 5
B cells alone	48 ± 2[b)]	169 ± 6[b)]	124 ± 4[b)]	82 ± 3[b)]
B cells + T cells	48 ± 1[b)]	153 ± 4[b)]	21 ± 3[b)]	7 ± 1[b)]
% growth inhibition	0	9	83	91
λ2 (ng/ml)	61 ± 4[)c]	290 ± 101[c)]	64 ± 2[c)]	252 ± 28[c)]

a) F9 transfectant was cultured in inside wells of Costar Transwell™ plates (F9 T.W.), parental A20/46 T.W. in outside wells. After 4 and 5 days, $\lambda 2^{315}$ concentrations were measured in supernatants (see c) and growth-inhibition experiments (1) were set up. 5×10^3 B lymphoma cells and 1×10^4 2000 rad-irradiated T cells were cultured in triplicate for 48 hours before a 10-18 hour pulse with 2 μCi ^{3}HTdR. Similar results were obtained with other A20 sublines and transfectants and at days 10 and 11 of cocultivation (not shown).

b) $cpm \times 10^{-3}$ incorporated into B lymphoma cells ± SD.

c) When the growth-inhibition assays were set up, the concentration of total λ2 chains in B lymphoma cell supernatants ± SD was measured by a soluble radio-immuno-assay.

REFERENCES

1. Bogen,B., Malissen,B. and Haas,W. Eur. J. Immunol. 16:1373 (1986).
2. Bogen,B. et al. Eur. J. Immunol. 16:1379 (1986).
3. Wu,G.E. et al. Nuc. Ac. Res. 10:3831 (1982).
4. Banerji,J., Olson,L. and Schaffner,W. Cell 33:729 (1983).
5. Kim,K.J. et al. J. Immunol. 122:549 (1979).
6. Shimonkevitz,R. et al. J. Exp. Med. 158:303 (1983).
7. Germaine,R.N. Nature 322:687 (1986)
8. Townsend,A.R.M. et al. Cell 44:959 (1986).
9. Jerne,N.K. Ann. Immunol (Inst. Pasteur) 125c:373 (1974).
10. JØrgensen,T., Bogen,B. and Hannestad,K. In Immunoglobulin idiotypes, ICN-UCLA symposia on molecular and cellular biology, eds. Janeway,C.A. et al., Academic Press, New York 1981 p.573.
11. Bogen,B. Thesis 1984, TromsØ. Available from author.
12. Leserman,L. Immunology Today. 6:352 (1985).
13. Kourilsky,P. et al. Proc. Natl. Acad. Sci. (U.S.A.) 84:3400 (1987).
14. Bikoff,E. and Birshtein, B.K. J. Immunol. 137:28 (1986).
15. Bogen,B. JØrgensen,T. and Hannestad,K. Eur. J. Immunol. 15:278 (1985).
16. JØrgensen,T. and Hannestad,K. Nature 288:396 (1980).
17. Bogen,B. JØrgensen,T. and Hannestad,K. Eur. J. Immunol. 13:353 (1983).
18. JØrgensen,T. and Hannestad,K. Eur. J. Immunol. 7:426 (1977)
19. Hannestad,K., Kristoffersen,G. and Briand,J.P. Eur. J. Immunol. 16:889 (1986).

GERMINAL CENTER B CELLS PRESENT ANTIGEN OBTAINED *IN VIVO* TO T CELLS *IN VITRO* AND STIMULATE MIXED LYMPHOCYTE REACTIONS

Marie H. Kosco, Alan P. Monfalcone, Andras K. Szakal and John G. Tew

Deparment of Microbiology and Immunology, Medical College of Virginia, VCU, Richmond, VA 23298, USA

Department of Anatomy and Immunobiology, Medical College of Virginia, VCU, Richmond, VA 23298, USA

INTRODUCTION

Shortly after secondary immunization, germinal center (GC) B cells have been observed in ultrastructural studies to obtain antigen from follicular dendritic cells (FDCs) in the form of immune complexes (1). This antigen appears to be degraded by GC B cells and may be processed for presentation to T cells. The present study was undertaken to determine if GC B cells can process and present antigen obtained from FDCs in vivo to appropriate T cells in vitro. In addition, the stimulatory capacity of GC B cells on T cells was evaluated in primary mixed lymphocyte reactions (MLR). The findings presented here demonstrate that antigen obtained in vivo by GC B cells can be processed and presented to T cells. The ability of GC B cells to interact with T cells is indicated by the fact that they served as potent stimulators of allogeneic T cells. In contrast, the small high density (HD) B cells are weak to negative stimulators under the same conditions. In addition, the GC B cells were even able to stimulate detectable syngeneic MLRs. These data suggest that in vivo, GC B cells are capable of interacting with T cells, in turn inducing the T cell help needed for the germinal center reaction, generating B memory cells, and influencing the high titers of antibody associated with the secondary antibody response.

MATERIALS AND METHODS

GC B cells were isolated from immune mice by methods described elsewhere (2). Briefly, after enzymatic digestion of lymph nodes, GC B cells were obtained using Percoll density centrifugation followed by a panning procedure. The panning technique utilizes the ability of the plant lectin, peanut agglutinin (PNA), to selectively bind to GC B cells (3,4). The resulting enriched population was approximately 80% positive for PNA, 97% positive for Ia and surface IgM, but less than 0.01% positive for Thy 1.2 or esterase. In some experiments, the PNA positive fraction was further purified to near 100% using fluorescein conjugated PNA and a fluorescence activated cell sorter. Antigen present tion by GC B cells was assessed by the ability of these cells to elicit IL 2 production from the 3DO-54.8 cells an OVA-specific T hybridoma. IL 2 production was measured using [^{3}H]-thymidine incorporation by the IL 2 dependent cell line, CTLL.

Mixed lymphocyte reactions (MLR) were carried out with three isolated cell types: 1. T cells were obtained from spleens of unstimulated C3H mice for the allogeneic MLR, and Balb/c spleens for the syngeneic MLR. The T cell population was purified by an adherence step and passage through nylon wool columns; 2. GC B cells were obtained from immune Balb/c mice 5-6 days after challenge with OVA as outlined above; 3. High density (HD) B cells (i.e. small, resting lymphocytes, Ref. 5). were used as comparison for the stimulatory capacity of lower density GC B cells. The HD B cells were obtained from the spleens of actively immunized Balb/c mice. HD B cells were purified by passage through two Sephadex G-10 columns, treatment with anti-Thy 1.2 and rabbit low-Tox compliment, and collected from the 60-65% interface of discontinuous Percoll gradients (density>1.080). For the MLR, cell cultures were set up in medium plus 10% fetal calf serum containing $4x10^5$ T cells plus various amounts of a B cell population.

RESULTS AND DISCUSSION

Initially, the ability of GC B cells to present antigen was tested by adding ovalbumin (OVA) to cultures containing GC B cells from OVA immune mice and the OVA-specific T cell hybridoma, 3DO-54.8. GC B cells were obtained about a month after a second challenge with OVA (i.e. not recently stimulated) and 10^6 of these cells were added to each well in the presence or absence of the 3DO-54.8 cells and OVA. GC B cells together with exogenous OVA did not produce any IL 2 (300 cpm). Upon addition of the 3DO-54.8 cells, IL 2 production was obtained (53,300 cpm). As a positive control, lymph node macrophages from the same set of mice were included. These macrophages elicited a comparable amount of IL 2 production from the 3DO-54.8 cells upon the addition of OVA (52,500 cpm).

Based upon ultrastructural observations reported in the article by Szakal et al. (this volume and Ref. 1), we next set out to determine if GC B cells could process and present antigen they obtained in vivo from FDCs. GC B cells were obtained 5 days after antigenic challenge, which corresponded to the time when they were observed in situ at the ultrastructural level to possess the greatest amount of ingested antigen (1). The results demonstrated that without adding OVA to the cell cultures, OVA-specific Balb/c GC B cells isolated 5 days after secondary immunization stimulated the 3DO-54.8 cells in a dose dependent fashion. Controls using GC B cells from an inappropriate strain of mouse or challenging with an anotherantigen resulted in no IL 2 production. These results indicated that the appropriate genetic restrictions applied and that the immunological response was specific and not attributable to polyclonal or nonspecific events.

Next, we investigated the kinetic pattern of GC B cells to process and present antigen obtained in vivo at various times after challenge with OVA (Table 1).

Table 1. Kinetics of IL 2 production by 3DO-54.8 cells in response to presentation by GC B cells of antigen obtained from FDCs in vivo.

Days After Antigenic Challenge with OVA	^{3}H-Thymidine Incorporation* *cpm ± SE*
Day 0	626 ± 170
Day 1	2,257 ± 300
Day 3	4,244 ± 285
Day 5	10,068 ± 940
Day 8	7,150 ± 520
Day 11	704 ± 160

*- IL 2 production was a measurement of proliferation of the IL 2 dependent CTLL cell line.

The results shown above were consistant with the prediction that maximum antigen presentation should occur around Day 5 when peak levels of antigen were observed in GC B cells (1). At Day 0 (i.e. before boosting the mice), when no iccosomes (i.e. FDC-derived immune complex coated bodies; see Szakal et al., this volume for complete description) were present, the GC B cells did not stimulate IL 2 production. One day after the boost, when the first iccosomes were being formed by the FDCs, the GC B cells elicited a slight, but statistically significant, IL 2 response. Three days following a boost, paralleling the appearance of iccosomes in germinal centers in peak numbers, GC B cells were able to elicit a greater IL 2 response then at Day 1. By Day 5, when the maximum amount of antigen uptake by GC B cells was observed (1), the isolated GC B cells were able to elicit the greatest IL 2 response. Finally, at later times when a decline in the amount of antigen endocytosed by B cells was observed, there was a concommittant decline in the ability of the isolated GC B cells to present the OVA obtained in vivo.

To be sure that the antigen being presented was associated with the PNA positive fraction of the PNA enriched population, a cell sorting experiment was conducted. Mice were boosted with ^{125}I-labeled OVA and five days later, lymph node cells were obtained. Following the panning step, the cells eluted from the PNA plates were incubated with fluorescein conjugated-PNA, washed and separated into PNA high and PNA low-to-negative fractions using the FACS. Seventy-two percent of the cells were sorted into a highly PNA positive fraction and 28% into a PNA low-to-negative fraction. When the amount of ^{125}I-labeled antigen in each population was measured, over 90% of the activity was in the highly PNA positive fraction, thus confirming that antigen was associated with the PNA positive GC B cells and not the contaminating fraction. Furthermore, the addition of the sorted PNA positive cell fraction to 3DO-54.8 cells induced the production of IL 2.

Next, we determined the amount of OVA present in GC B cells isolated at Day 5 and the minimum amount of antigen necessary to stimulate IL 2 production. The mice were immunized initially intraperitoneally with unlabeled OVA and then ^{125}I-labeled OVA each time they were challenged in the footpads. The amount of OVA present in 10^6 GC B cells 5 days after challenge with OVA was remarkably small, 4.5 picograms. Nevertheless, this amount of antigen was sufficient to elicit a significant level of IL 2 production (5,700 cpm above backgound). Decreasing the amount of OVA in each well by adding only half the number of GC B cells per culture still resulted in significant production of IL 2. Dose response studies done by adding exogenous OVA to GC B cells obtained 3 weeks or more after a secondary immunization demonstrated that 100 to 1000 times more free antigen was required to obtain a similar IL 2 response. This data indicates that the FDC-derived antigen obtained by the GC B cells _in vivo_ was remarkably immunogenic as compared to the antigen added to the cell cultures.

In order to evaluate the ability of GC B cells to interact with and stimulate T cells, mixed lymphocyte reactions were studied. For comparative purposes, B cells from the high density fraction of the Percoll gradients (HD B cells) were also purified and assayed in the MLR. (GC B cells are found in the low density fraction.) The results from the two B cell populations are tabulated below in Table 2. The data show that GC B cells are potent stimulators of an allogeneic MLR requiring as little as 15,000 GC B cells to stimulate a significant proliferative response of the T cells. In comparison, the HD B cells were not able to stimulate the MLR at even the highest cellular concentrations. Subsequent studies conducted with HD B cells cultured first in neuraminidase (which increases the number PNA-binding receptors exposed on the cell surface), then assayed in the MLR, show a dramatic increase in their ability to stimulate the MLR.

In addition, the ability of GC B cells from OVA-primed Balb/c lymph nodes to stimulate a syngeneic MLR with T cells from unprimed Balb/c spleens was tested. The data showed that the mitomycin C treated GC B cells caused a significant proliferation of the naive T cells (1.25×10^5 GC B cells stimulated over 20,000 cpm above background). These results from the mixed lymphocyte reactions demonstrate that GC B cells express the appropriate stimulatory molecules to efficiently activate T cells.

SUMMARY AND CONCLUSIONS

In the present study we sought to test the ability of GC B cells to activate T cells by antigen presentation and mixed lymphocyte reactions. The first set of experiments demonstrated that GC B cells obtained three or more weeks after immunization could induce IL 2 production by T cells only by adding antigen to the cell cultures. In the next

Table 2. Ability of GC B cells[1] versus High Density B cells[2] to stimulate allogeneic mixed lymphocyte reactions[3].

Number of B cells[4] added	Allogeneic T cells[5] plus GC B cells	Allogeneic T cells[5] plus HD B cells
	(cpm ± SE)[6]	
250,000	171,000 ± 11,000	2,000 ± 150
125,000	96,000 ± 10,000	1,300 ± 300
60,000	75,000 ± 7,500	1,500 ± 200
30,000	24,000 ± 7,000	1,100 ± 100
15,000	5,000 ± 1,300	1,400 ± 100
B or T cells alone	1,300 ± 800	------

1- Source of GC B cells - lymph nodes of mice challenged 5 days prior in foot pads with OVA;
2- Source of high density (HD) B cells - spleens of Balb/c mice challenged 5 days prior in foot pads with OVA;
3- Measured after 4 days in culture;
4- B cells were treated with Mitomycin C to inhibit B cell proliferation;
5- Source of T cells - nylon wool enriched from spleens of unstimulated C3H mice;
6- Measurement of 3H-Thymidine incorporation.

series, it was demonstrated that GC B cells isolated as early as 1 day and for over a week following an antigenic challenge with OVA (i.e. where they obtained antigen from the FDC-derived iccosomes in vivo), were able to stimulate high levels of IL 2 production in the absence of exogenous OVA. This response was maximal on Day 5 which corresponded precisely with the keinetics of the ultrastructural studies which document the uptake of antigen by GC B cells in vivo (see Szakal et. al, this volume). Furthermore, it was shown that the FDC-derived antigen was remarkably immunogenic when compared with exogenous antigen. The data from the MLRs demonstrated that GC B cells, which are low density cells, are potent stimulators of T cells compared to the resting, low density B cells. In addition, GC B cells even stimulated a significant syngeneic response. These results indicate that GC B cells can present antigen and interact with T cells efficiently. This is believed to be important in the germinal center reaction and in the induction and maintenance of the immune responses.

ACKNOWLEDGEMENTS

This work is supported by NIH Grant #AI-17142.

REFERENCES

1. Szakal, A.K., M.H. Kosco and J.G. Tew. 1987. A novel in vivo follicular dendritic cell dependent iccosome mediated mechanism for delivery of antigen to antigen processing cells. J. Immunol. (In Press).

2. Kosco, M.H., A.K. Szakal and J.G. Tew. 1987. In vivo obtained antigen presented by germinal center B cells to T cells in vitro. J. Immunol. (In Press).

3. Butcher, E.C., R.V. Rouse, R.L. Coffman, C.N. Nottenburg, R.R. Hardy, and R.L. Weissman. 1982. Surface phenotype of Peyer's patch germinal center cells: implications for the role of germinal centers in B cell differentiation. J. Immunol. 129:2698.

4. Coico, R.F., B.S. Bhogal, and G.J. Thorbecke. 1983. Relationship of germinal centers in lymphoid tissue to immunologic memory. VI. Transfer of B cell memory with lymph node cells fractionated according to their receptors fro peanut agglutinin. J. Immunol. 131:2254.

5. Krieger, J.I., S.F. Grammar, H.M. Grey and R.W. Chestnut. 1985. Antigen presentation by splenic B cells: Resting B cells are effective accessory cells for T cells responses. J. Immunol.135:2937.

CHARACTERIZATION OF CELL POPULATIONS, REAPPEARING IN THE MOUSE SPLEEN AFTER ELIMINATION BY LIPOSOME ENCAPSULATED DICHLOROMETHYLENE DIPHOSPHONATE: REAPPEARANCE OF MARGINAL ZONE LYMPHOCYTES IS INDEPENDENT OF MARGINAL ZONE MACROPHAGES

Nico van Rooijen, Nel Kors and Georg Kraal

Immunocytochemistry Unit, Department of Histology, Medical Faculty, Free University, P.O. Box 7161, 1007 MC Amsterdam, The Netherlands

Macrophages in the spleen can be eliminated by treatment with liposome encapsulated dichloromethylene diphosphonate (DMDP; reviewed in ref. 1. These macrophages disappeared during the first day after intravenous injection with liposome-encapsulated DMDP, remained absent for one week and reappeared gradually in the next weeks, as studied with the aid of enzyme histochemical techniques (2). Macrophages in different compartments of the mouse spleen can now be characterized by various monoclonal antibodies (3-6).

We also showed that in addition to the macrophages the marginal zone lymphocytes disappeared from this compartment after treatment with liposome encapsulated DMDP, in spite of the fact that they do not ingest and digest the liposomes (7). This phenomenon may be explained by the release of enzymes from dying macrophages, affecting the marginal zone lymphocytes. Apart from elimination of marginal zone lymphocytes, their disappearance from the marginal zone may also be caused by migration due to an altered microenvironment in that part of the spleen following the removal of macrophages (7). However, it is shown in the present study that marginal zone B lymphocytes reach their normal concentration amply before the reappearance of marginal zone macrophages in the spleen. It is concluded that marginal zone macrophages are not obligatory for the homing and/or retention of marginal zone B lymphocytes.

MATERIALS AND METHODS

Preparation of liposomes with entrapped DMDP

Multilamellar liposomes were prepared as described earlier (8). In brief, 75 mg phosphatidylcholine and 11 mg cholesterol (Sigma Chemical Co., St. Louis, Missouri, U.S.A.) were dissolved in chloroform in a round-bottom flask. The thin film that formed on the interior of the flask after low vacuum rotary evaporation at 37°C was dispersed by gentle rotation for 10 min in 10 ml phosphate-buffered saline (PBS; 0.15 M NaCl/10 mM phosphate buffer, pH 7.4). For DMDP containing liposomes 1.89 DMDP was added to the 10 ml PBS. More DMDP could not be dissolved in the buffer. DMDP was a kind gift of Procter and Gamble, U.S.A. Liposome encapsulated DMDP was separated from free DMDP by washing of the liposomes. The DMDP containing liposomes were finally resuspended in 5 ml PBS. 0.2 Ml of the liposome suspension (containing 5 mg lipid and 2 mg DMDP) was intravenously injected in 21 mice.

Characterization of macrophages and lymphoid cells in the marginal zone of the spleen

Blocks of fresh splenic tissue were frozen in liquid nitrogen and stored at -20°C. Cryostat sections of 8-10 µm thickness were fixed in acetone for 10 min and air-dried for at least 30 min. After washing in 0.01 M phosphate-buffered saline (PBS, pH 7.4) the sections were incubated with culture supernatants of the monoclonal rat anti-mouse antibodies directed to marginal zone macrophages (ERTR9; 3, 4) at saturating concentrations in PBS containing 0.1% bovine serum albumin (BSA, Poviet, Tilburg, The Netherlands) for 30 min. After washing thoroughly in PBS, the slides were incubated with a 1:100 dilution of peroxidase-conjugated rabbit anti-rat IgG (Dako, Glostrup, Denmark) in PBS/BSA containing 1% normal mouse serum for 30 min. After washing in PBS again, the peroxidase activity was visualized with 3,3'-diaminibenzidine-tetrahydrochloride (DAB, Sigma, St. Louis, MO, USA), 0.5 mg/ml Tris-HCl buffer (pH 7.6) containing 0.01% H_2O_2. The sections were stained for 10-15 min at room temperature.

B-lymphocytes in the marginal zone were stained with peroxidase conjugated goat IgG anti-mouse IgM (µ-chain specific; Cappel Labs. Cochranville, PA, USA).

Experimental design

Twenty-seven (C3D2)F1 female mice were used for the experiments and kept under routine laboratory conditions throughout the experiment. Six mice were killed and their spleens taken, the remaining mice received a single intravenous injection with 0.2 ml of the DMDP liposomes suspension at day 0. Three mice were killed at each of the following time intervals after injection: 1, 2, 4, 8, 16, 32, 64 days. Their spleens were taken and prepared for histology as described above.

RESULTS

In the normal mouse spleen, ERTR9 positive cells are found exclusively in the marginal zone surrounding the white pulp (Fig. 1A). ERTR9 positive cells have long branched cell processes and have been characterized as a specialized subset of spleen macrophages (3, 4). IgM positive cells are found in the normal spleen in all B lymphocyte compartments of the white pulp, i.e. in the follicles, in the periphery of the periarteriolar lymphocyte sheaths (PALS) and in the marginal zone. Marginal zone lymphocytes stain more intensely than B lymphocytes in the adjacent corona of the follicles (Fig. 1B). At one day after a single intravenous injection with DMDP liposomes, ERTR9 positive cells have disappeared from the spleen, and only few IgM positive marginal zone lymphocytes remained present. This picture did not differ from that at day 2, 4 (Fig. 2A, B) and 8 after injection. At day 16 after injection with DMDP liposomes, about normal numbers of IgM positive marginal zone lymphocytes were found again (Fig. 3A, 3B). However, the first ERTR9 positive macrophages reappearing in the spleen were not found until day 32 and by day 64 they had reached about their normal concentration, i.e. long after the marginal zone lymphocytes.

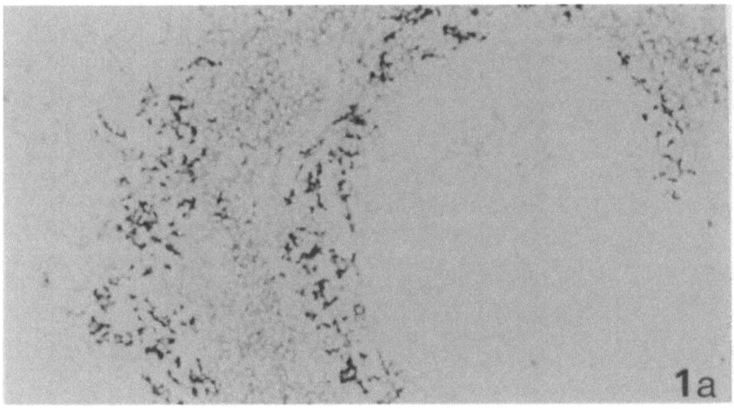

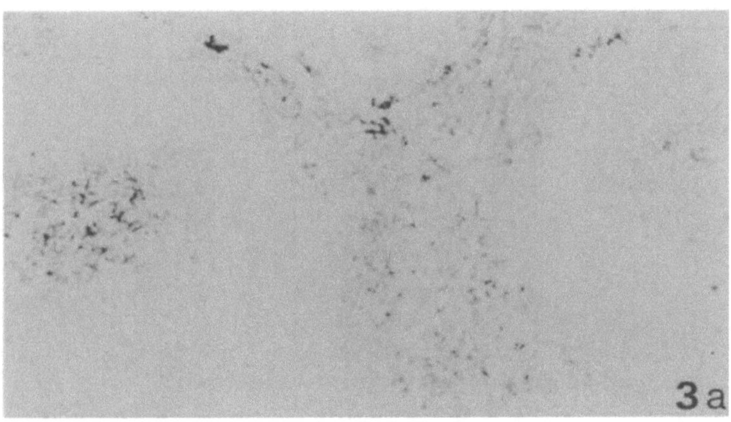

Figs. 1a, 2a and 3a: ERTR9 positive cells (marginal zone macrophages) in the spleens of normal mice (1a) and mice injected with DMDP liposomes 4 days earlier (2a) or 16 days earlier (3a).

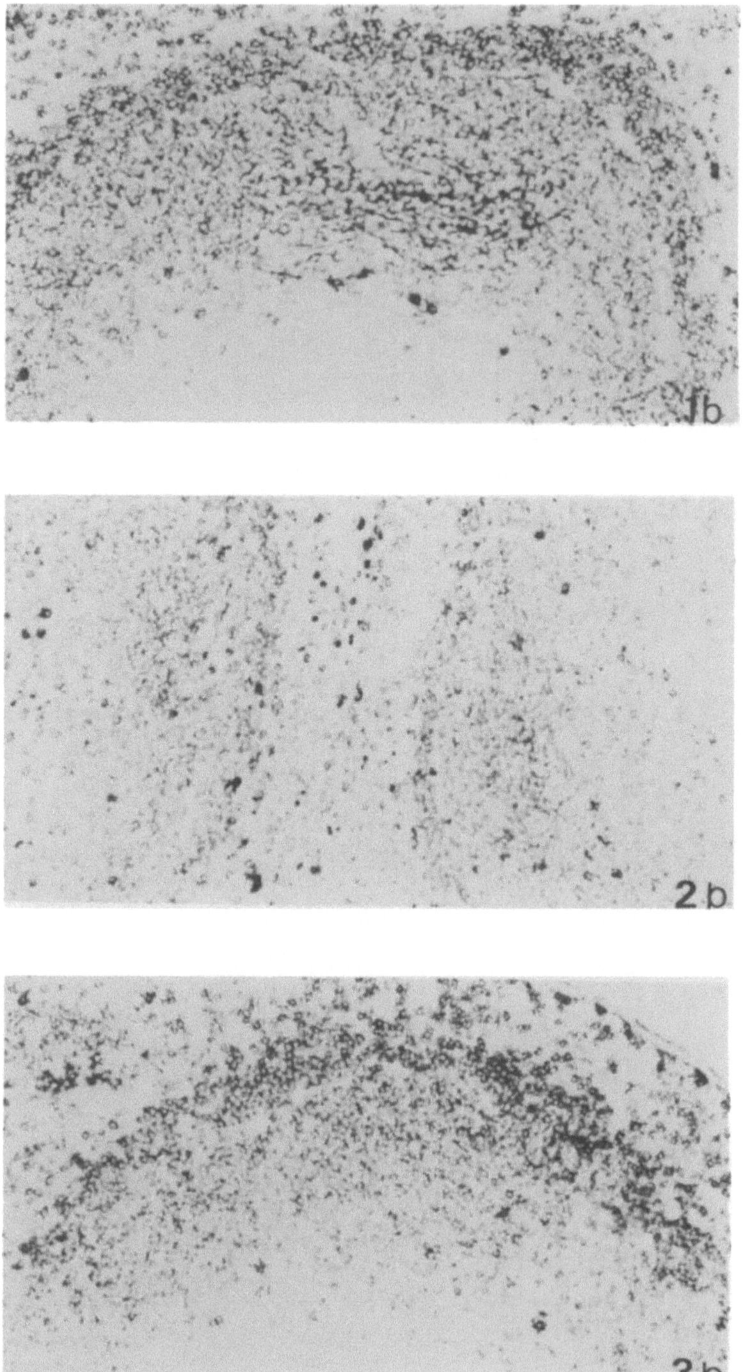

Figs. 1b, 2b and 3b: IgM positive cells (B lymphocytes) in the spleens of normal mice (1b) and mice injected with DMDP liposomes 4 days earlier (2b) or 16 days earlier (3b).

DISCUSSION

It is shown in the present study that both marginal zone macrophages and marginal zone lymphocytes are affected by intravenous administration of liposome encapsulated DMDP. Marginal zone cells were characterized using a monoclonal antibody, ERTR9, directed to a surface antigen of marginal zone macrophages (3, 4) and an anti-IgM antibody, recognizing the surface IgM of marginal zone lymphocytes.

That these marginal zone cells not only lost their surface antigens used for their characterization, but were virtually eliminated by treatment with DMDP liposomes has been shown at ultrastructural level in earlier studies (7). Our conclusion that marginal zone macrophages are not obligatory for the retention and/or homing of marginal zone lymphocytes, is based on two findings:

1. Complete disappearance of all marginal zone macrophages at day 2-8 after treatment with DMDP liposomes, does not result in complete disappearance of all marginal zone lymphocytes in the same period.
2. Marginal zone lymphocytes reappeared earlier than did marginal zone macrophages. At day 16 after treatment marginal zone macrophages were still absent, but marginal zone lymphocytes had reached about their normal concentration.

In an earlier study (9) we demonstrated that elimination of marginal zone macrophages did not inhibit the migration of labelled lymphocytes from the marginal zone into the PALS. It is concluded in the present study that marginal zone macrophages have no role in the retention and/or homing of the marginal zone lymphocytes themselves. That marginal zone macrophages are not involved in the migration or retention of lymphocytes does not implicate that they cannot be involved in the presentation of antigens to these lymphocytes as has been suggested by Humphrey (10).

REFERENCES

1. N. van Rooijen, and E. Claassen, <u>in</u>: 'Liposomes as drug carriers: Trends and progress', G. Gregoriadis, ed., John Wiley, Chichester, U.K. (1987).
2. N. van Rooijen, and R. van Nieuwmegen. Cell Tissue Res. 238:355 (1984).
3. E. van Vliet, M. Melis, and W. van Ewijk. J. Histochem. Cytochem. 33:40 (1985).
4. C.D. Dijkstra, E. van Vliet, E.A. Dopp, A.A. van der Lelij, and G. Kraal. Immunology 55:23 (1985).
5. G. Kraal, and M. Janse. Immunology 58:665 (1986).
6. G. Kraal, M. Breel, E.M. Janse, and G. Bruin. J. Exp. Med. 163:981 (1986).
7. N. van Rooijen, R. van Nieuwmegen, and E.W.A. Kamperdijk. Virchows Arch. (Cell Pathol.) 49:375 (1985).
8. N. van Rooijen, and R. van Nieuwmegen. Meth. Enzymol. 93:83 (1983).
9. E. Claassen, and N. van Rooijen. Immunology 56:689 (1985).
10. J.H. Humphrey. Immunol. Lett. 11:149 (1985).

LOCALIZATION OF TI2 ANTIGEN (FICOLL-FITC) AND CORRESPONDING SPECIFIC ANTI-FITC ANTIBODY FORMING CELLS IN THE RAT SPLEEN

Kenjiro Matsuno, Taichi Ezaki and Masahiko Kotani

Department of Anatomy, Kumamoto University Medical School, Kumamoto 860, Japan

INTRODUCTION

Specific antibody forming cells (AFC) to various antigens can be readily stained on tissue sections (1,2) using enzyme-labelled antigens. But the antigen itself was not observed on the same sections by these authors.

Ficoll-FITC (Fic-F,TI2 antigen) can be visualized on the spleen section by fluorescence microscopy after iv injection (3,4). When Fic-F is administrated into rats, Ficoll behaves as a TI2-type carrier (5) and FITC is recognized as a hapten (6), so that the antibody production to FITC can be expected. By a modification of indirect immuno-alkaline phosphatase technique, we have newly established a method to stain specific AFC to FITC determinants (aF-AFC).

In this work, time kinetic studies on the intrasplenic localization of both Fic-F and aF-AFC were performed after iv injection of Fic-F into rats to know the relationship between antigen-laden macrophages and corresponding specific AFC.

MATERIALS AND METHODS

Preparation of sera and tissues

Fic-F, 400 μg/ rat, was injected iv into male DA rats of 2 to 5 months old. After time interval from 1 day to 3 wk, sera were collected and spleens were removed. Two to four rats at each time point and total 30 rats were used. The tissues were fresh frozen and cryosections of 6 μm thick were made.

Titration of serum anti-FITC antibodies in the immunized rats

This was examined by the passive hemoagglutination of FITC-erythrocytes. Sheep red cells were labelled with FITC according to Möller and Coutinho (6). Sera were heat-inactivated for 30 min at 56°C and each 25 μl was serially diluted with phosphate buffered saline. Fifty μl of 0.5 % FITC-erythrocyte suspension was added to the above solution.

After incubation for 30 min at 37°C, the titer was read as a final dilution of positive hemoagglutinations.

Localization of Fic-F in the spleen

Cryosections were fixed in acetone and examined under a fluorescent microscope (epi-illumination, Nikon Fluophoto, Japan).

Specific anti-FITC AFC staining (Indirect immuno-alkaline phosphatase technique)

Spleen cryosections were fixed in acetone and FITC-labelled Avidin (Avidin-F, 5 µg/ml, Vector, USA) were overlayed for 3 h at room temperature (RT). After washing, the sections were reacted with biotinylated alkaline phosphatase (2 units/ml; Vector, USA) for 30 min at RT and coloured with the substrate solution consisting of Naphthol AS-MX phosphate and Fast Red TR salt for 10 min at RT (7). Endogenous alkaline phosphatase activity was blocked by adding 1 mM Levamisole in the substrate solution. Then sections were counterstained with hematoxylin and examined under a light microscope.

The AFC staining was proved to be specific since no staining was obtained when sections were reacted under following conditions:

1. where the target section was that of unimmunized control.
2. where the preincubation of sections was done with either FITC or FITC-BSA solution.
3. without Avidin-F.
4. with unlabelled Avidin instead of Avidin-F.

Since the red colour product of the specific staining in this method emitted orange-red fluorescence at a wavelength exciting FITC (490nm) and did not interfere Fic-F fluorescence, the localization of both antigen and corresponding specific AFC in the spleen section became simultaneously detectable under a fluorescence microscopy, ie, Fic-F antigen emitted green fluorescence and aF-AFC orange-red.

RESULTS

Fic-F antigen localization

Most of Fic-F in the spleen was concentrated into marginal zone (MZ) macrophages and marginal metallophils (MM) located at the outer border of the periarterial lymphatic sheath (PALS) and at the MZ bridging channel (MZ-BC). But MM located around the follicular area did not ingest Fic-F (Fig.1). This pattern was stable throughout the period examined (day 1 to 3 wk). In addition, a small amount of antigen was detected in the germinal center(GC) on day 4 after the injection and persisted till 3 wk.

Kinetics of serum anti-FITC antibodies

The serum antibodies became detectable on day 4 (titer:4X), reached its peak on day 7 (32X) and declined by 2 wk (4X), but persisted till 3 wk (2X).

Kinetics of aF-AFC

AF-AFC first appeared in the outer PALS, at the border between GC and PALS, and in MZ-BC on day 1 to day 2 after the immunization. The

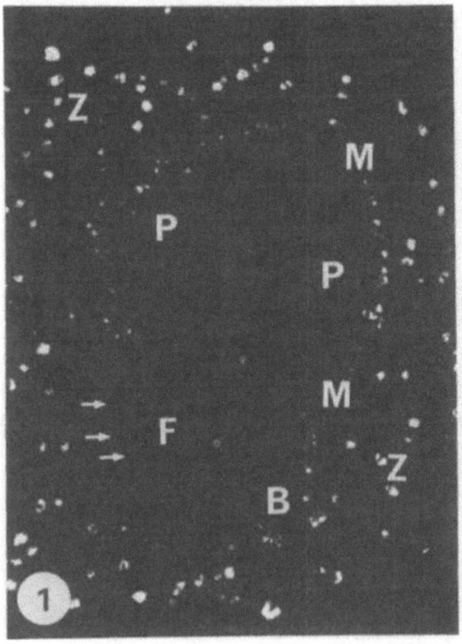

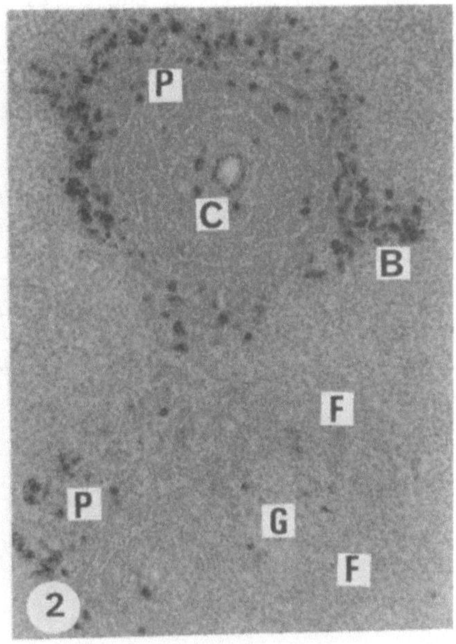

Fig.1. Localization of Fic-F on day 4. 145x. P:PALS, M:MM, B:MZ-BC, F:follicular area, Z:MZ macrophage. Note MM around follicular area (arrows) were not fluorescent.

Fig.2. AF-AFC on day 7. 145x. G:GC, C:central artery. Note aF-AFC are mainly located at the outer PALS and MZ-BC, while few aF-AFC exists near MM that located around the follicular area. Granular deposits of aF antibody are seen in GC.

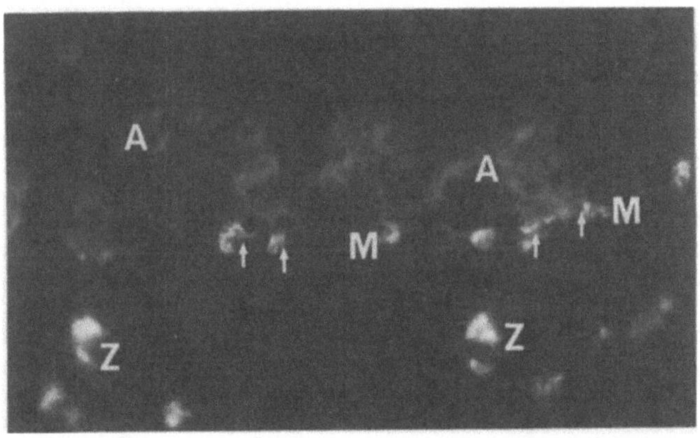

Fig.3. Double fluorescence of Fic-F-laden MM (white) and aF-AFC (grey) on day 4. 580x. A:aF-AFC, Z:MZ macrophages. Note several white cells are superimposing on grey cells (arrows).

number of aF-AFC reached its peak between day 4 and day 6 (Fig.2). AF-AFC were found dominantly in the outer PALS, and noteworthly, some aF-AFC in the outer PALS and MZ-BC were observed in the immediate vicinity of MM. On the other hand, aF-AFC was only occasionally seen at the periphery of the follicular area where MM were located (Fig.2). A few aF-AFC were also found in MZ. After 1 wk, number of aF-AFC decreased in PALS, while many aF-AFC were found in the MZ-BC and the red pulp. This suggested migration of these cells from the white pulp. The number of aF-AFC reduced considerably on day 10 and almost disappeared by 2 wk. Specific granular anti-FITC staining also appeared in GC on day 4 and its intensity was markedly decreased on day 10 but detectable till 3 wk.

Double fluorescence microscopy of both Fic-F and aF-AFC

An intimate relationship between Fic-F-laden MM and aF-AFC was further confirmed by this method. This was constantly observed from day 2 till day 10 after immunization. In outer PALS and MZ-BC, some green cells were superimposing on orange-red cells (Fig.3) which suggested a direct contact of MM with aF-AFC.

DISCUSSION

In the present study, by using Fic-F as an immunogenic tracer and indirect immuno-alkaline phosphatase staining for anti-FITC antibody, it has become possible for the first time to demonstrate simultaneously localization of both antigen-laden cells and corresponding antibodies on the same spleen secion. It was proved that both Fic-F fluorescence and aF-AFC staining were highly specific with low back ground and easily recognized. Even extra-cellular Fic-F and aF antibody in GC could be identified as granular deposits which strongly suggested formation of specific immune complexes.

It was shown that in rats MM except for the follicular area did ingest Fic-F, as well as MZ macrophages as reported before (4). The localization of Fic-F had not changed throughout the period studied except that granular fluorescence of antigens appeared in GC from day 4 to 3 wk. The numbers of aF-AFC corresponded to the titers of aF antibody in the serum and the kinetics of AF-AFC showed the similar pattern of localization to the study of van Rooijen et al (1), though they used rabbits and thymus-dependent protein antigen.

As a novel finding, in the outer PALS and MZ-BC, the close association of MM with aF-AFC was constantly observed by the double fluorescence study. On the other hand, at the periphery of the follicular area, where MM did not ingest Fic-F, only a few aF-AFC was seen. These results suggest that MM around the outer PALS and MZ-BC play an important role in controlling the generation of specific AFC. In other words, MM may regulate the proliferation or differentiation of AFC by forming direct contact with AFC. Our preliminary experiments using Bromo-deoxy-Uridine (BrdU) and anti-BrdU immunohistochemistry in addition to aF-AFC staining showed that cells which incorporated BrdU increased significantly in PALS regions on day 4 after Fic-F immunization. And there, number of BrdU-positive aF-AFC reached up to 70 % of total aF-AFC. This supports the above speculation. Classen et al (2) did not find the close relationship between MZ macrophages and MM and anti-TNP-AFC when rats were immunized with TNP-Ficoll. The difference between our results and theirs may be due to the experimental system involved including difference between the haptens, but it should be further studied. Concerning TI2 antigen presentation by MZ macrophages to MZ type B cells (8), we did not find the evidence though a few aF-AFC were seen in MZ.

In conclusion, by using Ficoll-FITC as an immunogenic tracer, we could study the intrasplenic kinetics of both antigen-laden cells and corresponding AFC simultaneously on the single cryostat sections. This can be used as one of the most suitable models for analysing the immune response in vivo.

REFERENCES

1. N. Van Rooijien, N. Kors, R. van Nieuwmegen, and P. Eikelenboom. Anat. Rec. 206:189 (1983).

2. E. Classen, N. Kors, C.D. Dijkstra, and N. van Rooijen. Immunology 57:399 (1986).

3. J.H. Humphrey, and D. Grennan D. Eur. J. Immunol. 11:221 (1981).

4. K. Matsuno, H. Fujii, and M. Kotani. Cell. Tissue Res. 246:263 (1986).

5. D.E. Mosier, and B. Subbarao. Immunology Today 3:217 (1982).

6. G. Möller, amd A. Coutinho. J. Exp. Med. 141:647 (1975).

7. D.Y. Mason, in: "Techniques in immunocytochemistry, Vol.3", p. 25, G.R. Bullock, and P. Petrusz, eds., Academic Press, London (1985).

8. D. Gray, I.C.M. MacLennan, B. Platteau, H. Bazin, J. Lortan, and G.D. Johnson, in: "Microenvironments in the lymphoid system", p. 437, G.G.B. Klaus, ed., Plenum, New York, London (1985).

IN VITRO INFECTION OF HUMAN EPIDERMAL LANGERHANS CELLS WITH HIV

G. Ramirez, L.R. Braathen*, R.O.F. Kunze, and H. Gelderblom

Dept. Virology, Robert Koch Institute
Bundesgesundheitsamt, Berlin
*Dept. Dermatol., Rikshospitalet
University of Oslow

INTRODUCTION

Human epidermal Langerhans cells (Lc) play an important role in antigen presentation and immunoregulation in the skin (for review see ref.(1)).They carry antigens and receptors which are identical to those of classical macrophages, i.e. FcIgG, C3 and the MHC class II HLA-D/DR antigens (2,3),but also the T-cell markers CD1(T6) and CD4(T4) (4-6). Remarkably, Lc exhibit 50-fold higher amounts of HLA-DR than other cells of the monocyte-macrophage lineage (7).
Human immunodeficiency virus (HIV) seems to have a special affinity for CD4 positive helper/inducer T-cells resulting in functional damage to these cells (8). It is believed that the CD4 molecules serve as a binding site for HIV (9-12). When mononuclear cells from peripheral blood of healthy individuals are inoculated with HIV, virus replication occurs predominantly in CD4 positive lymphocytes (13). However,HIV can also be encountered in follicular dendritic cells (14) and certain CD4 positive mononuclear phagocytes (4,15-17).
It has been demonstrated that ARC/AIDS patients have decreased numbers of epidermal Lc (18,19). A recent report shows Lc reaction with a monoclonal antibody specific for the HTLV-III core protein p17 in seven out of 40 HIV infected patients, and electronmicroscopy revealed Lc-associated viral particles characteristic of HIV and cytopathic changes in about one third of the Lc population in one skin biopsy (20).
We therefore asked the question whether Lc can be infected by HIV directly, and here we demonstrate that epidermal Lc from non-HIV-infected individuals infected in vitro with HIV after some days in culture react with a mixture of monoclonal antibodies directed against the p15, p24 and gp120 HIV specific proteins.

MATERIALS AND METHODS

Donors of epidermal cells

Epidermal cells were obtained from non-affected abdominal skin of 6 patients with various skin diseases. Antibody tests for HIV infection were negative, and the patients did not belong to any risk groups. They came from rural areas of Norway where HIV infections have not yet occured.

Epidermal cells

Epidermis was separated from dermis at the basement membrane in vivo by means of a suction blister device as previously described (21). The sheets of epidermis obtained were then dissociated by means of trypsin and DNA'se. The cells were then, after washings, resuspended in medium RPMI-1640 with L-glutamine (BIO-Cult, Gibco, Glasgow, Scotland) supplemented with penicillin, streptomycin and 20% autologous serum. Viability was determined by trypan blue exclusion and was more than 90%. The cell suspensions contained 2-3% Lc as judged after staining of the cells with monoclonal anti-HLA-DR and OKT6 antibodies in fluorescence technique.
The Langerhans cells were enriched by Hypaque-Ficoll gradient centrifugation, and flow cytometry using anti-HLA-DR antibody demonstrated 5-10% positive i.e. Lc. The cells were then incubated with HIV.

Infection of epidermal cell suspensions with HIV

One ml epidermal cell suspension (2.5x10 cells) and one ml supernatant of the HIV-1(strain HTLV-IIIB) infected cell line H9 were mixed and incubated for 1 h at 37'C. The cells were then washed three times in RPMI-1640 before being resuspended in RPMI-1640 with L-glutamine supplemented with penicillin, streptomycin and 20% autologous serum. The cell suspensions were cultured in the wells of Linbro plates (Costar) in a humid 5% CO2 atmosphere at 37'C.
From the 3rd to the 8th day in culture aliquots of cells were cytocentrifuged onto slides and stained for light microscopic evaluation with a mixture of monoclonal antibodies directed against the p15, p24 and gp120 HIV specific proteins in the APAAP technique. Noninfected cells and cells infected with heat-inactivated HIV-1 (56 C, 30') served as controls.

Infection of allogenic peripheral blood mononuclear cells(PBMC) with supernatants from the in vitro infected epidermal cells

Supernatants (0,5 ml stored at -20 C) obtained at days 5-8 from two different in vitro infected epidermal cell suspensions were added to Lymphoprep-isolated PBMC (0,5 ml of 2x10 cells/ml) stimulated with PHA (10ug/ml) two days earlier from two other HIV negative healthy donors.Recombinant Interleukin-2 (20 units) was also added to the cultures. At day 7 additional PHA prestimulated PBMC from the same donor and Interleukin-2 was added. The cells were cultured in medium RPMI-1640 with L-glutamine, penicillin, streptomycin and 20% fetal calf serum. From day 3 until day 14 aliquots of the mononuclear cells were stained with the monoclonal antibodies directed against the HIV specific proteins in the APAAP technique. Noninfected PHA-prestimulated PBMC served as controls.

Monoclonal antibodies

Anti-HLA-DR was obtained from Becton Dickinson and used in a dilution of 1:400. T-6A (anti-CD1) used diluted 1:100 was from Coulter Immunology. Anti-p15, -p24 and -gp120 were obtained from DuPont and mixed to a cocktail with a final dilution of 1:50.

APAAP staining technique

The APAAP staining technique was performed as described by Cordell et al.(22).

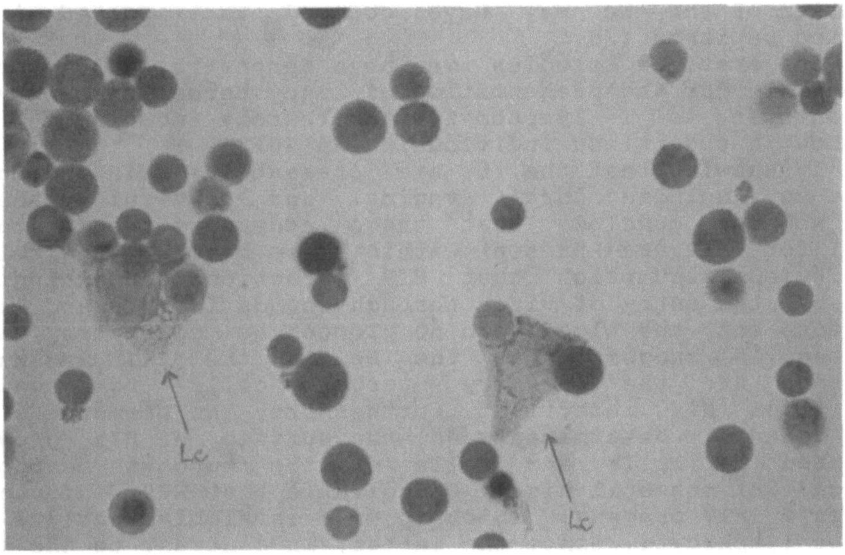

Fig. 1
The micrograph demonstrate HIV positive Langerhans cells in epidermal cell suspension 6 days after the in vitro infection. The cells are stained with a mixture of monoclonal antibodies directed against the p15, p24 and gp120 HIV specific proteins in the APAAP technique. The positive cells appear with red granules. Keratinocytes are brown coloured.

RESULTS

4-5 days after the infection in all 6 of the epidermal cell suspensions the dendritic Lc were stained with the mixture of monoclonal antibodies against the HIV specific proteins p15, p24 and gp120. The staining appeared granular in the cytoplasm of the cells (Fig 1). Noninfected cells and

cells infected with heat-inactivated HIV did not demonstrate staining.
After 11 days in culture the PBMC which were obtained from two heterologous donors and incubated with supernatants from one of two different in vitro infected epidermal cell suspensions demonstrated staining for the HIV specific proteins. Noninfected PHA prestimulated PBMC did not demonstrate positive staining.

DISCUSSION

The present results demonstrate that human epidermal Lc can be directly and easily infected in vitro with HIV, and that they after a few days in culture produce and release HIV which then are capable of infecting allogenic PBMC.
Furthermore, the Lc seem to be much easier infected than PBMC since the PBMC required prestimulation with PHA and in addition Interleukin-2.

Lc originate from bone marrow precursors and migrate to the epidermis (23). There are numerous sites at which HIV infection of the Lc may have occured as reported in HIV infected patients (20).

In previous studies we have demonstrated that Lc are responsible for the presentation of live herpes simplex virus and antigen to T lymphocytes and thereby inducing a T-cell response in sensitized individuals (24-26).

In knowing that the Lc are present in skin and mucous membranes including oral, vaginal and cervical epithelium (27), we must conclude that these accessory cells (target cells) for HIV are present within these barriers themselves. The previous assumption that HIV infection can exclusively occur by the entry of virus through wounds in skin and mucous membranes into the blood can no longer be considered valid. Our results suggest that the Lc in the skin and mucous membranes are the primary target cells for sexually transmitted HIV infection. Furthermore, the presence of MHC class I and II determinants on the surface of HIV grown in activated T cells and cells of the monocyte-macrophage lineage (28) makes it highly conceivable that HIV infected Lc produce a HIV progeny, which in vivo is highly reactive with the CD4 lymphocyte. Since the latter, in contrast to the cells of the macrophage-monocyte lineage, is highly sensitive to the cytopathic action of HIV, the pathognomotic continuous loss of CD4 positive T helper cells in HIV infected individuals may be readily initiated. Thus the Lc under the condition of natural infection probably serve as the initial cell to be infected and remain, since being relatively resistant to the cytopathic effect of HIV, a longlasting source of infectious virus.

ACKNOWLEDGEMENTS

We are grateful for the excellent technical assistance of Hilde Børufsen. This work was supported by The Norwegian Research Council for Science and the Humanities.

REFERENCES

1. L.R. Braathen, S. Bjercke, and E. Thorsby. Immunbiol. 168:301 (1984).

2. G. Stingl, E.C. Wolff-Schreiner, W.J. Pichler, F. Gschnait, W. Knapp, and K. Wolff. Nature 268:245 (1977).
3. L. Klareskog, U.M. Tjernlund, U. Forsum, and P.A. Peterson. Nature 268:247 (1977).
4. G.S. Wood, N.L. Warner, and R.A. Warnke. J. Immunol. 131:212 (1983).
5. K. Wolff, and G. Stingl. J. Invest. Dermatol. 80 (suppl):17s (1983).
6. V. Groh, M. Tani, A. Harrer, K. Wolff, and G. Stingl. J. Invest. Dermatol. 86:115 (1986).
7. S. Bjercke, G. Gaudernack, and L.R. Braathen. Scand. J. Immunol. 21:489 (1985).
8. F. Wong-Staal, R.C. Gallo. Nature 317:395 (1985).
9. G. Dalgleish, P.L.C. Beverley, P.R. Clapham, D.H. Crawford, M.F. Greaves, R.A. Weiss. Nature 312:763 (1984).
10. D. Klatzmann, E. Champagne, S. Chamaret, J. Gruest, D. Guetard, T. Hercend, J.C. Gluckman, L. Montagnier. Nature 312:767 (1984).
11. J.S. McDougal, M.S. Kennedy, J.M. Sligh, S.P. Cort, A. Mawle, and J.K.A. Nicholson. Science 231:382 (1986).
12. S.M. Lederman, J. Barasch, M. Godfrey, D. Sherman, M. Gershon, and L. Chess. Clin. Res. 34:670A (1986)
13. D. Klatzmann, F. Barre-Sinoussi, M.T. Nugeyre, C. Dauguet, E. Vilmer, C. Griscelle, F. Brun-Vezinet, C. Rouzioux, J.C. Gluckman, J-C. Chermann, and L. Montagnier. Science 225:59 (1984).
14. J.A. Armstrong, R.L. Dawkins, and R. Horne. Immunology Today 6:121 (1985).
15. D.D. Ho, T.R. Rota, M.S. Hirsch. J. Clin. Invest. 77:1712 (1986).
16. S. Gartner, P. Markovitz, D.M. Markovitz, R.F. Betts, and M. Popovic. JAMA 256:2365 (1986).
17. G.S. Wood, B.F. Burns, R.F. Dorfman, and R.A. Warnke. Am. J. Pathol. 120:371 (1985).
18. D.V. Belsito, M.R. Sanchez, R.L. Baer, F. Valentine, G.J. Thorbecke. N. Engl. J. Med. 310:1279 (1984).
19. D.V. Belsito, and G.J. Thorbecke. N. Engl. J. Med. 311:857 (1984).
20. E. Tschachler, V. Groh, M. Popovic, D.L. Mann, K. Konrad, B. Safai, L. Eron, F.D. Veronese, K. Wolff, and G. Stingl. J. Invest. Dermatol. 88:233 (1987).
21. L.R. Braathen, and E. Thorsby. Scand. J. Immunol. 11:401 (1980).
22. J.L. Cordell, B. Falini, W.N. Erder, A.K. Ghosh, Z. Abdulaziz, S. MacDonald, K.A. Pulford, H. Stein, and D.Y. Mason. J. Histochem. Cytochem. 32:219 (1984).
23. B. Volc-Platzer, G. Stingl, K. Wolff, W. Hinterberger, and W. Schnedl. N. Engl. J. Med. 310:1123 (1984).
24. L.R. Braathen, E. Berle, U. Mobech-Hanssen, and E. Thorsby.Acta Dermatovener (Stockh). 60:381 (1980)
25. L.R. Braathen, and E. Thorsby. Br. J. Dermatol. 108:139 (1983).
26. L.R. Braathen, and E. Thorsby. Scand. J. Immunol. 15:55 (1982).
27. S. Bjercke, H. Scott, L.R. Braathen, and E. Thorsby. Acta Obstet. Gynecol. Scand. 62:585 (1983).
28. H. Gelderblom, H. Reupke, T. Winkel, R. Kunze, and G. Pauli. Submitted.

ACCESSORY CELL NEOPLASIA AS A RESULT OF THE BREAKDOWN OF IMMUNOLOGICAL REGULATION?

Valerie Wallis, Monisha Chaudhuri and Mark Brittel

Institute of Cancer Research
Haddow Laboratories, Clifton Avenue
Sutton, Surrey SM2 5PX, England

INTRODUCTION

We have previously shown that the injection of spleen cells from old individual donors into young syngeneic recipients has two unexpected consequences (1). Firstly, in most recipients there is a gradual increase in the proportion of donor T cells and, secondly, the majority of the recipients develop tumours. These also derive from the donor inoculum. According to the classification of Dunn (2) they are type B reticulum cell neoplasms (RCN-B). The tumour cells have neither T nor B cell characteristics, most constitutively express class-II antigens, and those that have so far been tested function as accessory cells in in vitro assays (3-5). We have therefore suggested that they are the neoplastic counterpart of normal lymphoid dendritic cells (6). In the present paper the relationship between the two consequences of the injection of old cells is explored, and a hypothesis proposed to account for the finding that similar tumours, also of donor origin, can be induced in recipients of young syngeneic lymphocytes if they are subjected to repeated extrinsic antigenic stimulation.

MATERIALS AND METHODS

Mice. Male mice of the syngeneic strains, CBA/Ca and CBA/T6T6, which have a pair of chromosome markers, and the F_1 hybrid between them were used throughout. Young recipient mice were 12-16w of age; donor mice were either 12-16w or >18m old.

Preparation of cells for injection. Suspensions of spleen cells were prepared by passing through a sieve, washed twice in TC199, and injected intravenously. Enriched T cell preparations were made by incubating spleen cells on nylon wool columns (7). Suspensions prepared in this way had >80% Thy1+ cells and were functionally depleted of accessory cells (4).

Peripheral blood cultures and chromosome preparations. Samples of peripheral blood (approx. 0.3ml) were taken from the retro-orbital sinus at intervals and, after defribination and separation on plasmagel, 10^6 lymphocytes were cultured in 0.5ml RPMI 1640 containing 10% foetal calf serum, antibiotic, L-glutamine, 10^{-4}M 2-mercaptoethanol and phytohaemagglutinin (PHA) at 2.5ug per ml. They were incubated in 10% O_2, 86% N_2 and 4% CO_2 for 2d. 4-5h before harvesting colcemid (0.004ug/ml) was added to arrest dividing cells at metaphase. Chromosome spreads were made by the method of Ford (8); where possible 200 cells on each of 2 slides were analysed by two independent observers and scored as being of host or donor origin.

Antigenic stimulation. Allostimulation was achieved by injection of 2.10^7 spleen cells from $(CBA/CaxC57BL)F_1$ mice intravenously at 2w intervals.

RESULTS

In the first experiment CBA/Ca mice were injected with 5.10^7 pooled spleen cells from either age-matched or old CBA/T6T6 mice. Peripheral blood samples were taken at intervals and stimulated by the specific T cell mitogen, PHA. The percent of donor cells scored in 3 of the mice is shown in Figure 1. Although all three showed the same rise in T cell chimaerism, only two of them developed tumour. In the experiment as a whole all the recipients of old cells showed similar donor T cell chimaerism and 73% developed tumour with a mean latent period of 87 ± 13 weeks. All the tumours analysed were of donor origin. Recipients of young spleen cells did not show any increase in chimaerism, nor did any develop tumour. Analysis of the tumour cells has shown that they lack both T and B cell surface markers, are strongly Ia positive, and function well in assays for accessory function for both T cell mitogenesis and antigen (SRBC) presentation. For these reasons, as well as their morphology (Figure 2), we have postulated that the target cell for tumorigenesis might be a lymphoid dendritic cell (3,4).

The donor T cell hyperplasia was in fact preceded by a general hyperplasia of the spleen, as demonstrated by a steady increase in spleen weight following injection of old spleen cells into young mice. Both T and B cells were involved since their relative proportions remained unchanged.

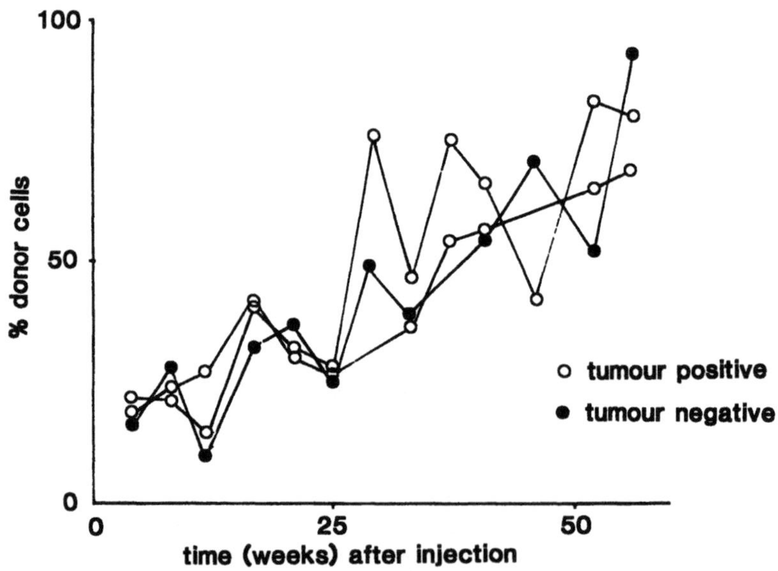

Fig. 1. Percent donor cells in PHA-stimulated cultures of peripheral blood lymphocytes of 3 mice which had been injected intravenously with 5.10^7 spleen cells from old syngeneic donors.

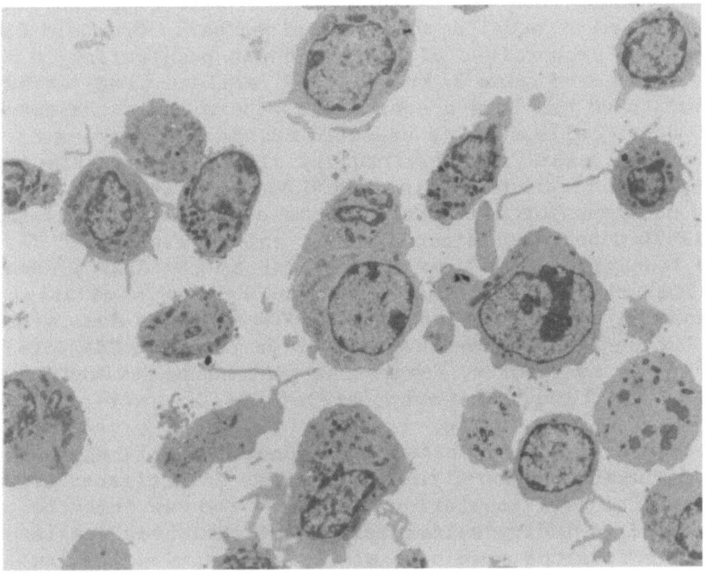

Fig. 2. Ultrathin section through RCN-B tumour cells enriched for the predominant non-adherent, low-density population (x2500).

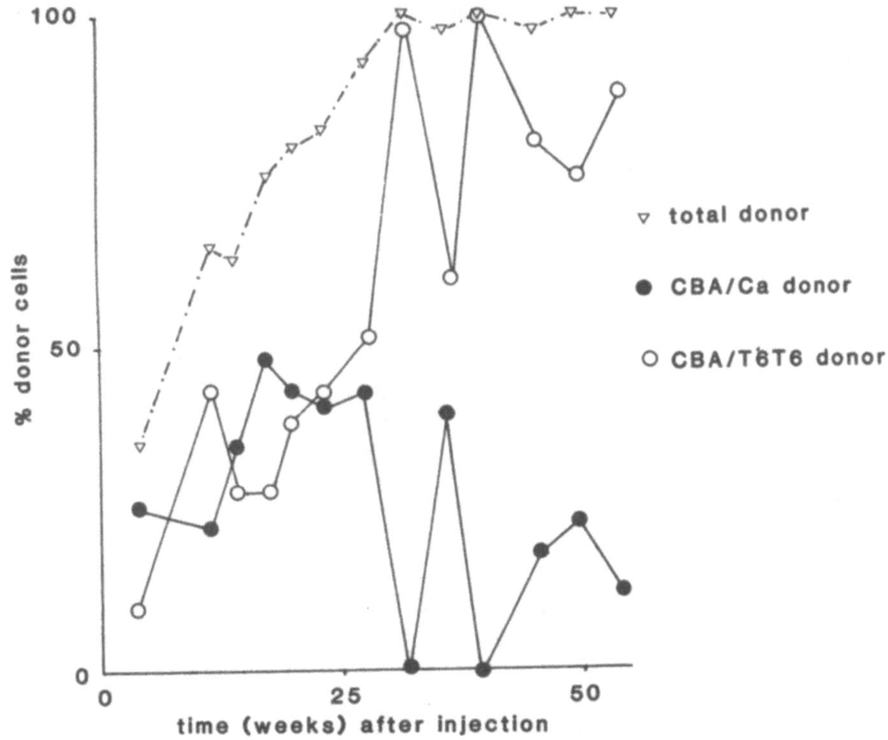

Fig. 3. Percent donor cells in PHA-stimulated cultures of peripheral blood lymphocytes from a $(CBA/CaxCBA/T6T6)F_1$ recipient of equal numbers of spleen cells from old CBA/Ca and old CBA/T6T6 mice.

In order to investigate whether the donor T cell hyperplasia was poly- or oligo-clonal, young $(CBA/CaxCBA/T6T6)F_1$ mice were injected with 5.10^7 cells from a mixture of equal numbers of spleen cells from old CBA/Ca and CBA/T6T6 mice. The proportions of the two donor populations in a typical recipient are shown in Figure 3. Even at the earliest time tested in this experiment (4w) there had been a substantial increase in the total proportion of donor cells from an expected initial chimaerism of about 10% and this increased steadily until virtually all the cells stimulated into mitosis by PHA were of donor origin. For about 6m after injection approximately equal proportions derived from each donor; after this cells from the CBA/T6T6 donor predominated. This animal was killed 57w after injection and found to have a tumour derived from the CBA/Ca donor population. The other mice in the experiment behaved similarly, i.e. the initial increase in the proportion of donor T cells was due to both donor populations, but at later times the population from the CBA/T6T6 donor predominated in all recipients, suggesting that these derived from very few clones. Most of the tumours however were of CBA/Ca donor origin.

The relationship, if any, between the donor T cell hyperplasia and development of tumours of non-T cell origin was investigated by injecting enriched or depleted cell populations. Figure 4 shows that the percent of donor cells scored in individual recipients of enriched T cells or unseparated cells from the same pool of old spleens was the same. However, the tumour incidence was 90% in the recipients of unseparated cells and only 22% in the mice which received enriched T cells. In a further experiment, spleens from individual old T cell deprived mice were used as a source of cells depleted of T cells but possibly enriched for the putative target cells for tumorigenesis. They produced tumour in only 1/10 recipients.

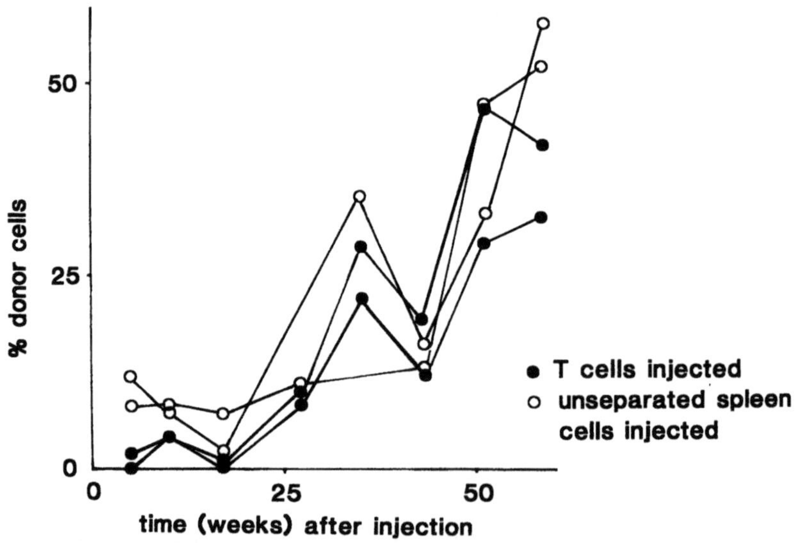

Fig. 4. Percent donor cells in PHA-stimulated cultures of peripheral blood lymphocytes from two recipients of unseparated spleen cells from old donors and two recipients of enriched T cells from the same spleens.

Table 1. Effect of continued allostimulation with (CBA/CaxC57BL)F_1 spleen cells on tumour induction in CBA/Ca mice previously injected with 10^6 or 5.10^6 syngeneic spleen cells.

Cells injected	Allostimulation	n	% tumour incidence
old spleen	-	4	0
	+	4	25
age-matched spleen	-	20	5
	+	20	70
none	+	4	0

It appeared from the previous experiments that although donor T cell hyperplasia occurred independently of the presence of donor accessory cells, tumorigenesis only occurred when both were present. It seemed possible that a similar environment conducive to malignant transformation might be provided by extrinsic alloantigenic challenge. Accordingly, young CBA/Ca mice were injected with 10^7 CBA/T6T6 spleen cells from either young or old donors and were then challenged intravenously every 2w with spleen cells from (CBA/CaxC57BL)F_1 mice. The tumour incidence is recorded in Table 1. Those groups of mice which received both syngeneic spleen cells and continued antigenic challenge developed tumours. It was notable that the tumour incidence was highest in the group which received age-matched syngeneic cells. The latent period was short (4-6m) compared with average latent periods following the injection of old spleen cells alone of 12-18m. The tumours which have been analysed have been found to be of syngeneic donor origin.

DISCUSSION

The experiments described in this paper suggest that there are two requirements for the induction of RCN-B following transfer of syngeneic cells. Firstly, as the tumours are almost invariably of donor origin, even though the donor cells must be greatly outnumbered by host cells of similar type, the transfer of cells must have greatly increased their risk of malignant transformation. Experiments are in progress, in which autologous spleen cells have been transferred, to see whether the procedure of making cell suspensions and injecting them intravenously is sufficient to put the cells at risk, or whether it is the transfer to a different immunological network which is responsible. It may well be that following intravenous injection the target cells locate ectopically and are thereafter unable to interact normally with other cells of the lymphoid system.

The second component in tumorigenesis in this model appears to be continued T cell stimulation, which can be provided either by transfer of cells from old mice into young, or extrinsically by continued alloantigenic challenge to provide a strong stimulus for T cell proliferation. It is not at present known why transfer of T cells from old mice into young should result in their selective proliferation, but it appears to be independent of the presence of donor accessory cells.

The changes consequent upon injection of old spleen cells probably proceed steadily:- firstly, an increase in spleen weight which appears to be due to increase in both T and B lymphocytes of both host and donor origin; secondly, selective proliferation of T lymphocytes derived from the old donor; thirdly, a gradual outgrowth of oligoclonal populations of donor T cells.

It is well documented that chronic graft-versus-host disease in mice can result in a high incidence of lymphoreticular neoplasms (9). However, the majority of tumours were of host origin, suggesting that the mechanism of induction might be different from the present system in which the tumours are almost invariably of donor origin, even when as few as 1 million donor cells are transferred. There is no evidence of any histoincompatibility between the two substrains of CBA mice used in the present experiments, and in fact the same incidence of tumours has been obtained when spleen cells from old CBA/T6T6 mice were transferred into either CBA/T6T6 or CBA/Ca recipients. It has also proved possible to induce RCN-B by transfer of lymphocytes from old to young mice in both DBA/2 and C57BL strains (5).

The experiments described here seem to demonstrate the consequences of cellular interactions which may be under imperfect control. Transfer of old T cells into young recipients appears to release them from homeostatic control so that at least some of them continue to expand, and eventually very few donor clones take over. Transfer of old accessory cells alone probably has few consequences but puts them at risk when a lymphoproliferative environment is provided, either by the continued proliferation of old donor T cells, or by continued extrinsic challenge. The fact that even a few young accessory cells transferred to animals which are subsequently subjected to extrinsic antigenic challenge can lead to tumours of donor origin, whereas the host's accessory cells are not put at similar risk, suggests that transfer may have altered the control exerted by interactions at a cellular level.

ACKNOWLEDGMENTS

This work was supported by grants to the Institute of Cancer Research: Royal Cancer Hospital from the Cancer Research Campaign and the Medical Research Council.

The authors are grateful to Prof. A.J.S. Davies for helpful discussions and to Ken Gomer and Richard Goucher for expert technical assistance.

REFERENCES

1. V.J. Wallis, M. Chaudhuri, M.C. Jacob, B.A. Valkova, and A.J.S. Davies. Immunology 53:769 (1984).
2. T.B Dunn. J. Natl. Cancer Inst. 14:1281 (1954).
3. M.P. Brittle, M.C. Jacob, K. Gomer, and D. Robertson. Immunology 54:541 (1985).
4. M.P. Brittle, M.C. Jacob, and K.J. Gomer. Immunology 55:663 (1985).
5. M.P. Brittle, V.J. Wallis, M. Chaudhuri, R.A. Goucher, and K.J. Gomer. Submitted to British Journal of Cancer.
6. R.M. Steinman, and M.C. Nussenzweig. Immunol. Rev. 53:127 (1980).
7 M. Julius, E. Simpson, and L.A. Herzenberg. J. Immunology 3:645 (1973).
8. C.E. Ford, <u>in</u>: "Tissue Grafting and Radiation", Appendix I, H.S. Micklem, and J.F. Loutit, eds., Academic Press, New York (1966).
9. E. Gleichmann, H. Gleichmann, and W. Wilke. Transplant. Rev. 31:156 (1976).

INDEX

MIX
Papier aus verantwortungsvollen Quellen
Paper from responsible sources
FSC® C105338

If you have any concerns about our products,
you can contact us on
ProductSafety@springernature.com

In case Publisher is established outside the EU,
the EU authorized representative is:
Springer Nature Customer Service Center GmbH
Europaplatz 3, 69115 Heidelberg, Germany

Printed by Libri Plureos GmbH
in Hamburg, Germany